AIR-CONDITIONING AND REFRIGERATION INSTITUTE

1501 Wilson Boulevard, 6th floor
Arlington, VA 22209–2403

SECOND EDITION

REFRIGERATION AND AIR-CONDITIONING

PRENTICE-HALL, INC., ENGLEWOOD CLIFFS, NJ 07632

Library of Congress Cataloging-in-Publication Data

Air-Conditioning and Refrigeration Institute.
  Refrigeration and air-conditioning.

  Includes index.
    1. Refrigeration and refrigerating machinery.
2. Air conditioning. I. Title.
TP492.A27 1987   621.5′6   86–9420
ISBN 0–13–770181–0

Editorial/production supervision: **Colleen Brosnan**
Interior design: **Linda Conway**
Cover design: **Network Graphics**
Manufacturing buyer: **Rhett Conklin**
Page layout: **Martin J. Behan**

Printed in the United States of America

10 9 8 7 6

ISBN 0-13-770181-0 025

PRENTICE-HALL INTERNATIONAL (UK) LIMITED, *London*
PRENTICE-HALL OF AUSTRALIA PTY. LIMITED, *Sydney*
PRENTICE-HALL CANADA INC., *Toronto*
PRENTICE-HALL HISPANOAMERICANA, S.A., *Mexico*
PRENTICE-HALL OF INDIA PRIVATE LIMITED, *New Delhi*
PRENTICE-HALL OF JAPAN, INC., *Tokyo*
PRENTICE-HALL OF SOUTHEAST ASIA PTE. LTD., *Singapore*
EDITORA PRENTICE-HALL DO BRASIL, LTDA., *Rio de Janeiro*

Contents

Contents vii

viii   Contents

Contents   ix

Preface

The First Edition of *Refrigeration and Air-Conditioning* was written and edited by the Manpower Development Committee of the Air-Conditioning and Refrigeration Institute (ARI), the national industry trade association whose members manufacture more than 90% of all air-conditioning and refrigeration equipment produced in the United States. The original textbook was based on course outlines developed in cooperation with the U.S. Office of Education's Vocational Education Branch and, subsequently, was approved by that agency.

Refrigeration and Air-Conditioning is the culmination of the ARI's long-standing involvement in the promotion and development of training programs designed to prepare beginning service technicians to install, service, and maintain air-conditioning and refrigeration equipment. Over 100,000 beginning service technicians have already learned their first skills with the help of this book.

This Second Edition reflects the results of a survey of countless instructors and their students who have used *Refrigeration and Air Conditioning* at nearly 200 institutions nationwide and, indeed, around the world. In order to make this text an even better learning tool, several basic changes were made in organization, content matter, and pedagogy to reflect the recommendations and suggestions of these concerned educators.

To provide a step-by-step and uniform sequence of study, chapters have been rearranged to progress from basic theory to trouble analysis in each category of Refrigeration, Air-Conditioning, and Heat Pumps. The subject of Heat Pumps has been extracted and expanded into its own separate section. Additionally, a chapter on Solid-State Controls has been included in the Air-Conditioning section.

This new Second Edition of *Refrigeration and Air-Conditioning* is now supported with a complete teaching and learning package. In addition to the new design and modern look of the parent textbook, instructors will now have access to a new and complete Instructor's Manual as well as a new set of Transparency Masters for classroom use. Additionally, a new Student Lab Manual/Workbook is available to enhance and improve study skills and student performance.

Due to the rapid changes that have been occurring in the equipment sizing and application of air-conditioning and heat pumps, the sections on Heat Gain–Heat Loss Calculation and Air Distribution Systems have been deleted from the textbook and now appear in the new Student Lab Manual/Workbook. This information is published on a continuously updated basis by the Air Conditioning Contractors of America (ACCA), 1228 17th Ave. N.W., Washington, DC 20036. We recommend that ACCA's Environmental Systems Library be used in conjunction with this text for study of this portion of the industry.

As before, it should be noted that refrigeration, air-conditioning, and the environmental control industry, in general, all have dramatically expanded in the variety of available equipment and applications to a level beyond the scope of a single text. This text, is therefore, limited to general principles and applications. We recommend that the use of specific product information, as well as installation and service information from individual manufacturers, become part of the learning process.

Acknowledgments

The Air-Conditioning and Refrigeration Institute wishes to convey its utmost appreciation to Lee Miles, the revising author, for his pen and patient prose, and for his untiring efforts in reorganizing and updating the wealth of material for this Second Edition.

We are also indebted to the many instructors who reviewed the manuscript and made valuable contributions to this improved Second Edition, particularly: A. Blaustein, St. Petersburg College; D. Giovanni, Long Beach City College; J. Haag, ITT Technical Institute; W. Harris, Thomas Nelson Community College; A.J. Knight, Bessemer State Tech.; E. Morgan, Pinellas Institute; R.G. Reichlmayr, Trident Technical College; D.R. Smith, SUNY; Alfred J. Smoker, Climate Control Institute; T. Sumners, Indiana Vocational Technical College; and J. Tokar, The Refrigeration School.

The preparation of this text would not be possible without the cooperation of all the related trade and professional organizations and manufacturers within our industry. Special thanks go to the following who so generously provided material:

A C & R Components
Addison Products Company
Air Conditioning Contractors of America (ACCA)
Aeroquip Corporation
Airserco Manufacturing Company
Allied Chemical Corporation
Alnor Instrument Company
Amana Refrigeration, Inc.
American Air Filter Company, Inc.
American Gas Association (AGA)
American Society of Heating, Refrigeration and Air Conditioning Engineers (ASHRAE)
A.W. Sperry Instruments, Inc.
Bacharach–United Technologies
Bard Manufacturing Company
Bally Case and Cooler Company
BDP Company–United Technologies
Beckett Corporation R.W.
The Black & Decker Manufacturing Co.
Bohn Heat Transfer Group, Gulf & Western Manufacturing Co.
Borg-Warner Central Environmental Systems, Inc.
Bussman Manufacturing Company
Carrier Corporation–United Technologies
Climate Control
Conco Inc.
Copeland Corporation
Davis Instruments Manufacturing Co.
Dole Refrigerating Company
Duro Metal Products Company
Dwyer Instruments, Inc.
E.I. du Pont de Nemours & Co.
E-Tech Inc.
Fedders Air Conditioning–USA
Federal Pacific Electric Company
Fisher Controls International Inc.
Furnas Electric Company
General Electric Company
General Controls Company
Gould, Inc.
Heat Exchangers, Inc.
Hoffman Controls Corporation
Honeywell Inc.
Hussman Corporation
Imperial Eastman Company
Johnson Controls, Inc.
Kramer Trenton Company
Koldwave, Inc.

Lennox Industries, Inc.
Liquid Carbonic Corporation
Marley Cooling Tower Company
Marsh Instrument Company
Marshalltown Instruments, Inc.
Monarch Manufacturing Works
Mueller Brass Company
National Fire Protection Association (NFPA)
Nicholson File Company
Ranco, Inc.
Refrigeration Engineering Corporation
Ridge Tool Company
Robertshaw Controls Company
Robinair Division, Sealed Power Corp.
Rotorex Company
Skuttle Manufacturing Company
Sporlan Valve Company
Standard Refrigeration Company
Suntec Industries, Inc.
Tecumseh Products Company
The Delafield Company, Division of Alco
Thermal Engineering Company
Thermal Industries Company, Inc.
TIF Instruments, Inc.

Tyler Refrigeration Corporation
Underwriters Laboratories (UL)
Vendo Company
Wallace–Murray Corporation
Walton Laboratories
Wayne Home Equipment
Weil McLain, Division of Wylain, Inc.
White-Rodgers Division, Emerson Electric Co.
York International Corporation

Also, recognition is due to the many officers of ARI—past and present—for their support of ARI's programs and especially to those who have served and continue to serve on ARI's Education and Training Committee. The scope of activity and successes of ARI's education program is due in large measure to the direction and hard work of this important committee. Appreciation is also due to Ed Moura and Colleen Brosnan of Prentice-Hall for their work in publishing this edition.

AIR-CONDITIONING
AND REFRIGERATION INSTITUTE
Arlington, VA

REFRIGERATION
AND
AIR-CONDITIONING

REFRIGERATION

R1

Introduction to Refrigeration

R1-1
REFRIGERATION THROUGH THE AGES

Although man's earliest forebears probably knew about, and observed, the effects of cold, ice, and snow on their bodies and on things around them—such as the meat they brought home from the hunt—it is not until we reach the early history of China that we find any reference to the use of these natural refrigeration phenomena to improve the lives of people, and then only for the cooling of beverages. But other uses were developed; the Chinese are the first known civilization to harvest winter's ice and store it, packed in straw or dried weeds, for use in the summer months.

Natural ice and snow provided the only means of refrigeration for many centuries. The early Egyptians discovered that evaporation could cause cooling, so they learned to put their wine and other liquids into porous earthenware containers, placing them on the rooftops at night so that cool breezes caused evaporation and cooled the contents (Fig. R1-1).

Some of our own early colonists developed methods of preserving perishable food and drink with snow and ice. They built storage buildings (ice houses) in which they could keep ice, harvested during the cold New England winters, for use in the hot months (Fig. R1-2).

During colonial days and well into the nineteenth century ice was an important commodity in trade with foreign countries that produced no natural ice.

By the early 1900s industrial refrigeration using the mechanical cycle had been developed, and meat packers, butcher shops, breweries, and other industries were beginning to make full use of mechanical refrigeration.

With the growth of the electric industry and wiring of homes, household refrigerators became popular and replaced window and standing iceboxes, which required a block of ice daily.

This increased interest in household refrigerators was aided by the design of fractional horsepower electric motors to operate the compressors in mechanical "iceboxes." Since the early 1920s these appliances have been produced in large numbers and have become a necessity for all rather than a luxury for a few.

Not only is food preserved in our homes today, but commercial preservation of food is one of the most important current applications of refrigeration. Commercial preservation and transport of food is so common that it would be difficult to imagine an unrefrigerated America.

FIGURE R1-1 Even the early Egyptians knew that evaporation of a liquid cooled the surrounding air or substance around it. These porous jars are typical of those used by the Egyptians for cooling wine, beer, and other liquids by placing them on the rooftops at night. They are among the earliest "refrigerators."

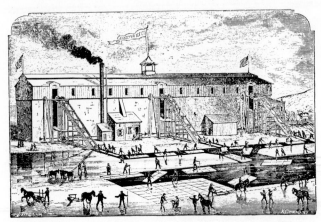

FIGURE R1-2 Throughout the ages, and almost up until the beginning of the twentieth century, natural ice, cut from lakes, ponds, and rivers, was the sole source of refrigeration . . . and scenes like this were common each winter in all parts of the country.

More than three-fourths of the food that appears on American tables every day is produced, packaged, shipped, stored, and preserved by refrigeration. Millions of tons of food are stored in refrigerated warehouses; millions more are in frozen food warehouses, in private lockers, and in packing and processing plants.

Without the many types of modern refrigeration in stores, warehouses, aircraft, railway cars (Fig. R1-3), trucks, and ships, the storage and transportation of all types of perishables would be impossible.

We are not limited to the enjoyment of fruits, vegetables, and other produce grown locally at any given time of the year; we can have food products from other parts of the country and even from distant lands the year round.

Refrigeration has improved the economy of many areas by providing a means of preserving their products en route to remote consumers. It has aided in the development of agricultural regions through greater demand for the products, and it has helped the dairy and livestock-producing areas similarly.

Prior to 1941, most of the tires for automobiles, trucks, airplanes, and the like were made of natural rubber produced on plantations in southeast Asia. When the shipment of latex from the foreign rubber plantations to the United States was curtailed following the outbreak of World War II, American industry and the federal government established a cooperative synthetic rubber program based upon previous research. Scientists learned that some manufacturing processes could make the artificial rubber more durable and wear resistant through use of low temperatures; thus refrigeration became vital to still another industry.

There has been a rapid increase of new products since World War II. The petrochemical industry (plastics), textile plants, and data processing industry are heavy users of process refrigeration. Without refrigeration (and air conditioning) many of these new products could not be manufactured and/or used.

One of the more recent developments associated with the energy situation is the importation of liquefied natural gas from foreign sources. Temperatures of $-270°F$ are required to change the gas into a liquid, which is then loaded onto a refrigerated tanker for shipment to a U.S. receiving port. All the while the liquid must be maintained at $-270°F$ until it is ready for vaporization back into a gas.

Solar heating and solar refrigeration, along with air-conditioning, are expected to provide many opportunities for new products and for qualified application, installation, and service personnel.

In 1980, U.S. homes, plants, and commercial buildings added new refrigeration and air-conditioning equipment valued at more than $8 billion. The dollar total was actually much larger, since the cost of many accessory products, such as ductwork, grilles, insulation, and controls that go into a complete system was not included. Thus, refrigeration and air-conditioning constitute one of the major domestic industries in the United States.

World markets in this field are also experiencing rapid growth. Canada, Japan, West Germany, the United Kingdom, France, Mexico, Iran, and Venezuela are among the major users of refrigeration and air-conditioning. And this market will continue to expand as

FIGURE R1-3 Trains made up of refrigerator cars carry refrigerated food from one part of the country to another.

foreign countries improve their standard of living and engage in more industrial development.

R1-2
MANPOWER REQUIREMENTS

As a result of the expanding markets for refrigeration and air-conditioning products and changing technology, there will be a dramatic increase in the need for qualified personnel both at home and abroad.

In 1973 ARI published a *Manpower Survey Report* that estimated a great number of new jobs to be filled through the 1980s and 1990s. Of the 1973 estimate, most were for technician/mechanically skilled personnel. It was estimated that for every $1 million in installed value of equipment, the following employees are needed:

1 graduate engineer

2 technicians/master mechanics

11 refrigeration, air-conditioning, and heating mechanics

2 refrigeration, air-conditioning, and heating helpers

7 sheet-metal mechanics

1 sheet-metal helper

2 salespeople

In actual practice it is common for a technician/mechanic to be qualified or required to work on refrigeration, air-conditioning, or heating—or all three. Thus you can readily see the types of jobs available in this broad category.

These job opportunities not only result from a continuously expanding market for refrigerating machinery, but because of the fact that many installers and service mechanics who have been active for years will be retiring or moving up to supervisory positions.

R1-3
OCCUPATIONAL OPPORTUNITIES

Where are the skills noted above used in the refrigeration industry (and this includes air-conditioning, since both achieve their purpose through application of the "refrigeration cycle")?

Fields in which these employment opportunities exist include high-temperature cooling (usually comfort or process air-conditioning) as well as applications involving medium temperature (product storage above freezing), low temperature (quick-freezing processes as well as storage below freezing), and extreme low temperature (utilized in medical and chemical applications and some industrial and scientific applications).

Equipment manufacturers hire a number of trained technicians for various jobs. Laboratory technicians may be involved in actually building prototypes of new products, or they may conduct tests on products for performance ratings, life-cycle testing, or compliance to ARI certification or UL safety standards or sound evaluation.

Factory service personnel prepare installation instructions, training material, renewal parts information, etc. Some equipment manufacturers employ large numbers of qualified service technicians for field assignment, which involves installation and startup situations and rendering actual customer service.

The largest employer classifications are the wholesale/contracting trades. The refrigeration wholesaler must have technical personnel capable of servicing a dealer/contractor organization on the proper application, installation, and maintenance of the product lines they carry.

The dealer/contractor installs and services the product for the end user. Depending on the extent of the contractor's business, a technician may be involved in installation only or in startup or perhaps customer service. In smaller organizations he may do all three.

In the larger metropolitan areas there are refrigeration service companies that are involved in nothing but service. They specialize in "maintenance contracts" with the user, to provide scheduled inspections as well as emergency repairs.

It is common for large manufacturing or industrial plants or institutions that utilize considerable refrigeration or air-conditioning products to have in-house maintenance departments that do most of the routine service work.

Large food chain companies such as Kroger, A & P, Safeway, etc., employ skilled technicians to assist in or direct the installation and service of the refrigeration equipment. The modern supermarket cannot afford down time on either its food preservation equipment or its comfort air-conditioning. Other areas of employment include the electric utilities, schools, and certain industries.

Knowledge, training, and experience are the key ingredients to being a successful technician. This text is designed to help prepare you to take advantage of these broad opportunities.

R1-4
DEFINITIONS

Before getting started on the fundamentals of a refrigeration system a few definitions should be considered since they form the groundwork for any discussion of this subject.

R1-4.1
Heat

Heat is a form of energy transferred by a difference in temperature or (by a carrying means) a difference in pressure. Heat exists everywhere to a greater or lesser degree. As a form of energy it can be neither created (except by nuclear energy release) nor destroyed, although it can be tampered with and converted into other forms of energy and ultimately end up again as heat. Heat energy can be caused to travel only by the creation of a difference in energy levels called heat intensity or temperature.

R1-4.2
Cold

Cold is a relative term to describe the energy level or temperature of an object or area compared to a known energy level or temperature. A 50°F outdoor temperature would be considered cold to a person living at the equator but warm to someone living at the arctic circle. Some definitions describe it as an "absence of heat." This implies that there is no heat at all, but, in fact, there is nothing known in the world today from which heat is totally absent. No process yet devised has been capable of achieving "absolute zero," the removal of all heat energy (molecular movement) from a space or object. Theoretically, this zero point would be 459.69 degrees below zero (−459.69°F) on the Fahrenheit thermometer scale or 273.16 degrees below zero (−273.16°C) on the Celsius thermometer scale.

R1-4.3
Refrigeration

Refrigeration is the transfer of heat from one place to another by a change in state of a liquid. Heat can also be transferred in many other ways as long as a difference in level of heat energy (temperature) exists. Thus, cold liquid can absorb heat from a hot object, or hot liquid can give up heat to a cold object, depending on the desired results. Without a change in state, however, the results are not because of the action of a "refrigeration system" or because of a "refrigeration effect."

R1-4.4
Mechanical Refrigeration

Mechanical refrigeration is the use of work energy in mechanical components as arranged in a "refrigeration system" to obtain the desired results. Mechanical articles are needed only when the heat energy is being transferred from an area of lower heat concentration (temperature) to an area of higher heat concentration (temperature), such as from the interior of the refrig-

erator cabinet to the surrounding area of the kitchen in the case of the household refrigerator. When transferring heat from a higher temperature to a lower temperature, mechanical compressor units are not needed, as in the example of the steam heating system (high-temperature fire to the room air) in the home.

R1-4.5
Refrigerants

Refrigerants are chemical compounds that are used to absorb heat by evaporation or boiling from the liquid state to the vapor state and reject heat from the vapor state to the liquid state by condensation. Many different refrigerants are in use and the selection of a particular refrigerant depends on the conditions under which it is intended to operate. The subject of refrigerants is discussed in detail throughout many sections of the book.

R1-4.6
Absorption Refrigeration

Absorption refrigeration is the use of heat energy to produce the conditions necessary to transfer heat energy from one place to another. The heat energy is converted to work energy and the desired results are then obtained, in the same manner as in a mechanical refrigeration system. Conditions under which the refrigerants operate are discussed in Chapter R15.

R1-5
SYSTEM COMPONENTS— SOME FUNDAMENTALS

As pointed out above, the job of the refrigeration cycle is to remove unwanted heat from one place and discharge it into another. To accomplish this, the refrigerant is pumped through a completely closed system. If the system were not closed, it would be using up refrigerant by dissipating it into the air. Because it is closed, the same refrigerant is used over again, each time it passes through the cycle removing some heat and discharging it (Fig. R1-4).

Imagine as a comparable situation a boat that is filled with water. We want to remove the water, so we use a pail. We scoop out the unwanted water and transfer it to the outside of the boat, but we hold onto the pail, and use it over and over again to continue bailing.

The closed cycle serves other purposes as well: it keeps the refrigerant from becoming contaminated and controls its flow, for it is a liquid in some parts of the cycle and a gas or vapor in other phases.

Let's take a look at what happens in a simple refrigeration cycle, and the major components of which it is made (this fundamental approach will be enlarged upon in subsequent chapters; see Fig. R1-5).

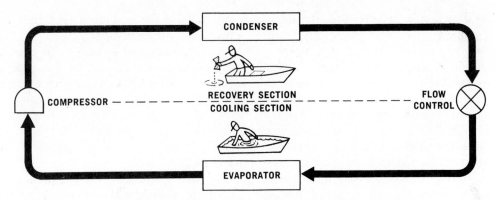

FIGURE R1-4   (*Courtesy* Allied Chemical Corporation)

Two different pressures exist in the cycle—the evaporating or low pressure in the "low side," and the condensing, or high pressure, in the "high side." These pressure areas are separated by two dividing points; one is the pressure-reducing device, where the refrigerant flow is controlled and the boiling point lowered, and the other is at the compressor, where the vapor is compressed and the condensing temperature raised.

The *pressure-reducing device* is a good place to start the trip through the cycle. This may be an expansion valve, a capillary tube, or other device to control the flow of the refrigerant into the *evaporator,* or cooling coil, as a low-pressure, low-temperature liquid refrigerant. With the boiling point lowered, the expanding refrigerant evaporates (changes state) as it goes through the cooling coil, where it removes the heat from the space in which the evaporator is located.

Heat will travel from the warmer air to the coils cooled by the evaporation of the refrigerant within the system, causing the liquid refrigerant to "boil" or evaporate, changing it to a vapor. This is similar to the change that occurs when a pan of water is boiled on

the stove and the water changes to steam—except that the refrigerant boils at a much lower temperature.

This low-pressure, low-temperature vapor is moved to the *compressor,* where it is compressed into a high-temperature, high-pressure vapor. The compressor discharges it to the *condenser,* so that it can give up the heat that it picked up in the cooling coil or evaporator. The refrigerant vapor is at a higher temperature than is the air passing across the condenser (air-cooled type); therefore, heat is transferred from the warmer refrigerant vapor to the cooler air.

In this process, as heat is removed from the vapor, a change of state takes place and the vapor is condensed back into a liquid, at a high pressure and a high temperature.

The liquid refrigerant now travels to the pressure-reducing device, where it passes through a small opening or orifice where a drop in pressure, (boiling point), and temperature occurs. Then it enters into the evaporator or cooling coil. As the refrigerant makes its way into the larger opening of the tubing or coil, it vaporizes, ready to start another cycle through the system.

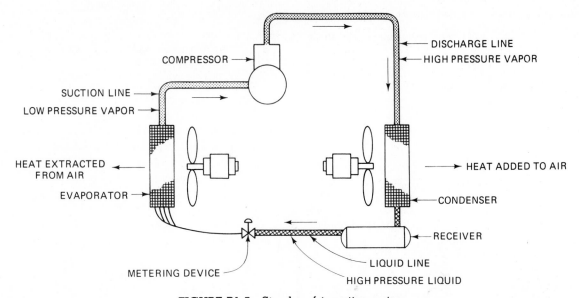

FIGURE R1-5   Simple refrigeration system.

The refrigeration system requires some means of connecting the major components—evaporator, compressor, condenser, and metering device—just as roads connect communities. Tubing or "lines" make the system complete so that the refrigerant will not leak out into the atmosphere. The *suction line* connects the evaporator or cooling coil to the compressor, the *hot-gas* or *discharge line* connects the compressor to the condenser, and the *liquid line* is the connecting tubing between the condenser and the pressure-reducing device. Some systems will have a *receiver* or storage tank immediately after the condenser and before the metering device, where the liquid refrigerant remains until it is needed for heat removal in the evaporator. The line between the condenser and the receiver is the *liquid return line*.

There are many different kinds and variations of the refrigerant cycle components. For example, there are at least a half dozen different types of compressors, from the reciprocating piston through a centrifugal impeller design, but the function is the same in all cases—that of compressing the heat-laden vapor into a high-temperature, high-pressure vapor.

The same can be said of the condenser and evaporator surfaces. They can be bare pipe, or they can be finned condensers and evaporators with electrically driven fans to pass the air through them.

There are a number of different types of metering devices to meter the liquid refrigerant into the evaporator, depending on size of equipment, refrigerant used, and its application.

The mechanical refrigeration system described above is essentially the same whether the system be a domestic refrigerator, a low-temperature freezer, or a comfort air-conditioning system. Refrigerants will be different and size of equipment will vary greatly, but the principle of operation and the refrigeration cycle remain the same. Thus, once a student understands the simple actions that are taking place within the mechanical cycle, he or she has gone a long way toward understanding how a refrigeration system works.

PROBLEMS

R1-1. The country of _____ is the earliest known to use ice for refrigeration.

R1-2. The mechanical refrigeration cycle was first used in the early _____s.

R1-3. Refrigeration and air-conditioning systems represent one of the major domestic and world industries. True or False?

R1-4. For every $1,000,000 in installed value of equipment, are two, six, or 13 refrigeration and air-conditioning technicians needed?

R1-5. Heat is the absence of cold. True or False?

R1-6. "Absolute zero" is 0°F on the Fahrenheit thermometer. True or False?

R1-7. Name the four major components of a mechanical refrigeration cycle.

R1-8. Name the four connecting lines in a refrigeration system.

R1-9. Define the refrigerating effect.

R1-10. The line between the evaporator and the compressor is called the _____.

R1-11. The line between the compressor and the condenser is called the _____.

R1-12. The line between the condenser and the pressure-reducing device is called the _____.

R1-13. If a receiver is used in the system, the line from the condenser to the receiver is called the _____.

R1-14. What two types of coils are used in refrigeration systems?

R2

Matter and Heat Behavior

R2-1

STATES OF MATTER

All known matter exists in one of three physical forms or states: solid, liquid, or gaseous (vapor). There are distinct differences among these physical states:

1. *Solid State:* Matter in a "solid" state will retain its shape and physical dimensions. Solids of sufficient density will retain their size and weight. Solids of light density will lose molecular quantity under certain conditions and lose weight and quantity. Carbon dioxide (CO_2) in solid state (known as dry ice) will pass from the solid state into the gaseous state under certain conditions (see Section R2-9).

2. *Liquid State:* Matter in a "liquid" state will retain its quantity and weight if of sufficient density but must depend on a containing vessel for shape. If a cubic foot of water in a container measuring 1 foot on each side is transferred to a container of different dimensions, the quantity and weight of the water will remain the same, although the dimension will change (see Fig. R2-1). In time, liquids of light density, such as water, will gradually lose quantity and weight by molecular loss to the gaseous state (see Section R2-9).

3. *Gaseous State:* Matter in a "gaseous" state, commonly called the "vapor" state or just vapor, will not hold its dimension or density but must depend on a container to accomplish this. If a 1-cubic-foot cylinder containing gaseous state water, called "steam," or some other vapor, is connected to a 2-cubic-foot cylinder of which, theoretically, a perfect vacuum has been drawn, the vapor will expand to occupy the volume of the larger cylinder as well as the original. Other changes will occur in the vapor which will be discussed later (see Fig. R2-2).

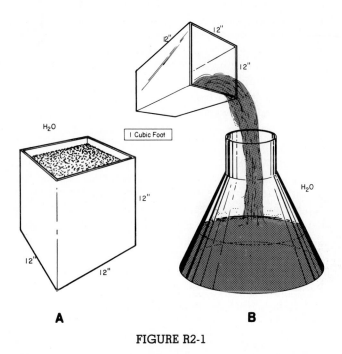

H_2O

1 Cubic Foot

12"

12"

12"

12"

12"

12"

H_2O

A   **B**

FIGURE R2-1

Although these specific differences exist in the three states of matter, quite frequently, under changing conditions of pressure and temperature, the same substance may exist in any one of the three states: for example, water could exist as a solid—ice, a liquid—water, or a gas—steam. Solids always have a definite shape, whereas liquids and gases have no definite shape of their own and so will conform to the shape of their container.

R2-2

MOLECULAR MOVEMENT

All matter is composed of small particles known as molecules, and the molecular structure of matter (as studied in chemistry) can be further broken down into atoms.

9

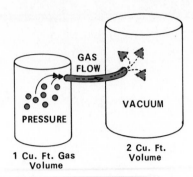

FIGURE R2-2

In Chapter R16 you will learn how the atom is further divided into electrons, protons, and neutrons. In that area you will study the electron theory, which concerns electric current flow—the movement of electrons through a conductor.

For the present we concern ourselves only with the molecule, the smallest particle into which any matter or substance can be broken down and still retain its identity. For example, a molecule of water (H_2O) is made up of two atoms of hydrogen and one atom of oxygen. If this molecule of water were to be broken down more, or divided further, into subatomic particles, it would no longer be water.

Molecules vary in shape, size, and weight. In physics we learn that molecules have a tendency to cling together. The character of the substance of matter itself is dependent on the shape, size, and weight of the individual molecules of which it is made and also the space or the distance between them, for they are, to a large degree, capable of moving about.

When heat energy is applied to a substance (Fig. R2-3), it increases the internal energy of the molecules,

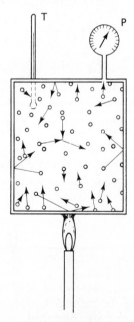

FIGURE R2-3

which increase their motion or velocity of movement. With this increase in the movement of the molecules, there is also a rise or increase in the temperature of the substance.

When heat is removed from a substance, it follows that the velocity of the molecular movement will decrease and also that there will be a decrease or lowering of the internal temperature of this substance.

R2-3
FIRST LAW OF THERMODYNAMICS

The first law of thermodynamics (the branch of science dealing with mechanical action of heat) states that "energy can neither be created nor destroyed, it can only be converted from one form to another." (With the development of nuclear energy, this is no longer correct. However, for the purposes of this text, the first law of thermodynamics applies totally.)

Energy itself is defined as the ability to do work and heat is one form of energy. It is also the final form, as ultimately all forms of energy end up as heat. There are other common forms of energy: mechanical, electrical, and chemical, which may be converted easily from one form to another. The steam-driven turbine generator of a power plant is a device that converts heat energy into electrical energy. Chemical energy may be converted into electrical energy by the use of a battery. Electrical energy is converted into mechanical energy through the use of an electric magnetic coil to produce a push-pull motion or the use of an electric motor to create rotary motions. Electrical energy may be changed directly to heat energy by means of heating resistance wires such as in an electric toaster, grill, furnace, anticipator, and the like.

R2-4
SECOND LAW OF THERMODYNAMICS

The second law of thermodynamics states that "to cause heat energy to travel, a temperature difference must be established and maintained." Heat energy travels downward on the intensity scale. Heat from a higher-temperature (intensity) material will travel to a lower-temperature (intensity) material, and this process will continue as long as the temperature difference exists. The rate of travel varies directly with the temperature difference. The higher the temperature difference (commonly called the delta temperature or ΔT), the greater the rate of heat travel. Commonly, the lower the ΔT, the lower the rate of heat travel. Heat and heat transfer are expressed in different forms and terms that are important to the refrigeration industry.

R2-5
TEMPERATURE MEASUREMENT

We are all acquainted with common measurements such as those referring to length, width, volume, etc. We must also acquaint ourselves with the methods of measuring heat energy.

The quantity of heat energy in a substance starts with the size of the substance as well as the intensity or level of heat energy in the substance. The level of heat energy is measurable on a comparison basis by means of a thermometer. The thermometer was developed using the principle of the expansion and contraction of a liquid, such as mercury, in a tube of small interior diameter which includes a reservoir for the liquid. Upon being subjected to a temperature (heat intensity) change, the liquid will rise up the line upon increase of temperature, or fall upon decrease of temperature.

The two most common standards of temperature (intensity) measurement are the Fahrenheit and Celsius (formerly centigrade) scales. Figure R2-4 shows a direct comparison of the scales of a Fahrenheit and a Celsius thermometer.

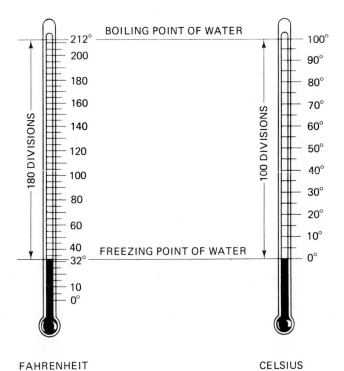

FIGURE R2-4   Comparison of the Fahrenheit and Celsius temperature scales.

R2-6
TEMPERATURE CONVERSION

Most frequently a conversion from one temperature scale to the other is made by the use of a conversion table, but if one is not available, the conversion can be done easily by using these formulas:

$$°F = 1.8°C + 32$$
$$°F = \tfrac{9}{5}°C + 32 \tag{R2-1}$$

$$°C = \frac{°F - 32}{1.8}$$
$$°C = \tfrac{5}{9}(°F - 32) \tag{R2-2}$$

EXAMPLE

To convert a room temperature of 77°F to its equivalent on the Celsius scale:

(a) $°C = \dfrac{77 - 32}{1.8} = \dfrac{45}{1.8} = 25°C.$

(b) $°C = \tfrac{5}{9}(77 - 32) = \tfrac{5}{9}(45) = \tfrac{225}{9} = 25°C.$

EXAMPLE

To convert to the Fahrenheit scale a temperature of 30°C:

(a) $°F = 1.8 \times 30 + 32 = 54 + 32 = 86°F.$

(b) $°F = \tfrac{9}{5} + 30 + 32 = 54 + 32 = 86°F.$

Thus far, in the measurement of heat intensity, we have located two definite reference points—the freezing point and the boiling point of water on both the Fahrenheit and Celsius scales. We must now locate still a third definite reference point—absolute zero. This is the point where, it is believed, all molecular action ceases. As already noted, on the Fahrenheit temperature scale this is about 460° below zero, $-460°$ F, while on the Celsius scale it is about 273° below zero, or $-273°C$.

Certain basic laws, to be discussed in later chapters, are based on the use of absolute temperatures. If a Fahrenheit reading is given, the addition of 460° to this reading will convert it to degrees Rankine or °R; whereas if the reading is from the Celsius scale, the addition of 273° will convert it to degrees Kelvin, °K. These conversions are shown in Fig. R2-5.

EXAMPLE

What is the freezing point of water in degrees Rankine (°R)? Since the freezing point of water is 32°F, adding 460° makes the freezing point of water *492° Rankine* (32° + 460° = 492°).

R2-7
HEAT QUANTITY

Heat quantity is different from heat intensity, because it takes into consideration not only the tempera-

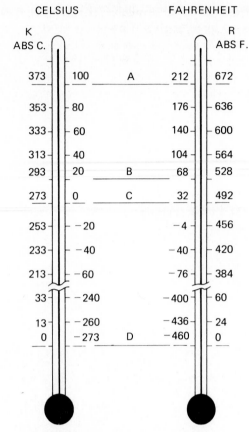

CELSIUS   FAHRENHEIT

K ABS C.				R ABS F.
373	100	A	212	672
353	80		176	636
333	60		140	600
313	40		104	564
293	20	B	68	528
273	0	C	32	492
253	−20		−4	456
233	−40		−40	420
213	−60		−76	384
33	−240		−400	60
13	−260		−436	24
0	−273	D	−460	0

FIGURE R2-5   Fahrenheit, Celsius, Rankine, and Kelvin thermometer scales: (a) boiling temperature of water; (b) standard conditions temperature; (c) freezing temperature of water; (d) absolute zero.

ture of the fluid or substance being measured but also its weight. The unit of heat quantity is the *British thermal unit* (Btu). Water is used as a standard for this unit of heat quantity; a Btu is the amount of heat required to change the temperature of 1 pound of water 1 degree Fahrenheit at sea level.

Two Btu will cause a change in temperature of 2 degrees Fahrenheit of 1 pound of water; or it will cause a change in temperature of 1 degree Fahrenheit of 2 pounds of water. Therefore, when considering a change in temperature of water, the following equation may be utilized:

$$\text{Btu} = W \times \Delta T \qquad \text{(R2-3)}$$

where

Change in heat (in Btu)
= weight (in pounds) × temperature difference

EXAMPLE

Calculate the amount of heat necessary to increase 10 lb of water from 50°F to 100°F.

SOLUTION

Since the temperature difference (ΔT) = 50 °, then

$$\text{heat} = \text{Btu} = W \times \Delta T = 10 \times 50 = 500 \text{ Btu}$$

In the example above, heat is added to the quantity of water, but the same equation is also used if heat is to be removed.

EXAMPLE

Calculate the amount of heat removed if 20 lb of water is cooled from 80°F to 40°F.

SOLUTION

$$\text{Btu} = W \times \Delta T = 20 \times 40 = 800 \text{ Btu}$$

R2-8
SPECIFIC HEAT

The specific heat of a substance is the quantity of heat in Btu required to change the temperature of 1 pound of the substance 1 degree Fahrenheit. Earlier we were presented with the information that 1 Btu was the amount of heat necessary to change the temperature of 1 pound of water 1 degree Fahrenheit, or to change the temperature of the same weight of water by the same unit of measurement on a thermometer.

Therefore, the specific heat of water is 1.0; and water is the basis for the specific heat table in Fig. R2-6. You will see that different substances vary in their capacity to absorb or to give up heat. The specific heat

FIGURE R2-6
Specific heats of common substances (Btu/lb/°F).

Water	1.00
Ice	0.50
Air (dry)	0.24
Steam	0.48
Aluminum	0.22
Brass	0.09
Lead	0.03
Iron	0.10
Mercury	0.03
Copper	0.09
Alcohol	0.60
Kerosene	0.50
Olive oil	0.47
Glass	0.20
Pine	0.67
Marble	0.21

values of most substances will vary with a change in temperature; some vary only a slight amount, whereas others can change considerably.

Suppose that two containers are placed on a heating element or burner side by side, one containing water and the other an equal amount, by weight, of olive oil. You would soon find that the temperature of the olive oil increases at a more rapid rate than that of the water, demonstrating that olive oil absorbs heat more rapidly than does water.

If the rate of temperature increase of the olive oil were approximately twice that of the water, it could be said that olive oil required only half as much heat as water to increase its temperature 1 degree Fahrenheit. Based on the value of 1.0 for the specific heat of water, it would show that the specific heat of olive oil must be approximately 0.5, or half that of water. (The table of specific heats of substances shows that olive oil has a value of 0.47.)

Equation (R2-3) can now be stated as

$$\text{Btu} = W \times c \times \Delta T \qquad \text{(R2-4)}$$

where

$$c = \text{specific heat of a substance}$$

EXAMPLE

Calculate the amount of heat required to raise the temperature of 1 lb of olive oil from 70°F to 385°F.

SOLUTION

Since $\Delta T = 315°$ and c of olive oil = 0.47, then

$$\text{heat} = \text{Btu} = W \times c \times \Delta T$$
$$= 1 \times 0.47 \times 315 = 148 \text{ Btu}$$

The specific heat of a substance will also change with a change in the state of the substance. Water is a very good example of this variation in specific heat. We have learned that, as a liquid, its specific heat is 1.0; but as a solid—ice—its specific heat approximates 0.5, and this same value is applied to steam—the gaseous state.

Within the refrigeration circuit we will be interested, primarily, with substances in a liquid or a gaseous form, and their ability to absorb or give up heat. Also, in the distribution of air for the purpose of cooling or heating a given area, we will be interested in the possible changes in the values for specific heat—more about this later.

Air, when heated and free or allowed to expand at a constant pressure, will have a specific heat of 0.24. Refrigerant 12 (R-12) vapor at approximately 70°F and at constant pressure, has a specific heat value of 0.148, whereas the specific heat of R-12 liquid is 0.24 at 86°F.

When dealing with metals and the temperature of a mixture, a combination of weights, specific heats, and temperature differences must be considered in the overall calculations of heat transfer. In order to calculate the total heat transfer of a combination of substances it is necessary to add the individual rates as follows:

$$\text{Btu}, = (W_1 \times c_1 \times \Delta T_1)$$
$$+ (W_2 \times c_2 \times \Delta T_2) \qquad \text{(R2-5)}$$
$$+ (W_3 \times c_3 \times \Delta T_3) \dots \text{ etc.}$$

EXAMPLE

How much heat must be added to a 10-lb copper vessel, holding 30 lb of water at 70°F to reach 185°F if the specific heat of copper is 0.095 F?

SOLUTION

Equation (R2-5) may be used as follows:

$$\text{Btu} = (W_1 \times c_1 \times \Delta T_1) + (W_2 \times c_2 \times \Delta T_2)$$

where W_1, c_1, and ΔT_1 pertain to the copper vessel, and W_2, c_2, and ΔT_2 pertain to the water. Therefore,

$$\text{Btu} = (10 \times 0.095 \times 115) + (30 \times 1.0 \times 115)$$
$$= 109.2 + 3450$$
$$= 3559.2 \text{ Btu}$$

R2-9
CHANGE OF STATE

We are now ready to look at the five principal changes of state:

Solidification A change from a liquid to a solid

Liquefaction A change from a solid to a liquid

Vaporization A change from a liquid to a vapor

Condensation A change from a vapor to a liquid

Sublimation A change from a solid to a vapor without passing through the liquid state.

When a solid substance is heated, the molecular motion is chiefly in the form of rapid motion back and forth, the molecules never moving far from their normal or original position. But at some given temperature for that particular substance, further addition of heat will not necessarily increase the molecular motion within the substance; instead, the additional heat will cause some solids to liquefy (change into a liquid). Thus the additional heat causes a change of state in the material.

The temperature at which this change of state in a substance takes place is called its *melting point*. Let us assume that a container of water at 70°F, in which a thermometer has been placed, is left in a freezer for hours (Fig. R2-7). When it is taken from the freezer, it has become a block of ice—*solidification* has taken place.

Let us further assume that the thermometer in the block indicates a temperature of 20°F. If it is allowed to stand at room temperature, heat from the room air will be absorbed by the ice until the thermometer indicates a temperature of 32°F, when some of the ice will begin to change into water.

With heat continuing to transfer from the room air to the ice, more ice will change back into water; but the thermometer will continue to indicate a temperature of 32°F until all the ice has melted. *Liquefaction* has now taken place. (This change of state without a change in the temperature will be discussed later.)

As mentioned, when all the ice is melted the thermometer will indicate a temperature of 32°F, but the temperature of the water will continue to rise until it reaches or equals room temperature.

If sufficient heat is added to the container of water through outside means (Fig. R2-8), such as a burner or a torch, the temperature of the water will increase until it reaches 212°F. At this temperature, and under "standard" atmospheric pressure, another change of state will take place—*vaporization*. Some of the water will turn into steam and, with the addition of more heat, all of the water will vaporize into steam; yet the temperature of the water will not increase above 212°F. This change of state—without a change in temperature—will also be discussed later.

If the steam vapor could be contained within a closed vessel, and if the source of heat were removed, the steam would give up heat to the surrounding air, and it would condense back into a liquid form—water. What has now taken place is *condensation*—the reverse process from vaporization.

Oxygen is a gas above −297°F, a liquid between that temperature and −324°F, and a solid below that

FIGURE R2-8 Convection currents caused by temperature differential.

point. Iron is a solid until it is heated to 2800°F, and it vaporizes at a temperature approximating 4950°F.

Thus far we have learned, with some examples, how a solid can change into a liquid, and how a liquid can change into a vapor. But it is possible for a substance to undergo a *physical* change through which a solid will change directly into a gaseous state without first melting into a liquid. This is known as *sublimation*.

All of us probably have seen this physical change take place without fully recognizing the process. Damp or wet clothes, hanging outside in the freezing temperature, will speedily become dry through sublimation; just as dry ice (solid carbon dioxide, or CO_2) sublimes into a vapor under normal temperature and pressure.

R2-10
SENSIBLE HEAT

Heat that can be felt or measured is called *sensible heat*. It is the heat that causes a change in temperature of a substance, but not a change in state. Substances, whether in a solid, liquid, or gaseous state, contain sensible heat to some degree, as long as their temperatures are above absolute zero. Equations used for solutions of heat quantity, and those used in conjunction with specific heats, might be classified as being *sensible* heat equations, since none of them involve any change in state.

As mentioned earlier, a substance may exist as a solid, liquid, or as a gas or vapor. The substance as a solid will contain some sensible heat, as it will in the other states of matter. The total amount of heat needed to bring it from a solid state to a vapor state is dependent on (1) its initial temperature as a solid, (2) the temperature at which it changes from a solid to a liquid, (3)

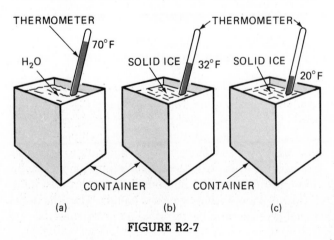

FIGURE R2-7

the temperature at which it changes from a liquid to a vapor, and (4) its final temperature as a vapor. Also included is the heat that is required to effect the two changes in state.

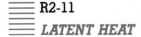

LATENT HEAT

Under a *change of state,* most substances will have a melting point at which they will change from a solid to a liquid without any increase in temperature. At this point, if the substance is in a liquid state and heat is removed from it, the substance will solidify without a change in its temperature. The heat involved in either of these processes (changing from a solid to a liquid, or from a liquid to a solid), without a change in temperature, is known as the *latent heat of fusion.*

Figure R2-9 shows the relationship between temperature in Fahrenheit degrees and both sensible and latent heat in Btu.

As pointed out earlier, the specific heat of water is 1.0 and that of ice is 0.5, which is the reason for the difference in the slopes of the lines denoting the solid (ice) and the liquid (water). To increase the temperature of a pound of ice from 0°F to 32°F requires only 16 Btu of heat; the other line shows that it takes only 8 Btu to increase the temperature of a pound of water 8°F (60°F to 68°F), from 188 to 196 Btu/lb.

Figure R2-9 also shows that a total of 52 Btu of sensible heat is involved in the 196 Btu necessary for converting a pound of 0°F ice to 68°F water. This leaves a difference of 144 Btu, which is the latent heat of fusion of water or ice, depending on whether heat is being removed or added.

The derivation of the word *latent* is from the Latin word for *hidden.* This is hidden heat, which does not register on a thermometer, nor can it be felt. Needless to say, there is no increase or decrease in the molecular motion within the substance, for it would show up in a change in temperature on a thermometer.

$$
\begin{aligned}
\text{Btu} &= (W_1 \times c_1 \times \Delta T_1) \\
&\quad + (W_1 \times \text{latent heat}) \\
&\quad + W_2 \times c_2 \times \Delta T_2)
\end{aligned}
\qquad \text{(R2-6)}
$$

EXAMPLE

Calculate the amount of heat needed to change 10 lb of ice at 20°F to water at 50°F.

SOLUTION

Utilizing Eq. (R2-6), we have

$$
\text{Btu} = (10 \times 0.5 \times 12) + (10 \times 144)
$$
$$
\downarrow
$$
$$
(32° - 20°)
$$

$$
+ (10 \times 1.0 \times 18)
$$
$$
\downarrow
$$
$$
(50° - 32°)
$$
$$
= 60 + 1440 + 180
$$
$$
= 1680
$$

Another type of latent heat that must be taken into consideration when total heat calculations are necessary is called the *latent heat of vaporization.* This is the heat that 1 pound of a liquid absorbs while being changed into the vapor state. Or it can be classified as the *latent heat of condensation;* for when sensible heat is removed from the vapor to the extent that it reaches the condensing point, the vapor condenses back into the liquid form.

The latent heat of vaporization of water as 1 pound is boiled or evaporated into steam at sea level is 970 Btu. That amount also is the heat that 1 pound of steam must release or give up when it condenses into water. Figure R2-9 also shows the relationship between temperature and both sensible heat and latent heat of vaporization.

The absorption of the amount of heat necessary for the change of state from a liquid to a vapor by evaporation, and the release of that amount of heat necessary for the change of state from a vapor back to a liquid by condensation are the main principles of the refrigeration process, or cycle. Refrigeration is the transfer of heat by the change in state of the refrigerant.

Figure R2-9 shows the total heat in Btu necessary to convert 1 pound of ice at 0° to superheated steam at 230°F under atmospheric pressure. This total amounts to 1319 Btu. Only 205 Btu is sensible heat; the remainder is made up of 144 Btu of latent heat of fusion and also 970 Btu of latent heat of vaporization.

R2-12

HEAT-TRANSFER METHODS

The second law of thermodynamics, discussed earlier, states that heat transfers (flows) in one direction only, from a high temperature (intensity) to a lower temperature (intensity). This transfer will take place using one or more of the following basic methods of transfer:

1. Conduction
2. Convection
3. Radiation

R2-12.1
Conduction

Conduction is described as the transfer of heat between the closely packed molecules of a substance, or between substances that are touching or in good contact

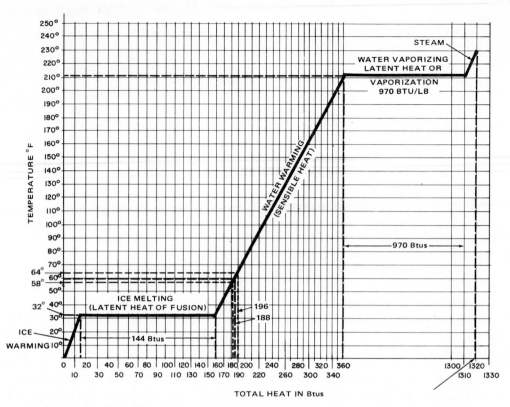

FIGURE R2-9 Chart demonstrating sensible and latent heat relationships in melting ice, changing ice to water and water to steam.

with one another. When the transfer of heat occurs in a single substance, such as a metal rod with one end in a fire or flame, movement of heat continues until there is a temperature balance throughout the length of the rod.

If the rod is immersed in water, the rapidly moving molecules on the surface of the rod will transmit some heat to the molecules of water, and still another transfer of heat by conduction takes place. As the outer surface of the rod cools off, there is still some heat within the rod, and this will continue to transfer to the outer surfaces of the rod and then to the water, until a temperature balance is reached.

The speed with which heat will transfer by means of conduction will vary with different substances or materials if the substances or materials are of the same dimensions. The rate of heat transfer will vary according to the ability of the material or substances to conduct heat. Solids, on the whole, are much better conductors than liquids; in turn, liquids conduct heat better than gases or vapors.

Most metals, such as silver, copper, steel, and iron, conduct heat fairly rapidly, whereas other solids, such as glass, wood, or other building materials, transfer heat at a much slower rate and therefore are used as insulators.

Copper is an excellent conductor of heat, as is alu-

minum. These substances are ordinarily used in the evaporators, condensers, and refrigerant pipes connecting the various components of a refrigerant system, although iron is occasionally used with some refrigerants.

The rate at which heat may be conducted through various materials is dependent on such factors as (1) the thickness of the material, (2) its cross-sectional area, (3) the temperature difference between the two sides of the material, (4) the heat conductivity (k factor) of the material, and (5) the time duration of the heat flow. Figure R2-10 is a table of heat conductivities (k factors) of some common materials.

Note: The k factors are given in Btu/hr/ft²/°F/in. of thickness of the material. These factors may be utilized correctly through the use of this equation:

$$\text{Btu} = \frac{A \times k \times \Delta T}{X} \qquad (\text{R2-7})$$

where

A = cross-sectional area, ft²

k = heat conductivity, Btu/hr

ΔT = temperature difference between the two sides

X = thickness of material, in.

16   Refrigeration

FIGURE R2-10

Conductivities for common building and insulating materials. *k* values expressed in Btu/hr/ft²/°F/inch thickness of material.

Material	Conductivity *k*
Plywood	0.80
Glass fiber—organic bonded	0.25
Expanded polystyrene insulation	0.25
Expanded polyurethane insulation	0.16
Cement mortar	5.0
Stucco	5.0
Brick (common)	5.0
Hard woods (maple, oak)	1.10
Soft woods (fir, pine)	0.80
Gypsum plaster (sand aggregate)	5.6

Metals with a high conductivity are used within the refrigeration system itself because it is desirable that rapid transfer of heat occur in both evaporator and condenser. The evaporator is where heat is removed from the conditioned space or substance or from air that has been in direct contact with the substance; the condenser dissipates this heat to another medium or space.

In the case of the evaporator, the product or air is at a higher temperature than the refrigerant within the tubing and there is a transfer of heat downhill; whereas in the condenser the refrigerant vapor is at a higher temperature than the cooling medium traveling through or around the condenser, and here again there is a downhill transfer of heat.

Plain tubing, whether copper, aluminum, or another metal, will transfer heat according to its conductivity or *k* factor, but this heat transfer can be increased through the addition of fins on the tubing. They will increase the area of heat transfer surface, thereby increasing the overall efficiency of the system. If the addition of fins doubles the surface area, it can be shown by the use of Eq. (R2-7) that the overall heat transfer should itself be doubled compared to that of plain tubing.

R2-12.2
Convection

Another means of heat transfer is by motion of the heated material itself and is limited to liquid or gas. When a material is heated, convection currents are set up within it, and the warmer portions of it rise, since heat brings about the decrease of a fluid's density and an increase in its specific volume.

Air within a refrigerator and water being heated in a pan are prime examples of the results of convection currents (see Fig. R2-11). The air in contact with the cooling coil of a refrigerator becomes cool and therefore more dense, and begins to fall to the bottom of the re-

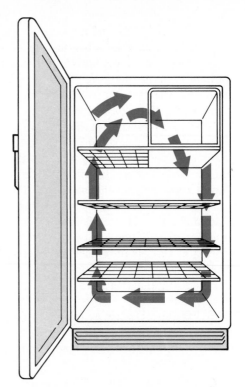

FIGURE R2-11 Convection currents caused by temperature differential.

frigerator. In doing so, it absorbs heat from the food and the walls of the refrigerator, which, through conduction, has picked up heat from the room.

After heat has been absorbed by the air it expands, becoming lighter, and rises until it again reaches the cooling coil where heat is removed from it. The convection cycle repeats as long as there is a temperature difference between the air and the coil. In commercial-type units, baffles may be constructed within the box in order that the convection currents will be directed to take the desired patterns of airflow around the coil.

Water heated in a pan will be affected by the convection currents set up within it through the application of heat. The water nearest the heat source, in absorbing heat, becomes warmer and expands. As it becomes lighter, it rises and is replaced by the other water, which is cooler and more dense. This process will continue until all of the water is at the same temperature.

Convection currents as explained and shown here are natural (passive), and, as in the case of the refrigerator, a natural (passive) flow is a slow flow. In some cases, convection must be increased through the use of fans or blowers and, in the case of liquids, pumps are used for forced (active) circulation to transfer heat from one place to another.

R2-12.3
Radiation

A third means of heat transfer is through radiation by waves similar to light or sound waves. The sun's rays heat the earth by means of radiant heat waves, which travel in a straight path without heating the intervening matter or air. The heat from a light bulb or from a hot stove is radiant in nature and is felt by those near them, although the air between the source and the object, which the rays pass through, is not heated.

If you have been relaxing in the shade of a building or a tree on a hot sunny day and move into direct sunlight, the direct impact of the heat waves will hit like a sledgehammer even though the air temperature in the shade is approximately the same as in the sunlight.

At low temperatures there is only a small amount of radiation, and only minor temperature differences are noticed; therefore, radiation has very little effect in the actual process of refrigeration itself. But results of radiation from direct solar rays can cause an increased refrigeration load in a building in the path of these rays.

Radiant heat is readily absorbed by dark or dull materials or substances, whereas light-colored surfaces or materials will reflect radiant heat waves, just as they do light rays. Wearing-apparel designers and manufacturers make use of this proven fact by supplying light-colored materials for summer clothes.

This principle is also carried over into the summer air-conditioning field where, with light-colored roofs and walls, less of the solar heat will penetrate into the conditioned space, thus reducing the size of the overall cooling equipment required. Radiant heat also readily penetrates clear glass in windows, but will be absorbed by translucent or opaque glass.

When radiant heat or energy (since all heat is energy) is absorbed by a material or substance it is converted into sensible heat—that which can be felt or measured. Every body or substance absorbs radiant energy to some extent, depending on the temperature difference between the specific body or substance and other substances. Every substance will radiate energy as long as its temperature is above absolute zero and another substance within its proximity is at a lower temperature.

If an automobile has been left out in the hot sun with the windows closed for a long period of time, the temperature inside the car will be much greater than the ambient air temperature surrounding it. This demonstrates that radiant energy absorbed by the materials of which the car is constructed is converted to measurable sensible heat.

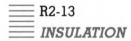
R2-13
INSULATION

In Section R2-12.1 it was pointed out that certain substances are excellent conductors of heat, while others are poor conductors, which may be classified as insulators. Therefore, any material that deters or helps to prevent the transfer of heat by any means is called and may be used as insulation. Of course, no material will stop the flow of heat completely. If there were such a substance, it would be very easy to cool a given space down to a desired temperature and keep it there.

Such substances as cork, glass fibers, mineral wool, and polyurethane foams are good examples of insulating materials; but numerous other substances are used in insulating refrigerated spaces or buildings. The compressible materials, such as fibrous substances, offer better insulation if installed loosely packed or in blanket or batt form than if they are compressed or tightly packed.

The thermal conductivity of materials, the temperature to be maintained in the refrigerated space, the ambient temperature surrounding the enclosed space, permissible wall thicknesses of insulating materials, and the cost of the various types of insulation, are all points to consider in selecting the proper material for a given project. Most service personnel are not involved in the selection or the installation of insulating material in a refrigeration application, but they may come in contact with different types of insulation, and under various conditions.

Insulation should be fire and moisture resistant, and also vermin-proof. Large refrigeration boxes or walk-in types of coolers are usually insulated with a rigid type of insulation such as corkboard, fiberglass, foam blocks, and the like; whereas smaller boxes or receptacles might be filled or insulated with a foam type that flows like a liquid and expands to fill up the available cavity with foam. Low-temperature boxes require an insulation that is also vapor resistant, such as unicellular foam, if the walls of the refrigerated enclosure are not made of metal on the outside, so that water vapor will not readily penetrate through into the insulation and condense there, reducing the insulating efficiency.

R2-14
REFRIGERATION EFFECT—"TON"

A common term that has been used in refrigeration work to define and measure capacity or refrigeration effect is called a *ton* or *ton of refrigeration*. It is the amount of heat absorbed in melting a ton of ice (2000 lb) over a 24-hour period.

The *ton of refrigeration* is equal to 288,000 Btu. This may be calculated by multiplying the weight of ice (2000 lb) by the latent heat of fusion (melting) of ice (144 Btu/lb). Thus

$$2000 \text{ lb} \times 144 \text{ Btu/lb} = 288,000 \text{ Btu}$$

in 24 hours or 12,000 Btu per hour (288,000 ÷ 24). Therefore, *one ton of refrigeration* = 12,000 Btu/hr.

A 10-ton refrigerating system will have a capacity of 10 × 12,000 Btu/hr = 120,000 Btu/hr.

R2-15
SUMMARY

The change of state of matter can be effected by adding or taking away heat. Heat effect or intensity can be measured by the use of thermometers. Heat always travels from a warmer condition to a colder condition. Substances have different capacities to absorb heat. Heat exists in two forms: *sensible* and *latent*. The unit of measure to express heat quantity is the Btu. Heat can be transferred by several methods: conduction, radiation, and convection. An insulator is a substance that will retard the flow of heat.

In the refrigeration cycle we work with an enclosed system where the effects of heat and pressure are highly related, and in Chapter R3 we examine the behavior of fluids and pressures.

PROBLEMS

R2-1. Matter exists in one of three forms. Name them.

R2-2. What is the smallest particle of matter?

R2-3. Is temperature a measure of heat quantity or intensity?

R2-4. What is meant by the term "absolute zero"?

R2-5. Define the first law of thermodynamics.

R2-6. Define the second law of thermodynamics.

R2-7. Define a British thermal unit (Btu).

R2-8. Convert 68°F to the Celsius scale.

R2-9. How much heat is required to raise the temperature of 100 lb of water from 70°F to 120°F?

R2-10. If 750 Btu is supplied to 15 lb of water at 72°F, what will be the resulting temperature?

R2-11. What is the specific heat value of water?

R2-12. Latent heat can be measured on a thermometer. True or False?

R2-13. Heat energy travels by one or more of three ways. Name them.

R2-14. The heat that changes the temperature of a substance is called _____.

R2-15. The heat that changes the state of a substance is called _____.

R2-16. A substance that can change from a solid directly into a vapor is called a _____ substance.

R2-17. The heat that changes a liquid to a vapor is called the _____ heat of _____.

R2-18. Define a "ton" of refrigerating effect.

R2-19. How many Btu are there in a standard ton of refrigerating effect?

R2-20. Is insulation a good or a poor conductor of heat?

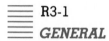

Fluids and Pressure

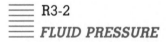 **R3-1**
GENERAL

The dictionary describes a fluid as "any substance that can flow, liquid or gas." Therefore, a refrigerant may be classified as a fluid, since within the refrigeration cycle, it exists both as a liquid and as a vapor or gas. Although, as previously mentioned, ice—a solid—is also used in heat removal, its usage in refrigeration has been overshadowed somewhat by the discovery of the versatility of the chemicals and chemical combinations used as refrigerants today.

R3-2
FLUID PRESSURE

The weight of a block of wood or any other solid material acts as a force downward on whatever is supporting it. The force of this solid object is the overall weight of the object, and the total weight is distributed over the area on which it lies.

The weight of a given volume of water, however, acts not only as a force downward on the bottom of the container holding it, but also as a force laterally on the sides of the container. If a hole is made in the side of the container below the water level (Fig. R3-1), the water above the hole will be forced out because of its force acting downward and sideways.

Fluid pressure is the force per unit area that is exerted by a gas or a liquid. It is usually expressed in terms of psi (pounds per square inch). It varies directly with the density and the depth of the liquid, and at the same depth below the surface, the pressure is equal in all directions. Notice the difference between the terms used: force and pressure. *Force* means the total weight of the substance; *pressure* means the unit force or pressure per square inch.

If the tank in Fig. R3-1 measures 1 foot in all dimensions, and it is filled with water, we have a cubic foot of water, the weight of which approximates 62.4 lb. Therefore, we would have a total force of 62.4 lb being exerted on the bottom of the tank, the area of which equals 144 square inches (12 in. × 12 in. = 144 in.²). Using the equation

$$\text{pressure} = \frac{\text{force}}{\text{area}} \qquad \text{(R3-1)}$$

or

$$P = \frac{F}{A}$$

the unit pressure of the water will be

$$\text{unit pressure} = \frac{\text{force}}{\text{area}} = \frac{62.4}{144} = 0.433 \text{ psi}$$

This pressure of 0.433 psi is exerted downward and also sideways. If a tank is constructed as in Figure R3-2, the pressure of the water would cause the tube to be filled with water to the same level as it is in the tank. Fluid pressure is the same on each square inch of the walls of the tank at the same depth, and it will act at right angles to the surface of the tank.

EXAMPLE

A tank is 4 ft square and is filled with water to its depth of 3 ft. Find (a) the volume of water, (b) the weight of the water, (c) the force on the bottom of the tank, and (d) the pressure on the bottom of the tank.

SOLUTION

(a) Volume = 4 ft × 4 ft × 3 ft = 48 ft³

(b) Weight = 48 ft³ × 62.4 lb/ft³ = 2995 lb

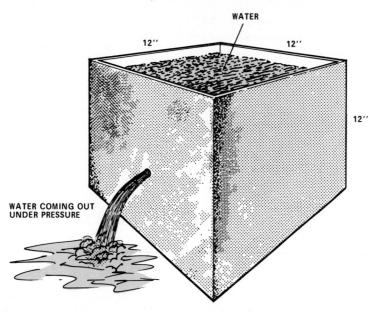

FIGURE R3-1   Water pressure in a container exerts pressure in all directions.

(c) Force = weight = 2995 lb

(d) Pressure = weight ÷ area = 2995 ÷ 2304
    = 1.3 psi

(area = 4 ft × 4 ft × 144 in.²/ft² = 2304 in.²)

 R3-3

HEAD

Pressure and depth have a close relationship when a fluid is involved. In hydraulics, a branch of physics that has to do with properties of liquids, the depth of a

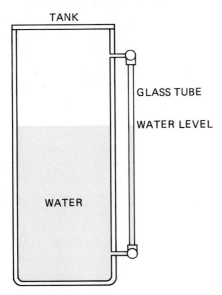

FIGURE R3-2   Fluid pressure same in tube as in tank.

body of water is called the *head* of water. *Water pressure varies directly with its depth.* As an example, if the tank in Figure R3-1 was 2 ft high and filled with water, it would contain a volume of 2 ft³ of water and would weigh 2 × 62.4, or 124.8 lb. Now the force of the water on the bottom of the tank would still be 144 in.², and the unit pressure would be 0.866 psi (124.8 ÷ 144). This is twice the amount of pressure that was exerted when the head of water was only 1 ft. Therefore, in an open-top container, the pressure of the water will equal 0.433 psi for each foot of head.

If there is a decrease or increase in the head of a body of water, there will be a corresponding decrease or increase in the pressure involved, as well as in the weight of the water, provided that the other dimensions stay the same. If the head of water was only 6 in. ($\frac{1}{2}$ ft), the pressure would equal 0.433 × $\frac{1}{2}$, or 0.217 psi; and if there were 10 ft of water in an open tank, the pressure would be 0.433 × 10 or 4.33 psi.

This relationship can be expressed in the equation

$$p = 0.433 \times h \qquad \text{(R3-2)}$$

where

p = pressure, psi
h = head, feet of water

The tank in Fig. R3-1 has an area of 1 ft² with 1 ft of head; therefore, the pressure on the bottom of the tank is 0.433 psi. If there is a fish pond covering an area of 50 ft², and the depth of water in it is 1 ft, pressure on the bottom of the pond will still be just 0.433 psi, even though there is a larger volume of water. This points up the relationship between pressure and depth and demonstrates that there is not necessarily a relationship between pressure and volume.

Fluids and Pressure   21

With this relationship between pressure and depth established, we can transpose the equation so that the depth of water in a tank can be found if we know the pressure reading at the bottom of the tank.

If $p = 0.433 \times h$, then $h = p/0.433$ is true, by transposition.

EXAMPLE

If a pressure gauge located at the bottom of a 50-ft-high water tower showed a reading of 13 psi, (a) what is the depth of the water in the tank, and (b) what would the gauge indicate if the tower were filled with water?

SOLUTION

(a) $h = \dfrac{p}{0.433} = \dfrac{13}{0.433} = 30$ ft of water

(b) $p = 0.433h = 0.433 \times 50 = 21.65$ psi

R3-4
PASCAL'S PRINCIPLE

In the middle of the seventeenth century, a French mathematician and scientist named Blaise Pascal was experimenting with water and air pressure. His scientific experiments led to the formulation of what is known as Pascal's principle, namely, that pressures applied to a confined liquid are transmitted equally throughout the liquid, irrespective of the area over which the pressure is applied. The application of this principle enabled Pascal to invent the hydraulic press, which is capable of a large multiplication of force. In Fig. R3-3, the unit pressure is equal in all vessels regardless of their shape, for pressure is independent of the shape of the container.

Figure R3-4, illustrating this principle, shows a vessel containing a fluid such as oil; the vessel has a small and a large cylinder connected by a pipe or tubing, with tight-fitting pistons in each cylinder. If the cross-sectional area of the small piston is 1 in.² and the area of the large one is 30 in.², a force of 1 lb when applied to

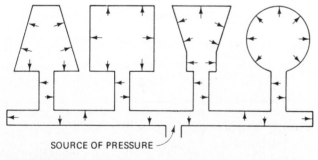

FIGURE R3-3 Pascal's principle of pressure equalization in various-shaped containers.

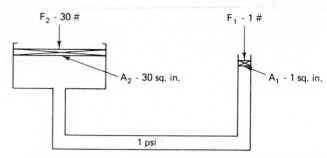

FIGURE R3-4 Pascal's principle of pressure transmission by hydraulic action.

the smaller piston will support a weight of 30 lb on the larger piston, because a pressure of 1 psi throughout the fluid will be exerted.

Since we found earlier that pressure equals force divided by area, a force of 1 lb applied to an area of 1 in.² will create a pressure of 1 psi. By transposition, force equals pressure multiplied by the area; therefore, a weight of 30 lb will be supported when a pressure of 1 psi is applied over an area of 30 in.²

R3-5
DENSITY

From a scientific or physics viewpoint, *density* is the weight per unit volume of a substance, and it may be expressed in any convenient combination of units of weight and volume used, such as *pounds per cubic inch* or *pounds per cubic foot*. An equation can be formulated which expresses this relationship:

$$D = \frac{W}{V} \qquad \text{(R3-3)}$$

where

D = density

W = weight

V = volume

As mentioned previously, the weight, or density, of water is approximately 62.4 lb/ft³, and it can be expressed as 0.0361 lb/in.³ (1 ft³ contains 1728 in.³, and $62.4 \div 1728 = 0.0361$). The density of some other common substances are listed in Fig. R3-5.

The *specific gravity* of any substance is the ratio of the weight of a given volume of the substance to the weight of the same volume of a given substance (where solids or liquids are concerned, water is used as a basis for specific gravity calculations that may have to be made, and air or hydrogen is used as a standard for gases).

density (solid or liquid)

= specific gravity × density of water (lb/ft³)

The specific gravity of water is considered as 1.0 and values for other substances are listed in Fig. R3-5.

Pressure within a fluid is directly proportional to the density of the fluid. Consider the tank in Fig. R3-1, which, if filled with water weighing 62.4 lb, has a force on the bottom of the tank of 62.4 lb and a unit pressure of 0.433 psi. If this tank were filled instead with gasoline, which has a specific gravity of 0.66, the force on the bottom would be only 66% as great as when water is in the tank, and the pressure of the gasoline would be only 66% as great. Therefore, the relationship can be expressed as

$$\text{pressure} = \text{head} \times \text{density}$$
$$p = h \times D \qquad \text{(R3-4)}$$

where

p = pressure, lb/ft²

h = head or depth below the surface, ft

D = density, lb/ft³

or where

p = pressure, psi

h = head or depth below the surface, in.

D = density, lb/in.³

FIGURE R3-5
Density and specific gravity of some common substances.

Substance	Density (lb/ft³)	Specific Gravity
Water (pure)	62.4	1
Aluminum	168	2.7
Ammonia (liquid, 60°F)	38.5	0.62
Brass	530	8.5
Brick (common)	112	1.8
Copper	560	8.98
Cork (average board)	15	0.24
Gasoline	41.2	0.66
Glass (average)	175	2.8
Iron (cast)	448	7.2
Lead	705	11.3
Mercury	848	13.6
Oil (fuel) (average)	48.6	0.78
Steel (average)	486	7.8
Woods		
Oak	50	0.8
Pine	34.2	0.55

Care must be taken whenever this equation is used to make sure that the proper units of weight and measurement are placed in the equation.

R3-6
SPECIFIC VOLUME

The specific volume of a substance is usually expressed as the number of cubic feet occupied by 1 lb of the substance. In the case of liquids, it will vary with temperature and pressure. The volume of a liquid will be affected by a change in its temperature; but since it is practically impossible to compress liquids, the volume is not affected by a change in pressure.

The volume of a gas or vapor is definitely affected by any change in either its temperature or the pressure to which it is subjected. In refrigeration, the volume of the vapor under the varying conditions involved is most important in the selection of the proper refrigerant lines.

The appropriate specific volumes and pressures for refrigerants will be covered later in Chapter R7, but as an example of the effect temperature has on a refrigerant vapor, refer to Fig. R3-6 for the properties of Refrigerant 12 (R-12).

Note that at +5°F the specific *volume of vapor* is 1.46 ft³/lb, whereas at 86°F it is only 0.38 ft³/lb. Correspondingly, there is an increase in pressure (psia) from 26.48 psia to 108.04 psia. Note also the change in density from 90.14 lb/ft³ to 80.67 lb/ft³.

Under the pressure columns there are two sets of data shown: psia for absolute pressure and psig for gauge pressure. The following discussion explains the difference between them.

R3-7
ATMOSPHERIC PRESSURE

The earth is surrounded by a blanket of air called the atmosphere, which extends 50 or more miles upward from the surface of the earth. Air has weight and also exerts a pressure known as *atmospheric pressure*. It has been computed that a column of air with a cross-sectional area of one square inch and extending from the earth's surface at sea level to the limits of the atmosphere, would weigh approximately 14.7 lb. As was pointed out earlier in the chapter, force means also the weight of a substance, and pressure means unit force per square inch; therefore, standard atmospheric pressure is considered to be 14.7 psi at sea level.

This pressure is not constant; it will vary with altitude or elevation above sea level, and there will be variations due to changes in temperature as well as water vapor content of the air. This atmospheric pressure can be demonstrated by construction of a simple barometer,

Properties of liquid and saturated vapor of R-12. Note pressures corresponding to standard evaporating temperature of 5°F and condensing temperature of 86°F.

Temp °F	PRESSURE		VOLUME (VAPOR)	DENSITY (LIQUID)	HEAT CONTENT (BTU/LB)	
	psia	psig	(ft/lb)	(lb/ft³)	Liquid	Vapor
−150	0.154	29.61ᵃ	178.65	104.36	−22.70	60.8
−125	0.516	28.67ᵃ	57.28	102.29	−17.59	83.5
−100	1.428	27.01ᵃ	22.16	100.15	−12.47	66.2
−75	3.388	23.02ᵃ	9.92	97.93	−7.31	69.0
−50	7.117	15.43ᵃ	4.97	95.62	−2.10	71.8
−25	13.556	2.32ᵃ	2.73	93.20	3.17	74.56
−15	17.141	2.45	2.19	92.20	5.30	75.65
−10	19.189	4.49	1.97	91.70	6.37	76.2
−5	21.422	6.73	1.78	91.18	7.44	76.73
0	23.849	9.15	1.61	90.66	8.52	77.27
5	26.483	11.79	1.46	90.14	9.60	77.80
10	29.335	14.64	1.32	89.61	10.68	78.335
25	39.310	24.61	1.00	87.98	13.96	79.9
50	61.394	46.70	0.66	85.14	19.50	82.43
75	91.682	76.99	0.44	82.09	25.20	84.82
86	108.04	93.34	0.38	80.67	27.77	85.82
100	131.86	117.16	0.31	78.79	31.10	87.63
125	183.76	169.06	0.22	75.15	37.28	88.97
150	249.31	234.61	0.16	71.04	43.85	90.53
175	330.64	315.94	0.11	66.20	51.03	91.48
200	430.09	415.39	0.08	60.03	59.20	91.28

ᵃInches of mercury below 1 atm.

Source: The DuPont Company

as shown in Fig. R3-7, using a glass tube about 36 in. long and closed at one end, an open dish or bowl, and a supply of mercury. Fill the tube with mercury and invert it in the bowl of mercury, holding a finger at the open end of the tube so that the mercury will not spill out while the tube is being inverted. Upon removal of the finger, the level of the mercury in the tube will drop somewhat, leaving a vacuum at the closed end of the tube. The atmospheric pressure bearing down on the open dish or bowl of mercury will force the mercury in the tube to stand up to a height determined by the pressure being exerted on the open surface of the mercury, which will be approximately 30 in. at sea level.

As previously shown in Fig R3-5, the specific gravity of mercury is 13.6—that is, mercury weighs and exerts pressure 13.6 times that of an equal volume of water. Since a 1-in.-high column of water exerts pressure of 0.0361 psi, a similar column of mercury will exert pressure that is 0.0361 × 13.6 = 0.491 psi. A 30-in. column of mercury will exert pressure of 0.491 × 30 = 14.7 psi, or an amount equal to atmospheric pressure. Conversely, atmospheric pressure at sea level (14.7) bearing down on the open dish of mercury divided by the unit pressure that is exerted by a 1-in. column of mercury (0.491) should cause the column of

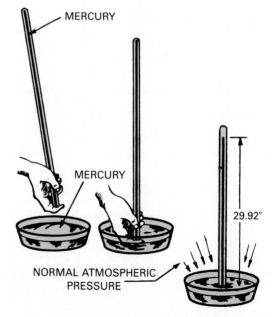

MERCURY

MERCURY

NORMAL ATMOSPHERIC PRESSURE

29.92″

COLUMN OF MERCURY SUPPORTED BY NORMAL ATMOSPHERIC PRESSURE

FIGURE R3-7   Simple barometer.

mercury in the tube to stand approximately at a height of 30 in. (actually, 29.92 in.).

$$\frac{14.7}{0.491} = 30\text{-in. column of mercury (Hg)}$$

Of course, water could possibly be used in this barometer instead of mercury; but the tube would have to be about 34 ft high, and this is not too practical.

$$\frac{14.7}{0.036} = 407\text{ in. of water (33.9 ft)}$$

R3-8
MEASUREMENT OF PRESSURE

A *manometer* is one type of device utilized in the refrigeration and air-conditioning field for the measurement of pressure. This type of pressure gauge utilizes a liquid, usually mercury, water, or gauge oil, as an indicator of the amount of pressure involved. The water manometer or water gauge is customarily used when measuring air pressures, because of the lightness of the fluid being measured.

A simple open-arm manometer is shown in Fig. R3-8. The U-shaped glass tube is partially filled with water, as shown in Fig. 3-8a, and is open at both ends. The water is at the same level in both arms of the ma-

nometer; because both arms are open to the atmosphere, and there is no external pressure being exerted on them.

Figure R3-8b shows the manometer in use with one arm connected to a source of positive air pressure that is being measured. The water is at different levels in the arms, and the difference denotes the amount of pressure being applied.

A space that is void, or lacking any pressure, is described as having a *perfect vacuum*. If the space has pressure less than atmospheric pressure, it is defined as being a *partial vacuum*. It is customary to express this partial vacuum in *inches of mercury*, and not as negative pressure. In some instances it is also referred to as a given amount of absolute pressure, expressed in psia, and this will be covered a little later in this chapter.

If a partial vacuum has been drawn on the left arm of the manometer by means of a vacuum pump, as shown in Fig. R3-9, the mercury in the right arm will be lower, and the difference in levels will designate the partial vacuum in inches of mercury.

Pressure gauges most commonly used in the field by service technicians, to determine what is going on within the refrigeration system, are of the Bourdon tube type. As is shown in Fig. R3-10, an internal view, the essential element of this type of gauge is the Bourdon tube. This oval metal tube is curved along its length and forms an almost complete circle. One end of the tube is closed, and the other end is connected to the equipment or component being tested.

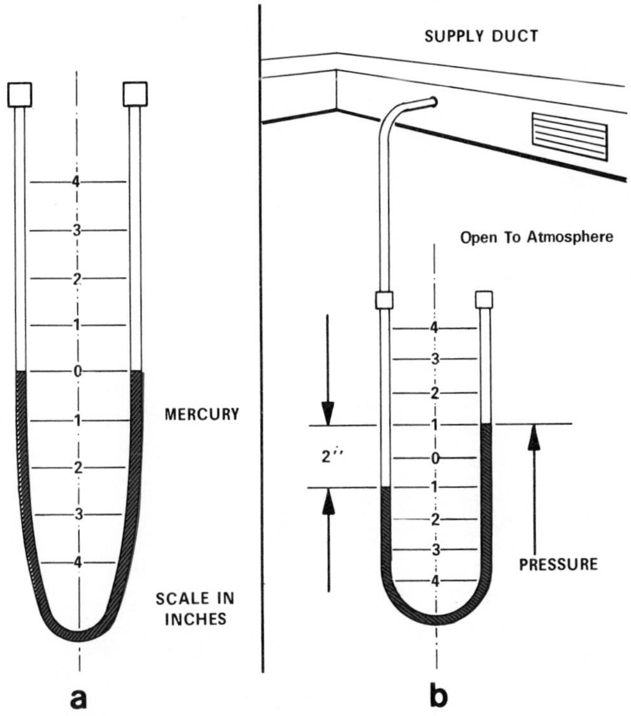

FIGURE R3-8   Water fill manometer used to measure air pressure.

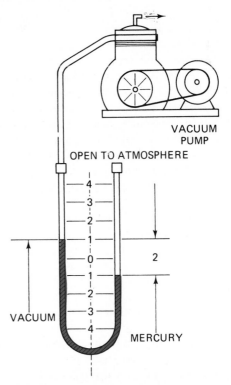

FIGURE R3-9   Mercury fill manometer measuring "vacuum" pressure.

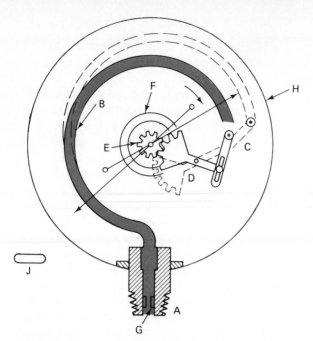

FIGURE R3-10   Internal construction of a pressure gauge: (a) adapter fitting, usually an $\frac{1}{8}$-in. pipe thread; (b) Bourdon tube; (c) link; (d) gear sector; (e) Pinter shaft gear; (f) calibrating spring; (g) restricter; (h) case; (j) cross section of the Bourdon tube. The dashed lines indicate how the pressure in the Bourdon tube causes it to straighten and operate the gauge.

As shown in Fig. R3-11, the gauges are preset at 0 lb, which represents the atmospheric pressure of 14.7 psi. Therefore, any additional pressure applied when the gauge is connected to a piece of equipment will tend to straighten out the Bourdon tube, thereby moving the needle or pointer and its mechanical linkage, thus indicating the amount of pressure being applied. Figure R3-11a is an example of a pressure gauge to indicate the amount of pressures above that of the atmosphere. Figure R3-11b is a compound pressure gauge, which has a dual function: that of registering a pressure above atmospheric pressure, and registering one that may be below that of the atmosphere.

Pressures below atmospheric are customarily expressed in inches of mercury. There is an indication of the range between 0 gauge and 30 in. of mercury (Hg) on the compound gauge.

R3-9
ABSOLUTE PRESSURE

Figure R3-12 shows a definite relationship among absolute, atmospheric, and gauge pressures. For many problems, atmospheric pressure does not need to be considered, so the customary pressure gauge is calibrated and graduated to read zero under normal at-

FIGURE R3-11   Refrigeration gauges with temperature scales. (*Courtesy Marsh Refrigeration*)

mospheric conditions. Yet when gases are contained within an enclosure away from the atmosphere, such as in a refrigeration unit, it is necessary to take atmospheric pressure into consideration, and mathematical calculations must be in terms of the absolute pressures involved.

FIGURE R3-12

Relationship between absolute, atmospheric, and gauge pressure.

Gauge Pressure	Absolute Pressure
40	54.7
Pressure	
30	44.7
above	
20	34.7
atmospheric in psi	
10	24.7
atmospheric	
0	14.7
pressure	(30)
10	20
Pressure below atmospheric	
20	10
in in. Hg	
30	0
(29.92)	

R3-10
PRESSURE OF GAS

The volume of a gas is affected by a change in either the pressure or temperature, or both. There are laws that govern the mathematical calculations in computing these variables.

Boyle's law states that the volume of a gas varies inversely as its pressure if the temperature of the gas remains constant. This means that the product of the pressure times the volume remains constant, or that if the pressure of a gas doubles the new volume will be one-half of the original volume. Or it may be considered that, if the volume is doubled, the absolute pressure will be reduced to one-half of what it was originally.

This concept may be expressed as

$$p_1 V_1 = p_2 V_2 \qquad \text{(R3-5)}$$

where

p_1 = original pressure

V_1 = original volume

p_2 = new pressure

V_2 = new volume

It must be remembered that p_1 and p_2 have to be expressed in the *absolute pressure* terms for Eq. (R3-5) to be used correctly.

EXAMPLE

If the gauge pressure on 2 ft³ of gas is increased from 20 psig to 50 psig while the temperature of the vapor remains constant, what will be the new volume?

SOLUTION

p_1 = 20 + 14.7   Since

  = 34.7 psia $p_1 V_1 = p_2 V_2$

p_2 = 50 + 14.7   then

  = 64.7 psia $V_2 = \dfrac{p_1 V_1}{p_2}$

V_1 = 2 ft³   Therefore,

V_2 = ? $V_2 = \dfrac{34.7 \times 2}{64.7} = 1.072 \text{ ft}_3$

The same basic equation may be used to determine the given new pressure.

EXAMPLE

If additional pressure is applied to a volume of 2 ft³ of gas at 20 psig so that the volume is lessened to 1.072 ft³ and the temperature of the gas remains constant, what is the new pressure in psig?

p_1 = 34.7 psia $p_2 = \dfrac{p_1 V_1}{V_2}$
  (20 + 14.7)

V_1 = 2 ft³ $p_2 = \dfrac{34.7 \times 2}{1.072} = 64.7$ psia

V_2 = 1.072 ft³ $p_2 = 64.7 - 14.7 = 50$ psig

R3-11
EXPANSION OF GAS

Most gases will expand in volume at practically the same rate with an increase in temperature, provided that the pressure does not change. If the gas is confined so that its volume will remain the same, the pressure in the container will increase at about the same rate as an increase in temperature.

Theoretically, if the pressure remains constant, a gas vapor will expand or contract at the rate of $\frac{1}{492}$ for each degree of temperature change. The result of this theory would be a zero volume at a temperature of $-460°F$, or at $0°$ absolute.

Charles' law states that the volume of a gas is in direct proportion to its absolute temperature, provided that the pressure is kept constant; and the absolute pressure of a gas is in direct proportion to its absolute temperature, provided that the volume is kept constant. That is,

$$\frac{V_1}{V_2} = \frac{T_1}{T_2} \qquad \text{(R3-6)}$$

and

$$\frac{P_1}{P_2} = \frac{T_1}{T_2} \qquad \begin{array}{l} T = \text{absolute temperature} \\ P = \text{absolute pressure} \end{array} \text{(R3-7)}$$

To clear the fractions, these may also be expressed as

$$V_1 T_2 = V_2 T_1 \quad \text{and} \quad p_1 T_2 = p_1 T_2 = p_2 T_1$$

EXAMPLE

If the temperature of 2 ft³ of gas is increased from 40°F to 120°F, what would be the new volume if there was no change in pressure?

SOLUTION

$$V_2 = \frac{V_1 T_2}{T_1} = \frac{2 \times (120 + 460)}{40 + 460}$$

$$= \frac{1160}{500} = 2.32 \text{ ft}^3$$

EXAMPLE

If a container holds 2 ft³ of gas at 20 psig, what will be the new pressure in psig if the temperature is increased from 40°F to 120°F?

SOLUTION

$$p_2 = p_1 T_2 = \frac{(20 + 14.7) \times (120 + 460)}{40 + 460}$$

$$= 40.25 \text{ psig}$$

$$40.25 - 14.7 = 25.55 \text{ psig}$$

In numerous cases dealing with refrigerant vapor, none of the three possible variables remains constant, and a combination of these laws must be utilized, namely the *general law of perfect gas:*

$$\frac{p_1 V_1}{T_2} = \frac{p_2 V_2}{T_2} \quad \text{or} \quad p_1 V_1 T_2 = p_2 V_2 T_1 \qquad \text{(R3-8)}$$

in which the units of p and T are always used in the absolute.

EXAMPLE

If a volume of 4 ft³ of gas at a temperature of 70°F and at atmospheric pressure is compressed to one-half its original volume and increased in temperature to 120°F, what will be its new pressure?

SOLUTION

Transposing Eq. (R3-8), we have

$$p_2 = \frac{p_1 V_1 T_2}{V_2 T_1}$$

Therefore,

$$p_2 = \frac{14.7 \times 4(120 + 460)}{2 \times (70 + 460)} = \frac{34{,}104}{1{,}060} = 32.17 \text{ psia}$$

$$32.17 - 14.7 = 17.47 \text{ psig}$$

R3-12
BOILING POINT

The most important point to understand when dealing with the action in a refrigeration system is the "boiling point" of the liquid (refrigerant) in the system. It is lowering the boiling point that causes the refrigerant to absorb heat and vaporize or "boil," and conversely by raising the "boiling point" the vapor gives up the heat and condenses. Basically, the refrigeration system operates by "control of the boiling point."

In Chapter 2, boiling point was mentioned and defined as the temperature at which a liquid turns from a liquid to a vapor or condenses from a vapor to a liquid depending on the absorption or rejection of heat energy.

The chart used was based on using water at standard atmospheric pressure of 29.92 in. of mercury and 70°F. At these conditions water will boil at 212°F or 100°C with the addition of heat energy or condense at this same temperature with the removal of heat energy.

When referring to "boiling point," the pressure that the liquid is subjected to must also be considered. When referring to the boiling point of water as 212°F or 100°C, the assumption is made that the water is subjected to standard barometric pressure of 29.92 in. of mercury.

In reality, the boiling point of a liquid will change in the same direction as the pressure to which the liquid is subjected. This is a very important basic law of physics that must be remembered. In a later chapter this law will be applied to various refrigerants, but for discussion in this chapter water will be used as the refrigerant. Figure R3-13 gives examples of the boiling point of water at sea level (212°F) and at 14,100 ft above sea level (167°F). Obviously, this is because the boiling point of water drops as the atmospheric pressure drops. Again, the boiling point of a liquid will vary in the same direction as the pressure to which the liquid is subjected.

If the boiling point of water is determined at various pressures, both above (pressure) and below atmospheric pressure (vacuum) and the temperatures are plotted on a graph, the results would be as pictured in Fig. R3-14. Here we can see that the boiling point of water can be raised to 276°F with a pressure of 30 lb

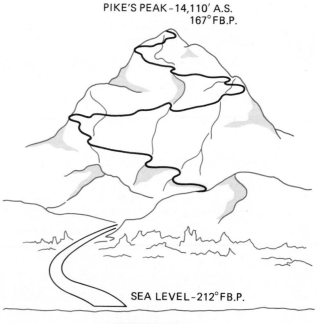

PIKE'S PEAK - 14,110' A.S.
167° F.B.P.

SEA LEVEL - 212° F.B.P.

FIGURE R3-13

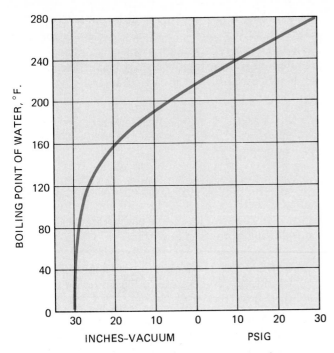

FIGURE R3-14 Pressure–temperature curve for water.

psig or lowered to 40°F at a pressure of 29.67233 in. Hg. Therefore, to obtain a desired boiling point of water it is only necessary to maintain an equivalent pressure based on the pressure–temperature curve of the liquid.

When talking boiling point, the automatic implication is that heat energy is added to the liquid trying to raise the sensible temperature above the temperature of the liquid; thus boiling takes place. For all practical purposes, at no time is it possible to lower the boiling point of a liquid below its sensible temperature without vaporization (boiling) taking place or raise the sensible temperature above the boiling point without vaporization (boiling) taking place. To remain as a liquid, the boiling point must always be higher than the sensible temperature.

R3-13
CONDENSING TEMPERATURE

With the liquid in a vapor state, to remain as a vapor, the sensible temperature must be higher than the condensing temperature. If the heat energy is removed from the vapor to the point where the sensible temperature is attempting to fall below the condensing temperature of the vapor, the vapor will liquefy or condense.

The boiling point and the condensing temperature for liquid are the same. Only a difference in the action taking place is implied. Boiling point—liquid to vapor; condensing temperature—vapor to liquid. In each case, the temperature at which the change takes place varies as the pressure to which the liquid is subjected. Low-

ering the pressure lowers the boiling point or condensing temperature. Raising the pressure raises the boiling point or condensing temperature.

R3-14
FUSION POINT

The fusion point (temperature at which solidification of liquid or melting of a solid takes place) is also affected by the pressure to which the solid is subjected. If the fusion point of a solid is raised by increasing the pressure on the solid until the fusion point is equal to the sensible temperature of the solid, any further attempt to raise the fusion point will cause the solid to liquefy.

This principle is what allows glaciers to move downhill. When the weight of the snow and ice in the glacier is high enough to raise the fusion or melting point of the base of the ice up to the sensible temperature of the ice, any further rise in pressure will cause the base of the ice to liquefy and the glacier moves on a layer of water.

Control of the fusion point is of little importance in refrigeration, as it is usually an undesirable result of temperature maintenance. In air-conditioning, formation of ice is a definite detriment and must be eliminated.

R3-15
SATURATION TEMPERATURE

In Section R3-12 the statement was made that the boiling point and condensing temperature of a liquid at a given pressure are the same. This means that the liquid has reached the point where it contains all the heat energy it can without changing to a vapor. This condition is described by referring to it as a "saturated liquid." This means that if any more heat energy is added, the liquid will boil. Commonly, if the vapor is cooled to a point where the vapor is so dense that any further reduction in heat energy causes it to condense to a liquid, the condition is referred to as a "saturated vapor."

In sections R2-9 and R2-10, where sensible heat and latent heat were discussed, sensible heat changes temperature and latent heat changes state. Therefore, liquid at the boiling point is saturated with sensible heat and any heat added would be latent heat to vaporize the liquid.

Vapor at the condensing temperature has been reduced to that temperature by removing the sensible heat until the density of the vapor is to a point where any further removal of heat will cause condensation of the vapor and removal of the latent heat of vaporization. At this point the vapor is said to be a "saturated vapor."

R3-16
SUPER HEAT

Superheat is "the heat added to a vapor after it becomes a vapor," a simple rise in temperature of the vapor above the boiling point. If, for example, water were to boil at 212°F and before leaving the passages in the boiler it were to take on more heat and the steam temperature rises to 220°F, the steam would be superheated 8°F, the difference between the boiling point of the liquid and the actual physical (sensible) temperature of the vapor. This is called superheated steam. To remove the heat and condense the steam to a liquid, the first action required is to de-superheat the steam to the saturation point (condensing temperature) and then remove the latent heat of vaporization to produce the liquid.

Superheat is very important in refrigeration systems to produce the maximum system capacity together with heat equipment life and will be referred to throughout the various chapters.

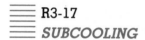
R3-17
SUBCOOLING

When a liquid is at a sensible temperature below its boiling point, it is said to be "subcooled." For example, water at standard atmospheric conditions with a

sensible temperature of 70°F will be subcooled 142°F (212°F − 70°F = 142°F subcooled). The subcooling of the liquid in the refrigeration or air-conditioning system is important for maximum capacity and efficiency. This will be discussed further in later chapters.

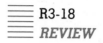
R3-18
REVIEW

To tie all these terms together, Fig. R2-9 has been reinserted here as Fig. R3-15 with the various terms added. Starting with ice at 0°F and adding heat energy, the following changes take place:

1. Sensible heat is added to the subcooled ice raising its sensible temperature to the saturated temperature of 32°F or 0°C.
2. Upon reaching the saturation temperature, additional heat is latent heat that changes the saturated ice to saturated liquid.
3. With all the solid (ice) now liquid (water), further heat energy addition is sensible heat that raises the temperature of the liquid (water) to the boiling point, where it is now a saturated liquid.
4. Additional heat (latent heat) changes the liquid from a saturated liquid to a saturated vapor.
5. After the liquid is completely vaporized, addi-

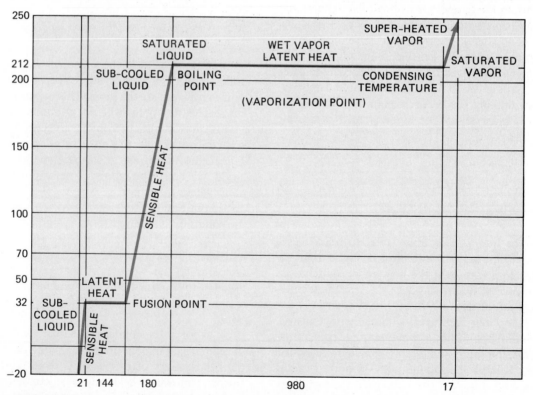

FIGURE R3-15 Chart demonstrating all terms concerning changes in physical conditions and temperature of solid, liquid, and vapor of water.

tional heat added will raise the temperature of the vapor (sensible heat) and superheat the vapor.

Starting with superheated steam and removing heat energy will produce the following changes:

1. Removing sensible heat from the vapor will lower the temperature and desuperheat the vapor to the saturated vapor point, referred to as the condensing temperature.
2. Further removed latent heat will reduce the saturated vapor to a saturated liquid still at the same temperature. Condensing temperature (saturated vapor containing latent heat) and boiling point (saturated liquid that does not contain the latent heat) are the same temperature.

3. Further removal of sensible heat from the liquid will lower the temperature of the liquid (subcool the liquid below the boiling point) until the temperature of the liquid reaches the fusion point.
4. Although never discussed, this liquid at the fusion point could also be classified as a saturated liquid based on the liquid-to-solid reaction.
5. Removal of latent heat from the liquid changes the liquid to a solid (latent heat of fusion).
6. Further removal of sensible heat will reduce the temperature or subcool the solid (ice) below the melting point.

All these terms must be understood before proceeding into discussions of refrigeration systems.

PROBLEMS

R3-1. Define "fluid pressure."

R3-2. What term is usually used to express fluid pressure?

R3-3. Define the term "force" when discussing fluid pressure.

R3-4. Find the total force and also the unit pressure exerted on the bottom of a tank filled with water if the tank measures 3 ft × 3 ft × 1 ft high.

R3-5. Define the term "head."

R3-6. If a tank with a flat base is filled with water to a level of 8 ft, what pressure is exerted on the bottom of the tank?

R3-7. What pressure is exerted halfway between the bottom of the tank and the surface of the water?

R3-8. Find the unit pressure exerted on the roof of a building on which is located a cooling tower. The tower weighs 1580 lb when filled with water and operating. The size of the base of the tower is 3 ft × 4 ft.

R3-9. In a hydraulic press, what force must be exerted on the small piston having an area of 2 in.² if a 600-lb weight must be supported on a larger piston which has an area of 16 in.²?

R3-10. What would be the pressure of the liquid in the press in Problem R3-9?

R3-11. Define "density."

R3-12. The density of water is _____.

R3-13. Define "specific gravity."

R3-14. Define "specific volume."

R3-15. What factor affects the specific volume of a liquid?

R3-16. What factors affect the specific volume of a vapor?

R3-17. Define "Boyle's law."

R3-18. If the volume of 10 ft³ of gas at a pressure of 25 psi is to be compressed to 2 ft³ with no change in temperature, what will be the new pressure in psi?

R3-19. If the volume of 2 ft³ of gas at a pressure of 183.8 psi is to be compressed to 1 ft³ with no change in temperature, what will be the new pressure in psi?

R3-20. Define "Charles' law."

R3-21. With the pressure remaining constant, find the new volume of 4 ft³ of gas when its temperature is increased from 60°F to 250°F.

R3-22. Give the formula for the general law of perfect gas.

R3-23. If a volume of 10 ft³ of gas at a temperature of 60°F at a pressure of 57.7 psig were compressed to 5 ft³ and the temperature raised to 127 psig, what would be its new pressure?

R3-24. What would be the new volume if 20 ft³ of gas at 70°F and 70.2 psig were raised to 115°F and 146.8 psig?

R3-25. Define "superheat."

R3-26. Define "subcooling."

R3-27. What is the difference between a saturated liquid and a saturated vapor?

Refrigeration Technician's Hand Tools and Accessories

R4-1
TYPICAL HAND TOOLS

The development of the technician's toolkit will, of course, depend on the scope of his or her needs. These will vary somewhat, depending on whether the technician is involved in installation or service repair, or both, and the kind of work: refrigeration, air-conditioning, heating, or all three. Regardless of job requirements, the careful selection, care, and knowledge of the use of tools are important considerations for the technician. Poor workmanship or injury can frequently be traced to a lack, or improper use, of hand tools.

The term *toolkit* implies a toolbox of some sort, and of course this would be true for small hand tools; in reality the technician will most likely be working out of a truck or van that provides storage for hand tools, power tools, soldering equipment, piping materials, refrigerant cylinders, etc.

The following discussion separates common hand tools from measuring or testing equipment such as thermometers, pressure gauges, electric meters, leak detectors, and the like, which will be covered later. It is also assumed that special equipment such as the tube cutters, benders, flaring tools, and brazing equipment previously presented will be available as needed.

A partial list of common hand tools needed by a refrigeration technician follows. Some do not need explanation; others require reference to their scope or use.

- Wrenches—of several sorts (as reviewed below)
- Pliers—of several sorts (as reviewed below)
- Spirit level
- Tin snips
- Screwdrivers—of several sorts (as reviewed below)
- Hammers—ball peen and common claw

- Mallets—with nonmetallic heads (plastic, wood, rubber)
- Hacksaw—blades with 14, 18, and 32 teeth per inch
- Brushes—of several sorts (as reviewed below)
- Files—of several sorts (as reviewed below)
- Tape measures and hand rule—(as reviewed below)
- Micrometer and calipers—(as reviewed below)
- Punches—for marking drill point
- Chisel—$\frac{3}{4}$-in. flat cold chisel
- Drills—of several sorts (as reviewed later)
- Vises—machinist and pipe vise
- Pocket knife
- Flashlight
- 50-ft electrical extension cord
- Stopwatch

R4-2
WRENCHES

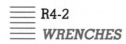

In the list above the wrenches are probably the most frequently used tool of all, so we will cover these in some detail. Several types of wrenches are listed, but not necessarily in general order of use or preference (see Figs. R4-1 through R4-10.

- Ratchet wrench
- Socket wrenches
- Box wrenches
- Open-end wrenches
- Adjustable wrenches
- Allen wrenches
- Nut driver

32

R4-2.1
Ratchet Wrench

The steel ratchet wrench (Fig. R4-1) is especially adapted for use on small refrigerant cylinders and shut-off valves. The ratchet permits rapid changing of direction so the operator can switch movement to open or close a valve, etc. The wrench handle openings vary from $\frac{1}{4}$ to $\frac{1}{16}$ in. Some also have a $\frac{1}{2}$-in. hex socket cast into one end.

R4-2.2
Socket Wrenches

Socket wrenches (Fig. R4-2) are used to slip over bolt heads. They are made of steel, and the array of sockets vary from square to 6-point hexagonal shape or up to 12-point double hexagonal shape. Common sizes run from $\frac{5}{32}$ to 2 in. in English measurement. Metric sizes are also readily available. Swivel sockets are particularly useful to reach hard-to-get-to nuts and bolts. The more points a socket has the easier it is to use in a restricted area.

The handle of a socket wrench may vary from a straight fixed tee drive to a ratchet operation to a special torque wrench handle that includes a gauge to measure the force being applied. In certain operations the manufacturer may specify torque requirements or limitations. The force or torque is measured in foot-pounds or inch-pounds.

R4-2.3
Box Wrenches

Box wrenches (Fig. R4-3) are useful in certain close-quarter situations. The ends are usually in the form of a 12-point double hexagonal shape, as illustrated. Ends may be of the same size or different sizes. The handle may be straight or offset. Common sizes range from $\frac{1}{4}$ to $1\frac{1}{2}$ in.

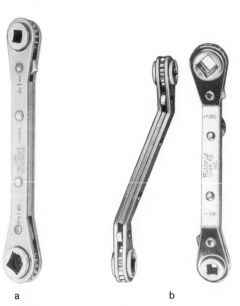

a   b

FIGURE R4-1   Ratchet wrenches. (*Courtesy* Ritchie Engineering Company, Inc.)

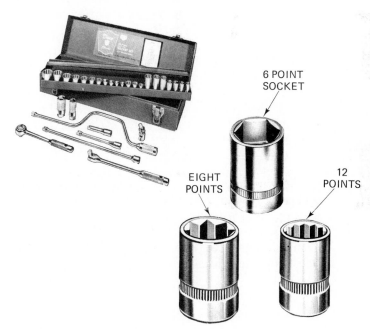

FIGURE R4-2   Socket wrenches. (*Courtesy* Duro Metal Products, Inc.)

FIGURE R4-3   Box wrenches. (*Courtesy* Duro Metal Products, Inc.)

The flare-nut wrench (Fig. R4-4) is a special variation of the box wrench in that the heads are slotted to allow the wrench to slip over the tubing and then onto a flare nut. After tightening, the wrench is removed in a reverse manner. The wrench heads are made to fit the standard SAE flare nuts. Several sizes are needed.

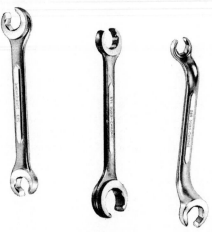

FIGURE R4-4 Flare-nut wrenches. (*Courtesy* Duro Metal Products, Inc.)

Open-end wrenches (Fig. R4-5) are needed where it is impossible to fit a socket or box wrench on a nut, bolt, or fitting from the top. The open-end wrench permits access to the object from the side. Of necessity the wrench has only two flats. The distance between these flats determines the size, which normally ranges from $\frac{1}{4}$ to $1\frac{5}{8}$ in.

FIGURE R4-5 Open-end wrench. (*Courtesy* Duro Metal Products, Inc.)

Some service mechanics prefer the combination box and open-end wrench (Fig. R4-6) to gain the advantage of each type. The ends are made to fit the same size nut or bolt so the technician can use the best selection.

FIGURE R4-6 Combination box and open-end wrench. (*Courtesy* Duro Metal Products, Inc.)

The familiar adjustable wrench as illustrated in Fig. R4-7 is useful where a regular open-end wrench could be used, but the adjustable screw permits fitting the flat to any size object within the maximum and minimum opening. The handle size indicates the general capacity. For example, a 4-in. size will take up to a $\frac{1}{2}$-in. nut. A 16-in. handle will take up to a $1\frac{7}{8}$-in. nut. Always use this type of wrench in a manner such that the force is in a down or clockwise direction when tightening a bolt; this keeps the force against the head. If the wrench were used in an opposite manner, it might suddenly loosen and injure the operator.

FIGURE R4-7 Adjustable wrench. (*Courtesy* Duro Metal Products, Inc.)

R4-2.8
Pipe Wrenches

The pipe wrench (Fig. R4-8) is a common tool used in refrigeration installation and service work to assemble or disassemble threaded pipe. It is tough and can withstand a great deal of punishment. At least two sizes are recommended—an 8-in. size, which can handle up to 3-in.-diameter pipe and a 14-in. size for up to 8-in.-diameter pipe.

Another form of adjustable pipe wrench is called the *chain wrench*. This wrench can make work easier in a confined area or on round, square, or irregular shapes.

FIGURE R4-8   Pipe wrenches. (*Courtesy* Ridge Tool Company)

R4-2.9
Allen Wrenches

Allen wrenches (Fig. R4-9) are necessary for removing or adjusting fan pulleys, fan blade hubs, and other components that are held in place or adjusted by allen set screws. The wrenches are tough alloy steel with 6-point flat faces. The wrench goes inside the set screw and may be used at either end. Sizes range from about $\frac{1}{16}$ to about $\frac{1}{2}$ in.

FIGURE R4-9   Allen wrenches. (*Courtesy* Duro Metal Products, Inc.)

R4-3
NUT DRIVERS

Nut drivers (Fig. R4-10) are actually a type of small socket wrench. Many equipment manufacturers have started using metal screws that have both a slot and a nut head; some have eliminated the slot. The nut driver consists of a plastic handle that has different drive sockets to fit over the screw or nut head. The nut driver is useful in tightening or removing sheet-metal screws that hold equipment panels in place or fasten control-box covers; it is also used to secure the control mechanism itself.

FIGURE R4-10   Nut drivers. (*Courtesy* Duro Metal Products, Inc.)

R4-4
PLIERS

Pliers are also one of the most frequently used tools and are available in many types (Fig. R4-11). The familiar slip-joint pliers (Fig. R4-11a) are handy for general use. At least two sizes are recommended. The curved-joint and/or arc-joint pliers (Fig. R4-11b) are necessary for working on larger objects and for holding pipe. Locking or vise grip pliers (Fig. R4-11c) are often needed to clamp objects during soldering, for example, thus freeing the hand of the operator.

In electrical work several different styles of pliers are needed. The diagonal cutting plier (Fig. R4-11d) is used to cut wire (or rope). Needle-nose pliers (Fig. R4-

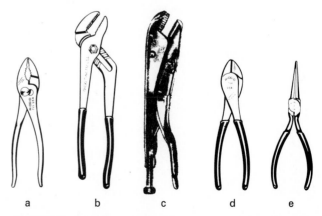

FIGURE R4-11   Pliers: (a) slip-joint; (b) curved joint; (c) vise grip; (d) diagonal cutting; (e) needle nose. (*Courtesy* Ridge Tool Company)

11e) are used to form wire loops and grip tiny pieces firmly; some also have built-in side cutters. Pliers in general are not made to tighten or unscrew heavy bolts and nuts, but are useful tools to start or hold the item until another tool takes over.

R4-5
SCREWDRIVERS

The well-equipped toolkit must have a variety of screwdrivers. The most common is the flat-blade design (Fig. R4-12a), and a complete set is recommended, from $\frac{1}{8}$-in. blades to the large $\frac{5}{16}$-in. size. Handle sizes vary with the blade dimension. Flat blades fit the common single slot screw and it is important to have a tight fit; otherwise, it is possible to strip the screw slot.

Phillips-tip screwdrivers (Fig. R4-12b) are needed where Phillips-head screws are used. These are more common in the electrical phases of refrigeration work. Again a complete set is recommended for accurate fitting to screw heads.

Starting with these two standard items the technician will customize his needs with specialized items like screwdrivers with magnetized blades or blades with a screw grasping clip. Screwdrivers, in addition to their primary purpose, can be used (with discretion) for light-duty prying, wedging, or scraping, *but never pound a screwdriver with a hammer.*

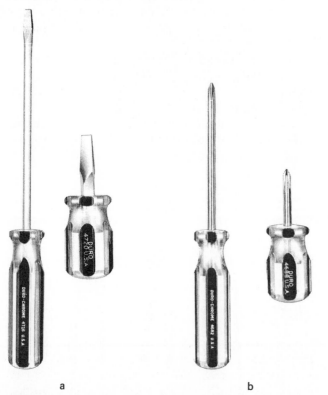

FIGURE R4-12   Screwdrivers: (a) square flat blade; (b) Phillips tip. (*Courtesy* Duro Metal Products, Inc.)

R4-6
BRUSHES

The use of a wire brush (Fig. R4-13) is recommended to clean the inside of tubing and fitting. These brushes come in sizes from $\frac{1}{4}$ to $2\frac{1}{8}$ in. to fit the outside diameter (OD) of the soldered fitting. The brush should have fine steel wire bristles, thickly set and firmly attached to the handle stem.

A solder flux brush is also recommended in applying paste. There is only one size and the brushes are very inexpensive. Frequently, they become too dirty to use and must be discarded. A paintbrush is useful to brush dirt or dust out of a control box, for example. Or it may be used to apply cleaning solvent to an object.

FIGURE R4-13   Brushes.

R4-7
FILES

Files come in several shapes: flat or rectangular, round, half-round, triangular, square, etc. In refrigeration work the common flat file and the half-round are used in preparing tubing for soldering—squaring the end or removing burrs.

The ability of a file to cut metal or other surfaces depends on the tooth size, shape, and number of cuts or directions of cut. Figure R4-14 illustrates single-cut and double-cut direction files.

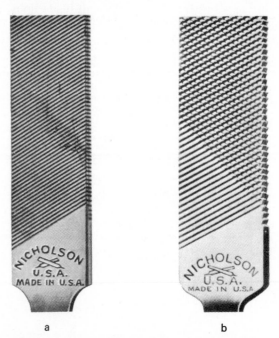

FIGURE R4-14   Files: (a) single cut; (b) double cut. (*Courtesy* Imperial Eastman Company)

A single-cut file is used for finishing a surface, for example, in preparing copper pipe for soldering. A double-cut file is more coarse and would be used where deeper and faster metal removal is needed. A rasp is an extremely coarse double-cut file intended for very rough work. The selection of files will vary with the need.

R4-8
VISES

A machinist's vise (Fig. R4-15a) is quite useful in holding parts for drilling, filing, and other operations. A medium-size portable model is recommended for field service work; larger models may be needed in shop work.

The pipe vise (Fig. R4-15b) is a requirement both in the field and in the shop to hold tubing or pipe while cutting or threading operations are taking place. *Caution:* When holding copper or other soft metal, be sure that the jaws are soft so as not to mar the surface. Also, do not overapply clamping pressure so as not to squeeze tubing out of round.

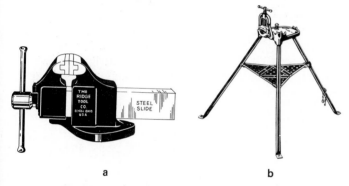

FIGURE R4-15   Vises: (a) machinist vise; (b) pipe. (*Courtesy* Ridge Tool Company)

R4-9
TAPES AND HAND RULES

Start with the familiar 6- or 8-ft flexible steel tape—it is invaluable for many measurements; tube diameters, short tube lengths, filter sizes, duct sizes, etc. Where long piping runs are to be assembled, a 50- or 100-ft steel or plastic tape is recommended for accurate measurements. A cloth tape is not recommended due to its stretch characteristics. Some technicians also like to have a 12-in. nonrusting steel ruler handy with graduations down to $\frac{1}{32}$ in. for more precise measurements.

R4-10
MICROMETERS AND CALIPERS

Sometimes a service technician must check dimensions of parts very accurately in thousandths of an inch.

Such measurements require special tools, which are, however, commonly available and reasonably inexpensive.

For measuring the outside diameter or width of an object such as a fan shaft or metal thickness the use of a micrometer (Fig. R4-16a) is required. The object is placed inside the micrometer frame and the thimble is tightened to force the object lightly against the anvil. One turn of the thimble advances the spindle $\frac{1}{40}$ in. or 0.025 in. (25 thousandths of an inch). The sleeve and spindle are so graduated as to permit the user to read measurements down to $\frac{1}{1000}$ (0.001) of an inch. Micrometers come in ranges, meaning the maximum permissible size that can be measured. A 1-in. micrometer measures from 0 to 1 in., a 2-in. micrometer measures from 1 to 2 in., a 3-in. measures from 2 to 3 in., and so on.

The caliper (Fig. R4-16b) is an instrument for measuring the inside diameter or dimension of an object such as a bearing opening or cylinder bore in a compressor casting. The jaw is placed inside the opening and the dimension is read from a vernier scale or dial, depending on the model. Again, these read to 0.001-in. accuracy. Some calipers can be used to measure both inside and outside dimensions.

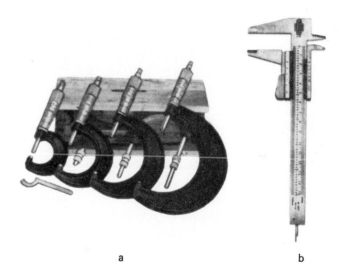

FIGURE R4-16   Typical micrometer and caliper measuring instruments: (a) micrometer sets; (b) caliper.

R4-11
DRILLS

Drills (Fig. R4-17) are a must and are frequently used by the refrigeration mechanic in the work of installation and repair. In the field it is assumed a hand-held portable electric drill will be used. A $\frac{1}{4}$-in. or $\frac{3}{8}$-in. chuck size is sufficient for small operations, such as drilling wood, plastic, thin metals, and light-duty masonry work.

FIGURE R4-17 Electric drill. (*Courtesy* The Black and Decker Manufacturing Company)

ACCESSORY ITEMS

In addition to hand tools, the refrigeration technician will need an array of accessory items. Most of these are expendable and will need periodic replacement. Following is a partial listing:

* Abrasive sand cloth, which comes in rolls or sheets (several grades will be needed)
* Steel wool and cleaning pads
* Cleaning solvent that is not dangerous to health or explosive (among those *not recommended* are gasoline and carbon tetrachloride; consult the refrigeration supply outlets for various acceptable solvents)
* Roll of friction tape
* Roll of rubber tape
* Pipe "dope" for sealing threads

Although it is not a tool or accessory as such, the availability and use of safety protection is one of the most important considerations in establishing minimum requirements for personal protection and/or to meet and comply with OSHA (Occupational Safety and Health Act) or local codes. Minimum safety equipment should include:

* Hard hat
* Safety glasses
* Safety shoes
* Gloves
* Fire extinguisher
* First-aid kits
* Checklist of actions to take in case of an emergency
* Common sense

Heavy-duty drilling in brick, cement, and thick steel is often required during installation work; for such operations a $\frac{1}{2}$-in. heavy-duty model drill is recommended. Such a drill can also double for rotary-hammer drills.

Either drill should have a variable-speed control plus a reversing switch to back out stuck bits. Bit selection will depend, of course, on the nature of the material, but as a minimum, a set of high-speed alloy steel bits for metal is a must. Such bits are also fine for drilling wood and plastic. Masonry bits and wood-boring bits may be added as job requirements demand.

PROBLEMS

R4-1. A special type of wrench used in refrigeration work is called a _____.

R4-2. Socket wrenches can be obtained with sockets for different points. True or False?

R4-3. What are the most common number of points?

R4-4. The more points a socket has, the harder it is to use in a restricted area. True or False?

R4-5. Setscrews on a fan shaft are turned by _____ wrenches.

R4-6. Name two common types of screwdriver blades.

R4-7. Files are made in what two types?

R4-8. A machinist's vice is also good for holding copper tubing. True or False?

R4-9. Common micrometers and calipers read to _____ of an inch.

R4-10. Gasoline and carbon tetrachloride are recommended cleaning solvents. True or False?

R4-11. What does "OSHA" stand for?

Refrigeration Piping Materials and Fabrication

R5-1
GENERAL

Chapter R1 introduced the major components of a mechanical refrigeration cycle: the compressor, evaporator, condenser, and metering device. It also defined the tubing or piping necessary to connect these elements and form a sealed system so that the refrigerant did not escape. This chapter reviews the materials, tools, and methods most commonly used by technicians to form and assemble the refrigeration piping system.

R5-2
REFRIGERANT PIPING MATERIALS

Most tubing used in refrigeration air-conditioning piping is made of copper. However, aluminum is now widely used for fabrication of the evaporator and condenser internal coil circuits, although it has not become popular for field fabrication of the connecting refrigerant lines, principally because it cannot be worked as easily as copper and is more difficult to solder.

Steel piping is used in assembly of the very large refrigeration systems where pipe sizes of 6-in. diameter and above are needed. Threaded steel pipe connections are not used in modern refrigeration work, since they cannot be made leakproof. These systems are welded, and couplings are bolted to the equipment and/or where service joints are needed.

The term *tubing* generally applies to thin-wall materials, which are joined together by means other than threads cut into the tube wall. *Piping,* on the other hand, is the term applied to thick pipe-wall material (e.g., iron and steel), into which threads can be cut into the wall and which are joined by fittings that screw onto the pipe. Piping can also be welded. Another distinction between tubing and piping is the method of size measurement

(see Fig. R5-1). Tubing sizes are expressed in terms of the outside diameter (OD), and pipe sizes are expressed as nominal inside diameters (ID). Thus in Fig. R5-1 the $\frac{1}{2}$-in.-OD (type *L*) copper tube will have an inside diameter of 0.43 in. The $\frac{1}{2}$-in.-ID nominal steel pipe will have an inside diameter of 0.50 in. and an outside diameter of 0.75 in.

Because of the special treatment of aluminum tubing and welded steel piping, the technique of fabrication will not be covered in this discussion.

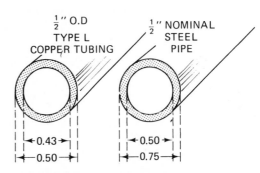

FIGURE R5-1 Method of sizing tubing and pipe.

R5-3
COPPER TUBING

The tubing used in all domestic refrigeration systems is specially annealed copper. Copper tubing when formed has a tendency to harden, and this hardening action could cause cracks in the tubing ends when they are flared or formed. The copper may be softened by heating to a blue surface color and allowing it to cool. This process is called *annealing* and is done at the factory.

Copper tubing used in refrigeration and air-conditioning work is known as *ACR tubing,* meaning that it is intended for use in air-conditioning and refrigeration work and has been specially manufactured and processed for this purpose. ACR tubing is pressurized with nitrogen gas to keep out air, moisture, and dirt, and also to provide maximum protection against the harmful oxides that are normally formed during brazing. The ends are plugged, and these plugs should be replaced after cutting a length of tubing.

R5-3.1
Copper Tubing Classification

Copper tubing has three classifications: K, L, and M, based on the wall thickness:

K—heavy wall—ACR approved

L—medium wall—ACR approved

M—thin wall—not used on refrigeration systems

Type M thin-wall tubing is not used on pressurized refrigerant lines, for it does not have the wall thickness to meet the safety codes. It is, however, used on water lines, condensate drains, and other associated system requirements.

Type K heavy-wall tubing is meant for special use where abnormal conditions of corrosion might be expected. Type L is most frequently used for normal refrigeration applications. Figure R5-2 provides a table of specifications for both type K and L tubing. Both K and L copper tubing are available in soft- or hard-drawn types.

R5-3.2
Soft-Drawn Copper Tubing

Soft-drawn copper tubing, as the name implies, is annealed to make the tubing more flexible and easier to bend and form. It is commercially available in sizes from $\frac{1}{8}$ to $1\frac{5}{8}$ in. OD and is usually sold in coils of 25, 50, and 100-ft lengths. The coils are dehydrated and sealed at the factory. Soft copper tubing may be soldered or used with flared or other mechanical-type fittings. Since it is easily bent or shaped it must be held by clamps or other hardware to support its own weight. The more frequent application is for line sizes from $\frac{1}{4}$ to $\frac{3}{4}$ in. OD. Above $\frac{3}{4}$ in. OD forming becomes rather difficult.

R5-3.3
Hard-Drawn Copper Tubing

Hard-drawn tubing is also used extensively in commercial refrigeration and air-conditioning systems. Unlike soft-drawn, it is hard and rigid and comes in straight lengths. It is intended for use with formed fittings to

FIGURE R5-2
Specifications of common copper tubing sizes.

Type	DIAMETER Outside (in.)	Inside (in.)	Wall Thickness (in.)	Weight per foot (lb)
K	$\frac{1}{2}$	0.402	0.049	0.2691
	$\frac{5}{8}$	0.527	0.049	0.3437
	$\frac{3}{4}$	0.652	0.049	0.4183
	$\frac{7}{8}$	0.745	0.065	0.6411
	$1\frac{1}{8}$	0.995	0.065	0.8390
	$1\frac{3}{8}$	1.245	0.065	1.037
	$1\frac{5}{8}$	1.481	0.072	1.362
	$2\frac{1}{8}$	1.959	0.083	2.064
	$2\frac{5}{8}$	2.435	0.095	2.927
	$3\frac{1}{8}$	2.907	0.109	4.003
	$3\frac{5}{8}$	3.385	0.120	5.122
L	$\frac{1}{2}$	0.430	0.035	0.1982
	$\frac{5}{8}$	0.545	0.040	0.2849
	$\frac{3}{4}$	0.666	0.042	0.3621
	$\frac{7}{8}$	0.785	0.045	0.4518
	$1\frac{1}{8}$	1.025	0.050	0.6545
	$1\frac{3}{8}$	1.265	0.055	0.8840
	$1\frac{5}{8}$	1.505	0.060	1.143
	$2\frac{1}{8}$	1.985	0.070	1.752
	$2\frac{5}{8}$	2.465	0.080	2.479
	$3\frac{1}{8}$	2.945	0.090	3.326
	$3\frac{5}{8}$	3.425	0.100	4.292

make the necessary bends or changes in direction. Because of its rigid construction it is more self-supporting and needs fewer supports. Sizes range from $\frac{3}{8}$ to over 6 in. OD. Hard-drawn tubing comes in standard 20-ft lengths that are dehydrated, charged with nitrogen, and plugged at each end to maintain a clean moisture-free internal condition. The use of hard-drawn tubing is most frequently associated with larger line sizes of $\frac{7}{8}$ in. OD and above.

R5-4
CUTTING COPPER TUBING

There are two methods of cutting copper tubing. The first uses the hand-held tube cutters shown in Fig. R5-3; they are suitable for cutting soft- or hard-drawn tubing. Hand-held cutters may be obtained in different models to cut from $\frac{1}{8}$ in. OD to as much as $4\frac{1}{8}$ in. OD.

The hand-held cutter is positioned on the tubing at the proper cut point. Tightening the knob forces the cutting wheel against the tube. Then by rotating the cut-

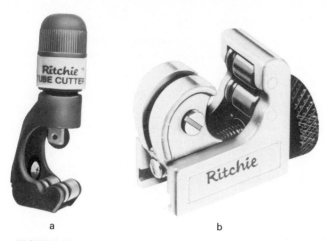

FIGURE R5-3 Typical tubing cutters. (*Courtesy* Yellow Jacket Division, Ritchie Engineering Company, Inc.)

ter around the tube and continual tightening of knob, the cut is made. A built-in reamer blade is used to remove burrs from inside the tube after cutting.

A second (but less desirable) method of cutting larger size hard-drawn tubing is the use of a hack saw and a sawing fixture to help square the end and make more accurate cuts (see Fig. R5-4).

The saw blade should have at least 32 teeth per inch to ensure a smooth cut. Try to keep saw filings from entering the tubing to be used. Some mechanics also file the end of the tube to provide a smooth surface. Wipe the inside of the tube with a clean cloth.

For contractors who do considerable refrigeration piping there are portable power-operated machines that use abrasive wheels to cut the tubing and buff both the inside and outside of the tube; however, such machines are not required for the average installation.

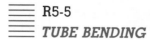

R5-5
TUBE BENDING

Where smaller sizes of soft-drawn tubing are used it is generally more convenient and economical to simply bend the tubing to fit the application requirements without using formed fittings. This can be done by hand without special tools—but it takes practice so as not to make too sharp or too tight bends and so flatten the tube.

As a rule of thumb, the minimum bending radius in which a smaller tube may be curved is about five times the tube diameter, as illustrated in Fig. R5-5. Larger tubing may require a radius of up to 10 times the diameter.

To make a hand bend, start with a larger radius and gradually work the tubing into the proper shape as you decrease the radius. *Never* try to make the first bend as small as the final radius desired.

Tube-bending springs as illustrated in Fig. R5-6 are available to insert inside or on the outside of the tube so as to prevent the tube from collapsing. These springs are relatively inexpensive and come in sizes to fit most common tubing requirements.

The most accurate and reliable method of tube bending is a lever-type tube-bending tool kit as illustrated in Fig. R5-7. It bends both hard and soft tubing. Various sizes of forming wheels and blocks are furnished up to about $\frac{7}{8}$ in. OD. Bends can be made at any angle up to 180°.

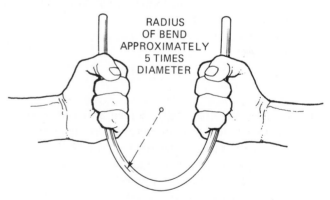

FIGURE R5-5 Recommended technique in bending tubing by hand. (*Courtesy* Imperial Eastman)

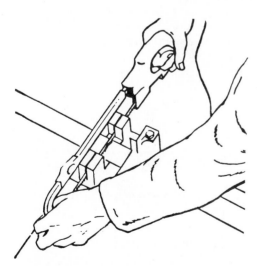

FIGURE R5-4 Cutting tube with hack saw and sawing fixture. (*Courtesy* Imperial Eastman)

FIGURE R5-6 Spring-type tube bender. (*Courtesy* Imperial Eastman)

Refrigeration Piping Materials and Fabrication 41

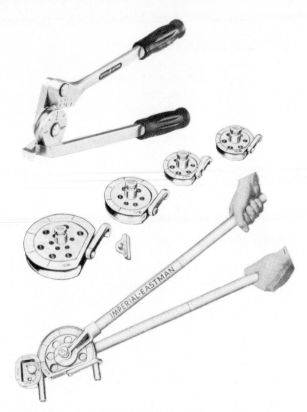

FIGURE R5-7 Lever-type tube bender. (*Courtesy* Imperial Eastman)

R5-6
METHODS OF JOINING TUBING

As mentioned earlier, the walls of copper tubing are too thin for threading, so other means must be used to connect the tubing. These may be divided into two broad catagories:

1. **Mechanical Couplings** Flared and compression fittings that are semi-permanent in that they can be mechanically taken apart
2. **Heat Bonding** Soldering and brazing, which means that they are permanent

R5-7
MECHANICAL COUPLINGS

R5-7.1
Flared Connections

Since about 1890 the flare connection has been one of the most widely used techniques to join soft-drawn copper tubing. A properly made flare is most important if leakproof joints are to be achieved—and this requires the right tools and practice.

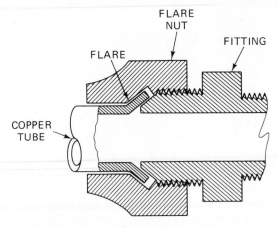

FIGURE R5-8 Cross section of 45° angle flared fitting.

Flares are made with special tools, which expand the end of the copper tubing into a cone shape as shown in Fig. R5-8. This cone is formed at a 45° angle that mates against the face of a flare fitting. The flare nut when tightened will press the soft copper against the machined fitting seat, thus forming a tight seal. The flare illustrated in Fig. R5-8 is called a *single-thickness flare*. Others are called *double-thickness flares*. Both forming techniques will be shown.

R5-7.2
Single-Thickness Flares

Figure R5-9 represents a typical flaring tool consisting of a flaring block or base and a slip-on yoke that holds the screw-driven flaring cone.

Note that the base-block dies slide to allow the tubing to be inserted into the particular size hole. The end wing nut is then tightened to hold the material in place during flaring.

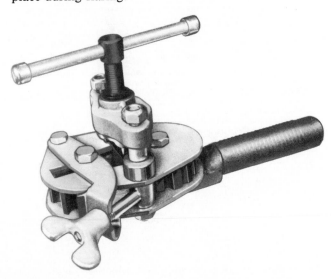

FIGURE R5-9 Typical flaring tool operation. (*Courtesy* Yellow Jacket Division, Ritchie Engineering Company, Inc.)

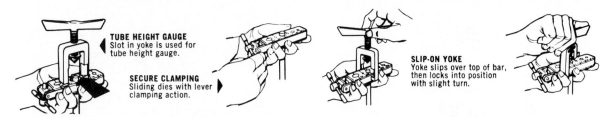

FIGURE R5-10   Typical flaring operation. (*Courtesy* Yellow Jacket Division, Ritchie Engineering Company, Inc.)

Figure R5-10 represents a typical flaring operation. The chamber in the block is at a 45° angle. The tubing should extend slightly above the flare block approximately one-third the height of the flare. This is necessary to provide enough material to fill the opening after the flaring cone is driven into the tubing and also to ensure the proper surface area against the fitting face. If it is too small, the flare nut may not hold the tubing; if too large, the flare nut may not fit over it. This operation takes practice.

Before making the flare it is important to prepare the tubing. Use a good grade of tubing (not all types of tubing are recommended for flaring). Follow proper cutting and deburring practices. Cut the tube squarely (use file if necessary) and remove internal and external burrs, using a deburring tool on the cutter.

To form the flare, first put the flare nut on the tubing. Insert the tubing into the proper size hole (adjusting the height above the block) and clamp with wing nuts. Put a drop of oil on the forming cone. Tighten the forming cone into the tubing using an initial one-half turn. Then back off one-quarter turn. Retighten three-quarters turn and then back off one-quarter turn. Continue this back and forth procedure until a flare is formed; with practice this motion becomes routine. Continuous turning is not recommended, since it does not give the operator a feel of the progress being made and may tend to harden the metal. Remove the flare from the block and carefully examine the completed flare to see that the sides have no splits or other imperfections.

Maintain tools in a clean and well-lubricated condition at all times for ease of operation and extended tool life. Never overtorque the feed mechanism when making a flare—flare washout will result.

Note: 45° flares are the standard of the refrigeration and air-conditioning industry. In other industries, such as automotive, steel or brass tubing is used. These metals do not form as easily as copper, so 37° angle flares are used. Therefore, flare fittings and tools are not interchangeable.

R5-7.3
Double-Thickness Flares

A double flare consists of a twin-wall sealing surface. Double-flare fabrication is accomplished through use of a conventional flaring tool provided with double-flare adapters.

As illustrated in Fig. R5-11, the tubing is clamped in the flaring bar and allowed to protrude a measured distance above the top of the bar. The adapter is positioned in the tube, the yoke is engaged with the bar, and the feed screw is advanced until a positive resistance is encountered. This completes the preform operation, which folds the tube inward. The adapter is removed and the forming cone is advanced until moderate resistance is encountered; then final forming is done as described for a single flare. This operation completes the two steps required in forming a double flare.

Why double flares? They are used mainly on larger-size tubing where a single flare may have a tendency to be weak due to excessive expansion. They offer greater resistance to fracture on installations that are subjected to excessive vibration, and they can be assembled and disassembled more frequently without flare washout.

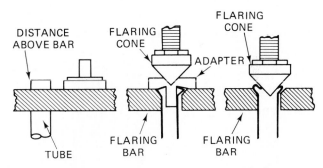

FIGURE R5-11   Method of making double flare.

R5-7.4
Flared Fittings

To accommodate the many types of flared fittings needed in refrigeration and air-conditioning systems, a variety of elbows, tees, unions, etc., are available to select from, as illustrated in Fig. R5-12. The fittings are usually drop-forged brass and are accurately machined to form the 45° flare face.

Threads for attaching the flare nuts are SAE National Fine Thread. Some fittings such as the flared half-union are made to join to national pipe threads. All fittings are based on the size of tubing to be used. The flare

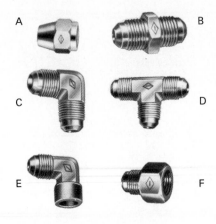

A = STANDARD FLARE NUT

B = FLARE COUPLING

C = FLARE 90° ELBOW

D = FLARE TEE

E = FLARE TO FEMALE PIPE ELBOW

F = FLARE TO FEMALE PIPE ADAPTER

FIGURE R5-12   Common flared fitting.
(*Courtesy* Imperial Eastman)

nuts are hexagon-shaped for easy wrench tightening. The fitting body also usually has a flat surface to take an open-end wrench attachment.

R5-7.5
Compression Fittings

In recent years there has been a trend toward using compression-type fittings for joining refrigerant tubing. This method has gained popularity in the residential air-conditioning field because it reduces the field labor requirement in making flared connections or soldered piping. Figure R5-13 represents a typical mechanical coupling concept as manufactured by the Aeroquip Corporation.

A gastight seal is obtained by simply connecting the tubing to the coupler with a coupling nut, crimp collar, and O-ring to assure a leakproof assembly. Once assembled, the joint may be disconnected and reassembled without lessening its sealing effectiveness.

As long as the tubing is straight and round in order to be inserted into the tube entrance, a proper connection can be made with minimum mechanical skills. This particular coupling comes in sizes from $\frac{1}{4}$ to $1\frac{1}{8}$ in. OD. Adapters are available to join these couplings to other fittings such as soldered pipe or threaded fittings.

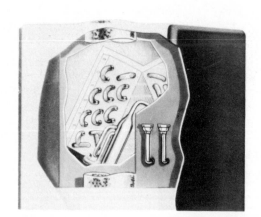

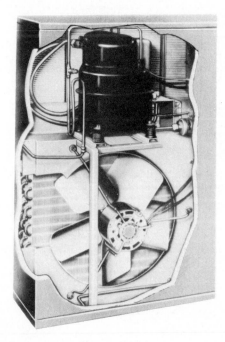

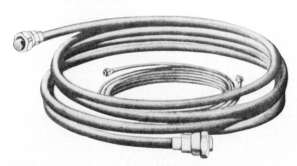

FIGURE R5-13   Precharged refrigerant lines. (*Courtesy* Aeroquip Corporation)

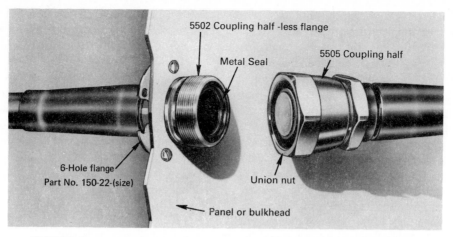

FIGURE R5-14   Line to unit coupling. (*Courtesy* Aeroquip Corporation)

R5-7.6
Diaphragm Fittings

The development of flexible, quick-connect, precharged refrigerant lines has played a major role in the growth and reliability of home air-conditioning. The liquid and suction lines are made of bendable copper or convoluted steel tubing. Suction lines are factory covered with a foam-rubber insulation material. Each end of the tubing is fitted with a coupling half that mates with a coupling half on the equipment (Fig. R5-14). Both coupling halves have diaphragms that provide a seal, which prevents refrigerant loss before connection. The male half (on the equipment) contains a cutter blade, the metal refrigerant sealing diaphragm, and an intermediate synthetic rubber seal to prevent loss of refrigerant while the coupling is being connected. The female half (on the tubing) contains a metal diaphragm, which

is a leakproof metal closure. Tightening the union nut draws the coupling halves together (Fig. R5-15), piercing and folding both metal diaphragms back and opening the fluid passage.

When fully coupled (Fig. R5-16) a metal seal forms a permanent leakproof joint between the two coupling halves. Note the service port for checking refrigerant pressure. The port is equipped with a valve (not shown), similar to a tire valve, that opens when depressed by gauge lines. When installing quick-connect precharged refrigerant lines, follow the manufacturer's installation directions on the radius of the bend for the size of tubing, lubrication of the coupling, and proper torque values when couplings are joined together. If excess tubing is present, form a loop or coil in a flat, horizontal manner; do not form vertical loops, which will create an oil trap. Precharged lines are manufactured in various sizes and lengths from 10 to 50 ft. It is necessary to plan the

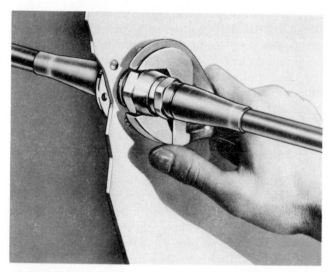

FIGURE R5-15   Tightening fitting. (*Courtesy* Aeroquip Corporation)

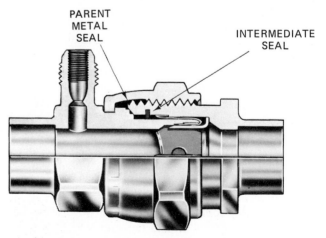

FIGURE R5-16   Completed fitting. (*Courtesy* Aeroquip Corporation)

Refrigeration Piping Materials and Fabrication   45

installation in order to have sufficient but not excessive tubing, which adds cost and pressure drop to the system.

The coupling described above is called a "one-shot connection," as it is designed to be coupled and remain coupled. It can be undone, however, to revise piping, change equipment location, etc. The only factor that must be considered is the loss of refrigerant from the system. Uncoupling and recoupling in demonstration use reveals the ability to recouple this type of connection several times when proper tools are used.

This coupling, as well as compression-type fittings, are subject to a high leak failure rate when the wrong tools are used. Only open-end type wrenches that fit snugly on the wrench pads of the nuts should be used. Adjustable-type open-end wrenches are permissible if the wrench is in good condition with little worm wear or jaw spring. Do not use pliers, electrician's pliers, pipe wrenchs, etc., on these fittings. Any tool that produces a squeezing pressure causes the nut to collapse into an egg shape. When this happens, the fitting is destroyed, cannot be made to hold, and must be replaced.

R5-8
HEAT BONDING

Heat bonding of refrigerant tubing is called *soldering* or *brazing* and consists of joining two pieces of metal together with a third metal (or solder), which melts at a lower temperature than the pieces to be joined. When melted, the solder flows between the two pieces. The molten metal adheres to the surfaces of the two metals and forms a good bond between them. The solder usually has less strength than the metals it joins, so for greatest strength of the solder joint the layer of solder must be very thin. The essential difference between soldering and brazing is the temperature at which the molten solder flows.

Welding differs from soldering or brazing in that it does not use another metal to act as the bonding material of the joint. The two pieces of material to be welded must be "puddled," that is, their edges or surfaces to be joined must be melted and allowed to mix together, so that when they cool they are a part of one another.

R5-8.1
Soft Soldering

The most common solder is a mixture of tin and lead. If it is one-half tin and one-half lead, it is called *fifty-fifty* (50–50) solder. Fifty-fifty solder starts to melt at 360°F but does not become fluid until further heated to 415°F. The 50–50 tin-lead solders are called *soft sol-*

ders and are primarily used on plumbing and heating systems with working temperatures up to 250°F. *They are not recommended for refrigeration work.*

Another form of soft solder, but one that is somewhat harder than tin-lead solders, is composed of 95% tin and 5% antimony and is commonly referred to as *ninety-five-five* (95–5). It starts to melt at 450°F and is fully liquid at 465°F. It is easily worked with a small hand-held propane or acetylene torch-kit. Ninety-five-five is the recommended soft solder to use on refrigeration work, particularly on small-diameter soft-drawn copper tubing.

R5-8.2
Hard Soldering

When larger hard-drawn copper tubing is used, or where local building codes require the use of hard solder, the hard solder joint makes a much stronger bond. Over the past twenty years or so, hard solders (or *silver solders* or *silver brazing alloys,* as they are now called) have been widely accepted by the refrigeration and air-conditioning industry for joining metal. The selection of these alloys over others was prompted by the need for high-strength, corrosion-resistant, vibration-proof, and leak-tight joints. Silver brazing satisfied all these requirements, and also offered the additional advantage of joining similar or dissimilar metals with greater ease of application and at low brazing temperatures.

Silver solders in general flow at a temperature of approximately 1,100 to 1,200°F. Copper melts at 1,981°F, so this alloy flows at some 800°F below the melting point of copper, making it a very safe alloy to use on copper tubing and fittings. When properly made, the silver-solder joint will have a tensile strength above the material it joins.

Some of the lower temperature silver-solders may be applied with air-acetylene or air-propane torches, but the oxygen-acetylene torch illustrated later is most commonly used. Techniques of proper soldering and brazing, and pointers on brazing equipment are also discussed later.

R5-8.3
Solder Fittings

For use with both soft- and hard-soldered systems, a number of *sweat fittings* are available, as illustrated in Fig. R5-17. Some sweat fittings are for connecting tubing to tubing, and others may have a connection on one end for National pipe threads.

These fittings are brass rod, brass forging, or wrought copper and are accurately manufactured to permit the tubing to be inserted into the fitting opening with a snug fit, leaving only a very thin *solder clearance*

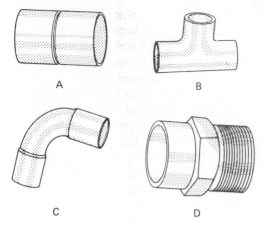

FIGURE R5-17 Some common soldered or brazed copper tube fittings: (a) coupling with rolled stop, sweat × sweat; (b) tee, sweat × sweat; (c) 90° elbow, sweat × sweat; (d) adapter, sweat × male pipe thread (m.p.t.). (*Courtesy* Mueller Brass Company)

for the flow of solder. If solder clearance is too large, the joint will be weakened.

R5-9
SWAGING COPPER TUBING

Sometimes in the assembly of copper tubing with the same diameter, some fabricators feel that it is more reliable to join the two pieces of tubing by making a *swaged connection* with only one solder joint, as illustrated in Fig. R5-18a. This is done with a swaging tool

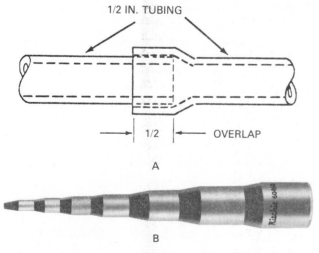

FIGURE R5-18 (a) Swaged connection. Two pieces of soft copper tubing are shown swaged and ready for soldering or brazing to make a joint connecting the two pieces of tubing. Note that both pieces of the tubing are of the same diameter. (b) Swaging tool. (*Courtesy* Yellow Jacket Division, Ritchie Engineering Company, Inc.)

(Fig. R5-18b), which is somewhat similar to the flaring block. The block holds the tube, and the correct-size punch is forced into the end of the tubing until the bell shape is produced. Some tools use the screw mechanism for force. Others require a blow by a hammer to force the punch into the tube.

Swaging, when properly done and soldered, reduces the number of soldered joints and thereby reduces leak hazards; however, it does take more time than using preformed fittings, and thus the option becomes one of individual choice.

R5-10
SOLDERING AND BRAZING

In the refrigeration industry, certain fundamental procedures of silver-alloy brazing have been arrived at through long experience. In many respects, these fundamentals are similar to those recommended for soft soldering, and a good soft-soldering operator generally makes a good silver-brazing operator.

There are six simple steps to follow in producing strong, leaktight joints that are basic to both soft soldering and silver brazing:

1. Good fit and proper clearance
2. Clean metal
3. Proper fluxing
4. Assembling and supporting
5. Heating and flowing the alloy
6. Final cleaning

R5-10.1
Four Steps of Preparation

1. Cutting and fitting the pipe or tubing (Fig. R5-19).
   a. Cut the tube to the proper length. Make sure that the ends are cut square; a tube cutter is by far the best tool. If a hacksaw is used, tilt the tube downward so that the cuttings fall out.
   b. Remove burrs with a reamer or half-round file, as shown in Fig. R5-20.
   c. Try the end of the tube in the fitting to be sure that it has the proper close fit. Clearance should be uniform all around. Where necessary, on soft-drawn copper, a sizing tool may be used to round out the tubing.
2. Cleaning the tubing and fittings (Figs. R5-21 and R5-22). This means that the surfaces to be joined must be *free of oil, grease, rust, or oxides*.

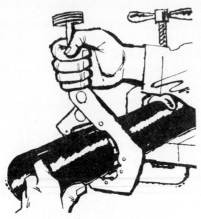

FIGURE R5-19 Cutting tube.
(*Courtesy* Mueller Brass Company)

FIGURE R5-21 Wire brushing.
(*Courtesy* Mueller Brass Company)

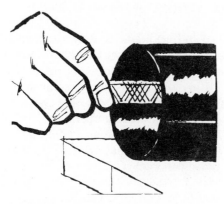

FIGURE R5-20 Deburring. (*Courtesy* Mueller Brass Company)

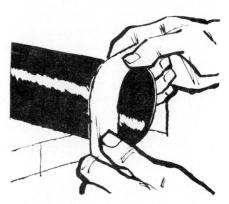

FIGURE R5-22 Sanding. (*Courtesy* Mueller Brass Company)

a. In those cases where the tubing or fittings have a coating of oil or grease, a liberal application of cleaning solvent with a brush or clean rag will effectively remove this contamination.

b. Clean socket of fitting and end of tubing with a clean wire brush and fine sand cloth. *Do not use steel wool* (shreds may adhere to the fitting and cause a void). *Do not use emery cloth,* since it often contains oils and undesirable abrasives that can cut too deeply.

c. Do not handle the surfaces after cleaning.

d. Cleaning should be done just before soldering so that oxidation is reduced to a minimum.

3. Proper fluxing. Flux does not clean the metal. It keeps the metal clean once it has been mechanically cleaned as mentioned above. Flux is available in both paste and liquid form, but in general the paste form is preferred by most refrigeration installers.

a. The first rule is to select the proper flux depending on whether the job is soft soldering or silver brazing. For brazing, use a good-quality low-temperature silver brazing flux. *This is very important* and requires special attention in re-

frigeration work; consult a local welding supply jobber.

b. Always stir the flux before using (see Fig. R5-23), since when flux stands the chemicals tend to settle to the bottom, especially in hot weather. Use a brush—never apply soldering paste with your fingers; perspiration and oils may prevent solder from sticking.

c. In most ordinary work, both the end of the tube and the inside of the fitting are fluxed (Fig. R5-24), but in refrigeration work the end of the tube is inserted partway into the fitting and the paste flux is brushed all around the outside of the joint.

d. The tube is then inserted to full depth in the socket (Fig. R5-25).

e. Where possible, revolve the fitting or tubing to spread the flux uniformly.

f. *Caution:* Too much flux can be harmful to the internal components and operation of the refrigeration system.

g. *Note:* With certain silver brazing alloys, copper-to-copper brazing can be done without

FIGURE R5-23   Stirring flux. (*Courtesy* Mueller Brass Company)

FIGURE R5-24   Applying flux. (*Courtesy* Mueller Brass Company)

FIGURE R5-25   Joining. (*Courtesy* Mueller Brass Company)

fluxing. On copper tube-to-brass fittings, flux is always required.

4. Supporting the assembly:

   a. Before soldering or brazing, the assembly should be carefully aligned and adequately supported.

   b. Arrange supports so that expansion and contraction will not be restricted.

   c. See that no strain is placed on the joint during brazing and cooling.

   d. A plumber's pipe strap makes an excellent temporary support until permanent arrangements are made.

R5-10.2
Heating and Flowing Soft Solder

To soft solder, apply the flame to the shoulder of the fitting so that the heat will flow in the direction of the tube (Fig. R5-26). Be careful not to allow the flame to enter directly into the opening where the solder will be drawn. With an ell or tee, heat the heaviest part of the fitting first and then move toward the opening where the solder will enter. In this way, heat is distributed uniformly to both the fitting and tube. Occasionally, remove the flame momentarily and touch the joint with solder to see whether the metal is hot enough to melt solder.

When applying the soft solder (Fig. R5-27), never push the solder into the joint. When the joint is hot enough, merely touch the solder, and capillary attraction will draw it into the clearance space between the surfaces to be joined. When a ring of solder appears all around the circumference, you have made a joint that is leakproof and tremendously strong. Wipe the joint clean with a clean cloth while the solder is still molten (Fig. R5-28). This is called *smoothing* the solder.

When working with large fittings (2 in. or over) it is helpful to use two torches or, better yet, a Y-shaped torch tip (Fig. R5-29), which gives a double or quadruple flame. This ensures a sufficient and even distribution of heat.

Also, when working with larger fittings 2 in. and over, each fitting should be rapped (with a small mallet)

FIGURE R5-26   Applying flame. (*Courtesy* Mueller Brass Company)

Refrigeration Piping Materials and Fabrication   **49**

FIGURE R5-27 Applying solder.
(*Courtesy* Mueller Brass Company)

FIGURE R5-28 Smoothing solder.
(*Courtesy* Mueller Brass Company)

FIGURE R5-29 Large-size work.
(*Courtesy* Mueller Brass Company)

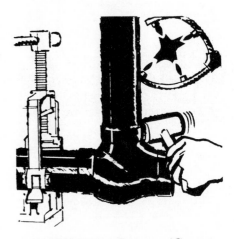

FIGURE R5-30 Tapping. (*Courtesy*
Mueller Brass Company)

at two or three points around its circumference while the solder is being fed (Fig. R5-30). This settles the joint and also releases any trapped gases that might hinder the flow of the solder.

 R5-10.3
Heating and Flowing Silver Brazing Alloy

- For rapid, efficient silver brazing, a soft bulbous oxyacetylene flame provides the best type of heat. Air-acetylene or air-propane torches have been used successfully for silver brazing fittings and tubing up to about 1 in. in size.
- Adjust the oxyacetylene for a slightly reducing (less oxygen) flame (*note*—brazing equipment and flames will be covered later).
- Start heating the tube about $\frac{1}{2}$ to 1 in. away from the end of the fitting (Fig. R5-31). Heat evenly all around to get uniform expansion of the tube and to carry the heat uniformly to the end inside the fitting.
- When the flux on the tube adjacent to the joint has melted to a clear liquid, transfer heat to the fitting.
- Sweep the flame steadily back and forth from fitting to tube, keeping it pointed toward the tube. Avoid letting the flame impinge on the face of the fitting, as this can easily cause overheating.

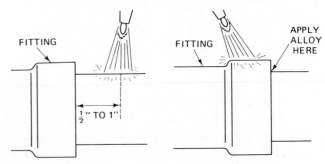

FIGURE R5-31 Brazing horizontal joints.

When the flux is a clear liquid on both fittings and tube, pull the flame back a little and apply alloy firmly against the tube and the fitting. With proper heating the alloy will flow freely into the joints.

The technique in making vertical joints (Fig. R5-32) is essentially the same: start with the preliminary heating of the tube, and then move to the fitting. (*Note:* This is slightly different from the soft-solder technique.) When the tube and fitting reach a black heat the flux will become viscous and milky in appearance. Continued heating will bring the material up to brazing temperature, and the flux will become clear; at that point apply the silver brazing alloy and sweat it.

Brazing larger-diameter pipe requires using the above technique but selecting only a 2-in. segment at a time and overlapping the braze from segment to segment. This operation requires a degree of practice once the basic points are learned.

Cleaning After Brazing: Do not quench. For cast fittings especially, allow to air cool until the brazing alloy has set; then apply a wet brush or swab to the joint to crack and wash off the flux. All flux must be removed before inspection and pressure testing. Use a wire brush if necessary.

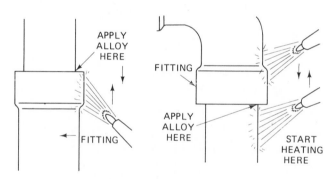

FIGURE R5-32   Brazing vertical joints.

R5-10.4
Safety Hints for Soldering and Brazing

- Never attempt to solder or braze while the system is under pressure or a vacuum.
- Many fluxes contain chloride or fluoride and should be handled carefully to prevent excessive skin contact or inhalation of fumes.
- Many silver solders contain cadmium in varying degrees. Cadmium fumes are very poisonous, so make sure that the work space is well ventilated.
- *Use safety glasses.*

R5-10.5
Brazing Equipment

As mentioned earlier, it is possible to achieve melting temperatures for some low-temperature silver solders by using air–acetylene or air–propane torches, but where extensive installation or repairing of refrigeration equipment is involved, the use of oxyacetylene brazing equipment has proven to be the most satisfactory. The introduction of pure oxygen along with the acetylene produces a very hot flame. Figure R5-33 illustrates the components involved in a typical oxyacetylene rig.

The efficient use of oxyacetylene equipment depends on a constant metered flow of oxygen and acetylene in correct proportions. Therefore, the operator must become thoroughly familiar with his particular equipment.

Both the oxygen tank and acetylene will have pressure regulators and two sets of gauges: one to register tank pressure and one to indicate pressure being furnished to the torch. The required torch pressure will vary with the particular torch and torch tip used.

While operating the oxyacetylene equipment never point the unlit torch toward any open flame or source of sparks. The acetylene is highly flammable—and the oxygen supports combustion very actively.

Lighting the torch calls for a torch lighter or sparker. *Do not use matches.* Open the torch acetylene valve approximately one-quarter turn. The torch oxygen valve is then "cracked" open as the spark or torch lighter is used to ignite the flame. Once ignition is achieved, the

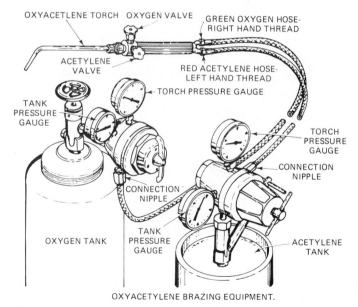

FIGURE R5-33   Oxyacetylene brazing equipment. (Reprinted from *Basic Air-Conditioning,* by G. Schweitzer and A. Ebling, copyright Hayden Book Company, Inc.; used by permission of the publisher)

acetylene valve is adjusted to obtain the desired flame size. The oxygen valve is turned *slowly* to establish the type of flame, as illustrated in Fig. R5-34.

The correct flame is called a *neutral flame;* it has a luminous blue cone with a touch of reddish purple at the tip. A carbonizing flame will be evidenced by a greenish flame caused by too much acetylene. It produces soot or carbon that will restrict the flow of solder. An oxidizing flame results from too much oxygen; it will cause pitting of the metal and tends to harden the joint, making it more susceptible to breakage from vibration.

R5-10.6
Soldering Is an Art

Beginners should test a few joints that they have soldered by reheating them and taking a joint apart to examine it. They can then determine whether some places in the joint were not soldered, and if this is true, what caused the failure and how to correct it.

There is probably no work that service technicians are called on to do that requires more skill than soldering. They may know all the principles of soldering, but will find that considerable experience will be required before they become adept at making consistently tight, strong, and neat soldered joints.

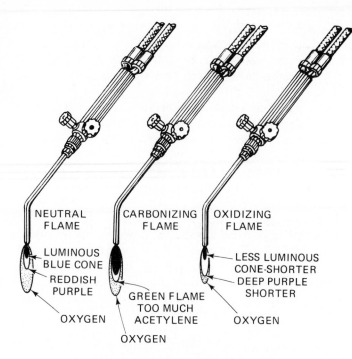

OXYACETYLENE FLAMES.

FIGURE R5-34   Oxyacetylene flames. (Reprinted from *Basic Air-Conditioning,* by G. Schweitzer and A. Ebling, copyright Hayden Book Company, Inc.; used by permission of the publisher)

PROBLEMS

R5-1. What is the most common material used in refrigeration tubing?

R5-2. Copper tubing differs from pipe in that it has a thicker wall. True or False?

R5-3. Copper tubing is marketed in three wall thickness classifications. What are they?

R5-4. Two of the three wall thickness classifications are approved for ACR installations. What are they?

R5-5. What are the two forms of copper tubing called?

R5-6. Name three methods of joining tubing.

R5-7. At what angle are flares used in refrigeration work formed?

R5-8. When bending tubing, what is the smallest radius of bend that can be used?

R5-9. What is the only type of wrench that should be used on compression fittings?

R5-10. What is the name of the procedure for fastening two tube ends without using a coupling?

R5-11. What are the four stages of preparation when brazing tubing?

R5-12. The correct oxyacetylene flame for brazing is called a _____ flame.

Compression Cycle Components

R6-1
GENERAL

The vapor compression refrigeration cycle as discussed in Chapter R1 is the most common method of heat energy transfer. There are four major components in the compression cycle: evaporator, compressor, condenser, and pressure-reducing device.

R6-2
EVAPORATORS

The evaporator or cooling coil is the part of the refrigeration system where heat is removed from the product: air, water, or whatever is to be cooled. As the refrigerant enters the passages of the evaporator it absorbs heat from the product being cooled, and as it absorbs heat from the load, it begins to boil and vaporizes. In this process, the evaporator accomplishes the overall purpose of the system—refrigeration.

Manufacturers develop and produce evaporators in several different designs and shapes to fill the needs of prospective users. The blower coil or forced-convec-

tion type of evaporator (Fig. R6-1) is the most common design; it is used in both refrigeration and air-conditioning installations.

Specific applications may require the use of flat plate surfaces for contact freezing. Continuous tubing is formed or placed between the two metal plates, which are welded together at the edges, and a vacuum is drawn on the space between the plates. These plates also may be assembled in groups arranged as shelving, utilizing refrigerant in a series flow pattern (see Fig. R6-2).

Other shapes of plate-type evaporators are shown in Fig. R6-3. They are widely used in small refrigerators, freezers, and soda fountains, where mass production is economical, and the plates can easily be formed into a variety of shapes.

Plate-type evaporators also are assembled in groups or banks for installation in low-temperature storage rooms and are mounted near the ceiling as shown in Fig. R6-4. This type may be connected for either series or parallel refrigerant flow, depending on usage requirements. Plate-type coils are also used in refrigerated

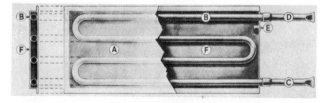

FIGURE R6-2 The Dole vacuum plate: (a) Outside jacket of plate. Heavy, electrically welded steel. Smooth surface. (b) Continuous steel tubing through which refrigerant passes. (c) Inlet from compressor. (d) Outlet to compressor. Copper connections for all refrigerants except ammonia where steel connections are used. (e) Fitting where vacuum is drawn and then permanently sealed. (f) Vacuum space in dry plate. Space in holdover plate contains eutectic solution under vacuum. No maintenance required due to sturdy, simple construction. No moving parts; nothing to wear or get out of order; no service necessary. (*Courtesy* Dole Refrigerating Company)

FIGURE R6-1 Blower coil. (*Courtesy* Kramer Trenton Company)

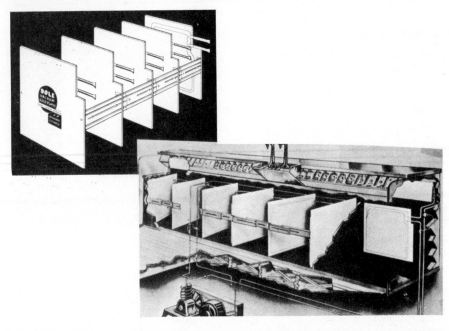

FIGURE R6-3   Plate evaporators in soda fountain. (*Courtesy* Dole Refrigerating Company)

trucks and railway cars for the transportation of refrigerated food and frozen-food products, and are of a design similar to that shown in Fig. R6-5. Frequently, the space between the plates is filled with a solution that retains its refrigeration if the unit is not in operation for short periods.

The bare-tube type of coil may be used for the cooling of either air or a liquid, with the smaller evaporators being constructed of copper tubing. Steel pipe is used for evaporators in systems using ammonia as the refrigerant and in the larger evaporators containing other refrigerants.

An air film adheres to the outside surface of a coil, acting as an insulator and slowing down the heat-transfer process, which is dependent primarily on surface area and temperature differential. One of the methods used

to overcome or compensate for the conduction loss due to the air film is to increase the surface area. This may be accomplished through the addition of fins to the evaporator pipe or tubing, as shown in Fig. R6-6. The addition of fins does not eliminate air film, for it furnishes more area to which air film will cling or adhere; but it does afford more surface area for heat transfer, without increasing the size of the coil to any great extent.

Another method of overcoming the heat-transfer loss caused by air film is through the addition of a fan or blower, which will cause rapid movement of air across the evaporator. Such a type of forced convection coil is shown in Fig. R6-7. Depending on the design and usage of the coil, the fan may be located for the movement of air across the coil either by means of an induced or

FIGURE R6-4   Plate evaporators for storage rooms. (*Courtesy* Dole Refrigerating Company)

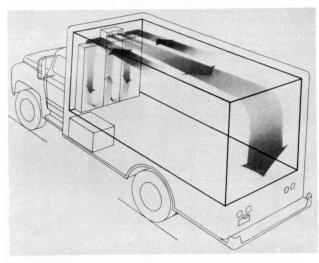

FIGURE R6-5   Refrigerated truck. (*Courtesy* Dole Refrigerating Company)

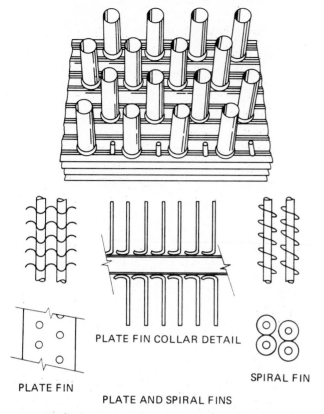

PLATE FIN COLLAR DETAIL

PLATE FIN

SPIRAL FIN

PLATE AND SPIRAL FINS

FIGURE R6-6 Finned tube evaporator. (Reprinted by permission from *ASHRAE Handbook and Product Directory*, 1975)

drawing action of the air or by a forced circulation or blowing action of the air across the evaporator coil.

The use of a fan improves the airflow and transfer of heat from the air to the refrigerant within the coil, since a greater proportion of the air will come in contact with the coil's surface area. Many coil manufacturers have designed their heat-transfer units with staggered rows of tubing, thus permitting, with the use of a blower, a large volume of air to come in contact with either the tubing surface or the fins connected to it. Forced or in-

FORCED AIR EVAPORATOR (FINNED)

FIGURE R6-7 Forced-air evaporator (finned). (*Courtesy Carrier Air-Conditioning Company*)

duced air motion across the coil usually will result in a greater portion of the air giving up its heat to the refrigerant within the coil over a specific period.

In the early days of mechanical refrigeration, cooling coils were constantly maintained at a temperature below freezing. Since these evaporators did not reach a temperature above 32°F, the frost accumulating on them did not have an opportunity to melt off while the equipment was in operation. The units had to be shut off and manually defrosted, since frost accumulating on the evaporator curtailed the amount of heat it could remove from the air passing across the coil.

In many of today's refrigeration applications, low temperatures must be maintained so that products may be kept in frozen storage condition. But defrosting of the cooling unit is performed by means other than manual, which is explained in Chapter R9. Frost accumulation on the cooling unit comes from moisture in the air and the products in the refrigerated space. When this moisture is removed from the air, the humidity is lowered.

Conditions may be such that an extremely low temperature or a low moisture content of the air surrounding the cooling unit is not desirable. If the temperature in the refrigerated space needs to be maintained at approximately 35°F, a coil or evaporator in which the refrigerant is at a temperature below this desired temperature must be used. As the air comes in contact with the cooling coil at a temperature below 32°F, some frost will form on the surface of the evaporator. When the desired temperature is achieved, the control mechanism will stop operation of the refrigeration unit. With the surrounding air temperature at 35°F, this warm air will melt the frost on the cooling unit and thereby defrost the cooling unit. This will occur naturally, particularly if it is a forced air coil and the warmer temperature air is forced across the evaporator surface.

The OFF-cycle period of the refrigeration unit should be long enough to assure complete defrosting of the cooling coil. If not, there is a possibility of only partial defrosting, which results in moisture collecting on the lower section of the unit. If this occurs, an icing condition on the coil may result. If this condition is allowed to continue, ice may cover the entire surface of the coil and may develop into a complete blockage of the coil.

Design conditions may be such that a high humidity must be maintained, to preserve freshness in the product being cooled, or to prevent a loss in weight, or deterioration. Examples of applications in which a high relative humidity is desired are a cold-storage room where fresh meat is kept, and a florist display cabinet with a moisture-laden atmosphere. Vegetable storage rooms or boxes also should be kept at a high humidity.

These conditions may be achieved through the use of nonfrosting evaporators, which are oversized coils used with thermostatic expansion valves as the metering devices. To maintain the refrigerated space temperature at approximately 35°F, the large-surface coil needs an internal refrigerant temperature of only 23° to 25°F. This permits an external coil temperature of about 30° to 32°F, allowing only a rare accumulation of frost, which will disappear rapidly when the space temperature requirement is satisfied and the compressor shuts off.

The condensate drain is shown in Fig. R6-8 beneath the nonfrosting coil, even though the drain is not put to any great use when the system is operating correctly. This type of coil is designed not to remove too much moisture from the air, so that a relative humidity of up to 85° may be maintained within the refrigerated space.

The evaporators described thus far have been of the *dry-expansion type,* as compared with the flooded type. The direct or dry-expansion type coil is designed for complete evaporation of the refrigerant in the coil itself, with only a vapor leaving the coil outlet. This vapor is usually superheated in the last part of the cooling coil. (*Superheating* means raising the temperature of the refrigerant vapor above that temperature required to change it from a liquid to a vapor.) It will reach the compressor in a superheated condition, picking up additional heat as it passes through the suction line. Figure R6-9 is a schematic showing a direct-expansion coil with a thermostatic expansion valve. The coil contains a mixture of liquid and gaseous refrigerant at all times when the unit is in operation. A constant superheat is maintained by the modulating of the valve, which is caused by the sensitivity of the thermal bulb to temperature changes at its location.

The characteristics of the dry or direct-expansion coil can be maintained by the automatic expansion valve, which maintains a constant pressure within the evaporator. This type of valve is usually used when a steady load is anticipated. Refrigerant controls are discussed more fully later in this chapter and again in Chapter R10.

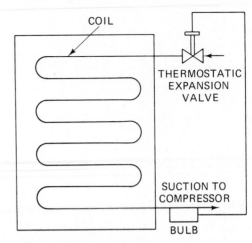

FIGURE R6-9   Dry expansion coil with thermostatic expansion valve. (Reprinted by permission from *ASHRAE Handbook and Product Directory,* 1975)

The *flooded* type of evaporator is filled with liquid refrigerant. It is designed so that the refrigerant (liquid) level is maintained by a float arrangement located in an accumulator that is situated outside the evaporator coil itself. A typical design is shown in Fig. R6-10. Part of the liquid refrigerant evaporates in the coil, and this vapor goes to the accumulator. From there the vapor is drawn from the top into the suction line and then to the compressor, while any liquid left in the accumulator is available for recirculation in the evaporator coil. When the equipment is properly calibrated, the remaining liquid is minimal.

As the refrigerant in the flooded coil evaporates as a result of the heat it has absorbed, the liquid level lowers in the coil. As the float lowers with the liquid level, it permits more refrigerant to flow into the accumulator so that a fairly constant liquid level is maintained. A flooded coil has excellent heat transmission efficiency because its interior surfaces are liquid-wetted instead of being vapor-wetted.

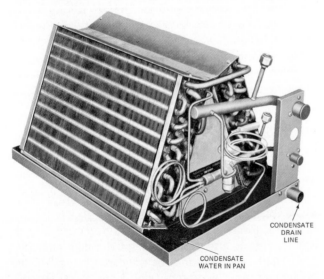

FIGURE R6-8   Condensate drain. (*Courtesy* Borg-Warner Central Environmental Systems, Inc.)

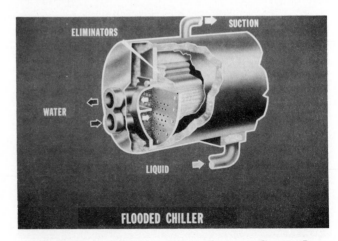

FIGURE R6-10   Flooded chiller. (*Courtesy* Carrier Air-Conditioning Company)

Liquid-cooling coils vary in their design depending on their application and usage, just as do air-cooling coils. Since there is a greater heat transfer between liquids and metals than between air and metals, a submerged coil has the capability of removing several times as many Btu's as an air-cooled coil under similar conditions. Submerged coils are used in a water-bath type of cooler, in which the "cold-holding" capacity is put to good use when cans filled with warm milk or other liquids are placed in the cooler.

Shell-and-tube and *shell-and-coil* are other types of arrangements for the cooling of one or more liquids, even in the cooling of brine solutions. A shell-and-coil water cooler is shown in Fig. R6-11. It is a direct-expansion type of system with the refrigerant circulated within the coil as the water is circulated within the shell

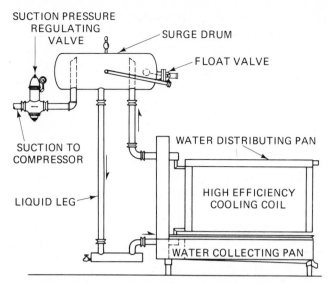

FIGURE R6-12   Boudelot cooler. (Reprinted by permission from *ASHRAE Handbook and Product Directory,* 1975)

at a temperature not much below 40°F to prevent freezing.

Tube-in-tube, sometimes classified as *double-pipe evaporator,* is a liquid-cooling coil that provides high heat-transfer rates between the refrigerant and the liquid being cooled. The path of refrigerant flow may be through either of the tubes, although usually the brine or liquid to be cooled is circulated through the inner tubing, and the refrigerant removing the heat is between the two tubes. This type of heating-exchange coil is also used in condenser design, described later in this chapter.

A *Baudelot cooler,* shown in Fig. R6-12, has several applications. It may be used for cooling water or other liquids for various industrial uses, and it is frequently used as a milk cooler. The evaporator tubing is arranged vertically, and the liquid to be cooled is circulated over the cooling coils by gravity flow from the trough type of arrangement located above the coils. The liquid gathers in a collector tray at the bottom of the coil, from which it may be recirculated over the Baudelot cooler or pumped to its destination in the industrial process.

R6-3
COMPRESSORS

After it has absorbed heat and vaporized in the cooling coil, the refrigerant passes through the suction line to the next major component in the refrigeration circuit, the compressor. This unit, which has two main functions within the cycle, is frequently classified as the *heart* of the system, for it circulates the refrigerant through the system. The functions it performs are:

1. Receiving or removing the refrigerant vapor from

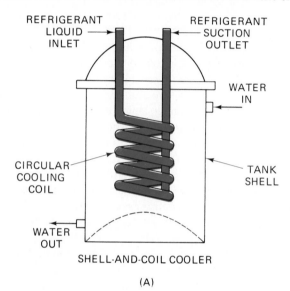

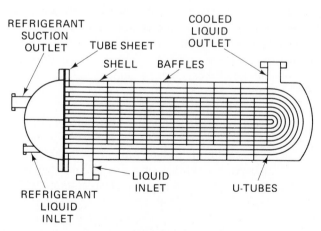

FIGURE R6-11   Shell-and-coil water cooler and direct (U-tube type) expansion liquid cooler. (Reprinted by permission from *ASHRAE Handbook and Product Directory,* 1975)

the evaporator, so that desired pressure and temperature can be maintained

2. Increasing the pressure of the refrigerant vapor through the process of compression, and simultaneously increasing the temperature of the vapor so that it will give up its heat to the condenser cooling medium

Compressors are usually classified into three major types: reciprocating, rotary, and centrifugal. The *reciprocating compressor* is used in the majority of domestic, small commercial and industrial condensing unit applications. This type of compressor can be further classified according to its construction, according to whether it is open and accessible for service in the field, or fully hermetic, not able to be serviced in the field.

Reciprocating compressors vary in size from that required for one cylinder and its operating piston to one large enough for 16 cylinders and pistons. The body of the compressor may be constructed of one or two pieces of cast iron, cast steel, or, in some cases, aluminum. The arrangement of the cylinders may be horizontal, radial, or vertical, and they may be in a straight line or arranged to form a V or a W.

Figure R6-13 shows an external view of a common type of reciprocating compressor used in commercial applications. As compressors differ in design and construction, so do the individual components within the compressors. But their main goal remains the same—the compression of the refrigerant vapor to high temperature and high pressure, so that its heat content can be reduced and it will condense into a liquid to be used over again in the cycle.

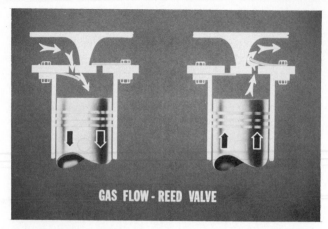

FIGURE R6-14   Gas flow reed valves. (*Courtesy* Carrier Air-Conditioning Company)

Pistons within the compressors may have the suction valve located in the top of the piston; this is classified as a *valve-in-head* type, or the piston may have a solid head, with the suction and discharge valves located in a valve plate or cylinder head. A typical valve plate, showing the suction and discharge internal valves of a two-cylinder reciprocating compressor is shown in Fig. R6-14.

Figure R6-15 presents sketches of a compressor piston and the internal suction and discharge valves in different stages of the compression cycle.

Figure R6-16 shows an assembly consisting of the piston, wrist pin, connecting rod, and crankshaft. All components of the reciprocating piston arrangement are finely machined, balanced carefully to eliminate vibration, and fitted with close tolerances to assure that the compressor will have a high efficiency in pumping the refrigerant vapor. A different type of crankshaft, one of an eccentric design, is shown in Fig. R6-17. The connecting rod is assembled on an offcenter eccentric fastened with balance weights. If the crankshaft is not almost completely machined, it should be dynamically balanced.

The internal valves of a compressor receive quite a bit of wear and tear in normal operation, since they must open and close hundreds of times each minute the compressor is running. Small commercial units usually have a high-grade steel disk or reed type of valve, both of which are quieter operating, efficient, simpler in construction, and longer lasting than the nonflexing ring-plate type of valve. Figure R6-18 shows some of the various designs of internal compressor valves. The proper operation of the valves is very important to the overall efficiency of the compressor.

If the suction valves do not seat properly and allow refrigerant vapor to escape from the cylinder, the piston cannot pump out all of the compressed vapor into the hot-gas line. If the suction valve leaks, the compressed

FIGURE R6-13   Typical reciprocating compressor. (*Courtesy* Borg-Warner Air Conditioning, Inc.)

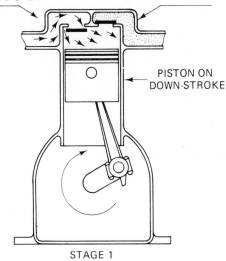

DOWN STROKE OF PISTON CREATES VACUUM IN CYL-INDER. PRESSURE IN SUCTION LINE FORCES SUCTION VALVE OPEN

PRESSURE IN DISCHARGE LINE HOLDS DISCHARGE VALVE CLOSED.

PISTON ON DOWN-STROKE

STAGE 1

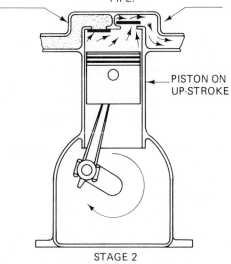

PRESSURE IN CYLINDER HOLDS SUCTION VALVE CLOSED

PRESSURE IN CYLINDER RAISES DISCHARGE VALVE GAS FLOWS INTO DISCHARGE PIPE.

PISTON ON UP-STROKE

STAGE 2

FIGURE R6-15   How differential pressures work the valves of the reciprocating compressor.

vapor, or part of it, will go into the suction line and heat up the low-pressure, low-temperature vapor there. If the discharge valve leaks some of the high-pressure, high-temperature vapor in the hot gas line will leak back into the cylinder on the down stroke of the piston, limiting the volume of suction vapor entering the cylinder.

In an open-type compressor, one end of the crank-shaft extends through the crankcase housing for connection directly to an outside drive motor, or it may have a pulley attached for belt drive by an external motor.

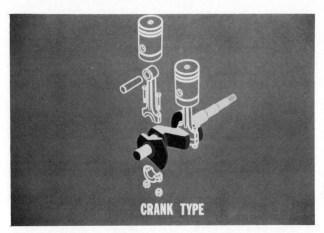

FIGURE R6-16   Crank-type assembly. (*Courtesy* Carrier Air-Conditioning Company)

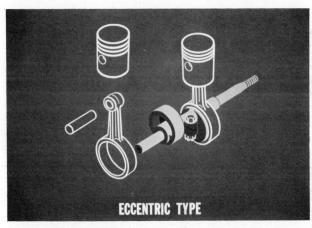

FIGURE R6-17   Eccentric-type crankshaft. (*Courtesy* Carrier Air-Conditioning Company)

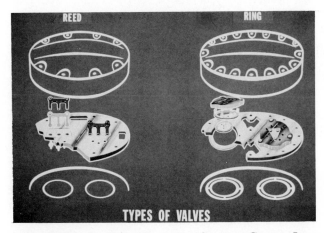

FIGURE R6-18   Types of valves. (*Courtesy* Carrier Air-Conditioning Company)

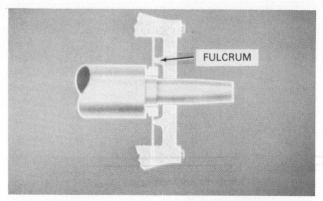

FIGURE R6-19 Crankshaft seal (diaphragm type). (*Courtesy* Carrier Air-Conditioning Company)

Some provision must be made to prevent the leakage of gas and oil around the crankshaft where it extends through the compressor shell; this is accomplished through the addition of a shaft seal.

One type of shaft seal is pictured in Fig. R6-19. The type of crankshaft shown has a seal shoulder built into it, against which a neoprene washer and self-lubricating seal ring are held stationary by the seal cover plate. A gas- and oil-tight seal is maintained between the seal ring and the crankshaft shoulder seal by the neoprene washer, which fits tightly on the shaft. Seals on reciprocating compressors are on the low-pressure or suction side. It is desirable that as nearly perfect a seal as possible be maintained, since if conditions require that the low-pressure side of the system operate in a vacuum, a leak at the seal or elsewhere in the low side would draw air and moisture into the system.

In most reciprocating compressors gaskets are utilized between mating parts to assure leakproof conditions, because most surfaces are not so finely machined as to provide metal-to-metal leakproof joints. Primarily, gaskets are used between the cylinder head and the valve plate, between the valve plate and the compressor housing, between the compressor body and the bottom plate (if there is one), and also between the exterior service valves and their mounting bases.

When the mating parts are tightly secured, they impress their form and outline on the gasket material, which is usually soft and resilient enough to take the impression and thus seal off any gas or oil from possibly leaking to the atmosphere and prevent taking in air. The gasket material must be such that there will be no chemical reaction when it comes in contact with the oil and refrigerant in the system. When gaskets need replacing after some component has been removed and possibly replaced, replacement gaskets should be of the same material originally used by the manufacturer and of the same thickness as those removed, whether they are of aluminum, cork, rubber, asbestos, or composition. A variation in thickness will affect the efficiency and

operation of the compressor. Too thick a gasket between the compressor housing and the valve plate will increase the clearance space above the piston and cause a loss in the volumetric efficiency. Too thin a gasket may permit the piston to hit against the valve plate, damaging the compressor.

As already mentioned, reciprocating compressors of the open type need externally driven motors, which may be connected directly through the use of couplings. This causes the compressors to operate at the same speed as the driving motors. Or a compressor may have a flywheel on the end of the crankshaft, which is turned by means of one or more V-belts between the flywheel and a pulley mounted on the motor shaft. The speed at which the compressor will turn depends on the ratio of the diameters of the flywheel and the motor pulley. The speed of the motor may be calculated as follows:

$$\text{compressor rpm} = \frac{\text{motor rpm} \times \text{pulley diameter}}{\text{flywheel diameter}}$$

EXAMPLE

At what speed will a compressor turn if it has a 10-in. flywheel and is driven by a 1725-rpm motor with a pulley 4 in. in diameter?

SOLUTION

$$\text{Compressor rpm} = \frac{1725 \times 4}{10} = 690 \text{ rpm}$$

The purpose of the hermetic compressor is the same as that of the open compressor, to pump and compress the vapor, but it differs in construction in that the motor is sealed in the same housing as the compressor. A typical fully hermetic compressor cutaway is shown in Fig. R6-20. Note the vertical crankshaft, with the connecting rod and piston in a horizontal position. The fully hermetic unit has an advantage in that there is no projecting crankshaft; therefore no seal is necessary, and there is no possibility of leakage of refrigerant from the compressor or of air being drawn in when the system is operating in a vacuum. A compressor of this design cannot be serviced in the field; internal repairs must be made in a region or area repair station or at the factory where it was manufactured.

Some hermetic compressors are constructed with internally mounted springs to absorb vibration caused by the pulsation of the refrigerant vapor being pumped by the pistons. Some hermetic compressors also have springs or hard-rubber vibration mounts located on the outside to absorb shock and vibration.

The bottom portion of the hermetic compressor acts as an oil sump, like the crankcase of an open-type compressor. As the oil circulates and lubricates the internal moving parts, it picks up some of the compressor heat caused by friction of the moving parts. The oil

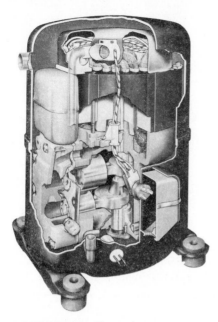

FIGURE R6-20 Typical reciprocating compressor. (*Courtesy* Tecumseh Manufacturing, Inc.)

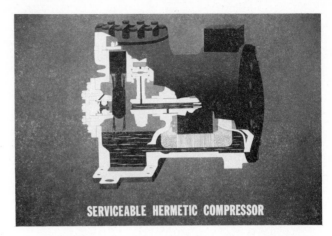

SERVICEABLE HERMETIC COMPRESSOR

FIGURE R6-22 Serviceable hermetic compressor. (*Courtesy* Carrier Air-Conditioning Company)

transfers some of this heat to the external shell of the compressor.

Most hermetic compressors are constructed so that the suction vapor is drawn across the motor windings before it is taken into the cylinder or cylinders. This, of course, helps to remove some of the heat from the motor windings and also helps to evaporate any liquid refrigerant that may have entered the compressor.

Suction and discharge mufflers are built into some of the smaller hermetic compressors, to absorb or lessen the sound caused by the pulsing vapor as it is pumped through the compressor. A suction muffler can be seen in the cutaway view shown in Fig. R6-20.

Figure R6-21 shows a typical external-type discharge muffler arrangement used on some compressors.

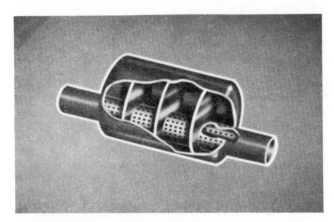

FIGURE R6-21 External discharge muffler. (*Courtesy* Carrier Air-Conditioning Company)

Since the crankshaft and motor shaft are the same unit, high-speed operation causes considerably more noise than the slower operating open compressors, thus requiring mufflers.

Another type of compressor is shown in Fig. R6-22. It combines the motor in the same shell as the compressor, but, unlike the fully hermetic unit, this type provides access to the compressor for repair in the field. This unit is called by several names such as *semihermetic, accessible,* and *serviceable hermetic.*

Rotary compressors are so classified because they operate through application of a rotary, or circular, motion, instead of the reciprocating operation previously described. A rotary compressor is a positive displacement unit, and usually can be used to pump a deeper vacuum than a reciprocating compressor.

There are two primary types of rotary compressors used in the refrigeration field: the *rolling piston* type, with a stationary blade, and the *rotating blades* or *vanes* type. Both are similar in capacity, variety of applications, physical size, and stability, but they differ in manner of operation.

The rolling piston type, as shown in Fig. R6-23, has the roller mounted on an eccentric shaft. The blade is located in a keeper in the housing of the compressor. As the rolling piston rotates, vapor is drawn into the space ahead of the spring-loaded blade, as shown, and is compressed by the roller into a continually smaller space until it is forced out the discharge port, and the compression cycle begins again. The roller does not make a metal-to-metal contact with the cylinder, because a film of oil, in normal operation, provides a clearance between the two surfaces.

Figure R6-24 shows the other primary type of rotary compressor. This unit consists of a cylinder and an eccentric roller or rotor having several blades, held in place by either springs or centrifugal force. As the roller turns in the cylinder, suction vapor is trapped in the

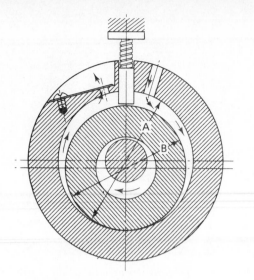

FIGURE R6-23 Rolling piston rotary compressor. (*Courtesy* ASHRAE)

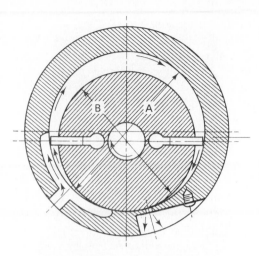

FIGURE R6-24 Rotating vane rotary compressor. (*Courtesy* ASHRAE)

crescent-shaped space between two of the blades. As the roller continues to turn, the suction gas is compressed in volume, and its pressure and temperature are increased until it is discharged from the cylinder.

As mentioned earlier, a film of oil prevents leakage of vapor from the cylinder, or between the spaces separated by the blades, while the unit is in operation. To prevent hot gas from leaking back into the cylinder from the discharge port when it is uncovered by the roller, a check valve is usually placed in the discharge line. During the OFF cycle or period of shutdown, warm vapor

is prevented from leaking back into the evaporator by the check valve.

Rotary compressors are well-balanced units, and those enclosed in a hermetic shell are usually spring-supported or spring-mounted. Usually, they are very quiet in operation. Since hermetic rotary compressors are direct-driven, they operate at the motor speed—usually 3450 rpm—and, although the sound level is in direct ratio to speed and horsepower, they operate comparatively quietly by any standards. Discharge mufflers are widely used to prevent the pulsations of gas being discharged

FIGURE R6-25 Complete centrifugal refrigeration unit. (*Courtesy* Carrier Air-Conditioning Company)

from causing vibrations to be carried over to the discharge line and condenser. The principles of design and operation given here pertain to household and small commercial units, although, in general, the same principles also apply to larger rotary compressors, some of which have primary usage in the low-temperature field where they are used as low-stage or booster compressors.

A complete centrifugal refrigeration unit is shown in Fig. R6-25. Like the other types of compressors, it compresses the refrigeration vapor, as its name implies, through centrifugal action or force. This action is performed mainly by the impeller or rotor as shown in Fig. R6-26. Vapor is drawn in at the intake near the shaft of the rotor and discharged from the exhaust openings at the outer edge of the rotor. With the rotation of the impeller, suction vapor is drawn rapidly into the impeller chambers, where it is forced to the outside of the housing sections through centrifugal action. To maintain the centrifugal force, the impeller is operated at a high rate of speed by an outside driving force, such as an electric motor, gasoline engine, or steam turbine. The pressure differential between inlet and outlet vapor is small. Therefore, it is not a positive displacement unit like the types described previously, and it is not capable of building up pressure against a closed valve in the system.

A centrifugal compressor may have one or more impellers. Compressors with several stages are constructed so that the discharge of one impeller or stage enters the suction inlet of the next. If the speed of the driving motor does not provide the desired operating speed of the compressor, speed-increasing gears or fluid couplings may be used to obtain optimum operating conditions. Since there are no pistons or internal suction and discharge valves, there is little wear and tear on the unit. The main bearings in the housing, supporting the drive shaft, are the components most subject to wear.

R6-4
CONDENSERS

The next major component in the refrigeration system, following the compression stage, is the condenser. Basically, the condenser is another heat-exchange unit in which the heat picked up by the refrigerant in the evaporator—as well as that added to the vapor in the compression phase—is dissipated to some condensing medium. High-pressure, high-temperature vapor leaving the compressor is superheated, and this superheat customarily is removed in the hot-gas discharge line and in the first portion of the condenser. As the temperature of the refrigerant is lowered to its saturation point, the vapor condenses into a liquid for reuse in the cycle.

Condensers may be air-cooled, water-cooled, or cooled by evaporation. Domestic refrigerators usually have an air-cooled condenser, which depends on the gravity flow of air circulated over it. Other air-cooled units use fans to blow or draw large volumes of air across the condenser coils.

Figure R6-27 depicts a typical small commercial condensing unit using an air-cooled condenser. It is dependent on an ample supply of relatively "cool" air, for, in order to have a heat transfer from the refrigerant in the condenser to the coolant, the air must be at a lower temperature than the refrigerant. Even when the surrounding temperature is above 100°F, the air is still cooler than the refrigerant in the condenser, which must give up some heat to return to its liquid state.

FIGURE R6-26 Impeller or rotor. (*Courtesy* Carrier Air-Conditioning Company)

FIGURE R6-27 Small air-cooled condensing unit. (*Courtesy* Copeland Corporation)

Air-cooled condensers are constructed somewhat like other types of heat exchangers, with coils of copper or aluminum tubing equipped with fins. Evaporators usually have filters in front to reduce clogging by dust, lint, and other matter, but condensers are not so equipped, and so must be cleaned frequently to prevent reduction of their capacities.

Remote air-cooled condensers usually have wider fin spacing to prevent clogging as quickly as in those directly mounted on the condensing unit. Also, they can be located away from the compressor, which is a distinct advantage. Occasionally, a complete condensing unit is placed somewhere inside the building in which it is to be used, where the heat dissipated from the condenser and motor can cause an increase in temperature within the storage or mechanical equipment room. As a result, the unit might have a higher operating discharge temperature and pressure, which would decrease its efficiency.

Figure R6-28 shows a remote air-cooled condenser, which may be located outdoors—beside a building or on a flat roof. In such an open, outdoor location, an adequate supply of air as a coolant is readily available at the ambient outdoor temperature, thus avoiding undesirable temperatures in the building. The air movement across the coil is created by either a belt-driven centrifugal fan or a direct-drive propeller-type fan. The slow-speed wide-blade propeller fan moves the required volume of air without creating unreasonable noise.

This type of condenser may be assembled in any combination of units that may be required for the necessary heat removal. The air may either be drawn through or blown through the coils. In another design, a single condenser may have more than one circuit in its coil arrangement, so that it may be used with several separate evaporators and compressors.

FIGURE R6-28 Remote air-cooled condenser. (*Courtesy* Carrier Air-Conditioning Company)

In most installations of remote air-cooled condensers in this country, the difference between the ambient air temperature and the condensing temperature of the refrigerant is approximately 30°F. Therefore, if the outdoor temperature is 95°F, the refrigerant will condense at approximately 125°F.

Some difficulty may arise with remote air-cooled condensers when they are operated in low ambient temperatures, unless proper precautions are taken to maintain head pressures that are normal for the unit. Earlier in this chapter it was stated that a too-high condensing temperature and pressure lessens the overall operating efficiency of the unit. Conversely, a too-low condensing temperature and pressure will affect the efficient operation of the system by causing a reduction in pressure difference across the metering device, thus resulting in a loss of refrigerant flow into the cooling coil. Later in this chapter the effect of pressure drop through a metering device on the overall capacity and efficiency of a system will be discussed.

Some remote air-cooled condensers equipped with multiple fans have controls for the cycling of one or more of the fans during periods of low ambient temperatures. The flow of air across other types of condensers may be controlled by adjustable louvers. Still other manufacturers install controls to allow partial flooding of the condenser with liquid which, in turn, will lessen the condensing capacity. This is another means of keeping the head pressure within allowable limits.

Water-cooled condensers permit lower condensing temperatures and pressures, and also afford better control of the head pressure of the operating units. They may be classified as follows:

1. Shell-and-tube
2. Shell-and-coil
3. Tube-in-a-tube

As pointed out in an earlier chapter, water usually is an efficient medium for transferring heat, since the specific heat of water is 1 Btu per pound per °F change in temperature. If 25 lb of water increases 20°F, and if this heat is removed in 1 minute, 500 Btu is removed from the source of heat each minute. Also, if this rate of heat transfer continues for 1 hour (60 minutes), this means that the water is absorbing 30,000 Btu/hr.

If the example above involved a water-cooled condenser, it would mean that 3 gallons of water was being circulated per minute. Also, if the heat of compression amounted to 6000 Btu/hr, it would mean that the evaporator load was 24,000 Btu/hr—a 2-ton refrigeration load.

Water coming from a well or other underground source will be quite a bit cooler than the ambient outdoor air. If cooling-tower water is used, its temperature can be lowered in the cooling tower, after it has picked up heat in the condenser, to within 5 to 8°F of the outdoor wet-bulb temperature. The use of a cooling tower and circulating pump permits the reuse of the water, except for a slight loss due to evaporation, and keeps the consumption and cost of water to a minimum.

The shell-and-tube type of water-cooled condenser consists of a cylindrical steel shell containing several copper tubes running parallel with the shell. Water is pumped through the tubes by means of the inlet and outlet connections on the end plates. The hot refrigerant vapor enters the shell at the top of the condenser, as shown in Fig. R6-29, and the liquid refrigerant flows as it is needed from the outlet at the bottom of this combination condenser-receiver.

The end plates are bolted to the shell of the condenser for easy removal to permit the rodding or cleaning of the water tubes of minerals that may be deposited on the inside of the tubes, causing a restricted water flow, a reduction in the rate of heat transfer, or both. A control of the water flow, that is, the number of times it travels the length of the condenser or the number of passes it makes, is built into the end plates of the condenser. If the water enters one end plate, passes through all the tubes once, and leaves the condenser at the other end plate, it is called a *one-pass condenser*. If the water inlet and outlet are both in the same end plate, it is a *two-pass* or some other even-numbered type of pass condenser.

If, instead of a number of tubes within the condenser shell, there are one or more continuous or assembled coils through which water flows to remove heat from the condensing vapor, it is classified as a *shell-and-coil* type of condenser. Figure R6-30 shows a condenser of this type. It is a compact unit and usually serves as a combination condenser-receiver within the circuit. Usually, this type of condenser is used only on small-capacity units and when there is an assurance of rea-

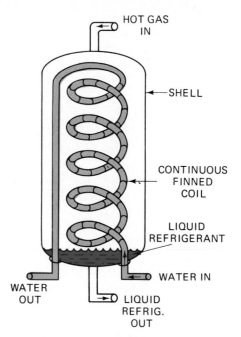

FIGURE R6-30   Shell-and-coil water condenser. (*Courtesy* Carrier Air-Conditioning Company)

sonably clean water, for the only means of cleaning it is by flushing with a chemical cleaner.

The *tube-in-a-tube,* or *double-tube* as it is also known, may be classified as a combination air-and-water-cooled type of condenser. As pictured in Fig. R6-31, it has the refrigerant flowing through the outer tubing where it is exposed to the cooling effect of air flowing naturally over the outside of the outer tubes while water is being circulated through the inner tubes. Generally, water enters the bottom tubes of water-cooled condensers and leaves at the top. In this manner peak efficiency is obtained, for the coolest water is capable of removing some heat from the refrigerant in a liquid state, thereby subcooling it. Then the warmer water still is able to absorb heat from the vapor, assisting in the condensation process.

When the ambient temperature is such that a satisfactory condensing temperature cannot be obtained with an air-cooled condenser, and when the water supply is inadequate for heavy usage, an *evaporative condenser* may be used to advantage. A diagram of this type of condenser is shown in Fig. R6-32 which shows the combined use of air and water for the purpose of heat removal from the refrigerant vapor within the condenser coil.

There is actually a double heat transfer in this unit: The heat of the vapor and the coil containing it is transferred to the water, wetting the outer surface of the coil, and then is transferred to the air as the water evaporates. The air can either be forced or drawn through the spray water.

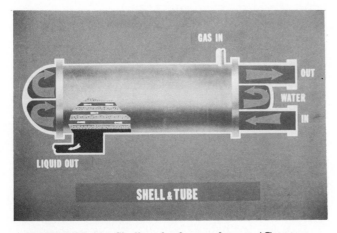

FIGURE R6-29   Shell-and-tube condenser. (*Courtesy* Carrier Air-Conditioning Company)

Compression Cycle Components   65

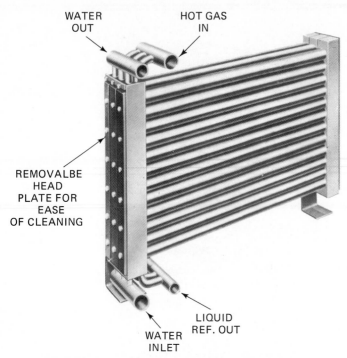

FIGURE R6-31 Water-cooled tube-in-tube condenser. (*Courtesy* Halstead & Mitchell, Division of Climate Control)

When the air is blown through the unit, the fan and motor are in the dry entering airstream. When the system has a draw-through fan, it is essential that eliminators be installed before the air mover. Otherwise, there would quickly be an accumulation of scale on all of the air-mover components. Even with blow-through units, there is a possibility of some of the spray water blowing out of the evaporative condenser, and a set of eliminator plates should be installed to prevent this.

R6-5
RECEIVERS

As mentioned earlier, some water-cooled shell-and-tube condensers also act as receivers, with liquid refrigerant occupying the space in the bottom of the condenser where there are no water tubes. If there is too much liquid in this type of condenser-receiver, some of the water tubes may be covered by the liquid level. This reduces the area of heat transfer surface in the condenser.

In systems other than those having condenser-receivers and those operating with a critical refrigerant charge, a receiver is needed, which is actually a storage container for the refrigerant not in circulation within the system. Receivers that are part of small, self-contained commercial units usually are large enough to hold the complete operating charge of refrigerant in the systems. This applies to a number of larger systems as well. Yet, in some cases, the receiver may not be large enough to hold the entire refrigerant charge if a pump-down becomes necessary for repair or replacement of a component. An auxiliary receiver would be necessary to provide pump-down capacity. If this is not provided, the surplus refrigerant would have to be pumped into an empty refrigerant drum or wasted to the atmosphere.

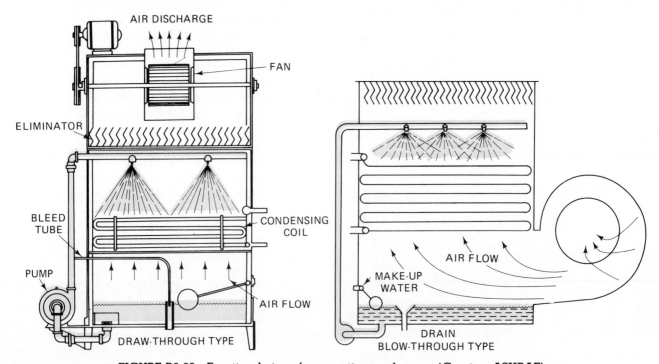

FIGURE R6-32 Functional view of evaporative condensers. (*Courtesy* ASHRAE)

Precautionary measures are usually taken by manufacturers of receivers against the possibility of too much pressure or too high a temperature build-up in the receiver or in a combination condenser-receiver. These safety measures usually include the installation of pressure relief valves, customarily spring-loaded, that will open if excessive pressure should build up within the receiver. A fusible-plug type of relief valve may be installed, which is designed to melt at a preselected temperature and thus release the refrigerant if, for any reason whatsoever, that temperature is reached within the receiver.

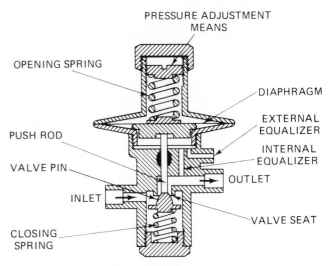

*VALVE IS USED WITH EITHER INTERNAL OR EXTERNAL EQUALIZER, BUT NOT WITH BOTH

FIGURE R6-33   Constant-pressure expansion valve. (*Courtesy* ASHRAE)

R6-6
REFRIGERANT FLOW CONTROLS

A fundamental and indispensable component of any refrigeration system is the *flow control,* or pressure-reducing device. Its main purposes are:

1. To maintain the proper pressure and boiling point in the evaporator to handle the desired heat load
2. To permit the flow of refrigerant into the evaporator at the rate needed to remove the heat of the load

The pressure-reducing device is one of the dividing points in the system.

The principal means of refrigerant flow control, in the early stages of refrigeration, was a basic hand valve. Knowing their work and their equipment, early operators of units in ice plants and similar operations having constant loads knew how far to open the hand valve for the work to be performed. However, in modern applications that have frequently varying loads, this is impractical because the hand-valve setting would have to be changed as the load changed.

The five main types of pressure-reducing devices now used in various phases of refrigeration are:

1. Automatic expansion valve
2. Thermostatic expansion valve
3. Capillary tube
4. Low-side float
5. High-side float

All are used to reduce the liquid refrigerant pressure and in some cases, to control the volume of flow.

The hand valve is obviously not suited for automatic operation, since any variation in requirements needs manual adjustment, so the automatic expansion valve came into being.

The *automatic expansion* or *constant-pressure valve,* shown in Fig. R6-33, maintains a constant pressure in the cooling coil while the compressor is in operation. In this diaphragm type of constant pressure expansion valve, the pressure in the evaporator effects the movement of the diaphragm, to which the needle assembly is attached.

A condition of stability in refrigerant flow and evaporation is necessary for the correct operation of the constant-pressure expansion valve. Like the hand valve, its use is limited to conditions of more or less constant loads on the evaporator, a situation that applies to the automatic expansion valve as well. In either valve there is a screw that applies pressure to the spring above the bellows or diaphragm. When the screw is adjusted clockwise, it causes more pressure on the bellows or diaphragm, forcing the valve to open more, admitting additional refrigerant to the evaporator, and resulting in higher operating pressure. If a lower operating pressure in the cooling coil is desired, the screw is turned counterclockwise, releasing pressure on the spring and therefore on the bellows or diaphragm. This allows the valve to close and curtails the flow of refrigerant. Following any adjustment, ample time should be allowed for a controlling device to settle down before any further change is made in its setting. For a given load on the evaporator coil being fed refrigerant, there is only one correct setting of the automatic expansion valve: when the coil is completely frosted. If the pressure is lowered, there will be a curtailment in the refrigerant flow, and the heat absorption capability of the coil will be lessened. If the pressure is raised, the flow of refrigerant will increase, with the possibility of liquid refrigerant flooding into the suction line, from which the refrigerant might reach the compressor and damage it.

Because all refrigeration loads do not remain constant, and someone cannot always be present at every installation to make compensating adjustments, another type of valve, the *thermostatic expansion valve,* was developed. Like the automatic expansion valve, the thermostatic expansion valve may be either the bellows type or the diaphragm type shown in Fig. R6-34. Both are equipped with a capillary tube and feeler bulb assembly, which transmits to the valve the pressure relationship of the temperature of the suction vapor at the outlet of the evaporator coil, where the feeler bulb is attached.

The basic purpose of the thermostatic expansion valve is to maintain an ample supply of refrigerant in the evaporator, without allowing liquid refrigerant to pass into the suction line and the compressor. When the metering device is a thermostatic expansion valve, its operation will depend on superheated vapor leaving the evaporator, since a portion of the evaporator is used for the superheating of the vapor about 5 to 10°F above the temperature corresponding to the evaporative pressure.

The capillary tube, which is based on the principle described above, is the simplest form of refrigerant control or metering device and generally the least expensive. There are no moving parts to wear out or require replacing, since it is a small-diameter tube of the right length for the refrigeration load it is designed to handle. This pressure-reducing device, like any other, is located between the condenser and evaporator, at the end of the liquid line or instead of a liquid line. One type of capillary refrigerant control is shown in Fig. R6-35. The advantages of this control have just been discussed; however, its disadvantages are that it is subject to clog-

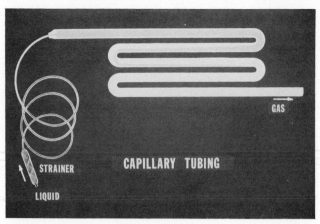

FIGURE R6-35 Capillary tubing. (*Courtesy* Carrier Air-Conditioning Company)

ging, requires an exact refrigerant charge, and is not as sensitive to load changes as other metering devices. Its internal cross-sectional area is so small that it takes only a minute dirt particle to plug the tube, or a small amount of moisture to freeze in it. A dryer and filter or strainer should be installed at the inlet to the capillary tube to prevent this clogging possibility.

Another type of refrigerant control device is the *float arrangement,* which also meters the flow of refrigerant into the evaporator. The float itself is made of metal that will not cause a reaction with the refrigerant used in the system. It is constructed in the shape of a ball or an enclosed pan, which will rise or fall within the float chamber with the level of the refrigerant. It is connected through an arm and linkage to a needle valve, which opens and closes against a seat, allowing and curtailing the flow of refrigerant into the chamber.

A *high-side float,* as its name indicates, is located in the high-pressure side of the system. It may be of a vertical or horizontal design and construction, and it may be located near either the condenser or the evaporator. A typical high-side float is shown in Fig. R6-36. Its design is such that, as the float chamber fills with refrigerant, the buoyancy of the float lifts it and raises the valve pin away from the seat. This permits the refrigerant to flow or be metered to the low-pressure side of the system and into the evaporator.

Since the float, through the pivot arrangement, is set to open at a given level, only a small amount of liquid stays in the high-side float chamber: most of the refrigerant in the system is in the evaporator. Therefore, the system refrigerant charge is critical, to the extent that only enough refrigerant to maintain the proper level in the flooded evaporator is desirable, with no liquid flooding over into the suction line and compressor. If there is an overcharge, flooding will result, whereas a shortage of refrigerant will cause the evaporator to be starved and the system will be inefficient.

A *low-side float* metering device is one in which

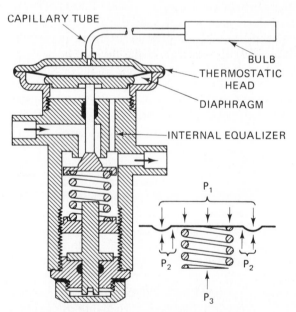

P₁—THERMOSTATIC ELEMENT'S VAPOR PRESSURE
P₂—EVAPORATOR PRESSURE
P₃—PRESSURE EQUIVALENT OF THE SUPERHEAT SPRING FORCE

FIGURE R6-34 Diaphragm-type thermostatic expansion valve. (*Courtesy* ASHRAE)

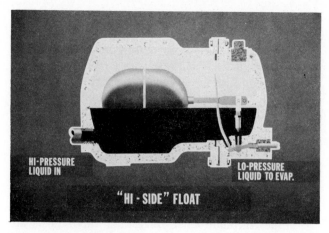

FIGURE R6-36 High-side float. (*Courtesy* Carrier Air-Conditioning Company)

the float is located in the evaporator, or in a chamber adjacent to the cooling coil that is flooded, maintaining a definite liquid level within the evaporator. It is constructed somewhat like the high-side float, with the exception that as the float rises it closes off the flow of the refrigerant. Its action is shown in Fig. R6-37: the high-pressure liquid is at the inlet to the float chamber, and the float assembly itself is in the low-pressure liquid, maintaining a definite level within the float chamber. As the load on the evaporator increases, liquid is evaporated and the liquid level in the evaporator and the float chamber drops. As the float lowers, the needle is pulled away from the seat, allowing additional refrigerant to enter until the desired level is reached.

If the load on the evaporator decreases, less evaporation will take place, and the liquid level will be maintained, the float causing the needle to close against the seat. In this manner the low-side float assembly can maintain the correct flow of refrigerant as it is needed, by a fluctuating load condition. Usually, the same float arrangement cannot be used if a different refrigerant is

desired in the system because of its operating characteristics, for the refrigerants will have different specific gravities as other characteristics change. A float with the correct buoyancy must be used.

A *check valve,* as shown in Fig. R6-38, is sometimes used to prevent the high pressure of one system or evaporator from backing up into a lower-pressure evaporator, when there are multiple coils in a system operating at different pressures. It may also be used to prevent an equalization of pressures when a system is not in operation. This type of valve opens easily if the flow is in the proper direction, but it will shut off the flow in the reverse direction, with the aid of a spring pressure or the weight of the internal valve itself.

A *solenoid,* or *magnetically operated, valve,* the operating principles of which will be explained in a later chapter, is frequently used in refrigeration lines. Figure R6-39 shows an exterior view and a cross section of the internal construction of such a valve. When a positive shutoff of liquid flow is desired, for a pump-down cycle

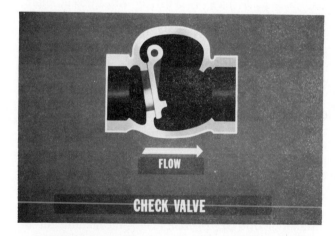

FIGURE R6-38 Check valve. (*Courtesy* Carrier Air-Conditioning Company)

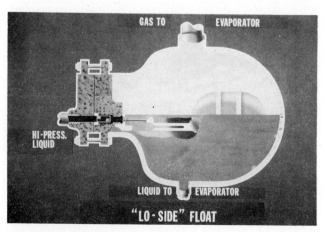

FIGURE R6-37 Low-side float. (*Courtesy* Carrier Air-Conditioning Company)

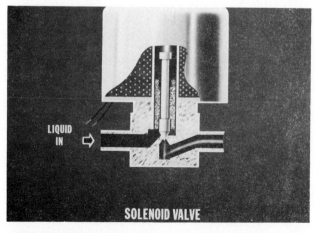

FIGURE R6-39 Solenoid valve. (*Courtesy* Carrier Air-Conditioning Company)

Compression Cycle Components 69

permitting the compressor to start up unloaded, a solenoid valve is installed in the liquid line ahead of the metering device. Its operation is controlled by a thermostat and an electrical circuit or by some other means of actuating the valve. Most of these valves close by gravity and the weight of the plunger and valve, when the electrical circuit or other means of holding the valve open is broken.

R6-7
REFRIGERANT LINES

Previous portions of this chapter have been devoted to the proper selection, balancing, and operation of the major components of a refrigeration system. But regardless of how well these have been selected and balanced, system operation is dependent on the means of moving the refrigerant, both liquid and vapor, from one component to another in the refrigeration circuit.

Just as a highway must be constructed, maintained, and kept open between communities to provide adequate access for the vehicles that must use it, so must the piping in a refrigeration system be properly sized and installed so that there will be no restrictions to the flow of the refrigerant.

Refrigerant oil, necessary for the proper lubrication of the moving parts in the compressor and the metering device, must be readily miscible with the refrigerant in the liquid state. The oil will travel with the liquid as long as the liquid line is sized so that the refrigerant will travel through its entire length at a proper velocity.

If the system is self-contained, proper sizing and installation of the refrigerant lines is the manufacturer's responsibility. But a system built up in the field, using products of several manufacturers, becomes the problem and responsibility of the person designing the complete system, as well as those who install and connect the components. A liquid line that is sized too small or that has too many restrictions (fittings and bends) could easily cause too great a pressure drop in the line, which could result in a loss in the capacity of the pressure-reducing device when compared to the capacity required in the cooling coil.

Proper sizing of all refrigerant lines and the tables to be used in the selection of liquid, suction, and hot gas lines is discussed in Chapter R13.

Lines through which refrigerant vapor flows are the most critical, and these are the suction and hot gas or discharge lines. The velocity of the vapor should be at least 750 ft/min in horizontal lines and above 1500 ft/min in vertical lines, so that the refrigerant oil will be entrained with the vapor and returned to the compressor. If the lines are too large, the desired velocity cannot be maintained and the oil may not return, and the compressor may become short of its proper oil charge.

If the evaporator is located above the compressor, the oil usually will return to the compressor through gravity flow, provided that no risers or possible traps are built into the line.

If the evaporator is situated below the compressor, or when the condenser is located a distance above the compressor, the vapor must have the proper velocity in order to carry the oil droplets with it. Oil traps may have to be built into the piping, or it may be necessary to use dual pipes, if the capacity control of the compressor varies because of changing load conditions. These measures ensure that even in the case of minimum load conditions of 10 to 25% of full capacity, the refrigerant vapor will have adequate velocity to carry the oil along with it.

PROBLEMS

R6-1. Name the four major components in the compression-type refrigeration cycle.

R6-2. What are the most frequently used methods for compensating for conduction loss due to an air film around the evaporator coil tubing?

R6-3. Under what circumstances should nonfrosting evaporator coils be installed?

R6-4. What are some types of cooling coils used for cooling liquids?

R6-5. What are the three major classifications (by method of compression) of compressors?

R6-6. What are the two types of reciprocating compressor valve reeds?

R6-7. What is the difference between a hermetic and an open compressor?

R6-8. What are the two primary types of rotary compressors used in the refrigeration field?

R6-9. How many valves are used in the compression operation of a centrifugal compressor?

R6-10. What is the basic purpose of the condenser?

R6-11. What are the primary types of condensers?

R6-12. What are the principal types of pressure-reducing devices used to control the flow of refrigerant to the evaporator?

R6-13. Which type of pressure-reducing device is the simplest? Why?

R6-14. What is the purpose of a check valve?

R6-15. What is the purpose of a solenoid valve?

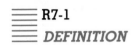

Refrigerants

R7-1
DEFINITION

Refrigerants are the vital fluids in both compression- and absorbtion-type refrigeration systems. They absorb heat from the place where it is not wanted and expel it elsewhere. The evaporation or boiling of the liquid refrigerant absorbs the heat that is released by the condensation of the vapor.

Any substance that undergoes phase change from liquid to vapor, and vice versa, may function as the refrigerant in vapor–compression-type systems. However, only those substances that undergo these changes at commercially useful temperature and pressure levels are of practical value.

R7-2
PRESSURE-TEMPERATURE RELATIONSHIP

This subject was discussed in Section R3-11 but because of its importance when discussing various refrigerants, we will also discuss it here. All liquids have pressure–temperature relationships, but they are different for the different liquids. To be more specific, at standard atmospheric conditions, water boils at 212°F, Refrigerant 12 at −21°F, Refrigerant 22 at −40°F, etc. For purposes of discussion on the principles of pressure–temperature relationships, we will start by using the most common liquid—water.

R7-3
WATER AS A REFRIGERANT

Water is the most efficient refrigerant from a heat-absorbing standpoint—980 Btu per pound of water evaporated. It can also be made to boil and produce re-

frigeration temperatures as low as 40°F and at one time was used extensively in large refrigeration systems producing 40°F water for air-conditioning and process work. To understand how this is done, you need only remember that a refrigeration system operates by "control of the boiling point" of the refrigerant. To cause water to absorb heat and boil, it is only necessary to lower the boiling point of the water far enough below the temperature from where the heat will be removed (the heat source) to obtain the rate of heat transfer desired. How this is done is discussed in later chapters.

If, for example, we want to remove heat from 50°F oil and the equipment design requires a 10°F ΔT (temperature difference) between the temperature of the oil

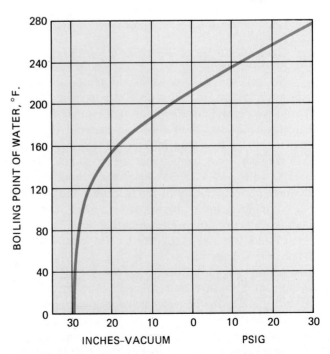

FIGURE R7-1 Pressure–temperature curve for water.

71

and the refrigerant (water), we need a boiling point of the water of 40°F. What pressure must we maintain on the water to obtain this boiling point?

R7-3.1
Pressure–Temperature Curve

Figure R7-1 shows a pressure–temperature curve for water. This curve is a plotting on a graph of the boiling points of water at various applied pressures. Using this curve, we can determine that to obtain a 40°F boiling point, finding the 40°F point on the left "boiling point of water" scale and moving to the right along this line we reach the pressure–temperature curve. At this point, if we move downward along the vertical pressure lines we come to what appears to be 29 in. Hg in the vacuum range.

The fault of using a curve of this type for water is the inaccuracy of the curve in the lower range of temperatures because of the wide change in boiling points over a very small change in applied pressure.

R7-3.2
Pressure–Temperature Chart

A more accurate way of presenting the pressure–temperature relationship is via a pressure–temperature chart. Figure R7-2 shows such a chart for water. Moving down the temperature column to the desired boiling point of 40°F, and moving horizontally to the pressure column, we find the desired applied pressure to be 29.67 in. Hg in the vacuum range.

R7-3.3
Disadvantages of Water as a Refrigerant

To reclaim the water vapor and reuse it in the refrigeration system, it is now necessary to raise the condensing temperature of the vapor high enough above the temperature of the heat-absorbing material (the heat sink) to get the heat-transfer rate necessary to accomplish the task. If we assume that we are to put the heat in 75°F water and a 25°F ΔT (temperature difference) between the temperature of the heat sink and the condensing temperature of the water is needed, what condensing pressure must we maintain? A 75°F heat sink plus 25° ΔT is a 100°F condensing temperature. From the pressure–temperature table we find that 28 in. Hg in the vacuum range is needed.

From this we conclude that the entire refrigeration system operates in the vacuum or below-atmospheric range. This operating pressure range makes it very difficult to keep the system leaktight and requires high maintenance. As a result, water as a refrigerant was short-lived.

FIGURE R7-2
Standard steam table.

Temperature (°F)	Vacuum (in. Hg) Referred to a 30-in. Bar (Hg at 58.4°F)	Temperature (°F)	Vacuum (in. Hg) Referred to a 30-in. Bar (Hg at 58.4°F)	Temperature (°F)	Vacuum (in. Hg) Referred to a 30-in. Bar (Hg at 58.4°F)
32	29.8191	81	28.934	114	27.088
40	29.7516	82	28.899		
		83	28.863	115	27.005
50	29.6365	84	28.826	116	26.919
51	29.6228			117	26.830
52	29.6087	85	28.788	118	26.739
53	29.5940	86	28.749	119	26.647
54	29.5788	87	28.708		
		88	28.666	120	26.553
55	29.5631	89	28.624	121	26.455
56	29.5470			122	26.355
57	29.5303	90	28.580	123	26.253
58	29.5131	91	28.535	124	26.149
59	29.4953	92	28.489		
		93	28.441	125	26.040
60	29.477	94	28.392	126	25.931
61	29.458			127	25.820
62	29.439	95	28.341	128	25.706
63	29.419	96	28.290	129	25.589
64	29.398	97	28.237		
		98	28.183	130	25.48
65	29.376	99	28.127	131	25.35
66	29.354			132	25.23
67	29.331	100	28.070	133	25.10
68	29.308	101	28.011	134	24.97
69	29.284	102	27.951	135	24.83
		103	27.889	140	24.11
70	29.259				
71	29.234	104	27.825	150	22.42
72	29.208	105	27.759	160	20.32
73	29.181	106	27.692	170	17.77
74	29.153	107	27.623	180	14.67
				190	10.93
75	29.125	108	27.550		
76	29.095	109	27.478	200	6.47
77	29.065			210	1.16
78	29.034	110	27.404	212	0.00
79	29.002	111	27.328		
		112	27.250		
80	28.968	113	27.170		

R7-4
GENERAL REFRIGERANTS

It can be said that there is no "universal" refrigerant. Since mechanical refrigeration is used over a wide range of the temperatures, some refrigerants are more suitable for high-temperature refrigeration, such as comfort cooling; others operate at lower temperature ranges, such as used for commodity storage, freezing process, and applications requiring even lower temperatures.

The choice of a refrigerant for a particular application frequently depends on properties not related to its ability to remove heat, for example, its toxicity, flammability, density, viscosity, and availability. Thus, the selection of a refrigerant for a particular purpose may be a compromise among conflicting properties.

Many different refrigerants have been used since the early days of refrigeration. Experimentation, research, and testing are still going on with various chemicals, or compounds and mixtures of chemicals. At one time or another air, butane, chloroform, ether, propane, water, and other organic and inorganic compounds have been used.

Ammonia is one of the old standby refrigerants. It was used in some of the earliest equipment, and its use continues in some of the larger commercial and industrial units. Ammonia requires the use of much heavier equipment than that needed by some of the other refrigerants in use today. Sulfur dioxide and methyl chloride were once widely used in household refrigerators and in small commercial condensing units. Because of the toxicity and flammability of these substances, most systems using these refrigerants have been replaced, or are being replaced as they wear out.

Sulfur dioxide was, and still is, a fairly stable refrigerant and is nonexplosive and nonflammable. It has an extremely irritating and unpleasant odor, which, should a leak occur in the system, is easily discernible and readily located. Because of its toxicity, however, even a slight leak can be hazardous to anyone in the vicinity of the equipment, and the fumes have been known to injure shrubbery and other plants.

Methyl chloride gives off a sweet-smelling vapor when it is exposed to the atmosphere, quite different from that of sulfur dioxide, but it is also toxic. The inhalation of too much of the vapor results in symptoms similar to those caused by an anesthetic. Although it is not considered harmful to flowers or other plants, it is a moderately flammable refrigerant. According to Underwriters' Laboratory Report MH-2375, its explosive limits in air are 8.1 to 17.2% by volume.

With the discovery of new chemicals and compounds, the practical advantages and disadvantages of each of the refrigerants in use were carefully appraised. Consideration is now given to the characteristics of the various refrigerants, from both a chemical and a physical standpoint. The availability and cost of each refrigerant is also of great importance. The characteristics and properties of several refrigerants will be found in the Appendix, under the *Refrigerant Table and Charts*. This data is presented by permission of the Freon Products Division, E. I. du Pont de Nemours & Co., Inc. In this chapter, pertinent and selected data of the refrigerants

R-12, R-22, and R-502 only will be shown in charts and tables, since they are the most commonly used.

The now universally used numbering system for refrigerants was developed by the Du Pont Company, the first to market many of the new breed of refrigerants. At the outset the letter *F* (for *Freon,* a registered Du Pont trademark) preceded the numbers. Later, as other producers of halocarbon refrigerants came into the picture, Du Pont made its numbering system available to the entire industry. As a result, refrigerants now are known by *R* or *refrigerant* numbers, such as R-12, R-22, etc.

The numbering system is described in ASHRAE (American Society of Heating, Refrigerating, and Air-Conditioning Engineers) Standard 34-67, which has been adopted by the American National Standards Institute as ANSI Standard B79-67.

For the purposes of this chapter, only that portion of the system covering the most commonly used refrigerants (designated in the previous text) will be used. Students seeking further information on different types of refrigerants may refer to the standard cited previously.

In the selection and use of a refrigerant in a specific and specialized project, the following characteristics must be taken into consideration:

1. Chemical properties
   a. Flammability
   b. Explosiveness
   c. Toxicity
   d. Stability
2. Physical properties
   a. Boiling point
   b. Freezing point
   c. Specific volume
   d. Density
   e. Critical pressure
   f. Critical temperature
   g. Latent heat content
   h. Oil miscibility
   i. Leak detection

R7-5
FLAMMABILITY AND EXPLOSIVENESS

When a system is operating satisfactorily, there is no need to worry about the refrigerant within it. However, should a fire due to an outside source occur in the vicinity of any refrigeration component, there is the danger that the fire might spread, and the possibility of an explosion exists if a questionable refrigerant escapes from ruptured lines or tubes.

If a leak occurs, or if repairs must be made to some component, the flammability and possible explosiveness and toxicity of the refrigerant must be taken into consideration. Even though the system may be evacuated and all of the refrigerant contained in the receiver or condenser-receiver, the brazing, soldering, or welding that may be necessary in the repair or replacement of a component may present hazards because enough vapor may remain in the component or connecting lines to cause danger from the heat or flame of the repair equipment.

The hydrocarbon refrigerants, among which are butane (R-600), ethane (R-170), and propane (R-290), are all highly flammable and explosive. Proper care must be taken with an open flame around a system using any of these refrigerants. The halocarbon refrigerants are considered to be nonflammable, but some become toxic when exposed to flame. Among these are trichlorofluoromethane (R-11), dichlorodifluoromethane (R-12), and monochlorodifluoromethane (R-22). R-502, which is an azeotrope refrigerant, is a mixture of 48.8% R-22 and 51.2% R-115 (chloropentafluoroethane), and is nonflammable.

There are some "tongue-twisters" in the chemical names of some refrigerants, and it was a relief when a numbering system was developed, which is more commonly used today instead of the chemical names. The number of the refrigerant is customarily preceded by the trade name of the manufacturer, but in general practice the number is preceded by *Refrigerant* or just *R*.

R7-6
TOXICITY

Underwriters' Laboratories has made extensive tests of the toxicity of all refrigerants. Human beings were not used in these tests; guinea pigs usually quickly show the effects of inhaling toxic gases and vapors.

Air is the exception to the rule that all gaseous substances are, to some degree, toxic. There are, of course, various degrees of toxicity. Carbon dioxide is exhaled whenever people breathe and is therefore harmless to human beings up to a certain level in the surrounding atmosphere. However, a person exposed to an atmosphere containing about 8 to 10% carbon dioxide or more would soon become unconscious.

The tests made by Underwriters' Laboratories pertain to the type of chemical or compound, the percent of the vapor being tested in the given atmosphere, and the duration of time that the subject is exposed to or breathes the mixture. Underwriters' Laboratories has devised a number classification that ranges from 1 to 6, the lowest number designating the most toxic and dan-

FIGURE R7-3
Hazard-to-life classification of gases and vapors by Underwriters' Laboratories.

Group	Limitations	Refrigerants
1	Concentration of gases or vapors about 0.5–1% for about 5 min is capable of producing serious injury or death.	Sulphur dioxide
2	Concentration of gases or vapors about 0.5–1% for about 30 min is capable of producing serious injury or death.	Ammonia
3	Concentration of gases or vapors about 2–2.5% for about 1 hr is capable of producing serious injury or death.	Chloroform
4	Concentration of gases or vapors about 2–2.5% for about 2 hr is capable of producing serious injury or death.	Methyl chloride
5	Gases or vapors that are less toxic than group 4 yet more toxic than those in group 6.	Refrigerants 11, 22, 502; butane, propane
6	Concentration of gases or vapors about 20% for periods of about 2 hr has apparently no injury.	Refrigerants 12, 114

gerous refrigerant and the highest number being the least toxic refrigerant. Fig. R7-3 lists the classification of refrigerants and the concentration and length of exposure that lead to serious injury.

R7-7
CRITICAL PRESSURE
AND TEMPERATURE

Every refrigerant, whether a single element or a mixture of elements (a compound), has among its characteristics a pressure above which it will remain as a liquid, even though more heat is added. Each element or compound also has a temperature above which it cannot remain or exist in a liquid state, regardless of the pressure exerted on it. These points are the *vapor pressure* and the *critical temperature,* respectively, of the refrigerant, some of which are shown in Fig. R7-4.

The pressures and temperatures shown in Fig. R7-4 are well above the pressures that can be expected to occur in the condenser, and also above the temperatures that accompany these operating pressures. These

FIGURE R7-4

Critical pressures and temperatures of some refrigerants.

Refrigerant	Critical Pressure (psia)	Critical Temperature (°F)
R-12	596.9	233.6
R-22	721.9	204.8
R-502	591	179.9

refrigerants are chemically stable in the ranges specified. They do not have a tendency, under normal conditions, to react with any material used in the manufacture of the components in the refrigeration system, nor do they ordinarily break down chemically. If the refrigerant in a system is unstable, it quickly becomes useless, since a chemical reaction will change the original characteristics or properties of the refrigerant. Decomposition of the refrigerant will add noncondensable gases, causing abnormal pressures and temperatures within the system, and the resulting sludge may cause mechanical troubles.

R7-8
PHYSICAL PROPERTIES

It has been stated earlier that a refrigerant absorbs much more heat when it changes state from a liquid to a vapor than when it absorbs heat as a liquid or as a vapor. Therefore, the boiling point of a refrigerant is of great importance, for it must readily evaporate below the temperature of the product or space which is to be cooled. Refrigerants that do not have a relatively low boiling point require that the compressor be operated with a deep vacuum. This could cause a lowering of the system's capacity and efficiency.

The freezing point of a refrigerant is another important property, particularly in extreme low-temperature applications, for this point should be sufficiently lower than any anticipated evaporator temperature.

The latent heat content of the selected refrigerant usually is high, which is a desirable characteristic for large-capacity systems. With a high latent heat content, less refrigerant is circulated for each ton of refrigeration effect produced, and a compressor of lower horsepower could be used. In systems of low capacity, the use of a refrigerant with a low latent heat will require the circulation of more refrigerant than that of one with a higher latent heat. Such a refrigerant will also make the control of the system easier, since less sensitive control devices are needed in a system that circulates a large amount of refrigerant.

The density and specific volume of a refrigerant are two properties that have a reciprocal relationship. In comparing refrigerants for different applications, advantages and disadvantages must be taken into consideration. In order to utilize smaller lines, tubes, or refrigerant pipes (these terms are used interchangeably) a high-density, low-specific-volume refrigerant should be used. Such a system costs less to construct and install. In some commercial installations where there is a considerable vertical distance between the major components of the system, it may be desirable to use a refrigerant with a low density and a high specific volume. In this type of system, less pressure is required to circulate the refrigerant through the components and piping.

Substances that are compared must have a mutual relationship at some point. The three refrigerants in Fig. R7-5 are compared in the vapor stage at 5°F.

FIGURE R7-5

Comparison of refrigerant vapors at 5°F.

Refrigerant	SPECIFIC VOLUME (ft³/lb)	DENSITY (lb/ft³)
R-12	1.46	0.685
R-22	1.24	0.80
R-502	0.825	1.21

Figure R7-6 furnishes a comparison of the specific volume and density of the same three refrigerants in a liquid state at 86°F. The temperatures of 5°F for the vapor and 86°F for the liquid used in Figs. R7-5 and R7-6 are classified as the standard cycle temperatures in the *ASHRAE Handbook of Fundamentals*.

Low oil solubility or miscibility is also an important characteristic of a refrigerant, but it can cause oil-return problems if it is too low. Solubility, or miscibility, is the capability of the refrigerant in its liquid state to mix with the oil necessary for the lubrication of the

FIGURE R7-6

Comparison of refrigerant liquids at 86°F.

Refrigerant	SPECIFIC VOLUME (ft³/lb)	DENSITY (lb/ft³)
R-12	0.0124	80.671
R-22	0.0136	73.278
R-502	0.0131	76.13

moving parts in the compressor. The oil will circulate with a highly soluble refrigerant in a liquid state from the condenser and/or receiver, through the liquid line and the metering device, and into the evaporator.

But since the refrigerant vaporizes in the evaporator, a different situation exists, for oil and refrigerant vapor do not mix readily. The oil circulating through the system can continue on its way to the compressor only if the vapor is moving rapidly enough to entrain the oil and carry it along with it through the suction line to the compressor crankcase. If the vapor lines, either suction or hot gas, run in a horizontal path, it is customary to size them so that the vapor travels at a velocity rate not less than 750 ft/min. When the vapor lines are run vertically upward, the velocity of the vapor should be not less than 1,500 ft/min to assure that the oil will travel up the vertical line. An example of this would occur when the evaporator is located somewhere beneath the compressor or condenser unit.

Leak detection was and is fairly easy when ammonia or sulphur dioxide is being used, since their peculiar odors are discernible from all others. Yet, depending strictly on odor for leak detection of these refrigerants may be deceiving as well as dangerous.

Ammonia leaks in a water-cooled condenser may be confirmed by the use of litmus or other test paper dipped in the outlet water from the condenser. Leaks in the rest of the system may be located by the use of a sulphur candle which, when lit, gives off a white cloud of smoke when exposed to ammonia fumes or vapor.

Conversely, leaks in a system using sulphur dioxide as the refrigerant are located by the use of a cloth or taper dipped in an ammonia and water solution of approximately 25% strength. A white cloud of smoke will appear when the ammonia vapor from the solution nears the leak in the sulphur dioxide system. Frequently, in order to pinpoint a leak when there is quite a bit of vapor in the air, a solution of soap and water is swabbed or wiped over the suspected area. When a system is still under pressure, bubbles will form in the immediate area of a leak. This means of leak detection should not be used if a section of the system is under a vacuum, for the soap and water solution might be drawn into the system.

Figure R7-7 shows a halide leak detector, which has been used successfully on systems with halogenated refrigerants for a number of years. This leak detector consists of two major segments: the cylinder containing gas and the detector unit. The principle involved is that air is drawn through the search hose and across a copper reactor plate, which has been heated red hot from the gas flame. When the search hose is manipulated so that it comes in contact with the leaking refrigerant vapor, gas is drawn through the hose. When the gas or vapor comes in contact with the flame and the reactor plate, the flame changes to a bluish green or violet in color.

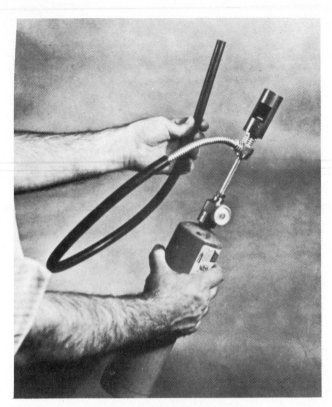

FIGURE R7-7 Halide leak detector.

For most leaks this type of detector works satisfactorily. However, it is not as sensitive as the electronic leak detector shown in Fig. R7-8 or R7-9. This type of detector is capable of detecting leaks as small as 0.5 oz per year. Its use is growing in the field because of its sensitivity; it is also available as a battery-operated unit. Some units have a dual type of control setting—a single HI-LOW selector—in addition to a balance-control dial.

This detector contains an internal pump that draws air into the probe and hose or tube. If there is any halogenated gas in the air drawn across the electrodes in

FIGURE R7-8 Electronic leak detector. (*Courtesy* Robinair Manufacturing Corporation)

(a)

(b)

FIGURE R7-9 Electronic leak detectors. (*Courtesy* Ritchie Engineering Co., Inc.)

the sensing element, a built-in signal light will flash frequently. The rate of the flashing will increase depending on the amount of refrigerant vapor in the air. If there is a heavy concentration of refrigerant, the source of the leak can be fairly accurately pinpointed by correctly using the HI-LOW selector and also the balance control. This type of leak detector can be modified to give an audible signal of a leak.

At present, most units are either case-mounted pump style in ac or dc or are hand-held units. Many units are available in either automatic or manual calibration. The units shown in Fig. R7-8 have a new tip that minimizes the biggest headache with leak detectors. The tip no longer requires filter materials. If a drop of water puts out the corona in the tip, just shake it out and the unit will operate again. Tip dirty? Wash with vaporizable solvent and dry.

Figure R7-3 listed some refrigerants grouped according to their toxic limitations by the Underwriters' Laboratories. Refrigerants also are classified in the National Refrigeration Safety Code under ANSI Standard B-9, in which they are rated in three groups as to their toxicity and fire hazard characteristics. On the basis of low fire hazard or low toxicity, R-12, R-22, and R-502 are placed in group I—the safest of the refrigerants. Group II refrigerants, which are toxic and somewhat flammable, includes such refrigerants as ammonia (R-717) and methyl chloride (R-40). Group III, which contains the flammable refrigerants, has among others butane (R-600) and propane (R-290).

For personal safety and to avoid possible damage to equipment, certain precautions should be taken. R-12, which is considered a safe refrigerant, may result in the formation of pungent acid vapors when it comes in contact with an open flame. Proper ventilation will prevent the possibility of breathing any minute quantities of phosgene vapor that might be produced. Therefore, when a halide leak detector is used on or around a system, the user should be careful that fumes from the torch are not inhaled.

If the refrigeration unit is located in a confined space, that area should be well ventilated before any time is spent there. Even a refrigerant of low toxicity can cause suffocation if the oxygen in the air is replaced by refrigerant vapor that has escaped from a unit. Liquid refrigerant spilled onto the skin can cause severe frostbite, because the liquid vaporizes rapidly at atmospheric pressure, removing heat from the skin. When refrigerant is handled, protective goggles should be worn because of the possibility of permanent damage should any liquid refrigerant get into the eyes.

Refrigerants should not be mixed in the field, either in a system or outside. To prevent the accidental mixture of refrigerants, the cylinders are color-coded for easy identification of contents, such as:

R-12—White   R-22—Green   R-502—Orchid

However, the words on the label should be the final means of identifying the contents.

The U.S. Department of Transportation has regulations pertaining to the manufacture, testing, filling, and shipment of all compressed gas containers subject to interstate transportation. The cylinders or containers must have identification markings stamped in the metal; they are also hydrostatically tested initially and, if returnable, are inspected at regular intervals. Most cylinders have built-in pressure relief valves, so that the pressure can be lowered, should the container be ex-

FIGURE R7-10   Liquid–vapor valve. (*Courtesy* E. I. du Pont de Nemours and Company)

FIGURE R7-11   Disposable Freon cylinder. (*Courtesy* E. I. du Pont de Nemours and Company)

posed to extremely high temperatures and in danger of exploding.

Many refrigerant cylinders used for storage or in the field are equipped with liquid–vapor valves, which permit withdrawal of either liquid or vapor. This type of valve makes it possible to charge a system or a service drum without inverting the storage cylinder, a necessary procedure before the combined valve came into use. One of the valves is depicted in Fig. R7-10, showing the two handwheels. The vapor handwheel is on the top of the valve in the customary position; the liquid handwheel is located on the side and is attached to a dip tube, which extends to the bottom of the cylinder. The handwheels are clearly marked to show whether they control the flow of refrigerant in a liquid or vapor state, and they also are color-coded: blue for vapor and red for liquid.

In addition to the larger, rechargeable cylinders, there are also smaller, "disposable" cylinders, as illustrated in Fig. R7-11, which are portable for field use. *Never* try to refill these containers.

Refrigerant containers should not be heated to a temperature greater than 125°F, and a direct flame should never be applied to the cylinder. Containers should not be stored near hot pipes or other heating devices, nor should they be left in a closed service vehicle in the sun, since temperatures under such conditions can easily reach 160°F or more. If the container, filled with liquid, is allowed to become heated excessively, its walls will be stretched by the expanding liquid. No matter how strong the walls are, they can stretch just so much before they must burst under the excessive pressure. *No cylinder should ever be more than 80% full.*

R7-10
PRESSURE-TEMPERATURE CHARACTERISTICS

With many different refrigerants used throughout the refrigeration field, when these are plotted on the same graph (Fig. R7-12), a comparison of the operating range possibilities can be seen. Here we see that in the 40°F boiling-point range, R-12, R-22, R-502, and even R-40, R-717, or R-764 could be used. However, the flammability and toxicity of R-764 (sulphur dioxide) and the attack of copper and aluminum by R-717 (ammonia) rule out those refrigerants. In the boiling-point range −20 to 20°F the refrigerants noted above can be used and selection is done on the same basis of efficiency, toxicity, flammability, etc. In this text we limit discussion to the three most popular refrigerants: R-12, R-22, and R-502.

Instead of using the curve-type presentation of

pressure-temperature relationships, the popular packet-type pressure-temperature chart that is put out by many manufacturers in the industry will be used. Figure R7-13 shows a typical pressure-temperature chart.

Using the same conditions of heat source and heat sink that we did for water—50°F heat source with 40°F boiling point of refrigerant and 75°F heat sink with 100°F condensing temperature of the refrigerant—we get the following pressure requirements. For R-12—37 psig for the 40°F boiling point and 117.2 psig for the con-

densing temperature. For R-22—68.5 psig for the 40°F boiling point and 195.9 psig for the 100°F condensing temperature. For R-502—80.2 psig for the 40°F boiling point and 214.4 psig for the 100°F condensing temperature.

From this we can see that different refrigerants have different pressures. Therefore, they are not interchangeable, nor are they designed to be mixed except in specific pieces of equipment where the manufacturers exact specifications must be met.

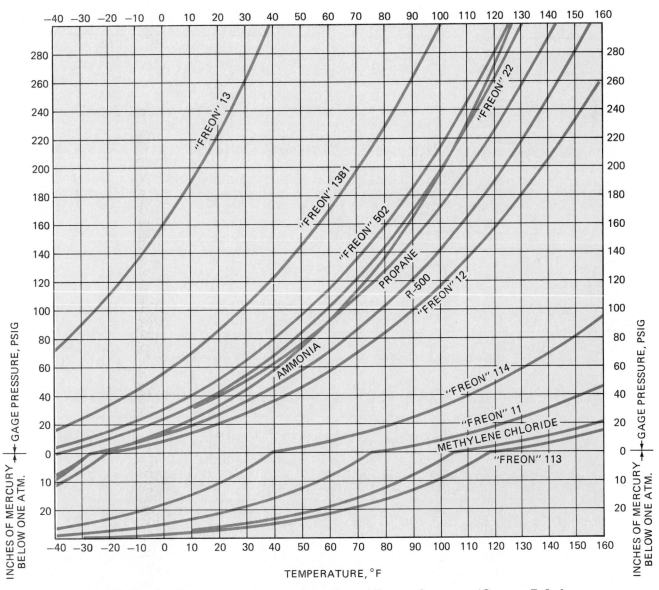

FIGURE R7-12   Pressure-temperature relationships of Freon refrigerants. (*Courtesy* E. I. du Pont de Nemours and Company)

FIGURE R7-13
Pressure–temperature chart.

Temp F	\multicolumn{8}{c}{GAGE PRESSURE, PSIG}							
	11	12	13	22	113	114	500	502
−50	28.9*	15.4*	57.0	6.2*	—	27.1*	12.8*	0.0
−48	28.8*	14.6*	60.0	4.8*	—	26.9*	11.9*	0.7
−46	28.7*	13.8*	63.1	3.4*	—	26.7*	10.9*	1.5
−44	28.6*	12.9*	66.2	2.0*	—	26.5*	9.8*	2.3
−42	28.5*	11.9*	69.4	0.5*	—	26.3*	8.8*	3.2
−40	28.4*	11.0*	72.7	0.5	—	26.0*	7.6*	4.1
−38	29.3*	10.0*	76.1	1.3	—	25.8*	6.4*	5.1
−36	28.2*	8.9*	79.7	2.2	—	25.5*	5.2*	6.0
−34	28.1*	7.8*	83.3	3.0	—	25.2*	3.9*	7.0
−32	27.9*	6.7*	87.0	3.9	—	25.0*	2.6*	8.1
−30	27.8*	5.5*	90.9	4.9	29.3*	24.6*	1.2*	9.2
−28	27.7*	4.3*	94.9	5.9	29.3*	24.3*	0.1	10.3
−26	27.5*	3.0*	98.9	6.9	29.2*	24.0*	0.9	11.5
−24	27.4*	1.6*	103.0	7.9	29.2*	23.6*	1.6	12.7
−22	27.2*	0.3*	107.3	9.0	29.1*	23.2*	2.4	14.0
−20	27.0*	0.6	111.7	10.1	29.1*	22.9*	3.2	15.3
−18	26.8*	1.3	116.2	11.3	29.0*	22.4*	4.1	16.7
−16	26.6*	2.1	120.8	12.5	28.9*	22.0*	5.0	18.1
−14	26.4*	2.8	125.5	13.8	28.9*	21.6*	5.9	19.5
−12	26.2*	3.7	130.4	15.1	28.8*	21.1*	6.8	21.0
−10	26.0*	4.5	135.4	16.5	28.7*	20.6*	7.8	22.6
− 8	25.8*	5.4	140.5	17.9	28.6*	20.1*	8.8	24.2
− 6	25.5*	6.3	145.8	19.3	28.5*	19.6*	9.9	25.8
− 4	25.3*	7.2	151.1	20.8	28.4*	19.0*	11.0	27.5
− 2	25.0*	8.2	156.5	22.4	28.3*	18.4*	12.1	29.3
0	24.7*	9.2	162.1	24.0	28.2*	17.8*	13.3	31.1
2	24.4*	10.2	167.8	25.6	28.1*	17.2*	14.5	32.9
4	24.1*	11.2	173.7	27.3	28.0*	16.5*	15.7	34.8
6	23.8*	12.3	179.7	29.1	27.9*	15.8*	17.0	36.9
8	23.4*	13.5	185.8	30.9	27.7*	15.1*	18.4	38.9
10	23.1*	14.6	192.1	32.8	27.6*	14.4*	19.7	41.0
12	22.7*	15.8	198.5	34.7	27.5*	13.6*	21.2	43.2
14	22.3*	17.1	205.7	36.7	27.3*	12.8*	22.6	45.4
16	21.9*	18.4	211.9	38.7	27.1*	12.0*	24.1	47.7
18	21.5*	19.7	218.7	40.9	27.0*	11.1*	25.7	50.0
20	21.1*	21.0	225.7	43.0	26.8*	10.2*	27.3	52.5
22	20.6*	22.4	232.9	45.3	26.6*	9.3*	28.9	54.9
24	20.1*	23.9	240.2	47.6	26.4*	8.3*	30.6	57.5
26	19.7*	25.4	247.7	49.9	26.2*	7.3*	32.4	60.1
28	19.1*	26.9	255.4	52.4	26.0*	6.3*	34.3	62.8
30	18.6*	28.5	263.2	54.9	25.8*	5.2*	36.0	65.6
32	18.1*	30.1	271.2	57.5	25.6*	4.1*	37.9	68.4
34	17.5*	31.7	279.4	60.1	25.3*	2.9*	39.9	71.3
36	16.9*	33.4	287.7	62.8	25.1*	1.7*	41.9	74.3
38	16.3*	35.2	296.2	65.6	24.8*	0.6*	43.9	77.4
40	15.6*	37.0	304.9	68.5	24.5*	0.4	46.1	80.5
42	15.0*	38.8	313.9	71.5	24.2*	1.0	48.2	83.8
44	14.3*	40.7	322.9	74.5	23.9*	1.7	50.5	87.0
46	13.6*	42.7	332.2	77.6	23.6*	2.4	52.8	90.4
48	12.8*	44.7	341.5	80.8	23.3*	3.1	55.1	93.9

Temp F	GAGE PRESSURE, PSIG							
	11	12	13	22	113	114	500	502
50	12.0*	46.7	351.2	84.0	22.9*	3.8	57.6	97.4
52	11.2*	48.8	360.9	87.4	22.6*	4.6	60.0	101.1
54	10.4*	51.0	371.0	90.8	22.2*	5.4	62.6	104.8
56	9.6*	53.2	381.2	94.3	21.8*	6.2	65.2	108.6
58	8.7*	55.4	391.6	97.9	21.4*	7.0	67.9	112.4
60	7.8*	57.7	402.3	101.6	21.0*	7.9	70.6	116.4
62	6.8*	60.1	413.3	105.4	20.6*	8.8	73.5	120.5
64	5.9*	62.5	424.3	109.3	20.1*	9.7	76.3	124.6
66	4.9*	65.0	435.4	113.2	19.7*	10.6	79.3	128.9
68	3.8*	67.6	446.9	117.3	19.2*	11.6	82.3	133.2
70	2.8*	70.2	458.7	121.4	18.7*	12.6	85.4	137.6
72	1.6*	72.9	470.6	125.7	18.2*	13.6	88.6	142.2
74	0.5*	75.6	482.9	130.0	17.6*	14.6	91.8	146.8
76	0.3	78.4	495.3	134.5	17.1*	15.7	95.1	151.5
78	0.9	81.3	508.0	139.0	16.5*	16.8	98.5	156.3
80	1.5	84.2	520.8	143.6	15.9*	18.0	102.0	161.2
82	2.2	87.2	534.0	148.4	15.3*	19.1	105.6	166.2
84	2.8	90.2	—	153.2	14.6*	20.3	109.2	171.4
86	3.5	93.3	—	158.2	13.9*	21.6	112.9	176.6
88	4.2	96.5	—	163.2	13.2*	22.8	116.7	181.9
90	4.9	99.8	—	168.4	12.5*	24.1	120.6	187.4
92	5.6	103.1	—	173.7	11.8*	25.5	124.5	192.9
94	6.4	106.5	—	179.1	11.0*	26.8	128.6	198.6
96	7.1	110.0	—	184.6	10.2*	28.2	132.7	204.3
98	7.9	113.5	—	190.2	9.4*	29.7	136.9	210.2
100	8.8	117.2	—	195.9	8.6*	31.2	141.2	216.2
102	9.6	120.9	—	201.8	7.7*	32.7	145.6	222.3
104	10.5	124.6	—	207.7	6.8*	34.2	150.1	228.5
106	11.3	128.5	—	213.8	5.9*	35.8	154.7	234.9
108	12.3	132.4	—	220.0	4.9*	37.4	159.4	241.3
110	13.1	136.4	—	226.4	4.0*	39.1	164.1	247.9
112	14.2	140.5	—	232.8	3.0*	40.8	169.0	254.6
114	15.1	144.7	—	239.4	1.9*	42.5	173.9	261.5
116	16.1	148.9	—	246.1	0.8*	44.3	179.0	268.4
118	17.2	153.2	—	252.9	0.1	46.1	184.2	275.5
120	18.2	157.7	—	259.9	0.7	48.0	189.4	282.7
122	19.3	162.2	—	267.0	1.3	49.9	194.8	290.1
124	20.5	166.7	—	274.3	1.9	51.9	200.2	297.6
126	21.6	171.4	—	281.6	2.5	53.8	205.8	305.2
128	22.8	176.2	—	289.1	3.1	55.9	211.5	312.9
130	24.0	181.0	—	296.8	3.7	58.0	217.2	320.8
132	25.2	185.9	—	304.6	4.4	60.1	223.1	328.9
134	26.5	191.0	—	312.5	5.1	62.3	229.1	337.1
136	27.8	196.1	—	320.6	5.8	64.5	235.2	345.4
138	29.1	201.3	—	328.9	6.5	66.7	241.4	353.9

*Inches of Mercury Below One Atmosphere.

Source: The DuPont Company

R7-1. Name the primary purpose of a refrigerant.

R7-2. The heat a vaporizing refrigerant absorbs is called _____.

R7-3. In what two categories may refrigerant properties be classified?

R7-4. The flame of a halide torch will burn what color in the presence of halogenated refrigerants?

R7-5. What leak detectors may be used in place of a halide torch?

R7-6. Refrigerant cylinders should never be filled more than _____% of the cylinder volume.

R7-7. Refrigerant cylinders should not be heated to a temperature greater than _____°F.

R7-8. What are the color codes for R-12, R-22, and R-502?

R7-9. Use your pocket pressure–temperature chart to find the pressure required to maintain a 45°F boiling point in an evaporator using R-22.

R7-10. What is the condensing temperature of a refrigeration system using R-12 and operating at a discharge pressure of 146.8 psig?

R8

Pressure–Enthalpy Diagrams

R8-1
REFRIGERATION EFFECT

If a specific job is to be done in a refrigeration system or cycle, each pound of refrigerant circulating in the system must do its share of the work. It must absorb an amount of heat in the evaporator or cooling coil, and it must dissipate this heat—plus some that is added in the compressor—out through the condenser, whether air-cooled, water-cooled, or evaporatively cooled. The work done by each pound of the refrigerant as it goes through the evaporator is reflected by the amount of heat it picks up from the refrigeration load, chiefly when the refrigerant undergoes a change of state from a liquid to a vapor.

As mentioned in Chapter R2, for a liquid to be able to change to a vapor, heat must be added to or absorbed by it. This is what happens—or should happen—in the cooling coil. The refrigerant enters the metering device as a liquid and passes through the device into the evaporator, where it absorbs heat as it evaporates into a vapor. As a vapor, it makes its way through the suction tube or pipe to the compressor. Here it is compressed from a low-temperature, low-pressure vapor to a high-temperature, high-pressure vapor; then it passes through the high-pressure or discharge pipe to the condenser, where it undergoes another change of state—from a vapor to a liquid—in which state it flows out into the liquid pipe and again makes its way to the metering device for another trip through the evaporator. Shown in Fig. R8-1 is a schematic of a simple refrigeration cycle, describing this process.

When the refrigerant, as a liquid, leaves the condenser it may go to a receiver until it is needed in the evaporator; or it may go directly into the liquid line to the metering device and then into the evaporator coil. The liquid entering the metering devide just ahead of the evaporator coil will have a certain heat content (*enthalpy*), which is dependent on its temperature when it enters the coil, as shown in the refrigerant tables in the Appendix. The vapor leaving the evaporator will also have a given heat content (enthalpy) according to its temperature, as shown in the refrigerant tables.

The difference between these two amounts of heat content is the amount of work being done by each pound of refrigerant as it passes through the evaporator and picks up heat. The amount of heat absorbed by each pound of refrigerant is known as the *refrigerating effect* of the system, or of the refrigerant within the system.

This refrigerating effect is rated in Btu per pound of refrigerant (Btu/lb); if the total heat load is known (given in Btu/hr), we can find the total number of pounds of refrigerant that must be circulated each hour of operation of the system. This figure can be broken down further to the amount that must be circulated each minute, by dividing the amount circulated per hour by 60.

EXAMPLE

If the total heat to be removed from the load is 60,000 Btu/hr and the refrigerating effect in the evaporator amounts to 50 Btu/lb, then

$$\frac{60,000 \text{ Btu/hr}}{50 \text{ Btu/lb}} = 1200 \text{ lb/hr or } 20 \text{ lb/min}$$

Since 12,000 Btu/hour equals the rate of 1 ton of refrigeration, the 60,000 Btu/hr in the example above amounts to 5 tons of refrigeration, and the 20 lb of refrigerant that must be circulated each minute is the equivalent of 4 lb/min/ton of refrigeration. (One ton of refrigeration for 24 hours equals 288,000 Btu.)

In this example, where 20 lb of refrigerant having a refrigerating effect of 50 Btu/lb is required to take care of the specified load of 60,000 Btu/hr, the results can also be obtained in another manner. As mentioned previously, it takes 12,000 Btu to equal 1 ton of refrigeration, which is equal to 200 Btu/min/ton.

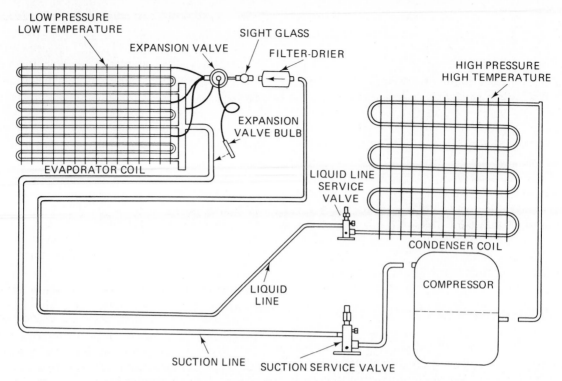

FIGURE R8-1 Schematic diagram of simple refrigeration cycle.

Therefore, 200 Btu/min, when divided by the refrigerating effect of 50 Btu/lb, amounts to 4 lb/min. This computation can be shown by the equation

$$W = \frac{200}{NRE} \qquad \text{(R8-1)}$$

where

    W = weight of refrigerant circulated per minute, lb/min

    200 = 200 Btu/min—the equivalent of 1 ton

    NRE = net refrigerating effect, Btu/lb of refrigerant

Because of the small orifice in the metering device, a fact that will be discussed more thoroughly in a later chapter, when the compressed refrigerant passes from the smaller opening in the metering device to the larger tubing in the evaporator, a change in pressure occurs together with a change in temperature. This change in temperature occurs because of the vaporization of a small portion of the refrigerant (about 13%) and, in the process of this vaporization, the heat that is involved is taken from the remainder of the refrigerant.

From the table of saturated R-12 in the Appendix, it can be seen that the heat content of 100°F liquid is 31.10 Btu/lb and that of 40°F liquid is 17.27 Btu/lb; this indicates that 13.83 Btu/lb has to be removed from each pound of refrigerant entering the evaporator. The

latent heat of vaporization of 40°F R-12 (from the appendix tables) is 64.17 Btu/lb, and the difference between this amount and that which is given up by each pound of refrigerant when its liquid temperature is lowered from 100°F to 40°F (13.83 Btu/lb) is 50.34 Btu/lb. This is another method of calculating the refrigerating effect—or work being done—by each pound of refrigerant under the conditions given.

The capacity of the compressor must be such that it will remove from the evaporator that amount of refrigerant which has vaporized in the evaporator and in the metering device in order to get the necessary work done. The previous examples have shown that 3.07 ft³/min must be removed from the evaporator to perform 1 ton of refrigeration under the conditions specified; therefore, when the total load amounts to 5 tons, then 5 times 3.07 ft³/min, or 15.35 ft³/min, must be removed from the evaporator. The compressor must be able to remove and send on to the condenser the same weight of refrigerant vapor, so that it can be condensed back into a liquid and so continue in the refrigeration circuit or cycle to perform additional work.

If the compressor, because of design or speed, is unable to move this weight, some of the vapor will remain in the evaporator. This, in turn, will cause an increase in pressure inside the evaporator, accompanied by an increase in temperature and a decrease in the work being done by the refrigerant, and design conditions within the refrigerated space cannot be maintained.

A compressor that is too large will withdraw the refrigerant from the evaporator too rapidly, causing a lowering of the temperature inside the evaporator, so that design conditions will not be maintained in this situation either.

In order for design conditions to be maintained within a refrigeration circuit, there must be a balance between the requirements of the evaporator coil and the capacity of the compressor. This capacity is dependent on its displacement and on its volumetric efficiency. The measured displacement of a compressor depends on the number of cylinders, their bore and stroke, and the speed at which the compressor is turning. Volumetric efficiency depends on the absolute suction and discharge pressures under which the compressor is operating. A thorough and elaborate presentation of these facts concerning displacement, as well as the variables pertaining to volumetric efficiency, will be offered in a later chapter, together with equations and other data.

R8-2
CYCLE DIAGRAMS

Figure R8-1 shows a schematic flow diagram of a basic cycle in refrigeration, denoting changes in phases or processes. First the refrigerant passes from the liquid stage into the vapor stage as it absorbs heat in the evaporator coil. The compression stage, where the refrigerant vapor is increased in temperature and pressure, comes next; then the refrigerant gives off its heat in the condenser to the ambient cooling medium, and the refrigerant vapor condenses back to its liquid state where it is ready for use again in the cycle.

Figure R8-2 is a reproduction of a Mollier diagram (commonly known as a *P-î chart*) of R-12, which shows the pressure, heat, and temperature characteristics of this refrigerant. Pressure–enthalpy diagrams may be utilized for the plotting of the cycle shown in Fig. R8-1, but a basic or skeleton chart as shown in Fig. R8-3 might be used as a preliminary illustration of the various phases of the refrigerant circuit. There are three basic areas on the chart denoting changes in state between the saturated liquid line and saturated vapor line in the center of the chart. The area to the left of the saturated liquid line is the subcooled area, where the refrigerant liquid has been cooled below the temperature corresponding to its pressure; whereas the area to the right of the saturated vapor line is the area of superheat, where the refrigerant vapor has been heated beyond the vaporization temperature corresponding to its pressure.

The construction of the diagram, or rather a knowledge and understanding of it, may bring about a clearer interpretation of what happens to the refrigerant at the various stages within the refrigeration cycle. If the state and any two properties of a refrigerant are known and this point can be located on the chart, the other properties can easily be determined from the chart.

If the point is situated anywhere between the saturated liquid and vapor lines, the refrigerant will be in the form of a mixture of liquid and vapor. If the location is closer to the saturated liquid line, the mixture will be more liquid than vapor, and a point located in the center of the area at a particular pressure would indicate a 50% liquid–50% vapor situation.

Referring to Fig. R8-3, the change in state from a vapor to a liquid—the condensing process—occurs as the path of the cycle develops from right to left; whereas the change in state from a liquid to a vapor—the evaporating process—travels from left to right. Absolute pressure is indicated on the vertical axis at the left, and the horizontal axis indicates heat content, or enthalpy, in Btu/lb.

The distance between the two saturated lines at a given pressure, as indicated on the heat content line, amounts to the latent heat of vaporization of the refrigerant at the given absolute pressure. The distance between the two lines of saturation is not the same at all pressures, for they do not follow parallel curves. Therefore, there are variations in the latent heat of vaporization of the refrigerant, depending on the absolute pressure. There are also variations in pressure–enthalpy charts of different refrigerants and the variations depend on the various properties of the individual refrigerants.

R8-3
REFRIGERATION PROCESSES

Based on the examples presented earlier in the chapter, it will be assumed that there will be no changes in the temperature of the condensed refrigeration liquid after it leaves the condenser and travels through the liquid pipe on its way to the expansion or metering device, or in the temperature of the refrigerant vapor after it leaves the evaporator and passes through the suction pipe to the compressor.

Figure R8-4 shows the phases of the simple saturated cycle with appropriate labeling of pressures, temperatures, and heat content or enthalpy. A starting point must be chosen in the refrigerant cycle; let it be point *A* on the saturated liquid line where all of the refrigerant vapor at 100°F has condensed into liquid at 100°F and is at the inlet to the metering device. What occurs between points *A* and *B* is the expansion process as the refrigerant passes through the metering device; and the refrigerant temperature is lowered from the condensation temperature of 100°F to the evaporating temperature of 40°F.

When the vertical line *A–B* (the expansion process) is extended downward to the bottom axis, a reading of 31.10 Btu/lb is indicated, which is the heat content of 100°F liquid. To the left of point *B* at the saturated liq-

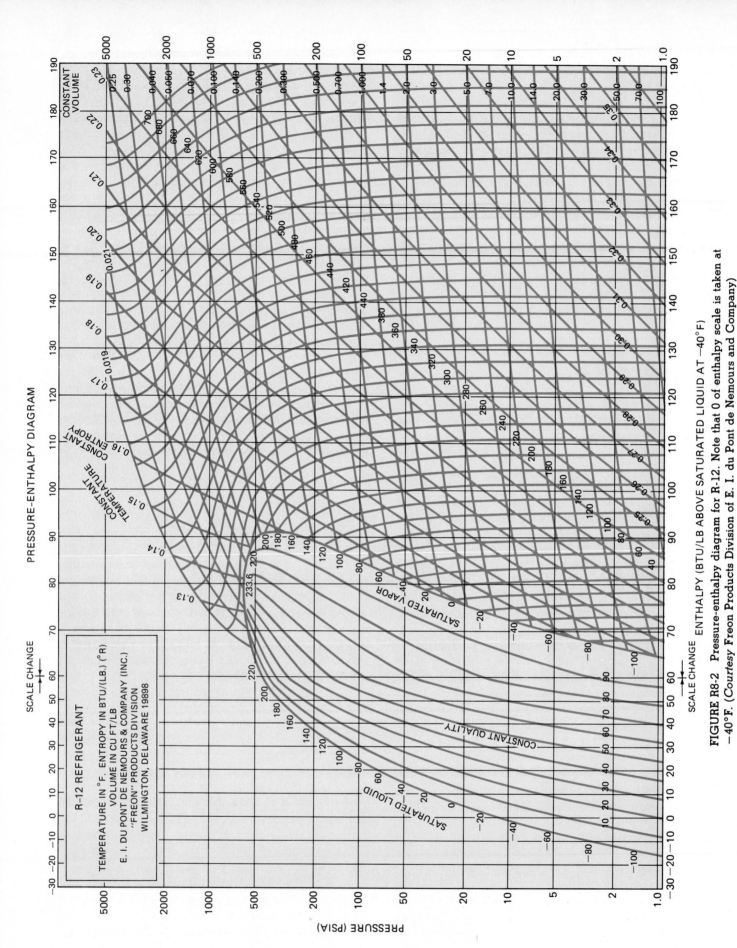

PRESSURE-ENTHALPY DIAGRAM

CONSTANT VOLUME

CONSTANT TEMPERATURE

CONSTANT ENTROPY

SCALE CHANGE

R-12 REFRIGERANT

TEMPERATURE IN °F. ENTROPY IN BTU/(LB.) (°R)
VOLUME IN CU FT/LB
E. I. DU PONT DE NEMOURS & COMPANY (INC.)
"FREON" PRODUCTS DIVISION
WILMINGTON, DELAWARE 19898

SATURATED LIQUID

SATURATED VAPOR

CONSTANT QUALITY

PRESSURE (PSIA)

ENTHALPY (BTU/LB ABOVE SATURATED LIQUID AT −40°F)

FIGURE R8-2 Pressure-enthalpy diagram for R-12. Note that 0 of enthalpy scale is taken at −40°F. (*Courtesy* Freon Products Division of E. I. du Pont de Nemours and Company)

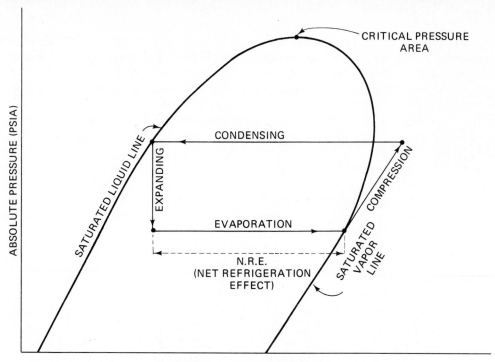

FIGURE R8-3 Enthalpy (Btu/lb).

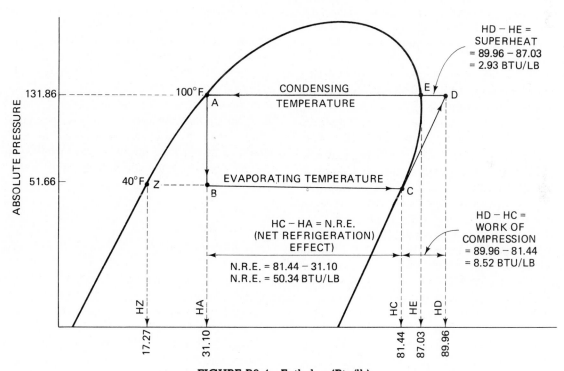

HD − HE =
SUPERHEAT
= 89.96 − 87.03
= 2.93 BTU/LB

HD − HC =
WORK OF
COMPRESSION
= 89.96 − 81.44
= 8.52 BTU/LB

HC − HA = N.R.E.
(NET REFRIGERATION)
EFFECT)

N.R.E. = 81.44 − 31.10
N.R.E. = 50.34 BTU/LB

FIGURE R8-4 Enthalpy (Btu/lb).

uid line is point Z, which is also at the 40°F temperature line. Taking a vertical path downward from point Z to the heat content line, a reading of 17.27 Btu/lb is indicated, which is the heat content of 40°F liquid; this area between points Z and B is covered later in the chapter.

The horizontal line between points B and C indicates the vaporization process in the evaporator, where the 40°F liquid absorbs enough heat to completely va-

porize the refrigerant. Point C is at the saturated vapor line, indicating that the refrigerant has completely vaporized and is ready for the compression process. A line drawn vertically downward to where it joins the enthalpy line indicates that the heat content, shown at h_c is 81.44 Btu/lb, and the difference between h_a and h_c is 50.34 Btu/lb, which is the refrigerating effect, as shown in an earlier example.

The difference between points h_a and h_c on the en-

Pressure–Enthalpy Diagrams   87

thalpy line amounts to 64.17 Btu/lb, which is the latent heat of vaporization of 1 lb of R-12 at 40°F. This amount would also exhibit the refrigerating effect, but some of the refrigerant at 100°F must evaporate or vaporize in order that the remaining portion of each pound of R-12 can be lowered in temperature from 100°F to 40°F.

The various properties of refrigerants, mentioned earlier, should probably be elaborated on before we proceed with the discussion of the compression process. All refrigerants exhibit certain properties when in a gaseous state; some of them are: *volume, temperature, pressure, enthalpy* or *heat content,* and *entropy.* The last property—entropy—is really the most difficult to describe or define. It is the ratio of the heat content of the gas to its absolute temperature in degrees Rankine, and it relates to internal energy of the gas.

The Mollier chart plots the line of constant entropy, which stays the same providing the gas is compressed and no outside heat is added or taken away. When the entropy is constant, the compression process is called *adiabatic;* which means that the gas changes its condition without the absorption or rejection of heat either from or to an external body or source. It is common practice, in the study of cycles of refrigeration, to plot the compression line either along or parallel to a line of constant entropy.

In Fig. R8-4, line *C–D* denotes the compression process, in which the pressure and temperature of the vapor are increased from that in the evaporator to that in the condenser, with the assumption that there has been no pickup of heat in the suction line between the evaporator and the compressor. For a condensing temperature of 100°F, a pressure gauge would read approximately 117 psig; but the *P–h* chart is rated in absolute pressure and the atmospheric pressure of 14.7 must be added to the psig, making it actually 131.86 psia.

Point *D* on the absolute pressure line is equivalent to the 100°F condensing temperature; it is not on the saturated vapor line, it is to the right in the superheat area, at a junction of the 131.86 psia line, the line of constant entropy of 40°F, and the temperature line of approximately 116°F. A line drawn vertically downward from point *D* intersects the heat content line at 89.96 Btu/lb, which is h_d; the difference between h_c and h_d is 8.52 Btu/lb—the heat of compression that has been added to the vapor. This amount of heat is the heat energy equivalent of the work done during the refrigeration compression cycle. This is the theoretical discharge temperature, assuming that saturated vapor enters the cycle; in actual operation, the discharge temperature may be 20 to 35° higher than that predicted theoretically. This can be checked in an operating system by strapping a thermometer or a thermocouple to the outlet of the discharge service valve on the compressor.

During the compression process the heat that is absorbed by the vapor is a result of friction caused by the action of the pistons in the cylinders and by the vapor itself passing through the small openings of the internal suction and discharge valves. Of course, the vapor is also heated by the action of its molecules being pushed or compressed closer together, commonly called *heat of compression.* Some of this overall additional heat is lost through the walls of the compressor. A lot depends, therefore, on the design of the compressor, the conditions under which it must operate and the balance between the heat gain and heat loss to keep the refrigerant at a constant entropy.

Line *D–E* denotes the amount of superheat that must be removed from the vapor before it can commence the condensation process. A line drawn vertically downward from point *E* to point h_e on the heat content line indicates the distance h_d–h_e, or heat amounting to 2.93 Btu/lb, since the heat content of 100°F vapor is 87.03 Btu/lb. This superheat is usually removed in the hot-gas discharge line or in the upper portion of the condenser. During this process the temperature of the vapor is lowered to the condensing temperature.

Line *E–A* represents the condensation process that takes place in the condenser. At point *E* the refrigerant is a saturated vapor at the condensing temperature of 100°F and an absolute pressure of 131.86 psia; the same temperature and pressure prevail at point *A,* but the refrigerant is now in a liquid state. At any other point on line *E–A* the refrigerant is in the phase of a liquid–vapor combination; the closer the point is to *A,* the greater the amount of the refrigerant that has condensed into its liquid stage. At point *A,* each pound of refrigerant is ready to go through the refrigerant cycle again as it is needed for heat removal from the load in the evaporator.

R8-4
COEFFICIENT OF PERFORMANCE

Two factors mentioned earlier in this chapter are of the greatest importance in deciding which refrigerant should be used for a given project of heat removal. Ordinarily, this decision is reached during the design aspect of the refrigeration and air-conditioning system, but we will explain it briefly now, and elaborate later.

The two factors that determine the coefficient of performance (COP) of a refrigerant are refrigerating effect and heat of compression. The equation may be written as

$$\text{COP} = \frac{\text{refrigerating effect}}{\text{heat of compression}} \qquad \text{(R8-2)}$$

Substituting values from the *P–h* diagram of the simple saturated cycle previously presented, the equation would be

$$COP = \frac{h_c - h_a}{h_d - h_c} = \frac{50.34}{8.52} = 5.91$$

The COP is therefore a rate or a measure of the efficiency of a refrigeration cycle in the utilization of expended energy during the compression process in ratio to the energy that is absorbed in the evaporation process. As can be seen from Eq. (R8-2), the less energy expended in the compression process, the larger will be the COP of the refrigeration system. Therefore, the refrigerant having the highest COP would probably be selected—provided other qualities and factors are equal.

R8-5
EFFECTS ON CAPACITY

The pressure–enthalpy diagrams in Figs. R8-4 and R8-5 show a comparison of two simple saturated cycles having different evaporating temperatures, to bring out various differences in other aspects of the cycle. In order that an approximate mathematical calculation comparison may be made, the cycles shown in Figs. R8-4 and R8-5 will have the same condensing temperature, but the evaporating temperature will be lowered 20°F. Data can

either be obtained or verified from the table for R-12 in the Appendix; but we will take the values of *A, B, C, D,* and *E* from Fig. R8-4 as the cycle to be compared to that in Fig. R8-5 (with a 20° evaporator). The refrigerating effect, heat of compression, and the heat dissipated at the condenser in each of the refrigeration cycles will be compared. The comparison will be based on data about the heat content or enthalpy line, rated in Btu/lb.

For the 20°F evaporating temperature cycle shown in Fig. R8-5:

net refrigerating effect $(h_{c'} - h_a)$ = 48.28 Btu/lb

heat of compression $(h_{d'} - h_{c'})$ = 10.58 Btu/lb

In comparing the data above with those of the cycle with the 40°F evaporating temperature (Fig. R8-4), we find that there is a decrease in the NRE of 4% and an increase in the heat of compression of 28%. There will be some increase in superheat, which should be removed either in the discharge pipe or the upper portion of the condenser. This is the result of a lowering in the suction temperature, the condensing temperature remaining the same.

By utilizing Eq. (R8-1), it will be found that the weight of refrigerant to be circulated per ton of cooling,

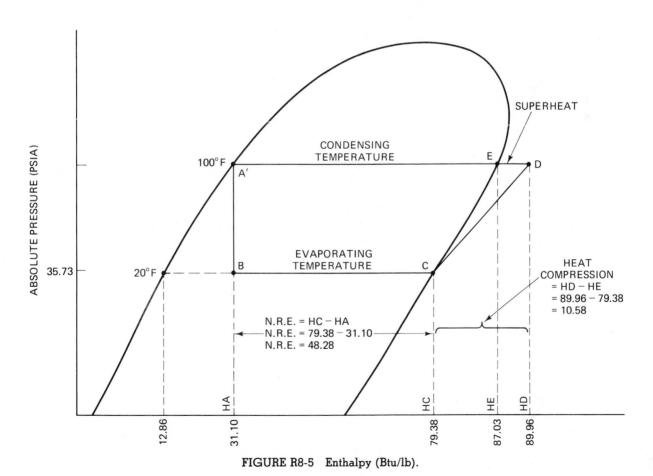

FIGURE R8-5 Enthalpy (Btu/lb).

in a cycle with a 20°F evaporating temperature and a 100°F condensing temperature, is 4.14 lb/min/ton:

$$W = \frac{200 \ (\text{Btu/min})}{\text{NRE (Btu/lb)}}$$

$$= \frac{200 \ \text{Btu/min}}{48.29 \ \text{Btu/lb}}$$

$$= 4.14 \ \text{lb/min}$$

This of course, would also necessitate either a larger compressor, or the same size of compressor operating at a higher rpm.

Figure R8-6 shows the original cycle with a 40°F evaporating temperature, but the condensing temperature has been increased to 120°F.

Again taking the specific data from the heat content or enthalpy line, we now find for the 120°F condensing temperature cycle that $h_a = 36.01$, $h_c = 81.43$, $h_d = 91.33$, and $h_e = 88.61$.

net refrigerating effect $(h_c - h_{a'}) = 45.4$ Btu/lb

heat of compression $(h_{d'} - h_c) = 9.90$ Btu/lb

condenser superheat $(h_{d'} - h_{e'}) = 2.72$ Btu/lb

In comparison with the cycle having the 100°F condensing temperature, it can be calculated that by allowing the temperature of the condensing process to increase 20°F, there is a decrease in the NRE of 9.8%, an increase of heat compression of 16.2%, and a decrease of superheat to be removed either in the discharge line or in the upper portion of the condenser of 7.1%.

Through the use of Eq. (R8-1) it is found that with a 40°F evaporating temperature and a 120°F condensing temperature the weight of refrigerant to be circulated will be 4.4 lb/min/ton. This indicates that approximately 11% more refrigerant must be circulated to do the same amount of work as when the condensing temperature was 100°F.

Both of these examples show that for the best efficiency of a system, the suction temperature should be as high as feasible, and the condensing temperature should be as low as feasible. Of course, there are limitations as to the extremes under which systems may operate satisfactorily, and other means of increasing efficiency must then be considered. Economics of equipment (cost + operating performance) ultimately determine the feasibility range.

Referring to Fig. R8-7, after the condensing process has been completed and all of the refrigerant vapor at 120°F is in the liquid state, if the liquid can be sub-

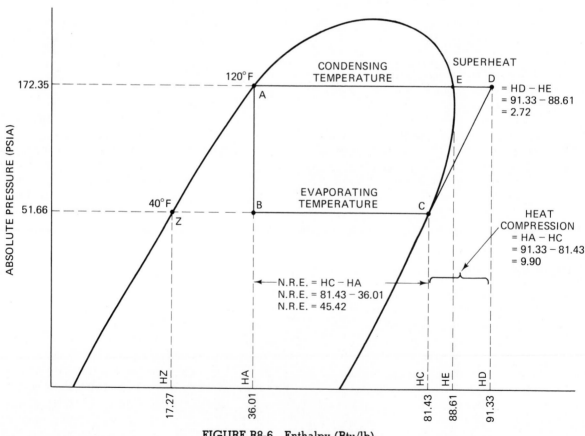

FIGURE R8-6   Enthalpy (Btu/lb).

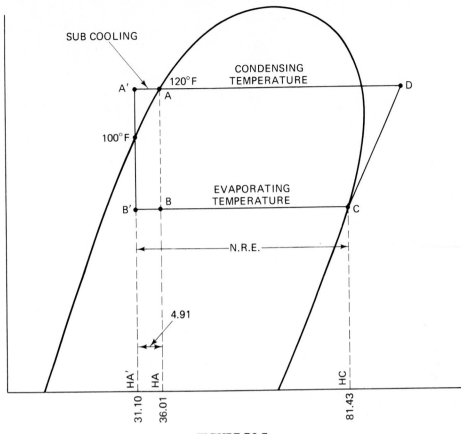

FIGURE R8-7

cooled to point A' on the 100°F line (a difference of 20°F), the NRE ($h_c - h_{a'}$) will be increased 4.91 Btu/lb. This increase in the amount of heat absorbed in the evaporator without an increase in the heat of compression will increase the COP of the cycle, since there is no increase in the energy input to the compressor.

This subcooling may take place while the liquid is temporarily in storage in the condenser or receiver, or some of the liquid's heat may be dissipated to the ambient temperature as it passes through the liquid pipe on its way to the metering device. Subcooling may also take place in a commercial-type water-cooled system through the use of a liquid subcooler, which, in a low-temperature application, may well pay for itself through the resulting increase in capacity and efficiency of the overall refrigeration system.

Another method of subcooling the liquid is by means of a heat exchanger between the liquid and suction lines, whereby heat from the liquid may be transferred to the cooler suction vapor traveling from the evaporator to the compressor. This type is shown in Fig. R8-8, a refrigeration cycle flowchart, using a liquid-suction heat exchanger. True, heat cannot be removed from the liquid and then added to the suction vapor without some detrimental effects to the overall refrigeration cycle; for example, the vapor would become superheated, which would in turn cause an increase in the spe-

cific volume of each pound of refrigerant vapor and consequently a decrease in its density. Thus, any advantage of subcooling in a saturated cycle would be negated; but, in an actual cycle, the conditions of a simple saturated cycle do not exist.

In any normally operating cycle, the suction vapor does not arrive at the compressor in a saturated condition. Superheat is added to the vapor after the evaporating process has been completed, in the evaporator and/or in the suction line, as well as in the compressor. If this superheat is added only in the evaporator, it is doing some useful cooling; for it too is removing heat from the load or product, in addition to the heat that was removed during the evaporating process. But if the vapor is superheated in the suction line located outside of the conditioned space, no useful cooling is accomplished; yet this is what takes place in the majority—if not all—of refrigeration systems.

Now, were some of this superheating in the suction pipe curtailed through the use of a liquid-suction heat exchanger, this heat added to the vapor would be beneficial, for it would be coming from the process of subcooling the liquid. As an example, suppose that the suction temperature in the evaporator is at 40°F; the superheated vapor coming out of the evaporator may be about 50°F, and the temperature of the vapor reaching the compressor may be 75°F or above, depending on the

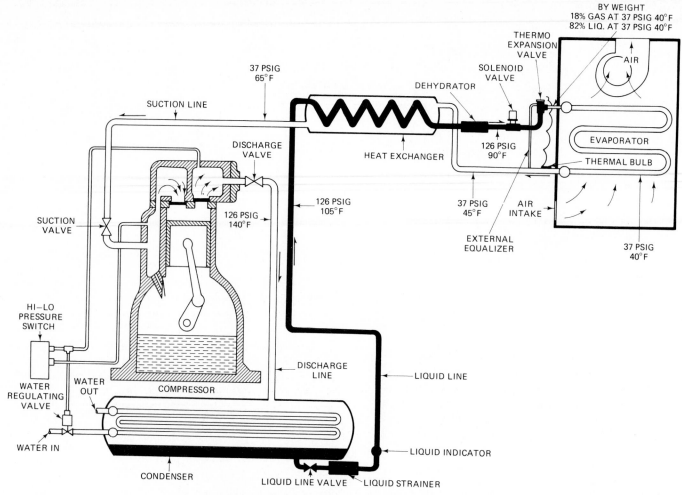

FIGURE R8-8 Flow diagram of R-12 refrigeration system.

ambient temperature around the suction. This means that the temperature of the vapor has been increased 25°F, without doing any useful cooling or work, because this heat has been absorbed from the ambient air outside of the space to be cooled.

If some or most of this 25°F increase in the vapor temperature were the result of heat absorbed from the refrigerant liquid, it would be performing useful cool-ing, since the subcooling of the liquid will result in a refrigerating effect higher than it would be if the refrig-erant reached the metering device without any subcool-ing. It is possible to reach an approximate balance between the amount of heat in Btu/lb removed by sub-cooling the liquid and the amount of heat added to the refrigerant vapor in the suction pipe without the heat exchanger.

PROBLEMS

R8-1. What is meant by "net refrigerating effect"?

R8-2. What two factors are involved in finding the NRE?

R8-3. What is the equation used to find the weight of refrig-erant to be circulated for a given load?

R8-4. Define "superheat."

R8-5. What is the difference between saturated and super-heated vapor?

R8-6. Upon what factors does the capacity of a compressor depend?

R8-7. What is a $P-e$ chart of a system?

R8-8. What are the divisions shown in a $P-e$ chart or dia-gram?

R8-9. List the refrigerant properties that can be determined from a $P-e$ chart or diagram.

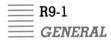

Evaporators

R9-1
GENERAL

In Chapter R6 an evaporator was described as that part of a refrigeration system in which the refrigerant is converted from a liquid to a vapor through the process of evaporation. This takes place as the heat from the produce or load is absorbed by the refrigerant in the evaporator. As discussed previously, evaporators fall into three types:

1. Bare tube
2. Finned-tube
3. Plate

Most commercial display cases, walk-in coolers, reach-in boxes, and florists' refrigerators utilize the finned-tube design of coil. An evaporator of this type has a definite advantage over the bare-tube type of evaporator, shown in Fig. R9-1. The heat load handled by the evaporator reaches the cooling coil by one or more of the three means of heat transfer: conduction, convection, or radiation. But this heat is transferred to the refrigerant through only one of these means—conduction. The extra area of the fins in addition to that of the bare tube permits a higher degree of heat transfer from the air surrounding the coil, as can be seen in Fig. R9-2. Therefore, the greater the surface area for the conduction of the heat from the product to the refrigerant in the evaporator, the greater the possible heat transfer. If the heat from the product reaches the evaporator but is not absorbed by the refrigerant within the coil, the box or area will reach a higher temperature than is desired. Increasing the surface area of the evaporator increases the capacity of the evaporator.

Many commercial evaporators are designed to use natural convection or flow of air through the coil. Of course, the capacity of this type of coil is based primarily on unrestricted flow of air. If the coil is improp-

erly located, or if its design and/or installation is such that air circulation around and through it is restricted, the coil cannot operate at peak efficiency.

As the air passes through the cooling coil it gives up its heat and is therefore cooled. As it is cooled, it contracts into a smaller volume, which weighs more than an equal volume of warmer air. In this way convection currents of air are set up which carry away the heat from the product being cooled.

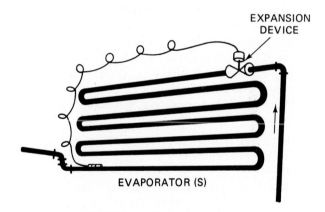

FIGURE R9-1   Bare tube evaporator.

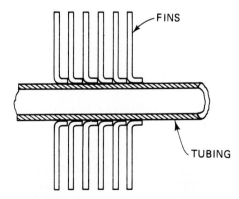

FIGURE R9-2   Flanged fins added to tubing.

93

FIGURE R9-3 Typical commercial evaporator with air-circulating fan. (*Courtesy* Kramer Trenton Company)

The circulation of this convection air can be, and frequently is, assisted by various means, one of which is the use of metal or composition baffles. These baffles are placed so that the air currents are forced to move in certain paths to give the optimum heat transfer.

If there is any doubt that the convection currents can produce the required flow of air across the evaporator coil to remove satisfactory amounts of heat from the air and product, forced circulation of air across the cooling coil is indicated. Evaporators utilizing forced circulation of air usually have finned tubes, supplemented by one or more axial (propeller-type) fans to accomplish the circulation of air. In such a case, baffles are not absolutely necessary to provide a standard pattern for required circulation, but are often used nonetheless. Figure R9-3 shows a typical commercial evaporator with an air-circulating fan.

With forced-air circulation, it is necessary that the fan be capable of:

1. Circulating enough air to remove the heat from the product
2. Distributing air at a satisfactory velocity around the room or conditioned space
3. Assuring that there are no "dead spaces"

If the velocity of air across some products is greater than recommended, there is a tendency toward too rapid dehydration, as related in an earlier chapter.

R9-2
REFRIGERATING EFFECT

Transfer of heat from the product to the air and then to the refrigerant within the cooling coil has been described. The amount of heat each pound of refrigerant picks up from the product and the air as it travels through the evaporator is called the *refrigerating effect*.

The liquid refrigerant entering the metering device and cooling coil has a certain heat content at its given temperature, as does the refrigerant vapor leaving the cooling coil at its lower temperature. The difference in heat content of these two stages is the amount of heat absorbed by the pound of refrigerant as it circulates through the cooling coil. Therefore, the refrigerating effect (RE) is rated in the terms of Btu per pound of refrigerant circulated.

The heat absorbed by the refrigerant depends on two main conditions of the refrigerant and the temperatures at these conditions:

1. The temperature of the liquid refrigerant entering the refrigerant control
2. The evaporating temperature, or the temperature of the refrigerant vapor leaving the evaporator

Under the following conditions, what would be the refrigerating effect?

The refrigerant enters the metering device in a liquid state at the condensing temperature of 100°F, and it leaves the evaporator in a vapor state at 40°F.

Referring to the refrigerant table in the Appendix, which lists the properties of R-12 in a saturated vapor state, one finds that the enthalpy (heat content) of the liquid refrigerant at 100°F is 31.16 Btu/lb, whereas the enthalpy of 40° refrigerant vapor leaving the cooling coil 82.71 Btu/lb. Therefore, the difference between these two figures amounts to 51.55 Btu/lb, or the amount of heat that each pound of refrigerant absorbs from the product and air under the given conditions.

Using the example above, it can be seen that if the amount of heat to be removed from a project amounts to 30,930 Btu, 10 lb of refrigerant must be circulated each minute. For under the conditions given in the example, the heat-removal capacity of the refrigerant amounts to 51.55 Btu/lb circulated through the evaporator. When this capacity of 1 lb of refrigerant is divided into the total Btu to be removed, we find this equation:

$$\frac{30.930 \text{ Btu/hr}}{51.55 \text{ Btu/lb}} = 600 \text{ lb/hr}$$

Then

$$\frac{600 \text{ lb/hr}}{60 \text{ min/hr}} = \frac{10 \text{ lb/min of refrigerant}}{\text{to be circulated}}$$

As already pointed out, the two variables that will alter the refrigerating effect per pound of refrigerant circulated involve the *entering liquid temperature* and the *leaving vapor temperature*. Therefore, by lowering the entering liquid temperature, the refrigerating effect will be increased. This means that fewer pounds of refrigerant will have to be circulated to do the work required. Also, by raising the evaporating temperature while the condensing temperature remains the same, the amount of refrigerant needed to be circulated will be lowered. This can be proved by the following:

If the condensing temperature—the temperature of the liquid refrigerant entering the evaporator—can be lowered from 100°F to 86°F, the enthalpy will be decreased to 27.72 Btu/lb. With the enthalpy of the leaving vapor remaining the same, there will be an increase of 3.44 Btu/lb in the refrigerating effect, producing a new RE value of 54.99 Btu/lb.

The same required load of 30,930 Btu as in the preceding example divided by the new RE of 54.99 Btu/lb results in a requirement of 562 lb/hr. When we divide by the 60 minutes in an hour, we find that approximately 9.37 lb of refrigerant must be circulated each minute.

Conversely, if the evaporator temperature can be increased so that the vapor leaving the cooling coil is at 50°F instead of 40°F, there will be an increase of 1.07 Btu/lb in the enthalpy of the vapor. This bears out that with the variables previously mentioned, the lower the temperature of the liquid entering the metering device and coil, the greater will be the refrigerating effect. It follows that the higher the evaporating temperature, the greater the RE.

We are studying the evaporator first in the commercial refrigeration cycle primarily because that is where the heat is removed from the product. The heat is absorbed by the refrigerant, which vaporizes in the coil and then proceeds to the compressor, which compresses the refrigerant vapor before it continues in the cycle to the condenser.

R9-3
DIRECT-EXPANSION COIL CAPACITY

The capacity of any direct-expansion (DX) coil is dependent on:

1. The temperature of the refrigerant circulated
2. The temperatures (dry and wet bulb) of the air being circulated through the coil
3. The volume of air being circulated

As shown earlier, if the temperature of the liquid entering the cooling coil is varied, the refrigerating effect is also varied. This affects the capacity of the cooling coil, and, if the temperature of the air on the coil

remains the same, any variation in the suction temperature will also change the temperature difference between the refrigerant and the air. If this temperature difference decreases, the rate at which the refrigerant evaporates also will decrease.

The same decrease in refrigerant evaporation will occur if the quantity of air across the cooling coil is decreased, since the lesser amount of air will cool to a lower temperature and reduce the temperature difference between the refrigerant and the air.

Figure R9-4 is a direct-expansion coil capacity table. It shows that as the suction temperature is increased, the coil capacity is lowered. This is caused by the decrease in the temperature difference between the refrigerant and the air.

As will be pointed out later, the identical condition affects the cooling coil and the compressor differently: as the capacity of the coil increases, the capacity of the compressor decreases. Therefore, components that are selected and installed in the field (not a prepackaged condensing unit) require careful balancing of the evaporator and the compressor components (with regard to their individual capacities) to find out the point or points at which each will have the same capacity.

Heat energy is removed from a substance, be that substance air, water, or any other material, by means of a refrigeration system. The heat is transferred into a refrigerant at a boiling point *far enough below the temperature of the substance to be cooled to obtain the desired rate of heat transfer.* This means that the rate of heat transfer, which is measured in Btu/h (Btu transferred in an hour's period of time) depends on the size of the DX coil. That is the amount of surface exposed to substance to be cooled as well as the temperature difference (ΔT°F) between the substance and the boiling point of the refrigerant.

A third consideration is made at this point because the type of evaporator surface configuration will directly affect the amount of expanded surface available to absorb heat.

The Btu/h capacity of an evaporator, therefore, depends on three things:

1. Surface
2. *U* factor
3. Mean Temperature Difference (MTD)

R9-3.1
Surface

As discussed in Chapter R6, evaporators fall into three types:

1. Bare tube
2. Finned tube
3. Plate

Coil capacity table for DX coil ratings.

75° EDB   64° EWB

Suct Temp. (°F)	Row	Fin No.	AIR VELOCITY (FT/MIN)											
			400			500			600			700		
			MBH	LDB	LWB	MBH	LDB	LWB	MBH	LDB	LWB	MBH	LDB	LWB
35°	4	8	18.0	49.0	48.0	20.4	51.0	49.8	22.3	52.7	51.3	23.9	54.0	52.4
	5		20.4	46.0	45.4	23.4	48.0	47.3	25.9	49.7	48.8	28.0	51.1	50.1
	6		22.3	43.8	43.4	25.8	45.6	45.2	28.9	47.2	46.7	31.6	48.6	48.0
	8		25.0	40.5	40.3	29.5	42.1	41.9	33.4	43.6	43.4	36.9	45.0	44.8
	4	10	19.6	47.0	46.4	22.2	49.0	48.3	24.3	50.8	49.9	26.1	52.2	51.2
	5		21.9	44.1	43.8	25.2	46.1	45.7	28.1	47.8	47.3	30.3	49.3	48.7
	6		23.7	41.9	41.8	27.6	43.8	43.6	30.9	45.5	45.3	33.8	47.0	46.7
	8		26.1	39.1	38.9	31.0	40.8	40.6	35.3	42.1	42.0	39.1	43.6	43.4
	4	12	20.7	45.5	45.1	23.6	47.6	47.1	26.0	49.3	48.8	27.9	50.7	50.1
	5		23.0	42.9	42.6	26.6	44.8	44.6	29.6	46.5	46.2	32.2	48.0	37.6
	6		24.7	40.9	40.7	28.9	42.7	42.5	32.5	44.4	44.2	35.6	45.9	45.7
	8		26.8	38.2	38.0	32.0	39.7	39.5	36.6	41.2	41.0	40.7	42.5	42.3
40°	4	8	15.2	52.0	50.9	17.2	53.7	52.3	18.8	55.0	53.5	20.1	56.2	54.5
	5		17.3	49.6	48.8	19.7	51.1	50.3	21.8	52.5	51.5	23.5	53.7	52.6
	6		18.9	47.5	47.1	21.9	49.2	48.6	24.4	50.4	49.8	26.5	51.7	51.0
	8		21.3	44.8	44.6	25.0	46.1	45.9	28.3	47.5	47.1	31.2	48.5	48.2
	4	10	16.5	50.2	49.5	18.7	52.0	51.1	20.5	53.4	52.4	22.0	54.6	53.4
	5		18.6	47.8	47.4	21.3	49.5	49.0	23.6	50.9	50.4	25.5	52.1	51.5
	6		20.1	46.0	45.8	23.4	47.7	47.3	26.1	48.9	48.6	28.5	50.2	49.8
	8		22.2	43.6	43.4	26.3	45.0	44.8	30.0	46.3	46.1	33.1	47.3	47.1
	4	12	17.5	49.1	48.6	19.9	50.7	50.2	21.8	52.2	51.5	23.4	53.4	52.7
	5		19.5	46.7	46.4	22.5	48.5	48.1	24.9	49.8	49.5	27.0	51.0	50.6
	6		21.0	45.1	44.9	24.5	46.5	46.4	27.4	48.0	47.7	30.0	49.1	48.9
	8		22.9	42.9	42.7	27.3	44.1	43.9	31.1	45.3	45.1	34.5	46.5	46.3
45°	4	8	12.3	55.0	53.7	13.8	56.3	54.8	15.1	57.4	55.7	16.1	58.4	56.5
	5		14.0	52.8	52.0	15.9	54.2	53.2	17.6	55.3	54.2	18.9	56.2	55.0
	6		15.4	51.2	50.7	17.7	52.5	51.9	19.6	53.7	52.9	21.3	54.5	53.8
	8		17.3	49.0	48.8	20.3	50.0	49.8	22.9	51.0	50.8	25.2	52.0	51.6
	4	10	13.3	53.5	52.7	15.0	54.8	53.9	16.4	56.0	54.9	17.6	56.9	55.7
	5		15.0	51.4	51.0	17.2	52.9	52.3	19.0	54.1	53.3	20.5	54.9	54.2
	6		16.4	49.9	49.7	18.9	51.2	50.9	21.1	52.4	52.0	22.9	53.5	52.9
	8		18.2	48.0	47.8	21.5	49.1	48.9	24.3	50.2	50.0	26.8	51.1	50.9
	4	12	14.1	52.4	51.9	16.0	53.9	53.2	17.5	55.0	54.2	18.8	55.9	55.1
	5		15.8	50.6	50.2	18.1	51.9	51.5	20.0	53.1	52.6	21.7	53.9	53.5
	6		17.0	49.3	49.1	19.8	50.4	50.2	22.1	51.5	51.3	24.2	52.5	52.2
	8		18.7	47.4	47.2	22.2	48.4	48.2	25.3	49.4	49.2	28.0	50.3	50.1

Each type of coil will have the same heat-absorbing ability *per square inch of material surface*. The amount of tube involved will vary between the different types because the ability of each length of tube to absorb heat will be different. The finned tube will have much more surface per inch of length than the bare tube, which in turn will have more surface than the tube enclosed in a plate. The standard calculation of all the types of tubes is based on each square inch of the basic tube surface, whether bare, finned, or plate, and the difference in the heat-absorbing effect is expressed by the U value of the particular type.

U value is called the overall heat-transfer coefficient and is defined as the amount of heat energy in Btu/h that will be absorbed by 1 square foot of surface for each degree of mean temperature difference between the heat source and the boiling refrigerant. The *U* value applies only to basic surface of the tube and is adjusted to allow for additional surface in the form of fins attached to the basic tube surface or if the *U* value of the basic surface is affected by enclosing it between plates.

Bare pipe coils have a *U* value of 0.5 to 1.0 Btu/h per square foot per $\Delta T°$F. Finned coils will produce *U* values of 2.0 to 3.5 Btu/h per square foot per $\Delta T°$F, depending on the number of fins per inch of tube. The greater number of fins, the higher the *U* value. Of course, the higher the number of fins, the harder it is to get the air to be cooled through the fins in order to remove the heat. In addition, if the coil is intended to operate at a temperature below 32°F (0°C) any moisture condensed out of the air will deposit between the fins and rapidly reduce the capacity of the coil.

Coils in freezing or below 30°F applications are bare pipe where frost formation is excessive and continuous with widely spaced defrost periods. They are in the range of three to six fins per inch where defrosting occurs during each operation OFF cycle of the system. Air-conditioning coils are usually in the range of 10 to 13 fins per inch for high heat-transfer capacity but still have good condensate drain ability.

R9-3.3
Mean Temperature Difference

Mean temperature difference (MTD) is the mean or average temperature difference between the substance to be cooled and the boiling point of the refrigerant. Since the difference in temperature causes heat to flow from the air to the refrigerant and since both the air and the refrigerant change in temperature as they pass through an evaporator, it is necessary to determine the mean difference to obtain the true temperature difference.

To properly check out the refrigeration or air-conditioning system to determine if it has the correct load and is operating in the correct boiling point of refrigerant range (suction pressure range), it is necessary to establish the proper MTD over the coil. This establishes the heat input into the evaporator.

In refrigeration cases, such as dairy, vegetable, floral, etc., the manufacturer will usually establish and label the equipment for proper MTD of air through the coil. On liquid cooling devices, the temperature drop of the liquid is usually given. In air-conditioning units, however, the temperature drop desired for best results is determined by the conditions of air surrounding the conditioned area. These conditions are referred to as *outside design conditions*. In areas of high humidity, where the day and night temperatures remain close (low-daily-range areas), the mean temperature difference is calculated by measuring the temperature of the air entering the DX coil, the temperature of the air leaving the DX coil, and the boiling point of the refrigerant in the coil, and applying these figures in the following:

$$MTD = \frac{(EAT - BP) + (LAT - BP)}{2}$$

The terms used in this equation are defined as follows:

EAT—Entering Air Temperature This is measured on the entering side of the DX coil. The reading should be taken as close to the coil face as possible but not closer than 2 in. to prevent radiation loss from the thermometer to the cold surface of the coil.

LAT—Leaving Air Temperature The leaving air temperature should be measured on the leaving side of each coil circuit to make sure that the coil is evenly loaded over all the circuits. On an A-type coil, the leaving air temperature should be taken on each side of the A and averaged. Air temperatures leaving various parts of the coil should not vary more than 2°.

BP—Boiling Point of Refrigerant This is derived from the suction pressure converted from a pressure–temperature curve or table for the refrigerant involved. This pressure should be measured at the coil outlet. If not possible, and measurement is done at the condensing unit, suction-line pressure drop must be added to the gauge reading to obtain as accurate a boiling point as possible.

This method of MTD calculation is not 100% accurate because of the varying cooling effect on the air as it passes through the coil. It is, however, close enough for general application and service of refrigeration and air-conditioning systems.

The best MTD (in low-daily-range areas) through an air-conditioning coil is the range 25°F +. In average areas (medium daily range) the MTD is in the range 20 to 25°F. In hot dry climates (high daily range) the MTD is in the range 15 to 20°F.

A more accurate MTD setting is required for peak operating efficiency in air-conditioning units. This is discussed in Chapters A3 and A17.

Whenever moisture is removed from the air or other product being cooled or frozen, frost will accumulate on the cooling element and must be removed periodically. Frost acts as an insulator, reducing the heat transfer between the air and the refrigerant in the cooling coil.

In the early days of refrigeration, when bare-tube or plate-coil evaporators were used in storage areas in locker plants, defrosting of these coils had to be done manually. It was not uncommon, particularly when the coils were located overhead at a height of 10 ft or more, for an employee to put on a raincoat and hat and use a long stick to knock down the accumulation of frost on the pipes or plate-coils. When he had finished, he would use a shovel or a broom to finish the job of defrosting by picking up the fallen frost or ice. This manner of defrosting was necessary because it was desired to continue the operation of the refrigeration equipment during the defrost period.

This type of manual defrosting is still being used in some small home or farm freezers, or even in some small commercial refrigeration units having plate-type evaporators. In these situations this method is satisfactory, because there are no fins or other obstructions on this type of cooling coil. These units cannot be shut off in order to permit the defrosting of the coils unless the contents or products are removed from the freezer to prevent melting.

In commercial installations where the room temperature is above 35°F, the finned-tube type of cooling coils does not present too great a problem. It is necessary to shut off the operation of the refrigeration unit temporarily and, with the continued circulation of room air at 35°F or above, the frost will be removed from the fins and cooling coil at a fairly rapid rate. This temporary shutdown of the refrigeration unit can be done manually or with some type of controlling device. This is called *OFF-cycle defrost,* for the refrigeration cycle has been temporarily stopped.

When the room air temperature is below 32°F, the accumulation of frost will not melt off the evaporator during the OFF cycle, and some means of artificial defrosting becomes necessary. This may be accomplished through the application of warm air brought in through ductwork, or by warm water, brine, or heat applied by some other means, such as an electrical heating element or a system of hot-gas defrost. Unless there is ductwork already being utilized for the distribution and circulation of air within the space being cooled, it is not feasible to use the warm-air method, since it necessitates the installation of complicated ductwork and accessories.

Of course, each of the methods of defrosting requires that the individual system be out of operation long enough to permit the defrosting of the coil. This usually takes a coil temperature between 36 and 40°F. Therefore, it is not practical to shut down the unit for any length of time in an application where the room temperature must be 32°F or below.

If two or more evaporators are connected to the same condensing unit but are at different locations, the condensing unit itself does not need to be out of operation in order to defrost the coils. The coils can be shut down individually and defrosted one at a time while the remaining coil or coils and the condensing unit continue to operate at the desired temperature. This may be accomplished by manually closing a valve in the liquid line of the coil being defrosted. When the defrosting of the individual coil is completed, the valve is opened, permitting the normal operation of that cooling coil. The other coil or coils are defrosted in the same manner until all coils in the system are defrosted.

This process can be done automatically by the use of a clock-timer to shut down part or all of a system for a given period of time at regular intervals. The length of the defrost period and the frequency of defrosting depend on the individual installation. It is recommended that the length of the defrost cycle be adjusted carefully so that the refrigeration system is put back into operation as soon as possible after defrosting. The more frequent the defrost cycles the smaller the accumulation of frost, thus shortening the defrosting period.

Water or brine may be used to defrost a cooling coil. Water is usually used when the evaporator temperature is above −30°F. The method is to spray the water over the coils of the evaporator unit. If the evaporator temperature is below −30°F, brine or an antifreeze solution should be sprayed over the coil instead of water. When brine is used, precautions must be taken to prevent corrosion of the coils. A water spray defrost system also requires precautions, so that the water does not spill from the drain pan onto the floor and freeze. Care must be taken to drain the lines properly, so that there will be no residue of condensate water to freeze when the unit is again put into operation.

A number of manufacturers are producing cooling coils equipped with electric heating coils, either inserted into the copper tubing or placed in contact with each row of finned tubing in a forced convection system. Where an adequate supply of water is not readily available or the temperature of the available water is too low for satisfactory defrosting, an electrical defrosting system is frequently used.

The principle of hot-gas defrosting is the use of the hot discharge refrigerant gas from the compressor as a medium for defrosting the cooling coil. A schematic of a basic hot-gas defrost system is shown in Fig. R9-5. An auxiliary or bypass line is installed between the hot-gas

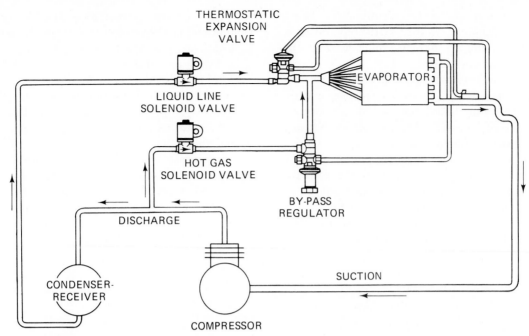

FIGURE R9-5 Typical schematic diagram of hot gas bypass from compressor discharge to evaporator inlet. (*Courtesy* ASHRAE)

line and the refrigerant line entering the cooling coil after the metering device. The valve installed in this line can be operated either manually or by a solenoid. If hand operated, the defrosting takes place only when an operator is present. If it is operated by a solenoid, a timer is installed in the electrical circuit, and the system is automatically defrosted periodically at the selected times.

This system has some weak points, which make its use impractical if there is only one cooling coil in the system. Theoretically, its method of operation is that the compressor continues to run when the defrosting is accomplished. When the valve in the bypass line is opened, hot gas should flow from the discharge line out of the compressor to the evaporator. This hot gas going to the coil will cause the frost to melt off its exterior surface. But when the refrigerant gas gives up its heat, it condenses into a liquid, and there is a possibility that a large slug of liquid refrigerant will flow down the suction line and damage the compressor.

Also, since no liquid enters the metering device and is vaporized in the evaporator while the system is on the defrost cycle, less refrigerant vapor is compressed by the compressor, and therefore less hot refrigerant vapor will be available for defrosting purposes.

Figure R9-6 shows a refrigerating system utilizing multiple evaporators in which two are illustrated. In this case the hot-gas defrost method is feasible since, while one coil is being defrosted, the other continues to operate at its normal temperature. As mentioned previously, when the hot gas passes through the evaporator on the defrost cycle, it gives up its heat to melt the frost. In so doing, it condenses into a liquid state and—since only one evaporator coil is being used—it is quite possible that liquid slugs will damage the compressor.

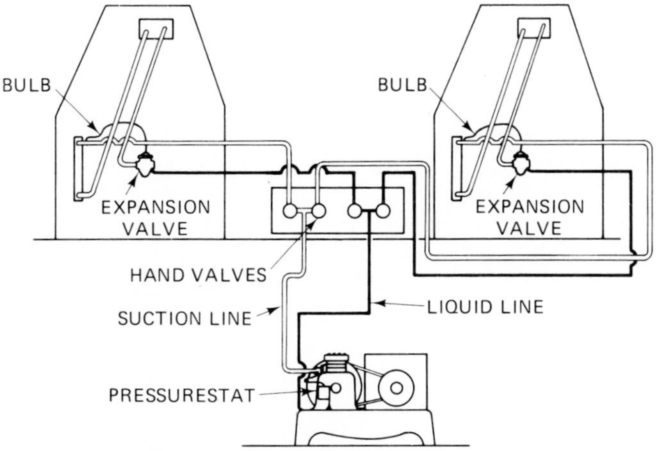

R9-5
OIL CIRCULATION

Oil travels in the refrigeration cycle, together with the refrigerant, for the lubrication of moving parts. Since the compressor must be lubricated, the flow of oil will be considered from that initial point. The refrigerant vapor comes in direct contact with the oil clinging

FIGURE R9-6 Refrigerating system with multiple evaporators. (*Courtesy* ASHRAE)

to the cylinder walls as the pistons are lubricated. Some of this oil is carried along with the refrigerant vapor as it passes into the hot-gas discharge line between the compressor and the condenser.

If only a small amount of oil travels along with the refrigerant, it will pass into the condenser and receiver and through the liquid line into the evaporator, returning to the compressor crankcase before that segment of the system is pumped short of oil.

But some compressors normally pump relatively large volumes of oil, and unless provisions are made for its speedy return to the crankcase, serious damage can occur to the compressor. A precautionary measure is the correct installation of refrigerant lines having the proper pitch to correctly sized lines, with oil loops provided in the piping system where needed. These factors are considered and elaborated on in Chapter R13. In Chapter R14 oil traps and oil separators are reviewed.

Small amounts of refrigeration oil will not be harmful to the cooling coil, but large amounts collecting in the circuits or passages of the evaporator will cause an increase in the cooling coil temperature. Such an increase means less work will be done in the evaporator. There will be less total cooling if the suction pressure remains constant, so the entire system will be less efficient.

If the oil is allowed to remain in the evaporator, it will take up space in the coils that should be used for the vaporization of the refrigerant, and refrigeration will decrease in efficiency.

PROBLEMS

R9-1. What are the major types of evaporators?

R9-2. What is the advantage of a finned-tube evaporator over a bare-tube type?

R9-3. Which of the three means of heat transfer—radiation, convection, or conduction—is the one by which heat is transferred to the refrigerant in an evaporator?

R9-4. When it is desirable to use a fan or blower to provide forced circulation of air over or through an evaporator coil?

R9-5. Upon what three factors does the capacity of a direct-expansion (DX) coil depend?

R9-6. In what applications are bare pipe coils used?

R9-7. Why must frost be removed when it collects on cooling coils?

R9-8. What is the disadvantage of using a hot-gas defrost system in an application having only one cooling coil?

R9-9. Why is it essential that refrigeration oil traveling with the refrigerant in the system be returned to the compressor crank case?

R9-10. What is the result of too much oil remaining in the evaporator?

R10

Refrigerant Control Devices

R10-1
GENERAL

In Chapter R9 the following statement was made and is well worth repeating: "The heat is transferred into a refrigerant at a boiling point far enough below the temperature of the substance to be cooled to obtain the desired rate of heat transfer." What this means is that the pressure on the liquid refrigerant must be lowered—before it enters the DX coil—to a pressure where the corresponding boiling point is far enough below the temperature of the substance to be cooled ($\Delta T°F$) to obtain the desired rate of heat transfer. At the same time, the quantity of refrigerant flow must match the evaporation rate of the refrigerant in the evaporator—no more or no less.

If the flow rate is too high, the liquid refrigerant will flow out of the coil before it is vaporized. If this should enter the compressor, damage could result. Even if the excess quantity is not sufficient to reach the compressor, any refrigeration effect (heat pickup) in the suction line before it reaches the compressor is heat energy that must be handled and represents a loss in efficiency and increase in operating cost. To prevent this loss, the proper refrigerant flow rate will usually result in a vapor superheat leaving the coil of 5 to 7°F.

If the refrigerant flow rate is too low, the DX coil will be starved and lose capacity, the suction pressure to the compressor will drop, the compressor works harder at lower efficiency, and the operating cost will rise. It is, therefore, necessary to have the correct size of refrigerant control device and the correct charge of refrigerant in the system.

R10-2
TYPES OF CONTROL DEVICES

The five main types of pressure-reducing devices used in various phases of refrigeration systems are:

1. Automatic expansion valve
2. Thermostatic expansion valve
3. Low-side float
4. High-side float
5. Capillary tube

As pointed out earlier, the use of hand valves in early refrigeration systems was a means of metering, but it is not suitable to application in modern automatic equipment and thus will not be covered.

R10-3
AUTOMATIC EXPANSION VALVES

The first important development after manual valve operation was the *automatic expansion valve*. This is not a good descriptive name, for some other types of expansion valves are also automatic. It could more accurately be called a *constant evaporator pressure expansion valve,* for it maintains constant outlet pressure regardless of changes in inlet liquid pressure, load, or other conditions.

Figure R10-1 is an illustration of an automatic expansion valve. This metering device is designed to maintain a constant pressure in the evaporator. Its primary motivating force is evaporator pressure, which exerts a force against the bottom of the diaphragm. An adjustable spring exerts a pressure on the top of the diaphragm. As the evaporator pressure increases, it overcomes the spring pressure and moves the diaphragm up, thus closing the valve. As the evaporator pressure decreases, the spring pressure overcomes the evaporator pressure and pushes the valve open.

As this valve maintains a constant evaporator pressure it also attempts to maintain a constant evaporator temperature. It is important to understand that this valve has a reverse action according to varying load conditions. As the load on an evaporator increases, the back

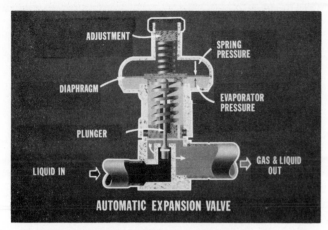

FIGURE R10-1 Automatic expansion valve. (*Courtesy* Carrier Air-Conditioning Company)

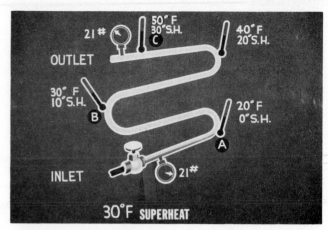

FIGURE R10-2 30°F superheat. (*Courtesy* Carrier Air-Conditioning Company)

pressure normally rises due to the increased rate of evaporation. To pick up the load this rise in back pressure should be accompanied by an increased rate of flow of liquid refrigerant to the evaporator. Where there is an automatic expansion valve, increased pressure closes the valve. By closing the valve when an increase in load is present, the supply of refrigerant is shut off rather than increased.

This means that this valve should only be used where the load is relatively constant. It is found primarily on such items as household refrigerators, small air-conditioners, water chillers, and the like.

≡ R10-4
≡ *THERMOSTATIC EXPANSION VALVES*

Before we explain the thermostatic expansion valve a brief review of *superheat* will be helpful. Figure R10-2 shows a bare-tube evaporator using a hand valve. In this evaporator no pressure drop is shown; in other words, the pressure is the same throughout the evaporator. When the valve is opened slightly, a small amount of refrigerant enters the evaporator. Heat passes into the refrigerant through the tube, causing the refrigerant to boil. If only a small quantity of refrigerant is flowing, it will all be boiled away at some point, such as *A*. (Throughout this example we assume the evaporator pressure to be 21 psig; the saturation temperature at 21 psig is 20°F.) From point *A*, the refrigerant is a gas, and any heat absorbed results in superheat. Superheat is the difference between the actual gas temperature and the saturation temperature (i.e., the boiling temperature corresponding to the gas pressure). At *B*, the gas temperature is 30°F and we have 10°F of superheat. At the last return bend the gas temperature is 40°F and the gas has 20°F of superheat. At point *C*, the difference be-

tween the actual gas temperature of 50°F and the saturation temperature of 20°F is 30°F superheat.

When the valve is opened wider (Fig. R10-3), the refrigerant flow increases and the point at which the last liquid boils will move to *B*. The reduced evaporator surface available to superheat the refrigerant results in a lower gas superheat. The gas is shown here leaving at 40°F, which represents 20°F of superheat.

The ideal condition would be that shown in Fig. R10-4, where the valve is opened sufficiently to give 0°F, or *no* superheat, in the refrigerant leaving the coil. This chart shows the last of the liquid being boiled at point *C*.

If the hand valve were opened wide enough, it would be possible for the refrigerant flow to be so great that liquid could flood back to the compressor. This is a dangerous condition, since the compressor may be damaged.

The practical application of the thermostatic expansion valve approaches the condition shown in Fig.

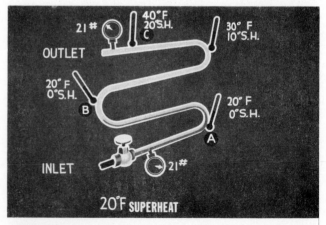

FIGURE R10-3 20°F superheat. (*Courtesy* Carrier Air-Conditioning Company)

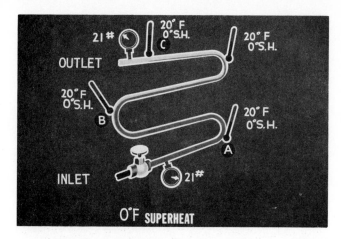

FIGURE R10-4   0°F superheat. (*Courtesy* Carrier Air-Conditioning Company)

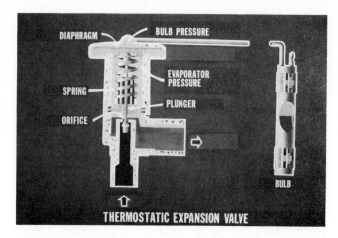

FIGURE R10-6   Thermostatic expansion valve. (*Courtesy* Carrier Air-Conditioning Company)

R10-5. In this example, the point of complete vaporization is at the last return bend of the evaporator, which allows the gas to pick up 10°F of superheat before it leaves the coil.

Because of variations in loads and a lag in the control of the metering devices, 10°F is the minimum practical superheat for air conditioning. In refrigeration and special applications lower superheat settings are often used.

By far the most widely used of all metering devices is the TXV (thermostatic expansion valve). This valve controls the flow of refrigerant by maintaining a relatively constant superheat at the end of the evaporator coil. Though widely used, this valve is the most difficult to understand of all metering devices.

Figure R10-6 shows a cutaway of the thermostatic expansion valve, with the major components labeled. The forces operating on the plunger are emphasized in this diagram. They are spring pressure and evaporator pressure on the bottom of the diaphragm and bulb pres-

sure on the top of the diaphragm. The evaporator pressure is admitted to the bottom of the diaphragm through the internal port in the valve. This is referred to an *internally equalized valve.*

Figure R10-7 is a schematic representation of the thermostatic expansion valve. Here again the three operating pressures are emphasized. The bulb pressure is on the top of the diaphragm and the spring pressure and evaporator pressure are on the bottom. When the bulb pressure is greater than the sum of the spring pressure and evaporator pressure, the plunger will be pushed down, opening the orifice. When the bulb pressure is less than the sum of the spring pressure and evaporator pressure, the plunger will be pushed up, closing the orifice.

Figure R10-8 is a pictorial representation of a thermostatic expansion valve and pressure and temperatures added. Shown is a thermostatic expansion valve set for 10°F superheat with the pressure above the diaphragm equal to the sum of the two pressures below the dia-

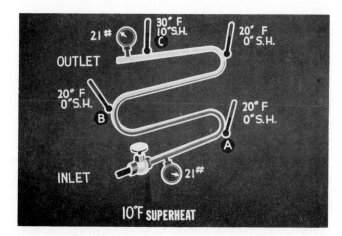

FIGURE R10-5   10°F superheat. (*Courtesy* Carrier Air-Conditioning Company)

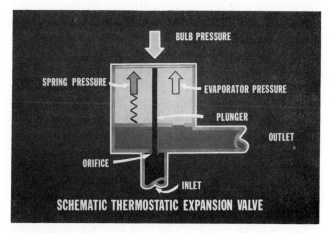

FIGURE R10-7   Schematic thermostatic expansion valve. (*Courtesy* Carrier Air-Conditioning Company)

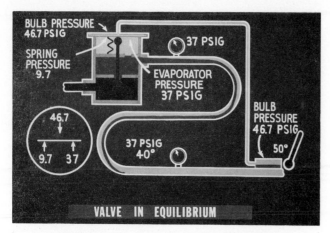

FIGURE R10-8 Schematic valve in equilibrium. (*Courtesy* Carrier Air-Conditioning Company)

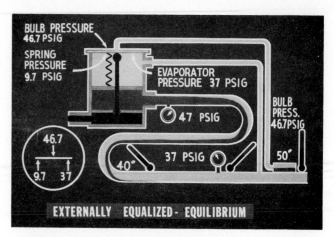

FIGURE R10-9 Valve externally equalized. (*Courtesy* Carrier Air-Conditioning Company)

phragm; that is, spring pressure plus evaporator pressure equals bulb pressure. R-12 is used in this illustration. The evaporator pressure is 37 psig as shown by the gauge. Assuming no pressure drop, this pressure is on the bottom of the diaphragm. The hand-set spring pressure is also exerted on the bottom of the diaphragm and is equal to 9.7 lb as shown in the inset. The total pressure on the bottom of the diaphragm is 9.7 + 37 or 46.7 lb.

The gas leaving the evaporator has a 10°F superheat. This indicates that the gas temperature at the end of the evaporator will be 10°F higher than the temperature corresponding to the evaporator pressure. The evaporator pressure is 37 psig, which corresponds to 40°F. Since the gas leaving the evaporator has a 10°F superheat, the temperature at the end of the evaporator will be 40°F plus 10°F, or 50°F, as shown by the thermometer. The temperature of the thermostatic expansion valve bulb will therefore be 50°F. If there is liquid R-12 in the bulb at 50°F, its pressure will be 46.7 psig. This pressure will be transmitted to the top of the diaphragm through the capillary. Because the pressure on the top and the bottom of the diaphragm are equal, the valve is in equilibrium, and a continuous, steady rate of refrigerant flow to the evaporator is the result. The valve would remain in this position as long as there was no change in the rate of heat flow to the evaporator.

To compensate for pressure drop in an evaporator, the externally equalized valve is used (Fig. R10-9). In this valve the internal equalizing port is eliminated and the pressure under the diaphragm is being taken from the end of the coil near the thermal bulb as shown. All other conditions are the same, but the evaporator pressure under the diaphragm has dropped to 37 psig, giving a total pressure under the diaphragm of 37 + 9.7, or 46.7 psig. With 46.7 psig above the diaphragm, as shown in the inset, the valve will be in equilibrium at 10°F superheat.

In this valve operation the pressure drop in the coil has been ignored. The externally equalized valve obtains its operating pressures from the point on the coil at which the coil superheat is measured and is completely unaffected by the pressure at the head of the coil. Whenever a pressure drop of several pounds is encountered, an externally equalized thermostatic expansion valve should be used.

The thermostatic expansion valve is the most versatile of all metering devices. It can be used either as a primary metering device or as a pilot device for evaporator control on almost any application. Because of its complexity, however, it must be thoroughly understood by both application and service engineers in order that satisfactory results be obtained.

Three important points to remember are shown in Fig. R10-10. The first is that the touching surface between the thermostatic expansion valve bulb and the suction line must be as clean and tight as possible to

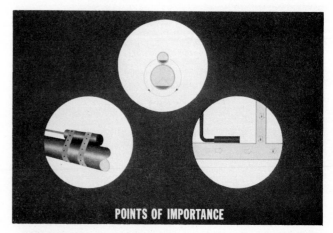

FIGURE R10-10 Points of importance. (*Courtesy* Carrier Air-Conditioning Company)

assure good heat transfer. The temperature of the thermal bulb must be as close as possible to the leaving gas temperature.

The center chart in Fig. R10-10 shows the importance of the bulb position on the suction line. Oil or liquid refrigerant may flow in the bottom of the line, giving a false temperature reading. The bulb must be placed so as to get a gas temperature reading, not an oil or liquid refrigerant temperature reading. Application or size of line will dictate the position of this valve on the line, but *it should never be placed on the bottom*.

The chart on the right in Fig. R10-10 shows the position of the equalizing line. It should be downstream from the bulb so that a slight leakage through the packing will not affect the temperature at the bulb. These are only three of many important points relative to the application of the thermostatic expansion valve. The thermostatic expansion valve is an excellent metering device, but requires proper understanding and handling.

R10-5
ADJUSTING THERMOSTATIC EXPANSION VALVES

The best way to adjust the TXV is first to open it a little too much (too low superheat), so as to frost out beyond the bulb. Then gradually close the TXV (increase the superheat) until the frost is barely back to the bulb. Change the adjustment in small steps and wait at least 15 minutes of running time before making another adjustment.

The main purpose in using a TXV is to get all the refrigeration possible from an evaporator by keeping it fully active. If the evaporator is not fully active, do not hesitate to adjust the TXV for lower superheat (open the valve), should that seem to be the cause of the partial activity; that is the purpose of the adjustment screw on the TXV. The TXV should be readjusted to a higher superheat (closed a little) if overfeeding of the evaporator, a frosted suction line, or liquid slugging by the compressor appears to be causing a low superheat.

As a rule, the TXV should rarely need adjustment after it has been properly installed. If at that time it is set at the correct superheat to make the coil fully active, and later the coil becomes only partially active, it is highly unlikely that readjusting the TXV is the proper procedure, since it is probably something other than the TXV superheat that has changed.

R10-6
FLOAT CONTROLS

Figure R10-11 shows a *low-pressure float*. The name is derived from the fact that the float ball is located in the low-pressure side or *low side* of the system.

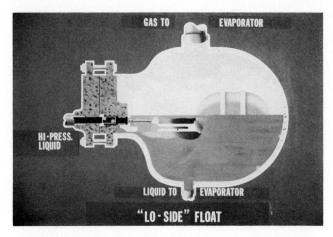

FIGURE R10-11   Low-side float. (*Courtesy* Carrier Air-Conditioning Company)

The primary means of control is the level of the liquid in the float chamber as shown.

This type of control is always used with a flooded evaporator. The float ball may be located directly in the evaporator or it may be in a float chamber adjacent to the evaporator. If the float chamber is used, both the top and bottom of the chamber must be connected to the evaporator so that the liquid level in both remains the same at all times.

As the load on the evaporator increases, more liquid is boiled away and the liquid level in the evaporator and float chamber drops. The float drops with it until the orifice opens and admits more liquid from the high-pressure side. As the load on the evaporator decreases less liquid is boiled away and the float will rise until the orifice is closed. The simplest type of float mechanism is shown here, but float valve construction can take many forms.

The low-pressure float is considered one of the best metering devices available for the flooded system. It gives excellent control and its simplicity makes it almost trouble free. It can be found in any flooded application, large or small, and it can be used with any refrigerant. On the larger systems it is generally used as a pilot device.

Figure R10-12 is an illustration of a *high-side float*. This float is located on the high side of the system and is immersed in high-pressure liquid, which is the primary control. This type of metering device can only be used in a system that has a "critical" charge of refrigerant. As fast as the hot gas is condensed it flows to the metering device. As the liquid level in the float chamber rises, the float opens and allows the refrigerant to flow to the evaporator. This control allows liquid to flow to the evaporator at the same rate that it is condensed; therefore, there can be no provision in the system for the automatic storage of liquid refrigerant other than in the evaporator. Thus an overcharge of refrigerant would

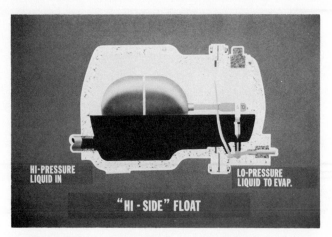

HI-PRESSURE
LIQUID IN

LO-PRESSURE
LIQUID TO EVAP.

"HI - SIDE" FLOAT

FIGURE R10-12 High-side float. (*Courtesy* Carrier Air-Conditioning Company)

result in liquid flooding back to the compressor and an undercharge of refrigerant would result in a starved evaporator. Frequently, suction accumulators are used to make the refrigerant charge less critical.

R10-7
CAPILLARY TUBES

Capillary tubes are placed in the same classification as expansion valves, flow valves, etc., and are called refrigerant control devices. In the case of the capillary tube, this is a serious error because a capillary tube does not control the flow of refrigerant into the coil. A capillary tube is only a very small diameter section of pipe, usually in the 0.030 to 0.085 in. range (inside-diameter size and of sufficient length, depending on the inside diameter) to produce the desired pressure drop. The capillary tube does not change its size and/or length and must depend on the difference between the high and low pressures in the system to produce the desired results (Fig. R10–13).

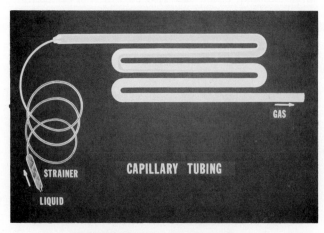

GAS

STRAINER   CAPILLARY TUBING

LIQUID

FIGURE R10-13 Capillary tube.

The sizing of capillary tubes must therefore, be at established design conditions for the particular refrigeration system. Operation at other than design conditions adversely affects the operation of the system. Capillary tubes are best used where the load is fairly constant, such as refrigerators, freezers, small air-conditioners, and even residential and small commercial air-conditioning systems.

If the system is expected to operate over a wide range of load, a more positive means of pressure reduction and refrigerant volume control is required. The recommended device in this situation is the thermostatic expansion valve.

R10-7.1
Advantages of Using Capillary Tubes

There are, however, several advantages to using a capillary tube for pressure reduction.

1. First and foremost is cost. The primary reason for development of this use was price reduction by reducing material cost.

2. A second reason for use of the capillary tube was reduction in starting-torque requirements. It is during the OFF cycle that this advantage comes into play. A float valve or expansion valve retains a seal between the high- and low-pressure sides of the system when the compressor shuts off. The rising pressure in the evaporator closes off the expansion valve and the reduction in boiling of the refrigerant in the low-side float prevents the float from opening. Thus, head pressure remains high and suction pressure low. When the compressor attempts to cycle on, a pressure difference exists which the compressor must start against, and extra starting torque is required.

The capillary tube, on the other hand, does not maintain a seal because it is an open pipe. When the compressor stops, the refrigerant continues to flow through the capillary tubes until the high and low pressures balance. The compressor is now in an unloaded condition and comparatively little torque is required to start it again. This usually permits the use of a low-cost, low-torque hermetic motor-compressor assembly which has been designed for capillary tube application. When the unit cycles on, the high pressure begins to build immediately and liquid begins to flow through the capillary tube. The flow capacity of the capillary tube depends more on the head pressure than on the suction pressure. (Actually, the Δpsig across the capillary tube is the deciding factor.) Because the head pressure does not build very rapidly, the compressor will pump the refrigerant vapor out of the coil at a higher-than-normal rate and suction pressure will drop very rapidly and below-normal readings will result. This will continue until the condenser becomes thermally saturated to its normal operating temperature and pressure.

Do not attempt to judge the performance of a capillary tube until a sufficient operating time has occurred to stabilize the head pressure.

R10-7.2
Disadvantages of Using Capillary Tubes

There are also several disadvantages to the use of capillary tubes:

1. As explained previously, the capillary tube does not have the ability to regulate refrigerant flow. Therefore, when the load on the unit decreases or increases, the efficiency of the system falls at a higher rate than in systems using a thermostatic expansion valve.

2. The capillary tube does not stop the flow of refrigerant when the unit cycles off. This is an advantage to the compressor starting-torque requirement but possibly a disadvantage to the mechanical life of the compressor. The amount of refrigerant in the system can seriously reduce the life of the compressor. The system must not contain any more liquid refrigerant than the DX coil can hold during the OFF cycle. The coil must be as full of liquid refrigerant during the ON cycle for maximum capacity so that a balance must be maintained between these quantities.

The system volumes are designed to accomplish this by the manufacturer, and when the refrigerant charge quantity is specified, the quantity of refrigerant put in the system *must not exceed this amount*. Figures R10-14 through R10-16 show the effect of refrigerant charge quantity on the operation of a 2-hp air-conditioning unit. The curve in Fig. R10-14 shows the net coil capacity (the room cooling capacity) of a unit operating at 90°F outside ambient and 75°F inside dry bulb and 63°F wet bulb—50% relative humidity.

With the refrigerant charge at 100% of the required amount, the net capacity of the unit was 26,400 Btu/h. When the refrigerant charge was increased 5% (3 oz), the capacity dropped to 24,600 Btu/h; with an increase of another 5% (3 oz), the capacity dropped to 19,000 Btu/h. A total overcharge of 9 oz reduced the capacity to 13,000 Btu/h.

Working the other way, from the correct charge, when the quantity was reduced 5% (3 oz), the net capacity dropped to 25,000 Btu/h, another 5% (2.5 oz) reduced the capacity to 22,000 Btu/h. A further reduction of 5% (2.5 oz) reduced the capacity to 18,000 Btu/h. From this it can be concluded that the correct charge results in the best net capacity.

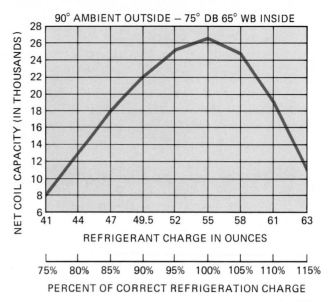

FIGURE R10-14 Net coil capacity versus % refrigerant charge.

Figure R10-15 shows the kilowatt requirement of the unit to handle the load. At 100% of charge the kilowatt required was 3.195 kW; at 5% overcharge, 3.45 kW; at a 110% overcharge, 3.97 kW; and at a 115% overcharge, 4.8 kW. With a reduced charge at 95% overcharge it was 2.97 kW; at 90% overcharge, 2.77 kW; and at 85% overcharge, 2.57 kW.

The true comparison of the system is the operating efficiency, the energy efficiency ratio (EER). The EER is determined by dividing the net capacity in Btu/h by

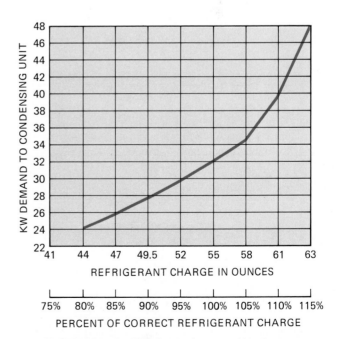

FIGURE R10-15 KW demand versus % refrigerant charge.

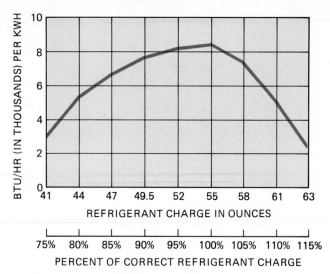

FIGURE R10-16 EER rating versus % refrigerant charge.

the watts of electricity needed to produce the capacity. Figure R10-16 shows the curve of the EER rating of the example unit at the various refrigerant quantities. At 100% of refrigerant charge the EER rating was 8.4; at 105%, 7.45; at 110%, 5.1; and at 115%, 2.4. With the undercharge at 95% of charge, an 8.2 EER resulted; at 90%, 7.7; and at 85%, 6.75. This points out that with refrigeration systems using capillary tubes, the refrigerant quantity in the system must be accurate. The charge tolerance is plus zero–minus 1 oz.

R10-8
REPLACEMENT OF CAPILLARY TUBES

Do not attempt to change the size or length of capillary tubes in the system. Using a capillary tube gauge as shown in Fig. R10-17, determine the inside diameter of the tube. An accurate rule to determine the physical length is to allow 1 in. on each end for insertion in the condenser outlet and coil inlet. Replace the capillary tube with the *exact size and length of the original. Do not attempt to substitute a different size or length*. If you do not have the correct size, *get it*.

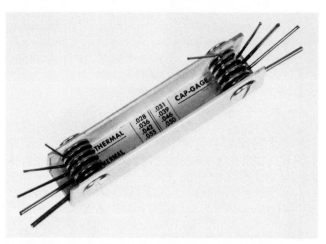

FIGURE R10-17 Capillary tube sizing gauge. (*Courtesy* Thermal Engineering Company)

PROBLEMS

R10-1. Name the five major pressure-reducing devices used in refrigeration systems.

R10-2. What is the disadvantage of an automatic expansion valve?

R10-3. A thermostatic expansion valve regulates the refrigerant boiling point according to the load and the coil by maintaining a fairly constant _____ .

R10-4. What three pressures are used to operate a thermostatic expansion valve?

R10-5. Of the three pressures that operate a thermostatic expansion valve, which two pressures oppose the third?

R10-6. The feeler bulb of the TXV should always be placed on the suction line between the 4 o'clock and 8 o'clock positions. True or False?

R10-7. What are the two principal advantages of using capillary tubes as pressure-reducing devices?

R10-8. What are the two disadvantages of using capillary tubes as pressure-reducing devices?

Compressors

Early models of refrigeration compressors were largely the vertical, single-acting compressor typical of the ammonia machine shown in Fig. R11-1. Since ammonia was the most popular refrigerant in the old days, these compressors were very heavy in order to withstand very high pressures, and, in comparison to modern compressor designs, the early ones ran at relatively low speeds. Advances in valve design, compressor shaft seals, bearings, and lubrication systems resulted in a gradual increase in the design speed. This allowed the compressors to become smaller for a given horsepower, since increased displacement was obtained from higher-speed operation.

The introduction of new refrigerants also had considerable bearing on the designs and development of

FIGURE R11-1 Single-acting ammonia compressor. (*Courtesy* Borg-Warner Air Conditioning, Inc.)

compressors. For example, when using ammonia, all those parts of the system exposed to refrigerant had to be constructed of steel. The introduction of sulphur dioxide and methyl chloride as refrigerants did make the use of nonferrous metals possible in some cases. The advent of the halogenated hydrocarbon refrigerants, however, had perhaps the greatest effect on compressor design. It became possible to use nonferrous metals such as aluminum. Simultaneously with the introduction of R-12, the hermetically sealed type of compressor became popular.

The development of compressors for commercial refrigeration and air-conditioning applications has been influenced considerably by the use of compressors in household refrigerators. Hermetically sealed compressors and capillary tube refrigerant feed devices were first introduced and field-proven in household refrigerator applications. In the early thirties hermetically sealed compressors began to become the standard of the household refrigerator manufacturers. Within a few years, belt-driven compressors practically disappeared from the household refrigerator field. The manufacturers of ice cream cabinets, beverage coolers, water coolers, etc., were next to adopt hermetically sealed compressors.

In 1935 the first hermetic compressor for air-conditioning service was introduced and by the early 1940s most air-conditioning manufacturers had switched to the hermetic compressor for their products. The trend toward the use of the hermetic compressor for both commercial refrigeration and air-conditioning has continued.

≡ R11-2
≡ *DESIGNS*

The compressor is often called the heart of any refrigeration system. Chapter R6 noted the four designs of compressors in general terms; let us restate those as follows:

- Positive displacement compressors
  - —Reciprocating
  - —Rotary
  - —Helical (screw)
- Kinetic compressors
  - —Centrifugal

Positive-displacement compressors such as illustrated in Fig. R11-2 are so classified because the maximum capacity is a function of the speed and volume of cylinder displacement. Since the speed is normally fixed (i.e., 1750 or 3500 rpm for typical hermetic reciprocating compressors), the volume or weight of gas (refrigerant) pumped becomes a mechanical ratio of strokes per minute times the cylinder(s) volume.

The kinetic compressor (centrifugal), shown in Fig. R11-3 and sometimes called the *turbocompressor,* is a member of a family of turbo machines including fans, propellers, and turbines, where the pumping force is subject to the speed of the impeller and the angular moment between the rotating impeller and the flowing fluid (refrigerant). Because their flows are continuous, turbo machines have greater volumetric capacities, size for size, than do positive-displacement machines. However, at present, the design and cost of such compressors do not lend themselves to smaller applications (50 tons and less). Centrifugal machines currently start at the 80- to 100-ton range and extend upward to 8000 tons and above.

Among the positive-displacement compressors, the reciprocating compressors have gained the widest acceptance and application from fractional horsepower size up to the 100- to 150-ton range. At that point the crossover to centrifugals becomes apparent.

As described in Chapter R6, the rotary compressor (Fig. R11-4) was used mainly in small, fractional-

FIGURE R11-3   Centrifugal compressor. (*Courtesy* Borg-Warner Air Conditioning, Inc.)

horsepower sizes associated with refrigerators, etc. More recently, however, the rotary has gained popularity in the nominal $1\frac{1}{2}$- to 5-ton residential air-conditioning field. The rotary compressor has not been adapted to commercial refrigeration duty, perhaps because it is usually inefficient in pumping against very high discharge pressures, especially when operating at low suction pressure.

The helical (screw) compressor (Fig. R11-5) is also a positive-displacement design and will perform satis-

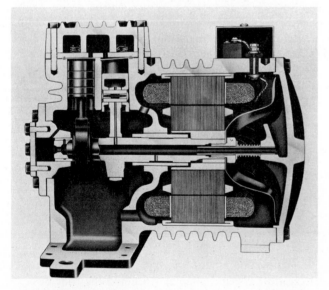

FIGURE R11-2   Reciprocating compressor. (*Courtesy* Copeland Corporation)

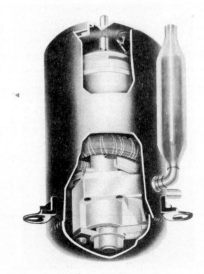

FIGURE R11-4   Rotary compressor. (*Courtesy* Rotorex Company)

FIGURE R11-5   Screw compressor. (*Courtesy* Borg-Warner Air Conditioning, Inc.)

factorily over a wide range of condensing temperatures. The screw compressor has been used for refrigeration duty in the United States since about 1950. The original design was invented and patented in Sweden in the early 1930s.

Figure R11-6 demonstrates the compression cycle of a screw compressor:

1. Gas is drawn in to fill the interlobe space between adjacent lobes.
2. As the rotors rotate the interlobe space moves past the inlet port, which seals the interlobe space. Continued rotation progressively reduces the space occupied by the gas, causing compression.
3. When the interlobe space becomes exposed to the outlet port, the gas is discharged.

Capacity control is achieved by internal gas recirculation, thus providing smooth capacity down to as low as 10% of design capacity. Current sizes of screw machines in operation range from 100 tons up to 700 tons, based on nominal ARI conditions for chilled-water systems. Like the centrifugal equipment, screw machines are not currently being used in small-tonnage refrigeration or air-conditioning. Thus, the reciprocating compressor is used in over 90% of those units, ranging from fractional horsepower through 100 tons and therefore should be the starting point for any novice learning about compressors. The balance of this chapter is devoted to this particular design.

R11-3
TYPES OF RECIPROCATING COMPRESSORS

As mentioned previously, the distinction in reciprocating machines is between the open type of compressor and the hermetic. By *open type* we mean a compressor driven by an external motor (Fig. R11-7), either belt-driven or direct-connected. This type requires a shaft extending through the crankcase of the compressor and, thus, a shaft seal. Opposed to this type of compressor is the hermetically sealed compressor (Fig. R11-8), in which the motor is incorporated in the same housing as the compressor. Thus, the hermetically sealed compressor has no shaft extending through the crank-

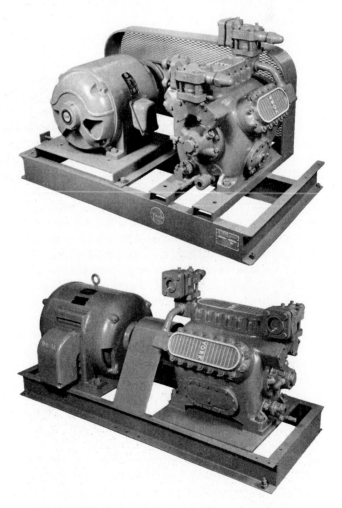

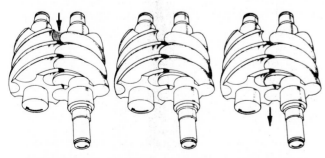

FIGURE R11-6   Screw compressor compression cycle. (*Courtesy* Borg-Warner Air Conditioning, Inc.)

FIGURE R11-7   Open-drive compressors. (*Courtesy* Borg-Warner Air Conditioning, Inc.)

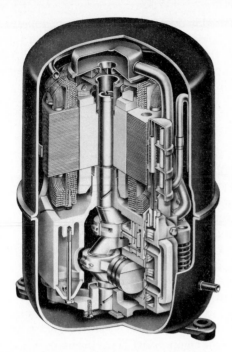

FIGURE R11-8 Hermetically sealed compressor. (*Courtesy* Tecumseh Products Company)

case and therefore requires no seal. Each type has some advantages over the other.

A belt-driven open-type compressor is quite flexible. Its speed may be varied so that a single compressor can often be used for two or three different-size units. By merely changing the size of the motor pulley and compressor valve clearance, in most instances this same compressor may be used not only with different sizes of motors but also for high-, medium-, or low-temperature applications. This is the most outstanding advantage of an open-type compressor as compared to a hermetic type. Other advantages are that this type of compressor may be used with motors available for odd voltages and frequencies, for which hermetic-type compressors are not produced. This has been an important factor in the overseas markets where voltages and phases differ from those in the United States, for example, where 50-cycle (hertz) or direct current is the only power available. Hermetic compressors are not available in dc, but open motors are.

Open compressors are always field-serviceable, which is not true of all hermetics. In the event of a motor burnout, it is probably easier to replace the motor on an open-type system than on a hermetic system, where the motor is exposed to the refrigerant. In the event that a motor burns out in a hermetically sealed system, the entire compressor must be replaced and returned to a repair shop or factory to be dismantled and reconditioned. The refrigerant must be discharged and the system cleaned to eliminate any possibility that acid, which might have resulted from the burning of insulation, can circulate through the system.

There are instances, however, where hermetically sealed compressors have distinct advantages compared to the open type. Perhaps the greatest of these is the elimination of the shaft seal. Shaft seals are vulnerable to dirt, temporary failure of lubrication, anything abrasive that might accumulate in the system (such as scale), and physical damage due to rough handling, etc. Although seals produced today are much improved over those of 15 to 20 years ago, they are still a potential source of trouble, especially on a low-temperature system where the low-side pressure may be at considerable vacuum. In such a case, a leak at the seal allows air and moisture vapor to enter the refrigerant system, which is more serious than the loss of refrigerant from the system.

Other advantages of the hermetic type of compressor are that it is smaller, more compact, more free of vibration, and has its motor continuously cooled and positively lubricated. There are no belts requiring frequent adjusting and eventual replacement.

To achieve its continuous cooling, the open compressor motor is cooled by the air surrounding the motor housing; if the motor is located in an area of high ambient temperature or poor ventilation, proper dissipation of heat may be a problem. In the hermetic design motor, heat is dissipated by passing cool suction gas through or around the stator winding to pick up the motor heat, and the refrigerant thus carries this heat to the condenser, where it is dissipated. Another advantage in this process is that suction gas is further superheated to help prevent any liquid refrigerant from entering the suction openings, and this is important in low-temperature work to dry the incoming gas.

This discussion of the relative merits of open and hermetic compressors may lead the reader to assume that the open compressors are still widely used in the industry. They actually represent a small part of the total number sold, and they are concentrated in the area of commercial refrigeration or process cooling, where refrigerant temperatures and conditions are different and more varied than found in the comfort air-conditioning field. There are a great number of open machines still operating, and in the course of service and maintenance it is quite possible for a technician to encounter one in need of repair or replacement, so it is important to understand the background of the open-type compressor.

R11-4
HERMETIC FIELD-SERVICEABLE VERSUS TOTALLY SEALED

Most early hermetic compressors were of the field-serviceable design illustrated in Fig. R11-9. Frequently called the *bolted hermetic,* it can be almost totally torn

FIGURE R11-9 Bolted hermetic compressor. (*Courtesy* Copeland Corporation)

down on the job and refitted with new parts, a characteristic that is a distinct advantage in higher-tonnage machines where sheer weight involved in removing a complete unit makes such actions physically and economically undesirable.

On the other hand, the completely sealed hermetic, or *welded shell hermetic* (Fig. R11-10) is not field serviceable and, whatever the internal problem, be it motor failure, valve breakage or anything else, the unit must be returned to a repair station or factory. An exchange compressor is then installed.

FIGURE R11-10 Welded hermetic compressor. (*Courtesy* Tecumseh Products Company)

The range of welded hermetics starts in fractional horsepower sizes and generally goes through the nominal $7\frac{1}{2}$-ton sizes; however, there are compressors on the market of up to 20 tons in a single shell.

Does it make sense to replace a complete compressor? The answer is "yes," and it is not necessarily a technical decision. When the growth explosion in residential air-conditioning came about in the late 1950s it became apparent that there could never be enough skilled technicians in the industry to handle the field repair of millions of compressor installations. Nor did it make sense for manufacturers to stock locally all the parts needed to make field repairs; the financial impact would have been staggering.

Mass production and standardization of compressors brought improved quality to the industry. At the same time there were also technical changes in system design from field-fabricated refrigerant lines to the precharged variety. System reliability and compressor life expectancy improved rapidly, to the point where failure rates represented a very small percentage, and economically it made sense to replace complete compressors in sizes up to the nominal $7\frac{1}{2}$-ton capacity. Between $1\frac{1}{2}$ and 5 tons, which is the bulk of the residential air-conditioning market and many commercial refrigeration applications, it is apparent the *welded hermetic* in reciprocating and rotary will dominate the market. Therefore, from a service technician's point of view it is not totally necessary to be able to tear down and reassemble a compressor. It is more important that the technician concentrate on techniques of proper application, installation, and trouble-shooting, so as to minimize failures and maintenance requirements.

The mechanic who becomes involved in heavier commercial work and larger air-conditioning units that do require field repair can advance his or her knowledge as the situation requires. At this stage, the ability to tear down and reassemble a compressor teardown becomes more of an art than a science, and interpreting the manufacturers' procedures and design is important.

Chapter R6 detailed the internal working parts of a typical reciprocating compressor as well as types of motors and associated starting equipment used with this compressor. However, several factors affecting the operation and life expectancy of compressors were not covered.

Nearly all motors used for refrigeration applications are *induction motors*, so named because the current in the moving part of the motor is induced, since the moving component has no connection to the source of current. The stationary part of an induction motor is called the *stator*, and the moving part, the *rotor*. The stator windings are connected to the power source, while the rotor is mounted on the motor shaft, the rotation of the rotor providing the motor with its driving power source.

In refrigeration systems, moving parts in several components create friction, which can be destructive to metal surfaces. In addition, friction results in an increase in the temperature of the moving parts involved. Because proper lubrication reduces possible damage resulting from friction, it is an important aspect in the maintenance of the mechanical components. The compressor requires proper lubrication for the bearings, pistons, and gears.

In the case of a reciprocating compressor, the space between the piston and the cylinder wall must be sealed off so that all refrigerant vapor will be forced out of the cylinder and into the hot-gas discharge line. This sealing off is accomplished by the refrigerant oil as it is forced to travel along with the compressed refrigerant vapor. If the oil film does not seal off the space as the piston moves back and forth, some of the vapor leaks back into the compressor crankcase, resulting in a loss of efficiency.

As mentioned earlier, oil used in refrigerating systems mixes with and travels along with most refrigerants in the liquid state. It is imperative that oil from the crankcase be forced out of the compressor and into the condenser through the hot gas line. To maintain proper lubrication of moving parts and to keep the correct oil level in the compressor crankcase, the oil must complete the circuit along with the refrigerant, and then make its way back to the compressor.

Traveling with the liquid refrigerant, the oil reaches the evaporator, which is one of the components in which moving the oil presents a problem. If the oil does not travel from the evaporator to the suction line, the evaporator can become oil-logged, decreasing the heat-transfer surface of the cooling coil.

Suction lines must be sized correctly to maintain the velocity of the vapor, in order to entrain the oil with it through the circuit back to the crankcase of the compressor. If the oil is not returned to the compressor, this component could soon be operating in a "dry" condition. When this happens, with no oil being pumped through the cylinder, the vapor seal will be gone and the compressor will lose efficiency. If this situation continues uncorrected for a lengthy period, there will be damage to the compressor.

Two principal methods are used for proper lubrication of compressors: (1) the splash system and (2) the force-feed or pressure system (Figure R11-11). In the first method, lubrication is initiated by the revolving of the crankshaft in the oil within the crankcase. Fingers or throws on the crankshaft dip into the oil and throw it on the bearings or small grooves that lead to the bearings and seal. Oil is also thrown onto the pistons and cylinder walls, thus maintaining the vapor seal between

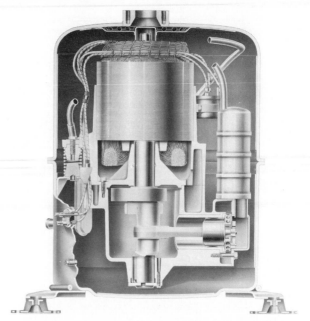

FIGURE R11-11   Lubricating pressure system for compressor. (*Courtesy* Carrier Air-Conditioning Company)

these components. The importance of keeping the proper oil level in the crankcase cannot be stressed too strongly, along with the need for keeping the oil moving through the system together with the refrigerant.

A small pump is used in the pressure system to force the oil to the bearings, seal, piston pins, pistons, and the cylinder walls. A compressor with this type of lubricating system is, of course, more costly than one using the splash system, but the former supplies more protection and assurance of proper lubrication of the compressor, as long as there is an adequate supply of oil in the crankcase.

Some compressors are inherent "oil pumpers." That is, they pump oil out along with the refrigerant vapor at a rate faster than it can be returned through the system to the crankcase. Often, the manufacturer includes an oil separator on the condensing unit. If the compressor is to be used in a built-up system, the manufacturer recommends that such an oil separator be included in the installation.

Figure R11-12 shows a cutaway of an oil separator installed in a refrigeration circuit. It is important that the oil be returned to the compressor as soon as possible, so the oil separator is located between the compressor and the condenser. The high-temperature, high-pressure vapor, together with the oil forced out of the compressor, travel through the discharge line from the compressor until it reaches the oil separator. There its direction of flow is changed, and its rate of flow lessens, since the separator has a larger volume and cross-sectional area than does the discharge line. Depending on the design of the separator, it may contain impingement screens or other devices that will force the oil to drop

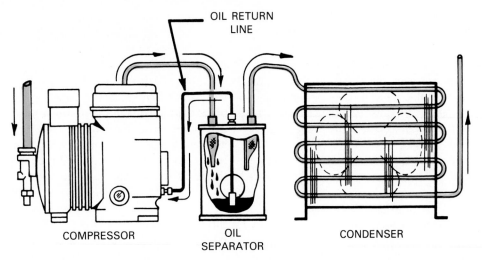

OIL RETURN LINE

COMPRESSOR   OIL SEPARATOR   CONDENSER

FIGURE R11-12   Cutaway of oil separator. (*Courtesy* E. I. du Pont de Nemours and Company)

into the reservoir of the separator while the refrigerant vapor continues on its path through the separator.

As shown in Fig. R11-12, most separators contain float and valve assemblies for the release of the oil back into the compressor. When a given amount of oil has accumulated in the reservoir chamber of the separator, the buoyancy of the oil will raise the float and the valve will open. The pressure of the refrigerant discharge vapor is greater than the pressure in the compressor crankcase, and this difference in pressure forces the oil to return to the crankcase. As the oil level in the separator lowers, so does the float, causing the needle valve to close and permitting the accumulation of more oil in the separator.

The separator is usually insulated so that it is kept warm; otherwise, refrigerant vapor might condense in the separator when the unit is not in operation. If the unit is part of a system that tends to have long off-cycle periods, it might be advisable to install an electric heater either on or in the separator to keep the refrigerant in a vapor state. Several oil separators are shown in Fig. R11-13.

R11-6
REFRIGERANT OIL REQUIREMENTS

As mentioned earlier, a refrigeration oil must have good lubricating qualities and the capability to seal off the low side from the high side in the compressor. While the oil lubricates the bearings in the compressor, it also

FIGURE R11-13   Oil separators. (*Courtesy* AC & R Components, Inc.)

acts as a cooling medium, removing from these bearings the heat caused by the friction of the moving components when the compressor is in operation.

The perfect oil for use with all refrigerants and under all conditions has not been developed as yet. Each of the refrigerant oils available has its good and not-so-good characteristics, and these must be balanced against the requirements of the installation and the use to which the particular systems is to be put.

Here are some of the qualities of oil that are essential:

1. It must remain fluid at low temperatures.
2. It must remain stable at high temperatures.
3. It must not react chemically with the refrigerant, metals, motor insulation (when used in hermetic compressors), air, or other contaminants.
4. It must not decompose into carbon under expected operating conditions.
5. It must not deposit wax when subjected to the low operating temperatures that must be met.
6. It must be as free of moisture as possible.

For all practical purposes, oils available for refrigeration systems are of mineral origin. They can be separated into three main categories: (1) paraffin base, (2) naphthene base, or (3) a mixture of (1) and (2) called a mixed base. The different categories are derived from the crude oil found in different parts of the world. Proper refining processes remove the heavier paraffins and naphthenes from the crude oil.

Some of the characteristics of refrigerant oils are listed below, not necessarily in their order of importance:

1. Viscosity
2. Pour point
3. Floc point
4. Flash point
5. Dielectric strength
6. Fire point
7. Corrosion tendency
8. Oxidation resistance
9. Color

The *viscosity* of a refrigerant oil, or any other liquid, is a measurement of its resistance to flow, or simply how thin or thick it is, under a given set of conditions. A measured sample of the liquid, at a specific temperature, flows through a calibrated orifice; the time it takes, in seconds, expresses its viscosity.

The *pour point* of an oil is the lowest temperature at which the oil will flow. Usually the temperature of the oil will be lowered to the point where it will no longer

flow, and then 5°F is added to this temperature. A low pour point is an indication that the oil will not congeal at the lowest temperatures reached in the system, at its design operating conditions.

It has been found that all refrigerant oils contain wax in varying degrees. This wax will separate from other components in the oil when the temperature of the oil is lowered enough. When a refrigerant oil is dewaxed as much as possible (for the wax cannot be completely removed) tests are run to find the temperature at which the remainder of the wax will separate from the oil.

When the wax separates from the oil the mixture of oil and refrigerant becomes cloudy. As the temperature of the mixture is lowered further, fine suspended particles of wax will form into small balls or clusters. The temperature at which this formation is visible is called the *floc point* of the oil.

Since wax will collect at the colder areas within the refrigeration system (the expansion valve and the evaporator), there will be a loss of heat-transfer efficiency in the evaporator, and the expansion valve or other type of metering device may very easily become restricted or clogged.

A particular oil may be used in high-temperature refrigeration or comfort conditioning, but it may not be satisfactory for use in low-temperature applications. Therefore, the floc point is an important property to consider when choosing a refrigerant oil for a specific use.

Although refrigerant oils usually present no danger or fire hazard within various systems, it is important to know the *flash point* of the particular oil. This is the temperature at which oil vapor, when exposed to a flame, will flash into fire. This occurs at a specific temperature, and the oil becomes unstable and some of its components tend to separate. Therefore, the flash point must be avoided.

Many compressors and motors are hermetically sealed together in housings or sheels, and the refrigerant vapor from the evaporator passes across the insulated motor windings. In such cases the refrigerant oil must have a resistance to the flow of electric current, and the *dielectric strength* of a refrigeration lubricating oil is the measurement of this resistance.

The *fire point* of a refrigerant oil is associated with the flash point of the fluid, which was previously described. When the temperature is increased beyond the flash point of the oil vapor and the oil continues to burn during the test, the oil's fire point has been reached.

Sulfur compounds in a refrigerant lubricating oil are undesirable. Sulfurous acid forms when moisture mixes with a sulfur compound. This acid, which is not considered much of a factor in oils today, can be very corrosive to the metal components of a refrigeration system. A good lubricating oil should show a minimum of corrosive tendency when a strip of highly polished copper is immersed in a sample of the oil and subjected to temperatures above 200°F. After a period of about 3 or 4 hours, the copper strip is removed from the oil sample. If it is pitted or more than slightly discolored, this is evidence that the oil contains too much sulfur.

Stability of the refrigerant lubricating oil was discussed in connection with the flash point of refrigerant oils. Still another indication of an oil's stability is its resistance to chemical reaction.

Oil to be used in most lubricating processes must be refined to remove unsaturated hydrocarbons. But the more an oil is refined, the lower its lubricating quality. In the early days of refrigeration, oil used in this process was continuously refined until it was almost colorless. The color of a good refrigeration oil ordinarily is light yellow, indicating that most of the hydrocarbons have been refined, without loss of its lubricating qualities.

R11-7
MOTOR TEMPERATURE

The first law of thermodynamics states that energy can be neither created or destroyed, but may be converted from one form into another. The motor receives electrical energy from the power source, but because of friction, not all the energy can be turned into mechanical output energy. The balance of the input energy is converted to heat energy, and unless this heat is dissipated, the temperature in the motor windings will rise until the insulation is destroyed. If a motor is kept free from contamination and physical damage, heat is practically the only enemy that can damage the windings.

The amount of heat produced in the motor depends on both the load and the motor efficiency. As the load is increased, the electrical input to the motor increases. The percent of the power input converted to heat in the motor depends on motor efficiency, the heat decreasing with an increase in efficiency, and increasing as motor efficiency decreases.

The temperature level that a motor can tolerate depends largely on the type of motor insulation and the basic motor design, but the actual motor life is determined by the conditions to which it is subjected during use. If operated in a proper environment, at loads within its design capabilities, a well-designed motor should have an indefinite life. Continuous overloading of a motor resulting in consistently high operating temperatures will materially shorten its life.

Since heat is the worst enemy of motors, the hermetic compressor enjoys a great advantage by utilizing suction gas to dissipate it effectively. By designing a hermetic motor for a specific application and controlling the motor temperature closely, a motor may be matched to a given load, and the motor output can be at its max-

imum capacity while maintaining a generous safety factor considerably above that available with standard open-type motors.

R11-8
NAMEPLATE AMPERAGE

On open-type standard motors, NEMA (National Electrical Manufacturers Association) horsepower ratings are used to identify a motor's power output capability. Because of industry practice, this nominal horsepower classification has carried over into hermetic motor identification, but it may be misleading when applied to this type of motor. With controlled cooling and motor protection sized for the exact load, a hermetic motor may be operated much closer to its maximum ability, so a particular motor may be capable of much greater power output as a hermetic motor than an equivalent open motor. The amperage draw and watts of power required are the best indicators of hermetic motor operation.

Most industry hermetic compressors carry on their nameplates ratings for both locked-rotor and full-load amperes. The designation *full-load amperage* persists because of long industry precedent, but in reality a much better term is *nameplate amperage*. On most industry compressors with inherent protection or internal thermostats, nameplate amperage has been arbitrarily established as 80% of the current drawn when the motor protector trips. The 80% figure is derived from standard industry practice in sizing motor protection devices at 125% of the current drawn at rated load conditions.

This does not mean that every hermetic compressor may be operated continuously at a load greatly in excess of its nameplate rating without fear of failure. Motor amperage is only one factor in determining a compressor's operating limitations. Discharge pressure and temperatures, motor cooling, and torque requirements are equally critical. Safe operating limits have been established by manufacturers for each compressor and are published on compressor or unit specifications.

R11-9
VOLTAGE AND FREQUENCY

Although electricity distributed in the United States is 60 cycles (hertz), distribution voltages are not standardized. Single-phase voltages may be 115, 208, 220, 230, or 240, and most utilities reserve the right to vary the supply voltage plus or minus 10% from the nominal rating. Three-phase voltage may be 208, 220, 240, 440, 460, and 480, again plus or minus 10%. Unless motors are specifically designed for the voltage range in which they are operated, overheating may result.

Most open motors, whether used on refrigeration compressor, fans, etc., may be operated at the voltage on the motor nameplate, plus or minus 10%, without danger of overheating; however, this is not the case on single-phase hermetic compressor motors, where the windings have been more critically sized to the exact load and operating conditions. A case in point is the PSC (permanent split capacitor) motor, which does not have a starting capacitor and does have a relatively low starting torque. The PSC motor is used where it is assumed refrigerant pressures will equalize between the off and on cycles. Low-voltage starting of PSC motors can be very detrimental. It is common practice to design hermetic compressors with only a 5% undervoltage and 10% overvoltage rating. So when the nameplate states 208–230 dual voltage, the minimum operating voltage permitted is 5% off 208 or 197 V. On the high side the maximum is 230 plus 10% or 253 V. Much too frequently electrical contractors who do not understand refrigeration compressors assume the 10% plus and minus rating. Should the actual voltage be a 208-V supply and the power company drops the voltage 10% to 187 V, the hermetic compressor will be in trouble. The point here is to make sure what the power company supply voltage is before installing the compressor. Three-phase compressors are not as critical in this respect as single-phase, because each winding is 120° out of phase with the other windings, which results in a motor with a very high starting torque.

In many parts of the world, the electrical power supply is 50 cycle (hertz) rather than 60 cycle (hertz). If both the voltage and frequency supplied to a motor vary at the same rate, the operation (within narrow limits) of a given motor at the lower-frequency condition is all right, in some cases. For example, a 440-V three-phase 60-Hz motor will operate satisfactorily on 380-V three-phase 50-Hz power. In the export of domestic compressors to foreign countries it is common practice for the manufacturer to approve standard 60-Hz machines for 50-Hz application, but with reduced capacity. In some cases, specially wound 50-Hz motors are required to meet local codes or specifications.

R11-10.
COMPRESSOR WIRE SIZES

Low or under-voltage conditions are not always because the power company permits variations from standard rating. Too frequently the power source wiring to the compressor, condensing unit, etc., may be improperly sized, so that line losses exceed normal limits and the terminal voltage to the compressor is too low. This situation is further complicated by the use of copper and aluminum conductors, which have different rating characteristics. Most manufacturers list recommended wire sizes to the unit and the maximum length between the unit and main power source so as not to exceed a 3% voltage drop. Should excessive line loss be coupled with minimum power supply voltage, the resulting impact on the compressor can be significant. The point here is to ensure that proper electrical data are furnished to the electric installer—if wiring is not a part of the refrigeration installation contract.

R11-11
REDUCED-VOLTAGE STARTING

Previous discussions introduced common starters such as starting relays, capacitors, contactors, and across-the-line starters. These devices are meant to be used with full voltage, which is the least expensive method of starting a compressor designed for full-voltage starts. However, because of some power company limitations on starting current, a means of reducing the inrush starting current on higher-horsepower motors is occasionally necessary (particularly in other countries and also in some sections of the United States and Canada), to prevent light flicker, television interference, and undesirable side effects on the equipment because of the momentary voltage dip. The reduced-voltage start allows the power company voltage regulator to pick up the line voltage after part of the load is imposed, and thus avoids the sharper voltage dip that would occur if the whole load were thrown across the line. Some electrical utilities may limit the inrush current drawn from their lines to a given amount for a specified period of time. Others may limit the current drawn on startup to a given percent of locked rotor current.

Unloading the compressor can be helpful in reducing the starting the pull-up torque requirement and will enable the motor to accelerate quickly (unloading is covered later in the chapter). Whether the compressor is loaded or unloaded, however, the motor will still draw full starting amperage for a small fraction of a second. Since the principal objection is usually to the momentary inrush current drawn under locked-rotor conditions when starting, unloading the compressor will not always solve the problem. In such cases some type of starting arrangement is necessary that will reduce the starting current requirement of the motor.

Starters that accomplish this are commonly known as *reduced-voltage starters,* although in two of the most common methods the line voltage to the motor is not actually reduced. Since manual starting is not feasible for refrigeration compressors, only magnetic starters will be considered.

There are five types of magnetic reduced-voltage starters, each of which is suited for specific applications:

- Part winding
- Star-delta
- Autotransformer
- Primary resistor
- Reduced voltage step-starting accessory

As the starting current is decreased, the starting torque also drops, and the selection of the proper starter may be limited by the compressor torque requirement. The maximum torque available with reduced voltage starting is 64% of full-voltage torque, which can be obtained with an autotransformer starter, 45% for part winding, and 33% for star-delta—which means that for star-delta an unloaded start is essential if the compressor is to start under reduced voltage.

R11-12
COMPRESSOR PERFORMANCE

The performance of a machine is an evaluation of the ability of the machine to accomplish its assigned task. Compressor performance is the result of design compromises involving certain physical limitations of the refrigerant, compressor, and motor, while attempting to provide:

1. The greatest trouble-free life expectancy
2. The most refrigeration effect for the least power input
3. The lowest cost
4. A wide range of operating conditions
5. A suitable vibration and sound level

Two useful measures of compressor performance are capacity, which is related to compressor displacement, and performance factor.

System capacity is the refrigeration effect that can be achieved by a compressor. It is equal to the difference in total enthalpy between the refrigerant liquid at a temperature corresponding to the pressure of the vapor leaving the compressor and the refrigerant vapor entering the compressor. It is measured in Btu/lb.

The *performance factor* for a hermetic compressor indicates combined operation efficiency of motor and compressor:

$$\text{performance factor (hermetic)}$$
$$= \frac{\text{capacity (Btu/hr)}}{\text{power input (watts)}}$$
$$= \text{Btu/watt}$$

Recently, the performance factor has become important to the industry because of the spotlight on energy conservation. It is now termed *EER* (*energy efficiency ra-*

tio), and the actual performance of refrigeration and air-conditioning units is being certified and listed in ARI directories so that users, specifiers, installers, and power companies can evaluate the relative efficiency of various machines.

There are three other definitions and measurements of performance for a compressor that are used primarily by compressor design engineers and are generally not of practical use to the refrigeration technician; however, it is well to be familiar with them.

- *Compressor Efficiency* Considers only what occurs within the cylinder. It is a measure of the deviation of the actual compression from the perfect compression cycle and is defined as work done within the cylinder.
- *Volumetric Efficiency* Defined as the volume of fresh vapor entering the cylinder per stroke divided by the piston displacement.
- *Actual Capacity* A function of the ideal capacity and the overall volumetric efficiency.
- *Brake Horsepower* A function of the power input to the ideal compressor and to the compression, mechanical, and volumetric efficiencies of the compressor.

This text and the functions of a refrigeration technician are more concerned with real or actual capacity and power input of the compressor or condensing unit over a defined range of operating conditions.

Manufacturers of compressors perform exhaustive tests (Fig. R11-14) on their compressors for ratings, which must either be in accordance with ASHRAE and/ or ARI conditions. There are two types of tests for compressors. The first determines capacity, efficiency, sound level, motor temperature, etc. The second, equally necessary, determines the probable life expectancy of the

FIGURE R11-14 Compressor test room. (*Courtesy* Carrier Air-Conditioning Company)

machine. Life-expectancy testing must be conducted under conditions simulating those under which the compressor must operate for years. Safety and adherence to codes are major factors in all this work.

From this information the manufacturer can present or publish the performance and application data needed to use the product properly.

Capacity ratings are published in either tabular form or curves which include:

1. Compressor identification—the number of cylinders, bore and stroke, etc.
2. Degrees of subcooling, or a statement that data have been corrected to zero degrees subcooling
3. Compressor speed
4. Type of refrigerant
5. Suction gas superheat
6. Compressor ambient
7. External cooling requirements (if necessary)
8. Maximum power or maximum operating conditions
9. Minimum operating conditions under full-load and unloaded operation

Figure R11-15 represents a typical capacity and power input curve for a hermetic reciprocating compressor. First note the stated facts: refrigerant 22, 10°F liquid subcooling, 20°F superheated vapor, and 1750-rpm compressor speed. Capacity is shown on the left vertical axis and stated in Btu/hr. Power input in kilowatts is shown on the right vertical axis. On the bottom is a range of evaporating temperatures. Condensing temperatures are plotted on the diagonal curves. (*Note:* This is the refrigerant condensing temperature, not to be confused with air- or water-cooled condenser nomenclature. The compressor does not know what kind of condenser is being used—it only knows what condensing temperature and pressures it must produce.)

To determine the refrigerating capacity, let us assume a 25°F evaporator temperature and 115°F condensing temperature. Read up from 25°F on the bottom to the 115°F capacity curve to point *A*, then across to the left; the final reading is about 105,000 Btu/hr.

To determine the power input, read up from 25°F to the power curves at the intersection of 115°F and point *B*, then across to the right. The reading is approximately 11.5 kW power input.

Observe at a constant condensing temperature the rapid falloff capacity with lowering evaporator temperatures, which of course is due to the lower density of the gas being pumped by a constant displacement machine. Note, however, the power input curves do not fall as rapidly, demonstrating the high levels of work required to raise lower pressure vapors to proper condensing pressures. Thus, the relative requirements for

commercial refrigeration and air-conditioning are quite different.

Obviously, it is not practical, or perhaps even technically possible, to use this particular compressor over a wide range of conditions. Thus, the industry offers units with variations in speed (belt-driven models), or bore sizes, or length of stroke, and/or larger motors to meet specific ranges of application.

The manufacturers will offer capacity rating tables similar to Fig. R11-16. This example is a hermetic single-stage compressor with $2\frac{5}{8}$-in. bore used on R-22. Assume that the required load is 26.5 tons at 115°F discharge (condensing temperature) and a 40°F saturated suction temperature (evaporator temperature less refrigerant piping losses). Note that model H61 can produce only 23.5 tons. The H61 (from the manufacturer's data) is a six-cylinder compressor with a short stroke. The H62 model, which is also a six-cylinder model but with a longer stroke, at the same conditions can produce 28.2 tons, *but* the higher power input requires a motor change (see the footnote). Note the shaded arrow marked *Q*.

The heat rejection column specifies the amount of Btu/hr the condenser must handle (437,000), be it water-cooled, air-cooled, or evaporative. At this point the compressor does not know what condenser will be uti-

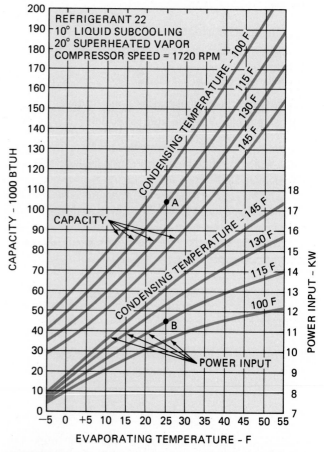

FIGURE R11-15 Typical capacity and power curves for a hermetic reciprocating compressor. (*Courtesy* ASHRAE)

FIGURE R11-16

Capacity ratings for compressor units—R-22; $2\frac{5}{8}$-in. bore hermetic single-stage compressor and condensing units.

UNIT MODEL		H32SM-22E			H61SN-22E H61SP-22E			H62SP-22E H62SQ-22e			H92SQ-22E H92SR-22E H92SS-22E		
Sat. Disch. Temp. and Pressure	Saturated Suction Temp. F	Tons	KW	Heat Rej. MBH	Tons	KW	Heat Rej. MBH	Tons	KW	Heat Rej. MBH	Tons	KW	Heat Rej. MBH
(226.4 PSIG)	15	8.0	11.8	136	13.3	19.3	244	16.0	23.0	269	24.0	34.8	405
	20	9.1	12.4	151	15.2	20.3	251	18.3	24.2	301	27.4	36.8	452
	25	10.4	13.0	168	17.3	21.3	279	20.7	25.4	334	31.1	38.5	503
	30	11.7	13.5	186	19.5	22.3	309	23.4	26.5	370	35.1	40.0	556
	35	13.1	13.9	204	21.9	23.0	340	26.2	27.4	406	39.4	41.3	612
	40	14.6	14.2	223	24.4	23.5	372	29.3	27.9	445	43.9	42.0	668
	45	16.3	14.4	244	27.2	23.8	406	32.6	28.4	487	48.9	42.7	730
	50	18.1	14.6	266	30.2	24.1	443	36.2	28.7	531	54.3	43.1 ⓡ	796
115 F (242.7 PSIG)	20	8.7	12.7	147	14.5	20.8	244	17.4	24.8	292	26.2	37.6	440
	25	9.9	13.3	163	16.5	21.9	272	19.8	26.1	325	29.8	39.4	490
	30	11.2	13.9	181	18.7	23.0	320	22.4	27.3	361	33.6	41.2	542
	35	12.6	14.4	200	21.0	23.8	332	25.2	28.4	398	37.8	42.7	597
	40	14.1	14.8	219	(23.5)	24.5	364	(28.2)	29.2	(437)	42.3	43.9	655
	45	15.7	15.1	239	26.2	25.0	398	31.5	29.9	478	47.2	44.8	717
	50	17.5	15.3	261	29.1	25.4	434	35.0	29.8	520	52.5	45.4	783
120 F (259.9 PSIG)	20	8.3	12.9	143	13.8	21.2	237	16.6	25.2	284	24.9	37.9	426
	25	9.5	13.6	160	15.8	22.5	265	18.9	26.7	317	28.4	40.0	475
	30	10.7	14.3	176	17.9	23.6	294	21.5	28.1	352	32.2	41.9	527
	35	12.1	14.9	195	20.2	24.6	325	24.2	29.3	389	36.3	43.7	582
	40	13.5	15.3	213	22.5	25.4	355	27.1	29.9	426	40.6	45.1	639
	45	15.1	15.8	234	25.2	25.6	388	30.3	30.7	467	45.4	46.4	701
	50	16.7	16.1	254	28.1	26.1	425	33.7	31.2	510	50.6	47.2	766
130 F (296.8 PSIG)	30	9.7	14.9	166	16.2	24.2	276	19.5	29.0	331	29.2	43.8	498
	35	11.0	15.6	184	18.4	25.4	306	22.1	30.4	367	33.1	45.9	551
	40	12.4	16.3	204	20.6	26.5	336	24.8	31.7	404	37.1	47.8	606
	45	13.9	16.9	224	23.2	27.6	371	27.8	32.9	444	41.7	49.7	667
	50	—	—	—	25.9	28.5	407	31.1	34.0	487	16.6	51.3	732
135 F (316.6 PSIB)	30	9.3	15.1	162	15.4	24.5	267	18.5	29.4	321	27.8	44.4	483
	35	10.5	16.0	180	17.5	26.0	297	21.1	31.7	360	31.6	47.0	537
	40	11.9	16.7	199	19.8	27.2	329	23.7	32.5	394	35.6	50.0	595
	45	—	—	—	22.2	28.5	362	26.6	33.9	443	40.0	51.1	652
	50	—	—	—	24.8	29.6	397	29.8	35.3	476	44.7	53.1	715

Important: Larger-size motors are required for certain operating temperatures as indicated by the shaded motor identification letter, P, Q, R or S. If future operation at these temperatures is anticipated, order the compressor with the large motor. All ratings are applicable to units with the larger motors.

lized; it only knows that it is pumping against a 242.7-psig condensing pressure corresponding to 115°F temperature for R-22.

Remember that these curves and tables reflect only compressor performance. Future discussions deal with the condenser options and will illustrate curves and/or tables for water, air, and evaporative condensing mediums.

R11-13
COMPRESSOR BALANCING

Commercially available components will seldom exactly match the design requirements of a given system, and, since system design is normally based on estimated peak loads, the system may often have to operate at other than design conditions. More than one combination of components may meet the performance requirements. The efficiency of the system usually depends on the point at which the system reaches stabilized conditions or balances under operating conditions. The capacities of the three major system components, the compressor, the evaporator, and the condenser, are each variable but interrelated. Because of the many variables involved, the calculation of system balance points is extremely complicated. A simple, accurate, and convenient method of forecasting system performance from readily available manufacturers' catalog data can be graphically displayed in a component balance chart similar to that in Fig. R11-17, which is for an air-cooled condenser system. Evaporator capacity curves from the coil manufacturer are based on entering air temperatures with variable evaporating temperatures. Similar heavy lines can be drawn for balance conditions for the

condenser–compressor balance points at various ambient temperatures. Assume the evaporator loading is 55,000 Btu/hr at 40°F entering air; this requires a 26°F evaporating temperature or a 14°F temperature difference between entering air and the refrigerant evaporating temperature (point *A*). Also assume that the compressor/condenser is operating at 100°F outdoor ambient temperature; at the same 26°F evaporating temperature its capacity would be approximately 59,000 Btu/hr (point *B*), so there is a difference of 4000 Btu/hr. If the outdoor ambient temperature remains constant at 100°F and the evaporator load also remains constant, the result will be a depression in evaporating temperature and thus an increase in evaporator TD and capacity, while the compressor capacity falls correspondingly until the system reaches stabilized conditions at about 57,000 Btu/hr and 24° evaporator temperature (point *C*).

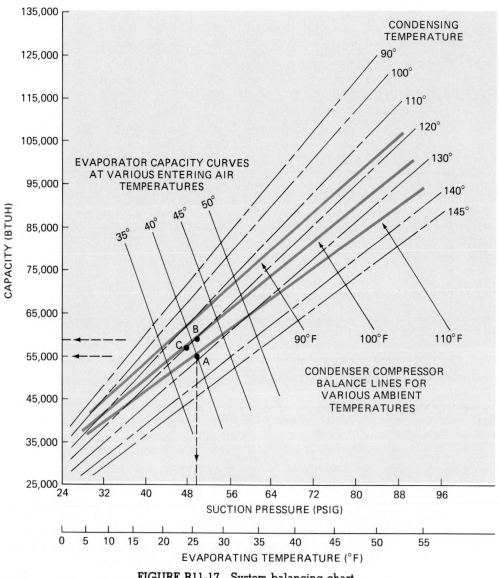

FIGURE R11-17 System balancing chart.

This example ignores refrigerant line losses and other influences but does provide a good illustration nonetheless. The point is that *if* the original equipment selected requires a major shift in balance points, the selection is not a good one and must be reexamined. A severe change in evaporating coil temperatures can measurably affect the coil's latent and sensible capacity and frosting and defrosting relationship to a point where the food or products may be affected by over- or undercooling, or too much humidity or too much dehydration.

The example also assumes a constant evaporator load, which seldom happens, and as the food, produce, or whatever is cooled, the load drops and with it the balance point will also drop. So where wide variations in load are anticipated, it is necessary to provide some kind of capacity modulation of the system.

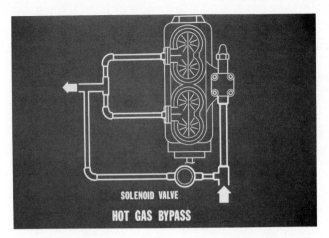

FIGURE R11-19 Hot-gas bypass. (*Courtesy* Carrier Air-Conditioning Company)

R11-14
CAPACITY CONTROL

To provide a means for changing compressor capacity under fluctuating load conditions, larger compressors are frequently equipped with compressor unloaders. Unloaders on reciprocating compressors are of two general types. In the first (Fig. R11-18), suction valves on one or more cylinders are held open by some mechanical means in response to a pressure-control device. With the refrigerant suction valves open, refrigerant is forced back into the suction chamber during the compressor stroke, and the cylinder performs no pumping action.

A second means of unloading is to bypass internally a portion of the discharge gas into the compressor

suction chamber. Care must be taken to avoid excessive discharge temperature when this is done.

A hot-gas bypass may also be achieved outside the compressor (Fig. R11-19). The solenoid in the bypass line can be controlled by temperature or pressure, depending on the nature of the application. As the controller calls for capacity reduction, the solenoid opens, allowing some hot gas to go directly to the suction line.

Specific techniques recommended by compressor manufacturers employ different hardware, but the end result is essentially the same. The stages of unloading naturally depend on the size of machine, the number of cylinders, and the application requirements. *One point that must be considered is the decreased amount of suction vapor and entrained oil returning from the system. In hermetic compressors those quantities must be sufficient for proper lubrication and prevention of motor overheating.*

R11-15
TWO-STAGE COMPRESSORS

The discussion on reciprocating compressors so far has concentrated on single-stage units, but because of the high compression ratio encountered in ultra-low-temperature applications, two-stage compressors (Fig. R11-20) have been developed for increased efficiency when evaporating temperatures are in the range −30 to −80°F. Two-stage compressors are divided internally in low (or first) and high (or second) stages. Suction gas enters the low-stage cyclinders directly from the suction line and is discharged and metered into the interstage manifold so that it can provide adequate motor cooling and prevent excessive temperatures. The superheated refrigerant vapor at interstage pressure enters the suction ports of the high-stage cylinder and is then discharged to the condenser at the condensing pressure.

ELECTRIC UNLOADER MECHANISM

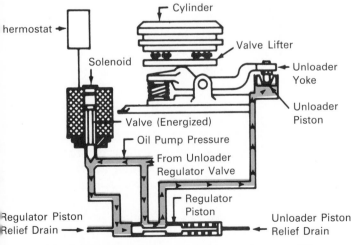

FIGURE R11-18 Electric unloader mechanism. (*Courtesy* Airtemp Corporation)

FIGURE R11-20 Two-stage compressor. (*Courtesy* Copeland Corporation)

R11-16
COMPRESSOR MAINTENANCE

Long life is, of course, desirable in any product. Compressors built today are expected to provide many years of constant trouble-free and quiet operation. In many applications the compressors are called upon to run 24 hours a day, 365 days per year. Such continuous operation, however, is often not as hard on a compressor as is a cycling operation, where temperatures constantly change and the oil is not maintained at a constant viscosity.

The compressor must not only be designed to withstand normal operating conditions but also, at times, some periodic abnormal conditions such as liquid slugging, excessive discharge pressure, etc. The industry's compressor manufacturers have done a creditable job in designing and producing machines that take extra punishment, *but* no manufacturer has ever shipped a complete system, and this is where the professional refrigeration technician plays such an important role, since most compressor failures stem from system faults and not from operating fatigue. The degree of technical skill and plain old common sense technicians use to install, operate, and maintain the equipment will ultimately determine the actual life expectancy of a system, particularly that of the compressor. Although the technician may not actually select the equipment, he or she can call attention to deficiencies before they become serious problems.

Subsequent chapters will deal with the installation, startup, and servicing of the complete system, but it will be helpful at this point to review some of the factors

that shorten the life of a compressor, reduce its efficiency, or create undesirable conditions, factors that are related to the reciprocating design. The extent to which a unit can be corrected on the job depends on the particular compressor (whether it is serviceable or sealed).

R11-16.1
Loss of Efficiency

This can result from a number of things:

- If liquid refrigerant enters the compressor, the efficiency and resulting capacity will be seriously affected. Besides physical damage, a shortage in capacity can also result from liquid slugging.
- Leaking discharge valves reduce pumping efficiency and cause the crankcase pressure to rise rapidly.
- Leaking suction valves seriously affect compressor efficiency (and capacity) especially of lower temperature applications.
- Loose pistons cause excessive blow-by and lack of compression.
- Worn-bearings—especially loose connecting rods or wrist pins—prevent the pistons from coming up as far as they should on the compression stroke. This has the effect of increasing the clearance volume and results in excessive reexpansion.
- Belt slippage on belt-driven units.

R11-16.2
Motor Overloading

When a compressor is not performing satisfactorily, the motor load sometimes provides a clue to the trouble. Either an exceptionally high or exceptionally low motor load is an indication of improper operation.

- Mechanical problems such as loose pistons, improper suction valve operation, or excessive clearance volume usually lead to a reduction in motor load.
- Another common problem is a restricted suction chamber or inlet screen (caused by system contaminants). The result is a much lower actual pressure in the cylinders at the end of the suction stroke than the pressure in the suction line as registered on the suction gauge. If so, an abnormally low motor load will also result.
- Improper discharge valve operation, partially restricted ports in the valve plate (which do not show up on the discharge pressure gauge), and tight pis-

tons will usually be accompanied by high motor load.

- Abnormally high suction temperatures created by an excess load or other problem will cause a high motor load.
- Abnormally high condensing temperatures, created by problems associated with the condenser, will also lead to a high motor load.
- Low voltage at the compressor, whether the source is the power supply or excessive line loss, will contribute to high motor loading, particularly if any of the other problems are also present.

R11-16.3
Noisy Operation

This condition usually indicates that something is wrong. There might be some abnormal condition outside the compressor or something defective or badly worn in the compressor itself. Obviously, if it is due to some cause outside the compressor, nothing is gained by changing the compressor. Therefore, before changing a compressor one should first check for these possible causes:

- *Liquid Slugging.* Make sure that only superheated vapor enters compressor.
- *Oil Slugging.* Possibly oil is being trapped in the evaporator or suction line and is intermittently coming back in slugs to the compressor.
- *Loose Flywheel.* (on belt-driven units).
- *Improperly Adjusted Compressor Mountings.* In externally mounted hermetic-type compressors, the feet of the compressor may be bumping the studs, the hold-down nuts may not be backed off sufficiently, or springs may be too weak, thus allowing the compressor to bump against the base.

R11-16.4
Compressor Noises

Noises coming from internal sources could be any of the following:

- *Insufficient Lubrication.* The oil level may be too low for adequate lubrication of all bearings. If an oil pump is incorporated, it may not be operating properly, or it may have failed entirely. Oil ports may be plugged by foreign matter or oil sludged from moisture and acid in system.
- *Excessive Oil Level.* The oil level may be high enough to cause excessive oil pumping or slugging.
- *Tight Piston or Bearing.* A tight piston or bearing can cause another bearing to knock—even though it has proper clearance. Sometimes in a new compressor such a condition will "wear in" after a few hours of running. In a compressor that has been in operation for some time, a tight piston or bearing may be due to copper plating, resulting from moisture in the system.
- *Defective Internal Mounting.* In an internally spring-mounted compressor the mountings may be bent, causing the compressor body to bump against the shell.
- *Loose Bearings.* A loose connecting rod, wrist pin, or main bearing will naturally create excessive noise. Misalignment of main bearings, shaft to crankpins or eccentrics, main bearings to cylinder walls, etc., can also cause noise and rapid wear.
- *Broken Valves.* A broken suction or discharge valve may lode in the top of a piston and hit the valve plate at the end of each compressor stroke. Chips, scale, or any foreign material lying on a piston head can cause the same result.
- *Loose Rotor or Eccentric.* In hermetic compressors a loose rotor on the shaft can cause play between the key and the keyway, resulting in noisy operation. If the shaft and eccentric are not integral, a loose locking device can be the cause of knocking.
- *Vibrating Discharge Valves.* Some compressors, under certain conditions, especially at low suction pressure, have inherent noise, which is due to vibration of the discharge reed or disc on the compression stroke. No damage will result, but if the noise is objectionable, some modification of the discharge valve may be available from the compressor manufacturer.
- *Gas Pulsation.* Under certain conditions noise may be emitted from the evaporator, condenser, or suction line. It might appear that a knock and/or a whistling noise is being transmitted and amplified through the suction line or discharge tube. Actually, there may be no mechanical knock, but merely a pulsation caused by the intermittent suction and compression strokes, coupled with certain phenomena associated with the size and length of refrigerant lines, the number of bends, and other factors.

The recognition of compressor problems and corrective options comes with the experience, skill, and knowledge of the specific product's operating characteristics, which come with the development of a matured and professional refrigeration technician.

R11-1. Name four designs of compressors.

R11-2. A centrifugal compressor is a positive-displacement design. True or False?

R11-3. Name two types of reciprocating compressors.

R11-4. Sealed hermetic compressors require less air for ventilation than do open compressors. True or False?

R11-5. Name two types of hermetic compressors.

R11-6. Nearly all motors used in compressors for refrigeration application are of which type: split phase, shaded pole, or induction?

R11-7. What is the purpose of an oil separator?

R11-8. What does the term "viscosity" of lubricating oil mean?

R11-9. What are the two principal methods of compressor lubrication?

R11-10. Name the six essential qualities of refrigeration oil.

R11-11. What is the "pour point" of oil?

R11-12. Why is the dielectric strength of an oil important?

R11-13. What is the greatest enemy of a motor winding?

R11-14. Compressor wire must be sized to limit voltage drop to _____ %.

R11-15. The operating voltage tolerance of a single voltage motor is plus _____ % minus _____ %.

R11-16. The operating voltage tolerance of a dual-voltage motor is plus _____ % minus _____ %.

R11-17. "EER" stands for _____ .

R11-18. Unloaders on reciprocating compressors control _____ .

R11-19. Are two-stage compressors needed for the high or the low compression ratio of low-temperature applications?

R11-20. Compressor maintenance is an important function of its _____ .

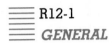

Condensers

GENERAL

The condenser is another component in the system; it transfers heat from a place where it is not wanted to a place where it is unobjectionable; it transfers the heat from the refrigerant to a medium that can absorb it and move it to a final disposal point.

We have already discussed three other major components of a refrigeration system: the refrigerant, the evaporator, and the compressor. By adding the condenser we complete a typical system, as illustrated in Fig. R12-1. The receiver is associated with the condenser and will be included in this discussion.

In Chapter R6 it was pointed out that the media used in the condenser heat-transfer process were (1) air, (2) water, or (3) a combination of air and water. Let's reidentify and restate them in terms of types of condensers (Fig. R12-2):

- Air-cooled
- Water-cooled
- Evaporative

There are four basic types of water-cooled condensers:

1. Double-pipe
2. Open vertical shell-and-tube
3. Horizontal shell-and-tube
4. Shell-and-coil

The first two types are discussed only briefly because they are rarely used. The horizontal shell-and-tube and the shell-and-coil condensers are discussed at length, because they are used extensively and represent by far the greatest percentage of today's installations.

The diagram on the right in Fig. R12-3 is one form of double-pipe condenser. Water flows through the inner pipe and the refrigerant flows in the opposite direction between the inner and outer tubes. This arrangement provides some air cooling in addition to water cooling. Its advantages are its high efficiency and its flexibility as to size and its adaptability in arrangement. Its construction, although efficient, leaves much to be desired because the great number of flanges and gaskets makes leakage possible. Double-pipe condensers may be constructed according to designs other than the ones shown. In some, flanges and gaskets are eliminated.

The diagram at the left in Fig. R12-3 illustrates an open-type vertical shell-and-tube condenser. Water is distributed over the head of the condenser, enters each tube, and flows down the inner surface. This condenser is normally found in medium-size or large ammonia plants. Its advantages are its low maintenance cost and the accessibility of its tubes for cleaning during operation. This condenser is the least efficient of the four types, requiring a larger unit for equivalent capacity.

Figure R12-4 shows the horizontal shell-and-tube condenser. This is similar in construction to the vertical shell and tube condenser except that water heads have been added, making a closed water circuit. This allows water to be redirected through the condenser more than once. This illustration shows the water making four passes through the condenser. By being able to redirect the water through the condenser, it is possible to obtain increased efficiency. A further increase in efficiency can be obtained by using finned condenser tubing in place of the bare tubes shown here. This type of condenser is used extensively in ammonia installations of all sizes and in medium-size and large installations using other refrigerants. The horizontal shell-and-tube condenser is very efficient and can be constructed at a reasonable cost. It is easily cleaned, and normal maintenance costs are relatively low.

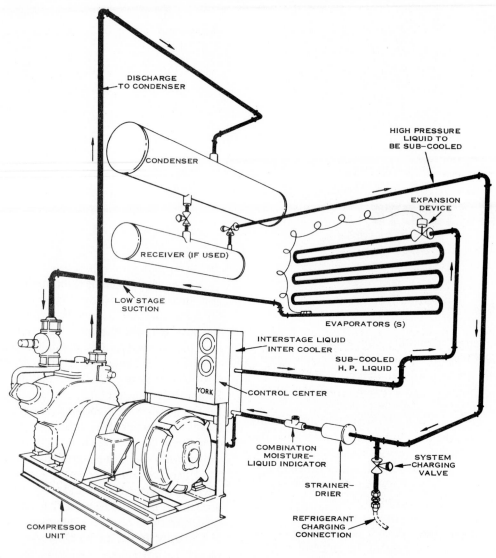

FIGURE R12-1   Refrigeration system. (*Courtesy* Borg-Warner Air Conditioning, Inc.)

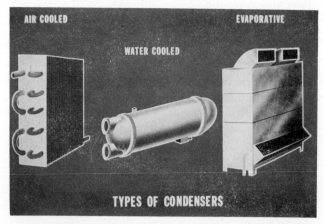

FIGURE R12-2   Types of condensers (*Courtesy* Carrier Air-Conditioning Company)

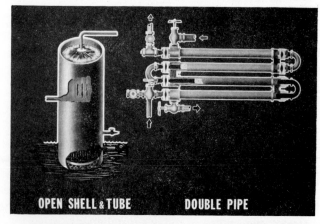

FIGURE R12-3   Double pipe, open shell and tube. (*Courtesy* Carrier Air-Conditioning Company)

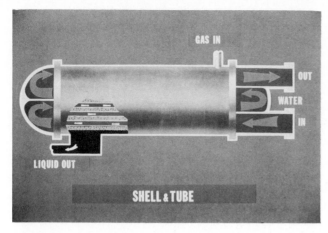

FIGURE R12-4   Shell and tube. (*Courtesy* Carrier Air-Conditioning Company)

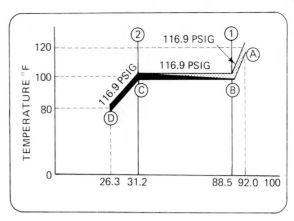

FIGURE R12-6   Heat content (Btu/lb). (*Courtesy* Carrier Air-Conditioning Company)

Figure R12-5 shows a vertical shell-and-coil condenser. It consists of a welded shell containing a coil of finned water tubing. The coil within the shell is continuous and without joints. This condenser can also be made in a horizontal type. Either type may be of flanged, rather than welded, construction. This is one of the most efficient, compact, and cheapest of all condensers; consequently, it is used extensively on small packaged equipment. However, it must be cleaned chemically, and when leaks develop, the cost of a single leak repair may exceed the cost of a new condenser.

Figure R12-6 shows what happens to the refrigerant in the condenser. It is also known as a *temperature Btu chart,* since the vertical scale is in degrees Fahrenheit and the horizontal scale is in Btu heat content of the refrigerant. On this chart, all the area to the right of line 1 represents the refrigerant in a gaseous state. The area to the left of line 2 represents the refrigerant in a liquid state. The area between lines 1 and 2 represents a gas–liquid mixture. For this illustration R-12 at 116.9

psig is used. The corresponding condensing temperature is 100°F.

Hot refrigerant gas from the compressor enters the condenser at 120°F, represented by point *A*. The gas at this point has a 20°F superheat. As it comes in contact with the tubes and/or fins, which are cooled by the cooling medium, the hot gas begins to give up its heat. Since the gas is superheated, this results in a reduction in temperature only. The temperature drops from 120°F to 100°F, which is represented by the line from *A* to *B*. The gas has reached the saturation temperature corresponding to its pressure and is ready to condense. Heat removal continues, and all the gas condenses to a liquid at point *C*. Note that the temperature has not changed between *B* and *C*. All heat removal in this area is latent heat; therefore, there is no temperature change. Since all the refrigerant is now a liquid, further cooling will subcool it below the condensing temperature of 100°F. In normally operating condensers, there would be some degree of subcooling. In this example, the liquid is subcooled 20°F to point *D* and leaves the condenser at 80°F.

An interesting point shown in this chart is the small amount of sensible heat removed compared to the latent heat. The sensible heat removal from *A* to *B* is only 3.5 Btu/lb of refrigerant. From *C* to *D* only 4.9 Btu of sensible heat has been removed. This is a total of 8.4 Btu of sensible heat removed compared to 57.3 Btu of latent heat removed between points *B* and *C*. It is thus easily seen that most of the heat removed in the condenser is latent heat. To further clarify the operation of the condenser, actual quantities of water used by a condenser are shown in Fig. R12-7.

A basic concept of condenser theory is that the heat given up by the refrigerant must equal the heat gained by the cooling medium. While 10 lb of refrigerant and 43.8 lb of water are passing through the condenser, the heat removed from the refrigerant results in a water temperature rise of 15°F, from 70°F to 85°F. The chart in Fig. R12-8 illustrates how these figures are used.

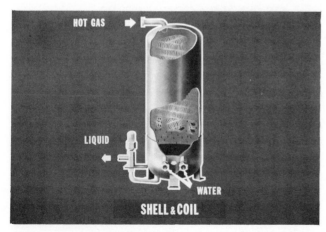

FIGURE R12-5   Vertical shell and coil. (*Courtesy* Carrier Air-Conditioning Company)

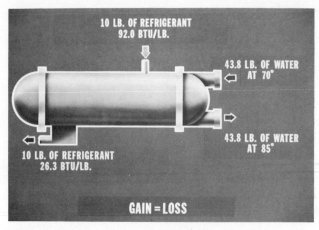

FIGURE R12-7 Gain-loss water condenser. (*Courtesy* Carrier Air-Conditioning Company)

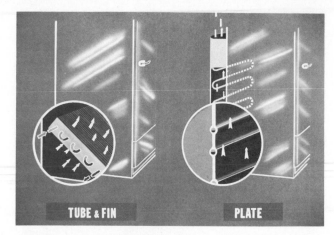

FIGURE R12-9 Tube-and-fin plate condenser. (*Courtesy* Carrier Air-Conditioning Company)

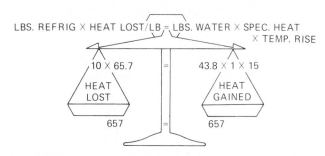

FIGURE R12-8 Heat balance of water-cooled condenser. (*Courtesy* Carrier Air-Conditioning Company)

Figure R12-8 shows the mathematics used in checking the heat lost by the refrigerant against the heat gained by the water. Each pound of refrigerant passing through the condenser loses 65.7 Btu. As there were 10 lb of refrigerant passing through the condenser, the total heat lost by the refrigerant should be 10 times 65.7, or 657 Btu. During the same period of time, 43.8 lb of water passes through the condenser with a temperature rise of 15°F. The heat gained by the water is then 43.8 times the specific heat, which is 1, times the 15° temperature rise. This equals 657 Btu. In actual operation some heat is lost through radiation, but this quantity of heat is insignificant and is usually not considered in calculations. The concept of heat balance is true for all condensers, including those which are air-cooled and evaporative-cooled.

R12-2
NATURAL-DRAFT (PASSIVE) AIR-COOLED CONDENSERS

In natural-draft air-cooled condensers (Fig. R12-9) air circulates over the condenser by convection. As the air comes in contact with the warm condenser it ab-

sorbs heat and rises. This allows the cooler air underneath to rise to where it may, in turn, also absorb condenser heat.

The natural-draft air-cooled condenser comes in one of two types. The tube-and-fin type is shown on the left in Fig. R12-9. On the right is a typical plate-type condenser, in which the plates are pressed into an outline of the condenser coil and seam-welded together. This leaves an interior space in the form of tubes through which the hot refrigerant gas passes.

The natural-draft air-cooled condenser has a very limited use. Because the air moves very slowly it is not capable of removing heat rapidly from the condenser. Therefore, relatively large surfaces are required. One of its most common uses is in household refrigerators. It is cheap, easy to construct, and requires very little maintenance.

R12-3
FORCED-DRAFT (ACTIVE) AIR-COOLED CONDENSERS

Condenser capacity can be increased by forcing air over the surfaces (Fig. R12-10). This illustration shows a forced-air condenser. A fan has been added to increase the airflow.

Some of the earlier condensers of this type were of bare-tube construction. However, condensers are now usually of the tube-and-fin construction shown. Unlike the natural-draft condenser, the forced-air type is more practical for larger cooling loads. The major limiting factors are economics and available space.

Either a propeller-type fan (as shown) or a centrifugal-type fan is used on air-cooled condensers. The selection of the fan depends on design factors such as air resistance, noise level, space requirements, etc.

FIGURE R12-10 Condenser capacity, gain or loss. (*Courtesy* Carrier Air-Conditioning Company)

R12-4
HEAT BALANCE

Actual quantities and temperatures of air and refrigerant are indicated in the illustration for the purpose of showing heat balance. The refrigerant quantity and heat loss per pound are the same as for the water-cooled condenser. However, air is the condensing medium. Ten pounds of refrigerant enters with a heat content of 92 Btu/lb and leaves with a heat content of 26.3 Btu/lb. During the same period of time 3910 ft³ of air pass over the condenser with a temperature rise of 10°F, since the air enters at a temperature of 85°F and leaves at a temperature of 95°F.

The chart in Fig. R12-11 shows the actual heat balance. The amount heat given up by the refrigerant is determined by multiplying pounds of refrigerant times the heat loss per pound. The heat gained by the air is calculated by multiplying the pounds of air by the specific heat of air, times its temperature difference. Through substitution we find that the refrigerant gives up 657 Btu of heat. It is necessary to convert cubic feet of air to pounds by dividing by the specific volume of air; the result is 14.3 ft³/lb. The weight of air is multiplied by 0.24 (its specific heat), and then by 10 (the temperature rise in degrees). As the calculations show, this equals 657 Btu, again indicating heat lost by the refrigerant must equal heat gained by the condenser medium.

R12-5
EVAPORATIVE CONDENSERS

Figure R12-12 illustrates the typical evaporative condenser. Note that it has some of the features of both air- and water-cooled condensers. In the evaporative condenser, heat is absorbed from the coil by the evaporation of water. In the case of both air- and water-cooled condensers, no evaporation takes place.

In operation, water is pumped from a pan in the base of the unit through a series of spray nozzles and then flows over the refrigerant coil. At the same time air enters through the inlet at the base, passes up through the coil and water spray, and then through eliminators to remove free water, next through the fans, and is then discharged from the unit. Water lost through evaporation is replaced through a water supply line, and the water level in the pan is controlled by a float valve. As impurities always remain behind when water is evaporated, a small continuous bleed-off is used to reduce this concentration.

As mentioned in previous chapters, large quantities of heat are required to change a liquid to a vapor. Water is no exception. When the spray water contacts the warmer refrigerant coil, evaporation takes place. The heat necessary to evaporate this water comes from the coil. At the same time an equal amount of heat is given up by the hot refrigerant gas inside the coil, and the refrigerant condenses. While this process is going on, the fan is removing the moisture-laden air from around the wetted coil and replacing it with air that has the ability to absorb more moisture.

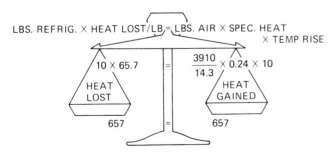

FIGURE R12-11 Heat balance of air-cooled condenser. (*Courtesy* Carrier Air-Conditioning Company)

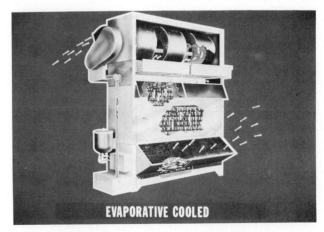

FIGURE R12-12 Typical evaporative condenser. (*Courtesy* Carrier Air-Conditioning Company)

Condensers   **131**

The capacity of an evaporative condenser is determined by the amount of heat the entering air is capable of absorbing. Therefore, the more heat contained in the entering air the lower the capacity; the less heat in the entering air the higher the capacity. From this it is seen that the capacity of the evaporative condenser depends on the heat content of the entering air. The heat content of air is calculated from its wet-bulb temperature. This temperature is easily determined by placing a wetted wick over the bulb of an ordinary thermometer and holding the thermometer over the entering air stream. The wet-bulb temperature is the lowest reading noted. After the wick begins to dry the temperature will rise to the dry-bulb temperature.

Since the wet-bulb temperature can be used to calculate the heat content of the air, it follows that the difference between the wet-bulb temperature of the entering air and that of the air leaving, along with the air quantity, determines the capacity of any evaporative condenser.

In Fig. R12-13 actual air and refrigerant quantities and conditions have been added. Ten pounds of refrigerant enters with a heat content of 92 Btu/lb and leaves with a heat content of 26.3 Btu/lb. During the same period 610 ft³ of air passes over the condenser. From an entering air wet-bulb temperature of 70°F, it has been determined that the total heat content of this entering air is 34.2 Btu/lb. From the wet-bulb temperature of the leaving air, it has been determined that the total heat content is 49.6 Btu/lb.

Figure R12-14 then represents the actual heat balance through the evaporative condenser. The heat given up by the refrigerant is shown as pounds of refrigerant multiplied by the heat loss per pound. Therefore, the heat loss is 10 lb of refrigerant times the loss of 65.7 Btu/lb, or 657 Btu lost through the condenser. The heat gained by the air is calculated by multiplying the pounds of air by the heat gained per pound. By dividing the

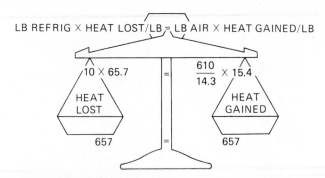

FIGURE R12-14 Heat balance scale. (*Courtesy* Carrier Air-Conditioning Company)

cubic feet of air by its volume, 14.3 ft³/lb, the pounds of air can be determined. If this quantity is multiplied by 15.4 Btu, which is the difference between the heat in the air leaving and the air entering, we derive the total heat loss. The calculation shows that 657 Btu are picked up by the air, which is equal to the heat given up by the refrigerant.

Evaporative condensers are of course efficient water conservation devices, but they do have a disadvantage in that extensive on-site refrigeration piping and a receiver are needed. This tends to increase job costs and reliability risks subject to the skills of the installers. It also precludes the manufacturers' assembling, testing, and shipping complete refrigeration units with minimum installation requirements. The result has been a trend toward more packaged air-cooled equipment for smaller systems and to the use of water towers where large quantities of heat must be dissipated.

R12-6
WATER-COOLED CONDENSERS

In earlier applications of water-cooled condensers to refrigeration and air-conditioning it was common practice to pipe the condensers to tap the city water supply and then waste the discharge water to a drain connection as illustrated in Fig. R12-15. An automatic water valve was placed in the line and the flow of incoming water was controlled by the condenser operating head pressure through a pressure tap. The temperature of the incoming water would naturally affect the condenser performance and flow rate of any heat load. Water temperatures in city water mains rarely rise above 50 to 60°F even in summer and frequently drop to much lower temperatures in winter. Condensers piped for city water flow were always arranged for series flow, circuited for several water passes so as to achieve maximum heat rejection to the water, which is then wasted. Thus condensers drawing on city water could use only 1 to 2 gal per minute per ton of refrigeration. The multipass circuit created high water-pressure drops (P_1-P_2) of 20 lb or more; however, most city pressures were able to supply the

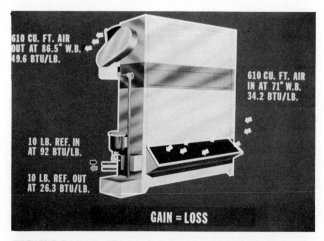

FIGURE R12-13 Heat balance of evaporative condenser. (*Courtesy* Carrier Air-Conditioning Company)

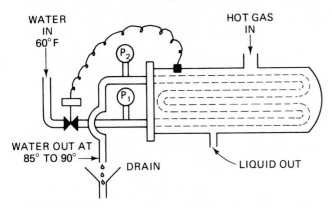

RATE 1 TO 2 GAL/MIN/TON OF REF.

FIGURE R12-15   Series flow condenser on city water.

minimum pressure requirement (usually 25 psig). In time the cost and scarcity of using city water (unless drawn from a lake or wells and returned) became prohibitive and even outlawed by many local city codes. Ordinances restricted the use of water for refrigeration and air-conditioning to the point where such activities were forced to use all air-cooled equipment, water-saving devices such as the evaporative condenser described above, or the so-called water tower. At this point we will concentrate on the water-cooled condenser and its application to water towers. First, the condenser (Fig. R12-16), when used with a recirculated flow such as in a water tower system, is usually designed for parallel tubes with fewer water passes to accommodate a greater water quantity (3 to 4 gal/min/ton) and lower pressure drop (P_1-P_2) at 8 to 10 psi (18 to 23 ft of head of water) pressure drop. The nominal cooling tower application will involve a water temperature rise of 10°F through the condenser, with a condensing temperature approximately 10°F above the water outlet temperature.

Another consideration in using the water-cooled condenser for open recirculated flow is the fouling factor, which affects heat transfer and water pressure drop.

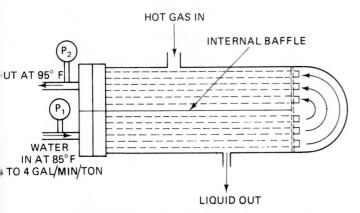

FIGURE R12-16   Condenser on tower operation.

Fouling is essentially the result of a scale buildup on the inside of the water tubes, some of which comes from chemical solids (calcium) but most from biological contaminants (algae, fungi, etc.) and entrained dirt and dust from the atmosphere. The progressive buildup of scale creates an insulating effect that retards heat flow from the refrigerant to the water. As the internal diameter of the pipe is reduced, so is the water flow, unless more pressure is applied. Reduced water flow naturally cannot absorb as much heat and thus condensing temperatures rise—as do operating costs.

In selecting condensers the application engineers will usually allow for the results of fouling so that the condenser will have sufficient excess tube surface to maintain satisfactory performance in normal operation, with a reasonable period of service between cleanings. For conditions of extreme fouling and poor maintenance, higher fouling factors are used. Proper maintenance depends on the type of condenser, meaning that mechanical or chemical cleaning or both may be needed or employed to remove scale deposits.

R12-7
COOLING TOWERS

The function of the water tower is to pick up the heat rejected by the condenser and discharge it into the atmosphere, which it does by evaporation.

Why is the temperature of the water in a lake cooler than the surrounding air during a hot summer season? *Surface evaporation* is the reason, and surface evaporation is also what takes place when one stands in front of an electric fan and feels the cooling effects of moisture evaporating from the surface of the skin.

The evaporation of water from any surface removes heat in the water vapor that is produced. This heat is called the *latent heat of vaporization.* When absorbing heat from water in this manner, air is capable of cooling water below the ambient dry-bulb temperature. When 1 lb of water is evaporated it takes with it approximately 1000 Btu in the form of latent heat. And this removal of latent heat by air is the cooling effect that makes it possible to cool the water in a cooling tower to a temperature below that of the surrounding air as read on an ordinary thermometer.

In air-conditioning and refrigeration work, the *dry-bulb* and *wet-bulb* temperatures are the basis for system design. Closely related to wet-bulb and dry-bulb temperature is relative humidity.

Relative humidity is the ratio of the quantity of water vapor actually present in a cubic foot of air to the greatest amount of vapor that air could hold if it were saturated. When the relative humidity is 100%, the air cannot hold anymore water, and therefore water will not evaporate from an object in 100% humid air. But when

Condensers   133

the relative humidity of the air is less than 100%, water will evaporate from the surface of an object, be it a lake, a wet sponge, or the *drops of falling water in a cooling tower.*

When the relative humidity is 100%, the wet-bulb temperature is the same as the dry-bulb temperature, because when the wet-bulb thermometer, whose bulb is covered with a wet cloth, is whirled, water cannot evaporate from the cloth. But when the humidity is less than 100%, the wet-bulb will be less than the dry-bulb temperature, since water will evaporate from the cloth and will carry latent heat away from the bulb, thus cooling it below the dry-bulb temperature. The drier the air, the greater will be the difference between the dry-bulb and wet-bulb temperature and the easier it will be for water to evaporate.

It follows, then, that cooling-tower operation does not depend on the dry-bulb temperature. The ability of a cooling tower to cool water is a measure of how close the tower can bring the water temperature to the wet-bulb temperature of the surrounding air. The lower the wet-bulb temperature (which indicates either cool air, low humidity, or a combination of the two), the lower the tower can cool the water. It is important to remember that no cooling tower can ever cool water below the wet-bulb temperature of the incoming air. In actual practice the final water temperature will always be at least a few degrees above the wet-bulb temperature, depending on the design conditions. The wet-bulb temperature selected in designing cooling towers for refrigeration and air-conditioning service is usually close

to the average maximum wet-bulb for the summer months at the given location.

By way of summary, the reason cooling towers cool is because air, passing over exposed water surfaces, picks up small amounts of water vapor. The small amount of water that evaporates (about 1% for each 10°F of cooling) removes a large amount of heat from the water left behind.

How do cooling towers cool? There are two types of cooling towers: *mechanical draft* and *atmospheric draft* (Fig. R12-17). A mechanical draft tower utilizes a motor-driven fan to move air through the tower, the fan being an integral part of the tower. An atmospheric draft tower is one in which the operating efficiency of the tower is dependent on atmospheric conditions, predominantly wind movement.

When the water to be cooled arrives at the mechanical draft cooling tower it contains heat that has been picked up in the condenser of the refrigeration or air-conditioning unit. This heat usually amounts to about 250 Btu/min for each ton of refrigeration. Typically, the water enters the tower at the top or upper distribution basin. It then flows through holes in the distribution basin and into the tower filling, which retards the fall of the water and increases its surface exposure. The concept of splash-type filling is as shown in Fig. R12-18. Meanwhile, the fan is pulling air through the filling. This air passes over and intimately contacts the water, and the resulting evaporation transfers heat from the warm water into the air. Finally, the falling water is cooled and collects in the lower (cold water)

(a)

(b)

FIGURE R12-17   (a) Mechanical draft cooling tower. (b) Atmospheric draft cooling tower. (*Courtesy* Marley Cooling Tower Company)

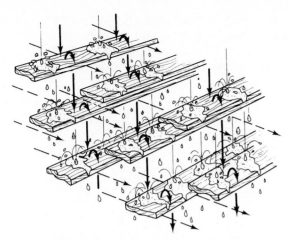

FIGURE R12-18 Typical filled section of a splash-filled cooling tower; filling provides cooling surface and water breakup; air moves across water drops and filling surfaces. (*Courtesy Marley Cooling Tower Company*)

basin of the tower. It is then pumped back to the water-cooled condenser to pick up more heat.

When the cooling water arrives at the atmospheric draft cooling tower, it is piped to the top, where it enters spray nozzles. These nozzles break the water into fine droplets, and induce air movement by the aspirating force of the sprays. Cooled by the time it reaches the bottom of the tower, the water is collected in a basin and pumped back to perform its cooling job again.

The atmospheric draft tower depends on the spray nozzles to break up the water and effect air movement. This tower has no filling or fan, and it will be seen that its size, weight, and location requirements (compared to mechanical draft towers) have reduced its use consid-

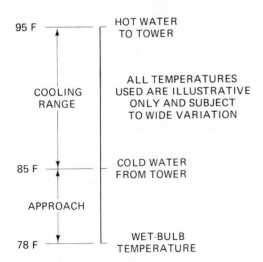

FIGURE R12-19 Diagram showing definition of "cooling range" and "approach." (*Courtesy Marley Cooling Tower Company*)

erably. However, in normal refrigeration service work there may be occasions when the technician is called upon for service or maintenance on such units, and thus it is important to be familiar with the operation of natural-draft towers.

The following terms and definitions apply to cooling towers (see Fig. R12-19):

- *Cooling Range* The number of degrees Fahrenheit through which the water is cooled in the tower. It is the difference between the temperature of the hot water entering the tower and the temperature of the cold water leaving the tower.
- *Approach* The difference in degrees Fahrenheit between the temperature of the cold water leaving the cooling tower and the wet-bulb temperature of the air entering the tower.
- *Heat Load* The amount of heat "thrown away" by the cooling tower in Btu per hour (or per minute). It is equal to the pounds of water circulated multiplied by the cooling range. For example, a tower circulating 18 gal/per min with a 10° cooling range will have a capacity of

$$18 \text{ gal/min} \times 60 \text{ min} \times 8.33 \text{ lb/gal}$$
$$\times 10°\text{F} = 90,000$$

Btu/hr = 6.0 cooling tower tons

Figure R12-20 represents a typical mechanical draft tower piping arrangement to the condenser of a packaged refrigeration or air-conditioning unit.

- *Cooling Tower Pump Head* The pressure required to lift the returning hot water from the cold water basin operating level to the top of the tower and force it through the distribution system. These data are found in the manufacturer's specifications and are usually expressed in feet of head (1 lb of pressure = 2.31 ft of head).
- *Drift* The small amount of water lost in the form of fine droplets retained by the circulating air. It is independent of, and in addition to, evaporation loss.
- *Bleed-off* The continuous or intermittent wasting of a small fraction of circulating water to prevent the buildup and concentration of scale-forming chemicals in the water.
- *Makeup* The water required to replace the water that is lost by evaporation, drift, and bleed-off.

We will now consider the piping system design. The first step is to determine the water flow to be circulated, based on the heat load given above. Normally, towers run between 3.0 and 4.0 gal/min/ton. Water supply lines should be as short as conditions permit. Standard-weight steel pipe (galvanized), type L copper tubing, and CPVC plastic pipe are among the satisfactory materials, subject to job conditions and local codes.

Piping should be sized so that water velocity does not exceed 5 ft/sec. Figure R12-21 lists approximate friction losses in standard steel pipe and type L copper tubing. Plastic pipe will have the same general friction loss as copper. The data are based on clear water, reasonable corrosion and scaling, and velocity flow at or below the 5 ft/sec range.

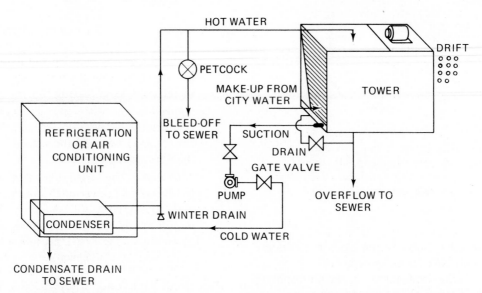

FIGURE R12-20 Typical cooling tower hookup. (*Courtesy* Marley Cooling Tower Company)

FIGURE R12-21

Approximate friction losses in standard steel pipe and type L copper tubing (figures are shown as head loss in feet per 100 ft of pipe).

Water Flow (gal/ min)	Type of Pipe or Tubing	¾ in.		1 in.		1¼ in.		1½ in.		2 in.	
		Velocity (ft/sec)	Head Loss (ft/100ft)	Velocity (ft/sec)	Head Loss (ft/100ft)	Velocity (ft/sec)	Head Loss (ft/100ft)	Velocity (ft/sec)	Head Loss (ft/100ft)	Velocity (ft/sec)	Head Loss (ft/100ft)
6	Std. steel	3.61	14.7	2.23	4.54						
	Copper type L	3.98	11.5	2.34	3.13						
9	Std. steel	5.42	31.1	3.34	9.72	1.93	2.75				
	Copper type L	5.96	24.2	3.50	6.63	2.30	2.38				
12	Std. steel			4.46	16.4	2.57	4.31	1.89	2.04		
	Copper type L			4.67	11.3	3.06	4.04	2.16	1.73		
15	Std. steel			5.57	24.9	3.22	6.35	2.36	3.22		
	Copper type L			5.84	17.1	3.83	6.12	2.70	2.62		
22	Std. steel					4.72	13.2	3.47	6.25	2.10	1.85
	Copper type L					5.21	12.5	3.96	5.57	2.28	1.40
30	Std. steel							4.73	11.1	287	3.29
	Copper type L							5.41	9.44	3.11	2.45
45	Std. steel									4.30	6.96
	Copper type L									4.66	5.20

Note: Data on friction losses based on information published in *Cameron Hydraulic Data* by Ingersoll Rand Company. Printed by permission. Data based on clear water and reasonable corrosion and scaling.

EXAMPLE

100 ft of $1\frac{1}{4}$ in. standard steel pipe would have a pressure loss of 4.31 ft at 12 gal/min, 50 ft would have a pressure loss of $4.31 \times 50/100 = 2.15$ ft, and 200 ft would have a pressure loss of $4.31 \times 200/100 = 8.62$ ft.

In using the table, select the smallest pipe size that will provide proper flow and velocity, to keep installation costs to a minimum. Friction loss is expressed in feet of head per 100 ft of straight pipe length (see the example above).

The entire piping circuit should be analyzed to establish any need for proper valves for operation and maintenance of the system. A means of adjusting water flow is desirable; shut-off valves should be placed so that each piece of equipment can be isolated for maintenance.

Valves and fittings (elbows, tees, etc.) create added friction loss and pumping head. Figure R12-22 lists the approximate friction loss expressed in equivalent feet of pipe.

Following is an example of calculating pipe sizing.

EXAMPLE

Determine the total pump head required for a 5-ton installation requiring 75 ft of steel pipe, 10 standard elbows, four gate valves, and a net tower static lift of 60 in. Water circulation will be 15 gal/min and the pressure drop across the condenser is 13 psi (data obtained from the manufacturer).

SOLUTION

Assume $1\frac{1}{4}$-in. pipe, since the velocity of 15 gal/min is less than 5 ft/sec.

Equivalent
$\frac{1}{4}$-*in.*
Pipe Length

75 ft of $1\frac{1}{4}$-in. standard steel pipe	=	75.0 ft
(10) $1\frac{1}{4}$-in. standard elbows × 3.5	=	35.0 ft[1]
(4) $1\frac{1}{4}$-in. open gate valve × 0.74	=	2.96 ft[1]
		112.96 ft

[1] See Fig. R12-2.

From Fig. R12-21, we find that for 100 ft of $1\frac{1}{4}$-in. pipe, the loss is 6.35 ft; and for 112.96 ft,

$$\text{loss} = \frac{112.96}{100} \times 6.35 = 7.17 \text{ ft}$$

Pressure loss due to piping and fittings	= 7.17 ft
Pressure loss due to condenser[2] = 13×2.31	= 30.00 ft
Pressure loss due to static lift-cooling tower pump head	= 5.00 ft
Total head	42.17 ft

Pipe size is adequate, since velocity is less than 5 ft/sec.

[2] To convert pounds per square inch pressure to feet of head, multiply by 2.31.

How is the pump selected? The selection is based on the gallons of water per minute (e.g., 15 gal/min) and the total head (42.17), or 43 ft of head. The pump manufacturer's catalog will rate the pump capacity in gal/min versus feet of head and the horsepower size needed to do the job. There are many types of pumps on the market, but generally a slower-speed unit (1750 rpm) is recommended for quiet operation. Units operating at 3450 rpm usually cost less, but are noisier.

The pump generally should be installed as follows:

- The pump should be located between the tower and the refrigeration or air-conditioning unit so that the water is "pulled" from the tower and "pushed" through the condenser. See Fig. R12-20 for a typical piping diagram. It is good practice to place a flow-control valve (a gate valve is satisfactory) in the pump discharge line.
- The pump should be installed so that the pump suction level is lower than the water level in the cold-water basin of the tower. This assures pump priming.

FIGURE R12-22

Approximate friction losses in fittings and valves (equivalent feet of pipe).

Pipe Size (in.)	Gate Valve Full Open	45° Elbow	Long Sweep Elbow or Run of Std. Tee	Std. Elbow or Run of Tee Reduced One-Half	Std. Tee through Side Outlet	Close Return Bend	Swing Check Valve Full Open	Angle Valve Full Open	Globe Valve Full Open
$\frac{3}{4}$	0.44	0.97	1.4	2.1	4.2	5.1	5.3	11.5	23.1
1	0.56	1.23	1.8	2.6	5.3	6.5	6.8	14.7	29.4
$1\frac{1}{4}$	0.74	1.6	2.3	3.5	7.0	8.5	8.9	19.3	38.6
$1\frac{1}{2}$	0.86	1.9	2.7	4.1	8.1	9.9	10.4	22.6	45.2
2	1.10	2.4	3.5	5.2	10.4	12.8	13.4	29.0	58.0

Note: Data on fittings and valves based on information published by Crane Company. Printed by permission.

- If noise is not objectionable, the pump may be located indoors so as to eliminate outside wiring and to permit use of an *open* or *drip-proof* motor.
- If circumstances require the use of an open or drip-proof motor outdoors, units should be sheltered by adequate housing covers.
- The pump should be accessible for maintenance and should be installed so as to permit complete drainage for winter shutdown.

What are some conventional methods of wiring a cooling tower? The most desirable wiring and control arrangement varies depending on the characteristics of the available power supply and the size of the equipment being installed. In every case, the objective is to provide the specified results with optimum operating economy and protection to the equipment involved.

For small refrigeration and air-conditioning equipment, the ideal arrangement is based on a sequence beginning with the cooling-tower fan and pump. The starter controlling the fan and pump would then activate the compressor motor starter through an interlock. This method, illustrated in Fig. R12-23, assures sufficient condenser water flow so that compressor short-cycling is eliminated in the event of pump motor failure. In other words, the compressor cannot run unless the tower is operating.

There are other, more economical methods based on using the compressor starter to activate the pump and fan, but water-temperature or flow-sensing devices should be incorporated as protection for the compressor. Where multiple refrigeration units are used on a common tower, the first unit that is turned on activates the cooling tower.

Winter operation or low-ambient operation of a cooling tower is subject to special treatment for temperatures above freezing. Water that comes off the tower too cold may cause thermal shock to the condenser and result in a very low condensing temperature. One method of raising water-OFF temperature is to cycle OFF or reduce the speed of the tower fan (assuming that it is a mechanical draft tower). This will reduce the tower's thermal capability, causing the water temperature to rise.

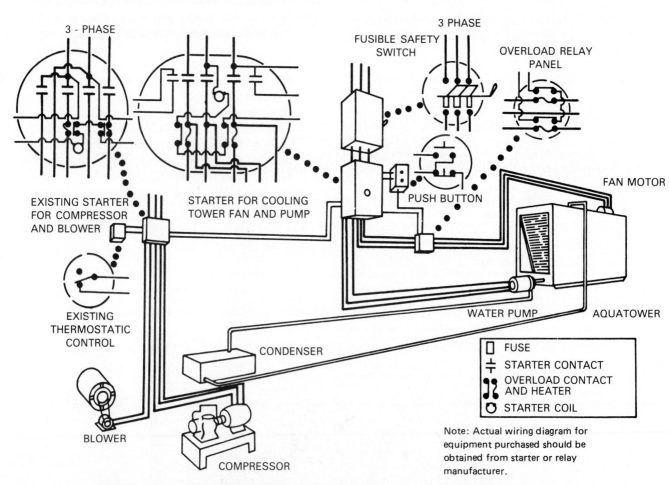

FIGURE R12-23 Typical wiring diagram showing separate starter for 220-V 3 phase electric motors driving cooling tower fan and pump. Compressor will not start until cooling tower fan and pump are operating. (*Courtesy* Marley Cooling Tower Company)

If this is not sufficient, a bypass valve can be placed in the tower piping to permit dumping warm water directly into the tower base, thus bypassing the spray or water distribution system. A combination of both fan and total bypass may be needed to assure stable water-OFF temperatures during cold weather, but reducing the water flow will contribute to tower freeze-up.

Prolonged below freezing conditions usually requires total shutdown and draining of the tower sump and all exposed piping.

R12-8
PERFORMANCE RATINGS

The foregoing discussion on water-cooled condensers was presented as if most refrigeration and/or air-conditioning equipment had separate condensers and compressors connected as in Fig. 18-1. True, separate components are sold for built-up systems, but these are relatively few. Most condensers are assembled at the factory as part of a water-cooled condensing unit (Fig. R12-24) or as part of a complete water-chilling unit. Therefore, the performance rating of the equipment becomes a system rating similar to the capacity rating table (Fig. R12-25), wherein the temperature of the water leaving the condenser and the saturated suction temperature to the compressor determines the unit tonnage, kilowatt power consumption, resulting condensing temperature, and condenser flow in gal/min. As stated previously, 95°F water off the condenser is normal for tower operation.

Additionally, the manufacturer's data will list the water pressure drop across the condenser (in feet of head) for various flow rates and number of passes; the refrigerant charges in pounds (maximum and minimum) when the condenser is used as a receiver or with an external receiver; and the pump-down capacity, which is usually equal to about 80% of the net condenser volume but does not exceed a level above the top row of tubes.

Pump-down is used to store or contain all the system refrigerant in the condenser, so as to be able to perform service or maintenance operations on other components and not lose the refrigerant. Shutoff valves on the condenser permit this operation.

Water-cooled condensers are usually equipped with a purge connection to vent noncondensable gases and have provisions for a relief valve to meet certain national and/or local code requirements. Setting of the relief valve varies somewhat with the refrigerant being used.

Peak performance of a water-cooled condenser and cooling tower system depends heavily on regular maintenance. Growths of slime or algae, which reduce heat transfer and clog the system, should be prevented. Sodium pentachlorophenolate, available under various trade names, is effective in killing these organisms. Cop-

FIGURE R12-24 Condensing unit. (*Courtesy* Borg-Warner Air Conditioning, Inc.)

Condenser Leaving Water Temp. (F)	Saturated Suction Temp. (F)	CONDENSING UNIT MODEL NUMBER											
		JS43L-12W413						JS53M-12W523					
		TONS CAP.	KW	Cond. Temp. (F)	Heat Rej. MBH	Cond. GPM	Cond. △P	TONS CAP.	KW	Cond. Temp. (F)	Heat Rej. MBH	Cond. GPM	Cond. △P
80	20	27.0	25.0	86.8	409	81.8	9.5	31.6	30.9	86.4	485	97.0	8.6
	30	33.5	26.6	88.3	493	98.6	13.3	39.5	32.8	87.8	586	117.2	12.2
	40	41.0	28.0	90.2	588	117.6	18.3	48.1	34.5	89.6	695	139.0	16.6
	50	—	—	—	—	—	—	58.3	36.2	91.8	824	164.8	22.7
85	20	26.0	26.0	91.6	401	80.2	9.1	30.5	32.1	91.2	476	95.2	8.3
	30	32.4	27.8	93.1	484	96.8	12.9	38.1	34.4	92.7	574	114.8	11.7
	40	39.8	29.4	94.9	578	115.6	17.8	46.7	36.3	94.4	684	136.8	16.1
	50	48.1	31.1	97.1	683	136.6	24.3	56.7	38.2	96.4	810	162.0	22.0
90	20	25.1	26.8	96.4	393	78.6	8.8	29.5	33.2	96.0	467	93.4	8.0
	30	31.5	29.0	97.9	477	95.4	12.6	37.0	35.9	97.5	567	113.4	11.4
	40	38.5	30.9	99.6	568	113.6	17.2	45.3	38.1	99.1	674	134.8	15.7
	50	46.8	32.8	101.7	673	134.6	23.6	55.0	40.2	101.1	797	159.4	21.4
95	20	24.2	27.6	101.1	384	76.8	8.5	28.5	34.2	100.9	459	91.8	7.8
	30	30.4	30.1	102.5	468	93.6	12.2	35.7	37.2	102.2	555	111.0	11.0
	40	37.4	32.3	104.3	559	111.8	16.7	44.0	39.9	103.8	664	132.8	15.3
	50	45.5	34.5	106.4	664	132.8	23.0	53.5	42.3	105.8	786	157.2	20.8
100	20	23.3	28.3	106.0	377	75.4	8.2	27.4	35.1	105.7	449	89.8	7.5
	30	29.4	31.2	107.3	460	92.0	11.8	34.5	37.6	107.0	542	108.4	10.5
	40	36.0	33.7	109.0	547	109.4	16.1	42.5	41.6	108.6	652	130.4	14.8
	50	44.0	36.1	111.0	651	130.2	22.1	51.9	44.5	110.4	775	155.0	20.3
105	20	22.3	28.9	110.8	367	73.4	7.8	26.3	35.8	110.5	438	87.6	7.1
	30	28.3	32.2	112.2	450	90.0	11.3	33.2	39.8	111.8	534	106.8	10.3
	40	35.0	35.0	113.7	540	108.0	15.7	41.2	43.3	113.3	642	128.4	14.4
	50	42.7	37.8	115.7	641	128.2	21.5	50.5	46.7	115.1	765	153.0	19.8

△P = Pressure Drop (Ft. H₂O)

FIGURE R12-25 Ratings and engineering data. (*Courtesy* Borg-Warner Air Conditioning, Inc.)

per sulfate is also satisfactory, but it is corrosive if too concentrated. Chlorine and potassium permanganate are also satisfactory. We recommend that you familarize yourself with the companies, chemicals, and cleaning techniques recommended. The success of any water treatment lies in beginning it early and using it regularly. Once scale deposits have formed it can be costly to remove them.

In addition to chemical treatment, regular draining and flushing of the tower basin is recommended. Also, the float valve should be adjusted to cause a small amount of overflow that trickles down the drain; this is called *bleed-off,* or sometimes *blow-down.* It is the continuous or intermittent removal of a small amount of water (1% or less) from the system. Dissolved concentrates are continually diluted and flushed away.

The water problems mentioned above are also common in the evaporative condenser but are not quite as critical, since all the water is contained in the sump.

There is no water-cooled condenser or extended water lines. Scale builds up on the exterior surface of the evaporative condenser refrigerant coil. Regular mechanical cleaning, chemical treatment, and bleedoff are also recommended.

Water treatment and maintenance can be an expensive and time-consuming process; as water restrictions have become widespread the refrigeration and air-conditioning industry has turned to alternate systems.

R12-9
AIR-COOLED CONDENSERS

In addition to being free of water-maintenance problems, the use of air-cooled equipment has other advantage: low ambient or winter operation can be accomplished with relative ease; location of equipment is flexible; the elimination of extensive water lines reduce the need for a plumbing tradesman and means the refrigeration technician can do all the mechanical work.

On the other hand, air-cooled condensing has some disadvantages too. First, an air-cooled condensing system is not as efficient as water-cooled in terms of Btu/W (EER) ratios. Second, if the compressor and condenser are separated, the need for extended on-site refrigeration piping adds to the cost of, and technical skills required in, the installation, as well as the need for increased reliability where prevention of leaks, oil-return problems, vibration, and noise must be considered.

It is apparent, however, that the advantages outweigh the disadvantages, because the bulk of commercial refrigeration and air-conditioning, from fractional-horsepower sizes to 100 tons and more, are using air-cooled condensing.

Note the expression *air-cooled condensing;* there is a great difference in hardware between an air-cooled condenser and an air-cooled condensing unit. Earlier, it was pointed out that in a remote air-cooled condenser installation such as shown in Fig. R12-26 the compressor is located inside the building, serving one or more evaporators or fan coil units, and the air-cooled condenser is on the roof. Figure R12-26 shows the use of a separate receiver; this subject will be covered later. The condenser itself as illustrated in Fig. R12-27 is nothing more than a fin-tubed coil within a cabinet or frame. The fan may be a propeller (as shown) or a centrifugal blower, which moves air across the coil face. This type

FIGURE R12-27   Air-cooled condenser. (*Courtesy* Carrier Air-Conditioning Company)

of installation can and is still being done today, but several factors should be considered when choosing such a system:

- The need for on-site assembly of refrigeration lines with associated evacuation, leak testing, charging, etc.
- The separation of the compressor motor and condenser fan motor adds to the cost of power and control wiring.
- The compressor unit takes up building space that may be needed for other purposes.
- The compressor unit noise is contained within the building.
- Separate components are more expensive than complete factory packages.

These factors are not critical disadvantages, but they have led to the development of more and more self-contained air-cooled condensing units. The remote air-cooled condenser is an excellent solution for converting existing water-cooled equipment to air-cooled operation.

In selecting a remote air-cooled condenser the following information is necessary:

1. Design suction temperature
2. Compressor capacity at design suction temperature
3. Refrigerant being used
4. Geographic location of installation
5. Type of condenser required

Items 1, 2, and 3 have already been discussed and are based on evaporator duty and compressor performance ratings. Item 4 is concerned with summer dry-bulb design conditions relative to the suction temperature; this information will help determine the proper operating temperature difference in selecting a condenser. Item 5 deals with a number of factors, such as type of fan (propeller or centrifugal); type of airflow

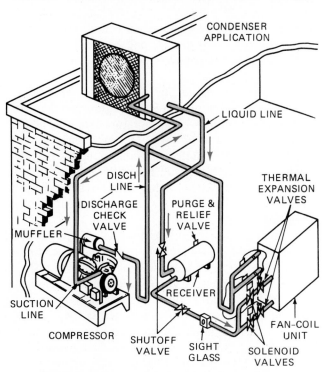

FIGURE R12-26   Remote air-cooled condenser installation. (*Courtesy* Carrier Air-Conditioning Company)

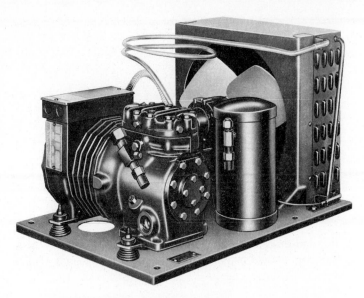

FIGURE R12-28 Small self-contained condensing unit. (*Courtesy* Copeland Corporation)

(vertical or horizontal); type of coil (single or split circuit); and accessories such as wind deflectors, winterstats, low ambient fan controls, etc. Manufacturers' catalogs contain ratings based on specifications for the information above as well as selection methods.

The self-contained air-cooled condensing unit varies from small tonnage commercial refrigeration units (Fig. R12-28) used on meat cases, dairy cases, commercial refrigerators, etc., to units for large refrigeration loads (walk-in coolers and freezers, process freezing plants, and air-conditioning duty).

The physical characteristics of larger self-contained air-cooled condensing units vary widely from one manufacturer to another. Figure R12-29 represents a typical unit used in commercial refrigeration work. It consists of a complete cabinet or enclosure to contain the condenser coil(s) and air-movement fan(s); a compressor and control compartment and in many cases an

FIGURE R12-29 Large self-contained condensing unit. (*Courtesy* Kramer Trenton Company)

external receiver are suspended underneath. Some of the advantages of this arrangement of the equipment are:

- All the heavy-duty mechanical equipment is outside (on the roof or at ground level) and does not take up valuable floor space. Noise and vibration are removed from the area inside.
- All major internal power and controls are preselected and prewired to a weatherproof panel.
- Refrigeration piping is reduced—the discharge line is prepiped at the factory.
- The complete assembly can be factory tested for leaks and operation run-in to improve reliability.

Figure R12-30 is a schematic piping diagram of a multiple-evaporator system on one air-cooled condensing unit. Instead of the evaporators commonly used in commercial refrigeration, it could be a fan coil unit for air-conditioning purposes. The essential difference would be the condensing unit selection procedure and the necessary controls to accomplish the application requirement.

R12-10
LOW-AMBIENT CONTROL

Condensing units for commercial refrigeration of rooms purposes must have an approximate operating range from +35°F down to −20°F when matched with appropriate evaporators. Outdoor conditions vary from 115°F down to zero degrees and below. Refrigeration piping for low-temperature work also requires extra care.

Condensing units for comfort air-conditioning duty are nominally rated at 95°F outdoor ambient temperature to function with evaporators operating at 40°F with incoming air at 67° WB. However, there is an increasing need for comfort air-conditioning to operate when outside ambients fall below 75°F. High internal heat loads from people, lights, and mechanical equipment may demand cooling when outside conditions go down to 35°F and below.

So in the case of both commercial refrigeration and air-conditioning, low-ambient operation is an important need. In earlier discussions we learned that in order to have proper refrigerant vaporization in a direct-expansion (DX) evaporator coil it was necessary to maintain a reasonable pressure differential across the expansion device. In a normal air-cooled condenser operating between 80 and 115°F ambient, condensing pressures are sufficiently high, but in winter they can fall off 100 psi or more; thus the pressure across the expansion device will be insufficient to maintain control of liquid flow. Evaporator operation becomes erratic. The thermal ex-

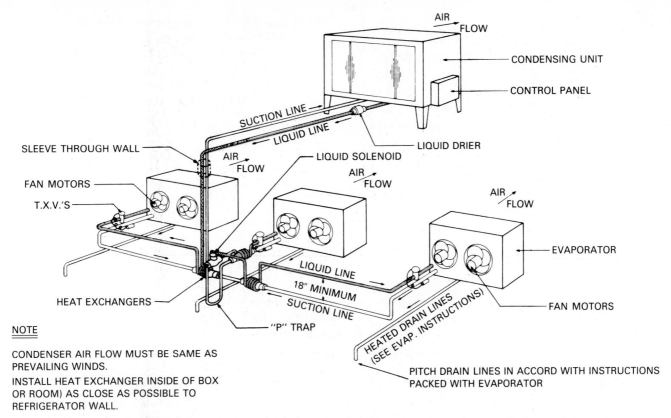

AIR FLOW

CONDENSING UNIT

CONTROL PANEL

SUCTION LINE

LIQUID LINE

LIQUID DRIER

SLEEVE THROUGH WALL

AIR FLOW

LIQUID SOLENOID

AIR FLOW

AIR FLOW

FAN MOTORS

T.X.V.'S

EVAPORATOR

LIQUID LINE

18" MINIMUM

SUCTION LINE

HEAT EXCHANGERS

FAN MOTORS

"P" TRAP

HEATED DRAIN LINES (SEE EVAP. INSTRUCTIONS)

NOTE

CONDENSER AIR FLOW MUST BE SAME AS
PREVAILING WINDS.

INSTALL HEAT EXCHANGER INSIDE OF BOX
OR ROOM) AS CLOSE AS POSSIBLE TO
REFRIGERATOR WALL.

PITCH DRAIN LINES IN ACCORD WITH INSTRUCTIONS
PACKED WITH EVAPORATOR

FIGURE R12-30   Schematic piping diagram for condensing units using two or more evaporators. (*Courtesy* Kramer Trenton Company)

pansion valve will alternately open and close, first causing flooding back of liquid refrigerant and then starving the coil, and the valve closes. The capillary tube (if used) is worse, because it is a fixed metering device, and as pressure difference falls the flow of refrigerant is severely reduced. Below 65° outdoor air a capillary tube system is in real trouble.

The solution to low-ambient operation is to raise the head pressure artificially in order to maintain proper expansion across the evaporator. There are several different ways of doing this.

Where multiple condenser fans are employed on a single coil, the control system of the condensing unit can be equipped with devices to switch or cycle "off" the fans in stages. These controls are usually air stats that sense outdoor ambient temperature or pressure controls that sense actual head pressure. As fans are turned off the airflow across the coil is reduced, and thus its heat rejection is also reduced. Condensing temperatures therefore rise. All fans can be turned off, and then the coil acts as a static condenser with only atmospheric air movement. Its capacity at this point is usually ample to continue operating below freezing conditions. Where only one condenser fan is used, air capacity can be reduced by employing a two-speed motor or solid-state speed controls, which have infinite speed control.

Another technique of restricting airflow across the condenser coil is the use of dampers on the fan discharge. These are used on nonoverloading centrifugal-type fans, not propeller fans. Dampers modulate from a head pressure controller to some minimum position, at which time the fan motor is shut off.

One common and important characteristic of the airflow restriction methods is that the full charge of refrigerant and entrained oil is in motion at all times, ensuring positive motor cooling and oil lubrication to the compressor.

The other technique of artificially raising the head pressure is to back the liquid refrigerant up into the condenser tubes. The normal free-draining condenser has very little liquid in the coil; it is mostly all vapor. But if a control valve is placed in the liquid outlet of the condenser (Fig. R12-31) and is actuated by ambient temperature, some of the discharge gas is allowed to bypass the condenser and enter the liquid drain. This restricts drainage of the liquid refrigerant from the condenser, flooding it exactly enough to maintain the head and receiver pressure. This method is not as common as restricting the air flow, because it is associated with systems that use an external receiver. Critically charged DX systems for air-conditioning normally do not employ receivers.

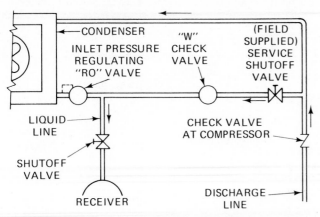

FIGURE R12-31 Head pressure control system for low ambient operation. (*Courtesy* Carrier Air-Conditioning Company)

Receivers, as mentioned in Chapter R6 and shown in Fig. R12-1, are pressure vessels used to store refrigerant. They are most often used in commercial refrigeration systems, where the amount of refrigerant circulated varies widely and as a result can flood a water-cooled or air-cooled condenser to the point where its efficiency is impaired. So an external storage vessel is needed. Receivers are also needed on evaporative condensers, which likewise have limited storage capacity in their condenser surfaces.

Condensers and receivers should never be filled to more than 80% of their volume. The remaining 20% must be left empty to allow for expansion. The total system charge should always be checked against the storage capacity.

R12-11
MAINTENANCE

Maintenance of air-cooled condenser coils, as compared to water towers or evaporative condensers, is relatively easy. The coil face should be kept clean of leaves, sticks, blowing paper, etc. If the coil is in an industrial area it should be inspected for accumulations of grease and dust or chemical substances. Remove lint with compressed air or a vacuum cleaner. Grease or chemicals should be removed with nonflammable, nonpoisonous solvent that will not attack aluminum or copper. Condenser fins can be corroded by salt air and over a period of time become less efficient. Fin damage from flying objects can usually be repaired with a fin comb.

PROBLEMS

R12-1. Name three types of condensers.

R12-2. Name four types of water-cooled condensers.

R12-3. Name two types of air-cooled condensers.

R12-4. In the evaporative condenser, is most of the heat removal from sensible heat or latent heat?

R12-5. Is the capacity for cooling tower basically a function of the wet-bulb or the dry-bulb temperature?

R12-6. Name two types of cooling towers.

R12-7. Bleed-off is used to control _____ .

R12-8. The amount of makeup water to a tower depends on three factors. Name them.

R12-9. In a cooling tower, what is meant by the "cooling range"?

R12-10. In a cooling tower, what is meant by the "approach"?

R12-11. The normal quantity of water supplied to the tower per 12,000 Btu/hr load is _____ to _____ gallons per minute.

R12-12. Water velocity in tower piping should not exceed _____ feet per second.

R12-13. The lowest outside temperature an air-cooled system using capillary tubes is _____ °F.

R12-14. Receivers can be added to systems using capillary tubes. True or False?

R12-15. The liquid holding capacity is what percent of the total volume?

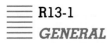

Refrigeration Piping

R13-1
GENERAL

The reliability of a field-assembled refrigeration or air-conditioning system is greatly influenced by proper design and installation of the various portions of the piping system. Proper refrigeration piping is as essential to the successful operation of the system as your veins and arteries are to your body. Improper layout or sizing can change the efficiency of the various components, and thus effect the system's capacity and efficiency.

The piping layout is usually made by an application engineer, but the refrigeration technician who installs and services the system is also concerned in this layout because of the possibility of difficulties and system faults. Also, the application engineer's layout may be diagrammatic only, with little regard for distances involved, either horizontal and/or vertical; therefore, the technician is frequently in the position of having to interpret the engineer's intent and then apply sound technical modification to properly complete the installation.

R13-2
FUNCTION OF REFRIGERANT PIPING

The piping that connects the four major components of the system (Fig. R13-1) has two major functions: (1) it provides a passageway for the circulation of refrigerant, either in liquid or vapor form, depending on the portion of the system involved; and (2) it provides a passageway through which oil is returned to the compressor. Each section of the piping should fulfill these two requirements with a minimum of pressure drop of the refrigerant, yet good oil flow for maximum protection of the compressor.

R13-3
OIL IN THE PIPING

The second function, the return of oil to the compressor, is usually considered of secondary importance, but experience has proven that it is of equal importance to the function of carrying refrigerant. Reciprocating compressors, as well as rotary and centrifugal types, use a force-feed method of providing lubrication to rotating parts. For example, in the reciprocating type, some of the oil in the crankcase gets on the cylinder walls during the down or intake stroke of the piston and is blown out with the compressed gaseous refrigerant through the discharge valve parts on the up or compression stroke (Fig. R13-2). Some compressors pump much less oil than others, depending on the design and manufacturing methods. However, there is no way to design a compressor so that none of the oil escapes into the refrigerant piping. This oil serves no useful purpose except to lubricate the compressor. Presence of oil in the heat exchangers (evaporator and condenser) can reduce the capacity of the heat exchange surfaces as much as 20%. Therefore, the presence of oil in the piping must be taken into con-

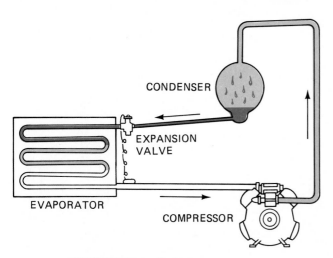

FIGURE R13-1   Refrigerant piping.

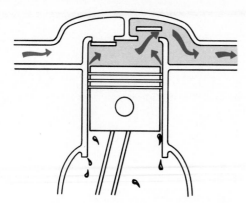

FIGURE R13-2 Compressor oil pumping.

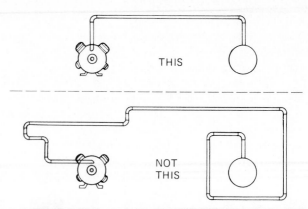

FIGURE R13-3 Improper piping arrangement.

sideration when installing piping. A piping system that is not correctly selected or installed can cause the following:

1. Burned-out compressor bearings due to lack of oil returning to the compressor for lubrication.
2. Broken compressor valves, valve plates, pistons, and/or connecting rods due to liquid refrigerant and/or large quantities (slugs) of oil entering the compressor. This is especially true of a high-efficiency-type compressor that does not have the rotating oil separator to break up the quantities before they enter the cylinder input. Remember that the compressor is designed to pump vapor and will not pump liquid.
3. Loss of capacity by occupying portions of the evaporator, thus reducing the amount of effective surface and the overall system capacity.

R13-4
BASIC PIPING PRECAUTIONS

A refrigeration technician must learn this subject well enough to avoid costly mistakes. There are five basic rules to keep in mind when piping a system:

1. *Keep it clean.* Cleanliness is a key factor in the actual installation. Dirt, sludge, and/or moisture will cause breakdown in the system *and must be avoided.* Neat, clean work will save many service difficulties.
2. *Proper sizing.* Each section of the piping system must be properly sized to ensure proper oil return as well as maintain system capacity and efficiency as high as possible. In those installations where one must be sacrificed, proper oil return takes precident. Proper sizing and limits are given later in the chapter.
3. *Use as few fittings as possible* (see Fig. R13-3).

Fewer fittings mean less chance for leaks, but more important, less needless pressure drop.

4. *Take special precautions in making every connection.* Use the right material and follow the method recommended by the equipment manufacturer.
5. *Pitch horizontal lines in the direction of refrigerant flow* (see Fig. R13-4). To aid in forcing oil to travel through lines that contain vapor (suction line, hot gas line, and liquid return line), horizontal lines should be pitched in the direction of refrigerant flow. This pitch, which helps the oil to flow in the right direction, should be $\frac{1}{2}$ in. or more for each 10 ft of run. Pitch also helps to prevent backflow of the oil during shutdown.

R13-5
SUCTION LINE

As noted in Section R13-3, oil circulates throughout the system and must be returned to the compressor to prevent damage. The most critical line in performing this function is the suction line.

The two refrigerants most commonly used in refrigeration and in conditioning are R-12 and R-22. In

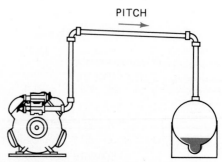

PITCH HORIZONTAL LINES 1/2 INCH
PER TEN FEET

FIGURE R13-4 Proper pitch.

their liquid form these refrigerants will mix with oil and carry it along the piping with ease. Therefore, few, if any, oil problems exist in the liquid line. However, in their gaseous state they are poor carriers of oil.

Oil in the suction line is, however, at a lower temperature than the rest of the system and therefore has higher viscosity—it is harder to cause to flow over pipe surface. Also, the refrigerant is in vapor form and has only a mechanical effect on the oil, similar to wind blowing over the surface of a lake. To make the water move in the lake, the wind must move against the water with sufficient velocity to cause waves. The suction line must, therefore, be carefully designed to ensure a uniform return of dry refrigerant gas and oil to the compressor.

The minimum load gas velocities within the suction line must be maintained at a minimum of 500 ft/min in horizontal runs and 1000 ft/min through vertical rises with upward gas flow. The minimum recommended pressure drop is 2 psig for R-12 and 3 psig for R-22 and R-502 systems. This results in the equivalent of no more than a 2°F change in saturated refrigerant temperatures. The reason the suction pressure and temperature drop are so important relates to the capacity of the compressor lines on the ratio between the discharge and suction pressures. Any increase in this ratio reduces the capacity of the compressor to pump refrigerant vapor and also increases the power required.

R13-5.1
Thermostatic Expansion Valve Systems

To learn how to prevent liquid refrigerant and oil slugging as well as oil retained in the coil, causing oil logging, we first consider what happens in a simple system when the compressor stops operating because controls are satisfied. The evaporator still contains refrigerant, part liquid and part vapor. There will also be oil in the refrigerant. The liquid refrigeration will draw by gravity to points where, when the compressor starts again, the liquid will be drawn into the compressor and could cause liquid slugging. The piping design must prevent liquid refrigerant and/or oil from draining to the compressor during shutdown.

When the compressor is above the evaporator, liquid drawing to the compressor is not a problem. In all TX Valve/DX coil installations, especially in those where the compressor is at the same level or below the evaporator, a trap and riser to at least the top of the coil must be placed in the suction line (see Fig. R13-5). This trap is to prevent liquid flowing from the DX coil into the compressor during shutdown. The trap and riser combination also promote free drainage of liquid refrigerant away from the thermostatic expansion valve bulb, thus permitting the blub to sense suction gas superheat instead of evaporating liquid temperature.

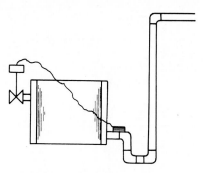

FIGURE R13-5 Suction line trap.

Although it is important to prevent liquid refrigerant from draining from the evaporator to the compressor during shutdown, it is just as important to avoid unnecessary traps in the suction line near the compressor. Such traps would collect oil, which on startup might be carried to the compressor in the form of slugs, thereby causing serious damage.

Where the system capacity is variable because of capacity control or some other arrangement, a short riser will usually be sized smaller than the remainder of the suction line (Fig. R13-6) for a velocity of not less than 1500 ft/min to ensure oil return up the riser. Although this smaller pipe has a higher friction, its short length adds a relatively small amount to the overall suction line friction loss.

In general, the pressure drop for the total suction line should be a maximum of 2°F or 2 psi, to avoid loss of system capacity. The compressor cannot pump or draw gas nearly as effectively as pushing or compressing.

If the system employs capacity control, it will be necessary to provide a double riser in the suction line, as illustrated in Fig. R13-7. Its operation is similar to the hot-gas double riser. When maximum cooling is required, the system will run at full capacity, and both risers will carry refrigerant and oil. On part load, as the amount of refrigerant being evaporated decreases, the gas velocity will also decrease to a point where it will not carry oil upward through the vertical risers. The oil trap, which is located at the bottom of the large riser, will fill with oil. All the refrigerant vapor will then pass

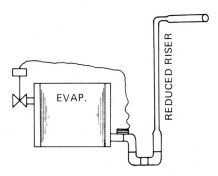

FIGURE R13-6 Suction line trap and riser.

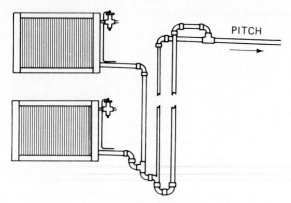

FIGURE R13-7 Double suction line trap and riser for capacity control.

up through the smaller riser, carrying oil with it. As the system load increases and more refrigerant is passed through the evaporator, this increased pressure will break the oil seal in the trap and carry oil upward through both risers. *Velocity through both risers should be sized for not less than 1500 ft/min at minimum load.*

Figure R13-7 also illustrates the use of multiple evaporators installed below the compressor. Notice that the piping is arranged so that refrigerant cannot flow from the upper evaporator into the lower evaporator.

Where a number of vertical risers are necessary on either suction or discharge lines, as illustrated in Fig. R13-8, it is recommended that line traps be installed approximately every 15 ft so that the storage and lifting of oil can be done in smaller stages.

How are refrigerant lines sized to give sufficient velocity and still avoid excessive pressure drop? To begin with, the choice and internal dimensions of refrigerant tubing must be determined. Copper tubing is available in three standard weights, known as K, L, and M. K is heavy duty, L is average duty, and M is light duty. K has the thickest wall, L is next, etc. L is recommended for ordinary refrigeration work and will be used in this discussion. Figure R13-9 is a table of nominal outside sizes and their inside diameters for type L tubing.

FIGURE R13-9

Dimensions of type L copper tubing.

DIAMETERS		TRANSVERSE AREA OF BORE	
Outside (in.)	Inside (in.)	in.²	ft²
$\frac{1}{2}$	0.430	0.1452	0.001008
$\frac{5}{8}$	0.545	0.2333	0.001620
$\frac{3}{4}$	0.666	0.3484	0.002419
$\frac{7}{8}$	0.785	0.4840	0.003361
$1\frac{1}{8}$	1.025	0.8252	0.005730
$1\frac{3}{8}$	1.265	1.257	0.008728
$1\frac{5}{8}$	1.505	1.779	0.01235
$2\frac{1}{8}$	1.985	3.095	0.02149
$2\frac{5}{8}$	2.465	4.772	0.03314
$3\frac{1}{8}$	2.945	6.812	0.04730
$3\frac{5}{8}$	3.425	9.213	0.06398

To learn how to prevent liquid slugging, we first consider what happens in a simple system when the compressor stops operating when its cooling requirements are satisfied. The evaporator is still filled with refrigerant, part liquid and part gas. There will also be some oil present. The liquid refrigerant and oil may drain by gravity to points where, when the compressor starts running again, the liquid will be drawn into the compressor and cause liquid slugging. The piping design must prevent liquid refrigerant or oil from draining to the compressor during shutdown.

When the compressor is above the evaporator there is no problem. If the compressor is on the same level or below the evaporator, as in Fig. R13-10, a rise to at least the top of the evaporator must be placed in the suction line. This inverted loop is to prevent liquid draining from the evaporator into the compressor during shutdown. The sump at the bottom of the riser promotes free drainage of liquid refrigerant away from the thermostatic expansion valve bulb, thus permitting the bulb to sense suction gas superheat instead of evaporating liquid temperature.

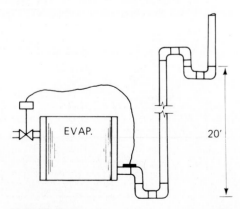

FIGURE R13-8 Multiple suction line traps.

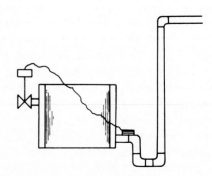

FIGURE R13-10 Proper TX valve bulb location.

Figure R13-11 is a chart for R-12, published by the Trane Company, that reflects tubing sizes, suction gas velocity, and load in tons of refrigeration (at 40°F suction and 105°F condensing temperatures, standard conditions for compressor ratings). To save time we will assume that the load is 5 tons and the suction tubing connection is $2\frac{5}{8}$ in. OD (refer to Fig. R13-11). Follow the vertical line up from 5 tons to where it intersects the diagonal line representing $2\frac{5}{8}$ in. OD. At that point read to the left and note the velocity is slightly less than 500 ft/min; obviously, the tubing is too large. What would the velocity be at 10 tons with the same-size tube? It would be nearly 1000 ft/min, and at 15 tons it would be approximately 1500 ft/min.

In most cases the condensing unit suction connection has been factory sized based on average job con-ditions. However, it is important to recheck the velocity since local conditions sometimes require an increase or decrease in size. Suppose that you had a load of 15 tons at 40°F suction and 105°F condensing temperature using R-12. What size tubing will give sufficient velocity in horizontal piping? Moving up the vertical line in the chart, at 15 tons, you find that to provide a minimum of 500 ft/min—in horizontal runs—$3\frac{5}{8}$-in. OD tube would be used. In vertical runs, the minimum velocity in 1500 ft/min to ensure proper oil carry. The maximum tube size would, therefore, be $2\frac{5}{8}$ in. OD.

This chart is not of sufficient accuracy to size the suction, as it is based on an equivalent length of the pipe (all fittings plus actual pipe length of 100 ft E.L.). For other suction and condensing temperatures, apply the correction factors from Fig. R13-11b.

FIGURE R13-11
(a) Suction gas velocities, R-12. (b) Suction gas velocity correction factors, R-12.

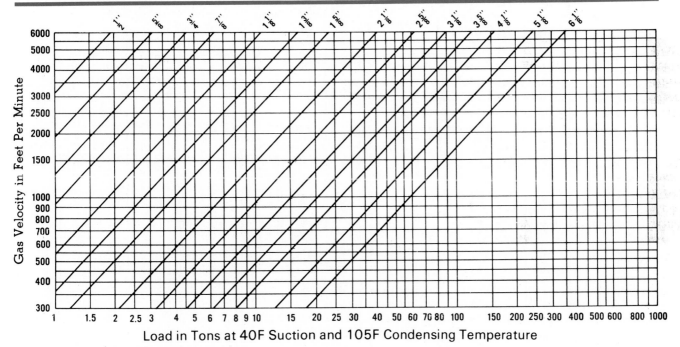

Outside Diameter-Type L Copper Tubing

(a)

COND. TEMP.	SUCTION TEMPERATURE																
	-30	-25	-20	-15	-10	-5	0	5	10	15	20	25	30	35	40	45	50
85	4.18	3.68	3.26	2.89	2.57	2.30	2.06	1.84	1.71	1.49	1.35	1.22	1.10	1.00	0.91	0.83	0.76
90	4.29	3.77	3.34	2.97	2.64	2.35	2.11	1.89	1.75	1.53	1.38	1.25	1.13	1.03	0.93	0.85	0.78
95	4.41	3.88	3.43	3.04	2.71	2.41	2.16	1.93	1.80	1.56	1.41	1.27	1.15	1.05	0.95	0.87	0.79
100	4.52	3.98	3.52	3.12	2.78	2.48	2.21	1.98	1.84	1.60	1.45	1.31	1.18	1.07	0.98	0.89	0.81
105	4.66	4.10	3.63	3.21	2.85	2.54	2.28	2.04	1.89	1.64	1.48	1.34	1.21	1.10	1.00	0.91	0.83
110	4.80	4.22	3.73	3.31	2.94	2.62	2.34	2.09	1.94	1.69	1.52	1.37	1.25	1.13	1.03	0.93	0.85
115	4.95	4.35	3.84	3.40	3.02	2.69	2.40	2.15	2.00	1.74	1.56	1.41	1.28	1.16	1.05	0.96	0.88
120	5.11	4.49	3.97	3.51	3.12	2.77	2.48	2.22	2.06	1.79	1.61	1.45	1.31	1.19	1.08	0.99	0.90
125	5.29	4.64	4.10	3.63	3.22	2.86	2.55	2.28	2.12	1.84	1.66	1.50	1.35	1.23	1.11	1.02	0.92
130	5.48	4.81	4.24	3.75	3.33	2.96	2.64	2.36	2.19	1.90	1.71	1.54	1.39	1.26	1.15	1.04	0.95
135	5.69	4.99	4.40	3.89	3.45	3.06	2.73	2.44	2.26	1.96	1.77	1.59	1.44	1.30	1.18	1.07	0.98
140	5.92	5.18	4.57	4.04	3.58	3.18	2.83	2.53	2.34	2.03	1.83	1.65	1.49	1.35	1.22	1.11	1.01
145	6.16	5.40	4.76	4.20	3.72	3.30	2.94	2.62	2.43	2.10	1.89	1.70	1.54	1.39	1.26	1.15	1.04

Source: Trane Company

(b)

Figure R13-12 gives the tonnage capacity of a range of tube sizes and total equivalent lengths. Also included is the total pressure drop in psig that will be encountered at standard conditions. Note that 2 psig pressure drop and the tonnage capacities at this pressure drop are in boldface. These are the *maximum* recommended tonnages calculated to minimize suction-line temperature penalty.

Figure R13-12 is based on standard conditions—40°F suction and 105°F condensing temperatures. For

FIGURE R13-12

Tonnage capacities of suction lines delivering vapor from evaporation to compressor.

EQUIV. LENGTH IN FEET	TOTAL PRESS. DROP PSI	½	⅝	¾	⅞	1⅛	1⅜	1⅝	2⅛	2⅝	3⅛	3⅝	4⅛	5⅛	6⅛
10	0.3	0.4	0.7	1.2	1.9	3.7	6.6	10.7	21.8	39.4	61.5	93	130	234	376
	0.6	0.6	1.1	1.8	2.8	5.6	9.6	15.5	31.7	57.0	89.0	132	186	339	540
	1.0	0.7	1.4	2.4	3.7	7.4	12.5	20.2	41.3	74.0	116.0	172	242	440	700
	2.0	1.1	2.1	3.5	5.4	10.7	18.0	29.0	59.0	108.0	167.0	244	349	631	1010
	3.0	1.4	2.6	4.3	6.6	13.2	22.1	35.9	73.0	133.0	206.0	302	427	779	1240
20	0.3	0.3	0.5	0.9	1.3	2.6	4.6	7.4	15.2	27.4	42.7	65.5	90	162	262
	0.6	0.4	0.7	1.2	1.9	3.7	6.6	10.7	21.8	39.3	61.5	93.0	130	234	376
	1.0	0.5	1.0	1.6	2.5	5.1	8.7	14.1	28.8	51.5	81.0	120.0	169	307	491
	2.0	0.7	1.4	2.4	3.7	7.4	12.5	20.2	41.3	74.0	116.0	172.0	242	440	700
	3.0	1.0	1.8	3.0	4.7	9.2	15.5	24.9	51.0	92.5	144.0	212.0	301	544	870
30	0.3		0.4	0.7	1.1	2.1	3.7	6.0	12.2	22.0	34.4	52.0	72.5	130	209
	0.6		0.6	1.0	1.5	3.1	5.3	8.6	17.5	31.7	49.6	74.5	105.0	188	302
	1.0	0.4	0.8	1.3	2.0	4.1	7.1	11.3	23.1	41.5	65.5	97.5	136.0	246	396
	2.0	0.6	1.2	1.9	3.0	6.0	10.1	16.3	33.6	60.5	94.0	139.0	197.0	358	570
	3.0	0.7	1.4	2.4	3.7	7.4	12.5	20.2	41.3	74.0	116.0	172.0	242.0	440	700
40	0.3		0.4	0.6	0.9	1.8	3.2	5.1	10.5	19.0	29.7	44.5	63	112	180
	0.6	0.3	0.5	0.9	1.3	2.6	4.6	7.4	15.2	27.4	42.7	64.5	90	162	262
	1.0	0.4	0.7	1.1	1.7	3.5	6.1	9.7	19.8	35.6	56.0	83.5	117	211	340
	2.0	0.5	1.0	1.6	2.5	5.1	8.7	14.1	28.8	51.5	81.0	120.0	169	307	491
	3.0	0.7	1.2	2.0	3.1	6.3	10.8	17.4	35.4	64.0	100.0	148.0	209	378	606
50	0.3		0.3	0.5	0.8	1.6	2.8	4.5	9.4	16.8	26.0	39.5	55.5	100	160
	0.6		0.5	0.8	1.2	2.3	4.1	6.6	13.5	24.2	37.6	57.0	80.0	144	230
	1.0	0.3	0.6	1.0	1.5	3.1	5.3	8.6	17.5	31.7	49.6	74.5	105.0	188	302
	2.0	0.5	0.9	1.4	2.2	4.5	7.7	12.5	25.3	45.9	71.5	107.0	150.0	272	437
	3.0	0.6	1.1	1.8	2.8	5.6	9.6	15.5	31.7	57.0	89.0	132.0	187.0	339	540
75	0.3		0.3	0.4	0.6	1.3	2.3	3.6	7.5	13.6	21.0	31.9	44.7	80.5	128
	0.6		0.4	0.6	0.9	1.9	3.3	5.3	10.9	19.6	30.5	46.6	64.5	116.0	186
	1.0	0.2	0.5	0.8	1.2	2.5	4.3	6.9	14.2	25.3	39.8	60.0	84.0	152.0	242
	2.0	0.4	0.7	1.2	1.8	3.6	6.2	10.0	20.2	36.8	57.5	86.0	121.0	219.0	352
	3.0	0.5	0.9	1.4	2.2	4.5	7.7	12.5	25.3	45.9	71.5	107.0	150.0	272.0	437
100	0.3			0.4	0.5	1.1	2.0	3.1	6.4	11.5	18.0	27.3	40.5	68.5	111
	0.6		0.3	0.5	0.8	1.6	2.8	4.5	9.4	16.8	26.0	39.6	55.5	100.0	160
	1.0		0.4	0.7	1.1	2.1	3.7	6.0	12.2	22.0	34.4	52.0	72.5	130.0	209
	2.0	0.3	0.6	1.0	1.5	3.1	5.3	8.6	17.5	31.7	49.6	74.5	105.0	188.0	302
	3.0	0.4	0.7	1.2	1.9	3.9	6.6	10.7	21.8	39.4	61.5	93.0	130.0	234.0	376
125	0.3			0.3	0.48	1.0	1.7	2.7	5.6	10.2	15.8	24.0	33.7	60.5	96.5
	0.6		0.3	0.5	0.71	1.4	2.5	4.0	8.2	14.9	23.1	34.9	49.1	88.0	142.0
	1.0		0.4	0.6	0.9	1.9	3.3	5.3	10.9	19.6	30.5	46.6	64.5	116.0	186.0
	2.0	0.3	0.5	0.9	1.4	2.8	4.7	7.7	15.6	28.0	44.0	66.5	93.0	167.0	270.0
	3.0	0.3	0.7	1.1	1.7	3.4	5.9	9.5	19.4	34.9	54.5	82.0	115.0	207.0	332.0
150	0.3			0.3	0.4	0.9	1.6	2.5	5.1	9.2	14.3	21.6	30.4	55.0	88
	0.6		0.3	0.4	0.6	1.3	2.3	3.6	7.5	13.6	21.0	31.9	44.7	80.5	128
	1.0		0.3	0.6	0.9	1.7	3.0	4.8	9.9	17.8	30.0	42.0	58.5	106.0	169
	2.0	0.3	0.5	0.8	1.2	2.5	4.3	6.9	14.2	25.3	39.8	60.0	84.0	152.0	242
	3.0	0.3	0.6	1.0	1.5	3.1	5.3	8.6	17.5	31.7	49.6	74.5	105.0	188.0	302
175	0.3			0.3	0.4	0.8	1.4	2.3	4.7	8.4	13.3	20.1	28.0	51.0	81
	0.6			0.4	0.6	1.2	2.1	3.3	6.8	12.4	19.3	29.2	41.0	73.5	118
	1.0		0.3	0.5	0.8	1.6	2.8	4.4	9.1	16.3	25.3	38.6	54.0	97.0	156
	2.0	0.3	0.4	0.7	1.2	2.3	4.0	6.4	13.1	23.6	37.1	55.5	78.0	140.0	222
	3.0		0.6	0.9	1.4	2.9	5.0	8.0	16.3	29.4	45.9	69.0	96.5	174.0	280
200	0.3				0.4	0.8	1.3	2.1	4.3	7.9	12.3	18.6	26.0	47.1	75
	0.6			0.4	0.5	1.1	2.0	3.1	6.4	11.5	18.0	27.3	40.5	68.5	111
	1.0		0.3	0.5	0.7	1.5	2.6	4.1	8.5	15.2	23.7	36.2	50.5	90.5	145
	2.0		0.4	0.7	1.1	2.1	3.7	6.0	12.2	22.0	34.4	52.0	72.5	130.0	209
	3.0	0.3	0.5	0.9	1.3	2.7	4.6	7.4	15.2	27.4	42.7	64.5	90.0	162.0	262
250	0.3				0.3	0.6	1.2	1.9	3.9	6.9	10.8	16.4	23.0	41.8	66.5
	0.6			0.3	0.5	1.0	1.7	2.7	5.6	10.2	15.8	24.0	33.7	60.5	96.3
	1.0		0.3	0.4	0.6	1.3	2.3	3.6	7.5	13.6	21.0	32.0	44.7	80.5	128.0
	2.0		0.4	0.6	0.9	1.9	3.3	5.3	10.9	19.6	30.5	46.6	64.5	116.0	186.0
	3.0		0.5	0.8	1.2	2.4	4.1	6.6	13.5	24.2	37.6	57.0	80.0	144.0	230.0
300	0.3					0.6	1.1	1.7	3.5	6.3	9.8	14.9	20.8	37.8	60.5
	0.6				0.3	0.9	1.6	2.5	5.1	9.2	14.3	21.6	30.4	55.0	88.0
	1.0			0.4	0.6	1.2	2.1	3.3	6.8	12.2	19.0	28.8	40.3	73.0	117.0
	2.0		0.3	0.6	0.9	1.7	3.0	4.8	9.9	17.8	30.0	42.0	58.5	106.0	169.0
	3.0		0.4	0.7	1.1	2.1	3.7	6.0	12.2	22.0	34.4	52.0	72.5	130.0	209.0
350	0.3					0.6	1.0	1.6	3.2	5.8	9.1	13.8	19.2	34.9	55.5
	0.6			0.3	0.4	0.8	1.4	2.3	4.7	8.4	13.3	20.1	28.0	51.0	81.0
	1.0			0.3	0.5	1.1	1.9	3.0	6.2	11.2	17.5	26.5	37.1	67.0	106.0
	2.0		0.3	0.5	0.8	1.6	2.8	4.4	9.1	16.3	25.3	38.6	54.0	97.0	156.0
	3.0		0.4	0.6	1.0	2.0	3.4	5.5	11.3	20.3	31.6	47.9	67.0	121.0	194.0
400	0.3					0.5	0.9	1.5	3.0	5.4	8.4	12.8	17.8	32.2	51.5
	0.6				0.4	0.7	1.3	2.1	4.3	7.9	12.3	18.6	26.0	47.1	75.0
	1.0			0.3	0.5	1.0	1.8	2.8	5.8	10.5	16.3	24.6	34.6	62.0	100.0
	2.0		0.3	0.5	0.7	1.5	2.6	4.1	8.5	15.2	23.7	36.2	50.5	90.5	145.0
	3.0			0.6	0.9	1.8	3.2	5.1	10.5	19.0	29.8	44.7	63.0	112.0	180.0
450	0.3				0.2	0.5	0.9	1.4	2.8	5.0	7.9	10.8	16.8	32.4	48.6
	0.6				0.4	0.7	1.3	2.0	4.1	7.4	11.6	17.5	24.5	44.5	71.0
	1.0			0.3	0.5	0.9	1.7	2.7	5.4	9.8	15.3	23.1	32.2	58.5	93.5
	2.0		0.3	0.4	0.7	1.4	2.4	3.9	7.9	14.3	22.2	33.9	47.1	85.0	137.0
	3.0		0.3	0.5	0.9	1.7	3.0	4.8	9.9	17.8	30.0	42.0	58.5	106.0	169.0
500	0.3					0.5	0.8	1.3	2.6	4.7	7.4	11.3	15.8	29.0	45.7
	0.6				0.3	0.6	1.2	1.9	3.9	6.9	10.8	16.4	23.0	41.8	66.5
	1.0			0.3	0.4	0.9	1.6	2.5	5.1	9.2	14.3	21.6	30.4	55.0	88.0
	2.0		0.3	0.4	0.6	1.3	2.3	3.6	7.5	13.6	21.0	32.0	44.7	80.5	128.0
	3.0		0.3	0.5	0.8	1.6	2.8	4.5	9.4	16.8	26.1	39.6	55.5	100.0	160.0

Based on 40°F suction and 105°F condensing. For other conditions, apply the correction factor from Fig. R13-13 to the design tons before entering this table.

Note: Figures in boldface type are maximum recommended tonnages at pressure drops calculated to minimize suction line temperature penalty.

Source: Trane Company

≡ Suction-line tonnage correction factors for R-12.

COND. TEMP.	EVAPORATOR TEMPERATURE																
	-30	-25	-20	-15	-10	-5	0	5	10	15	20	25	30	35	40	45	50
85	2.20	2.04	1.90	1.77	1.66	1.55	1.46	1.36	1.29	1.21	1.13	1.07	1.02	0.97	0.91	0.87	0.82
90	2.25	2.09	1.95	1.81	1.70	1.59	1.50	1.39	1.32	1.23	1.16	1.10	1.04	0.99	0.93	0.89	0.84
95	2.32	2.15	2.00	1.86	1.74	1.63	1.53	1.43	1.35	1.26	1.19	1.12	1.07	1.01	0.95	0.91	0.86
100	2.38	2.21	2.05	1.91	1.79	1.67	1.57	1.47	1.38	1.30	1.22	1.15	1.10	1.04	0.98	0.93	0.88
105	2.45	2.27	2.11	1.96	1.84	1.72	1.62	1.50	1.42	1.33	1.25	1.18	1.12	1.06	1.00	0.95	0.90
110	2.52	2.34	2.17	2.02	1.89	1.76	1.66	1.55	1.46	1.37	1.28	1.21	1.15	1.09	1.03	0.98	0.92
115	2.60	2.41	2.24	2.08	1.95	1.82	1.71	1.59	1.50	1.40	1.32	1.24	1.18	1.12	1.05	1.00	0.95
120	2.69	2.49	2.31	2.14	2.01	1.87	1.76	1.64	1.54	1.44	1.36	1.28	1.21	1.15	1.08	1.03	0.97
125	2.78	2.57	2.39	2.21	2.07	1.93	1.82	1.69	1.59	1.49	1.40	1.31	1.25	1.18	1.11	1.06	1.00
130	2.88	2.66	2.47	2.29	2.14	2.00	1.88	1.74	1.64	1.54	1.44	1.36	1.29	1.22	1.15	1.09	1.03
135	2.99	2.76	2.56	2.37	2.22	2.07	1.94	1.80	1.70	1.59	1.49	1.40	1.33	1.26	1.18	1.12	1.06
140	3.11	2.87	2.66	2.46	2.30	2.14	2.01	1.87	1.76	1.64	1.54	1.45	1.37	1.30	1.22	1.16	1.09
145	3.24	2.99	2.77	2.56	2.40	2.23	2.09	1.94	1.82	1.70	1.59	1.50	1.42	1.34	1.26	1.20	1.13

Multiply the design by the factor in this table to obtain the corrected tons for use in Fig. R13-12.

Source: Trane Company

other conditions a correction factor table (Fig. R13-13) must be used. For a system operating at 30°F suction and 94°F condensing, the table shows that the load in tons must be adjusted by a factor of 1.07. You then use the corrected tonnage to calculate the tube size in Fig. R13-11. In this case, a tonnage of 5.35 tons would be used. In the case of an air-cooled air-conditioning unit operating at 45°F suction and 125°F condensing, the connection factor would be 1.06 and a design tonnage of 5.35 tons would be used to size the suction tube.

This discussion is based on the suction-line requirements of single-compressor systems of single capacity and on material supplied by the Trane Company. On installation that will result in a wide variation of suction gas quantities due to compressor capacity control, compressor staging, etc., we recommend the use of the Trane Company's manual *Reciprocating Refrigeration* or consulting the specific equipment manufacturer's recommendation since there are some differences in techniques from one manufacturer to another.

≡ R13-5.2
≡ Capillary Tube Systems

Sizing of the suction line depends on the amount of suction vapor to be handled. Therefore, suction lines in capillary tube systems are sized the same way as in TX valve systems. The difference in suction-line design between the two types of systems is in the use of traps. Traps are not recommended in capillary tube installations, as they only add resistance to the flow of refrigerant. Remember, the trap is composed of three 90° bends, and each bend is equal to 5 ft of straight pipe. Adding a trap is the same as adding 15 ft of straight pipe. Practically all D.X. coils using capillary tubes are bottom feed, which reduces the possibility of gravity drain during the OFF cycle of the system. Also, the refrigerant charge is limited to what the coil will hold dur-

ing the OFF cycle. When properly charged with refrigerant, the capillary tube system has less chance of liquid slugging. Also, because the coil fills with refrigerant when the pressures balance, oil in the coil is lifted out into the suction leader of the coil on each OFF cycle. This prevents oil accumulation in the coil.

Traps are also not required in vertical rises because of the high gas velocities on startup. Standing idle, the DX coil, full of liquid refrigerant, will acquire considerable thermal energy at the balance pressure. Immediately on startup of the compressor, when the suction pressure drops, a high quantity of vapor is produced in the DX coil, resulting in suction-line velocities in excess of 6000 ft/min. This velocity is sufficient to lift oil in vertical riser suction lines up to 65 ft. It is not recommended that installations be made of capillary tube systems where a rise in the suction line in excess of 65 ft is required. In these installations, TX valves and suction-line traps should be used.

≡ R13-5.3
≡ Suction-Line Insulation

Insulation on the suction line is an absolute requirement. This eliminates the following:

1. Sweating of the suction line. Water condensing on the suction line can fall on occupants, which is objectionable and can cause damage to ceilings, floors, etc.
2. High-suction temperature gain, since hermetic and hermetic motor compressor assemblies are suction gas cooled. Therefore, the lower the returning suction gas temperature, the better the heat removal from the motor and the lower the motor operating temperature; also, the lower the oil temperature, resulting in better lubrication and bearing heat removal. This promotes longer compressor life.

FIGURE R13-14

Tonnage capacities of discharge lines delivering hot vapor from compressor to condenser.

EQUIV. LENGTH IN FEET	TOTAL PRESS. DROP PSI	COPPER TUBE SIZE — OD — TYPE L												
		1/2	5/8	3/4	7/8	1 1/8	1 3/8	1 5/8	2 1/8	2 5/8	3 1/8	3 5/8	4 1/8	5 1/8
20	1	1.1	2.1	3.6	5.5	11.2	19.9	31.0	64.2	116.0	181.0	266.0	370.0	675.0
	2	1.7	3.1	5.3	8.1	16.8	29.2	45.8	95.0	170.0	265.0	390.0	542.0	1000.0
	3	2.1	3.9	6.7	10.2	21.0	36.5	57.5	120.0	211.0	332.0	484.0	688.0	1270.0
	4	2.5	4.6	7.8	12.0	24.8	41.9	67.8	141.0	250.0	393.0	578.0	810.0	1485.0
	5	2.8	5.1	8.9	13.6	28.1	48.8	76.2	160.0	282.0	445.0	657.0	920.0	1680.0
	6	3.1	5.7	9.8	15.1	30.2	54.0	84.4	178.0	314.0	490.0	725.0	1010.0	1860.0
	7	3.4	6.2	10.7	16.6	33.8	58.9	92.0	193.0	340.0	538.0	798.0	1100.0	2010.0
30	1	0.93	1.7	2.9	4.5	9.2	16.0	25.0	53.0	94.0	148.0	217.0	302.0	557.0
	2	1.4	2.5	4.3	6.6	13.8	23.8	37.2	77.5	140.0	216.0	320.0	445.0	808.0
	3	1.7	3.1	5.2	8.3	17.0	29.9	46.2	97.5	173.0	270.0	402.0	565.0	1035.0
	4	2.0	3.7	6.3	9.7	20.1	35.0	54.0	115.0	205.0	318.0	468.0	659.0	1200.0
	5	2.3	4.2	7.2	11.1	22.9	39.7	61.5	130.0	229.0	360.0	536.0	742.0	1370.0
	6	2.5	4.6	7.9	12.3	25.0	44.1	69.0	144.5	251.0	400.0	592.0	829.0	1515.0
	7	2.7	5.0	8.6	13.2	27.4	47.8	74.0	158.0	278.0	435.0	641.0	902.0	1625.0
40	1	0.8	1.4	2.5	3.8	7.8	13.8	21.3	44.8	79.0	125.0	185.0	260.0	478.0
	2	1.2	2.1	3.7	5.6	11.6	20.0	31.1	65.6	118.0	183.0	270.0	380.0	696.0
	3	1.5	2.7	4.6	7.1	14.5	25.1	39.2	82.5	148.0	228.0	339.0	476.0	875.0
	4	1.7	3.1	5.4	8.3	17.0	29.5	46.0	97.0	174.0	268.0	400.0	556.0	1025.0
	5	1.9	3.5	6.2	9.4	19.2	33.5	52.3	110.0	196.0	305.0	451.0	635.0	1160.0
	6	2.1	4.0	6.8	10.3	21.3	37.2	58.1	128.0	218.0	338.0	502.0	702.0	1280.0
	7	2.3	4.3	7.4	11.3	23.0	40.1	62.4	134.5	237.0	363.0	545.0	760.0	1390.0
50	1	0.7	1.3	2.2	3.4	6.9	12.2	19.0	39.8	70.0	110.0	165.0	230.0	419.0
	2	1.0	1.9	3.2	5.0	10.2	18.1	28.0	58.0	114.0	162.0	240.0	336.0	608.0
	3	1.3	2.4	4.1	6.3	12.9	22.6	34.8	73.6	131.0	205.0	304.0	420.0	775.0
	4	1.5	2.8	4.8	7.4	15.1	26.5	40.9	86.0	154.0	241.0	358.0	495.0	905.0
	5	1.7	3.2	5.4	8.3	17.0	29.9	47.0	98.2	175.0	283.0	404.0	562.0	1020.0
	6	1.9	3.5	6.0	9.2	18.9	33.0	51.6	109.5	193.0	302.0	446.0	624.0	1140.0
	7	2.1	3.8	6.5	10.0	20.7	36.2	55.8	117.5	211.0	328.0	489.0	678.0	1245.0
75	1	0.5	1.0	1.8	2.7	5.6	9.6	15.0	31.8	56.8	88.0	130.0	180.0	333.0
	2	0.8	1.5	2.6	4.0	8.2	14.2	22.0	46.5	82.4	130.0	191.0	265.0	497.0
	3	1.0	1.9	3.3	5.0	10.3	17.9	28.0	58.6	104.0	162.0	241.0	332.0	629.0
	4	1.2	2.2	3.9	5.9	12.1	21.0	32.6	69.1	122.0	193.0	283.0	391.0	732.0
	5	1.4	2.5	4.4	6.7	13.8	23.9	37.3	78.2	139.0	218.0	320.0	448.0	829.0
	6	1.5	2.8	4.9	7.4	15.2	26.5	41.1	86.1	152.0	241.0	355.0	494.0	920.0
	7	1.7	3.1	5.3	8.0	16.8	29.9	44.5	94.0	167.0	262.0	388.0	539.0	998.0
100	1	0.5	0.9	1.5	2.3	4.8	8.4	13.0	27.2	48.0	75.0	112.0	158.0	288.0
	2	0.7	1.3	2.2	3.4	7.1	12.3	19.1	40.0	70.8	112.0	167.0	230.0	420.0
	3	0.9	1.6	2.8	4.3	8.8	15.4	24.0	50.1	89.0	140.0	209.0	291.0	531.0
	4	1.0	1.9	3.3	5.0	10.5	18.1	28.3	59.2	104.0	165.0	245.0	340.0	622.0
	5	1.2	2.2	3.8	5.8	11.8	20.8	32.0	66.8	118.0	187.0	277.0	384.0	710.0
	6	1.3	2.4	4.2	6.4	13.0	22.9	35.5	74.1	131.0	206.0	308.0	429.0	780.0
	7	1.4	2.6	4.5	6.9	14.1	24.8	39.1	80.3	142.0	224.0	332.0	464.0	842.0
125	3	0.8	1.4	2.5	3.8	7.8	13.7	20.9	44.1	78.0	122.0	182.0	251.0	468.0
	4	0.9	1.7	2.9	4.4	9.1	15.8	24.6	51.8	92.0	142.0	212.0	299.8	546.0
	5	1.1	1.9	3.3	5.0	10.4	17.9	27.8	58.6	104.0	162.0	232.0	335.0	620.0
	6	1.2	2.1	3.6	5.6	11.5	20.0	30.8	64.5	115.0	180.0	268.0	370.0	685.0
	7	1.3	2.3	3.9	6.0	12.5	21.3	33.2	70.2	126.0	194.0	291.0	402.0	748.0
150	3	0.7	1.3	2.2	3.4	7.0	12.2	19.1	40.0	70.0	110.0	165.0	228.0	420.0
	4	0.8	1.5	2.6	4.0	8.3	14.3	22.2	47.1	83.0	130.0	194.0	267.0	492.0
	5	0.9	1.7	3.0	4.5	9.4	16.2	25.1	53.1	94.0	149.0	220.0	302.0	559.0
	6	1.0	1.9	3.3	5.0	10.4	18.0	28.1	59.5	104.0	163.0	242.0	337.0	621.0
	7	1.1	2.1	3.6	5.4	11.2	19.7	30.2	64.0	113.0	178.0	264.0	368.0	676.0
175	3	0.6	1.2	2.0	3.1	6.4	11.2	17.2	36.1	63.9	100.0	150.0	209.0	388.0
	4	0.8	1.4	2.4	3.7	7.6	13.2	20.3	42.8	75.4	119.0	177.0	244.0	450.0
	5	0.9	1.6	2.7	4.2	8.6	15.4	22.2	48.8	85.2	135.0	200.0	278.0	516.0
	6	1.0	1.7	3.0	4.6	9.5	16.6	25.5	53.9	94.8	150.0	220.0	308.0	569.0
	7	1.0	1.9	3.3	5.0	10.2	18.0	27.9	58.1	103.0	162.0	240.0	332.0	622.0
200	3	0.6	1.1	2.0	3.0	6.2	10.8	16.8	35.0	63.0	96.8	146.0	201.0	366.0
	4	0.7	1.3	2.3	3.5	7.3	12.7	19.8	41.2	74.0	116.0	170.0	238.0	430.0
	5	0.8	1.5	2.6	4.0	8.3	14.3	22.2	47.0	84.0	130.0	193.0	270.0	490.0
	6	0.9	1.7	2.9	4.4	9.1	15.9	24.8	52.0	93.0	143.0	210.0	299.0	542.0
	7	1.0	1.8	3.2	4.8	9.9	16.2	26.8	56.0	100.0	157.0	231.0	323.0	591.0
250	3	0.5	1.0	1.6	2.5	5.2	9.1	14.1	29.5	52.0	82.0	122.0	170.0	315.0
	4	0.6	1.1	1.9	3.0	6.1	10.7	16.7	34.2	61.0	96.0	143.0	200.0	370.0
	5	0.7	1.3	2.2	3.4	6.9	12.2	18.9	39.4	69.4	110.0	162.0	226.0	425.0
	6	0.8	1.4	2.4	3.8	7.6	13.4	21.0	43.5	76.3	121.0	180.0	250.0	470.0
	7	0.9	1.5	2.7	4.1	8.3	14.6	23.9	47.3	83.6	133.0	198.0	272.0	516.0
300	3	0.5	0.9	1.5	2.4	4.9	8.5	13.1	27.9	49.5	76.0	114.0	158.0	286.0
	4	0.6	1.0	1.8	2.8	5.8	10.0	15.4	32.6	57.9	90.5	134.0	185.0	337.0
	5	0.7	1.2	2.1	3.2	6.5	11.2	17.7	37.0	65.4	101.0	152.0	211.0	385.0
	6	0.7	1.3	2.3	3.5	7.3	12.6	19.7	40.9	72.8	112.0	169.0	232.0	424.0
	7	0.8	1.4	2.5	3.8	7.9	13.8	21.1	44.9	79.5	123.0	183.0	254.0	460.0
350	3	0.4	0.8	1.4	2.1	4.3	7.6	11.8	24.6	43.5	68.0	100.0	141.0	266.0
	4	0.5	0.9	1.6	2.5	5.1	8.9	13.9	29.0	51.0	80.0	119.0	165.0	312.0
	5	0.6	1.1	1.8	2.8	5.8	10.1	15.8	33.1	58.1	90.0	134.0	189.0	355.0
	6	0.6	1.2	2.0	3.1	6.4	11.2	17.3	36.4	64.0	100.0	148.0	209.0	393.0
	7	0.7	1.3	2.2	3.4	7.0	12.1	18.9	39.6	70.0	109.0	161.0	227.0	428.0
400	3	0.4	0.8	1.4	2.1	4.3	7.6	11.7	24.4	43.2	67.0	98.0	135.0	252.0
	4	0.5	0.9	1.6	2.5	5.1	8.9	13.8	28.8	49.6	78.0	117.0	160.0	295.0
	5	0.6	1.1	1.8	2.8	5.8	10.1	15.7	32.9	57.9	88.0	132.0	175.0	335.0
	6	0.6	1.2	2.0	3.1	6.4	11.2	17.3	36.2	63.7	98.0	147.0	200.0	365.0
	7	0.7	1.3	2.2	3.4	7.0	12.1	18.8	39.4	69.7	107.0	159.0	210.0	401.0

Based on 40°F suction and 105°F condensing. For other conditions, apply the correction factor from Fig. R13-15 to the design tons before entering this table.

Source: Trane Company

R13-6
HOT-GAS LINE

In sizing and arranging hot-gas lines, select tubing with a diameter small enough to provide the velocity to carry the hot vaporized oil to the condenser. On the other hand, the diameter must be large enough to prevent excessive pressure drop. In suction lines, the maximum pressure drop is 2 psig, but in hot-gas lines, a maximum of 6 psig is permitted.

If a higher pressure drop is used, the velocity of gas flow through the line can be excessive, to the point of causing noise, vibration, and serious reduction in system capacity and increase in operating cost due to the higher compressor discharge pressure required.

In Fig. R13-14 the tonnage capacities of hot-gas lines carrying R-22 are given at standard conditions. The 5-ton unit used in the example in the discussion of suction lines would require a $\frac{7}{8}$-in.-OD line if the total equivalent length were 100 ft. If, however, the total equivalent length were 20 ft, a $\frac{5}{8}$-in.-OD tube could be used. As in calculating suction-line sizes, when other than standard conditions are encountered, an adjustment must be made.

Figure R13-15 gives the correction factors to be used for other-than-standard conditions. In the example of the air-conditioning unit operating at 45°F suction and 125°F condensing, the correction factor is 0.95. Multiply the original 5 tons by 0.95, giving a corrected design capacity of 4.75 tons.

If the total equivalent length of the hot-gas line were 75 ft at standard conditions, the line size would be $\frac{7}{8}$ in. OD. Corrected for the actual conditions, the line size would be $\frac{3}{4}$ in. OD. In condensing units, the hot-gas line (from compressor to condenser) is usually very short and factory sized. Only where field-builtup systems are applied or where units are converted to air-cooled by the addition of a remote air-cooled condenser does the previous discussion have field application.

When larger built-up systems with compressor capacity control are encountered, special rules for multiplexing hot-gas lines apply. Material on this subject is published by the Trane Company in their manual *Reciprocating Refrigeration* or consult the manufacturer's literature covering the particular equipment when making these installations.

R13-6.1
Hot-Gas-Line Insulation

In package units and condensing units with the short hot-gas line between compressor and condenser,

no insulation should be used on the hot-gas line. On remote condensers, if the unit is expected to operate in low outside ambiancy, it is possible to reach the condensing temperature of the discharge refrigerant before the refrigerant reaches the condenser. This can cause liquid slugs to fall backwards down the hot-gas line into the superheated vapor from the compressor. Violent expansion of the slug vaporizing can cause "steam hammer" and could result in noise and vibration, even to the point of line breakage. Insulate the hot-gas line to prevent this action.

R13-7
LIQUID LINE

Liquid lines do not present a problem from an oil standpoint because liquid refrigerant and oil mix easily and the oil is carried through the liquid line with ease. Liquid lines, however, are critical of pressure loss both from a pressure drop due to pipe size and, in the case of the vertical upflow line, vertical lift of the refrigerant.

R13-7.1
Pressure Drop

In the table of tonnage capacities of liquid lines in Fig. R13-16, you will note that the maximum pressure drop due to line size is 4 psig. Using this table for the previous example of the 5-ton system (at standard conditions) the smallest line given is $\frac{1}{2}$ in. and has a tonnage capacity of 7 tons with 3 psig pressure drop for 20 equivalent feet of length.

To limit the amount of refrigerant in the system, most manufacturers use $\frac{3}{8}$-in. and even $\frac{1}{4}$-in. liquid lines in remote systems. This will vary with tonnage and line length. *Do not attempt to change the liquid line size from the original specifications,* as this will seriously affect

FIGURE R13-15
Discharge gas line tonnage correction factors for R-22.

COND. TEMP.	SATURATED EVAPORATOR TEMPERATURE °F																
	-30	-25	-20	-15	-10	-5	0	5	10	15	20	25	30	35	40	45	50
85	1.21	1.20	1.19	1.18	1.17	1 16	1.15	1.14	1.13	1.12	1.11	1.11	1.10	1.09	1.08	1.08	1.07
90	1.18	1.17	1.16	1.15	1.14	1.13	1.12	1.11	1.10	1.10	1.09	1.08	1.07	1.07	1.06	1.05	1.05
95	1.16	1.15	1.14	1.13	1.12	1.11	1.10	1.09	1.08	1.07	1.07	1.06	1.05	1.04	1.04	1.03	1.00
100	1.14	1.13	1.12	1.11	1.10	1.09	1.08	1.07	1.06	1.05	1.05	1.04	1.03	1.02	1.02	1.01	1.00
105	1.13	1.12	1.11	1.09	1.08	1.08	1.07	1.06	1.05	1.04	1.03	1.02	1.02	1.01	1.00	0.99	0.99
110	1 12	1.10	1.09	1.08	1.07	1.06	1.05	1.04	1.03	1.03	1 02	1.01	1.00	0.99	0.99	0.98	0.97
115	1.11	1.09	1.08	1.07	1.06	1.05	1.04	1.03	1.02	1.01	1.01	1.00	0.99	0.98	0.98	0.97	0.96
120	1.10	1.09	1.08	1.06	1.05	1.04	1.03	1.02	1.01	1.01	1.00	0.99	0.98	0.97	0.97	0.96	0.95
125	1.10	1.08	1.07	1.06	1.05	1.04	1.03	1.02	1.01	1.00	0.99	0.98	0.97	0.97	0.96	0.95	0.94
130	1.09	1.08	1.07	1.06	1.05	1.04	1.02	1.01	1.00	1.00	0.99	0.98	0.97	0.96	0.95	0.95	0.94
135	1.10	1.08	1.07	1.06	1.05	1.04	1.03	1.02	1.01	1.00	0.99	0.98	0.97	0.96	0.95	0.94	0.94
140	1.10	1.09	1.08	1.06	1.05	1.04	1.03	1.02	1.01	1.00	1.00	0.99	0.98	0.97	0.96	0.95	0.94	0.93
145	1.12	1.10	1.09	1.07	1.06	1.05	1.04	1.02	1.01	1.00	0.99	0.98	0.97	0.96	0.96	0.95	0.94

Apply to Fig. R13-14.
Source: Trane Company

FIGURE R13-16

Tonnage capacities of lines delivering liquid R-12 from receiver to evaporator.

EQUIV. LENGTH IN FEET	TOTAL PRESS. DROP PSI	COPPER TUBE SIZE—OD—TYPE L											
		1/2	5/8	3/4	7/8	1 1/8	1 3/8	1 5/8	2 1/8	2 5/8	3 1/8	3 5/8	4 1/8
20	0.3	1.9	3.6	6.1	9.5	19.6	33.5	53.0	112.0	202.0	318.0	475.0	700.0
	0.6	2.8	5.4	9.1	14.1	29.0	49.5	78.0	166.0	301.0	470.0	708.0	1030.0
	1.0	3.7	7.2	12.2	19.0	39.0	67.0	105.0	223.0	405.0	619.0	950.0	1400.0
	2.0	5.6	10.7	18.1	28.1	58.2	100.0	138.0	330.0	600.0	950.0	1410.0	2060.0
	3.0	7.0	13.4	23.0	35.4	73.5	127.0	198.0	420.0	760.0	1200.0	1790.0	2600.0
	4.0	8.3	15.8	27.0	41.5	86.1	150.0	230.0	495.0	895.0	1410.0	2100.0	3050.0
	5.0	9.4	18.0	30.8	47.4	99.0	170.0	262.0	560.0	1010.0	1605.0	2395.0	3490.0
	6.0	10.5	20.0	34.0	52.2	110.0	188.0	290.0	625.0	1130.0	1800.0	2620.0	3850.0
30	0.3	1.5	2.9	5.0	7.6	15.9	27.1	42.0	91.0	162.0	260.0	380.0	560.0
	0.6	2.2	4.6	7.4	11.3	23.1	40.0	62.0	134.0	240.0	380.0	560.0	820.0
	1.0	3.0	5.8	9.9	15.1	31.0	54.0	84.0	180.0	325.0	515.0	760.0	1100.0
	2.0	4.5	8.6	14.8	22.6	46.5	80.0	126.0	268.0	480.0	760.0	1130.0	1600.0
	3.0	5.6	10.8	18.6	28.2	69.0	102.0	159.0	338.0	610.0	880.0	1440.0	2010.0
	4.0	6.7	12.9	22.0	33.5	58.5	120.0	185.0	400.0	720.0	1145.0	1690.0	2380.0
	5.0	7.6	14.5	25.0	38.0	79.0	137.0	210.0	450.0	815.0	1300.0	1910.0	2700.0
	6.0	8.4	16.0	27.6	42.0	86.5	150.0	230.0	500.0	900.0	1450.0	2120.0	2960.0
40	0.3	1.3	2.4	4.2	6.5	13.2	23.1	35.5	76.0	140.0	220.0	325.0	470.0
	0.6	1.9	3.6	6.2	9.6	19.9	34.5	53.0	115.0	208.0	328.0	480.0	708.0
	1.0	2.6	4.9	8.4	13.0	26.6	46.5	72.0	155.0	278.0	440.0	650.0	950.0
	2.0	3.8	7.3	12.6	19.5	40.0	69.5	108.0	230.0	412.0	659.0	980.0	1420.0
	3.0	4.8	9.2	15.9	24.4	50.0	88.0	137.0	292.0	520.0	835.0	1230.0	1800.0
	4.0	5.7	11.0	18.9	29.0	60.0	103.0	160.0	345.0	620.0	985.0	1460.0	2150.0
	5.0	6.5	12.4	21.2	33.0	68.0	119.0	181.0	395.0	702.0	1115.0	1660.0	2440.0
	6.0	7.2	13.8	23.5	36.2	75.0	131.0	202.0	435.0	780.0	1245.0	1830.0	2700.0
50	0.3	1.1	2.2	3.8	5.8	12.0	20.5	32.0	69.0	125.0	195.0	290.0	420.0
	0.6	1.7	3.2	5.6	8.6	17.8	30.5	48.0	102.0	182.0	290.0	430.0	620.0
	1.0	2.2	4.3	7.4	11.4	23.6	41.0	64.0	138.0	245.0	388.0	575.0	830.0
	2.0	3.4	6.4	11.1	17.0	35.0	61.5	94.0	204.0	360.0	575.0	845.0	1240.0
	3.0	4.3	8.2	14.0	21.4	44.5	77.0	120.0	259.0	458.0	730.0	1090.0	1560.0
	4.0	5.0	9.6	16.6	25.0	52.0	91.0	140.0	302.0	540.0	860.0	1180.0	1810.0
	5.0	5.8	11.0	18.9	28.5	59.5	104.0	160.0	340.0	610.0	979.0	1450.0	2080.0
	6.0	6.4	12.0	20.8	31.5	66.0	116.0	177.0	380.0	682.0	1085.0	1600.0	2290.0
75	0.3	0.9	1.7	2.9	4.6	9.4	16.0	25.0	54.0	95.0	150.0	221.0	330.0
	0.6	1.3	2.6	4.3	6.7	13.8	23.8	37.0	79.0	141.0	220.0	330.0	485.0
	1.0	1.8	3.4	5.8	9.0	18.6	32.0	50.0	106.0	190.0	300.0	445.0	650.0
	2.0	2.7	5.1	8.7	13.3	27.5	47.0	73.0	158.0	280.0	445.0	659.0	960.0
	3.0	3.3	6.4	11.0	16.9	34.8	60.0	93.0	200.0	350.0	560.0	835.0	1205.0
	4.0	3.9	7.6	13.0	20.0	41.0	70.0	110.0	232.0	415.0	660.0	990.0	1415.0
	5.0	4.5	8.6	14.8	22.4	46.0	80.0	124.0	262.0	470.0	750.0	1110.0	1600.0
	6.0	5.0	9.5	16.2	25.0	51.5	89.0	137.0	290.0	525.0	835.0	1230.0	1795.0
100	0.3	0.8	1.5	2.5	3.9	8.0	14.0	21.7	46.0	83.0	130.0	195.0	282.0
	0.6	1.1	2.2	3.7	5.8	11.9	20.6	32.0	68.0	124.0	193.0	290.0	420.0
	1.0	1.5	2.9	5.0	7.7	16.0	28.0	42.5	91.0	165.0	260.0	385.0	560.0
	2.0	2.3	3.4	7.4	11.6	23.9	41.0	64.0	137.0	246.0	390.0	578.0	840.0
	3.0	2.9	5.6	9.4	14.8	30.0	52.0	80.0	172.0	310.0	490.0	730.0	1060.0
	4.0	3.4	6.5	11.1	17.0	35.0	62.0	95.0	204.0	368.0	582.0	860.0	1260.0
	5.0	3.9	7.4	12.7	19.5	40.0	70.0	108.0	231.0	417.0	660.0	980.0	1410.0
	6.0	4.3	8.2	14.0	21.4	45.0	78.0	120.0	259.0	466.0	740.0	1090.0	1595.0
125	1.0	1.4	2.6	4.5	7.0	14.3	18.5	38.5	83.0	150.0	238.0	350.0	502.0
	2.0	2.1	3.9	6.8	10.4	21.2	25.0	58.0	124.0	220.0	350.0	520.0	742.0
	3.0	2.6	4.9	8.6	13.1	27.0	37.0	72.0	158.0	280.0	445.0	655.0	940.0
	4.0	3.1	5.8	10.0	15.5	32.0	55.0	85.0	184.0	330.0	528.0	770.0	1102.0
	5.0	3.5	6.6	11.4	17.6	36.2	63.0	97.0	210.0	375.0	600.0	875.0	1265.0
	6.0	3.9	7.3	12.7	19.5	40.2	70.0	108.0	231.0	416.0	660.0	978.0	1400.0
150	1.0	1.2	2.3	4.0	6.0	12.6	21.8	33.0	72.0	131.0	205.0	304.0	440.0
	2.0	1.8	3.4	5.9	9.1	18.8	34.5	50.0	108.0	196.0	308.0	450.0	660.0
	3.0	2.3	4.3	7.4	11.5	23.7	41.0	63.0	137.0	248.0	389.0	575.0	830.0
	4.0	2.7	5.2	8.8	13.6	28.5	48.5	74.0	160.0	290.0	460.0	680.0	980.0
	5.0	3.1	5.8	10.0	15.3	31.5	55.0	84.0	181.0	330.0	520.0	765.0	1105.0
	6.0	3.4	6.5	11.1	17.0	35.0	61.0	93.0	202.0	365.0	580.0	860.0	1235.0
175	1.0	1.2	2.2	3.8	5.9	12.1	21.0	32.0	65.0	128.0	200.0	295.0	429.0
	2.0	1.7	3.3	5.7	8.8	18.0	31.0	49.0	104.0	189.0	299.0	440.0	638.0
	3.0	2.2	4.2	7.2	11.0	23.0	39.5	61.0	131.0	238.0	378.0	555.0	800.0
	4.0	2.6	4.9	8.5	13.0	27.0	47.0	72.0	154.0	280.0	442.0	650.0	940.0
	5.0	2.9	5.6	9.6	14.8	30.5	53.0	82.0	176.0	319.0	500.0	740.0	1075.0
	6.0	3.3	6.2	10.7	16.3	34.0	58.0	91.0	195.0	350.0	560.0	825.0	1195.0
200	1.0	1.0	1.9	3.3	5.0	10.3	18.0	28.0	59.5	109.0	170.0	250.0	370.0
	2.0	1.5	2.9	4.9	7.4	15.6	27.0	41.6	89.0	161.0	255.0	375.0	555.0
	3.0	1.9	3.6	6.2	9.4	19.6	34.0	52.0	112.0	204.0	323.0	470.0	700.0
	4.0	2.2	4.2	7.2	11.0	23.0	40.0	62.0	132.0	241.0	381.0	560.0	825.0
	5.0	2.5	4.8	8.2	12.7	26.0	46.0	70.0	151.0	272.0	438.0	640.0	942.0
	6.0	2.8	5.4	9.2	14.0	29.0	51.0	78.0	169.0	302.0	485.0	710.0	1050.0
250	1.0	0.9	1.8	3.1	4.8	9.8	17.0	26.5	57.0	101.0	160.0	240.0	350.0
	2.0	1.4	2.7	4.6	7.1	14.8	25.6	39.5	84.0	152.0	241.0	360.0	520.0
	3.0	1.8	3.5	5.8	9.0	18.4	32.0	50.0	107.0	194.0	310.0	455.0	658.0
	4.0	2.1	4.1	7.0	10.6	22.0	38.0	59.0	128.0	230.0	360.0	538.0	770.0
	5.0	2.4	4.6	7.9	12.0	25.0	43.5	67.5	143.0	260.0	415.0	612.0	880.0
	6.0	2.7	5.2	8.8	13.4	27.8	49.0	75.0	160.0	290.0	460.0	686.0	978.0
300	1.0	0.8	1.5	2.6	4.0	8.3	14.2	22.0	47.0	86.0	137.0	200.0	290.0
	2.0	1.2	2.3	3.9	6.0	12.4	21.1	33.0	71.0	129.0	203.0	300.0	438.0
	3.0	1.5	2.9	4.9	7.6	15.7	27.0	41.5	89.0	161.0	256.0	380.0	550.0
	4.0	1.8	3.4	5.8	8.9	18.2	32.0	49.0	105.0	190.0	300.0	445.0	650.0
	5.0	2.0	3.9	6.6	10.1	21.0	36.0	56.0	120.0	216.0	342.0	510.0	740.0
	6.0	2.2	4.3	7.4	11.3	23.1	40.5	62.0	132.0	241.0	380.0	560.0	825.0
350	1.0	0.8	1.5	2.6	4.0	8.2	14.1	21.8	46.0	84.0	134.0	198.0	285.0
	2.0	1.2	2.2	3.8	5.9	12.2	21.0	32.8	70.0	128.0	200.0	298.0	428.0
	3.0	1.5	2.8	4.8	7.5	15.5	26.8	41.0	88.5	160.0	251.0	372.0	540.0
	4.0	1.8	3.3	5.7	8.8	18.0	31.6	48.5	104.0	189.0	298.0	440.0	640.0
	5.0	2.0	3.8	6.5	10.0	20.8	35.8	55.0	119.0	214.0	339.0	500.0	730.0
	6.0	2.2	4.2	7.2	11.1	23.0	40.0	61.0	131.0	240.0	376.0	554.0	805.0
400	1.0	0.7	1.3	2.3	3.5	7.2	12.2	19.4	42.0	74.0	119.0	177.0	260.0
	2.0	1.0	2.0	3.4	5.2	11.0	18.6	29.0	62.0	102.0	179.0	262.0	386.0
	3.0	1.3	2.5	4.3	6.6	13.7	23.6	36.5	78.0	141.0	228.0	330.0	490.0
	4.0	1.5	3.0	5.0	7.8	16.1	28.0	43.0	83.0	168.0	270.0	395.0	580.0
	5.0	1.7	3.4	5.7	8.8	18.4	31.8	49.0	107.0	190.0	305.0	445.0	650.0
	6.0	1.9	3.7	6.4	9.8	20.5	35.5	55.0	119.0	211.0	340.0	500.0	735.0

NOTE: BOLD FACE FIGURES ARE MAXIMUM RECOMMENDED TOTAL PRESSURE DROPS. SHADED AREAS ARE FOR GENERAL INFORMATION ONLY.

Based on 40°F suction and 105°F condensing. For other conditions, apply the correction factor from Fig. R13-18 to the design tons before entering this table.

Source: Trane Company

FIGURE R13-17
Liquid line tonnage correction factors for R-12.

COND. TEMP.	EVAPORATOR TEMPERATURE																
	-30	-25	-20	-15	-10	-5	0	5	10	15	20	25	30	35	40	45	50
85	1.06	1.05	1.03	1.02	1.01	1.00	0.99	0.98	0.97	0.96	0.95	0.94	0.93	0.92	0.91	0.90	0.90
90	1.08	1.07	1.06	1.05	1.03	1.02	1.01	1.00	0.99	0.98	0.97	0.96	0.95	0.94	0.93	0.92	0.92
95	1.11	1.10	1.09	1.07	1.06	1.05	1.04	1.03	1.01	1.00	0.99	0.98	0.97	0.96	0.95	0.95	0.94
100	1.14	1.13	1.12	1.10	1.09	1.08	1.06	1.05	1.04	1.03	1.02	1.01	1.00	0.99	0.98	0.97	0.96
105	1.18	1.16	1.15	1.13	1.12	1.11	1.09	1.08	1.07	1.06	1.04	1.03	1.02	1.01	1.00	0.99	0.98
110	1.21	1.20	1.18	1.17	1.15	1.14	1.12	1.11	1.10	1.08	1.07	1.06	1.05	1.04	1.03	1.02	1.01
115	1.25	1.23	1.22	1.20	1.19	1.17	1.16	1.14	1.13	1.11	1.10	1.09	1.08	1.06	1.05	1.04	1.03
120	1.29	1.27	1.26	1.24	1.22	1.21	1.19	1.18	1.16	1.15	1.13	1.12	1.11	1.09	1.08	1.07	1.06
125	1.34	1.32	1.30	1.28	1.26	1.25	1.23	1.21	1.20	1.18	1.17	1.15	1.14	1.13	1.11	1.10	1.09
130	1.39	1.36	1.34	1.32	1.31	1.29	1.27	1.25	1.24	1.22	1.20	1.19	1.17	1.16	1.15	1.13	1.12
135	1.44	1.42	1.39	1.37	1.35	1.33	1.31	1.29	1.29	1.26	1.24	1.23	1.21	1.20	1.18	1.17	1.15
140	1.50	1.47	1.45	1.42	1.40	1.38	1.36	1.34	1.32	1.30	1.29	1.27	1.25	1.24	1.22	1.21	1.19
145	1.56	1.53	1.51	1.48	1.46	1.44	1.41	1.39	1.37	1.35	1.33	1.31	1.30	1.28	1.26	1.25	1.23

Multiply the design tons by the factor in this table to obtain the corrected tons for use in Fig. R13-16.
Source: Trane Company

performance and equipment life. Some systems also depend on the pressure drop in the liquid line to add to the pressure drop when using a capillary tube for the pressure-reduction device. In these cases, both the size and the length of the liquid line are important for peak performance. *Do not attempt to change either one.*

Figure R13-17 lists the correction factors (using R-12) when the liquid line must be sized for conditions other than standard. For the system operating at 45°F suction temperature and 125°F condensing temperature, the correction factor would be 1.06. Down the column under 45°F, evaporator temperature, the horizontal line from the 125°F condensing temperature, where the line and column cross, 1.06 is shown. The 5-ton load must then be multiplied by the 1.06 and the line sized on the basis of 5.3 tons.

Figure R13-18 lists the line capacity for lines carrying R-22. This table is used in the same fashion as Fig. R13-6. Figure R13-19 is the correction factor table for the R-22 liquid line at other than standard conditions.

R13-7.2
Vertical Lift

Liquid lines can present a problem if the design causes a radical temperature change or pressure drop, since refrigerant leaving the condenser remains a liquid only as long as its condensing temperature is higher than its sensible temperature. If a drop in pressure (drop in boiling point) or a rise in the sensible temperature were to reverse (boiling point fall below sensible temperature or sensible temperature rise above boiling point), preexpansion of the liquid would occur before it passes through the pressure-reducing device. The vapor formed

is called *preexpansion flash gas.* As discussed previously, pressured flash gas within the liquid line is very undesirable, since it displaces liquid at the port of the expansion valve, reducing its capacity. It also affects the capacity of capillary tubes very seriously, as the ability of the tube to carry vapor is considerably less than the ability to carry liquid. Liquid leaving the condenser is generally at a higher boiling point than the temperature of the surrounding air. However, as the liquid line, friction, and static losses do occur in properly sized lines, static loss is the most frequent cause of the creation of preexpansion flash gas.

For example, a standing column of liquid R-22 at 100°F and 210.6 psig exerts, due to its weight, a pressure of approximately 0.50 psi for every foot of height of the column (R-12 exerts 0.55 psi per foot of height); therefore, the pressure at the bottom of a 10-ft column of R-22 is 5 psi greater than the pressure at the top of the column. Conversely, for every 10 ft that R-22 is lifted in a vertical riser, the pressure at the top of the column is reduced by 5 psi.

Assume that the liquid R-22 leaves the condenser at standard conditions of 105°F condensing temperature with 3°F of subcooling. The pressure leaving the condenser would be 210.765 psig. Therefore, considering the 3°F of subcooling, the liquid leaves the condenser at 102°F. If the liquid-line riser has a 30-ft lift, the pressure at the top of the riser would be 15 psi less due to the weight of the refrigerant or 210.765 psig less 15 psig or 195.765 psig minus the line loss of 6 psig (if properly sized) for a final pressure of 189.765 psig. This will produce a boiling point of 97.86°F. Since the liquid left the condenser at 102°F, enough liquid will vaporize to cool the remaining liquid to the 97.86°F. To avoid the formation of flash gas in this riser, a liquid subcooling leav-

Tonnage capacities of lines delivering liquid R-22 from receiver to evaporator.

EQUIV. LENGTH IN FEET	TOTAL PRESS. DROP PSI	COPPER TUBE SIZE—OD—TYPE L												
		3/8	1/2	5/8	3/4	7/8	1 1/8	1 3/8	1 5/8	2 1/8	2 5/8	3 1/8	3 5/8	4 1/8
20	0.5	1.5	3.4	6.5	11.0	17.6	35.5	61.0	99.0	204.0	350.0	565.0	890.0	1155.0
	1	2.2	5.1	9.6	16.1	26.0	52.0	90.0	146.0	298.0	520.0	835.0	1310.0	1698.0
	2	3.2	7.6	14.2	23.9	38.0	76.0	133.0	212.0	436.0	768.0	1225.0	1920.0	2490.0
	3	4.1	9.7	18.0	30.0	48.0	96.0	166.0	268.0	548.0	970.0	1540.0	2408.0	3125.0
	4	4.7	11.4	21.1	35.0	56.0	112.0	197.0	322.0	640.0	1145.0	1810.0	2840.0	3670.0
	5	5.4	13.0	24.0	40.0	64.0	128.0	220.0	354.0	730.0	1300.0	2070.0	3225.0	4180.0
	6	5.9	14.3	26.8	44.5	71.0	141.0	245.0	385.0	810.0	1425.0	2300.0	3550.0	4650.0
	7	6.5	15.8	29.0	48.6	77.5	153.0	265.0	430.0	885.0	1560.0	2505.0	3890.0	5000.0
30	0.5	1.1	2.5	4.9	8.2	13.2	26.4	46.0	74.0	152.0	267.0	420.0	660.0	860.0
	1	1.7	3.8	7.3	12.1	19.9	39.0	68.0	110.0	222.0	395.0	625.0	970.0	1285.0
	2	2.5	5.7	10.9	17.9	29.1	56.8	100.0	160.0	330.0	578.0	915.0	1440.0	1880.0
	3	3.1	7.2	13.7	22.4	36.0	71.0	127.0	201.0	415.0	736.0	1160.0	1800.0	2360.0
	4	3.6	8.5	16.0	26.3	42.5	84.0	148.0	236.0	487.0	860.0	1370.0	2115.0	2780.0
	5	4.1	9.6	18.0	30.0	48.2	94.0	168.0	268.0	550.0	985.0	1530.0	2400.0	3110.0
	6	4.5	10.8	20.1	33.0	53.5	104.0	186.0	297.0	610.0	1080.0	1700.0	2670.0	3470.0
	7	5.0	11.8	22.0	36.0	58.2	114.0	202.0	324.0	665.0	1180.0	1870.0	2900.0	3800.0
40	0.5	1.0	2.3	4.3	7.3	11.6	23.1	40.1	65.0	136.0	235.0	376.0	590.0	778.0
	1	1.5	3.4	6.4	10.9	17.1	34.0	59.8	95.0	200.0	345.0	558.0	877.0	1140.0
	2	2.2	5.1	9.6	16.0	25.3	50.0	88.5	140.0	291.0	510.0	820.0	1290.0	1685.0
	3	2.8	6.4	12.1	20.0	32.0	63.0	110.0	178.0	366.0	640.0	1025.0	1600.0	2100.0
	4	3.3	7.6	14.2	23.4	37.5	74.0	130.0	210.0	430.0	755.0	1200.0	1900.0	2500.0
	5	3.7	8.6	16.1	26.5	42.0	84.0	148.0	236.0	487.0	860.0	1380.0	2140.0	2810.0
	6	4.1	9.5	18.0	29.6	47.0	93.0	163.0	260.0	540.0	950.0	1505.0	2380.0	3105.0
	7	4.5	10.5	19.6	32.1	51.0	100.0	179.0	281.0	589.0	1020.0	1650.0	2595.0	3400.0
50	1	1.2	2.8	5.4	8.8	14.2	28.0	49.0	79.0	161.0	283.0	460.0	720.0	950.0
	2	1.8	4.2	7.9	13.0	21.0	41.0	73.0	116.0	240.0	420.0	678.0	1065.0	1200.0
	3	2.2	5.4	9.9	16.4	26.3	52.0	91.0	146.0	300.0	530.0	845.0	1320.0	1400.0
	4	2.7	6.3	11.8	19.3	31.0	61.0	108.0	170.0	350.0	625.0	995.0	1570.0	1760.0
	5	3.0	7.2	13.2	21.9	35.0	69.0	121.0	192.0	400.0	710.0	1130.0	1798.0	2080.0
	6	3.3	8.0	14.9	24.0	39.0	76.0	136.0	213.0	440.0	795.0	1270.0	1990.0	2350.0
	7	3.6	8.8	16.0	26.2	42.5	83.0	148.0	232.0	479.0	860.0	1380.0	2125.0	2820.0
75	1	1.0	2.4	4.4	7.4	12.0	23.4	41.0	66.0	133.0	233.0	370.0	587.0	770.0
	2	1.5	3.5	5.6	10.9	17.7	34.6	60.0	97.0	198.0	348.0	545.0	860.0	1125.0
	3	1.9	4.5	8.2	13.7	22.0	43.0	75.0	121.0	249.0	435.0	690.0	1100.0	1405.0
	4	2.3	5.3	9.7	16.0	25.8	51.0	88.0	142.0	289.0	510.0	810.0	1290.0	1670.0
	5	2.6	6.0	11.0	18.1	29.3	57.5	100.0	160.0	330.0	580.0	920.0	1460.0	1900.0
	6	2.8	6.6	12.1	20.0	32.4	63.8	111.0	179.0	360.0	642.0	1008.0	1600.0	2100.0
	7	3.1	7.3	13.2	21.9	35.2	69.0	121.0	195.0	394.0	700.0	1105.0	1730.0	2300.0
100	2	1.3	2.9	5.6	9.2	14.9	29.0	51.0	81.0	168.0	295.0	470.0	740.0	980.0
	3	1.6	3.7	7.0	11.6	18.7	36.5	64.0	102.0	210.0	365.0	590.0	925.0	1210.0
	4	1.9	4.4	8.2	13.5	22.0	43.0	75.0	121.0	250.0	435.0	695.0	1095.0	1420.0
	5	2.1	5.0	9.3	15.4	25.0	49.0	85.0	138.0	280.0	495.0	790.0	1220.0	1605.0
	6	2.4	5.6	10.3	17.0	27.4	54.0	94.0	151.0	310.0	542.0	865.0	1360.0	1800.0
	7	2.6	6.1	11.1	18.6	30.0	59.0	102.0	165.0	340.0	600.0	955.0	1499.0	1985.0
125	2	1.1	2.6	4.8	8.0	12.9	25.0	43.5	70.0	147.0	256.0	410.0	640.0	840.0
	3	1.4	3.2	6.0	10.0	16.0	31.4	55.0	88.0	182.0	320.0	510.0	800.0	1055.0
	4	1.6	3.8	7.2	11.9	19.0	37.0	64.0	103.0	213.0	375.0	600.0	940.0	1220.0
	5	1.8	4.4	8.1	12.4	21.2	42.0	74.0	118.0	241.0	430.0	680.0	1060.0	1400.0
	6	2.0	4.9	9.0	14.9	23.8	46.5	81.0	165.0	270.0	475.0	750.0	1185.0	1540.0
	7	2.2	5.3	9.8	16.1	25.9	50.0	89.0	141.0	290.0	520.0	820.0	1290.0	1690.0
150	2	1.0	2.4	4.4	7.4	11.9	23.0	41.0	65.0	133.0	233.0	375.0	588.0	780.0
	3	1.3	3.0	5.6	9.4	15.0	29.0	51.0	81.5	168.0	298.0	470.0	740.0	985.0
	4	1.5	3.5	6.6	11.0	17.4	34.0	60.0	96.0	197.0	346.0	558.0	867.0	1150.0
	5	1.7	4.1	7.5	12.6	20.0	38.8	68.5	110.0	222.0	395.0	630.0	980.0	1300.0
	6	2.0	4.5	8.3	13.9	21.9	43.0	76.0	120.0	247.0	435.0	700.0	1095.0	1445.0
	7	2.1	4.9	9.0	15.0	24.0	47.0	82.0	131.0	269.0	470.0	760.0	1185.0	1580.0
175	2	0.9	2.1	3.9	6.4	10.4	20.6	35.5	57.0	118.0	208.0	330.0	518.0	680.0
	3	1.1	2.6	4.9	8.2	13.0	25.8	45.0	72.0	149.0	260.0	415.0	642.0	850.0
	4	1.3	3.1	5.8	9.6	15.3	30.0	53.0	84.0	171.0	305.0	486.0	760.0	1000.0
	5	1.5	3.5	6.6	10.9	17.2	34.0	60.0	95.0	195.0	350.0	546.0	855.0	1120.0
	6	1.6	3.9	6.6	12.0	19.1	38.0	66.0	105.0	216.0	385.0	610.0	950.0	1260.0
	7	1.8	4.3	8.0	14.2	21.0	41.0	72.0	115.0	238.0	420.0	660.0	1030.0	1370.0
200	2	0.8	2.0	3.7	6.2	10.0	20.0	35.0	55.0	115.0	200.0	320.0	500.0	660.0
	3	1.1	2.5	4.7	7.8	12.8	25.0	43.0	69.0	142.0	250.0	400.0	620.0	820.0
	4	1.3	3.0	5.6	9.2	15.0	29.4	51.0	82.0	169.0	300.0	470.0	735.0	960.0
	5	1.4	3.4	6.3	10.5	17.0	33.0	58.0	93.0	190.0	336.0	540.0	830.0	1100.0
	6	1.6	3.8	7.0	11.6	18.9	37.0	64.0	101.0	210.0	370.0	600.0	925.0	1210.0
	7	1.7	4.2	7.6	12.7	20.4	40.0	70.0	111.0	230.0	405.0	640.0	1000.0	1325.0
250	2	0.8	1.8	3.4	5.6	9.0	18.0	31.5	50.0	102.0	180.0	290.0	450.0	590.0
	3	0.9	2.3	4.2	7.0	11.3	22.5	39.0	63.0	129.0	228.0	360.0	560.0	740.0
	4	1.1	2.7	5.0	8.3	13.4	26.0	46.0	74.0	151.0	267.0	425.0	660.0	860.0
	5	1.3	3.1	5.6	9.4	15.1	30.0	52.0	84.0	172.0	302.0	480.0	750.0	990.0
	6	1.4	3.4	6.2	10.4	16.9	33.0	58.0	93.0	190.0	335.0	540.0	840.0	1100.0
	7	1.5	3.8	6.8	11.3	18.2	35.7	63.0	101.0	209.0	365.0	580.0	915.0	1200.0
	8	1.6	4.1	7.4	12.2	19.9	38.5	68.5	110.0	225.0	396.0	630.0	988.0	1300.0
300	2	0.7	1.6	3.0	5.0	8.0	16.0	28.0	44.0	91.5	160.0	227.0	400.0	525.0
	3	0.9	2.0	3.8	6.3	10.0	20.0	35.0	55.0	115.0	200.0	320.0	500.0	660.0
	4	1.0	2.4	4.5	7.4	12.0	23.1	41.0	65.0	134.0	238.0	380.0	595.0	765.0
	5	1.2	2.7	5.1	8.4	13.4	26.0	46.0	74.0	151.0	268.0	430.0	664.0	865.0
	6	1.3	3.0	5.6	9.3	14.9	29.0	51.0	82.0	168.0	299.0	470.0	740.0	960.0
	7	1.4	3.4	6.1	10.0	16.1	32.0	56.0	88.0	180.0	320.0	510.0	800.0	1050.0
	8	1.5	3.6	6.6	11.0	17.3	34.0	60.0	96.0	197.0	348.0	558.0	867.0	1135.0
350	2	0.6	1.5	2.8	4.7	7.4	14.9	25.8	40.5	83.0	148.0	232.0	363.0	480.0
	3	0.8	1.9	3.5	5.9	9.4	18.3	32.0	51.0	105.0	185.0	295.0	460.0	600.0
	4	0.9	2.2	4.2	6.9	11.0	21.5	38.0	60.0	123.0	218.0	347.0	540.0	708.0
	5	1.1	2.5	4.8	7.8	12.4	24.3	42.0	67.8	140.0	247.0	390.0	610.0	802.0
	6	1.2	2.8	5.2	8.6	13.8	27.0	47.0	75.0	152.0	270.0	430.0	678.0	899.0
	7	1.3	3.1	5.6	9.4	15.0	29.2	51.0	81.0	169.0	298.0	470.0	740.0	990.0
	8	1.4	3.3	6.1	10.0	16.1	31.6	54.8	88.0	180.0	318.0	508.0	897.0	1045.0
400	2	0.6	1.4	2.5	4.2	6.9	13.5	23.3	38.0	77.0	136.0	219.0	340.0	450.0
	3	0.7	1.7	3.2	5.3	8.6	17.0	29.5	47.0	98.0	171.0	272.0	430.0	560.0
	4	0.9	2.0	3.8	6.3	10.1	20.0	34.6	53.0	115.0	202.0	320.0	500.0	658.0
	5	1.0	2.3	4.3	7.1	11.5	22.3	39.5	63.0	130.0	230.0	360.0	570.0	740.0
	6	1.1	2.6	4.8	7.9	12.7	25.0	43.0	70.0	144.0	252.0	400.0	635.0	820.0
	7	1.2	2.8	5.2	8.6	13.9	27.0	47.0	76.0	158.0	276.0	440.0	690.0	895.0
	8	1.3	3.1	5.6	9.2	15.0	29.2	51.0	82.0	170.0	299.0	472.0	747.0	962.0

NOTE: BOLD FACE FIGURES ARE MAXIMUM RECOMMENDED TOTAL PRESSURE DROPS. SHADED AREAS ARE FOR GENERAL INFORMATION ONLY.

Based on 40°F suction and 105°F condensing. For other conditions, apply the correction factor from Fig. R13-19 design tons before entering this table.

Source: Trane Company

COND. TEMP.	SUCTION TEMPERATURE																
	-30	-25	-20	-15	-10	-5	0	5	10	15	20	25	30	35	40	45	50
85	1.01	1.00	0.99	0.99	0.98	0.97	0.96	0.95	0.95	0.94	0.93	0.93	0.92	0.91	0.91	0.90	0.90
90	1.04	1.03	1.02	1.01	1.00	0.99	0.99	0.98	0.97	0.96	0.96	0.95	0.94	0.94	0.93	0.92	0.92
95	1.07	1.06	1.05	1.04	1.03	1.02	1.01	1.00	1.00	0.99	0.98	0.97	0.97	0.96	0.95	0.95	0.94
100	1.09	1.09	1.08	1.07	1.06	1.05	1.04	1.03	1.02	1.01	1.00	1.00	0.99	0.98	0.98	0.97	0.96
105	1.13	1.12	1.11	1.09	1.08	1.08	1.07	1.06	1.05	1.04	1.03	1.02	1.02	1.01	1.00	0.99	0.99
110	1.16	1.15	1.14	1.13	1.12	1.11	1.10	1.09	1.08	1.07	1.06	1.05	1.04	1.03	1.03	1.02	1.01
115	1.20	1.18	1.17	1.16	1.15	1.14	1.13	1.12	1.11	1.10	1.09	1.08	1.07	1.06	1.06	1.05	1.04
120	1.24	1.22	1.21	1.20	1.19	1.17	1.16	1.15	1.14	1.13	1.12	1.11	1.10	1.09	1.09	1.08	1.07
125	1.28	1.27	1.25	1.24	1.23	1.22	1.20	1.19	1.18	1.17	1.16	1.15	1.14	1.13	1.12	1.11	1.10
130	1.33	1.31	1.30	1.28	1.27	1.25	1.24	1.23	1.22	1.21	1.19	1.18	1.17	1.16	1.15	1.15	1.14
135	1.38	1.36	1.35	1.33	1.32	1.30	1.29	1.28	1.26	1.25	1.24	1.23	1.22	1.20	1.19	1.19	1.18
140	1.43	1.42	1.40	1.38	1.37	1.35	1.34	1.32	1.31	1.30	1.28	1.27	1.26	1.25	1.24	1.23	1.22
145	1.50	1.48	1.46	1.44	1.43	1.41	1.39	1.38	1.36	1.35	1.33	1.32	1.31	1.30	1.28	1.27	1.26

Applicable to Fig. R13-11.

Source: Trane Company

ing the condenser of 8°F or more is required instead of 3°F to compensate for lift and presure drop in the line.

Again, assume that the same conditions are applied to a typical air-conditioning unit which works at peak efficiency with 19°F of subcooling. Instead of leaving the condenser at 102°F the liquid would leave at 86°F. Again, assuming 6 psig line loss for a properly sized line, the 30 ft of vertical lift would not produce flash gas. The vertical lift could increase to 105 ft before flash gas would occur.

This demonstrates the value of a proper amount of subcooling. It not only provides latitude for the designer when laying out the liquid line, but the proper amount of subcooling provides the maximum system capacity at the lowest operating cost.

R13-7.3
Liquid-Line Insulation

Normally, no insulation is used on the liquid line because the greater the heat loss from the line, the lower the temperature of the liquid entering the pressure-reducing device, and the less system flash gas produced. However, when the line is installed in a higher-than-normal ambient, such as carrying liquid refrigerant to an evaporator in an air handler located in an attic space, the temperature gain of the liquid could produce liquid temperatures in excess of the condensing temperature, if this were possible. Because it is not possible for the sensible temperature of a liquid to exceed the boiling point, vaporization or boiling takes place and produces preexpansion. Insulation on both the suction and liquid lines are required.

Another very common cause of preexpansion flash gas production is laying a liquid line on the roof of a building, where it is on tar that has been heated by the sun. Refrigerant lines should be raised at least 18 in. above such a roof to minimize the effects of the sun.

R13-8
CONDENSER DRAIN LINE

The line between a condenser and a liquid receiver, called the condenser drain line, must be carefully sized. Although it is almost impossible to oversize the line (0 psig pressure drop is desirable), undersizing or too long a line is to be avoided. An undersized or excessively long line can restrict the flow of refrigerant to the thermostatic expansion valve to the extent that some will be held back in the condenser, which reduces the effective condenser surface and reduces condenser capacity. This causes the head pressure to rise, decreasing the overall system capacity. At the same time, power requirements and operating costs increase.

For proper operation the length of the line between the condenser must be kept as short as possible, preferably not more than 18 in. The top of the receiver should be level with or below the bottom of the condenser. This eliminates the need for the liquid to be forced uphill to enter the condenser.

These rules apply to standard installations of receivers. Where special requirements exist for capacity control, low ambient operation, etc., consult the Trane Company manual *Reciprocating Refrigeration* or the literature furnished by the equipment manufacturer.

R13-8.1
Condenser Drain Line Insulation

Except in very special applications, the condenser drain line is not insulated.

Refrigeration Piping   157

R13-1. Piping that connects the four major components of the system has two major functions. What are they?

R13-2. What are the five basic rules for proper piping practice?

R13-3. Is refrigerant in vapor form a good or a poor carrier of oil?

R13-4. The minimum gas velocities through horizontal and vertical suction lines are ——————— and ———————.

R13-5. What are the maximum recommended pressure drops in the suction line for systems using R-12, R-22, and R-504?

R13-6. Suction line traps are required on all DX coils using TX valves. True of False?

R13-7. In what situation are additional traps required in the suction line?

R13-8. In what types of systems are double trap and riser systems required?

R13-9. Traps should also be used in systems using capillary tubes. True or False?

R13-10. Suction-line insulation is an absolute requirement. True or False?

R13-11. The maximum pressure drop allowable in hot-gas lines is ——————— psig.

R13-12. Is insulation ever required on the hot-gas line?

R13-13. What is the maximum pressure drop allowable in the liquid line?

R13-14. Define "preexpansion flash gas."

R13-15. What causes preexpansion flash gas?

R13-16. When is insulation required on the liquid line?

R14

Accessories

R14-1
GENERAL

Earlier, the term *accessory* was used to describe an oil separator, which is only one of many devices used in refrigeration work. A refrigeration accessory is an article or device that adds to the convenience or effectiveness of the system. The essential items of the basic refrigeration system are the compressor, condenser, metering device, and evaporator. An accessory gives the basic system certain conveniences or allows it to attain a degree of performance that is impractical or virtually impossible with commercially available basic system components. These accessories come in many shapes and sizes and serve many functions, but their primary purpose is to add to the convenience or effectiveness of the refrigeration system.

Following is a list of common accessories (there are others, but they are so specialized or so infrequently used that they will not be listed).

Oil separator	Moisture indicator
Muffler	Water valve
Heat exchanger	Solenoid valve
Strainer-drier	Check valve
Suction accumulator	Evaporator pressure regulator
Crankcase heaters	Relief valve
Sight glass	Fusible plugs

R14-2
OIL SEPARATOR

The purpose of the oil separator (Fig. R14-1) is to reduce the amount of oil in circulation through the sys-

tem and thereby increase efficiency. All refrigeration systems will have some oil passing through them. In some cases the amount of oil in circulation can affect the evaporator heat-transfer characteristics, create false float action, or even affect expansion valve operation. In these cases, an oil separator, by reducing the oil circulating within the system, can improve evaporator efficiency or reduce float and valve problems.

Systems with improperly sized or trapped lines quite frequently do not return oil to the compressor, thus creating compressor lubrication problems. The insertion of an oil separator into such a system will not correct these problems. An oil separator is not 100% efficient and some oil will pass through it. The installation of an oil separator on a system that is trapping oil will merely delay the problem; it will not in any way solve the problem.

In the separator illustrated, the hot gas–oil mixture from the compressor enters at the left and flows down and out through the perforated pipe. The mixture strikes against the screen, where the oil usually separates from the gas. The oil drains down the screen into the small

FIGURE R14-1 Oil separator. (*Courtesy* Carrier Air-Conditioning Company)

159

sump at the bottom of the separator. The gas passes through the screen and leaves the separator at the upper right. When the oil level rises in the sump, the float ball is raised and the oil is returned through an orifice to the crankcase.

The separator is normally placed in the discharge line as close to the compressor as practical. The oil return line leads directly to the crankcase. The internal construction of the separator will vary greatly; however, its external appearance and its location in the system make the oil separator comparatively easy to identify.

When an oil separator is used, certain precautions must be taken. A cool separator will condense refrigerant gas into liquid, which, if allowed to return to the compressor crankcase, can cause damage to the compressor. Care should also be taken to keep the float orifice clean, as this orifice is subject to any oil sludge that might leave the compressor discharge. If the float were to stick open, hot gas might flow into the compressor crankcase. If the float were to stick closed, no oil would be returned to the compressor.

R14-3
MUFFLER

The purpose of the muffler (Fig. R14-2) is to dampen or remove the hot-gas pulsations set up by a reciprocating compressor. Every reciprocating compressor will create some hot-gas pulsations. Although a great effort is made to minimize pulsations in system and compressor design, gas pulsation can be severe enough to create two closely related problems. The first is noise, which, although irritating to users of the equipment, does not necessarily have a harmful effect on the system. The second problem is vibration, which can result in line breakage. Frequently, these problems occur simultaneously.

The illustration is a good example of muffler construction. It is designed to eliminate gas pulsations by allowing them to dissipate within the muffler itself. The pulsations force the hot gas through the holes in the pipe and into the chambers, where they are absorbed.

The muffler is inserted in the discharge line as close to the compressor as is practical. In welded hermetic compressors, the muffler is frequently within the compressor shell itself. Because it is usually constructed within the shell, the muffler is a natural trap. It will easily trap oil and may also trap liquid refrigerant. The muffler should be installed either in a downward flow or a horizontal line.

Selecting the proper size and location for a muffler can be a difficult engineering problem. Improper field selection of a muffler will sometimes lead to a condition of increased rather than decreased vibration.

R14-4
HEAT EXCHANGER

The heat exchanger (Fig. R14-3) is a device used to transfer heat from the liquid refrigerant to the suction gas. The exchanger has two principal uses. The first is to reduce the temperature or subcool the liquid refrigerant flowing from the condenser to the metering device. This reduction in temperature is necessary in systems that have high pressure drops to prevent flashing in the liquid line. These pressure drops may be caused by excessively long lines or by lengthy liquid risers.

The second use of the exchanger is to assure that the suction gas flowing into the compressor is dry. On systems with rapid load fluctuations, it is not unusual to find liquid "slopping" over from the evaporator. The heat interchanger will allow lower evaporator superheat settings, since some liquid carryover need not be considered dangerous.

FIGURE R14-2 Muffler. (*Courtesy* Carrier Air-Conditioning Company)

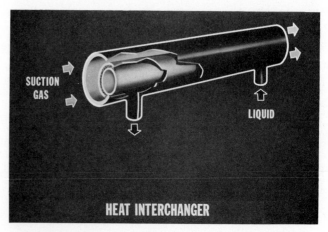

FIGURE R14-3 Heat interchanger. (*Courtesy* Carrier Air-Conditioning Company)

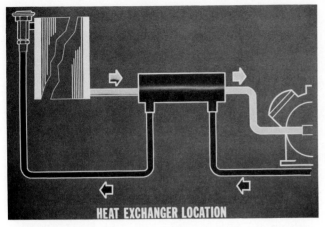

FIGURE R14-4 Heat interchanger location. (*Courtesy* Carrier Air-Conditioning Company)

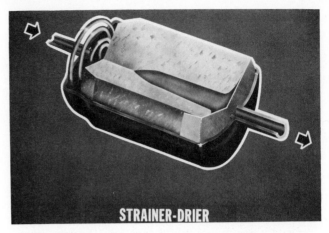

FIGURE R14-5 Strainer-drier. (*Courtesy* Carrier Air-Conditioning Company)

The triple-pipe heat exchanger illustrates the principle of the heat interchanger. It is a counterflow device with warm liquid entering at the right and wet or cool gas entering at the left. As the liquid is at a higher temperature than the gas, the heat will flow out of the liquid and into the gas. This subcools the liquid to reduce flashing and superheats the gas to prevent flooding. Other forms of heat interchangers are the double-pipe and the shell-and-coil.

The location of the heat exchanger (Fig. R14-4) will depend on its intended usage. If the heat exchanger is being used for the purpose of subcooling the liquid, it will be found as close to the condenser as practical piping practices will allow. A heat exchanger that is being used as protection against slopover may be mounted in the suction line quite close to the evaporator. Because both liquid and suction lines must be brought to the heat exchanger, equipment layout may have more bearing on the position of the heat exchanger than either of the other two factors.

Although heat exchangers serve many useful purposes, care must be exercised in their use, particularly with hermetic compressors. Hermetic compressor motors are cooled by suction gas and frequently have well-defined upper limits on suction gas temperatures. Because heat exchangers tend to increase suction gas temperatures, indiscriminate field use without proper engineering investigation can be dangerous.

R14-5
STRAINER-DRIER

The dangers of moisture within a refrigeration system have already been discussed. It will simply be repeated that no moisture should be present, particularly in those systems using halogenated hydrocarbon refrigerants. However, if moisture does enter the refrigeration system, it must be removed. One method of removing moisture is the strainer-drier shown in Fig. R14-5. This accessory consists of a shell through which the liquid refrigerant will pass. Inside the shell is a material known as a desiccant. As the moisture-laden refrigerant passes through the drier, the desiccant removes a portion of the moisture. Each passage through the drier removes additional moisture until the refrigerant is sufficiently dry or until the drier has reached its moisture-holding capacity. When this happens, the drier must be replaced.

The strainer-drier also performs a second service by filtering any solid particles from the flowing liquid refrigerant. These particles are filtered out by the desiccant core.

The strainer-drier is almost always found in the liquid line of the refrigeration system, as in Fig. R14-6. Because the volume of the liquid is much smaller than that of the gas, a smaller drier may be used. This, of course, results in lower cost. Also, the metering device

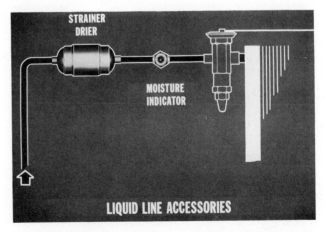

FIGURE R14-6 Liquid-line accessories. (*Courtesy* Carrier Air-Conditioning Company)

is protected from solid particles when the strainer is in this location.

Two important facts should be noted about driers. The first is that they should be replaced when the desiccant is saturated. The second is that they create some pressure drop; therefore, they must be properly sized to prevent excessive pressure drop with resulting flash gas.

R14-6
SUCTION ACCUMULATOR

The suction accumulator (Fig. R14-7) is a simple device serving a very useful function. On some evaporators, the metering device action is not rapid enough to maintain pace with load changes. Also, capillary tubes are not designed to "shut off" under light evaporator loads. In both cases some liquid will occasionally leave the evaporator through the suction line. This could damage the compressor. The accumulator is nothing more than a trap to catch this liquid before it can reach the compressor. This surplus liquid is boiled or evaporated in the trap and returned to the compressor as a gas.

Sometimes space limitations require that the whole evaporator be used for boiling liquid. Here the suction accumulator can be used to advantage, as it allows full use of the evaporator without fear of spilling liquid into the compressor suction part. Its construction is not complicated. It consists of a container to collect and evaporate liquid refrigerant. The use of the liquid line to obtain heat to evaporate the liquid and the oil-return line are optional features.

The suction accumulator is generally found quite close to the evaporator from which the liquid comes (Fig. R14-8). Occasionally, it will be found in the main suction line of a multievaporator system, thereby protecting against slopover from all evaporators.

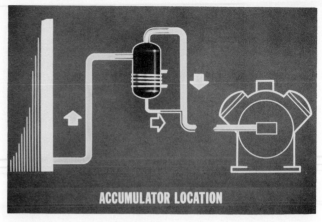

FIGURE R14-8 Accumulator location. (*Courtesy* Carrier Air-Conditioning Company)

The suction accumulator must be properly sized or it may fill with liquid and cause compressor damage. There must also be provisions for oil return; this is essential, as it is a natural oil trap. The suction accumulator is used primarily in packaged equipment using a capillary feed.

R14-7
CRANKCASE HEATER

The crankcase heater (Fig. R14-9) is used to prevent the accumulation of liquid refrigerant in the compressor crankcase during shutdown. Due to the affinity of oil for refrigerant, refrigerant may migrate to the crankcase during shutdown. Condensation of refrigerant in the compressor crankcase may also occur when the temperature of the compressor is lower than that of the rest of the system.

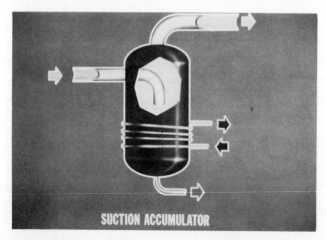

FIGURE R14-7 Suction accumulator. (*Courtesy* Carrier Air-Conditioning Company)

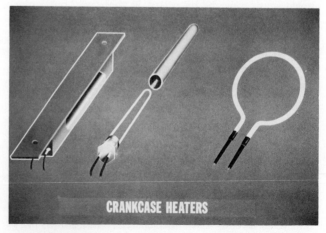

FIGURE R14-9 Crankcase heaters. (*Courtesy* Carrier Air-Conditioning Company)

If refrigerant does gather in the crankcase, a reduction in suction pressure at startup will cause the liquid to evaporate or boil. This boiling causes the oil-refrigerant mixture to foam, and some of this foam leaves the crankcase and passes into the cylinder. There it may create liquid slugs, which will damage the valves. The refrigerant also reduces the lubricating value of the oil.

Accumulation of liquid in the crankcase during shutdown can be minimized by keeping the crankcase warmer than the rest of the system. The crankcase heater is designed for that purpose. Figure R14-9 shows three heaters designed for three different applications. The heater on the left is fastened to the bottom of the crankcase, the heater in the center may be inserted into the crankcase, and the one on the right is designed to be wrapped around the crankcase.

Compressors that perform well in low-temperature work, low-ambient operation, or heat pump application where cold oil conditions are encountered, will most likely be shipped from the factory with the crankcase heat installed. Or, the compressor will have a plugged opening (threaded) into which a heater may be inserted during installation.

R14-8
SIGHT GLASS

The sight glass (Fig. R14-10) in a refrigeration system permits the installer or serviceman to observe the condition of the refrigerant at that particular point. The sight glass usually consists of a glass opening in the liquid line of the system. Frequently, a glass on each side of the line is used to assure illumination of the line.

When the line is completely filled with liquid refrigerant, there is almost no obstruction when looking through the line. However, if any gas is in the liquid

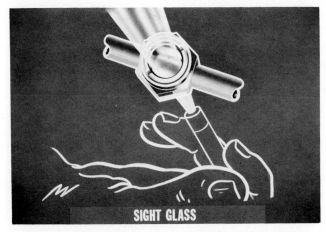

FIGURE R14-10 Sight glass. (*Courtesy* Carrier Air-Conditioning Company)

FIGURE R14-11 See-All moisture and liquid indicator. (*Courtesy* Sporlan Valve Company)

line, it will show immediately in the form of bubbles passing the sight glass. It should be pointed out that in a sight glass of this type, the glass will also show clear when only gas is present.

At first glance, the moisture indicator (Fig. R14-10) may appear to be a simple sight glass, a natural assumption since the indicator is normally a part of the sight glass. Contained within the glass, but exposed to the liquid refrigerant, is a small colored dot as shown in Fig. R14-11. This dot is the moisture indicator. It is of a special chemical composition that will change color depending on the amount of moisture in the refrigerant. When the amount of moisture is within the limits set by the manufacturer, the dot will be one color. However, if too much moisture is present, the dot will change color. When a maintenance or serviceperson sees that the color of the dot has changed, he or she knows that steps must be taken to remove the moisture from the system before harm is done. The moisture indicator will not be accurate unless it is completely filled with liquid refrigerant.

Depending on the purpose for which it is intended, the sight glass or moisture indicator may be found in more than one location. If the glass is to be used as an aid in determining whether the system is properly charged, it will be found at the condenser or receiver outlet. If it is to be used to assure that no flash gas is present prior to the liquid entering the metering device, it will be found immediately before the metering device, as shown.

R14-9
WATER REGULATING VALVE

The water valve (Fig. R14-12) is designed to control head pressure by governing the flow of water to a water-cooled condenser. The action of the water valve is controlled by the refrigerant pressure within the condenser. The gas in the condenser enters the valve at the bottom and exerts a pressure on the bellows. As the pressure increases it opens the valve to allow more water to flow. As pressure decreases the bellows contracts,

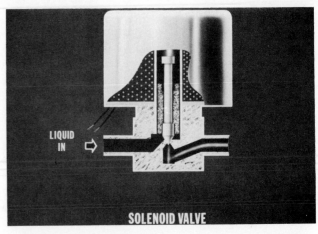

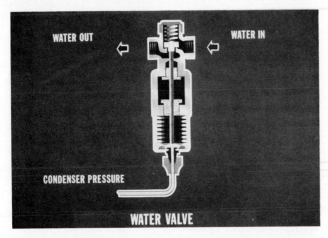

FIGURE R14-12 Water valve. (*Courtesy* Carrier Air-Conditioning Company)

FIGURE R14-13 Solenoid valve. (*Courtesy* Carrier Air-Conditioning Company)

closing the valve port. The proper water flow and condenser pressure are adjusted by setting the spring tension in the valve head.

Installation is simple but care should be taken that the direction of water flow through the valve is correct. The water valve is usually virtually trouble free; an occasional change of seat is all that should be required.

R14-10
SOLENOID VALVE

The solenoid valve may be used to control the flow of a gas or a liquid. It is frequently used in refrigerant liquid lines to control the flow of refrigerant to the evaporator. It is sometimes used in suction lines to isolate the evaporators of multievaporator, two-temperature systems. Another important use of the solenoid is as a pilot for a much larger valve.

Solenoid valves, like the one in Fig. R14-13, are used in many ways to control fluid flow, but one of the most common is a refrigerant liquid-line valve. Here a solenoid valve is used to stop the flow of liquid to the evaporator when refrigeration requirements are satisfied. When refrigeration is again required, the solenoid valve is opened and the liquid flows to the evaporator.

The operating principle of a solenoid valve is simple. A coil of wire is placed around a tube containing a movable plunger. When current flows through the wire a magnetic field is created. This pulls the plunger up into the tube. As the plunger rises, it opens the valve and allows liquid to flow.

The valve shown in Fig. R14-13 is a normally closed valve, because the valve is in the closed position unless current is flowing. Solenoid valves are also made in the normally open type. These valves will close only when the coil is energized.

The solenoid valve is an extremely dependable device when installed as recommended by the manufacturer. Poor installation practices such as failing to set the valve up right, poor wiring practices, and warping the valve body by excessive heat during brazing can make a solenoid valve a source of constant trouble. Proper installation is thus most important for this device.

R14-11
CHECK VALVE

The check valve (Fig. R14-14) is designed to allow the flow of liquid or gas in one direction only. When the fluid flows in the direction of the arrow, the force of the fluid will lift the gate from its seat and allow the fluid through. When the fluid tries to flow in the opposite direction, the gate closes, thereby stopping the flow. This device is useful in preventing the return of liquid to a

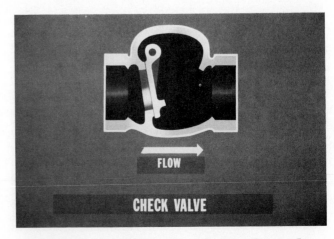

FIGURE R14-14 Check valve. (*Courtesy* Carrier Air-Conditioning Company)

compressor through the discharge line during shutdown. It is also found in reverse-cycle systems and is frequently inserted in the suction lines on two-temperature systems to prevent equalizing of pressures during shutdown.

The check valve has many uses and is considered a reasonably trouble-free device. By its very nature, however, it creates a pressure drop within the line in which it is used. Therefore, it should not be used indiscriminately.

R14-12
EVAPORATOR PRESSURE REGULATOR

The evaporator pressure regulator is frequently referred to as a *back-pressure regulator*. It is designed to maintain a constant pressure or temperature in the evaporator regardless of how low the compressor suction pressure falls. It is always found in the suction line as close to the evaporator as is practical.

Figure R14-15 is an illustration of a back-pressure valve or evaporator pressure regulator. This valve is designed to maintain a constant pressure or temperature in the evaporator regardless of the compressor suction pressure. Evaporator pressure enters the port at the left. As the evaporator pressure increases the pressure under the bellows increases. This pressure overcomes the spring tension and raises the valve plunger from the seat. This allows evaporator gases to flow to the suction line. As the gas leaves the evaporator the pressure in the evaporator is reduced. As the pressure under the bellows decreases, the spring pressure closes the valve. The operating pressure at the valve may be regulated by changing the spring pressure.

This type of valve is useful where control of temperature and humidity is important. It assures that the evaporator will not drop below a specified temperature.

It is also used where two or more evaporators on the same system are operating at different temperatures.

Another type of back-pressure valve frequently found in larger systems is the pilot-operated valve shown in Fig. R14-16. Though it works on the same principle as the valve previously discussed, the control point is not at the valve itself. In this control, the control pressure enters through the external pilot connection. As the pressure increases, it forces the seat of the pressure pilot valve up away from the pilot port. This allows the gas to flow through the pilot port down to the top surface of the piston, creating sufficient pressure to lower the piston and open the valve. As the pressure at the external pilot decreases, the pressure pilot adjusting spring closes the valve and shuts off the pilot port. The gas pressure on top of the piston then bleeds through the piston orifice, and the main spring forces the piston up, closing the valve. The pressure to operate this valve

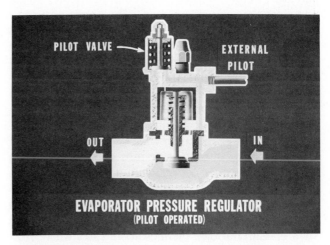

FIGURE R14-16 Evaporative pressure regulator—pilot operation. (*Courtesy* Carrier Air-Conditioning Company)

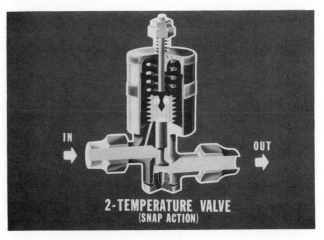

FIGURE R14-17 Two-temperature valve. (*Courtesy* Carrier Air-Conditioning Company)

FIGURE R14-15 Evaporative pressure regulator. (*Courtesy* Carrier Air-Conditioning Company)

should be taken from the evaporator because it gives better performance than the suction line.

The third type of evaporator pressure regulator is the *snap-action* or *two-temperature valve* (Fig. R14-17). This valve operates on the same principle as the two valves previously discussed. However, the mechanical action is designed so that the valve will be fully opened or fully closed.

The gas at evaporator pressure flows through the body orifice into the bellows. As the pressure increases, the bellows expands and moves the diamond-shaped finger up. As the finger moves up, the spring-loaded wheels ride up on the finger to the maximum finger width. If the finger continues to tilt, the wheels go over the wide point and the spring pressure tends to force them together. This force causes the wheels to ride down the diamond slope to push the valve plunger down and open the valve. The valve will stay open until the evaporator pressure is reduced enough to let the finger-wide point drop below the wheels, at which time the spring pressure, by forcing the wheels together, will lift the plunger and close the valve.

The valve is used in two-temperature applications where there are two or more evaporators operating on the same compressor. The two-temperature valve is always placed in the suction line of the highest temperature evaporator. It is so set that it will maintain a higher average evaporator pressure or temperature in this evaporator than will be maintained in the evaporator connected directly to the compressor suction.

The most important feature of the snap-action valve is that it does not maintain a constant pressure. Since the valve is wide open or closed, it cycles the evaporator in the same manner as it would were it directly connected to its individual condensing unit, controlled by a pressurestat or a thermostat. In cases where frost may be a problem, the high temperature in the evaporator can be adjusted so as to be above freezing, thus providing automatic defrost.

R14-13
RELIEF VALVE

The relief valve is designed to protect the system against refrigerant pressures great enough to do physical damage. The valve shown in Fig. R14-18 is held closed by a spring. The valve will automatically close when the pressure has been reduced to within the required limits, and will remain closed until high pressure again requires relief. If the valve does open there will naturally be a loss of refrigerant, and the system subsequently must be checked. Since these valves are often required by law, they often have some type of seal device so that they cannot be tampered with by unauthorized persons.

R14-14
SAFETY FITTINGS

The fusible plug or rupture disk fittings, illustrated in Fig. R14-19, are not literally controls, but are safety devices designed to protect the system against extreme pressures. The fusible-plug fitting on the left contains a core of soft metal with a low melting point. In case of fire, the soft metal would melt and allow the gas to escape to the atmosphere before hazardous pressures could be built up.

The rupture-disk fitting on the right in Fig. R14-19 is designed to serve the same purpose as the fusible plug, but in a different manner. The rupture disk consists of a thin piece of metal that is meant to break or rupture at a pressure below that which might create a dangerous condition.

Many shell vessels, such as water-cooled condensers and receivers, are factory equipped with fusible plugs or relief devices, since the systems are located indoors, where there are people. Air-cooled systems are not normally so equipped, and these devices would be installed at or near the outside condenser to "blow" into the open atmosphere.

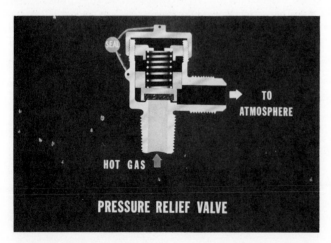

FIGURE R14-18 Pressure relief valve. (*Courtesy* Carrier Air-Conditioning Company)

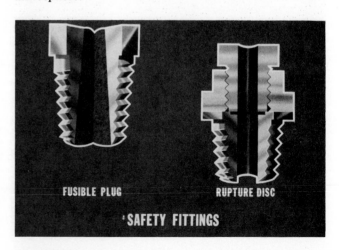

FIGURE R14-19 Safety fittings. (*Courtesy* Carrier Air-Conditioning Company)

R14-1. An oil separator will cure all oil-return problems. True or False?

R14-2. Mufflers may some times be required on what type of compressor?

R14-3. Mufflers must be installed in only two positions. Name them.

R14-4. Name the two principal uses of a heat exchanger.

R14-5. What are the two purposes of installing a strainer-driver?

R14-6. Name two types of strainer-driers.

R14-7. What is the primary purpose of an accumulator?

R14-8. Name three types of resistance-type crankcase heaters?

R14-9. What is the purpose of a crankcase heater?

R14-10. A moisture indicator reveals the presence of water in the system by changing ＿＿＿＿＿＿＿ .

R14-11. What controls the water regulating valve?

R14-12. A solenoid valve can be installed in any convenient position. True or False?

R14-13. A back-pressure regulator is used to (select one):

(a) Protect the compressor from too low a suction pressure.

(b) Regulate the unit head pressure.

(c) Prevent the coil operating temperature from going too high.

(d) Prevent the coil operating temperature from going too low.

R14-14. What is the difference between a back-pressure regulator and an evaporator pressure regulator?

R14-15. What is the purpose of a relief valve?

R14-16. What is the purpose of a fusible plug?

R14-17. What is the difference in operation between a pressure relief valve and a fusible plug?

R14-18. What is the difference between a pressure relief valve and a rupture disk?

Absorption Refrigeration

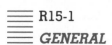

R15-1
GENERAL

The mechanical refrigeration cycle, reviewed previously, is based on the use of a compressor as a source of energy to transfer heat from one source to another. A different method of moving heat (or refrigerating) is called the *absorption refrigeration cycle*.

Absorption liquid chillers, illustrated in Fig. R15-1, utilize heat as the energy source. Steam or hot water are the usual heat media, and they originate from several sources, such as:

1. Existing heating boilers that are now used only during the winter
2. New boilers installed for heating and refrigeration air-conditioning
3. Low-pressure steam or hot water used in an industrial plant for process work
4. Waste heat recovered from the exhaust gases of gas engines or gas turbines
5. Low-pressure steam from steam turbine exhaust

Utilizing waste heat to power the absorption chiller results in a very inexpensive system to own and operate. Already in existence are systems that use solar energy to heat water. The results of these applications may cause even greater interest in the use of absorption equipment as part of energy conservation methods.

Other advantages are sound and vibration levels that are quite low compared to mechanical systems, making them ideally suited for installation in almost any part of a building or on the roof. Commercial units are available in capacities ranging from 25 to well over 1000 tons. Residential air-conditioning absorption units are available in the range 3 to 10 tons.

R15-2
OPERATION

The operation of the absorption system depends on two factors: a refrigerant (water) that boils or evaporates at a temperature below that of the liquid being cooled, and an absorbent (lithium bromide) having great affinity for the refrigerant.

The refrigerant (water) in an open pan will boil or evaporate at 212°F at sea level (14.7 psia) when heat is applied to it. By placing a tight lid on the pan we can cause the pressure and the evaporation temperature to increase. If, on the other hand, we create a vacuum in the closed pan of water, the evaporation will take place at a lower pressure and temperature (refer back to Chapter R3). Now we have only to replace the closed pan with an adsorption chiller and the desired evaporator temperature will dictate the vacuum to be maintained.

For the absorbent, many types of salts could be used. Common table salt (sodium chloride) is an absorbent. You know what happens to a salt shaker during humid weather; it clogs up because the salt has absorbed moisture from the air.

Calcium chloride is another salt, frequently used on dirt roads or clay tennis courts to settle the dust. The calcium chloride absorbs moisture from the air, which keeps the surface of the road or court moist.

Lithium bromide is also a salt and it is in crystal form when it is dry. Tests were made on many types of salts before lithium bromide was selected as having the best overall characteristics for use in large-capacity absorption chillers.

For use in the absorption chiller, lithium bromide crystals are dissolved in water; the mixture is known as *lithium bromide solution*. The amounts of lithium bromide and water in a solution are measured by weight

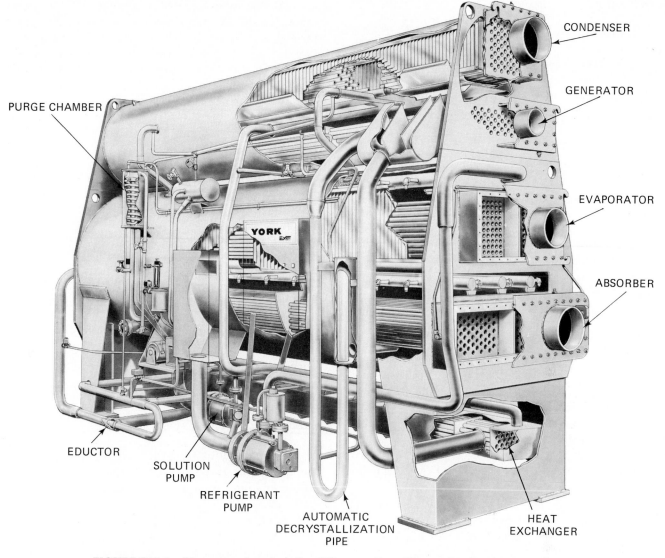

CONDENSER

GENERATOR

EVAPORATOR

ABSORBER

PURGE CHAMBER

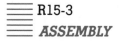

EDUCTOR

SOLUTION PUMP

REFRIGERANT PUMP

AUTOMATIC DECRYSTALLIZATION PIPE

HEAT EXCHANGER

FIGURE R15-1   Absorption liquid chiller. (*Courtesy* Borg-Warner Air Conditioning, Inc.)

and not by volume. The concentration of the solution is stated in percent of lithium bromide in the total solution. A 65% concentration means that 65% of the total weight is lithium bromide.

EXAMPLE

For each 100 lb of solution at 65% concentration there are 65 lb of lithium bromide and 35 lb of water.

$$100 \text{ lb (solution)} \times \frac{65}{100} = 65 \text{ lb lithium bromide}$$

$$100 \text{ lb (solution)} - 65 \text{ lb} = 35 \text{ lb water}$$

The absorbent (lithium bromide) serves as a vehicle to absorb and transport the refrigerant (water)

from a part of the system where it is not needed to another part of the system where the refrigerant can be recovered to be used again.

R15-3
ASSEMBLY

The absorption liquid chiller, shown in Fig. R15-1, consists of two main shells. The upper shell contains the generator and condenser and is maintained at a vacuum of approximately one-tenth of an atmosphere (1.470 psia).

The lower shell contains the evaporator and absorber and is maintained at a vacuum of approximately one-hundredth of an atmosphere (0.147 psia).

R15-4
COMPONENTS

All absorption chillers include four basic heat-exchange components, which, when properly balanced, will cool a liquid to the desired temperature. The components are:

1. Evaporator
2. Absorber
3. Generator
4. Condenser

In addition, there are auxiliary items which assist the four basic components to perform their functions. Typically, they are:

1. A heat exchanger
2. Two fluid pumps
3. A purge unit
4. A vacuum pump
5. An automatic decrystallization device
6. A solution control valve
7. A steam or hot-water valve
8. An eductor
9. A control center

All this sounds fairly complicated until you take it one simple step at a time.

Throughout this discussion certain temperatures and pressures are mentioned to permit a better understanding of the work done in each part of the system. These example conditions are approximate and might vary for other leaving chilled water temperatures, cooling water temperatures, steam conditions, etc.

R15-4.1
Evaporator

The purpose of the evaporator is to cool a liquid for use in process refrigeration work or in an air-conditioning system. Let's consider a typical application (see Fig. R15-2). Chilled water enters the evaporator at 56°F and is to be cooled to 44°F. To accomplish this the lower shell is maintained at a pressure of 6 mmHg (0.117 psia). Under these conditions the refrigerant (water) will evaporate at 39°F, thereby providing a large enough temperature difference to cool the chilled water to 44°F.

Since the refrigerant (water) can evaporate more easily if it is broken up into small droplets, a pump recirculating system is used. The refrigerant (water) enters the top of the lower shell, part of it is evaporated as it comes in contact with the relatively warm tubes, and the liquid that is not evaporated is collected under the evaporator tubes. A pump recirculates this refrigerant through a spray header over the evaporator tubes. This system makes maximum use of the refrigerant and improves heat transfer by keeping the tube surface wetted at all times.

R15-4.2
Absorber

Refrigerant vapor from the evaporator passes through eliminators, which remove any entrained refrigerant liquid. Lithium bromide can absorb water vapor more easily it its surface area is increased; therefore, a pump is used to circulate solution from the bottom of the absorber to a spray header at the top of the absorber (see Fig. R15-3). The vapor is absorbed by the lithium bromide solution, which is flowing over the outside of the absorber tubes. The mixture of lithium bromide and refrigerant water is called the *dilute solution*. Heat generated in this process is called *heat of absorption*. It is removed by condenser water, flowing through the absorber tubes.

R15-4.3
Generator

Dilute solution from the bottom of the absorber is pumped to the generator located in the upper shell. The dilute solution flows over the outside of the hot generator tubes. Steam or hot water in the generator raises the temperature of the solution to the boiling point and vaporizes a portion of the refrigerant. Once again there are two substances, a concentrated lithium bromide solution and refrigerant water vapor (see Fig. R15-4). The refrigerant vapor moves onto the condenser section and the concentrated solution of lithium bromide returns to the absorber to be reused.

R15-4.4
Condenser

The refrigerant vapor released in the generator passes through eliminators, which remove any entrained lithium bromide solution. A pressure of 70 mmHg (1.346 psia) is maintained in the upper shell and causes the refrigerant to condense at 112°F on the condenser tubes (see Fig. R15-5). Condenser water is used, which after passing through the absorber tubes flows inside the condenser tubes. Condensed refrigerant flows by gravity and pressure differential through an orifice to the evaporator. This refrigerant plus that recirculated by the refrigerant pump is distributed over the evaporator tubes to complete the refrigerant cycle.

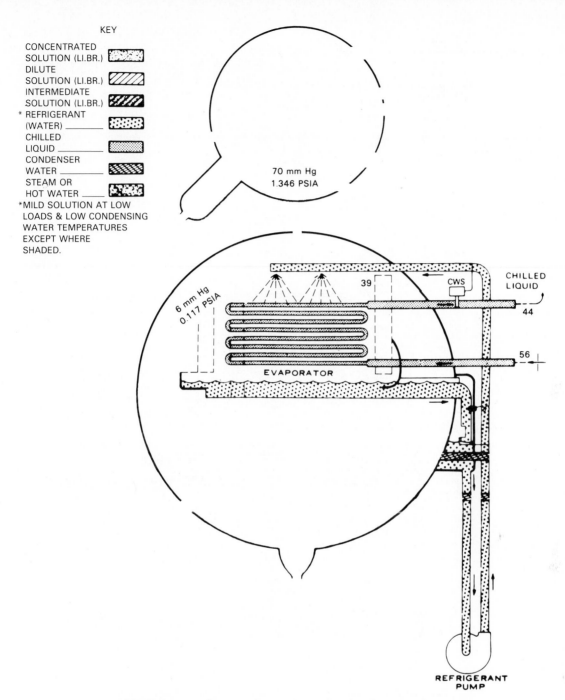

KEY

CONCENTRATED
SOLUTION (LI.BR.)
DILUTE
SOLUTION (LI.BR.)
INTERMEDIATE
SOLUTION (LI.BR.)
* REFRIGERANT
(WATER)
CHILLED
LIQUID
CONDENSER
WATER
STEAM OR
HOT WATER

*MILD SOLUTION AT LOW
LOADS & LOW CONDENSING
WATER TEMPERATURES
EXCEPT WHERE
SHADED.

70 mm Hg
1.346 PSIA

6 mm Hg
0.117 PSIA

39

CWS

CHILLED
LIQUID

44

56

EVAPORATOR

REFRIGERANT
PUMP

FIGURE R15-2   (*Courtesy* Borg-Warner Air Conditioning, Inc.)

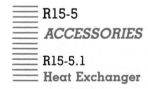

R15-5
ACCESSORIES

R15-5.1
Heat Exchanger

To make the absorption cycle more efficient, several accessory items are normally added; the first is a *heater exchanger* in the line between the absorber and generator (see Fig. R15-6). The heat exchanger brings the warm, concentrated lithium bromide solution coming from the generator in contact with the relatively cool dilute solution coming from the absorber.

The dilute solution leaves the absorber at a temperature of 102°F, and the concentrated lithium bromide solution is at a temperature of 214°F as it comes from the generator.

The introduction of a heat exchanger improves the efficiency of the cycle by reducing the amount of steam or hot water required in the generator. The effect of the heat exchanger on both solutions is shown by the temperatures in Fig. R15-6.

Absorption Refrigeration   171

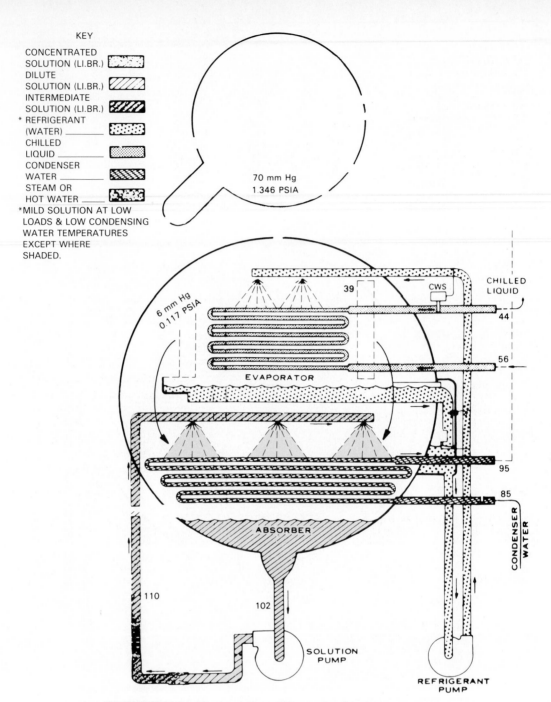

KEY

CONCENTRATED
SOLUTION (LI.BR.)
DILUTE
SOLUTION (LI.BR.)
INTERMEDIATE
SOLUTION (LI.BR.)
* REFRIGERANT
(WATER)
CHILLED
LIQUID
CONDENSER
WATER
STEAM OR
HOT WATER

*MILD SOLUTION AT LOW
LOADS & LOW CONDENSING
WATER TEMPERATURES
EXCEPT WHERE
SHADED.

70 mm Hg
1.346 PSIA

6 mm Hg
0.117 PSIA

39

CWS

CHILLED
LIQUID

44

56

EVAPORATOR

95

85

ABSORBER

CONDENSER WATER

110

102

SOLUTION
PUMP

REFRIGERANT
PUMP

FIGURE R15-3   (*Courtesy* Borg-Warner Air Conditioning, Inc.)

R15-5.2
Eductor

A second accessory item is the *eductor,* which provides for circulation of the lithium bromide solution over the absorber tubes (see Fig. R15-6). This increases the efficiency of absorbing water vapor. A portion of the dilute solution leaving the solution pump at the bottom of the absorber is directed through an eductor, which induces the concentrated solution to mix with the dilute solution from the pump. This mixture is delivered to the spray headers over the absorption tube bundle.

R15-5.3
Automatic Decrystallization Pipe

A third accessory item is the *automatic decrystallization pipe,* which is designed to prevent crystallization of the lithium bromide solution. Crystallization can occur when the lithium bromide solution becomes to concentrated (see Fig. R15-7).

Should the absorption chiller be shut down due to a prolonged power failure, the refrigerant and the lithium bromide solution temperatures would eventually reach the equipment room temperature, possibly caus-

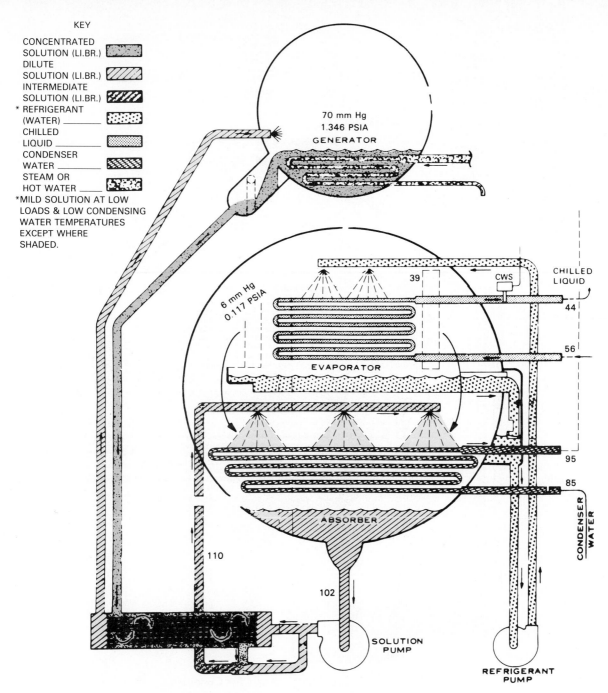

CONCENTRATED
SOLUTION (LI.BR.)
DILUTE
SOLUTION (LI.BR.)
INTERMEDIATE
SOLUTION (LI.BR.)
* REFRIGERANT
(WATER)
CHILLED
LIQUID
CONDENSER
WATER
STEAM OR
HOT WATER
*MILD SOLUTION AT LOW
LOADS & LOW CONDENSING
WATER TEMPERATURES
EXCEPT WHERE
SHADED.

70 mm Hg
1.346 PSIA
GENERATOR

6 mm Hg
0.117 PSIA

EVAPORATOR

39

CWS

CHILLED
LIQUID

44

56

95

85

CONDENSER WATER

ABSORBER

110

102

SOLUTION
PUMP

REFRIGERANT
PUMP

FIGURE R15-4   (*Courtesy* Borg-Warner Air Conditioning, Inc.)

ing the concentrated solution in the generator, the heat exchanger, and the connecting piping to crystallize.

When power is restored, heat from the steam or hot water will increase the temperature of the concentrated solution in the generator, causing it to liquefy. However, the concentrated solution in the piping and the heat exchanger will remain in crystalline form and the addition of heat is required to liquefy it.

The liquid solution from the generator will back up in the pipe on the generator side of the heat exchanger until it overflows into the automatic decrystalliza-

tion pipe and then into the absorber. The solution temperature in the absorber will increase to about 214°F instead of 102°F. The solution pump will then pump 214°F concentrated solution through the heat exchanger. This additional heat will liquefy the solution in the heat exchanger and the piping.

When the flow from the generator to the absorber returns to normal, the flow through the automatic decrystallization pipe will cease and the system will return to normal operation.

In normal operation the automatic decrystalliza-

Absorption Refrigeration   173

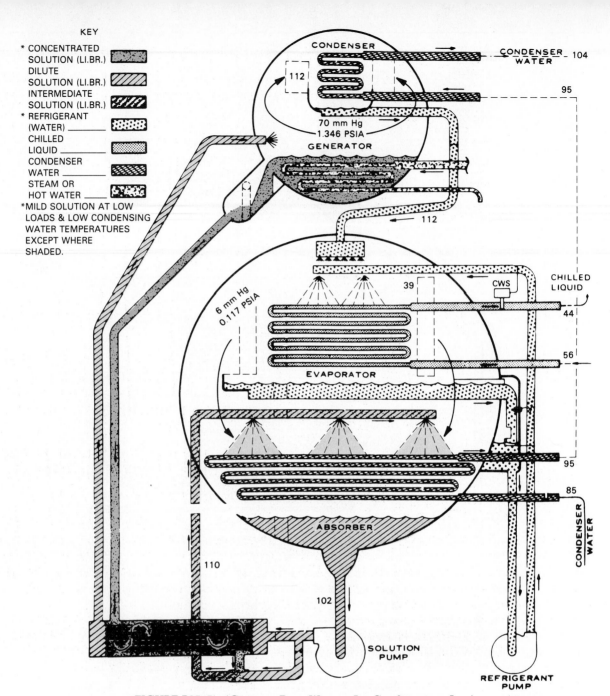

FIGURE R15–5 (*Courtesy* Borg-Warner Air Conditioning, Inc.)

tion pipe is designed so that there is a seal between the upper and lower shells to prevent an equalization of pressure.

To keep the decrystallization pipe ready for use at all times, a small amount of dilute solution is constantly flushed through it.

R15-5.4
Purge System

A fourth accessory item is the *purge system,* which is designed to remove noncondensables. The noncondensable gases are collected in the water-cooled purge

chamber, and they are removed by periodic operation of an electric motor compressor type purge unit. Manual operation of the purge unit assures that the operator will know the amount of noncondensables in the system. If the unit was purged automatically, the unit might have a large leak that would go undetected until extensive damage occurred.

R15-5.5
Control Valve

The operation of the absorption chiller is controlled by a fifth accessory, a *steam* or *hot water control*

174 Refrigeration

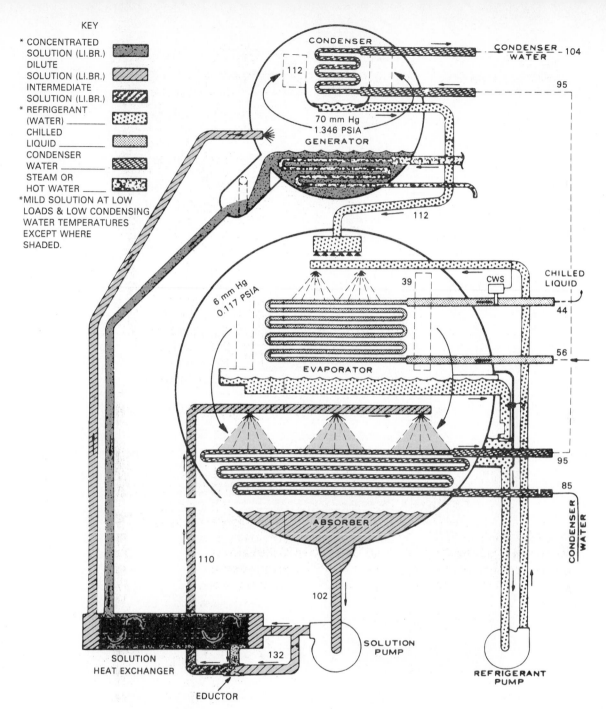

KEY

* CONCENTRATED SOLUTION (LI.BR.)	
DILUTE SOLUTION (LI.BR.)	
INTERMEDIATE SOLUTION (LI.BR.)	
* REFRIGERANT (WATER)	
CHILLED LIQUID	
CONDENSER WATER	
STEAM OR HOT WATER	

*MILD SOLUTION AT LOW LOADS & LOW CONDENSING WATER TEMPERATURES EXCEPT WHERE SHADED.

FIGURE R15–6 (*Courtesy* Borg-Warner Air Conditioning, Inc.)

valve (see Fig. R15-7). This valve modulates to control the flow to the generator tubes. It is activated by a sensing element in the chilled water line leaving the evaporator. In this way the energy supplied to the generator is only the amount required to produce a sufficient quantity of refrigerant to maintain the system chilled water temperature.

R15-5.6
Solution Valve

The basic system includes the evaporator, absorber, generator, and condenser, as well as the heat ex-

changer, the eductor decrystallization pipe, the purge unit, and the control valve, all of which are standard equipment on most units. They are all needed for efficient operation of the unit. A valuable optional accessory item is the *solution control valve* (see Fig. R15-7). It is designed to provide the maximum in economical operation at part load.

Under normal conditions, at full load, 12 lb of dilute lithium bromide solution is supplied to the generator for every pound of refrigerant that is boiled off. At 25% load, without a solution valve, 12 lb of solution is still circulated to the generator but only $\frac{1}{4}$ lb of refrigerant is boiled off.

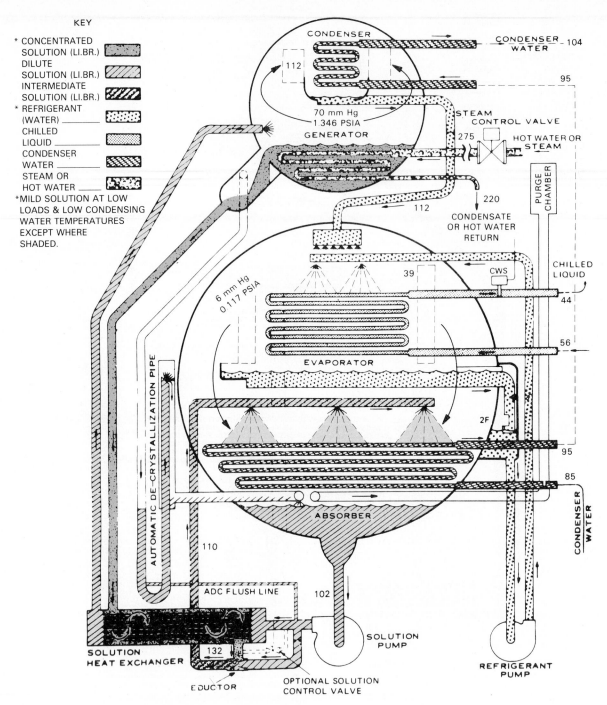

KEY

* CONCENTRATED SOLUTION (LI.BR.)
DILUTE SOLUTION (LI.BR.)
INTERMEDIATE SOLUTION (LI.BR.)
* REFRIGERANT (WATER)
CHILLED LIQUID
CONDENSER WATER
STEAM OR HOT WATER

*MILD SOLUTION AT LOW LOADS & LOW CONDENSING WATER TEMPERATURES EXCEPT WHERE SHADED.

CONDENSER

CONDENSER WATER — 104

95

112

70 mm Hg
1.346 PSIA

GENERATOR

STEAM CONTROL VALVE

275

HOT WATER OR STEAM

112

220

CONDENSATE OR HOT WATER RETURN

PURGE CHAMBER

CHILLED LIQUID

6 mm Hg
0.117 PSIA

39

CWS

44

EVAPORATOR

56

2F

95

85

ABSORBER

CONDENSER WATER

AUTOMATIC DE-CRYSTALLIZATION PIPE

110

102

ADC FLUSH LINE

SOLUTION HEAT EXCHANGER

132

EDUCTOR

OPTIONAL SOLUTION CONTROL VALVE

SOLUTION PUMP

REFRIGERANT PUMP

FIGURE R15-7   (*Courtesy* Borg-Warner Air Conditioning, Inc.)

The addition of a solution valve restricts the flow of dilute solution to the generator in accordance with the reduced load requirements. The solution valve is actuated by a sensing element in the generator outlet box, which maintains a constant temperature of the concentrated solution regardless of the load. The solution valve greatly improves the efficiency of the system at part load.

R15-1. Absorption refrigeration units use _____ as their energy source.

R15-2. Are sound and vibration levels of absorption equipment higher or lower than in conventional systems?

R15-3. What refrigerant is used in an absorption unit?

R15-4. The most common absorbant is _____ .

R15-5. Name the four main components of an absorption chiller.

R15-6. Does the absorber contain the evaporator or the condenser?

R15-7. The condenser is located within the _____ .

R15-8. The function of the purge unit is to remove _____ gases.

R15-9. Is the purge unit normally an automatic or a manual operation?

R15-10. The control valve modulates the flow of steam or hot water to the generator. True or False?

R16

Basic Electricity

R16-1
GENERAL

Electricity alone is a subject that would require several books to cover even the basics. However, in the next several chapters we discuss electricity as it relates to the refrigeration, heating, and air-conditioning industry.

R16-2
THE ELECTRON THEORY

All matter is composed of atoms—the smallest or basic components of molecules. Atoms, in turn, are composed of a heavy, dense nucleus containing *protons* and sometimes *neutrons*, surrounded by lighter particles called *electrons*. The proton carries a positive charge (+), the electron carries a negative charge (−), and the neutron (if present) is neutral. The attraction between the positively charged protons in the nucleus and the surrounding negatively charged electrons tend to hold them together in the unit we call the *atom* (see Fig. R16-1). The number of protons in the nucleus determines the

type of element. The hydrogen atom, illustrated in Fig. R16-1a, is the smallest, having one proton in the nucleus and one electron circling in orbit around it. Copper (Fig. R16-1b), on the other hand, has 29 protons and 34 neutrons.

Returning to the definition of positive or negative charges—a positive charge does not indicate an excess of protons; it means a deficiency of electrons. Thus, the terms *negative* and *positive*, simply mean more or less electrons are involved.

Materials that are charged with static electricity either attract or repel each other. Attraction takes place between unlike charges, because the excess electrons of a negative charge seek out a positive charge, which has a shortage of electrons (Fig. R16-2). Unlike charges (+ and −) attract. Like charges (− and −) or (+ and +) will repel each other.

Static electricity is the condition when the electrons are at rest but have a potential to move. *Dynamic electricity* is electrons in motion. The movement of electrons is called *current*.

Before the acceptance of the electron theory as the basis of electrical behavior, it was thought that current flowed from positive to negative. Now it has been established that current (the flow of electrons) is actually from negative to positive.

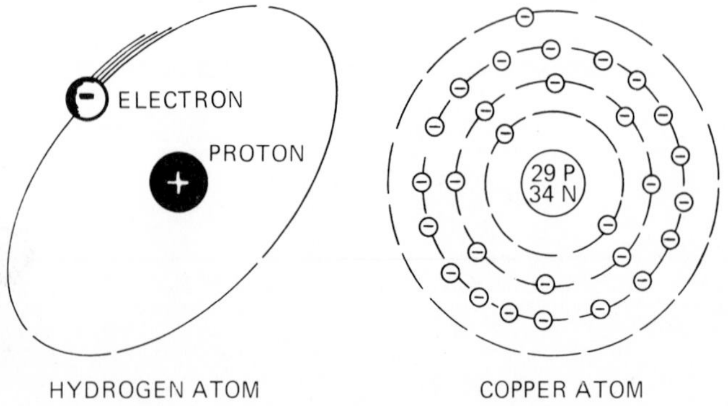

HYDROGEN ATOM   COPPER ATOM

FIGURE R16-1   (a) Hydrogen atom. (b) Copper atom.

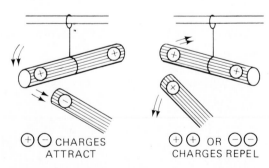

ATTRACTING AND REPELLING OF CHARGES

FIGURE R16-2   Attracting and repelling of charges.

SOURCES OF ELECTRICAL ENERGY

Energy is defined as the ability to do work, and since energy cannot be created or destroyed it must be converted from one form to another. Sources of electrical energy may be:

1. Chemical action
2. Friction
3. Heat
4. Light action
5. Pressure
6. Mechanical action
7. Nuclear action
8. Magnetism

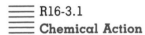
R16-3.1
Chemical Action

The prime example of converting chemical action to electrical energy is the common battery we use in cars, flashlights, test instruments, communicating equipment, radios, calculators, etc. Batteries are divided into two types: Rechargeable, called *secondary* type, such as those used in automobiles, and single use, called *primary* type, such as those used in flashlights, test instruments, etc.

Rechargeable batteries were the first type developed and use a nonporous case of hard rubber or plastic containing a solution of sulfuric acid and water. This solution is called the *electrolyte*. In this solution plates of sponge lead and lead peroxide are suspended in the solution. These plates are called the *electrodes*. The chemical action of the sulfuric acid/water solution (the electrolyte) causes electrons to be drawn from the lead peroxide electrode, leaving this electrode with a shortage of free electrons. This creates a positive charge in the electrode.

The free electrons are then deposited on the sponge lead electrode. This creates an excess of free electrons in this electrode and the electrode becomes negatively charged, as compared to the positive charge in the lead peroxide plate. A voltage (potential) difference is built up between the plates.

Figure R16-3 shows a single lead–acid secondary battery with an external circuit of a lamp controlled by a switch. With the switch open, no electrical circuit exists and thus no electrical flow. When the switch is closed, the electrical path is completed. Electricity (free electrons) flows from the negatively charged spongy lead electrode to the positively charged lead peroxide electrode through the electrical circuit and the load (the lamp). In the lamp, electrical energy is changed to heat

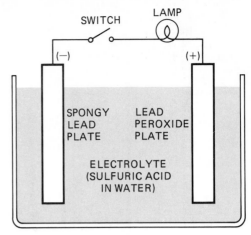

FIGURE R16-3 Simple lead–acid chemical cell (battery). (*Courtesy* American Gas Association)

energy, causing the element to glow and light is emitted.

Chemical reactions between the electrodes and the electrolyte continue to take place. These reactions constantly supply free electrons to the negative electrode, removing them from the positive electrode, and electrical energy continues to flow through the circuit. Because the electrical flow is constantly in one direction, from the negative electrode to the positive electrode, through the connected load circuit, this type of electrical energy is known as *direct current* (dc).

Electrical energy can be drawn from the cell only as long as the supply of free electrons exist in the positive electrode. Removal of the electrons gradually changes the electrode from lead peroxide to lead sulfate. Then the electrode is completely changed to lead sulfate. The action stops and the cell is "dead."

Before the cell is completely dead, it is possible to "recharge" the cell with an external dc power source. This could be the alternator and diodes in the automobile engine, or a battery charger, or even another charged cell. Use of another cell or battery is called "jumping the battery." To recharge the cell, electrical energy is forced backward through the cell and electrons are driven back from the negative to the positive electrode through the electrolyte. Thus, the positive electrode is restored to lead peroxide and the negative electrode to spongy lead. The cell is in its original condition and is still to be recharged.

The ability of these cells to store energy in chemical form to be used on demand as electrical energy led to the name "storage battery." Remember, there is no "electrical energy" in a storage battery, only chemical energy that is converted to electrical energy upon demand.

A single cell can only produce 1.5 V of electricity pressure, so two or more cells are combined for higher voltage (potential). A 6-V motorcycle battery has four

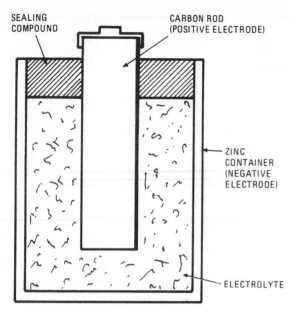

FIGURE R16-4　Construction of a dry cell. (*Courtesy* American Gas Association)

cells—4 cells × 1.5 V per cell = 6 V. A 12-V automobile battery uses eight cells—8 cells × 1.5 V per cell = 12 V.

Unlike the storage battery, which is rechargeable, cells used in flashlights and test instruments are the throwaway, single-charge type. Fig. R16-4 shows the construction of a throwaway or "dry cell." This type of battery is referred to as a dry type because the electrolyte is in a paste form rather than a liquid as in the lead-acid cell. The action is the same in that free electrons are removed from the positive (carbon rod) electrode and transferred to the negative (zinc container) electrode whenever electrical energy is demanded. This action continues until all free electrons leave the carbon rod and the battery is dead.

Because of the high energy loss in the form of heat as well as gas formation during the charging process, a temperature high enough to produce vaporization of the paste is possible and explosion could result from the pressure produced. When the cell is specifically marked as not rechargeable, this process should not be done, as it can be extremely dangerous.

≡≡ R16-3.2
≡≡ Heat Action

When two wires of different materials are fused together at the ends and a temperature difference is created between the ends by heating one end, an electrical voltage (potential) is created between the hot and cold ends. The heated end is called the *hot junction* and the cooled end, the *cold junction*. The greater the $\Delta T°F$ between the ends, the higher the voltage. This effect is called the *thermo* (temperature)-*electric effect* and a sin-

gle pair of wires used for this purpose is called a *thermocouple*. Fig. R16-5 illustrates the thermoelectric effect in a heated thermocouple. The voltage produced by a single thermocouple is very small (0.15 to 0.35 V), but even at this small voltage, the thermocouple has been a very important device in the development of the gas industry.

A thermocouple can be used as a very sensitive temperature-measuring device because of the fact that the voltage developed depends on the $\Delta T°F$ between the hot and cold junctions. Also, different metal combinations will produce different voltages at a given temperature—or, in reverse, the same voltage and a variety of temperatures. Thus tables can be prepared for use of a single instrument (voltmeter calibrated with a temperature scale) to indicate different temperature ranges by using the appropriate metal combinations.

Of even greater importance in the gas industry is the widespread use of thermocouples in gas controls, such as in automatic pilot safety devices. Figure R16-6 shows the use of a thermocouple in an automatic pilot device. The hot junction of the thermocouple is heated by the flame of a gas pilot. The cold junction is left cool by some of the air for combustion to the main burner passing over the thermocouple body. The temperature difference between the hot and cold junctions is thus maintained.

Even though only 0.15 to 0.35 V is produced by the thermocouple, it is sufficient voltage to produce enough electrical flow through the electromagnet to hold the keeper on the magnet pole faces. Thus, the gas valve safety pilot valve is held open as long as the gas pilot

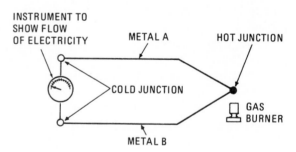

A. HOT JUNCTION NOT HEATED—NO ELECTRICITY PRODUCED

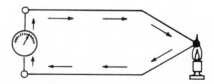

B. HOT JUNCTION HEATED—ELECTRICITY FLOWS IN CIRCUIT

FIGURE R16-5　Thermoelectric effect. (*Courtesy* American Gas Association)

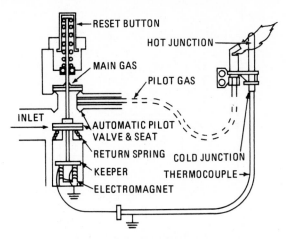

FIGURE R16-6 Use of thermocouple in automatic pilot device. (*Courtesy* American Gas Association)

flame is "heating the hot junction." If for some reason the hot portion of the pilot flame leaves the hot junction, because of not enough pilot gas, too much pilot gas, or excessive draft blowing the flame away or blowing out the pilot, the hot junction cools. The thermocouple output drops and the current flow decreases until the electromagnet power drops below the strength of the spring, the keeper is forced off the magnet poles, and the gas safety valve closes.

In those applications where higher thermocouple voltage is required, the voltage can be increased by using a number of thermocouples in the same heated device. Such an item is called a *thermopile*. Figure R16-7 shows five thermocouples connected in series to produce an output voltage five times the output of a single couple. Such a device is used in "self-generating" control systems where the gas valve is controlled by a thermostat using the power from the thermopile, and no outside power source is required. Thermocouples and thermopiles will be discussed in more detail in future chapters.

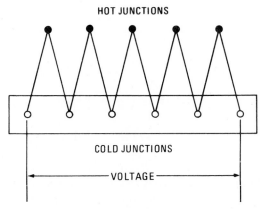

FIGURE R16-7 A thermopile is composed of many thermocouples connected together so that the electrical effect is additive. (*Courtesy* American Gas Association)

For a complete discussion of gas controls, it is suggested that material published by the American Gas Association and the Gas Appliance Manufacturers Association (Catalog XH0175) be used.

R16-3.3
Light Action

Light has energy and when it strikes the surface of a substance it will dislodge free electrons from the stricken material (positive electrode) and pass to the backup material (negative electrode), creating an electrical voltage. This cell, called a *photovoltaic cell*, is connected in series and parallel with many more cells to produce dc current from solar energy.

Such cells have little use in environmental control systems. The photocell in an oil burner is sometimes used as an illustration of such a cell. The photocell in the oil burner control system, however, does not produce electrical potential. The presence of light affects only the electrical resistance of the cell, thus affecting its ability to carry electrical energy. This feature is then used in the oil burner control to affect the safety shut down feature.

Another very popular example of resistance change is the use of a photocell to control relays which in turn control structure area lighting, automatically turning lights on at dusk and off at dawn.

R16-3.4
Pressure

If certain materials such as quartz or barium titanate are compressed in certain directions, electrons in the material will shift and create a voltage difference across the two surfaces of the material. This is called the *piezoelectric effect* (pressure-electronic effect) and materials that have this quality are called *piezoelectric materials*.

The amount of voltage created depends on the change of pressure applied and the change time involved. The highest pressure applied in the shortest time possible produces the highest voltage. Thus, a steady surge of high pressure with a gradual increase in the pressure will not produce a piezoelectric effect. Rather, rapid squeezing or striking of the material will produce sufficient voltages to produce a spark across wires leading from the material's surfaces. Such piezoelectric materials are used in camping stoves and lanterns as well as hand-held "matchless" pilot lighters.

R16-3.5
Nuclear Action

Nuclear action does not produce usable electrical energy directly but by the use of energy conversion to heat, which in turn produces mechanical pressure or energy, which in turn is converted to electrical energy through the electromechanical process.

R16-3.6
Mechanical Action

Through the electromechanical process, large amounts of electrical energy are produced to serve needs in residential, commercial, and industrial needs. Mechanical energy is converted to electrical energy using rotating units called generators, which in turn are driven by various means, such as steam turbines whose steam is produced by fossil-fuel combustion, as nuclear-powered water turbines in hydroelectric plants, and in both reciprocating and rotary thermal combustion engines using fossil fuels. Thus, the mechanical energy of the turbine and driver combination is converted to electrical energy.

R16-3.7
Magnetism

Earlier it was pointed out that objects of like charges repel each other, and unlike charges attract. You may recall from your study of physics that a natural magnet, such as a lodestone, is surrounded by a force field. The classic experiment is to place a bar magnet under a sheet of paper or glass on which rest iron filings (Fig. R16-8). The filings will align themselves along the lines of force that leave one end of the magnet and return to the other. One end is called the *north* (N) *pole* and the other the *south* (S) *pole*.

Now if we take this bar magnet and pass it inside a coil of copper wire as illustrated in Fig. R16-9, we observe, on a sensitive galvanometer, the development of a current as the copper wire (conductor) cuts the force field of the magnet.

In position *a*, the motion of the magnet is in a downward direction and the galvanometer needle will move to the right. When the magnet is at rest (position

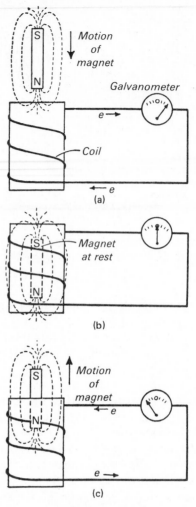

FIGURE R16-9

b), no current is established. With the magnet moving in an upward direction (position *c*) current again flows, but in the opposite direction to that in position *a*.

This phenomenon is fundamental to all practical power generation and electric motors. *When an electrical conductor cuts through a magnetic field, an electromotive force is set up between the ends of the conductor.* Later we discuss the principles of electromagnetism as applied to generators and motors.

R16-4
SIMPLE ELECTRICAL CIRCUIT

With a source of potential electrical energy established, let's examine a simple electric circuit. The term *circuit* is to be taken literally, for unless electricity is able to follow a complete path from the source to a load and return to its source, it will not accomplish any work—such as lighting a lamp or running a motor. The two wires at the bottom of Fig. R16-10, marked L_1 and L_2, go to the source of electricity for this circuit.

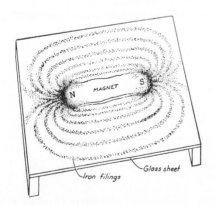

FIGURE R16-8   Magnet and iron filings.

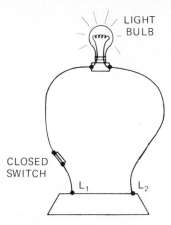

FIGURE R16-10   Simple electric circuit.

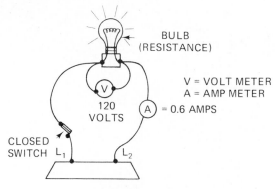

FIGURE R16-11

The electricity flows from and to the source, passing through the switch and through the bulb (load) on its way. We say that electricity flows. This popular conception of an electrical current as a flow of electrons reflects the common attempt to make electrical phenomena understandable by direct comparison to flowing water. (Note the term *current*, which again slips into our discussion.) For our purpose, it is sufficient to say that when electrons move through a material, we have electricity.

For electricity to flow, there must be, at the source, buildup of electrons sufficient to create a pressure or force. This electron buildup may be accomplished chemically, as in a battery; magnetically, as in a generator; or by heat, as in a thermocouple. The proper term for this pressure is *electromotive force* (emf). Since the unit of measurement for emf is the *volt*, the term *voltage* is synonymous with electromotive force in practical work.

As is true for all motion, the movement of electrons through a material is opposed by friction. Depending on the material, this opposition to electron flow, called *resistance*, may be great, moderate, or low.

Copper, for example, offers low resistance to the flow of electrons; therefore this metal is by far the most common electrical conductor used, whether in the form of small wires, large cables, or giant busbars. Aluminum is also a good conductor and is being used increasingly in electrical work. Tungsten, used in the filaments of incandescent lamps, offers a very high resistance to electron flow.

The unit of measurement for electrical resistance is the *ohm*. Devices, such as bulbs, with high resistance are called *resistors* and are denoted by a wavy line in wiring diagrams.

If the total resistance in a circuit is great enough to prevent any flow of electrons at a particular emf, no current can pass. When the voltage is high enough to overcome the resistance, electrons flow through the conductor and the resistor. For practical purposes, we are interested in knowing how many electrons flow past any point in a circuit during a given time. It is this *rate of flow* of electrons that is properly called *current*.

The unit of measurement for the rate of electron flow is the *ampere*, almost always called the *amp*. Summarizing:

Electromotive force is expressed in volts.

Electrical resistance is expressed in ohms.

Current, the electrical rate of flow, is expressed in amperes.

We apply this information to the simple electrical circuit in Fig. R16-11. A voltmeter shows the electromotive force is 120 V; an ammeter indicates the current is 0.6 A. Obviously, the emf is sufficient to cause current to flow through the total resistance offered by the wires and the bulb. This total resistance may be measured with an ohmmeter or it may be easily calculated; the calculations will be described later.

With the same current flowing through the wire conductors, the switch, and the bulb, why do they not all get hot enough to emit light? Recall that different substances offer different resistances to the flow of electrons. Stated another way, a copper wire and a tungsten wire of the same diameter and length can both pass 0.6 A under a pressure of 120 V. But the electrons must work harder to get through the high-resistance tungsten than through the low-resistance copper. The harder work in this case results in heat and light.

An ohmmeter will show the resistance of the copper wire conductors to be practically zero; the resistance of a 25-W bulb is about 570 Ω, when lighted. Obviously, the current can more readily flow through the copper than through the tungsten filament of the lamp.

If the resistor (light bulb) were not in the circuit, and only copper wire was connected to the emf, too much current—too many amperes—would flow, and the copper wire would overheat and burn out (or a fuse would blow; the reasons for this will be explained later). We would have, in effect, a short circuit.

A short occurs when the current takes an accidental path that does not pass through the circuit's normal loads, causing the total resistance to drop near zero. If

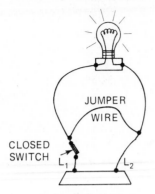

FIGURE R16-12

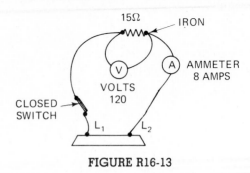

FIGURE R16-13

L_1 and L_2 were bare and accidentally touched, for example, or if a copper jumper wire were installed ahead of the resistor, as illustrated in Fig. R16-12, the current would take the path of least resistance and never touch the bulb. Since the resistance of the copper wire alone is low, so many amperes would flow that the highly conductive copper would soon overheat.

An *open circuit* occurs when, for any reason, the path of the current flow is broken. Pulling the switch opens the circuit. A broken wire or broken bulb filament has the same effect.

Although some substances are nonconductors in a practical sense, any material may be made to pass some current if a sufficient emf (voltage) is applied. Thus, if the switch was open and a higher and higher voltage were supplied through L_1 and L_2, at some point a spark would jump across the open switch contacts, indicating that current has passed through the air gap. Although air is normally a nonconductor, it will pass current if the voltage is high enough. (Artificial lighting can be made to travel hundreds of feet through the air.)

A circuit is said to be *closed* when all necessary contacts are joined to permit normal current flow through conductors and loads. The term must not be misunderstood to mean that the circuit is closed to current flow, as a valve might close off or shut off water flow. The circuit in Fig. R16-10 is closed, and current flows through properly made contacts and properly connected wires and load (resistor).

R16-5
OHM'S LAW

The relationship, indeed the interdependence, among electromotive force, resistance, and current is clearly indicated by the formal definition of a volt: the amount of electromotive force required to cause 1 ampere of current to flow through a resistance of 1 ohm.

The relationship is even more clearly defined by *Ohm's law*, named after the German scientist Georg

Simon Ohm, who worked out these mathematical relationships early in the nineteenth century. Stated practically, Ohm's law shows that the greater the emf, the greater the current; the greater the resistance, the less the current.

Mathematically, the relationship is stated as: I equals E over R:

$$I = \frac{E}{R}$$

where I is the symbol for current in amperes, E is the symbol for emf in volts, and R is the symbol for resistance in ohms.

Suppose, in the simple circuit in Fig. R16-13, the resistance symbol represents an electric iron, rather than a lamp bulb as it did in our previous discussion. With an ohmmeter, we determine the resistance to be 15 Ω. Either by taking the power company's word for it or by using a voltmeter, we determine the voltage is 120 V. How many amps will the iron draw?

$$I = \frac{E}{R} = \frac{120}{15} = 8 \text{ A}$$

If the iron is replaced with a 100-W bulb, we might find the bulb's resistance to be in order of 145 Ω. With the same emf, how many amperes will the light bulb draw?

$$I = \frac{E}{R} = \frac{120}{145} = 0.83 \text{ A}$$

It is obvious that the lower the resistance, at a given voltage, the higher the current that will be drawn. Theoretically, as the resistance approaches zero, the current approaches infinity.

If, in our simple circuit, we short the resistance out of the line, 120 V will be pushing against almost zero resistance and hundreds of amperes will flow—until a fuse or a burned wire opens the circuit. Assume the wire in our circuit has 0.2 Ω resistance; the current flow will be

$$I = \frac{E}{R} = \frac{120}{0.2} = 600 \text{ A}$$

The basic *E*-over-*R* formula may be altered, or transposed, to calculate any one factor, if the other two are known. For example, a volt-ammeter had been used to measure 1000 A at only 2 V. What is the resistance?

$$R = \frac{E}{I} = \frac{2}{1,000} = 0.002 \ \Omega$$

Returning to the electric iron, assume that the readings were 120 V and 8 A. What is *R*?

$$R = \frac{E}{I} = \frac{120}{8} = 15 \ \Omega$$

The formula may be transposed in one other way: $E = IR$. In the 100-W bulb example, suppose that we had measured the resistance and the current but had no voltmeter. What is the emf in volts? (The current was 0.8 A and the resistance 145 Ω.)

$$E = IR = 0.83 \times 145 = 120 \ V$$

In electrical talk, the amount of voltage used up by each resistance in a circuit is called the *voltage drop* or the *IR drop* through that resistance.

There are several "helpful" tricks in remembering the formula for Ohm's law and its transpositions. We do not recommend such memory aids. If Ohm's law is used frequently enough in his or her work, the service technician has no need for a memory aid. No good refrigeration technician trusts memory when making a calculation involving an almost forgotten formula; he or she looks it up.

Actually, most of the difficulty lies in remembering the transpositions, since the statement of the law as "*I* equals *E*-over-*R*" gives the first formula,

$$I = \frac{E}{R}$$

To determine how to transpose, plus simple numbers into the basic formula. Substitute, for instance, 4 for *I*, 8 for *E*, and 2 for *R*, thus:

$$4 = \frac{8}{2}$$

From this equation it is easy to determine that 8 will equal 2 times 4 and 2 will equal 8 over 4.

Ohm's law, then, allows us to determine mathematically the third factor when any two are know by measurement or by reference to a specification sheet.

Although it is theoretically possible to get an increasing, even an unlimited number of amperes from any voltage source—as resistance approaches zero—practical considerations, such as the heat needed to light the lamp or operate the iron, limit the current to a level that can be carried safely by the conductors and the devices on the line.

Here are some practical applications of the information so far:

1. A 100-W bulb manufactured for 120-V operation will glow extra brightly on 240 V for a few seconds, and then burn out. The resistance of the filament was taken into consideration in the design of its length and diameter, to make it compatible with 120 V. At 240 V this same resistance passes twice as much current—more than it is designed to carry. The filament overheats and burns out, opening the circuit.

2. For a given material, resistance increases as the length of the conductor increases. This accounts for the voltage drop, the *IR* drop, on long extension lines. At a wall outlet, an electric iron may draw 9 A at 120 V. At the end of a 12-ft household extension wire, it draws 8 A at 105 V. In rural areas, it is not uncommon for the customer farthest from a pole transformer to have low voltage because of the voltage drop from the transformer through the long line to his home.

   Ideally, *IR* drop should be kept under 3% of rated voltage. It is common to have a 5% drop; a voltage drop over 10% can cause malfunction and failure of motors, relays, and similar electrical devices.

3. For a given material, resistance increases as the diameter of the conductor decreases. A lamp circuit in a home may call for No. 14 wire. Install a central air-conditioner or electric range, and you must use No. 8 wire for the circuit. (The diameter of wires gets larger as the number gets smaller.) A larger-diameter wire can carry more amperes (45 to 60 A for a range) than can a smaller wire (15 A for a lighting circuit) without overheating. (Wire sizes are discussed more thoroughly later.)

R16-6
SERIES CIRCUITS

When more than one resistance device is placed in a circuit, the current can flow through more than one type of path on its way from and to the emf source. Depending on how the path is designed, the circuit will fall into one of three categories:

1. Series
2. Parallel
3. Series–parallel

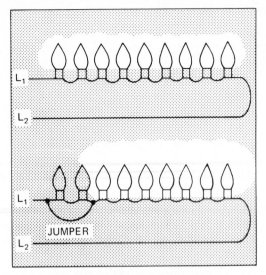

FIGURE R16-14   Series circuit.

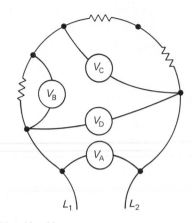

$V_b + V_c = V_d$
$V_d = V_a$ (WIRE RESISTANCE NEGLIGIBLE)

FIGURE R16-16   Resistances in series.

Simplest to understand is the series circuit. It is familiar to most through the older type of Christmas tree lights; when any one of them burns out, they all go out because of an open circuit. (We can see why all the lights go out by looking at Fig. R16-14.)

To travel from and to the emf source, the current must pass through each resistance in succession. Resistors in series are connected end to end across a voltage source. Remove one lamp and the current stops because the circuit is no longer complete. Place a jumper around the defective bulb(s) and the rest will light.

Such a current stoppage can be useful. All switches, for example, must be in series with the devices they control. Protective devices such as fuses and overload protectors, are wired in series so equipment cannot operate when the safety device is electrically opened for any reason.

In Fig. R16-15, taken from a full circuit diagram for a refrigeration unit, the fan motor overload protec-

tor (OL) is in series with the power supply. Should the protector open due to excessive current or overheating, no electricity can reach the fan motor windings. B and BR are color codes for the wires. What about volts, amperes and ohms in a series circuit? Remember that voltage may be read between any two points on a conductor or resistor and that the amount of voltage used up between any two points is termed the *voltage drop*.

If three equal resistances, such as three 25-W bulbs, are wired in series as in Fig. R16-16, the voltage read by a meter at V_A will be the supply voltage, 120 V. A voltmeter across a single resistance, V_B, will show 40 V. V_C, across two 25-W bulbs, will show 80 V, and V_D, across all three resistors, will again show 120 V. (The negligible resistance of the wires is ignored.)

The practical effect of voltage drop in a series circuit can be seen in Fig. R16-17 that a simple circuit with a motor designed to operate off 120 V. A 25-W bulb placed in series with the motor causes enough voltage drop to prevent the motor from starting. Actual measurements on the illustrated circuit showed a 95-V drop through the bulb; therefore, only a 28-V supply to the motor remains. Twenty-eight volts is not enough to operate a device designed for 120-V application.

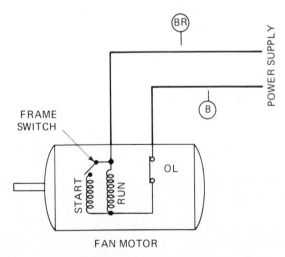

FIGURE R16-15   Fan motor.

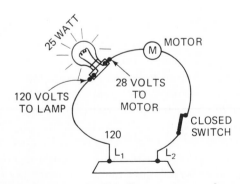

FIGURE R16-17   25-W-bulb-motor series diagram.

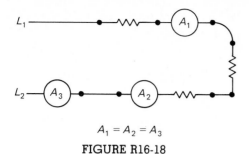

$$A_1 = A_2 = A_3$$

FIGURE R16-18

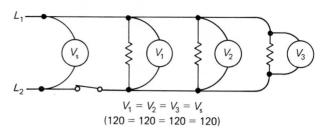

$$V_1 = V_2 = V_3 = V_s$$
$$(120 = 120 = 120 = 120)$$

FIGURE R16-19 Resistances in parallel.

If one 25-W and one 75-W bulb are wired in series, the 25-W bulb glows almost at full brilliance, with a 110-V drop, while the 75-W bulb takes the remaining 10 V without even glowing.

In a series circuit, then, the voltage drop through, or across, each resistance is only part of the total voltage and depends on the ohm-value of each resistor. The sum of these individual *IR* drops always equals the total applied voltage.

In a series circuit, the total resistance to current flow is the sum of all the individual resistances. The entire circuit may be treated as a simple circuit in Ohm's law calculations. For example, if a 75-W bulb with 194 Ω resistance and a 25-W bulb with 570 Ω resistance are in series, the total resistance of the circuit is 194 + 570 = 764 Ω.

For the entire circuit

$$I = \frac{E}{R} = \frac{764}{120} = 0.16 \text{ A}$$

Thus, 0.16 A will flow through each resistor. This amount of current is enough to heat the filament of the 25-W bulb, but not enough to warm the filament of the 75-W bulb or to start the motor in the example above. Using the same method and reasoning, it can be seen why a single 25-W bulb will glow brilliantly at 120 V, but three 25-W bulbs in series will each glow only dimly.

The foregoing discussion has clearly indicated that the current, the ampere draw, is the same in all parts of a series circuit. An ammeter placed anywhere in the circuit will read the same current (*A*) as at any other point. In Fig. R16-18, A_1, equals A_2 equals A_3 no matter how diverse the ohm values of the separate resistors.

R16-7
PARALLEL CIRCUITS

Just as a series circuit is described as one in which the resistors are connected end to end, a parallel circuit is one in which the resistors are connected side by side across the voltage source.

The three bulbs in Fig. R16-19 are wired in parallel; this is a conventional line drawing for such a cir-

cuit, which clearly depicts the side-by-side arrangement of the resistances.

Utility wiring in dwellings and most commercial-industrial structures is designed in parallel. The newer Christmas tree lights, those that do not all go out when one bulb fails, are in parallel.

Several characteristics of parallel circuits have immediate practical significance. You would notice first the full brilliance of the three 25-W bulbs in parallel. The same three bulbs in series, across the same 120-V source, would be considerably dimmer. A discussion of volts, ohms, and amperes in a parallel circuit will indicate why this is so.

The voltage drop across each resistance in a parallel circuit is the same and each IR drop equals the emf of the source. A voltmeter across any of the resistors will read 120 V, the same as directly across L_1 and L_2. Mathematically, V_1 equals V_2 equals V_3 equals V_S.

Another parallel circuit is shown in Fig. R16-20. This diagram indicates that electrons (*e*) from the power source may flow through more than one path on their way from and to the emf source. All of the current must flow in L_1 and L_2, but the total electron flow is divided among the three resistances. If the resistance of each device is equal, the current flow through each will be equal. If the resistances are unequal, the current flow through each will be unequal. In both cases, the total current will equal the sum of the individual currents in each branch, or leg, of the parallel circuit.

Assume, for example, a household living-room circuit, fused for 15 A (Fig. R16-21). All devices are in parallel when plugged into wall outlets, with three lamps, the TV, and an air-conditioner operating simultaneously. Let us assume the lamps, all together, are drawing 1.5 A, the TV is pulling 1 A, and the air-conditioner is drawing 7.5 A.

Plug in a 10-A iron so that the housewife can iron while watching TV in the air-conditioned living room and the 15-A fuse will open under the 20-A load.

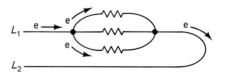

FIGURE R16-20 Resistances in parallel.

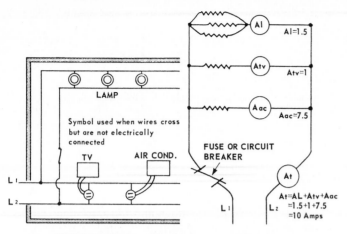

FIGURE R16-21   Combination parallel circuits.

Each leg of the circuit draws its portion of the total current, but all the electrons must come from and return to the emf source, through the fuse for the circuit. An ammeter placed at A_T will record this total ampere draw of all the devices on the circuit. Remember, the ampere draw through branches of a parallel circuit are equal only if the resistances in the several legs happen to have the same ohm value.

Total resistance, in a parallel circuit, is *not the* sum of individual resistances. Instead, *total resistance in a parallel circuit is always less than that of the lowest individual resistance.*

Assume a two-device parallel circuit carrying an iron with a 10-Ω resistance and a toaster with a resistance of 23 Ω (Fig. R16-22). Using Ohm's law as in a series circuit, we would add the resistances and divide them into the voltage to get amperes:

$$I = \frac{E}{R} = \frac{120}{33} = 3.6 \text{ A}$$

Actual measurement, however, shows that the toaster draws 5 A and the iron draws 11 A—a total of 16 A. Using the Ohm's law transposition:

$$R = \frac{E}{I} = \frac{120}{16} = 7.5 \ \Omega$$

7.5 ohms is the actual total resistance in this circuit, and this ohm value is less than the lowest individual resistance, 10 Ω.

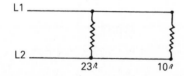

FIGURE R16-22   Parallel resistances of different value.

The formula for finding the total resistance (R_r) in a parallel circuit is

$$\frac{1}{R_T} = \frac{1}{R_1} + \frac{1}{R_2} + \frac{1}{R_3} + \cdots$$

In our example,

$$\frac{1}{R_T} = \frac{1}{23} + \frac{1}{10}$$

Usually, the easiest way to calculate is to convert the fractions to decimals (divide 23 into 1 and 10 into 1):

$$\frac{1}{R_T} = 0.043 + 0.10$$

$$= 0.143$$

$$0.143 R_T = 1$$

$$R_T = \frac{0.143}{1} = 7 + \text{ ohms}$$

Seven ohms approximates the 7.5 Ω obtained by calculation from field-measured amps and volts.

The calculated R_T can then be used in any calculations using Ohm's law.

Summary: In a parallel circuit:

Voltage across each resistor is the same and equals the applied voltage.

Total ampere draw divides among the parallel branches in the circuit and may differ in each branch.

Total resistance is always less than that of the lowest individual resistance.

R16-8
COMPARISON OF SERIES AND PARALLEL CIRCUITS

There is no "best" circuit. It is meaningless to dispute which type of circuit—series or parallel—is more useful, since each is indispensable to modern technology.

All but the most simple electrical circuits, indeed, make use of both types in what are logically called *series–parallel* circuits. On a regrigerated display case, the supply voltage is used to operate the compressor, the fans, the lighting circuit, and the defrost heaters. Depending on the manufacturer's design, each component may be wired to operate independently or together with any one or more of the other devices. The choice is made, in effect, by the electrical circuitry—by proper use of series and parallel circuits either alone and in combination.

R16-9
DIRECT AND ALTERNATING CURRENTS

All the information so far presented is valid for both direct current circuits and for alternating current circuits, the latter having resistance-type devices only on the lines. (The reason for this qualification about ac circuits will be explained later; it involves added resistance that is encountered in magnetic fields and is called *inductive reactance*.)

The rest of this discussion, dealing with such items as transformers, power distribution, solenoids, relays, and motors, will require an understanding of the distinction between ac and dc circuits and of differences in behavior between the two. Further, most of the remaining material will pertain to alternating current.

As defined earlier, electricity is said to flow whenever electrons move through a conductor. *If the electrons always move in the same direction through the conductor, the flow is called direct current (dc). If the electrons alternately move first in one direction, then in opposite direction through the conductor, the flow is called alternating current (ac).*

Electrical current produced by chemical action, as in a battery, is always dc. this is also true of current produced by thermocouples. Either alternating or direct current may be obtained from the mechanical action of steam or hydroelectric turbine generators.

R16-10
INDUCTANCE-CAPACITANCE-REACTANCE

Earlier it was stated that the discussion of Ohm's law and other such factors up to that point were valid for both direct current circuits and for *alternating current circuits with resistance-type devices only in the line.* The more perceptive reader may have noticed that the specific term *resistance* was applied to such things as incandescent lamps and heating elements but a coined term, like *ohm value*, is used to refer to the opposition to current flow offered by a motor or other induction device.

An ohmmeter applied to the leads of a small shaded-pole motor may indicate that the pure resistance of the motor windings and leads is some 35 Ω. Using

Ohm's law on an application across a 120-V line yields

$$I = \frac{E}{R} = \frac{120}{35} = 3.4 \text{ A}$$

A small motor like this does not draw 3.4 A. In fact, field measurement with a snap-around ammeter shows almost no deflection; clearly, the current is under 1 A.

The pure resistance of a solenoid coil was found to be 4 Ω. At 120 V, this would operate at 30 A theoretically. In actual operation, the coil draws 11.5 A.

Obviously, something other than pure resistance must be opposing current flow in these instances. The increased opposition is due to *inductance*. In a later discussion of splitting the phase to enable a single-phase motor to start, it will be shown that merely creating more inductance in one coil than in the other drops current in the first far enough behind that in the second to get the shift of magnetic poles needed to obtain starting torque.

It is difficult at this point not to become too complex for the beginner and for those who need only an awareness, rather than a full understanding, of what follows. Even simplified explanations of alternating current circuits call for more mathematics than is advisable in a discussion at this level.

For now, it is accurate enough to say that in alternating current circuits, several factors combine to affect current flow in one way or another; in direct current circuits, only pure resistance is normally encountered.

Inductive reactance is that opposition to current flow offered by induction devices. Since inductance can only affect current flow while the current is changing (current changes generate an induced emf) it does not occur in dc circuits except at the moments of closing and opening the circuit.

If an ac circuit has only pure resistance, the current rises and falls at the same time as the voltage, and the two waveforms are said to be *in phase* with each other as illustrated in Fig. R16-23.

Although it is only theoretically possible, if an ac circuit had only pure inductance, the opposed current

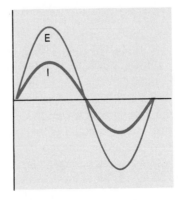

FIGURE R16-23 Voltage–amperage sine wave for resistance load.

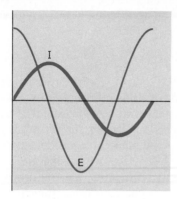

FIGURE R16-24   Voltage–amperage sine wave for inductive load.

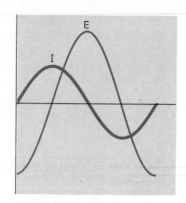

FIGURE R16-25   Voltage–amperage sine wave for pure capacitance load.

lags behind the voltage a full quarter of a cycle, or 90° as shown in Fig. R16-24. In this instance, the voltage and current are said to be 90° out of phase. The unit of measurement for inductance, symbol L, is the *henry*.

In a practical circuit containing both inductive reactance and resistance, the current wave will lag the voltage by an amount between 0° and 90°. Equal resistance and inductive reactance would produce a *45° phase angle*.

Just as inductive reactance opposes any change in current in an ac circuit, *capacitance* opposes any change in *voltage*. (Again, this phenomenon affects dc circuits only at the moments the current is turned on and off.) When the voltage increases, capacitance tries to hold it down; when the voltage decreases, capacitance tries to hold it up.

The electrical devices used to add capacitance to a line are, of course, the familiar *capacitors*, frequently called *condensers* by electronic and automotive people.

The actual action of capacitance in a circuit is to store a charge and to increase its charge if the voltage rises, discharge if the voltage falls.

Capacitance also offers an opposition to current flow—termed *capacitive reactance*—but the capacitive reactance (opposition to current flow) *decreases* as the capacitance (opposition to voltage change) *increases*.

As a result, in a theoretical circuit containing only pure capacitance, and no resistance, voltage can only exist *after* current flows. In such a theoretical circuit, the current wave leads the voltage wave by 90°, as illustrated in Fig. R16-25. The unit of measurement for capacitance, symbol C, is the *farad*, more practically the *microfarad*—one millionth of a farad.

In a practical circuit containing both resistance and capacitance, the phase angle is between 0 and 90°, and the current always leads the voltage. This is exactly opposite of what occurs in an inductive circuit where the current lags the voltage.

The fact that inductance and capacitance have opposing effects in a circuit is the explanation for the use

of capacitors with a motor or on a power line to improve power factor, an item that will be discussed later.

In an alternating current circuit, then, the total opposition to current flow offered by resistance, inductive reactance, and capacitive reactance is termed *impedance*, symbol Z, and is expressed in ohms. The term *total opposition* here is *not* synonymous to an arithmetic sum of the several factors; we are ignoring the mathematics involved in converting henries and farads to ohms and the calculation of the total effect of varying inductances and capacitances.

Z, after it is determined, can be substituted for R in the normal Ohm's law statement:

$$I = \frac{E}{R} \quad \text{or} \quad I = \frac{E}{Z}$$

R16-11
POWER FACTOR

In direct-current circuits, and in alternating-current circuits containing pure resistance only, power—in watts—is equal to the product of volts and amps.

$$P = EI$$

Thus, an incandescent lamp drawing 0.8 A on a 120-V line would be

$$P = EI = 120 \times 0.8 = 96 \text{ W}$$

or, practically, a 100-W bulb.

A 120-V circuit containing several resistance devices with a total ampere draw of 15:

$$P = EI = 1800 \text{ W or } 1.8 \text{ kW} = 120 \times 15$$

In this type of circuit, power factor is practically 100%; that is, the circuit actually expends very close to the calculated 1800 W. *Power factor* may be defined as the ratio of consumed power to supplied power, or as the percentage of time that the product of volts and amperes equals actual power.

A 100% PF can only exist when the voltage and current are in phase as explained earlier. When either inductive reactance or capacitive reactance causes the current to lag or lead the voltage (get out of phase) the product of volts and amperes gives only the *apparent power*, rather than the *actual power*.

Again ignoring the fundamental mathematics of the phenomena for the practical, the point is that for a typical commercial circuit, high in inductance, the utility must supply some power which does not actually perform work and which is not measured on the normal wattmeter, thus depriving the company of revenue.

If, for example, the calculation of measured volts and amperes equals 2000 W, but the device on the line actually shows a 1600-W consumption, the power factor is:

$$PF = \frac{1600}{2000} \times 100 = 80\%$$

This means that only 80% of the supplied power is doing measurable work. The remaining 20% is *magnetizing current*, which makes possible the functioning of induction devices but which does the work itself and therefore is not normally recorded.

Induction devices, such as motors and fluorescent lights, never have a 100% power factor. In fact, fluorescent lighting uses so much magnetizing current that utilities tend to look with disfavor on this type of illumination. However, since induction devices are necessary to our technology, power companies in general have established a 90% PF as a practical minimum which must be maintained on their lines.

Depending on design and application, motors may have an inherent power factor as low as 60%. In terms of waveforms, this means that the inductive reactance is causing the current to lag considerably behind the voltage. Previous discussion showed that capacitive reactance tends to make the current lead the voltage. Logically, then, a low power factor due to inductance can be raised by placing capacitance in the line.

In practice this is what happens. A running capacitor is placed on a motor; a bank of capacitors is installed in an industrial plant; pole capacitors are strategically located on distribution lines by utility companies. The resulting capacitance, acting in opposition to the inductance, establishes a more favorable power factor than would be possible with only inductance acting on the line.

PROBLEMS

R16-1. The atom is composed of _____ and _____ .

R16-2. The neutron has a _____ charge.

R16-3. Current (the flow of electrons) is from negative to positive. True or False?

R16-4. Define "static electricity."

R16-5. Define "dynamic electricity."

R16-6. Fill in the following statement: _____ charges attract, _____ charges repel.

R16-7. Batteries are divided into two types: _____ and _____ . Give an example of each.

R16-8. Do batteries produce an ac current or dc current?

R16-9. A battery produces electricity by _____ action.

R16-10. Two different wires fused together, with one end heated to produce electrical energy, is called a _____ .

R16-11. In what range is the voltage produced by a single thermocouple?

R16-12. Several thermocouples connected in series to produce a higher voltage output is called a _____ .

R16-13. A unit that converts light to electrical energy is called a _____ .

R16-14. The compression of some materials such as quartz or barium titanate will create a voltage difference across the material. The result is called the _____ .

R16-15. The generation of electrical energy in a turbine-driven generator is called the _____ process.

R16-16. What does "emf" stand for?

R16-17. What is the common term used instead of "emf"?

R16-18. Resistance opposes the flow of electrons. True or False?

R16-19. The unit of measure for resistance is the _____ .

R16-20. The rate of flow of electrons (current) is called the _____ .

R16-21. Name the two basic types of electrical circuits.

R16-22. All electrical circuits must have three basic elements. Name them.

R16-23. Give the formula for Ohm's law.

R16-24. The heat element in an electric furnace uses 20 A when 240 V is applied across it. What is the resistance of the element?

R16-25. In series circuits the total resistance is the sum of all individual resistances. True or False?

R16-26. In parallel circuits, the total resistance is the sum of all the individual resistances. True or False?

R16-27. Resistance to current flow in an induction device (motor) is called _____ .

R16-28. The unit of measurement for inductance is called a _____ .

R16-29. Capacitance opposes any change in _____ .

R16-30. A device that adds capacitance to a line is called a _____ .

R16-31. The measurement for capacitance is called a _____ .

R16-32. Define "power factor."

R16-33. The average power factor for refrigeration motor compressor assemblies is _____ %.

R17

Electrical Generation and Distribution

R17-1
ALTERNATING-CURRENT GENERATION

Although batteries and thermocouples are called chemical and thermal generators of electricity, respectively, the term *generator,* used alone, always refers to a mechanical-magnetic machine, such as a hydroelectric or steam turbine, which is used to generate almost all the electrical power consumed for domestic and industrial purposes.

Essential to this last type of power generation is induced electromotive force and the resulting induced current; the voltage and current are caused by the relative motion of a magnet and an electrical conductor.

Assume that the single coil or loop of wire shown in Fig. R17-1 is rotating in a magnetic field in the direction of the arrow. The following events occur as the loop is rotated (see Fig. R17-2).

1. In position 1, two sides of the loop are parallel to the magnetic field, no lines of force are being cut, and no current is produced in loop. Emf (volts) is zero.

2. In position 2, the coil has moved through 90°, the wires are now at right angles to the magnetic field, and maximum number of lines are being cut. This means that the emf is maximum in one direction.

3. At position 3, the loop has completed half of a complete turn, and again motion is parallel to the field and the emf again is zero.

4. At position 4, the loop again is cutting maximum lines, and the emf again is maximum but in the opposite direction.

5. The loop now is returned to position 1 and the cycle is completed.

These five steps describe the fundamental operation in the generating of alternating current. The smooth curve in Fig. R17-3, which traces the rise and fall of generated emf, illustrates one complete cycle. During the generating of typical household voltage, say, the curve traces the increase from zero to 120 V in one direction, the drop from 120 V to zero, the increase from zero to 120 V in the opposite direction, and the drop again to zero. A cycle that takes place in 1/60 of a second is called 60-hertz (cycle) current, meaning 60 cycles per second. The curve is also referred to as a *sine wave* or an *alter-*

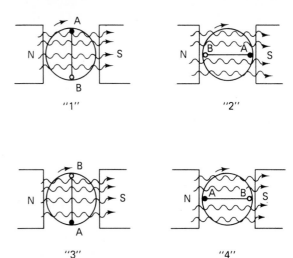

"1"   "2"

"3"   "4"

FIGURE R17-2   Generating one sine wave.

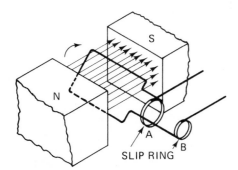

FIGURE R17-1   Rotating wire loop.

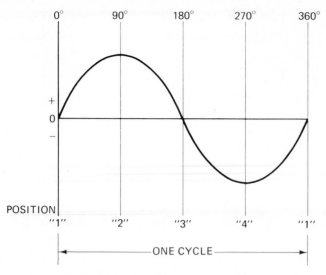

FIGURE R17-3   1 phase sine wave.

nating current wave form—these are mathematical terms that define the shapes of curves.

The current wave, as illustrated in Fig. R17-3, is said to be a *single-phase current,* meaning that only one loop is cutting the magnetic field. Addition of another loop at right angles to the first loop produces two separate voltages, one in each loop. Voltages produced are 90° apart; this is termed a *two-phase current.* Addition of a third separate loop results in a *three-phase current,* illustrated in Fig. R17-4.

Each loop is at a 120° angle to the other. Single- and three-phase systems of alternating current are the most popular.

Depending on the size of wires in the loop the number of loops in a coil, and the speed of rotation, several to many thousand volts can be generated from this mechanical-magnetic induction apparatus. The speed of rotation also bears directly on the number of cycles (hertz) generated per unit of time. Almost universal in this country is 60-Hz ac. To generate 60-Hz current, modern generators rotate at 3600 rpm. To produce the 25-Hz current, used until recently in most of Canada, the generator rotates at 1500 rpm. In most parts of the world 50-Hz current is common, and in aircraft 400-Hz equipment is used to get the same work out of lighter and smaller machines than would be possible with 60-Hz apparatus.

R17-2
DIRECT-CURRENT GENERATION

If the voltage or current curve of a battery were plotted against time, as was the ac waveform, it would look like Fig. R17-5. The slanted line leading to the flat line denotes the moment of time required for direct current to build up to full value when the circuit is closed. Since chemical generation always causes electrons to build up on one of the materials involved and thus flow only in one direction at a constant rate (battery is full strength), the dc curve is really a straight line very unlike the ac waveform.

Such steady dc is frequently termed *pure dc* or *true dc* to differentiate it from the pulsating direct current that is obtained from a mechanical-magnetic generator. The curve for such generated direct current looks like Fig. R17-6.

Since a rotating coil in a magnetic field always generates alternating current, whence the direct current?

The transition from ac to pulsating dc is accomplished by altering the ac slip-ring arrangement to the split-ring setup illustrated in Fig. R17-7. The effect of the split ring is to cut off the induced emf to the con-

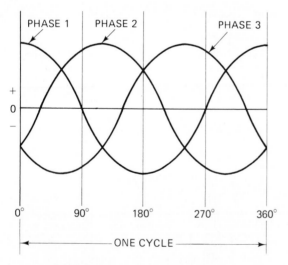

FIGURE R17-4   3 phase sine wave.

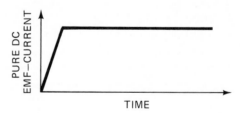

FIGURE R17-5   Current flow—true DC.

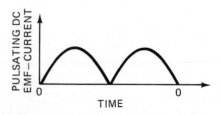

FIGURE R17-6   Current flow—pulsating DC.

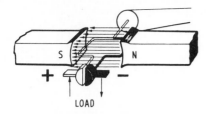

FIGURE R17-7 DC generator using slip rings.

ductors just before it reverses direction on each cycle. As a result, the current can build up to two peaks per cycle, but both in the same direction—no curve below the zero line as in the case of alternating current. A single-loop conductor causes wide peaks or pulsations, so in actual practice many loops are used to smooth out the current.

Although a few cities still have some streets (usually downtown) on direct current service, the most widely used type of electrical power used in the United States is alternating current, 60 Hz. The chief reason for the predominance of ac over dc is the greater ease and economy with which ac may be transmitted from relatively remote generating plants and distributed to consumers.

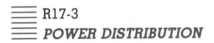

R17-3
POWER DISTRIBUTION

In the beginning dc current was used to supply lighting and motor loads to the fledgling electrical industry. However, the major disadvantage of dc, its inability to be transmitted for any great distance without excessive power loss and expensive voltage control apparatus, made ac current the better choice.

Briefly, this segment of the discussion on electricity will describe the transmission and distribution steps involved in getting electrical power from the generating station to the consumer.

Turbines, driven by hydropower or by gas, oil, or coal-fired steam, produce the mechanical rotation needed to cause relative motion between a magnetic field and electrical conductors. The induced electricity is fed outside into giant step-up transformers. Transformers are electrical devices that either raise (step up), or lower (step down), the voltage supplied to them. Their operation will be discussed later.

In the step-up transformer, the 18,000 V generated by the turbines is raised to 120,000 V, or whatever voltage the system is designed for. (Specifics in the following discussion apply to a typical power system. Although transmission and distribution voltages may differ in certain areas, the basic principles of generating and handling power are universal.)

The next stage after the step-up transformer is the mat—a type of shipping department. In its maze of steel towers, cables, and relay equipment, the electricity begins its journey to the customers along the transmission lines. The term *transmission lines* refers to the conductors that carry the electricity from the generating station through utility-owned switching stations to substations.

In the substation the transmission voltage is lowered to 4800 V in a step-down transformer. From the substation, distribution lines carry the electricity either directly to industrial customers who purchase power at a primary rate and do their own stepping down or to strategically located step-down transformers, each of which serves a group of industrial or domestic customers.

An ac network can reduce the 4800 V from the substation to the proper utilization voltages and deliver those to individual customers in a downtown or commercial district. In residential zones, the step-down is usually accomplished by pole transformers, and the power enters the home through the service entrance (meter, fuse box arrangement).

Figure R17-8 shows how different types of power are obtained. The primary wires, running along the tops of the utility poles, carry the current at distribution voltages, say 4800 V. Where three-phase power is to be taken off, all three wires are carried on the poles; when single phase will be used, either three or only two primaries may be used.

At the far left in Fig. R17-8, two wires enter the pole transformer where the secondary is tapped so as to give single-phase 120/240-V power over three wires, one neutral. The three secondaries from a single pole may travel from pole to pole and serve several houses—as many as the capacity of the transformer will allow.

Next, three wires are run from the primary into two transformers (which may be in a single shell). From the four wires issuing from the transformers, both single-phase and three-phase power are obtained at 120/240 V. Again, secondaries can run from pole to pole so that one transformer serves several consumers.

The third sketch shows another connection for obtaining the same secondary service as in the setup just discussed. The power company chooses a method according to its overall need to balance its distribution system. At the far right in Fig. R17-8 is shown the manner in which the 120/208 V or the 277/480 V network is obtained.

From this brief review of electrical generation and distribution, it is obvious that the several utility voltages (120/240 V, 120/208 V, 265/460 V) and single- and three-phase power must all be obtained from the generating station and transmission lines which generally produce and transmit only one emf, not a multitude of voltages. Normally, three-phase alternating current is generated.

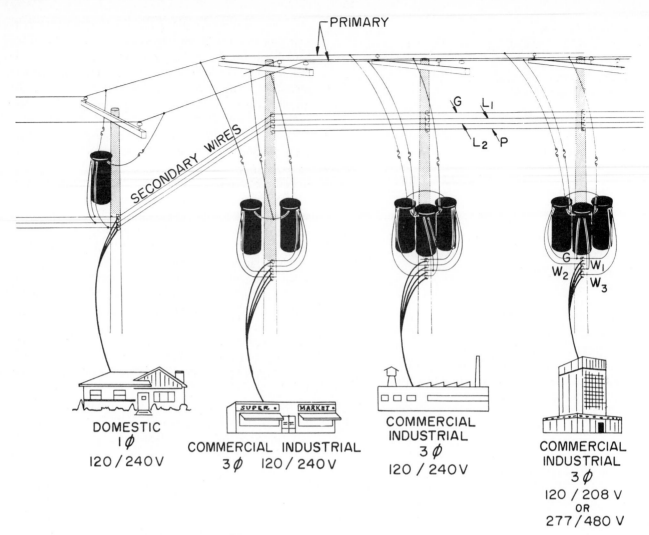

PRIMARY

SECONDARY WIRES

G L₁
L₂ P

G W₁
W₂ W₃

DOMESTIC
1φ
120/240V

COMMERCIAL INDUSTRIAL
3φ 120/240V

COMMERCIAL
INDUSTRIAL
3φ
120/240V

COMMERCIAL
INDUSTRIAL
3φ
120/208V
OR
277/480V

FIGURE R17-8 Electrical distribution system.

R17-4
VOLTAGES UTILIZED

From the three-phase (3φ) and three-wire primary lines (wires) the following voltage and phase combinations can be obtained (see Fig. R17-9).

These are domestic 60-Hz voltages. Outside the United States both the voltage and cycles may be different. The term *nominal voltage* indicates the desired voltage and not the exact line voltage at any given time. It is impossible for the power company to hold exactly to a given voltage; they are allowed a plus or minus variation, in some areas as high as 10%. This point is extremely important in the application of hermetic refrigeration compressors, as will be seen later.

The trend is toward higher and higher voltages and for good reason.

1. All normal distribution and utilization wiring is rated at 600 V, so the insulation is the same for specific classes, whether the voltage is 120 or 480 V.

2. There is little difference in the switches, relays, and other distribution and utilization details.

3. Motors that are wound for higher voltages, although the geometry of their windings may be different, will not cost very much more than regular motors.

4. Most important—in larger installations the customer may realize savings up to 35% in wiring costs over a 120-V base system.

The explanation of this last factor lies in the fact that when voltage is doubled, current is cut in half. At the lower ampere draw, smaller-diameter conductors may be used. For a house, the total savings would be insignificant, but for a large industrial or commercial establishment, the savings can come to hundreds, or even thousands, of dollars.

(Although the ampere draw is less, the customer

Nominal Utilization Voltage	60-Hz Service
120 and 240-1ϕ	120/240: 3-wire
120 and 240-1ϕ	120/240: 4-wire
240-3ϕ	
120 and 208-1ϕ	120/208: 4-wire
208-3ϕ	
120 and 240-1ϕ	240: 2 wire
240-1ϕ	240: 2 wire
120 and 240-1ϕ	240/416: 4-wire
416-3ϕ	
120 and 240-1ϕ	265/460: 4-wire[a]
460-3ϕ	

[a]Also termed 227/480.

still consumes and pays for the same amount of electrical power. Power, in watts, is equal to volts-times-amperes. The product of this multipliction is the same when volts are doubled and amperes are halved.)

A major consideration in the move toward higher voltages is the very great increase in the use of fluorescent lighting, which operates off 277-V ballasts. In a large installation supplied with 277/480-V power, the 480-V 3ϕ supply is for equipment, the 277-V 1ϕ power is for fluorescent lighting, and a small transformer is used on the premises to get the 120-V 1ϕ needed for small appliances, office machines, and control systems.

R17-5
CONDUCTORS

A *conductor* is the path through which the electrical charges are transferred from one point to another. *Insulation,* such as the rubber or plastic coverings on the surface of conductors, confines the flow of electrical charges along a desired path.

Both conductors and insulators conduct current, but in vastly different amounts. The current flow in an insulator is so small that for all practical purposes it is zero.

Materials that readily conduct current, such as copper or aluminum wires, busbars, and connectors, are the primary components of electrical equipment, circuits, and systems. The use of conductors and their insulation is governed primarily by the *National Electrical Code®* (NEC). The NEC lists the minimum safety precautions needed to safeguard persons and buildings and their contents from hazards arising from the use of electricity for light, heat, refrigeration, air-conditioning, and other purposes. Compliance with the NEC and proper installation and maintenance result in a system that is more likely to be free from electrical hazards.

Wires and cables are the most common conductors, and the NEC specifies that the conductor size be selected to limit the voltage drop and also the maximum current that is safe for conductors of different sizes having different insulations and wired under different conditions.

R17-6
CONDUCTOR SIZES

The system of measurement used to size a conductor is the *mil* or *circular mil,* as illustrated in Fig. R17-10. A mil is one-thousandth (0.001) of an inch. A circular mil is the area of a circle 1 mil in diameter. Expressing wire sizes in mils is much simpler than in fractions of an inch. For example, the diameter of a wire is said to be 25 mils instead of 0.025 in.

The mil-foot as illustrated in Fig. R17-11 is a convenient unit for comparing the resistance and size, called *resistivity,* of one conductor material with another. The resistivity valves are used to rate the current-carrying capacity of various conductor sizes as specified in the NEC.

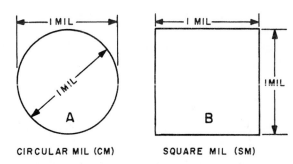

CIRCULAR MIL (CM)   SQUARE MIL (SM)

FIGURE R17-10 Numbering units for conductors.

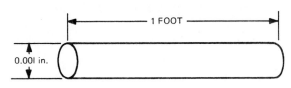

FIGURE R17-11 Circular mil-foot.

R17-7
WIRE SIZES

Wire sizes as published in the NEC are based on American Wire Gauge (AWG) numbers. This is a convenient way of comparing the diameter of wires. A wire gauge is shown in Fig. R17-12. It measures wires using a range calibrated from 36 to 0. To measure wire size, insert the *bare wire* into the smallest slot that will just accommodate it. The gauge number corresponding to that slot indicates the wire size. Lamp cord is commonly

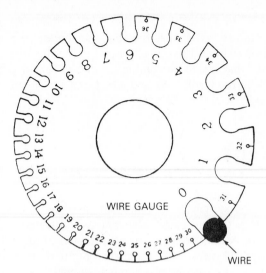

FIGURE R17-12   Wire gauge.

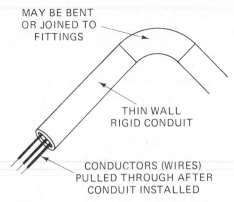

FIGURE R17-13   Rigid conduit (thin wall).

a No. 18 AWG. No. 14 copper wire is the size commonly used in household lighting circuits that carry 15 A current. (*Note:* Wire size numbers do *not* indicate the ampere capacity. This value is obtained from the NEC book and depends on the type of conductor, the type of insulation, and surrounding conditions.)

R17-8
CONDUCTOR CLASSIFICATION

A basic classification of wires and cables can be made according to the degree of covering used:

- *Bare conductors* have no covering. The most common use of bare conductors is in high-voltage electrical overhead transmission lines.
- *Covered conductors* are not insulated, but provide protection against weather and resistance to heat.
- *Insulated conductors* have a insulating covering over the metallic conductor that electrically isolates the conductor and allows grouping of conductors.
- *Stranded conductors* are composed of a group of wires in combination, usually twisted together.
- A *cable* can be either two or more bare conductors or insulated conductors. In general, the term *cable* is applied to the large conductors.

R17-9
CONDUCTOR PROTECTION

In residential wiring (inside the house) the wire covering or insulation is usually sufficient to protect the wire from damage and from being grounded or shorted.

However, in commercial and industrial buildings there is considerably greater risk of physical damage. Therefore, most local electrical codes require the wires or cable to be installed in a *raceway*, which is simply a channel for holding wires, cables or busbars, etc. The raceway may be metal or an insulating material.

Rigid metal conduit is one of the best known raceways (Fig. R17-13). Rigid conduit is made of steel or aluminum pipe that incloses all wires. A lubricant is used to help pull the wires through the conduit during installation. Such conduits are used for concealed wiring in buildings where the wiring is buried in concrete or masonry, since they provide good protection for wiring subject to mechanical damage or in hazardous locations. Rigid conduit bends can be made with a tube bender much like that used on copper refrigerant tubing.

Flexible conduit (Fig. R17-14), often called *Greenfield,* is used where rigid conduit is impractical. Flexible conduit is formed with a single strip of galvanized steel, spirally wound on itself and interlocked to provide strength and flexibility. It can be easily formed into numerous bends and turns. It absorbs vibration and can be used where adjustments in length (as when shifting a motor) are needed. Greenfield is not approved for wet or hazardous locations.

Liquid-tight, flexible metal conduit, sometimes called *sealtight,* is suitable where oil, water, and certain chemicals and corrosive atmospheres might present problems. Essentially it is a flexible metal conduit with an outer liquid-tight jacket.

Armored cable (Fig. R17-15), generally known as *BX,* consists of insulated wire covered with an armor that looks like flexible conduit. The armor must be stripped back about 8 in. from the end before it is attached to a junction box.

FIGURE R17-14   Flexible conduit (Greenfield).

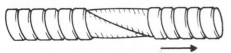

CUT ARMOR AND SLIDE OFF

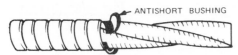

ANTISHORT BUSHING

INSERT BUSHING BETWEEN
PAPER WRAP AND ARMOR

ANTISHORT BUSHING

UNTWIST CONDUCTORS AND TEAR
OFF PAPER CLOSE TO BUSHING

FIGURE R17-15   B.X. cable.

Removing armor from BX cable.

a high-resistance conductor like tungsten (lamp filaments) or nichrome (heating elements), temperature rise is more pronounced.

The point to remember is that for each conductor of a certain material and a certain dimension, the temperature rise limits the amount of current that can flow safely. A $\frac{1}{4}$-in. piece of copper 1 ft long will conduct more electricity without overheating than a piece of tungsten of the same dimensions. But the $\frac{1}{4}$-in. piece of copper 1 ft long cannot carry as much current without overheating as a $\frac{1}{2}$-in. piece of copper of the same length.

Similarly, the 15-A branch circuits in a home are wired with copper conductors capable of passing 15 A without overheating. Should a short develop, resistance falls practically to zero and hundreds or thousands of amperes flow through the wire. Without a fuse, the temperature rise caused by this excessive current will overheat the 15-A conductors, at last burning them out, possibly causing a fire.

R17-10
FUSES AND PROTECTIVE DEVICES

Why fuses? For protection, yes. But why the need for protection? Simply that whenever electrons flow through a material, some heat is generated, and the temperature of the conductor rises. In a low-resistance conductor like copper, temperature rise is relatively low; in

R17-11
BASIC PRINCIPLE OF FUSE OPERATION

Essentially, a fuse is a strip of metal that will melt at a comparatively low temperature, placed in series with the circuit or devices it is designed to protect (see Fig. R17-16). When, for any reason, more current passes than the circuit is designed to carry, the metal strip melts

Plug Fuses

Ordinary Plug Fuse   FUSETRON dual-element plug fuse   Fustat

Cartridge Fuses

SUPER-LAG Renewable Fuse (Cut-away view)

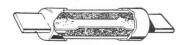

Ordinary One-time Fuse (Cut-away view)

How a Fuse Works

If a current of more than rated load is continued sufficiently long, the fuse link becomes overheated, causing center portion to melt. The melted portion drops away but—

Due to the short gap, the circuit is not immediately broken and an arc continues, and—

burns back the metal at each end, until

the arc is stopped because of the very high increase in resistance, and because the material surrounding the link tends

to mechanically break the arc. The center portion melts first because it is farthest removed from the terminals which have the highest heat conductivity.

On a short-circuit, the entire center section is instantly heated to an extremely high temperature, causing it to volatilize (turn into a vapor). This vapor has a high current-carrying capacity, which would permit the arc to continue, but the arc-extinguishing filler cools and condenses the vapor and stops the arc. The degree or extent of the volatilization of metal is dependent upon the capacity and resistance of the circuit and upon the design of the fuse.

FIGURE R17-16   Fuse. (*Courtesy* Bussman Manufacturing, a McGraw-Edison Company Division)

Electrical Generation and Distribution   199

apart, opening the circuit before the excessive current is able to overheat the conductors dangerously or burn out devices designed to operate at a much lower ampere draw.

This is the basic principle of operation for all thermally operated fuses, whether of the plug or cartridge type. Where the entire electrical load on a circuit is resistance load—incandescent lamps, heaters, toasters—a simple strip fuse, plug, or cartridge is sufficient protection. On such circuits, the ampere rating of the fuse must not be greater than that of the conductors.

When motors are on the line, the situation is a bit different. When a refrigeration unit or air-conditioner starts, for example, the ampere draw on startup may be five to seven times what it is while in operation. For the part of a second that the motor requires to reach running speed, the conductors can carry the increased current safely. But a simple strip fuse, say a 15-A fuse—designed to protect a motor with a 10-A running current—may blow in less than half a second on a 60-A starting current, causing an annoying current stoppage.

If a large fuse (60-A) is used instead of a 15-A fuse, the new fuse will still protect against a short in the circuit, but it will not protect against an overload. An overload, electrically speaking, is any condition that causes an electrical device to draw more current than normal, although it continues to operate. A bad bearing that results in a drag on a motor can cause the motor to pull higher than normal amperes in operation.

An oversized fuse in the circuit would make it possible to overheat the conductors dangerously, even though not enough current was drawn to blow the fuse chosen to protect against starting current.

R17-12
FUSE DEVICES

Time-delay fuses are the answer to such problems. Supplied in both plug and cartridge types, the time-delay fuse has a dual element. One element is the simple metal link that will open the circuit instantly in the event of a short. The second element makes use of a very small solder pot and a spring to make normal electrical contact. If current draw is excessive for a long enough period to melt the solder, the spring action breaks electrical contact, opening the circuit.

The pot and spring assembly may be designed for different protection ranges. One, for example, is so designed that it will allow overloads as high as 500% for 10 seconds without opening the circuit. Another will take a 200% overload for 110 seconds. This delay gives the motor ample time to start.

When a fuse blows, one of the following is the cause:

A short

An overload

Insufficient time delay in the fuse

Poor contacts on, in, or near the fuse

Overheating of the fuse (high ambients or too many fuses in a panel)

Wrong fuse size

Vibration

A Cardinal Rule: Never replace a blown fuse without determining and correcting the cause of the open circuit.

R17-3
OVERCURRENT CUTOUT DEVICES

Also operating on the principle of heat generation due to current flow, but applying the principle mechanically, is the *thermal overload protector,* used on hermetic compressors. Its basic component is a bimetallic strip. This is a thin strip of metal actually fabricated from two thinner strips of unlike metals pressed together. One metal has a higher rate of expansion than the other when heated (see Fig. R17-17).

Exposed to heat, the two metals expand at different rates, causing the strip to warp or buckle, since one or both ends are held fast. The bimetallic strip can be designed and shaped so as to warp enough to open an electrical circuit—with which it is in series—at a predetermined temperature. (The temperature rise may be due to current flow or to high internal heat in the compressor.)

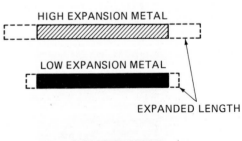

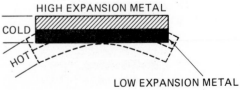

FIGURE R17-17 Bimetallic strips.

When the bimetallic element cools off, it reverts to its original size and configuration and closes the circuit again (automatic reset). If the cause of the first open was transient, the unit will operate normally. If an existing condition is causing chronic high temperature or current, the overload protector will repeatedly open the circuit until the condition is corrected.

Bimetallic elements are also applied in circuit breakers, motor starters, and relays, all of which are discussed later.

PROBLEMS

R17-1. Induced current results from rotating a wire coil in a magnetic field. True or False?

R17-2. Does alternating current produce a voltage curve of a straight line or a sine curve?

R17-3. Sixty-hertz current means producing one cycle every 60 seconds. True or False?

R17-4. Single- and three-phase currents are the most popular. True or False?

R17-5. Common residential voltages are _____.

R17-6. Three-phase power is normally the voltage used by _____.

R17-7. A material that readily conducts current is called a _____.

R17-8. The measurement used to size a conductor is called a _____.

R17-9. "NEC" stands for _____.

R17-10. A channel for holding wires, cables, etc., is commonly called a _____.

R17-11. Are fuses installed to protect the electrical appliances (lights, motors, etc.) or the electrical conductors (wires)?

R17-12. The type of fuse used on circuits to motors is called a _____ type.

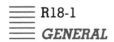

Electrical Components

R18-1
GENERAL

In Chapter R16, basic electricity was discussed with mention of the various components, such as battery transformers and generators as well as types of loads such as light bulbs. In Chapter R17, electrical generation via mechanical/magnetic means as well as conductors for distribution were discussed. In the field of refrigeration and air-conditioning, where the greatest number of potential problems are in the electrical portion of the equipment, we must go beyond simple circuitry, for the application and service technician must be acquainted with such components as transformers, relays, contactors, motor starters, motors, motor protectors, etc.

R18-2
BASIC CIRCUIT COMPONENTS

Every electrical circuit, where electrical energy is obtained, moved, and used, consists of three basic parts:

1. *Power source* A source of electrical energy is required, be it a battery, generator, piezo crystal, or thermocouple.
2. *Load* Something has to be able to convert the electrical energy into another form to be usable for a certain purpose or purposes.
3. *Conductors* A means must be provided to conduct the electrical energy from the power source to the load. Conductors were discussed in Chapter R17, so little further discussion is needed at this point.

Sometimes considered to be a basic part of the electrical circuit is controls. This is an error, however,

because not all electrical circuits have controls. In fact, the load would not operate properly, in the case of an electrical time clock or the control transformer in the air-conditioning unit, if an on/off control were put into the circuit to control the clock or power transformer. As a general rule, however, controls are necessary in the refrigeration and air-conditioning field to get the desired results.

The most basic circuit of all is a simple flashlight (Fig. R18-1). In this circuit, the batteries are the power source, the light is the load, and the shell is the conductor. Because on/off operation is desired, a switch has been added as a control device. This circuit is the same as a simple light circuit in a home (Fig. R18-2), where the transformer on the power pole feeding electricity into the house is the power source, the light bulb is the load, and wires from the transformer, through the service entrance, fuse panel, and switch to the light are the conductors. A control device (on/off switch) is used to provide operation whenever desired and fuses are also provided for protection.

Electrical circuits can become confusing, however, because many loads may be connected to a single power source, each controlled by a control device; a single load may be controlled by many control devices; or many loads may be controlled by many control devices. *Electrical circuits must have only one power source.*

Therefore, as stated previously, all electrical circuits consist of the following:

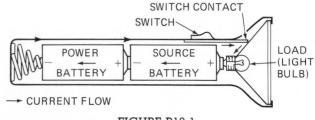

FIGURE R18-1

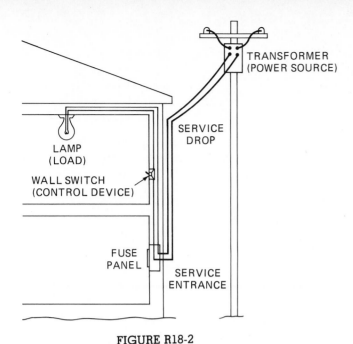

FIGURE R18-2

1. Power source
2. Load
3. Conductors
4. Controls (optional)

R18-3
POWER SOURCE

In Chapter R17 the generation of ac electrical power by mechanical means as well as ac power distribution were discussed. From this we learned that the generator was the load on the mechanical source of power, be it a turbine, internal combustion engine, etc. In turn, the transformer substation at the generator is the load on the generator, the transformers feeding the secondary system are the load on the substation transformers, and in turn the electrical devices in the domestic, commercial, and industrial unit all at once are the loads on the secondary transformers.

From this we conclude that transformers are loads when they receive energy and power sources when they give it off. The discussion here will center around the transformer as a power source, as this is its main function. To be more explicit, we should state that the main function of the transformer is to change electrical pressure (voltage) up or down to provide the proper voltage for the circuit in question. The greatest potential for electrical failure in refrigeration or air-conditioning units centers around the control transformer supplying the 24-V control system and the load it is expected to carry.

Transformers used in power work consist of two coils of wire wound together around a laminated inner core and are capable of changing the electrical pressure

FIGURE R18-3 Low voltage control transformer. (*Courtesy* Honeywell, Inc.)

or voltage in direct proportion to the number of turns in the two coils (see Fig. R18-3).

To understand the operation of a transformer we must go back to the wire in a magnetic field. When discussing power generation, we passed a wire through a magnetic field, which produced an electrical potential (voltage) in the wire. In transformers we use the principle that a magnetic field is produced around a wire when electrical current is passed through it as the source of the magnetic field for generation in the second wire. By placing wires in parallel (Fig. R18-4) and applying voltage to points *A* and *B* of wire 1, a current flow will occur through wire 1 and a magnetic field will build around wire 1. The amount of magnetic field will depend on the amount of current flowing through the wire. As the magnetic field builds around wire 1 it will cut through wire 2 and induce a voltage in wire 2. Again, the amount of voltage induced in wire 2 will depend on the amount of magnetic field cutting wire 2.

Because it is the action of the magnetic field passing through the second wire that produces the induced voltage, transformers must be supplied with a varying voltage (ac) and will not work on a fixed voltage (dc). The amount of voltage that is produced in the second wire will be in direct proportion to the length of the wires that are exposed to each other. If the wires are of equal length, the voltage into wire 1 (this is called the *primary*

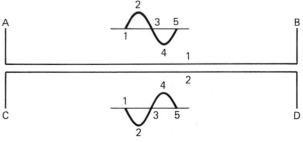

FIGURE R18-4

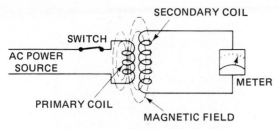

SECONDARY COIL

SWITCH

AC POWER
SOURCE

PRIMARY COIL

MAGNETIC FIELD

METER

FIGURE R18-5 (*Courtesy* ACCA)

side or coil) will be equal to the voltage induced in wire 2 (this is called the *secondary side or coil*). If the secondary wire were only 10% as long as the primary, if it were only exposed to 10% of the primary wire, the voltage induced in the secondary wire would be 10% of the primary voltage (see Fig. R18-5).

Also, for more efficiency and smaller volume, the wires are in coils around laminated magnetic iron cores. The voltage change is then determined by the number of turns in each coil rather than by the length of the wire. This process can be stated as a rule: A secondary voltage (E_s) is equal to the primary voltage (E_p) when it is multiplied by the ratio of the secondary turns (N_s) to the primary turns (N_p):

$$E_s = E_p \ \frac{N_s}{N_p} \qquad \text{(R18-1)}$$

EXAMPLE

A given transformer has 250 turns in the primary winding and 125 turns in its secondary winding. If a source emf of 240 V is applied to the primary coil, what will be the value of the voltage available in the secondary coil?

SOLUTION

Using Eq. (R18-1), the solution would be

$$E_s = 240\left(\frac{125}{250}\right) = 240(0.5) = 120 \text{ V}$$

It will be noted that the number of turns in the secondary coil are only-half the number of turns in the primary coil; therefore, the secondary voltage is half the primary voltage. Another equation that might be set up is

$$\frac{\text{primary source voltage}}{\text{secondary induced voltage}} = \frac{\substack{\text{total number of} \\ \text{primary turns}}}{\substack{\text{total number of} \\ \text{secondary turns}}}$$

or it can be simplified to

$$\frac{E_p}{E_s} = \frac{N_p}{N_s} \qquad \text{(R18-2)}$$

If, as shown in Eq. (R18-2), the source voltage to the primary coil is 120 V and there are 250 turns in the primary coil with only 50 turns in the secondary coil, the resulting secondary

voltage would be 24 V. The equation would be

$$E_s = 120\left(\frac{50}{250}\right) = 120 \ (0.20) = 24 \text{ V}$$

A type of transformer in which the secondary coil has a smaller number of turns—and therefore less voltage—than the primary coil is classified as a step-down transformer. This is the type of transformer utilized in low-voltage control circuits used in refrigeration, air-conditioning, and heating. A step-up type of transformer is one in which the secondary coil has a larger number of turns than the primary. The step-up transformer is utilized when a higher voltage is desired but is not readily obtainable from the primary source. If the primary coil has 100 turns and the secondary coil has 1000 turns, the secondary voltage will be 10 times the primary voltage (see Fig. R18-6).

Since power is the result of the effective voltage multiplied by effective current, or $W = EI$, the relationship of power in the secondary coil is closely maintained to that in the primary coil. This relationship is expressed as

$$E_p I_p = E_s I_s \qquad \text{(R18-3)}$$

where

E_p = primary voltage

I_p = primary current

E_s = secondary voltage

I_s = secondary current

EXAMPLE

If the primary voltage in a transformer is 120 V and has a current of 1 A, and the secondary coil has a value of 24 V, what current is available in the secondary?

SOLUTION

Utilizing Eq. (R18-3), we have

$$I_s = \frac{E_p I_p}{E_s} = \frac{120 \times 1}{24} = \frac{120}{24} = 5\text{A}$$

In both the step-up and step-down transformers, another equation is used. The current in amperes flowing through a primary coil, multiplied by the number of turns in the coil, will equal the ampere-turns of the secondary coil. This is expressed as

$$I_p N_p = I_s N_s \qquad \text{(R18-4)}$$

Therefore, in a step-down transformer, the secondary current will be greater than the current carried in the

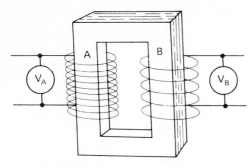

FIGURE R18-6 (*Courtesy* ACCA)

primary, since the number of turns are fewer in the secondary coil. If, in the example above, there are 250 turns in the primary coil and 50 turns in the secondary coil, Eq. (R18-4) shows that

$$1 \times 250 = 5 \times 50$$
$$= 250 \text{ ampere-turns in each coil}$$

The secondary winding of a transformer is always connected to the load being handled. The two principal types of transformer construction—the core type and the shell type—are shown in Fig. R18-7. Figure R18-7a shows the core-type transformer, in which the copper winding surrounds the laminated iron core. Figure R18-7b pictures

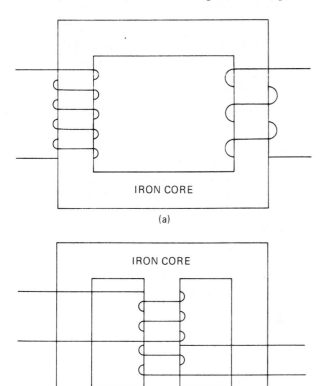

FIGURE R18-7 Transformers: (a) core type; (b) shell type. (*Courtesy* Trane Company)

the shell-type transformer, in which the iron core surrounds the copper winding of the two coils.

In summary, the voltage relationship between the primary and secondary windings of a transformer is in direct ratio to the number of turns of wire in each winding. If one winding has twice as many turns as the other winding, it will naturally have twice the voltage. Conversely, the current circulation in the two windings will be inversely proportional to the ratio of the two coils. The designation of the ratio in the transformer, such as that used in the example explaining Eq. (R18-1), would be classified as a two-to-one step-down transformer, since the primary voltage is higher.

Because a transformer will work in either direction, if the 24-V coil were connected to a 24-V power source, 115 V would be obtained from the 115-V coil. In this instance, the 24-V coil would be the primary and the 115-V coil would be the secondary. Caution must be used in wiring controls together where several tranformers are involved, to prevent this reverse action. High voltage could result, which could be dangerous, as well as cause erratic control operation and control failure.

Care must also be used to make sure that the transformer coils are connected to the electrical pressure or voltage they are designed to handle. If the voltage applied is higher than the coil rating, the coil will not handle the higher voltage and will burn out very quickly. If the voltage is lower than the coil rating, the transformer will not deliver as designed and will overheat and eventually burn out. Always check the coil voltage rating and make sure it is supplied the rated voltage.

Transformers, like batteries, have a maximum rate capacity. Batteries are rated in volt-ampere-hours and transformers are rated in volt-amperes. A battery will put out rated voltage for a period of time depending on the connected load. If the battery is rated at 1 volt-ampere-hour it will deliver 1 ampere for 1 hour at 1 volt or any multiple thereof before the voltage drops: for example, $\frac{1}{2}$ A (ampere) for 2 hours, 2 A for $\frac{1}{2}$ hour, etc.

Because a transformer does not depend on stored energy but draws its energy from a power source, the time rating does not apply. Therefore, the transformer is rated in volt-amperes only. For example, if the transformer is rated at 40 VA at 24 V, its maximum capacity is 40 VA divided by 24 V, or 1.665 A. As long as the amperage load connected to the transformer does not exceed this maximum amount, the transformer will deliver full voltage.

If the maximum amount is exceeded, however, the transformer output voltage will drop in approximately the same ratio as the connected amperage load increases. This results in reduced voltage to the relay coils, solenoid coils, and other connected loads. Also, as the excessive load on the transformer increases, the heat generated in the transformer increases, the life of the transformer decreases, and burnout occurs.

Electrical Components 205

All alternating-current devices using magnetic coils, such as relays, solenoids, motors, etc., have a voltage tolerance of plus or minus 10% of rated voltage regardless of the rated voltage. A 115-V relay will operate at plus or minus 10% of 115 V—maximum of 115 V plus 10%, or 126.5 V, minimum of 115 V minus 10%, or 103.5 V. This is common knowledge.

What is not generally realized, however, is that 24-V devices also have the same tolerance. A 24-V relay will operate satisfactorily at a maximum of 24 V plus 10% or 26.4 V and a minimum of 24 V minus 10%, or 21.6 V. High voltage on 24-V devices is seldom a problem, as it rarely occurs. To exceed the 26.4-V output from the secondary of a 115-V to 24-V transformer, the primary voltage (115 V rating) must exceed 126.72 V.

The reverse situation is common, however, in large complex control systems. If, for example, the control system connected to the air-conditioning system used damper motors for individual zone control, a fan relay liquid-line solenoid for automatic pump down, a damper motor for seasonal changeover, and a compressor contactor for control of the motor compressor assembly, and all these devices were connected to the 40-VA transformer in the condensing unit, the following could happen.

1. Setting the season switch on the thermostat to cooling energizes the season changeover damper motor. This places a load on the transformer of an average of 0.625 A, depending on the damper motor used.

2. Setting the fan switch on the thermostat subbase energizes the fan relay, which draws an average of 0.229 A. A total transformer load is up to 0.85 A.

3. A rise in room temperature causes the cooling thermostat to close, energizing the damper motor, placing another load of 0.625 A of the transformer. Transformer load is now up to 1.475 A.

4. The zone damper motor opens the damper and closes a switch to energize the liquid-line solenoid valve. This places an additional load on the transformer of approximately 0.5 A. The transformer load is up to 1.975 A, an excess over the rated capacity of the transformer of 0.32 A. Although the output voltage of the transformer does not drop in direct proportion with the amount of overload, it can be expected that the output voltage will be approximately 40 VA divided by the 1.975-A load, or 21.3 V. Because 21.3 V is below the minus 10% voltage tolerance of the liquid-line solenoid, the solenoid will be slow opening, will hum loudly, and will slowly overheat. The transformer will also slowly overheat at this load and eventually burn out.

5. However, this is not the final step. The liquid-line solenoid will open and allow refrigerant flow. This supply of refrigerant to the cooling coil will create suction pressure buildup and cause the low-pressure control to close. Closing the contacts in the low-pressure control energizes the coil in the compressor motor contactor, placing an additional load of approximately 0.229 A on the transformer. The total load on the transformer is now up to 2.204 A and the output voltage of the transformer can be expected to drop to approximately 18 V. Because this voltage is well below the minimum voltage of 21.6 V, the relay will not pull in properly, the relay will clatter excessively, the contacts will bounce instead of closing solidly, the contacts will burn, the electrical system to the compressor motor will be subjected to fast make and break of electrical supply, and starting capacitors, relays, and compressor motors will burn out. All this is because of extremely low voltage due to excessive load on the 24-V control circuit transformer. Therefore, *always make sure that the total connected load never exceeds the output capacity of the transformer.*

R18-4
LOADS

The load portion of the electrical circuit is that portion that connects the electrical energy to another form of usable energy.

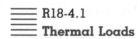

R18-4.1
Thermal Loads

If the conversion is to heat, the load is called a *thermal load.* Thermal loads are divided into two types:

1. Primary type
2. Parasitic type

Primary Type: The primary-type resistance load is a resistance sized to convert the total electrical pressure or voltage to heat. This resistance is connected across the power source with or without control devices and is sized to give off a certain amount of heat and draw whatever amperage is necessary to produce the necessary heat. A typical primary resistance load in an air-conditioning unit would be:

1. Crankcase heater
2. Strip heater
3. Heat-motor-type damper motor
4. Heat-motor-type valve

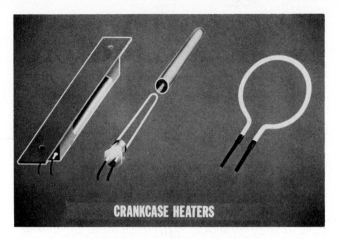

CRANKCASE HEATERS

FIGURE R18-8   Primary loads.

Crankcase Heater Crankcase heaters are used to prevent the accumulation of liquid refrigerant in the motor compressor assembly during long OFF-cycle periods or over the winter. Because of the necessity of operation when the unit is not running, they are wired to the line side of the motor starter or contactor and are energized at all times. They are generally a continuous tube type (a resistance wire within a metal shell). Open-resistance wire is never used because of the lack of protection from the elements and the necessity for insulation to prevent grounding of the wire and burning out. (See Fig. R18–8.)

Strip Heaters Strip heaters are primary-type resistance loads generally used to convert electrical energy to heat energy for space or comfort conditioning or as supplemental heat used in conjunction with heat pumps. Because of their enclosure in a metal cabinet and suspension in insulating frames, open wire resistances as well as shielded resistances may be used. Heaters of this type are generally controlled so as to maintain comfort conditions. (See Fig. R18–9.)

FIGURE R18-9   Ducted resistance heater—primary load. (*Courtesy* Lennox Industries)

FIGURE R18-10   Primary load.

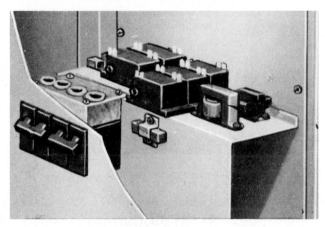

FIGURE R18-11   Primary load.

Heat-Motor-Type Damper Motor In an application of this type, the load resistor is used to produce the heat required to warp a bimetallic-element to produce the necessary motion to operate dampers. The resistor is designed to be placed across the full voltage of the power source and is generally controlled by a switch device. (See Fig. R18–10.)

Heat-Motor-Type Relay In this application, like the damper motor, heat from the load resistor is used to warp a bimetallic element. Instead of the element producing a rotary motion, single-direction motion is used to open and close the relay contact. This is also an across-the-line (primary) type of resistance and is controlled by switching devices. (See Fig. R18–11.)

Parasite Type: A parasite resistor uses a small portion of the amperage drawn by another load device (generally inductance type) so as to reduce the required amount of heat. The greater the amperage passed through the parasite heater by the demands of the load, the greater the amount of heat given off by the heater.

This device is used to open a control circuit or to open contacts if the heat given off by the parasite heater

Electrical Components   207

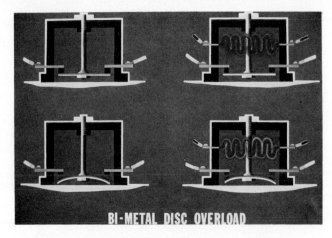

FIGURE R18-12 Parasitic load.

becomes excessive due to unusual amperage demands by the main load. This excessive heat then causes a bimetallic element to warp and open the contacts. The opening of the contacts then cuts the main load off from the electrical power source by opening a separate control circuit as in Fig. R18-12 (right side) or by open in the main line to the load as in Fig. R18-12 (left side).

Another example of a parasite heater is the heat anticipator resistor in a thermostat. In a heating thermostat, the heat anticipator resistor uses a small proportion of the amperage through the heating control circuit to produce the heat necessary to give the anticipated operation of the bimetallic strip in the thermostat. Thus, because the amount of heat necessary to activate the thermostat is always the same regardless of the amperage requirements of the heating controls, it is necessary that the heat anticipator resistor be matched to the amperage requirements of the heating controls. If

FIGURE R18-13 Heat anticipator—parasitic load. (*Courtesy* Control Products Division, Johnson Controls, Inc.)

the heating control required 1 A to activate properly, a smaller parasite resistor would be used as a heat anticipator if the controls only required $\frac{1}{2}$ A to activate. Failure to check the size of the heat anticipator resistance has caused a high percentage of thermostat replacements in the field. (See Fig. R18-13.)

R18-4.2
Inductive (Magnetic) Loads

Inductance loads are devices that use the electrical energy by converting this energy into magnetic energy for the creation of a magnetic field. Such magnetic fields are then used to produce motion of an armature, such as in a relay or contactor, or rotary motion of a rotor, as in a motor.

In a resistance load device, the amperage draw depends directly on the resistance of the resistor and the voltage applied. In an inductance load device, however, the amperage drawn is limited by the amount of opposing electric pressure or voltage generated in the coil in the device. This counter voltage, called counter electromotive force (cemf), is generated when the magnetic field created by the electrical energy flowing through the coil changes due to the rise and fall of the electrical pressure or voltage. Because alternating current reverses itself or alternates (starts at zero, builds to maximum in one direction, falls to zero, builds to maximum in the reverse direction, and returns to zero to create an alternation or cycle) the magnetic field also follows this pattern. The field starts at zero, builds to maximum intensity in one direction, falls to zero, builds to maximum in the reverse direction, and returns to zero. As the magnetic field builds up or collapses in the device, it cuts through other turns of the coil. This produces an electrical pressure or voltage which is in the opposite direction to the original electrical pressure or voltage.

Naturally, the less the resistance to flow the magnetic field encounters, the greater the intensity of the field and the greater the generating effect. If the magnetic field encountered no resistance and was able to generate full counter voltage, the counter voltage would equal the applied voltage. The two voltages would then counteract each other and no current flow would take place. However, the magnetic field is expected to do some work, such as moving an armature in a relay, pulling up a plunger in a solenoid, or turning a rotor in a motor. This work makes up the resistance to flow of the magnetic field and determines the ability of the coil to generate the counter voltage.

Because an open iron core (Fig. R18-14) has more resistance to flow of the magnetic field than a closed iron core, the amount of counter voltage generation is less than when the core is closed. This is the reason that a relay coil draws more current to close the relay than to keep it closed or to open a solenoid valve than to keep

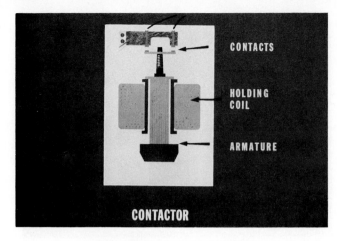

FIGURE R18-14 Primary load.

it open. This is also the reason why a motor draws more current as the load on the motor is increased. Because the magnetic field in the motor is expected to do more work as the load is increased, more of the magnetic energy is converted to work energy, there is less magnetic energy left to generate the counter voltage, the voltage difference between applied and counter voltage increases, and current flow through the windings increases. Also, as the current flow through the winding increases, the more the resistance of the wire in the winding converts the electrical energy to heat energy and the hotter the wire gets. This means that if the armature of the relay does not close, the solenoid plunger does not pull up, or the motor is overloaded, the winding wire will heat to the point of insulation failure and burn up.

In the section on transformers, the statement is made that "alternating-current appliances will tolerate a voltage of plus or minus 10% of rated voltage." In order to have the proper magnetic field quantity to do the required work, the electrical pressure or voltage must be high enough to force sufficient current through the coil. Therefore, when the voltage is too low, the coil does not develop the required "pull," and the device will not do the required work. In the case of a relay or solenoid, the armature or plunger will not pull up, the resistance to the magnetic field remains high, the counter voltage generated remains low, and the coil draws excessive current. This overheats and burns out the coil. In the case of a motor, the magnetic fields are not sufficient to get the rotor turning, again the counter voltage generated is very low, and the motor draws excessive current and quickly burns out. Of course, the motor has overload cutouts to protect against burnout, but even these have their limitations.

Excessive voltage also causes burnouts, because the applied voltage remains higher than the generated counter voltage, keeping the voltage difference higher than the device is designed to handle. As a result, the windings are forced to carry more current than normal, they overheat, the insulation breaks down, and the coils

burn out. The speed of the burnout process is slower from high voltage than from low voltage, but the end result, burn out, is inevitable. Therefore, when the unit is trying to start, always make sure that the applied voltage is at the rated voltage of the unit.

R18-5
CONTROLS

Controls are divided into two types:

1. Total interruption
2. Characteristic change

A total interruption control totally makes or breaks the circuit, as, for example, the light switch.

A characteristic change control affects the performance of the connected load but does not discontinue the power supply, as, for example, the dimmer control portion of a light dimmer switch or the modulating speed control on the ceiling fan.

R18-5.1
Total Interruption

There are two categories of total interruption or cycle controls, primary and secondary. A primary control actually starts and stops the cycle, either directly or indirectly, as dictated by temperature or humidity requirements. Secondary controls regulate and/or protect the cycle when required to do so by either a primary control or by conditions within the cycle. Many controls can serve both functions in the same system, so the control cannot be defined as primary or secondary until its main purpose is known. The first group of controls discussed include those most frequently used as primary controls.

Primary Controls: There are three types of primary controls. The first is operated by temperature and is called a *thermostat*. The second is operated by pressure and is called a *pressurestat*. The third is operated by moisture or humidity and is called a *humidistat*. Each of these controls can be used to regulate cycle operation. For example, when a home refrigerator becomes too warm, that is, the temperature becomes too high for food storage, the thermostat senses this and starts the compressor.

Pressurestats are often used to control temperature conditions in a display case by controlling evaporator pressure. When the evaporator pressure and corresponding temperature become high, the pressurestat operates and starts the compressor.

In storage rooms where humidity is all-important, the humidistat is designed to start the refrigeration cycle

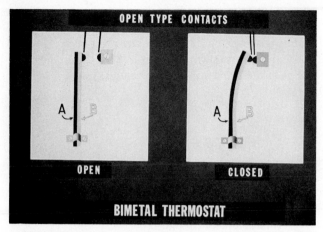

FIGURE R18-15 Bimetallic thermostat. (*Courtesy* Carrier Air-Conditioning Company)

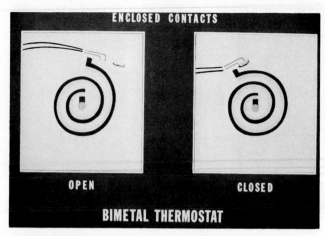

FIGURE R18-16 Bimetallic mercury bulb thermostat. (*Courtesy* Carrier Air-Conditioning Company)

when the humidity rises to a predetermined point. Conversely, all the above controls will stop cycle operation when the conditions are satisfied.

Thermostats respond to temperature; they can do this because of the warping effect of a bimetallic strip or of fluid pressure. The thermostat illustrated in Fig. R18-15 is a bimetallic thermostat. The bimetallic element is composed of two different metals bonded together. As the temperature surrounding the element changes, the metals will expand or contract. As they are dissimilar metals, having different coefficients of expansion, one will expand or contract faster than the other. In this drawing the open contacts are illustrated on the left. No current is flowing. If the temperature surrounding the bimetallic element is raised, metal *A* and *B* will both start to expand. However, metal *A* is chosen because it will expand faster than metal *B*. This will cause the bimetallic strip to bend and close the contacts as shown on the right. As the temperature drops, *A* would contract faster than *B,* thereby straightening the element and opening the contacts. This thermostat is widely used, especially in residential heating and cooling units, as well as in refrigeration systems that work above freezing conditions, where the stat is actually located in the controlled space. It is simple and cheap to construct, yet is dependable and easily serviced.

One of the common variations of the bimetallic thermostat is a mercury bulb thermostat (Fig. R18-16). Although the bimetallic principle is used, the contacts are enclosed in an airtight glass bulb containing a small amount of mercury as shown. The action of the bimetallic element tilts the bulb. By tilting the bulb to the left, mercury in the bulb will roll to the left and complete the electrical circuit. By tilting the bulb to the right, the mercury will flow to the other end of the bulb and break the electrical circuit. This thermostat is used extensively in space heating and cooling. Although a little more expensive than the bimetallic open-contact ther-

mostat, its operation is more dependable because dirt cannot collect on the contacts.

The second method of thermostat control is the fluid pressure type shown in Fig. R18-17. With a liquid and gas in the bulb, the pressure in the bellows will increase or decrease as the temperature at the bulb varies. The thermostat illustrated is a heating thermostat. As the pressure in the bulb increases with the rise in bulb temperature, the bellows expands and, through a mechanical linkage, opens the electrical contacts. As the pressure in the bellows decreases with a lowering of bulb temperature, the bellows contacts and closes the electrical contacts. This thermostat is sometimes referred to as a *remote bulb thermostat.* The controlling bulb can be placed in a location other than that used for the operating switch mechanism. For example, with a cooling thermostat in a refrigerated room, the bulb can be placed inside the room and the capillary run through the wall to the operating switch mechanism located outside the room. Thermostat adjustments can be made without en-

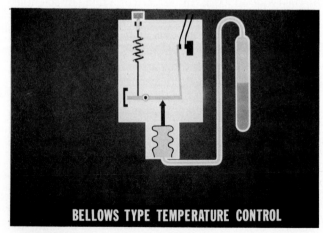

FIGURE R18-17 Bellows-type thermostat. (*Courtesy* Carrier Air-Conditioning Company)

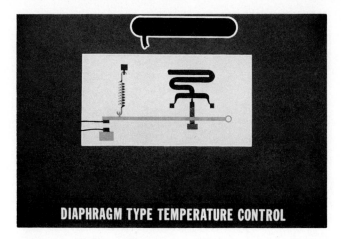

FIGURE R18-18 Diaphragm-type thermostat. (*Courtesy* Carrier Air-Conditioning Company)

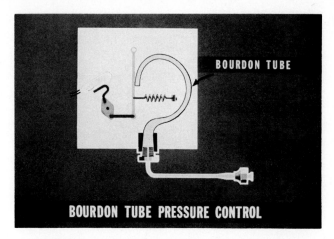

FIGURE R18-20 Bourdon tube pressure control. (*Courtesy* Carrier Air-Conditioning Company)

tering the refrigerated room, and the switch mechanism is not exposed to the extreme conditions inside the room.

Another version of the pressure-actuated thermostat is the diaphragm type (Fig. R18-18). In this control the diaphragm is completely filled with liquid. The liquid expands and contracts with a change in temperature. The movement of the diaphragm is very slight, but the pressure that can be exerted are tremendous. Because the coefficient of expansion of liquid is small, relatively large volume bulbs are used on the control. This results in sufficient diaphragm movement and also ensures positive control from the bulb.

Pressure controls can also be divided into two categories: the bellows type and the Bourdon-tube type. By far the most common is the *bellows type,* illustrated in Fig. R18-19. The bellows is connected directly into the refrigerant system through a capillary tube. As the pressure within the system changes, so does the pressure within the bellows, causing the bellows to move in and out in conjunction with pressure variation. As shown,

the electrical connections are made as the pressure rises. This type of control is used normally as both a low-and a high-pressure control. That is, it can be connected into the high or low side of the system. Because of its simplicity, dependability, and adaptability, this control is found on almost every air-conditioning or refrigeration system.

Figure R18-20 is a *Bourdon-tube-type* pressure control. Notice that a mercury bulb switch is shown here. The Bourdon-tube control is ideally suited for mercury bulb operation and is frequently found in applications requiring enclosed contacts. As pressure inside the tubing increase, the tube will tend to straighten out. This in turn moves the linkage attached to the mercury bulb, causing it to move over center, thus moving the mercury from one end of the bulb to the other and making electrical contacts.

The third type of primary control is the humidity control or *humidistat* (Fig. R18-21). Hygroscopic elements are used on these controls, the most common

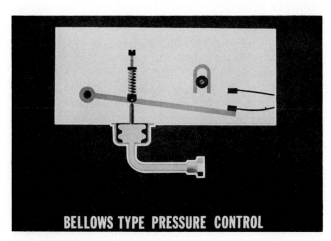

FIGURE R18-19 Bellows-type pressure control. (*Courtesy* Carrier Air-Conditioning Company)

FIGURE R18-21 Human hair humidistat. (*Courtesy* Carrier Air-Conditioning Company)

Electrical Components 211

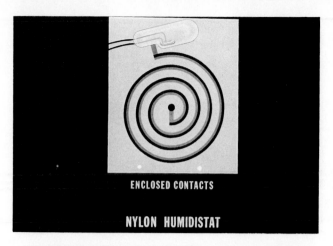

ENCLOSED CONTACTS

NYLON HUMIDISTAT

FIGURE R18-22 Nylon humidistat. (*Courtesy* Carrier Air-Conditioning Company)

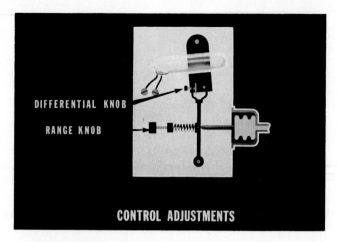

DIFFERENTIAL KNOB

RANGE KNOB

CONTROL ADJUSTMENTS

FIGURE R18-23 Range control. (*Courtesy* Carrier Air-Conditioning Company)

being human hair. As the air becomes more moist and the humidity rises, the hair expands and allows the electrical contacts to close. As the hair dries it contracts, thus opening the electrical contacts. This type of control is susceptible to dirt and dust in the air. Although accurate, it must be carefully maintained.

Another form of humidistat, using a nylon element, is shown in Fig. R18-22. The nylon is bonded to a light metal in the shape of a coil spring. the expanding and contracting of the nylon creates the same effect as that found in the spiral bimetallic strip used in thermostats.

There are almost as many possibilities for correct settings on controls as there are applications. Therefore, all controls have some way of being adjusted to compensate for the different conditions under which they may be required to operate. These may be field adjustable or fixed at the factory. The first such adjustment is range. *Range* is the difference between the minimum and the maximum operating points within which the control will function accurately.

For example, a high-pressure control may be used to stop the compressor when head pressure becomes too high. This control may have an adjustment that will allow a maximum cutout, or stopping point, of 300 psig and a minimum cutout of 100 psig. Therefore, the control would have a range of 200 psig. It may be set at any cutout point within this range. A control should never be set outside its range, as it will always be inaccurate and frequently will not even function.

Figure R18-23 shows one of the simpler methods of range control. As the pressure in the bellows increase, the lever is pushed to the left, thus opening the contacts. By varying the spring pressure, we can increase or decrease the bellows pressure necessary to open the contacts, thus raising or lowering the system pressure at which the control will operate. Many controls used this principle and usually have some external adjustment for field use.

If a control cuts out, or breaks the circuit, it is just as important that it cuts in, or remakes the circuit. The cutout and cut-in point cannot be the same. The difference between these points is the differential. *Differential* can be defined as the difference between the cutout and cut-in points of the control. For example, if the high-pressure control discussed previously cuts out or breaks the electrical circuit at 250 psig and cuts in or remakes it at 200 psig, the pressure-stat differential is 50 psig.

Referring again to Fig. R18-23, the control also has a differential adjustment. By opening or closing the effective distance between the prongs of the operating fork, the pressure at which the control will cut in can be varied. As the pressure on the bellows increase, the operating fork moves to the right, and the left prong of the fork will tilt the bulb and remake the electrical circuit. When the adjustable stop on the left prong of the fork is moved away from the prong on the right, the pressure in the bellows must decrease further before the bulb will be tilted and the electrical distance between the prong and stop, the difference between the cutout and cut-in points can be varied.

The preceding discussion explained one way to regulate both range and differential adjustment. There are, however, many ways in which each can be adjusted, depending on the application, size, and manufacturer's preference.

There is another feature of electrical controls that should be mentioned briefly—*detent* or *snap action*. For technical reasons, all electrical contacts should be opened and closed quickly and cleanly. Detent or snap action is built into most controls to accomplish this purpose. Figure R18-24 shows four common examples. The magnet in the upper left should require little explanation. The pull of the magnet merely accelerates the closing and opening rate of the contacts. The closer the magnet is to the bimetallic strip in the closed position, the more positive the snap action.

The bimetallic disk, as shown in the upper right,

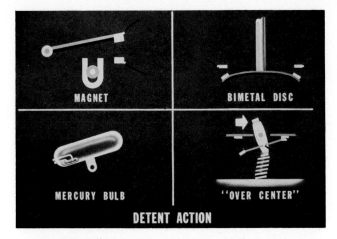

FIGURE R18-24 Four-control illustration. (*Courtesy* Carrier Air-Conditioning Company)

is normally in a concave position. As the temperature rises, the two metals expand at different rates until the disk snaps to a convex position, sharply breaking the electrical contacts.

The mercury bulb in the lower left is another method of obtaining detent action. As the element moves the bulb over the top center position, the heavy mercury runs from one end of the bulb to the other. This shifting of weight causes the bulb to move quickly from one side of the center to the other, thereby rapidly making or breaking contacts.

Shown in the lower right is a fourth method of obtaining detent action. Here the action is induced by using a compressed spring, such as is commonly found in a household toggle switch. Force is applied to the operating arm in the direction of the arrow. This rotates the toggle plate, which starts to compress the spring. As the switch approaches the center position the spring is at maximum compression. The moment the switch passes dead center position the compressed spring will force rapid completion of the switching action.

Secondary Controls: Secondary controls can be divided into two categories: operating and safety controls.

Operating Controls Operating controls can be expected to regulate the cycle during normal operation.

1. *Relays.* As mentioned previously, a relay is a magnetically operated switch used to control an electrical circuit or circuits, depending on the switching arrangement, according to the action produced by a magnetic coil and armature arrangement.

Such a device is required when:

• The voltage of the controlling device and controlled device are not the same.
• The controlling device is not capable of handling

the current demand of the controlled device.

More than one circuit must be controlled by a control device.

An example of (1) would be a fan relay controlling a 115- or 220-V fan motor by means of a 24-V coil and power supply. An example of (2) would be a fan relay controlling a 230-V fan motor by means of a 230-V coil and power supply where the current demand of the fan motor exceeds the capacity of the contacts in the control device. An example of (3) would be a season-changeover relay, where heating circuits are controlled separately from cooling circuits at the demand of a single-contact-type control device.

Because of the many types of arrangements possible, it would not be possible to explain each type. They all have the common function of controlling circuits by electrically operated actuating means.

2. *Contactors.* Contactors are relays capable of handling larger amounts of current or may be defined as extra-heavy-duty relays. The contactor performs the same function, that of controlling an electrical circuit by means of a magnetic coil and linkage assembly opening and closing contact, but because of its heavier construction and large surface contacts, it will make and break higher current demands. Where the relay contact capacity stops at 20 A, the contactor capacity starts at 20 A and goes to 600 A or more (see Fig. R18-25).

3. *Starters.* "Starter" is the term used to describe a contactor that has the overload cutout protecting the controlled device in the contactor. Therefore, a starter is nothing more than a contactor with built-in overloads.

All the secondary controls just discussed are operating controls, that is, controls that regulate the cycle during normal operation.

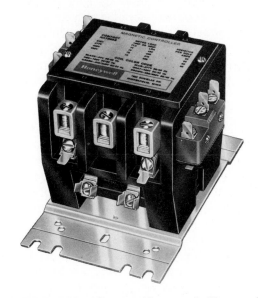

FIGURE R18-25 (*Courtesy* Minneapolis Honeywell)

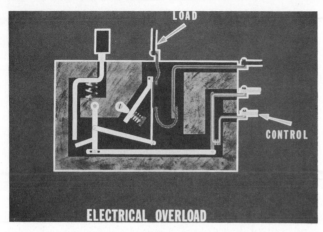

FIGURE R18-26 Electric heating overload. (*Courtesy* Carrier Air-Conditioning Company)

FIGURE R18-27 Bimetallic disk overload. (*Courtesy* Carrier Air-Conditioning Company)

Safety Controls The electrical heating overload control (Fig. R18-26) is the first of the secondary safety controls to be discussed. The safety control protects cycle components and is not expected to operate until a malfunction of the system occurs.

In all electrical circuits, some kind of protection against excessive current must be provided. The common household fuse is a good example of this type of protection. In motor circuits, however, a reset type of protector is frequently used. The overload shown is one method by which this is accomplished. It is used in conjunction with the contactor previously discussed. When a contactor is used, there is both a control circuit, in which the primary control is inserted, and a load circuit, which is opened and closed by the contactor. When excessive current is drawn in the load circuit, this device will break the control circuit.

The control circuit to the contactor coil passes through the controls at the lower right. The load circuit is passing through the bimetallic element shown in the center. If the current through the bimetallic element becomes too high, the element bends to the left. This action forces the contact arm to the left and breaks the contacts. This breaks the control circuit and allows the contactor to open, thereby interrupting the power to the load. The bimetallic element will now cool and return to its original position, but because of the slot in the bottom arm, the contacts will not be remade. The control circuit will remain broken until the reset button is pushed. This moves the arm to the right and closes the control circuit contacts. Because this control must be reset by hand, it is usually referred to as a manual reset overload.

A second type of electrical overload protector is the bimetallic disk (Fig. R18-27). When the temperature is raised, the disk will warp and break the electrical circuit. On the left is shown the simplest device, in which heat from an external source activates the disk. On the

right, external heat is supplemented by resistance heat from an electrical load. The device on the right will react much faster on electrical overload than the device on the left because the supplemental heat is within the device itself. The bimetallic disk overload protector is widely used in the protection of electrical motors.

The current relay (Fig. R18-28) is another type of electrical overload that can be manually or automatically reset. The greater advantage of this type of relay is that it is only slightly affected by ambient temperatures, thereby avoiding nuisance trip-outs.

The current relay is made up of a sealed tube completely filled with a fluid and holding a movable iron core. When an overload occurs, the movable core is drawn into the magnetic field, but the fluid slows its travel. This provides a necessary time delay to allow for the starting in-rush of momentary overloads. When the core approaches the pole piece, the magnetic force increases and the armature is actuated, thus breaking the control circuit.

FIGURE R18-28 Current relay. (*Courtesy* Carrier Air-Conditioning Company)

On short circuits, or extreme overloads, the movable core is not a factor, because the strength of the magnetic field of the coil is sufficient to move the armature without waiting for the core to move. The time-delay characteristics are built into the relay and are a function of the core design and fluid selection. The only difference between the thermostatic primary control previously discussed and the safety thermostat involves the applications in which they are used; the controls themselves may be identical.

A good illustration of a thermostat used as a safety control can be found in most water-chilling refrigeration systems (Fig. R18-29). In water-chilling systems it is important that the water not be allowed to freeze, because it could then do physical damage to the equipment. In such systems a thermostat may be used with the temperature-sensing element immersed in the water at the coldest point. The thermostat is set so that it will break the control circuit at some temperature above freezing. This will stop the compressor and prevent a further lowering of water temperature and possible freezing. The pressure control used for safety purposes is physically the same as that used for operational purposes.

The high-pressure cutout is probably the best example of a pressure control used for safety purposes (Fig. R18-30). It can be set to stop the compressor before excessive pressures are reached. Such conditions might occur because of a water supply failure on water-cooled condensers or because of a fan motor stoppage on air-cooled condensers.

The operation of the low-pressure control is the same as discussed under the subject of primary controls. Mechanically, there is no difference between the low-pressure control used as a primary control and one used as a safety control. The low-pressure control is used to stop the compressor at a predetermined minimum operating pressure. As a safety device, the low-pressure control can protect against loss of charge, high com-

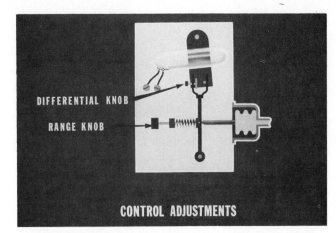

FIGURE R18-30 Control adjustments. (*Courtesy* Carrier Air-Conditioning Company)

pression ratios, evaporator freeze-ups, and entrance of air into the system through low-side leaks.

The oil safety switch is activated by pressure. However, it is operated by a pressure differential rather than by straight system pressure. It is designed to protect against the loss of oil pressure. Inasmuch as the lubrication system is contained within the crankcase of the compressor, a pressure reading at the oil pump discharge will be the sum of the actual oil pressure plus the suction pressure. The oil safety switch measures the pressure difference between the oil pump discharge pressure and suction pressure and shuts down the compressor if the oil pump does not maintain an oil pressure as prescribed by the compressor manufacturer. For example, if a manufacturer indicated that a 15-lb net oil pressure were required, a compressor with a 40-lb back pressure must have at least a 55-lb pump discharge pressure. At any lower discharge pressure from the pump this switch would stop the compressor. An examination of the operating force inset on the right of Fig. R18-31

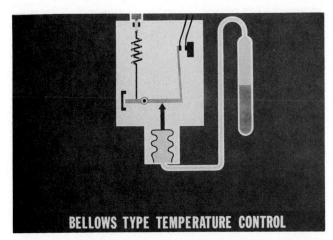

FIGURE R18-29 Bellows-type thermostat. (*Courtesy* Carrier Air-Conditioning Company)

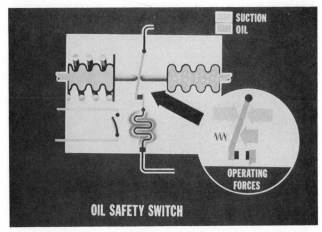

FIGURE R18-31 Oil safety switch. (*Courtesy* Carrier Air-Conditioning Company)

Electrical Components   215

will show that since suction pressure is present on both sides, its effect is canceled. The control function depends on the net oil pressure overcoming the predetermined spring pressure. The spring pressure must equal the minimum oil pressure allowed by the manufacturer.

When a compressor starts, there is no oil pressure. Full oil pressure is not obtained until the compressor is up to speed. Therefore, a time-delay device is built into this type of control to allow the compressor enough time to start. One of the methods used to create this time delay is a small resistance heater. When the compressor is started, because the pressure switch is normally closed, the resistance heater is energized. If the oil pressure does not build up to the cutout setting of the control, the resistance heater warps the bimetallic element, breaks the control circuit, and stops the compressor. The switch in Fig. R18-31 is automatically reset; however, these switches are normally reset manually.

If the oil pressure rises to the cutout setting of the oil safety switch within the required time after the compressor starts, the control switch is opened and deenergizes the resistance heater, and the compressor continues to operate normally.

If oil pressure should drop below the cut-in setting during the running cycle, the resistance heater is energized and, unless oil pressure returns to cutout pressure within the time-delay period, the compressor will be shut down. The compressor can never be run longer than the predetermined time on subnormal oil pressure. The time-delay setting on most of these controls is adjustable from approximately 50 to 125 seconds.

There are as many control methods for operating refrigeration systems as there are engineers to design them, but there are two simple methods that are standard and should be discussed briefly. The first, illustrated in Fig. R18-32, is the simple thermostat start–stop method. As the thermostat calls for cooling and closes

the contacts, the control circuit is completed through the contactor coil. This control circuit includes a klixon, a pressure control, and an overload electrical heater as secondary controls. When the contactor coil is energized it closes the contacts and completes the power circuit to the motor. The thermostat is the primary control and the motor cycles as the thermostat dictates. Any of the secondary controls will break the control circuit and stop the compressor in the same manner as the thermostat.

This is one of the simplest of all control methods and is therefore widely used. Some small refrigeration systems and almost all small air-conditioning systems use this principle of control.

A slightly more complex but still widely used control method is shown in Fig. R18-33. This is known as pump-down control. In this method the thermostat operates a solenoid valve. When the thermostat calls for cooling it completes a circuit through the solenoid valve, which is in the liquid line ahead of the metering device. By opening the valve, high-pressure liquid flows to the evaporator.

A pressure control is connected to the low side of the system. As the pressure rises in the evaporator coil, the pressure control closes and completes the circuit through the safety device to the contractor coil. The compressor starts and the system operates normally. When the thermostat is satisfied, the solenoid coil is deenergized and the flow of liquid is cut off to the evaporator. Because the compressor is still running, the evaporator pressure is reduced to the cutout point of the low-pressure control. The control contacts open and the compressor stops.

This system is designed to keep liquid refrigerant from filling the evaporator during shutdown. Regardless of the position of the thermostat, the low-pressure control will operate the compressor upon sensing a rise in pressure in the vaporator.

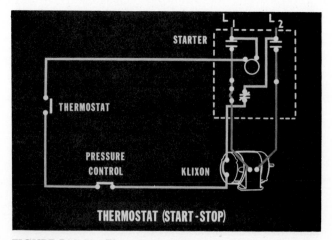

FIGURE R18-32 Thermostat start–stop. (*Courtesy* Carrier Air-Conditioning Company)

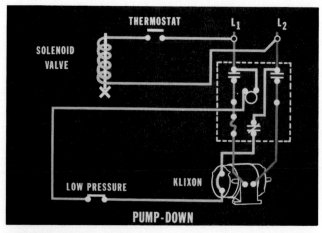

FIGURE R18-33 Pump-down control. (*Courtesy* Carrier Air-Conditioning Company)

Occasionally, a control is needed that does not interrupt the electrical supply to the load but changes the characteristics of the power supply to produce the desired results. The only control of this type used in a summer air-conditioning unit would be the capacitor used in conjunction with the motor compressor assembly.

Capacitor: The capacitor start–capacitor run type motors used in large-horsepower summer air-conditioning units are not single-phase motors even though connected to a single-phase line. They are actually two-phase motors using both the starting and running windings to both start and run the motor.

In order to apply two-phase electricity to the second winding of the motor, it is necessary to use some type of control to delay the voltage change in this winding. This is done by means of an electrical storage device called a capacitor (see Fig. R18–34). A capacitor can be defined as "two plates separated by an insulator where, when a positive voltage is induced on one plate, a negative voltage will be created on the other."

This action may be technically correct but would be difficult to apply to the application of an air-conditioning unit. Instead, the ability of a capacitor to receive and store electrical energy up to the amount of electrical pressure or voltage supplied by the power source and to give up this stored energy on demand would more closely describe its function.

This action can best be described as follows. Suppose that we construct a Y-pipe arrangement as shown in Figure R18-35. The Y arrangement is composed of a tee, two short nipples, two 90° elbows, and two lengths

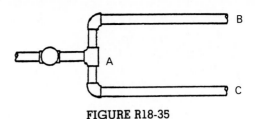

FIGURE R18-35

of pipe of equal length. This arrangement is connected, through a solenoid valve, to a source of water under pressure. With the Y-pipe arrangement in a perfectly level position so that gravity has no effect, and assuming equal flow resistance through each pipe (resistance from point *A* to point *B* equals the resistance from point *A* to point *C*), if the solenoid is energized and opened, from which pipe would the water emerge first? Because of equal resistance in the pipes and because the assembly is level, water would emerge from both pipes at the same time.

Referring now to the starting and running windings in a standard motor (Fig. R18-36), if the resistance (composed of the total of the resistance, reactance, reluctance, and impedance of the coil) from point *A* to point *B* through the running winding were exactly the same as from point *B* through the starting winding, when the applied voltage builds up (point 1 to point 2), in which winding does the magnetic field build up first? Assuming equal resistance, the field builds in both windings at the same time. When the applied voltage falls (point 2 to point 3), in which winding does the magnetic field fall first? Again, both at the same time. When the applied voltage builds up in the opposite direction (point 3 to point 4) and falls (point 4 to point 5), the reaction in both coils is at the same time.

This motor, then, depends on the mechanical magnetic difference between the two coils to produce rotating power. The mechanical magnetic difference is produced by placing the starting and running windings at different positions in the stator of the motor (see Fig. R18-37).

The mechanical magnetic difference may be sufficient to start an ordinary blower or burner motor, where light starting loads are encountered, but is not sufficient to start against the heavier-starting loads of refrigeration compressors. Therefore, a greater magnetic difference must be created between the two windings (see Fig. R18-38).

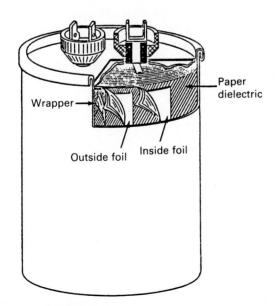

FIGURE R18-34 Capacitor construction.

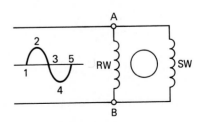

FIGURE R18-36

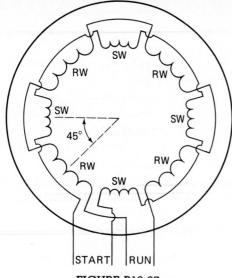

FIGURE R18-37

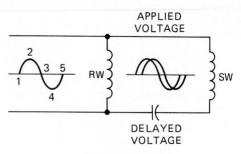

FIGURE R18-40

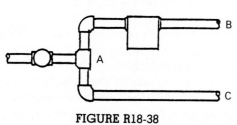

FIGURE R18-38

Returning to the water pipe arrangement, if a storage tank were placed in the pipe from point *A* to point *B* and the solenoid valve were opened, from which pipe would the water flow first? Naturally, pipe *A–C* would have water flowing from it first, as it would take time to fill the storage bucket in pipe *A–B*. Suppose, now, that the storage bucket were piped with a drain solenoid, as in Fig. R18-39, to drain the bucket when the main water valve were closed. From which pipe would the water stop flowing first? Pipe *A–C* would stop immediately upon the closing of the main valve and pipe *A–B* would continue to flow until the bucket was empty. Therefore, because of the delaying action of the bucket, the water flow start and stop from pipe *A–B* always lags behind the water flow start and stop from pipe *A–C*.

Referring to Fig. R18-40, in order to produce a greater delay between the buildup of the magnetic fields in the starting winding and the running winding, it is only necessary to insert an electrical storage bucket in the starting winding circuit.

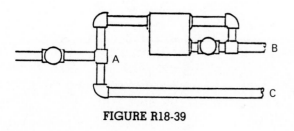

FIGURE R18-39

These storage buckets are called "capacitors." Therefore, upon the rise of applied voltage (point 1 to point 2) the rise in voltage in the starting winding is delayed behind the rise in voltage in the running winding, accomplishing the delay in the rise of the magnetic field of the starting winding behind the rise in magnetic field of the running winding. Upon a drop in voltage (point 2 to point 3), because the capacitor has an amount of stored power that must be used by the starting winding, the magnetic field of the running winding is reduced by the drop in the applied voltage, but the magnetic field is held up by the extra power in the bucket. This causes the change in the starting winding magnetic field to lag behind the field of the running winding on both buildup and fall. Therefore, even though single-phase power is applied to the motor, the motor actually feels two-phase power (see Fig. R18-41).

Motors used in air-conditioning compressors also use the two-phase power produced by capacitors to reduce the amperage draw of the motor for a given load. These motors are classified as "capacitor start–capacitor run." Like any other motor, more torque (power) is required to start the motor than to run it. This means that more capacitors are required to start than to run. Therefore, it is necessary to remove part of the storage capacity from the circuit after the motor comes up to approximately 80% of normal running speed. The capacitors are then classified as to the length of time they are in the circuit. Those that are across the line at all times are called "running" capacitors. Those that are

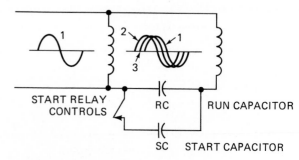

1. APPLIED VOLTAGE
2. DELAYED VOLTAGE (START)
3. DELAYED VOLTAGE (RUN)

FIGURE R18-41

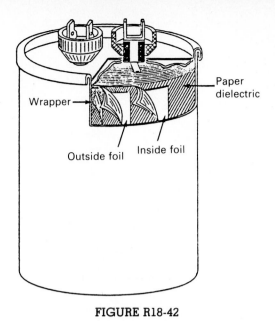

FIGURE R18-42

used only for starting the motor are called "starting" capacitors.

Running Capacitor: The running capacitor is a metal can oil-filled type that is designed for continuous duty (Fig. R18-42). In other words, this capacitor is built sturdily enough to be in use the entire time the motor is running. Rated in microfarads, which is a unit of electrical storage capacity similar to gallons as units of water quantity, this capacitor has a predetermined filling and discharging time for the electrical pressure or voltage to which it is connected. Therefore, it will give a predetermined time delay to the magnetic buildup and fall in the start winding. As each type or size of motor has definite time delay requirements, the same microfarad capacity must be installed if the capacitor is replaced. If a 40-μF capacitor is required, a 40-μF capacitor is used as replacement. *Do not use larger or smaller capacitors as replacements.*

The construction of a capacitor is also like a barrel. If the barrel is of a certain metal thickness, it will hold up to a given pressure without bursting. If the barrel is connected to higher pressure, it will burst. However, it will hold any pressure lower than that which it is designed to hold. A capacitor also has a normal working voltage or electrical pressure to which it is designed to be connected. If the voltage is higher than the capacitor is designed to hold (its normal working voltage), the capacitor will start to leak, heat up, and explode. It will, however, stand voltages under the normal working voltage. There are voltage tolerances that are taken into account in capacitor design for long working life, but these should never be considered when making replacements.

Start Capacitor: Running capacitors can be used as start capacitors if their microfarad capacity is sufficient.

However, because start capacitors are expected to perform on start only, usually a maximum of 2 seconds and not over an average of four times per hour, their construction is considerably lighter. As a result, the basic cost of the capacitor is considerably less. Also, because start capacitors are required to have more storage capacity (higher microfarad rating) and yet keep their physical size to a minimum, they are of the electrolytic filled type.

Although the electrolytic used has a higher storage ability, it also heats under use. This heating is so rapid that if the capacitor is left in use, it quickly heats to the boiling point, steam pressure forms, and the capacitor blows up. Because of this short operation requirement, it is very important that the proper microfarad and working voltage ratings be adhered to when changes are made. This will ensure proper starting time of the motor and removal of the starting capacitors from the circuit in the shortest possible time.

Because the shorting out or rupturing of the starting capacitor can be caused by so many things, such as low voltage, excessive load, bad start relay, etc., replacement of just the start capacitor generally does not cure the cause and repeated failures occur. *Always check the entire circuit in cases where start capacitors fail.*

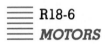

R18-6
MOTORS

Motors convert electrical energy into mechanical energy. Motors, like electrical circuits, are divided into two general categories, depending on the type of current involved—direct or alternating. Since most applications in refrigeration involve alternating current, this presentation will deal primarily with ac motors.

Alternating-current motors are classified in two main divisions (depending on the type of power used):

1. Single-phase
2. Polyphase (three-phase)

Single-phase motors are less efficient than three-phase motors. They are used chiefly where the demand is for fractional-horsepower units or where single-phase electrical service is the only type available. In some types of special applications, single-phase motors ranging up to several horsepower may be used, but generally the single-phase field stops at around 5 hp.

Naturally, all types of motors are not alike, since they are designed for different jobs. It is possible to turn over a small motor with a touch of a finger, whereas a motor used with a belt-driven compressor will need considerably more than the pressure of a finger to turn its shaft.

There seems to be more difficulty in understanding the operation of small, single-phase motors than that of three-phase motors. Learning why there is a need for different kinds of motors will help toward an understanding of the composition and operation of single-phase motors. Single-phase motors may be categorized as:

1. Split-phase
2. Capacitor-start
3. Permanent-split capacitor
4. Capacitor-start–capacitor-run
5. Shaded-pole

According to the dictionary, *torque* is a force, or combination of forces, that produces or tends to produce rotation. In the case of an electric motor it is the ability to exert power, resulting in the turning of an object "load," such as a compressor flywheel. The greater the capability a motor has to turn a heavy object, the greater the torque of the motor. Of course, it costs more to construct a motor with a high starting torque than one with a low starting torque, which accounts for the many types of motors produced. For example, it would be uneconomical to produce a high-torque motor to turn a household fan.

The two main parts of a motor are called the rotor and the stator, as shown in Fig. R18-43. The *rotor* is the rotating or revolving part, and is sometimes called an *armature*. The stationary part is known as the *frame* or the *stator*. Keep in mind the principles of magnetism: "Like" magnetic poles repel and "unlike" poles attract. Figure R18-43 shows two magnetized poles mounted on the outside of the motor shell or stator, while between these magnetic poles is a permanent magnet placed on a shaft. This permanent magnet in the center corre-

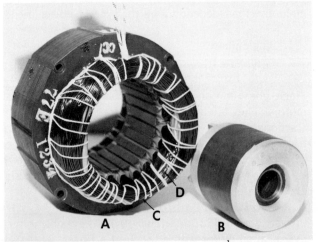

FIGURE R18-43 Stator and rotor from a $\frac{1}{4}$-hp split-phase hermetic motor: A, stator; B, rotor; C, running winding; D, starting winding. (*Courtesy* General Electric Company, Hermetic Motor Department)

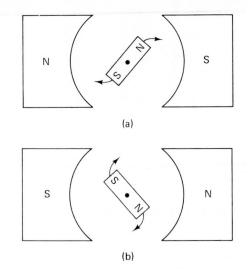

FIGURE R18-44 (*Courtesy* The Trane Company)

sponds to the rotor. The north magnetic pole of the stator attracts the south pole of the permanent magnet or the rotor, and the south magnetic pole of the stator attracts the rotor's magnetic pole. This attraction occurs because, when current passes through the coils making up the magnetic poles of the stator, magnetic fields are set up in the coils.

Since Fig. R18-43 is a diagram of a simple alternating-current motor, the direction of the flow of current will reverse completely and reverse the polarity of the stator coils as the flow of current alternates. Therefore, the north coil in Fig. R18-44a will soon become the south pole of the stator as depicted in Fig. R18-44b. When this occurs, with the south pole of the rotor pivoting on the shaft, and with the alternating current reversing the polarity of the stator, the two south poles will repel one another. The rotor will continue to pivot on the shaft as long as the current flow continues to reverse in the coils of the stator. This briefly explains the operation of the motor, or the rotation of the permanent magnet that is pivoted on the shaft. The motor will operate in this way until the current is shut off, at which time the demonstration motor will reach the condition shown in Fig. R18-45: As you will see, the north pole of the stator attracts the south pole of the rotor and the permanent magnet or magnetized bar will remain as it is. Even when the current flow is reversed due to alternating current, the rotor will stay where it is because it is not at an angle comparable to that in Fig.

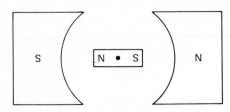

FIGURE R18-45 (*Courtesy* The Trane Company)

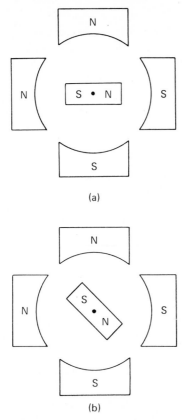

(a)

(b)

FIGURE R18-46   (a) (*Courtesy* The Trane Company) (b) (*Courtesy* ACCA)

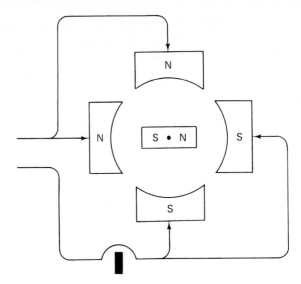

FIGURE R18-47   (*Courtesy* The Trane Company)

R18-44a or b. The attraction or repulsion of the poles cannot cause rotation when the bar is in the position shown in Fig. R18-45.

If we insert another set of poles in the stator or shell, as shown in Fig. R18-46a, the new set of magnetic poles will offer enough attraction to start rotation of the permanent magnet again. But difficulty may be encountered if the rotor should stop in a position midway between the two poles, as shown in Fig. R18-46b. There would be an equal attraction between the south pole in the rotor and the two north poles in the stator as well as between the rotor north pole and the stator south poles. This problem must be overcome, since corresponding poles in the stator have the same magnetic strength at the same time. Each is using the same source of electrical current, for we are dealing with a single-phase source. If we put a resistance in one of the paths, temporarily obstructing the flow of electrons to that pole in the stator (as shown in Fig. R18-47), the electrons will not arrive at their destination in the poles at the same time. This will cause the electrons flowing to one pole to be out of step with those flowing to the other, because a second phase has developed. We now have a two-phase current that will help to start the motor operating again by aiding in the pulling or pushing of the rotor. The second phase is needed by all single-phase motors so that they can start rotating. The main differences among sin-

gle-phase motors is the method used for producing and controlling this second phase of electrical current.

This second phase may be better understood by a study of Fig. R18-48. In Fig. R18-48a, two electrons are running together up and down the "hills" that range between the maximum north polarity and the maximum south polarity. When the electrons keep pace with one another, we have what is called single-phase current. If they travel up and down these hills 60 times a second,

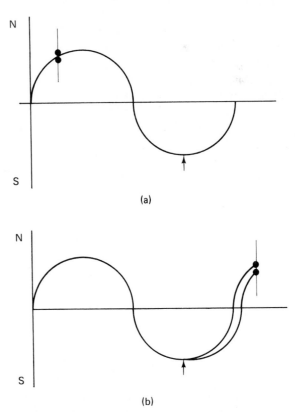

FIGURE R18-48   (*Courtesy* The Trane Company)

the current is a 60-Hz current. This is the condition mentioned earlier, in which the rotor stopped midway between poles, as shown in Fig. R18-48b.

In Fig. R18-47, with a resistance placed in one of the paths, the equally capable electrons have been placed in an unbalanced condition. The electrons are in corresponding positions on different hills, with one trailing behind the other. Because of the insertion of a resistance into one of the paths, the single-phase current has been changed to a current having two phases.

All of the five categories of single-phase motors mentioned earlier need this second phase of electrical current to start. The main difference among the categories is the way in which the second phase is produced. (Conceivably, since all of these motors have a second phase, they could be classified as two-phase motors instead of single-phase.)

R18-6.1
Split-Phase Motors

If the single-phase motor had but one set of windings of the stator, the rotor would not always start to rotate, but would sometimes just sit there and hum when energized. If turned by hand when it fails to start, the rotor will pick up speed until it reaches its operating rpm. This hand-turning will begin the rotation of the magnetic field and set up the necessary induction in the motor.

In the split-phase type of motor, the self-starting capability is brought about through the addition of a second set of windings on the stator. This is known as the *starting winding;* it is made up of considerable more turns than the primary or running winding. Because of the greater number of turns in the starting winding, the current flowing in it lags behind the current in the running winding. The added number of turns in the starting winding provides the resistance placed in one of the paths, as shown in Fig. R18-47. It also duplicates the condition of the electrons flowing in an unbalanced condition, as shown in Fig. R18-48b.

The starting winding in a split-phase motor stays in the circuit during the starting period or until the motor reaches about 75% of its full-rated speed. At this point, the starting winding is disconnected from the circuit by a centrifugal switch or a relay (thermal, current, or potential) to disconnect the starting winding.

Split-phase motors have low starting torques and are used on small, fractional-horsepower refrigeration units. This type of motor is often used in a system that has a capillary tube for refrigerant control, because the pressures in the system equalize on the OFF cycle of the unit. Figure R18-49 shows the arrangement of the starting and running windings in a split-phase motor, in conjunction with a centrifugal switch.

R18-6.2
Capacitor-Start Motor
(Capacitor-Start–Induction Run Motor)

The capacitor-start motor shown in Fig. R18-50 has a motor winding arrangement similar to that of a split-phase motor. An additional component, a capacitor, is wired in series with the centrifugal switch and the starting winding. In this motor the capacitor causes a "leading" current, and thus the starting winding is out of phase with the running winding. Again, in this type of motor, when the rotor reaches approximately 75% of its rated speed, the centrifugal switch will take the starting winding and the capacitor out of the electrical circuit.

R18-6.3
Permanent-Split Motor Capacitor

If a capacitor is used in the starting winding circuit, it is simpler if the capacitor can be left in the circuit

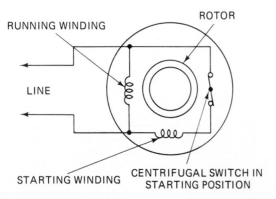

FIGURE R18-49   Split-phase motor. (*Courtesy* ACCA)

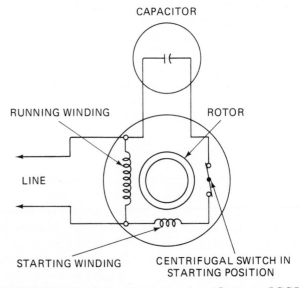

FIGURE R18-50   Capacitor-start motor. (*Courtesy* ACCA)

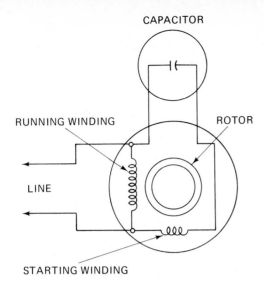

FIGURE R18-51 Permanent-split capacitor. (*Courtesy* ACCA)

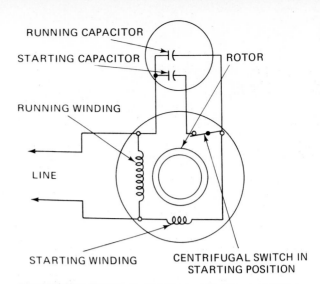

FIGURE R18-52 Capacitor-start–capacitor-run motor. (*Courtesy* ACCA)

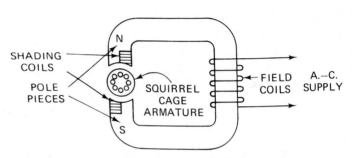

FIGURE R18-53 Shaded-pole motor. (*Courtesy* The Trane Company)

at all times. Of course, in such a case, the correct capacitor must be selected for use with the starting winding. It must be able to pass enough current to the starting winding to provide adequate torque, but it must not allow enough current to overheat the starting winding while the motor is in operation. A diagram of this type of motor, which has a good starting torque, is shown in Fig. R18-51. The diagram shows a capacitor connected in series with the starting winding, which is energized at all times when the motor is in operation. Consequently, no centrifugal switch is needed with this type of motor: which is classified as a permanent-split capacitor (PSC) motor.

R18-6.4
Capacitor-Start-Capacitor-Run Motor
(Two-Value Capacitor Motor)

The capacitor-start–capacitor-run type of motor is quite similar in design and operation to the capacitor-start–induction-run motor. Figure R18-52 shows this type of motor using two capacitors. In this case, a centrifugal switch is wired in series with the large-capacity type of capacitor, and this series circuit is parallel to the running capacitor. In this motor, like others utilizing a centrifugal switch, when the motor reaches approximately 75% of its rated speed, the centrifugal switch will break the circuit, putting the larger capacitor out of use. This type of motor is adapted to conditions in which a strong starting torque is necessary, such as in a compressor that must start under full-load conditions.

R18-6.5
Shaded-Pole Motor

A shaded-pole motor, shown in Fig. R18-53, does not have a starting winding but is otherwise similar to a split-phase induction motor. In the motors previously described, some effort must be used to get the motor turning, since it is only in the turning of the rotor that the lines of force are cut and set up their own magnetic field to maintain rotation. Figure R18-53 shows that the poles of the motor are split or grooved so that a smaller pole replaces part of the main pole. Around this smaller pole is a copper loop or strap called a *shading coil*. Instead of depending on a separate starting winding to create a second phase, the shading coils produce the rotation of the rotor through a shifting of a portion of the magnetic field. Figure R18-54 suggests the principles involved in shifting this magnetic field.

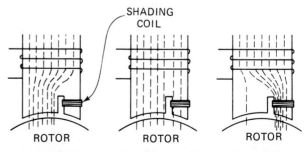

FIGURE R18-54 (*Courtesy* The Trane Company)

Built-in motor protectors are provided to permit the motor to perform beyond its normal duty to a predetermined safe temperature limit without cutting off its power supply. Most regulations and codes are strict about overload protectors, since danger to persons or property could arise from a malfunctioning motor or equipment connected with it.

There are two general types of built-in protectors. The first is activated primarily by temperature, and the second is activated by motor current or temperature, or both. The temperature-responsive protectors (Fig. R18-55a) may be of either bimetallic or thermistor type, both of which operate at a fixed motor temperature, or of the rod-and-tube type, which has an anticipatory or "rate-of-rise" temperature sensitivity.

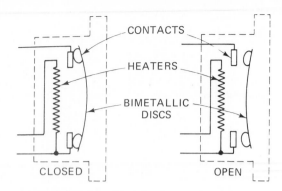

FIGURE R18-56   Thermal protector (dual-responsive) for end bracket mounting. (*Courtesy* Gould, Inc., Electric Motor Division)

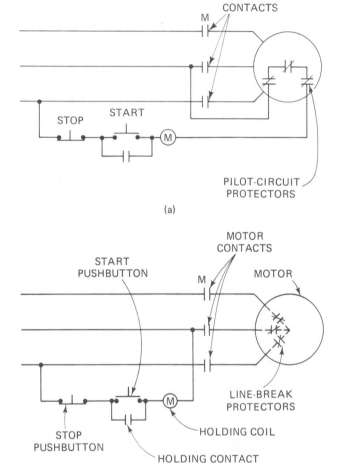

(a)

(b)

FIGURE R18-55   (a) Temperature-responsive (pilot circuit). (b) Dual-responsive (line break) thermal protector. (*Courtesy* ACCA)

Bimetallic protectors use bimetallic strips or disks to control a normally closed switch, and are designed to require minimum space so that in most cases they can be placed on the motor windings. Thermistors are semiconductors, the electrical resistance of which varies with temperature. They are extremely sensitive to relatively minute temperature changes, but the resistance variation must be amplified through an electronic circuit or mechanical relay to actuate a holding coil.

Rate-of-rise protectors are almost always of a rod-and-tube type and operate through a pilot circuit. An external metal tube and an internal metal rod are connected so that they operate as a differential-expansion thermal element to actuate a self-contained snap-action switch. The metal of the tube has a higher coefficient of expansion than that of the rod. The element is installed in the motor winding, where it will be affected by winding heating. On slow temperature changes, both tube and rod heat at the same rate, and the switch is set to open when the tube reaches a predetermined temperature. When the temperature change is rapid, the tube heats at a faster rate than the rod, so that the differential-expansion movement required to operate the switch is reached at a lower tube temperature than for a slow rate of change. This temperature anticipation makes it possible for the tube-and-rod protector to provide protection for both running and stalled conditions.

Dual-responsive protectors (the location of one type of which is shown in Fig. R18-55b) are used to protect against both slow and fast temperature increases, and are thus used for both running overloads and locked-rotor conditions. These protectors are essentially bimetallic thermostats with built-in heating elements in series with the motor winding. They may be meant for "on-winding" attachment or for attachment within the end bracket (Fig. R18-56).

The heating element may be the bimetallic strip or disk itself, a separate heater, or both. Heating of the bimetallic element is a function of both the temperature around the protector and the internal heat generated by the motor current through the element.

Resistance of the disk and heater are coordinated with the temperature setting of the disk to establish an operating point suited to the winding temperature limit. In the case of running overload, most of the heat reaching the thermal element is produced by the motor windings. During locked-rotor conditions, the rate of rise is too fast to cause the protector to operate quickly enough in response to motor heat alone. In this situation, the heat generated within the protector augments the winding heat, resulting in a response that is rapid enough to prevent damage.

For single-phase motors up to about 5 hp, thermal protectors are used. For three-phase motors, a single protector with three heating elements connected in series with the motor windings at the neutral point is used. The protector opens all three phases, as shown in Fig. R18-55b.

PROBLEMS

R18-1. Every electrical circuit consists of three parts. Name them.

R18-2. Most circuits (though not all) have a fourth item. What is it?

R18-3. What are the two types of transformers?

R18-4. What is the difference between the two types of transformers?

R18-5. How are amperage and voltage affected by the relationship of the number of turns on primary and secondary coils?

R18-6. If a 240- to 24-V control transformer were to be connected in reverse (output across the 240 V), what would be the output voltage?

R18-7. Loads are divided into two types. Name and describe each.

R18-8. Name the most common parasitic-type load used in the refrigeration and air-conditioning field.

R18-9. Name the two types of primary loads.

R18-10. Name the two types of controls.

R18-11. Most room thermostat sensing elements are made of _____ .

R18-12. Many pressure-sensitive controls use the _____ -tube principle.

R18-13. What three situations require the use of a relay?

R18-14. An oil pressure control stops the compressor immediately if it senses no oil pressure. True or False?

R18-15. Define a capacitor.

R18-16. Some controls are equipped for either _____ or _____ reset.

R18-17. There are two types of capacitors used in refrigeration units. What are they?

R18-18. Describe the difference in construction of the two types.

R18-19. What is torque?

R18-20. Name the five different types of single-phase motors.

R18-21. In a motor using both starting and running capacitors, the starting capacitor is to start the motor and the running capacitor keeps the motor running. True or False?

R18-22. What is the purpose of a built-in motor protector?

R19

Electrical Testing Devices

R19-1
GENERAL

Most service problems in the field of refrigeration and air-conditioning are found in the electrical power and control circuitry. Knowledge of the electrical system and electrical testing instruments enables the service technician to locate the trouble easily.

Many service technicians have an aversion to using test instruments and do not understand how they work or the need for their proper care. If the service technician makes a sincere effort to learn the proper use of test instruments, these valuable tools can work for his or her benefit as well as that of the customers. Test instruments should be just as familiar to the service technician as the conventional tools in his or her toolbox. These test instruments are delicate in nature and proper care should be taken in their handling and use. If the voltage is to be checked, the instrument to use is a voltmeter. If the electrical current or the flow of electrons, measured in amperes, is to be checked, an ammeter must be used. If the system or its components are to be checked for electrical resistance, for shorts in the circuit, or for continuity of the electrical circuit, the instrument to use is an ohmmeter. If the amount of power in watts must be determined, a wattmeter is used.

When measuring the resistance of insulation such as found in an electrical motor, a high-voltage resistance meter called a *megger* is used. Each of these instruments is discussed in the following sections.

R19-2
VOLTMETER

The voltmeter is the basic meter used in all the previously listed meters. The difference is where the amperage that actuates the meter is obtained directly from the power source to be measured (voltmeter), the voltage drop across a known resistance in the circuit (ammeter), or the voltage drop across the measured resistance using a fixed-voltage power source (ohmmeter).

In all cases, the voltmeter measures current flow. For purposes of this discussion, the moving pointer on scale-type instruments will be discussed. A moving pointer on a scale meter has a meter coil that moves a pointer across a calibrated scale to a mark that indicates the measurement to be taken. The moving meter coil is a coil of wire that is free to rotate between the north- and south-seeking poles of a permanent magnet. When voltage is applied to the moving coil, the resulting current flow through the coil sets up a magnetic field. This field reacts with the field existing between the poles of the magnet and causes the coil to rotate. A pointer attached to the coil moves to a position on the meter scale depending on the amount of current passing through the coil. Figure R19-1 shows the rotating coil and permanent-magnet relationship.

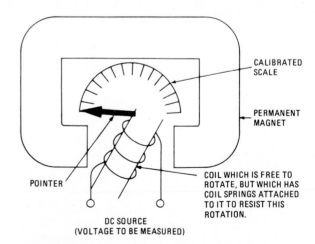

FIGURE R19-1 Movement components of a voltmeter. (*Courtesy* American Gas Association)

When a small current passes through the coil, a low voltage has been applied across the coil, and a weak magnetic field is produced. This causes a small turning force, called *meter torque,* to be created between the coil field and the permanent-magnet field. Thus, the coil and pointer turn a small amount against the tension of the return springs. A higher voltage will produce a larger current through the coil, producing a stronger magnetic field around the coil, a greater amount of meter torque, and more rotation of the coil and pointer.

Because the voltmeter is actually a current-reading instrument, the electrical values of its circuit component must be fairly accurate. The circuit design of the instrument must provide for all the voltage ranges to be measurable by the instrument. The meter shown in Fig. R19-1 can be used to measure ac voltages by adding a diode in the circuit to change the ac to dc.

Most meters, to be accurate, operate on a very small amount of coil amperage. This amperage, called *current sensitivity,* depends on the number of turns in the meter coil and the strength of the permanent-magnet field. Current sensitivity is expressed as the number of milliamperes or microamperes required for full-scale reading of the dial. Typical meter movements have a current sensitivity of 1 mA or 50 μA.

The meter sensitivity may also be expressed on an "ohms-per-volt" basis. This is determined by the total resistance that must be in series in the meter circuit to obtain full-scale reading when 1 V is applied across the meter. Using Ohm's law to determine the resistance, the meter with a 1-mA rating would require 1000 Ω/V in the circuit and the 50-μA meter would require 20,000 Ω/V in the circuit.

If the meter were to be able to measure a range of voltage between 0 and 10 V full scale, the 1-mA meter would require 10,000 Ω in the circuit to produce the proper scale results. If the meter were to be able to measure from 0 to 250 V full scale, a resistance of 250,000 Ω must be in the circuit. Therefore, to be able to use the same meter over a multirange of voltages, it is only necessary to be able to change the amount of resistance in the circuit to change the voltage range of the meter.

Figure R19-2 shows a combination-type volt-ohm-meter with two voltage ranges, 150 and 300 V. If the basic meter has a current sensitivity of 1 mA, the 150-V range would require a circuit resistance of 150,000 Ω and the 300-V range a resistance of 300,000 Ω. By means of a selector switch, the various resistances are cut into or out of the circuit to permit the use of a single meter over multiple ranges. When the ac voltages are selected, the resistances for the various ranges are the same. Only a diode is connected into the circuit to convert the ac to dc for the meter action.

Because the meter will not take higher current flow than design, subjecting the meter to a higher voltage than the range of the meter could cause burnout of the coil

FIGURE R19-2 Combination ac–dc voltmeter–ohmmeter. (*Courtesy* Robinair Manufacturing Corporation)

and/or circuit resistances or could cause bending of the pointer arm from the heavy movement of the pointer against the pointer stop. Some meters have air damper chambers to slow the action of the pointer, but these have their limitations. When using a multirange meter, always start to measure voltage using the highest range on the meter. When the approximate voltage is read, the meter range can then be reduced to the proper range for greater reading accuracy.

R19-3
AMMETER

Ammeters measure the electrical current flowing through the circuit (Fig. R19-3) and involve the same operating principles as voltmeters. Current measurements require that the ammeter be placed in series in the circuit being measured rather than parallel as when using a voltmeter.

Because it is impractical to build an ammeter with a coil large enough to carry all the current in the elec-

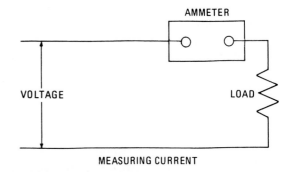

FIGURE R19-3 Measuring current in a circuit with an ammeter. (*Courtesy* American Gas Association)

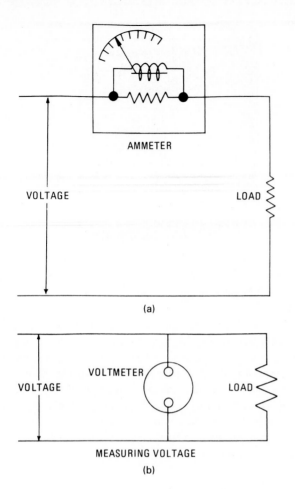

(a)

MEASURING VOLTAGE

(b)

FIGURE R19-4   A voltmeter measures voltage drop across a resistance. (*Courtesy* American Gas Association)

trical circuit, the ammeter is actually a sensitive voltmeter that measures the voltage drop across a small resistance in series in the circuit (see Fig. R19-4). This type of ammeter requires that the meter be inserted into the circuit to measure the amount of current flowing. This is practical when measuring current flow in the low-amperage range, such as control circuits using 24 V or the microampere range of thermocouples. In high-amperage ranges, however, it is more practical to use a clamp-on split-armature ammeter.

Figure R19-5 shows a clamp-on ammeter with ranges from 0 to 600 A. This type of instrument needs no actual contact with wires or terminals. It is closed around a wire carrying the current to be measured. The current through the wire produces a magnetic field that induces a magnetic effect in the clamp (yoke) of the instrument, causing it to work like a step-up transformer and generating a voltage in a secondary winding on the meter armature. A voltmeter measures the voltage drop over a resistance in series with the secondary winding. Remember—this instrument works only on ac current, because it depends on transformer effect.

FIGURE R19-5   Clamp-on ammeter. (*Courtesy* A. W. Sperry Instruments, Inc.)

To produce the multiple range of current measurement, a switch arrangement is used to vary the resistances in the circuit.

Figure R19-6 shows the internal circuitry of a typical clamp-type ammeter. The size of the various resistances must be balanced against the sensitivity of the meter, and a bridge of rectifiers is used to connect the ac current to dc current for the meter.

R19-4
OHMMETER

In the ohmmeter, the basic meter movement of a voltmeter is also used for measuring the resistance of a component. However, the ohmmeter has its own voltage source within the instrument (usually a single battery or

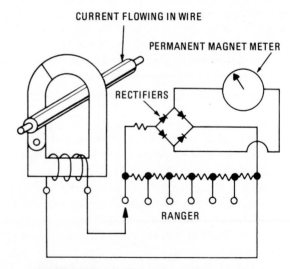

FIGURE R19-6   Clamp-type split-armature current meter. (*Courtesy* American Gas Association)

combination of batteries, depending on the range settings available with the instrument). *The ohmmeter should never be connected to any other source of power. If it is, the meter will certainly burn out.*

The simple ohmmeter circuit in Fig. R19-7 consists of a voltage source (battery), a basic meter (rm), a variable resistance (rh) to zero the meter, and a fixed resistance to limit the current flow to the meter coil. The unknown resistance to be measured is placed between terminals *A* and *B*. Note that the scale of the ohmmeter is reversed from that of the voltmeter. With the voltmeter the 0 is at the left end (no current flow position of the pointer) and maximum voltage is at the right end. With the ohmmeter, full travel of the pointer to the right is a reading of maximum current flow or zero (0) ohms (a dead short), while the left end of the scale indicates no current flow and infinite resistance (shown by the symbol ∞).

Before measuring a resistance, terminals *A* and *B* are connected together (by touching the lead terminals together) so that the resistance between the leads is zero. The needle should indicate zero ohms. If not, the resistance Rh is adjusted up or down to position the pointer on the zero mark. This adjustment is especially important when measuring resistances in the range 0 to 5 Ω, such as measuring the start and run windings in a motor. The unknown resistance is then connected between *A* and *B* and the pointer indicates the resistance in ohms on the scale. The meter is actually performing the function of Ohm's law with a known voltage and indicating the result.

The ohmmeter is the basic test instrument in trouble analysis. A voltmeter will measure the voltage available, the ammeter the current flow, but the ohmmeter is used to check quality of the parts, condition of the

FIGURE R19-8   Volt-wattmeter. (*Courtesy* Robinair Manufacturing Corporation)

circuit, as well as condition of individual parts. For example, the conditions of a contact in a relay, overload, contractor, etc., can be checked only with an ohmmeter. If the contact is a high-voltage type (above 50 V), the resistance must not be over 1 Ω. Over the 1-Ω reading, the contact requires replacement or further electrical damage could be done. If the contact is of the low-voltage type (50 V or less), the maximum allowable resistance is 0.5 Ω.

R19-5
WATTMETER

On thermal-type loads (resistance type) the watts the load draws equals the volts applied times the amperage drawn. On magnetic-type loads, power factor must also be considered because the voltage and amperage are out of phase, due to the inductance and reactance developed in the load.

A wattmeter uses two coils to affect the movement of the needle: one powered by the voltage directly and the other driven by the voltage drop across a short resistor in the circuit. Each coil reacting to its prime voltage source will affect the meter to produce the true watts in the circuit, taking the power factor into account.

The volt-wattmeter (Fig. R19-8) enables the service technician to obtain simultaneous readings of the voltage and true wattage (the power factor has been compensated for) when checking and servicing an electrical appliance. The refrigeration technician may have to check appliances other than those directly related to refrigeration, and he or she may find appliances that have ratings in watts rather than amperage. In such a case where a rating in amperes is desired, the watt reading can be divided by the voltage of the circuit to give the amperage.

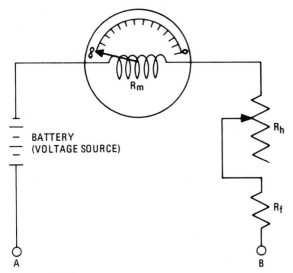

FIGURE R19-7   Simple ohmmeter circuit. (*Courtesy* American Gas Association)

FIGURE R19-9 Combination meter. (*Courtesy* A. W. Sperry Instruments, Inc.)

FIGURE 19-10 Megohmmeter-insulation tester. (*Courtesy* TIF Instruments, Inc.)

R19-6
MILLIVOLTMETER AND MILLIAMMETER

The technician may be called on to check an electrical circuit having low voltage and current measurements. In such cases, he or she would use test instruments known as millivoltmeter and milliammeters, which record smaller values than the more generally used voltmeter and ammeter. For example, many gas appliances use thermocouples generating dc voltage in the millivolt range to operate safety-cutoff devices that use amperage in the milliampere range. A standard voltmeter and ammeter could be used but the reading would be so low with so little pointer movement that the reading would mean very little. Figure R19-9 shows a combination meter that measures in the millivolt range as well as the milliampere and microampere range.

R19-7
MEGGER

When checking high-voltage insulation the ohmmeter used needs a higher voltage than the 1.5 V usually found in the standard ohmmeter. A high-voltage-type ohmmeter using a voltage of approximately 500 V is used for this application. This instrument, commonly called a *megger* because it is designed to work in the megohm resistance range, is pictured in Fig. R19-10. Originally, the megger had a power source using a hand-operated generator to supply the 500-V power supply. With the development of solid-state voltage boosters, a 6-V power supply (batteries) is raised to the 500 V for the resistance test.

PROBLEMS

R19-1. Are the greatest number of service problems in refrigeration and air-conditioning equipment in the refrigeration circuit or the electrical circuit?

R19-2. What instruments would be used to check the following?
(a) Applied voltage _____
(b) Motor terminal to ground resistance _____
(c) Current draw _____
(d) Coil resistance _____
(e) Total power drawn _____
(f) Continuity of a circuit _____
(g) Shorts in a circuit _____

R19-3. When using a multirange meter, should the starting point be the lowest or the highest range?

R19-4. A clamp-type ammeter will work equally well on both dc and ac current. True or False?

R19-5. On an ohmmeter, does travel of the needle to the top of the scale mean that the item tested has maximum resistance or minimum resistance?

R19-6. No movement of the needle when using an ohmmeter on maximum range to check a relay coil means that the coil is _____ .

R19-7. When using an ohmmeter to check the condition of relay contacts, the maximum resistance for high-voltage contacts is _____ and for low-voltage contacts is _____ .

R19-8. What is the purpose of millivoltmeters and milliammeters?

R19-9. Name a widely used device that would require millivoltmeters and milliammeters for checking.

R19-10. What is the voltage developed by a megger?

Electrical Circuits and Controls

R20-1
BASIC COMPONENTS

In Chapter R18 electrical components were discussed as individual items. It is now necessary to put the various items together to obtain the end product desired from the electrical sources. Whether the end product is heat (thermal load) or work (mechanical load), each circuit must contain the necessary basic items: power source, load, and conductors. Unless continuous operation is required, as in the case of clock motors, some type of control is required.

Further, some components may have multiple uses, the most common being the power source. The transformer on the pole outside a residence is a single power source but is connected to many loads through many control devices. *Remember: Each circuit supplying power from the common transformer to an individual load possibly through a control device is an individual circuit and is not necessarily related to any other circuit.* The light in the dining room controlled by a wall switch is connected to the same power source as the lamp in the living room. However, they operate independently of each other. On the other hand, it may be necessary to be able to operate the lights from a single location in such an order that either lamp may operate but not both at the same time. A single control or switching device would be used to accomplish this.

R20-2
SWITCHING ACTION

All controls of the total interruption type (switching devices) have a common factor in that they contain contacts or switching devices for the control of electrical loads. Because mechanical switching devices (with the exception of the toggle switch with center off) are ca-

pable of two positions only, selection of the contact arrangement will depend on whether the contact must be closed or open at the desired conditions. In the case of the motor starter, relay, or contactor, the controls must close when the operating coil is energized. In the case of a high-pressure control, the contacts must open if the head pressure exceeds the required amount. In the case of a suction pressure control, the contacts must open if the suction pressure is reduced below the set point of the control.

When picturing control devices in wiring diagrams, the contacts are shown in the position they would be in at standard conditions. Standard conditions are as follows:

1. **Temperature Controls:** When the control is at 70°F or 21.1°C
2. **Pressure Controls:** When the pressure is at atmospheric (0 psig)
3. **Electrical Controls:** When the power is off or the power source has been disconnected

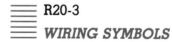

R20-3
WIRING SYMBOLS

Figure R20-1 is the recommended simplified graphic electrical diagram symbols as recommended by the Educational Assistance Committee of the Refrigeration Service Engineers Society. In the switches section, the various switch applications were given. It should be noted that in some instances, the switch is shown as an open circuit and in others as a closed circuit. When the switch contact is shown in an open position at standard conditions, the switch is classified as *normally open.* When the switch or contact is shown in a closed position at standard conditions, the switch is classified as *normally closed.*

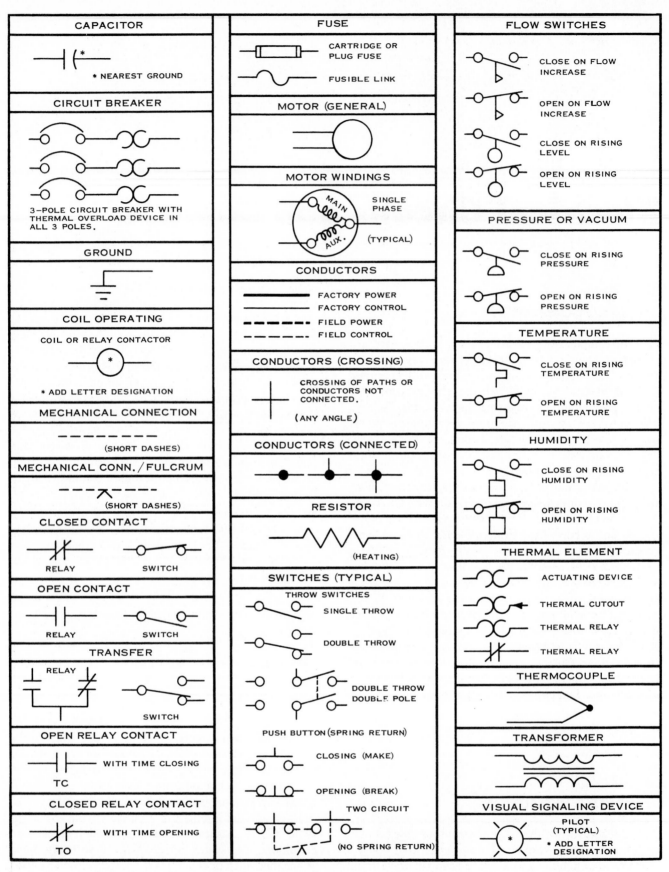

FIGURE R20-1 Recommended simplified graphic electrical diagram symbols. (*Courtesy* ACCA)

FIGURE R20-2   Normally open contacts.

Normally open contacts are contacts that rest in the open position when the control device is inactive. For example, when the magnetic coil of the relay in Fig. R20-2 is deenergized, the contacts open and rest in the open position.

Normally closed contacts are contacts that rest in the closed position when the control device is inactive. For example, when the magnetic coil of the relay in Fig. R20-3 is deenergized, the contacts close and rest in the closed position.

It is also possible to have different combinations of contact arrangement, depending on the control functions desired. For example, the relay in Fig. R20-4 has a single pole–double throw arrangement, which means it is made up of one normally open contact and one normally closed contact, with one common terminal. Contacts 1 and 2 are made in the closed position, and contacts 3 and 4 are made in the open position. This would be used when one power source is used with two load devices: one load device operating with the relay deenergized, and one operating when the relay is energized.

FIGURE R20-3   Normally closed contacts. (*Courtesy* Honeywell, Inc.)

FIGURE R20-4   (*Courtesy* Honeywell, Inc.)

Another arrangement would also use one normally open contact and one normally closed contact but without a common terminal. This arrangement would be used to control two circuits from two separate power sources, possibly at two different voltages, where one circuit operates when the relay is energized and the other operates when the relay is deenergized. The most common contact arrangements are as follows:

1. **SPST—Single Pole, Single Throw:** This arrangement incorporates one set of contacts that control a given load from a single power source (Fig. R20-5).

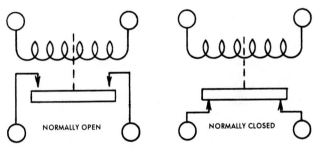

FIGURE R20-5   (*Courtesy* Honeywell, Inc.)

2. **SPDT—Single Pole, Double Throw:** This arrangement incorporates two sets of contacts that control two loads from a common power source but only one load connected at a time (Fig. R20-6).

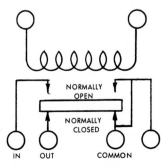

FIGURE R20-6   (*Courtesy* Honeywell, Inc.)

Electrical Circuits and Controls   233

3. **DPST—Double Pole, Single Throw:** This arrangement incorporates two sets of contacts that control two loads from two separate power sources. Actual control could be as follows:

   a. *Both contacts normally open.* Both loads are disconnected from their respective power source when the relay is deenergized (Fig. R20-7).

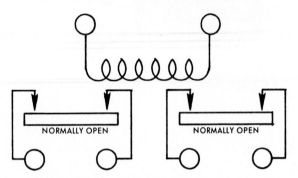

FIGURE R20-7 (*Courtesy* Honeywell, Inc.)

   b. *Both contacts normally closed.* Both loads are connected to their respective power source when the relay is deenergized (Fig. R20-8).

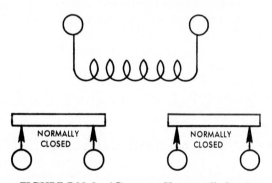

FIGURE R20-8 (*Courtesy* Honeywell, Inc.)

   c. *One contact normally open—one contact normally closed.* One load connected to its power source and the other load is disconnected from its power source when the relay is deenergized (Fig. R20-9).

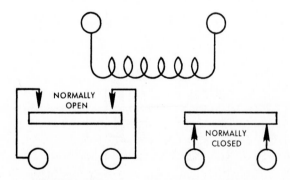

FIGURE R20-9 (*Courtesy* Honeywell, Inc.)

4. **DPDT—Double Pole, Double Throw:** This arrangement would use four sets of contacts arranged to control a load from two separate power sources. Each set would have the one normally open contact and one normally closed contact with a common terminal. Thus, a load would be connected to a power source in the open position (relay deenergized) and the other loads connected to the power sources in the closed position (relay energized) (Fig. R20-10).

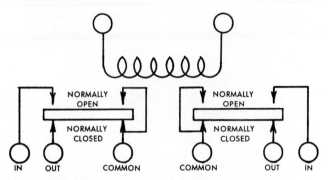

FIGURE R20-10 (*Courtesy* Honeywell, Inc.)

Any number of combinations are possible but they would all be multiples of these basic switching arrangements.

In contactors and starters, a greater number of contact sets are required to accomplish the desired control. For example, in a three-phase contactor (Fig. R20-11), three single-pole, single-throw contacts are required to disconnect all three lines from the power source when the unit is disconnected. These would be three normally open contacts.

It is also possible to raise the amperage capacity of the control without using heavier contacts using four sets of contacts wired so as to give the effect of two sets;

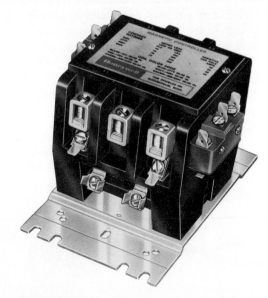

FIGURE R20-11 (*Courtesy* Honeywell, Inc.)

such a contact carries one-half the total load and the overall capacity of the control is doubled. Thus, four sets of single-pole, single-throw contacts are wired in parallel to make up two sets of single-pole, single-throw contacts.

R20-4
CONTROL SEQUENCE

About the only time that single control is encountered is in the case of a single light controlled by a single wall toggle switch. Usually, a sequence or series of controls are used. A typical case would be the three-level table lamp with an on/off cord switch. The cord switch supplies the main power control to the lamp while the base switch selects the amount of light to be available by selecting the bulb filaments to be used. Figure R20-12 shows the wiring diagram for such a light arrangement. The cord switch is the "master controller." This control decides if the light is active or off. The selector switch selects only which bulb filament will be active. Therefore, to control the action of all elements in the load (the three-way bulb), the master controller must be installed ahead of the selector or operation control.

A simple control circuit in refrigeration work would be the basic temperature control system in a domestic refrigerator along with the interior lighting circuit. In Fig. R20-13 we see two circuits being supplied through the 120-V male plug with a third-wire ground for required protection. Proper control of 120-V circuits requires the control to be in the hot or black lead (although this is not always done in unitary equipment). The black or hot lead supplies two circuits:

1. The interior light through the door switch. Opening the door completes the circuit and the light glows. The door switch is a normally closed switch maintained in an open position by the pressure of the door frame when the door is closed.

2. The power to operate the compressor motor and condenser fan to produce the desired refrigeration effect in the cabinet; here we see that the temperature control is the master controller.

The closing of the master controller feeds power to two circuits in parallel: the condenser fan circuit and the

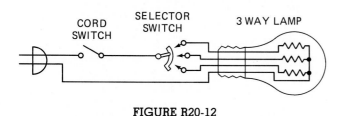

FIGURE R20-12

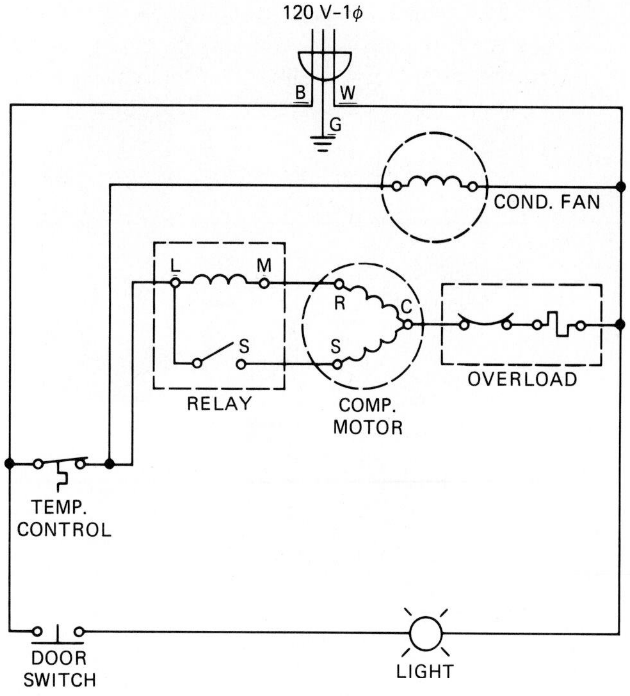

FIGURE R20-13

compressor motor circuit. The condenser fan motor is a shaded-pole type of motor and does not need additional control to operate. Also, because of this type of construction, it depends on the 120-V line fuse for protection.

Because the compressor motor is an induction start–induction run type of motor, a starting winding is required to provide the necessary starting torque for operation of the motor. Also, upon start the starting winding must be deenergized when the motor reaches approximately 80% of full-load speed. Because the motor compressor assembly is a hermetic type, a separate current type relay is used to accomplish this (Fig. R20-14).

When the temperature control closes, power flows through the relay, the run winding of the motor, and the overload out to the other side of the line. The high current draw of the motor (the inrush or loaded rotor current) is great enough to pull the relay closed, connecting the start or auxiliary winding into the circuit.

As the motor accelerates, the current draw decreases until the magnetic pull of the relay is not sufficient to keep the relay armature in, the kick-off spring forces the relay armature to drop away, opening the start or auxiliary winding circuit, and the motor continues to run. Because the contact in the relay is open when no power is applied, this is a normally open contact.

At standard air temperature (70°F or 21.1°C) the overload cutout is closed to permit current flow for normal operation; therefore, this contact is a normally closed type. Should the compressor motor not start and

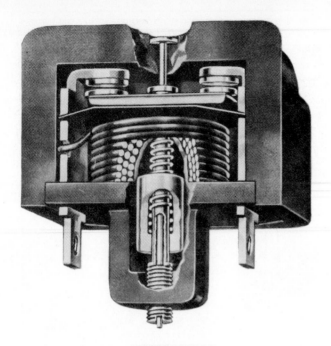

FIGURE R20-14

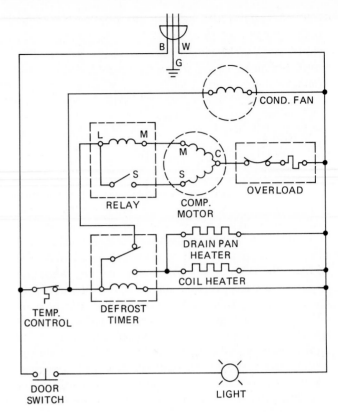

FIGURE R20-15

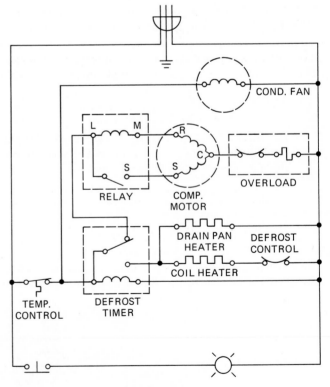

FIGURE R20-16

continue to require the high inrush current, the parasite-type heater in the overload quickly heats up to a temperature level high enough to cause the bimetallic total contact assembly to operate and open the contact. This then stops the current flow through the motor and presents damage. From the diagram we see that we have a master control in each circuit. The compressor circuit also has two additional controls: one normally open (the starting relay) and one normally closed (the overload circuit).

In the setup shown in Fig. R20-13, when the evaporator coil accumulated excessive frost and/or ice, it was necessary to stop operation of the system until the frost had melted and the coil was clear. The demand to eliminate this nuisance led to the inclusion of automatic defrost of the coil. This was especially required in refrigerators having a separate freezer compartment. When the unit was shut off for a sufficient period of time to remove the frost, food warm-up and possible spoilage could occur. A heater was added to the evaporator coil for fast warm-up and a heater to the drain pan and drain tube to prevent ice forming and causing the condensate water to back up in the pan (Fig. R20-15).

The operation of the unit then became refrigeration or defrost under the control of a defrost timer. To operate the unit after a given number of hours of compressor operation (usually 12 hours) the defrost timer motor becomes an additional load under the control of the temperature control. The temperature control now controls the condenser fan, the compressor motor, and the timer motor. Power is fed from the hot side of the power supply through the temperature control to the compressor motor through the normally closed contact of the defrost timer and normal operation results. The refrigerator temperature is maintained at the set point of the thermostat (Fig. R20-16).

After the required period of compressor operating time (actually, timer motor operating time—both operate simultaneously) the timer normally closed contact is broken (the compressor stops) and the normally open contact is closed (the heaters are energized). Because the refrigerant pressure is built up in the evaporator because of the added heat, hard starting of the compressor motor could result. To bring the discharge pressure to as close to room temperature as possible, the condenser fan motor continues to operate.

After a predetermined period of time, the timer motor reverses the contacts, shuts off the heaters, and puts the unit back in operation. Under normal operating conditions the preset time is sufficient for complete defrost of the coil with a minimum of product temperature rise. In areas of extremely high humidity or under higher than normal usage, it is sometimes necessary to exchange the timer motor for more frequent defrost cycles. This is an individual application problem that is best discussed with the equipment manufacturer.

In some applications where the manufacturer designs the defrost system for peak-humidity conditions and/or high service usage (frequent door openings) a limiting device is put in the defrost heater circuit to limit the heat to the coil during low-humidity or light service loads. This reduces the energy requirement as well as the load on the compressor motor. To limit the control of the defrost thermostat to the coil heater only, the defrost thermostat is in the heater circuit only. Note that for convenience of production technique the manufacturer has placed this control in the downstream or dead side of the circuit. Therefore, even with this control open, *the heater is hot to ground—be careful.*

In all this discussion, we see that we have added simple series circuits that can be traced from the power source (the 120-V plug) through control devices, through a load either magnetic (work) or thermal (heat), and then to the other side of the power source—in one case from the load through another control and to the other side of the power source. If the service technician will form the habit of tracing each circuit in such a manner, control circuits are much easier to follow and problems are easier to analyze.

══ *PROBLEMS* ══

R20-1. Control contacts that are open when the control is not actuated are called _____ .

R20-2. Control contacts that are closed when the control is not actuated are called _____ .

R20-3. Standard conditions for picturing controls in wiring diagrams are as follows:

  (a) For temperature controls _____

  (b) For pressure controls _____

  (c) For electrical controls _____

R20-4. Are the contacts in a current-type motor-starting relay normally open or normally closed?

R20-5. Are the contacts in a voltage-type motor-starting relay normally open or normally closed?

R20-6. Would a set of contacts that move in only one direction and control one circuit be called DPDT, DPST, SPST, or SPDT?

R20-7. Would a set of contacts that make one of two circuits in each direction of movement be called DPDT, DPST, SPST, or SPDT?

R20-8. Would the contacts be both normally open, normally closed, or one of each?

R20-9. A contactor breaking all legs of a three-phase motor would have what kind of contact arrangement?

R20-10. Is is possible to use a four-PST contactor of 20-A rating on a 40-A load? If so, how?

R20-11. Is the switch that controls the light in a refrigerator a normally closed or normally open type?

R21

Solid-State Electronics

R21-1
GENERAL

The use of solid-state material for electrical use has been in the refrigeration and air-conditioning field for many years. The first electrostatic air filters used vacuum tubes as rectifiers to convert the high-voltage alternating current to direct current to charge the filter cells. With the development of the selium rectifier, the vacuum tube was eliminated and the product greatly simplified.

The so-called "energy crunch" of the late 1970s brought out the spark ignition system for gas appliances, and this was the forerunner of the electronic or solid-state temperature control and the rest of the controls systems that went with it. The use of solid-state systems is not new, but the development has accelerated very rapidly.

The subject of electronics is so vast that several books would be required to cover all applications in the radio, TV, communications, control, etc., fields. For the purpose of this chapter only those topics that pertain to the control systems in the refrigerating and air-conditioning fields will be discussed.

R21-2
THE ELECTRON THEORY

The electron theory was covered in Chapter R16, but it is necessary to discuss it here in more detail to provide background for the use of solid-state material.

All matter is composed of atoms, the smallest or basic components of molecules. Atoms, in turn, are composed of a heavy, dense nucleus containing "protons" and sometimes "neutrons," surrounded by lighter particles called "electrons."

R21-3
PROTONS

The proton carries a positive electrical charge (+), the electron carries a negative electrical charge (−), and the neutron, if present, carries no electrical charge—it is neutral. The attraction between the positively charged protons in the nucleus and the surrounding negatively charged electrons tend to hold them together in the unit we call the atom (see Fig. R21-1).

The number of protons in the nucleus will determine the type of or atomic weight of the element. Figure R21-1a pictures the hydrogen atom with one proton in the nucleus. Hydrogen has the atomic number of 1, the least of all the basic elements. Figure R21-1b pictures the copper atom. It has 29 protons in the nucleus and is assigned the atomic number of 29.

Figure R21-2 lists the atomic numbers for the known natural elements (92) and the 11 elements that have been developed. Each is listed according to the number of protons in the nucleus—1 for hydrogen, 29 for copper, 32 for germanium, 34 for selium, 42 for silver, 79 for gold, etc.

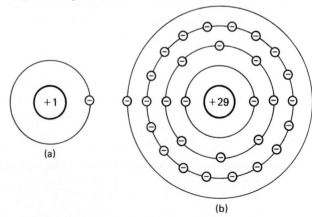

FIGURE R21-1   (a) Hydrogen atom. (b) Copper atom.

R21-4

ELECTRONS

The number of electrons in an element equals the number of protons, so that the element is in electrical balance. The electrons are also orbiting around the nucleus at a speed that balances the centrifugal force of the electron movement against the electrical attraction between the proton and the electron. The path that the electron takes around the proton is called the *shell*. For example, the path the moon takes around the earth or the earth around the sun is called its "orbit" but could also be referred to as its "shell" if compared to the path of the electron around the proton.

FIGURE R21-2
Atomic weights of basic elements.

THE NATURAL ELEMENTS

Atomic Number	Name	Symbol	Atomic Number	Name	Symbol	Atomic Number	Name	Symbol
1	Hydrogen	H	32	Germanium	Ge	62	Samarium	Sm
2	Helium	He	33	Arsenic	As	63	Europium	Eu
3	Lithium	Li	34	Selenium	Se	64	Gadolinium	Gd
4	Beryllium	Be	35	Bromine	Br	65	Terbium	Tb
5	Boron	B	36	Krypton	Kr	66	Dysprosium	Dy
6	Carbon	C	37	Rubidium	Rb	67	Holmium	Ho
7	Nitrogen	N	38	Strontium	Sr	68	Erbium	Er
8	Oxygen	O	39	Yttrium	Y	69	Thulium	Tm
9	Fluorine	F	40	Zirconium	Zr	70	Ytterbium	Yb
10	Neon	Ne	41	Niobium	Nb	71	Lutetium	Lu
11	Sodium	Na		(Columbium)		72	Hafnium	Hf
12	Magnesium	Mg	42	Molybdenum	Mo	73	Tantalum	Ta
13	Aluminum	Al	43	Technetium	Tc	74	Tungsten	W
14	Silicon	Si	44	Ruthenium	Ru	75	Rhenium	Re
15	Phosphorus	P	45	Rhodium	Rh	76	Osmium	Os
16	Sulfur	S	46	Palladium	Pd	77	Iridium	Ir
17	Chlorine	Cl	47	Silver	Ag	78	Platinum	Pt
18	Argon	A	48	Cadmium	Cd	79	Gold	Au
19	Potassium	K	49	Indium	In	80	Mercury	Hg
20	Calcium	Ca	50	Tin	Sn	81	Thallium	Tl
21	Scandium	Sc	51	Antimony	Sb	82	Lead	PB
22	Titanium	Ti	52	Tellurium	Te	83	Bismuth	Bi
23	Vanadium	V	53	Iodine	I	84	Polonium	Po
24	Chromium	Cr	54	Xenon	Xe	85	Astatine	At
25	Manganese	Mn	55	Cesium	Cs	86	Radon	Rn
26	Iron	Fe	56	Barium	Ba	87	Francium	Fr
27	Cobalt	Co	57	Lanthanum	La	88	Radium	Ra
28	Nickel	Ni	58	Cerium	Ce	89	Actinium	Ac
29	Copper	Cu	59	Praseodymium	Pr	90	Thorium	Th
30	Zinc	Zn	60	Neodymium	Nd	91	Protactinium	Pa
31	Gallium	Ga	61	Promethium	Pm	92	Uranium	U

THE ARTIFICIAL ELEMENTS

Atomic Number	Name	Symbol	Atomic Number	Name	Symbol	Atomic Number	Name	Symbol
93	Neptunium	Np	97	Berkelium	Bk	101	Mendelevium	Mv
94	Plutonium	Pu	98	Californium	Cf	102	Nobelium	No
95	Americium	Am	99	Einsteinium	E	103	Lawrencium	Lw
96	Curium	Cm	100	Fermium	Fm			

Solid-State Electronics 239

Atomic No.	Element	ELECTRONS PER SHELL				
		1	2	3	4	5
1	Hydrogen, H	1				
2	Helium, He	2				
3	Lithium, Li	2	1			
4	Beryllium, Be	2	2			
5	Boron, B	2	3			
6	Carbon, C	2	4			
7	Nitrogne, N	2	5			
8	Oxygen, O	2	6			
9	Fluorine, F	2	7			
10	Neon, Ne	2	8			
11	Sodium, Na	2	8	1		
12	Magnesium, Mg	2	8	2		
13	Aluminum, Al	2	8	3		
14	Silicon, Si	2	8	4		
15	Phosphorus, P	2	8	5		
16	Sulfur, S	2	8	6		
17	Chlorine, Cl	2	8	7		
18	Argon, A	2	8	8		
19	Potassium, K	2	8	8	1	
20	Calcium, Ca	2	8	8	2	
21	Scandium, Sc	2	8	9	2	
22	Titanium, Ti	2	8	10	2	
23	Vanadium, V	2	8	11	2	
24	Chromium, Cr	2	8	13	1	
25	Manganese, Mn	2	8	13	2	
26	Iron, Fe	2	8	14	2	
27	Cobalt, Co	2	8	15	2	
28	Nickel, Ni	2	8	16	2	
29	Copper, Cu	2	8	18	1	
30	Zinc, Zn	2	8	18	2	
31	Gallium, Ga	2	8	18	3	
32	Germanium, Ge	2	8	18	4	
33	Arsenic As	2	8	18	5	
34	Selenium, Se	2	8	18	6	
35	Bromine, Br	2	8	18	7	
36	Krypton, Kr	2	8	18	8	
37	Rubidium, Rb	2	8	18	8	1
38	Strontium, Sr	2	8	18	8	2
39	Yttrium, Y	2	8	18	9	2
40	Zirconium, Zr	2	8	18	10	2
41	Niobium, Nb	2	8	18	12	1
42	Molybdenum, Mo	2	8	18	13	1
43	Technetium, Te	2	8	18	14	1
44	Ruthenium, Ru	2	8	18	15	1
45	Rhodium, Rh	2	8	18	16	1
46	Palladium, Pd	2	8	18	18	0
47	Silver, Ag	2	8	18	18	1
48	Cadmium, Cd	2	8	18	18	2
49	Indium, In	2	8	18	18	3
50	Tin, Sn	2	8	18	18	4
51	Antimony, Sb	2	8	18	18	5
52	Tellurium, Te	2	8	18	18	6

Atomic No.	Element	ELECTRONS PER SHELL						
		1	2	3	4	5	6	7
53	Iodine, I	2	8	18	18	7		
54	Xenon, Xe	2	8	18	18	8		
55	Cesium, Cs	2	8	18	18	8	1	
56	Barium, Ba	2	8	18	18	8	2	
57	Lanthanum, La	2	8	18	18	9	2	
58	Cerium, Ce	2	8	18	19	9	2	
59	Praseodymium, Pr	2	8	19	20	9	2	
60	Neodymium, Nd	2	8	19	21	9	2	
61	Promethium, Pm	2	8	18	22	9	2	
62	Samarium, Sm	2	8	18	23	9	2	
63	Europium, Eu	2	8	18	24	9	2	
64	Gadolinium, Gd	2	8	18	25	9	2	
65	Terbium, Tb	2	8	18	26	9	2	
66	Dysprosium, Dy	2	8	18	27	9	2	
67	Holmium, Ho	2	8	18	28	9	2	
68	Erbium, Er	2	8	18	29	9	2	
69	Thulium, Tm	2	8	18	30	9	2	
70	Ytterbium, Yb	2	8	18	31	9	2	
71	Lutetium, Lu	2	8	18	32	9	2	
72	Hafnium, Hf	2	8	18	32	10	2	
73	Tantalum, Ta	2	8	18	32	11	2	
74	Tungsten, W	2	8	18	32	12	2	
75	Rhenium, Re	2	8	18	32	13	2	
76	Osmium, Os	2	8	18	32	14	2	
77	Iridium, Ir	2	8	18	32	15	2	
78	Platinum, Pt	2	8	18	32	16	2	
79	Gold, Au	2	8	18	32	18	1	
80	Mercury, Hg	2	8	18	32	18	2	
81	Thallium, Ti	2	8	18	32	18	3	
82	Lead, Pb	2	8	18	32	18	4	
83	Bismuth, Bi	2	8	18	32	18	5	
84	Polonium, Po	2	8	18	32	18	6	
85	Astatine, At	2	8	18	32	18	7	
86	Radon, Rn	2	8	18	32	18	8	
87	Francium, Fr	2	8	18	32	18	8	1
88	Radium, Ra	2	8	18	32	18	8	2
89	Actinium, Ac	2	8	18	32	18	9	2
90	Thorium, Th	2	8	18	32	19	9	2
91	Protactinium, Pa	2	8	18	32	20	9	2
92	Uranium, U	2	8	18	32	21	9	2
93	Neptunium, Np	2	8	18	32	22	9	2
94	Plutonium, Pu	2	8	18	32	23	9	2
95	Americium, Am	2	8	18	32	24	9	2
96	Curium, Cm	2	8	18	32	25	9	2
97	Berkelium, Bk	2	8	18	32	26	9	2
98	Californium, Cf	2	8	18	32	27	9	2
99	Einsteinium, E	2	8	18	32	28	9	2
100	Fermium, Fm	2	8	18	32	29	9	2
101	Mendelevium, Mv	2	8	18	32	30	9	2
102	Nobelium, No	2	8	18	32	31	9	2
103	Lawrencium, Lw	2	8	18	32	32	9	2

R21-5

SHELL

Some elements have more electrons than others; therefore, a greater number of orbits take place. Also, each neutron follows its own path rather than being in line with others, so that the neutrons traveling the same distance from the proton form a ball-type or spherical shell around the proton. There is also a limit to the number of electrons that can travel in each sphere or orbital path.

The path closest to the proton can contain only

two electrons. This is because the unlike charge attraction is greater due to the shorter distance and the electron speed is greater to produce the centrifugal force necessary to counteract the attraction to keep the electron in orbit. Similarly, the second shell cannot hold more than 8 electrons, the third a maximum of 18 electrons, the fourth shell a maximum of 32 electrons, and the fifth also a maximum of 32. We do not know of the capacity of the sixth and seventh shells, as we have not developed elements with atomic weights in the range where the sixth and seventh shells have been filled. It is assumed because of the 32-maximum repeat between the fourth and fifth shells that this applies to all succeeding shells. Figure R21-3 shows the number of electrons and the quantity in each shell for the various basic elements, both natural and artificial.

R21-6
VALENCE SHELL

The outside shell of an atom is called the *valence shell* and the electrons that orbit in the outer shell are known as *valence electrons*. In Fig. R21-1a the hydrogen has only one electron in one shell, so it has one valence electron; the copper atom in Fig. R21-1b has one electron in its valence shell (the fourth shell), so it also has one valence electron.

In reviewing the electron shells listed in Fig. R21-3, you will see that the third, fourth, fifth, and sixth shells have a peculiar trait in that they will not hold more than 8 electrons as a valence shell. Only when an additional shell is started and is no longer a valence shell will it hold more electrons. A rule can thus be stated: The outer or valence shell of an atom cannot hold more than 8 electrons. This rule is important because it indicates which atoms make good conductors of electrical energy, which are good insulations, and which are semiconductors.

R21-7
ENERGY LEVEL

Each electron in the atom has the same negative charge, but not all the electrons have the same *energy level*. Those traveling in the outer shells, because they travel a greater distance, require a higher energy level. If we could add energy to the electrons traveling in the inner shells, we could move them to the outer shells. If enough energy is added to a valence electron, it can be forced out of its orbital path and since there is no orbit it can enter, it becomes a "free" electron. When energy, thermal, electrical, light, etc., are added to the atom,

the valence shell receives the energy first. Therefore, the valence electrons are the easiest to "free" from the atom.

R21-8
STABLE AND UNSTABLE ATOMS

The ease with which an atom gives up its valence electrons depends on the number of electrons in the valence shell. If the shell is less than half full, contains one to three electrons, it tends to empty its valence shell. This allows the next inner shell, which is full, to be the valence shell. The atoms with this tendency are the best conductors of electrical energy: for example, copper, 1 valence electron; aluminum, 3 valence electrons; silver, 1 valence electron; gold, 1 valence electron. If the valence shell is more than half full, contains 5 or more valence electrons, the atom resists electron loss and is a good insulator: For example, the very stable gases: neon, 8 valence electrons; argon, 8 valence electrons; etc.

R21-9
BONDING

R21-9.1
Electrovalent Bonding

To use the atoms as conductors, insulators, or semiconductors, the atoms combine with other atoms to make compounds. When this occurs they combine in such proportions as to produce stable valence shells. For example: two hydrogen atoms, each with one valence electron, will combine with one atom of oxygen, with six valence electrons, to produce a compound (water, H_2O) with eight valence electrons in the oxygen shell.

Of more importance to the refrigeration technician is the fact that clean copper connections carry electricity well but connections coated with copper oxide have high resistance due to the insulating value of the copper oxide. Pure copper has only one valence electron, easily set free to carry electrical energy. However, two copper atoms plus one oxygen atom combine to produce a copper oxide molecule with eight valence electrons. As a result, copper oxide has a high resistance to electrical flow. It is classified as an insulator. When different atoms combine to form stable molecules or filled valence shells, they are said to be *electrovalent bonds* because they are bonded by sharing their valence electrons. They are also held together because of the positive and negative charges between the atoms.

When an atom has an equal number of protons and electrons it is in balance. If it either gives up or takes on electrons and the balance is eliminated, it is called an *ion.* If the atom gives up an electron, it has a surplus of protons; it is now a positive ion. If it takes on any electrons, it has a surplus of electrons and is now a negative ion. Because opposites attract, the copper ions (positive) and the oxygen ion (negative) attract and combine to form the molecule copper oxide. This process of combining opposite ions is called *ionic bonding.*

R21-9.2
Metallic Bonding

Metallic bonding occurs mostly in materials that are good conductors, such as copper, aluminum, silver, gold, etc. These all have a minimum of valence electrons which are easily ejected and free to roam from atom to atom in the material. As a result, the valence electron leaves its orbit at random but encounters another orbit immediately, where it joins. This frees the electron in the new orbit to seek an orbit in the next atom. This ease of travel allows the transfer of energy from atom to atom with little outside energy required. As a result, these materials are excellent conductors of electrical energy.

R21-9.3
Covalent Bonding

In electrovalent bonding, one of the atoms involved must be unstable so as to be able to give up the electron and produce the ion difference. In metallic bonding, the atoms must be unstable so as to have free electron action. If two atoms that resist either giving up or taking electrons are combined, they must share electrons to be bonded. When two such atoms meet, each atom allows one of its valence electrons to be shared by the other. For example, if we have two atoms of germanium, each with four valence electrons, each will allow one to be shared. They each keep three in their own personal valence shell and two more alternately from shell to shell. In this way neither actually gives up an electron, but instead the orbital path of the electron is extended to include the other atoms. With atoms doing this in combination and sharing electrons with atoms in different directions, each atom will acquire an additional four electrons traveling through its valence shell and become stable. This bonding is called *electron-pair bonding,* although the established term is *covalent bonding.*

Germanium and silicon materials used in most semiconductor electronics are joined by the electron-pair or covalent bonding. In these materials, the covalent bonding is between several atoms in groups that are combined with other groups into larger groups, with each atom in bond with four other atoms. This arrangement is called a *creptal lattice* structure. This term is used because the basic pattern of one atom bonded to form others, repeats throughout the molecular structure of the material. This arrangement holds true throughout the material except at the surface, where atoms at the surface do not have atoms on one side to bond to. This is why the surface of such materials have different electrical characteristics than the interior material.

R21-10
HOLES

A pure semiconductor material allows each of its atoms to see eight valence electrons, three for its personal use only and five on a shared basis with four other atoms. Therefore, each atom tends to be stable and the material acts as an insulator because there are no free electrons to carry energy through the material.

This would be true if no excess heat energy were involved. Even at room temperature enough heat energy is in the material to allow some of the unshared valence electrons to raise their energy level high enough to leave their valence shell. These electrons are free and can wander from one valence shell to another. Because the electron-pair bonds between the individual atoms made use of all the valence electrons, both personal and shared, the existence of any free electrons causes an electron-pair bond to be broken. Therefore, when the electron leaves its position in the valence shell it leaves an empty space called a *hole.* There was an electron in the space but it was set free and a hole created by heat or "thermal" agitation. Because of this action, therefore, a pure semiconductor or any other compound class has some free electrons that can take part in current flow. In addition, the material has holes in the covalent bonds that will readily accept electrons.

If the energy level of the electrons is raised by any energy addition (heat, electrical, light, etc.), many electrons will have their energy level increased to the point where, although they are not free to wander, they can jump from one valence shell to fill a hole in the next valence shell. When this happens, the hole that is left jumps to the covalent bond that lost the electron. That hole, in turn, will be filled by another valence electron, so the hole appears to move again. As a result, a pure semiconductor, in a practical sense, has free electrons and holes moving about in a random way.

Because of the fact that most of the atoms are stable, the pure semiconductor has only a few free electrons and only a slight current flows. This shows that the semiconductor has a high resistance. This resistance is, however, affected by temperature; the higher the temperature, the higher the electron activity, the higher the transfer of electrons and holes, the greater the current flow, and the lower the resistance. This action is the basis of the "thermister."

Even though a pure semiconductor trys to produce stable atoms, it still conducts current with some free electrons the same way a conductor does. Because the semiconductor has only a few free electrons, only a slight current flows.

Compared with a good conductor, which has an abundance of free electrons, the good conductor has high current-flow ability. Also, the good conductor electrons do not leave holes because they are from unstable valence shells.

In the semiconductor, the current flow is because of rupture of electron bond pairs and holes are left when the electron leaves. The holes allow the other valence electrons to move about because the holes continually try to be filled. The movement of the valence electrons does not require high energy because they are only moving to adjacent holes and the attraction of the holes causes them to move. The space between the electron and the hole acts like a positive–negative charge attraction, so the hole is considered to be positively charged.

If a dc voltage were applied across the semiconductor, the resulting voltage would draw the negative electrons toward the positive voltage. This would make the holes appear to move to the negative voltage. The free electrons would naturally be drawn to the positive voltage.

You can see that there are two types of current flow: electrons to the positive voltage and holes to the negative voltage. Even though the holes are only an evident movement, they are the ones that are used to describe the current movement in the "valence" area because they go in the opposite direction of the free electron flow. The two currents can, therefore, be easily identified. A semiconductor has a negative free electron current and a positive hole current.

Electron current flow continues because a free electron enters the semiconductor from the conductor on the negative side for each one that leaves on the positive side. Also, for each hole that is filled on the negative side, a new one is created on the positive side when the electron leaves and enters the conductor attached thereto. The holes do not leave or enter the semiconductor. They cannot because there are no covalent bonds or holes in the conductors.

The current flow in the wire is the total of the free electron flow and hole flow in the semiconductor. It must be remembered that in a semiconductor the current flow takes place at two different energy levels; one current is in the conduction bond of free electrons and the other is in the valence bond with valence electrons or holes. In a conductor, current flow takes place only in the conduction bond of free electrons.

So far we have assumed that the hole current and the electron current through the semiconductor are equal. This may be true for pure semiconductors with only a slight energy level applied. As applied energy increases, however, the electron current rises in a linear or straight line increase per increase in voltage. The hole current, however, does not rise appreciably until the applied energy level is appreciable. Because of the different reactions in current flow to applied energy (voltage) changes, the ohm resistance of a semiconductor does not follow Ohm's law.

R21-12
DOPING

To improve the energy-carrying capacity of semiconductors, foreign atoms are mixed with the atoms of the semiconductor. The type of foreign atom used will depend on the current carrying action desired in the semiconductor. Although the electrons and holes in the atom are not the actual current, they are the means of carrying the current. Electrons move from negative to positive, so they are classified as *negative current carriers*. The holes appear to move from positive to negative, so they are called *positive current carriers*.

If a material added to the semiconductor has five valence electrons (such as arsenic) and the basic semiconductor is silicon, which has four valence electrons, the arsonic atom will form an electron-pair bond with the silicon atom but one electron will be left over. This releases a free electron that can move freely between atoms. This combination will have a higher ability to carry negative current.

If the material added to the basic material of the semiconductor (silicon) has a lower number of valence electrons than the basic material, for instance, boron with only three valence electrons, the combination will leave the electronic-pair bond short an electron. In other words, it will leave excessive holes in the atom combinations. This combination has a higher ability to carry positive current.

This process of controlling the electron and hole makeup of the atomic combinations is called *doping*. Also, if the foreign material used has five valence electrons to produce free electrons, atoms of this material are called *pentavalent*. If the material used has three valent electrons, the atoms are called *trivalent*. Remember, pentavalent impurities provide excess free electrons or negative current carriers; trivalent impurities provide excess holes or positive current carriers.

Semiconductors doped for excess free electrons are predominantly negative current carriers and are classified *N-type*. Semiconductors doped for excess holes for heavier positive current carrying capacity are called

P-type. P-type semiconductors will carry current easier than pure semiconductors but not as easily as the N-type semiconductors.

R21-13
SOLID-STATE DIODES

In solid-state diodes, current will pass through the diode in one direction but not in the reverse direction, the same as in a vacuum tube. There are two general types of diodes with this ability: metallic rectifier and semiconductor diode.

R21-13.1
Metallic Rectifiers

Two types of metallic rectifiers have been predominantly used, although others are known. These two are the copper oxide rectifier and the selenium rectifier.

The copper oxide rectifier is made of a thick disk of copper on which a thin layer of copper oxide has been deposited. When a negative voltage is applied to the copper and a positive voltage to the copper oxide, the resistance to flow is very low and current passes easily. When the voltages are reversed, however, the resistance becomes very high and the current flow almost disappears. The selenium–iron rectifier, which is composed of an iron plate on which a thin layer of selenium has been deposited, has the same characteristics.

These rectifiers have high current-carrying abilities as well as the ability to handle high voltages. The selenium–iron combination is the most durable and is used in the majority of applications. In the air-conditioning field, the chief application is in electronic air cleaners when the current flow is in the 0.25-A range but the voltage involved is upwards of 6000 V. This application is explained in more detail in Sections R21-6 and R21-6.1. In controls such as thermostats, load centers, etc., the rectifiers are of much lighter duty and voltages are considerably lower; therefore, semi-conductors of the N-type and P-type are used.

R21-13.2
Semiconductor Diodes

When N-type and P-type semiconductors are combined into a single unit, they are called a *P-N* or *semiconductor diode.* Because each half of the *P-N* unit has opposite majority and minority currents, the resistance of the unit to current flow going in one direction is much higher than the resistance to current going in the other direction.

Remember, in the P-type the majority current carriers on the holes are positive current carriers and the free electrons are the minority current carriers. In the N-type the free electrons are the majority current carriers and the holes are the minority current carriers. When the two types of semiconductors are bonded automatically together, when the bond occurs, a "depletion region" develops.

R21-14
DEPLETION REGION

Before bonding took place the opposite changes in each N-type and P-type semiconductor were neutral, with the N-type having free electrons and the P-type having free holes. When atomic bonding of the two types take place, at the connection, the free electrons in the N-type will cross the bond to attempt to fill the holes in the P-type. As a result the positive ion changes in the N section outnumber the negative free electrons and the N section takes a positive charge in the bond region. At the same time, the P section takes on a negative charge as more and more free electrons cross the junction to fill the holes. As the holes fill, the negative charge becomes greater and greater until the overall negative charge in the P section becomes sufficient to repel the free electrons from the N section. This stops the action. At the same time the positive charge built up in the N section attracts the free electrons on the N side of the bond. The electrical charge buildup stops the electron-hole combination formation at the bond and limits the depth into each section that these changes can occur. The section of each semiconductor on each side of the bond is called the *depletion region.* Because of the formation of the depletion region, the P and N sections of the diode have equal and opposite charges. Like a battery, a voltage difference exists between the two sections. The region is called a *potential barrier* ("potential" referring to voltage).

R21-15
FORWARD CURRENT FLOW

If an outside energy source, a battery, is connected with the negative of the battery connected to the N side of the semiconductor and the positive side to the P side of the semiconductor, the electron from the battery through the wire will enter the semiconductor and cause an increase in the electron quality in the N side (see Fig. R21-4). At the same time, the positive side of the battery will draw the electrons out of the P side of the semiconductor and leave an excess of holes in the P side. With an excess of holes on the P side and an excess of electrons on the N side, the transfer of electrons through the potential barrier will increase and we have electron (current) flow through the diode.

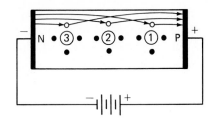

FIGURE R21-4   Forward current flow.

R21-16
REVERSE CURRENT FLOW

If the battery were to be reversed, with the negative side of the battery connected to the *P* side and the positive side to the *N* side (see Fig. R21-5), the positive voltage on the *N* side will pull the electrons away from the potential barrier. The electrons will not be able to combine with the holes in the *P* side and the majority current flow will stop. The minority current flow between the two sides will still exist, so some current flow will still take place, although of minor consequences compared to the majority current flow.

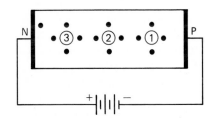

FIGURE R21-5   Reverse current flow.

R21-17
P-N DIODE RECTIFIER

In as much as the *P-N* diode will carry current more readily in one direction than in the other, it can be used to convert an alternating current to a pulsating direct current. Figure R21-6 shows a diode connected to an alternating current source and supplying a resistance (thermal)-type load. Pictured also is the resulting waveform that is produced across the resistor.

Diodes are pictured as an arrow point against a surface and the arrow point represents the *P* section of the diode. The arrow points in the direction of each flow if the power source is converted with power positive to

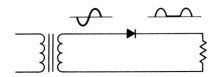

FIGURE R21-6   AC circuit using P-N diode.

diode positive and power negative to diode negative. The diode resistance in this direction would be low, maybe in the hundreds of ohms. With the battery reversed, however, the diode would show a very high resistance to current flow, in the 500,000+ Ω range.

When using an ohmmeter to check diodes, if the ohmmeter battery power supply is larger than the potential capacity of the diode, the diode may reach the breakdown point, called the *avalanche-breakdown point,* and be destroyed. Normal rectifier and operating-type diodes are rated with applied voltage limits to prevent application where avalanche breakdown or runaway voltages are encountered.

Figure R21-6 shows the wave pattern resulting from the use of a single diode to rectify the ac voltage. Notice that due to the minority currents that flow through the diode, the OFF cycle is not completely cut off. Also, a timespace exists between the voltages that are passed because the half of the ac cycle has been reduced. The resulting dc power is "half-wave dc." To produce a more even power supply, a full wave dc voltage can be produced by the use of two diodes connected as shown in Fig. R21-7. In this case the power source consists of a double transformer to supply the same power to each side of the diode circuit, depending on the direction of current flow. The resulting wave pattern shows a positive flow to the resistor using each side of the ac cycles. This type of power is called *full-wave dc.* The wave pattern is also marked to show that alternate cycles are conducted by the respective diodes.

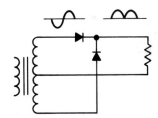

FIGURE R21-7   AC circuit using 2 P-N diodes.

Figure R21-8 shows a *bridge rectifier circuit,* which does not require a double or center-tapped transformer. It may, in some instances, be placed directly across the line without a transformer if the desired dc voltage is within 10% of the source voltage. In this circuit, four diodes are used. On the positive half of the cycle, current flows through diode 3, through the load, and back through diode 2. Diodes D_1 and D_4 have reverse voltage and they have high resistance, like an open switch.

On the negative side of the cycle, diodes 2 and 3 have the high resistance and diodes 1 and 4 have the low resistance. Current flows through diode 4, through the load, and back through diode 1. This again produces full-wave pulsating dc current. This arrangement is called a *full-wave rectifier.*

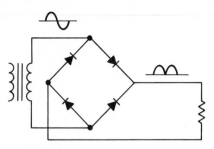

FIGURE R21-8 Bridge rectifier circuit.

R21-18
FILTERS

The dc voltage produced by semiconductor rectifiers is a pulsating dc voltage, either half wave or full wave, depending on the type of rectifier circuit used. Because of the reaction speed of solid-state-type control, the 60-pulse or 120-pulse-per-second voltage wave pattern would cause rapid on and off cycling of any control system. It is necessary, therefore, to filter out those pulsations and produce a smooth dc voltage.

To accomplish this, an arrangement of capacitors, resistors, and inductors in various combinations are used. Two general arrangements are an L section, using a capacitor and two resistors, or a capacitor and an inductor; or the pi (Π) section, using two capacitors and a resistor, or two capacitors and an inductor.

R21-18.1
Capacitative Filters

The capacitative filter is simply a capacitor connected across the power source. As the pulsating dc voltage from the half-wave or full-wave rectifier circuit is applied across the capacitor, it charges the capacitor to the peak voltage of the voltage wave. If there were no load on the circuit, the capacitor would take on and retain this high voltage. With a load connected, however, the current draw of the load causes the capacitor to discharge and the voltage to drop during the low-voltage peak of the cycle.

Figure R21-9 shows the difference in the voltage cycle entering the capacitor circuit versus the cycle after the capacitor circuit. From this you can see that the capacitor, taking time to fill and time to empty, causes the

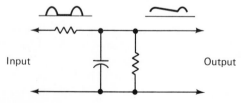

FIGURE R21-9 Single capacitor filter circuit.

peaks of the input cycle to be decreased and the valleys to fill. The range of the pulsation of the dc voltage has been reduced but the remaining ripple is still too extensive for proper application.

R21-18.2
Section Filters

The ability of the capacitor to take on voltage causes high surge current when the capacitor is charging. This can cause overload breakdown in *P-N* semiconductor rectifiers. To eliminate this possibility, a resistor is placed ahead of the capacitor to limit the surge current. Remember, the amount of resistance a resistor develops depends on the amount of current trying to pass through: a large amount of current, high resistance to flow; a small amount of current, small resistance to flow. Another way to say this would be: high current flow, high voltage drop across the resistor; low current flow, low voltage drop across the resistor. The result is a closer range of the voltage applied to the capacitor, allowing the capacitor to produce less ripple voltage. The limiting resistor controls the surge current by introducing a resistive time period to slow the charging of the capacitor.

This resistor–capacitor arrangement is called an *L-section filter*. An inductor commonly called a "choke" coil may be used instead of the resistor. An inductor is a number of turns of wire around an iron core. When a voltage is applied and current starts to flow through the coil, an induced counter voltage develops as long as the current is building. This was covered in Section 16-9. This inductive reactance also provides a time factor for the voltage buildup, to limit the surge current. Inductors also react with the capacitor to produce a more complete leveling of the voltage—more so than the resistor. In larger power filters, the inductor is used predominantly. Its production cost, however, is higher, which promotes the use of the resistor-type L-section filter in small, light-duty systems.

R21-18.3
Pi-Section Filters

When a smooth dc voltage is required, another capacitor is installed ahead of the surge-limiting resistor or inductor to provide a first-stage ripple effect. This further reduces the voltage variation into the rest of the filter system and allows the inductor or resistor and the final capacitor to emit a smooth dc voltage into the connected load. Figure R21-10 shows the two types of pi-section filters, the one on the top using a resistor and the one on the bottom using an induction or choke.

By the use of the semiconductor rectifier, we can convert ac to dc at whatever dc voltage we want from the 6000 V dc used across the filter cells of the electronic air cleaner to the 0.5-V dc used in microcircuits of load

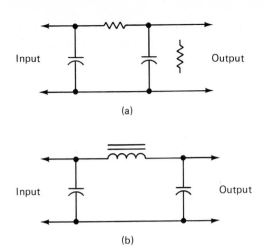

(a)

(b)

FIGURE R21-10 Dual capacitor filter circuit.

controls, all this without the use of batteries that wear out or run down and of such a bulk size as to be totally impractical.

R21-19
APPLICATIONS OF SEMICONDUCTOR DIODES

It would be impossible to list in this text all applications of the semiconductor diode without creating many volumes. For a broader knowledge of particular applications, it is suggested that manufacturers' published material be used, such as that published by Honeywell, Penn Control, Ranco, White Rodgers, Robertshaw, etc. The information in this chapter has been primarily through the courtesy of Honeywell.

We will attempt, however, to include a few common examples of the use of *P-N* semiconductors in commonly used equipment.

R21-19.1
Electronic Air Filters

As stated previously, the electronic air cleaner was one of the first solid-state applications. The filtering ac-

tion is covered in Chapter A6. The electrical operation is as follows. Figure R21-11 shows the electrical wiring diagram of a typical electronic air cleaner. Some cleaners use meters to indicate performance levels; others use indicating lights; still others have voltage variations for "ozone control"; etc. These have been left out of the discussion to limit it to *P-N* semiconductors. Following the diagram, power travels from the hot lead of the 120-V power supply through the control switch, the door safety switch, to the primary of the step-up transformer and back to the other side (white) of the 120-V supply. The transformer increases the 120-V ac to 5000-V ac. This 3000-V ac is across the selenium rectifier or capacitor bridge circuit. Notice that two capacitors are used in place of two rectifiers to produce a voltage-doubling effect of pulsating full-wave dc. Because we do not require a smooth dc voltage for filter action, no filters are used. The capacitors are charged to full capacity for peak output voltage.

Figure R21-12 shows the current flow during the positive side of the applied ac power. Electrons flow from terminal 1 of the transformer 3000-V output side into the capacitor producing the negative charge. A positive charge is produced on the other side of the capacitor because of electrons from the capacitor through the selenium rectifier to terminal 2 of the transformer output. This charges capacitor C_1 with 3000-V.

Figure R21-13 shows the current flow during the negative side of the cycle of the applied ac power. Here the polarities reverse. Electrons now flow from terminal 2 of the transformer output through the selenium rectifier 2 into capacitor 2 and produce a 3000-V negative charge in the capacitor. Electrons flow from the other side of the capacitor to the other side of the power source on terminal 1. The voltage applied at this moment to selenium rectifier 1 is reversed. The resistance of the rectifier is now high and it does not allow the current to pass. When the ac power has completed a cycle (60 per second in 60-Hz power), we have stored 3000 V dc on each of the two capacitors (see Fig. R21-14). Since these capacitors are in series between the ionizer section of the filter cell and ground, the two voltages add and produce a voltage of 6000 V dc across the filter.

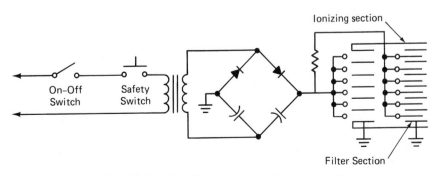

FIGURE R21-11 Electronic air cleaner circuit.

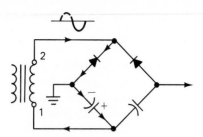

FIGURE R21-12 Voltage doubler—positive half of cycle.

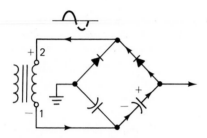

FIGURE R21-13 Voltage doubler—negative half of cycle.

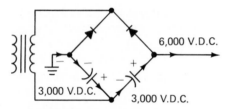

6,000 V.D.C.

3,000 V.D.C.

3,000 V.D.C.

FIGURE R21-14 Voltage doubler—complete cycle.

R21-19.2
Electronic Ignition

The spark that lights the gas flame in the electronic ignition system is also produced by *P-N* semiconductor action as a total interuption switch acting upon a transformer to produce a high voltage across an ignitor. This system is compared to the ignition system in an automobile, where a breaker point makes and charges the coil, then breaks; a capacitor charging causes a rapid

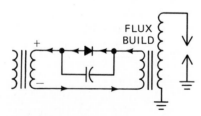

FLUX BUILD

FIGURE R21-15 Ignition system—charge half of cycle.

field collapse in the coil and a high-voltage impulse across an ignitor (spark plug) fires the engine cylinder (Fig. R21-15).

Figure R21-16 shows the basic diagram of the ignitor system showing the *P-N* semiconductor with a capacitor in parallel connected to a step-up transformer called a "pulse coil," which in turn has an ignitor mounted on a burner connected to the pulse coil output. With 24-V ac applied to the system, on the negative side of the cycle, electron flow occurs through the *P-N* diode, and the flow through the primary coil of the pulse transformer produces a magnetic saturation in the coil. This

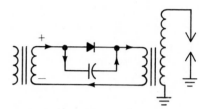

FIGURE R21-16 Ignition system—ignition half of cycle.

is a high quantity of magnetic flux in the coil. At the same time, the closing action of the *P-N* diode shorts out and empties the capacitor.

When the action reverses, the positive side of the cycle, the reverse action of the voltage causes a collapse of the fields in the coil. At the same time because the *P-N* diode, acting as a switch, opens, the capacitor no longer is shorted and takes on a charge. The action of the voltage charge accelerated by the drawing action of the capacitor causes a very rapid collapse of the magnetic field in the coil. Because voltage produced in a coil is equal to the speed of the lines of force cutting through the coils, the rapid increase in this speed induces an extremely high voltage (10,000+ V), enough to cause a jump from the ignitor to ground, and a spark is produced. This action is repeated each cycle.

R21-19.3
Control Systems

The state of the art in control work is that these controls are manufactured in encapsulated units to prevent moisture damage, which also prevents repair of the individual unit. Service therefore consists of replacement of the individual components. The service procedure also varies with each individual item or system. The same type of sytem produced by different manufactures may use different service procedures. The only policy for proper and efficient service on electronic controls is to *follow the manufacturer's instructions to the letter*.

R21-1. The atom is composed of _____ and _____ .

R21-2. Protons carry a _____ electrical charge.

R21-3. Electrons carry a _____ electrical charge.

R21-4. What determines the atomic weight of an element?

R21-5. If copper has 29 protons in the nucleus, how many electrons are in its electrical field?

R21-6. The path the electron takes around the proton is called the _____ .

R21-7. The number of electrons in the first shell is _____ .

R21-8. The shell farthest from the proton is called the _____ .

R21-9. The outer shell will not hold more than _____ electrons.

R21-10. Which has a greater energy level, electrons traveling in the inner shell or those traveling in the outer shell?

R21-11. Define a "free" electron.

R21-12. Copper, silver, and aluminum are good conductors of electricity. Why?

R21-13. If copper is a good conductor of electricity, why is copper oxide an insulator?

R21-14. Define an atom in balance.

R21-15. If a balanced atom gives up an electron, it is called an _____ .

R21-16. An atom with a surplus of protons is called a _____ .

R21-17. An atom with a surplus of electrons is called a _____ .

R21-18. Describe "ionic" bonding.

R21-19. Good conductors of electricity depend on what condition in the atoms?

R21-20. Describe covalent bonding.

R21-21. Why are covalent bonded atoms poor conductors of electricity?

R21-22. Describe a hole.

R21-23. Describe the action of the thermister.

R21-24. A semiconductor has two current flows through it. What are they?

R21-25. In a conductor only one current flow takes place. Name it.

R21-26. The process of controlling the electron and hole makeup of atomic combinations is called _____ .

R21-27. Negative current carriers are produced by adding impurities called _____ .

R21-28. Positive current carriers are produced by adding impurities called _____ .

R21-29. Semiconductors doped to carry negative current are classified as _____ types.

R21-30. Semiconductors doped to carry positive current are classified as _____ types.

R21-31. Define a solid-state diode.

R21-32. Name the two general types of diodes.

R21-33. The two most predominately used metallic rectifiers are _____ and _____ .

R21-34. Describe the depletion region in a semiconductor.

R21-35. The use of a *P-N* diode and ac transformer will produce a _____ direct current.

R21-36. Describe the Avalanche breakdown point of a diode.

R21-37. A bridge rectifier circuit uses how many diodes?

R21-38. When can a bridge rectifier circuit be used without a power supply transformer?

R21-39. Why are resisters used in the circuit between the diode output and the filter section of an ac/dc power supply?

R21-40. An inductor called a _____ may be used instead of a resister.

R21-41. The one resistor–one capacitor filter arrangement is called an _____ filter.

R21-42. The use of two capacitors with a limiting resistor or choke coil is called a _____ filter.

R21-43. The use of two capacitors and two rectifiers in the bridge circuit is called a _____ .

R21-44. In the electronic system, what part performs the same function as the breaker points in the distributor of an automobile?

R21-45. What is the best policy to follow in servicing electronic controls?

Refrigeration Measuring and Testing Equipment

R22-1
GENERAL

In refrigeration (and air-conditioning) work you will use many different kinds of instruments and test equipment. Previous chapters have detailed basic hand tools and certain test equipment such as refrigerant-leak testing devices as well as measuring and testing equipment associated with the electrical phases of refrigeration systems. This chapter reviews measuring and testing equipment needed to determine temperature and pressure conditions.

R22-2
TEMPERATURE MEASUREMENTS

When analyzing a refrigeration system accurate temperature readings are important. The most common temperture measuring device is the pocket glass thermometer, illustrated in Fig. R22-1. Note that it fits into a protective metal case. The thermometer head has a ring for attaching a string to suspend it if needed. The ranges of glass thermometers vary but −30° to +120°F is a common scale for refrigeration systems, and the thermometer is calibrated in 2° marks. Some have a mercury fill, but others use a red fill that is easier to read.

To check the calibration of a pocket glass thermometer insert it into a glass of ice water for several minutes. It should read 32°F plus or minus 1°F. Should the fill separate, place the thermometer in a freezer, and the resulting contraction will probably rejoin the separated column of fluid. Another way to connect the separated fill is to carefully heat the stem, *not the bulb*. In most cases the liquid will coalesce as it expands.

Another form of pocket thermometer is the dial type, shown in Fig. R22-2. It too has a carrying case with pocket clip. The dial thermometer is more convenient or more practical in measuring air temperatures in a duct. The stem is inserted into the duct, but the dial remains visible. Again several ranges are available, depending on the accuracy needed and the nature of application. A common refrigeration range is −40° to +160°F. A different type of dial thermometer is the superheat thermometer, illustrated in Fig. R22-3.

The highly accurate expansion-bulb thermometer is used to measure suction-line temperature(s) in order to calculate, check, and adjust superheat. A common range is −40 to 65°F. The sensing bulb is strapped or clamped to the refrigerant line and covered with insulating material (such as a foam rubber sheet) to prevent air circulation over the bulb while you take a reading. Of course, the superheat bulb thermometer can be used to measure air or water temperatures as well.

The thermometers just described have been the basic temperature-measuring tools for many years. However, they do have certain limitations; for example, the operator is required to actually be in the immediate area during the time of reading. An example would be trying to measure the inside temperature of your home refrigerator without opening the door.

FIGURE R22-1   Glass thermometer. (*Courtesy* Robinair Manufacturing Corporation)

FIGURE R22-2   Pocket dial thermometer. (*Courtesy* Robinair Manufacturing Corporation)

FIGURE R22-3 Superheat thermometer.
(*Courtesy* Marsh Instrument Company)

FIGURE R22-5 Electronic thermometer.
(*Courtesy* Robinair Manufacturing
Corporation.

The thermometer illustrated in Fig. R22-4 is also a hand-held device, but it has a probe length of 30 in. to allow a degree of remote temperature measurement.

Different probes are available to measure surface of air temperatures. In refrigeration use, the surface probe determines superheat settings of expansion valves, motor temperature, condensing temperature, and water temperature. This is a useful instrument for many applications, but it too is limited, since only one reading in one place at a time can be taken.

As a result of the rapid development of low-cost electronic devices the availability and use of electronic thermometers is now very common. The electronic thermometer as illustrated in Fig. R22-5 consists of a tester with provision to attach one or several (three to six) sensing leads. The sensing lead tip is actually a thermistor element, which, when subjected to heat or cold, will vary the electrical current in the test circuit because its resistance changes with temperature change. The tester converts changes in electrical current to temperature readings. The sensing leads vary in length depending on

the make of the unit, but extensions can be used to allow for remote testing.

Once the operator places the sensing probe(s) in the spots chosen to be tested he or she may switch from position to position and record temperatures without actually going into each test area, such as the refrigerator, walk-in cooler, freezer, or air duct. The sensing probe can also be used to check superheat.

Sometimes it may be necessary to record temperatures over long periods of time such as a day or even

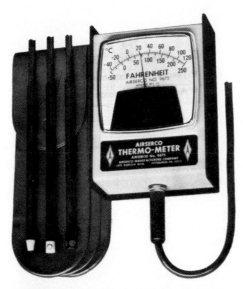

FIGURE R22-4 Thermometer. (*Courtesy* Airserco Manufacturing Company)

FIGURE R22-6 Portable recording thermometer.
(*Courtesy* Airserco Manufacturing Company)

Refrigeration Measuring and Testing Equipment 251

a week, in order to examine the changes in system conditions. Very expensive recording thermometers for highly sophisticated jobs are available, but for the average installation there are reasonably inexpensive, compact, and portable recording thermometers, shown in Fig. R22-6, which consist of a hand-wound chart-driving unit. A recording thermometer takes the guesswork out of setting the system operating conditions or diagnosing and locating trouble areas—and it provides a permanent record of the results.

The final selection of temperature-measuring instruments will depend on the scope of work with which a technician is involved. If servicing refrigeration, plus air-conditioning and heating, the technician will need a wide variety of thermometers. It is important to remember that these are sensitive devices and require consistent care and calibration to provide accuracy and reliability.

R22-3
PRESSURE MEASUREMENTS

Temperature measurements are usually taken outside the operating system. But it is also necessary for the service technician to know what is going on inside the system, and this must basically be learned from pressure measurements.

Chapter R3 discussed two pressure gauges as the measuring instruments necessary to obtain these readings (see Fig. R22-7). On the right in Fig. R22-7 is the *high-pressure gauge*, which measures high-side or condensing pressures. It is normally graduated from 0 to 500 psi in 5-lb graduations. The *compound gauge* (left) is used on the low side (suction pressures) and is normally graduated from a 30-in. vacuum to 120 psi; thus, it can measure pressure above and below atmospheric pressure. This gauge is calibrated in 1-lb graduations. Other pressure ranges are available for both gauges, but these two are the most common.

Note that on the dial faces there is an intra scale, which gives the corresponding saturated refrigerant temperatures at a particular pressure. (Remember from Chap. R7 that there is a definite relationship between pressure and saturated temperature for a given refrigerant). In Fig. R22-7 the dials are marked for R-12 and R-22. Gauges for other refrigerants and for metric measurements are also available.

A service device that includes both the high and compound gauges is called a *gauge manifold*. It enables the serviceman to check system operating pressures, add or remove refrigerant, add oil, purge noncondensibles, bypass the compressor, analyze system conditions, and perform many other operations without replacing gauges or trying to operate service connections in inaccessible places.

FIGURE R22-7 Pressure gauges. (*Courtesy* Robinair Manufacturing Corporation)

The testing manifold as illustrated in Fig. R22-8 consists of a service manifold containing service valves. On the left the compound gauge (suction) is mounted and on the right the high-pressure gauge (discharge). On the bottom of the manifold are hoses which lead to the equipment suction service valve (left), refrigerant drum (middle), and the equipment discharge, or liquid-line valve (right).

Many equipment manufacturers color code the low-side gauge casing and hose *blue* and the high-side gauge and hose *red*. The center or refrigerant hose is colored *white*. This system is very helpful to avoid crossing hoses and damaging gauges. A hook is provided to hang the assembly and free the operator from holding it.

By opening and closing the refrigerant valves on gauge manifold *A* and *B* (Fig. R22-9) we can obtain different refrigerant flow patterns. The valving is so arranged that when the valves are closed (front-seated) the center port on the manifold is closed to the gauges (Fig. R22-9). When the valves are in the closed position, gauge ports 1 and 2 are still open to the gauges, permitting the gauges to register system pressures.

With the low-side valve (1) open and the high-side valve (2) closed (Fig. R22-9), the refrigerant is allowed

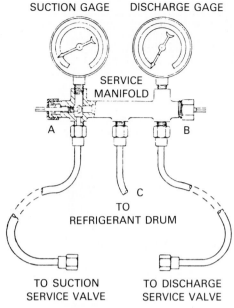

SUCTION GAGE   DISCHARGE GAGE

SERVICE
MANIFOLD

A   B

C
TO
REFRIGERANT DRUM

TO SUCTION   TO DISCHARGE
SERVICE VALVE   SERVICE VALVE

FIGURE R22-8   Testing manifold.

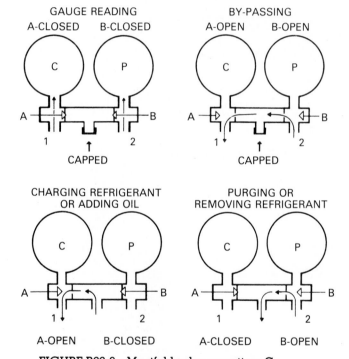

GAUGE READING   BY-PASSING
A-CLOSED   B-CLOSED   A-OPEN   B-OPEN

C   P   C   P

A   B   A   B
1   2   1   2

CAPPED   CAPPED

CHARGING REFRIGERANT   PURGING OR
OR ADDING OIL   REMOVING REFRIGERANT

C   P   C   P

A   B   A   B
1   2   1   2

A-OPEN   B-CLOSED   A-CLOSED   B-OPEN

FIGURE R22-9   Manifold valve operation: C,
compound gauge; P, pressure gauge; 1, gauge line
to suction service S.O.V.; 2, gauge line to discharge
service S.O.V. (*Courtesy* BDP Company)

to pass through the low side of the manifold and the center port connection. This arrangement might be used when refrigerant or oil is added to the system.

Figure R22-9 illustrates the procedure for bypassing refrigerant from the high side to the low side. Both valves are open and the center port is capped. Refrig-

erant will always flow from the high-pressure area to a lower-pressure area.

Figure R22-9 shows the valving arrangement for purging or removing refrigerant. The low-side valve is closed. The center port is open to the atmosphere or connected to an empty refrigerant drum. The high-side valve is opened, permitting a flow of high pressure out of the center port.

Note: Purging large quantities of flurocarbon refrigerants to the open atmosphere is not recommended unless absolutely necessary.

The method of connecting the gauge manifold to a refrigerant system depends on the state of the system, that is, whether the system is operating or just being installed. For example, let's assume that the system is operating and equipped with back-seating in-line service valves (Fig. R22-10).

The first step is to purge the gauge manifold of contaminants before connecting it to the system:

1. Remove the valve stem caps from the equipment service valves and check to be sure that both service valves are back-seated.
2. Remove the gauge port caps from both service valves.
3. Connect the center hose from the gauge manifold to a refrigerant cylinder, using the same type of refrigerant that is in the system, and open both valves on gauge manifold.
4. Open the valve on the refrigerant cylinder for about 2 seconds, and then close it. This will purge any contaminants for the gauge manifold and hoses.
5. Next, connect the gauge manifold hoses to the gauge ports—the low-pressure compound gauge to the suction service valve and the high-pressure gauge to the liquid-line service valve, as illustrated in Fig. R22-11.
6. Front-seat or close both valves on the gauge manifold. Crack (turn clockwise) both service valves one turn off the back seat. The system is now allowed to register on each gauge. With the gauge manifold and hoses purged and connected to the system, we are free to perform whatever service function is necessary within the refrigeration cycle.

To remove the gauge manifold from the system, follow this procedure:

1. Back-seat (counterclockwise) both the liquid and suction service valves on the in-line type.
2. Remove hoses from gauge ports and seal ends of hoses with $\frac{1}{4}$-in. flare plugs to prevent hoses from being contaminated. (Some manifold assemblies have built-in hose seal fittings).

Refrigeration Measuring and Testing Equipment   253

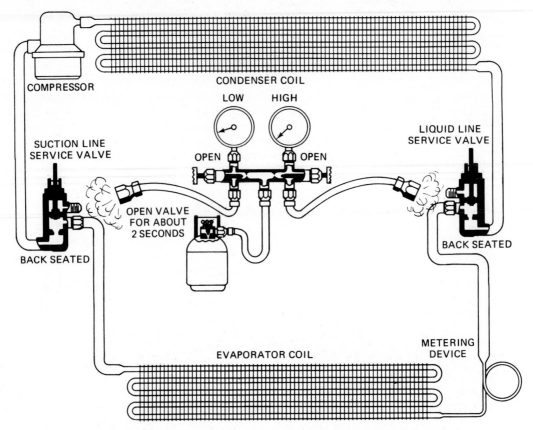

FIGURE R22-10   Purging gauge manifold. (*Courtesy* BDP Company)

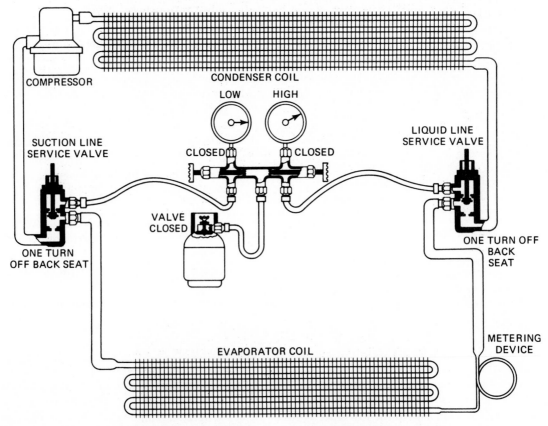

FIGURE R22-11   Connecting manifold. (*Courtesy* BDP Company)

3. Replace all gauge port and valve stem caps. Make sure that all caps *contain the gaskets provided with them* and are tight.

The manifold and gauges are necessary tools to perform many system operations. Once the system has been completed and cleaned of most of the air by purging, it must be tested for leaks. Or, whenever a component has been repaired or replaced, it is imperative that the entire system be checked for leaks.

R22-4
LEAK TESTING

In most cases a low-pressure refrigerant may be used to build sufficient pressure in the system to check for leaks such as illustrated in Fig. R22-12. Install the gauge manifold as described previously. Open both valves on the gauge manifold. Open the service valves on system. Open the valve on the refrigerant cylinder and pressurize the system to the cylinder pressure with refrigerant vapor (keep cylinder in upright position). Use any one of the three methods of leak testing described in Chapter R7. Where the application or requirement of

local codes requires a pressure test above the refrigerant vapor pressure, some other gas may be used for testing, for example, dry nitrogen. *Under no circumstances should oxygen be used.* Nitrogen would be introduced into the system through the center hose connection after disconnecting the refrigerant cylinder.

Caution: The test gas cylinder must be equipped with a pressure gauge and regulator so that system test pressures do not exceed maximum allowable limits as prescribed by national or local codes or by the equipment manufacturer.

R22-5
PURGING

Whenever a system is exposed to atmospheric conditions for a short period of time (less than 5 minutes, for example) during component replacement, it is necessary to purge the system to remove any contaminants that may have entered the system. Similarly, during installation, if the refrigerant lines are left open for more than 5 minutes, the system should be purged.

The theory behind purging is to use a high-velocity charge of gaseous refrigerant to blow any contaminants

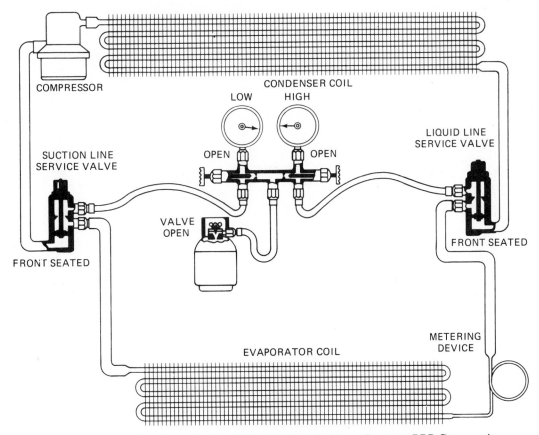

FIGURE R22-12 Using test manifold while leak testing. (*Courtesy* BDP Company)

Refrigeration Measuring and Testing Equipment 255

from the system. To purge a system that has just been installed, proceed as follows:

Install the gauge manifold as illustrated in Fig. R22-13, with the low-side valve closed and not connected to the suction service valve. Connect the center hose to the refrigerant drum. Connect the high-side hose to the liquid-line valve. Front-seat both service valves and open high-side manifold valve. Open the valve (wide) on the refrigerant cylinder and allow a high-velocity charge of refrigerant vapor ($\frac{1}{2}$ to 1 lb or more depending on the size of the equipment) to enter the system. The refrigerant will push any contaminants through the system to the suction service valve, where they will be purged through the gauge port.

Whenever a defective component such as an expansion valve is to be removed, the system should be pumped down and that part of the system should be isolated by means of the service valves. Then when the new component is installed the lines should be purged from both sides. *Pumping down* means to store the refrigerant in the receiver or condenser.

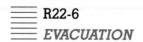
Proper evacuation of a unit will remove noncondensibles (mainly air, water, and inert gases) and assure a tight, dry system before charging. There are generally two methods used to evacuate a system: the *deep vacuum* method and the *triple evacuation* method. Each has its advantages and disadvantages. The choice depends on several factors: type of vacuum pump available, the time that can be spent on the job, and whether there is liquid water in the system.

In refrigeration work, especially those systems that operate at very low suction pressures, the deep vacuum method is recommended. In higher temperature refrigeration systems and in air-conditioning work, triple evacuation is practiced. Both methods will be discussed.

The tools needed to evacuate a system properly depend on the method used. A good vacuum pump and vacuum indicator are needed for the deep vacuum method, and a good vacuum pump and compound gauge are needed for the triple evacuation method.

Figure R22-14 shows a tool used to remove and replace the valve core from Schrader or Dill valves. An-

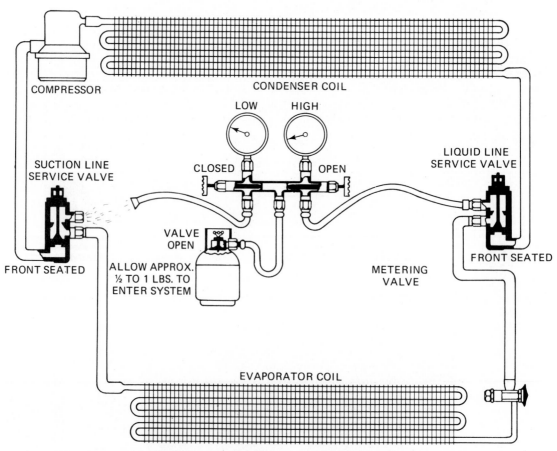

FIGURE R22-13   Purging. (*Courtesy* BDP Company)

FIGURE R22-14   Valve core remover.
(*Courtesy* Ritchie Engineering Co., Inc.)

other use is as a vacuum charge valve. By removing the core, a vacuum can be pulled much quicker. The valve core is a high restriction to gas flow and is best removed. This can be done without opening the system to atmosphere with this tool.

R22-7
VACUUM PUMP

A vacuum pump, illustrated in Fig. R22-15, is somewhat like an air compressor in reverse. Most are driven by a direct- or belt-driven electric motor but gasoline engine-driven pumps are also available. The pump may be single or two stage depending on the design. Most pumps for normal field service are portable; they have carrying handles or are mounted on wheel dollies. Vacuum pump sizes are rated according to the free air displacement in cubic feet per minute or liters per minute metric. Specifications may also include a statement as to the degree of vacuum the pump can achieve, expressed in terms of microns.

FIGURE R22-15   Vacuum pump. (*Courtesy* Robinair Manufacturing Corporation)

What is a micron? When the vacuum pressure approaches 29.5 to 30 in. on the compound gauge, the gauge is working within the last half inch of pressure, and the readout beyond 29.5 in. is not reliable for the single deep vacuum method. The industry has therefore adopted another measurement, called the *micron*. The micron is a unit of linear measure equal to 1/25,400 of an inch and is based on measurement above total absolute pressure, as opposed to gauge pressure, which can be affected by atmospheric pressure changes. Figure R22-16 is a comparison of measurements starting at a standard atmospheric conditions and extending to a deep vacuum.

Figure R22-16 not only demonstrates the comparison in units of measure but dramatically shows the changes in the boiling point of water as the evacuation approaches the perfect vacuum. This is the main purpose of evacuation—to reduce the pressure or vacuum enough to boil or vaporize the water and then pump it out of the system. It will be noted the compound gauge could not possibly be read to such minute changes in inches of mercury.

R22-8
HIGH-VACUUM INDICATORS

To measure these high vacuums the industry developed electronic instruments, such as is shown in Fig. R22-17. In general these are heat-sensing devices in that the sensing element, which is mechanically connected to the system being evacuated, generates heat. The rate at which heat is carried off changes as the surrounding gases and vapors are removed. Thus the output of the sensing element (either thermocouple or thermistor) changes as the heat dissipation rate changes, and this

FIGURE R22-16
Absolute pressure/temperature relationship.

BOILING POINT OF WATER		UNIT OF ABSOLUTE PRESSURE		Units of Vacuum (in. Hg)
°F	°C	psia	Microns of Mercury	
212	100	4.7	—	0
79	26	0.5	25,400	29.0
72	22	0.4	20,080	29.8
32	0	0.09	4,579	29.99
−25	−31	0.005	250	29.99
−40	−40	0.002	97	29.996
−60	−51	0.0005	25	29.999

Refrigeration Measuring and Testing Equipment   257

FIGURE R22-17 Electronic high-vacuum gauge.
(*Courtesy* Robinair Manufacturing Corporation)

change in output is indicated on a meter that is calibrated in microns of mercury.

The degree of accuracy of these instruments is approximately 10 microns, thereby approaching a perfect vacuum as shown in Fig. R22-16.

R22-9
DEEP VACUUM METHOD OF EVACUATION

The single deep vacuum method is the most positive method of assuring a system free of air and water. It takes longer but the results are far more positive. Select a vacuum pump capable of pulling at least 500 microns and a reliable electronic vacuum indicator. The procedure is illustrated in Fig. R22-18 and described below.

1. Install the gauge manifold as described earlier.
2. Connect the center hose to the vacuum manifold assembly. This is simply a three-valve operation allowing you to attach the vacuum pump and vacuum indicator and a cylinder of refrigerant, each with a shutoff valve.
3. Open the valves to the pump and indicator. Close the refrigerant valve. Follow the pump manufacturer's instructions for pump suction line size, oil, indicator location, and calibration.
4. Open (wide) both valves on the gauge manifold and mid-seat both equipment service valves.
5. Start the vacuum pump and evacuate the system until a vacuum of at least 500 microns is achieved.

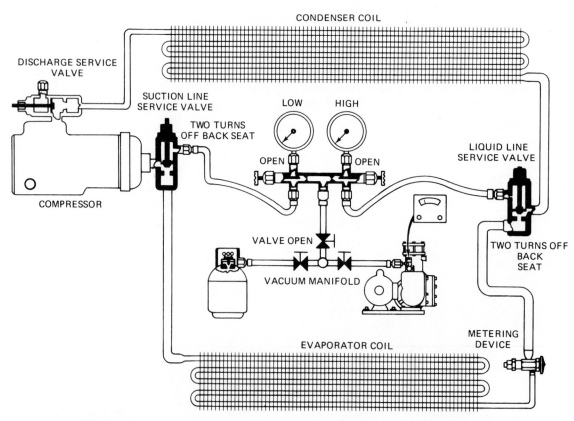

FIGURE R22-18 Deep vacuum evacuation assembly. (*Courtesy* BDP Company)

6. Close the pump valve and isolate the system. Stop the pump for 5 minutes and observe the vacuum indicator to see if the system has actually reached 500 microns and is holding. If the system fails to hold, check all connections for tight fit and repeat evacuation until the system does hold.

7. Close the valve to the indicator.

8. Open the valve to the refrigerant cylinder and raise the pressure to at least 10 psig or charge the system to the proper level (covered later).

9. Disconnect the pump and indicator.

R22-10
TRIPLE EVACUATION

The triple evacuation method requires no specialized high-vacuum equipment. However, this method should not be used if liquid water is suspected in the system. An evacuation pump of sufficient capacity to pull 500 microns of mercury vacuum will be needed. Good-quality refrigeration service gauges are important.

This method of evacuation is based on the principle of diluting the noncondensables and moisture with clean, dry refrigerant vapor. This vapor is then removed from the system, carrying with it a portion of the en-

trained contaminants. As the procedure is repeated, the remaining contaminants are proportionately reduced until the system is contaminant-free. Figure R22-19 illustrates the assembly procedure, described below.

1. Install the gauge manifold as described earlier.

2. Connect the center hose to the vacuum manifold valves.

3. Connect the pump and the refrigerant cylinder to manifold valves. Purge the lines with refrigerant.

4. Close the refrigerant cylinder valve and open the pump valve.

5. Open (wide) both valves on the gauge manifold and mid-seat both service valves.

6. Start the evacuation pump and evacuate the system until a 29.5+ -in. mercury vacuum is reached on the compound gauge. Let the pump operate for 10 minutes per unit horsepower (e.g., 3 hp unit, 30 minutes).

7. Close the pump valve and stop the pump.

8. Open the refrigerant valve. Allow the pressure to rise to 20 psig. Then close the refrigerant valve. Allow the refrigerant to diffuse through the system and absorb moisture for 5 minutes before the next evacuation.

9. Close the refrigerant valve. Open the pump valve and repeat the evacuation steps to again reach

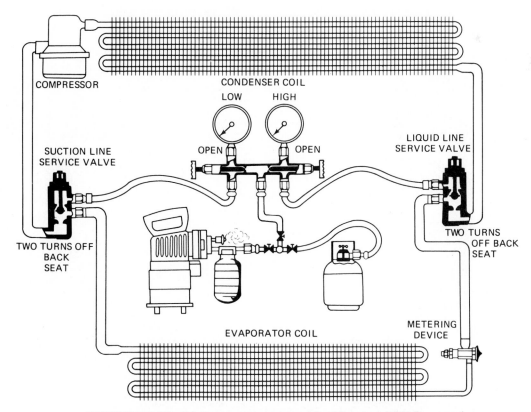

FIGURE R22-19   Triple evacuation assembly. (*Courtesy* BDP Company)

Refrigeration Measuring and Testing Equipment   259

29.5+ -in. mercury vacuum and hold for 10 minutes per horsepower.

10. Close the pump valve and turn off the pump. Open the refrigerant valve and charge to 20 psig, again holding for 5 minutes.

11. Close the refrigerant valve. Open the pump valve. Start the pump and evacuate again to 29.5+ in. vacuum and hold 20 minutes per unit horsepower (e.g., 3 hp unit, 60 minutes).

12. Stop the pump and break the vacuum, this time charging the system to the proper pressure.

R22-11
CHARGING THE SYSTEM

The quantity of refrigerant to be added to the system for initial charge or recharging depends on the size of the equipment and the amount of refrigerant to be circulated. In very large systems it is common practice to simply weigh the charge by placing the refrigerant cylinder or drum on a suitable scale and observing the reduction in weight in pounds. This method is fine for systems that have receivers or condenser volume ample enough to take a slight overcharge.

On smaller systems, and particularly those that are self-contained packaged units without receivers, the system refrigerant charge is *critical* to ounces, rather than whole pounds. In this case a "charging cylinder" is recommended as illustrated in Fig. R22-20. Refrigerant from the refrigerant drum is transferred to the charging cylinder. The charging cylinder has a scale that is visible to the operator so that he may precisely measure the quantity of a specific refrigerant and compensate for temperature and pressure conditions. These cylinders are accurate to one-quarter of an ounce. Optional electric heaters are available to speed the charging operation.

Where considerable installation and service work is involved many contractors use a mobile evacuation and charging station, as illustrated in Fig. R22-21. It contains a vacuum pump, charging cylinder, and service manifold and gauges. More elaborate models may also include a vacuum indicator and space for refrigerant cylinder.

R22-12
CHARGING TECHNIQUES

Refrigerant may be added in either liquid or vapor form. Refrigerant is added in the vapor form, when the unit is operating, through the suction valve. Refrigerant may be added in the liquid form, when the unit is off and in an evacuated condition, through the liquid-line service valve only.

Figure R22-22 illustrates the charging procedure for the vapor form (unit running). For simplicity we show only a refrigerant cylinder and assume that the charge is weighed in during operation.

1. Install the gauge manifold.
2. Attach the refrigerant cylinder to the center con-

FIGURE R22-20 Charging cylinder. (*Courtesy* Robinair Manufacturing Corporation)

FIGURE R22-21 Mobile vacuum and charging station. (*Courtesy* Robinair Manufacturing Corporation)

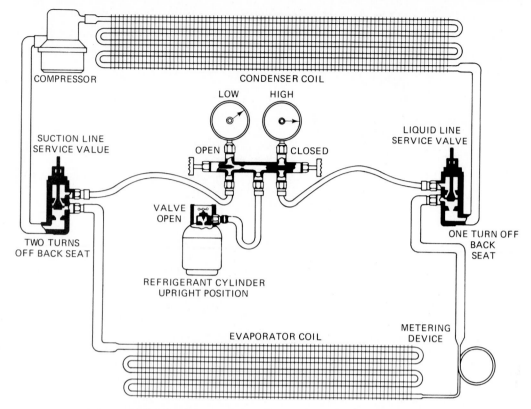

FIGURE R22-22 Vapor-form charging. (*Courtesy* Imperial Eastman)

nection hose and open the valve on the low side of the manifold.

3. Place the cylinder in the upright position.
4. Crack the suction service valve two turns off back-seat.
5. Open the valve on the refrigerant cylinder and weigh in the desired charge.
6. When the correct charge has been added, close the lowside manifold valve and the valve on the refrigerant cylinder.
7. Back-seat both the suction- and liquid-line service valves; remove the hoses and cap ports.

The charging procedure for the liquid form (unit not operating and evacuated) is detailed below and shown in Fig. R22-23.

1. Install the gauge manifold.
2. Attach the refrigerant cylinder. Invert the cylinder unless it is equipped with a liquid-vapor valve, which permits liquid withdrawal in the upright position.
3. Open both suction and liquid service valves one turn off back-seat.
4. Open the valve on the high side of the gauge manifold.
5. Open the valve on the refrigerant cylinder and add refrigerant.

6. After the correct charge is introduced, close the valve on the high side of the manifold and close the valve on the refrigerant cylinder. Back-seat both suction and liquid service valves.
7. Remove the gauge manifold.

In both descriptions above the use of a charging cylinder would be recommended on smaller, critically charged systems where more accuracy is needed.

R22-13
CHECKING THE CHARGE

Checking the charge of a new installation or of an existing unit is another function of the service manifold gauges. For example, the following procedure would be used for an air-cooled unit.

1. Install the gauge manifold.
2. Allow the system to operate until the pressure gauge readings stabilize (approximately 15 minutes).
3. While the unit is operating, record the following information.
   a. High-pressure gauge reading.
   b. Dry-bulb temperature of air entering the condenser coil.
   c. Wet-bulb temperature of air entering the evap-

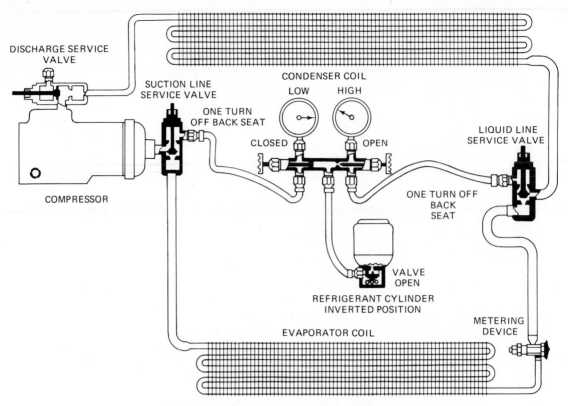

FIGURE R22-23 (*Courtesy* Imperial Eastman)

orator coil. (This is done with a wet-wick thermometer.)

4. A comparison measuring of the measurements above with the head pressure charging table supplied with the unit will indicate if the system is adequately charged and operating properly.

The refrigeration measuring and testing equipment and their proper use as discussed above are the most basic tools for field service and installation. With experience or by need they may be supplemented with special devices to improve skills or save time.

PROBLEMS

R22-1. Name two common pocket thermometers used by refrigeration technicians.

R22-2. The electronic thermometer sensing probe makes use of the _____ element to register changes in temperature.

R22-3. A compound gauge measures pressure _____ and _____ atmospheric pressure.

R22-4. The refrigerant scales on the pressure gauge dials indicate the _____ for the stated refrigerant.

R22-5. The device that includes both a pressure gauge and a service valve is called a _____ .

R22-6. The end valve on a gauge manifold closes off the gauge from its connecting hose. True or False?

R22-7. Removing contaminants from manifold bases is called _____ .

R22-8. The act of removing the contaminants from each hose on the manifold should not take more than _____ seconds.

R22-9. The use of oxygen for leak-detection pressure testing could result in _____ .

R22-10. Name the two methods of evacuation.

R22-11. A micron is equal to _____ of an inch.

R22-12. For evacuation by either method, a pump capable of _____ microns is recommended.

R22-13. Triple evacuation requires the use of an electronic high-vacuum gauge. True or False?

R22-14. A device that can be used to critically charge a system is called a _____ .

R22-15. To charge refrigerant into the suction side of the system, the refrigerant must always be in _____ form.

Installation and Startup

R23-1
GENERAL

The importance of this phase in the successful application of a refrigeration system cannot be overemphasized. No matter how well the equipment has been designed and manufactured or how good the application engineer's planning, poor installation can easily ruin the best system.

A good installation requires more than just technical or mechanical skills; it requires personal integrity

to do the best job possible for the user and the development of an attitude called "pride in workmanship." In other words, it demands a professional approach.

The installation of the equipment depends of course on the type of product and system involved. The manufacturer's instructions prescribe the specific procedures to be followed, but there are some factors that are common to almost any situation.

R23-2
CODES, ORDINANCES, AND STANDARDS

In earlier discussions we have occasionally referred to "compliance to national and local codes and/or regulations." This statement really has several meanings. Some codes or standards are related to the design and performance of the product, its application, or safety considerations. Other codes are directed at the installation phases. The most important national standards for refrigeration products are established by the organizations listed in Fig. R23-1.

R23-3
ARI

The ARI (Air-Conditioning and Refrigeration Institute) has its headquarters in Arlington, Virginia. As mentioned in the introduction of this text, the ARI is an association of manufacturers of refrigeration and air-conditioning equipment and allied products. Although it is a public relations and information center for industry data, one of the institute's most important functions is to establish product or application standards by which the associate members can design, rate, and apply their hardware. In some cases, the products are sub-

FIGURE R23-1

mitted for test and are subject to certification and listing in nationally published directories. The intent is to provide the user with equipment that meets a recognized standard.

R23-4
ASHRAE

ASHRAE (American Society of Heating, Refrigeration, and Air-Conditioning Engineers) is an organization started in 1904 as the American Society of Refrigeration Engineers with some 70 members. Today its membership is composed of thousands of professional engineers and technicians from all phases of the industry. ASHRAE also creates equipment standards for the industry, but its most important contribution probably has been the publication of a series of books that have become the reference bibles of the industry. These include the *Guide and Data Books for Equipment, Fundamentals, Applications, and Systems.*

R23-5
ASME

ASME (American Society of Mechanical Engineers) is concerned primarily with codes and standards related to the safety aspects of pressure vessels.

R23-6
UL

UL (Underwriters' Laboratories) is a testing and code agency which is concerned primarily with the safety aspects of electrical products, although its scope sometimes also includes an overall review of the product. Most of us are familiar with the UL seal on household appliances (irons, toasters, etc.), but it also approves refrigeration and air-conditioning equipment. Its scope of activity has now even expanded into large centrifugal refrigeration machines.

UL approval for certain types of refrigeration and air-conditioning products is almost mandatory in order for it to be locally accepted by electrical inspectors. Compliance with UL is the responsibility of the manufacturer, and the approved products are listed in a directory sent to all local agencies. Installation in accordance with the approved standards is the responsibility of the installer, and violation of these standards can obviously cause a safety hazard as well as possibly voiding the user's insurance coverage should an accident or fire result. So rule number 1 of installation procedures—*conform to UL approval standards.*

R23-7
NFPA

Closely associated with the work of UL in terms of electrical safety is the *National Electrical Code®* (Fig. R23-2), sponsored by the NFPA (National Fire Protection Association). The original code was developed in 1897 as a united effort of various insurance, electrical, architectural, and allied interests. Although it is called the National Electrical Code, its intent is to guide local parties in the proper application and installation of electrical devices. It is the backbone of most state or local electrical codes and ordinances, so rule number 2 of installation—*get a copy of the National Electrical Code Book and become thoroughly familiar with its scope and where to find any information it contains.*

There are other national agencies involved in the industry and their names and activities will be learned through exposure and experience.

Compliance with *local codes* is the second phase of providing good installation. Most states, cities, and counties have or are in the process of adopting local codes, which may be based on suggested national standards and also supplemented by local interpretation. These ordinances are usually divided into (1) electrical, (2) plumbing and refrigeration, and (3) other codes such as sound control. Become familiar with the requirements so the chances of violations and changes are minimized. Corrections are usually expensive and are the responsibility of the installer.

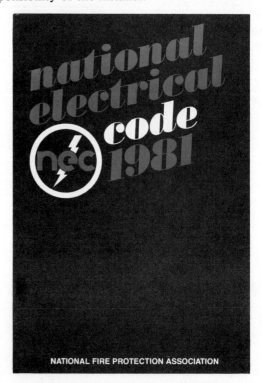

FIGURE R23-2 (*Courtesy* NFPA)

Local codes, ordinances, or regulations are not always administered by civil agencies, but can represent special situations such as power company restrictions on current draw. So rule number 3 of providing a good installation is: *become familiar with all applicable local codes and regulations.* If you are not sure, find out. Do not wait until the inspector puts a red tag on the installation.

Remember! *Standards* serve as guidelines to improve the performance and reliability of a system, but *codes and ordinances* are specific and mandatory rules to be complied with.

R23-8

EQUIPMENT PLACEMENT

Despite what may seem to be a great many possibilities for positioning major cycle components during any installations three factors must be considered in the placement of equipment. First, when air-cooled condensing equipment is installed, ample space must be provided for air movement to and from the condenser. Second, all major components must be installed so that they may be serviced readily. When an assembly is not easily accessible for service, the cost of service becomes excessive. Third, vibration isolation must always be considered, not only in regard to the equipment itself, but also in relation to the interconnecting piping and sheet-metal duct-work. When selecting the placement of all the major components of a refrigeration system, these factors must be considered if a satisfactory installation and proper operation is to be assured. All manufacturers supply recommendations of the space required; these recommendations should be followed.

Figure R23-3 shows a properly placed air-cooled condensing unit. Although the unit has been installed in an inside corner, sufficient room has been left for the passage of incoming air around and over the unit.

Noise is also an important factor in the placement

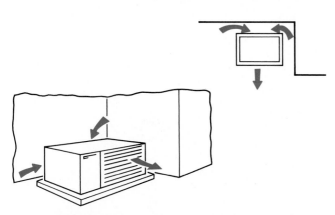

FIGURE R23-3   Air movement. (*Courtesy* Carrier Air-Conditioning Company)

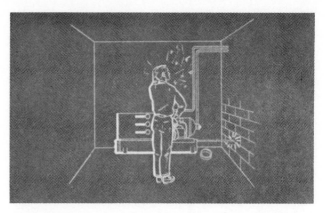

FIGURE R23-4   Service accessibility. (*Courtesy* Carrier Air-Conditioning Company)

of air-cooled condensing equipment. Noise generated within the unit will be carried out by the discharge air. It is poor practice to "aim" the condenser discharge air in a direction where noise may be disturbing, such as neighboring office windows.

When positioning major system components, care must be taken to assure service accessibility. Figure R23-4 is an example of a common error. A shell-and-tube water-cooled condenser has been placed in such a position that the entire condensing unit must be moved to replace a single condenser tube. So be sure to allow room for replacement of items such as compressors, fan motors, fans, and filters. High service costs are often attributable to the poor placement of system components.

When installing major components, it is important to control vibrations set up by rotating assemblies such as compressors, fans, and fan motors. These vibrations can break refrigerant lines, cause structural building damage, and create noise.

Vibration isolation is required on all refrigeration and air-conditioning equipment where noise or vibration may be disturbing. Almost all manufacturers use some form of vibration isolation in the production of their equipment. This is usually enough for the average installation; however, unusual circumstances may require that more stringent measures be taken.

The compressor is considered to be the chief source of system vibration. Since it is good practice to isolate vibration at its source, compressor vibration isolation is essential during installation. One method of minimizing compressor vibration at the source is to bolt the compressor firmly to a solid foundation. An example of this is the use of poured concrete pads. Stud bolts are positioned in the concrete when it is poured and the compressor or condensing unit is bolted firmly to the pad, as shown in Fig. R23-5.

When the compressor or condensing unit is to be installed on the roof or in the upper stories of a multistory building, vibration isolators as shown at the upper right in Fig. R23-5 may be used. This type of isolator is

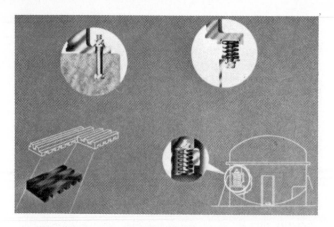

FIGURE R23-5   Vibration-isolation compressor. (*Courtesy* Carrier Air-Conditioning Company)

usually available from the manufacturer of the unit and, in many cases, it is standard equipment.

An isolation pad designed specifically for vibration dampening is shown at the lower left in Fig. R23-5. This type of material is designed to dampen the vibration from a given amount of weight per square inch of area. As the right amount and type of this material can be properly selected only when both weight and vibration cycle are taken into consideration, a competent engineer should be consulted.

A final method of vibration isolation, used with small hermetic compressors, is shown at the lower right in Fig. R23-5. In this case the compressor is spring-mounted within the hermetic shell.

Figure R23-6 illustrates a vibration isolator that is inserted in the discharge line. This isolator is designed to absorb compressor discharge line pulsation before it creates noise or breaks refrigerant lines. The isolator consists of a flexible material banded and held in place by a metal mesh. The mesh will allow some linear movement of the flexible material, but no expansion. This type of isolator is normally placed in the compressor discharge line as close to the compressor as practical. It is particularly effective on installations where the compressor and condenser are on different bases, yet quite close together.

The eliminator should be placed in the line so that the movement it absorbs is in a plane at right angles to the device. Do not place this unit in a position that will put tension on it, as its useful life will be shortened.

Some system components, such as the evaporator, may be suspended from the ceiling. As the air-handling unit containing the evaporator will also normally contain a fan and fan motor, it too is a possible source of vibration. Most manufacturers isolate the fan and fan motor inside the unit with rubber mounts. If this supplies sufficient vibration isolation, the unit may be directly connected to the ceiling, as shown in Fig. R23-7. When more isolation is required, the same method used with compressors is effective. Coil springs support the air-handling unit for additional isolation.

In addition to the placement of major system components, most refrigeration system installations have three other major phases. They are:

1. Erection of piping (both refrigerant and water)
2. Making electrical connections
3. Erection of ductwork

The erection of refrigerant piping is a primary responsibility of the refrigeration installer, along with the placement of major system components.

Although not always the responsibility of the refrigeration installer, electrical and ductwork are an important part of most installations. The refrigeration installer should be familiar with good electrical and air duct installation techniques, since quite often, particularly with small equipment, the refrigeration installer is called upon to do the electrical wiring and make connections to ductwork.

The art of making flared and soldered connections in copper tubing has been discussed, as well as the procedures for sizing lines, installing traps, etc., but several important points should be remembered while actually erecting the system piping.

When hard copper tubing is selected for refrigerant piping, it is recommended that low-temperature silver alloy brazing materials be used. These alloys have flow points ranging from 1100 to 1300°F.

To attain these temperatures, oxyacetylene welding equipment is required. Figure R23-8 shows the equipment necessary for this type of brazing. Both oxygen and acetylene tanks with gauges and reducing valves are

FIGURE R23-6   Vibration-isolation discharge line. (*Courtesy* Anaconda Metal Hose)

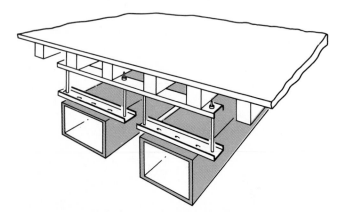

FIGURE R23-7   Vibration-isolation evaporators. (*Courtesy* Carrier Air-Conditioning Company)

needed. Also shown is the torch used with the tanks. At the right is a bottle of dry nitrogen also equipped with gauges and reducing valves. The use of nitrogen is recommended, since it serves to keep the interior of the pipe clean during the brazing process. During low-temperature brazing the surface of the copper will reach a temperature at which the metal will react with oxygen in the air to form a copper oxide scale. If this scale forms on the interior surface of the pipe, it might be washed off by the refrigerant and circulated in the system. The scale can clog strainers or capillary tubes and will damage orifices.

This scaling can be prevented by replacing the air in the pipe by nitrogen. As nitrogen is an inert gas and will not combine with copper even under high-temperature conditions, the pipe interior will remain clean during brazing and no scale is formed, though discoloration may occur with overheating.

Figure R23-9 shows a method by which nitrogen may be introduced into the refrigerant tubing during the brazing operation. Instead of connecting the liquid line to the "king" valve or liquid-line stop valve, connect the liquid line directly to the nitrogen bottle as shown. By use of the reducing valve on the nitrogen bottle, a slight pressure is admitted to the tubing. This is just enough pressure to assure that air will be forced from the pipe. As shown in the insert, if the nitrogen flow can be felt on the palm of the hand it is sufficient. This nitrogen pressure is kept in the tubing throughout the entire brazing operation, thus assuring an oxygen-free pipe.

Clean pipe is essential in refrigeration installation; therefore, nitrogen is an extremely important part of the brazing operation, as it assures scale-free interior pipe walls.

The temperatures during brazing operations can warp metals and burn or distort plastic valve seats. It is important that brazing heat does not reach metal or plastic components that might be damaged. Figure R23-10 shows the results of brazing heat damage and also

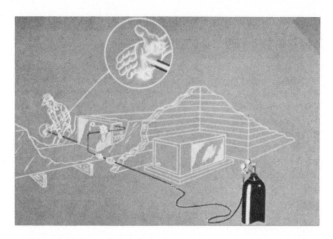

FIGURE R23-9 Brazing with nitrogen. (*Courtesy* Carrier Air-Conditioning Company)

one method of assuring this. The valve at the upper left was not protected from brazing heat; the plastic seat has been damaged. Also, the seat holder has been warped, as shown in the insert. This valve obviously will not operate properly and would require immediate replacement. However, the valve at the bottom right has a wet rag wrapped around the body. The rag absorbs the heat that flows to the valve body during the brazing operation. By keeping the rag wet, the valve and its component parts are protected from heat damage.

In many cases the suction line of the refrigerant cycle installation will run through a nonrefrigerated or a non-air-conditioned area. The outside surface temperature of this pipe is frequently below the dew point of the surrounding air. In this case the moisture in the air will condense on the exterior of the pipe. This will not only create problems where this continuous moisture drips, such as rotting wood, separating floors, etc., but it can be annoying. When this condensation is expected or does occur, the pipe should be insulated; thus the condensation is prevented.

The insulation must be of good quality, so that the temperature of its exterior surface will never drop below

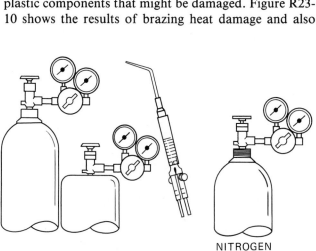

FIGURE R23-8 Brazing equipment. (*Courtesy* Carrier Air-Conditioning Company)

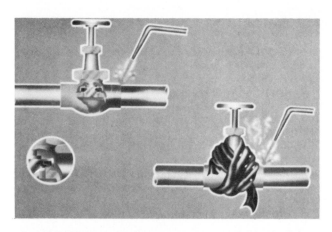

FIGURE R23-10 Overheat protection. (*Courtesy* Carrier Air-Conditioning Company)

the dew point of the surrounding area. It must also be well sealed, so that the air and moisture it contains cannot flow through the insulation to the pipe, thus causing condensation underneath the insulation.

There are many different types of insulation available; some are designed for specific uses. The insulation for a specific refrigeration installation should be chosen by a competent technician.

The primary purpose of the pipe hanger is to support the pipe properly, but it may also serve as a vibration isolator. If vibration problems are not anticipated, ordinary plumbing practices such as shown in Fig. R23-11 may be used. The only purpose of this hanger is to support the pipe firmly.

The hanger shown on the left also has a height-adjustment feature. This enables the installer to level or pitch the pipe if required for oil flow. This type of hanger can be used with or without insulation.

At the right in Fig. R23-11 is an example of a typical hanger that might be found on a small commercial-refrigeration installation. This hanger will serve as a vibration isolator as well as a pipe supporter. This example is being used on an insulated line. A short length of light-gauge rustproof metal has been wrapped around the insulated pipe. A metal strap has been attached to this length of metal; in some cases it is merely wrapped around the length of sheet metal. The free end of the strap is then fastened to the joist or ceiling.

The insulation in such a hanger will act as the vibration isolator. The purpose of the short length of metal is to prevent the thin strap from cutting the insulation.

The service technician who installs the refrigeration cycle is sometimes responsible for the final wiring connections between the installed unit and the fused disconnect switch shown in Fig. R23-12. All electrical power to the refrigeration unit must pass through this switch. When this switch is pulled, or opened, all electrical power to the unit must be disconnected. This same disconnect switch also contains fuses, which will inter-

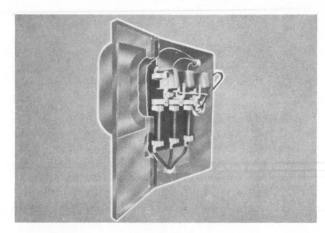

FIGURE R23-12   Fused disconnect switch. (*Courtesy* Carrier Air-Conditioning Company)

rupt the flow of current whenever a severe electrical overload occurs. This mechanism is a protection against fires and explosions and also against electrical shocks to people.

Electrical codes, both national and local, are made to protect property and life; they should always be followed. All refrigeration cycle installations having electrical connections are governed by national or local electrical codes. For example, electrical codes require that the fused disconnect switch in Fig. R23-12 always be placed within sight of the unit that receives the power passing through the switch.

When electrical circuits are to be connected by the refrigeration installer, good wiring practices should be followed and every effort should be made to assure good electrical contacts. Lug-type connectors are recommended.

When braided wire is used, a single wire may separate and create a potential electrical hazard. Loose wires might contact other wires or "ground," causing electrical problems. Braided wire should be looped into the size required and soft solder should be allowed to flow over and coat the braided loop. This will assure good contact and eliminate the hazard of wire separation.

The refrigeration system installer is frequently required to make the final connection between the evaporator or air-handling unit and the ductwork (Fig. R23-13). This final connection is in the form of a canvas connection, which will eliminate any vibration carryover from the air-handling unit to the ductwork. Good weatherproof canvas is required and must be installed correctly. If, as shown in the upper left in the illustration, it is too loose, it will drop into the airstream and obstruct normal air flow. It may also flap and create noise. If it is too tight it will stretch, harden, and disintegrate over a period of time, causing leaks. If the canvas is damp and installed too tightly, the ductwork might be pulled out of alignment. The center of Fig. R23-13

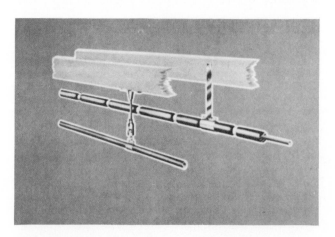

FIGURE R23-11   Pipe hangers. (*Courtesy* Carrier Air-Conditioning Company)

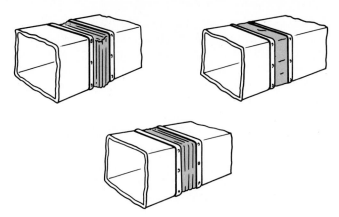

FIGURE R23-13 Canvas duct connection. (*Courtesy* Carrier Air-Conditioning Company)

shows the proper application: the canvas is loose enough to absorb vibration and not loose enough to interfere with airflow.

When the refrigeration system has been completely assembled and all electrical and duct connections completed, there are still a number of important steps that must be taken before the equipment is started.

The unit must be leak-tested and charged, and all belts must be checked for tension and alignment. There must be an electrical motor check of the direction of rotation, and the bearings must be oiled. The power source must be checked to be absolutely sure that the correct power will be applied to the unit. None of these steps is time consuming or difficult to perform, but they are all extremely important and cannot be neglected. Field leak detection of the halogenated hydrocarbons is generally done using the halide torch or electronic leak detectors.

After all interconnecting tubing has been assembled, some refrigerant is introduced into the system as a gas. Although a leak test could be made at this time, some refrigerants do not exert sufficient pressure at room temperature to assure trustworthy results. By using nitrogen, the pressure in the system may be built up

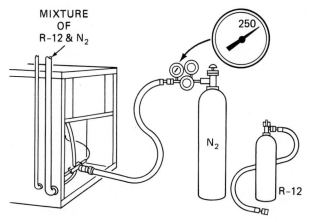

MIXTURE OF R-12 & N₂

250

N₂

R-12

FIGURE R23-14 Leak test. (*Courtesy* Carrier Air-Conditioning Company)

to approximately 250 psig, at which pressure a true leak test can be made (Fig. R23-14). The mixture of refrigerant and nitrogen inside the unit will cause a reaction on the detector if a leak is present. Since nitrogen is an inert gas, the system should be purged with refrigerant gas after it has been determined to be leak-free.

Following the leak test the system is ready for charging or dehydration. In a small system operating at medium evaporator temperatures, elaborate dehydration procedures will not be required; a simple refrigerant gas purge of the system will be sufficient. If some moisture has inadvertently entered the system, it will show on the moisture indicator as previously described. A drier replacement program should then be started and continued until the system is dry.

If moisture is known to be in the system or if the system is designed for low-temperature application, dehydration by evacuation before charging is recommended. When in doubt, dehydrate. Dehydration by evacuation is accomplished by the use of a vacuum pump as detailed in Chapter R22.

R23-9
CHARGING THE SYSTEM

The system may be charged with refrigerant in its normal state, or as a liquid or gas. Although a number of factors may affect the method of charging, the most important is the quantity of refrigerant involved.

Relatively speaking, most refrigerant containers, regardless of size, have a small outlet. In small units, sufficient gas vapor may be passed through this outlet to complete the full charging of the unit in a reasonable length of time. On large units, however, the time required for the proper amount of vapor to pass through this small outlet may take so long that gas charging is impractical. When this is the case, liquid charging is used. The small orifice will pass a much greater weight of liquid refrigerant in any given amount of time.

The point at which the refrigerant will actually enter the system is determined by one basic consideration: whether the unit is to be charged with refrigerant in a gas or liquid form.

Under normal charging conditions the refrigerant cylinder will be at ambient temperature and corresponding pressure. As most field refrigerant charging is done with the unit in operation, the refrigerant cylinder pressure will usually be below the head pressure of the condenser and above the back pressure of the evaporator.

The metering device orifice is not large enough to make gas charging on the high side of the device practical because of the time consumed; therefore, gas is normally charged into the system after the metering device. This charging can be done at any convenient point between the metering device and the compressor.

If the refrigerant is to be charged as a liquid, the charging should take place ahead of the metering device to protect the compressor from damage due to liquid flooding.

With open-type and serviceable hermetic systems that are small enough to make gas charging practical or where only a small amount of refrigerant is required, charging is usually done through the suction service valve on the compressor. Since welded hermetic systems are normally small enough so that gas charging is practical, pinch-off tubes are made available at the compressor for this purpose. These pinch-off tubes are connected into the suction side of the compressor and are designed to allow the compressor to pump gas directly from a refrigerant drum.

The original charge in units of this type is placed in the unit at the factory during production, but occasionally the charge must be replaced in the field. When this is necessary, the pinch-off tube may be cut and a connection placed on the tube. The unit is then charged through the tube, which is repinched and brazed. Occasionally, valves are placed permanently on this pinch-off line. This is practical where the pinch-off tube is either very short or inaccessible.

There are also devices that are designed to puncture a refrigerant line for charging purposes. The device then remains on the line and serves as a valve for either future charging or as a gauge connection.

Another device used for charging the hermetic circuit is the *tire* type of valve. This valve contains a plunger, which, when depressed, opens the circuit. A special adapter on the charging hose will depress the plunger when the hose is firmly attached to the valve. Refrigerant will then flow through the valve into the system. When the charging hose is removed, the plunger returns to its original position and reseals the refrigerant circuit. Always replace the valve cap after servicing. There are four ways to determine whether the proper amount of refrigerant is introduced into the system. These methods are: charging by sight glass; charging by pressures; weighing the charge in; and using a frost line.

All four methods are used extensively in commercial refrigeration work, and the choice of method depends on the size and type of system involved. "Critical-charge" methods, common to small residential air-conditioning systems, *are not included;* they are critical to ounces of refrigerant charge.

The first method to be discussed will be charging by sight glass. In a properly charged system there should always be a solid flow of liquid refrigerant (no bubbles) to the metering device. A sight glass will indicate the presence of gas by bubbles in the glass.

To charge properly by the sight-glass method, refrigerant is added in the usual manner. Following the addition of a reasonably large percentage of the estimated charge of gas or liquid, the refrigerant cylinder is shut off and the system is allowed to "settle down" for a period of time. When the operating pressures have stabilized, the sight glass is again observed. If bubbles are still present, additional charge is added slowly until the bubbles disappear. If, after stabilizing the pressures, no bubbles are present when the unit is operating under maximum load conditions, the unit is properly charged.

A common method of determining whether the proper amount of charge is being introduced into the system is by "weighing" the charge. This is the most accurate method of adding a full charge when the required charge is known. The method is quite simple, as shown in Fig. R23-15.

With the system evacuated and ready to receive the refrigerant, the refrigerant bottle is weighed. On the left the scales read 190 lb, indicating that the bottle and refrigerant weigh a total of 190 lb. If the system requires a charge of 40 lb, as determined by the manufacturer or design engineer, gas or liquid is released from the bottle into the system. When 40 lb of refrigerant have been removed, the total weight of refrigerant and bottle should be 190 − 40, or 150 lb, as shown at the right in Fig. R23-15.

This method is used mostly on packaged equipment and then only when it requires a complete charge; the method is of little use when only a portion of the charge is required. Rarely is the exact charge known under such conditions.

A third way of charging, which may be used with factory-designed and balanced packages, is the head pressure method. From test data, the factory determines the proper head pressures under various evaporator loads or temperatures. This information is furnished to the installer either as a graph or a table, which consists of a list of possible low-side pressures along with the proper head pressures supplied by the test data of the unit.

After charging with an estimated amount as directed by the factory, the unit is run for enough time to allow the pressures to settle and stabilize. Pressure readings are then compared to the manufacturer's chart.

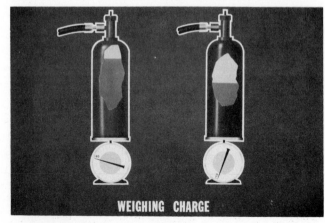

FIGURE R23-15 Weighing charge. (*Courtesy* Carrier Air-Conditioning Company)

If the actual head pressure is below the pressure that the chart indicates is correct, refrigerant is added. If the actual head pressure is above that indicated by the chart, refrigerant should be removed.

Figure R23-16 shows how the head pressure method is used. An original charge as suggested by the factory has been introduced into the unit. The unit is run long enough for pressure stabilization, and the back pressure is found to be 35 psig. This point is on the horizontal axis at point *A*. The chart is then read upward vertically until the intersection with the outdoor temperature line at *B*. From *B*, read horizontally to the left until point *C* is reached. This is the head pressure under which the unit should be operating when the back pressure is 35 psig. The chart shows this figure to be 244. If the discharge pressure gauge reading had been below this figure, refrigerant would have to be added until the gauge pressure is brought up to 244 psig. If the discharge pressure gauge reading had been above this figure, refrigerant would have to be removed until the pressure was brought down to 244 psig.

The fourth and final way of charging to be discussed is the frost-line method. This method can only be used in small hermetic systems that use capillary tubes; such systems are not too common in commercial refrigeration work. When a system of this type is operated without evaporator load, the back pressures will normally drop below the freezing temperature and frost will form on the coil.

In Figure R23-17 the evaporator load has been removed by placing a piece of cardboard over the coil face, thus shutting off the airflow. Since the load has been removed from the evaporator, the refrigerant will not evaporate as rapidly, and some will pass through the evaporator and evaporate in the suction line. Tests have shown that under these conditions a properly charged unit will normally frost to within a few inches of the compressor. By factory testing, this final frost point can be determined, and given to the installer.

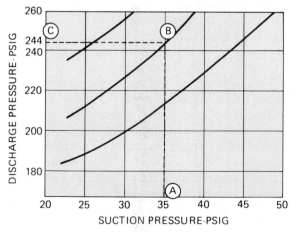

FIGURE R23-16 Charging chart. (*Courtesy* Carrier Air-Conditioning Company)

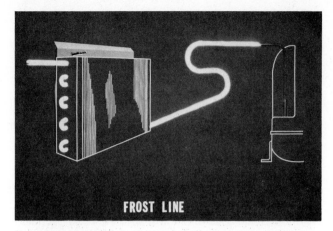

FIGURE R23-17 Frost line. (*Courtesy* Carrier Air-Conditioning Company)

By recreating the frost line on the suction line, the installer can determine the proper charge. If the frost line does not reach the point designated by the factory tests, more refrigerant should be added. If the frost goes beyond the point designated by the factory, refrigerant should be removed.

In most systems some overcharging can be tolerated, but an undercharge is rarely acceptable. Overcharging will create high head pressure and high temperature, with all the resultant problems, such as motor overloading, sludge formation, and compressor valve failure. High head pressures can also result in poor load control, with liquid refrigerant flooding to the compressor.

Although the biggest problem with undercharging is that of capacity, it may also create frost conditions on the evaporator in higher temperature refrigeration equipment, and may also cause high evaporator superheats. As some hermetic compressor motors depend on suction gas for cooling, they can be damaged by overheating due to high suction gas temperatures.

Both overcharging and undercharging should be avoided, since either condition can do serious harm or destroy system components.

R23-10
OIL CHARGING

There are several ways by which the proper amount of oil can be measured into the system. In a new system it can be measured or weighed in as is the refrigerant itself. Unit installation instructions include the compressor oil requirements, in either weight or liquid measurements. This method is also applicable following a compressor overhaul, when all the oil has been removed from the compressor; however, it should be used only when the system has no oil in it.

A second method of determining proper oil charge is the dip stick. This is used primarily with small vertical-shafted hermetic compressors, but some larger, open types of compressors may have openings designed for the use of a dip stick. The manufacturer's recommendations of the correct level should always be followed.

The third way to arrive at the correct oil charge is by using the compressor crankcase sight glass. When determining the correct oil charge by this method, the system should be allowed to operate for a period of time under normal conditions before the final determination of the proper oil level is made. This procedure will assure proper oil return to the crankcase; it will also allow the oil lines and reservoirs to fill and, where halogenated hydrocarbons are used, give the refrigerant an opportunity to absorb its normal operating oil content.

When a compressor is replaced, the new unit should be charged with the same amount of oil as the old unit.

Oil is normally introduced into a refrigerant system by one of two methods (Fig. R23-18). It may be poured in as shown on the left, providing the compressor crankcase is at atmospheric pressure. This method is normally used prior to dehydration, since it will expose the compressor crankcase interior to air and the moisture the air contains.

On the right in Fig. R23-18 the method normally used with an operating unit is shown. In this case the crankcase is pumped down below atmospheric pressure and the oil is drawn in. When this method is used, the tube in the container subjected to air pressure should never be allowed to get close enough to the surface of the oil to draw air. As shown, the tube is well below the level of the oil in the container.

Oil charging of replacement welded hermetic compressors should be done according to the manufacturer's recommendations and depends on whether the replacement compressor has been shipped with or without an oil charge.

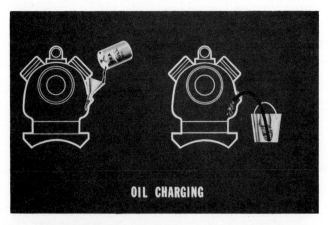

OIL CHARGING

FIGURE R23-18 Oil charging. (*Courtesy* Carrier Air-Conditioning Company)

There are three precautions to take in charging or removing oil. The first is to use clean, dry oil. Hermetically sealed oil containers are available and should be used. Second, pressure must be controlled when the crankcase is opened to the atmosphere. Too much pressure can force oil out through the opening rapidly and create quite a mess.

Third, system overcharging should be avoided. Not only will this create the possibility of oil slugs damaging the compressor, but it also can hinder the performance of the refrigerant in the evaporator. Oil overcharging will also cause liquid refrigerant to return to the compressor from the evaporator.

R23-11
INITIAL STARTUP

Care must be taken to prevent damage to the refrigeration system during initial startup (Fig. R23-19). Valve positions must be checked to make certain that only proper and safe pressures will occur at the compressor. The compressor discharge valve must always be opened prior to start. Low-side valves must be adjusted in such a way as to assure neither excessively high nor low pressures in the compressor suction. Pressures should be carefully observed and regulated until the unit is operating normally. When starting a new system, try to keep the operating pressures as close to normal as practical.

When the installed equipment is belt-driven, both the alignment and tension of the belt must be checked prior to startup. Improperly aligned belts show excessive wear and have an extremely short life. Loose belts wear rapidly, slap, and frequently slip. Belts that are too tight may cause excessive motor-bearing and fan-bearing wear. Belt alignment is simple to check; a straightedge laid along the side of the pulley and flywheel will show any misalignment immediately.

The belt tension should also be checked. The belt stretches very little, so the belt should be set at the right tension during initial startup. The correct tension allows about one inch of deflection on each side of the belt when squeezed by the fingers.

Many parts of newly installed equipment will require simple maintenance prior to original startup. A good example of this type of maintenance is motor lubrication. Every motor that is not permanently lubricated should receive the proper amount of oil as recommended by the manufacturer. Instructions enclosed with all equipment indicate what maintenance of this type is required.

It is good installation practice to check electric motor nameplates against the available voltage prior to initial startup. Figure R23-20 shows the voltage at the disconnect switch being checked for comparison with the

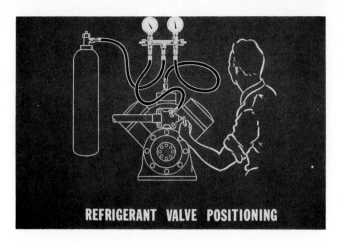

FIGURE R23-19 Refrigerant valve positioning. (*Courtesy* Carrier Air-Conditioning Company)

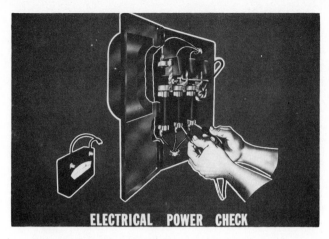

FIGURE R23-20 Electric power check. (*Courtesy* Carrier Air-Conditioning)

electric motor nameplate. It is also good practice to determine if a phase unbalance exists in polyphase units prior to the final wiring connection.

When the system has been thoroughly checked for leaks, properly charged, valves positioned for startup, belts aligned, motors oiled, and wiring checked, then the unit can be started.

Following startup the unit must be checked to make certain that it is doing the job it was designed to do efficiently. This includes determining electrical load characteristics. No installation should ever be considered complete until it has been proven it can do the job it was designed to do.

Following the completion of tests to determine that the unit is performing satisfactorily, the final cleanup should be started. All excess tubing, wire, scrap metal,

etc. should be removed and the unit left in a condition that will indicate the installer's pride of workmanship. The equipment should be an asset to the area in which it is located.

The final step in any installation is to instruct the customer in the operation of the equipment. There are probably as many service complaints because of a lack of customer understanding of the new equipment as from any other reason. Customers should be clearly and concisely shown how their equipment can be made to operate properly and efficiently. Maintenance requirements should be explained, including such things as filter changes and motor oiling. Customers who are thoroughly familiar with the way their equipment should operate and what they can do to keep it that way are going to be satisfied customers.

PROBLEMS

R23-1. The National Electrical Code Book is published by _____ .

R23-2. UL approvals are not generally required on refrigeration or air-conditioning equipment. True or False?

R23-3. What is the difference between codes and standards?

R23-4. When installing equipment, placement is important to allow for _____ .

R23-5. Vibration of equipment has been eliminated by the manufacturer and is no problem in the field. True or False?

R23-6. Dry nitrogen is used in the brazing process to reduce _____ .

R23-7. Suction lines running outside the refrigerated area will generally require insulation to eliminate _____ .

R23-8. To prevent vibration transmission in metal ducts, a _____ connection is used.

R23-9. A charging connection on welded hermetic compressors is called a _____ .

R23-10. The frost-line method of checking the refrigerant charge applies to all systems. True or False?

R23-11. The head-pressure method of checking the refrigerant charge can be used regardless of the ambient outside temperature. True or False?

R23-12. What is the most accurate method of refrigerant charging the system?

R23-13. Which is more critical, overcharging or undercharging the system?

R23-14. Checking the oil level by the sight-glass method does not require the compressor be operated before checking. True or False?

R23-15. Oil overcharging is not critical to the system. True or False?

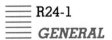

Troubleshooting

R24-1
GENERAL

In previous chapters we have learned the major components of the refrigeration cycle, the theoretical operation of a cycle, the materials and tools needed to fabricate a system, and the instruments needed to test and check the operation. We will now put these elements to work by reviewing what might be expected, first during normal operation and then under abnormal conditions.

First, field service technicians do not usually carry with them a pressure–enthalpy diagram for each refrigerant. They will, however, carry a pocket-size pressure–temperature chart, as illustrated in Fig. R24-1, which lists the saturated refrigerant pressure temperatures for various refrigerants. R-12, R-22, R-500, and R-502 are listed; they are the most common systems refrigerants. At any given saturated liquid temperature the technician can determine the equivalent pressure (or vacuum, if it is a low-temperature application). Now let's look at three systems and see what actually happens.

Figure R24-2 represents an air-cooled refrigeration cycle used in a freezer operation where the room storage temperature is to be held at 0°F and the outdoor (ambient) temperature is 95°F.

First identify the system components: compressor, condenser, evaporator, and expansion valve metering device. Next identify the glass (or dial) thermometers used to measure air temperature to evaporator and condenser. A six-station electronic thermometer is used to check temperatures at points *A–F*. Manifold gauges are attached to the respective service valves in the system. The system refrigerant for this example is R-22.

Let's first consider the evaporator conditions. Since the freezer room is to be held at 0°F it will be necessary to maintain a temperature differential between the air across the evaporator and the refrigerant temperature; this differential varies with the particular evaporator design. For this example we will assume a 10°F differential. The saturated refrigerant temperature at point *A* just as it enters the evaporator is −10°F. From the *P–T* chart in Fig. R24-1 note that the corresponding pressure of R = 22 is 16.4 psig. As the evaporator picks up heat and the refrigerant boils, it will leave the evaporator at 0°F, meaning that it has gained 10°F of superheat in terms of temperature rise, but the vapor pressure will remain essentially the same if we assume that there is no pressure drop in the coil. The feeler bulb from the thermostatic expansion valve reacts to changes in temperature at point *B* and will regulate refrigerant flow in an attempt to maintain a constant 10°F superheat.

Depending on the size and length of the suction line, two things will occur. As the suction line runs outside the freezer room into a warmer location, it will gain some additional superheat (2°F is shown for this example). And because of friction loss within the pipe, there will be some pressure reduction at point *C* as the gas enters the compressor suction valve. Pressure at *C* can be read from the compound low-pressure gauge on our manifold.

The low-temperature, low-pressure vapor enters the compressor and is elevated to a high-pressure, high-temperature vapor as indicated at point *D*. Due to the heat of compression the temperature will be approximately 100°F above the condenser saturation temperature. The pressure at points *D* and *E* are shown to be the same, because the distance between the compressor and condenser in this example is assumed to be close coupled. This may not always be true and pressure loss in the discharge line would have to be considered accordingly.

The pressure at point *D* is really determined by the condenser design and application. In this example it is air cooled with outdoor air over the coil at 95°F. For the condenser to reject heat it must operate above 95°F. Generally, this will be around 30 to 40°F above the entering air temperature. In our example the average condensing temperature in the condenser as the hot vapor becomes a saturated liquid will be 125°F at point *E.* For R-22 the corresponding pressure (refer to Fig. R24-1) is 277.9 or 278 psig. Therefore, the compressor must be capable of raising the pressure to that level.

Most of the superheat is removed from the vapor by the upper rows of the condenser, so that the bottom two or three rows are then full of liquid refrigerant. This fulfills two functions. First, a liquid seal is formed to prevent vapor from getting into the liquid line. Second, it permits the liquid refrigerant to be subcooled below its saturation temperature. A 10°F subcooling as shown in our example between points *E* and *F* is a common condenser design. The pressure (278 psig) remains constant and can be measured at the liquid-line service valve and read on the manifold high-pressure gauge.

Liquid refrigerant then flows from point *F* by way of the liquid line back to the expansion valve where there is a pressure differential of over 250 psig. In actual practice there are pressure losses in the liquid line due to friction and also when the liquid must be raised to a higher level. If pressure loss is too great the refrigerant will flash into a gas prior to entering the expansion valve. Chapter R13 covers in detail procedures for determining refrigerant line sizes so as to minimize pressure loss, ensure proper oil return to the compressor, and eliminate objectionable noise.

As a contrast to the low-temperature freezer application, let's now examine a high-temperature application as would be found in a comfort air-conditioning system.

Figure R24-3 shows a refrigerant cycle diagram for a typical air-cooled system employing an expansion valve and using R-500 refrigerant. The temperature-pressure (*T–P*) relationships shown are considered normal for a system operating under the following conditions: the entering evaporator coil air temperature is 80°F DB, 67°F WB; condenser air entering temperature is 95°F DB. Since the pressures shown will be affected by equipment design, always consult the manufacturer's pressure-temperature charts for each specific model.

Starting at point *A,* the liquid refrigerant is subcooled approximately 16 to 114°F and exerts a pressure of 218 psig. Assuming the liquid refrigerant line 25 ft long and properly sized, the pressure at the entering side of the expansion valve will be approximately 218 psig with the temperature of the refrigerant still about the same as at *A* (i.e., 114°F).

Downstream from the expansion valve at point *B,* the *T–P* readings will be approximately 44°F and 50.7 psig. With the expansion valve regulating the refrigerant flow for 12° of superheat, the *T–P* readings, at point *C* will be 56°F and 50.7 psig. This pressure reading neglects the nominal refrigerant pressure drop of the vaporizing refrigerant through the evaporator coil.

At the suction gas intake at the compressor (point *D*), the *T–P* relationship of 61°F and 50.7 psig represents the temperature pickup of additional superheat caused by the suction line. No pressure loss is shown because we assume that the cooling coil and compressor are reasonably close-coupled. At the hot-gas discharge from the compressor (at *E*), these readings become 195°F and 218 psig. The condenser average *T–P* is 130°F and 218 psig at point *F.* The subcooling (16°F) then takes place at the last two and three bottom rows of the condenser and the cycle is completed at point *A.*

The two previous systems employed a thermostatic

SPORLAN **TEMPERATURE PRESSURE CHART**

Vacuum-Inches of Mercury – Italic Figures   Pressure-Pounds Per Square Inch Bold Figures

TEMPER-ATURE °F.	12-F	22-V	500-D	502-R	717-A	TEMPER-ATURE °F.	12-F	22-V	500-D	502-R	717-A	TEMPER-ATURE °F.	12-F	22-V	500-D	502-R	717-A
−60	19.0	12.0	17.0	7.2	18.6	12	15.8	34.7	21.2	43.2	25.6	42	38.8	71.5	48.2	83.8	61.6
−55	17.3	9.2	15.0	3.9	16.6	13	16.5	35.7	21.8	44.3	26.5	43	39.8	73.0	49.4	85.4	63.1
−50	15.4	6.2	12.8	0.2	14.3	14	17.1	36.7	22.6	45.4	27.5	44	40.7	74.5	50.5	87.0	64.7
−45	13.3	2.7	10.4	1.9	11.7	15	17.7	37.7	23.4	46.5	28.4	45	41.7	76.0	51.6	88.7	66.3
−40	11.0	0.5	7.6	4.1	8.7	16	18.4	38.7	24.1	47.7	29.4	46	42.7	77.6	52.8	90.4	67.9
−35	8.4	2.6	4.6	6.5	5.4	17	19.0	39.8	24.8	48.9	30.4	47	43.6	79.2	54.0	92.1	69.5
−30	5.5	4.9	1.2	9.2	1.6	18	19.7	40.9	25.7	50.0	31.4	48	44.7	80.8	55.1	93.9	71.1
−25	2.3	7.4	1.2	12.1	1.3	19	20.4	41.9	26.5	51.2	32.5	49	45.7	82.4	56.3	95.6	72.8
−20	0.6	10.1	3.2	15.3	3.6	20	21.0	43.0	27.3	52.5	33.5	50	46.7	84.0	57.6	97.4	74.5
−18	1.3	11.3	4.1	16.7	4.6	21	21.7	44.1	28.1	53.7	34.6	55	52.0	92.6	63.9	106.6	83.4
−16	2.1	12.5	5.0	18.1	5.6	22	22.4	45.3	28.9	54.9	35.7	60	57.7	101.6	70.6	116.4	92.9
−14	2.8	13.8	5.9	19.5	6.7	23	23.2	46.4	29.8	56.2	36.8	65	63.8	111.2	77.8	126.7	103.1
−12	3.7	15.1	6.8	21.0	7.9	24	23.9	47.6	30.6	57.5	37.9	70	70.2	121.4	85.4	137.6	114.1
−10	4.5	16.5	7.8	22.6	9.0	25	24.6	48.8	31.5	58.8	39.0	75	77.0	132.2	93.5	149.1	125.8
− 8	5.4	17.9	8.8	24.2	10.3	26	25.4	49.9	32.4	60.1	40.2	80	84.2	143.6	102.0	161.2	138.3
− 6	6.3	19.3	9.9	25.8	11.6	27	26.1	51.2	33.3	61.5	41.4	85	91.8	155.7	111.0	174.0	151.7
− 4	7.2	20.8	11.0	27.5	12.9	28	26.9	52.4	34.2	62.8	42.6	90	99.8	168.4	120.6	187.4	165.9
− 2	8.2	22.4	12.1	29.3	14.3	29	27.7	53.6	35.1	64.2	43.8	95	108.3	181.8	130.6	201.4	181.1
0	9.2	24.0	13.3	31.1	15.7	30	28.5	54.9	36.0	65.6	45.0	100	117.2	195.9	141.2	216.2	197.2
1	9.7	24.8	13.9	32.0	16.5	31	29.3	56.2	37.0	67.0	46.3	105	126.6	210.8	152.4	231.7	214.2
2	10.2	25.6	14.5	32.9	17.2	32	30.1	57.5	37.9	68.4	47.6	110	136.4	226.4	164.1	247.9	232.3
3	10.7	26.5	15.1	33.9	18.0	33	30.9	58.8	38.9	69.9	48.9	115	146.8	242.7	176.5	264.9	251.5
4	11.2	27.3	15.7	34.9	18.8	34	31.7	60.1	39.9	71.3	50.2	120	157.7	259.9	189.4	282.7	271.7
5	11.8	28.2	16.4	35.9	19.6	35	32.6	61.5	40.9	72.8	51.6	125	169.1	277.9	203.0	301.4	293.1
6	12.3	29.1	17.0	36.9	20.4	36	33.4	62.8	41.9	74.3	52.9	130	181.0	296.8	217.2	320.8	—
7	12.9	30.0	17.7	37.9	21.2	37	34.3	64.2	42.9	75.9	54.3	135	193.5	316.6	232.1	341.2	—
8	13.5	30.9	18.4	38.9	22.1	38	35.2	65.6	43.9	77.4	55.7	140	206.6	337.3	247.7	362.6	—
9	14.1	31.8	19.0	39.9	22.9	39	36.1	67.1	45.0	79.0	57.2	145	220.3	358.9	264.0	385.0	—
10	14.6	32.8	19.7	41.0	23.8	40	37.0	68.5	46.1	80.5	58.6	150	234.6	381.5	281.1	408.4	—
11	15.2	33.7	20.4	42.1	24.7	41	37.9	70.0	47.1	82.1	60.1	155	249.5	405.1	298.9	432.9	—

FIGURE R24-1 Pocket-size *P-T* table.

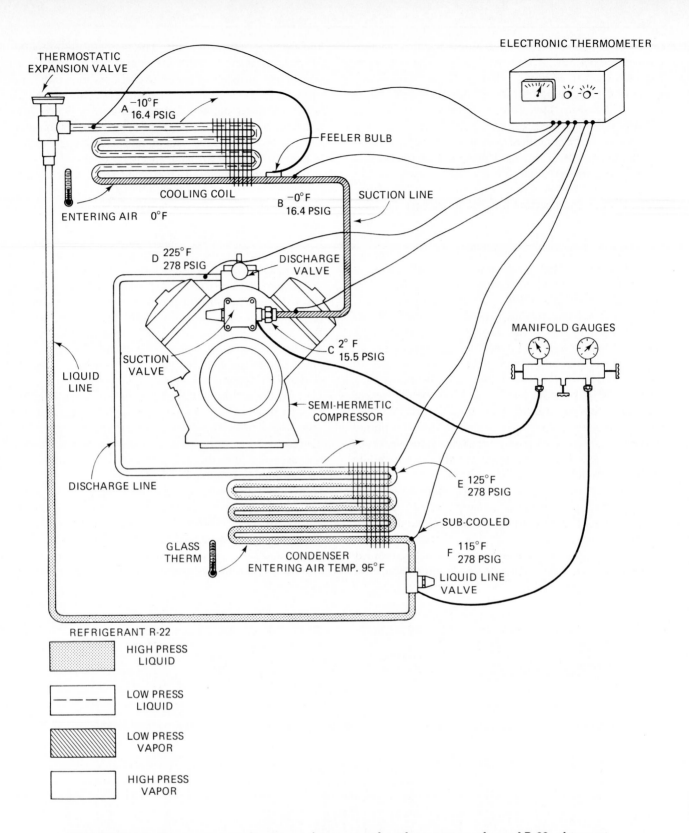

FIGURE R24-2 Freezer room application—refrigerant cycle with expansion valve and R-22 refrigerant.

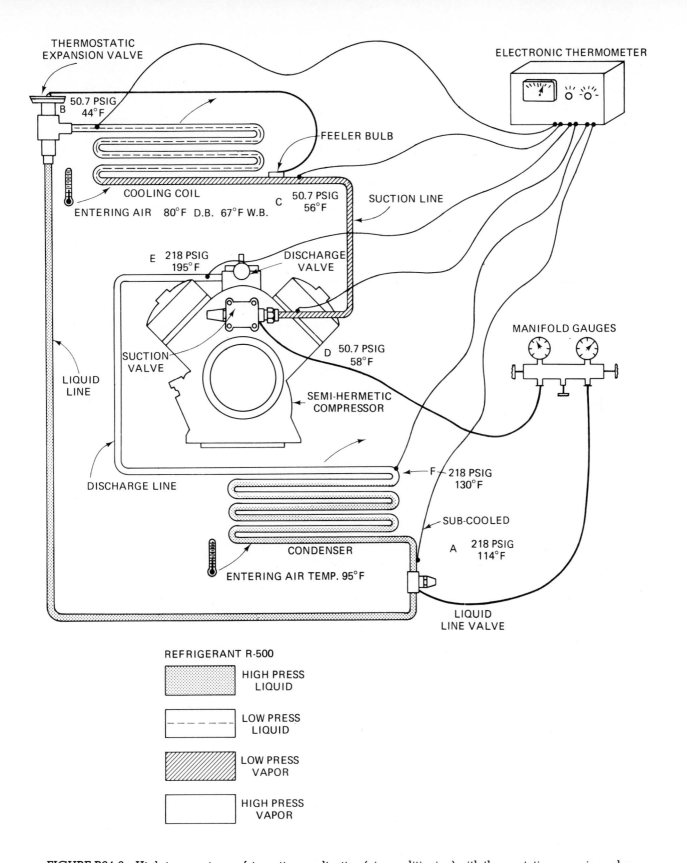

THERMOSTATIC
EXPANSION VALVE

ELECTRONIC THERMOMETER

50.7 PSIG
44°F
B

FEELER BULB

COOLING COIL

ENTERING AIR 80°F D.B. 67°F W.B.

C 50.7 PSIG
56°F

SUCTION LINE

E 218 PSIG
195°F

DISCHARGE
VALVE

SUCTION
VALVE

D 50.7 PSIG
58°F

MANIFOLD GAUGES

LIQUID
LINE

SEMI-HERMETIC
COMPRESSOR

DISCHARGE LINE

F 218 PSIG
130°F

SUB-COOLED

A 218 PSIG
114°F

CONDENSER

ENTERING AIR TEMP. 95°F

LIQUID
LINE VALVE

REFRIGERANT R-500

HIGH PRESS
LIQUID

LOW PRESS
LIQUID

LOW PRESS
VAPOR

HIGH PRESS
VAPOR

FIGURE R24-3 High-temperature refrigeration application (air-conditioning) with thermostatic expansion valve.

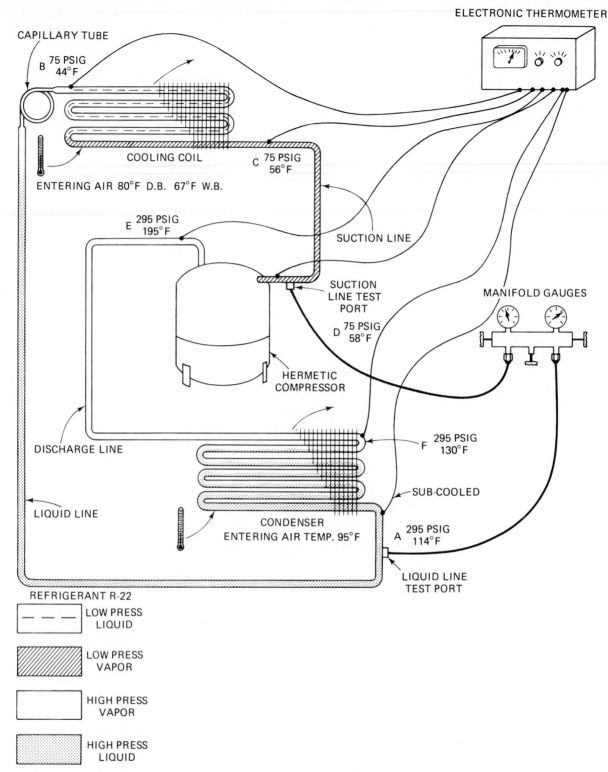

CAPILLARY TUBE

B 75 PSIG 44°F

ELECTRONIC THERMOMETER

COOLING COIL

C 75 PSIG 56°F

ENTERING AIR 80°F D.B. 67°F W.B.

E 295 PSIG 195°F

SUCTION LINE

SUCTION LINE TEST PORT

MANIFOLD GAUGES

D 75 PSIG 58°F

HERMETIC COMPRESSOR

DISCHARGE LINE

F 295 PSIG 130°F

SUB-COOLED

LIQUID LINE

CONDENSER ENTERING AIR TEMP. 95°F

A 295 PSIG 114°F

LIQUID LINE TEST PORT

REFRIGERANT R-22

- - - - LOW PRESS LIQUID

LOW PRESS VAPOR

HIGH PRESS VAPOR

HIGH PRESS LIQUID

FIGURE R24-4   Refrigerant cycle with capillary tube.

expansion valve metering device. Now let's compare these to a system using a capillary tube metering device.

Figure R24-4 shows a refrigerant cycle diagram for a typical air-cooled system employing a capillary tube and using R-22 refrigerant. The temperature–pressure relationships shown are considered normal for a system operating under the following conditions: entering evaporator coil air temperature is 80°F DB, 67°F WB; condenser air entering temperature is 95°F DB. Since these pressures will be affected by equipment design, always consult the manufacturer's pressure–temperature charts for each specific model. The system shown has a sealed hermetic compressor. Pressure-test connections (Schrader valves) are shown in the suction and liquid lines.

Starting at point *A*, the liquid refrigerant is subcooled approximately 19 to 111°F and exerts a pressure of 295 psig. Assuming the liquid refrigerant line is 25 ft long and properly sized, the pressure at the entering side of the capillary tube will be approximately 195 psig with the temperature of the refrigerant still about the same as at *A* (i.e., 111°F).

Downstream from the capillary tube at *B*, the *T–P* readings will be approximately 44°F and 75 psig. With the capillary tube restricting the refrigerant flow for 12°F of superheat, the *T–P* readings at *C* will be 56°F and 75 psig. This pressure reading neglects the nominal refrigerant pressure drop of the vaporizing refrigerant through the evaporator coil.

Moving down to the suction gas intake at the compressor (point *D*), the *T–P* relationship of 61°F and 75 psig represents the temperature pickup of additional superheat caused by the suction line. At *E*, these readings become 195°F and 295 psig. The condenser average *T–P* is 130°F and 295 psig at *F*. The subcooling (16°F) then takes place at the last two and three bottom rows of the condenser and the cycle is completed at point *A*.

The examples above show what might be expected during normal system operation. Of course, this varies with the refrigerant, the condensing medium (air or water or evaporative), the type of application, and the make of equipment. Naturally, the entire scope of differences cannot be presented here, but these variations are discussed in later chapters and will become familiar through experience.

Regardless of the type of system, there will be some common operating problems encountered, and the service technician must, like a doctor, be able to recognize the symptoms, diagnose the cause, and take corrective action. In most cases, the medical doctor is able to prescribe medicine or treatment immediately to relieve the patient. The refrigeration serviceman may have to arrive at a satisfactory diagnosis through the process of elimination of several possible causes, each of which may be the source of the complaint or the problem in the refrigeration system.

Craftsmen must be conscientious in their attempts to put the system back in proper operating condition and ethical in their dealings with the customer. There have been complaints from the field that some service persons do not always measure up to the highest standards. For example, one may have added refrigerant to a system when there were indications of a shortage and, when this action did not correct the trouble, the service person was negligent and did not remove the excess refrigerant. (This, in itself, might be the cause of a future service complaint.) Some manufacturers of components such as expansion valves have had parts returned that were not defective. Rather, the strainers in some of the valves were merely dirty or clogged, but the service person would replace the valve, blaming the trouble on its operation.

R24-2
PROBLEM CATEGORIES

Remember, a refrigeration system is a *heat-energy-transfer device*. It is only capable of absorbing heat from a *heat source* and rejecting the heat into a *heat sink*. The rate at which it does this is dependent on the amount of heat available and the rate of transfer. The rate of transfer will depend on maintaining the proper temperature difference between the refrigerant and the material from which the heat is to be extracted from or rejected into.

R24-2.1
Heat Source

Is there sufficient heat source to satisfy the capacity of the system? If it is a system capable of extracting 12,000 Btu of heat in an hour's time with 20°F ΔT, this amount of heat must be available. If air is the means of carrying the heat from the product to be cooled to the evaporator for extraction, the correct amount of air must pass through the coil. If insufficient air is being supplied because of:

1. Dirty air filters
2. Slow pumping fan or blower
3. Dirty coil fins
4. Improperly placed product

or any other reason for reduction in the air quantity, the amount of heat absorbed is reduced. The coil operates at a lower temperature, the refrigerant boiling point is lower, and the system capacity is lost.

This also applies if the heat is transferred by means of a liquid. Again, the same problem can apply:

1. Dirty liquid filters—plugged drains
2. Slow-running pump—damaged inpuller
3. Limed-up exchanger tubes

A reduction in the quantity of the liquid through the heat exchanger reduces heat absorbed, lowers coil boiling point, and lowers suction pressure and system capacity. To check the amount of load on the evaporator, the quantity of air or liquid through the evaporator must be measured, and both entering and leaving temperatures of the air or liquid must be taken.

The measurement of quantity of air through the coil of a dairy case for example is almost impractical, but thermometers in the supply air off the coil and return air to the coil can usually be inserted to measure air temperature. If the air temperature of the coil drops, the air through the coil is higher than factory specifications, the fault is in the air quantity, because a refrigeration system has no ability to suddenly increase in capacity. If higher than normal, however, the fault is usually in the refrigeration system, as the air supply does not usually increase in quantity over design. In this case, the refrigeration system must be checked.

R24-2.2
Heat Sink

Problems in the heat-sink area are usually easier to diagnose because the change in the system becomes more radical with a change in the operating conditions of the heat sink. When the air through the air-cooled condenser is reduced, head pressures and compressor amperage draw go up and system capacity drops. When the liquid through the liquid condenser is reduced, head pressures rise together with amperage draw of the compressor and reduction in capacity results. The effect on capacity is not as great, however, as a change in load on the evaporator, so this problem usually exists and grows until a radical departure from normal occurs.

The most frequent causes of problems in air-cooled condensers are:

Dirt, leaves, etc., on coil surface

Restricted air inlet or outlet

Dirty blower blades

Motor and/or blower bearings tight

Incorrect fan or blower rotation (especially on three-phase blower motors)

Incorrect fan or blower speeds

Prevailing winds

Unit located in restricted airflow area

The most frequent causes of problems in liquid-cooled condensers are:

Limed-up tubes in condensers.

Insufficient liquid pressure.

Insufficient liquid supply.

Improperly set or malfunctioning regulating valve.

Inability to dispose of liquid leaving condenser.

In all cases, the measurement of the air or liquid quantity through the condenser as well as the entering and leaving temperatures are important for trouble diagrams.

In all instances, the head pressure will be too high if the fault is condenser performance.

If the head pressure is too low, it is usually due to the refrigeration system; the one exception is low ambient temperatures in the case of air-cooled condensers when trying to operate in ambients of 65°F or below without head pressure control.

R24-3
REFRIGERATION SYSTEM PROBLEMS

In the refrigeration system, problems can be divided into two categories: refrigerant quantity and refrigerant flow rate. If the system is correctly charged to the point where the evaporator is fully active and the condenser contains a sufficient quantity to produce the desired subcooling and the compressor is pumping the required amount of vapor, the system has to perform as designed. Any deviation from these requirements will show in the test pressures and temperatures that are taken on the system.

The first test to be determined is if the problem in the refrigeration system is electrical or mechanical. Does it operate, and if so, does it operate correctly?

R24-3.1
Electrical Problems

In the electrical category the problems are usually that the unit will not start, will not continue to run, or will run erratically. Following is a list of common complaints or problems that apply to the electrical portion of the system. Under each problem are listed possible causes. These causes are listed and explained in Chapters R25 and R26.

Motor Compressor Assembly

1. Will not run—no hum
   - E6. Open control contacts
   - E4. Overload cutout tripped
   - E8. Improper wiring
   - E5. Overload cutout B.O.
   - R25-12. Refrigerant shortage
2. Will not run—hums—cycles on overload
   - E8. Improper wiring
   - E12. Voltage low
   - E19. Start capacitor B.O. or wrong one
   - E20. Run capacitor B.O. or wrong one
   - E18. Start relay B.O. or wrong one

R25-4. High head pressure

R25-22. Compressor B.O.

3. Starts—start capacitor remains in

    E8. Improper wiring

    E12. Voltage low

    E19. Start capacitor B.O. or wrong one

    E18. Start relay B.O. or wrong one

    E20. Run capacitor B.O. or wrong one

    R25-4. High head pressure

    R25-10. Refrigerant pressure-reducing device B.O. or misadjusted

    R25-22. Compressor B.O.

4. Starts and runs—cycles on overload cutout

    E8. Improper wiring

    E20. Run capacitor B.O. or wrong one

    R25-4. High head pressure

    E12. Voltage low

    E11. Voltage high

    E5. Overload cutout B.O.

    R25-10. Refrigerant pressure-reducing device B.O. or misadjusted

    R25-22. Compressor B.O.

5. Starts and runs after several attempts

    E12. Voltage low

    E19. Start capacitor B.O. or wrong one

    E18. Start relay B.O. or wrong one

    R25-4. High head pressure

    R25-9. Restricted or plugged liquid line

    R25-10. Refrigerant pressure-reducing device B.O. or misadjusted

    R25-21. Low ambient temperature

    R25-6. Improper load on evaporator

    R25-11. Undersized refrigerant line

6. Starts—cuts out on main overload protection

    E8. Improper wiring

    E5. Overload cutout B.O.

    E12. Voltage lower

    E11. Voltage high

    E19. Start capacitor B.O. or wrong one

    E18. Start relay B.O. or wrong one

    R25-10. Refrigerant pressure-reducing device B.O. or misadjusted

    R25-22. Compressor B.O.

7. Runs but noisy

    R25-4. High head pressure

    R25-10. Refrigerant pressure-reducing device B.O. or misadjusted

    R25-13. Refrigerant overcharge

Start Relay

1. Contacts do not open

    E12. Voltage low

    E19. Start capacitor B.O. or wrong one

    E20. Run capacitor B.O. or wrong one

    R25-4. High head pressure

    E18. Start relay B.O. or wrong one

    R25-22. Compressor B.O.

2. Contacts will not close

    E18. Start relay B.O. or wrong one

3. Burns out

    E8. Improper wiring

    E11. Voltage low

    R25-4. High head pressure

    E19. Start capacitor B.O. or wrong one

    E20. Run capacitor B.O. or wrong one

    E11. Voltage high

    R25-2. Low suction pressure

    E18. Start relay B.O. or wrong one

    R25-22. Compressor B.O.

Start Capacitor

1. Burns out

    E8. Improper wiring

    E12. Voltage low

    E19. Start capacitor B.O. or wrong one

    R25-22. Compressor B.O.

    E20. Run capacitor B.O. or wrong one

    R25-2. Low suction pressure

    R25-4. High head pressure

2. Capacitor tests OK—no effect

    E8. Improper wiring

    E7. Loose connection or broken wire

Run Capacitor

1. Burns out

    E11. Voltage high

2. Tests OK—no effect

    E8. Improper wiring

Fan Motor

1. Will not run

    E7. Loose connection or broken wire

    E5. Overload cut out B.O.

2. Will not run—burns—cuts out on overload

    E20. Run capacitor B.O. or wrong size

    E17. Motor B.O.

3. Runs—cycles on overload
    E16. Motor overloaded
    E12. Voltage low
    E5. Overload cutout B.O.
    E17. Motor B.O.

Relay or Contactor

1. Will not pull in—no hum
    E6. Open control contacts
    E8. Improper wiring
    E7. Open circuit
    E15. Burned-out coil
2. Will not pull in—hums
    E12. Voltage low
    E14. Shorted coil
    E13. Armature sticking

R24-3.2
Refrigeration Problems

In the refrigeration category, the most common problem is refrigerant shortage, which results in failure of the refrigerating affect. As in the electrical category, however, problems of refrigeration system malfunction are varied and can arise from the refrigeration system itself or from an electrical cause or a combination thereof. Following is a list of common complaints or problems that can be encountered in the refrigeration system. Under each problem are listed possible causes. These causes are listed and explained in Chapters R25 and R26.

Entire System

1. Will not run—no hum
    R26-2. Disconnect switch open
    R26-4. Blown fuse
    R26-3. Main overload tripped
    R26-6. Overload cutout B.O.
    R26-5. Overload cutout tripped
    R26-8. Loose connection or broken wire
    R26-9. Improper wiring
2. Runs but short cycles
    R26-10. Improper control settings
    R25-4. High head pressure
    R25-2. Low suction pressure
    R25-6. Improper load on evaporator
    R25-12. Refrigerant shortage
    R25-10. Refrigerant control device B.O. or misadjusted

R25-20. Oil-logged coil
R25-9. Restricted or plugged liquid line

3. Runs continuously
    R26-9. Improper wiring
    R25-6. Improper load on evaporator
    R25-10. Refrigerant pressure-reducing device B.O. or misadjusted
    R25-21. Low ambient temperature
    R25-12. Refrigerant shortage
    R25-4. High head pressure
    R25-22. Compressor B.O.
    R26-11. Control contacts overloaded
    R25-20. Oil-logged coil
    R25-23. Undersized unit
4. Noisy
    R25-24. Vibration
    R25-4. High head pressure
    R25-6. Improper load on evaporator
    R25-13. Refrigerant overcharge
    R25-10. Refrigerant pressure-reducing device B.O. or misadjusted
    R25-22. Compressor B.O.
    R25-11. Undersized refrigerant lines
5. Evaporator temperature too high
    R25-6. Improper load on evaporator
    R25-12. Refrigerant shortage
    R25-21. Low ambient temperature
    R25-5. Low head pressure
    R25-10. Refrigerant pressure-reducing device B.O. or misadjusted
    R25-20. Oil-logged coil
    R25-11. Undersized refrigerant lines
    R25-23. Undersized unit
6. Evaporator temperature too low
    R25-6. Improper load on evaporator
    R26-9. Improper wiring
    R26-11. Control contacts overloaded
    R25-12. Refrigerant shortage
    R25-10. Refrigerant pressure-reducing device B.O. or misadjusted
    R25-21. Low ambient temperatures
    R25-20. Oil-logged coil
    R25-16. Unit location
7. Suction line sweats heavily or ices up
    R25-6. Improper load on evaporator
    R26-9. Improper wiring
    R26-11. Control contacts stuck
    R25-10. Refrigerant pressure-reducing device B.O. or misadjusted

R25-13. Refrigerant overcharge

R25-4. High head pressure

8. Liquid line extremely hot

R25-4. High head pressure

R25-12. Refrigerant shortage

9. Liquid line sweats or frosts

R25-9. Restricted or plugged liquid line

10. Occasional operation on OFF cycle

R25-19. Refrigerant control device leaking

R25-22. Compressor B.O.

R24-4
SERVICE CHECK SHEETS

Practically all manufacturers of refrigeration equipment publish check sheets designed to supply the necessary information to determine the cause of the problem that has been encountered in their product, regardless of the particular arrangement on the form. The information asked for always covers the three steps of heat energy transfer that is required. The steps are: (1) transfer of heat energy from the heat source to boiling refrigerant, (2) flow of heat from evaporator to condenser, and (3) rejection of heat from refrigerant to the heat sink. Taking each step in turn, the information required would be:

1. a. Temperature of air or liquid entering the evaporator
   b. Temperature of air or liquid leaving the evaporator
   c. Quantity of air or liquid through the evaporator (this would be very difficult to measure in the case of gravity-type-coil cooling air)

2. a. Make and model number of unit
   b. Serial number of unit
   c. Btu/h capacity of unit
   d. Type of refrigerant in system
   e. Line voltage to unit with unit off
   f. Voltage at compressor common and run terminals when trying to start
   g. Starting or inrush current to compressor
   h. Amperage draw of compressor through:
      (1) Common terminal
      (2) Run winding
      (3) Start winding
   i. Type of pressure-reducing device
      (1) Automatic expansion valve
      (2) Thermostatic expansion valve
      (3) Capillary tube
      (4) Other
   j. Compressor suction pressure (psig) and boiling point temperature equivalent (°F)
   k. Suction-line temperature
      (1) At outlet of evaporator
      (2) At compressor
   l. Compressor discharge pressure (psig) and condensing temperature equivalent (°F)
   m. Temperature of liquid leaving condenser (°F)

3. a. Temperature of air or liquid entering the condenser (°F)
   b. Temperature of air or liquid leaving the condenser (°F)
   c. Quantity of air or liquid through the condenser in cubic feet per minute (ft³/min) or gallons per minute (gal/min)

With this information, problem causes in Chapters R25 and R26 become usable.

PROBLEMS

R24-1. In the suction line, is it normal to expect a drop in pressure or in temperature?

R24-2. For an air-cooled condenser to reject heat to the atmosphere, must the condensing temperature be above or below the outdoor ambient temperature?

R24-3. The difference between the condensing temperature and the temperature of the air entering the condenser is called the _____ .

R24-4. Does subcooling in the air-cooled condenser take place in the top or the bottom rows?

R24-5. Usually, subcooling of the liquid will occur in the last _____ rows of the condenser.

R24-6. Problems in a refrigeration system may occur in two classifications: _____ and _____ .

R24-7. Will a dirty condenser produce high or low head pressure?

R24-8. What is the lowest outdoor ambient temperature a standard air-cooled unit will operate at satisfactorily?

R24-9. Problems in the refrigeration system are divided into two categories: _____ and _____ .

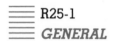

Troubleshooting: Refrigeration

R25-1
GENERAL

As stated several times throughout the various chapters, a refrigeration system is a device that transfers heat from a heat source to a heat sink by the vaporization and condensation of a refrigerant. Also, the refrigeration system operates by control of a boiling point of the liquid refrigerant.

With this in mind, if the heat energy available in the heat source is sufficient, the refrigeration system has the proper amount of refrigerant, the refrigerant is flowing at the desired rate, and the heat sink is capable of absorbing the heat rejected at the desired rate, the system has to work. Any deviation from these conditions will show up in temperature and pressure measurements as well as electrical current needed to operate the unit.

In Chapter R24 various problems in the electrical and refrigeration systems are listed together with possible solutions. Following are the suggested causes that pertain to the refrigeration system.

- R25-2. Low suction pressure
- R25-3. High suction pressure
- R25-4. High head pressure
- R25-5. Low head pressure
- R25-6. Improper load on evaporator
- R25-7. Poor load distribution
- R25-8. Plugged distributor or coil circuits
- R25-9. Restricted or plugged liquid line
- R25-10. Refrigerant pressure-reducing device B.O. or misadjusted
- R25-11. Undersized refrigerant lines
- R25-12. Refrigerant shortage
- R25-13. Refrigerant overcharge
- R25-14. Clogged condenser

- R25-15. Defective condenser fan motor or drive
- R25-16. Unit location
- R25-17. Air in the system
- R25-18. Restricted hot-gas line
- R25-19. Refrigerant control device leaking
- R25-20. Oil-logged coil
- R25-21. Low ambient temperature
- R25-22. Compressor B.O.
- R25-23. Undersized unit
- R25-24. Vibration/noise

R25-2
LOW SUCTION PRESSURE

The operating suction pressure of a refrigeration system, be it an air-conditioning unit (high temperature), refrigerator (medium temperature), or freezer cabinet (low temperature), varies over a range of pressures depending on the heat load on the evaporator. In turn, because the heat load on the evaporator varies with the amount of air or liquid and/or the temperature of the air or liquid entering the evaporator, it is not possible to establish definite operating suction pressures. Therefore, the operating suction pressure of a refrigeration system has no significant value unless it is unusually high or low.

Normally, suction pressures will operate in the range 35 to 65°F boiling point of the refrigerant in the high-temperature range, −10 to +5°F in the medium-temperature range, and −25 to −5°F in the low-temperature range. Suction pressure equivalants much lower than these ranges indicate that the gas is not returning to the compressor as fast as the compressor is handling the gas. In the case of too high a suction pressure, the compressor is not handling the gas as fast as it is being returned from the evaporator.

This situation can be compared with filling a leaking tank from an air pressure supply. If the air enters the tank faster than the air leaves from the tank, the tank pressure will increase. If the air supply is less than the air leaking out, the pressure will decrease. In regard to the refrigeration system, the problem is to determine the reason for the reduced volume of vapor returning to the compressor or the compressor's inability to handle the refrigerant vapor.

The efficiency of a motor compressor depends on the pressure of the gas entering the compressor (suction side) and the pressure of the gas leaving the compressor (discharge). This difference in pressure is known as the *compression differential*. If the suction pressure drops and/or the discharge pressure rises, the pressure differential increases and the compressor volumetric efficiency (ability of the compressor to pump gas) decreases. In reverse, if the suction pressure increases and/or the discharge pressure decreases, which results in a decrease in compression differential, the compression volumetric efficiency increases. This means that in a normally operating refrigeration system, the amount of refrigerant developed by the evaporator and the volumetric efficiency of the compressor reach a balance point such as when the amount of air into the tank equals the amount of air leaking out of the tank and the tank pressure remains constant. Therefore, the suction gas balance point pressure becomes important only when this balance point is below or above the normal range.

The following could be causes for low suction pressure:

- R25-6. Improper load on evaporator
- R25-7. Poor load distribution
- R25-8. Plugged distributor or coil circuits
- R25-9. Restricted or plugged liquid line
- R25-10. Refrigerant pressure-reducing device B.O. or misadjusted
- R25-11. Undersized refrigerant line
- R25-12. Refrigerant shortage
- R25-20. Oil-logged coil
- R25-5. Low head pressure

R25-3
HIGH SUCTION PRESSURE

As explained in Section R25-2, the suction pressure is a balance between the evaporator producing suction vapor by absorbing heat and boiling off liquid refrigerant and the compressor compressing the vapor and discharging the hot high-pressure vapor into the condenser.

If for any reason the evaporator increases the rate of vaporization or the compressor loses the ability to handle the vapor, the suction pressure will balance at a higher reading. Naturally, when a system first cycles ON, the suction pressure will be high because of the accumulated heat load in the evaporator. After a reasonable operating time, however, the suction pressure should come down to normal. If it continues to be too high, it may be because of an excessive vaporization rate in the evaporator, or reduction in pumping rate of the compressor, or a combination of both.

Excessive vaporization rate in the evaporator may be caused by:

- R25-6. Improper load on evaporator
- R25-10. Refrigerant pressure-reducing device B.O. or misadjusted
- R25-13. Refrigerant overcharge

Reduction in pumping rate in the compressor could be caused by:

- R25-4. High head pressure
- R25-13. Refrigerant overcharge
- R25-14. Clogged condenser
- R25-15. Defective condenser fan motor or drive
- R25-16. Unit location
- R25-17. Air in the system
- R25-18. Restricted hot-gas line
- R25-22. Compressor B.O.

R25-4
HIGH HEAD PRESSURE

High head pressure in the condensing unit is caused by the inability of the condenser to remove the heat from the compressed gas from the compressor as fast as the gas is compressed by the compressor. Inasmuch as the ability of the condenser to transfer heat is determined by the transfer rate of each square inch of fin and tube surface, the quantity of air through the condenser and the temperature of air (unit ambient) entering the condenser, any change in either factor will affect the overall transfer capacity of the condenser. Following are items that will affect one or more of the condenser transfer factors:

- R25-13. Refrigerant overcharge
- R25-14. Clogged condenser
- R25-15. Defective condenser fan motor or drive
- R25-16. Unit location
- R25-17. Air in the system
- R25-18. Restricted hot-gas line

R25-5
LOW HEAD PRESSURE

Low head pressure is caused by a higher-than-normal heat rejection rate by the condenser or a decrease in the vapor pumping rate of the compressor. As long as the condenser is removing the heat from the vapor at the required rate, the condensing temperature and pressure to maintain it will be within the desired range.

If the low head pressure is due to decreased pumping rate of the compressor, the problem could be:

- R25-2. Low suction pressure
- R25-6. Improper load on evaporator
- R25-7. Poor load distribution
- R25-8. Plugged distributor or coil circuits
- R25-9. Restricted or plugged liquid line
- R25-10. Refrigerant pressure-reducing device B.O. or misadjusted
- R25-11. Undersized refrigerant line
- R25-12. Refrigerant shortage
- R25-20. Oil-logged coil
- R25-22. Compressor B.O.

If the low head pressure is due to an increase in the heat removal rate of the condenser, the problem could be:

- R25-21. Low ambient temperature

R25-6
IMPROPER LOAD ON EVAPORATOR

The most common cause of low suction pressure is insufficient load on the evaporator.

R25-6.1
Air

This problem, when air is the transfer medium, is usually due to improper blower speed, high duct or air passage resistance (improper product placement in the case of medium- and low-temperature applications), dirty air filters, or a clogged coil, with dirty air filters the principal cause. Frequently, it is a combination of two or more of these causes. Because of the predominance of trouble with dirty air filters:

1. Check and clean or replace the air filters before any further checks are made. All temperature drop tests through the coil as well as suction pressure checks must be made with clean air filters.

2. Check to make sure that volume dampers or air control damper settings have not been changed. Make sure that airways are not blocked with product. Frequently, changes in the air distribution system are caused by people who lack the proper knowledge, closing air dampers or controls without regard to possible results.

3. Check the blower motor and blower and drive in the case of belt driven blowers to make sure that:
   a. The blower motor is properly lubricated and operating freely.
   b. The blower wheel is clean—the blades could be loaded with dust and dirt or other debris. If the wheel is dirty, *it must be removed and completely cleaned.* Do not try brushing because a poor cleaning job will only cause an unbalance to occur in the wheel and extreme vibration in the wheel and noise will result. This could result in disintegration of the wheel.
   c. On belt-driven blowers, blower bearings must be lubricated and operating freely.
   d. The blower drive belt must be in good condition and properly adjusted (see Fig. R25-1). Cracked or heavily glazed belts must be replaced. Heavy glazing of the belt is caused by running the belt too tight. Proper adjustment requires the ability to depress the belt. Midway between the pulleys, approximately 1 in. for each 12-in. distance between shaft centers.
   e. The blower speed is properly set. The blower speed is set by measuring the temperature drop of the air as it passes through the evaporator.

Generally, high-temperature applications (air conditioning) operate with a 15 to 25°F ΔT through the coil, medium-temperature applications with a 5 to 15°F ΔT through the coil, and low-temperature application with a 5 to 10°F ΔT through the coil.

Because of the vast variety of refrigeration application, the manufacturer of the equipment should be contacted for proper ΔT for their particular piece of

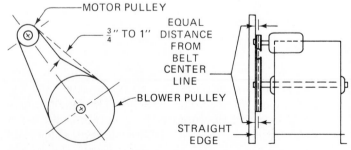

CHECKING BELT TENSION — CHECKING PULLEY ALIGNMENT

MOTOR PULLEY
$\frac{3}{4}''$ TO 1''
EQUAL DISTANCE FROM BELT CENTER LINE
BLOWER PULLEY
STRAIGHT EDGE

FIGURE R25-1   Blower drive belt adjustment.

equipment. Air-conditioning applications will be discussed in more detail in the air-conditioning section of this text.

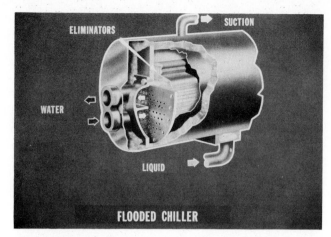

FIGURE R25-2 Flooded chiller. (*Courtesy* Carrier Air-Conditioning Company)

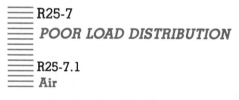

This problem, when liquid is the heat-transfer means, is usually caused by deposits on the walls of the liquid sections of the liquid cooling evaporator. This can be lime deposits, if water is used, or sediment materials of the liquid to be cooled.

The best way to determine if such deposits have accumulated to excess is to determine the loss or drop in pressure of the liquid leaving the coil versus the entering pressure. The difference will be the flow resistance through the coil liquid circuit.

Figure R25-2 shows a flooded chiller with a removable head in order to make the liquid tubes available for rod-type cleaning. This arrangement is usually used in large-tonnage units. In smaller systems, to reduce cost, tubes in a tube-type evaporator are used. These heat exchangers are known as coaxial type. Because of the configuration of the tube in this unit, chemical cleaning is the only method available. There are many types of cleaners available and the chemical manufacturers' recommended methods of use for their chemicals should be closely followed.

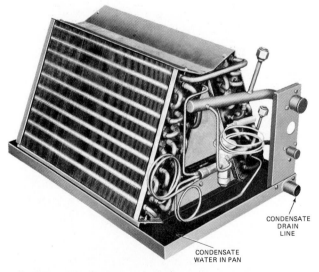

R25-7
POOR LOAD DISTRIBUTION

R25-7.1
Air

With coil cooling air, it is very important that each circuit of the coil receive equal amounts of cooling load (air ft³/min) so as to handle its proportionate share of the cooling load. If proper balance of air over the coil is not obtained, the capacity and efficiency of the coil is reduced. The action of this unbalance, however, is different if the pressure-reducing device is a thermostatic expansion valve than if multiple capillary tubes are used.

TX Valves: In multiple-circuit coils using a TX valve, the distributor and feeder tubes will divide the refrigerant flow through the TX valve equally through all the circuits. Figure R25-3 shows a flat coil mounted in a vertical position for horizontal airflow through the coil. The coil has four circuits fed by the TX valve and distributor. Therefore, each circuit, which consists of a horizontal quarter section of the coil, must receive the same amount of air (one-fourth of the total) to vaporize the refrigerant flowing through the circuit and prevent liquid runout.

The situation where the greatest potential for unbalance exists is where the air must make a 90° turn before entering the coil. When air turns a corner, the centrifugal force produced, when the air changes direction, will cause the air to pack at the outer radius of the curve. As a result, the coil sections on the outer part receive more air than the coil sections in the inner part.

Because the sections receive equal amounts of refrigerant, the liquid refrigerant in the sections of the coil nearest the inner radius will not be completely vaporized. It will pass from the coil in liquid form and will cool the gas leaving the entire coil assembly to a temperature colder than normal. This lower gas temperature will cause the TX valve to close down according to the requirements of the sections carrying the lighter load and thus will rob the rest of the coil sections of the re-

FIGURE R25-3 Coil with T.X. valve. (*Courtesy* Borg-Warner Central Environmental Systems, Inc.)

quired refrigerant to handle the load. The capacity and efficiency of the entire coil is greatly reduced, the suction pressure will balance at a lower than normal pressure, and the overall capacity of the system will be reduced.

Capillary Tubes: In coils using capillary tubes, unbalanced load on the coil circuits will cause liquid run-out and possible liquid return to the compressor, with possible compressor liquid slugging and compressor failure. Figure R25-4 shows an A-type coil using six capillary tubes each feeding a separate circuit. Because the flow resistance of each circuit is equal to all the others, each circuit gets an equal amount of refrigerant. In the case of the coil in Figure R25-4, each circuit gets one-sixth of the total.

If any of the circuits do not receive sufficient load to boil the refrigerant in the coil, any liquid left at the end of the circuit will leave the coil, enter the suction line, and travel this line until sufficient heat is acquired to complete the boiling process. With insulated suction lines the heat available is usually not enough to prevent liquid refrigerant from entering and possibly damaging the compressor.

Regardless of the pressure-reducing device used, the coil should be checked to make certain that the temperature drop of the air through each circuit cross section is approximately the same and that the temperature rise of the refrigerant (superheat) in each circuit is the same.

Occasionally, it is necessary to install turning vanes or splitter dampers to ensure the necessary air distribution. These are always installed on the entering side of the coil. Air distribution leaving the coil is never any problem from a coil performance standpoint.

FIGURE R25-4 Coil with capillary tubes. (*Courtesy Addison Products*)

The previous discussion on air unbalance also applied to liquid cooling coils. Unbalanced loads are seldom a problem unless some repair mechanic has replaced head plate gaskets with the wrong type or has put the gasket in improperly and causes liquid bypass between circuits. Unfortunately, in the case of water coolers, this unbalanced load not only shows up as low suction pressure but usually causes coil freeze-up of the water, with resulting damage of coil tubes and coil leaks. Drying out the refrigerant circuit in a 150-ton water chiller because it filled with water due to a freeze-up resulting from an improperly installed water head gasket is a time-consuming and expensive task.

R25-8
*PLUGGED DISTRIBUTOR
OR COIL CIRCUITS*

Another cause of low suction pressure, especially on new installations, is plugged distributor port, coil feeder tubes, or plugged capillary tubes. This causes the same reactions as poor air distribution of air over the coil because the coil is not able to produce refrigerant vapor fast enough to maintain the suction pressure in the proper range.

Test results of the leaving-air-temperature test will be the same as for improper air temperature drop through the section of the coil except that the temperature of the gas leaving the section will be considerably higher than the rest of the coil section and the temperature drop of the air through the particular coil section will be very low. The result is that the compressor has more ability to handle the vapor than the coil has the ability to produce the vapor, and the system operates at a lower-than-normal coil and system capacity.

Repair of coils with plugs is a strictly mechanical procedure. However, the process of cutting out coil tubes, bypassing coil tubes, or cutting out sections of circuit feeder tubes or capillary tubes in order to bypass or eliminate the plug should never be done. It is better to replace the coil if the block cannot be removed. Bypassing coil tubes or shortening feeder tubes results in an unbalance in capacity of the coil sections and results in lower coil and system capacity. Shortening the length of capillary tubes will only reduce the pressure drop through the tube, raise the coil operating pressure, reduce the temperature difference between the refrigerant and the load, and reduce the coil capacity. It can also cause extensive liquid refrigerant runout and system damage. Plugged capillary tubes should be replaced with tubes of like size and length to retain the design performance of the system.

R25-9
RESTRICTED OR PLUGGED LIQUID LINE

For the evaporator to produce sufficient refrigerant vapor to satisfy the compressor capacity at a normal suction pressure balance, the coil must receive an unrestricted flow of liquid refrigerant. This means that the liquid line must be free of restrictions from the point of pickup of the refrigerant at the outlet of the condenser, through the receiver (if used), through driers, sight glass, refrigerant control (such as a liquid line solenoid valve), and the liquid line, to and through the pressure-reducing device at the inlet of the coil. Restrictions or plugs, if serious enough to affect operation, will usually produce sufficient pressure drop to cause a reduction in the boiling point of the liquid to a point where refrigerant expansion takes place. This will produce a detectable temperature drop across the restriction which can be detected by means of surface-temperature thermometers.

Figure R25-5 shows a four-rack–dual-range thermometer with tape-on leads that can be used for such a purpose. A sensitive, accurate thermometer must be used. The fingers are not sensitive enough.

Figure R25-6 shows a digital version of the same instrument which is easier to read and more accurate. Instruments such as these are an essential part of service technicians' test equipment.

When extra pressure drop in the liquid line is suspected, but normal means of detection do not produce

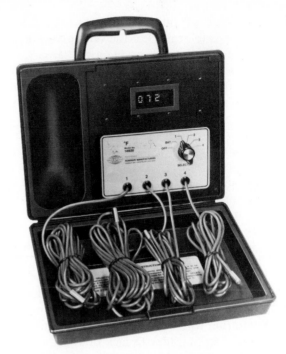

FIGURE R25-6 Robinair model 14820. (*Courtesy* Robinair Manufacturing Corporation)

results, it may be necessary to install a pressure gauge immediately ahead of the pressure-reducing device to read actual liquid pressure at this point. This liquid pressure should never be more than 10 psig lower than the compressor discharge pressure. If greater than 10 psig pressure drop is found, a systematic search of the line must be conducted. There are exceptions to this rule when the manufacturer (especially when using capillary tubes or liquid-line restrictors) uses small liquid lines as part of the pressure-reduction function. In these cases, the liquid-line size and length must not be changed (see Section R13-5).

R25-10
REFRIGERANT PRESSURE-REDUCING DEVICE B.O. OR MISADJUSTED

R25-10.1
Thermostatic Expansion Valve

Operation of the thermostatic/expansion valve is covered in Section R10-4. Thermostatic expansion valves shipped on units are factory set and the adjustment should never be changed except after an accurate superheat check.

Because the TX valve acts to maintain a constant superheat of the gas as it passes through the coil, changing the adjustment of the valve only floods the coil if the valve is opened or reduces the capacity of the coil if the valve is closed. Therefore, check all other possibilities of trouble first before the TX valve is adjusted or

FIGURE R25-5 Robinair model 14840. (*Courtesy* Robinair Manufacturing Corporation)

Troubleshooting: Refrigeration   289

replaced. Even if all other possibilities have been exhausted, disassemble and clear the valve body and inlet screen before adjustment is made on the valve.

It is possible for the valve to stick in a nearly closed position, a fully closed position, or a fully open position. Sometimes dirt, solder, or other material from poor installation practice or frozen moisture from improper evacuation will restrict the flow of refrigerant liquid through the valve or stop the flow of any liquid at all into the evaporator. In that case, if the unit has a low-pressure control, the compressor will short cycle (i.e., start and stop at frequent intervals) when the expansion valve is partially plugged and insufficient liquid is entering the evaporator.

When the expansion valve is completely plugged, the compressor will lower the pressure in the evaporator down to the cutout point of the low-pressure control, which will stop the compressor. If there is no low-pressure control switch in the system, the compressor will continue to run, with no work being done until the motor windings heat up (due to lack of vapor for motor cooling) and will cut off on the overheat protection. The only actual failure that can occur in a TX valve is loss of charge in the power element. This loss of pressure above the diaphragm in the valve will allow the internal spring to drive the needle into the seat and close the valve off. It will then act the same as a plugged valve. To adjust the superheat setting of valves, see Section R10-5.

R25-10.2
Capillary Tubes

Capillary tubes are only small-diameter liquid lines of sufficient resistance to liquid refrigerant flow to produce the pressure drop and boiling point drop to produce the desired results. As the resistance of the capillary tube is set by tube length and inside diameter of the tube, the only trouble that can be encountered would be plugging. Because of the small inside diameter of the tube, cleaning is very difficult and complete replacement is recommended. For best results, the replacement tube must be of the same inside diameter and length (see Section R10-7.2).

R25-11
UNDERSIZED REFRIGERANT LINES

R25-11.1
Suction Line

Even though the evaporator is able to produce sufficient refrigerant to balance the compressor capacity within the normal suction pressure range, if the suction line is undersized and has too high a flow resistance, the suction pressure measured at the suction service valve connection will be low. This also means low compressor efficiencies, higher evaporator operating temperatures, etc., with the attended reduction in system performance. Proper sizing of the suction line is a must. For sizing, see Section R13-5.

R25-11.2
Liquid Line

There is also the possibility that the liquid line does not carry sufficient liquid to satisfy the evaporator requirements, not from restrictions as explained in Section R25-2.4, but because the line is undersized. In the liquid line, it is also very important that the line be sized for minimum pressure drop. See Section R13-7 for proper sizing.

R25-12
REFRIGERANT SHORTAGE

The most common cause for system failure and demand for service is loss of refrigerant from the system. Factory-built systems are pressure tested with presure buildup to 250 to 300 psig in the system and leak tested with electronic leak detectors capable of detecting leaks of $\frac{1}{2}$ oz of refrigerant per year. In turn, the field-installed portion of the system should be checked in the same way.

If at any time the system is found to be low on refrigerant, the leak should be found and repaired and the system evacuated before the system is recharged. When checking for leaks, check all surfaces of each device in the refrigeration system. This includes all tubes in the evaporator and condenser, all factory- as well as field-made joints, all runs of pipe or tube, all joints and fittings, as well as electrical terminals on the compressor and the weld joints of the compressor.

Shut metal mounting screws have been known to pierce tubing runs, coil bends, etc.; children have been known to pierce condenser tubes with nails, wires, etc. *Never assume that a part of a system is gastight even though factory made or installed.* For evacuation and charging the system, see Sections R22-6 through R22-13.

When liquid refrigerant is vaporized, the space occupied by the refrigerant is increased many times. For example, a cubic inch of liquid R-22 expands to 52 in.[3] of vapor and 68 in.[3] if it is R-12. Therefore, any quantity of R-22 vapor *entering* the evaporator (not produced in the evaporator) reduces the refrigerating effect by 52 times the amount of entering vapor. If 1 in.[3] of vapor is formed in the liquid line or enters the liquid line from the condenser because the refrigerant charge is low, this vapor takes the place of the liquid necessary to produce 52 in.[3] of vapor in the evaporator. Along with the reduction in vapor produced, the cooling capacity of the coil is also reduced because of the loss of liquid capacity to absorb the necessary heat. Thus, it is very important that sufficient refrigerant is in the system at all times to supply the full capacity of the cooling coil for maximum operating capacity.

R25-12.1
Determining Amount of Refrigerant

The amount of refrigerant in the system for proper operation will depend on the type of pressure-reducing device used and if a receiver is used.

Systems with a Receiver: In systems using receivers such as expansion valve, low-side float, or liquid-level control devices, the amount of refrigerant charge is not critical as long as there is sufficient liquid in the receiver to keep a liquid seal at the pickup tube or quill in the receiver. Measurement of proper charge in this system would be one-third to one-half the height of the receiver when the system is operating at maximum capacity. As the load decreases the surplus refrigerant would then be stored in the receiver and cause liquid heights of one-half to three-fourths of the receiver height. It is only necessary to add sufficient refrigerant to obtain the proper liquid height in the receiver.

Systems without a Receiver: Systems without receivers depend on the lower tubes of the condenser to act as storage for the surplus refrigerant when the system is under light loads. Therefore, the amount of refrigerant charge in the system is much more critical than if a receiver is used. When charging these systems, the correct amount of charge will produce the correct amount of subcooling of the liquid before it leaves the condenser. For low-temperature and medium-temperature applications where the load is fairly constant, 10 to 12°F subcooling will usually produce the best capacity and unit efficiency. Air-conditioning units, because of the wider variation of load, usually operate at 10 to 20°F of subcooling. Subcooling tests must be conducted in the normal outside ambient range of 65°F minimum to 115°F maximum.

These subcooling readings will usually mean that the bottom two tubes of the condenser will contain liquid and be at a lower temperature than the rest of the condenser surface. This is only at or near standard or design conditions; therefore, this is not an accurate check for the quantity of liquid in the system. It can be a rough check because if the temperature break between condensing refrigerant and liquid refrigerant is much higher than the bottom two or three tubes, the unit is definitely overcharged. If no temperature break is felt in the lower tubes, the unit is short on refrigerant. In either case, gauges and thermometers must be attached and tests made to determine existing conditions.

R25-12.2
Charging System with Refrigerant

Systems with a Receiver: Should the system be evacuated and recharged or refrigerant simply added until the proper level of refrigerant is obtained in the receiver? The answer to this is: "Is there any liquid refrigerant in the system or is there only refrigerant vapor?"

A quick check for this is to use a sling psychrometer to determine the ambient temperature of the condensing unit with the unit off for a sufficient period of time for the unit temperature and ambient temperature to equalize. When the unit temperature is at ambient, the pressure in the idle system should equal the pressure equivalent of the ambient temperature for the type of refrigerant in the system. If the pressure is lower than this equivalent, the system has only vapor and should be evacuated and charged starting from deep vacuum. If the pressure is higher than this equivalent, the system could have air or other noncondensibles in the system. Here again, the system must be evacuated and charged, starting from a deep vacuum. For proper evacuation methods, see Chapter R23.

Systems without a Receiver: Systems that do not have a receiver are critical on charge to within 4 oz of refrigerant on systems using a thermostatic expansion valve and $\frac{1}{2}$ oz of refrigerant using capillary tubes. In all cases, because the weight of the refrigerant in the system is not known, it is best to evacuate the system and start to charge from a deep vacuum. For evacuation and charging methods, see Sections R22-6 through R22-13.

Figure R25-7 shows an electronic charging unit

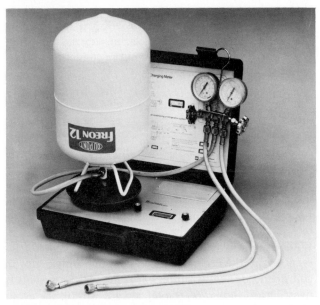

FIGURE R25-7 Electronic charging meter. (*Courtesy* TIF Instruments, Inc.)

with a digital readout that will permit charging a unit within 0.5 oz of correct charge. Use of such a meter is highly recommended for best performance of the unit with lowest operation cost and longest system life.

R25-13
REFRIGERANT OVERCHARGE

An overcharge of refrigerant in the system will give symptoms that are the same as a clogged condenser, fan motor, or blower motor and drive B.O. or air in the system. The reason for the higher head pressure is because the liquid refrigerant occupies a higher percentage of the condenser tubes, reducing the heat transfer capacity of the condenser. As a result, the amount of subcooling of the liquid refrigerant leaving the condenser will be higher than normal. The only thing that can raise the amount of subcooling is excessive refrigerant in the system. *Subcooling is affected only by liquid quantity in the condenser.* This can be caused by restrictions to flow in the liquid line (see Section R25-9) or an overcharge of refrigerant in the system. Because of the overcharge of refrigerant, further symptoms would depend on the type of system.

R25-13.1
Thermostatic Expansion Valve Systems

The thermostatic expansion valve is designed to limit the refrigerant flow into the evaporator according to the load requirements of the evaporator. Therefore, the total amount of the refrigerant in the system will not affect the action of the thermostatic expansion valve unless the system is short on refrigerant or so overcharged that the head pressure is extremely high.

Also, because the thermostatic expansion valve closes off during the OFF cycle, any excess of refrigerant in the system will remain in the condenser and receiver (if one is used).

R25-13.2
Capillary Tube Systems

The capillary tube used in this type of system will pass liquid refrigerant according to the pressure drop through the tube. Therefore, if the pressure on the liquid entering the tube rises for any reason, the flow capacity of the tube increases. This will cause the refrigerant quantity in the coil to be higher than normal. If the system is properly charged, only enough refrigerant to fill the coil is in the system and only a rise in coil operating temperature will result. If the system is overcharged, however, the refrigerant will fill the evaporator and flow out of the evaporator and possibly to the compressor in liquid form. This reduces the capacity

of the system because of the vapor formed by heat picked up from other than the evaporator. If sufficient liquid refrigerant returns to the compressor, the compressor can be robbed of oil or oil floating sufficiently to enter the compressor cylinders and cause slugging and compressor damage.

On the OFF cycle, the capillary tube continues to flow refrigerant until the pressures in the system are balanced. This will mean that the excess refrigerant will flow out of the coil and through the suction line to the compressor. Here the liquid refrigerant floats the oil and settles in the bottom of the compressor. If sufficient refrigerant is in the system to float the oil above the inlet to the oil pump in the compressor assembly, the pump will pick up liquid refrigerant instead of oil. As the liquid refrigerant is an excellent solvent and contains no lubricating qualities, the liquid refrigerant washes all the oil from the pump, oil tubes, and bearing surfaces. The compressor is quickly ruined from this lack of lubrication.

Also, as the pressure drops in the compressor above the oil level, the liquid refrigerant vaporizes due to the drop in boiling point below the temperature of the oil. The refrigerant vapor passes up through the oil, causing foaming. The foam fills the compressor housing, then enters the suction intake to the cylinders (high-efficiency motor compressor assemblies do not have oil separators on top of the motor shaft) and the cylinders themselves. The hydrostatic pressure that results when the pistons of the compressor try to compress the oil causes valve rupture, rod or valve plate breakage, etc. This is always accompanied by excessive slugging noise and vibration.

Therefore, on capillary tube systems, the charge of refrigerant is very critical. If there is doubt as to the correct charge in the system, discharge and evacuate the entire system and recharge with a charge as given on the rating plate of the unit. It is not possible to accurately measure the amount of charge and add refrigerant to bring it to the correct charge. Evacuate and start over.

R25-14
CLOGGED CONDENSER

If the radiator of an automobile becomes plugged with insects, leaves, grass, dirt, or other debris, the engine will overheat. Similarly, if the condenser of the refrigerator or air-conditioning unit becomes clogged with insects, leaves, grass, and other debris, the heat-transfer capacity of the condenser is reduced, head pressure rises, current draw goes up, etc., until the unit overheats sufficiently to cut the unit out on overload or cuts off on the high-pressure control. Because of the gradual buildup of condenser clogging, it is possible for the unit to operate under an overloaded condition without high-pressure cutoff for some time. This can result in high operating cost as well as the attendant electrical problems, because of overload on the electrical parts. Refrigeration and air-conditioning units should be thoroughly cleaned at the start of each peak-requirement season, usually in the spring of the year going into the high-temperature season.

In the case of heat pumps, the "outside coil" (the condenser) should be cleaned at the start of the cooling season (spring of the year) *but must be cleaned at the start of the heating season* (fall of the year). Because of the lower temperatures of air through the outside coil, when it becomes the evaporator during the heating cycle, it is important to keep the coil capacity as high as possible. The condenser can generally be cleaned with a stiff brush or broom. A stiff round brush with a wire handle, such as a bottle brush with the head of the brush bent to a 90° angle, can be used with vertical strokes between the fins. *Be careful not to bend the fins.*

Heavy layers of dust and dirt can be removed by wetting the condenser surface with a strong solution of Electro-Sol dishwasher detergent and water or commercially available coil cleaners and flushing with a garden hose. Always flush in the opposite direction of the operating airflow through the condenser. Pressure-type sprayer cleaners are also available for this task. Periodic cleaning will keep the job a simple procedure. If let go until a heavy dirt crust accumulates, the cleaning job can become difficult and expensive and in extreme cases, require replacement of the condenser.

R25-15
DEFECTIVE CONDENSER FAN MOTOR OR DRIVE

R25-15.1
Belt-Driven Motors

Heavy drag or binding in the condenser fan motor or other electrical troubles that slow the motor, a bad belt, too tight a belt setting (see R25-6.1 step 3), or dragging or binding blower bearings will also cause high head pressure. Any dragging, sticking, or binding that reduces the speed of the blower will drastically cut the air quantity through the condenser, reduce the condenser efficiency, and raise the head pressure.

R25-15.2
Propeller-Type Condenser Fans

The quantity of air through the condenser in units using propeller type fans is factory set and cannot be changed in the field. Because an increase in resistance to airflow through the fan very seriously reduces the output of the propeller-type fan, it is not possible to add resistance to the condenser air circuit. No ductwork is allowed on the entering or leaving side of a unit, with the exception of factory-designed packages, where the unit is designed accordingly.

Unbalanced air entry into the condenser can cause serious fan failure. On large air-cooled condensers where air is taken in on opposite sides and discharges out the top of the unit, setting the unit against the building or any other vertical surface will reduce the air going into one side of the unit. This reduction of the air will cause unbalanced pressure on the inlet of the fan blades, extensive flexing of the blades, and a high incidence of blade fracture. Blades have been thrown through condenser surfaces and out the top of units, endangering property and people.

There is also the possibility of reduction in air quantity through the condenser by a slowdown in speed of the propeller fan motor. Improper lubrication of the fan motor or mechanical drag or binding of the motor are the most common causes of this reduction in fan speed.

Check the amperage draw of the fan motor; the current draw of the motor should be given on the rating plate of the motor on the unit. Current draws in excess of the nameplate rating should be checked. If the condenser fan motor is a permanent-split-capacitor type (has a running capacitor in the start winding circuit), check this capacitor before further work is done on the motor. If the capacitor is shorted, the motor will run at a slower speed and draw excessive current. Generally, the excess current draw is enough to cause the motor to cycle on the built-in automatic reset overload cutout. If the capacitor is open, the motor will start and run in either direction. If wind is causing the fan to run backwards, it will continue to run in this direction.

A quick check of the capacitor is, with the unit off, to spin the fan in the reverse direction. While the fan is spinning in the reverse direction, start the unit. The fan should stop, reverse, and run in the proper direction. If it continues to run in the reverse direction, the capacitor is open or wired wrong.

Most propeller-type fan motors use semipermanently lubricated bearings. No oiling is required for the first 2 years. After the 2-year period, $\frac{1}{2}$ teaspoon of No. 10 marine oil, special electric motor oil, or pure mineral oil should be added to the motor not more than once each year. Do not use rust-preventative oils such as Three-in-One, Finol, or automobile oils. All automobile oils contain detergents, soaps, or other additives.

R25-16
UNIT LOCATION

The location of an air-cooled condensing unit or the high-side section of a self-contained unit is very important from an operating standpoint. The condenser can be compared with the radiator of an automobile. Both are designed to transfer heat to the air passing through the fin area. Operating an automobile with an obstruction in front of the radiator will cause the engine to overheat; this obstruction can be merely idling the engine when the front bumper is up against the garage wall. Idling the engine with the automobile parked so that the rear of the automobile is into a heavy wind will cause overheating; the wind will blow the hot air discharged under the automobile back to the front of the automobile and mix it with the air into the radiator.

The same principle that governs the operation of the automobile radiator applies to the condenser of the condensing unit. Never locate the condensing unit where prevailing winds will blow the discharge air toward the intake of the unit. Also, where the discharge of a second unit will enter the intake of the first unit, the higher-temperature air of the second unit will cause the first unit to overheat.

Never locate a condensing unit where the discharge air will be trapped, such as discharging the air into the corner of a building or locating the unit in an areaway or between adjacent buildings with less than 10 ft between them. Keep decorative bushes and shrubbery away from the unit. They look pretty but are costly from an operating and maintenance standpoint.

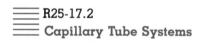

R25-17
AIR IN THE SYSTEM

A common cause of high head pressure is air or other noncondensibles in the system. A poor evacuation job for moisture removal or just bleeding the system for moisture removal can result in air being left in the system. Also, if dry nitrogen or dry carbon dioxide is used as an aid for building pressure for pressure testing, these noncondensible gases will accumulate in the condenser. Because they occupy space in the condenser, it is necessary for the system refrigerant to be compressed further to occupy the space that is left. This results in the necessity for the compressor to create higher head pressure to perform the necessary heat transfer. Also, with the noncondensibles occupying part of the condenser volume, that portion of the condenser so occupied is removed from the effective heat-transfer area of the condenser. This forces the remaining portion of the condenser to develop a higher $\Delta T°F$ between the refrigerant vapor and the heat sink, resulting in higher head pressures. It is necessary, if noncondensibles exist in the system, to purge them off to reduce the operating head pressure.

R25-17.1
TX Valve Systems

To determine if noncondensibles exist in the system:

1. Turn off the condensing unit.
2. Jump the condenser fan relay contacts or connect the condenser fan motor to the hot side of the compressor contactor and operate the condenser fan motor until the head pressure has reached minimum pressure.
3. This pressure should be within 5 psig of the pressure equivalent of the refrigerant at the condenser entering air temperature (the ambient temperature of the unit). For example, assuming a 90° ambient on a unit using R-22 refrigerant, the minimum head pressure will be 170.1 psig; therefore, if the minimum head pressure is 175 psig or less, the system is clear of noncondensibles.

If the head pressure does not drop sufficiently, it is necessary to purge off the noncondensible gases. This is done by opening the discharge gauge side of the gauge manifold with the condensing unit off and the condenser fan running by means of jumping or reconnecting until the head pressure comes down to the proper amount. Purging should be done in small amounts with a short period of time between purging spurts so as to allow the air in the condenser to accumulate in the condenser intake manifold so as to be purged with a minimum waste of refrigerant. Accumulation of noncondensible gases should be prevented by a complete evacuation of the system by means of a good vacuum pump before the system is charged with refrigerant (see Section R22-7).

R25-17.2
Capillary Tube Systems

It is not possible to determine the presence of noncondensible gases in the capillary tube systems because of the fact that the pressures in the system balance out during the OFF cycle. Therefore, if it is suspected that the head pressure is high due to noncondensibles, discharge the unit, evacuate the system, and recharge with refrigerant. Because of the inability to detect noncondensibles in this type of system, a thorough evacuation is an absolute necessity.

R25-18
RESTRICTED HOT-GAS LINE

In systems using remote air-cooled condensers, the possibilities of restrictiveness from excess solder in pipe joints, etc., represent a source of high-head-pressure troubles. This will be indicated by a high compressor discharge pressure with a rapid pulsation of the pressure caused by the loading of the section of the pipe between the compressor and the restriction. Also, when the compressor stops, the discharge pressure drops very rapidly to the normal pressure equivalent of the refrigerant condensing temperature. Generally, a restriction of this type can be detected by the sound of the high-pressure gas going through the restriction, producing a whistling or singing noise like air from a leaking tire. In a situation of this type, it is necessary to discharge the entire system, open the plugged joint, clean out and resolder, evacuate the system, and discharge.

R25-19
REFRIGERANT CONTROL DEVICE LEAKING

When liquid-line solenoids, pilot-operated thermostatic expansion valves, or other liquid flow control devices are used in conjunction with a suction pressure control to cycle the unit, it is possible for the suction pressure to rise above the cut-in set point of the suction pressure control and cause it to close and start the unit. The compressor will very rapidly reduce the suction pressure to the cutout set point of the suction pressure control and shut the unit down. This short cycling action is detrimental to the electrical system, especially to the start capacitor, will cause capacitor failure, and possibly compressor motor burnout.

If the device is leaking refrigerant in the OFF cycle, this can usually be picked up by means of electric thermometers. Any temperature drop between the inlet and outlet of the valve is a good indication of expansion of liquid refrigerant through the device. The device will have to be removed from the system and cleaned or replaced and the system evacuated and recharged for proper operations.

It is possible that the suction pressure rise is not due to the refrigerant control device leaking but is due to the refrigerant reversing through the idle compressor (see Section 25-22).

R25-20
OIL-LOGGED COIL

Chapter R13 covers the proper installation of suction lines to prevent oil accumulation in the coil; therefore, this problem should be a rare occurrence. The problem can exist, however, and cause low suction pressure, low evaporator capacity, and serious system capacity loss because it removes part of the evaporator capacity. The circuits in the evaporator coil will fill with oil to a point where the pressure drop in the coil is not high enough to clear the circuit of the oil: This applies to thermostatic expansion valve-type evaporators. The only cure is system pump-down or discharge, installation of proper traps and pipe slopes in the suction line, evacuation, and recharge of the system. In properly charged capillary tube systems, oil logging will not occur.

R25-21
LOW AMBIENT TEMPERATURE

Air-cooled refrigeration and air-conditioning systems are designed to operate at a minimum condenser ambient temperature of 65°F. At temperatures below this the head pressure will drop too low for the pressure-reducing device to feed sufficient liquid refrigerant to the evaporator to maintain the suction pressure in the proper range. As a result, system capacity loss is extensive, operating time increases, and if the system is controlled by a low-pressure control, short cycling will result (see Section R12-8).

R25-22
COMPRESSOR B.O.

See Section R11-16.

R25-23
UNDERSIZED UNIT

With the system operating with suction and discharge pressures in the normal range, $\Delta T°F$ drop of the air through the evaporator in the required range, the $\Delta T°F$ rise of the air through the condenser in the required range, and the current draw of the motor-compressor assembly appropriate, the unit is working as expected. If the results in the conditioned area are still not satisfactory, it is possible that the unit capacity is not sufficient to handle the load. For estimating refrigeration loads, see Chapter R27.

For estimating air-conditioning loads, the manuals published by the Air Conditioning Contractors of America are the best source of this information. The procedure for calculating heating and cooling loads has been eliminated from this test because of the rapidly changing nature of the procedure. The Air Conditioning Contractors of America manuals are updated as pertinent information is developed and are an excellent source of this information.

R25-24
VIBRATION

The chief source of vibration in the condensing unit or package unit is the motor compressor assembly. With reciprocative action of piston and rod in the compressor, this action is transmitted to the compressor shell and thus to the unit structure. Normally, the vibration of the internal compressor assembly is handled and dampened sufficiently by the compressor assembly mounting means. When the load on the compressor is excessive, however, the heavier pulsations of the compressor can overcome the mounting means and cause unit vibration.

Some vibration is inevitable, so do not mount the condensing unit or package unit in such a location that the vibration is transmitted to walls or ceilings and thus into the occupied area. Keep the unit off light wooden floors, off the attic floor, etc. If necessary to locate in such locations, the use of vibration dampening pad or material is required; however, follow the manufacturer's directions exactly. The overuse of dampening material can sometimes be worse than its lack.

Vibration from blower assemblies is not possible with properly balanced blower wheels, drive, and driven pulleys and properly adjusted belt in good condition. Unbalanced or damaged components can only be replaced.

The chief source of blower wheel unbalance is improper cleaning. When cleaning of the blower blades is required, a thorough cleaning is required. Merely brushing the blades will not remove all the accumulation from each blade. Unless this is done, severe imbalance and vibration will result, possibly to the extent of wheel disintegration.

Propeller fans will also provide vibration if the blades are bent and do not track in the same place. Viewing the blade rotation from the side should show all the blades moving in the same plane. If this is not done, the amount of discharge pressure from each blade will be different and the resulting difference can set up vibrations in the motor-blade assembly. This vibration can be extensive enough to cause blade failure. If air entry into the unit is not balanced, this will also cause blade pulsation and vibration in the unit (see Section R25-16).

R25-25
NOISE

Noise in refrigeration or air-conditioning systems can be classified into three categories:

1. Air
2. Mechanical
3. Refrigerant circuit

R25-25.1
Air Noise

Usually, air noises are the result of too much air traveling at an excessive speed of misdirected air. A certain amount of air noise has to be expected: the rush of air from the discharge of the forced-air blower in the walk-in cooler, the sound of the condenser fan in the condensing unit section of the refrigerator, the sound of air from the supply grill of the air-conditioning unit. Only when the sound of air movement becomes excessive is it classified as noise. The maximum recommended velocity of air from a fan coil unit is 550 ft/min; from a supply register in the air-conditioning system, 350 to 400 ft/min, depending on the grill design. Noise problems of this type are best solved in conjunction with information obtained for the manufacturer of the grill or register.

Air noises encountered in the field are usually from the air-cooled condensing units outside the conditioned area, usually located adjacent to other property and situated where the noise from the unit affects persons other than the occupants of the conditioned area. With the air-cooled unit in an areaway between buildings less than 10 ft apart, the sound waves of the unit can expect to rebound between the vertical surfaces and be objectionable. This is especially so if the condensing unit is a horizontal discharge type, where air is forced out in the direction of the opposite building. This situation has been so prevalent that practically all manufacturers have gone to vertical discharge units to force the major portion of the sound upward and try to minimize the horizontal flow.

The best cure for air noise problems is careful location of the equipment to prevent sound trapping. If not possible, sound-absorbing buffles between the unit and complainant may have to be used.

R25-25.2
Mechanical Noise

Excessive mechanical noise in the condensing unit is usually caused by excessive vibration (see Section R25-24). There are certain mechanical noises in refrigeration and air-conditioning units that are inherent and are not possible to eliminate; they can only be contained. If an air handler is located next to an opening in the walk-in cooler wall, expect to hear the mechanical noise of the blower assembly through the opening. If the furnace or air handler of the air-conditioning system is located in a closet so that when looking through the return air opening one can see the blower wheel, expect to hear the mechanical noise of the blower.

To dampen mechanical noise, the air must make at least two 90° turns from the time it leaves the observer until it enters the blower assembly. In an air-conditioning system, there should be two 90° turns between the return air grill and the blower compartment of the air handler. In extreme cases, the air transfer means (the return air duct) must be lined with acoustical material. This is one of the reasons that fiberglass sheets have risen in popularity as air duct material.

Mechanical noises may also be transmitted via the refrigerant lines from the motor compressor assembly to the means of support of the pipe and cause vibration and noise. Practically all refrigeration and/or air-conditioning units use vibration loops in the connecting lines to the compressor if compressor vibration is excessive, such as in an externally spring-mounted type. In large units, however, this is not practical. In such applications, a line vibration damper section can be used to eliminate the transmission of the vibration through the tube. Figure R25-8 shows such a device installed parallel with the compressor crank shaft to absorb the side thrust of the vibration. These devices do an excellent job of noise reduction. However, if installed in such a position that the vibration stretches and compresses the vibration elimination, the elimination is very quickly destroyed

FIGURE R25-8   Vibration-isolation discharge line. (*Courtesy* Anamet, Inc./Anaconda Metal Hose)

and produces refrigerant loss. The manufacturer's instructions for proper installation should be followed closely.

R25-25.3
Refrigerant Circuit Noise

Occasionally, the pulsation of the hot high-pressure vapor from the compressor cylinders will set up a high-frequency vibration through the hot-gas line to the condenser. This is more prevalent on refrigeration systems using a remote air-cooled condenser and in heat pumps on the heating cycle where the compressor is remote from the indoor coil acting as the condenser. In this case, a muffler is desired in the hot-gas line of the compressor (see Section R14-3).

PROBLEMS

R25-1. The basic principle on which a refrigeration system operates is _____ of _____ .

R25-2. Does reduced evaporator airflow cause high or low suction pressure?

R25-3. On an evaporator using a TX valve, will reducing the airflow cause the coil superheat to increase, decrease, or stay the same?

R25-4. On an evaporator using a capillary tube, will reducing the airflow cause the coil superheat to increase, decrease, or stay the same?

R25-5. If an expansion valve fails by loosing the feeder bulb charge, will the suction pressure rise or fall?

R25-6. Location of the expansion valve feeder bulb has no effect on valve operation. True or False?

R25-7. What is the most common cause of high head pressure?

R25-8. What is the most common cause of low suction pressure?

R25-9. What is the easiest way to check for a clogged capillary tube on multitube coils?

R25-10. The proper refrigerant charge in a TX valve system using a receiver is _____ the height of the receiver.

R25-11. With the unit off and system at ambient temperature, the pressure in the system should equal the _____ .

R25-12. Excessive subcooling of the liquid off the condenser can be caused by _____ or _____ .

R26

Troubleshooting: Electrical

R26-1
GENERAL

As explained in Chapter R18, all electrical circuits consist simply of power source, load, and conductor to transmit the power from the source to the load and control contacts for on/off operation or characteristic change type to accomplish the necessary operation.

R26-1.1
Power Source

The voltage available to the refrigeration unit must be within ±10% of the unit rating for single-phase units and +10%, −5% for three-phase units. Also, on three-phase units, the voltage between all pairs of legs of the power source must be within 3% of each other if the motor connected to this power source is expected to run with reasonable life. The ±10% of maximum voltage variation applys to 24-V devices as well as to the higher voltages of 120V, 240V, 480V, 560V, etc.

R26-1.2
Conductors

The system must be wired correctly and all connections must be tight. When a piece of equipment is received from a manufacturer, it is a good idea to go over all major connections of a screw pressure type to make sure that they have not relieved with time. Copper will tend to cold-creep under pressure and connections such as the main power leads from the contactor or motor starter to the compressor motor may loosen. Check all connections regardless.

In Chapter R24 various problems are listed in the electrical and refrigeration systems together with possible solutions. Following are the suggested causes that pertain to electrical problems.

- R26-2. Disconnect switch open
- R26-3. Main overloads tripped
- R26-4. Blown fuse
- R26-5. Overload cutout tripped
- R26-6. Overload cutout B.O. (defective)
- R26-7. Open control contacts
- R26-8. Lose connection or broken wire
- R26-9. Improper wiring
- R26-10. Improper control settings
- R26-11. Control contacts overloaded
- R26-12. Voltage high
- R26-13. Voltage low
- R26-14. Relay armature sticking
- R26-15. Shorted coil
- R26-16. Burned-out coil
- R26-17. Condenser fan overloaded
- R26-18. Condenser fan or blower motor B.O.
- R26-19. Start relay B.O. or wrong one
- R26-20. Start capacitor B.O. or wrong part
- R26-21. Run capacitor B.O. or wrong one

R26-2
DISCONNECT SWITCH OPEN

Lack of power supply to a unit is not always due to failure of any particular part. Before removing any part or conductor, check the power supply. Many times the source of trouble is only an open disconnect switch or control switch, open because of the owner's lack of knowledge of what switches to close when operation of the system is desired. Furthermore, closing of the switch does not always mean that electrical power will be made available. Check to make sure that power is available at the load (unit) side of the switch. Use of a voltmeter, using the correct voltage range of the meter for the voltage tested, will determine if both contacts in the control are closing. When in doubt, start with the highest voltage range of the meter and reduce the range for accurate reading if the meter indicates voltage is present. Figure R26-1 shows a swinging needle, clamp-type volt-ammeter. Figure R26-2 shows a digital clamp-type volt-ammeter. The digital meter is easier to read and will measure in smaller increments for better accuracy.

Measurement of voltage from terminal to ground will not always indicate a good switch. The load terminal of a bad contact may indicate voltage to ground because of a circuit (usually the primary side of the control transformer) feeding voltage backwards to the terminal. Always measure the voltage across the load terminals to pick up both sides of the power circuit through the switch.

If no voltage indication is obtained across the load terminals, measure across the line side to make sure that there is power to the disconnect switch. If no voltage is indicated across the line side, the branch circuit fuses in the power distribution panel could be blown or circuit breakers tripped (see Section R26-3). If the branch circuit fuses test OK or circuit breakers when reset test OK, refer this problem to the local utility. Do not attempt work on power distribution panels unless you are a certified licensed electrician. Figure R26-3 shows the contact arrangement of a typical disconnect.

If voltage is found across the line terminals and no voltage on the load side, it can be assured that the contacts are not closing. This can be confirmed by measuring the resistance of the closed contacts with an ohmmeter *with the branch circuit fuses removed or circuit breakers open* (voltage removed from the line side). Measure the resistance of each set of contacts, L_1 to T_1, L_2 to T_2 and L_3 to T_3. Any resistance reading over 1 Ω for high-voltage terminals (50 V or higher) or $\frac{1}{2}$ Ω for low-voltage terminals (less than 50 V) indicate burned contact and require replacement. Figure R26-4 shows a typical resistance meter (ohmmeter).

FIGURE R26-2   Digital readout—clamp-type meter. (*Courtesy* A. W. Sperry Instrument Company)

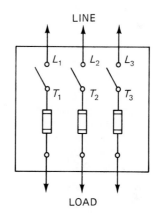

FIGURE R26-3

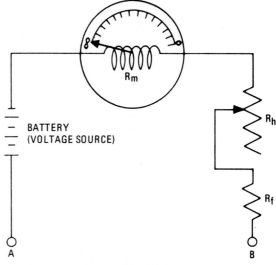

FIGURE R26-4   Simple ohmmeter circuit. (*Courtesy* American Gas Association)

FIGURE R26-1   Clamp-on ammeter. (*Courtesy* A.W. Sperry Instruments, Inc.)

MAIN OVERLOADS TRIPPED

In addition to the main fuse disconnect, the unit may contain manual reset overloads which act as additional protectors for the unit. The unit may have been subjected to unusual voltage conditions, causing the overloads to trip. Resetting the thermal trip levers should bring the unit on. If not, check the voltage across the load side of the overload to make sure that the overload circuits are good and contacts are closing. If no voltage is available with the main supply power off, check the quality of the overload contacts using the ohmmeter (see Section R26-2).

The overloads used may also be of the auxiliary contact type, where adverse current draw through the main circuit causes the opening of the contact in the control circuit of the unit. Figure R26-5 shows a typical motor starter circuit of this type.

Resetting of the overload closes the control circuit, which should cause the controls to actuate. The condition of the auxiliary contact can be checked by measuring the resistance through the terminals (*with the power off*) by means of the ohmmeter. The limitation of 1 Ω on high-voltage terminals and $\frac{1}{2}$ Ω on lower-voltage terminals applies.

Removing the leads from the auxiliary switch terminal (to remove the possibility of other circuits in parallel) and measuring the resistance between the terminals will give the condition of the terminal contact. Again, the maximum resistance rule is 1 Ω for high-voltage contacts and $\frac{1}{2}$ Ω for low-voltage contacts.

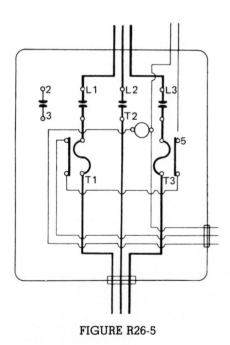

FIGURE R26-5

BLOWN FUSE

Instant-blow or link-type fuses are sometimes used as main fuses. Because they will not carry current above their rating for any length of time, they must be sized to carry the maximum starting current of the unit. This starting current could be more than 150% of the operating current of the circuit. Therefore, time-delay fuse or circuit breaker protection should always be used for refrigeration or air-conditioning systems. This will allow closer sizing of the fuse or circuit breaker to the running current of the unit for maximum protection.

Checking of fuses is best done by means of an ohmmeter. *With the disconnect switch open,* measure the resistance of each fuse. A reading of zero resistance on the ohmmeter indicates that the fuse is good. Infinite resistance reading indicates an open or blown fuse. A resistance reading of anything above 0.1 Ω indicates a burned fuse that should also be replaced.

Partial burning of the fuse or time blowing can be caused by poor contact between the fuse and the fuse clip. Fuses must enter the clip with force required and a definite snap of the clip around a cylindrical fuse or force required to send a bayonet-type fuse into the knife clips. Clips or fuse holder blades should be clean and not burned. Any sign of clip or blade deterioration requires replacement of the fuse holder.

OVERLOAD CUTOUT TRIPPED

Overload cutouts on motor compressor assemblies are usually automatically reset and will close after cooling. In blower motors, both types of overloads are available, although the automatic reset type is predominant. This does not prevent the replacement of the motor with an automatic reset type with a motor with a manual type. This practice is not recommended, but the possibility exists.

Resetting the overload should start the motor provided that power is up to the motor. Check the voltage on the line side of the overload to make sure that power is up to the motor. Check the voltage on the load side of the overload to make sure that voltage is available through the overload. Do not check the voltage to ground. Checking the voltage to ground will not indicate the condition of the overload contacts, as it is possible to get a voltage indication to ground through the motor circuits. On circuits of 240 V or higher, always check between the hot leads. If test results show voltage on the leaving or load side of the overload, check the motor (see Section R26-6).

OVERLOAD CUTOUT B.O. (DEFECTIVE)

Overload cutouts used in most units on motor compressor assemblies and fan motors are of the automatic reset type. These overloads reset themselves after cooling. Figure R26-6 shows a typical automatic type. The overload consists of a bimetallic disk across two terminals (1 and 2) carrying the full (starting and running) current draw of the compressor motor. If the running current or the temperature of the location of the overload becomes excessive, the bimetallic disk warps and snaps the contacts open, stopping the compressor. When the disk cools, the contacts snap closed, restarting the compressor. Because this action is slow, additional protection is needed to cut the compressor off the line, in case of failure of the compressor motor to start. Therefore, an additional heater, carrying the starting current only, is inserted under the bimetallic disk, terminals 2 and 3. The additional heat supplied by the heater causes quicker action of the bimetallic element by causing it to heat up faster. This reduces the time of the unit across the line when in a stalled condition.

Repeated operation of the overload when the unit is stalled and the main overloads are oversized will cause gradual weakening of the auxiliary heater and result in burnout. This will remove the starting circuit from the compressor motor and lead to eventual failure of the motor windings. Checking the overload cutout and auxiliary heater is done by removing the overload from the circuit and checking the circuits through the overload with the ohmmeter. The circuit from 1 to 2 (the main contacts) should show less than 1 Ω on high voltage or less then $\frac{1}{2}$ Ω on low voltage. The circuit from 2 to 3 (the auxiliary heater) should show $\frac{1}{2}$ Ω on larger units to $\frac{1}{5}$ Ω for smaller units. If any high resistance is shown, the part should be replaced with another with the same part number or competitive equivalent. Overloads are not

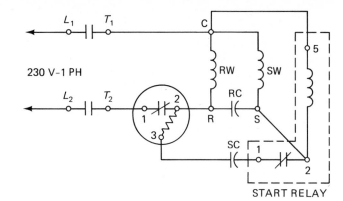

FIGURE R26-7

universal and changing types or sizes could result in extensive damage to the system.

The development of motor compressor protection has changed the type of overload protection that has been used. The original overload is the heater-assisted overload mounted on the outside of the compressor housing. This type is still used on small units under 1 hp. Figure R26-7 shows how this overload is wired into the circuit.

On large motor compressor assemblies, when changes in design resulted in high motor heat as well as smaller housings, the external overload did not react fast enough against high winding temperature. To overcome this problem, a thermostat was mounted in the motor winding inside the shell with two leads brought out of the shell. Used in conjunction with an external high current overload, Fig. R26-8 shows the wiring arrangement of overload system 2.

The external high current overload had the heater element in the common conductor to the motor so that it felt both starting and running current. The bimetallic contact circuit was in the control circuit to the motor

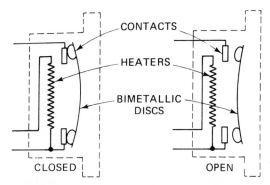

FIGURE R26-6 Thermal protector (dual-responsive) for end bracket mounting. (*Courtesy* Gould, Inc., Electric Motor Division)

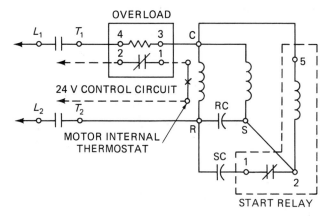

FIGURE R26-8

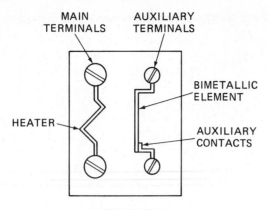

FIGURE R26-9

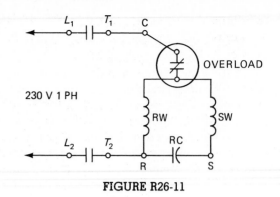

230 V 1 PH

FIGURE R26-11

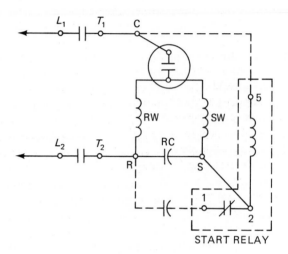

START RELAY

FIGURE R26-12

contactor. This usually is a 24-V circuit. Figure R26-9 shows the interval circuit of this overload.

To reduce the leak possibility of terminals through compressor housings and to give maximum protection from a high temperature–high current combination that the internal thermostat–external high current combination could not provide, a combination internal line break overload has been incorporated into the motor winding. Figure R26-10 shows a typical internal line break overload and the mounting means and location.

Figures R26-11 and R26-12 show the circuit arrangement when the internal line break overload is used. Figure 26.11 shows the circuit arrangement of a motor compressor assembly in a capillary tube refrigeration system. Because the suction and discharge pressure equalize on the OFF cycle, a low starting torque is required and a permanent-split capacitor (PSC) motor using only a running capacitor is satisfactory for starting and running torque requirements.

In systems using thermostatic expansion valves or any other pressure-reducing device that prevents equal-

izization of pressure during the OFF cycle, a higher starting torque is required. In these applications a capacitor start–capacitor run type of motor using a start capacitor and capacitor (start) relay must be used. Figure R26-12 shows in dashed lines the circuitry of adding a start capacitor and capacitor relay to a PSC motor of low starting torque to convert it into a CSCR motor with high starting torque. Most manufacturers supply these two parts together with the necessary wires and wiring diagram in kit form to make this conversion. A popular title for this arrangement is a "hard-start kit."

R26-7
OPEN CONTROL CONTACTS

Because of the many controls in the circuits of refrigeration and air-conditioning systems, operating controls, safety controls, defrost controls, etc., the electrical circuit may be open in more than one control. For example, it is possible to have a defective overload circuit which results in a compressor failure which results in a loss of refrigerant charge which opens a low-pressure control. When checking circuits, all control contacts should be checked.

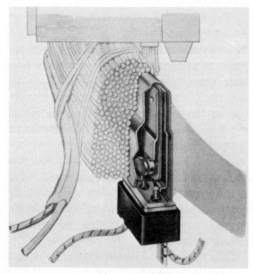

FIGURE R26-10

The best instrument to use would be the ohmmeter, checking each set of contacts for proper resistance readings. Infinite readings indicate open contact. Intermediate readings indicated burned contact. Contacts with readings above 1 Ω on high voltage or $\frac{1}{2}$ Ω on low voltage should be discarded. Parts with open contact should be checked to determine the cause of the contact opening. Low-pressure controls open on reduction in pressure from low suction pressure (see Section R25-2) refrigerant shortage (see Section R25-12), etc. High-pressure control contacts open because of unusually high head pressure (see Section R25-4). Check the purpose of each control and judge accordingly.

R26-8
LOOSE CONNECTION OR BROKEN WIRE

The hardest problem to find is a loose connection or broken wire. After checking all controls and contacts, if the circuit still does not operate, it is necessary to perform a systematic check of the entire circuit. For example, in the low-voltage control circuit, start at the low-voltage transformer and proceed as follows:

1. Measure the output of the transformer to make sure that the proper voltage is available. This measurement can be taken across the R and C terminals of the transformer or between the designated leads from the transformer body.
2. Turn off all the power to the unit.
3. Remove the lead from the R terminals of the transformer or one of the leads to the transformer. Connect one of the leads of the ohmmeter to the lead that was removed, using a pressure clip to ensure a good tight connection.
4. Using the other lead from the ohmmeter, on the medium-resistance range, touch the lead to the other side of the transformer. If there is no reading on the ohmmeter, proceed to the other end of this conductor. If a reading is obtained, the conductor is open. If there is no reading, proceed to the other side of the device to which the lead is connected. If a reading is obtained, the device is open. If no reading is present, the other end of the conductor is connected to this terminal. Repeat this checking of conductor and connected device until a reading is obtained. When the reading is obtained, the last device checked has the opening.

If the break appears to be in a conductor, a common pin inserted through the insulation will act as a probe when clamped in the ohmmeter lead clips.

Because checking circuits for breaks is the most time consuming, this procedure should be followed only after all controls, coils, contacts, etc., have been checked.

R26-9
IMPROPER WIRING

As in Section R26-8, it may be necessary to check out the wiring system of the unit. In this procedure it is necessary to have a wiring diagram of the particular unit. All manufacturers furnish wiring diagrams for their equipment. They are required by UL, AGA, and ARI to have wiring diagrams permanently secured to the unit in a place protected from the weather. *Leave the wiring diagram on the unit. Do not mutilate it.* You may need it again on future service calls.

If working on a particular type or brand of equipment, collect the same wiring diagrams in a reference binder. Most manufacturers will furnish wiring diagrams if you will furnish the model number, serial number, and bill of material number (if used) (see Chapter R20).

R26-10
IMPROPER CONTROL SETTINGS

Each control in a refrigeration or air-conditioning system has a definite function and has specific settings according to the individual unit requirement. Also, the type of refrigerant used in the unit has been taken into consideration.

Most safety controls, such as high-pressure cutouts, are factory set and are not adjustable. Usually, they are set at maximum of 150°F condensing temperature (235 lb on R-12, 380 lb on R-22, 400 lb on R-502, etc.). *Do not attempt to raise the settings of high-pressure controls*—dangerous pressures could result.

When the system is controlled by a low-pressure control, the control settings will depend on the coil temperatures that must be maintained for proper results. In specialized equipment, such as food display cases, dairy cases, etc. each manufacturer has their own desired settings. For example, Hussman Corporation's dairy, delicatessen, and produce case model BHDB8&12, if used as a dairy case, requires a suction pressure control setting of 10 to 14 psig cutout and 25 to 27 psig cut-in. If it is used as a delicatessen case, the low-pressure control settings are 8 to 12 psig cutout and 25 to 27 psig cut-in. The same unit used as a produce case requires a setting of 14 to 16 psig cutout and 38 to 40 psig cut-in. If in doubt about the settings required, *do not guess;* contact the manufacturer involved, giving them the model number, serial number, and usage involved for the particular unit. They will very gladly supply the correct information.

| | REFRIGERANT[a] | | | | | | | |
| Application | 12 | | 22 | | 502 | | 717 | |
	Out	In	Out	In	Out	In	Out	In
Ice cube maker—dry type coil	4	17	16	37	22	45	—	—
Sweet water bath—soda fountain	21	29	43	56	52	66	33	45
Beer, water, milk cooler, wet type	19	29	40	56	48	66	—	—
Ice cream trucks, hardening rooms	2	15	13	34	18	41	5	24
Eutectic plates, ice cream truck	1	4	11	16	16	22	4	8
Walk in, defrost cycle	14	34	32	64	40	75	23	55
Reach in, defrost cycle	19	36	40	68	48	78	30	57
Vegetable display; defrost cycle	13	35	30	66	38	77	—	—
Vegetable display case—open type	16	42	35	77	44	89	—	—
Beverage cooler, blower dry type	15	34	34	64	42	75	24	55
Retail florist—blower coil	28	42	55	77	65	89	44	67
Meat display case, defrost cycle	17	35	37	66	45	77	—	—
Meat display case—open type	11	27	27	53	35	63	—	—
Dairy case—open type	10	35	26	66	33	77	—	—
Frozen food—open type	7	5	4	17	8	24	—	—
Frozen food—open type—thermostat	2°F.	10°F.	—	—	—	—	—	—
Frozen food—closed type	1	8	11	22	16	29	—	—

[a]Vacuum - inches of mercury

Pressure - pounds per square inch gage

There are general rules for pressure settings that can be used as a stop gap *until the correct settings are obtained.* Figure R26-13 shows a typical approximate pressure control setting table that was published by Sporlan Valve company.

R26-11
CONTROL CONTACTS OVERLOADED

This is the most common cause of contact failure. Generally, overloading is caused by low voltage (see Section R26-13), high voltage (see Section R26-12), or the addition of auxiliary equipment to the extent of adding more electrical amperage load than the contacts will handle.

It is not advisable to add additional load to the compressor contactor, as it is sized to handle only the motor compressor assembly and the condenser fan motor or motors. If additional motors must be operated together with the condensing unit, it is better to operate the additional motors with their own contactors and overload protection. Even though only the original designed equipment is connected to the contacts, high or low voltage will cause contact failure because of the additional current draw because of the incorrect voltage. Single-phase 240-V units are designed to operate within a voltage range of 10% above or below design. Therefore, on a 240-V unit, the voltage must not drop below 216 V or rise above 264 V. Remember, this is the "operating" voltage tolerance. This does not imply normal operating life. Motors operate at their best efficiency and lowest winding temperature at design voltage. Any deviation raises winding temperatures and shortens life.

On three-phase equipment and on single-phase equipment of wide-range design voltage, the tolerance is plus 10%, minus 5%. This means that on a three-phase 208/230-V unit, the maximum voltage is 230 V plus 10% or 253 V. The minimum voltage is 208 V minus 5%, or 198 V.

A very common cause of contact arching and freezing closed is low voltage. Not only does low voltage increase the amperage draw of the unit, but it also reduces the magnetic pulling power of the coil in the contactor. As a result, the contacts do not close as fast or as cleanly as they should. The loss of pulling power also causes the contact to bounce when they pull in. The bounce produces an arc as the contacts momentarily separate. The arc softens the contact metal and when contact is made, the contacts weld together. On three-phase units, if two of the three contacts weld together, when the contactor tries to open, the compressor motor will single-phase because only one contact opens. This mechanical lockup and resulting single phasing produces a rapidly burned-up motor. Because the trouble is mechanical lockup of the contactor, the electrical protection of the unit cannot function to protect the unit from burnout. Therefore, it is very important that proper voltage be supplied to the unit at all times that the unit is trying to start as well as when running.

R26-12
VOLTAGE HIGH

At no time should the voltage supplied to a condensing unit be higher than 10% over the normal voltage rating of the unit. Therefore, on a 240-V unit, the maximum applied voltage should not exceed 264 V. Higher voltage than normal subjects the unit to higher current draw, with attendance damage to motor windings, control contacts, and relay coils, and shortens the life of running capacitors.

High-voltage problems are usually more difficult to detect. Remember that the refrigeration unit or air-conditioning unit has to operate 24 hours per day, 7 days per week. High voltage usually results during nighttime hours and/or on weekends when the commercial and/or industrial loads on the same power distribution system are at a minimum. As high voltage problems can be handled only by the electric utilities, they should be contacted when this problem occurs.

R26-13
VOLTAGE LOW

Low voltage is the most common voltage problem. Single-phase units are designed to start and run at a minimum voltage of 10% below rating. This means that on single-phase 240-V units, the minimum "applied" voltage at the unit is 216 V. On three-phase units the minimum is 5% below rated voltage. This means that the minimum "applied" voltage for a three-phase 208/230-V unit at the moment it is attempting to start is 197 V.

If the applied voltage is under these minimums, the unit either fails to start or delays long enough to blow start capacitors and start relays, burn up contacts and overloads, and burn compressor motors. Because considerable damage can be done due to low applied voltage, damage of this type is usually excluded from manufacturers' warranty coverage.

Throughout this section, the term "applied" voltage has been used. This means the voltage at the motor terminals at the time the unit attempts to start and run. Electricity is like water in that the amount of electrical volts (pressure) at the motor terminals will depend on the voltage (pressure) at the feed end of the circuit and the amount of current (gallons of water per hour) going through the conductors (pipes). Like electricity, the amount of water through a pipe (amperes) will depend on the size of the pipe (resistance) as well as the amount of pressure (volts) on the water as it enters the pipe. Therefore, to get the amount of water necessary through the pipe without too much loss in pressure, it is necessary to have adequate pipe size. Also, if the water-driven appliance requires a definite water pressure to operate, the pressure must be measured at the appliance and not where the water enters the pipe.

On an electrical unit, the applied voltage must be measured at the unit and not at the power distribution transformer where the electricity enters the circuit. The wire size from the power transformer, through all feeders and branch circuits, etc., must be adequate size to maintain the proper voltage at the unit when the unit tries to start.

If low starting voltage is due to inadequate power transformer capacity, this problem must be referred to the electric utility. This can be detected by the voltage drop measured at the transformer secondary terminals during the attempt to start. If the drop in voltage is due to inadequate wire size throughout the power distribution system, the problem should be referred to a licensed electrician.

Regardless of the situation, the condition of low voltage should not be allowed to continue or the unit permitted to operate until the situation is cleared up. To do so would only cause damage to the equipment.

R26-14
RELAY ARMATURE STICKING

Like all other facets of service work, it is not possible to say that each trouble is caused by a specific fault. Because contactors and relays contain mechanical movement, it is possible for a sticking armature or slide bar to cause burned contacts, stuck contacts, etc., with resulting damage to the unit.

If a test shows that the applied voltage is within the allowable range and the amperage load in the contacts is normal, it is possible that the mechanical drag or sticking is the trouble. In cases of this type, it is necessary to check all linkages for binding for excessive play as well as iron particles on pole faces, etc. Rarely do the mechanical parts of a relay stick, but cases have been known where washers, nuts, etc., have been dropped in the control panel when mounting in place. The fastenings, or iron chips, will stick to the armature because of the slight magnetism in the iron core of the magnet, and cause malfunction of the contact pull-in linkage. Disassembly and brushing of the parts are necessary to correct the trouble.

R26-15
SHORTED COIL

Excessively low or high voltage applied to a relay or contactor coil will cause excessive current draw the same as a motor. This excessive current draw will cause overheating of the coil, break down the insulation between winding, and result in shorts between the wire turns in the coil. Shorts between coils reduce the magnetic power of the coil as well as increase its current draw.

A noisy coil due to shorted turns will quickly develop into a burned-out coil because the current drain and heating are accumulative until burnout occurs. Therefore, coils that hum excessively should be checked and replaced before burnout occurs.

R26-16
BURNED-OUT COIL

Burned-out coils in relays or contactors are the result of adverse operating conditions. It is probable that burnout can be due to a defect in the coil itself, but this probability is remote. The principal reason for coil failure is improper voltage (see Sections R26-12 and R26-13). Coils, like other electrical appliances, are designed to operate at an applied voltage of plus or minus 10% of the design voltage. 24-V coils will operate with a maximum voltage of 26.4 V and a minimum of 21.6 V. High-voltage coils in a motor starter or solenoid valve rated at 120 V will stand a maximum of 132 V and minimum of 108 V. 240-V appliances will stand 264 V maximum and 216 V minimum.

On units with 24-V control systems, the major source of low control voltage is the overloading of the low-voltage transformer by the addition of damper motors, relays, solenoid valves, etc., in the auxiliary control system of the refrigeration or air-conditioning system. Because the control system is energized before the condensing unit starts, including zoning relays, damper motors, solenoid valves, etc., by the time the coil in the compressor contactor is energized, the low-voltage transformer is overloaded and its output voltage drops below minimum limits. This causes the compressor contactor armature to pull at a slower rate than normal, reduces the holding-in power of the armature, which permits the contacts to bounce and arc, resulting in burning and sticking, and causes the coil to draw additional current, which eventually burns out the coil.

On all control systems, check to make sure that the control circuit voltage stays within design limits with all the connected load operating. If control voltages do not remain within limits, the control system must be split with relays, additional transformer power must be phased in, or the initial low-voltage transformer must be replaced with one of larger VA capacity. (See the electrical wiring design for the particular unit.)

R26-17
CONDENSER FAN OVERLOADED

Cutout of the condensing unit on the high-pressure cutout is always due to excessive head pressure. If all other sources of high head pressure (see Section R25-4) have been checked and found satisfactory, it is well to check the current draw of the condenser fan motor.

On propeller-type condensing units, the addition of airflow resistance over and above that built into the unit will cause the fan motor to draw considerably more current and cut out on the automatic reset overload in the motor. Lack of condenser air will cause a rapid rise in unit head pressure and cutout of the unit on the manual reset high pressure cutout. Cooling the fan motor will bring the condenser fan on, but the condensing unit will remain off. Improper settings of air speed on blower-type condensing units or other troubles inherent with belt-driven blowers, such as belts too tight, wrong type of lubrication oil causing tight bearings, etc., will also cause excessive blower motor current draw, which produces the same reaction. On those units that seem to cut off on high head pressure for no apparent reason, check the current draw of the blower or fan motor to ensure that it is under the maximum nameplate rating.

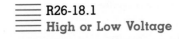

R26-18
CONDENSER FAN
OR BLOWER MOTOR B.O.

Burnout of condenser fan or blower motors can be caused by any of the following:

1. High voltage (see Section R26-12)
2. Low voltage (see Section R26-13)
3. Overloading
4. Improper lubrication

R26-18.1
High or Low Voltage

Like all other electrical appliances, condenser fan and blower motors are designed to operate with a voltage tolerance of plus or minus 10% of rated voltage. 240-V motors will operate at applied voltages up to 264 V and down to 216 V. At higher or lower voltages, excessive current draw will cause motor burn-out.

Practically all condenser fan and blower motors

are protected by automatic reset overload cutouts. These cutouts must be sized, however, with operating tolerances to prevent nuisance cutoffs when the voltage takes momentary changes; therefore, there is a range of motor load between normal and cutoff that will overheat and burn the motor if the load extends over a period of time. For best performance of the unit, make sure that the voltage applied to the motor is in the correct range.

R26-18.2
Overloading

Overloading the condenser fan motor is the most common cause of burnout. On propeller-fan-type condensing units, any increase in the resistance to the airflow will cause an increase in current draw of the motor. Therefore, it is necessary that no additional resistance be added to the unit in the form of air ducts, louvers, grills, etc. If these are added to act as head pressure control, the fan motor may have to be resized to handle the increased resistance. Also, it is very important that condenser surfaces be cleaned regularly for complete air passage. Allowances are designed into the unit when selecting motor sizes, etc., for normal accumulation of condenser dirt. Accumulation of leaves, paper, grass, and other debris has not been allowed for, however, and should be removed regularly and at intervals depending on the rate of accumulating.

Blower-type condensing units using centrifugal-type blowers react differently to increases in a resistance. As the resistance increases, the air quantity handled decreases. Because the load on the motor decreases as the air quantity handled decreases, the current draw of the motor decreases. Therefore, the danger of motor burnout on centrifugal-type units due to condenser plugging does not exist.

This does not eliminate the possibility of overloading of the motor, as all these units have an adjustable drive to provide the current amount of air through the condenser with different static resistances. Because it is possible to add ducts to the unit to handle condenser air, it must be possible to adjust blower speed accordingly. As the duct resistance is increased, the blower speed must be increased accordingly.

If the resistance in the condenser air-handling system is changed, such as removal of ductwork, louvers, grills, access panels, etc., it can be expected that the motor will overload. Therefore, all cabinet parts and panels, as well as duct access doors, etc., must be in place when the unit is in operation.

R26-18.3
Lubrication

Electric motors that require periodic lubrication should be checked and oiled with care and common sense. Because $\frac{1}{2}$ teaspoon of oil once a year is sufficient for lubrication for most standard motors does not mean that more than $\frac{1}{2}$ teaspoon of oil once a year or more than once a year is better. *More motors are damaged from over-oiling than for any other reason.*

When a motor is over-oiled, the oil only runs out of the bearing into the inside of the housing and is sprayed off the rotor throughout the interior of the motor. This coats the windings, terminals, and starting switch, causing gumming of moving parts, carboning of points, and contact and motor failure. It is better that the motor be oiled every 2 years than more often than required.

The type of oil is also very important. The oil should be heavy enough to provide proper lubrication but must not contain soaps, detergents, or any other additives. Three-in-One, Finol, or any other rust-preventive oil should never be used in blower or fan motors. Also, ordinary motor oil should never be used, regardless of the make or grade. Use electric motor oil or refined mineral oil of a weight or grade specified on the motor. Generally, this is a No. 10 weight.

R26-19
START RELAY B.O. OR WRONG ONE

Starting relays used in refrigeration and air-conditioning units are of two types: current type and potential type.

R26-19.1
Current-Type Relay

The method of operation of the current type is the same as that of any other relay. This relay is composed of a magnetic coil winding and a set of contacts to control the starting winding circuit of the motor. Figure R26-14 shows a typical cutaway of a current-type starting relay. The magnetic field produced by the coil pulls against a spring-loaded armature to which the contact points are connected. The contacts are normally open type—open when the coil is denergized. The magnetic coil is also a parasite-load type connected so as to use a small portion of energy supplied to the motor to operate the relay. To understand its function, a review of the starting function of the motor is in order.

When an electric motor is connected across a power source, only the straight resistance of the winding limits the amount of current through the motor. As the motor starts, this resistance climbs rapidly due to counter voltage generated in the motors. Therefore the initial amount of current, called *inrush current* or *locked rotor current* (LRA) can be two or four times as high as the normal running current. Figure R26-15 shows this amount of current draw in an electric motor from the

FIGURE R26-14

instant voltage if applied until the motor reaches full-load speed 1.5 seconds later. As you will note, the inrush current is 21.6 A and the normal running current is 7.2 A, three times as much.

By passing all of the current the motor uses through the magnetic coil of the relay, the relay will develop considerably more magnetic pull on the inrush current than on the normal running current. By having the armature in the relay spring loaded, the spring tension can be set so as to open the cutout when the amount of current reaches 80% of full-load spread. Thus, the starting winding is cut out at the required speed and the motor runs normally.

Because the contacts are normally open and the relay is usually of a sealed type, it is not possible to check the conditions of the contacts with an ohmmeter. Therefore, the only test is by substitution with a new part.

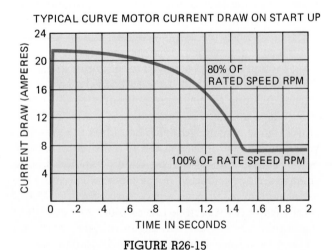

TYPICAL CURVE MOTOR CURRENT DRAW ON START UP

FIGURE R26-15

However, the new part must be the exact equivalent of the old part. Because the inrush versus full-load current ratio is different for different types of motors, an exact replacement must be used to assure proper starting of the motor.

R26-19.2
Potential-Type Relay

The start relay used in large-horsepower single-phase units must be wound with such a large wire if current type relays were used that the overall size of the coil would be impractical. Also, the fact that capacitor start–capacitor run motors used in the units generate higher voltages across the winding in series with the run capacitor permits the use of a high-voltage coil-type relay called a "potential" type. This relay, shown in Fig. R26-16, utilizes a coil of a great number of turns capable of handling voltages as high as 600 V depending on the motor it is designed to control. It also has a set of single pole–single throw normally closed contacts connected in the starting capacitor circuit. Thus, when the motor stands idle, the contacts are closed, and the starting capacitors are in the circuit ready to start the motor whenever voltage is applied to the circuit.

As the motor comes up to speed, the voltage across the auxilary or start winding in the motor builds up over and above the applied voltage. When the motor reaches approximately 80% of normal running speed, this voltage buildup reaches the actuative voltage of the relay and the relay pulls the starting contacts open. The starting capacitor is removed from the circuit and the motor then runs as a capacitor run motor.

FIGURE R26-16

Figure R26-17 shows a voltage curve of the voltage developed across the auxilary winding as the motor reaches full-load speed. Remember—the voltage across the coil is always considerably higher than the applied voltage. *Never place your hands or a tool around the coil or relay terminals when the unit is operating.* Because of mechanical delay time from the time the coil is activated to pull the contacts open and the actual opening of the contacts, the coils are usually sized to pull the contacts open before the desired speed point. The mechanical delay then furnishes the additional time to allow the motor to reach the 80% of normal running speed before the actual opening of the contacts.

If the relay is changed to one having a higher operating voltage of the coil or longer contact opening time, the motor will run up to too high a speed before the contacts open. This will cause a very noisy motor, possible motor failure, as well as rupture of the start capacitor and auxilary winding burnout.

If the relay is changed to one having a lower operating voltage, the relay contacts will open before the motor reaches the proper change over speed, and the motor will stall and drop back in speed; this will cause a very rapid cycling of the motor speed, rapid opening and closing of the relay contacts—burning them out—and burnout of the start capacitors. Therefore, never replace a relay with one of different characteristics. Use factory replacement or parts specifically designed to replace the defective part.

Because the actuation voltage is made up of the voltage applied to the unit (240 V) plus the generated voltage (100 to 300 V, depending on the unit motor), the actuation voltage for the coil can be anywhere from 330 to 530 V. Testing a coil is best done with an ohmmeter after the relay has been removed from the unit. The coil will be either OK, open, or shorted. If the coil is open, the ohmmeter needle will indicate infinite resistance (no movement of the needle). If the coil is shorted, the ohmmeter needle will move to the zero-resistance point. A good coil will indicate a resistance of 50 to 10,000 Ω, depending on the horsepower of the motor compressor assembly in the unit. Make sure that the proper terminals are used for checking coil resistances. These are usually terminals 2 and 5. The relay may not be numbered using the industry standard, so 2 and 5 may not apply.

At the time the coil resistance is checked it is wise to check the contact resistance. A contact resistance of 1 Ω or less is desired. Any higher resistance indicates burned contacts and the relay should be replaced. A reading of infinite resistance (no movement of the ohmmeter needle) indicates a broken contact lead, burned open contact, or stuck armature, and the relay must be replaced. *Under no circumstances should the relay be repaired or adjusted, as the activation voltages are critical.* Changing the settings of the relay or use of a relay other than the original part number or certified part suitable for substitution will only lead to failure of other electrical parts and eventual burnout of the compressor motor.

R26-20
START CAPACITOR B.O. OR WRONG PART

Loss of the action of the start capacitors in the starting circuit of CSCR compressor motors results with loss of the electrical difference between the starting and running windings of the motor. As a result, the motor will not produce the desired rotation in order to run; the current draw remains high because of the lack of counter voltage produced by the motor rotation.

Continued draw of this high current quickly causes the compressor overload to open or the main overload cutouts to open. Without these devices, the compressor motor would quickly overheat and burn out. Because of the type of construction of the start capacitor and because it is designed to remain in the circuit for a very short period of time, if the unit short cycles or stalls or if the start relay cutouts refuse to open, the start capacitor quickly heats up and ruptures. Theoretically, the pressure relief vent in the top of the capacitor is supposed to rupture and relieve the steam pressure developed. Occasionally, the capacitor shell will explode, and if the leads ground out, the starting winding in the compressor will burn.

Capacitors will seldom rupture because of defect. Invariably, rupture is the result of malfunction of some other electrical part. Therefore, if start capacitor trouble is encountered, look for trouble in other parts of the electrical system such as high voltage (Section R26-12, low voltage (Section R26-13), running capacitor (Section R26-21), starting relay (Section R26-19), or the motor compressor assembly (Section R25-22). There is also

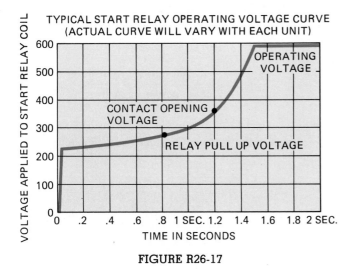

TYPICAL START RELAY OPERATING VOLTAGE CURVE
(ACTUAL CURVE WILL VARY WITH EACH UNIT)

FIGURE R26-17

the possibility that the wrong part has been used which does not have the proper microfarad capacity or is designed for a working voltage other than that required.

All capacitors are rated on a maximum working voltage basis as well as on a microfarad basis. Replacement capacitors should be of the same microfarad rating (with plus or minus 10%) as well as equal or up to 10% high working voltage rating for proper operation.

Capacitors, both starting and running type, can be tested very easily by means of an ohmmeter. Capacitors are in reality electrical storage tanks and it is only necessary to measure the ability of the capacitor to receive the store electrical energy. Therefore, the power supply of the ohmmeter can be used as the electrical energy source and the ohmmeter the indicator of the flow rate.

A good capacitor will receive electrical energy as fast as the power source will give it out until the capacitor is full and will not take any more. Therefore, when an ohmmeter is connected across the capacitor, the electrical flow rate will be high enough for the ohmmeter to indicate a low resistance; as the capacitor fills, however, the ability of the capacitor to take electrical energy decreases and the current flow decreases. This causes the ohmmeter to show an increase in resistance because of the decrease in current flow. The decrease in current flow will continue until the capacitor is filled, at which point the current flow stops and the needle of the ohmmeter indicates the infinite resistance. When using the ohmmeter to check capacitors, be sure to leave the ohmmeter connected for a sufficient length of time to fill the capacitor. On small capacitors, the action is very rapid, but on large capacitors this may take 10 seconds or more. If the test results show no motion of the needle on the ohmmeter from the zero point after connecting the leads, regardless of the time invoked, the capacitor is shorted. If the needle does not leave the infinite resistance position on the ohmmeter, the capacitor has an open internal circuit.

A starting capacitor should have a drain resistor across the terminals. This resistor is used to drain off any charge in the capacitor after the starting contacts in the starting relay open upon startup of the compressor motor. This drain-off of the capacitor charge eliminates burning of the starting relay contacts when the unit stops and the contacts close and cut the capacitor back into the circuit. The capacitor shorting resistor is usually of a $\frac{1}{2}$ W, 15,000 to 50,000 Ω rating. If the starting capacitor does not have a resistor on it, one should be installed to reduce possibility of future failure.

Remember, for accurate testing of the capacitor, one lead of the resistor should be disconnected from the capacitor lead so as to be able to check the capacitor with only the capacitor in the circuit. This test will check the capacitor only on a good/no good basis. It will not determine the capacity. The best way to determine the capacity is by the use of a capacitor tester, as shown in

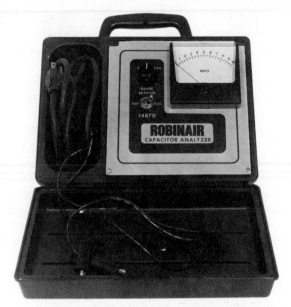

FIGURE R26-18 (*Courtesy* Robinair Manufacturing Corporation)

Fig. R26-18. The test instrument will accurately measure the capacity and power factor from 0 to 1000 μF capacity. As long as the test results show a capacitor capacity within plus or minus 10% of the capacitor rating, the capacitor may be used. If a wider range is indicated, the capacitor should be discarded.

In the absence of a capacitor tester, the use of a voltmeter and ammeter can be used to determine the capacity. Figure R26-19 shows the hook-up of the voltmeter and ampmeter to the capacitor using a 120-V power source. *Remember—This test is done only after determining, by means of the ohmmeter, if the capacitor is not open or shorted.* Connecting an ammeter in series with a shorted capacitor could mean purchasing another meter.

After connecting the meters and capacitor as shown in Fig. R26-19, the voltage across the capacitor and the amperage draw of the capacitor is read and recorded. These readings, used in the following formula, will result in the capacity rating of the capacitor:

$$\mu F = \frac{\text{amperage} \times 2650}{\text{voltage across capacitor}}$$

For example, if 120 V is applied across the capacitor and the capacitor draws 5 A, the microforad capacity is

$$\frac{5 \times 2650}{120} = 110.41666 \ \mu F$$

The capacitor is "tested" at 110 μF. If the rating is 100 to 120 μF, the capacitor is usable. If higher than 120 μF or lower than 100 μF, the capacitor should be discarded.

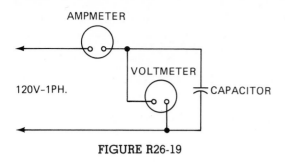

FIGURE R26-19

R26-21
RUN CAPACITOR B.O. OR WRONG ONE

Running capacitors have two purposes: to supply electrical difference between the starting and running windings or start of the motor as well as to use the start or auxilary winding as a second winding while the unit is running. Use of the run capacitor and auxilary winding for power creation while the motor is running provides additional power and reduces the amperage draw of the motor.

If the motor is a CSCR type, the running capacitor does not supply enough electrical difference between the windings for starting the unit, so a start capacitor and start relay are needed. If the running capacitor is open or out of the circuit, the total capacity available for starting the compressor is reduced and the motor will stall if attempting to start against a difference in suction and head pressure. If the pressures are balanced, the motor may start, but the high current draw will cause it to cut out on overload.

On PSC compressor motors where only the running capacitor is used to start and run the motor, the capacitor has been sized on a compromise basis between starting and running requirements, and therefore a start capacitor and start relay are not required to start the motor.

Loss of the capacitor from the circuit will cause the motor to fail to start if the capacitor is open, or start and run with extremely high current draw if the capacitor is shorted. Practically all running capacitors have internal fuses so that if a short develops in the capacitor, the fuse will blow to protect the compressor motors windings.

The main point to understand here is *that running capacitors are as necessary to start the compressor motor as are starting capacitors. Without the correct amount of capacity in the circuit, faulty starting and running will result.*

It is absolutely necessary that capacitors with the proper microfarad rating and working voltage be used in the unit. Never replace a running capacitor with capacitors of different microfarad capacity, lower working voltage, or different type.

Running capacitors are checked in the same manner as are starting capacitors (see Section R26-20). The difference in internal circuiting between running and starting capacitors is that the run capacitor has an internal fuse. If the common practice of shorting the capacitor with a screwdriver or jumpwire is done, the internal fuse could blow because of the high current surge and the capacitor is B.O. To safely discharge a capacitor, use a bleed resistor such as that found on start capacitors, a $\frac{1}{2}$-W 15,000-Ω resister with two insulated leads attached.

PROBLEMS

R26-1. The voltage tolerance for single-phase refrigeration systems is +_____ % and −_____ %.

R26-2. The voltage tolerance for three-phase refrigeration systems is +_____ and −_____ %.

R26-3. On three-phase units the voltage on all three pairs of legs of the power source must be within _____ % of each other.

R26-4. The resistance limit for closed contacts is _____ Ω for high-voltage contacts and _____ Ω for low-voltage contacts.

R26-5. The fuse or circuit breaker for motors must be of what type?

R26-6. What would be a possible cause of fuse failure even when the connected load never exceeds the fuse rating?

R26-7. Three different automatic reset overload arrangements have been used on motor compressor assemblies. What are they?

R26-8. A "hard-start kit" consists of what two parts?

R26-9. What is the high and low input voltage tolerance of a 240-V one-phase unit?

R26-10. What is the high and low input voltage tolerance of a 208/230-V three-phase unit?

R26-11. Where should the voltage be measured when starting and running a unit?

R26-12. What is the most common cause of relay coil noise and burnout?

R26-13. Will increasing the resistance against the propeller fan increase or decrease the current draw of the motor?

R26-14. Will increasing the resistance against a blower increase or decrease the current draw of the motor?

R26-15. A duct system can be used with a unit using a propeller fan. True or False?

R26-16. Automobile oil can satisfactorily be used to oil electric motors. True or False?

R26-17. Starting relays used in refrigeration and air-conditioning units are of what two types?

R26-18. The operating voltage of a current-type relay is in the range _____ to _____ .

R26-19. Starting relays are easily adjusted. True or False?

R26-20. The replacement tolerance of the microfarad rating of a capacitor is + _____ % to − _____ %.

R26-21. The replacement tolerance of the working voltage of a capacitor is + _____ % to − _____ %.

R26-22. When using an ammeter and a voltmeter to determine the capacity of a capacitor, what formula would be used?

R26-23. All running capacitors have internal fuses. True or False?

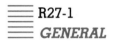

Refrigeration Load

GENERAL

The total refrigeration load of the system as expressed in Btu/hr comes from many heat sources. Figure R27-1 represents a cutaway view of a refrigerated storage room in a supermarket. Note the sources of heat caused by:

1. Heat transmission
   a. The temperature difference of 60°F between the 95°F outside air and the room temperature of 35°F, which can cause much heat conduction.
   b. The sun effect on the roof and walls results in radiation heat buildup.
2. Air infiltration
   a. Air that enters a room as a result of opening and closing the doors during normal working procedures.
   b. Air that enters a room through cracks in the construction or around door seals.
   c. Air that may be purposely introduced for ventilation reasons.
3. Product loads, which are the results of heat(s) contained within the product being stored. In some cases it is purely dry or sensible heat, such as cooling a can of juice from room temperature down to 35°F, or it may be a combination of dry heat (sensible) and moisture (latent heat) from produce; should the product be frozen, there are additional requirements connected with the latent heat of freezing. Some heat is also the result of chemical changes such as the ripening of fruit.
4. Supplementary loads are caused by such things as electric lights, motors, tools, and also arises from human beings.

Although the refrigeration design engineers are basically responsible for estimating the loads and for the construction planning of the installation and application of equipment, refrigeration technicians should understand how these heat sources affect the operation of the system so that they can adjust the equipment to perform in a manner consistent with the design engineer's recommendations.

R27-2
HEAT TRANSMISSION

The heat gain through walls, floors, and ceilings will vary with the type of construction, the area that is exposed to a different temperature, the type and thickness of insulation, and the temperature difference between the refrigerated space and the ambient air.

Thermal conductivity varies directly with time, area, and temperature difference and is expressed in Btu/hr, per square foot of area, per degrees Fahrenheit of temperature difference, per inch of thickness.

It is readily apparent that in order to reduce heat transfer, the thermal conductivity factor (based on material composition) should be as small as possible and the material as thick as economically feasible.

Heat transfer through any material is also subject to or affected by the surface resistance to heat flow, which is determined by the type of surface (rough or smooth); its position (vertical or horizontal); its reflective properties; and the rate of airflow over the surface.

Extensive testing has been done by many laboratories to determine accurate values for heat transfer through all common building and structural material. Certain materials (like insulation) have a high resistance to flow of heat; others are not so good.

To simplify the task of calculating heat loss the industry has developed a measuring term called *resistance*

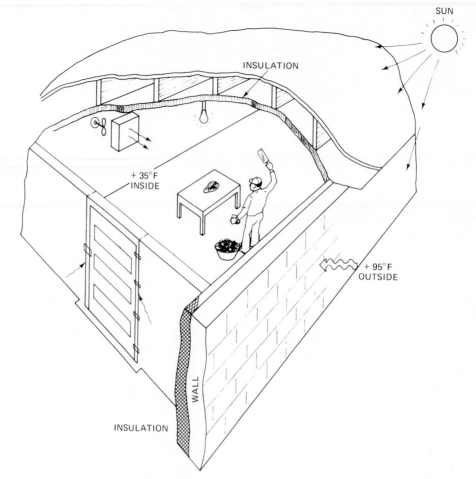

FIGURE R27-1   Refrigerated storeroom.

(R), which is the resistance to heat flow either of 1 in. of material, a specified thickness, an air space, an air film, or an entire assembly. Its value is expressed as degrees Fahrenheit temperature difference per Btu per hour per square foot. A high R value indicates low heat-flow rates. The resistances of several components of a wall may be added together to obtain the total resistance:

$$RT = R_1 + R_2 + R_3 + \ldots$$

Figure R27-2 lists some R values for common building materials in order to illustrate the differences in heat-flow characteristics. For more extensive listings of R values, refer to the ACCA Manual J.

The actual quantity of heat transmission (Q) through a substance or material is then calculated by the formula

$$Q = U \times A \times TD$$

where

Q = heat transfer, Btu/hr

U = overall heat transfer coefficient, (Btu/hr)(ft²)(°F ΔT)

$U = 1/R_t$ value   (if an assembly $R_t = R_1 + R_2$ $+ R_3$ for various components)

A = area, ft²

TD = temperature difference between inside and outside design temperature and refrigerated space design temperature

For example, calculate the heat flow through an 8-in. concrete block (cinder) wall (Fig. R27-3), 100 ft² in area, having a 60°F temperature difference between the inside and the outside.

R value for an 8-in. concrete block wall (see Table 21A) is 1.72.

U therefore is = 1/R = 1/1.72
                = 0.58 Btu/hr/°F/ft².

$Q = 0.58 \times 100$ ft² $\times 60°F = 3480$ Btu/hr heat flow into the space.

FIGURE R27-2
Typical heat transmission coefficients.

Material	Density (lb/ft³)	Mean Temp. (°F)	Conductivity, k	Conductance, C	Resistance, R Per inch	Resistance, R Overall
Insulating materials						
Mineral wool blanket	0.5	75	0.32		3.12	
Fiberglass blanket	0.5	75	0.32		3.12	
Corkboard	6.5−8.0	0	0.25		4.0	
Glass fiberboard	9.5−11.0	−16	0.21		4.76	
Expanded urethane, R11		0	0.17		5.88	
Expanded polystyrene	1.0	0	0.24		4.17	
Mineral wool board	15.0	0	0.25		4.0	
Insulating roof deck, 2 in.		75		0.18		5.56
Mineral wool, loose fill	2.0−5.0	0	0.23		4.35	
Perlite, expanded	5.0−8.0	0	0.32		3.12	
Masonry materials						
Concrete, sand and gravel	140		12.0		0.08	
Brick, common	120	75	5.0		0.20	
Brick, face	130	75	9.0		0.11	
Hollow tile, two-cell, 6 in.		75		0.66		1.52
Concrete block, sand and gravel, 8 in.		75		0.90		1.11
Concrete block, cinder, 8 in.		75		0.58		1.72
Gypsum plaster, sand	105	75	5.6		0.18	

Now add 6 in. of fiberglass insulation to the wall (Fig. R27-4) and recalculate the transmission load.

U factor $= 1/R_t$

$R_t = R_1$ (concrete block) $+ R_2$ (6-in. insulation)

$= 1.72 + 18.72$ (R value for 1-in. insulation $= 3.12$) (R_2 then equals 6×3.12 or 18.72)

$= 20.44$

Therefore,

$U = 1/20.44 = 0.049$

$Q = 0.049 \times 100\ \text{ft}^2 \times 60°\text{F} = 294\ \text{Btu/hr}$

The example above demonstrates the cumulative effect of R values on determining the total wall resistance, but it also dramatically shows the heat reduction that can be achieved through proper insulation, in this case, from 3480 Btu/hr down to 294 Btu/hr, which would drastically reduce the size of the refrigeration equipment needed and the resulting operating cost.

Insulation is the most effective method of reducing heat transmission. There are types made of material to meet the requirements of every application, although some are better than others. General classifications of the available forms of insulation are (1) loose fill, (2) flexible, (3) rigid or semirigid, (4) reflective, and (5) foamed-in-place.

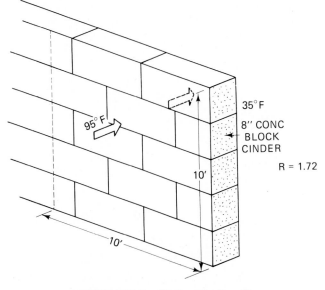

FIGURE R27-3 Cinder block wall.

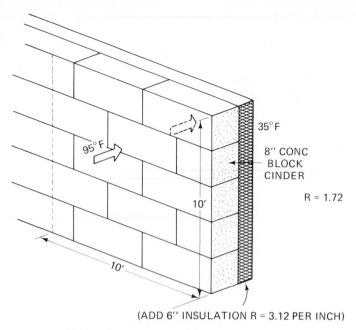

(a)

(b)

FIGURE R27-6   Batt or roll type insulation.

FIGURE R27-4   Cinder block wall with insulation.

Loose fill or "blown" insulation (Fig. R27-5) is used primarily in residential structures for reducing heat flow. Flexible insulations such as fiberglass, in batts or roll type (Fig. R27-6), are also common in new residential and commercial construction and come with a material such as kraft paper to act as a vapor barrier. Some are available with reflective surfaces to help reduce radiant-heat effects.

Rigid and semirigid insulations (Fig. R27-7) are made of such materials as corkboard, polystyrene, foam glass, and polyurethane, which are manufactured in various convenient dimensions and forms, such as boards, sheets, or blocks. Some have a degree of structural strength; others do not. It is in this category that we find

more widespread application to commercial refrigeration: walk-in coolers, freezers, display cases, etc. Because of their density and cell composition, they offer a built-in vapor barrier to moisture penetration.

Foamed-in-place insulation (Fig. R27-8) is widely used for filling cavities that are hard to insulate and also to cover drain pans, etc., where effective temperature control and water seal are needed. Foamed insulation is also used in on-site, built-up refrigerated rooms in connection with the rigid insulations mentioned above.

Regardless of what type of insulation is used, moisture control is most important. Figure R27-9 represents the gradual temperature changes within an insulating material from 90°F on the warm outside side to 40°F on the cold inside. At 90°F the warm-air side

FIGURE R27-5   Loose fill insulation.

FIGURE R27-7 Installing rigid insulation board.

will have a dew-point temperature of 83° (*dew-point temperature* is the temperature where condensation from the vapor to a liquid occurs). As illustrated, when there is no effective vapor seal (barrier) on the warm side, water will start to condense inside the insulation. Water is a good conductor of heat, about 15 times as fast as fiberglass. Thus if water gets into the insulation its insulating value is greatly reduced, not to mention the physical problems it causes in the construction.

Therefore, insulation must remain dry when first installed and be sealed perfectly so that it stays dry.

Vapor seals can be formed from various materials: metal casings, metal foil, plastic film, asphalt coverings, etc. Some are more effective than others and the selection will depend on the application. The ability of a material to resist water vapor transmission is measured in *perms*, a term related to permeability. Rating tables for various materials are available from industry sources. In general, vapor barriers of 1 perm or less have been found satisfactory for residential comfort heating and cooling work. But in low-temperature commercial refrigeration applications such as freezers, perm ratings of 0.10 and below may be needed. As with insulation and heat flow, the resistance to vapor flow is a function of the composite of all the materials, as constructed, not just the rating of the vapor barrier itself.

The effectiveness of both insulation and vapor barriers will be greatly reduced if any openings, however small, exist. Such openings may be caused by poor workmanship during construction and application, but they may also result from negligence in sealing around openings for refrigerant lines, drain lines, electrical wiring, etc., all of which are part of the refrigeration technician's responsibility.

FIGURE R27-8 Application of foam-type insulation.

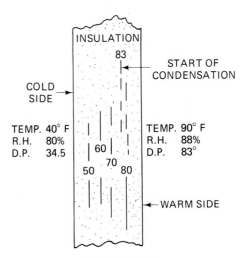

FIGURE R27-9

R27-3
SUN EFFECT

The primary radiation factor involved in the refrigeration load is the heat gain from the sun's rays. If the walls of the refrigerated space are exposed to the sun, additional heat will be added to the heat load. For ease in calculation or estimating the load, an allowance can be made by increasing the temperature differential by the factors in Figure R27-10 which are the degrees Fahrenheit to be added to the normal temperature difference between indoor and outdoor design conditions.

FIGURE R27-10
Allowance for sun effect.

Type of Structure	East Wall	South Wall	West Wall	Flat Roof
Dark-colored surfaces Slate, tar, black asphalt shingles, and black paint	8	5	8	20
Medium-colored surfaces Brick, red tile, unpainted wood, dark cement	6	4	6	15
Light-colored surfaces White stone, light-colored cement, white asphalt shingles, white paint	4	2	4	9

R27-4
DESIGN TEMPERATURES

Recommended outside design conditions are the results of extensive studies by the National Weather Service. For air-conditioning and refrigeration applications the maximum load occurs during the hottest weather. However, it is neither economical nor practical to design equipment for the hottest temperature that might possibly occur, since the peak temperature might last for only a few hours over a span of several years. Therefore, the design temperature chosen is less than that. Figure R27-11 is a segment of the ACCA outdoor design data chart, which lists the recommended design dry-bulb and wet-bulb temperatures for states and major cities.

R27-5
AIR INFILTRATION

Any outside air entering the refrigerated space must be reduced to the storage temperature, thus increasing the refrigeration load. In addition, if the moisture content of the entering air is above that of the refrigerated space, the excess moisture will condense out of the air, and the latent heat of condensation will add to the refrigeration load.

Because of the many variables involved, it is difficult to calculate the additional heat gain due to air infiltration. The traffic in and out of a refrigerator usually varies with its size and volume. Therefore, the number of times doors are opened will be related to the volume rather than the number of doors.

Some engineers use the air-change method of estimating infiltration; this method is based on the average number of air changes in a 24-hour period compared to the refrigerator volume, as illustrated in Fig. R27-12. Note that this is for rooms above 32°F average use. Heavier use would require increasing the values by 2. For storage at 0°F or below the usage will normally be less, and the values are reduced.

Another means of computing infiltration is by means of the velocity of airflow through an open door. Charts are available that list average infiltration velocity depending on door height and temperature difference. If the average time the door is opened each hour can be determined, the average hourly infiltration can be calculated. Once the rate of infiltration in cubic feet per hour has been determined by either method, the heat load can be calculated from the heat gain per cubic foot, given in the chart in Fig. R21-11.

Assume that the volume of a refrigerated room is 1000 ft³ and the storage temperature is 40°F, with an outside temperature of 95°F and 60% relative humidity. From Fig. R27-12 note that a 1000-ft³ volume would average 17.5 air changes per 24 hours, or an infiltration of 17,500 ft³ in 24 hours (1000 × 17.5). From Fig. R27-13 note that for a room at 40°F with 95°F outside temperature and 60% relative humidity the Btu/ft³ of heat is 2.62. Therefore, in 24 hours the load is 2.62 × 17,500, or 45,850 Btu. In 1 hour it is 45,850 divided by 24, or 1910 Btu/h.

If in some systems positive ventilation is provided by means of supply or exhaust fans, the ventilation load will replace the infiltration load (if greater), and the heat gain may be calculated on the basis of the ventilating equipment air volume.

R27-6
PRODUCT LOAD

The product load is any heat gain due to the product in the refrigerated space. The load may be the result of a product coming into the refrigerator at a higher temperature than that of the storage area, from a chilling or freezing process, or from the heat of respiration of perishable products. The total product load is the sum of the various types of product load of the particular application.

Summer outdoor design conditions for the United States and Canada.

Location	2½% Design Dry Bulb	Coincident Design Wet Bulb
ALABAMA		
Birmingham AP	94	75
Mobile AP	93	77
Montgomery AP	95	76
ALASKA		
Anchorage AP	68	58
Barrow (S)	53	50
Juneau AP	70	58
ARIZONA		
Flagstaff AP	82	55
Phoenix AP (S)	107	71
Tucson AP (S)	102	66
ARKANSAS		
El Dorado AP	96	76
Fayetteville AP	94	73
Little Rock AP (S)	96	77
CALIFORNIA		
Bakersfield AP	101	69
Eureka/Arcata AP	65	59
Los Angeles CO (S)	89	70
Sacramento AP	98	70
San Bernadino, Norton AFB	99	69
San Diego, AP	80	69
San Francisco CO	71	62
San Jose AP	81	65
COLORADO		
Denver AP	91	59
Durango	87	59
Fort Collins	91	59
CONNECTICUT		
Hartford, Brainard Field	88	73
New Haven AP	84	73
DELAWARE		
Wilmington AP	89	74
DISTRICT OF COLUMBIA		
Washington National AP	91	74
FLORIDA		
Gainesville AP (S)	93	77
Jacksonville AP	94	77
Key West AP	90	78
Miami AP (S)	90	77
Orlando AP	93	76
Tallahassee AP (S)	92	76
Tampa AP (S)	91	77

Location	2½% Design Dry Bulb	Coincident Design Wet Bulb
GEORGIA		
Atlanta AP (S)	92	74
Savannah-Travis AP	93	77
Valdosta-Moody AFB	94	77
HAWAII		
Hilo AP (S)	83	72
Honolulu AP	86	73
IDAHO		
Boise AP (S)	94	64
Coeur D'Alene AP	86	61
ILLINOIS		
Carbondale	93	77
Chicago CO	91	74
Springfield AP	92	74
INDIANA		
Evansville AP	93	75
Indianapolis AP (S)	90	74
South Bend AP	89	73
IOWA		
Burlington AP	91	75
Des Moines AP	91	74
Mason City AP	88	74
KANSAS		
Salina	100	74
Topeka AP	96	75
Wichita AP	98	73
KENTUCKY		
Bowling Green AP	92	75
Lexington AP (S)	91	73
Louisville AP	93	74
LOUISIANA		
Baton Rouge AP	93	77
New Orleans AP	92	78
Shreveport AP (S)	96	76
MAINE		
Bangor, Dow AFB	83	68
Caribou AP (S)	81	67
Portland AP (S)	84	71
MARYLAND		
Baltimore CO	89	76
Cumberland	89	74
Salisbury (S)	91	75
MASSACHUSETTS		
Boston AP (S)	88	71
New Bedford	82	71

Location	2½% Design Dry Bulb	Coincident Design Wet Bulb
Springfield, Westover AFB	87	71
Worcester AP	84	70
MICHIGAN		
Detroit	88	72
Grand Rapids AP	88	72
Marquette CO	81	69
MINNESOTA		
International Falls AP	83	68
Minneapolis/St. Paul AP	89	73
Rochester AP	87	72
MISSISSIPPI		
Biloxi, Keesler AFB	92	79
Jackson AP	95	76
Tupelo	94	77
MISSOURI		
Kansas City AP	96	74
St. Louis CO	94	75
Springfield AP	93	74
MONTANA		
Billings AP	91	64
Butte AP	83	56
Great Falls AP (S)	88	60
NEBRASKA		
Lincoln CO (S)	95	74
Omaha AP	91	75
Scottsbluff AP	92	65
NEVADA		
Las Vegas AP (S)	106	65
Reno CO	93	60
NEW HAMPSHIRE		
Berlin	84	69
Concord AP	87	70
Manchester, Grenier AFB	88	71
NEW JERSEY		
Atlantic City CO	89	74
Newark AP	91	73
Trenton CO	88	74
NEW MEXICO		
Albuquerque AP (S)	94	61
Las Cruces	96	64
Santa Fe CO	88	61
NEW YORK		
Albany CO	88	72
Buffalo AP	85	70

FIGURE R27-11 (Continued)

Location	2½% Design Dry Bulb	Coincident Design Wet Bulb
NYC-Central Park (S)	89	73
Rochester AP	88	71
Syracuse AP	87	71
NORTH CAROLINA		
Charlotte AP	93	74
Raleigh/Durham AP (S)	92	75
Wilmington AP	91	78
NORTH DAKOTA		
Bismark AP (S)	91	68
Fargo AP	89	71
Grands Forks AP	87	70
OHIO		
Cincinnati CO	90	72
Cleveland AP (S)	88	72
Columbus AP (S)	90	73
OKLAHOMA		
Muskogee AP	98	75
Oklahoma City AP	97	74
Tulsa AP	98	75
OREGON		
Eugene AP	89	66
Medford AP (S)	94	67
Portland AP	85	67
PENNSYLVANIA		
Allentown AP	88	72
Philadelphia AP	90	74
Pittsburgh CO	88	71
RHODE ISLAND		
Providence AP	86	72
SOUTH CAROLINA		
Charleston CO	92	78
Columbia AP	95	75
Greenville AP	91	74
SOUTH DAKOTA		
Aberdeen AP	91	72
Pierre AP	95	71
Sioux Falls AP	91	72

Location	2½% Design Dry Bulb	Coincident Design Wet Bulb
TENNESSEE		
Chattanooga AP	93	74
Memphis AP	95	76
Nashville AP (S)	94	74
TEXAS		
Corpus Christi AP	94	78
Dallas AP	100	75
Houston CO	95	77
Lubbock AP	96	69
San Antonio AP (S)	97	73
UTAH		
Cedar City AP	91	60
Provo	96	62
Salt Lake City AP (S)	95	62
VERMONT		
Barre	81	69
Burlington AP (S)	85	70
Rutland	84	70
VIRGINIA		
Norfolk AP	91	76
Richmond AP	92	76
Roanoke AP	91	72
WASHINGTON		
Bellingham AP	77	65
Seattle-Tacoma AP (S)	80	64
Spokane AP (S)	90	63
WEST VIRGINIA		
Charleston AP	90	73
Huntington CO	91	74
Wheeling	86	71
WISCONSIN		
Green Bay AP	85	72
Madison AP (S)	88	73
Milwaukee AP	87	73
WYOMING		
Casper AP	90	57
Cheyenne AP	86	58

Location	2½% Design Dry Bulb	Coincident Design Wet Bulb
Sheridan AP	91	62
ALBERTA		
Calgary AP	81	61
Edmonton AP	82	65
BRITISH COLUMBIA		
Vancouver AP (S)	77	66
Victoria CO	73	62
MANITOBA		
Flin Flon	81	66
Winnipeg AP (S)	86	71
NEW BRUNSWICK		
Fredericton AP (S)	85	69
Saint John AP	77	65
NEWFOUNDLAND		
Gander AP	79	65
St. John's AP (S)	75	65
NORTHWEST TERR.		
Fort Smith AP (S)	81	64
Yellowknife AP	77	61
NOVA SCOTIA		
Halifax AP (S)	76	65
Yarmouth AP	72	64
ONTARIO		
Sudbury AP	83	67
Thunder Bay AP	83	68
Toronto AP (S)	87	72
PRINCE EDWARD ISLAND		
Charlottetown AP (S)	78	68
QUEBEC		
Chicoutimi	83	68
Montreal AP (S)	85	72
Quebec AP	84	70
SASKATCHEWAN		
Regina AP	88	68
Saskatoon AP (S)	86	66
YUKON TERRITORY		
Whitehorse AP (S)	77	58

AP—Airport
CO—City Office

Source: Reprinted by permission from 1967 *ASHRAE Handbook of Fundamentals.*

AVERAGE AIR CHANGES PER 24 HR. FOR STORAGE ROOMS DUE TO DOOR OPENINGS AND INFILTRATION

(above 32° F.)

Volume cu. ft.	Air Changes per 24 hr.	Volume cu. ft.	Air Changes per 24 hr.
200	44.0	6,000	6.5
300	34.5	8,000	5.5
400	29.5	10,000	4.9
500	26.0	15,000	3.9
600	23.0	20,000	3.5
800	20.0	25,000	3.0
1,000	17.5	30,000	2.7
1,500	14.0	40,000	2.3
2,000	12.0	50,000	2.0
3,000	9.5	75,000	1.6
4,000	8.2	100,000	1.4
5,000	7.2		

Note: For heavy usage multiply the above values by 2.
For long storage multiply the above values by 0.6.

FIGURE R27-12 Average air changes per 24 hours for storage rooms due to door openings and infiltration (above 32°F). (*Courtesy* ASHRAE)

To calculate the refrigeration product load for food products, solids, and liquids, it is essential to know their freezing points, specific heats, percent water, etc. Figure R27-14 is a sample of food products data taken from ASHRAE information.

R27-7
SENSIBLE HEAT ABOVE FREEZING

Most products are at a higher temperature than the storage temperature when placed in the refrigerator. Since may foods have a high water content, their reaction to a loss of heat is quite different above and below the freezing point. Above the freezing point the water exists in liquid form, while below the freezing point, the water has changed its state to ice.

As mentioned in earlier chapters, the specific heat of a product is defined as the Btu required to raise the temperature of 1 lb of the substance 1°F. The specific heats of various food products are listed in Fig. R27-14, both for above and below freezing temperatures.

HEAT REMOVED IN COOLING AIR TO STORAGE ROOM CONDITIONS
(BTU per cu. ft.)

Storage room temp F	Temperature of Outside Air, F							
	85		90		95		100	
	Relative Humidity, Percent							
	50	60	50	60	50	60	50	60
65	0.65	0.85	0.93	1.17	1.24	1.54	1.58	1.95
60	0.85	1.03	1.13	1.37	1.44	1.74	1.78	2.15
55	1.12	1.34	1.41	1.66	1.72	2.01	2.06	2.44
50	1.32	1.54	1.62	1.87	1.93	2.22	2.28	2.65
45	1.50	1.73	1.80	2.06	2.12	2.42	2.47	2.85
40	1.69	1.92	2.00	2.26	2.31	2.62	2.67	3.06
35	1.86	2.09	2.17	2.43	2.49	2.79	2.85	3.24
30	2.00	2.24	2.26	2.53	2.64	2.94	2.95	3.35

Storage room temp F	Temperature of Outside Air, F							
	40		50		90		100	
	Relative Humidity, Percent							
	70	80	70	80	50	60	50	60
30	0.24	0.29	0.58	0.66	2.26	2.53	2.95	3.35
25	0.41	0.45	0.75	0.83	2.44	2.71	3.14	3.54
20	0.56	0.61	0.91	0.99	2.62	2.90	3.33	3.73
15	0.71	0.75	1.06	1.14	2.80	3.07	3.51	3.92
10	0.85	0.89	1.19	1.27	2.93	3.20	3.64	4.04
5	0.98	1.03	1.34	1.42	3.12	3.40	3.84	4.27
0	1.12	1.17	1.48	1.56	3.28	3.56	4.01	4.43
— 5	1.23	1.28	1.59	1.67	3.41	3.69	4.15	4.57
—10	1.35	1.41	1.73	1.81	3.56	3.85	4.31	4.74
—15	1.50	1.53	1.85	1.92	3.67	3.96	4.42	4.86
—20	1.63	1.68	2.01	2.09	3.88	4.18	4.66	5.10
—25	1.77	1.80	2.12	2.21	4.00	4.30	4.78	5.21
—30	1.90	1.95	2.29	2.38	4.21	4.51	4.90	5.44

From 1967 ASHRAE Handbook of Fundamentals, Reprinted by Permission

FIGURE R27-13 (*Courtesy* ASHRAE)

FOOD PRODUCTS DATA

Product	Average Freezing Point F	Percent Water	SP ht, Btu/ (lb) (F deg)		Latent Heat of Fusion Btu/lb	Heat of Respiration Btu per (24 hr) (ton) at Temp. Indicated	
			Above Freezing	Below Freezing		°F	BTU
VEGETABLES							
Artichokes	29.1	83.7	0.87	0.45	120	40	10,140
Asparagus	29.8	93	0.94	0.48	134	40	11,700-23,100
Beans, string	29.7	88.9	0.91	0.47	128	40	9700-11400
Beans, Lima	30.1	66.5	0.73	0.40	94	40	4300-6100
Beans, dried		12.5	0.30	0.24	18		
Beets	31.1	87.6	0.90	0.46	126	32	2700
						40	4100
Broccoli	29.2	89.9	0.92	0.47	130	40	11,000-17,000
Brussels sprouts	31	84.9	0.88	0.46	122	40	6600-11,000
Cabbage	31.2	92.4	0.94	0.47	132	40	1700
Carrots	29.6	88.2	0.90	0.46	126	32	2100
						40	3500
Cauliflower	30.1	91.7	0.93	0.47	132	40	4500
Celery	29.7	93.7	0.95	0.48	135	32	1600
						40	2400
Corn (green)	28.9	75.5	0.79	0.42	106	32	7200-11,300
						40	10,600-13,200
Corn (dried)		10.5	0.28	0.23	15		
Cucumbers	30.5	96.1	0.97	0.49	137		
Eggplant	30.4	92.7	0.94	0.48	132		
Sausage (franks)	29	60	0.86	0.56	86		
Sausage (fresh)	26	65	0.89	0.56	93		
Sausage (smoked)	25	60	0.86	0.56	86		
Scallops	28	80.3	0.89	0.48	116		
Shrimp	28	70.8	0.83	0.45	119		
Veal	29	63	0.71	0.39	91	◄———	VEAL
MISCELLANEOUS							
Beer	28	92	1.0				
Bread		32-37	0.70	0.34	46-53		
Bread (dough)		58	0.75				
Butter	30-0	15	0.64	0.34	15		
Candy			0.93				
Caviar (tub)	20	55				40	3820
Cheese (American)	17	60	0.64	0.36	79	40	4680
Cheese (Camembert)	18	60	0.70	0.40	86	40	4920
Cheese (Limburger)	19	55	0.70	0.40	86	40	4920
Cheese (Roquefort)	3	55	0.65	0.32	79	45	4000
Cheese (Swiss)	15	55	0.64	0.36	79	40	4660
Chocolate (coating)	95-85	55	0.30	0.55	40		
Cream (40%)	28	73	0.85	0.40	90		
Eggs (crated)	27		0.76	0.40	100		
Eggs (frozen)	27			0.41	100		
Flour		13.5	0.38	0.28			
Flowers (cut)	32						480/sq. ft. Floor Area
Furs—Woolens				0.40			

FIGURE R27-14 (*Courtesy* ASHRAE)

The heat to be removed from a product to reduce its temperature above freezing may be calculated as follows:

$$Q = W \times C \times (T_1 - T_2)$$

where

Q = Btu to be removed

W = weight of the product in pounds

C = specific heat above freezing (Fig. R27-14)

T_1 = initial temperature, °F

T_2 = final temperature, °F (freezing or above)

For example, the heat to be removed in order to cool 1000 lb of veal (whose freezing point is 29°F) from 42°F to 29°F can be calculated as follows:

$Q = W \times C \times (T_1 - T_2)$

$= 1000$ lb $\times 0.71$ specific heat (veal)

$\times 13 (42° - 29°)$

$= 9230$ Btu

R27-8
LATENT HEAT OF FREEZING

The latent heat of freezing for water is 144 Btu/lb, as mentioned in earlier chapters. Most food products have a high percent of water content. To calculate the heat removal required to freeze the product, only the water need be considered. Water content is also shown in Fig. R27-14.

Since the latent heat of freezing for water is 144 Btu/lb, the latent heat of freezing for a given product can be calculated by multiplying 144 Btu/lb by the percentage of water content. To illustrate, veal is 63% water, and its latent heat is 91 Btu/lb (63% × 144 Btu/lb = 91 Btu/lb).

The heat to be removed from a product for the latent heat of freezing may be calculated as follows:

$$Q = W \times hof$$

where

Q = Btu to be removed

W = weight of product, lb

hof = latent heat of fusion, Btu/lb

The latent heat of freezing 1000 lb of veal at 29°F is

$$Q = W \times hof$$
$$= 1000 \text{ lb} \times 91 \text{ Btu/lb}$$
$$= 91,000 \text{ Btu}$$

R27-9
SENSIBLE HEAT BELOW FREEZING

Once the water content of a product has been frozen, sensible cooling can occur again in the same manner as above freezing, with the exception that the ice in the product causes the specific heat to change. Note in Fig. R27-14 that the specific heat of veal above freezing is 0.71, while the specific heat below freezing is 0.39.

The heat to be removed from a product to reduce its temperature below freezing may be calculated as follows:

$$Q = W \times C_i \times (T_f - T_3)$$

where

Q = Btu to be removed

W = weight of product, lb

C_i = specific heat below freezing (Fig. R21-12)

T_f = freezing temperature

T_3 = final temperature

For example, the heat to be removed in order to cool 1000 lb of veal from 29°F to 0°F can be calculated as follows:

$$Q = W \times C_i \times (T_f - T_3)$$
$$= 1000 \text{ lb} \times 0.39 \text{ specific heat} \times (29 - 0)$$
$$= 1000 \times 0.39 \times 29$$
$$= 11,310 \text{ Btu}$$

Total product load is the sum of the individual calculations for the sensible heat above freezing, the latent heat of freezing, and the sensible heat below freezing. In the preceding example of 1000 lb of veal cooled and frozen from 42°F to 0°F, the total product load would be:

Sensible heat above freezing =	9,230 Btu
Latent heat of freezing =	91,000 Btu
Sensible heat below freezing =	11,310 Btu
Total product load	111,540 Btu

If several different commodities, or crates or baskets, etc., are to be considered, a separate calculation must be made for each item for an accurate estimate of the total product load. Note that in all the calculations above the factor time is not considered, either for initial pull-down or storage life.

R27-10
STORAGE DATA

For most commodities there are temperature and relative humidity conditions at which their quality is best perserved and their storage life is at a maximum. Recommended storage conditions for such things as perishable food products, flowers, etc., are published in the *ASHRAE Guide*. Figure R27-15 is a sample illustration. Apples have a long storage life (2 to 6 months), while fresh figs can be stored for only several days. Note that no information is given for bananas, for which conditions vary widely due to the internal heat generated during the ripening process.

In the calculations above no allowance was made for the containers used to store the commodity. Some are in cartons, some are in wooden boxes, and some are stored loose, on pallets. Naturally, these containers are also cooled along with the product, so a storage container allowance factor is usually added to the total load.

STORAGE REQUIREMENTS AND PROPERTIES OF PERISHABLE PRODUCTS

Commodity	Storage Temp. F	Relative Humidity %	Approximate Storage Life	Commodity	Storage Temp. F	Relative Humidity %	Approximate Storage Life
Apples	30-32	85-90	2- 6 months	Endive (escarole)	32	90-95	2- 3 weeks
Apricots	31-32	85-90	1- 2 weeks	Figs			
Artichokes (Globe)	31-32	90-95	1- 2 weeks	Dried	32-40	50-60	9-12 months
Jerusalem	31-32	90-95	2- 5 months	Fresh	28-32	85-90	5- 7 days
Asparagus	32	90-95	2- 3 weeks	Fish			
Avocados	45-55	85-90	4 weeks	Fresh	33-35	90-95	5-15 days
Bananas	—	85-95	—	Frozen	—10-0	90-95	8-10 months
Beans (Green or snap)	45	85-90	8-10 days	Smoked	40-50	50-60	6- 8 months
Lima	32-40	85-90	10-15 days	Brine salted		90-95	10-12 months
Beer, barrelled	35-40		3-10 weeks	Mild cured	28-35	75-90	4- 8 months
Beets				Shellfish			
Bunch	32	90-95	10-14 days	Fresh	33	90-95	3- 7 days
Topped	32	90-95	1- 3 months	Frozen	0 to —20	90-95	3- 8 months
Blackberries	31 32	85-90	7 days	Frozen-pack fruits	—10-0	—	6-12 months
Blueberries	31 32	85-90	3- 6 weeks	Frozen-pack vegetables	—10-0	—	6-12 months
Bread	0		several weeks	Furs and Fabrics	34-40	45-55	several years
Broccoli, sprouting	32	90-95	7-10 days				
Brussels sprouts	32	90-95	3- 4 weeks	Garlic, dry	32	70-75	6- 8 months
Candy	0-34	40-65	—	Grapefruit	50	85-90	4- 8 weeks
Carrots				Grapes			
Prepackaged	32	80-90	3- 4 weeks	American type	31-32	85-90	3- 8 weeks
Topped	32	90-95	4- 5 months	European type	30-31	85-90	3- 6 months
Cauliflower	32	90-95	2- 3 weeks	Honey	—	—	1 year, plus
Celeriac	32	90-95	3- 4 months	Hops	29-32	50-60	several months
Celery	31-32	90-95	2- 4 months	Horseradish	32	90-95	10-12 weeks
Cherries	31-32	85-90	10-14 days	Kale	32	90-95	2- 3 weeks
Coconuts	32-35	80-85	1- 2 months	Kohlrabi	32	90-95	2- 4 weeks
Coffee (green)	35-37	80-85	2- 4 months	Lard (without antioxidant)	45	90-95	4- 8 months
Corn, sweet	31-32	85-90	4- 8 days	Lard (without antioxidant)	0	90-95	12-14 months
Cranberries	36-40	85-90	1- 4 months	Leeks, green	32	90-95	1- 3 months
Cucumbers	45-50	90-95	10-14 days	Lemons	32 or 50-58	85-90	1- 4 months
Currants	32	80-85	10-14 days	Lettuce	32	90-95	3- 4 weeks
Dairy products				Limes	48-50	85-90	6- 8 weeks
Cheese	30-45	65-70		Logan blackberries	31-32	85-90	5- 7 days
Butter	32-40	80-85	2 months	Meat			
Butter	0 to —10	80-85	1 year	Bacon—Frozen	—10-0	90-95	4- 6 months
Cream (sweetened)	—15	—	several months	Cured (Farm style)	60-65	85	4- 6 months
Ice cream	—15	—	several months	Cured (Packer style)	34-40	85	2- 6 weeks
Milk, fluid whole				Beef—Fresh	32-34	88-92	1- 6 weeks
Pasteurized Grade A	33	—	7 days	Frozen	—10-0	90-95	9-12 months
Condensed, sweetened	40	—	several months				
Evaporated	Room Temp	—	1 year, plus	Fat backs	34-36	85-90	0- 3 months
Milk, dried				Hams and shoulders—Fresh	32-34	85-90	7-12 days
Whole milk	45-55	low	few months	Frozen	—10-0	90-95	6- 8 months
Non-fat	45-55	low	several months	Cured	60-65	50-60	0- 3 years
Dewberries	31-32	85-90	7-10 days	Lamb—Fresh	32-34	85-90	5-12 days
Dried fruits	32	50-60	9-12 months	Frozen	—10-0	90-95	8-10 months
Eggplant	45-50	85-90	10 days	Livers—Frozen	—10-0	90-95	3- 4 months
Eggs				Pork—Fresh	32-34	85-90	3- 7 days
Shell	29-31	80-85	6- 9 months	Frozen	—10-0	90-95	4- 6 months
Shell, farm cooler	50-55	70-75		Smoked Sausage	40-45	85-90	6 months
Frozen, whole	0 or below	—	1 year, plus	Sausage Casings	40-45	85-90	
Frozen, yolk	0 or below	—	1 year, plus	Veal	32-34	90-95	5-10 days
Frozen, white	0 or below	—	1 year, plus				
Whole egg solids	35-40	low	6-12 months	Mangoes	50	85-90	2- 3 weeks
Yolk solids	35-40	low	6-12 months	Melons, Cantaloupe	32-40	85-90	5-15 days
Flake albumen solids	Room Temp	low	1 year, plus	Persian	45-50	85-90	1- 2 weeks
Dried spray albumen solids	Room Temp	low	1 year, plus	Honeydew and Honey Ball	45-50	85-90	2- 4 weeks
				Casaba	45-50	85-90	4- 6 weeks
				Watermelons	36-40	85-90	2- 3 weeks

FIGURE R27-15 (*Courtesy* ASHRAE)

R27-11
SUPPLEMENTARY LOAD

In addition to the heat transmitted to the refrigerated space through the walls, air infiltration, and the product load, any heat gain from other sources must be included in the total cooling load estimate.

Any electric energy dissipated in the refrigerated space through lights and heaters (defrost), for instance, is converted to heat and must be included in the heat load. One watt equals 3.41 Btu, and this conversion ratio is accurate for any amount of electric power.

Electric motors are another source of heat load. For a motor that is actually in the refrigerated space, the following table gives the approximate Btu per horsepower per hour of heat generated.

Motor Horsepower	Btu/hp/hr
$\frac{1}{8}$ to $\frac{1}{2}$	4250
$\frac{1}{2}$ to 3	3700
3 to 20	2950

Motors outside the refrigerated space but coupled to a fan or pump that is inside will produce a lesser load, but should also be considered.

People give off heat and moisture, and the resulting refrigeration load will vary depending on the duration of occupancy in the refrigerated space, the temperature, the type of work, and other factors. The following table lists the average heat load due to occupancy, but for stays of short duration the heat gain will be somewhat higher.

Cooler Temperature (°F)	Heat Equivalent/ Person/Btu/hr
50	720
40	840
30	950
20	1050
10	1200
0	1300
−10	1400

The total supplementary load is the sum of the individual factors contributing to it. For example, the total supplementary load in a refrigerated storeroom maintained at 0°F in which there are 300 W of electric lights, a 3-hp motor driving a fan, and two people working continuously would be as follows:

300 watts × 3.41 Btu/h	1,023 Btu/h
3 hp motor × 2950 Btu/h	8,850 Btu/h
2 people × 1300 Btu/h	2,600 Btu/h
Total supplementary load	12,473 Btu/h

R27-12
TOTAL HOURLY LOAD

For refrigerated fixtures, display cases and refrigerators, prefabricated coolers, and cold-storage boxes that are produced in quantity, the load is normally determined by testing by the manufacturer; the refrigeration equipment is preselected and sometimes already installed in the fixture.

If it must be estimated, the expected load should be calculated by determining the heat gain due to each of the factors contributing to the total load. There are many short methods of estimating heat loads for small walk-in storage boxes, and they vary in their degree of accuracy. The most accurate use forms and data available from the manufacturer for such purposes, and each factor is considered separately as well.

Refrigeration equipment is designed to function continuously, and normally the compressor operating time is determined by the requirements of the defrost system. The load is calculated on a 24-hour basis, and the required hourly compressor capacity is determined by dividing the 24-hour load by the desired number of hours of compressor operation during the 24-hour period. A reasonable safety factor must be provided to enable the unit to recover rapidly after a temperature rise and to allow for any load that might be larger than originally estimated.

In cases where the refrigerant evaporating temperature does not drop below 30°F, frost will not accumulate on the evaporator, and no defrost period is necessary. The compressor for such applications is chosen on the basis of an 18- to 20-hour period of operation.

For applications with storage temperatures of 35°F or higher and refrigerant temperatures low enough to cause frosting, it is common practice to defrost by stopping the compressor and allowing the return air to melt the ice from the coil. Compressors for such applications should be selected for 16 to 18 hours of operation.

On low-temperature applications, some positive means of defrosting must be provided. With normal defrost periods, 18-hour compressor operation is usually acceptable, although some systems are designed for continuous operation except during the defrost period.

An additional 5 to 10% safety factor is often added to load calculations as a conservative measure to be sure that the equipment will not be undersized. When data concerning the refrigeration load are uncertain, this practice may be desirable, but generally the fact that the compressor is sized on the basis of 16- or 18-hour operation in itself provides a sizable safety factor. The load should be calculated on the basis of the peak demand at design conditions, and usually the design conditions are selected on the assumption that peak demand will occur no more than 1% of the hours during the summer months. If the load calculations are reasonably accurate and the equipment is properly sized, an additional safety factor may actually result in the equipment being oversized during light-load conditions, a situation that could lead to operating difficulties.

PROBLEMS

R27-1. The refrigeration load generally comes from four sources. What are they?

R27-2. Product loads may be made up of either _____ or _____ heat or both.

R27-3. Does insulation have a high or a low resistance value?

R27-4. How much heat will be transmitted through 10 ft² of a 4-in.-thick common brick wall with no insulation if the temperature difference across the wall is 70°F?

R27-5. Insulation is available in several different forms. Name them.

R27-6. Define "design temperature."

R27-7. Opening and closing the refrigerator door creates an _____ load.

R27-8. The heat-removal rate of a product is determined by the _____ of the product.

R27-9. Is the specific heat of a product higher above or below the freezing point of the product?

R27-10. The latent heat load of a product is related to the percentage of its _____ content.

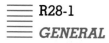

Refrigerated Storage

Commercial refrigeration is the broadest in scope of all the segments of the refrigeration industry, encompassing the range between domestic and industrial applications. At the lower end of the size scale we find such units as the physically compact, relatively maintenance-free household refrigerators and freezers. At the upper end of the commercial range are systems that cool, chill, or freeze millions of cubic feet of product and, due to their complexity, require the presence of an operating engineer and other skilled personnel. Between these two extremes are a multitude of hardware system types, both mobile and stationary, which play a part in the production, adaptation, or preservation of most of the commodities used throughout the world today.

Some of the first commercial storage refrigerators appeared in the early part of the twentieth century and were installed in large meat markets, breweries, and other similar storage facilities. Although of a semiautomatic design, they required full-time attendance by an operator and were expensive to maintain. Although not satisfactory by today's standards, this equipment did serve a definite need at the time. A few years later, smaller units were designed for storage of ice cream. These units utilized some of the early refrigerants, as described in Chap. 7. The newly designed ice cream storage used the old ice-and-salt mixture type of cooling.

Soon there was an influx of smaller commercial refrigeration units used as open or closed display cases, reach-in cabinets, and walk-in coolers, for fresh and frozen foods.

These units were used in grocery stores, meat markets, supermarkets, restaurants, drugstores, and the like. Different storage temperatures and conditions were required for various meats, poultry, fish, vegetables, fruits, candy, and ice cream, as well as medicines stored in drug stores.

Figure R28-1 shows a typical reach-in cabinet designed to store and refrigerate many different types of products. Figure R28-2 shows a typical walk-in cooler that may be used in many facilities such as grocery stores, markets, or restaurants.

Since numerous applications are covered by the broad field of commercial refrigeration, a wide variety of cabinets are constructed for this use. The reach-in cabinets used for keeping perishable produce and groceries at their required temperatures have one or more doors permitting access by the customers. Some cabinets are designed so the customer may have access to the product from the front of the cabinet while the store personnel have access from the rear of the cabinet to replenish the supply of fresh or frozen foods. Early boxes were constructed of wood and had limited glass area. Modern cabinets usually have fiberglass or por-

FIGURE R28-1 Typical reach-in refrigeration for commercial use. (*Courtesy* The Delfield Company, Division of ALCO)

FIGURE R28-2 Typical walk-in cooler. (*Courtesy* Bally Case & Cooler, Inc.)

FIGURE R28-3 Typical evaporator. (*Courtesy* Bohn Heat Transfer Division)

celain exteriors with thin layers of modern insulation material, and some have extensive glass display areas.

Some boxes depend on the gravity flow of air for the proper circulation in or around the product, while other units, particularly larger sizes, may use a circulating fan for air distribution within the cabinet. The refrigeration unit for this type of box may be inside the framework of the box itself or may be in another location.

When the cabinet is to be used for storage of meat or fresh produce, good control of temperature and humidity must be maintained, so that dehydration or deterioration of the product will not occur. In such an application, the temperature probably will be maintained between 34 and 40°F, and the relative humidity may be kept at 75% or higher. If the product is such that there is no concern about dehydration or deterioration, colder temperatures may be maintained and control of relative humidity will not be critical. Figure R28-3 shows a typical blower-type evaporator coil that may be used in a walk-in cabinet installation.

Figure R28-4 shows a display case often used in florist shops. In this type of application the air temperature may be kept higher than is customary in grocery or meat market boxes, since a temperature of about 55°F is best for storage of flowers.

Cabinets for frozen-food storage and display are usually maintained at or below 0°F. Proper precautions must be taken to assure that the access doors are not frozen in a closed position so that they cannot be opened by the customers. It is sometimes necessary to place an electric resistance heat coil around the door opening to prevent freezing.

Another operation requiring refrigeration is the cooling of bottled or canned beverages in automatic coin-operated dispensers. Some of these dispensers are designed to handle only one product, whereas others can dispense several different kinds of beverages. Figure R28-5 shows a unit that dispenses several kinds of beverages.

Figure R28-6 gives excerpts from the *Handbook & Product Directory 1974 Applications,* by courtesy of the American Society of Heating, Refrigerating and Air-Conditioning Engineers. This table lists many commodities and their recommended storage temperatures and other information pertinent to the products.

Since refrigeration plays such an important part in the preservation of food and other perishable products, one should recognize evidences of food spoilage and know recommended methods for its control. Some products may show effects of deterioration or food spoilage through a change in their appearance, odor, taste, loss of weight, or a change in the chemistry of the product itself.

FIGURE R28-4 Refrigerated floral display case. (*Courtesy* Bally Case & Cooler, Inc.)

FIGURE R28-5 Typical refrigerated coin-operated beverage cooler. (*Courtesy* Vendo Company)

If not properly cared for after being harvested, fruits eventually become overripe and soft because the chemical processes inside the fruits continue. Overripe fruit may still be fit for consumption by humans, but it cannot be commercially handled in a convenient manner without extensive damage to the fruit. Vegetables, like fruits, also must receive proper care after harvesting to prevent deterioration.

Eggs will lose weight through evaporation of the water content and emit an offensive odor when they spoil. Broken eggs must be salvaged immediately if they are to be processed into a frozen or powdered form that can be used by a bakery or other food industry.

Milk, if not refrigerated, will be "soured" in a very short time by growth of bacteria, producing a disagreeable odor and distinctively bitter taste. This obviously renders it unsuitable for direct consumption as a beverage. However, since milk that has been modified by bacterial growth may be used to produce other food products, the soured condition is not automatically categorized as spoilage. Other dairy products, such as butter or cheese, derived from milk or some of its components, will be decomposed by bacterial activity—even products that utilize other types of bacteria in their formation. The chemical change in this decomposition process will produce a sour odor and unpalatable taste in milk by-products as well as in unprocessed milk.

Beef, pork, and other types of meat products show very little apparent change in physical appearance when they first begin to deteriorate. As the deterioration process continues, however, they give off a slight odor, which increases in intensity as the deterioration progresses. When certain stages of deterioration are reached, a slimy covering or coating appears on the surface of the meat, which is a definite indication of spoilage.

All of the examples of food spoilage above are caused by a series of chemical changes within the products. These changes may be brought about through natural forces or by external contamination. Either way the result can be a complete breakdown or decay of the products.

Bacteria are found in the air, in the soil, and in water, as well as in almost every known substance on earth. For our purposes the two important types of bacteria are:

1. Those that cause food deterioration
2. Those used in the preservation of foods

The latter type was described in the paragraph concerning sour milk. It is also used in the processing of other foods such as wine, vinegar, and certain types of cheeses and butter. The first type of bacteria, however, is the most important to the refrigeration storage or processing plant operator.

Bacteria are subject to changes in temperature and moisture conditions and are even sensitive to normal lighting and changes in light conditions.

Since bacteria, under certain conditions, cause the decay or deterioration of food products, in any study of food preservation one must become conversant with the different types of bacteria that are found in food products. Each type has a minimum and maximum temperature range for ideal growth. Therefore the temperature to which bacteria are subjected must be closely controlled. Some become dormant or are dissipated if subjected to temperatures below 0°F, whereas other bacteria need the application of high-temperature steam under pressure before they are destroyed. Usually bacteria prefer darkness. As an example of the widespread presence of bacteria, more than 200 different types have been found in milk and milk products alone. Although the different types of bacteria are responsible for most of the spoilage of foodstuffs, another member of the plant kingdom—mold—also contributes.

Molds are the principal causes of deterioration and spoilage in citrus and other fruits. The different types of mold thrive particularly well in damp locations; darkness adds to their growth. As mentioned earlier, molds also are a form of plant life, and different types of molds cause the spoilage of citrus fruit and dairy products, soft rot of apples in storage, and deterioration of bread and grain.

It is obvious that control measures must be taken to suspend or prevent food from spoiling. Through experience and experimentation, different methods have

Commodity	Storage Temp. (°F)	Relative Humidity (%)	Approximate Storage Life[a]	Water Content (%)	Highest Freezing Point (°F)	Specific Heat Above Freezing[b] (Btu/lb/F)	Specific Heat Below Freezing[b] (Btu/lb/F)	Latent Heat (Calculated)[c] (Btu/lb)
Alfalfa meal	0 or below	70–75	1 year, plus	—	—	—	—	—
Apples	30–40	90	3–8 months	84.1	29.3	0.87	0.45	121
Apricots	31–32	90	1–2 weeks	85.4	30.1	0.88	0.46	122
Artichokes								
Globe	31–32	95	2 weeks	83.7	29.9	0.87	0.45	120
Jerusalem	31–32	90–95	5 months	79.5	27.5[d]	0.83	0.44	114
Asparagus	32–36	95	2–3 weeks	93.0	30.9	0.94	0.48	134
Avocados	45–55	85–90	2–4 weeks	65.4	31.5	0.72	0.40	94
Bananas	—	85–95	—	74.8	30.6	0.80	0.42	108
Beans								
Green or snap	40–45	90–95	7–10 days	88.9	30.7	0.91	0.47	128
Lima	32–40	90	1 week	66.5	31.0	0.73	0.40	94
Beer								
Keg	35–40	—	3–8 weeks	90.2	28.0[d]	0.92	—	129
Bottles, cans	35–40	65 or below	3–6 months	90.2	—	—	—	—
Beets								
Bunch	32	95	10–14 days	—	31.3	—	—	—
Topped	32	95–100	4–6 months	87.6	30.1	0.90	0.46	126
Blackberries	31–32	95	3 days	84.8	30.5	0.88	0.46	122
Blueberries	31–32	90–95	2 weeks	82.3	29.7	0.86	0.45	118
Bread	0	—	3 weeks to 3 months	32–37	—	0.70	0.34	46–53
Broccoli, sprouting	32	95	10–14 days	89.9	30.9	0.92	0.47	130
Brussels sprouts	32	95	3–5 weeks	84.9	30.5	0.88	0.46	122
Cabbage, late	32	95–100	3–4 months	92.4	30.4	0.94	0.47	132
Candy	0–34	40–65	—	—	—	—	—	—
Canned foods	32–60	70 or lower	1 year	—	—	—	—	—
Carrots								
Topped, immature	32	98–100	4–6 weeks	88.2	29.5	0.90	0.46	126
Topped, mature	32	98–100	5–9 months	88.2	29.5	0.90	0.46	126
Cauliflower	32	95	2–4 weeks	91.7	30.6	0.93	0.47	132
Celeriac	32	95–100	3–4 months	88.3	30.3	0.91	0.46	126
Celery	32	95	1–2 months	93.7	31.1	0.95	0.48	135
Cherries								
Sour	31–32	90–95	3–7 days	83.7	29.0	0.87	—	120
Frozen	0 to −10	—	1 year	—	—	—	0.45	—
Sweet	30–31	90–95	2–3 weeks	80.4	28.8	0.84	—	—
Cocoa	32–40	50–70	1 year, plus	—	—	—	—	—
Coconuts	32–35	80–85	1–2 months	46.9	30.4	0.58	0.34	67
Coffee (green)	35–37	80–85	2–4 months	10–15	—	0.30	0.24	14–21
Collards	32	95	10–14 days	86.9	30.6	0.90	—	—
Corn, sweet	32	95	4–8 days	73.9	30.9	0.79	0.42	106
Cranberries	36–40	90–95	2–4 months	87.4	30.4	0.90	0.46	124
Cucumbers	50–55	90–95	10–14 days	96.1	31.1	0.97	0.49	137
Currants	31–32	90–95	10–14 days	84.7	30.2	0.88	0.45	120
Dairy products								
Cheese	30–34	65–70	18 months	37.5	8.0	0.50	0.31	53
Cheddar	40	65–70	6 months	37.5	8.0	0.50	0.31	53

Commodity	Storage Temp. (°F)	Relative Humidity (%)	Approximate Storage Life[a]	Water Content (%)	Highest Freezing Point (°F)	Specific Heat Above Freezing[b] (Btu/lb/F)	Specific Heat Below Freezing[b] (Btu/lb/F)	Latent Heat (Calculated)[c] (Btu/lb)
Processed	40	65–70	12 months	39.0	19.0	0.50	0.31	56
Grated	40	60–70	12 months	31.0	—	0.45	0.29	44
Butter	40	75–85	1 month	16.0	−4 to 31	0.50	—	23
	−10	70–85	12 months	16.0	−4 to 31	—	0.25	23
Cream	−10 to −20	—	6–12 months	55–75	31.0	0.66–0.80	0.36–0.42	79–107
Ice cream	−20 to −15	—	3–12 months	58–63	21.0	0.66–0.70	0.37–0.39	86
Milk, fluid whole Pasteurized, grade A	32–34	—	2–4 months	87.0	31.0	0.93	—	125
	−15	—	3–4 months	87.0	31.0	—	0.46	125
Condensed, sweetened	40	—	15 months	28.0	5.0	0.42	0.28	40
Evaporated	70	—	12 months	74.0	29.5	0.79	0.42	106
	40	—	24 months	74.0	29.5	0.79	0.42	106
Milk, dried Whole	70	low	6–9 months	2–3	—	—	—	4
Nonfat	45–70	low	16 months	2–4.5	—	0.36	—	4
Whey, dried	70	low	12 months	3–4	—	0.36	—	4
Dates	0 or 32	75 or less	6–12 months	20.0	3.7	0.36	0.26	29
Dewberries	31–32	90–95	3 days	84.5	29.7	0.88	—	—
Dried foods	32–70	low	6 months to 1 year, plus	—	—	—	—	—
Dried fruits	32	50–60	9–12 months	14.0–26.0	—	0.31–0.41	0.26	20–37
Eggplant	45–50	90–95	7–10 days	92.7	30.6	0.94	0.48	132
Melons, Cantaloupe	36–40	90–95	5–15 days	92.0	29.9	0.93	0.48	132
Casaba	45–50	85–95	4–6 weeks	92.7	30.1	0.94	0.48	132
Honeydew and Honey Ball	45–50	90–95	3–4 weeks	92.6	30.3	0.94	0.48	132
Persian	45–50	90–95	2 weeks	92.7	30.5	0.94	0.48	132
Watermelons	40–50	80–90	2–3 weeks	91.1	31.3	0.97	0.48	132
Mushrooms[c]	32	90	3–4 days	91.1	30.4	0.93	0.47	130
Mushroom spawn Manure spawn	34	75–80	8 months	—	—	—	—	—
Grain spawn	32–40	75–80	2 weeks	—	—	—	—	—
Nectarines	31–32	90	2–4 weeks	81.8	30.4	—	—	—
Nursery stock	32–35	85–90	3–6 months	—	—	—	—	—
Nuts	32–50[c]	65–75	8–12 months	3–6	—[c]	0.22–0.25	0.21–0.22	4–8
Oil (vegetable salad)	70	—	1 year, plus	0	—	—	—	—
Okra	45–50	90–95	7–10 days	89.8	28.7	0.92	0.46	128
Oleomargarine	35	60–70	1 year, plus	15.5	—	0.32	0.25	22
Olives, fresh	45–50	85–90	4–6 weeks	75.2	29.4	0.80	0.42	108
Onions (dry) and onion sets	32	65–70	1–8 months	87.5	30.6	0.90	0.46	124
green	32	95	3–4 weeks	89.4	30.4	0.91	—	—

Commodity	Storage Temp. (°F)	Relative Humidity (%)	Approximate Storage Life[a]	Water Content (%)	Highest Freezing Point (°F)	Specific Heat Above Freezing[b] (Btu/lb/F)	Specific Heat Below Freezing[b] (Btu/lb/F)	Latent Heat (Calculated)[c] (Btu/lb)
Oranges	32–48	85–90	3–12 weeks	87.2	30.6	0.90	0.46	124
Orange juice, chilled	30–35	—	3–6 weeks	89.0	—	0.91	0.47	128
Papayas	45	85–90	1–3 weeks	90.8	30.4	0.82	0.47	130
Parsley	32	95	1–2 months	85.1	30.0	0.88	0.45	122
Parsnips	32	98–100	4–6 months	78.6	30.4	0.84	0.44	112
Peaches and nectarines	31–32	90	2–4 weeks	89.1	30.3	0.90	0.46	124
Pears	29–31	90–95	2–7 months	82.7	29.2	0.86	0.45	118
Peas, green	32	95	1–3 weeks	74.3	30.9	0.79	0.42	106
Peppers, Sweet[c]	45–50	90–95	2–3 weeks	92.4	30.7	0.94	0.47	132
Peppers, Chili (dry)[c]	32–50	60–70	6 months	12.0	—	0.30	0.24	17
Persimmons	30	90	3–4 months	78.2	28.1	0.84	0.43	112
Pineapples Mature green	50–55	85–90	3–4 weeks	—	30.2	—	—	—
Ripe	45	85–90	2–4 weeks	85.3	30.0	0.88	0.45	122
Plums, including fresh prunes	31–32	90–95	2–4 weeks[c]	82.3	30.5	0.88	0.45	118
Pomegranates	32	90	2–4 weeks	—	26.6	—	—	—
Popcorn, un-popped	32–40	85	4–6 months	13.5	—	0.31	0.24	19
Potatoes Early crop	50–55	90	—[c]	81.2	30.9	0.85	0.44	116
Late crop	38–50[c]	90	—[c]	77.8	30.9	0.82	0.43	111
Poultry Fresh	32	85–90	1 week	74.0	27.0[d]	0.79	—	106
Frozen, eviscerated	0 or below	90–95	8–12 months	—	—	—	0.42	—
Precooked, frozen								
Pumpkins[c]	50–55	70–75	2–3 months	90.5	30.5	0.92	0.47	130
Quinces	31–32	90	2–3 months	85.3	28.4	0.88	0.45	122
Radishes— Spring, pre-packaged	32	95	3–4 weeks	93.6	30.7	0.95	0.48	134
Winter	32	95–100	2–4 months	93.6	—	0.95	0.48	134
Rabbits Fresh	32–34	90–95	1–5 days	68.0	—	0.74	0.40	98
Frozen	−10–0	90–95	0–6 months	—	—	—	—	—
Raspberries Black	31–32	90–95	2–3 days	80.6	30.0	0.84	0.44	122
Red	31–32	90–95	2–3 days	84.1	30.9	0.87	0.45	121
Frozen (red or black)	−10–0	—	1 year	—	—	—	—	—
Rhubarb	32	95	2–4 weeks	94.9	30.3	0.96	0.48	134
Rutabagas	32	98–100	4–6 months	89.1	30.1	0.91	0.47	127
Salsify	32	98–100	2–4 months	79.1	30.0	0.83	0.44	113
Seed, vege-table[c]	32–50	50–65	—[c]	7.0–15.0	—	0.29	0.23	16
Spinach	32	95	10–14 days	92.7	31.5	0.94	0.48	132

Commodity	Storage Temp. (°F)	Relative Humidity (%)	Approximate Storage Life[a]	Water Content (%)	Highest Freezing Point (°F)	Specific Heat Above Freezing[b] (Btu/lb/F)	Specific Heat Below Freezing[b] (Btu/lb/F)	Latent Heat (Calculated)[c] (Btu/lb)
Squash[c]								
Acorn	45–50	70–75	5–8 weeks	—	30.5	—	—	—
Summer	32–50	85–95	5–14 days	94.0	31.1	0.95	—	135
Winter	50–55	70–75	4–6 months	88.6	30.3	0.91	—	127
Strawberries								
Fresh	31–32	90–95	5–7 days	89.9	30.6	0.92	—	129
Frozen	−10–0	—	1 year	72.0	—	—	0.42	103
Sweet potatoes	55–60	85–90	4–7 months	68.5	29.7	0.75	0.40	97
Tangerines	32–38	85–90	2–4 weeks	87.3	30.1	0.90	0.46	125
Tobacco								
Hogsheads	50–65	50–55	1 year	—	—	—	—	—
Bales	35–40	70–85	1–2 years	—	—	—	—	—
Cigarettes	35–46	50–55	6 months	—	—	—	—	—
Cigars	35–50	60–65	2 months	—	—	—	—	—
Tomatoes								
Mature green	55–70	85–90	1–3 weeks	93.0	31.0	0.95	0.48	134
Firm ripe	45–50	85–90	4–7 days	94.1	31.1	0.94	0.48	134
Turnips, roots	32	95	4–5 months	91.5	30.1	0.93	0.47	130
Vegetable seed	32–50	50–65	—	7.0–15.0	—	0.29	0.23	16
Yams	60	85–90	3–6 months	73.5	—	0.79	—	105
Yeast, compressed baker's	31–32	—	—	70.9	—	0.77	0.41	102

[a]Not based on maintaining nutritional value.

[b]Calculated by Siebel's formula. For values above freezing point $S = 0.008a + 0.20$. For values below freezing point $S = 0.003a + 0.20$. Recent work by H. E. Staph, B. E. Short, and others at the University of Texas has shown that Siebel's formula is not particularly accurate in the frozen region, because foods are not simple mixtures of solids and liquids and are not completely frozen even at −20 F.

[c]Values for latent heat (latent heat of fusion) in Btu per pound, calculated by multiplying the percentage of water content by the latent heat of fusion of water. 143.4 Btu.

[d]Average freezing point.

[e]Eggs with weak albumen freeze just below 30°F.

[f]Lemons stored in production areas for conditioning are held at 55 to 58°F; in terminal markets they are customarily stored at 50 to 55°F, but sometimes 32°F is used.

Note: Acknowledgment is due the following men for assistance with certain commodities: R. L. Hiner, L. Feinstein, and A. Kotula, meat and poultry; J. W. White and C. O. Willits, honey; E. B. Lambert, mushrooms; H. Landani, furs and fabrics; M. K. Veldhuis, orange juice; A. L. Ryall, plums and prunes; L. P. McColloch, tomatoes (all the former are United States Department of Agriculture staff members); J. W. Slavin, fish, United States Department of Commerce; and T. I. Hedrick, dairy products, Michigan State University.

been developed to preserve foodstuffs. Some of them are listed in Fig. R28-6; one must follow the recommended temperatures and conditions to preserve different types of foodstuffs. Figure R28-6 lists the *optimum* conditions and temperatures that will give the desired results. The variety or type of each product, the purpose for which it is being stored, and also the conditions under which it is being stored, dictate the conditions required for the most desirable storage environment.

If vegetables or fruits are refrigerated immediately upon harvesting, they will be preserved in storage for a longer period than if they are permitted to continue their ripening process after harvesting, before being refrigerated. This one difference in the preparation and storage of vegetables or fruits could easily result in variations of several months of continued satisfactory quality of the product. Some of the products listed in Figure R28-6 have a high moisture content; therefore conditions of high relative humidity are recommended for their proper storage.

Other areas where commercial refrigeration is important are in the manufacture of candy and the processing of food products—particularly those to be sold in a frozen state. Processing plants for name-brand frozen foods are often located in the areas where those foods are produced. Thus, it is natural that processing centers for fruits would be located in a fruit-producing area of the country. By the same token processing centers for different types of vegetables are located in truck-farming areas. Processors of frozen fish have plants located along the coast, from which they ship their products throughout the country.

Not too many years ago, before the widespread acceptance of home freezers, it was a common practice for communities to have locker plants which would process, freeze, and store various products either for the producers of these products or for the consumers.

Figure R28-7 shows the layout of a typical frozen food locker plant, in which these steps or phases of operation are followed:

1. *The Receiving Area* May or may not be refrigerated, where the incoming foods are weighed and listed.

2. *Chill and Age Areas* Must be refrigerated and maintained at approximately 35°F, where different types of meat, poultry, and even fish are prechilled before being cut into the desired size for packaging.

3. *Preparation Area* May or may not be refrigerated, where the meat, fowl, and fish are processed for distribution to either the freezer or cure room. This includes the washing and blanching of fruits and vegetables when these products are handled or processed in the locker plant.

4. *Cure or Smoke Area* Refrigerated and maintained at approximately 35°F. If some of the product is to be cured, smoked, or salted, such as bacon, hams, or sausage, the meat is taken into the cure room after it has been cut to size and/or processed.

5. *Packaging or Wrapping Area* May or may not be refrigerated. This is the area where, before freezing, all of the pieces or sections of meat and other products are packaged and/or wrapped. It is here also that the product is identified by labeling each package as to its content, weight, date, and owner.

6. *Quick-freeze Area* All processing plants make an effort to freeze food as rapidly as possible. This is done by one of several methods such as:

   a. Placing the food packages on cold-plate shelves, which lower the temperature of the

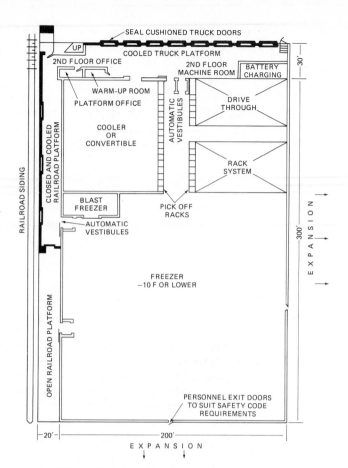

FIGURE R28-7   Typical frozen-food locker plant. (*Courtesy* ASHRAE)

product by the contact method. (An example of this type of cold-plate shelf arrangement is shown in Fig. R28-8.)

   b. Quick freezing may be achieved by placing the packages of food on shelves in a blast-freezer unit.

   c. Some foods are immersed in subzero liquids, while some are sprayed with a cooling liquid.

FIGURE R28-8   Cold-plate shelf freezer. (*Courtesy* Refrigeration Engineering Corporation)

In any case, the temperature of the product is brought down as quickly as possible to a range from minus 10°F to minus 40°F.

7. *Locker or Storage Area* Usually kept about 0°F. It contains individual lockers, which are rented. Also included in this locker room area is a storage space to accommodate food for customers who wish to store packages in their own home freezers.

Frozen-food locker plants still are operating in some communities. The function of these plants ranges from the storage of commodities prepared elsewhere to the complete local processing and storage of many different kinds of foods.

R28-3
TRANSPORTATION AND DISTRIBUTION

Prior to the use of mechanical refrigeration, producers of perishable products had to depend on the local distribution of their products, or rely on ice to keep them in salable condition. Railway cars used for the shipment of perishable commodities were equipped with bunkers at each end in which ice was packed to refrigerate products in the car during shipment. There were hatches or access openings to the bunkers on the roofs of the cars so that the ice could be replenished when needed. If the weather was hot and the distance was long, ice would be replenished several times before the product arrived at its destination. Floor drains were provided to permit drainage of the water from the melting ice.

In many cases, a blower was used for the circulation of cool air across the product in the railway car. The motor for the blower was driven either by a small gasoline engine or else there was an electric motor driven by a battery. A grill was located in the bottom of the bunker, which permitted the warmer return air from the product to pass across the ice and thereby give up its heat. The cold air was then forced down across the product directly or through ductwork for equal distribution.

The recent trend has been toward the installation of mechanical refrigeration in railway cars used for the transportation of perishable products. These refrigeration units are driven directly from the turning axles of the cars when they are in motion, or by electric motors or gasoline engines when the cars are in a freight yard for any length of time.

Refrigerated trucks and trailers today are more evident than they used to be on the highways. The mechanical refrigeration systems in these vehicles are quite similar to those used in other commercial applications, except that the compressors may be driven by electrical motors, in engine-motor combinations, or even by individual gasoline or diesel engines. In the type using an electric motor, the motor not only will operate from the engine-driven electric generator, but its voltage, cycle, and phase are such that it is capable of being plugged into a wall outlet in a garage when the truck engine is not operating. The engine-motor combination is such that, when the truck is in operation, the transmission power takeoff of the vehicle operates the refrigeration unit. But, when the truck is parked overnight with the engine shut off, a standard motor is utilized for its operation. The gasoline or diesel engine units operate automatically, as the system load requires. Some truck and trailer units utilize a dual refrigeration system with a standard refrigeration unit for standby purposes, when the truck engine is not operated for a long time.

Some truck and trailer units use eutectic plates as the evaporator within the storage compartment. This type of system can be cooled down to a low temperature during the night and maintain a "cooling" storage capacity for use the following day. Examples of this are milk delivery trucks, which may be cooled during the evening when the trucks are being loaded with their products, so they will stay cold the following day while deliveries are made. Some eutectic plates are capable of being charged or cooled continuously while the truck is being used.

Many of the air-freight lines have planes equipped to handle refrigerated products in transit. Marine refrigeration units are basically the same as other commercial units. An improvement is that fishing vessels now are equipped with automatic ice makers, which can keep the bunkers aboard the ship filled with ice, if the catch requires storage. In the past, fishing boats would load up with ice from land sources, in the anticipation that the day's catch would be large enough to utilize the ice that was brought aboard.

R28-4
ICE MANUFACTURE

As mentioned in Chapter R1, the use of ice to refrigerate food, beverages, and other products dates back to early civilizations. The shipment of ice by the clipper ships at times was a prosperous venture, but at times could be a disaster. These early ships used natural ice formed in rivers, lakes, and ponds in the cold areas of the country. With the advent of mechanical refrigeration, dependence on natural ice became a thing of the past. Since its inception, the manufacture of artificial ice has become an important part of the refrigeration industry. Large-scale production of ice is usually in block form, in sizes varying from 50 to 400 lbs. The variation in the sizes of the block depends on the size of the cans in which the ice or water is frozen.

Before an ice plant is established, a survey of the market determines the sizes of blocks of ice to be produced. The smaller-capacity freezing cans usually are approximately 2.5 ft high, with various inside dimensions. Cans accommodating larger blocks might have heights approximating 5 ft, with inside measurements about 1 by 2 ft. A typical current commercial operation coordinates the manufacture of ice with the use of refrigerated self-service ice-cube vending machines. The blocks are cut into convenient sizes of 10, 25, or 50 lb to be placed in the vending machines. These smaller blocks may be wrapped or unwrapped.

This ice, of course, must be maintained at proper storage temperatures, as must the packages of ice cubes dispensed in the same manner.

The demands for cubed, crushed, sized, or flake ice are great. To cool a beverage without too-rapid dilution, a smaller surface of ice is needed in the beverage, and ice cubes would be best.

If a beverage or other liquid is to be cooled quickly, such as a glass of water in a restaurant, flake ice will serve the purpose satisfactorily. The greater the area of the ice surfaces, the more quickly heat is removed from the liquid in which it is immersed.

If an ice cube is 1 in.^{2} on each side, it has 6 in.^{2} of surface in contact with the beverage for heat transfer. Therefore, if a glass contains four ice cubes, the total heat transfer area is 24 in.^{2}. If there is a hole in the center of each cube, the area is increased, but the volume of ice is less.

FIGURE R28-10   Typical flake ice machine. (*Courtesy* Liquid Carbonic Corporation)

Figure R28-9 shows a typical ice-cube maker capable of manufacturing about 8000 ice cubes in a 24-hour period, with a storage bin below it, which holds about 400 lb of ice cubes.

The manufacture of this ice is automatic, and its operation is controlled by a thermostat located in the storage bin, which is actuated by the level of ice remaining in the bin.

Flake ice machines, such as shown in Fig. R28-10, range in capacity from as small as 200 lb daily to those that produce approximately a ton of ice in a 24-hour period. The cylinder is sprayed with, or rotates in water, depending on its design—which also governs whether it rotates in a horizontal or a vertical position. A film of ice forms on the cylinder or drum and, when it becomes approximately $\frac{1}{16}$ of an inch in thickness, a cutting bar or roller breaks it up, allowing it to drop into the storage bin.

A glass filled with flake ice would probably hold at least 50 flakes measuring 1 in. by 1 in. by $\frac{1}{16}$ in. each. Even though they are thin, each flake would have 2 in.^{2} of heat-transfer area (one square inch on each side). The glass containing the flake ice would have approximately four times as much ice area in the beverage as the glass containing the ice cubes; therefore, the beverage with the flake ice would be cooled more quickly. Of course, the flake ice would melt much more quickly, too, thereby adding water to the liquid contents of the glass.

Ice for light commercial use is manufactured in many sizes and shapes, such as shell ice, tube ice, and even a thin film, which may be formed into small, briquette-type cubes. Some of the ice cube machines have attachments for crushing the cubes, if desired. Crushed or flake ice is used in restaurants and cafeterias for the chilling and display of salads and some desserts.

FIGURE R28-9   Typical ice cube maker. (*Courtesy* Liquid Carbonic Corporation)

R28-5
STORAGE LIFE

Figure R28-6 lists the approximate storage life of a number of perishable foods, particularly those that have been freshly picked or harvested. It also lists the storage life and temperatures of several types of frozen foods. It should be noted that there is a definite difference in storage temperatures, depending on the length of time the product is to be stored. The longer the storage time, the lower its storage temperature.

Of course, the condition or quality of the product has a definite relationship to the length of time it can be stored. If the product is to be presented for sale to the consumer immediately, it must be picked at its peak of ripeness. This also applies if the quality product is to be quick-frozen.

If they are to be shipped any appreciable distance, many vegetables and fruits are precooled before shipping and chilled while being transported. If this is not done, the respiration process will speed up in the fruits or vegetables and they will soon begin to decay. Meats, as pointed out earlier, are affected by the growth of bacteria, but they do not undergo the respiration process. Fruits and vegetables continue to absorb oxygen and release carbon dioxide even after harvesting.

FIGURE R28-11
Commodity storage requirements.
(Appropriate rates of evolution of heat by certain fresh fruits and vegetables when stored at the temperatures indicated.)[a]

Commodity	BTU/TON/24 HR			Commodity	BTU/TON/24 HR		
	32°F	40°F	60°F		32°F	40°F	60°F
Apples	500–900	1,100–1,600	3,000–6,800	Lettuce, leaf	4,500	6,400	14,400
Asparagus	5,900–13,200	11,700–23,100	22,000–51,500	Melons, cantaloupes	1,300	2,000	8,500
Avocados	—	—	13,200–30,700				
Bananas[b]	—	—	4,600–5,500	Melons, honeydews	—	900–1,100	2,400–3,300
Beans, green or snap	—	9,200–11,400	32,100–44,100	Mushrooms[d]	6,200–9,600	15,600	—
Beans, lima	2,300–3,200	4,300–6,100	22,000–27,400	Okra	—	12,100	31,600
Beets, topped	2,700	4,100	7,200	Onions	700–1,100	800	2,400
Blueberries[c]	1,300–2,200	2,000–2,700	7,500–13,000	Onions, green	2,300–4,900	3,800–15,000	14,500–21,400
Broccoli, sprouting	7,500	11,000–17,600	33,800–50,000	Oranges	400–1,000	1,300–1,600	3,700–5,200
Brussels sprouts	3,300–8,300	6,600–11,000	13,200–27,500	Peaches	900–1,400	1,400–2,000	7,300–9,300
				Pears	700–900	—	8,800–13,200
Cabbage	1,200	1,700	4,100	Peas, green	8,200–8,400	13,200–16,000	39,300–44,500
Carrots, topped	2,100	3,500	8,100	Peppers, sweet	2,700	4,700	8,500
Cauliflower	3,600–4,200	4,200–4,800	9,400–10,800	Plums	400–700	900–1,500	2,400–2,800
Celery	1,600	2,400	8,200	Potatoes, immature	—	2,600	2,900–6,800
Cherries	1,300–1,800	2,800–2,900	11,000–13,200	Potatoes, mature	—	1,300–1,800	1,500–2,600
Corn, sweet	7,200–11,300	10,600–13,200	38,400	Raspberries	3,900–5,500	6,800–8,500	18,100–22,300
Cranberries	600–700	900–1,000	—	Spinach	4,200–4,900	7,900–11,200	36,900–38,000
Cucumbers	—	—	3,300–7,300	Strawberries	2,700–3,800	3,600–6,800	15,600–20,300
Grapefruit	400–1,000	700–1,300	2,200–4,000				
Grapes, American	600	1,200	3,500	Sweet potatoes	—	—	4,300–6,300
				Tomatoes, mature green	—	1,100	6,200
Grapes, European	300–400	700–1,300	2,200–2,600	Tomatoes, ripe	1,000	1,300	5,600
Lemons	500–900	600–1,900	2,300–5,000	Turnips	1,900	2,200	5,300
Lettuce, head	2,300	2,700	7,900				

[a]Data largely from Table 1 of *USDA Handbook* No. 66, 1954. Acknowledgement is also due to the following for other commodities: Avocados, J. B. Biale; Brussels sprouts, J. M. Lyons and L. Rappaport; Cucumbers, I. L. Eaks and L. L. Morris; Honeydew melons, H. K. Pratt and L. L. Morris; Plums, L. L. Claypool and F. W. Allen; Potatoes, L. L. Morris; all from the Univ. of California. Asparagus, W. J. Lipton, USDA; Cauliflower, lettuce, okra, and onions, H. B. Johnson USDA; Sweet corn, S. Tewfik and L. E. Scott, Univ. of Maryland.

[b]Bananas at 68° F, 8,400–9,200.

[c]Blueberries at 50° F, 5,100–7,700; at 70° F, 11,400–15,000.

[d]Mushrooms at 50° F, 22,000; at 70° F, 58,000.

This process of absorbing oxygen and releasing carbon dioxide, through the chemical process of starch conversion into sugar within the living substance, generates heat. The increase of heat within the substance brings about an acceleration in the action of the enzymes in the product. This action should be reduced as much as possible so that the quality of the product will be maintained for presentation to the public.

Figure R28-11 lists the amounts of heat, rated in Btu/ton/day, given off by various fruits and vegetables at different temperatures. This table was used in Chapter R27, when calculations of heat loads, which occur under varying circumstances, are discussed. It is readily seen that this amount of heat is considerable, particularly at high temperatures. Therefore the product should be cooled to its lowest acceptable temperature as soon as possible.

The storage life of various products is important to fulfill the law of supply and demand. Usually, when there is a large supply of a product, even with an adequate demand, the price of such products may drop. This applies, for example, to fruits and vegetables "in season." When the product is not readily available or "out of season," the price demanded is usually somewhat higher. If people can have such things as strawberries, juicy apples, and even corn-on-the-cob during the winter months, they are often willing to pay a premium price for these delicacies.

Producers of many commodities may place some of their produce in storage to be available for sale during the "off season." An egg farmer may, when his hens are producing a surplus of eggs, place some of these products in storage so that they will be available for sale when production is lower.

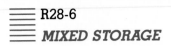

R28-6
MIXED STORAGE

To maintain quality merchandise under various storage conditions, separate storage areas or facilities for each of the products is desirable. This may not be feasible in many cases. It is possible to store several products in the same area if they have common recommended storage conditions. Usually, if mixed storage is used, it may call for compromise temperatures that may not be the optimum for some of the products. Some of the products may be stored at temperatures below point that is considered critical.

As already pointed out, higher storage temperatures may shorten the life of some of the products in mixed storage. Usually, this does not create too much of a problem, for the mixed storage is ordinarily only for a short period.

The big problem in mixed storage is the transfer or absorption of odors from one product to another. All dairy products, along with eggs, fruit, and even some kinds of nuts, easily pick up odors and flavors from other products. Onions and potatoes transfer or give off their odors and flavors to other products that may be stored with them in the same facility.

Customarily, larger storage facilities, such as those provided by wholesalers or jobbers, will be adequate to permit long-term storage of individual products in their own facilities; or facilities will be provided for the storage of vegetables in one area, fruit in another, and dairy products in still another.

PROBLEMS

R28-1. What refrigerant was used in the earliest commercial refrigerators?

R28-2. How is dehydration of food products in a refrigerator cabinet prevented?

R28-3. Approximately what temperature should be maintained in frozen-food storage or display cabinets?

R28-4. Generally speaking, there are two primary types of bacteria that affect foods in refrigerated storage. What are they?

R28-5. Refrigerating fruits or vegetables immediately upon harvesting preserves them for longer periods than if there is a time lag between the harvesting and refrigeration. Why?

R28-6. Before the use of refrigeration in railroad cars for fresh produce transportation, what method was used?

R28-7. What are eutectic plates?

R28-8. On what does the rate at which ice will cool water in a glass depend?

R28-9. Why is it necessary to store different types of produce under different conditions of temperature and relative humidity?

R28-10. The storage of different products in the same storage area is called _____ .

AIR-CONDITIONING

Introduction

A1-1
HOW OLD IS AIR-CONDITIONING?

Air-conditioning is as old as man himself. The primitive people shown in Fig. A1-1, who wore the skins of animals, were in a crude sense controlling the escape or containment of their own body heat and effecting a change in their personal comfort. Seeking shelter from the sun or finding refuge in caves from cold or heat were basically actions that changed their environment. The discovery and use of fire was perhaps the most important advance in that era.

Later history and artifacts show that the Egyptian ruling class used slaves (Fig. A1-2) equipped with palm branches to fan their masters. Thus, the art of evaporative cooling provided some relief from the desert heat. History also recalls the Romans, who engineered ventilation and panel heating into their famous baths. The Romans also brought ice from the northern mountains to chill wine, and possibly also to chill water for bathing.

FIGURE A1-1 Ancient couple around fire.

Moving into the Middle Ages, the remarkable Leonardo da Vinci built a water-driven fan to ventilate rooms of a house belonging to a patron friend. Other early innovations included rocking chairs with bellows action to produce spot ventilation for the occupant and clock mechanisms that activated fan devices above beds.

By today's standards, these examples of comfort conditioning seem rather crude, and perhaps, if fully explored, some would be humorous.

A1-2
EARLY TECHNICAL DEVELOPMENTS

The art of ventilation and central heating progressed rapidly during the nineteenth century. Fans, boilers, and radiators had been invented and were in fairly common use. Early warm-air furnaces were coal-fired cast iron with gravity air distribution. Soon mechanical fans were used for forced circulation of air through ducts. Modern concepts of furnaces bear little resemblance to some of those "iron monsters." Size, weight, and ventilation of combustion products have been drastically changed, but most important were the developments that led to the gradual coversion from coal to oil and gas, and from manual to automatic firing.

Early texts on refrigeration discussed the applications of using ice for preservation of food and the initial development of the concept of mechanical/chemical refrigeration in 1748 in Scotland by William Cullen.

It was in 1844 that John Gorrie (1803–1855), director of the U.S. Marine Hospital at Apalachicola, Florida, described his new refrigeration machine. In 1851 he was granted U.S. patent 8080. This was the first commercial machine in the world built and used for refrigeration and air-conditioning. Gorrie's machine received worldwide recognition and acceptance. Many improvements to Gorrie's work followed and by 1880 the development of reciprocating compressors with

FIGURE A1-2 Egyptian slave cooling master.

FIGURE A1-3 Willis Carrier. (*Courtesy* Carrier Air-Conditioning Company)

commercial application to ice making, brewing, meat packing, and fish processing were fairly common. Refrigeration engineering became a recognized profession and in 1904 some 70 members formed the ASRE (American Society of Refrigeration Engineers).

The real "father of air-conditioning" (Fig. A1-3) was Willis H. Carrier (1876–1950), as noted by many industry professionals and historians. Throughout his brilliant career, Carrier contributed more to the advancement of the developing industry than any other person. In 1911 he presented his epochmaking paper dealing with the properties of air. These assumptions and formulas formed the basis for the first psychrometric chart and became the authority for all fundamental calculations in the air-conditioning industry.

Carrier continued his work and invented the first centrifugal refrigeration machine in 1922 and later pioneered the use of induction systems for multiroom office buildings, hotels, apartments, and hospitals. During World War II, he supervised the design, installation, and startup of a system for the National Advisory Committee for Aeronautics (NACA) in Cleveland for cooling 10,000,000 ft³ of wind tunnel air down to −67°F (−19.45°C). Carrier died in 1950, having witnessed the real turning point in the industry's growth.

These were only a few of the steps along the way toward development of modern air-conditioning as we know it today.

A1-3
COMFORT COOLING COMES OF AGE

Comfort air conditioning had its first major use in motion picture theaters during the early 1920s. Famous New York City movie houses like the Rivoli, the Paramount, the Roxy, and Loew's Theaters in Times Square (Fig. A1-4) were among the first. By the late twenties several hundred theaters throughout the country had been air-conditioned. These were custom-designed, custom-manufactured, field-installed systems, which meant most of the assembly and erection of components was done on the job site.

Toward the end of the decade came the introduction of the first self-contained room air-conditioner. Not only was this an important technical achievement, but it became the industry's first attempt to "package" products that could be manufactured in so-called mass production and be factory tested and operated prior to shipment to the ultimate user.

The next milestone was the development of safe refrigerants. In 1930, Thomas Midgley of the Du Pont Company developed the now-famous fluorocarbon Freon refrigerant. In 1931 Freon-12 (Fig. A1-5) was introduced as a commercial refrigerant. Fluorocarbon refrigerants permit uses where more flammable or toxic material would be hazardous. Additionally, the operating characteristics of F-12 opened new vistas for compressor and system component designs. A whole family of Freon refrigerants soon followed where specific operating variations were needed. In 1955, other firms joined Du Pont in manufacturing the fluorocarbon refrigerants, and by 1956 a new numbering system was adopted to reflect the now-current R-12, R-22, etc., designations for these refrigerants.

About 1935 the industry introduced the first hermetic compressor for air-conditioning duty. Its size was considerably larger than today's equivalent capacities. Its motor speed was 1750 rpm compared to the common 3600 rpm of today. The outer shell was bolted rather than completely welded (Fig. A1-6). But the concept of passing cool suction gas to and over the motor windings started a trend in sizes up to $7\frac{1}{2}$ tons that has now become almost universal among compressor manufacturers. Full-welded hermetics (Fig. A1-7) are currently offered in capacities up to 20 tons in a single shell. The advantages of the hermetic concept include reduced size and weight; less cost to the manufacturer; no seal failure problems; less noise; no belt maintenance; location not critical insofar as ventilation requirements for motor heat dissipation; and the suction gas is superheated by

FIGURE A1-5 Freon 12. (*Courtesy* E. I. du Pont de Nemours and Company)

absorbing motor heat, so better oil separation is assured. Realiability reports from the industry support these technical advantages.

Post–World War II products consisted mainly of applied machinery systems (Fig. A1-8) for large buildings, store conditioners, and window air-conditioners. Window units (Fig. A1-9) were used extensively to cool residences, smaller offices, small stores, and just about any conceivable application where access to window or through-the-wall mounting could be achieved. However, the "main street" commercial comfort conditioning market, the drugstore, restaurants, dress shops, barber shops, grocery stores, etc., were handled by the *self-contained store conditioner* (Fig. A1-10). These units were primarily water cooled and were usually lo-

REMEMBER WHEN . . . summer meant straw hats and Saturday night talkies in a movie palace cooled with "frosted air". As air-conditioning caught on, this 1929 marquee was followed by more flamboyant signs featuring penguins and igloos. Temperatures in today's theatres are less of a novelty than in the past, a Loew's executive stated, because people are accustomed to air conditioning in their homes and places of work. In addition, industry product certification programs of the air-conditioning industry in effect for 13 years have helped air-conditioning engineers in assuring cool comfort in homes and public places.

FIGURE A1-4

FIGURE A1-6 Old bolted hermetic compressor. (*Courtesy* Copeland Corporation)

Introduction 343

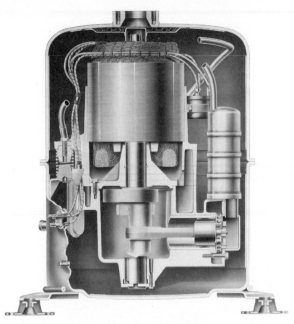

FIGURE A1-7   Three-ton compressor. (*Courtesy* Carrier Air-Conditioning Company)

FIGURE A1-9   Window air-conditioning unit.

cated in the conditioned space. Air distribution generally came from grilles on the discharge hood, although they could be ducted to some extent. Some disadvantages of these units included the need for a municipal water source and cost considerations, or cooling-tower application problems, excessive noise, high air velocity, and drafts; they also occupied valuable selling space in retail establishments. But even with these deficiencies, these units are credited with creating the public appreciation and awareness for central comfort air conditioning. Some manufacturers offered variations that could be adapted to residential use, but the number of installations were relatively few.

The next major breakthrough that really started the industry sales skyrocketing was the introduction in 1953 of air-cooled operation instead of water for condensing purposes. New technology in component and system design permitted pushing head pressures higher and higher so that machines could operate safely and reasonably efficiently up to 115°F outdoor ambient conditions. Early packaged units (Fig. A1-11) were mostly horizontal in configuration, for mounting in attics or on a slab at ground level. Installation consisted of mounting the unit, electrical hook-up, and a simple duct system to distribute air into the space. These units were equally adaptable to small commercial applications when mounted on the roof. Among the many advantages of these systems were: no refrigerant piping; little or no plumbing needed; a factory-charged, tested, and sealed refrigeration circuit; and minimum electrical wiring was

FIGURE A1-8   Applied machinery systems. (*Courtesy* Borg-Warner Air Conditioning, Inc.)

FIGURE A1-10   Typical water-cooled self-contained unit. (*Courtesy* Addison Products Company)

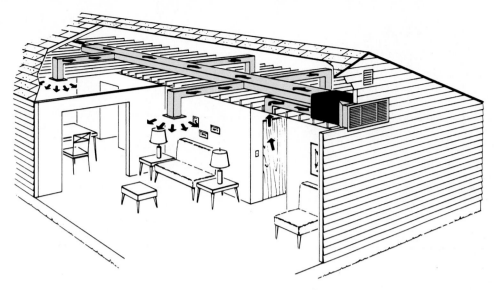

FIGURE A1-11 Horizontal packaged unit.

required; all of which reduced on-site labor and material costs and greatly increased reliability of operation. The main disadvantage associated with this product design was its lack of flexibility in adapting to all kinds of residential applications, particularly where heating and cooling were to be combined. As the expression goes, "necessity is the mother of invention"; the industry reacted quickly with the split-system concept.

The split system (Fig. A1-12), as the name implies, consists of two parts: an indoor cooling component and an outdoor condensing section. The two are connected by liquid and suction refrigerant lines. The indoor unit can be varied widely, first as an evaporator coil with upflow, downflow, or a horizontal airflow arrangement, depending on the heating system, or it can be equipped with a blower to provide its own air-handling capability. The adaptability of these systems made them an immediate success and there was extremely fast growth in their use. Early versions, however, required on-site fabrication and brazing of refrigerant lines and evacuating and refrigerant charging before becoming operational. This added measurably to installation costs and also required more highly trained installers. Reliability depended on the installer's technical skills. The industry responded with solutions and came up with the precharged refrigerant line and quick-connect couplings.

Precharged lines (Fig. A1-13) are cut and fabricated in various lengths, and the suction line is preinsulated with continuous covering. Both ends of the lines are equipped with mechanical couplings having a thin metal diaphragm to form a seal. The lines are dehydrated, evacuated, and charged with precise amounts of refrigerant. When installed at the job, the lines are mated to connections on both indoor and outdoor units. One part of the coupling has a cutting edge that pierces the diaphragm, folds it back, and opens the line. Thus, a complete refrigeration system is put together with a min-

imum of labor and minimum risk of contamination. The reliability of these systems also increased measurably. Chapter R5 defines other variations of mechanical fittings.

During this same period of development of the package and split-system cooling equipment, the industry also converted many of these cooling-only systems to reverse-cycle operation and called them *heat pumps* (Fig. A1-14), meaning that heat could be pumped into or out of the space, depending on which mode was needed. Unfortunately, because of pressure on electric utilities to build winter heating loads, early heat pumps were not fully developed and lacked reliability. To compound the matter, heat pumps were sold in extreme climates where operational performance was marginal. Finally, the number of qualified service personnel available was minimal or even nonexistent. The results were far from satisfying to all concerned. However, experience and technical improvements have restored confidence in the heat pump, and with the world's attention turning to energy conservation, it will probably gain more popularity in the future.

The final major innovation that occurred during the late fifties and early sixties and still accounts for the most dramatic growth rate of all packaged year-round conditioners was the *rooftop* combination gas-heating and electric-cooling unit (Fig. A1-15). It started with small 2-, 3-, 4- and 5-ton systems being installed on the rooftops of low-rise commercial structures of all descriptions, as well as residential applications. The trend rapidly spread. Equipment became more sophisticated and sizes changed rapidly; today's capacities range upwards of 100 tons of cooling. Both gas and electricity are available for heating. To coin a phrase, "Show me a shopping center and I'll show you a rooftop unit."

This review of the products that contributed to the coming-of-age for comfort air-conditioning by no means

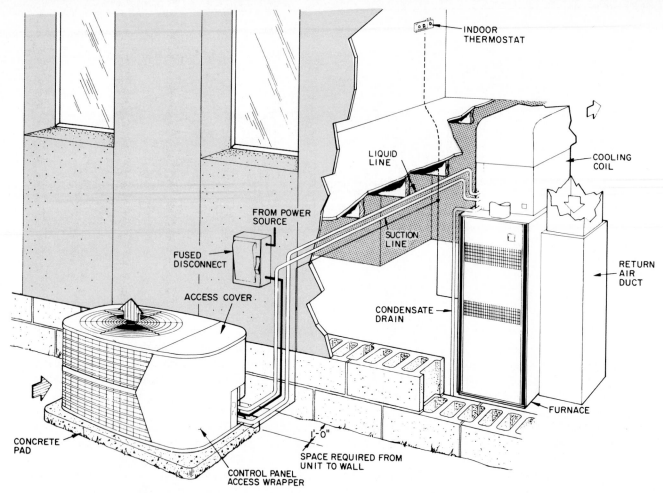

INDOOR
THERMOSTAT

LIQUID
LINE

COOLING
COIL

FROM POWER
SOURCE

SUCTION
LINE

FUSED
DISCONNECT

RETURN
AIR
DUCT

ACCESS COVER

CONDENSATE
DRAIN

CONCRETE
PAD

FURNACE

1'-0"

CONTROL PANEL
ACCESS WRAPPER

SPACE REQUIRED FROM
UNIT TO WALL

FIGURE A1-12   Split system. (*Courtesy* Carrier Air-Conditioning Company)

covers all the important technical advancements. In subsequent chapters, we explore further systems and products currently offered by the industry. Also, before leaving the history of the development of air-conditioning, we should note that *not all* the knowledge that led to such rapid growth came from manufacturers or suppliers of components. One of the most notable examples came from a trade association effort.

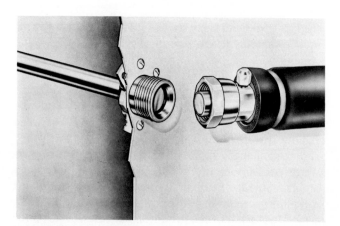

FIGURE A1-13   Precharged lines. (*Courtesy* Aeroquip)

In the late 1950s the Field Investigation Committee and the Research Advisory Committee of the National Warm Air Heating and Air-Conditioning Association, in cooperation with the University of Illinois, established the Engineering Experiment Station, which was in reality a residence on campus (Fig. A1-16). Many important studies were conducted on heat-flow characteristics, the effects of insulation, design, and the efficiency of air distribution systems, etc. The results of this effort helped greatly to advance the art of sizing selection and application of heating and air-conditioning and became the basis for a number of design manuals. The name of the association has now been changed to ACCA (Air Conditioning Contractors of America). Use of these manuals is highly recommended, as they are constantly updated.

A1-4
HISTORICAL MARKET GROWTH

Those who contemplate a career in air conditioning are naturally interested in the industry's past track record of growth and most frequently will ask, "Just

HEAT PUMP SYSTEM

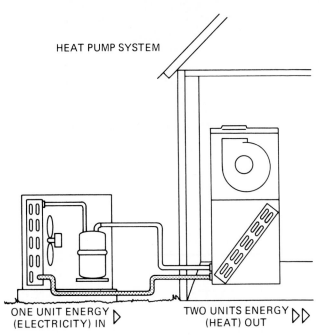

ONE UNIT ENERGY
(ELECTRICITY) IN ▷

TWO UNITS ENERGY ▷▷
(HEAT) OUT

FIGURE A1-14 Heat pump system. (*Courtesy* Carrier Air-Conditioning Company)

FIGURE A1-16 Engineering experiment station, University of Illinois campus. (*Courtesy* ACCA)

how big is this industry?" That's a difficult question to answer because of its many facets; however, we will attempt an explanation.

ARI keeps track of industry shipments and breaks them down into several classifications. First we will discuss *unitary* shipments, which for this report covers all of the packaged-type of air-conditioners for cooling only, heating and cooling, heat pumps, etc. Since 1950 the record of sales, represented by Fig. A1-17, shows a rise through 1983. Several observations can be made.

FIGURE A1-15 Rooftop combination gas heating and cooling unit. (*Courtesy* Borg-Warner Central Environmental Systems, Inc.)

First note the dramatic growth since 1958. Most will agree this must be related to the introduction of air-cooled systems into the residential market. By 1973 it is estimated that 75% of the total were residential units. The rate of growth demonstrated by the slant of the curve is even more dramatic when compared to growth curves of other industries or even the gross national product we hear so much about from economists or commentators. It will be noted that there was a slight dip in 1974; industry shipments totaled 2,490,000 units, down 13% from 1973. This can be directly related to the economy of 1974. Air-conditioning sales can be tied closely to the construction industry, residential and commercial, both of which were down measurably for that period. Nevertheless, the installed value of *unitary* equipment for 1974 was placed at over $3\frac{1}{2}$ billion. You will note the major dip in 1975. This recession was felt in all phases of the industry. It was short lived, however, as 1976 represented a 50% recovery as compared in 1974. 1978 was a complete recovery and equaled the peak year of 1973. 1980, 1981, and 1983 were down years compared to 1978, but 1983 has again equaled the peak sales years. The potential for future growth is only limited by the trained personnel available.

≡ A1-5
CENTRAL STATION EQUIPMENT

Sometimes also called *applied machinery* or *field-erected equipment,* this category of product includes water chillers, air handling equipment, ducts, etc., that together make up the larger tonnage central plant installation. In 1983, ARI estimated the installed value of this category to be approximately $5 billion, and since water chiller sales were up by 6.5% in 1984, it is reasonable to assume the total market increased accordingly.

Introduction   **347**

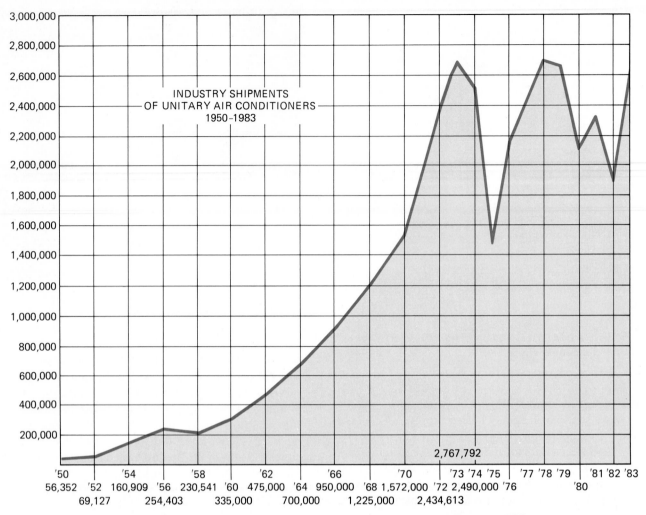

INDUSTRY SHIPMENTS
OF UNITARY AIR CONDITIONERS
1950-1983

2,767,792

| '50 | '52 | '54 | '56 | '58 | '60 | '62 | '64 | '66 | '68 | '70 | '72 | '73 '74 '75 | '76 | '77 '78 '79 | '80 | '81 '82 '83 |

56,352   '52   160,909   '56   230,541   '60   475,000   '64   950,000   '68 1,572,000 '72 2,490,000 '76

69,127   254,403   335,000   700,000   1,225,000   2,434,613

FIGURE A1-17   Industry shipments of unitary air-conditioners. (*Courtesy* ARI)

A1-6
OTHER PRODUCT MARKETS

Not included in the unitary or central station figures as reported by ARI are other products that can be classified in the broad, year-round air-conditioning field. Room-air-conditioner sales in 1983 were estimated at 5 million units. Package terminal units (through-the-wall units commonly used in motels) were estimated at 400,000 units. Sales of electronic air cleaners were slightly over 200,000 units.

Central heating equipment in 1983, as reported by GAMA (Gas Appliance Manufacturers Association), showed gas furnace sales at approximately 1.5 million units, oil furnaces at 275,000 units, electric furnaces at 400,000 units, and electrical duct heaters at 100,000 units.

Additionally, there is a group of products that are used in nonducted electric resistance heating, and estimates by the EEI (Edison Electric Institute) place this segment in excess of 400,000 units sold.

Finally, there are a great number of related accessories sold separately that are used to make up complete installations, such as grilles, registers, controls, filtering equipment, etc. There is no accurate method of determining or estimating their sales total, but it is probably in the millions of dollars.

Figure A1-18 shows the reported installed value of residential and commercial equipment installed in the United States, and since 1960 the amount exported.

In summary, the total annual installed value approaches $8 billion and that figure is underestimated, as not all manufacturers report to ARI. In addition, there is a developing foreign market.

A1-7
THE INTERNATIONAL MARKET

Of the 1983 unitary-air-conditioner shipments, 76,000, or 4%, went to foreign markets. U.S. manufacturers exported to some 153 countries. Major users

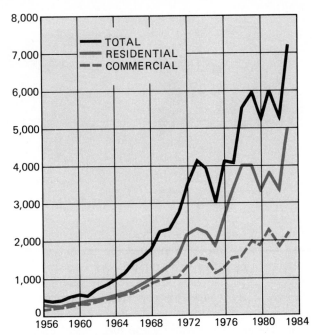

FIGURE A1-18 Estimated installed value of unitary air conditioners. (*Courtesy* ARI)

Schools	10.0%
Hotels	8.5%
Stores, supermarkets, shopping centers	14.0%
Hospitals	14.0%
Industrial plants	14.5%
Multiple dwellings	17.0%
Office buildings	22.0%

Of these commercial applications, some 57% were served by unitary products and 43% by central station systems.

A1-9
INDUSTRY MANPOWER NEEDS

In March 1973, ARI published a *Manpower Survey Report* (Fig. A1-19), which reflected projections of industry manpower demands through 1980. A survey was sent to air-conditioning contractors, mechanical contractors, service organizations, and commercial refrigeration companies, who were asked to estimate their requirements for *new* positions, not normal replacement. From this survey it was estimated that there would be over 50,000 new jobs to be filled. Of these, two-thirds were for technicians/mechanics in the fields of heating, air-conditioning, refrigeration, and sheet-metal work.

To put it another way, for every $1 million in installed value of equipment, the following employees would be needed:

  1 graduate engineer

  2 technicians/master mechanics

 11 air-conditioning, heating, and refrigeration mechanics

  2 air-conditioning, heating, and refrigeration helpers

  7 sheet-metal mechanics

  1 sheet-metal helper

  2 salespeople

were Canada, Japan, West Germany, Australia, the United Kingdom, Venezuela, France, Iran, Belgium, and Mexico. The United States and Japan produce about 90% of the world's HVAC (heating, ventilating, and air-conditioning) output, with Europe claiming the rest.

One major air-conditioning manufacturer estimates the total 1983 foreign consumption to be approximately $800 million, which includes the 76,000 units mentioned above. Factoring these out leaves a foreign manufactured supply of around $650 million.

By 1978 the same information source puts the total value at $1.6 billion and by 1983 at close to $3 billion. Obviously, this is an optimistic forecast, but it is reasonable in terms of the manufacturing facilities being erected in foreign lands.

A1-8
AIR-CONDITIONING USAGE BY TYPE OF MARKET

In 1973, U.S. manufacturers estimated 76% of unitary production went into residential consumption. The balance of 24% was spread across broad commercial uses. In 1974 these applications were estimated to be closer to a 70%–30% mix. Of the 70% residential use, approximately one-half went into new construction and the balance into existing home modernization.

Commercial applications were broken down into seven major groups:

A1-10
OCCUPATIONAL OPPORTUNITIES

Manufacturers hire a number of trained technicians for various jobs. Laboratory technicians actually build prototypes of new products, or they conduct tests on products for performance ratings, life-cycle testing, or compliance to ARI certification or UL safety standards or sound evaluation (Fig. A1-20). Factory service

1970-1980

MANPOWER
SURVEY
REPORT

FIGURE A1-19   Manpower survey report.

personnel prepare installation instructions, training material, parts information, etc. Some manufacturers employ a large number of service technicians for field assignment. They are active in installation and startup situations or in rendering actual customer service.

The contractor trades have the largest demand for personnel. To explain this, we will divide the contractor industry into two groups. Group I are heating, ventilating, and air-conditioning (HVAC) dealer/contractors who deal primarily with the residential and light commercial market. These businesses need personnel for installation and service (Fig. A1-21), as well as technically

FIGURE A1-20   Laboratory technician.

FIGURE A1-21   HVAC service personnel.

competent sales personnel who can estimate jobs, make duct layouts, and if so inclined, do the actual selling. Much of this type of contractor work is so-called self-designed, because there are no detailed plans and specifications; the contractor provides that service to the customer.

The other category is the mechanical contractor who works in the planning and specification market, which usually involves larger jobs that have been engineered by a consultant. This person must have qualified personnel to plan the work but, more important, must have skilled installers who possess a broad knowledge of refrigeration, steam fitting, general plumbing, sheetmetal, and control systems.

In the distribution function are HVAC and refrigeration wholesalers, distributors, supply houses, and factory branches where technicians are needed to help service the purchasing dealers. They act as counterpersonnel, service specialists, or salespeople, but must be well grounded in the technical arts.

Finally, there is a wide range of other opportunities, including operating and maintenance personnel for large institutions like universities, hospital complexes, office buildings, government buildings, military facilities, and industrial plants, all of which do their own in-house service and repair. This list is extensive and an excellent source of job opportunities.

Training and experience are the key ingredients to being a successful technician, and it starts with knowledge. Many programs are offered by vocational technical schools and community colleges for both the younger and the adult student. Manufacturers and trade associations are a continuing source of development training.

A1-11
TRADE ASSOCIATIONS

With the rapid growth and variety of interests, it was only natural that trade associations would evolve to represent specific groups. The list includes manufacturers, wholesalers, contractors, sheet-metal dealers, and service organizations. Each is important and makes a valuable contribution to the field. Space does not permit a detailed examination of all their activities, but throughout this book many of these associations will be acknowledged as specific subjects are covered. There is one organization, however, that is responsible for this book and related course outlines, and it is proper at this time to discuss that particular group.

The Air-Conditioning and Refrigeration Institute (ARI) is a trade association of manufacturing companies which produce residential, commercial, and industrial air-conditioning and refrigeration equipment, as well as machinery, parts, accessories, and allied products for use with such equipment.

ARI was formed in 1953 through a merger of two related trade associations. Since that time several other related trade associations have been merged into ARI, making it the strong association it is today. ARI traces its history back to 1903 when it started as the Ice Machine Builders Association of the United States. Today ARI has approximately 180 member companies.

ARI activities include:

1. Developing standards for testing and rating products.
2. Administering performance certification programs for industry products.

FIGURE A1-22 Company representatives at product section meeting.

3. Representing the industry to federal, state, and local government bodies in legislative and regulatory matters.
4. International trade research and analysis.
5. Public relations and consumer education programs.
6. Educational activities geared towards assisting vocational/technical schools.

FIGURE A1-23 ARI certification seals.

ARI is divided into separate product sections grouped according to product type. Member company representatives staff these sections, subsections, and committees (Fig. A1-22). Important decisions such as equipment standards and certification programs (Fig. A1-23) not only vitally affect equipment design and field application but also give assurance to customers, engineers, contractors, and specifiers.

ARI, in cosponsorship with ASHRAE, holds an annual International Air Conditioning–Heating–Refrigeration Exposition (Fig. A1-24), which may draw 30,000 to 40,000 people in the field, depending on key city location. Product exhibits and technical and business seminars highlight the event.

ARI has a full program of educational activities geared primarily toward helping the nation's vocational and technical schools improve and expand their education and training programs. Under the direction of its Education and Training Committee, ARI serves as the source of information from the manufacturers to school instructors, department heads, and guidance counselors. In addition to this textbook and its companion pieces, ARI produces the *Bibliography of Training Aids,* a career brochure, and a promotional videotape for schools to use to recruit students into HVACR pro-

grams. ARI's most recent efforts involve development of nationwide competency exams for students who graduate from HVACR programs.

FIGURE A1-24 International air-conditioning, heating, and refrigeration exposition.

PROBLEMS

A1-1. Name some ways in which early man tried to control the environment.

A1-2. Who was the first person to patent a "refrigeration machine"?

A1-3. Who is thought to be the first person to use the "refrigeration machine" to cool air?

A1-4. The Du Pont Company developed the first "safe" refrigeration, called _____ .

A1-5. Hermetic compressors for air conditioning were first introduced in about the year _____ .

A1-6. ARI stands for _____ .

A1-7. ARI is made up of _____ .

A1-8. "ACCA" stands for _____ .

A1-9. ACCA is made up of _____ .

A1-10. The industry growth potential is limited to _____ .

A1-11. For every $1,000,000 in installed equipment, how many air-conditioning and heating technicians are needed?

A2

Air-Conditioning Benefits

WHAT IS AIR-CONDITIONING?

That depends on what point of view is being considered. Ask the man on the street and he would most probably answer "keeping cool." Ask the owner of a printing plant and he would respond with a statement that might mean closely controlling temperature and humidity so that the behavior of the paper could be held within certain tolerances. One answer is from the standpoint of human comfort, the other is about a commercial consideration.

A dictionary definition might read "the process that heats, cools, cleans, and circulates air, and controls its moisture content on a continuous basis." So let's examine each definition as it affects the environment.

A2-2
HUMAN COMFORT CONTROL

The human body (Fig. A2-1) is a heat-generating device. Its normal temperature is 98.6° Fahrenheit. It can regulate or control this condition by four methods; convection, radiation, conduction, and evaporation. When in an environment where room conditions are too warm (but less than 98.6°), it will transfer heat to the air passing over the skin by convection. Simultaneously, it gives up heat by conduction to clothing, bedding, or whatever is in contact with the skin. Additionally, it throws off heat by means of radiation to the cooler surrounding objects. If these three are not sufficient, sweat glands will open, allowing skin moisture to evaporate. From previous discussion you noted this change of state from water to vapor absorbs much heat. Temperature change and air motion are important elements.

In colder surroundings, radiation, conduction, and convection take place more rapidly, thus requiring cloth-ing to insulate and hold body heat. Evaporation becomes minimal as the amount of skin perspiration decreases.

The body is also sensitive to impurities (Fig. A2-2). Dust, smoke, plant pollen, etc., cause irritation to the nose, lungs, and eyes, so this indicates another need for clean air.

Finally, the body requires "fresh air" (Fig. A2-3) to renew its oxygen supply or to dilute undesirable odors.

Stated simply, the body should have a comfortable and healthful atmosphere; and five properties of air must be treated:

1. Temperature (cooling or heating)
2. Moisture content (humidifying or dehumidifying)
3. Movement of the air (circulation)
4. Cleanliness of the air (filtering)
5. Ventilation (introduction of outside air)

The temperature of the air is indicated by the feeling of being hot or cold, and it can be measured by

FIGURE A2-1

FIGURE A2-2

FIGURE A2-3

means of an ordinary household thermometer (Fig. A2-4), commonly referred to as a *dry-bulb thermometer.*

The moisture content of the air is indicated by a feeling of dryness in winter or that muggy, sticky feeling in summer. Sometimes you can be uncomfortable from humidity in spite of the temperature. Humidity refers to the moisture that has evaporated into the air and exists as an invisible gas. To measure this feeling of humidity and express it in specific terms, a *wet-bulb thermometer* (Fig. A2-5) is used. Actually, it is nothing more than an ordinary thermometer with a wick or sock placed over the bulb. By wetting the sock and passing air over it, moisture will be evaporated until it balances with the moisture content of the air and no more evaporation takes place. Heat lost during the evaporation process lowers the temperature of the bulb; the lower reading is the *wet-bulb temperature.*

For the convenience of the technician, a *sling psychrometer* (Fig. A2-6) is used to take dry- and wet-bulb temperatures. It is simply two matching thermometers (scales and calibration the same) mounted on a common frame attached to a handle so the holder can rotate,

whirl, or sling the instrument around. It is rotated about two to three times per second until repeated readings become constant—perhaps a total of a minute.

By observing both the dry-bulb and wet-bulb thermometer readings, we have an indication of the *relative humidity,* meaning the actual amount of moisture in the air compared to the maximum amount that air could hold at that dry-bulb temperature. If the dry-bulb and wet-bulb readings were the same, the relative humidity would be 100%. The difference between dry-bulb and wet-bulb readings is called the *wet-bulb depression.*

Fig. A2-7 is a simple table that reflects the relative humidity in typical home conditions. For example, if the dry-bulb temperature were measured at 72°F and the wet-bulb at 61°F (a depression of 11°F), the relative humidity would be 53%.

Proper relative humidity is necessary so that the air will be dry enough in summer to absorb body perspiration for comfort. In winter the air should not be so dry that skin, nose, and throat have that dryness sensation. Also, too much humidity can cause mold, mildew, and rust.

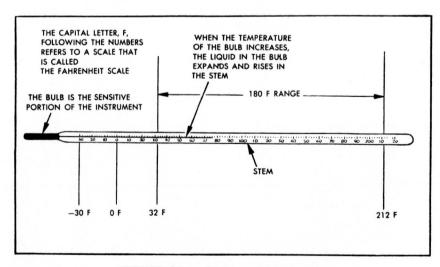

FIGURE A2-4   Ordinary thermometer.

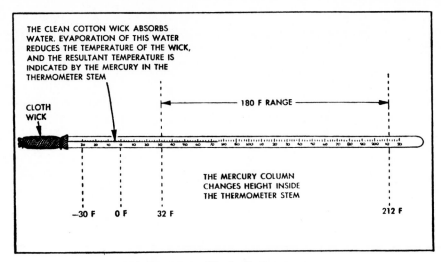

THE CLEAN COTTON WICK ABSORBS WATER. EVAPORATION OF THIS WATER REDUCES THE TEMPERATURE OF THE WICK, AND THE RESULTANT TEMPERATURE IS INDICATED BY THE MERCURY IN THE THERMOMETER STEM

CLOTH WICK

180 F RANGE

THE MERCURY COLUMN CHANGES HEIGHT INSIDE THE THERMOMETER STEM

−30 F   0 F   32 F   212 F

FIGURE A2-5   Wet-bulb thermometer.

A2-3
COMFORT ZONE

It is logical to ask what the desired temperature-humidity relationship is. The answer is that there is no one specific condition. People react differently to different conditions. ASHRAE (the American Society of Heating, Refrigerating, and Air-Conditioning Engineers) conducted a research study over many years, checking the reactions of large numbers of people to establish a range of combined temperatures, humidities, and air movement that provided the most comfort. This is known as the *comfort zone.* Each combination is known as an *effective temperature* (ET). It was found, for example, that with a given air velocity, a number of different combinations of dry-bulb temperatures and relative humidity readings would give the same feeling

of comfort to over 90% of the people involved. Thus, a comfort zone (Fig. A2-8) could be constructed. From the shaded zone of effective temperatures, it can be determined what dry-bulb temperature and relative humidity will produce that result. Note one obvious fact: the higher the humidity the lower the dry-bulb temperature can be.

The comfort zone chart is a good selling point with average people, since it explains how temperature and humidity should be controlled and thus shows the need for year-round air-conditioning. The chart is representative of the conditions found in homes, theaters, offices, etc., where periods of long occupancy occur. However, it is not completely accurate for conditions in retail stores, banks, drug stores, and similar situations, where short duration of occupancy coupled with rapid temperature changes and air motion will indeed change the experienced effective temperature. Therefore, when

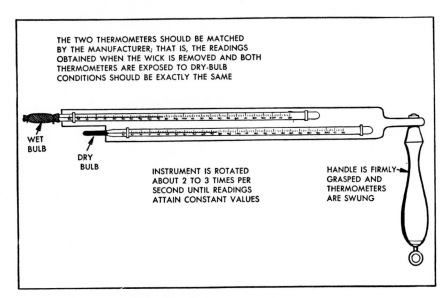

THE TWO THERMOMETERS SHOULD BE MATCHED BY THE MANUFACTURER; THAT IS, THE READINGS OBTAINED WHEN THE WICK IS REMOVED AND BOTH THERMOMETERS ARE EXPOSED TO DRY-BULB CONDITIONS SHOULD BE EXACTLY THE SAME

WET BULB

DRY BULB

INSTRUMENT IS ROTATED ABOUT 2 TO 3 TIMES PER SECOND UNTIL READINGS ATTAIN CONSTANT VALUES

HANDLE IS FIRMLY GRASPED AND THERMOMETERS ARE SWUNG

FIGURE A2-6   Sling psychrometer.

FIGURE A2-7

Psychrometric table: percent relative humidity from dry-bulb temperature and wet-bulb depression.

DB Temp. (°F)	1	2	3	4	5	6	7	8	9	10	11	12	13	14	15	16	17	18	19	20	21	22	23	24	25	26	27	28	29	30
												WB DEPRESSION																		
32	90	79	69	60	50	41	31	22	13	4																				
36	91	82	73	65	56	48	39	31	23	14	6																			
40	92	84	76	68	61	53	46	38	31	23	16	9	2																	
44	93	85	78	71	64	57	51	44	37	31	24	18	12	5																
48	93	87	80	73	67	60	54	48	42	36	34	25	19	14	8															
52	94	88	81	75	69	63	58	52	46	41	36	30	25	20	15	10	6	0												
56	94	88	82	77	71	66	61	55	50	45	40	35	34	26	24	17	12	8	4											
60	94	89	84	78	73	68	63	58	53	49	44	40	35	31	27	22	18	14	6	2										
64	95	90	85	79	75	70	66	61	56	52	48	43	39	35	34	27	23	20	16	12	9									
68	95	90	85	81	76	72	67	63	59	55	51	47	43	39	35	31	28	24	21	17	14									
72	95	91	86	82	78	73	69	65	61	57	53	49	46	42	39	35	32	28	25	22	19									
76	96	91	87	83	78	74	70	67	63	59	55	52	48	45	42	38	35	32	29	26	23									
80	96	91	87	83	79	76	72	68	64	61	57	54	54	47	44	41	38	35	32	29	27	24	21	18	16	13	11	8	6	1
84	96	92	88	84	80	77	73	70	66	63	59	56	53	50	47	44	41	38	35	32	30	27	25	22	20	17	15	12	10	8
88	96	92	88	85	81	78	74	71	57	64	61	58	55	52	49	46	43	41	38	35	33	30	28	25	23	21	18	16	14	12
92	96	92	89	85	82	78	75	72	69	65	62	59	57	54	51	48	45	43	40	38	35	33	30	28	26	24	22	19	17	15
96	96	93	89	86	82	79	76	73	70	67	74	61	58	55	53	50	47	45	42	40	37	35	33	31	29	26	24	22	20	18
100	96	93	90	86	83	80	77	74	71	68	65	62	59	57	54	52	49	47	44	42	40	37	35	33	31	29	27	25	23	21
104	97	93	90	87	84	80	77	74	72	69	66	63	61	58	56	53	51	48	46	44	41	39	37	35	33	31	29	27	25	24
108	97	93	90	87	84	81	78	75	72	70	67	64	62	59	57	54	52	50	47	45	43	41	39	37	35	33	31	29	28	26

designing systems you must consult the specific manufacturer, trade association, or local gas and electric utilities for recommendations. Until recently, it was general practice to design for the following indoor conditions; for winter, 72 to 75°F dry-bulb temperature with relative humidity at 35 to 40%; for summer, 75 to 78°F dry-bulb temperature and 50 to 55% relative humidity.

With the nation's attention on energy conservation and the ever-increasing cost of fuels, the above design conditions are subject to change. In 1974 the federal government recommended lowering the indoor winter dry-bulb temperature to 68°F and increasing the summer design temperature to 80°F for government buildings, and it encouraged private industry and homeowners to adopt similar practices.

A2-4
AIR MOTION

Air motion (Fig. A2-9) is another factor in comfort considerations. The comfort zone as presented above was based on an air movement of 15 to 25 ft/min velocity. The effective temperature drops sharply as velocity is increased. This would seem to be desirable for summer air-conditioning, but conditioned air being introduced into a room is usually 15 to 20° below room conditions; and if velocity should approach 100 ft/min, cold drafts would be noticeable.

Forced warm-air heating systems are even more subject to drafts, particularly when the blower first comes on. The skin seems to react more quickly to warm air currents, and a good rule of thumb is not to exceed 50 ft/min velocity in the comfort zone. Recent industry discussion would suggest limiting *year-round* velocity to 70 ft/min. Ways of planning for good air distribution will be covered more completely in later chapters.

Too little air circulation should also be avoided, because people tend to feel "closed in." This may be a disadvantage of the nonducted type of heating systems

COMFORT CHART

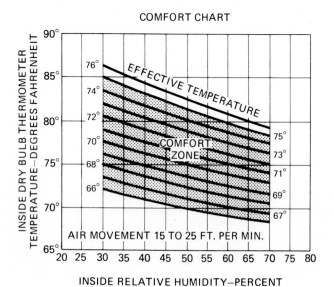

FIGURE A2-8 Comfort zone.

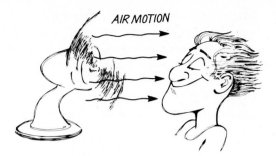

FIGURE A2-9 Air motion.

FIGURE A2-11 Ventilation.

that depend on gravity circulation and have no mechanical means of filtration. The situation becomes even more critical where the structure is well-insulated and infiltration of outdoor air very small.

Cleanliness and ventilation are the last two necessities for proper air treatment and are to some extent related and work against each other. We breathe 36 lb of air each day, compared to eating 3.8 lb of food and drinking 4.3 lb of water. Clean air is important to both our health and comfort. Ordinary air (Fig. A2-10) is contaminated with impurities such as dust, pollen, smoke, fumes, and chemicals. All of these should be filtered out of both the indoor air and the outdoor air that enters the structure.

The efficiency of filtration depends on the type of system, and we will review these options later. Some have the capability of removing more than 95% of all impurities. However, even with the finest filtering equipment a degree of ventilation air (Fig. A2-11) is needed to eliminate that stale, dead-air feeling, and also to dilute odors and supply oxygen for breathing and makeup air for vented appliances (exhaust fans, dryers, furnaces, water heaters, etc.). The amount of outside air needed depends on the conditioned space. Figure A2-12 lists some typical applications and the ASHRAE recommendations, based on cubic feet of air per minute per person. Outdoor air should always be provided and properly filtered before entering the conditioned space. Many local codes and ordinances specify minimum standards for schools and public buildings.

FIGURE A2-10 Cleanliness of air.

This review of the characteristics of air has not considered any economic impact. As with all human needs there must be compromise, depending on the individual's needs and his financial capability. A window air-conditioner basically performs the five functions, but cannot be compared to a true total comfort, year-round central air-conditioning system in function or form. Thus the industry offers a wide variety of products to suit many potential markets. In addition to the pure pleasure comforts, air-conditioning offers other important advantages.

A2-5
MEDICAL CONSIDERATIONS

Air-conditioning may contribute to better health as a result of controlling air temperature, humidity, cleanliness, ventilation, and motion. For example, it may help eliminate heat rash, particularly among infants. Some physicians believe that air-conditioning can provide an environment that is easier for people suffering from ailments such as heart trouble to accept. Just moving around in extreme heat may put an undue strain on vital organs, the same as the strain from heavy manual labor, like shoveling snow in cold weather. As a result, doctors sometimes recommend air conditioning for their patients.

FIGURE A2-12
ASHRAE ventilation table.

Application	Smoking	FT³/MIN PER PERSON	
		Optimum	Minimum
Bank	Some	10	7½
Cocktail bar	Heavy	40	25
Offices	Some	15	10
General	None	25	15
Private	Considerable	30	25
Residence	Occasional	20	10
Restaurant	Considerable	15	12
Retail store	Very little	10	7½
Theater	Some	15	10

An effective air filtering system may relieve the sufferings of asthmatic and allergic patients. Such a system might include an electronic air cleaner as part of the home central system. Ordinary house dust, in addition to dust in the outdoor air, can contribute to allergy problems. House dust is a complex mixture that develops from the breakdown and wearing out of material in the house such as cotton, linens, wool, furniture stuffings, and carpet. Pollen allergy irritants, however, come from the outdoors. In the height of the hayfever season the pollen count in a home with central air-conditioning may be reduced considerably if an effective air cleaning system is used.

The U.S. Public Health Service has called air pollution a contributing cause of cancer and a serious irritant to lung and respiratory tissue. In 1952, a heavy chemical-laden smog in London was reported to be the immediate cause of death for 4000 persons. The problem is also staggering throughout the United States, and thousands of specialists in industry and government are engaged in research looking for solutions. In the meantime, the homeowner and employer can protect themselves and their personnel with air-cleaning apparatus.

Although it has not been proven conclusively, there is some belief that proper humidity may also help protect health. Adequate humidity in the air can help the membranes in the nose and respiratory tract to remain moist. This can alleviate the effect of bacteria and viruses. Lack of humidity also promotes dryness of those household dust sources mentioned above and thereby keeps the dust airborne.

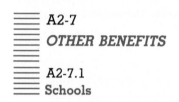

A2-6
HUMIDIFICATION PROTECTS PROPERTY

An atmosphere that is too dry can adversely affect furniture, clothing, shoes, books, documents, plaster, musical instruments—virtually everthing in the home, including structural members. Let's look at a few examples: Lack of adequate humidity causes glue to dry out in tables and chairs and other furniture, and joints separate and crack. Door panels shrink and expose unpainted surfaces. Plaster or drywall joints dry out, which may lead to unsightly cracks. Musical instruments may lose tone. Hardwood floors may shrink and separate. Rug fibers become brittle and break.

Still another phenomenon experienced with winter dryness is the presence of, or generation of, static electricity. Most of us have touched a door knob and jumped from a jolt of static electrical shock. Although not harmful, it is disconcerting to say the least. A properly sized humidifier installed in a forced-air duct system can minimize or eliminate all these conditions.

The security benefits are more or less a recent consideration but, with the ever-increasing crime rate, many homeowners are installing mechanical and electrical security protection devices that require all doors and windows to be closed and locked. Without central air-conditioning, these devices would not be practical.

The social benefits of having central air conditioning are perhaps the least definable, but there are a number of homeowners who do a considerable amount of entertaining at home because of their social position and also for business reasons. Such concentrations of many people, along with smoking at cocktail and dinner parties, demand full treatment for all five properties of air.

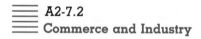

A2-7
OTHER BENEFITS

A2-7.1
Schools

Schools have been a prime market for year-round air conditioning, not only for the personal comfort it affords, but studies have demonstrated the learning process is definitely improved for the student and teacher. Interestingly, at the time this text was being prepared, two articles appeared in the *Air-Conditioning, Heating and Refrigeration News* trade paper which said in part, "Clearwater, Florida—More than 1000 teachers belonging to the Pinellas (County) Classroom Teachers' Association here have ranked air-conditioning as its top priority." From Fort Worth, Texas—" . . . air-conditioning for all schools here is the chief interest for a bond issue by the Certified Teachers' Association (CTA), which believes it is a necessity for a good learning environment." As population increases, the need for year-round school operation becomes apparent.

A2-7.2
Commerce and Industry

Commerce and industry have used air-conditioning in several ways: first to increase personal productivity and second to provide space process-cooling for specific needs.

Worker productivity in air-conditioned areas has improved, in terms of less absenteeism, less labor turnover, less noise distraction, less trips to the water fountain, more efficient production, fewer mistakes, and less time lost due to heat fatigue and accidents. In general, better morale and better relationships between employer and employee result. The degree of benefit, of course, is subject to the type of surroundings—office worker as compared to garment seamstress, as compared to assembly line operator. No one would think of constructing a modern office building today without air-conditioning. Trade unions have also been instrumental in the growth of comfort air-conditioning for their members, most notably in the textile industry.

The uses of process cooling would make a list almost as long as this book; however, just to name a few: electronic computer rooms must be controlled closely as to temperature and humidity. Tapes and card decks need uniform conditions. Current computer designs require that large amounts of heat be removed. Printing plants must have a controlled atmosphere to maintain tolerances of paper shrinkage, accurate register in color printing, and efficient paper feed through the presses. Clean rooms, for all types of tools and gauges and highly critical precision manufacturing, must maintain close tolerances of conditions to protect dimensional accuracy, fits, allowances, gauging, and manufacturing limits. Temperature, humidity, and cleanliness are very important. A telephone exchange building is something most of us accept as a part of telephone services. The myriad number of electrical contacts that are made and broken when phone numbers are dialed is fantastic. The slightest speck of dust or wide variations in temperature and humidity can cause an erratic or even a broken contact. The drug industry is one of the largest users of industrial air conditioning; the list goes on and on.

A2-8
MARKETING TOOL FOR BUILDER
AND DEVELOPER

In the early 1950s, residential builders were reluctant to install central air conditioning as standard equipment in new homes, and for good reason. The Federal Housing Administration and many local savings and loan companies were not favorably inclined toward increasing selling prices, which might result in mortgage payments that would possibly overextend the lower-income home buyers' ability to meet monthly payments. Air-conditioning was given a lower priority compared to certain other appliances in the new home. As a result, builders installed the heating system and offered cooling as a later option. Fortunately, there were builders like Levitt and Sons of Levittown, Pennsylvania, Ryan Homes of Pittsburgh, and Fox and Jacobs of Dallas, who built large tract housing developments with central air-conditioning as standard equipment. The systems were pre-engineered by personnel from the manufacturer and builder and installed by trained technicians. Most important, mass purchasing brought installation costs to a minimum. Thus, a trend was started. Other builders and the broad financial community jumped on the bandwagon, and soon all were making statements like "any house that is built without central air-conditioning is already obsolete."

The explosion of apartment and condominium townhouse development in today's housing market could not have reached the heights it has without central air-conditioning. Common walls and limited access to the outdoors makes indoor living a way of life and air-conditioning an absolute necessity, as well as a good sales tool.

This chapter has given an overview of the properties of air-conditioning from the physiological, sociological, and commercial standpoints. Now let's consider the mechanical aspects of air and its behavior: the art of psychrometrics.

PROBLEMS

A2-1. What are the five control functions of air conditioning?

A2-2. What is normal body temperature?

A2-3. Moisture in the air is called _____ .

A2-4. Define "comfort zone."

A2-5 What is the recommended indoor heating temperature?

A2-6. What is the recommended cooling temperature and humidity?

A2-7. Air motion is measured in _____ .

A2-8. The maximum air velocity to prevent drafts is _____ .

A2-9. The thermometer used to measure the temperature of air is called a _____ thermometer.

A2-10. The humidity level in air is measured by a thermometer called a _____ type.

A2-11. The common instrument used to measure temperature and humidity at the same time is called a _____ .

A2-12. The difference in readings between the two thermometers is called the _____ .

A2-13. A chart showing all the combinations of temperatures and humidity at which people are comfortable is called a _____ chart.

A2-14. A person breaths _____ lb of air each day.

A2-15. Common air pollution is made up of _____ .

A2-16. The recommended amount of fresh air that should be brought into a residence for proper ventilation is between _____ and _____ ft²/min per person.

A2-17. Electronic air cleaners are said to be able to remove an excess of _____ % of all impurities.

A3

Psychrometrics

A3-1
GENERAL

Who uses psychrometrics? Consulting engineers, research and product development engineers, and application engineers, for example, need an in-depth understanding and working knowledge of the art, for their professional existence requires practical application of the theories. Does a service technician really need to know all of the information? He or she needs to know enough to be able to determine if air problems are the cause of unsatisfactory results of system operation. A knowledge of the basic principles, definition of terms, the understanding of psychrometric charts, and the relationship of the elements involved are necessary to be able to analyze problems as well as report information on equipment performance.

For example, unitary packaged products, particularly the residential kind, are predesigned by factory engineers with specific performance characteristics. There is little the technician can do to change their operation except possibly to vary the fan speed. Performance tables are available that rate equipment based on entering dry-bulb and wet-bulb conditions, outdoor ambient conditions, sensible-to-latent-heat ratios, total output, etc. What the service person needs to know are the

terms and definitions being used and how to understand the information being presented. The following discussion is designed to provide these basic skills.

A3-2
BASIC TERMS

Atmosphere: The air (Fig. A3-1) around us is composed of a mixture of dry gases and water vapor. The gases contain approximately 77% nitrogen and 23% oxygen, with the other gases totaling less than 1%. Water vapor (Fig. A3-2) exists in very small quantity so it is either measured in *grains* or *pounds*. (It takes 7000 grains to make 1 lb.)

Dry-Bulb Temperature: The temperature as measured by an ordinary thermometer.

Wet-Bulb Temperature (see Chapter A2): The temperature resulting from the evaporation of water from a wetted sock on a standard thermometer.

Dew-Point Temperature: The saturation temperature at which the condensation of water vapor to visible water takes place. An example is the sweating on a glass of ice water. The cold glass reduces the air temperature below its dew-point, and the moisture that condenses forms beads on the glass surface.

Specific Humidity: The actual weight of water vapor in the air expressed in grains or pounds of water per pound of dry air, depending on which data are used.

Relative Humidity: The ratio of actual water vapor in the air compared to the maximum amount that could be present at the same temperature, expressed as a percent.

Specific Volume: The number of cubic feet (ft³) oc-

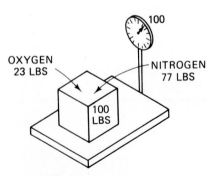

FIGURE A3-1 Atmosphere.

360

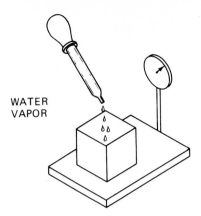

WATER
VAPOR

FIGURE A3-2 Water vapor.

cupied by 1 lb of the mixture of air and water vapor.

Sensible Heat: The amount of dry heat expressed in Btu per pound of air; it is reflected by the dry-bulb temperature.

Latent Heat: The heat required to evaporate the moisture a specific amount of air contains. This evaporation occurs at the wet-bulb temperature. It, too, is expressed in Btu per pound of air.

Total Heat (Enthalpy):   The total heat content of the air and water vapor mixture is also known as *enthalpy*. It is the sum of both sensible and latent heat values, expressed in Btu per pound of air.

A3-3
PSYCHROMETRIC CHART

The psychrometric chart in Fig. A3-3 is probably the best way of showing what happens to air and water vapor as these properties are changed. The chart is published by ASHRAE and is the one most commonly used in the industry. Enthalpy values at various temperatures are listed in Fig. A3-4. Some manufacturers have developed their own charts, which vary only in style and construction; the relationship of the air properties are all the same.

To make this chart all we do is start with the ordinary temperature scale called the *dry-bulb temperature*. Just extend the thermometer scale as shown in Fig. A3-5. Note on the actual chart that these lines are not truly perpendicular. This is done so that other lines will come out straight instead of curved.

Next, the vertical scale is set up according to the amount of water vapor mixed with each pound of dry air. This scale (Fig. A3-6), called the *humidity ratio,* is expressed in pounds of moisture per pound of dry air. We know that air can hold different amounts of mois-

ture depending on its temperature; if it is holding all the moisture it can (100%), it is termed *saturated*.

From the *ASHRAE Guide and Data Book* we can find out exactly how much moisture air can hold at saturated conditions. Following is a simple table taken from this reference book:

Saturated Temperature (°F DB)	Humidity Ratio (lb/lb dry air)
70	0.01582
72	0.01697
75	0.01882
78	0.02086
80	0.02233
82	0.02389
85	0.02642

Returning to the psychrometric chart construction, we can now plot saturation points (Fig. A3-7) for each condition of dry-bulb temperature, and when these are connected they form a curve or *saturation line*.

Assume an air sample (point *A*, Fig. A3-8) with a dry-bulb temperature of 80°F, holding 0.011 lb of moisture. If we were to heat the air without adding moisture, the point would move to the right on the horizontal line, showing an increasing dry-bulb temperature but an unchanging moisture content.

If we were to add moisture (humidify) without changing the dry-bulb temperature, the point would move vertically up. If the moisture were reduced (dehumidifying), it would move vertically down. If temperature and moisture were added, the point would move up and to the right, and if the air were cooled (without changing its moisture content), the point would move horizontally to the left.

Continuing the example, if the air sample is cooled, it eventually reaches the saturation line (point *B*, Fig. A3-9) where it cannot hold any more water vapor, and on further cooling some water would start to condense. That temperature is just below 60°F, or about 59.7°F. This is known as the *dew-point temperature* of the sample. It can be read from the vertical dry-bulb index temperature. In summary, at point *B*, we have a 59.7°F dry-bulb temperature, a 59.7° dew-point temperature, and a moisture content of 0.011 lb of moisture per pound of dry air.

Now if the sample is further cooled, for example to 50°F dry-bulb, moisture will condense out and follow along the saturation line to point *C* (Fig. A3-10), where it will have a dew point of 50°F and a humidity ratio of only 0.0076. Thus, the sample has lost 0.0034 lb of moisture. It has been cooled and dehumidified.

A practical example of this process is a cold supply air duct (as shown in Fig. A3-11) running through a

ASHRAE PSYCHROMETRIC CHART NO. 1

NORMAL TEMPERATURE
BAROMETRIC PRESSURE 29.921 INCHES OF MERCURY
COPYRIGHT 1963
AMERICAN SOCIETY OF HEATING, REFRIGERATING AND AIR-CONDITIONING ENGINEERS, INC.

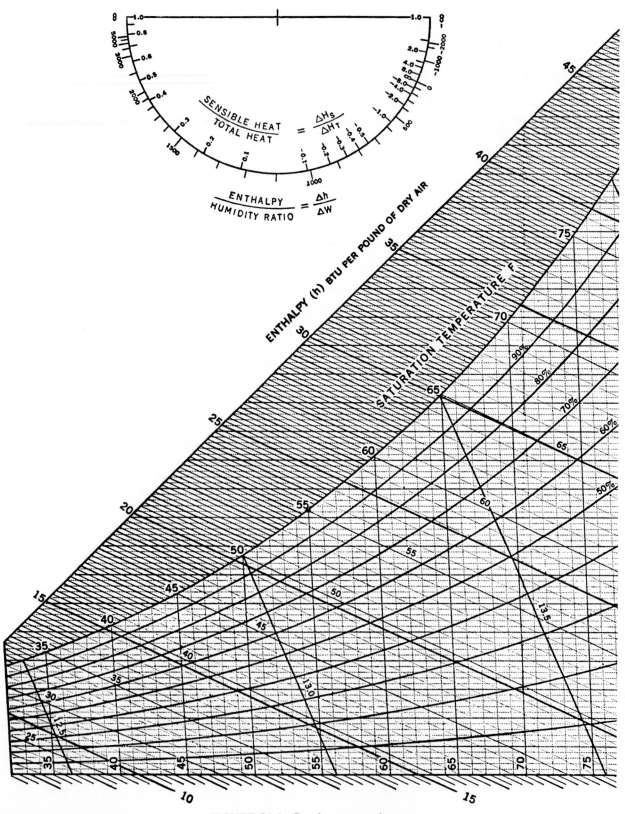

$$\frac{\text{SENSIBLE HEAT}}{\text{TOTAL HEAT}} = \frac{\Delta H_s}{\Delta H_T}$$

$$\frac{\text{ENTHALPY}}{\text{HUMIDITY RATIO}} = \frac{\Delta h}{\Delta W}$$

FIGURE A3-3 Psychrometric chart.

362 Air-Conditioning

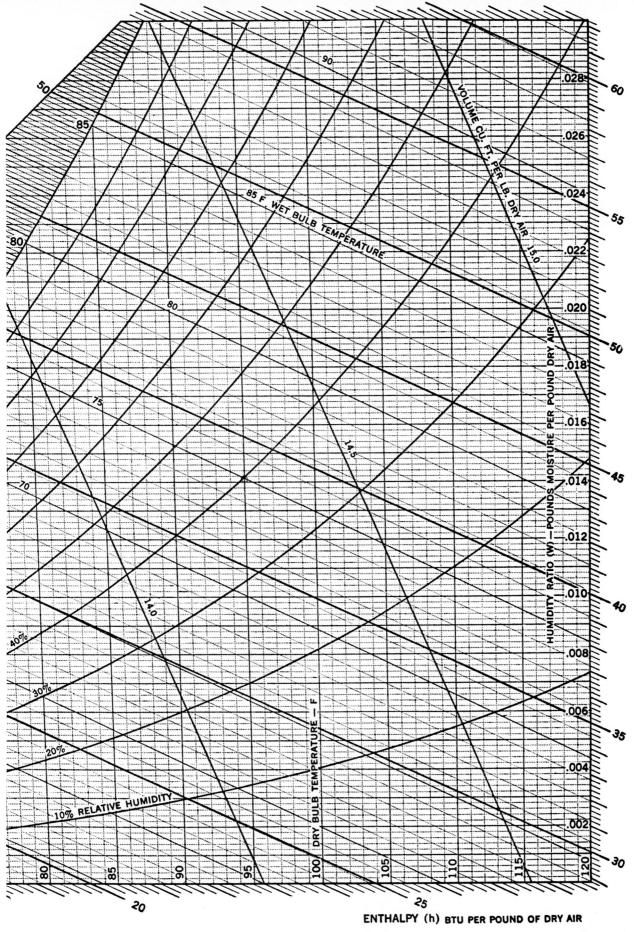

FIGURE A3-3   (Continued)

ENTHALPY* IN BTU PER POUND OF DRY AIR

Wet Bulb Temperature F	TENTHS OF A DEGREE									
	.0	.1	.2	.3	.4	.5	.6	.7	.8	.9
35	13.01	13.05	13.10	13.14	13.18	13.23	13.27	13.31	13.35	13.40
36	13.44	13.48	13.53	13.57	13.61	13.66	13.70	13.75	13.79	13.83
37	13.87	13.91	13.96	14.00	14.05	14.09	14.14	14.18	14.23	14.27
38	14.32	14.37	14.41	14.46	14.50	14.55	14.59	14.64	14.68	14.73
39	14.77	14.82	14.86	14.91	14.95	15.00	15.05	15.09	15.14	15.18
40	15.23	15.28	15.32	15.37	15.42	15.46	15.51	15.56	15.60	15.65
41	15.70	15.75	15.80	15.84	15.89	15.94	15.99	16.03	16.08	16.13
42	16.17	16.22	16.27	16.32	16.36	16.41	16.46	16.51	16.56	16.61
43	16.66	16.71	16.76	16.81	16.86	16.91	16.96	17.00	17.05	17.10
44	17.15	17.20	17.25	17.30	17.35	17.40	17.45	17.50	17.55	17.60
45	17.65	17.70	17.75	17.80	17.85	17.91	17.96	18.01	18.06	18.11
46	18.16	18.21	18.26	18.32	18.37	18.42	18.47	18.52	18.58	18.63
47	18.68	18.73	18.79	18.84	18.89	18.95	19.00	19.05	19.10	19.16
48	19.21	19.26	19.32	19.37	19.43	19.48	19.53	19.59	19.64	19.70
49	19.75	19.81	19.86	19.92	19.97	20.03	20.08	20.14	20.19	20.25
50	20.30	20.36	20.41	20.47	20.52	20.58	20.64	20.69	20.75	20.80
51	20.86	20.92	20.97	21.03	21.09	21.15	21.20	21.26	21.32	21.38
52	21.44	21.50	21.56	21.62	21.67	21.73	21.79	21.85	21.91	21.97
53	22.02	22.08	22.14	22.20	22.26	22.32	22.38	22.44	22.50	22.56
54	22.62	22.68	22.74	22.80	22.86	22.92	22.98	23.04	23.10	23.16
55	23.22	23.28	23.34	23.41	23.47	23.53	23.59	23.65	23.72	23.78
56	23.84	23.90	23.97	24.03	24.10	24.16	24.22	24.29	24.35	24.42
57	24.48	24.54	24.61	24.67	24.74	24.80	24.86	24.93	24.99	25.06
58	25.12	25.19	25.25	25.32	25.38	25.45	25.52	25.58	25.65	25.71
59	25.78	25.85	25.92	25.98	26.05	26.12	26.19	26.26	26.32	26.39
60	26.46	26.53	26.60	26.67	26.74	26.81	26.87	26.94	27.01	27.08
61	27.15	27.22	27.29	27.36	27.43	27.50	27.57	27.64	27.71	27.78
62	27.85	27.92	27.99	28.07	28.14	28.21	28.28	28.35	28.43	28.50
63	28.57	28.64	28.72	28.79	28.87	28.94	29.01	29.09	29.16	29.24
64	29.31	29.39	29.46	29.54	29.61	29.69	29.76	29.84	29.91	29.99
65	30.06	30.14	30.21	30.29	30.37	30.45	30.52	30.60	30.68	30.75
66	30.83	30.91	30.99	31.07	31.15	31.23	31.30	31.38	31.46	31.54
67	31.62	31.70	31.78	31.86	31.94	32.02	32.10	32.18	32.26	32.34
68	32.42	32.50	32.59	32.67	32.75	32.84	32.92	33.00	33.08	33.17
69	33.25	33.33	33.42	33.50	33.59	33.67	33.75	33.84	33.92	34.01
70	34.09	34.18	34.26	34.35	34.43	34.52	34.61	34.69	34.78	34.86
71	34.95	35.04	35.13	35.21	35.30	35.39	35.48	35.57	35.65	35.74
72	35.83	35.92	36.01	36.10	36.19	36.29	36.38	36.47	36.56	36.65
73	36.74	36.83	36.92	37.02	37.11	37.20	37.29	37.38	37.48	37.57
74	37.66	37.76	37.85	37.95	38.04	38.14	38.23	38.33	38.42	38.52
75	38.61	38.71	38.80	38.90	38.99	39.09	39.19	39.28	39.38	39.47
76	39.57	39.67	39.77	39.87	39.97	40.07	40.17	40.27	40.37	40.47
77	40.57	40.67	40.77	40.87	40.97	41.08	41.18	41.28	41.38	41.48
78	41.58	41.68	41.79	41.89	42.00	42.10	42.20	42.31	42.41	42.52
79	42.62	42.73	42.83	42.94	43.05	43.16	43.26	43.37	43.48	43.58
80	43.69	43.80	43.91	44.02	44.13	44.24	44.34	44.45	44.56	44.67
81	44.78	44.89	45.00	45.12	45.23	45.34	45.45	45.56	45.68	45.79
82	45.90	46.01	46.13	46.24	46.36	46.47	46.58	46.70	46.81	46.93
83	47.04	47.16	47.28	47.39	47.51	47.63	47.75	47.87	47.98	48.10
84	48.22	48.34	48.46	48.58	48.70	48.83	48.95	49.07	49.19	49.31
85	49.43	49.55	49.68	49.80	49.92	50.05	50.17	50.29	50.41	50.54

* Interpolated to tenths of degrees from 1963 edition ASHRAE Guide and Data Book. Published by the American Society of Heating, Refrigerating and Air Conditioning Engineers, Inc.

FIGURE A3-4 Enthalpy in Btu per pound of dry air. (*Courtesy* ASHRAE)

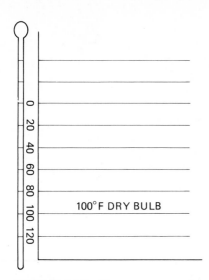

FIGURE A3-5   Thermometer scale.

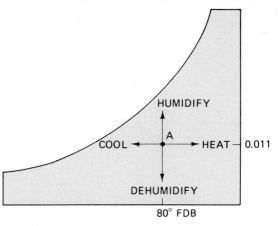

FIGURE A3-8   Air sample.

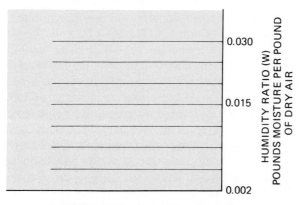

FIGURE A3-6   Humidity ratio scale.

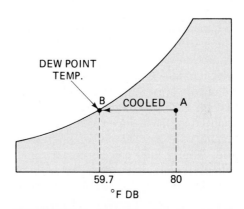

FIGURE A3-9   Saturation line for point *B*.

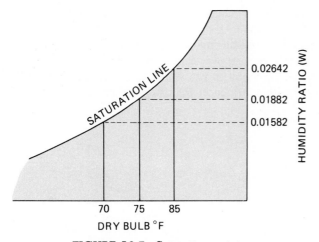

FIGURE A3-7   Saturation points.

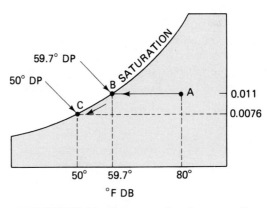

FIGURE A3-10   Saturation line for point *C*.

moist unconditioned area. Will the duct sweat and need to be insulated? Assume that the air temperature inside the duct is 55°F and the unconditioned air surrounding the duct is at 95°F with 0.0142 lb of moisture content. This condition means that the outside air would have a saturated (dew-point) temperature of 67°F. Thus, as the 55°F duct temperature cools the air touching its surface

to below the 67°F dew point, condensation will likely occur. Depending on conditions, it will be necessary to take some corrective action using appropriate insulation to prevent sweating.

The next element in our chart is the construction of relative humidity lines for partly saturated conditions (Fig. A3-12). We know the relative humidity is 100% at

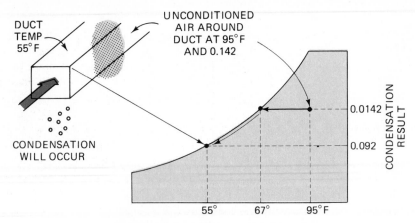

FIGURE A3-11 Cold-supply air duct.

the saturation line. Lines for 80%, 60%, 40%, etc., can be plotted, since we know specific moisture contents in relation to temperatures. As an example, 1 lb of air at 75°F dry bulb will hold 0.01882 lb of moisture (point A) at a saturation (100% relative humidity). Point B (50% relative humidity) can be located at approximately 0.0094 lb moisture ($\frac{1}{2}$ of 0.01882 lb). The same method can be used for each dry-bulb temperature, and eventually a connecting line is drawn that represents a 50% relative humidity for any chosen condition of dry-bulb temperature. Similar lines can be drawn for different relative humidity conditions. We already know from Chapter A2 how useful it is to be able to express relative humidity, since it affects human comfort.

Unfortunately, it is not practical or convenient to measure the amount of moisture content or dew point of the air except under laboratory conditions, so we need to plot another element that will give us an easier method. It was noted in Chapter A2 that the wet-bulb temperature also reflects the amount of moisture in the air. The rate of evaporation on the sling psychrometer determined the wet-bulb depression below the dry-bulb temperature. And from Fig. A2-7, we can determine the relative humidity.

For example, Fig. A2-7 showed that for an 80° dry-bulb temperature and an 11° wet-bulb depression (69°F actual measured WB), the relative humidity is 57%. Transferring this information to our psychrometric chart, we can plot point A (Fig. A3-13). If we were to cool the dry-bulb temperature to 76° and the wet-bulb temperature actually stayed at 69° on the sling psychrometer, we now have a WB (wet-bulb) depression of only 7°F and, from Fig. A2-7, a relative humidity of 70%. Point B can now be located. By connecting points A and B, we create a constant wet-bulb line. This process could be repeated over and over until a complete grid of wet-bulb lines fill the chart. Wet-bulb temperature is read at the saturation temperature line, because at that point it can hold no more moisture and becomes the same as the dry-bulb and dew-point temperatures.

This completes the construction of the simplified psychrometric chart (Fig. A3-14). Although it is not 100% accurate, this description should help you understand the relationship of the lines on the real chart. Fortunately, precise and accurate information has gone into the construction of the ASHRAE chart, and it may be used with confidence. *Remember,* if any two of the five

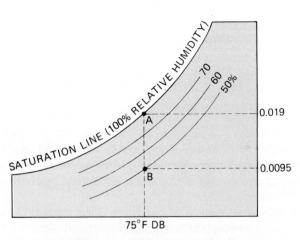

FIGURE A3-12 Relative humidity lines.

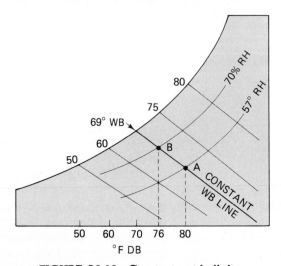

FIGURE A3-13 Constant wet bulb line.

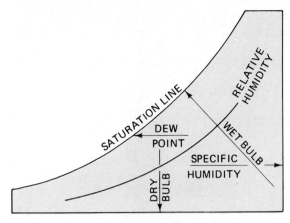

FIGURE A3-14 Simplified psychrometric chart.

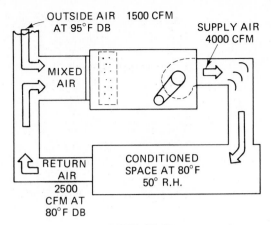

FIGURE A3-15

properties of air are known, the other three can be found on the psychrometric chart by locating the point of intersection of the lines representing the two known conditions.

Before going on, here are a few practice problems using the included ASHRAE chart.

1. The owner of a building wants to maintain an indoor condition of 80°F DB and 50% relative humidity. Find the wet-bulb and dew-point temperatures. (Answer: 66.7°F WB, 59.8°F DP.)

2. Assume a wet-bulb temperature of 60°F and a dry-bulb temperature of 72°F. What is the relative humidity? The humidity ratio? (Answer: 50% relative humidity, 0.0084 lb of moisture.)

3. Assume a DB of 68.6°F and a WB of 60.4°F. What is the relative humidity? (Answer: 63%.)

4. What is the WB of air that has a relative humidity of 80% and a DB of 70°F. (Answer: 65.7°F.)

5. The outdoor design conditions in Portland, Oregon, are 90° DB and 68° WB. Find the dew-point temperature and the relative humidity. (Answer: 56.0°F DB and 32% relative humidity.)

A3-4
AIR MIXTURES

Before plotting an air-conditioning problem on the psychrometric chart, we must first know the initial temperature of the air to be cooled (or heated, as the case may be). In most air-conditioning systems there will be ventilation air from outdoors mixed with room air returning to the unit. In Fig. A3-15 the system is handling 4000 ft³/min total air. Also, 1500 ft³/min of outside air at 95°F DB and 78°F WB is mixed with 2500 ft³/min return (room) air at 80°F DB and 67°F WB. What is the air mixture temperature? First determine the percent of ventilation air in the mixture. We can do this by dividing

1500 by 4000 = 0.375. Ventilation air is 37.5% and recirculated return air is 62.5%. The next step is very important! Multiply the dry-bulb temperature of each condition of the air by its percentage in the mixture. If the outdoor DB is 95°F and it is 37.5% of the mixture, it contributes 35.6 F DB degrees (95° × 0.375) to the mixture. The return air is 80° DB, so it contributes 80° × 0.625, or 50°F DB to the mixture. The total mixed air entering is thus 35.6° + 50°F or 85.6°F DB. Plotted on the psychrometric chart, it would be as represented by Fig. A3-16. The resulting air mixture wet-bulb temperature would be 71.2°F, the dew-point temperature would be 65°, the relative humidity 50%, and the humidity ratio 0.131 lb of moisture.

Let's plot another practice problem:

EXAMPLE

- Outdoor design   95°F DB,   78°F WB
- Indoor design   80°F DB,   67°F WB
- Total air   20,000 ft³/min
- Outside air   4000 ft³/min

Find the resulting conditions of the mixture.

SOLUTION

$$\frac{4000}{20,000} = 0.20 \qquad \text{(20\% outside air, thus 80\% return air)}$$

95° DB × 0.20 = 19.0°F

80° DB × 0.80 = 64.0°F

Mixed temp. = 83.0°F DB

Refer to Fig. A3-17. Plot point *A* (outdoor air) and point *B* (return air). Draw a straight line between them. Locate 83.0°F DB on this line; note as point *C*. From point *C*, record a wet-bulb temperature of 69.2°F, a dew-point temperature of 63.1°F, a relative humidity of 50%, and the humidity ratio as 0.012 lb of moisture.

Now let's use the chart for other air-conditioning processes. First we will look for latent heat and sensible

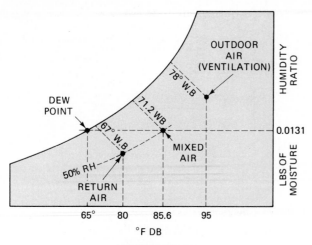

FIGURE A3-16

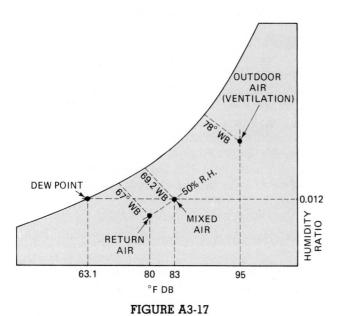

FIGURE A3-17

heat changes. A latent heat change occurs when water is evaporated or condensed and the dry-bulb temperature does not change. This shows as a vertical line on the chart, in Fig. A3-18. On the other hand, sensible heat shows as a change of temperature with no change in the amount of water vapor, as indicated by the horizontal line on the psychrometric chart in Fig. A3-19.

To illustrate a sensible heat process (Fig. A3-20), let's heat air by passing it over a heating coil. If the air starts out at point *A* (60°F dry-bulb and 56°F wet-bulb temperatures), its dew point is 53°F as obtained from the chart. After heating to point *B,* 75°F dry-bulb temperature (Fig. A3-21) the dew point remains the same, for no water vapor has been added. The wet-bulb temperature, however, has increased to 61.5°F. Also, note that the relative humidity has decreased. This explains why the relative humidity in the early morning is high, but decreases as the day gets warmer.

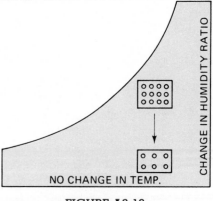

FIGURE A3-18

If the process is reversed and the 75°F dry-bulb temperature (a dew point of 53°F) is cooled back to 60°F DB, we have a sensible cooling process. The wet-bulb temperature drops, but the dew point remains the same (no moisture change).

However, if cooling is combined with dehumidification, the air tends to follow the sloping line to the left in Fig. A3-22. The amount of sensible heat and latent heat determines whether the line has a gentle slope or steep slope. This combination of sensible and latent cooling occurs so frequently in air-conditioning that the slope of the line has been named the *sensible heat factor* or the *S/T ratio,* which is the sensible heat divided by the sum of sensible plus latent heat, or total heat.

On the ASHRAE chart there is a circle at upper left. Note the sensible heat ratios inside the circle, with 1.0 on the horizontal baseline. This is used to estimate the slope of a line between two points on the chart and thus determine the *S/T* ratio by parallel lines rather than by calculation.

If no latent heat change occurs, the sensible heat factor is 1.0 and the line drawn between the two points is horizontal (Fig A3-23). If the sensible heat factor is 0.7, the line starts to slope measurably (Fig. A3-24). This means that 70% of the total heat is sensible and 30% latent. Putting it another way, if you had a 10-ton cooling unit, 7 tons would be for sensible and 3 tons for

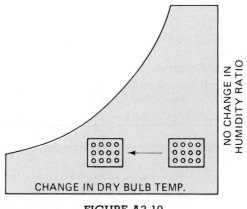

FIGURE A3-19

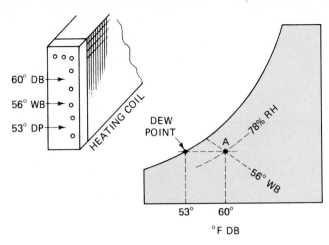

FIGURE A3-20 Heat process.

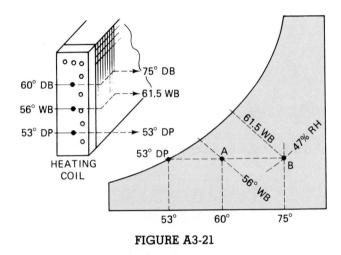

FIGURE A3-21

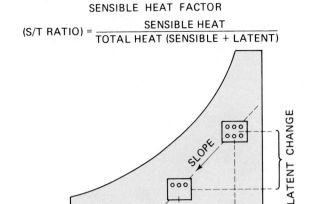

$$\text{SENSIBLE HEAT FACTOR}$$
$$(\text{S/T RATIO}) = \frac{\text{SENSIBLE HEAT}}{\text{TOTAL HEAT (SENSIBLE + LATENT)}}$$

FIGURE A3-22

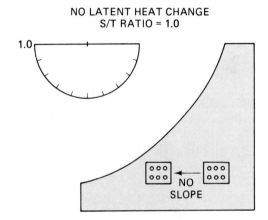

NO LATENT HEAT CHANGE
S/T RATIO = 1.0

FIGURE A3-23

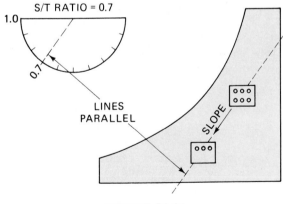

FIGURE A3-24

latent heat removal. Packaged air-conditioners are pre-designed with a fixed ratio usually of about 0.75 *S/T.* Large, built-up systems can be individually selected for a wider variety of conditions.

Another term that is used is the *total heat content,* or *enthalpy,* of the air and water vapor mixture. The enthalpy is very useful in determining the amount of heat that is added to or removed from air in a given process. It's found on the psychrometric chart in Fig. A3-25 by following along the wet-bulb temperature lines, past the saturation line, and out to the enthalpy scale. Point *A* represents air at 75°F DB and 0.0064 lb of moisture at about 35% relative humidity. Its enthalpy is 25.0 Btu per pound of dry air. Heat and humidify the air to point *B,* 95°F DB, 0.0156 lb moisture, and about 44% relative humidity, where the enthalpy is now recorded at 40 Btu per pound of dry air. The increase in total heat would thus be 40 − 25, or 15 Btu per pound of dry air.

If a triangle is drawn as shown in Fig. A3-26, the vertical distance represents the amount of moisture added; that is latent heat. The horizontal distance represents the sensible heating of air. The enthalpy at the intersection of the vertical and horizontal lines is 30 Btu/lb. Therefore, the amount of latent heat added is the difference between 40 and 30, or 10 Btu per lb. The sensible heat added is the difference between 25 and 30, which equals 5 Btu/lb.

The final property to be examined is *specific volume,* which is the number of cubic feet occupied by 1 lb of dry air. For example, in Fig. A3-27 of air at 75°F

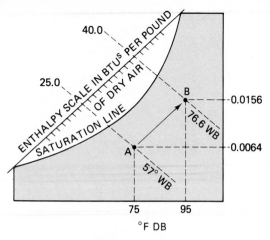

FIGURE A3-25

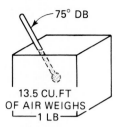

STANDARD BAROMETER (SEA LEVEL)

13.5 CU.FT OF AIR WEIGHS 1 LB

@ 95° IT OCCUPIES 14 CU.FT.
@ 55° IT OCCUPIES 13 CU.FT.

FIGURE A3-27

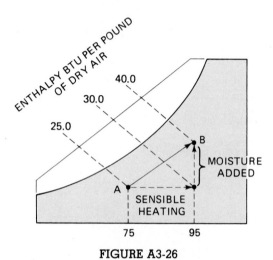

FIGURE A3-26

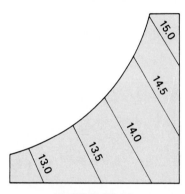

FIGURE A3-28

lines that slant to the left. Specific volume is used primarily for checking fan performance and for determining fan motor sizes for low and high-temperature applications.

As mentioned in the beginning of this chapter, the degree to which you will need and use psychrometrics depends on your particular activity. In most cases an air-conditioning technician will use only dry-bulb and wet-bulb information to determine system performance; however, should you need more information than was presented in this discussion, there are excellent materials available from industry sources for additional reading and understanding.

dry bulb has a volume of about $13\frac{1}{2}$ ft³ at sea level. If the air is heated to 95°F, it takes about 14 ft³, for the air is not as dense at the higher temperature. If cooled to 55°F, it would occupy only 13 ft³, for it is more dense at lower temperatures.

The lines for these specific volumes are shown on the psychrometric chart in Fig. A3-28 as almost vertical

=== *PROBLEMS* ===

A3-1. What is psychrometrics?

A3-2. The atmosphere is made up of _____ .

A3-3. Water vapor is measured in _____ or _____ .

A3-4. Dry-bulb temperature is measured by an _____ .

A3-5. Wet-bulb temperature measures the _____ from a wetted sock on an ordinary thermometer.

A3-6. A device that measures both dry-bulb and wet-bulb temperatures simultaneously is called a _____ .

A3-7. The difference between dry-bulb and wet-bulb readings is called the _____ .

A3-8. What is the dew-point temperature?

A3-9. Specific humidity is a reference of _____ .

A3-10. How many grains are there per pound of moisture?

A3-11. Define "relative humidity."

A3-12. Relative humidity at saturation is _____ %.

A3-13. The total heat in air is made up of _____ and _____ .

A3-14. Define "specific volume."

A3-15. Define "enthalpy."

A3-16. How is enthalpy expressed?

A3-17. What does the scale for the vertical lines of a psychrometric chart indicate?

A3-18. The horizontal lines of the psychrometric chart indicate _____ .

A3-19. Air that is holding all the moisture it can is called _____ .

A3-20. The curved line at the left edge of the psychrometric chart is called the _____ .

A3-21. The scale along the left-hand curved line indicates _____ temperatures.

A3-22. The curved lines that parallel the left-hand curved line indicate _____ .

A3-23. Changes in conditions along a horizontal line on the chart indicate a change in _____ heat.

A3-24. Changes in conditions along a vertical line on the chart indicate a change in _____ heat.

Basic Airflow Principles

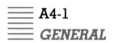

A4-1
GENERAL

In Chapter A3 the properties of air (psychrometrics) were discussed. Topics such as heat content and moisture content were reviewed. This tells us that the heat source for the air-conditioning system is the air moving through the evaporator or cooling coil. To accomplish this, a means of carrying the air from the space to be air-conditioned to the air-conditioning unit for specific treatment and returned to the conditioned space must be provided.

This system consists of duct or pipe to conduct the air, blowers to supply the power to move the air, as well as registers and grills to distribute the air throughout the conditioned area. The system must be quiet, distribute the air without uncomfortable drafts, and yet carry the proper amount of air to maintain the desired conditions in the occupied area.

The calculation of the heating and/or cooling requirements of a conditioned area, be it residential, commercial, etc., has been left out of this text because of its rapidly changing nature. Within the last decade so many new types and compositions of building material as well as insulations have been developed, it is not possible to remain current in a text of this type.

To cover the subject of load calculation, as well as duct design, the use of the air-conditioning application manuals published by the Air-Conditioning Contractors of America (ACCA) should be used. The nature of their application library is such that individual subject manuals are used and updated on a regular basis. Starting with Manual B, *Principles of Air-Conditioning*, duct design, load calculation, installation practice, and equipment selection are covered in 12 sections.

An additional source of information on duct design is the duct design standards published by the Sheet Metal and Air-Conditioning Contractors National Association. Formulated by sheet-metal contractors actually in the air-conditioning business, they publish a series of construction standards covering the construction of low-pressure ducts, high-pressure ducts, and fibrous glass ducts. They also publish a manual, *Heating and Air Conditioning Systems Installation Standards*, which is excellent reference material.

Therefore, because of the publication of this material, heating and cooling load calculation has been left out of this text. However, the service technician should have a basic knowledge of air handling and duct sizing. Following are some of the fundamentals that should be understood.

A4-2
FLUID FLOW

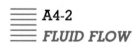

The flow of a fluid (Fig. A4-1) is caused either by a pressure difference due to an increase of pressure at some point in the path of flow because of a mechanical device like an air pump, or by a change in the density of the fluid, due to a temperature difference such as in an oil lamp. For purposes of this discussion, thermal

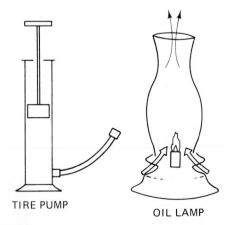

TIRE PUMP   OIL LAMP

FIGURE A4-1 Tire pump and an oil lamp.

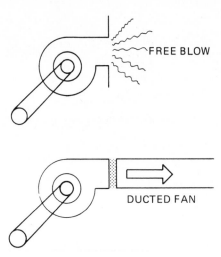

FIGURE A4-2   Ducted fan.

pipe. Air does not move along in a nice placid stream. Rather, it moves in what engineers call a *turbulent flow*, constantly churning and mixing. Metal, fiberglass, and flexible ducts each have a somewhat different friction effect. Constrictions or changes in shape require more pressure to speed up the velocity so that the volume of air flow is unchanged. An additional push is needed to make air change direction, such as a 90° turn. Restrictions such as filters, coils, dampers, grilles, and registers all add to the pressure requirement.

The total pressure (P_t, Fig. A4-4) required to move a desired amount of air through a duct is composed of two elements. Static pressure (P_s) is the pressure exerted against the side wall of the duct in all directions. Think of it as bursting pressure. It can be positive (+) on the push side of a fan or negative (−) on the suction side. Velocity pressure (P_v) is the pressure in the direction of flow. Think of it as the impact or push needed to bring the air up to speed. Total pressure is the sum of the static and velocity pressures at the point of measurement. For illustrative purposes the use of a U-tube manometer is shown.

motion is not involved. However, in precise system designs the engineers do consider the temperature of the fluid (air) because its density per cubic foot is a function of temperature and varies accordingly.

If the mechanical movement is further refined and a fan is substituted for the piston (Fig. A4-2), air can be set into motion. If the air is allowed to free-blow into the atmosphere, there is no resistance to flow, nor is there any direction or control. If a container (duct) is attached to provide a sense of direction and distribution, the energy input required by the fan increases due to the added resistance of the ducting system.

A4-3
RESISTANCE PRESSURES

Resistance to air movement in a duct system (Fig. A4-3) has several causes. First, friction is created by air moving against the duct wall surface, even in straight

A4-4
AIRFLOW MEASURING INSTRUMENTS

Pressures in a duct system are most frequently measured by an inclined manometer (Fig. A4-5), which is filled with a fluid that will be set in motion by pressures from either end. A sliding scale permits the calibration of the instrument at zero (0) pressure when it is exposed to atmospheric pressures only.

The scale will depend on the range of application needed. In residential work the common range is 0.10 to 1.0 in. of water column (W.C.). Commercial high-pressure duct systems may require a range as high as 6 in. of W.C. or more.

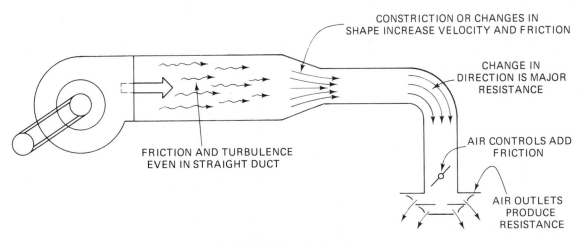

FIGURE A4-3   Duct system.

Basic Airflow Principles   373

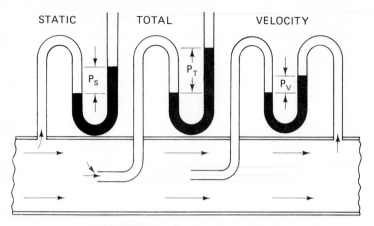

FIGURE A4-4   Duct pressure measurements.

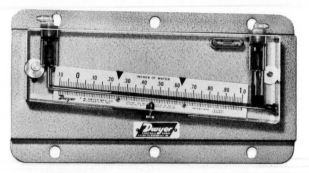

FIGURE A4-5   Inclined manometer. (*Courtesy* Dwyer Instruments)

The manometer can be used in several different ways. Putting positive pressure at connection *A* (Fig. A4-6) will cause the fluid to move to the right and register a positive numerical value. Similarly, putting negative or suction pressure at connection port *B* will register negative pressure on the scale. By putting positive pressure on both ports, a pressure difference can be measured across a coil or filter, as shown in the illustration. By placing a positive pressure at *A* and a negative pressure at *B*, the total static pressure of a system can be measured.

Going back to the pressures in a duct, the manometer can record static, velocity, and total pressures, as indicated by Fig. A4-4. In actual laboratory work the use of a pitot tube (Fig. A4-7) at the end of the air probe allows the engineer to measure total pressure and static pressure with only one probe. When connected as shown, it will indicate the difference, which is the velocity pressure.

In normal air-conditioning work the measurement of velocity pressure is not generally needed. Velocity in terms of speed or motion in feet per minute is important, since it is related to noise considerations and to balancing the systems.

Instrument manufacturers have modified the basic inclined manometer into a more compact, hand-held field instrument such as illustrated in Fig. A4-8a. An identical model can be used with a Pitot tube, and the scale will read velocity in feet per minute rather than in inches of water (Fig. A4-8b).

Another instrument used to measure the velocity of air from or into a grille is the *anemometer* (Fig. A4-9). It consists of propellers on a shaft that revolve when held in an airstream. In the center is a dial that reads velocity in (feet per minute). A stopwatch is used to clock the instrument at whatever time period is desired. Some situations might require the anemometer to run longer than 1 minute. Longer time measurements will provide averages that are more accurate. With the velocity known, the ft³/min can be determined by multiplying

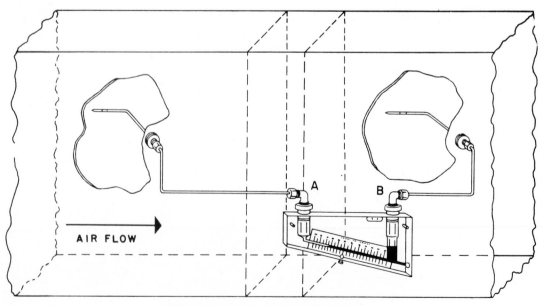

FIGURE A4-6   Measuring duct pressure.

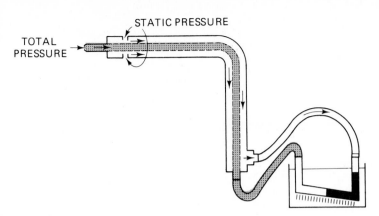

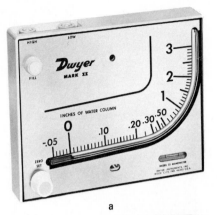

a

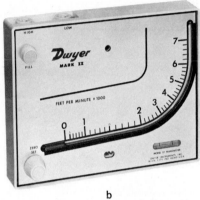

b

FIGURE A4-7 Pitot tube and manometer measuring velocity pressure.

FIGURE A4-8 (a) Manometer to measure inches of water; (b) Manometer to measure velocity in feet per minute. (*Courtesy* Dwyer Instruments, Inc.)

FIGURE A4-9 Anemometer. (*Courtesy* Davis Instruments Manufacturing Company)

the grille's net free area in square feet times the velocity in ft/min. As an example, assume that the grille in Fig. A4-9 had 2 ft^2 of net free area (data obtained from the manufacturer's catalog), and the velocity was recorded at 500 ft/min. The ft^3/min would thus be 500 ft/min times 2 ft^2 = 1000 ft^3/min.

The understanding of pressures, velocity, and the use of instruments can be valuable in designing, balancing, and servicing air distribution systems. The design of a system starts with the understanding of friction loss in straight ducts.

Instruments have been developed that eliminate the need for pressure readings and calculations by measuring actual air quantity from individual registers and/or grills. These can then be added to obtain the actual air quantity handled by the duct distribution system. Figure

A4-10 shows such an instrument, called a *balometer* (balance meter), that does this. Figure A4-11 shows the use of the instrument to measure the air quantity from a ceiling outlet (Fig. A4-11-1) or from a wall outlet (Fig. A4-11-2). Figure A4-11-3 shows the direct-reading dial in four different cubic feet permanent ranges.

The ASHRAE air friction chart (Fig. A4-12) is nothing more nor less than a graph on which the co-ordinates are the friction loss in inches of water column (W.C.) for each 100 ft of equivalent length of duct and the cubic feet of air per minute carried through the duct. As a result of laboratory studies and calculations, a separate line for each size of duct is plotted on the graph. Calculated duct air velocities are also plotted. Logarithmic graph paper is used in plotting the graph so that the lines will be almost straight. This makes the coordinates look a bit unusual, but the air friction chart is much easier to use. The chart illustrated covers up to 2000 ft³/min air volume for residential and light commercial work up to 5 tons. Figure A4-13 is for volumes up to 100,000 ft³/min. Both charts are reprints from the *ASHRAE Guide and Data Book* and are based on standard air conditions flowing through average, clean, round, galvanized metal ducts having approximately 40 joints per 100 ft.

An example of using Fig. A4-12 would be to assume that a 6-in. round duct has 200 ft³/min flowing. We wish to find its velocity and friction loss. Follow across from 200 ft³/min on the left vertical scale to the intersection of the 6-in. diagonal line marked *A*. Read down to the baseline and note a loss of 0.3 inches of water per 100 ft of length. Read the velocity from the other set of diagonal lines at just over 1000 ft/min. If the actual duct is only 75 ft long, the friction loss is cal-

FIGURE A4-10 (*Courtesy* Alnor Instrument Company)

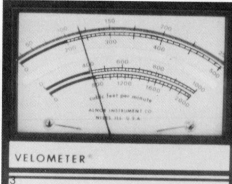

1. Measuring air from a ceiling diffuser
2. Measuring air from a grille
3. BALOMETER anti-parallax scale
4. BALOMETER and carrying case

FIGURE A4-11 (*Courtesy* Alnor Instrument Company)

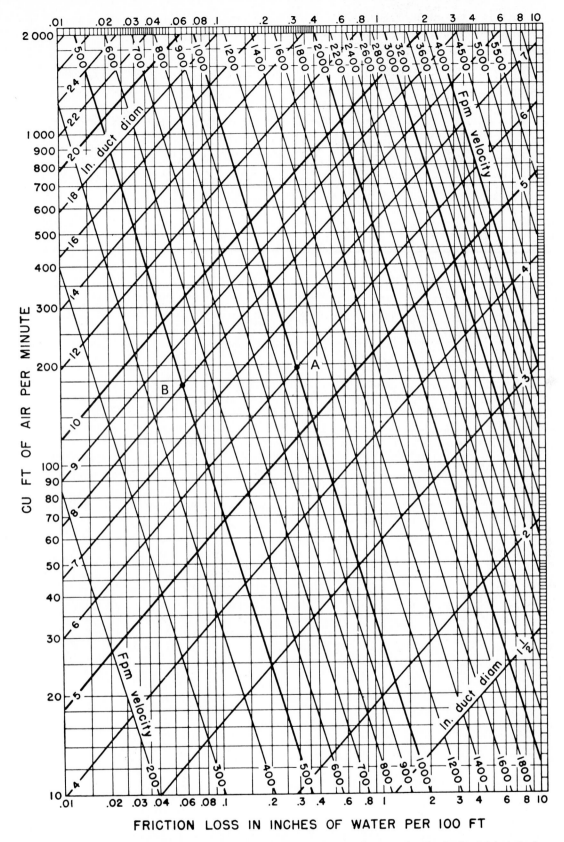

(Based on Standard Air of 0.075 lb per cu ft density flowing through average, clean, round, galvanized metal ducts having approximately 40 joints per 100 ft.) Caution: Do not extrapolate below chart.

FIGURE A4-12 Friction of air in straight ducts for volumes of 10 to 2000 ft³/min. (Reprinted by permission from ASHRAE Guide and Data Book)

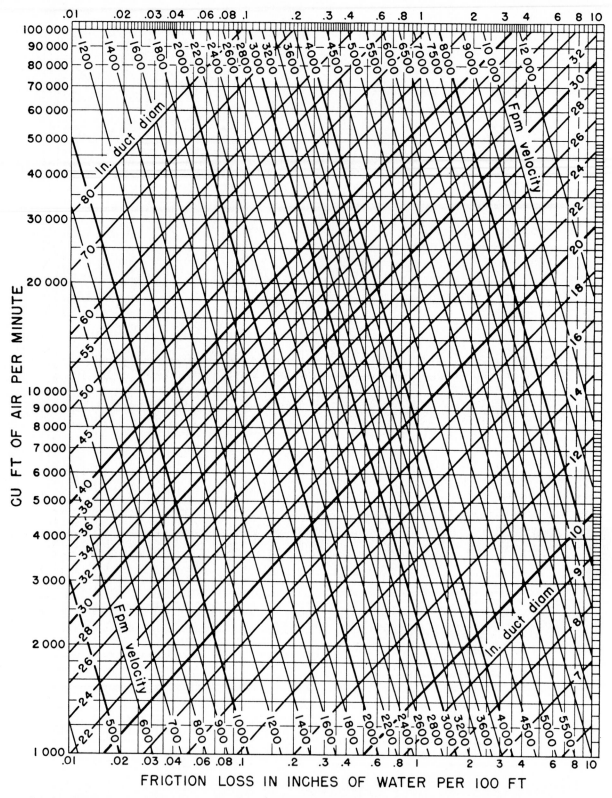

(Based on Standard Air of 0.075 lb per cu ft density flowing through average, clean, round, galvanized metal ducts having approximately 40 joints per 100 ft.)

FIGURE A4-13 Friction of air in straight ducts for volumes of 1000 to 100,000 ft³/min.
(Reprinted by permission from ASHRAE Guide and Data Book)

SIDE	6 Area sq ft	6 Diam in.	8 Area sq ft	8 Diam in.	10 Area sq ft	10 Diam in.	12 Area sq ft	12 Diam in.	14 Area sq ft	14 Diam in.	16 Area sq ft	16 Diam in.	18 Area sq ft	18 Diam in.	20 Area sq ft	20 Diam in.	22 Area sq ft	22 Diam in.
10	.39	8.4	.52	9.8	.65	10.9												
12	.45	9.1	.62	10.7	.77	11.9	.94	13.1										
14	.52	9.8	.72	11.5	.91	12.9	1.09	14.2	1.28	15.3								
16	.59	10.4	.81	12.2	1.02	13.7	1.24	15.1	1.45	16.3	1.67	17.5						
18	.66	11.0	.91	12.9	1.15	14.5	1.40	16.0	1.63	17.3	1.87	18.5	2.12	19.7				
20	.72	11.5	.99	13.5	1.26	15.2	1.54	16.8	1.81	18.2	2.07	19.5	2.34	20.7	2.61	21.9		
22	.78	12.0	1.08	14.1	1.38	15.9	1.69	17.6	1.99	19.1	2.27	20.4	2.57	21.7	2.86	22.9	3.17	24.1
24	.84	12.4	1.16	14.6	1.50	16.6	1.83	18.3	2.14	19.8	2.47	21.3	2.78	22.6	3.11	23.9	3.43	25.1
26	.89	12.8	1.26	15.2	1.61	17.2	1.97	19.0	2.31	20.6	2.66	22.1	3.01	23.5	3.35	24.8	3.71	26.1
28	.95	13.2	1.33	15.6	1.71	17.7	2.09	19.6	2.47	21.3	2.86	22.9	3.25	24.4	3.60	25.7	4.00	27.1
30	1.01	13.6	1.41	16.1	1.82	18.3	2.22	20.2	2.64	22.0	3.06	23.7	3.46	25.2	3.89	26.7	4.27	28.0
32	1.07	14.0	1.48	16.5	1.93	18.8	2.36	20.8	2.81	22.7	3.25	24.4	3.68	26.0	4.12	27.5	4.55	28.9
34	1.13	14.4	1.58	17.0	2.03	19.3	2.49	21.4	2.96	23.3	3.43	25.1	3.89	26.7	4.37	28.3	4.81	29.7
36	1.18	14.7	1.65	17.4	2.14	19.8	2.61	21.9	3.11	23.9	3.63	25.8	4.09	27.4	4.58	29.0	5.07	30.5
38	1.23	15.0	1.73	17.8	2.25	20.3	2.76	22.5	3.27	24.5	3.80	26.4	4.30	28.1	4.84	29.8	5.37	31.4
40	1.28	15.3	1.81	18.2	2.33	20.7	2.88	23.0	3.43	25.1	3.97	27.0	4.52	28.8	5.07	30.5	5.62	32.1
42	1.33	15.6	1.86	18.5	2.43	21.1	2.98	23.4	3.57	25.6	4.15	27.6	4.71	29.4	5.31	31.2	5.86	32.8
44	1.38	15.9	1.95	18.9	2.52	21.5	3.11	23.9	3.71	26.1	4.33	28.2	4.90	30.0	5.55	31.9	6.12	33.5
46	1.43	16.2	2.01	19.2	2.61	21.9	3.22	24.3	3.88	26.7	4.49	28.7	5.10	30.6	5.76	32.5	6.37	34.2
48	1.48	16.5	2.09	19.6	2.71	22.3	3.35	24.8	4.03	27.2	4.65	29.2	5.30	31.2	5.97	33.1	6.64	34.9
50			2.16	19.9	2.81	22.7	3.46	25.2	4.15	27.6	4.84	29.8	5.51	31.8	6.19	33.7	6.87	35.5
52			2.22	20.2	2.91	23.1	3.57	25.6	4.30	28.1	5.00	30.3	5.72	32.4	6.41	34.3	7.14	36.0
54			2.29	20.5	2.98	23.4	3.71	26.1	4.43	28.5	5.17	30.8	5.90	32.9	6.64	34.9	7.38	36.8
56			2.38	20.9	3.09	23.8	3.83	26.5	4.55	28.9	5.31	31.2	6.08	33.4	6.87	35.5	7.62	37.4
58			2.43	21.1	3.19	24.2	3.94	26.9	4.68	29.3	5.48	31.7	6.26	33.9	7.06	36.0	7.87	38.0
60			2.50	21.4	3.27	24.5	4.06	27.3	4.84	29.8	5.65	32.2	6.50	34.5	7.26	36.5	8.12	38.6
64			2.64	22.0	3.46	25.2	4.24	27.9	5.10	30.6	5.91	33.1	6.87	35.5	7.71	37.6	8.59	39.7
68					3.63	25.8	4.49	28.7	5.37	31.4	6.26	33.9	7.18	36.3	8.12	38.6	9.03	40.7
72					3.83	26.5	4.71	29.4	5.69	32.3	6.60	34.8	7.54	37.2	8.50	39.5	9.52	41.8
76					4.09	27.4	4.91	30.0	5.86	32.8	6.83	35.4	7.95	38.2	8.90	40.4	9.98	42.8
80					4.15	27.6	5.17	30.8	6.15	33.6	7.22	36.4	8.29	39.0	9.21	41.1	10.4	43.8
84							5.41	31.5	6.41	34.5	7.54	37.2	8.55	39.6	9.75	42.1	10.8	44.6
88							5.58	32.0	6.64	34.9	7.87	38.0	8.94	40.5	10.1	43.0	11.2	45.4
92							5.79	32.6	6.91	35.6	8.12	38.6	9.39	41.5	10.4	43.8	11.7	46.3
96							5.90	33.0	7.14	36.2	8.40	39.2	9.70	42.1	10.8	44.5	12.1	47.2
100									7.40	36.9	8.50	39.5	9.80	42.5	11.3	45.5	12.3	47.6
104									7.60	37.4	8.90	40.5	10.3	43.5	11.6	46.2	13.0	48.8
108									7.90	38.0	9.20	41.2	10.6	44.0	12.0	47.0	13.4	49.6
112									8.10	38.6	9.50	41.8	10.9	44.7	12.3	47.5	13.8	50.3
116											9.80	42.4	11.3	45.5	12.6	48.1	14.3	51.3
120											10.0	42.8	11.5	46.0	13.1	49.1	14.4	51.5
124											10.3	43.5	11.9	46.7	13.4	49.6	15.0	52.4
128											10.6	44.1	12.1	47.1	13.8	50.4	15.5	53.3
132													12.5	47.9	14.1	50.9	15.8	53.9
136													12.8	48.5	14.5	51.6	16.2	54.5
140													13.0	48.8	14.7	52.0	16.5	55.0
144													13.3	49.4	15.2	52.9	16.8	55.6

FIGURE A4-14 Duct dimensions, section area, circular equivalent diameter, and duct class. (*Courtesy* ASHRAE)

culated as 0.3 × 0.75 = 0.225 in. of W.C. friction loss.

If any two coordinates are known, the other values can be determined. Assuming that a maximum of 500 ft/min velocity is the important element to consider and the duct size cannot exceed 8 in., what would be the resultant ft³/min and friction loss? Plotting from the 500 ft/min diagonal line to the intersection with the 8-in. round duct (point B), read the volume as 175 ft³/min and the friction loss as 0.058 in. of W.C. per 100 ft of duct.

The significance of plotting friction loss per 100 ft of equivalent length cannot be overemphasized. The relationship between the friction loss and the design pressure must be understood if the friction loss chart is to be used correctly. If, for example, the total equivalent of a duct is 200 ft and the system design pressure is to be 0.20 in. of W.C., the total friction loss is 0.20 in. of W.C. For each 100 ft of equivalent length the friction loss cannot exceed 0.10 in. of W.C. In other words, if the duct were shortened to 100 ft of equivalent length, the same ft³/min of air would be delivered if the pressure were decreased from 0.20 to 0.10 in. of W.C.

A4-6
DUCT DIMENSIONS

The friction charts are based on the diameters of round ducts; to express the equivalent cross section in square inches of rectangular dimensions, refer to Fig. A4-14. Note the large numbers overprinted on the table. These are called *duct class numbers*, and are numerical representations of the initial cost of the ductwork. The larger the duct class, the more expensive the duct. Always use the smaller duct class wherever possible. The cross-sectional areas for equivalent round and rectangular ducts never agree exactly, so in using the chart go to the next higher rectangular value. For example, if you wish to convert a nominal 10-in. round duct to a rectangular duct with one side 8 in. long, you would select a 12 × 8-in. equivalent, using the chart. The 10 × 8-in. size would be slightly undersized.

With ductwork made of other materials than galvanized metal or aluminum, it is recommended that the manufacturer's application data be consulted for similar friction-loss tables of conversion factors that can be applied to the metal duct information. Fiberglass ducts and flexible ducting have somewhat different airflow characteristics and must be treated accordingly. Also, if metal ducts are lined with thermal or sound absorbing insulation, they not only have different friction values but also are smaller in the inside dimension by the thickness of the insulation, so do not forget that in the original size determination.

A4-7
DUCT CALCULATORS

For those who frequently work with duct design there are handy duct calculators (Fig. A4-15), which are based on the friction charts and provide the same information on ft³/min velocity, static pressure, etc. The setting also provides an instant conversion from round duct to its rectangular equivalent.

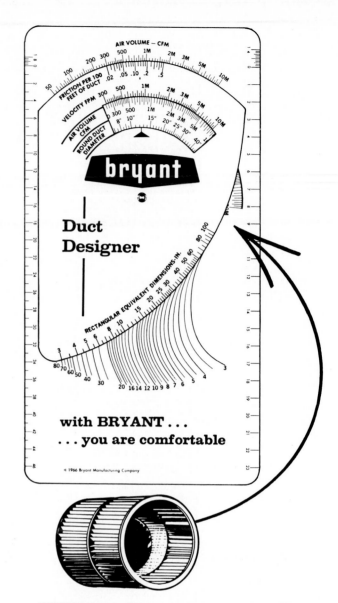

FIGURE A4-15   Duct calculator. (*Courtesy* BDP Company, Division of United Technologies, Inc.)

A4-8
DUCT FITTINGS

Fittings is the name given to the various parts of a duct system such as turning elbows, trunk duct takeoffs, boot connections, etc. These offer resistance to airflow and represent a major portion of duct loss; they also affect the sound characteristics of the system. The *ASHRAE Guide*, NESCA manuals, and duct manufacturers list conventional fittings (Fig. A4-16) and express the air losses in terms of equivalent length of straight duct. For example a type *B*, 90° elbow from group 2, with a trunk width of 12 to 21 in. will have the same or equivalent loss as 15 ft of the same size of straight duct. A glance at the chart will quickly reveal that it does not take many fittings to add an appreciable resistance.

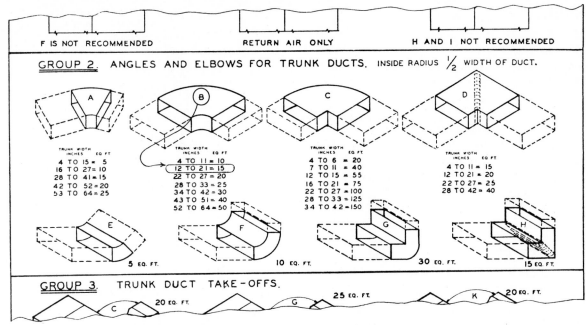

GROUP 2. ANGLES AND ELBOWS FOR TRUNK DUCTS. INSIDE RADIUS ½ WIDTH OF DUCT.

A

B

C

D

TRUNK WIDTH INCHES	EQ FT
4 TO 15 =	5
16 TO 27 =	10
28 TO 41 =	15
42 TO 52 =	20
53 TO 64 =	25

TRUNK WIDTH INCHES	EQ FT
4 TO 11 =	10
12 TO 21 =	15
22 TO 27 =	20
28 TO 33 =	25
34 TO 42 =	30
43 TO 51 =	40
52 TO 64 =	50

TRUNK WIDTH INCHES	EQ FT
4 TO 6 =	20
7 TO 11 =	40
12 TO 15 =	55
16 TO 21 =	75
22 TO 27 =	100
28 TO 33 =	125
34 TO 42 =	150

TRUNK WIDTH INCHES	EQ FT
4 TO 11 =	15
12 TO 21 =	20
22 TO 27 =	25
28 TO 42 =	40

E   F   G   H

5 EQ. FT.   10 EQ. FT.   30 EQ. FT.   15 EQ. FT.

GROUP 3. TRUNK DUCT TAKE-OFFS.

C   20 EQ. FT.   G   25 EQ. FT.   K   20 EQ. FT.

FIGURE A4-16 Duct fittings. (*Courtesy* ACCA)

Figure A4-17 is an example of several types of trunk duct takeoffs. Note method *G*, a round duct butt joint that has a 35-ft equivalent length as compared to *B*, which is a transition collar and is only 15 feet. Top takeoffs like *A, C,* and *F* are even higher in resistance because they actually require two 90° turns of the air as opposed to only one for a side outlet.

Filters, if installed in ductwork rather than in the rest of the equipment, plus air outlet and return air grilles, also present restrictions to airflow that must be considered in the overall system. The resistance of filters and grilles are given directly in terms of inches of water column static pressure taken from the manufacturers' catalogs. They vary widely, from low-pressure residential systems to higher-pressure commercial applications. The use of this information will be more fully covered under system designs.

A4-9
SOUND LEVEL

The noise or sound level of a duct system is the function of velocity, and the permissible limits for acceptable levels are tabulated in Fig. A4-18. These should not be exceeded, and if sound is a major consideration use branch duct values for the entire system design. Although most duct design is initially determined by the static loss method a careful check of velocity is advised so as to minimize the possibility of undesirable noise.

We have discussed the basic restrictions to airflow and velocity recommendations in a duct system, but what about the "push" needed to generate the movement: the fan or blower?

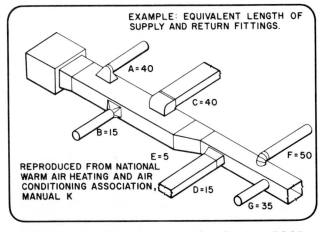

EXAMPLE: EQUIVALENT LENGTH OF SUPPLY AND RETURN FITTINGS.

A = 40
C = 40
B = 15
E = 5
F = 50
D = 15
G = 35

REPRODUCED FROM NATIONAL WARM AIR HEATING AND AIR CONDITIONING ASSOCIATION, MANUAL K

FIGURE A4-17 Trunk duct takeoffs. (*Courtesy* ACCA)

RECOMMENDED MAXIMUM DUCT VELOCITIES, FPM

APPLICATION	MAIN DUCTS*		BRANCH DUCTS	
	SUPPLY	RETURN	SUPPLY	RETURN
Apartments	1000	800	600	600
Auditoriums	1300	1100	1000	800
Banks	2000	1500	1600	1200
Hospital Rooms	1500	1300	1200	1000
Hotel Rooms	1500	1300	1200	1000
Industrial	3000	1800	2200	1500
Libraries	2000	1500	1600	1200
Meeting Rooms	2000	1500	1600	1200
Offices	2000	1500	1600	1200
Residences	1000	800	600	600
Restaurants	2000	1500	1600	1200
Retail Stores	2000	1500	1600	1200
Theaters	1300	1100	1000	800

* When noise control is critical use branch duct values.

FIGURE A4-18 (*Courtesy* ACCA)

Propeller fans, as illustrated in the review of air-conditioning products, are not used for air movement in ducts; they are fine, however, for situations requiring little static resistance, such as air-cooled condenser duty and for separate free-blow exhaust systems.

The centrifugal fan is used extensively in HVAC work because of its ability to move air efficiently and against pressure. The thrust or motion of a centrifugal fan wheel is illustrated in Fig. A4-19. The rotation is fixed and is at a 90° angle to the shaft. The motion imparted by the shape of the blade varies considerably depending on the design. A straight radial blade as illustrated in Fig. A4-19 is seldom used in HVAC work. It is efficient for handling abrasives and dust, and for exhausting fumes containing dirt, grease, or acids.

The typical heating and air-conditioning blower (Fig. A4-20) has a forward-curved multiblade wheel, with the resultant force as shown in the sketch. It moves large masses of air at low rpm and at medium pressure applications, which covers most residential and light commercial work.

Forward-curved blade fans are smaller and are considered relatively quiet and thus are universally used in the unitary type of heating and cooling equipment.

The backward-inclined wheel (Fig. A4-21) has blades inclined toward the rear, away from the fan wheel rotation. They may also be curved. It is sturdy, high speed, and has a low noise level with nonoverload horsepower characteristics. Its mechanical efficiency is high and it requires little power.

A variation of both the forward and backward-inclined blade is the airfoil wheel (Fig. A4-22) with its fewer but airfoil-shaped blades. Note that the force vector is more forward, which increases its static pressure capability for use in high-pressure systems but maintains a relatively low speed and power requirement. The airfoil blade and wheels are more expensive to manufacture and thus are only used on larger equipment.

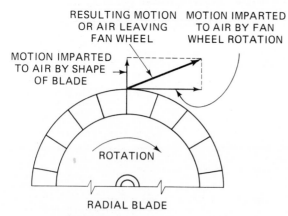

FIGURE A4-19   Radial blade fan.

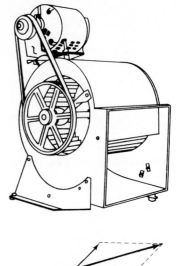

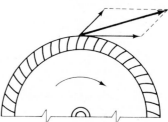

FORWARD CURVED BLADE

FIGURE A4-20   Forward-curved multiblade wheel.

A comparison of performance curves for backward and forward curved fans (Fig. A4-23) shows efficiency and brake horsepower compared to ft³/min at a constant rpm. One significant point can be observed from these curves. For the backward-curved fan, the horsepower rises to a maximum and then drops off and is said to have a *nonoverloading characteristic*. In contrast, the forward-curved fan horsepower continues to rise, and this type, therefore, is considered to be *overloading*. Selection is made within the range shown. Ideally it would be at the point of highest efficiency and minimum horsepower. When the efficiency of these two types of fans are compared, it can be seen that the backward-curved fan has a higher peak efficiency but it also has a wider efficiency variation. Thus for general-purpose application and size, the forward-curved multiblade wheel provides a better selection.

Of interest is the axial type of fan (Fig. A4-24), which has a tube configuration with blades or vanes on a center shaft, much like the aircraft engine turbine. It is versatile, simple in design, and relatively cheap for industrial ventilation and exhaust duty. Its high speed and the resulting noise preclude its use in ducted comfort air conditioning.

Fans are generally rated by the AMCA (Air-Moving and Conditioning Association) by class according to the operating limits of static pressure.

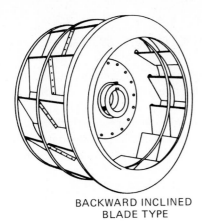

BACKWARD INCLINED
BLADE TYPE

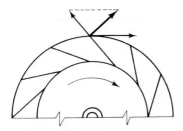

BACKWARD INCLINED BLADE

FIGURE A4-21   Backward-inclined blade wheel.

Class   I Fans— $3\frac{3}{4}$ in. maximum total pressure

Class   II Fans— $6\frac{3}{4}$ in. maximum total pressure

Class III Fans—$12\frac{1}{4}$ in. maximum total pressure

Class IV Fans—Greater than $12\frac{1}{4}$ in. total pressure

Most normal air-conditioning requirements fall in the class I or II rating.

A4-11
AIR DUCT DESIGN METHODS

The basic function of a duct system is to transmit air from the air-handling unit to the individual spaces to be conditioned. The designer must take into consid-

AIRFOIL BLADE
(FORWARD-INCLINED)

ROTATION

FIGURE A4-22   Air foil blade wheel.

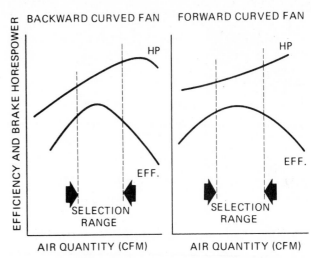

FIGURE A4-23   Fan performance curves.

eration the space available, sound levels, friction loss, initial cost, and heat gain or leakage factors. The prime objective is to consider all of them and determine which system can most nearly satisfy the collective requirements.

A simple duct system (Fig. A4-25) consists of a supply fan, ductwork, transition fitting, discharge grilles, and a return air duct from a room. The fan picks up the air for the room at an air velocity of about 25 ft/min, and after passing through the return air system, the fan discharges the air at a velocity of, say, 1500 ft/min. In this process, from point 1 to point 2 four things have been done to the air:

1. Velocity has been increased
2. Static pressure has been increased
3. Temperature has been increased
4. The air has been compressed slightly

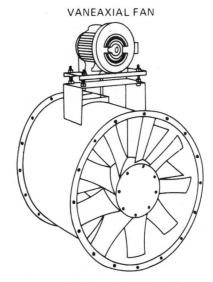

VANEAXIAL FAN

FIGURE A4-24   Axial-type fan.

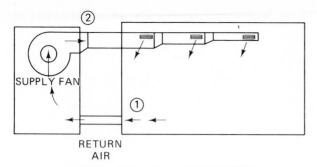

FIGURE A4-25 Duct system diagram.

These four items represent the work the fan has done.

On air-conditioning applications we do not concern ourselves with the compression of air because the pressures are so low. The temperature increase due to fan work has been accounted for in the heat load calculation. But the velocity and static pressure of the fan at point 2 are the two factors we must concern ourselves with in duct design.

Static pressure in the duct and behind the grille forces air through the grille (Fig. A4-26). The selection of these grilles is based on a given quantity of air at a given outlet velocity and static pressure. If the static pressure from point 2 to point 3 were constant, we would be able to get the vanes of each outlet in the same way and get the same amount of air and the same performance. But if we were to take static pressure readings as the air flowed down the duct, we would read perhaps 1.0 in. at point A and 0.9 in. at point B. So a loss of 0.1 in. of W.C. in static pressure has occurred between these two points. The outlets at the far end of the duct will deliver less air than those in the first part. We must be able to predict the losses in a duct system in order to design a system that will deliver the proper air quantities to all conditioned spaces.

There are three methods of sizing and predicting duct performance:

1. Velocity reduction
2. Equal friction
3. Static regain

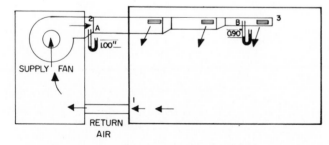

FIGURE A4-26 Duct system pressure measurements.

A4-12
VELOCITY REDUCTION

In the velocity reduction method (Fig. A4-27) a starting velocity is selected at the fan discharge, and then arbitrary reductions in velocity are made in the succeeding duct sections or runs. The assumed velocities are based on sound-related recommendations as reviewed in Fig. A4-18. But the reader will be quick to conclude that there is no constant way of telling what is happening to the pressure relationship. Each section, V1, V2, and V3, will be different, both in velocity and pressure. Therefore, since velocity reduction is an arbitrary method, it is recommended only for very sophisticated and experienced duct designers who have developed a "feel" for required velocity reductions versus the length of the duct section, the duct velocity, pressures, etc.

A4-13
EQUAL-FRICTION METHOD

In the equal-friction method each section of duct (Fig. A4-28) is designed to have the same friction loss per foot of duct. It matters not how long sections P_1, P_2, P_3, etc., actually are since their rate of loss per foot is constant. This method of sizing is the most popular today and is used to size both the supply and return duct systems. It is far superior to the velocity reduction method, since it requires less balancing for symmetrical duct layouts and results in more economical duct sizes.

A4-14
STATIC REGAIN

With duct systems designed on an equal friction basis, the calculated friction loss from one end to the other is not generally as great as calculated. As the air proceeds down the duct, the air velocity is reduced. This velocity represents energy, and since energy cannot be created or destroyed, the same total energy must still be available. Actually, every time the velocity is reduced there is a conversion of velocity pressure to static pressure. This pressure will partially offset the friction in the next duct section.

This is similar to what happens to a skier on a hill (Fig. A4-29). At the top of the slope all of the available energy is potential. Progressing down the hill the skier picks up speed, and at the bottom where he really speeds along, all energy is motion. This energy is sufficient to carry him partway up the next slope, but friction and wind resistance prevent him from reaching the top. This is comparable to what happens in a duct. Static pressure is the potential energy. If air progresses down a duct with

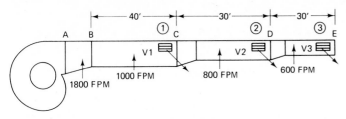

FIGURE A4-27

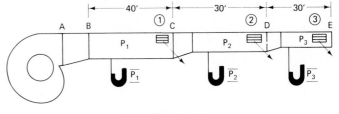

FIGURE A4-28

increasing velocity pressure, there would be a decrease in static pressure. Similarly, decreasing the velocity of air going through a duct increases the static pressure.

This carryover effect is called *static regain* in ducts. The regain efficiency is about 60%, and in high-velocity systems the use of this energy can be significant in reducing duct sizes and, therefore, initial costs. But for low-pressure duct systems as used in residential or light commercial work, the effect is not enough to be included in the design and its procedures.

For a more complete understanding of duct design and airhandling, we suggest that the application manuals published by the Air-Conditioning Contractors of America (ACCA) and the Sheet Metal and Air-Conditioning Contractors National Association Inc. (SMACNA) be used.

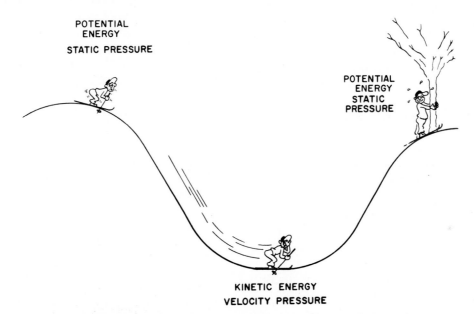

FIGURE A4-29   (*Courtesy* Carrier Air-Conditioning Company)

PROBLEMS

A4-1. Fluid flow may be caused by changes in _____ .

A4-2. Air friction in ducts is caused by _____ .

A4-3. Total pressure is made up of _____ and _____ .

A4-4. An instrument to measure air pressure is called a _____ .

A4-5. An instrument to measure air velocity is called a _____ .

A4-6. An instrument connected to a manometer to measure total pressure and static pressure is called a _____ .

A4-7. A chart that gives the resistance encountered in straight ducts depending on air quantity is called an _____ .

A4-8. The resistance to airflow of various duct fittings is based on their _____ .

A4-9. What are the recommended main and branch duct velocities for residential systems?

A4-10. Most packaged air conditioners use a _____ blower wheel.

A4-11. What are the three methods of sizing ductwork?

Basic Airflow Principles   385

A5

Winter Comfort

A5-1
FUELS

Heating requires the expenditure of some form of energy to raise the temperature of water or air, depending on the type of equipment used. Discounting solar heating, one can classify four types of energy sources (fuels) as the principal means of domestic heating; coal, oil, gas, and electricity. The choice of fuel is usually based on supply (availability), economy, operating requirements, dependability, cleanliness, and control. Major attention is now being focused on the cost of operation.

A5-1.1
Coal

Coal is a solid fuel consisting of carbon, hydrogen, oxygen, nitrogen, sulfur, and ash, and, depending on the proportions of each, we can classify two types—hard and soft (Fig. A5-1). Anthracite (hard) is a clean, dense, hard coal which creates little dust. It's hard to ignite but burns uniformly and with little smoke. Soft coal is classified as bituminous; it ignites easily and burns freely with a long flame, producing much smoke and soot if improperly fired. Both varieties have a heat content of around 14,000 Btu per pound but do differ in carbon and oxygen content, which accounts for the difference in burning characteristics.

COAL

HARD
ANTHRACITE

SOFT
BITUMINOUS

HEAT CONTENT
14,000 BTU PER POUND

FIGURE A5-1   Coal samples.

This description is an oversimplification of the coal classifications, but, because coal is no longer a major source of fuel for heating homes or commercial buildings, we need not go into greater detail on the subject. Coal, however, is still a vital energy source for power company generating plants and certain industrial applications. And with the focus on energy conservation and the high cost of fuels, it may well be that coal will return to the heating scene with new technologies and combustion equipment.

A5-1.2
Oils

Fuel oil is a major source of heating energy for homes and commercial buildings. It is very popular in the northern and eastern sections of the country and is found in many rural areas where the availability or desirability of gas or electrical energy are not alternate options. Fuel oils are mixtures of hydrocarbons derived from crude petroleum by various refining processes. They are classified by dividing them into grades according to their characteristics, mainly viscosity; however, other properties like flash point, pour point, water and sediment content, carbon residue, and ash, are important in the storage, handling, and types of burning equipment for oil. The viscosity determines whether the fuel oil can flow or be pumped through lines or if it can be atomized into small droplets.

For comfort heating applications, we are primarily interested in two grades of fuel oil: No. 1 and No. 2 (Fig. A5-2), which contain 84 to 86% carbon, up to 1% sulfur, and the remainder is mostly hydrogen. The heavier grades, No. 4 and No. 5, have even higher carbon contents, but also considerably more sulfur than is permissible in domestic grades. Number 1 grade fuel oil is considered premium quality and priced accordingly, and is used in room-type space heaters, which do not use high-pressure burners and depend on gravity flow, thus

FUEL OIL

85% CARBON

HEAT CONTENT

140,000 BTU PER GALLON

FIGURE A5-2 Fuel oil.

the need for the lower viscosity of No. 1. Number 2 grade is the standard heating oil sold by most oil supply firms. It weighs between 7.296 and 6.870 (lb/gal) as compared to water at 8.34 (lb/gal). Number 2 oil is used in equipment that has pressure-type atomizing, which covers most forced-warm-air furnaces and boilers. The heating value is approximately 135,000 to 142,000 Btu/gal. Thus the hourly heating effect is found by multiplying the gallons burned by the heat value per gallon.

A5-1.3
Gases

Fuel gases are employed for various heating and air-conditioning (cooling) processes and fall into three broad categories: natural, manufactured, and liquefied petroleum. Stories about the discovery and use of gas date back as far as 2000 B.C. History notes that the Chinese piped gas from shallow wells through bamboo poles and boiled seawater to obtain salt. Early explorers of America reported seeing "burning springs," which were probably jets of gas escaping from holes in the earth. Settlers often struck gas when drilling water wells.

Today gas has a thousand and one uses. In addition to the heating industry, the power generation industry, iron and steel mills, oil refineries, food service industries, and the glass, cement, and pottery industries are large users of gas. In 1983 AGA (American Gas Association) reported that 56 million homes had gas service.

Natural gas comes from wells. It is the product of the organic material of plants and animals that over mil-

lions of years were chemically converted into gas or oil. Most reserves of gas in the United States are not dissolved in or in contact with oil. Natural gas (Fig. A5-3) is mostly *methane*, which consists of one carbon atom linked to four hydrogen atoms, and *ethane*, which consists of two carbon and six hydrogen atoms. Liquid petroleum (LP) gases are *propane* and *butane* or a mixture of the two, and these fuel gases are obtained from natural gas or as a by-product of refining oil. From Fig. A5-3 you will note they contain more carbon and hydrogen atoms than natural gases, are thus heavier, and, as a result, have more heating value per cubic foot.

Manufactured gas, as the name states, is man made as a by-product from other manufacturing operations. For example, in ironmaking large amounts of gases are produced that can be used as fuel gases. The use of manufactured fuel gases has declined greatly in the United States. Today over 99% of sales by gas distribution and transmission companies is natural gas. However, the man-made kind is still a popular fuel in Europe.

Mixed gas, as the name implies, is also a man-made mixture of gases. A common example is a mixture of natural gas and manufactured gas.

It is important to know the specific gravity of a gas. Compared to standard air (Fig. A5-4), which has a specific gravity (sp gr) of 1.0, natural gas ranges from 0.4 to 0.08 and thus is lighter than air. On the other hand, of the liquid petroleum gases, the sp gr of propane is 1.5 and that of butane is 2.0, meaning they are heavier than air. The density of the gases is important because it affects the flow of the gas through orifices (small holes) and then to the burner. Should a leak develop in a gas pipe, natural gas will rise while LP gases will not and may drift to low spots and collect in pools, creating a hazard if open flames are present. Specific gravity also affects gas flow in supply pipes and the pressure needed to move the gas.

The *heating value* (or *heat value*) of a gas is the amount of heat released when 1 ft³ of the gas is com-

METHANE CH₄

ETHANE C₂H₆

PROPANE C₃H₈

BUTANE C₄H₁₀

FIGURE A5-3 Fuel gases.

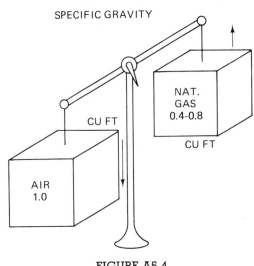

SPECIFIC GRAVITY

NAT. GAS 0.4-0.8

CU FT

CU FT

AIR 1.0

FIGURE A5-4

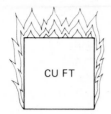

NATURAL GAS	950 TO 1150 BTU/FT^3
PROPANE	2,500 BTU/FT^3
BUTANE	3,200 BTU/FT^3

FIGURE A5-5   Heat content when burned.

1 TO 24

1 TO 10

METHANE GAS

PROPANE GAS

FIGURE A5-7

pletely burned (Fig. A5-5). Natural gas (which is largely methane) has a heating value of about 950 to 1,150 Btu/ft^3. Propane has a heat value of approximately 2500 Btu/ft^3 and butane, about 3200 Btu/ft^3. The hourly rate of a heating value is thus the number of cubic feet burned times the heating value. For example: If a furnace burned 75 ft^3 of natural gas in 1 hour and the heat value is 1000 Btu/ft^3, the total heating *input* is 75 × 1000 = 75,000 Btu/hr. The actual output depends on the operating efficiency of the furnace. The exact heating value of gases in your local area can be obtained from your gas company or LP gas distributor. Propane is the LP gas most used for domestic heating. Butane has more agricultural and industrial applications.

Combustion and ventilation are most important elements and should be understood by the technician. Combustion (Fig. A5-6) takes place when fuel bases are burned in the presence of air. Methane gas combines with the oxygen and nitrogen present in the air and the resulting combustion reaction produces heat, with by-products of carbon dioxide, water vapor, and nitrogen. *For each 1 cubic foot of methane gas, 10 cubic feet of air is needed for complete combustion* (Fig. A5-7). Although natural gas requires a 10-1 ratio of air, LP fuels require much more, due to the concentration of carbon and hydrogen atoms. Liquid petroleum combustion must have more than 24 ft^3 of air per cubic foot of gas to support proper combustion.

The by-products called *flue products* are vented to the outside. Insufficient makeup air can produce hazardous results. If too little oxygen is supplied, part of

the by-products will be dangerous carbon monoxide gas (CO) rather than harmless carbon dioxide gas (CO$_2$). Second, lack of makeup air causes poor flue action and spillage of combustion products into the room, and carbon monoxide in a living area is a serious problem. Specific recommendations on venting and makeup air will be covered later, but remember, combustion heating ventilation is most important.

There is an excellent manual entitled *Fundamentals of Gas Combustion* prepared by AGA (American Gas Association), 1515 Wilson Boulevard, Arlington, Virginia 22209, which covers combustion and the design of burners, their operation, and their maintenance. This author highly recommends it for all who wish to specialize in gas heating.

A5-1.4
Electrical Energy

The growth of electric heating came immediately after World War II and found its main use in areas of cheap public power such as the TVA (Tennessee Valley Authority) and in the Pacific Northwest. In both situations, the cost of power and the winter climate were favorable to the use of electric heating. In the late 1950s, when it became apparent that investor-owned private utilities were going to be faced with heavy summer cooling loads, electric utilities and heating manufacturers began promoting electric heating to build up their winter load. Special rates and promotion programs like "Live Better Electrically," "Total Electric Home," etc., were introduced in the sixties and the market developed rapidly. The situation in the mid-1970s changed somewhat due to higher and higher power-generating costs, which are being passed on to the consumer. Also, the federal government directed that the use of special promotional programs be stopped. However, even with these current obstacles, the use of electric heating continues to grow, due to availability, convenience, and other considerations. With more and more nuclear generating stations being built, the cost of power may be stabilized and may even start to decline in the future.

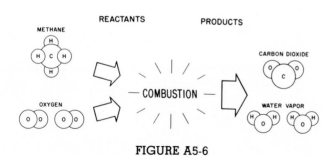

FIGURE A5-6

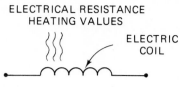

ELECTRICAL RESISTANCE
HEATING VALUES

ELECTRIC
COIL

1 WATT = 3.4 BTU
1 KILOWATT = 3400 BTU
1 KILOWATT/HOUR = 3400 BTU/HR

FIGURE A5-8

The heating value of electric resistance heat is easy to remember and calculate (Fig. A5-8). For each watt of power consumed, 3.4 Btu of heat output will be generated. Resistance heating is 100% efficient; there are no losses such as those experienced with oil and gas combustion processes. Electric rates are measured in kilowatts (1 kw is 1000 w). Therefore, consuming 1 kW in 1 hour is called a kilowatt/hour. And if the heat ratio is 3.4 Btu/W, then a kilowatt/hour of electric resistance heating will produce 3,400 Btu/hr.

This discussion of electric resistance heating does not include the all-electric heat pump. Its unique reverse refrigerant cycle can produce heating with efficiency ratios of 1 input to 3 output under ideal conditions, and 1 to 2 or 2.5 under normal operation. That means for each unit of electricity put in, 2 to 2.5 units of heat are produced, compared to the 1:1 ratio of straight resistance heat. The operation of heat pumps will be explained more fully in subsequent chapters, but because of their ability to produce all-electric heating and cooling, they are a vital factor in future electric energy developments.

Without engaging in a "battle of the fuels," we will cover the most common types of heating equipment that use gas, oil, and electric energy and attempt to establish the advantages and disadvantages as noted by manufacturers and other experts in the field.

Common to all types of heating apparatus sold domestically is the need to be tested, certified, or listed by the proper regulatory agencies.

The AGA (the American Gas Association) establishes the minimum construction safety and performance standards for gas heating equipment. The AGA maintains laboratories to examine and test furnaces, and also maintains a field inspection service. Furnaces submitted and found to be in compliance are listed in the

FIGURE A5-9   AGA seal. (*Courtesy* American Gas Association)

FIGURE A5-10   UL seal. (*Courtesy* Underwriters' Laboratories, Inc.)

AGA *Directory*; they also bear the Blue Star Seal of Certification (Fig. A5-9).

Oil heating equipment and electric heating products are subject to examination, testing, and approval by UL (Underwriters' Laboratories, Inc.). Most of us are familiar with the UL stamp (Fig. A5-10) appearing on everything from toasters to electric blankets, but UL also gets deeply involved in the approval and listing of heating and air-conditioning equipment, even dealing with large centrifugal machines one hundred tons and over. Local city codes and inspectors are guided by UL standards, and failure to comply with them may be costly to the manufacturer and installer. Underwriters' Laboratories maintains testing laboratories for certain types of products; however, they often perform the necessary tests at the manufacturer's plant.

FIGURE A5-11   CSA seal. (*Courtesy* Canadian Standards Association)

Manufacturers who actively sell their products in Canada seek CSA (Canadian Standards Association) approvals as well. CSA is the Canadian counterpart to our UL. So if you see the CSA symbol (Fig. A5-11) on a product, you will recognize its meaning.

A5-2
SPACE HEATERS

A5-2.1
Gas and Oil

Space heater is a general term applied to room heaters, wall furnaces, and floor furnaces that are fired by gas or oil combustion fuels.

A room heater (Fig. A5-12) is a self-contained, free-standing, nonrecessed, gas or oil-burning, air heating appliance intended for installation in the space being heated and not intended for duct connection. A room heater may be of the gravity or mechanical air circulation type, vented or unvented. If unvented, the input for

FIGURE A5-12 Vented room heater. (*Courtesy* Addison Products Company)

gas should not exceed 50,000 Btu/hr. Because unvented room heaters discharge their products of combustion into the space being heated and get their combustion air from the same space, it is essential that they be used in well-ventilated rooms. AGA and many local codes prohibit their use in public spaces like hotels, motels, or institutions such as nursing homes and sanitariums. This form of heating is low cost, but leaves much to be desired in comfort control.

Wall furnaces (Fig. A5-13) are also self-contained vented appliances complete with grilles or their equivalent, designed for incorporation into, or permanently attached to, a wall or partition. They furnish heated air circulated by gravity or by fan directly into the space to be heated. Venting is usually accomplished by an approved integral venting system to the outside wall, although some have optional vertical venting through the roof.

Floor furnaces (Fig. A5-14) are also completely self-contained units designed to be installed on the floor of the space being heated, taking combustion air from outside the living space. The user is able to observe the unit and its lighting; it also can be serviced from this same outside space. Floor furnaces have been very popular in the south for low-cost housing; the crawl space of the house is used as the outside air and service access source. The airflow may be by either gravity or fan-forced circulation into the room. Their width is designed so as to enable them to fit between joist spaces, with enough clearance to provide protection from combustible products. Venting of combustion products must be by proper flue arrangements to outdoors—*not* through

the crawl space. Floor furnaces are an improvement over in-space heaters in that they do not take up usable space; however, the circulation of warm air with this type of furnace leaves much to be desired.

This selection and application of oil and gas space heaters is not really a technical art, and in most cases the choice of heater is the result of a recommendation

FIGURE A5-13 Wall insert heater. (*Courtesy* Gould, Inc.)

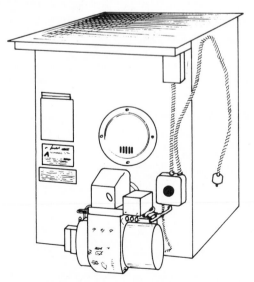

FIGURE A5-14 Floor furnace.

by a furniture or hardware dealer, or a retail store. The operation and service of space heaters is relatively simple; so a technician who understands the rest of this book, dealing with more complicated products, can readily handle this type of equipment.

A5-2.2
Electric Space Heating

Radiant heaters for residential use are available in several forms. Most popular is the wall-mounted variety (Fig. A5-15) that may be flush mounted (recessed) or surface mounted. Only about 2 in. deep, they have an enclosed heating element with a highly polished heat reflector behind. A built-in thermostat automatically controls temperature. These heaters are excellent for bathrooms and spot heating for workrooms, playrooms, etc. Sizes range from 500 to 1000 W (1700 to 3400 Btu/hr) and come suitable for 120-, 208-, or 240-V operation. Service is easy; it is usually a matter of changing a defective thermostat or heating element.

Another variety of heater is the cove radiant heater (Fig. A5-16), which takes advantage of the convective air movement that is produced as room temperatures warm from the radiant effect. Wattages range from 450 to 900 W, with a choice of voltage.

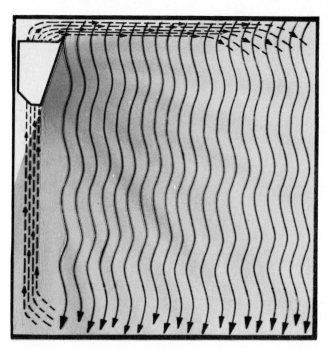

FIGURE A5-16 Ceiling cove radiant heater. (*Courtesy* Federal Pacific Electric Company)

Ceiling panel radiant heaters (Fig. A5-17) are equally useful in residential or commercial buildings. They attach easily to ceiling joists or hang in the T-bar suspended ceiling. Panel sizes conform to ceiling tile dimensions so they are easy to fit into standard building dimensions.

Another form of ceiling-mounted radiant heater is the tiered round grille design (Fig. A5-18). Frequently, this type is used in conjunction with a wall switch timer or wall-mounted thermostat. It adapts to a standard fixture outlet box.

Commercially, there are quite a range of infrared lamp heaters for spot heating for indoor or outdoor application (Fig. A5-19). Typically, these are used in industrial plants, aircraft hangars, garages, etc., where convective heating of the air may not be feasible. They may also be found under covered entrance ways to hospitals, stores, etc., to melt ice and snow. Control is by manual switch operation rather than by a thermostatic device.

Forced-air heaters add a measure of comfort to electric space heating. In homes and offices the recessed models (Fig. A5-20) combine style with the forced-air circulation, resulting in more uniform room temperature control. Most are wall mounted but models are also available for recessing into the ceiling if wall installation is not practical. There are even space heaters for use under kitchen counters. Wall-mounted units go up to 3000 W (10,240 Btu/hr), which can warm larger rooms or offices.

Suspended force-fan unit heaters (Fig. A5-21) are most functional, being used in homes (garages, work-

FIGURE A5-15 Radiant wall heater. (*Courtesy* Climate Control)

SURFACE MOUNTED

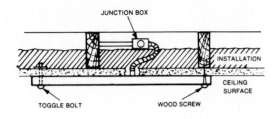

RECESS MOUNTED

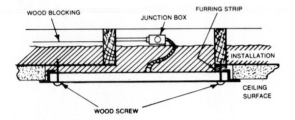

T—BAR CEILING MOUNTED

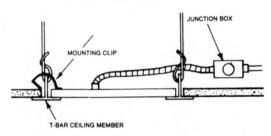

FIGURE A5-17   Ceiling panel radiant heaters. (*Courtesy* Federal Pacific Electric Company)

tion patterns. The heavy-duty construction of this heater category is geared for commercial applications.

Cabinet-type space heaters (Fig. A5-22) are found in classrooms, corridors, foyers, and similar areas of commercial and institutional buildings where cooling is not a consideration. Heating elements go up to 24 kW, and the fans are usually the centrifugal type for quiet but high volume airflow. Cabinets have provision for introducing outside air where use and/or local codes require minimum ventilation air. Outdoor air intake damper operation can be manual or automatic. Control

FIGURE A5-19   Infrared lamp heater. (*Courtesy* C.P.E.—Electric Heat)

rooms, playrooms, etc.), where large capacity and positive air circulation are important, as well as in commercial and industrial establishments of all kinds. They may be suspended in vertical or horizontal fashion with a number of control options such as wall switch only, thermostats, timers, and even night setback operation. Capacities generally range from 3 to 12 kW and above. Discharge louvers may be set to regulate air mo-

FIGURE A5-18   Ceiling heater. (*Courtesy* Standard Refrigeration Company)

FIGURE A5-20   Wall forced-air heater. (*Courtesy* Federal Pacific Electric Company)

392   Air-Conditioning

FIGURE A5-21 Suspended forced-air heater. (*Courtesy* Federal Pacific Electric Company

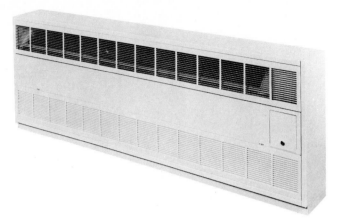

FIGURE A5-22 Cabinet unit heater. (*Courtesy* Federal Pacific/Reliance Electric)

systems can be a simple one-unit operation or multiunit zone control from a central control station.

The most frequently used electric space heater is the convective baseboard model (Fig. A5-23) that has been installed in millions of American homes as well as in countless commercial buildings. A cross section through the heater shows the convective air flow across the finned tube heating element. The contour of the casing provides warm air motion away from the walls thus keeping them cooler and cleaner.

It is important that drapes do not block the air flow and cause overheating. Similarly, it is important that carpeting does not block inlet air. Most baseboard units have a built-linear type of thermal protection that prevents overheating. If blockage occurs, the safety limit stops the flow of electrical current. It cycles off and on until the blockage is removed. Baseboard heaters come in lengths of 2 to 10 ft in the nominal 120-, 208-, 240-, and 277-V rating. Standard wattage per foot is 250 at the rated voltage. But lower heat output can be obtained by applying heaters on lower voltage. For example, a 4-

ft heater rated at 1,000 W on 240 V, when applied on 208 V, will produce about 750 W. Most manufacturers offer these reduced output (low-density) models by simply changing the heating element to 187 W ft.. Low density is generally preferred by engineers and utilities and is recommended for greatest comfort. Controls may be incorporated in the baseboard or through the use of wall-mounted thermostats. Accessories such as corner boxes, electrical outlets, and even plug-in outlets for room air-conditioners are available. Baseboards are finished painted, but can be repainted to match the room decor.

Electric radiant heating cable (Fig. A5-24) was one of the most popular methods of heating early in the development of this technology. It is an invisible source of heat and does not interfere with the placement of drapes or furniture. The source of warmth is evenly spread over the area of the room. It can be installed within either plaster or drywall ceilings. Staples hold the wire in place until the finished layer of plaster or drywall is applied. Wall-mounted thermostats control space temperature.

An important electric heating industry association is NEMA (National Electrical Manufacturers Association), which not only publishes manuals on electric heating load calculations (Fig. A5-25) but provides standards of operation for baseboard equipment.

This discussion on the room type of space heating covers a broad cross section of this type of winter heating equipment. We may not have mentioned all the individual models, but what we have discussed is representative of the industry. Sales of space heaters total many millions of dollars and as such, are important to certain phases of residential and commercial require-

FIGURE A5-23 Baseboard heater. (*Courtesy* Federal Pacific Electric Company)

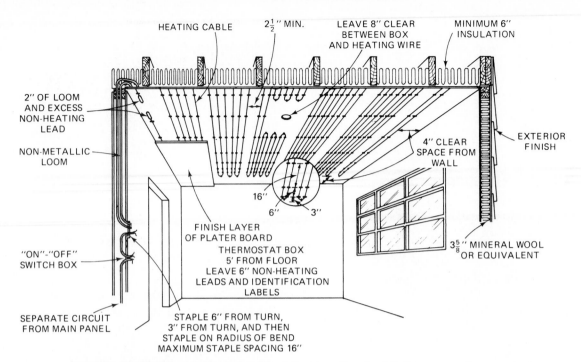

FIGURE A5-24 Electric heating cable. (*Courtesy* Standard Refrigeration Company)

STANDARDS PUBLICATION/NO. HE 1-1974

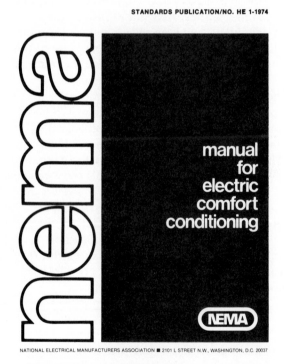

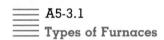

FIGURE A5-25 (*Courtesy* National Electrical Manufacturers Association)

ments. However, there is one important point to be considered—space heaters are not designed for the addition of cooling, and at best offer little or no filtering or humidity control and only limited air circulation. Therefore, in terms of their contribution to total comfort, we must conclude they provide a minimum objective: heating only.

A5-3
WINTER HEATING—GAS FORCED-AIR SYSTEMS

By far the most popular central system heating apparatus is the gas forced-air furnace; in 1979 over 2.3 million units were sold. Because of the difference in housing types, construction methods, and air distribution systems, several different forms of this system are needed.

A5-3.1
Types of Furnaces

Furnaces can be categorized into four styles: upflow highboy, lowboy, counterflow, and horizontal.

The upflow highboy (Fig. A5-26) is most popular. Its narrow width and depth allow for location in first-floor closets and/or utility rooms. However, it can still be used in most basement applications for heating only or with cooling coils where head room space permits. Blowers are usually direct drive multispeed. Air intake can be either from the sides or from the bottom.

Lowboy furnaces (Fig. A5-27) are built low in height and are the necessary choice where head room is minimum. These furnaces are a little over 4 ft high, providing easy installation in a basement. Supply and return ducts are on top for easy attachment. Blowers are most commonly belt driven. Many are sold as replacement equipment in older homes.

The counterflow or downflow (Fig. A5-28) is similar in design and style to the highboy, except that the

FIGURE A5-26   Upflow highboy furnace. (*Courtesy* Bard Manufacturing Company)

air intake and fan are at the top and the discharge is at the bottom. These are widely used where duct systems are set in concrete or in a crawl space beneath the floor. When mounted on a combustible floor, accessory bases are required. An extra safety limit control is also used.

The fourth type is the horizontal furnace (Fig. A5-29), which is adaptable to installation in crawl spaces, attics, or basements due to its low height. It requires no floor space. Intake air is at one end and discharged out the other. Burners are usually field changeable for left- or righthand application. Figure A5-30 is a composite drawing of typical furnace installation.

FIGURE A5-27   Lowboy furnace. (*Courtesy* Bard Manufacturing Company)

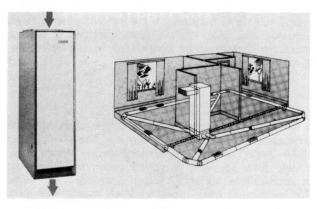

FIGURE A5-28   Counterflow furnace. (*Courtesy* Borg-Warner Central Environmental Systems, Inc.)

A5-3.2
Furnace Ratings

Gas furnaces and even gas-burning space heaters are rated on input Btu/hr capacity by AGA. The output is a function of the individual furnace design, but it is common practice to assume 80% efficiency. Therefore, a furnace rated at 100,000 Btu/hr input will have 80,000 Btu/hr output.

Naturally, in selecting a furnace to satisfy the heating load we would base our recommendation on the output as stated in the manufacturer's specifications. This is true for all gases: natural, manufactured, mixed, and LP. Oil furnaces, on the other hand, are tested and rated by UL for output.

A5-3.3
Basic Furnace Design

In its simplest form the cross section of a gas upflow warm-air furnace (Fig. A5-31) consists of a cabinet housing, a heat exchanger, a burner assembly, a blower to move air over the heat exchanger, an air filter, a flue vent system, and the necessary safety and operation controls. Let's examine each.

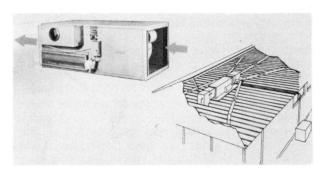

FIGURE A5-29   Horizontal furnace. (*Courtesy* Borg-Warner Central Environmental Systems, Inc.)

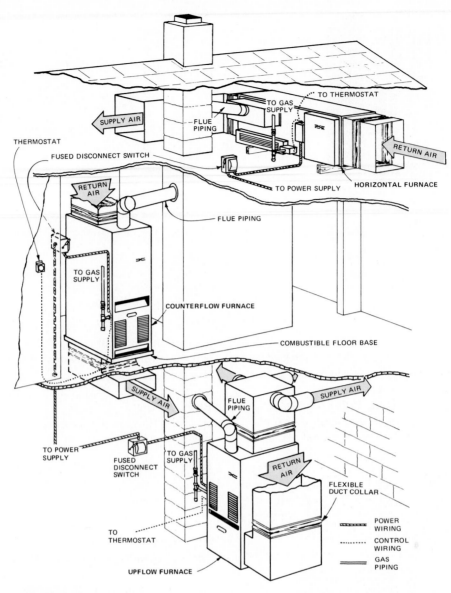

FIGURE A5-30   (*Courtesy* Borg-Warner Central Environmental Systems, Inc.)

Heat Exchanger: The typical gas furnace heat exchanger (Fig. A5-32) is made from two stamped or formed steel sheets which are then seam welded to make a section. These sections are then headered at top and bottom and welded into fixed position. The number of sections depends on the ultimate capacity. Most manufacturers estimate about 25,000 to 40,000 Btu/hr per section; thus, a 100,000 Btu/hr furnace will have four sections, and so on. The heat exchanger is of heavy-gauge steel and is then finished with protective coatings to help prevent rusting from condensation that occurs in the normal combustion process. Note the contours in the surface of the sections. This is to create air turbulence for better heat transfer, inside and outside.

Burner: The burner assembly that fits into the lower openings of the heat exchanger is sized and designed to produce the right amount of heat, flame pattern, and flue effect for efficient gas combustion. Figure A5-33 is a cross section of a typical atmospheric burner. Gas is piped from the meter to the furnace and enters the burner cavity through the spud, which has a precisely drilled orifice to meter the correct amount of gas. This gas ejects itself into the burner venturi throat, and from basic physics we learn that a negative pressure or vacuum is produced behind the gas jet. *Primary air* is thus drawn into the tube, where it mixes with gas and forms a positive pressure mixture ready for exhausting through the burner ports. As the mixture comes out of the ports, it is ignited by a pilot device and a flame is established. In a previous discussion we noted the importance of air for combustion. About half of the air is needed for primary air, but in order to expel the combustion by-products, a *secondary air* source is needed to help create a

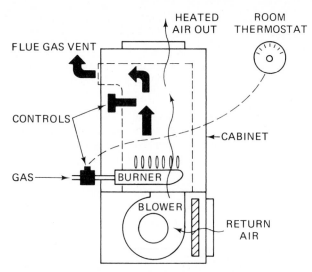

FIGURE A5-31   Cross section of gas upflow warm-air furnace.

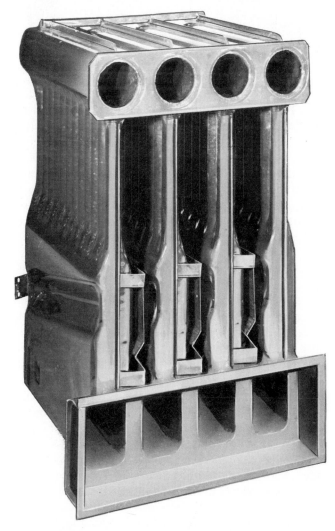

FIGURE A5-32   Heat exchanger. (*Courtesy* Addison Products Company)

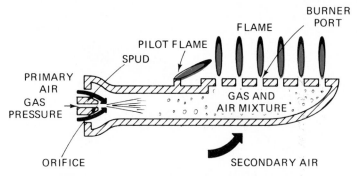

FIGURE A5-33   Burner cross section.

good flue effect. This secondary air through the burner pouch opening is controlled by restrictor baffles designed into the heat exchanger. The primary air is controlled by air shutters (Fig. A5-34), as would be seen from the burner end, if this was possible. Turning the shutters opens or closes the venturi and thus the amount of induced primary air. With a fixed gas jet, we can then vary the flame characteristics for most efficient burning. Burner construction, operation, and adjustment are covered in more detail in Chapter A14.

Pilot: Notice in Fig. A5-35 the existence of a pilot runner attached to the main burner tube. Its function is to carry burning gas to each of the main burners from the one which has the pilot lighter attached to it. The alignment of these runners is most important for smooth, positive ignition. If it is out of line, a delayed ignition can create noise, possibly with enough force to blow out the pilot.

The pilot lighter and flame (Fig. A5-36) are fed gas from the main gas valve (to be discussed later) by a small $\frac{1}{4}$-in. line. Attached to the bracket is a thermocouple sensor that, when heated, produces a small current of electricity (millivolts), which is directed back to the gas valve to actuate a magnetic operator. The flame adjustment should give a soft steady flame enveloping $\frac{3}{8}$ to $\frac{1}{2}$ in. of the tip of the thermocouple, which will glow a dull red. The flame should be correctly positioned to supply the pilot runners across all burners.

FIGURE A5-34   Adjusting primary air. (*Courtesy* Borg-Warner Central Environmental Systems, Inc.)

Winter Comfort   397

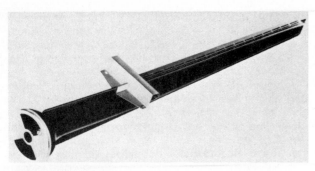

FIGURE A5-35 Slotted port burner. (*Courtesy* Borg-Warner Central Environmental Systems, Inc.)

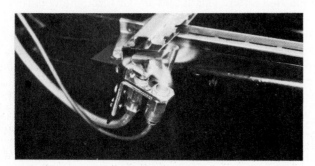

FIGURE A5-36 Pilot flame. (*Courtesy* Borg-Warner Central Environmental Systems, Inc.)

Main Gas Valve: Although control of the main flame is accomplished by adjusting the burner shutters, the gas supply or flow is controlled in the gas manifold section through the use of a combination gas valve (Fig. A5-37) that performs many functions. Valves from various manufacturers may differ, but they all accomplish essentially the same functions:

1. Manual control for ignition and normal operation
2. Pilot supply, adjustment, and safety shutoff
3. Pressure regulation of burner gas feed
4. On/off electric solenoid valve controlled by the room thermostat

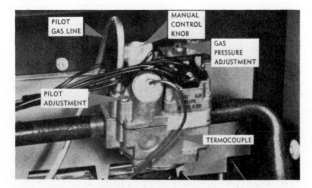

FIGURE A5-37 Combination gas valve. (*Courtesy* Borg-Warner Central Environmental Systems, Inc.)

The function of the gas control valves is covered in Chapter A10.

Blower: Another major component of our furnace is the blower (Fig. A5-38) which provides the energy to distribute conditioned air to the living space. Most residential furnaces use a double inlet centrifugal blower. Direct drives are possible where the motor is mounted inside the fan scroll and the shaft is directly connected to the fan wheel. Fan speed then is the same as the motor rpm. Most direct-drive blowers have multispeed motors for ft³/min adjustment. Larger furnace models often employ belt drives where the ft³/min adjustment can be varied over a wide range due to pulley diameter ratios.

FIGURE A5-38 Blower. (*Courtesy* Borg-Warner Central Environmental Systems, Inc.)

The multispeed direct-drive motors are wound with speed taps so that the rpm can be changed either at the motor itself or at a remote location. Some manufacturers use solid-state control devices to vary the ft³/min, depending on winter or summer needs or to compensate for additional duct pressures when cooling, humidifiers, air cleaners, and the like are added to the system. These solid-state devices vary motor speed by changing the electrical frequency sine waves of power input.

Cabinet: The final element of our built-up furnace is the cabinet itself (Fig. A5-39). Its primary purpose, of course, is to support all of the internal components rigidly, but it must also provide a reasonably attractive appearance. Many furnaces are installed in living areas, closets, utility rooms, etc., where an attractive appearance and sound control are desirable. The cabinet must also incorporate important safety features: to keep children away from the combustion area, to present a surface temperature suitable to the touch, and to provide proper clearances from combustible surfaces. The hot air passing over the outside heat exchanger shell surface is naturally in contact with the cabinet. The heated air, plus radiant heat, produce a surface temperature that must be considered in the safety aspects of a furnace.

This then leads to discussion of clearances. When gas furnaces are approved and listed by AGA (American Gas Association) testing laboratories, they are done so with approved minimum clearances from combustible

FIGURE A5-39 Cabinet. (*Courtesy* Bard Manufacturing Company)

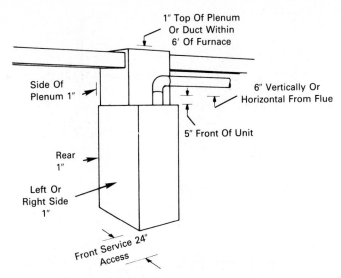

FIGURE A5-40 Typical furnace clearances. (*Courtesy* Borg-Warner Central Environmental Systems, Inc.)

surfaces. These clearances are listed in the installation instructions for each model and on a decal in the furnace cabinet. Figure A5-40 is a sample of the required clearances for an upflow highboy. The clearances shown in the following table are from any surface of the enclosure:

	Clearance (in.)
From top of plenum	1
From top of duct within 6 ft of unit	1
From front of unit	5
Horizontally from flue or draft hood	6
Vertically from flue or draft hood	6
From rear of unit	1
From left side	1
From right side	1
From side of plenum	1

A 24-in. service access must be provided at the front of the unit. It is necessary to comply with these minimum clearance standards. If they are not met, the risk of fire and voided insurance coverage may become liabilities for the dealer.

As previously mentioned, when a counterflow furnace is installed on a combustible floor, AGA requires the use of an approved floor base (Fig. A5-41). Sufficient clearance space is allowed for the passage of the plenum through the floor opening in the event of an overheating problem.

With a basic understanding of the furnace construction and the function of internal components, let's look at some related application data, for the furnace can be no better than the installation. Lack of knowl-

edge or attention to these elements can cause faulty or possible hazardous operation.

Venting: Another important application consideration is the matter of proper venting of flue gases. The basic function of the vent to which a gas furnace is connected is twofold:

1. To provide a safe and effective means for moving the products of combustion from the draft hood to the outside atmosphere without contaminating the room air.
2. To provide the mechanics for producing and maintaining a draft, which induces a supply of replacement air into the room where the furnace is situated.

For example, a 75,000-Btu LP furnace (Fig. A5-42) burns 30 ft³ of gas per hour. Remembering that LP also requires approximately 24 ft³ of air for each cubic

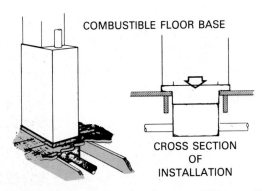

FIGURE A5-41 Counterflow furnace. (*Courtesy* Borg-Warner Central Environmental Systems, Inc.)

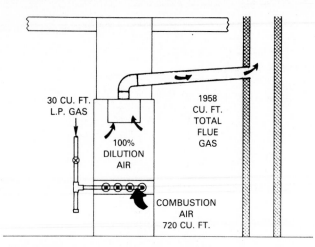

FIGURE A5-42   LP gas/air ratio. (*Courtesy* Borg-Warner Central Environmental Systems, Inc.)

foot of gas to support combustion, we then add 720 ft³ of air. When these two quantities are burned, the resultant release of combustion by-products, plus the 100% dilution of air entering the draft diverter, brings the total vent gases to 1440 ft³—more than 60 times the input gas capacity. A natural gas furnace would have about half that much, but this is why supply air and venting are so important on all gas appliances.

The draft hood on a furnace (Fig. A5-43) is the place where flue gases and dilution air are mixed. Under normal operation, furnace flue gases entering the draft hood are mixed with dilution air entering the relief opening and then flow out of the draft hood exit at a lower temperature than incoming flue gases.

Under strong updraft conditions, such as may be caused by vent terminal wind conditions, the low resistance to the flow of dilution air through the relief opening of a properly designed draft hood tends to neutralize the effect of excess stack action on the furnace.

If a downdraft occurs, the relief opening also provides a discharge opening for flue gases and downflowing air so that pilot outage and other possible effects are minimized.

Flue/Vents: The vent connector pipe between the furnace outlet and the vent or chimney is illustrated in Fig. A5-44. A large portion of the resistance to the flow of flue gases occurs in this vent connection. It is important to employ a minimum run and elbows and use a pipe at least equal in size to the draft hood outlet. Long horizontal runs and fittings increase flow resistance and also lower the flue gas temperature before it reaches the vertical vent. Always slope the vent connector up toward the chimney connector. The vent connector is then inserted into, but not beyond, the inside wall of the chimney flue liner.

Where two or more appliances (furnace and water heater) are vented into the same flue (Fig. A5-45), the connector should be at least the size of the largest flue, plus 50% of the other appliance. The chart shown gives venting combinations, and it is usually in the furnace installation bulletin for quick reference. The critical point occurs when one appliance is not operating and dilution air enters the common vent, lowering the draft-producing flue gas of the other appliance. For example, a water heater with a 3-in. vent and a furnace with 5-in. vent would require a common 6-in. flue connector.

The vertical vent (chimney or flue stack) must be properly designed and installed to create proper draft conditions. In some cases the heating and air-conditioning technician has little control over these conditions; in others, the installation of prefabricated vents may be in the heating contract from the builder or owner.

Masonry chimneys (Fig. A5-46) are field constructed and should be built in accordance with local and/or nationally recognized codes or standards. The size of the flue liner, offsets, terminal vent, and the chimney height can materially affect the draw or draft.

The chimney in Fig. A5-47 is factory built and listed by a nationally recognized agency. It is made of metal of an adequate thickness and is insulated, galvanized, and properly welded or riveted. It is commonly called a *type B gas vent* and is double walled. The double wall helps to conserve heat in the flue gas and thus promotes better draft as well as producing a lower sur-

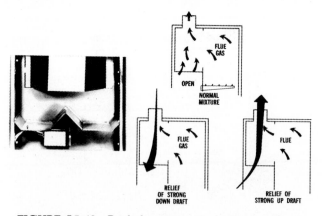

FIGURE A5-43   Draft diverter action. (*Courtesy* Borg-Warner Central Environmental Systems, Inc.)

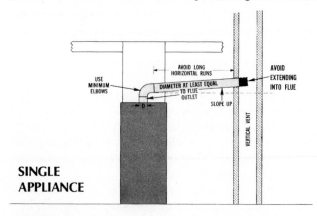

FIGURE A5-44   Vent connector. (*Courtesy* Borg-Warner Central Environmental Systems, Inc.)

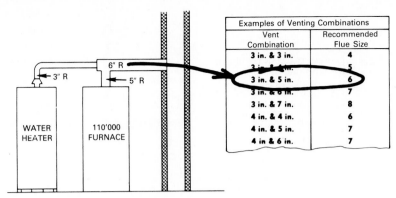

Examples of Venting Combinations	
Vent Combination	Recommended Flue Size
3 in. & 3 in.	4
3 in. & 4 in.	5
3 in. & 5 in.	6
3 in. & 6 in.	7
3 in. & 7 in.	8
4 in. & 4 in.	6
4 in. & 5 in.	7
4 in. & 6 in.	7

WATER HEATER   110'000 FURNACE

FIGURE A5-45   Vent connector multiple appliances. (*Courtesy* Borg-Warner Central Environmental Systems, Inc.)

face temperature on the outer surface; it also requires less clearance than the single-walled type.

Single-walled metal vents are not generally considered suitable for venting residential heating equipment and are prohibited by many local codes because they cannot be used in concealed spaces.

As mentioned above, although you, as a service technician, in most cases will not have a part in the type or design of the vertical flue, you can observe those poor practices that may reflect on the performance and operation of equipment or lead to serious service problems. This is important when you are replacing a furnace; you then must check the venting arrangement.

Most city venting codes are based on the standards published by AGA, the *National Fire Protection Association Bulletin,* or by the National Board of Fire Underwriters (NBFU) and Gas Vent Institute. Your local building inspector and gas company keep these publications on file.

AGA Report 1319 provides tables for the proper sizing of both masonry and metal flues based on heat input, the height of the chimney, horizontal run of connector, etc. The table shown in Fig. A5-48 for masonry

chimneys gives the maximum allowance of heat input based on the chimney height, vent connector size, and distance of horizontal run. For example, a chimney height of 15 ft with a 6-in. vent connection of 5 ft of horizontal run will handle 163,000 Btu/hr.

Type B double-wall gas vents are figured in a similar manner. Manufacturers' specifications will list the capacity based on height, horizontal run, and vent size. For example, in Fig. A5-49 an individual appliance with vent height of 20 ft, 6 in. round size and a 10-ft horizontal run will carry 228,000 Btu. This would be typical of a ranch-house application, with the furnace in the basement and a short stack to the roof. The specifica-

FIGURE A5-46   Vertical vents (masonry chimney). (*Courtesy* Borg-Warner Central Environmental Systems, Inc.)

FIGURE A5-47   Vertical vents (prefabricated chimney). (*Courtesy* Metalbestos Systems, Wallace Murray Corporation)

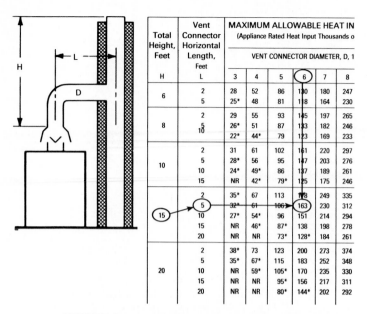

Total Height, Feet H	Vent Connector Horizontal Length, Feet L	MAXIMUM ALLOWABLE HEAT IN (Appliance Rated Heat Input Thousands o... VENT CONNECTOR DIAMETER, D, 1					
		3	4	5	6	7	8
6	2	28	52	86	130	180	247
	5	25*	48	81	118	164	230
8	2	29	55	93	145	197	265
	5	26*	51	87	133	182	246
	10	22*	44*	79	123	169	233
10	2	31	61	102	161	220	297
	5	28*	56	95	147	203	276
	10	24*	49*	86	137	189	261
	15	NR	42*	79*	125	175	246
15	2	35*	67	113	178	249	335
	5	32*	61	106	163	230	312
	10	27*	54*	96	151	214	294
	15	NR	46*	87*	138	198	278
	20	NR	NR	73*	128*	184	261
20	2	38*	73	123	200	273	374
	5	35*	67*	115	183	252	348
	10	NR	59*	105*	170	235	330
	15	NR	NR	95*	156	217	311
	20	NR	NR	80*	144*	202	292

FIGURE A5-48 Masonry chimney with single vent connector. (*Courtesy* American Gas Association)

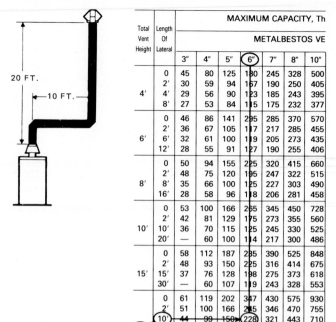

Total Vent Height	Length Of Lateral	MAXIMUM CAPACITY, Th METALBESTOS VE						
		3"	4"	5"	6"	7"	8"	10"
4'	0	45	80	125	180	245	328	500
	2'	30	59	94	167	190	250	405
	4'	29	56	90	123	185	243	395
	8'	27	53	84	115	175	232	377
6'	0	46	86	141	205	285	370	570
	2'	36	67	105	117	217	285	455
	6'	32	61	100	119	205	273	435
	12'	28	55	91	127	190	255	406
8'	0	50	94	155	225	320	415	660
	2'	48	75	120	195	247	322	515
	8'	35	66	100	125	227	303	490
	16'	28	58	96	118	206	281	458
10'	0	53	100	166	265	345	450	728
	2'	42	81	129	175	273	355	560
	10'	36	70	115	125	245	330	525
	20'	—	60	100	114	217	300	486
15'	0	58	112	187	285	390	525	848
	2'	48	93	150	225	316	414	675
	15'	37	76	128	198	275	373	618
	30'	—	60	107	119	243	328	553
20'	0	61	119	202	347	430	575	930
	2'	51	100	166	245	346	470	755
	10'	44	99	150	228	321	443	710
	20'	35	78	134	200	295	410	665
	30'	—	68	120	186	273	380	626
30'	0	64	128	220	336	475	650	1060
	2'	56	112	185	280	394	535	865
	20'	—	90	154	237	343	473	784

FIGURE A5-49 Type B gas vent selection table. (*Courtesy* American Gas Association)

tions also list the sizes for multiple appliance application (furnace and water heater).

In addition to the type and size of vertical flue, the terminal point in respect to the roof is also important (Fig. A5-50). Two specific rules apply:

1. If the vent terminates without a listed cap or similar device and depends on natural draft, it must extend at least 2 ft above the point where it passes through the roof, and be at least 2 ft higher than any portion of a building that is within 10 ft of the vent.

2. If the vent terminates with a listed or approved top, it must terminate in accordance with the terms under which such listing was approved.

Additionally, type B vents must be equipped with an approved cap having a capacity at least as great as the vent stack. Also, this cap must be at least five feet above the highest connected appliance draft hood.

Altitude also has a pronounced effect on combustion air and draft, and above 2000 ft the AGA calls for reducing the furnace input 4% for each 1000 ft above sea level.

Outside masonry chimneys or vents lose heat faster and take longer to heat up than do their indoor counterparts; thus prolonged priming and spilling of flue products from the draft hood relief will occur when the vented appliances are put into operation.

A simple test for proper vent operation can be made at the draft hood of the appliance (Fig. A5-51).

Light a match near the relief opening and move it around the entire perimeter of the hood. If vent gases are spilling out, the match will be extinguished because of a lack of oxygen, not because of the velocity of the air. The test should be made under actual (or even extreme) operating conditions.

A5-3.4
Combustion Air

Equally important as venting is the matter of sufficient supply air to maintain proper combustion air for diluting the flue gas. As previously pointed out, a 75,000-Btu LP furnace required 1440 ft³ of air per hour.

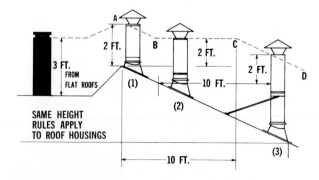

FIGURE A5-50 Vertical vent roof termination. (*Courtesy* American Gas Association)

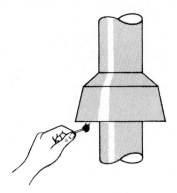

FIGURE A5-51 Vent test.

Any other gas-burning appliances within the structure add to the problem. With the tight construction of modern homes, equipped with storm doors and windows, normal infiltration cannot be depended on to provide such a volume of air, particularly if there is a kitchen ventilating system, a clothes dryer, or a fireplace competing for air to perform its function. The danger of a short supply of combustion air is greatest in small, single-floor houses or apartments where the furnace is located in the living area.

As a furnace installer, you do share the responsibility to insure that there are provisions to introduce make-up air to the equipment room. The following illustrations taken from the NFPA's (National Fire Protection Association) Manual 54 indicate various methods of introducing makeup air.

Method A (Fig. A5-52) is the most critical and least desirable because it depends on the infiltration air getting into the house and then through the supply air openings. This is typical of closet applications. The grilles are sized by dividing the total input Btu (of all appliances) by 1000. Remember that this is free area in square inches, and the particular grille must be selected from the manufcturer's tables to be at least that.

Method B (Fig. A5-53) provides all air from the outdoors.

Method C (Fig. A5-54) is good for single-floor residences where air may be taken from a ventilated attic. *Caution:* Make sure that louvers in the attic are adequate and open in winter and, of course, there should be no attic exhaust fan.

Finally, Method D (Fig. A5-55) may be used where there is a crawl space. Again vent louvers must be large enough and open during the winter.

Basement installations can usually use infiltration air from window openings, garage door leakage, etc., particularly if there is an abundance of open space.

Over the years closet installations have proven to be particularly dangerous where installers tried to use the equipment room as a plenum chamber for return air. Even though the installer calculated the theoretical air opening size or used a louvered door, there were, and still are, too many deaths reported from carbon mon-

METHOD A

ALL AIR FROM

INSIDE BUILDING

NOTE: Each opening shall have a free area of not less than one square inch per 1,000 BTU per hour of the total input rating of all appliances in the enclosure.

FIGURE A5-52 (*Courtesy* American Gas Association)

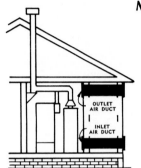

METHOD B

ALL AIR

FROM OUTDOORS

NOTE: Each air duct opening shall have a free area of not less than one square inch per 2,000 BTU per hour of the total input rating of all appliances in the enclosure.

FIGURE A5-53 (*Courtesy* American Gas Association)

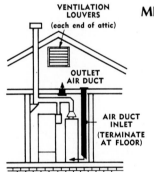

METHOD C

ALL AIR FROM

VENTILATED ATTIC

NOTE: The inlet and outlet air openings shall each have a free area of not less than one square inch per 4,000 BTU per hour of the total input rating of all appliances in the enclosure.

FIGURE A5-54 (*Courtesy* American Gas Association)

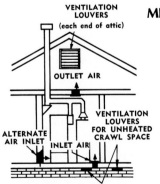

METHOD D

AIR, IN FROM CRAWL

SPACE, OUT INTO ATTIC

NOTE: The inlet and outlet air openings shall each have a free area of not less than one square inch per 4,000 BTU per hour of the total input rating of all appliances in the enclosure.

FIGURE A5-55 (*Courtesy* American Gas Association)

Winter Comfort **403**

oxide poisoning due to homeowners intentionally or unintentionally blocking openings and grilles and creating a negative pressure. Lack of fresh air causes poor combustion and produces excessive carbon monoxide, which is then introduced from the draft hood into the return air and circulated to the living area. *Always connect the furnace return air to a return duct from the living space.*

A5-4
WINTER HEATING—OIL FORCED-AIR SYSTEMS

As mentioned previously, oil has been a popular fuel in the northeastern part of the United States for some time. It is also widely used in rural areas where gas mains do not exist and/or where the use of electricity for heating is not practical. Vacation or second homes frequently use oil for primary heating.

A5-4.1
Types of Furnaces

Like its counterpart the gas furnace, oil forced-air furnaces are available in upflow (Fig. A5-56), lowboy, downflow (counterflow), and horizontal models. The type of general application, ductwork, and air distribution into the conditioned space are essentially the same. There are, however, diferences in the internal combustion chamber, burners, controls, clearances, and flue venting requirements.

A5-4.2
Heat Exchanger

The typical oil-fired heat exchanger (Fig. A5-57) is a cylindrical shell of heavy-gauge steel where combustion takes place; it offers additional surfaces for heat transfer from the products of combustion to the air over the outside of this heat exchanger. This type of heat exchanger is called a *drum and radiator*. The portion containing the flame is called the *primary surface*, and the additional section or sections are called the *secondary surface*.

Some manufacturers add baffles, flanges, fins, or ribs to the surfaces to provide faster heat transfer to the air passing over the surfaces.

The burner assembly is bolted to the heat exchanger with its firing assembly and blast tube extending into the primary surface in correct relationship to the combustion chamber or refractory. A flame inspection port is provided just above the upper edge of the refractory. This is essential to observe proper ignition, flame, and for measurement of over-fire draft for start-up and service operations.

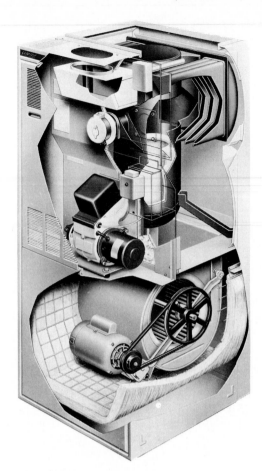

FIGURE A5-56 Oil upflow furnace. (*Courtesy* Lennox Industries)

A5-4.3
Refractory

To product maximum burning efficiency of the oil/air mixture, high temperatures are required in the flame area of the heat exchanger. To provide this increase in temperature, a reflective material is installed around the combustion area. This insulation type of material, called the *refractory*, is designed to obtain white-heat surface temperatures very quickly and to withstand high temperatures with minimum deterioration.

A5-4.4
Burner

The high-pressure atomizing gun burner (Fig. A5-58) is the type used on most residential and small commercial oil-fired heating systems. Viewed from the exterior, it has several functional components:

1. The oil pump
2. The air blower
3. Electric motor
4. Ignition transformer

FIGURE A5-57 Oil-fired heat exchanger. (*Courtesy* Bard Manufacturing Company)

5. The blast tube with included nozzles and ignition system

A cross section of the burner (Fig. A5-59) reveals a more detailed view of the operation. The pump and blower are driven from a common motor shaft. The oil pump is a gear arrangement, and by changing a bypass plug in the housing, the pump can operate as a single-stage or a two-stage system. Single-stage operation is used where the oil tank is above the burner and gravity oil feed to the burner is permitted. Two-stage operation

is needed where the oil tank is below the burner and the pump must lift the oil from the tank as well as furnish presssure to the nozzles. The pump supplies oil to the nozzle at 100 to 300 psi.

The burner blower is a centrifugal wheel also mounted on the common motor shaft. It furnishes air through the blast tube and turbulator for proper combustion. The amount of air is controlled by a rotating shutter band on the blower housing section.

The nozzle is mounted in an adapter and receives oil by means of a pipe from the pump. The orifice opening in the nozzle is factory bored to produce the correct firing rate. As a rule of thumb, the firing rate for No. 2 grade fuel oil will be approximately 0.8 gal/hr for each 100,000 Btu/hr output of the furnace. *Do not* attempt to change the firing rate by changing orifices or drilling an enlarged opening. Under high pressure the oil is atomized into fine droplets and mixes with primary air.

The air deflector vane ring (turbulator) serves an important function in getting the mixture of atomized oil and air to rotate or spin into the cavity of the heat exchanger. Ignition is established by a high-voltage electric spark, which may be continuous (on when the motor is on) or interrupted (on only to start combustion). The electric current is provided by an ignition transformer located in the burner control compartment.

Also located in the burner control compartment (not shown) is a flame-detecting device called a *cad cell* that is light sensitive. It has a hermetically sealed light-actuated cell that has low electrical resistance in the presence of light rays from the flame, thus permitting current flow, and high electrical resistance in the absence of light rays from the flame, thus prohibiting current flow. Therefore, if there is a flame failure for whatever

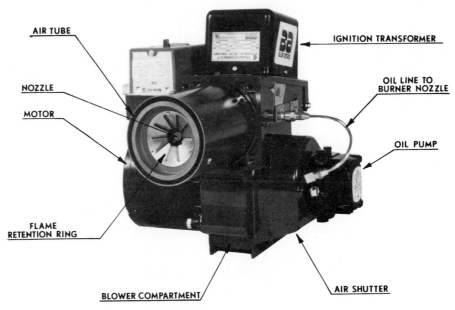

FIGURE A5-58 High-pressure atomizing gun type of burner. (*Courtesy* Wayne Home Equipment Company)

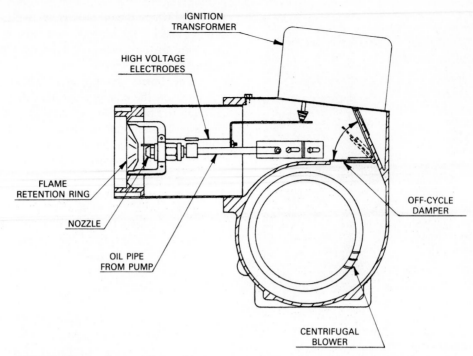

FIGURE A5-59   Cross section of burner. (*Courtesy* Wayne Home Equipment Company)

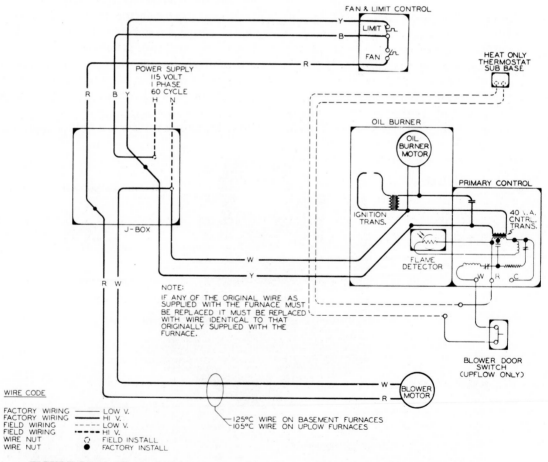

FIGURE A5-60   Typical oil furnace control. (*Courtesy* Borg-Warner Central Environmental Systems, Inc.)

406   Air-Conditioning

reason, the cad cell will stop the burner motor so that oil cannot flow into the heat exchanger. Some furnaces have a lockout system requiring the control to be reset manually before the burner can run again.

A5-4.5
Controls

External to the burner controls, the oil-fired forced-air furnace control system (Fig. A5-60) must also have a fan and limit control that performs the same function as on a gas furnace. The safety limit, if open, will stop the electrical current to the burner before overheating becomes excessive. The fan switch is set to cycle the furnace blower as desired. The space thermostat (24 V) feeds directly to burner low-voltage secondary control circuit, actuating the flame detector and relay to feed the primary line voltage to the burner motor and ignition transformer.

The operation of the controls, as well as startup, adjusting, and servicing the system are covered in subsequent chapters. Oil burner controls are discussed in more detail in Chapter A10.

A5-4.6
Flue Venting

Oil-fired furnaces must have an ample supply of makeup air for combustion, and the methods of introducing makeup air for a gas furnace will also be adequate for oil equipment.

Masonry chimneys used for oil-fired furnaces should be constructed as specified in the National Building Code of the National Board of Fire Underwriters. A masonry chimney shall have a minimum cross-sectional area equivalent to either an 8-in. round or an 8 × 8-in. square; it should never be less than the flue outlet of the furnace. Prefabricated lightweight metal chimneys (Fig. A5-61) are also available for use with oil. The double-wall type is filled with insulation. These vents are class A flues. Class B flues, made specifically for gas-fired equipment, are not suitable for use with solid or liquid fuels. Be sure that the class rating states "all fuels." The terminal vent heights above-the-roof recommendations outlined previously for gas should also be observed for oil.

Oil-fired furnaces operate on positive pressure from the burner blower, and it is most important to have a chimney that will develop a minimum draft of 0.01 to 0.02 in. of W.C. as measured at the burner flame inspection port. Consistency or stability of draft is also more critical, and the use of a barometric damper (Fig. A5-62) is required. The damper is usually installed in the horizontal vent pipe between the furnace and chimney. (*Note*: Some manufacturers attach them directly to the furnace flue outlet.) The damper has a movable weight

FIGURE A5-61 Factory-built "all fuel" chimney. (*Courtesy* Selkirk Metalbestos Chimney System)

so that it can be set to counterbalance the suction and to maintain reasonably constant flue operation. It is adjusted while the furnace is in operation and the chimney is hot. The overfire draft at the burner should be in accordance with the equipment installation instructions.

A5-4.7
Clearances

Cabinet temperatures and flue pipe temperatures run warmer for oil-burning equipment, and the clearances from combustible material should be adjusted accordingly. One-inch clearances are common for the sides and rear of the cabinet, as opposed to zero clearances for many gas units. Flue pipe clearances of 9 in. or more are needed, whereas only 6 in. of clearance is needed for gas.

Front clearance is generally determined by the space needed to remove the burner assembly. Combustible floor bases for oil are generally increased as compared to gas. Horizontal oil furnaces are particularly important in respect to attic installations and the installer must check the recommendations carefully.

Oil furnaces are rated and listed by Underwriters' Laboratories, and clearances are a vital part of this inspection and compliance procedure along with many other safety considerations.

FIGURE A5-62 Barometric damper. (*Courtesy* Field Control, Division of Conco, Inc.)

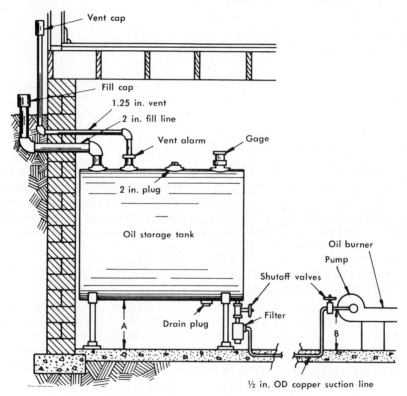

FIGURE A5-63 Indoor oil storage tank.

A5-4.8
Oil Storage

The installation of the fuel oil tank and connecting piping must conform to the standards of the National Fire Protection Association (Standard No. 31) and/or local code requirements. Regulations and space permitting, the oil tank can be located indoors as shown in Fig. A5-63. Note the proper use of shutoff valves and an ef-

fective filter to catch impurities. If the oil tank is placed outdoors above ground, firm footings must be provided. Exposed tanks and piping above ground are subject to more condensation of water vapor and the possibility of freezing during extremely low temperatures.

If a large tank is installed below ground (Fig. A5-64), it is important to keep it well filled with oil during periods of high water level (i.e., spring rains); otherwise,

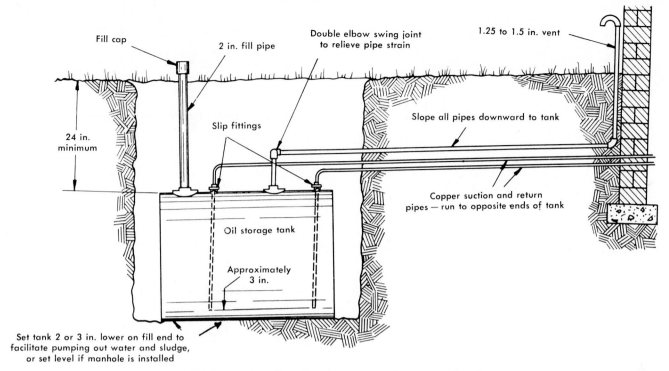

Fill cap

2 in. fill pipe

Double elbow swing joint
to relieve pipe strain

1.25 to 1.5 in. vent

24 in.
minimum

Slip fittings

Slope all pipes downward to tank

Oil storage tank

Copper suction and return
pipes — run to opposite ends of tank

Approximately
3 in.

Set tank 2 or 3 in. lower on fill end to
facilitate pumping out water and sludge,
or set level if manhole is installed

FIGURE A5-64   Outside oil storage tank two-pipe details.

groundwater may force it to float upward. Extra concrete for added weight over the tank is advised. Flexible copper lines are recommended so as to give with ground movement. Notice the use of suction and return lines; as mentioned previously, it is normal practice to use a two-stage fuel pump with a two-pipe system whenever it is necessary to lift oil from a tank that is below the level of the burner. The oil burner pump suction is measured in terms of inches of mercury vacuum. A two-stage pump should never exceed a 15-in. vacuum. Generally, there is 1 in. of vacuum for each foot of vertical oil lift and 1 in. of vacuum for each 10 ft of horizontal run of supply piping.

A5-5
WINTER HEATING—ELECTRIC FORCED-AIR SYSTEMS

At the end of 1973 the Electric Energy Association reported that over 5 million American residential dwellings were heated electrically. The growth rate is approximately 800,000 per year. Some of these systems are combination heating and cooling units (heat pumps or packaged terminal units), but the majority of installations (47.6% in 1973) were forced-air electric furnaces and all other forms of ducted resistance heating, and industry forecasts project this trend will continue.

A5-5.1
Electric Furnace

The typical electrical furnace (Fig. A5-65) is a most flexible and compact heating unit. It consists of a cabinet, blower compartment, filter, and resistance heating section. The cabinet size is generally more compact than the equivalent gas or oil furnaces, and due to cooler surface temperature, most units enjoy "zero" clearances all around from combustible materials. Thus, they may be located in very small closets. Additionally, since there is no combustion process involved, there is no requirement for venting pipes, chimney, or makeup air, thus simplifying installation and reducing building costs. Service access is the only dimensional consideration. Also, the absence of combustion permits mounting the units for up, down, or horizontal airflow application (Fig. A5-66). Some manufacturers provide space within the furnace cabinet for a cooling coil.

With the panel removed (Fig. A5-67) observe that the blower compartment usually houses a centrifugal, multispeed direct-drive fan or one of a belt-driven design where larger units are involved. Air flow through an electric furnace has less resistance and fan performance is more efficient. Mechanical filters of either the throwaway or cleanable variety are available.

The heating section consists of banks of resistance heater coils wound from nickel–chrome wire held in place by ceramic spacers. The heater resistance is de-

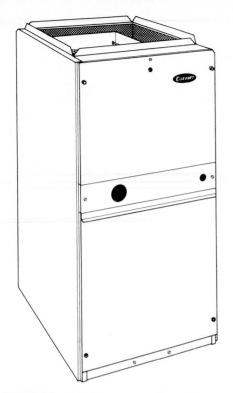

FIGURE A5-65 Electric furnace. (*Courtesy* Carrier Air-Conditioning Company)

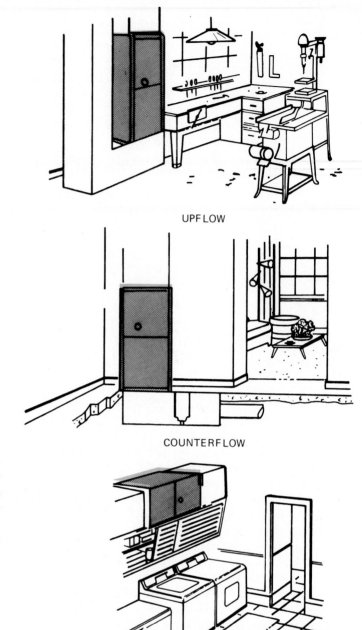

UPFLOW

COUNTERFLOW

HORIZONTAL

FIGURE A5-66 Airflow applications.

signed to operate on 208- to 240-V power with output according to the actual voltage used. The amount of heat per bank is a function of the amperage draw and/or staging. National and local electrical codes control the amount of current that can be put on the line in one surge, so it is necessary for manufacturers to limit this impact by the size (kW) of the heaters in a bank and the sequence of operation. Total furnace output capacities range from a low of 5 kW (17,000 Btu) to 35 kW (119,400) Btu/hr. At approximately 35 kW the amperage draw on 240 V approaches 150 A and, considering that 200 A is the total service to a residence, this only leaves some 50 A for other electrical uses. Seldom are all on continuously, but codes must assume so for safety purposes. Note, however, that well-insulated, electrically heated home requiring 35 kW will have considerable square feet. Zoning too is easier with two electric furnaces since location is not critical.

Electrical heating elements are protected from any overheating that may be caused by fan failure or a blocked filter. These are high-limit switch devices that sense air temperature and when overheated "open" the electrical circuit. Some furnaces also employ fusible links wired in series with the heater; these links melt at approximately 300°F and open the circuit. This is an auxiliary backup for the high-limit switches.

Built-in internal fusing or circuit breakers are provided by some manufacturers and are divided to comply with the *National Electric Code®*. This can be conve-

nient and may also mean installation savings for the contractor; external fuse boxes and associated on-site labor are eliminated.

The sequence control system is an important operation of the electric furnace. On a single-stage thermostat system there is an *electric sequencer control*, which contains a bimetallic strip. On a call for heat, the thermostat closes a circuit which applies 24 V across the heater terminals of the sequencer. As the bimetallic element heats up, the blower and first heater come on. The

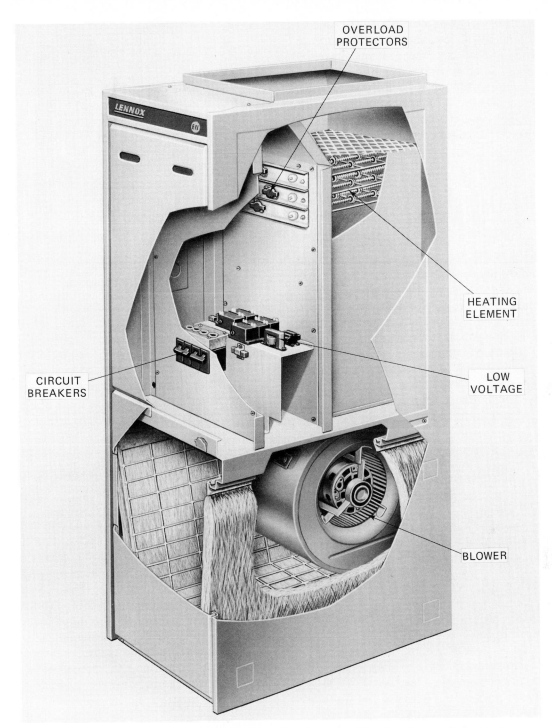

FIGURE A5-67   Lennox electric E-II upflow furnace. (*Courtesy* Lennox Industries)

operation characteristics of the sequencer are such that there is a time delay before each additional heater stage is energized. When the thermostat is satisfied, the sequencer is deenergized and the elements are turned off. The time delay (in seconds) is adequate to stagger power inrush and minimize the shock on the power system.

Where larger furnaces are installed, it is common to use two-stage thermostats in connection with the sequencer control. The first stage would operate as previously described and would bring on at least 50% of total capacity. The second stage of the thermostat would respond only when full heating capacity is needed. With this added control, wide variations of indoor temperature are avoided.

Previously, the text emphasized *electric* sequencer control as opposed to a *motor-driven* variety that may be used on furnaces having four to six elements. It fulfills the same function and, by a series of cams, times out the sequence of bringing heaters "on" and "off."

The air distribution system for an electric furnace

should receive extra care due to the normally lower air temperature of heated air coming off the furnace, as compared to gas and oil equipment. Temperatures of 120°F and below can create drafts if improperly introduced into the space. Additional air diffusers are recommended. Also, duct loss through unconditioned areas can be critical, so *well-insulated ducts are a must*, to maintain comfort and reduce operating cost.

A5-6
AIR-HANDLING UNITS WITH DUCT HEATERS

A variation of the electric furnace is the use of an air-handling unit with duct-type electric heaters (Fig. A5-68). The air handler consists of a blower housed in an insulated cabinet with openings for connections to supply and return ducts. Electric resistance heaters are installed either in the primary supply trunk or in branch ducts leading from the main trunk to the rooms in the dwelling. In terms of sales this system has not been as popular as the complete furnace package concept, primarily because it complicates installation requirements and adds cost. There is more comfort flexibility by zone control when heaters are installed in branch runs; rooms can be individually controlled.

Duct heaters (Fig. A5-69) are made to fit standard duct sizes and contain overheating protection. Electric duct heaters can also be utilized with other types of ducted heating systems to add heat in remote duct runs or to beef up the system if the house has been expanded. They may be interlocked to come on with the furnace blower and are controlled by a room thermostat.

A5-7
SUMMARY

Forced-air heating systems, be they gas, oil, or electric, offer the advantage of mechanical air movement whereby the air is not only heated but is cleaned by filtration, humidified, and "freshened" with an intake of outdoor air, which is circulated to the living area. Thus four of the five elements of total comfort air conditioning can be performed, with the fifth option of cooling as an easy addition.

A5-8
WINTER HEATING—HYDRONIC SYSTEMS

Another form of heating, and one that has existed for many years, is the hydronic method of conveying heat energy to the point of use by means of hot water or steam. The early steam system (Fig. A5-70) used a boiler partially filled with water. Heat from the fuel converts the water into steam (vapor). The steam passes through piping mains to the terminal units (radiators), where it is condensed upon giving up its heat. The condensate (liquid) returns to the boiler through the piping system, thus maintaining the proper water level within the boiler. The flow of heat is by pressure; no mechanical force is used.

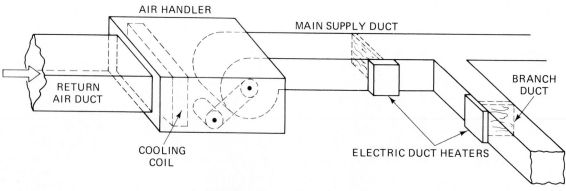

FIGURE A5-69   Duct heaters.

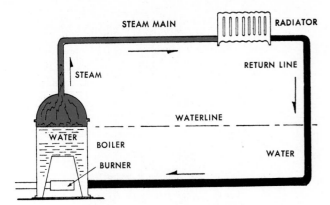

FIGURE A5-70 Steam system.

In contrast to the partially filled boiler of the steam heating system, the boiler, piping, and terminal units of a hot-water heating system (Fig. A5-71) are entirely filled with water. Heat produced by the fuel, be it gas, oil, or electric, is transferred to the water within the boiler or heat exchanger. The heated water is circulated through piping and terminal units of the system. The heat given off as the water passes through the terminal units causes the water to cool. The cooled water returns to the boiler and is reheated and recirculated. (*Note*: A similar but reverse process can be used for cooling by circulating chilled water; this technique will be covered later.)

By definition, hydronics includes all three systems: steam, hot water, and chilled water. However, since steam-heating systems are not frequently installed in residences or small commercial buildings today, this discussion will be confined to hot-water heating.

A5-8.1
Hot-water Heating

In Fig. A5-71 the flow of water is dependent on gravity circulation due to the difference in the density of water in the supply and return sides. Like steam heating, gravity-flow hot-water systems are pretty much obsolete; much more popular today is the forced circulation system using a pump to circulate the water from the boiler to terminal units and back again. Pipe sizes can be smaller and longer, and terminal units can be smaller and may be located below the boiler if needed. Forced circulation insures better distribution control.

Hot-water systems may be classified according to operating temperatures and pressures:

	Boiler Pressure (psi)	Maximum Hot-Water Temperature (°F)
LTW—low temp.	30	250
MTW—med. temp.	150	350
HTW—high temp.	300	400–500

Medium- and high-temperature systems are limited to large installations and are beyond the scope of this text.

Low-temperature hot-water heating systems are classified according to the piping layout. Four common types of systems are:

1. Series loop
2. One pipe
3. Two pipes
4. Hot-water panel

A5-8.2
Piping Systems

The series loop system (Fig. A5-72) is most commonly used in small buildings or in subcircuits of large systems. The terminal units, usually of the baseboard or fin tube-type radiation, serve as part of the distribution system. Water flows through each consecutive heating element in succession. The water temperature, therefore, is progressively reduced around the circuit, which may require that the length of each heating element be selected and adjusted for its actual operating tempera-

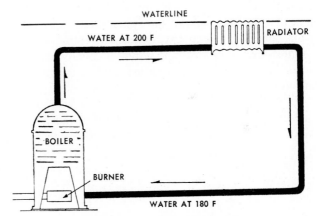

FIGURE A5-71 Hot-water system.

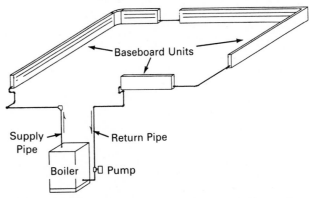

FIGURE A5-72 Series loop baseboard system (single circuit).

ture. The series loop system has the advantage of lower installed costs. Also, interconnecting piping can be above the floor, eliminating the need for pipe trenches, insulation, and furring. Its disadvantage lies in the fact that the water temperature to each unit cannot be regulated, and thus some form of manual air damper on the terminal is desirable. Also, circuit capacity is limited by the size of the tube(s) in the terminal element(s).

The one-pipe system (Fig. A5-73) is a variation of the loop, except that the terminals are connected to the pipe loop by means of branches. Individual terminals can be controlled, but special tees are needed, and these fittings are available only from a few manufacturers. A one-pipe system will generally cost more to install than a comparable series-loop system, because of the extra branch pipe, fittings, and special fittings themselves. The one-pipe system also shows a progressive temperature drop around the circuit with resulting limitations.

Two-pipe systems (Fig. A5-74) use one pipe (supply) to carry hot water to the terminal and a second to return cool water to the boiler. Known as a *two-pipe reverse return*, it is set up so that while terminal 1 is closest to the boiler on the hot water supply, it's the farthest on the return main. The reverse is true for terminal 5; it is farthest away on the hot-water supply and closest on the return main. This equalization of the distance the water travels through each unit provides even distribution of water throughout the system—and better control. Zones or several circuits may be employed with equal success. The two-pipe system is usually preferred when low-pressure-drop terminal units like baseboard are used.

Panel systems (Fig. A5-75) are installed in floors or ceilings and utilize a supply or return header where all of the coils can be brought together. Balancing valves

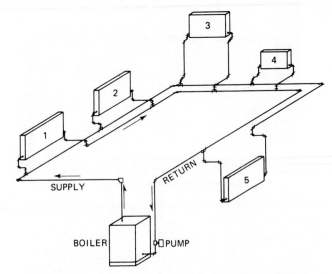

FIGURE A5-74 Two-pipe reverse hot-water heating system.

and vents on each of the coils can then be made accessible in a pit or in the ceiling of a closet.

The preceding discussions and illustrations were of single-circuit systems. Any of these hydronic systems may also be installed as multiple circuits where a main or mains form two or more parallel loops between the boiler supply and return.

Multiple circuits are employed for several reasons: to reduce the total length of circuits, to reduce the number of terminal units on one circuit, to reduce pipe size or the quantity of water circulated through a circuit, and for simplification of pipe design in certain types of buildings.

A5-8.3
Boilers

Steam boilers were in existence prior to the mid-1800s, and by the 1870s, hot water was beginning to replace steam. Modern boiler design includes the use of cast iron, steel, and copper materials.

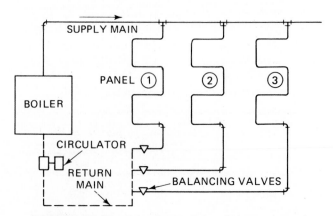

FIGURE A5-75 Forced circulation hot water panel system.

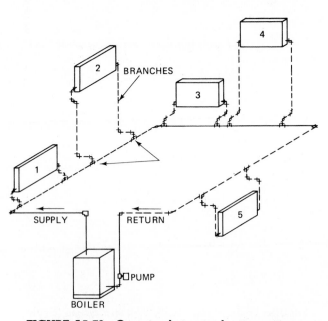

FIGURE A5-73 One-pipe hot-water heating system.

FIGURE A5-76 Gas boiler. (*Courtesy* Weil-McLain, Division of Wylain, Inc.)

Cast iron boilers are assembled from cast iron sections (Fig. A5-76). They can burn gas, oil, or coal. Steel boilers are designed with tubes inside a shell. Usually, the system water is in the tubes and the fire and hot gases surround the tubes and thus heat the water. This is called a *water-tube boiler* (Fig. A5-77). However, some boilers function in just the reverse fashion—the fire and hot gases pass through the tubes and heat the water in the shell. These are called *fire-tube boilers*. Residential-size steel boilers are sold as packages with pump, burner, and controls mounted and wired.

Flash boilers for gas or oil burning consist of copper coils which contain the system water surrounded by the fire and hot gases. These are very compact, and some models are designed to be used outdoors.

An electric boiler is essentially a conventional boiler with a large internal volume and electric heaters directly immersed in water. There are also instantaneous types of electric boilers small enough to "hang on the wall" (Fig. A5-78), which are fitted with either immersed electric heaters or electrical heating elements bonded to the outside of the water vessel. The pump and expansion tank are enclosed in the cabinet.

A5-8.4
Typical System

A typical low-temperature hot-water installation (Fig. A5-79) consists of:

1. The boiler, be it gas, oil, or electric, as described above and operated at 30 psi, with heating water from 180 to 220°F for normal supply.
2. A boiler water temperature control to cycle the burner or heaters to provide control of hot water temperature.
3. A circulating water pump large enough to allow the flow of sufficient water to heat the space. The control of the pump is from a space or zone thermostat.
4. A relief valve that will "open" if the boiler pressure exceeds 30 psi.

FIGURE A5-78 Wall-type electric boiler. (*Courtesy* Weil-McLain, Division of Wylain, Inc.)

FIGURE A5-77 Water-tube boiler.

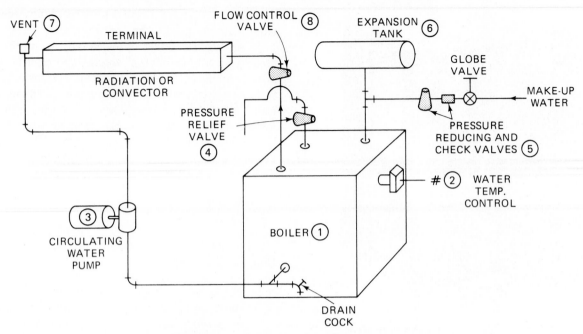

FIGURE A5-79 Typical residential hot-water system.

5. Pressure-reducing and check valves and an automatic water makeup valve that supplies water to the boiler for initial fill and also if pressure gets too low.

6. A closed air-cushion tank, which, on initial fill, has a pocket of trapped air within the tank. When water is heated it expands and compresses the trapped air as a cushion, thus providing space for extra water without creating excessive pressure.

7. Air vents at the terminals or high point of the piping system to purge unwanted trapped air.

8. A flow-control valve (check valve) to prevent gravity circulation of water when the pump is not in operation.

9. The necessary cocks, purge valves, and drain cocks for balancing, service, and maintenance adjustment.

Terminal units vary from cast iron radiators (seldom used today) to finned-tube residential baseboard radiation (Fig. A5-80), to cabinet-type convectors (Fig. A5-81), to hot-water coils for air-handling units or duct mounting (Fig. A5-82). The selection of the type of unit depends on whether it is used in residential, commercial, or institutional buildings, etc.

The selection of the terminal size is essentially a function of the temperature of the water, the flow rate, the surface area of radiation, and the type of space air circulation (natural or fan-forced). Manufacturers' cat-

alogs provide rating tables which prescribe the proper selection.

The overall design of hydronic heating systems is in itself a special art, and we would suggest that those who become involved in this area of heating take advantage of the information available through the Hydronic Institute, 35 Russo Place, Berkeley Heights, New Jersey 07922, and the North American Heating and Air-Conditioning Wholesalers Association, 1661 West Henderson Road, Columbus, Ohio 43220. Both of these associations offer excellent training data on the design, installation, and service of hydronic systems.

A5-9
WINTER HEATING—HUMIDIFICATION

In Chapter A2 the importance of maintaining proper humidity was discussed, and some of the harmful effects of poor humidity control on people and materials.

Room-type humidifiers are available that add moisture to the air, but these require constant attention in terms of manually filling with water. Their evaporation capacity is limited, and at best they are useful only for bedside medical purposes.

As mentioned previously, central forced-air heating and air-conditioning systems offer the ideal solution of adding moisture to the air at the furnace location and circulating it to the total living area, not to just one room.

How much moisture should be added to the air in a residence? This, of course, is subject to the outside

FIGURE A5-80 Finned-tube baseboard radiation.
(*Courtesy* Weil-McLain, Division of Wylain, Inc.)

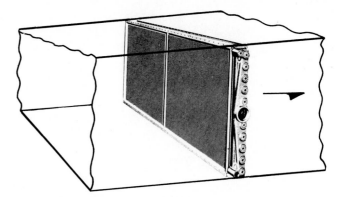

FIGURE A5-82 Hot-water coil in duct.

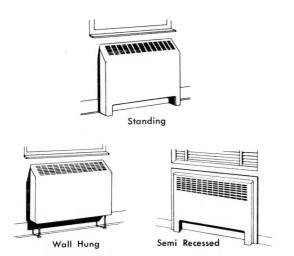

FIGURE A5-81 Convector radiation.

and inside temperature and moisture conditions, all of which are constantly changing. We could go through a rather lengthy discussion involving vapor pressure and moisture content, permeability of building materials, etc., but the answer would still only represent one sample and not be useful to practical application.

Fortunately, ARI developed a certification program known as *Standard 610*, open to all manufacturers, whereby a humidifier "size" is rated in terms of capacity—a measure in gallons of the amount of water the equipment can evaporate into the heated home in 24 hours. The size of the home and its air tightness are two important factors in determining the needed capacity required in a humidifier. Figure A5-83 is a sample chart published by one manufacturer of a specific model showing maximum moisture requirements. It is based on design conditions of: an outdoor dry-bulb temperature of 20°F, 80% relative humidity, an indoor dry-bulb temperature of 70°F and 40% relative humidity, and minimum moisture production from residential operations for an absolute humidity difference of 0.0049 lb/hr. The definition of a tight house is one that is well insulated, has vapor barriers, tight-fitting storm doors and windows with weather stripping, and its fireplace

will be dampered. An average house will be insulated, have vapor barriers, loose storm doors and windows, and a dampered fireplace.

With the volume of the residence known (in ft³—length × width × ceiling height) one can quickly determine how many pounds of water per hour or gallons per day will be needed to maintain the above conditions. This is an average guide but will apply to most of the country. Remember, the *average* outdoor winter temperature doesn't stay below 20°F for any extended period of time. Each ARI certified manufacturer will have similar information on capacity and/or home dimensions relative to particular units.

With a method of determining how much water to evaporate, the next question is what kind(s) of humidifiers are available. This, too, is a choice or compromise based on need, capacity, method of operation, price, and ease of maintenance.

Early attempts at humidification consisted of pan-type humidifiers (Fig. A5-84). These were limited in ca-

FIGURE A5-83

Maximum moisture requirements.

Volume of Residence (ft³)	TIGHT HOUSE		AVERAGE HOUSE	
	Pounds per Day	Gallons per Day	Pounds per Day	Gallons per Day
8,000	1.76	5.09	3.52	10.17
10,000	2.21	6.35	4.41	12.72
12,000	2.64	7.63	5.29	15.26
14,000	3.09	8.91	5.92	17.08
16,000	3.53	10.18	7.06	20.35
18,000	3.97	11.45	7.94	22.89
20,000	4.41	12.72	8.82	25.44
22,000	4.85	13.99	9.71	27.98
24,000	5.29	15.27	10.59	30.52
26,000	5.74	16.54	11.47	33.07
28,000	6.18	17.81	12.35	36.51
30,000	6.62	19.08	13.24	38.16

Evap. Rate .03 lbs./hr./ft.²
at Room Temperature and still air.
3 lbs./hr. Requires 100 ft.² Pan

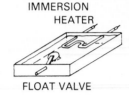

Pan with Plates or Discs
Mounted in Warm Air
Capacity Rating 1 to 3 lbs./hour

IMMERSION
HEATER

WATER SUPPLY

Capacity Rating with a:
1,000 Watt Element = 3 to 4 lbs./hr.
Hot Water Coil (160°F.)=4 to 6 lbs./hr.
Steam Coil (2 psig) = 10 to 20 lbs./hr.

FLOAT VALVE

FIGURE A5-84 Pan humidifiers.

pacity, fraught with scale problems when solids were precipitated out of hard water during evaporation, or required too much energy, such as the immersed electric heater.

Improvements have been made to the plate type by using rotating disks as represented in Fig. A5-85. The disks are made of a noncorrosive material and are rotated by a small motor (similar to the way a barbeque spit works). The disk constantly presents a wetted surface to the air. This type of unit is mounted under a warm-air duct and can evaporate up to 25 gallons of water per day. It is also relatively inexpensive to install and operate.

Another version is the atomizing humidifier (Fig. A5-86), which mechanically breaks the water into fine droplets to accelerate evaporation. Several methods are shown. Seldom is the high-pressure nozzle used in residential application. They are, however, used extensively in spray coil commercial systems that need large volumes of water. The centrifugal wheel design (Fig. A5-87) has been a popular unit for mounting on the furnace return plenum or suspended under a return air duct. A motor-driven centrifugal wheel throws the water against a set of slotted plates that break the water into a fine mist, which is then blown into the conditioned air. The limited capacity of this system is a disadvantage for larger homes.

The third and most effective type is called a *wetted element humidifier* (Fig. A5-88), which combines the advantages of a positive air stream passing over a large, wetted surface. The plenum power type of humidifier (Fig. A5-89) uses a small internal fan to draw the air through the wetted evaporating pad and then return it to the furnace airstream. Its capacity is high, but a significant amount of electrical power is consumed by the humidifier fan.

The bypass humidifier (Fig. A5-90) does not have a fan but depends on the air pressure differential between the warm-air plenum (+pressure) and the cold-air return (−pressure). Its capacity rating is slightly less

FIGURE A5-85 Rotating disk humidifier. (*Courtesy* BDP Company, Division of United Technologies Corporation)

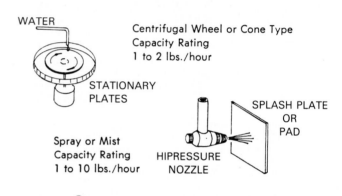
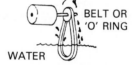

WATER

Centrifugal Wheel or Cone Type
Capacity Rating
1 to 2 lbs./hour

STATIONARY
PLATES

Spray or Mist
Capacity Rating
1 to 10 lbs./hour

HIPRESSURE
NOZZLE

SPLASH PLATE
OR
PAD

BELT OR
'O' RING

WATER

Splashing Ring Plenum Mounted
Capacity Rating
1 to 3 lbs./hour

FIGURE A5-86 Atomizing humidifiers.

than that of the fan model above, but no added electrical cost is introduced.

To recap, the decision as to which method, model, or brand to install is a function of need, application, price, and maintenance. And on the last point, maintenance is the key to effective humidifier operation. Regardless of the type, water is being evaporated, and any suspended solids such as calcium will collect and leave a crust on the surface that must be removed periodically. The question is, can or will a homeowner perform this task or must he depend on his installer? Some products are simple to service and require no special tools; others are not so easy to care for and should be attended to by competent service technicians.

The control of the humidifier depends on the degree of accuracy needed. Water make-up is introduced

FIGURE A5-87 Centrifugal wheel humidifier. (*Courtesy* Walton Laboratories)

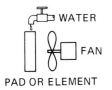

WATER

FAN

PAD OR ELEMENT

EVAPORATION RATE
1 TO 1.5 LBS./HR./100 CFM
WITH 70°F DRY BULB AND
35% RH ROOM AIR

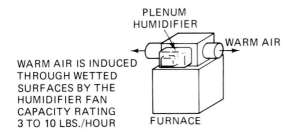

PLENUM
HUMIDIFIER

WARM AIR

WARM AIR IS INDUCED
THROUGH WETTED
SURFACES BY THE
HUMIDIFIER FAN
CAPACITY RATING
3 TO 10 LBS./HOUR

FURNACE

FIGURE A5-89 Plenum power humidifier. (*Courtesy* Skuttle Manufacturing Company)

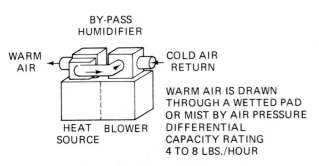

BY-PASS
HUMIDIFIER

WARM
AIR

COLD AIR
RETURN

WARM AIR IS DRAWN
THROUGH A WETTED PAD
OR MIST BY AIR PRESSURE
DIFFERENTIAL
CAPACITY RATING
4 TO 8 LBS./HOUR

HEAT   BLOWER
SOURCE

FIGURE A5-88 Wetted element humidifiers.

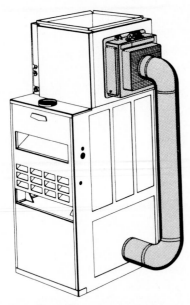

FIGURE A5-90 Bypass humidifier. (*Courtesy* BDP Company, Division of United Technologies, Inc.)

FIGURE A5-91 Humidistat. (*Courtesy* Honeywell, Inc.)

and controlled by a float valve that maintains a certain level or flow to the sump. If the humidifier has no internal power fan, such as in the bypass model, it will function as the furnace fan cycles on and off. But as long as the furnace fan runs, moisture will be introduced, so in order not to over-humidify, a shut-off water solenoid valve is employed.

And where close control of humidity is desired, a humidistat may be used to energize or deenergize the water solenoid and/or furnace fan relay. (This will be reviewed later under electrical controls systems.) A humidstat (Fig. A5-91) is a switch mechanism actuated by changes in moisture. The sensing or hygroscopic element of a typical residential humidistat is composed of tightly strung human hair or a Teflon strip that will contract or expand under varying relative humidity conditons. This movement is sufficient to open or close an electrical contact. Some are wall mounted, but others are made for duct mounting and may be equipped with a sail switch to sense air movement in the duct.

PROBLEMS

A5-1. Name four types of common fuel.

A5-2. What grades of fuel oil are used for residential heating? The approximate heating value is _____ .

A5-3. Natural gas is made up of _____ .

A5-4. Name the two most common types of standard LP fuel.

A5-5. Is LP gas heavier or lighter than air?

A5-6. How much combustion air (ratio) is needed to burn natural gas?

A5-7. How much combustion air (ratio) is needed to burn propane and butane?

A5-8. The products of properly burned gas, besides heat, are _____ and _____ .

A5-9. A product of improperly burned gas is _____ .

A5-10. In electric heat, 1 kilowatt equals _____ Btu.

A5-11. The agency that establishes the safety and construction standards for gas appliances is the _____ .

A5-12. The agency that sets the safety standards for oil-fired appliances is _____ .

A5-13. What is the most frequently used electric heating product?

A5-14. Name four types of forced-air furnaces.

A5-15. What is the basic difference between a highboy and a lowboy forced-air heating unit?

A5-16. If a gas furnace had an input rating of 120,000 Btu/hr, what would be the assumed output?

A5-17. Furnace clearances are distances from _____ .

A5-18. The vent for a gas furnace has two basic functions. What are they?

A5-19. The purpose of a draft hood or diverter on a furnace is to _____ .

A5-20. There are four methods of providing combustion to a residential heating unit. What are they?

A5-21. Is it permissible to use the equipment room as part of the return air system?

A5-22. Heat exchangers are built in two different styles. What are they?

A5-23. In an oil-fired heat exchanger, the portion containing the fire is called the _____ surface.

A5-24. The additional heat exchanger section or sections is called the _____ surface.

A5-25. The device in the combustion area that is provided to produce maximum combustion efficiency is called the _____ .

A5-26. Oil burners for residential forced-air furnaces are of the _____ type.

A5-27. A residential oil burner has five major parts. What are they?

A5-28. Electric furnace capacities are usually built in _____ kilowatt increments.

A5-29. Two types of protection are built into electric furnace heater assemblies. What are they?

A5-30. The time-delay control used with electric elements is called a _____ .

A5-31. The elements of an electric furnace are staggered on and off. Why?

A5-32. The term "hydronics" covers three types of heat-transfer systems. What are they?

A5-33. The maximum boiler pressure and water temperature for a residential hot water boiler are _____ psig and _____ °F.

A5-34. What are the four types of piping layout for low-temperature hot water systems?

A5-35. Describe a water tube boiler.

A5-36. Name three types of humidifiers.

A5-37. To control the moisture level in an area, a _____ is used.

Mechanical and Electronic Filtration

A6-1
MECHANICAL FILTRATION

In earlier discussions on forced-air heating equipment, mechanical filters were included as standard equipment. The term *mechanical* describes a typical fibrous material filter (Fig. A6-1) commonly called a *throwaway* or replacement variety. It consists of a cardboard frame with a metal grille to hold the filter material in place. The filter media material consists of continuous glass fibers packed and loosely woven that face the entering air side and more dense packing and weave on the leaving air side. Air direction is clearly marked. Filter thickness is usually one inch for residential equipment. Filters come in a wide variety of standard sizes (length and width). The recommended maximum air velocity across the face should be around 500 ft/min, with a maximum allowable pressure drop of about 0.5 in. W.C. Some makes also have the fiber media coated with an adhesive substance in order to attract and hold dust and dirt.

Many manufacturers and filter suppliers also offer what is known as a *permanent* or *washable* filter (Fig. A6-2). It consists of a metal frame with a washable viscous-impingement type of air-filter material supported by metal baffles with graduated openings or air passages. The filter material is coated with a thin spray of oil or adhesive in order to help the filtering effect. When needed, these filters may be removed and cleaned with detergent, dried, and recoated.

Another variation of dry filter is a pad or roll of material supported by a metal mesh frame of some type as shown in Figs. A6-3 and A6-4.

In commercial and industrial work there are variations of wet and dry filters for special use where larger volumes of air are involved, or the filtration of chemical substances such as in the removal of paint in spray booths, commercial laundries, hospital operating rooms, etc., but this is a matter for experts to handle.

The arrestance factor (efficiency) of standard residential and light commercial mechanical filters can vary from 25% for a typical window air-conditioner to perhaps 80 to 85% for the better heating and central air-conditioning equipment. This means they are effective in removing lint, hair, large dust particles, and somewhat effective on common ragweed pollen. They are relatively ineffective on smoke and staining particles. So for average use they do a creditable job. However, getting the homeowner to keep filters clean is not easy, and dirty filters are probably the main contributor to malfunctioning equipment. As the mechanical filter clogs with dirt, it can reduce airflow to a point where the cooling coil will freeze and perhaps lead to a compressor failure. Or, in a heating system, it can cause overheating and reduce the life of the heat exchanger, or it can cause nuisance tripping of the limit switch. Dirty filters increase operating costs as system efficiency goes down.

A6-2
ELECTRONIC FILTRATION

Where a high degree of filtering efficiency is desired (above 80%), the use of electronic air cleaners is highly recommended. Chapter A2 mentioned the many advantages to health, cleanliness, and savings resulting from having clean air. Industry product development has put these devices within the price range of most homeowners.

Electronic air cleaners use the principle of electrostatic precipitation in filtering air. This is not a recent discovery and has been used in commercial and industrial work such as smokestacks, clean rooms, etc. The adaptation to residential use, made possible by mass production and effective marketing, is more recent, within the last 10 to 15 years. There are two types of electrostatic cleaners: the ionizing type and the charged media.

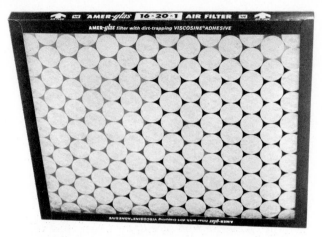

FIGURE A6-1 Fibrous material throwaway filter. (*Courtesy* American Air Filter, an Allis-Chalmers Company)

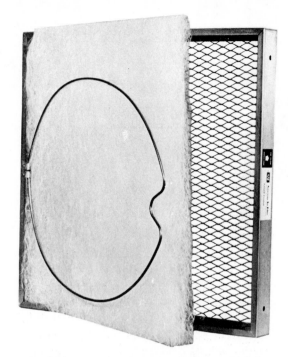

FIGURE A6-3 Dry filter pad. (*Courtesy* American Air Filter, an Allis-Chalmers Company)

FIGURE A6-2 Washable filter. (*Courtesy* American Air Filter, an Allis-Chalmers Company)

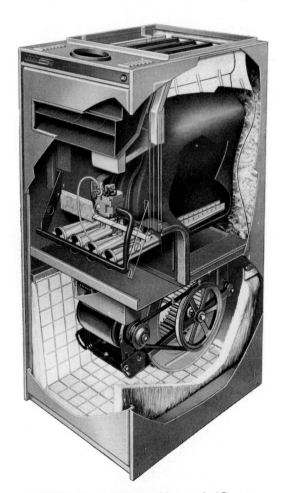

FIGURE A6-4 Bulk dry filter pad. (*Courtesy* Lennox Industries)

In electronic cleaners (Fig. A6-5), positive ions are generated in the ionizing section by the high-potential ionizer wires. The ions flow across the airstream, striking and adhering to any dust particles carried by the airstream. These particles then pass into the collecting cell, a system of charged and grounded plates. They are driven to the collecting plates by the force exerted by the electrical field on the charges they carry. The dust particles which reach the plates are thus removed from the airstream and held there until washed away. A prefil-

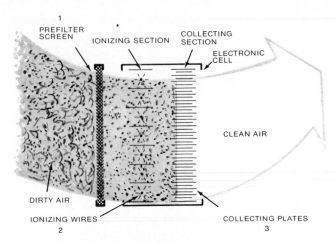

FIGURE A6-5 Ionizing-type cleaner.

ter/screen is usually employed to catch large particles such as lint and hair.

In a typical electronic air cleaner design (Fig. A6-6), a dc current potential of 12,000 V or more is used to create the ionizing field, and a similar or slightly lower voltage is maintained between the plates of the collector cell. These voltages necessitate safety precautions, which are designed into the equipment. The power pack boosts the 110-V supply to 12,000 V by means of a high-voltage transformer and converts ac current to dc, using solid-state rectifiers that contain silicon or selenium. Power consumption is relatively low, 50 W or less. The collector cell (Fig. A6-7), made of nonrusting aluminum plates, requires periodical removal for cleaning by soaking in a tub or any suitable container in detergent or liquid soap and then washing down with a hose. Some cells are small enough to be cleaned in an automatic dishwasher.

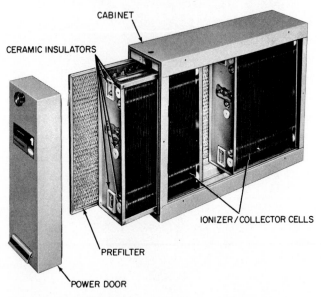

FIGURE A6-6 Electronic air cleaner. (*Courtesy* Fedders Air Conditioning–USA)

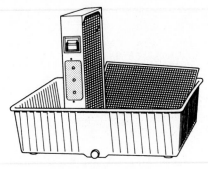

FIGURE A6-7 Cleaning collector cell.

The *charged-media type of air cleaner* had certain characteristics of both dry filters and electrostatic cleaners. It consists of a dielectric filtering medium, usually arranged in pleats or pads. Ionization is not used. The dielectric medium is in contact with a gridwork consisting of alternately charged members held at very high voltage potential. An intense electrostatic field was created, and airborne particles approaching the field were polarized and drawn into the filaments or fibers of the media. This type of cleaner offered much higher resistance to air flow, and the efficiency of the media was impaired by high humidity. At this time, charged-media air cleaners are no longer available.

How effective are electronic air cleaners? Most are capable of filtering airborne particles of dust, smoke, pollen, and bacteria ranging in size from diameters of 50 microns down to 0.03 micron. The micron is a unit of measure (length) equal to one millionth of a meter or 1/25,400 of an inch. The dot made by a sharp pencil is about 200 microns in diameter. It takes special electronic microscopes to examine particles of this size.

The operating efficiency of electronic air cleaners is a function of airflow. With a fixed number of ionizer wires and collector plates, a particular model may be used over a range of airflow quantities (ft^3/min). It will be rated at a nominal ft^3/min, and then for other efficiency ratings for the recommended span of operation. The higher the ft^3/min of air and the resultant velocity, the lower the efficiency. Residential electrostatic air cleaners come in sizes of 800 to 2000 ft^3. The number of models depends on the manufacturer. Most offer two sizes: 800 to 1400 ft^3 and 1400 to 2000 ft^3. Static pressure drop will run about 0.20 in. of W.C. at the nominal airflow rating point. Ratings are published in accordance with National Bureau of Standards Dust Spot Test and/or are certified under ARI standards.

The application of an air cleaner is most flexible (Fig. A6-8). When mounted at the furnace or air handler, it is installed in the return air duct for upflow, downflow, or horizontal airflow. Position does not affect its operation or performance. (*Note:* Some units are equipped with built-in washing, and, obviously, these must be installed vertically.)

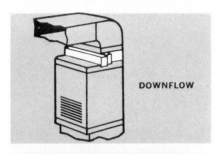
DOWNFLOW

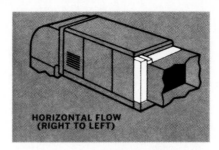
HORIZONTAL FLOW (RIGHT TO LEFT)

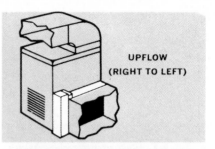
UPFLOW (RIGHT TO LEFT)

CLOSET INSTALLATION

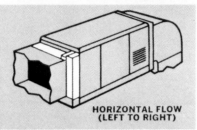
HORIZONTAL FLOW (LEFT TO RIGHT)

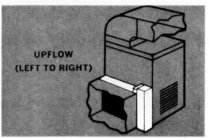
UPFLOW (LEFT TO RIGHT)

FIGURE A6-8   Air cleaner installation.

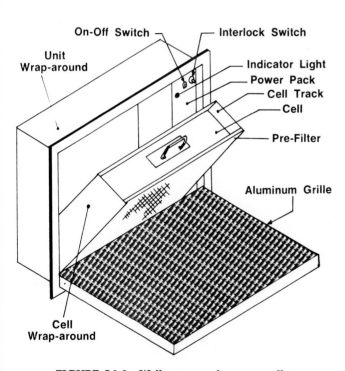

FIGURE A6-9   Wall-type air cleaner installation.

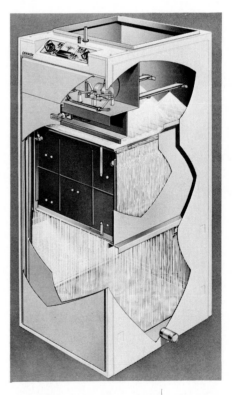

FIGURE A6-10   Electronic air cleaner.
(*Courtesy* Addison Products Company)

For the explanation of the electrical operating principles of the electronic air cleaner, see Section R21-6.1.

Where furnaces are installed in very small closets or utility rooms and no duct space is available, some manufacturers offer a model (Fig. A6-9) which will recess into a wall and actually provide a decorative return air louver that faces the living area. Service access is through the louver.

Some units, as shown in Fig. A6-10, have built-in washing without removing the cells. On an automatic control, washing action sprays water over the plates. The runoff is collected in the bottom of the unit, where it drains out.

Mechanical and Electronic Filtration   425

A6-1. What is the common name for noncleanable mechanical filters?

A6-2. Mechanical filters are generally effective up to _____ %.

A6-3. Filter maintenance is not important. True or False?

A6-4. Electronic air cleaners are made in two types. What are they?

A6-5. Operating voltage in an electronic filter can run as high as _____ .

A6-6. Electronic air cleaners are rated in _____ .

A6-7. What is a micron?

A6-8. The power consumption of a typical electronic air cleaner is about _____ watts.

A6-9. Position of the electronic air cleaner is not critical. True or False?

A6-10. The collect cell is usually cleaned by _____ .

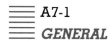

Unitary Packaged Cooling

A7-1
GENERAL

The chapters on refrigeration discussed the basics of a mechanical refrigeration cycle and reviewed each component in terms of its function in the system and its operating characteristics. This and subsequent chapters will build on that background and discuss the products that make up the broad range of cooling and combination heating and cooling systems that contain a mechanical refrigeration cycle. Chapter A7 deals with products essentially designed to provide cooling only, although it is sometimes difficult to make such a distinction because of the similar characteristics of and crossover between certain systems.

A7-2
ROOM AIR-CONDITIONERS

With annual sales of over 5 million units per year, the room unit (Fig. A7-1) obviously is meeting a need. It is installed in the window of a room and is meant to provide cooling for that immediate area, although in actual practice, the cooling effects may extend to other areas if there is good circulation. For example, a unit installed in the living room may offer some relief to bedrooms or kitchen. Not all units are window mounted. Models are available for permanent installation through the wall. This is common in low cost installations in motels, small offices, and apartments, but such units are not often mounted in the walls of single-family residential dwellings.

Room units have been used extensively in homes employing nonducted baseboard or ceiling cable heating systems.

The capacities of room units range from 5000 to over 30,000 Btu/hr. The larger units are more frequently associated with new residential construction, where the necessary electrical service and structural support is preplanned, and also for commercial applications. Most average bedrooms use a $\frac{1}{2}$- to $\frac{3}{4}$-ton unit. For existing homes the industry has developed what is known as a $7\frac{1}{2}$-A unit that will plug into an ordinary 15-A lighting circuit. Codes assume that the air conditioner will not exceed 50% of the circuit rating, so that breakers or fuses will not open when the compressor starts up. The maximum size of $7\frac{1}{2}$-A models are about 9000 Btu/hr. Larger units are wired independently of the lighting system for 208/230 V operation.

The sales and distribution of room units are predominately through appliance and chain store marketing channels rather than through professional air-conditioning contractors. Many installations are do-it-yourself situations. Capacity ratings and other technical standards are subject to the Association of Home Appliance Manufacturers (AHAM) rather than to the ARI for all other cooling apparatus. Also, in terms of war-

FIGURE A7-1 Room air-conditioning unit. (*Courtesy* Airtemp Corporation)

427

ranty and service, the room unit more closely follows the appliance industry warranties.

In terms of operation there is little or nothing that can be changed to control comfort except perhaps directional louvers on the room air outlet grille to minimize drafts. Filter replacement is the essential service requirement.

In summary, the room unit serves a definite need: spot cooling at a minimum installed cost and mobility of location, both of which preclude the consideration of central system cooling units.

A7-3
CENTRAL SYSTEM COOLING UNITS

A7-3.1
Packaged Air-Cooled Units

Basically developed as a residential central system cooling unit, the packaged air-cooled unit (Fig. A7-2) has also been broadly adapted to light commercial application and mobile home conditioning. The usual design configuration is a rectangular box with supply and return air connections on the front and provisions for the condensing section air (in and out) on the back and sides. The inside arrangement is relatively simple. Return air is drawn across the fin-tubed evaporator by a centrifugal blower and discharged out the front. On smaller sizes the blower is of the direct-drive variety. Larger units have adjustable belt drives. A condensate pan beneath the evaporator collects moisture runoff; it is connected to a permanent drain. The evaporator compartment is amply insulated to prevent heat gain and sweating. The filter is usually located in the return air duct. Separating the evaporator compartment and the condensing compartment is a barrier wall, which isolates air movements and insulates for minimum thermal and sound transmission to the conditioned air. The compressor and condenser coil form the *high-side* refrig-

FIGURE A7-3   Condensing unit control compartment. (*Courtesy* Addison Products Company)

erant circuit. Condenser air is drawn in from the sides and discharged through the condenser coil. This is called a *blow-through* arrangement. Some units use a *draw-through* arrangement and exhaust through the sides. The condenser fan is most frequently a propeller type. It moves large volumes of air where there is little resistance. Propeller fans are not intended for ducting. Refrigerant R-22 is almost universally used by all manufacturers, as is capillary expansion on smaller tonnage equipment.

The control box (Fig. A7-3) includes capacitors for fan motors and/or compressors, starting relays, and the terminal connection for a remote thermostat.

The capacities of packaged air-cooled cooling units range from $1\frac{1}{2}$ to $7\frac{1}{2}$ tons for residential use and up to 30 tons in commercial work. Most units are rated and certified in accordance with ARI Standard 210 rating conditions, which are 80°F dry-bulb and 67°F wet-bulb temperature of the entering air to the indoor evaporator coil and 95°F D.B. of the entering air to the outdoor condensing coil. ARI Standard 210 also requires that the unit must be capable of operating up to 115°F outdoor ambient without cutting off on high pressure or causing compressor overload cycling.

Diagrammatically, the system operating at ARI conditions has the characteristics shown in Fig. A7-4; at a dry-bulb temperature of 80°F, return air from the conditioned space at the rate of 400 to 450 ft³/min per ton goes through the filter and then through the evaporator where it is cooled and dehumidified. Air coming off the coil will be about 58 to 60° D.B. Thus, there is approximately a 20 to 22°F D.B. temperature drop across the coil. The ratio of sensible cooling to total cooling will be about 0.75. Suction pressure coming from R-22 off the coil would then be about 73 to 76 psig. The conditioned air is at 60°F, and, assuming that it picks up a small amount of heat in the ducts, it will enter the room

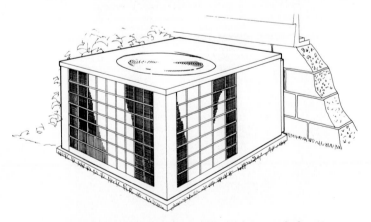

FIGURE A7-2   Two-ton packaged air-cooled unit. (*Courtesy* Addison Products Company)

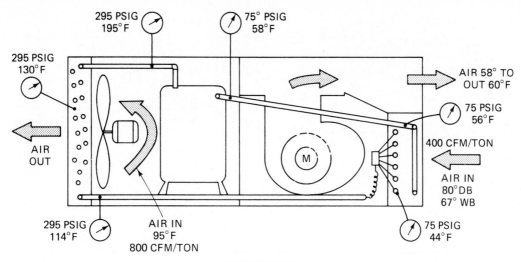

295 PSIG
195°F

75° PSIG
58°F

295 PSIG
130°F

AIR 58° TO
OUT 60°F

75 PSIG
56°F

AIR
OUT

400 CFM/TON

AIR IN
80°DB
67° WB

M

295 PSIG
114°F

AIR IN
95°F
800 CFM/TON

75 PSIG
44°F

FIGURE A7-4

at 62 to 65°F dry-bulb (15 to 18°F temperature difference, ΔT), which is a desirable temperature.

On the refrigerant high side, outdoor air for condensing will be introduced at 95°F across the coil. Depending on the coil design, airflow over the condenser coil will be a nominal 800 ft³/min per ton. The resulting pumping head of the compressor on R-22 will be in the range of 295 psig discharge pressure. The average condenser temperature will be 130°F, with approximately 16°F liquid subcooling to a liquid OFF temperature of 114°F.

In a single-package air-cooled unit as described

above, it is assumed that there will be a supply and return duct system. In a residence the unit can be mounted on a concrete slab at ground level (Fig. A7-5), with ducts into the basement or crawl space. This is a common method of adding cooling to an existing home that has hydronic or electric baseboard heating. Some units can be mounted through a concrete block foundation wall.

Attic, garage, or carport installations may be found in areas like Florida, where homes have concrete slab floors (Fig. A7-6). Structural support, adequate condensing air intake, and well-insulated ducts are important for proper operation.

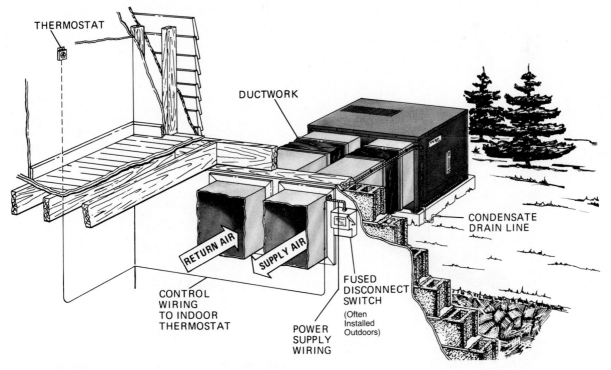

THERMOSTAT

DUCTWORK

RETURN AIR

SUPPLY AIR

CONTROL
WIRING
TO INDOOR
THERMOSTAT

POWER
SUPPLY
WIRING

FUSED
DISCONNECT
SWITCH
(Often
Installed
Outdoors)

CONDENSATE
DRAIN LINE

FIGURE A7-5 Typical ground-level installation. (*Courtesy* Borg-Warner Central Environmental Systems, Inc.)

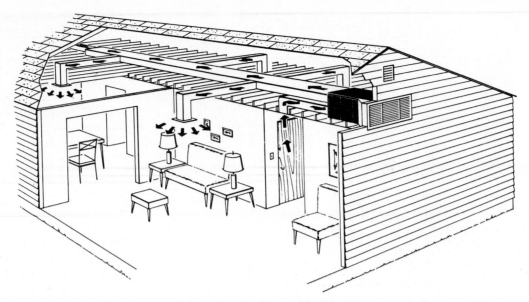

FIGURE A7-6   Attic installation

In light commercial applications (Fig. A7-7), the ability to mount the unit on a roof, with short supply and return ducts, allows for simple and minimum cost installations. The installer must make sure the roof is strong enough to support the weight and that a proper water seal exists around the ducts. On larger units, roof curbs are available to support the equipment.

Another commercial application approach is to install the unit through an outside wall (Fig. A7-8) and provide a simple air distribution plenum on the front of the unit. Conditioned air is free-blown into the space out of the top grille and returned through the bottom louvers; ductwork is thus eliminated.

As mentioned at the beginning of Chapter A7, the classification of a unit as "cooling only" is not really accurate, because most manufacturers have provided internal space for the option of adding electric heating resistance coils to provide winter heating. These coils function much as noted for an electric furnace. Other manufacturers offer ducted resistance heaters (Fig. A7-9) that are installed in the main trunk duct and/or in branch runs. Either method will convert the basic cooling only unit to a full year-round comfort system for either residential or commercial use.

Many manufacturers have also adapted the basic packaged air-cooled unit to mobile home air-conditioning (Fig. A7-10). An adapter box is placed on the front section. Flexible supply and return ductwork feeds under the structure to floor outlets and a return grille. The air-conditioner mounts on a slab on the paved surface

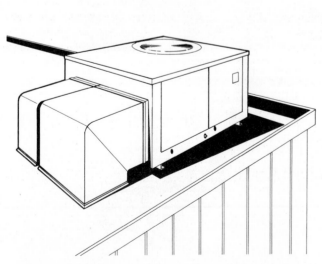

FIGURE A7-7   Rooftop installation. (*Courtesy* Addison Products Company)

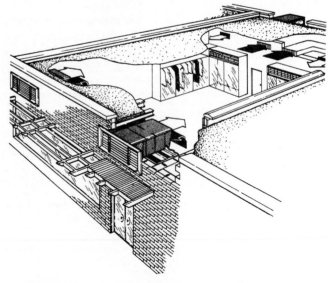

FIGURE A7-8   Commercial outside-wall installation. (*Courtesy* Borg-Warner Central Environmental Systems, Inc.)

FIGURE A7-9 Ducted resistance heater. (*Courtesy* Lennox Industries)

provided in a mobile home park. The system can be disconnected and transported to a new location without too much difficulty.

In commercial air-conditioning it is quite common that internal heat loads (due to lights, people, cooling, etc.), require cooling even when outdoor temperatures are at 35°F or lower. This presents problems for the standard refrigeration cycle. As outdoor temperatures fall, so does the refrigerant condensing pressure, until there may not be sufficient pressure drop across the expansion device (capillary tube or expansion valve) to feed the evaporator properly. The coil can freeze up, presenting a danger to the system and to compressor operation. The solution is to raise the head pressure. Several methods are used. Where only one condenser fan is present, its speed may be reduced by electrical speed controls, and thus airflow across the condenser coil is also reduced. The compressor head pressure rises as the condenser air quantity is reduced. Solid-state control devices are common.

Another popular method of low ambient control is the use of two or more condenser fans. Then it is only a matter of cutting off fan motors in stages until proper

pressures are established. Many units have multiple condenser fans and can be successfully operated down to 0°F and below outdoor ambient. (*Note:* It is assumed there is enough internal heat to keep evaporator coils above freezing.)

A third method of head-pressure control is to place dampers on the condenser air circuit, and so throttle the airflow across the coil. This system, too, is essentially automatically preset to maintain a minimum head pressure.

No matter which technique is used, it is important to understand that there is a difference between mild weather control (35°F and above) and true low ambient operation down to perhaps −20°F and below. Mild weather operation is standard on many units, while low ambient control is usually an optional extra.

The configuration of a horizontal rectangular cabinet is the most conventional design; however, there are variations that include vertical arrangements. Figure A7-11 depicts an air-cooled self-contained unit specifically designed for use in modernizing older high-rise buildings. The unit has been carefully designed to fit into elevators, to go through existing doors, and to be placed against outside window openings for the intake and exhausting of condenser air. With these limiting specifications, sizes range from $7\frac{1}{2}$ to 20 tons; but when used in multiples, these units can effectively condition large areas.

Whether horizontal or vertical, the operation is essentially the same, and packaged air-cooled units continue to offer many advantages. They are completely factory assembled and tested, are relatively easy to install with minimum electrical and plumbing work, and need little or no ductwork for simple air distribution.

A7-3.2
Packaged Water-Cooled Units

The self-contained vertical water-cooled packaged cooling unit (Fig. A7-12) is one of the earliest approaches to commercial air conditioning and is still of-

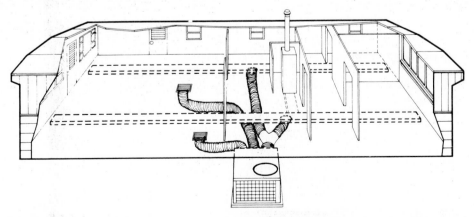

FIGURE A7-10 Mobile home air-conditioning. (*Courtesy* Addison Products Company)

FIGURE A7-11 Vertical air-cooled self-contained unit. (*Courtesy* Bard Manufacturing Company)

FIGURE A7-12 Packaged water-cooled unit. (*Courtesy* Addison Products Company)

fered by several major manufacturers. Capacities range in sizes from 3 to 60 tons, with the range of 3 to 15 tons for in-space, free-standing application utilizing discharge and return grilles. Larger units are generally located out of the conditioned space with necessary ducting for air distribution.

The original concept of operation (Fig. A7-13) is a water-cooled circuit, with the compressor and water-cooled condenser located in the bottom compartment. The center section houses filters and the fin-tubed cool-

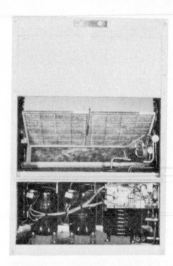

FIGURE A7-13 Packaged water-cooled unit. (*Courtesy* Addison Products Company)

ing coil. (*Note:* Some units allow space for the installation of a combination steam and hot-water coil.) The upper portion contains centrifugal blower(s) motor and the motor drive. Some units offer an optional fan discharge arrangement, depending on the application and space considerations.

The water-cooled condenser may be piped for series flow city-water circuit, controlled by an automatic head-pressure water valve, or piped for parallel flow water tower operation. The temperature rise and the pressure drop through the condenser vary with the flow arrangement. City water supply can mean water from the city mains or from a well or lake source. The automatic water valve is set to regulate the proper condensing pressure. Water tower operation (Fig. A7-14), as reviewed in the refrigeration chapters, is a function of tower selection, wet-bulb temperature, and flow rate. The normal design is to get water from the tower and to the condenser at about $7\frac{1}{2}°F$ approach to the wet bulb. So, if the design wet-bulb temperature was 78°F, the cold

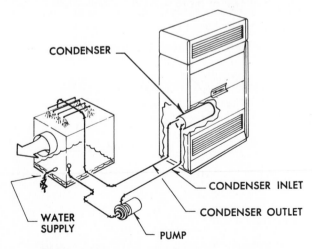

FIGURE A7-14 Typical water tower application.

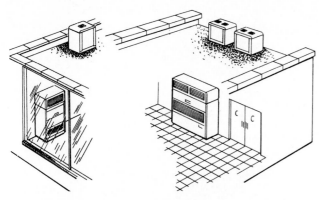

FIGURE A7-15 Typical air-cooled application. (*Courtesy Borg-Warner Central Environmental Systems, Inc.*)

water to the condenser would be 85°F. The water out of the condenser is usually based on a 10°F rise and thus would be 95°F. The gal/min flow rate at these conditions is determined from the tower performance data. Pump selection is a function of the water flow (gal/min) and the resistance head (expressed in feet) for the piping, condenser, and tower.

Whether on city water or water tower operation, the efficiency of the condenser operation is affected by water scale and other contamination. Tube scaling reduces the heat flow between water and refrigerant, and reduces the effective pipe diameter as well as cutting water flow and increasing the pumping head. Towers are also affected by dust, dirt, and algae and slime formations. Proper water treatment is necessary and regular maintenance is a must. Cleaning procedures and chemicals are suggested by the air-conditioning equipment manufacturers, as well as by water treatment companies.

Some localities do not allow the use of city water for air-conditioning; also, wastewater can be expensive. In addition, some users don't wish to have the service problems associated with water towers; however, they do like the design of a vertical self-contained air-conditioner. As a result, manufacturers offer models without the water-cooled condenser, permitting application of a remote air-cooled condenser (Fig. A7-15). The compressor stays with the air-conditioner. Liquid and discharge lines are run between indoor and outdoor units. This section functions like an air-cooled system, but the installer must be skilled in running extended refrigeration lines, installing proper oil traps and pipe insulation, and dealing with noise considerations.

A7-4
COMPUTER ROOM UNITS

Another form of single package cooling equipment is the so-called computer room unit (Fig. A7-16), for use in the precise environmental control of sophisticated EDP equipment. Its design is based on the fact that computer rooms have raised floors to accommodate the maze of electronic connections between equipment. This raised floor forms a natural plenum for the distribution of conditioned air or the return thereof, depending on the specific product design. The most obvious difference from other air-conditioning equipment is the attractive cabinet that must relate to the highly styled surrounding computer equipment. High-efficiency filters and built-in humidification are most important for the clean air and constant humidity conditions required. Equipment sizes range from 3 to 15 tons. At $7\frac{1}{2}$ tons and above, the units generally employ multiple compressor circuits for higher-capacity reduction and more accurate control.

Computer rooms must frequently operate on a 24-hour schedule all year long, and thus the air-conditioning equipment must also operate under all weather conditions. Thus the selection and operation of the condensing function is carefully planned. Some units use a remote air-cooled refrigerant condenser (as shown) that employs low ambient fan speed modulation for head-pressure control, along with the staging of multi-circuit compressor operation. Units are also available with conventional water-cooled condensers where city water or tower operation can be used without danger of freezing. Where freezing is a problem, some manufacturers offer a form of remote air-cooled condenser that uses a closed-loop system and circulates antifreeze liquid (glycol) instead of water. It is pumped to the computer unit condenser.

Since reliability and service are mandatory to minimize down time of EDP computers, these units are top shelf in terms of quality and ease of access. Operational and warning lights permit constant monitoring of the system.

A7-5
UNITARY SPLIT-SYSTEM COOLING

The development of split-system air-conditioning for both residential and commercial applications is related to certain limitations of the packaged units. In residential work there are millions of American homes that have existing furnaces and duct systems suitable for the addition of cooling, but to tie in a packaged cooling unit and to use the common ducts presents difficult and expensive dampering and control problems. This is particularly true where the furnace is centrally located in a closet or utility room.

In many apartments, it is very difficult to find outdoor space to put packaged units except on the roof; running ducts down several stories is not a practical and economical consideration when compared to running refrigerant lines. The same situation is true for many multifloor commercial buildings. So, to meet the many

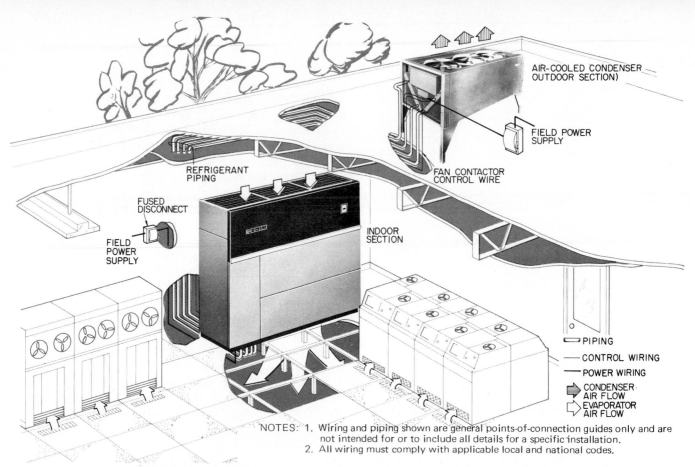

NOTES: 1. Wiring and piping shown are general points-of-connection guides only and are not intended for or to include all details for a specific installation.
2. All wiring must comply with applicable local and national codes.

FIGURE A7-16   Computer room unit. (*Courtesy* Carrier Air-Conditioning Company)

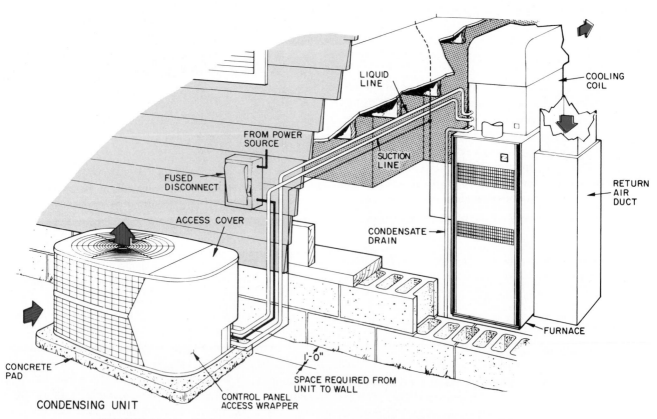

FIGURE A7-17   Add-on cooling system. (*Courtesy* Carrier Air-Conditioning Company)

434   Air-Conditioning

variations in construction techniques and to make use of existing ductwork, the industry developed the split system, which consists of an indoor cooling section (with or without a fan), an outdoor condensing unit, and interconnecting refrigeration piping.

A7-6

RESIDENTIAL SPLIT SYSTEMS

The so-called add-on cooling system (Fig. A7-17) accounts for a huge potential market in modernization work. Where the furnace exists (gas, oil, or electric) and the size of the ductwork is adequate, a cooling coil may be added to the discharge outlet of the furnace. An air-cooled condensing unit is located outdoors on a suitable base. The two are connected by properly sized liquid and suction refrigerant lines. Let's examine each component.

Add-on cooling coils (no fan) are offered in four designs. Upflow coils (Fig. A7-18) mount on top of upflow or lowboy furnaces, and most take the form of the letter A; however, there are flat (Fig. A7-19) and diagonal versions (Fig. A7-20), which may cut down on the coil height. Some are cased in a cabinet; others are not. The horizontal coil (Fig. A7-21) is for use with horizontal furnaces, or else it can be installed in a horizontal main trunk duct of any type of furnace. Counterflow coils (Fig. A7-22) install below the furnace and must offer a cabinet of sufficient strength to support the furnace. The counterflow coil casing must also have proper clearances from combustible surfaces, for it is used for heating air distribution as well.

In all four instances the coils of a particular manufacturer are usually designed to match their furnace outlets so as to simplify installation and eliminate costly ductwork transitions.

Coil cabinets are insulated to prevent sweating, and all models have adequate drain pans for collecting condensate runoff. Low-cost plastic hose may be used to pipe condensate drain water to the nearest sink or open plumbing drain in the house. If no nearby drain is available, a small condensate pump may be installed to pump the water to a drain or perhaps even to an outdoor disposal arrangement.

An important consideration in applying an add-on cooling coil to an existing furnace is the air resistance it adds to the furnace fan. During the heating cycle, the coil is inactive and thus is *dry*. In summer, when the unit is cooling and dehumidifying, the coil is termed *wet*. On the average, the wet coil will add 0.20 to 0.30 in. of W.C. of static pressure loss. This added resistance will be sufficient to require a change in furnace fan speed; also, an increase in motor horsepower will probably be needed. If the installation is new, then the air condi-

FIGURE A7-18   Upflow cooling coil. (*Courtesy* Addison Products Company)

FIGURE A7-19   Upflow coil. (*Courtesy* Borg-Warner Central Environmental Systems, Inc.)

FIGURE A7-20   Typical two-ton slant coil. (*Courtesy* Addison Products Company)

Unitary Packaged Cooling   435

FIGURE A7-21 Horizontal coil. (*Courtesy* Addison Products Company)

tioning contractor will select a furnace fan capacity capable of producing sufficient external static.

Continuing with the add-on concept, we next examine the outdoor air-cooled condensing unit (Fig. A7-23). It consists of a compressor, condenser coil, a condenser fan, and the necessary electrical control box assembly. On residential condensing units, the full hermetic compressor is universally used. It is sealed from dust and dirt and requires little ventilation. The condenser coil is a finned-tube arrangement. The rows of

depth and face area are a function of the particular manufacturer's design. On the condenser a large area is desirable, and thus you will see many units that offer almost a complete wraparound coil arrangement to gain maximum surface area.

The configuration of the condenser fan also depends on each manufacturer's design. Most utilize a propeller fan, because it can move large volumes of air where little resistance is offered. This is another reason why the condenser coil tube depth is limited so as not to create added resistance. Few air-cooled condensing units are designed for ducting, but some are available and these use centrifugal blowers. (Later we will see that these designs are meant to be used with heat pumps that require more positive air flow on defrosting.)

Airflow direction becomes a function of the cabinet and coil arrangement, and there is no one best arrangement. Most units, however, do use the draw-through operation over the condenser coil. Outlet air, in its direction and velocity, can have an effect on surrounding plant life. Top discharge is the most common arrangement. The fan motors are sealed or covered with suitable rain shields. Fan blades must be shielded so as to offer protection for prying hands and fingers.

The foregoing description of an add-on condensing unit applies to a free standing configuration, where there is walk space around the unit. But we must pause to acknowledge that there are residential condensing units (Fig. A7-24) that are primarily designed for in-wall mounting in new construction. An apartment may have a condensing unit installed in a balcony wall or a small utility closet, and obviously the air for condensing must be taken in and discharged through only one side. Oth-

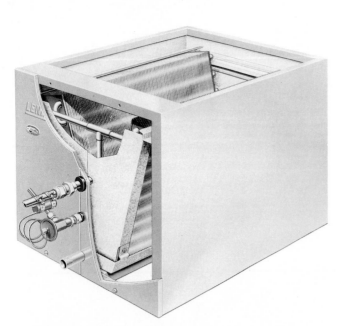

FIGURE A7-22 Counterflow coil. (*Courtesy* Lennox Industries)

FIGURE A7-23 Outdoor air-cooled condensing unit. (*Courtesy* Lennox Industries)

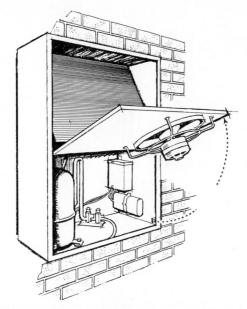

FIGURE A7-24 In-wall residential condensing unit. (*Courtesy* Trane Company, Dealer Products Group)

erwise, the internal components and operation are essentially the same as described above.

When installing a condensing unit at ground level, it is most important to provide a solid foundation (Fig. A7-25). A concrete slab over a fill of gravel is recommended to minimize the movement from ground heaving. In new construction, placing the slab on new back fill may not be wise, since the slab can settle several inches or more, thus placing stress on or even breaking the refrigerant lines.

Roof mounting small condensing units does not generally present structural problems, since the weight per square foot is relatively low. Larger commercial units will, however, require consideration of the support, and it may be necessary to provide steel rails spanning several bar joists.

Location of the outdoor condensing unit (Fig. A7-26) is a compromise among several factors: the actual available space, the length of run for refrigerant lines, the aesthetic effects on the home or landscaping, and sound considerations as they relate to the home or to neighbors. In general, avoid placement of the unit directly underneath a window or immediately adjacent to an outdoor patio. Do try to locate it where plants and shrubs may screen the unit visually, as well as provide some sound-absorbing assistance.

Noise pollution, along with air pollution, is gaining widespread attention. Outdoor air-cooled air-conditioning is a mechanical device and does produce sound levels that cannot be ignored. As a matter of fact, in the 1960s certain localities, such as Coral Gables, Florida, enacted local codes to help curb the compounding noise pollution of air conditioners, particularly from window units. Sound is a difficult subject to understand. It is hard to measure and even harder to evaluate and is frequently misunderstood by local code makers.

However, these early attempts to create local ordinances prompted action on the part of the ARI to establish an industry method of equipment sound rating and application standards (which local communities could adopt), setting forth acceptable sound standards.

Thus, ARI Standard 270 was created in 1971 with the publication of its first directory of sound-rated outdoor unitary equipment. Under the program, all participating manufacturers are required to rate the sound power levels of their outdoor units in accordance with rigid engineering specifications. They must submit test results on all their units to ARI engineers for review and evaluation and make any of their units available for sound testing by an independent laboratory. Units are sound-rated with a single number—the *sound rating number* (SRN). Most units will be rated between 14 to 24 on the ARI sound-rating scale. The ARI anticipates that the sound-rating program will guide and encourage manufacturers to produce quieter units in the future.

Along with equipment ratings, ARI Standard 270 sets forth recommended application procedures for using a sound-rated unit SRN to predict and, through the design of the system, control the sound level at a given point. The application procedures are worked out so that installation contractors can use them to properly predict and control the sound level. A technician involved in selling and installing outdoor units must become familiar with these standards and use the recommendations accordingly. These standards and recommendations apply to both packaged and split-system outdoor equipment.

Refrigerant piping between the indoor coil and the outdoor condensing unit can take several forms. On

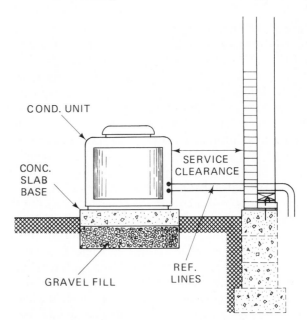

FIGURE A7-25 Condensing unit at ground level.

FIGURE A7-26 Outdoor condensing unit.

basically residential-type equipment, nonsoldered methods are used. As mentioned in Chapter A1, the development of flexible, quick-connect, precharged refrigerant lines (Fig. A7-27) played a major role in the growth and reliability of home air-conditioning. The liquid and suction lines are made of bendable copper or convoluted steel tubing. Suction lines are factory covered with a foam-rubber insulation material. Each end of the tubing is fitted with a coupling half that mates with a coupling half on the equipment (Fig. A7-28). Both coupling halves have diaphragms that provide a seal, which prevents refrigerant loss before connection. The male half (on the equipment) contains a cutter blade, the metal refrigerant sealing diaphragm, and an intermediate synthetic rubber seal to prevent a loss of refrigerant while the coupling is being connected. The female half (on the tubing) contains a metal diaphragm, which is a leakproof metal closure. Tightening the union nut draws the coupling halves together (Fig. A7-29), piercing and folding both metal diaphragms back, and opening the fluid passage. When

fully coupled (Fig. A7-30) a parent metal seal forms a permanent leakproof joint between the two coupling halves. Note the service port for checking refrigerant pressure. The port is equipped with a valve (not shown), similar to a tire valve, that opens when depressed by gauge lines. When installing quick-connect precharged refrigerant lines, follow the manufacturer's installation directions on the radius of the bend for the size of tubing, lubrication of the coupling, and proper torque values when couplings are joined together. If excess tubing is present, form a loop or coil in a flat, horizontal manner; do not form vertical loops, which will create an oil trap. Precharged lines are manufactured in various sizes and lengths from 10 to 50 ft. It is necessary to plan the installation in order to have sufficient but not excessive tubing, which adds cost and pressure drop to the system.

Another method of mechanically sealing refrigerant tubing is the compression fitting technique (Fig. A7-31). The refrigerant tube is first cut square and deburred. A coupling nut is slipped on the tubing, fol-

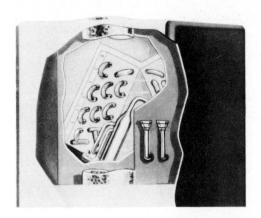

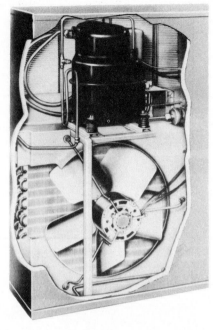

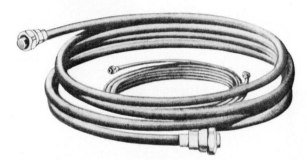

FIGURE A7-27 Precharged refrigerant lines. (*Courtesy* Aeroquip Corporation)

lowed by a compression O-ring. When mated with the male adapter, the O-ring is compressed to form a tight and leakproof seal. The male adapter on the air-conditioning unit is also part of a service valve assembly. Refrigerant tubing is shipped in coils that are clean and dehydrated. Removal of the seals and piping connection should be made as quickly as possible to minimize moisture condensation. Piping connections are first made hand-tight. By cracking open the service valves, refrigerant pressure is allowed to pressurize the tubing, thus purging air and moisture out through the hand-tight fittings. After purging, the compression fittings are securely tightened to form a leakproof joint.

The third form of mechanical coupling is the *flare connection,* which has been widely used on refrigeration systems for years. Flare connections are generally limited to tubing sizes of about $\frac{3}{4}$ in. outside diameter, and

the installer must be skilled in making the flare. The tubing kit is factory cleaned and capped at both ends to minimize dirt and moisture penetration, but it is not precharged with refrigerant. In the process of making the connections, if the system is open to the atmosphere for more than 5 minutes, a complete procedure of purging and leak testing and full evacuation and recharging will be needed.

In the case of quick-connect couplings, compression fittings, and flared fittings, no field soldering is needed, and pumpdown, evacuation, and refrigerant

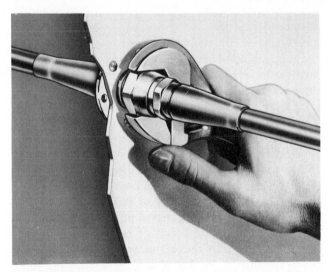

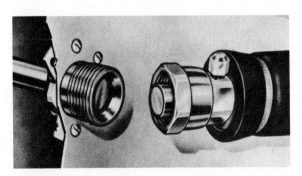

FIGURE A7-28 Diaphragm-type fittings. (*Courtesy* Aeroquip Corporation)

FIGURE A7-29 Tightening fitting. (*Courtesy* Aeroquip Corporation)

Unitary Packaged Cooling 439

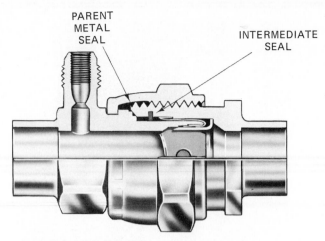

FIGURE A7-30 Diaphragm fitting cutaway. (*Courtesy* Aeroquip Corporation)

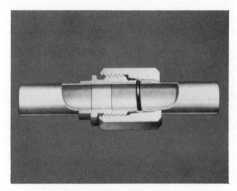

FIGURE A7-31 Compression fitting. (*Courtesy* Carrier Air-Conditioning Company)

charging are not usually required. The condensing unit and evaporators are factory sealed. The condensing unit is also factory charged with refrigerant for the complete system up to about 25 ft. On-site labor, tools, time, and mechanical skills are greatly reduced. All, however, have the common disadvantage of not being practical or economical in large pipe sizes. The limiting size of soft-drawn copper tubing that can be formed or bent is about $\frac{7}{8}$ in. OD. Thus, single-compressor precharged systems stop at the nominal range 5 to $7\frac{1}{2}$ tons. Unless dual circuits are employed, the large systems must go to the field-fabricated and soldered lines. All of these types of fittings and connections were discussed in Chapter R13.

In addition to the physical limitations of handling larger pipe sizes and because of refrigerant characteristics, split-system piping has limitations in both the length of line and the height of refrigerant lift which must be observed by the technician (Fig. A7-32). When the evaporator is located above the condensing unit, there is a pressure loss in the liquid line due to the weight of the column of liquid and the friction of the tubing

walls. The friction and weight of the refrigerant column may be interpreted in terms of pressure loss in pounds per square inch (psi), as shown in the following table for R-22.

Vertical lift in (ft)	5	10	15	20	25	30
Static pressure loss (psi)	$2\frac{1}{2}$	5	$7\frac{1}{2}$	10	$12\frac{1}{2}$	15

Thirty feet of vertical lift is considered the maximum limitation in normal use. If the pressure loss is great enough, vapor will form in the liquid line prior to normal vaporization in the evaporator, and operation and capacity are adversely affected.

Line lengths and the separation between a condenser and an evaporator using standard precharged refrigerant lines is 50 ft, or less if so noted by the manufacturer. Seventy-five feet is considered the maximum separation and usually requires the addition of refrigerant. Where the evaporator is above the condenser, pitch the suction line toward the condenser at least 1 in. in 10 ft of run to insure proper gravity oil return. (More information on refrigerant line application will be covered when we discuss commercial split-system cooling.)

Wiring and piping a typical single circuit add-on split system is relatively simple, as shown diagrammatically in Fig. A7-33. The electrical wiring consists of a line voltage (208/240 V) to the outdoor condensing unit through a properly fused outside disconnect switch. The line voltage (usually 110 V) is supplied to the furnace, which furnishes power to the fan as well as for the low-voltage (24-V) secondary OFF transformer. A 24-V combination heating and cooling room thermostat controls the on/off of cooling or heating through interlocking relays. The practice of using permanent split capacitor motors in small hermetic compressors for res-

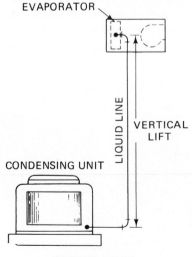

FIGURE A7-32

440 Air-Conditioning

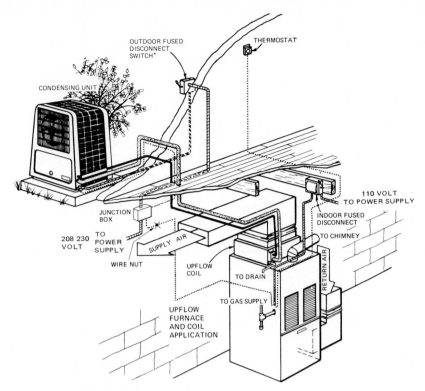

FIGURE A7-33   Single-circuit add-on split system. (*Courtesy* Borg-Warner Central Environmental Systems, Inc.)

idential work is included in the electrical review. It is assumed that refrigeration pressures equalize during the OFF cycle, and thus the compressor is not required to start against a load. The auxiliary winding remains energized at all times during motor operation. The running capacitor is added to provide additional torque both during starting and while the unit is running, but in cases where low-voltage fluctuation exists, it may be necessary to add a starting capacitor for "extra kick." Hard-start kits, which essentially consist of a starting capacitor to be field applied, are offered to overcome low-voltage starting problems and minimize light flicker.

The sizes of residential add-on split systems run from 1 ton to $7\frac{1}{2}$ tons, with the bulk of sales in the lower capacities, from 2 to 3 tons. Models are usually available in elements of 6000 Btu/hr (i.e., 12,000, 18,000, 24,000, 30,000, etc., Btu/hr). Most manufacturers rate and certify their equipment in accordance with ARI Standard 210, which is based on mating specific sizes condensing units and cooling coils. This combination rating is published in the ARI directory. Some manufacturers do publish mix-match ratings, which means selecting a different combination of condensing unit and cooling coil. This most notably occurs in the southwest part of the country, which has low humidity but high temperatures and thus needs a higher air flow and sensible cooling ratio, even though the condensing unit total capacity is not changed. These mix-match combinations may or may not be ARI-certified.

A7-7
FAN-COIL UNITS

Not all residential split systems are of the add-on variety for existing homes. Many are installed in new home construction. Many are also planned for apartments and motels, where the indoor element is a fan-coil unit rather than a furnace. Figure A7-34 is a ceiling evaporator blower designed to be installed in a dropped-ceiling application in furred areas above closets, hallways, bathrooms, etc., where room air is returned through a ceiling grille and conditioned air is discharged through a grille into a single room. These systems are basically not designed for ducting. Complete cabinets are eliminated to reduce weight and cost.

As noted previously, it is common to refer to a "cooling system" but then acknowledge that it can also be equipped for heating. Ceiling evaporator blowers in apartments and motels can be so supplied with electric resistance heaters. Cooling sizes of ceiling evaporator blowers range from 12,000 to 30,000 Btu/hr and are installed with quick-connect, precharged refrigerant lines. Condensate drain lines are treated with care, because an overflow could result in damage to the ceiling. Safety is recommended and required in some locations.

Larger evaporator blowers for homes and commercial use (Fig. A7-35) are available in size from $1\frac{1}{2}$ to $7\frac{1}{2}$ tons and are complete with total cabinet enclosure.

Unitary Packaged Cooling   441

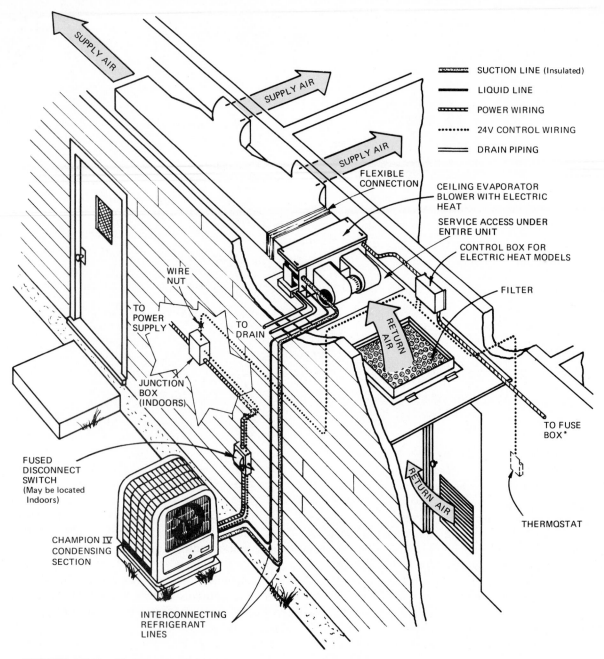

SUCTION LINE (Insulated)

LIQUID LINE

POWER WIRING

24V CONTROL WIRING

DRAIN PIPING

SUPPLY AIR

SUPPLY AIR

SUPPLY AIR

FLEXIBLE CONNECTION

CEILING EVAPORATOR BLOWER WITH ELECTRIC HEAT

SERVICE ACCESS UNDER ENTIRE UNIT

CONTROL BOX FOR ELECTRIC HEAT MODELS

FILTER

WIRE NUT

TO POWER SUPPLY

TO DRAIN

RETURN AIR

JUNCTION BOX (INDOORS)

TO FUSE BOX*

FUSED DISCONNECT SWITCH (May be located Indoors)

RETURN AIR

THERMOSTAT

CHAMPION IV CONDENSING SECTION

INTERCONNECTING REFRIGERANT LINES

FIGURE A7-34   Ceiling evaporator blower system. (*Courtesy* Borg-Warner Environmental Systems, Inc.)

These are also commonly called *air-handling units*. Models for horizontal application are suspended in a basement or floor-mounted in an attic. Vertical models for upflow or counterflow are easily adapted to closet installation. As mentioned above, these also can be equipped with electric resistance heaters for winter heating. The centrifugal blowers provide ample external static pressure for a complete air distribution system; optional motor speed control permits varying the airflow. Frequently these units are used in small commercial applications with ductwork or with an accessory discharge plenum for free-blow air distribution.

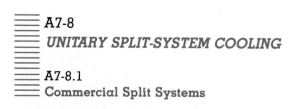

A7-8
UNITARY SPLIT-SYSTEM COOLING

A7-8.1
Commercial Split Systems

Up to $7\frac{1}{2}$ tons there is no distinct rule to classify a piece of equipment as commercial or residential, for there is a wide degree of application in both markets, using the same products. However, above $7\frac{1}{2}$ tons the application becomes distinctly commercial, and the product designs use different component standards.

The first notable difference other than size and capacity is the use of soldered refrigerant lines and expansion valves rather than capillary tubes. Capacity reduction methods and low ambient operation are improved by virtue of multiple compressors and condenser fans. Structural design becomes somewhat more functional and heavier, with less emphasis on appearance. Compressors all have three-phase electrical characteristics.

Commercial split systems range from $7\frac{1}{2}$ to 100 tons and above, and the condensing unit design most often used is similar to that in Fig. A7-36, where air discharge is vertical and inlet air is from the sides or bottom. Vertical discharge has several advantages. First, these units move large volumes of air and the best way to avoid blowing against obstructions or creating a disturbance is to blow straight up. Second, the horizontal arrangement of fan blades is least affected by windmilling if the unit is off and then tries to start with its blades rotating in the wrong direction. The use of direct-drive multiple propeller fans is almost universal in this type of equipment. Ducting is not needed when equipment is normally located on the roof or at ground level in open space. Capacity control is necessary because commercial air-conditioning loads are rarely constant. Capacity reduction is accomplished in several ways. When multiple compressors are used, as would be the case for higher tonnages, it then becomes a matter of staging or sequencing the operation by cutting compressors on and off to match the load. Where single compressors are used (and even in multiple arrangements), they can be

FIGURE A7-36 Typical condensing unit design. (*Courtesy* Borg-Warner Central Environmental Systems, Inc.)

equipped with cylinder unloaders to vary the pumping capacity down 25% or less. In other words, the machine starts almost unloaded, and the cylinders are cut in as the heat load demands. Unloaded starting reduces power demand.

An example of a cylinder unloader (Fig. A7-37) consists of a gas bypass valve in the cylinder head that is opened by a spring and closed by discharge gas pressure, and that is actuated by solenoid valves that are responsive to suction pressure. The compressor is always unloaded at startup for maximum power economy. The decrease in horsepower is almost directly proportional to capacity reduction. By the use of this method, the capacity in a compressor is limited only by the number of cylinders. Other unloaders use oil pressure to open the bypass valve.

Another method of capacity control is the hot gas bypass (Fig. A7-38). The hot-gas bypass valve in the bypass line can be controlled by temperature or gas pressure, depending upon the nature of the application. As the controller calls for capacity reduction, the bypass valve opens, allowing some hot gas to go directly to the coil distributor manifold. This reduces the effective capacity of the compressor by the amount of gas bypassed around the condenser. In the hot-gas bypass, there is no appreciable change in horsepower.

Low ambient control is most frequently provided by cycling the condenser fans as previously explained, except that the number of steps is increased by virtue of the multiple fans; closer control is thus achieved. There are other techniques, which include dampers to throttle the condenser air (Fig. A7-39). The choice is determined by the manufacturer's design. And since commercial equipment equipped for both mild weather and low ambient temperatures must operate under cold conditions, crankcase heaters in the compressor are needed to keep the lubrication oil warm. A reverse-acting switch will energize the heater as soon as the compressor is off.

FIGURE A7-35 Evaporator blower air-conditioning unit. (*Courtesy* Borg-Warner Central Environmental Systems, Inc.)

Unitary Packaged Cooling 443

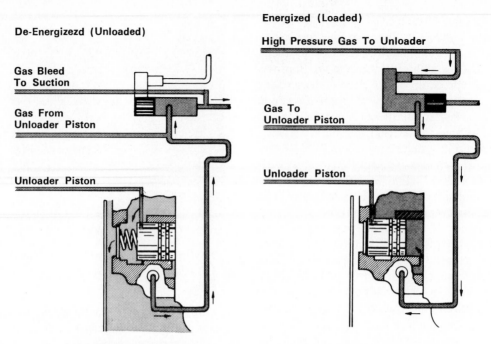

De-Energizezd (Unloaded)

Gas Bleed
To Suction

Gas From
Unloader Piston

Unloader Piston

Energized (Loaded)

High Pressure Gas To Unloader

Gas To
Unloader Piston

Unloader Piston

FIGURE A7-37   Cylinder unloader. (*Courtesy* Borg-Warner Central Environmental Systems, Inc.)

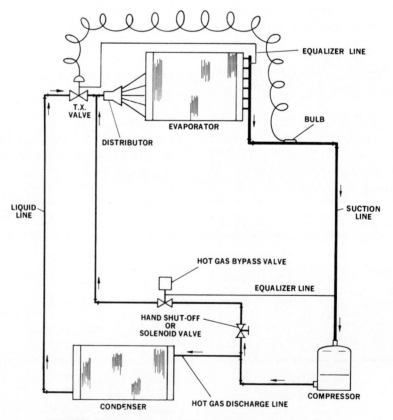

EQUALIZER LINE

T.X.
VALVE

DISTRIBUTOR

EVAPORATOR

BULB

LIQUID
LINE

SUCTION
LINE

HOT GAS BYPASS VALVE

EQUALIZER LINE

HAND SHUT-OFF
OR
SOLENOID VALVE

CONDENSER

HOT GAS DISCHARGE LINE

COMPRESSOR

FIGURE A7-38   Hot-gas bypass. (*Courtesy* Borg-Warner Central Environmental Systems, Inc.)

444   Air-Conditioning

FIGURE A7-39 Dampers to throttle the condenser air. (*Courtesy* Fedders Air Conditioning USA)

Crankcase heaters also vaporize liquid refrigerant in the crankcase, so that on cold weather starts liquid slugging doesn't harm the compressor valves.

Control systems that are more sophisticated include an oil-pressure failure switch, high and low-pressure refrigerant switches, lookout protector relays, time-delay relays between compressor starts and anticycling timers to prevent rapid or short-cycling of the compressors. Timing ranges from 3 to 5 minutes. The indoor air-handling equipment takes two main forms, as shown in Figs. A7-40 and A7-41. Horizontal evaporator blower sections (Fig. A7-40) are suspended from a ceiling either for in-space free-blow application with accessory plenum and discharge grille, or are remotely located with conditioned air ducted into the comfort space. Vertical models (Fig. A7-41) are free standing and can also be free blow or ducted. The choice of free blow or ducting is a funtion of cost, space, appearance, and type of facility. Large, open-space stores are typical applications for free blow, while an office with many partitions must be ducted.

The centrifugal blower must deliver a higher air flow and static pressure than are called for on residential work, and thus a V-belt drive with an optional variable pitch motor pulley and motor horsepower is needed. The fan scroll can frequently be rotated within the cabinet

to obtain alternate discharge arrangements for greater application flexibility.

The evaporator coil(s) will be circuited to match the number of compressors. Also, space is generally provided for the inclusion of either nonfreezable steam coils or hot water coils, for winter heating.

Ratings of *matched* condensing units with specific air handlers are published by the manufacturer and are certified under ARI Standard 210. An example of a specification is shown in Fig. A7-42. For a system combination (condensing unit and air handler) at a rated ft³/min (usually 400 ft³/min/ton), 95°F outdoor air on the condenser, and an indoor entering 67°F wet-bulb temperature to the evaporator coil it lists

1. The total capacity in MBtu/hr (93)
2. Leaving WB temperature off the evaporator (60.6°F)
3. Total power input in kW (9.9)
4. The sensible capacity in MBtu/hr

based on whatever entering dry-bulb exists in the return air from the conditioned space (80°F). In this example the sensible capacity would be 73 Mbh for an *S/T* ratio of 0.785. All these data have been precalculated, so the selection of a standard combination is very easy, and the alert application technician will always consider using standard equipment to keep cost and delivery time down. There are occasions when nonstandard combinations are desired, and so the art of selecting the air-handling

FIGURE A7-40 Horizontal model, evaporator blower section. (*Courtesy* Borg-Warner Central Environmental Systems, Inc.)

FIGURE A7-41 Vertical model, evaporator blower section. (*Courtesy* Borg-Warner Central Environmental Systems, Inc.)

Unitary Packaged Cooling   445

	AIR ON EVAPORATOR		Total Capacity (MBtu/hr)	WB Temp. off Evap. (°F)	Comp. and Cond. Fan Input (kW)	SENSIBLE CAPACITY (MBTU/HR) Dry-Bulb Temp. on Evaporator (°F)				
System	ft³ min	WB Temp. (°F)				70	75	80	85	90
CA 91 (EB92-B)	3300	72	101	65.9	10.4	—	40	56	69	89
(EBV92-B)		67	93	60.6	9.9	43	57	73	88	91
(C90UX⁴)		62	86	55.4	9.4	57	73	84	86	86
(92DX)		57	78	50.2	9.0	67	72	78	78	78
	4400	72	—	—	—	—	—	—	—	—
(EB122-B)		67	98	60.6	10.2	43	59	77	93	98
(EBV122-B)		62	91	55.4	9.7	61	79	87	91	91
(122DX)		57	84	50.2	9.3	83	84	84	84	84

Cooling capacity—95°F DB air on condenser (MBtu/hr).

equipment first and then cross plotting the coil capacity against the condensing unit is necessary, in order to determine the correct system balance in terms of refrigerant operating temperature. This procedure is presented by the manufacturer and varies with individual methods for system performance.

A knowledge of refrigerant piping for commercial split systems is important for the installation and servicing technician. The soldering skill needed to run hard-drawn copper piping has been discussed elsewhere; however, here we are concerned with system design in terms of proper oil return to the compressor, pressure loss in the lines, and the velocity of gas flow. Pressure drop in the suction line is more critical than on the liquid line, since it means a loss in system capacity, because it forces the compressor to operate at a lower suction pressure to maintain the desired evaporator temperature. Capacity falls off while the horsepower actually increases. The larger the suction line, the less the pressure drop, but oversizing also has its problems. First, the velocity of gas flow may not be sufficient to return entrained oil back to the compressor, and second, the cost of larger pipe becomes an economic factor. Therefore, as a technical and economical practice, suction lines are sized with nominal pressure drop corresponding to 2°F, or approximately 3 psi for R-22. Liquid-line sizes have nominal pressure drops corresponding to 1°F, or approximately 2 psi for R-22.

The size of the liquid and suction line connections on the equipment usually represents a compromise between maximum and minimum velocities at maximum and minimum loads with nominal pressure drops in pipes and fittings. It is impossible to specify one line size to cover all conditions; however, based on the volume above, standard line sizes on the equipment should fall within the suction lift of 30 ft, and the length of the liquid and suction lines should be from 50 to 75 ft. System applications exceeding these values should be checked by calculating the actual piping and fitting loss. Below is a procedure to follow.

First, determine the total pressure loss from pipe and fittings. This is done by the equivalent length method. Since every valve, fitting, and bend in a refrigerant line offers resistance to the smooth flow of refrigerant, each one also increases the pressure drop. To avoid the necessity of calculating the pressure drop of each individual fitting, an equivalent length of straight tube, or pipe, has been established for each fitting, as shown in Fig. A7-43. The line sizing tables shown are set up on the basis of the pressure drop per 100 ft of straight pipe or tubing. This permits consideration of the entire length of line, including all fittings, as an equivalent length of straight pipe and allows the data shown to be used directly.

Assume that you are calculating the suction pressure loss for the 10-ton system shown in Fig. A7-44. The actual separation between the condensing unit and the air handler is 75 ft for combined vertical and horizontal distances. The suction line layout includes an angle valve and six 90° elbows. From the manufacturer's catalog or by actually checking the line size on the equipment itself you determine the standard suction line connection is $1\frac{3}{8}$-in.-OD copper. Next, determine the fitting loss from Fig. A7-43.

$1\frac{3}{8}$-in.-OD angle valve	= 15.0 ft equiv. length
$1\frac{3}{8}$-in.-OD 90° elbows @ 2.4 =	14.4 ft equiv. length
Total fitting loss	29.4 ft equiv. length

Add this to the 75 ft of actual straight pipe for a total system value of 104.4 ft, or roughly 105 ft.

Line Size (in. OD)	Globe Valve	Angle Valve	90° Elbow	45° Elbow	Tee Line	Tee Branch
$\frac{1}{2}$	9	5	0.9	0.4	0.6	2.0
$\frac{5}{8}$	12	6	1.0	0.5	0.8	2.5
$\frac{7}{8}$	15	8	1.5	0.7	1.0	3.5
$1\frac{1}{8}$	22	12	1.8	0.9	1.5	4.5
$1\frac{3}{8}$	28	15	2.4	1.2	1.8	6.0
$1\frac{5}{8}$	35	17	2.8	1.4	2.0	7.0
$2\frac{1}{8}$	45	22	3.9	1.8	3.0	10.0
$2\frac{5}{8}$	51	26	4.6	2.2	3.5	12.0
$3\frac{1}{8}$	65	34	5.5	2.7	4.5	15.0
$3\frac{5}{8}$	80	40	6.5	3.0	5.0	17.0

Note: For complete information on equivalent lengths, refer to *ASHRAE Guide.*

Referring to the manufacturer's catalog, we find that $1\frac{3}{8}$-in.-OD tubing has a friction loss of 2.8 psi per 100 ft for R-22 at a 10-ton load, and, since we have 105 ft, the actual suction pressure loss would be 2.8 × 1.05 = 2.95 psi, which is just marginal in terms of the stated 2°F or approximately 3 psi recommended maximum. The liquid lines are calculated in the same way.

Most manufacturers publish line-size tables for various lengths of run, including the average number of fittings, and their recommendations are generally safe to follow in terms of leaving some margin for error; do not intentionally exceed these values.

The position of the condensing unit in relation to the air handler determines where to form the necessary suction line traps. In Fig. A7-44 the condensing unit is above the evaporator. In the suction riser, oil is carried upward by the refrigerant gas. A minimum gas velocity

must be maintained to entrain the oil. The sump or trap at the bottom of the riser promotes free drainage of liquid refrigerant away from the evaporator.

Where system capacity is variable, through compressor capacity control or some similar arrangement, a short riser will usually be sized smaller than the remainder of the suction line to ensure oil return up the riser. Some manufacturers also recommend a trap in the vertical riser at approximately 20 ft and at each additional 10 ft thereafter.

Where the system capacity is variable over a wide range, it may not be possible to find a pipe size for a single suction riser that will ensure oil return under minimum load conditions and still have a reasonable pressure drop during maximum conditions. A double suction riser as shown in Fig. A7-45 should be considered. In a double riser, the small pipe, *A,* on the lift is sized to return oil when the compressor is "unloaded" to minimum capacity. The second pipe, *B,* which is usually larger but not necessarily so, is sized so that the pressure drop through both pipes under maximum load is satisfactory. A trap is located between the two risers. During partial load operation the trap will fill up with oil until riser *E* is sealed off. The gas then travels up riser *A* only and has sufficient velocity to carry oil along with it back into the horizontal suction main. Both risers loop at the top before entering the horizontal suction line. This is to prevent oil drainage into an idle riser during partial load operation.

In Fig. A7-44 note the recommended slope of horizontal suction lines to either the traps or the condensing unit. Figure A7-46 shows how to form a line trap from standard fittings. Figure A7-47 illustrates the correct position for clamping the expansion valve bulb to the suction line. Between the nine o'clock and the three o'clock position the bulb feels only superheated gas temperature but will detect any excess liquid surge.

The choice of controls and coil circuiting deter-

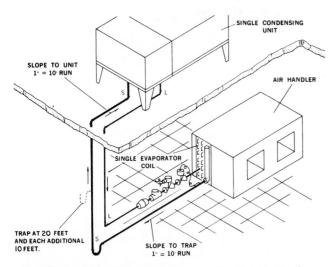

FIGURE A7-44 One condensing unit with one single circuit evaporator coil. (*Courtesy* Borg-Warner Central Environmental Systems, Inc.)

Unitary Packaged Cooling 447

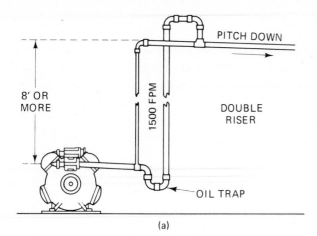

(a)

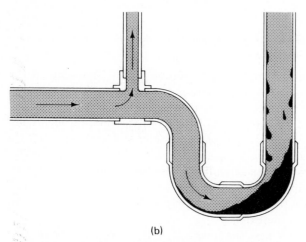

(b)

FIGURE A7-45　Double suction riser. (*Courtesy* Carrier Air-Conditioning Company)

mines the headers and the number of solenoid valves used, but in all cases it is recommended practice to install a filter/drier and sight glass in the main liquid line. Evaporator coils may be single circuit, dual circuit, face split, row split, or combinations thereof, and the proper mainfold piping is usually recommended by the manufacturer.

Although the liquid-line pressure loss from refrigerant flow is not as critical as the suction, the vertical lift is an important consideration, as explained earlier. (Figure A7-32 presented the static pressure loss for the vertical lift of R-22 refrigerant.) Remember that lifts over 30 ft may cause "flash gas," possibly resulting in damage to the expansion valve or at least erratic control of the evaporator.

Structural support of the refrigerant piping is important to hold the weight and also to isolate vibration

FIGURE A7-46　Typical line trap.

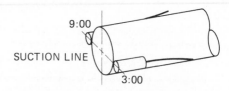

LOCATE BULB AT 9:00 or 3:00 O'CLOCK POSITION (NOT ON TOP AT 12:00 O'CLOCK)

FIGURE A7-47　Typical expansion valve bulb location.

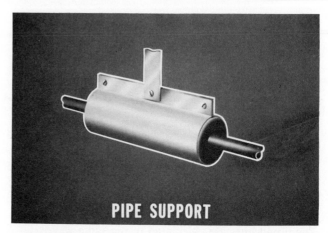

FIGURE A7-48　Pipe support. (*Courtesy* Carrier Air-Conditioning Company)

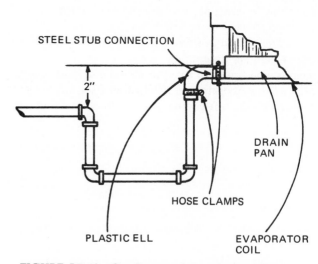

FIGURE A7-49　Condensing drain piping. (*Courtesy* Borg-Warner Central Environmental Systems, Inc.)

and minimize sound transmission. Figure A7-48 illustrates typical hanging methods. Supports should provide insulation between themselves and the pipe to absorb vibration, but it should not be tight enough to cut regular pipe insulation.

Thermal insulation of suction lines is necessary to prevent sweating and heat gain. The thickness of insulation will depend on the type of application and be subject to the specific manufacturer's recommendations. The correct thickness must raise the surface temperature above the surrounding dewpoint temperature. When exposed to the outdoors the insulation must also have an adequate vapor barrier. Some insulation, by virtue of its material, has a built-in or integral vapor barrier.

Evaporators are equipped with condensate drain connections, and the field-fabricated drain lines should remain full size (not reduced). A line trap similar to the one shown in Fig. A7-49 *must* be provided close to the unit to form an air lock. Otherwise, negative suction pressure from the fan on a draw-through arrangement could create improper draining. Drain lines must also be protected from freezing; they must empty into an open drain and not to a closed sanitary plumbing connection where sewer gas could be drawn into the air handler in the event the unit water trap went dry on the winter cycle.

PROBLEMS

A7-1. The capacity ratings and technical standards of window-type air-conditioners are under the control of _____ .

A7-2. What is the ARI standard operating conditions for rating air-cooled package units?

A7-3. Conventional *S/T* ratios for residential cooling units are around _____ %.

A7-4. ARI-listed air-cooled units must operate to _____ °F outdoor ambient.

A7-5. Operating an air-conditioning unit down to 55°F outdoor temperature is called _____ .

A7-6. Operating an air-conditioning unit down to 0°F outdoor ambient is called _____ .

A7-7. The most frequent method of low ambient control is _____ .

A7-8. Water-cooled condensers for water tower operation will have condenser water piping for _____ flow.

A7-9. Computer room units will frequently use remote air-cooled coils circulating _____ instead of water for year-round operation.

A7-10. Add-on evaporators are made in four types. What are they?

A7-11. Air-cooled condensing units are made in two configurations. What are they?

A7-12. Name three types of mechanical seals for joining refrigerant piping.

A7-13. Pressure loss in a liquid line because of height difference is called _____ .

A7-14. Fifty feet of lift in an R-22 liquid line would result in a liquid pressure loss of _____ psig.

A7-15. A blower coil assembly in a total cabinet enclosure is called an _____ .

A7-16. Capacity control can be accomplished by several means. Name two.

A7-17. Why are traps required in the suction line?

A7-18. Evaporator drain lines should have water traps to form an _____ .

A8

Unitary Combination Heating and Cooling Equipment

A8-1
GENERAL

The overall scope of heating and cooling products as previously mentioned is not clearly defined. Many packaged and split-system cooling products may offer heating as an option to provide year-round comfort. There are, however, three categories of unitary products designed to perform both heating and cooling, and in this chapter we will review their aspects.

A8-2
PACKAGED TERMINAL UNITS

Figure A8-1 portrays what are sometimes called *incremental systems,* which have shown rapid market growth in the past 10 years. This type of unit is geared to the new construction in motels, hotels, schools, offices, apartments, hospitals, nursing homes, etc., where the nature of the construction and the need are based on conditioning one room or one space. Ducted systems cannot provide individual control, and in some cases the mixing of room air such as in a hospital or nursing home is not desirable. Also, in an office building, each tenant can be on a separate meter. And finally, the very nature of the product lends itself to simple installation and ease of service. One of the largest installations of packaged terminal units is in the U.S. Department of Transportation building in Washington, D.C., with approximately 2000 units installed. Where great quantities of units are involved, it is common practice to have spares, so when major service is needed, it is done by exchanging the entire unit with minimum labor and disruption; this is one of the major design considerations. Packaged terminal units differ from through-the-wall window units; the former offer higher quality construction, reliability, and much broader application capability.

The typical design of a packaged terminal unit is usually in several components (Fig. A8-2). First a wall sleeve is permanently inserted and attached to the outside wall of the building. The wall sleeve also includes the outside condenser air grille. It must be attractive as well as functional, in terms of directing air intake and discharge, shedding rainwater, and above all, it must be noncorrosive.

The refrigeration chassis is a complete assembly that slides into the wall sleeve. It includes the entire cooling cycle and also space for electric resistance heaters or, in some cases, steam or hot water coils. Ninety percent of all systems use electric heating, but there are still areas where wet heat is available and is so specified. Within the refrigeration chassis components such as the fan deck, control console, etc., are the plug-in type and may easily be removed for quick service.

Facing the conditioned space is the room cabinet, which provides the discharge and air return functions. The air outlets are at the top and have directional control vanes to throw and spread the air pattern as needed. Return air comes back at the floor level and enters the

FIGURE A8-1 Packaged terminal unit, incremental system. (*Courtesy* Standard Refrigeration Company)

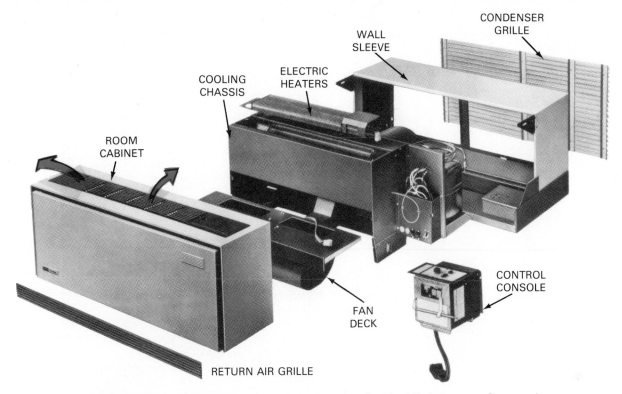

ROOM
CABINET

COOLING
CHASSIS

ELECTRIC
HEATERS

WALL
SLEEVE

CONDENSER
GRILLE

FAN
DECK

CONTROL
CONSOLE

RETURN AIR GRILLE

FIGURE A8-2   Packaged terminal unit. (*Courtesy* Standard Refrigeration Company)

unit through a concealed opening. The room cabinet must be decorative and sturdy. Projection into the room is a function of the individual manufacturer's model. Naturally the least possible extension is a desirable feature. Filter(s) are also included in the room cabinet and are easily accessible for routine replacement.

Several operational factors must be noted for this type of equipment. First, since all mechanical operation is contained in a comparatively small enclosure, accoustical control becomes vital to reduce the room-side sound level. Fan-speed control also helps to reduce air noise. Second, the condensate disposal system is accomplished without plumbing. Condensate water must be mechanically vaporized and exhausted in the condenser airstream. This also includes rainwater, should it collect during a storm; it cannot be allowed to drip down the exterior wall. Third, the wall-sleeve installation and cabinet mounting system must be properly sealed to prevent moisture or air infiltration. These units are frequently used in high-rise buildings where wind velocity, even in normal weather, can be high. Because of this, some manufacturers subject their units to a *hurricane test* (Fig. A8-3), which is conducted under conditions where water and air are blowing at 75 mph against the outside grille.

Another operating provision is the introduction of outside air. This is done through a small, dampered opening between the indoor and outdoor section. The air is filtered before entering the room. The damper is normally operated by a manual control on the control panel. The amount of outside air can sometimes be a code requirement if these units are installed in schools, hospitals, and nursing homes, etc.; 25% outside air would be a minimum need. In motels and offices it is left to personal preference to overcome smoke and stale air.

Since these units must operate on a year-round basis, it is entirely possible that mild weather or low ambient cooling may be needed. Therefore, the system must be capable of operating safely under low head pressure situations.

Control of packaged terminal units can range from a simple, manually operated fan, heating, cooling, and ventilation selection to very sophisticated central-point control for multiple unit operation. An example of the latter would be a motel where the desk clerk could activate a guestroom at check-in time. Offices too can be "zone controlled" from the building maintenance office. A night setback arrangement can also be included to reduce operating time during no-occupancy hours. Motorized outside air dampers can be controlled or programmed as a function of outdoor ambient temperature or in connection with night setback to minimize operating costs.

Cooling capacities of packaged terminal units go from 7000, 9000, 12,000 to 15,000 Btu/hr, with a few up to the 18,000 Btu/hr. Electric heating capacity usually ranges from 4000 to 16,000 Btu. Most manufacturers use two-stage heating arrangements to split electrical circuits and to offer closer room control without overheating.

Unitary Combination Heating and Cooling Equipment   451

FIGURE A8-3   Hurricane test.

The electrical installation is simple. 208/240-V units may be permanently wired or equipped for plug-in to wall-mounted receptacles: 15, 20, or 30 A, depending on the requirements. A power cord, or *pigtail,* with the proper ampere rating is prewired at the factory. The 277-V operation common to many offices is another consideration. Underwriters' Laboratories requires that 277-V electrical installations be protected by a raceway, so the use of a subbase with a built-in receptacle (Fig. A8-4) is needed. The subbase also serves another function, and that is to support the weight of the unit and wall sleeve when the unit is installed in a curtain wall. Most curtain-wall construction has no load-bearing capability and therefore cannot be used for that purpose.

Although a packaged terminal unit is basically designed to serve one room, the use of a duct kit (Fig. A8-5) allows conditioned air to be discharged into an adjoining room. Such might be the case in a hospital or nursing home room with an adjoining bathroom. Return air may not be brought back to the unit, and outside air is used to compensate for the loss.

Packaged terminal unit servicing is about half-way between that of a refrigerator and a packaged air-cooled central unit. All units use welded hermetic compressors and thus no on-the-spot compressor repairs can be made. Most units do not have service valves for checking operating pressures, and they rely on temperature measurements to monitor performance. Fan motors (evaporator and condenser) are direct drive and require little or no oiling. If a fan motor failure occurs, the motor simply unplugs for quick removal. Filter maintenance is a key to successful operation. Space simply does not permit large filters, so frequent replacement or cleaning is a must; fortunately, in larger, multiunit installations, this is a regularly scheduled activity. Control consoles are designed for quick access or complete removal by plug-in arrangement. Much repair and maintenance can be done on a bench in the maintenance shop, and because of the availability of spare replacement units, downtime and disruption are kept to a minimum.

Packaged terminal units are not usually used in single-family residences, where ducted systems offer a better approach for whole-house conditioning.

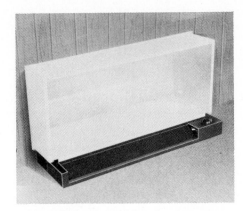

FIGURE A8-4 Subbase with built-in receptacle.

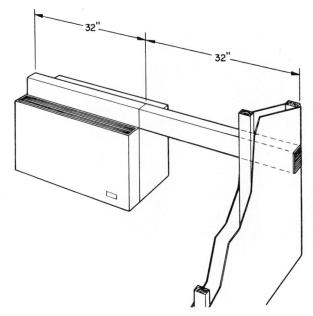

FIGURE A8-5 Duct kit. (*Courtesy* Carrier Air-Conditioning Company)

A8-3
ROOFTOP UNITS

Combination heating and cooling equipment for ducted installation fits into the description of a one-piece *rooftop unit* (Fig. A8-6). This unit is ideal for low-rise shopping centers and commercial and industrial buildings. This particular category of year-round air-conditioning has experienced the fastest growth rate among all those reported to the ARI.

The design of this unit has features similar to a large air-cooled packaged cooling conditioner, plus a heating section and an air distribution section. Figure A8-7 represents an oversimplified cross section of these components. Return air is drawn into the unit over fil-

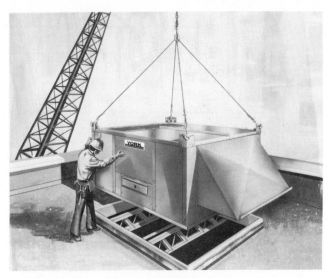

FIGURE A8-6 Rooftop combination gas heating and cooling unit. (*Courtesy* Borg-Warner Central Environmental Systems, Inc.)

ters and through the cooling coil by the centrifugal fan. It is then forced through the heat exchanger and then out through a supply duct to the conditioned space. The air-cooled condenser portion functions as described in previous products. Due to the weight involved and the need to provide a base that can be built in during building construction, a *roof curb* is needed; it actually becomes part of the roof itself. A cutaway picture of a single-zone unit is shown in Fig. A8-8, and the major components can be identified. This illustration depicts a gas heating system; however, the use of electric resistance heating is also very popular. Some models also burn oil, but such units are in the minority. The cooling capacities of standard models begin at around $1\frac{1}{2}$ tons and go up to 60 tons. Some manufacturers even offer custom-type units of up to 100 tons. The range of gas heating to go with standard cooling products runs from 45,000 Btu/hr to over 1 million Btu/hr in very large units. The Btu capacity ratio of gas heat to cooling is about 2 or 2.5:1. The ratio of electric heating is a bit lower, because of the assumption that tighter building construction, insulation, and internal heat loads will require less heating input.

The description of a rooftop unit really does not do justice to the versatility of this product line; it can also be installed at ground level with ductwork through the wall to the conditioned space (Fig. A8-9). This application is very popular in the smaller sizes for use in residential or light commercial work, because it is a complete factory-tested unit in one cabinet; it requires no chimney or flue; all mechanical sound is outside; and service is made easy by walk-around space at ground level.

Let's look more closely at certain features of this unit. *Single-zone units* as shown in Figs. A8-7 through

Unitary Combination Heating and Cooling Equipment   453

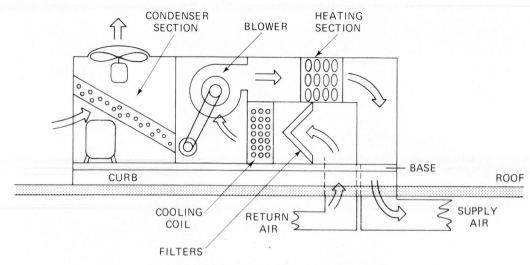

FIGURE A8-7 Cross section of rooftop unit.

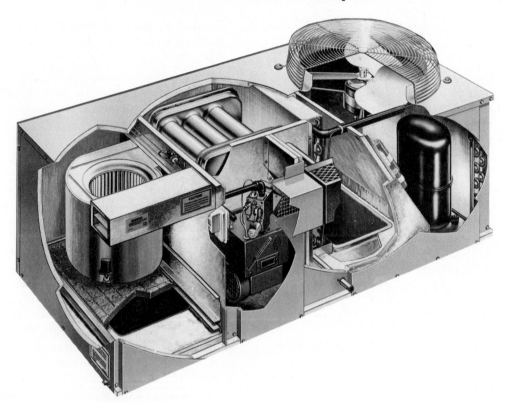

FIGURE A8-8 Single package gas heating and electric cooling unit. (*Courtesy* Lennox Industries)

A8-9 direct all of the conditioned air into one trunk duct, which then distributes air to branch runs as required. *Multizone units* (Fig. A8-10) differ in that the air is divided inside the unit and then is directed to several zone ducts that serve various parts of a building. Zone thermostats signal the proper mode (heating or cooling) in response to comfort needs. Multizone units do differ in their design concepts. Some manufacturers use constant air circulation and temper the heating, cooling, and humidity control, while others use a damper arrangement and premix the air to each zone. Still others use a double-duct system as shown in Fig. A8-11 (hot and cold),

and mix the air out in the conditioned space. In terms of sales, single-zone units are by far the most widely used, because they are the least complicated to apply, to install, and to control for the average need.

One important feature of rooftop units is the so-called *economizer cycle,* which uses filtered outdoor air for cooling rather than mechanical refrigeration when the outdoor ambient temperature is less than the indoor comfort zone temperatures. This is accomplished in the return air chamber by means of dampers on the return air and on an outside air inlet opening (Fig. A8-12). These dampers work in reverse, and as the outdoor air

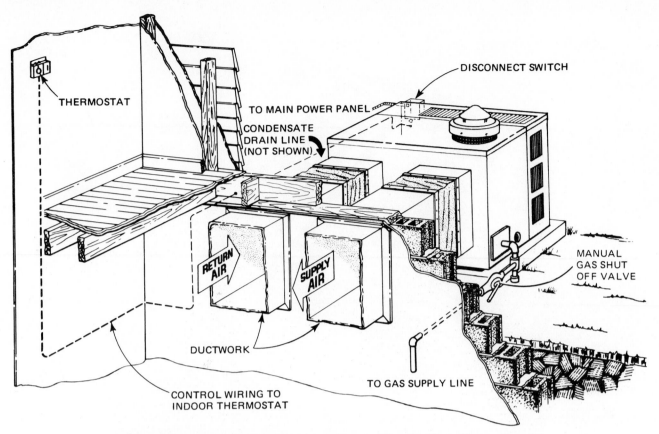

FIGURE A8-9   Typical ground-level installation. (*Courtesy* Borg-Warner Central Environmental Systems, Inc.)

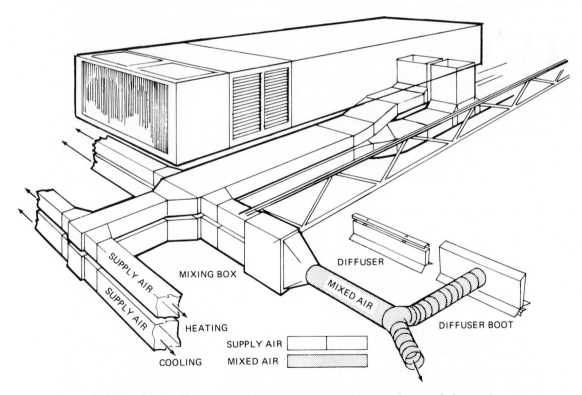

FIGURE A8-10   Direct multizone duct system. (*Courtesy* Lennox Industries)

Unitary Combination Heating and Cooling Equipment   455

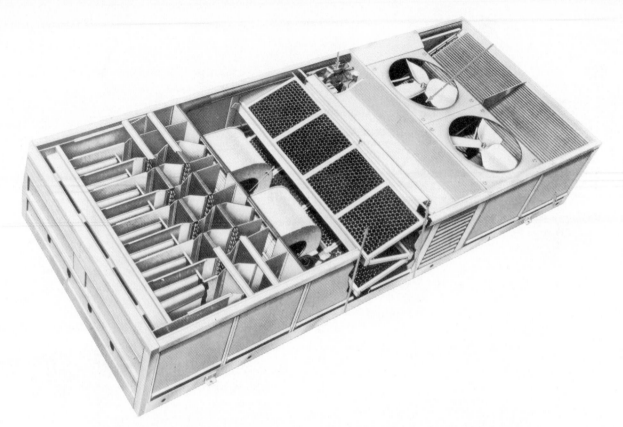

FIGURE A8-11   Double duct system. (*Courtesy* Carrier Air-Conditioning Company)

opens the return closes, mixing the two. Some units can take in 100% outside air, which means no return air comes back. This creates excessive building pressures, which must be exhausted somewhere to relieve pressure. Some units are equipped with built-in power exhaust fans as shown in Fig. A8-12. Others depend on remotely located exhaust fans that are energized by the economizer control. The resultant air mixture of outdoor and return air is controlled so as to provide a ducted air temperature to the conditioned space of not less than about 55°F. Below this the economizer will modulate to a closed or minimum outside air position; thereafter, heating will be needed.

In mild weather, low ambient operation of the cooling cycle is needed and is accomplished by the techniques of speed control, staging, or sequencing OFF the condenser fans. Some units also offer compressor unloading for capacity control. These devices must be programmed to work in conjunction with the economizer system.

The gas-fired heating system has several notable features. The most obvious is that there is no extended flue stack to create the needed draft for exhausting combustion products. Small units have short terminal vents. However, on larger units power venters (centrifugal fans, Fig. A8-13) are needed to provide mechanical draft through the heat exchanger and to discharge combustion products into the atmosphere. Because of this forced-air movement, heat exchanger design includes tu-

bular steel, as well as the conventional sectionalized type. Ignition is provided by a solid-state electrical spark. If the pilot light goes out, it will relight automatically. Pilot outage could be a real nuisance on rooftop equipment, but the spark method has solved this problem.

Due to the quantity of heat associated with large stores, the commercially sized rooftop equipment usually has a two-stage furnace operation controlled by a two-stage heating thermostat so as to match load vari-

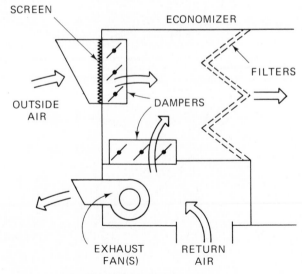

FIGURE A8-12   Economizer cycle.

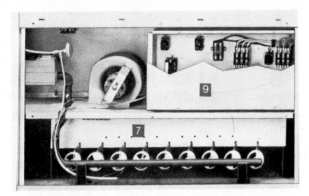

FIGURE A8-13 Power venter. (*Courtesy* Fedders Air Conditioning USA)

ations more closely. Natural gas and propane LP fuels are the standard gas-heating options.

Electric heating (Fig. A8-14) is accomplished by banks of nicrome wire elements supported by ceramic insulators. As described previously, they are protected from overheating by thermal cut-out switches and are backed up by fusible links. Some manufacturers install the heaters at the factory; others offer them as field-installed kits to be placed in the discharge air stream. Prewired fusing and contactors are contained in the equipment panel to simplify and reduce on-site electrical work.

Frequently, you will hear the expression "single-point electrical hook-up." This means that the total power requirement for the unit is brought to one point and then is split internally into its various functions. This is in contrast to individual external wiring to cooling, heating, fans, etc. Obviously this, too, reduces on-site labor and materials and simplifies the building electrical layout.

The roof curb (Fig. A8-15) varies for different manufacturers, but essentially it is a rigid galvanized

steel base that is attached to the roof joists prior to the roof installation. The positioning of the curb depends on the unit location. Leveling the curb is important to assure that the unit drains properly. Wood nailing strips are bolted to the curb and are used by the roofer to nail flashing and counterflashing. Insulation around the entire curb perimeter is needed to prevent sweating. When the unit is set over the curb, a gasket between the two prevents air leaks, and it is common practice to let the unit channels overlap the curb to form a rain shield. Roof leakage was a common problem in the early stages of this equipment's development, but that has been overcome with better equipment design and experienced installers.

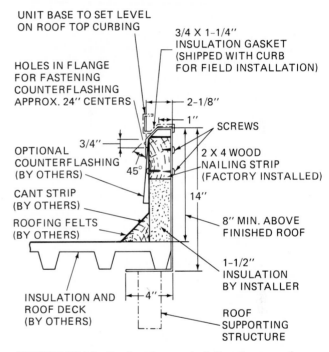

FIGURE A8-15 Roof curb mounted directly on roof supports. (*Courtesy* Borg-Warner Environmental Systems, Inc.)

FIGURE A8-16 Helicopter airlift.

FIGURE A8-14 Electric heating. (*Courtesy* Fedders Air Conditioning USA)

Unitary Combination Heating and Cooling Equipment 457

Lifting the unit onto the base is an interesting event. Large cranes with extension booms are usually used to spot the equipment on small buildings. But when many units are involved, it is faster to use a helicopter airlift, as shown in Fig. A8-16. The technique of maneuvering this size of equipment and setting it down on the curb indeed takes an expert pilot.

Two methods of air distribution are used with sin-gle-zone units when mounted on the roof of a commercial job. First, conventional trunk ductwork (supply and return) go down into the drop ceiling of the building as shown in Fig. A8-8. It then is routed to branch ducts and air diffusers as needed. The second method is to use a concentric duct connection (Fig. A8-17), consisting of a supply and return to a single ceiling diffuser below. This arrangement depends on the feasibility of spotting multiple units over the area so that complete air coverage is achieved. Use of this method can lower duct-work and installation costs considerably.

The indoor centrifugal fans on large equipment are belt driven and must be capable of producing much higher duct static pressure than is necessary for residential equipment. Pressures of $1\frac{1}{2}$, 2, and almost 3 in. of W.C. are needed, depending on the type of air distribution. Pulley and motor options provide a wide selection of speeds.

Since rooftop equipment is remotely located and not readily visible, the need for an in-space control panel (Fig. A8-18) is essential to permit the operator to select whichever mode (heat, cool, or ventilation) the system should be on. Panel lights may also be included that signal dirty filters or warn of malfunctions.

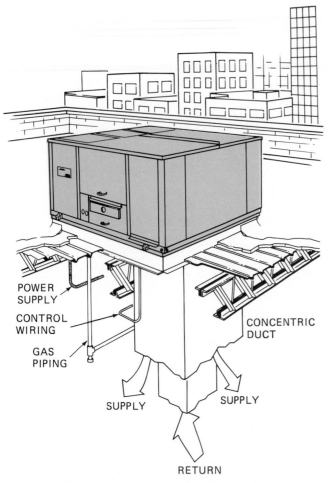

FIGURE A8-17   Typical rooftop installation with accessories. (*Courtesy* Borg-Warner Central Environmental Systems, Inc.)

FIGURE A8-18   In-space control panel. (*Courtesy* Borg-Warner Central Environmental Systems, Inc.)

═══ *PROBLEMS* ═══

A8-1. Packaged terminal units are also called _____ .

A8-2. Rooftop combination heating and cooling units also divide into two other descriptions based on airflow. What are they?

A8-3. The use of outdoor air for mild weather cooling is called the _____ cycle.

A8-4. Gas-fired rooftop units almost universally employ _____ ignition.

A8-5. What is meant by "single-point hook-up"?

A8-6. What does concentric duct connection on a rooftop unit permit?

A8-7. The indoor centrifugal blower on large equipment is capable of operating at dust static pressures as high as _____ in. of W.C.

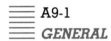

Central Station Systems

A9-1

GENERAL

The name *central station system* or *equipment* is also commonly called an *applied system, applied equipment, applied machinery,* and *engineered system.* Essentially, these are names adopted by the various companies that manufacture the equipment as a means of identifying the division or departments within their organizations. The end results are essentially the same.

Central station equipment is associated with installations where the cooling plant is located in the basement or in a penthouse on the roof of multistory buildings. It serves air-handling equipment and air distribution systems throughout the building. Although size is not necessarily the crossover point between unitary and central station, it is usually acknowledged that central station equipment starts at 25 to 50 tons and extends upward to multi-thousand-ton systems. Unitary equipment tapers off in the range 50 to 75 tons.

Another distinguishing difference is that central station systems use the medium of liquid (mostly water) to transfer heating and cooling to a space air terminal, while unitary systems are based on distributing conditioned air directly to the conditioned space. Unitary equipment makes use of factory packaged, balanced, and tested equipment, which requires a minimum of on-site labor and material to be operational. Central station systems are made up of separate components such as chillers, air handlers, water towers, controls, etc., which can become quite complex in terms of on-site installation and labor and related trades and crafts. Central station equipment is closely associated with the *plan and spec* engineered systems put out by consulting engineering firms. Equipment is then selected and/or built to order in compliance with these specifications. Delivery time can vary from a few months to even a year or more on very large apparatus.

The art of selecting and matching the components

is done by skilled professional engineers, and the servicing technician need not be concerned with that function. The service technician is, however, concerned with the system's installation, operation, and maintenance and, therefore, must learn the basics of the equipment being utilized and the system application. So we will now examine the scope of the systems available, beginning with an orientation sketch of a conventional central system (Fig. A9-1). First, identify the major components:

1. Water chiller
2. Boiler
3. Air-handling unit
4. Water-cooling tower
5. Control system

The water chiller will produce 40 to 45°F cold water, and by means of a pump circulate it to the cold-water coil in the air handler. Water off the coil will generally return at a 10°F rise.

Similarly, in winter the boiler will produce hot water at 180 to 200°F and pump it to the hot-water coil. Note that it is possible to have the boiler and chiller operating at the same time, because in large buildings there may be need for cooling and heating in different zones.

Condenser water off the chiller (95°F) is pumped to cooling-tower spray nozzles, where it is cooled within the tower to around 85°F and then is returned to the condenser. The tower bypass permits regulating return water temperature as a function of ambient temperature changes, but it is never less than 70°F.

The air-handling unit or units, depending on the number of floors or zones, generally contain:

- Chilled-water coils
- Main hot-water coil (can be steam)
- Humidifier
- Filters

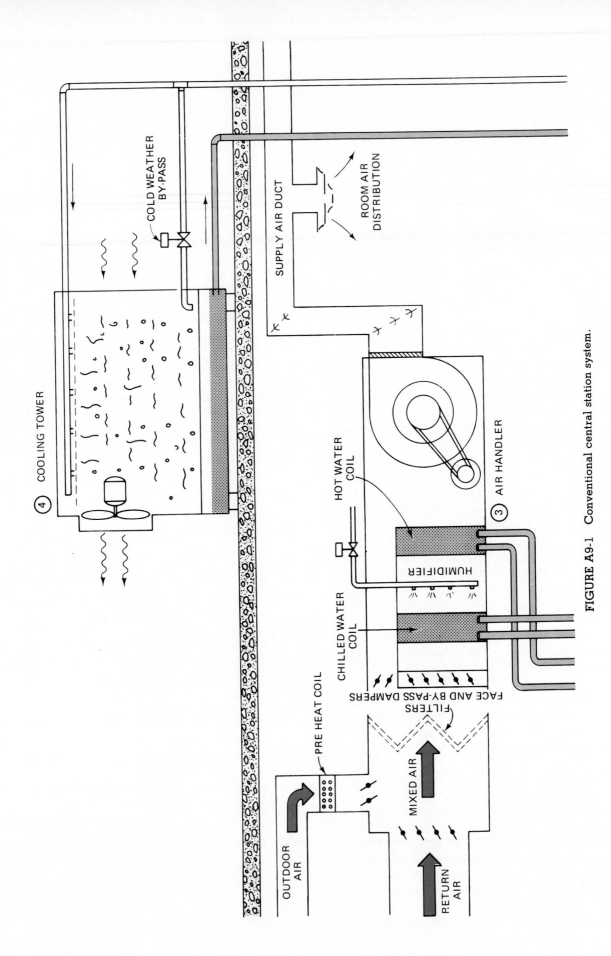

FIGURE A9-1   Conventional central station system.

COLD WEATHER BY-PASS

SUPPLY AIR DUCT

ROOM AIR DISTRIBUTION

④ COOLING TOWER

HOT WATER COIL

HUMIDIFIER

③ AIR HANDLER

PRE HEAT COIL

CHILLED WATER COIL

FACE AND BY-PASS DAMPERS

FILTERS

MIXED AIR

OUTDOOR AIR

RETURN AIR

- Face and bypass dampers
- Dampers for mixing return air and outside air
- Blower and motor

A preheat coil is frequently required where large amounts of outside air at or below 32°F are needed either by code or because of special application. It is installed in the outside duct and is controlled as a function of entering air temperature to the main heating coil.

The face and bypass dampers either permit all the air to go through the humidifiers and the heating and cooling coils or allow some of it to be bypassed, depending on the particular situation. All air to the conditioned space is always filtered or cleaned.

The control system must direct the operation of these elements and do it automatically. In large installations the control system is usually a separate installation job from the air-conditioning equipment. Controls may be electric, electronic, or pneumatic (air), or a combination of all three and must be included in the initial construction stages. Although it takes a thorough knowledge of controls to plan and install the system, it isn't so complicated when the elements are studied individually. Basically the water chiller operates by sensing the temperature of chilled water and controls it accordingly. The tower fan and condenser water pump come on when the chiller operates. The boiler is activated by hot-water temperature. The tower bypass valve controls the water by temperature. Space thermostats and humidistats control the functions within the air handler. Chilled-water and hot-water circulating pumps are a function of summer/winter operation and generally operate on a continuous basis.

Naturally, this is an oversimplification of the control system, and there are many details a technician must learn, but these come with exposure and experience and are also a function of the type of system. This illustration is only one approach; others may include air-cooled condensing.

With the system components identified, let's review the equipment hardware.

A9-2
WATER-CHILLING EQUIPMENT

A9-2.1
Packaged Chillers

The design and range of water chillers is somewhat related to the compressor and condensing medium (air or water). Starting at the lower capacity, the so-called packaged water chiller uses a reciprocating compressor (or multiples thereof). Figure A9-2 represents a small water-cooled chiller in the 20-ton range that has reciprocating compressors, liquid chillers and water-cooled condensers, compressor starters, controls, and refrigerant and oil pressure gauges neatly enclosed in a cabinet. This style of unit is ideal for motels, small office buildings, etc. Because it is packaged it can be factory operated and tested to job specifications. It must be used with a remote water tower or other sources of water for condensing.

Continuing with the reciprocating compressor, water-cooled models (Fig. A9-3), the larger packaged chillers range upward from 40 to 200 tons. These are characteristically multiple-compressor models that permit close control of capacity; they have, as well, standby

FIGURE A9-2 Small package chiller. (*Courtesy* Borg-Warner Air-Conditioning, Inc.)

FIGURE A9-3 Reciprocating large package chiller. (*Courtesy* Borg-Warner Air-Conditioning, Inc.)

compressor protection in the event of a malfunction. Starting inrush current is reduced. Where single compressors are used they are equipped with cylinder unloading to allow capacity reduction and minimum starting power requirements.

The basis of heat rejection for both units described above is a shell-and-tube condenser (Fig. A9-4). Water flows through the tubes, and refrigerant vapor fills the shell, condensing to a liquid that is collected at the bot-tom and is subcooled 10 to 15°F for greater cooling capacity. In the water circuit, the coldest condenser water enters the lower part of the shell and circulates through the tubes. The water will make two or three passes through the shell before it is discharged. This is arranged by circuit baffles in the condenser heads. The higher the number of passes the greater the pressure drop or pressure required to produce the required flow rate. There are also cross baffles within the shell which serve

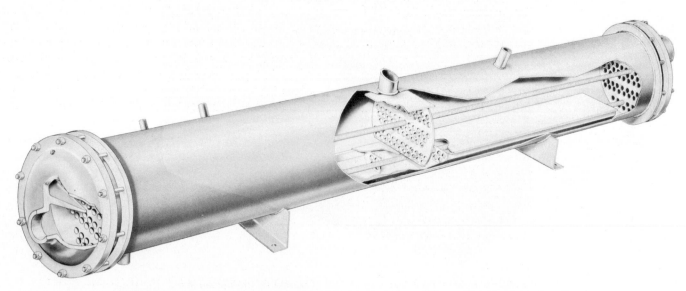

FIGURE A9-4 Shell-and-tube condenser. (*Courtesy* Borg-Warner Air-Conditioning, Inc.)

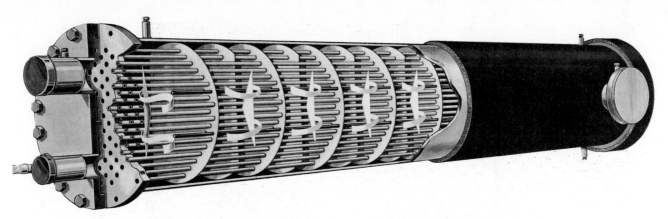

FIGURE A9-5 Chiller. (*Courtesy* Borg-Warner Air-Conditioning, Inc.)

to hold the tube bundles but also help to spread the refrigerant gas over the entire length of the shell. Condenser capacity is normally based on 85°F entering water temperature with a 10°F rise. Calculated into the performance rating is a *fouling factor,* which assumes there will be scale deposits collecting inside the water tubes. This amounts to about a 4% reduction from the performance of a clean condenser. Cleaning water-cooled condensers is influenced by the local water hardness conditions in addition to atmospheric contaminants that collect in the water tower operation. Chemical water treatment is a necessary routine maintenance activity, but eventually the condenser will need attention. Removable heads permit the mechanical cleaning of the tubes by reaming out the scale.

Most manufacturers offer a special sea-water condenser for marine duty where salt water is used as the cooling medium. These tubes are made of cupro-nickel steel to withstand the salt corrosion effects.

In the chiller size ranges associated with R-22 reciprocating compressors, the chiller (cooler) shown, in Fig. A9-5, is of the direct expansion design. Refrigerant flows through the tubes and will generally make two passes for standard operation, which gives a counterflow arrangement versus the water flow pattern. Water connections are on the sides instead of the ends. Cross baffles direct the chilled water back and forth over the refrigerant-filled tubes at the optimum velocity for best heat transfer. The water system within the building is usually a closed loop. That is, it is not open to evaporation and contamination as is the condenser circuit. Even so a fouling factor is assumed in the chiller ratings.

The cooler shell and suction lines must be properly insulated to prevent sweating; this is done at the factory with a layer of closed-cell foam insulation prior to painting.

Both condenser and cooler shells must comply with ANS B9.1 and the applicable ASME safety codes for pressure vessels.

Standard chiller rating points are based on ARI Standard 590: 44°F leaving water temperature off the cooler at 105°F and 120°F condensing temperatures;

also, 95°F leaving water off the condenser with a 10° rise. The 95°F condenser water rating will produce a condensing temperature near the 105°F figure. The rating point of 120°F condensing temperature is used for applications where remote air-cooled condensers (Fig. A9-6) are used instead of the water cooled type.

A more popular method of offering air-cooled water-chiller operation is the complete package (Fig. A9-7), which mounts on the roof and is not unlike a large air-cooled condensing unit, except that the cooler shell is suspended beneath the condenser coil and fan section. The cooler internal construction is as previously described, except that it must be protected against freezing. Electric heating elements are wrapped around the shell and then covered with a thick layer of insulation. Some manufacturers also add a final protective metal jacket that doubles as a good vapor barrier.

The sizes of reciprocating compressor air-cooled packaged water chillers range from 10 to over 100 tons. These are also rated in accordance with ARI Standard 590 at 44°F leaving chilled water temperature and 95°F DB condenser entering air temperature.

In addition to its space-saving advantages and the elimination of the water-cooled condenser and tower problems, the air-cooled packaged chiller has the ability to furnish chilled water at mild weather and low ambient conditions where water-cooled condensing methods would be subject to freezing problems. The techniques for low ambient control are as previously described for air-cooled condensing units and/or rooftop equipment.

A9-2.2
Centrifugal Chillers

Hermetic: For very large installations the industry offers a range of hermetic centrifugal compressor water chillers of up to 1300 tons in a single assembly; when used in multiples they handle applications of huge magnitude such as sports arenas, airports and high-rise office buildings. Interestingly, however, the comparatively new small centrifugal chillers, down to 100 tons and be-

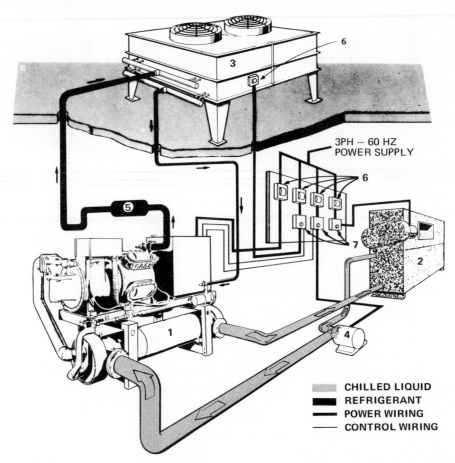

3PH — 60 HZ
POWER SUPPLY

CHILLED LIQUID
REFRIGERANT
POWER WIRING
CONTROL WIRING

FIGURE A9-6   Air-cooled condenser unit. (*Courtesy* Borg-Warner Air-Conditioning, Inc.)

low, are beginning to challenge the popularity of the reciprocating machines at lower tonnages.

Hermetic centrifugal compressors (Fig. A9-8) vary in design and refrigerant use. Some are single stage, others multistage. Some are direct drive while others are gear driven. However, as reviewed in the refrigeration chapters, the basic principle of all hermetic compressors is the same—and that is to use a rotating impeller to

FIGURE A9-7   Packaged air-cooled water chiller. (*Courtesy* Borg-Warner Air-Conditioning, Inc.)

draw suction gas from the chiller (cooler) and compress it through a voluted discharge passage into the condenser (water or air). The speed of the impeller is a function of the design, but some gear-driven units go up to 20,000 to 25,000 rpm. Capacity control is accomplished by a set of inlet vanes that throttle the suction gas to load or unload the impeller wheel.

A complete water-cooled hermetic centrifugal chiller assembly is illustrated in Fig. A9-9. This particular unit utilizes a combination chiller and condenser in one shell, although they are separated internally according to the respective function of each. Chillers of this size and design do not use expansion valves as such; they use either a float control or a metering orifice to flood the chiller shell with liquid refrigerant.

Hermetic centrifugal chillers are also being assembled in complete air-cooled units for roof mounting. Current product offerings range from 130 to 320 tons. The size and weight of these units are factors in their future development, as is the energy efficiency rating of air-cooled condensing methods in increasing tonnages.

Open Type: In addition to the hermetic centrifugal design, there are on the market open drive centrifugal units that produce up to 5000 tons. The choice of drive

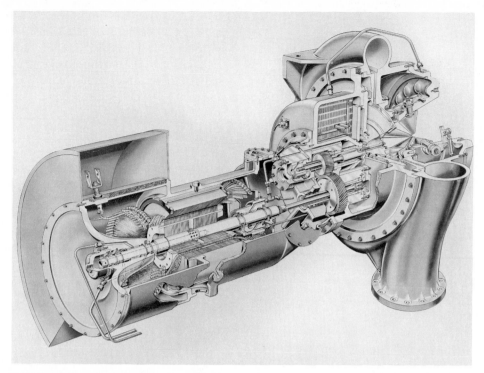

FIGURE A9-8   Hermetic centrifugal compressor and drive. (*Courtesy* Borg-Warner Air-Conditioning, Inc.)

can be a compatible gas, steam, diesel, or electric motor. This category of centrifugal unit is somewhat specialized and not as easily designed, selected, and installed as are the complete packaged units.

≣ A9-2.3
≣ Screw-Compressor Chillers

An important type of central station chiller is the screw-compressor design (Fig. A9-10), which utilizes mating sets of gear lobes that rotate; during rotation the space or mesh between the lobes becomes smaller and

smaller so that suction gas entering from the front is compressed and discharged at the end of the screw arrangement. The compressor shown in Fig. A9-11 is not hermetic but is driven by an external motor. The condenser and chiller (cooler) design is similar to the shell-and-tube system already reviewed. The development of screw-compressor chillers has been mostly useful for the refrigeration needs of the food and chemical industries, but recently this type of compressor is becoming popular in the comfort air-conditioning market.

FIGURE A9-9   Hermetic centrifugal liquid chiller cutaway. (*Courtesy* Borg-Warner Air-Conditioning, Inc.)

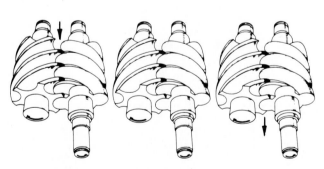

Gas drawn in to fill the interlobe space between adjacent lobes.

As the rotors rotate the interlobe space moves past the inlet port, which seals the interlobe space.

Continued rotation progressively reduces the space occupied by the gas causing compression.

When the interlobe space becomes exposed to the outlet port the gas is discharged.

FIGURE A9-10

Central Station Systems   465

FIGURE A9-11 Screw-type compressor. (*Courtesy* Borg-Warner Air-Conditioning, Inc.)

A9-2.4
Absorption Chillers

Unlike the conventional mechanical compression refrigeration cycle used in all the other equipment discussed, an absorption chiller (Fig. A9-12) uses either steam or hot water as an energy source to produce a pressure differential in a generator section. The absorption cycle substitutes physicochemical processes for the purely mechanical processes of the compression cycle. A complete thermodynamic analysis of an absorption cycle is relatively complex and is beyond the scope of this book. Also, it is a specialized field that should be considered by those who wish to become involved in that

type of work. For further information, consult the ASHRAE *Handbook of Fundamentals and Equipment*. For our purposes it is sufficient to note that these units are available from 100 tons to over 1500 tons. The weight and size of an absorption chiller is a major consideration in the selection of a system, its application, and its installation.

In the review of chilling equipment, we have worked with water as the fluid medium to be cooled and circulated, and so it is for most comfort air-conditioning applications. However, these machines will provide liquid cooling for other purposes, such as the circulation of low-temperature brine or glycol solutions for refrigerating in ice skating rinks, freezing plants, and the chemical, drug, and petrochemical industries.

With source of chilled water established, the next step is to review the terminal equipment that is used to transfer the cooling effect from the water to the conditioned space.

A9-3
AIR DISTRIBUTION EQUIPMENT

A9-3.1
Fan-Coil Units

Fan-coil units for combination heating and cooling come in a variety of designs. Perhaps the most familiar is the individual room conditioner (Fig. A9-13) so frequently seen in office buildings, apartments, dormitories, motels, and hotels. It has an attractive cabinet enclosure with an air return on the bottom and a set of discharge directional louvers on top. Inside it consists of a filter, direct-driven centrifugal fan(s) and a coil suitable for handling chilled or hot water. The size of the unit is based on its cooling ability. Heating then is usually more than adequate. For use with chilled water and hot water, a variety of water-flow control packages are available for manual, semiautomatic, or fully automatic motorized or solenoid valve operation. Airflow is controlled by fan speed adjustment (manual or optional automatic). Outside air is introduced through a dampered opening to the outside wall. The size of the cabinet is

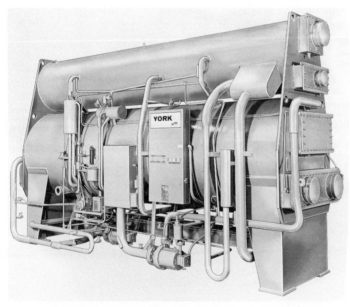

FIGURE A9-12 Absorption chiller. (*Courtesy* Borg-Warner Air-Conditioning, Inc.)

FIGURE A9-13 Fan-coil unit. (*Courtesy* Carrier Air-Conditioning Company)

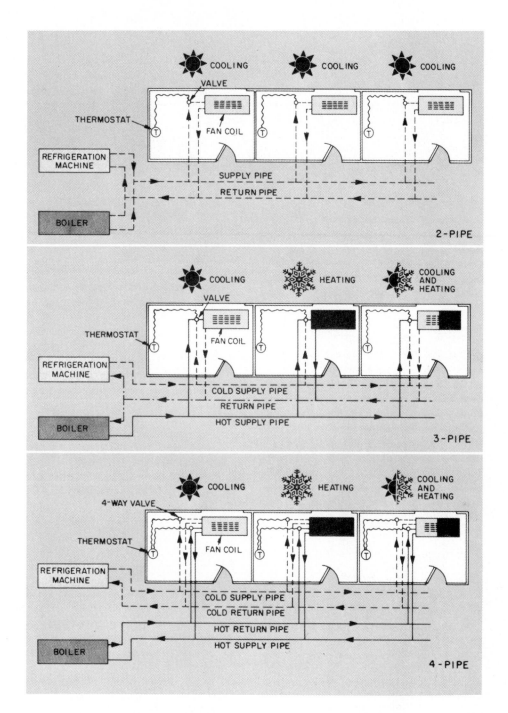

2-pipe system

Either hot or chilled water is piped throughout the building to a number of fan-coil units. One pipe supplies water and the other returns it. Cooling operation is illustrated here.

3-pipe system

Two supply pipes, one carrying hot and the other chilled water, make both heating and cooling available at any time needed. One common return pipe serves all fan-coil units.

4-pipe system

Two separate piping circuits — one for hot and one for chilled water. Modified fan-coil unit has a double or split coil. Part of this heats only; part cools only.

FIGURE A9-14   Fan-coil units. (*Courtesy* Carrier Air-Conditioning Company)

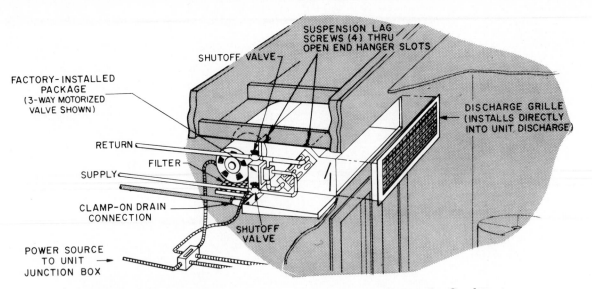

FACTORY-INSTALLED
PACKAGE
(3-WAY MOTORIZED
VALVE SHOWN)

SHUTOFF VALVE

SUSPENSION LAG
SCREWS (4) THRU
OPEN END HANGER SLOTS

DISCHARGE GRILLE
(INSTALLS DIRECTLY
INTO UNIT DISCHARGE)

RETURN

FILTER

SUPPLY

CLAMP-ON DRAIN
CONNECTION

SHUTOFF
VALVE

POWER SOURCE
TO UNIT
JUNCTION BOX

FIGURE A9-15 Ceiling-mounted fan-coil unit. (*Courtesy* Carrier Air-Conditioning Company)

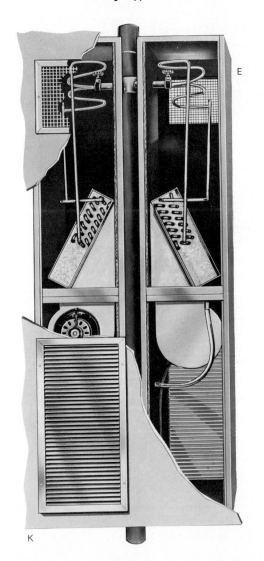

FIGURE A9-16 Room-type vertical column fan-coil unit. (*Courtesy* Borg-Warner Air-Conditioning, Inc.)

FIGURE A9-17 Ducted vertical fan-coil unit. (*Courtesy* Borg-Warner Air-Conditioning, Inc.)

FIGURE A9-18 Large air handler. (*Courtesy* Addison Products Company)

designated in cubic feet of airflow and ranges from 200 to 1200 ft³/min. Coils are then picked to match cooling conditions based on the number of rows of coil depth. These units may be installed on two-, three-, or four-pipe system designs as illustrated and described in Fig. A9-14.

With slight modifications in cabinetry, the same components are assembled in a horizontal ceiling-mounted version (Fig. A9-15). Where hot-water heating is not desirable, electric resistance heaters may be installed in the cabinet on the leaving air side of the cooling coil.

Another version of the room type of fan-coil unit is a vertical column design (Fig. A9-16), which may be installed exposed or concealed in the wall. These are placed in common walls between two apartments, motel rooms, etc. Water-piping risers are also included in the same wall cavity. They are not designed for ducting but can serve two rooms by adding supply grilles.

Small ducted fan-coil units (Fig. A9-17) with water coils may be installed in a drop ceiling or a closet, with ducts running to individual rooms in an apartment. These range in sizes from 800 to 2000 ft³. The unit's cooling capacity is selected based on the desired ft³/min and the number of rows of coil depth needed.

Larger fan-coil units (Fig. A9-18), used to condition offices, stores, etc., where common air distribution is feasible, are somewhat like the cabinets of self-contained store conditioners. They can be equipped with supply and return grilles for in-space application, or they may be remotely located for ducting. Sizes range from about 800 to 15,000 ft³/min. In larger sizes, cabinet and fan discharge arrangements permit flexible installations. These units approach the next category of equipment called *air handlers,* but in general, they do not have the size and functions available in central station air handlers.

A9-3.2
Central Station Air Handlers

The air handler illustrated in Fig. A9-1 is a single-zone, low-pressure, draw-through system, which is typ-

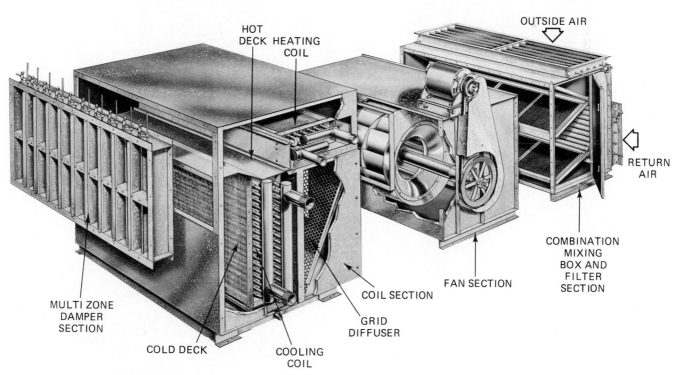

FIGURE A9-19 Blow-through central station air handler. (*Courtesy* Borg-Warner Air-Conditioning, Inc.)

Central Station Systems 469

FIGURE A9-20 Variable inlet vane dampers. (*Courtesy* Borg-Warner Air-Conditioning, Inc.)

ical of a small chiller application or where multiple air handlers are used to provide zone control. However, where large volumes of air are to be distributed and/or where multizone or double-duct applications are needed, the blow-through type unit as shown in Fig. A9-19 is used. First, the unit is made up of dimensionally match-

ing sections. Each section can then be selected for its own performance requirement.

Note the combination mixing box and filter section. The size of the outside air intake damper is such that 100% outside air may be taken in during intermediate seasons, just as was described for the economizer cycle of rooftop units. Filters may be of the throwaway or permanent (washable) type.

The blow-through fan section utilizes a large centrifugal fan, which is classified for either low-pressure or medium-pressure performance. Low-pressure wheels produce external static pressures from 0 to 3 in. of W.C. and usually have forward-curved blades. For medium static pressure duty from 3 to 6 in. of W.C., units may use the air-foil type of wheels as illustrated in Fig. A9-19. The larger fans may also be equipped with variable inlet vanes (Fig. A9-20) to control airflow performance. (Chapter A4 contains more information on basic fan designs and application.)

The coil section consists of a cooling coil and heating coil. Finned-tubed water coils for cooling (Fig. A9-

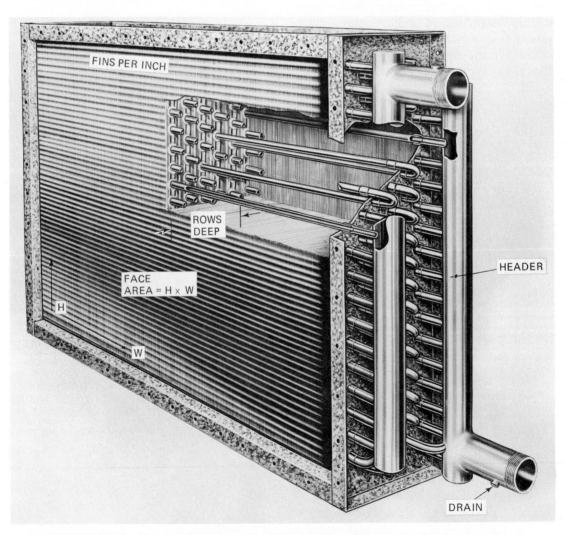

FIGURE A9-21 Finned-tube water coils. (*Courtesy* Borg-Warner Air-Conditioning, Inc.)

21) are selected on the basis of the square feet of face area needed for the most efficient use of space; the rows and fin spacing necessary to meet performance standards at the lowest cost; and the proper circuiting for the best transfer, within pressure-drop limitations. Actually, there are many possible options, and many manufacturers offer a computerized selection service to help the engineer in making the optimum selection.

Heating coils using hot water look identical to those shown for cooling, except for the face area and the number of rows of depth, which rarely exceeds three, since the temperature difference between the 180°F hot water and the 75°F conditioned air is so great, compared to the 44°F chilled water and 75°F air. Steam coils are also available and would be positioned in the area noted for the hot-water coil.

Note the grid diffuser (Fig. A9-19), which spreads the fan discharge air over the entire coil face. The position and separation of the two coils form a cold deck and a hot deck, so that air may be blended to provide the individual zone requirement. In many large buildings the perimeter (outside) zones may need heating while the interior core, because of lights, people, and equipment, may call for cooling. The multizone damper section will individually control each zone requirement.

A9-4
HUMIDIFIERS

Not shown in the air handler illustrated in Fig. A9-1 are humidifiers that can be used to add moisture to the air. This would be done in the hot deck as a function of heating need. Several types are used (Figs. A9-22 through A9-24), depending on need and application.

Water-spray types are used with hot water heating and provide optimum performance in applications where the humidity level is fairly low and the most precise control is not required.

The steam pan type is used when the introduction of steam directly into the airstream is undesirable. The vaporization of water from the pan provides moisture to the conditioned air.

The steam grid type is highly recommended because it offers simplicity in construction and operation, and humidification can be closely controlled. Obviously a source of steam must be available.

The duct systems into the conditioned space vary with the application and type of air handler. Some are relatively simple; others are complex due to the nature of the need.

A9-5
AIR DISTRIBUTION SYSTEMS

In terms of the principles of airflow, velocity, static pressure, etc., the air distribution systems for commercial work use the same basic design information as previously presented for residential work. The major differences lie in the fact that houses, their construction, and occupancy factors are really all quite similar to each other when compared to the huge variety of applications found in commercial and industrial comfort conditioning. The requirements for air conditioning a broadcasting studio are quite different than the air distribution in a textile plant. So to a great extent the application determines the design conditions.

In a high-rise office building the need to provide perimeter air distribution is a function of the external

SPRAY

FIGURE A9-22 Water-spray humidifier.

STEAM PAN

FIGURE A9-23 Steam pan humidifier.

STEAM GRID

FIGURE A9-24 Steam grid humidifier.

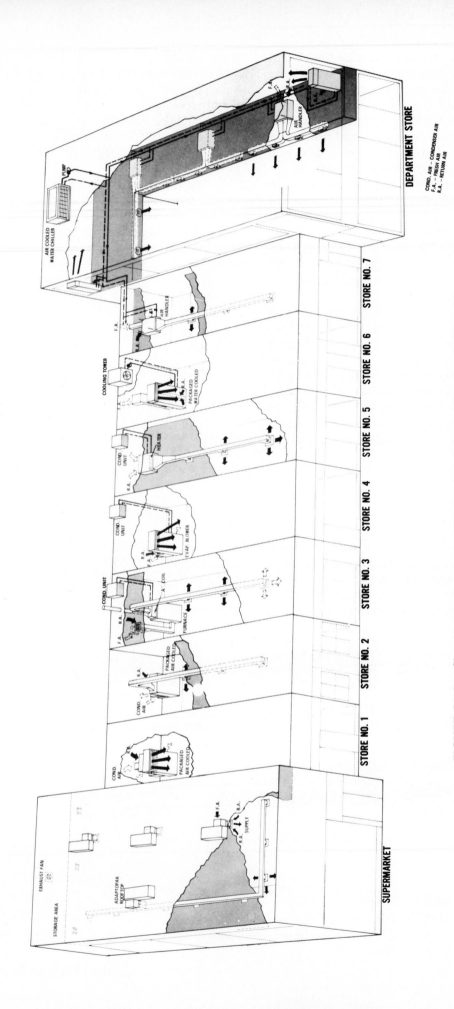

FIGURE A9-25 Shopping center air-conditioning installations. (*Courtesy Fedders Air Conditioning USA*)

COND. AIR — CONDENSER AIR
F.A. — FRESH AIR
R.A. — RETURN AIR

SUPERMARKET

STORE NO. 1 STORE NO. 2 STORE NO. 3 STORE NO. 4 STORE NO. 5 STORE NO. 6 STORE NO. 7

DEPARTMENT STORE

AIR COOLED WATER CHILLER

PUMP

AIR HANDLER

COOLING TOWER

AIR HANDLER

PACKAGED WATER COOLED

HEATER

COND UNIT

COND UNIT

EVAP. BLOWER

COND UNIT

A. COIL

FURNACE

PACKAGED AIR COOLED

COND AIR

PACKAGED AIR COOLER

EXHAUST FAN

STORAGE AREA

ADAPTOPAK ROOFTOP

SUPPLY

F.A.

R.A.

heat gain or loss through walls and windows; at the same time the center core air distribution is a function of the heat load from people, lighting, and machines. The perimeter loads are functions of time and would be affected by the sun and weather conditions. The core loads are a function of whether the building is occupied or not. So the air distribution systems of most commercial applications have to cope with wide variations in demand over a 24-hour period. Thus, we find the use of systems that are able to vary temperature and air volume to the space, and some even have provisions to transfer heat from where it is not needed to areas of need, thus conserving energy.

A9-5.1
Types of Systems

For simplicity, types of commercial air distribution may be classified as low-, medium-, or high-velocity systems.

Low-velocity systems are those associated with the application of smaller unitary packaged and split-system units as illustrated in Fig. A9-25. The use of limited ductwork or none at all (free blow) is typical of this classification. Where ductwork is used, the external static pressure is held down to the range 0.25 to 0.50 in. of W.C. The use of concentric supply and return ducts, as shown in the illustration of the supermarket, is a common application. Duct and air outlet velocities and thus resultant noise are somewhat higher than that of residential work; they should be kept within the recommendation noted in Chapter A4. The type of duct design is the same as that of the equal friction method, which permits the prediction or control of total static.

Medium-velocity systems are those associated with conditioners or air handlers that provide larger air volumes at external static pressures of up to 2.0 to 3.0 in. of W.C. Typical of this category are the larger rooftop and air-handling units for single and multizone application. It is in this category that the use of different air distribution techniques becomes necessary, depending on the application and economic considerations. The following discussions are presented to acquaint the reader with general systems, not necessarily the specific design illustrations.

A9-5.2
System Applications

Application in core areas may be handled in several ways.

1. Single-zone heating and cooling. If the interior core of a building is a large open area, a single-zone, constant-volume system (Fig. A9-26) can be used at a reasonable initial cost. A single heat/cool

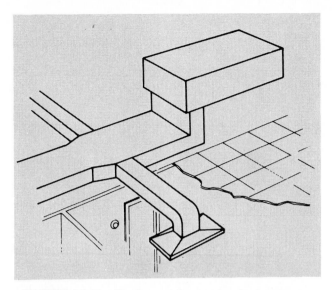

FIGURE A9-26 Single-zone heating/cooling constant volume.

thermostat provides year-round environmental control with automatic changeover; top floor and ground floor systems will have heating for morning warm-up. Intermediate floor systems usually omit the heat source. The design of the ducts follows conventional practices, and room air distribution will usually be from the ceiling.

2. For an area composed of numerous spaces subject to variations in lighting loads or people density, the variable-volume, single-zone system (Fig. A9-27) is a low-cost answer. Instead of delivering a constant flow of conditioned air to each space, the air quantity is thermostatically varied for each space to match fluctuations in space loads exactly. Under partial load conditions, excess air is simply bypassed back to the rooftop unit. The variable volume control terminal (Fig. A9-28) consists of a

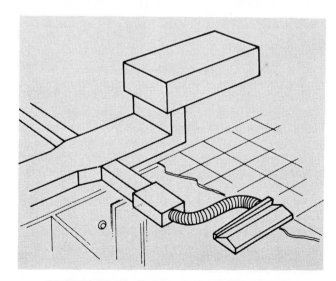

FIGURE A9-27 Single-zone variable volume.

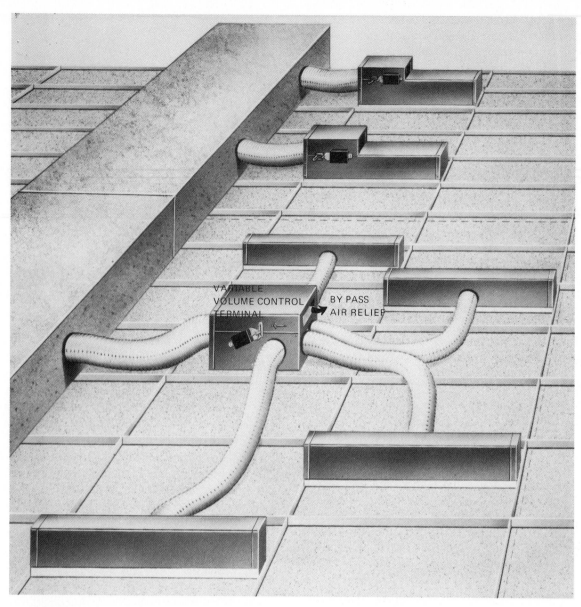

VARIABLE
VOLUME CONTROL
TERMINAL

BY PASS
AIR RELIEF

FIGURE A9-28   Variable-volume control terminal. (*Courtesy* Borg-Warner Air-Conditioning, Inc.)

main duct inlet and one or more branch outlets that feed the supply air diffusers. A damper operator modulates the air to each branch, and relieves or bypasses excess air to the ceiling plenum or the return air duct as well. Air diffusers are designed to fit into the T-bar suspended ceiling. Flexible round ducts feed the control terminal and air diffusers.

Variable-volume systems can be either low velocity or high velocity, depending on the type of controller and air diffusers used. The volume of primary air is automatically adjusted to the total cooling load of the building by pressure-controlled inlet vanes on both the supply and return air fans.

Heating and cooling requirements for various perimeter areas can easily be met with cooling-only single-zone equipment used in conjunction with individual, thermostatically controlled electric heaters or hot-water coils in the branch duct to the occupied space (Fig. A9-29). The duct coils provide tempering reheat in spring and fall and full heat in winter.

The single-zone, variable-volume system may also be applied to the perimeter areas, but air is introduced to the space at floor level on the outside wall rather than through ceiling outlets. Branch ducts are installed in the ceiling plenum of the floor below, and air diffusers are placed in the floor along outside walls under window exposures.

3. As a straightforward solution to perimeter air-conditioning problems, *multizone* heating and cooling systems are usually the best. In single-zone sys-

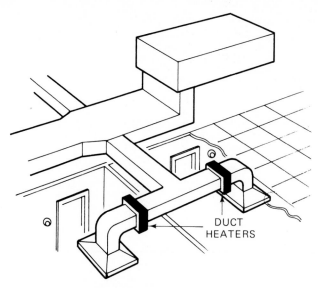

FIGURE A9-29 Constant-volume single-zone remote duct heaters.

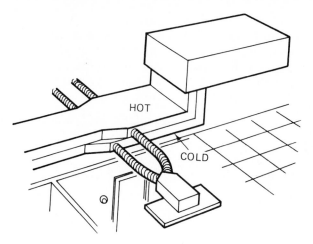

FIGURE A9-31 Variable-volume double-duct heating/cooling.

tems, a single main duct supplies all spaces. In the multizone system (Fig. A9-30), a small branch duct connects the rooftop unit or air handler with each space directly. A thermostat in each space modulates mixing dampers at the discharge of the multizone unit. This provides exactly the right supply air temperature to satisfy cooling and heating requirements. Since there is a separate thermostat, duct run, and zone mixing damper for each space, one area can receive heating while another receives cooling. Because a constant quantity of air is delivered to each space under all load conditions, better humidity, filtration, and odor control are possible.

4. Double-duct heating/cooling systems (Fig. A9-31) have the highest initial cost of all systems, but they offer distinct advantages for certain building re-

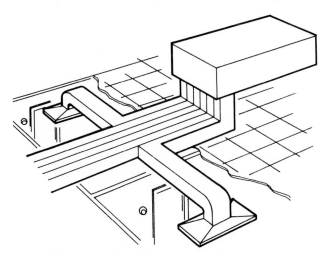

FIGURE A9-30 Constant-volume multizone heating/cooling.

quirements. The main benefit is maximum flexibility. This type of system is also advantageous where there are a large number of areas requiring less than 400 ft³/min. The duct system size is double that of a single-zone system because, as the name implies, two duct systems run to each space—one for warm air, the other for cool air. Both ducts connect to a mixing box serving each space, which is controlled by an individual thermostat; therefore, the supply air mixture temperature to the space is maintained at exactly the right level.

The double-duct system is a high-velocity and high-pressure type. The maximum velocity in the trunk supply ducts will range from 2500 to 6000 ft/min, depending on the air quantity. Static pressures can run in excess of 4 in. of W.C. Branch ducts feeding the mixing boxes will then be sized to help slow the air down; and when it enters the mixing box its velocities are further reduced, so the low-pressure ducts leaving the mixing box can be sized as any other conventional ductwork. The low-pressure ducts feeding the air diffusers are also lined with accoustical material to assure a quiet installation.

Return air ducts in a double-duct system and any other system should be sized at low velocity as in any conventional system. Attempts to pressurize the return duct have not as yet proved to be entirely successful.

When to use a combination of these systems is a function of building size and economics. Small buildings frequently cannot justify one system for the core and another for the perimeter. In these instances, any one of the three perimeter systems can be used for the entire building.

The systems are all of the type that use primary (supply) air from the air-conditioner, and the maximum quantity needed is based on the maximum load conditions of the area. Primary air temperatures will be approximately 20 to 22°F below the room air temperature. So if room air is held at 78°F, the entering primary air is at approximately 56°F. The quantity of air is a function of the sensible heat load and will be about 400 ft³/min per ton for cooling.

A9-5.3
Induction Air Distribution Systems

Since the cost of ductwork is a function of size, the smaller the ductwork, the lower the initial cost and the less labor is needed to install it. A reduction in the size of the ductwork means a reduction in the quantity of primary air, and to reduce the quantity of air it is necessary to lower the temperature so that the required sensible cooling capacity remains the same. This can be seen in the formula

LAT* (leaving air temperature)

$$= \text{room temperature}$$
$$- \frac{\text{room sensible heat capacity (RSHC)}}{1.1 \times \text{ft}^3/\text{min}}$$

If the room temperature is held constant (for example, 78°F DB) and the RSHC also remains constant, it becomes apparent that a reduction in ft³/min will require a reduction in the LAT or the primary air temperature. Thus, if the primary air is reduced to 38 to 40°F, there can be a considerable savings in the supply air ductwork. But supply air cannot be introduced into the room at 38°F, for this is 40° below the room conditions, which will cause uncomfortable cold drafts. Thus, induction air terminals, which blend cold primary air with warmer induced room air, were developed to temper the resultant discharge air.

An induction type of room terminal has an external physical appearance much like the fan-coil unit as described in Chapter A8; however, inside it is totally different. Figure A9-32 is an internal cutaway of an induction type of room terminal. Note there are no fans. High-pressure, high-velocity, primary air at 38 to 40°F is introduced at the bottom (or at the side) at 1500 to 2000 ft/min into a well-insulated plenum for sound and thermal control. An internal damper controls the primary air that feeds to the air nozzles where jets of air create a negative pressure and cause induced air to be drawn from the room. The resultant mixture is discharged at about 55 to 60°F DB in order to maintain

---

* LAT is the leaving air temperature off the coil and is assumed to be the same temperature as the primary ducted air without duct loss.

INDUCTION ROOM TERMINAL

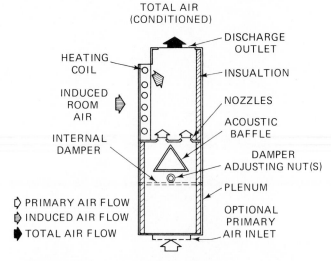

FIGURE A9-32 Induction room terminal. (*Courtesy* Carrier Air-Conditioning Company)

satisfactory room air motion without drafts. The air volume is constant. Note the heating coil on the induced room-air opening. It provides total heat in winter; it can also be used as a reheat coil during the cooling operation. Heating may be by steam, hot water, or electric, and may be selected on the basis of utility rates.

Room terminal induction units are most often used in schools, laboratories, hospitals, and single-floor or low-rise office buildings. The typical piping system is represented in Fig. A9-33. This system is also a high-velocity method. The main trunk duct will be sized for velocities of 2000 to 5000 ft/min. Branch takeoffs to terminals should be a maximum of 2000 ft/min. Round ducts are preferred to rectangular because of their greater rigidity. The ducts are sealed to prevent leakage of air, which may cause objectionable noise. A number of variations in equipment design are offered, including horizontal models that may be suspended in a dropped ceiling. Primary air at 38 to 40°F must be supplied from a chilled-water coil included in a central station air-handling apparatus that circulates a brine-cooling solution. Room-type induction units, even though they operate at high velocity, are considered medium-pressure applications, since the inlet static pressure to the terminal is normally 0.5 to 1.0 in. of W.C. at rated air flow; the maximum would be 2.0 in. of W.C. In the selection of room induction units, two parameters must be satisfied: The unit must supply air at an acceptable sound power level, and it must have enough unit capacity to maintain the proper room temperatures.

Ceiling induction systems use induction units installed in the primary supply duct and the feeding air diffusers. Figure A9-34 shows a single induction unit that draws air from the room and blends it with the cold

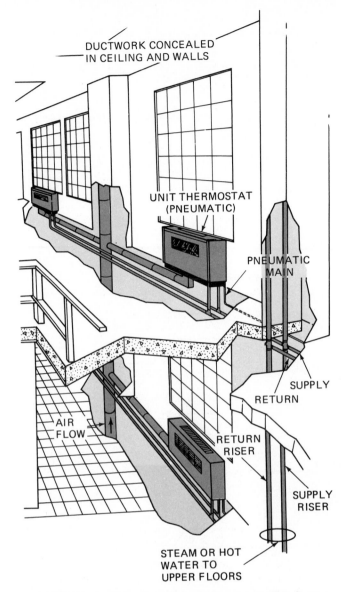

FIGURE A9-33 Piping system of room terminal induction unit. (*Courtesy* Carrier Air Conditioning Company)

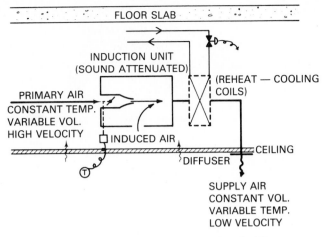

FIGURE A9-34 Single induction terminal. (*Courtesy* Borg-Warner Air-Conditioning, Inc.)

primary air. The high-velocity primary air is supplied to induction boxes in each zone in a variable volume at a constant temperature. The induction boxes are designed for both interior zones and perimeter zones. The temperature of the primary air can be regulated to meet seasonal variations in the total load, especially at the perimeter.

The low-velocity room supply air is discharged into a zone from an air induction box that induces air from the zone and mixes it with primary air. Room air is supplied at a constant volume. The temperature of the supply air is raised by increasing the amount of air induced and by decreasing the amount of primary air at the same time.

Additional heating or cooling may be required to cope with local variations that exceed the capacity of the

system. The limits of the system are established by the minimum primary air setting and the temperature of the primary air.

Because of the increasing emphasis on energy conservation, as well as the desire to reduce duct sizes, the use of double induction is becoming more popular. The double induction principle (Fig. A9-35) extends the range of control over space conditions. Double induction makes it possible to induce air from two sources—the room and the plenum—each at a different heat potential, thus taking advantage of the heat in the plenum that was created by the lighting. In winter the plenum air damper is open to take advantage of recirculating the heat given off by electric lighting. In summer it is mostly induced room air that is mixed with the cold primary air to maintain zone control. The positions of all the dampers are determined by automatically modulating operators in response to the zone controller.

The main return air ductwork is extended into the ceiling plenum from service ducts or shafts at appropriate locations to collect the return air from the plenum and transfer it back to the central plant. In most buildings the ceiling plenum is used as a void. Air is drawn from the space below into the plenum through grilles, registers, or slots in the lighting fixtures.

The double-induction method offers a more closely controlled room temperature; it reduces the total ft³/min delivery, the fan horsepower, and the air-handling apparatus size, and it results in lower building and operating costs. But, although it can use some of the heat produced by lighting and thus reduce the ft³/min requirement, it cannot reduce the refrigeration requirement.

Recently, this problem has been solved by the use of water-cooled lighting fixtures, or *luminaires* (Fig. A9-36). The object of all luminaire designs that remove light heat by return air is to minimize the amount of heat that

Central Station Systems   477

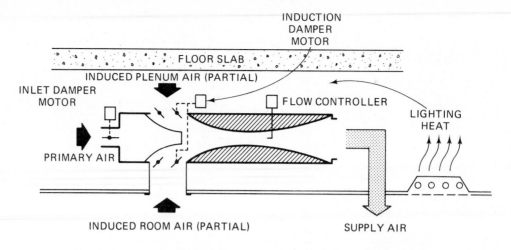

INDUCTION
DAMPER
MOTOR

FLOOR SLAB

INDUCED PLENUM AIR (PARTIAL)

INLET DAMPER
MOTOR

FLOW CONTROLLER

LIGHTING
HEAT

PRIMARY AIR

INDUCED ROOM AIR (PARTIAL)

SUPPLY AIR

DOUBLE INDUCTION TERMINAL

FIGURE A9-35   Double induction terminal. (*Courtesy* Borg-Warner Air-Conditioning, Inc.)

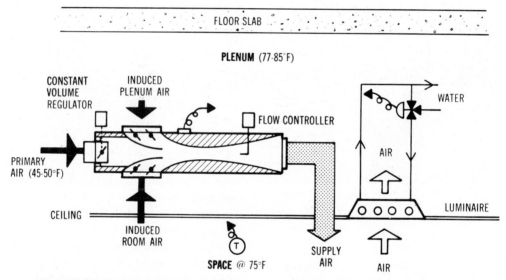

FLOOR SLAB

PLENUM (77-85°F)

CONSTANT
VOLUME
REGULATOR

INDUCED
PLENUM AIR

FLOW CONTROLLER

WATER

PRIMARY
AIR (45-50°F)

AIR

CEILING

INDUCED
ROOM AIR

LUMINAIRE

SPACE @ 75°F

SUPPLY
AIR

AIR

FIGURE A9-36   Luminaire design. (*Courtesy* Borg-Warner Air-Conditioning, Inc.)

becomes a space-conditioning load. However, the lighting heat (during the summer) must be removed from the return air before it is recirculated through the building. This means that the refrigeration plant must be large enough to cool the total heat generated by the light fixtures.

Therein lies the advantage in using water-cooled luminaires; they capture the maximum amount of lighting heat in a medium (water) that permits the heat to be either redistributed or rejected from the building. A water-cooled luminaire is a simple heat exchanger in the shape of a standard fluorescent luminaire fixture. Tubing is an integral part of the fixture housing. Water cir-culated through this tubing absorbs most of the lighting heat (60 to 80%) before it can increase the plenum temperatures and the heat regain in the area. The size of the central refrigeration plant can thus be reduced by the amount of heat removed from the fixture. The resulting hot water may be rejected by a cooling tower or used in the building domestic hot-water system. Lighting efficiency is improved by some 20% as a result of lower fixture temperatures. Ceiling induction air terminals are used as previously explained to blend the primary, room, and plenum air for space temperature control. However, the size of the terminal can be affected by the reduction in the heat removal requirement.

A9-1. Central station equipment is also called _____ .

A9-2. Central station systems are _____ fabricated.

A9-3. Heat transfer is by _____ to _____ to _____ .

A9-4. A conventional central system will generally include what five major components?

A9-5. Chilled-water systems for comfort application produce cold water in the range _____ to _____ °F.

A9-6. In the winter heating season the boiler water temperature is usually in the range _____ to _____ °F.

A9-7. What is the function of face and bypass dampers?

A9-8. Chillers up to a nominal 100 tons usually use _____ compressors.

A9-9. What is meant by "number of passes" in a chiller or condenser?

A9-10. What is meant by "fouling factor" of a condenser?

A9-11. The average fouling factor used in condenser size selection is _____ %.

A9-12. Condensers used for seawater (marine) applications are usually made of _____ .

A9-13. Centrifugal compressors may be _____ driven or _____ driven.

A9-14. Fan coil unit piping arrangements are of three types. What are they?

A9-15. Humidifiers used in large systems are of three types. What are they?

A9-16. The air distribution systems used with central station equipment are classified as _____ systems.

A9-17. On perimeter air-conditioning problems, the best system to use is usually a _____ heating and cooling system.

A9-18. What is the biggest advantage of a double-duct system?

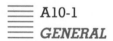

Controls

The development of controls and control systems has gone hand in hand with the development of overall heating and air-conditioning equipment. Particular controls have been developed to answer the need for one or more of several factors, including operation, safety, personal convenience, and economy. When designing a control system and selecting the type and quantity of controls to be used, all factors of operation, safety, personal convenience, and economy are taken into account.

Very simply, a control system checks or regulates within prescribed limits. Such a system consists of three major parts: (1) a source of power to operate the control system, (2) a load or loads to utilize the power to obtain the desired results, and (3) controllers to obtain the desired levels of the end results.

Controls and control circuits were discussed extensively in Chapter R20 as they apply to a refrigeration control system. It is the purpose of this chapter to discuss controls and control circuits as they apply to heating and air-conditioning systems. Controls for heat pumps are discussed in Chapter HP2.

Figure A10-1 shows a typical control system for a year-round air-conditioning system using gas for winter heating and refrigerated air for summer conditions. In addition, a power-type humidifier is included for winter humidification and an electronic air cleaner for year-round air filtration. A multispeed blower motor in the heating (air handling) unit is used to provide the best results in both the heating and cooling phase.

The operation of the system is under the master control of a room thermostat to maintain the desired conditions in the heating phase by controlling a gas valve, and in the cooling phase by controlling the condensing unit operation. In addition, the blower in the gas furnace must operate intermittently, depending on furnace supply plenum temperature in the heating phase

and continuously or with the operation of the condensing unit (depending on the conditioned area occupants pleasure) in the cooling phase. A low-voltage room thermostat is used to control a high-voltage blower motor (120 V) and a higher-voltage condensing unit (240 V).

As explained in Chapter R20, when a device of one voltage is controlled by a device of another voltage, intermediate controls are required. These could be relays, contactors, or motor starters, depending on the load characteristics encountered.

In this control circuit, the power source is a transformer which is part of a plate-mounted relay/transformer assembly. The other component of the assembly is a relay containing two sets of single-pole, double-throw contacts. One set of contacts controls the speed of the blower motor while the other set controls the operation of the blower motor either under the control of the furnace fan switch or directly for cooling operation.

The master control is the room thermostat for both the heating and cooling operation. Intermediate controls consist of the fan switch and limit switch for heating operation and a circulator switch for continuous blower operation with humidification in the heating season.

The loads in this control circuit consist of the gas valve for field input control to the heating unit, the condensing unit contactor for cooling operation, the blower relay coil for control of the blower motor, and the blower motor for air circulation.

The step-by-step operation of a control system is covered in Chapter A11. Discussion in this chapter will be limited to the various component parts.

Obviously, there must be a source of power to operate a control system. Automatic controls for residential and light commercial heating and air-conditioning are powered by electricity. The controls in engineered systems can be electrical, electronic, pneumatic (compressed air), or a combination of all three. Here we will concentrate on the electrical type of controls for residential and light commercial application.

We will first consider the power source. Heating and cooling control circuits can be designed to operate on either line voltage (115 to 120 V) or on low voltage, which is designated as a 24-V system. A low-voltage control circuit is superior to a line voltage circuit because (1) the wiring is simplified and safer, and (2) low-voltage thermostats provide closer temperature control than do line-voltage thermostats.

Essential to the development of low-voltage controls is the ac voltage transformer (Fig. A10-2). A step-down or low-voltage transformer is used in heating and air-conditioning control systems to reduce line voltage to operate the control components. Inside, a simple step-down transformer consists of two unconnected coils of insulated wire wound around a common iron core (Fig. A10-3). To go from 120 V (primary) to 24 V (secondary), there are five primary turns to one secondary turn. For a 240-V primary the ratio will be 10:1, etc. Thus,

the induction ratio is a direct proportion. Step-up transformers would be just the reverse.

In reducing the voltage, was any energy lost? Not really, because the value of the current on the secondary side would be five times higher than on the primary side, so that the power on both sides of the transformer remains the same—assuming a transformer that is 100% efficient. Thus the ratio can be expressed by the formula

$$\text{volts} \times \text{amperes} = \text{volts} \times \text{amperes}$$
$$\text{(primary)} \qquad \text{(secondary)}$$

So if we wanted a 40-VA capacity rating on the secondary, it would take a 0.334-A input on the primary at 120 V to produce 1.67 A on the 24-V secondary side:

$$120 \times 0.0334 = 24 \times 1.67 = 40 \text{ VA}$$

Transformers are available in a variety of voltages and capacities. The capacity refers to the amount of electrical current expressed in volt-amperes; a transformer for a control circuit must have a capacity rating sufficient to handle the current (amperage) requirements

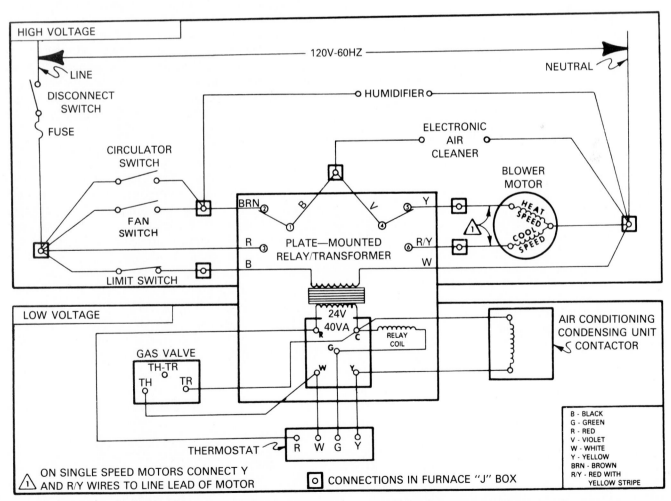

FIGURE A10-1 Total comfort wiring diagram. (*Courtesy* Addison Products Company)

FIGURE A10-2   Low voltage control transformer. (*Courtesy* Honeywell, Inc.)

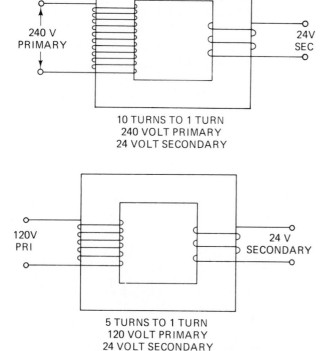

10 TURNS TO 1 TURN
240 VOLT PRIMARY
24 VOLT SECONDARY

5 TURNS TO 1 TURN
120 VOLT PRIMARY
24 VOLT SECONDARY

FIGURE A10-3   Step-down transformer. (*Courtesy* American Gas Association)

of the loads connected to the secondary. Twenty volt-ampere ratings are used only for heating forced-air furnaces. Heavier volt-ampere ratings are needed for air-conditioning duty, because electrical devices containing a coil and iron, such as solenoid valves and relays, have a power factor of approximately 50%—thus, for secondary circuits with such controls the capacity of a transformer must be equal to or greater than twice the total nameplate wattages of the connected loads.

The proper transformer rating for the electrical control circuits will have been selected by the equipment manufacturer. However, if accessory equipment is added, the additional power draw must be considered and might possibly result in an increased rating. This situation is common, for example, when cooling is added to an existing furnace and where the original transformer is too small. Also, when replacing a defective transformer, make sure that the rating is equal to or greater than the original equipment.

A10-3
THERMOSTATS

With a source of power established, let's turn our attention to the most familiar controller—the room thermostat. As stated previously, the low-voltage thermostat is much more accurate than line-voltage stats, but a number of line voltage stats are used in direct electrical resistant heating—wall, baseboard, room air-conditioners, etc.—and in order to appreciate the differences, let's first examine the line-voltage type.

Early typical line-voltage stats, sometimes called *snap-action thermostats,* consisted of bimetallic temperature-sensing elements (Fig. A10-4) made of two or more metallic alloys welded together, each having different coefficients of expansion when exposed to heat. One metal will expand more rapidly than the other and thus cause the bimetallic element to change its curvature when it "feels" a change in temperature. The bimetallic element may be a straight form (cantilever), U-shaped, or spiral. Line voltage usually flows through the bimetallic element, which has a moving contact that closes against a fixed contact point. The distance (h) the moving contact travels is the differential range between off and on, and in a simple, low-cost line stat this distance can be 3° or more. If a knob is attached to the fixed contact screw, the temperature sensing range can be adjusted within a fairly narrow range. The function of the permanent magnet is to snap the contacts closed when the moving contact comes within range, to help reduce the arcing of contacts. This type of control can be used for heating or cooling, depending on whether the contact is made on a rise or a drop in room temperature. Dual-contact models can thus perform both functions.

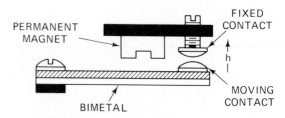

FIGURE A10-4   Snap-action thermostat.

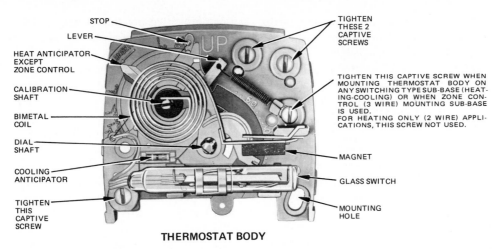

FIGURE A10-5   Low-voltage thermostat with cover removed. (*Courtesy* Robertshaw Controls Company)

Snap-action line-voltage thermostats are satisfactory for limited control situations where close temperature swings are not economically justified. Also, line-voltage stats can suffer from dirty contacts or pitting of the contacts due to arcing when making or breaking contact. The control of a snap-action stat is also a delayed reaction. This means that by the time the bimetallic element has traveled its range and the heating or cooling equipment is shut down, the actual room temperature will overshoot the desired space temperature. The anticipation method is a solution to the above (to be covered later), but it is harder to build into high-voltage controls than into low-voltage designs.

More recently, line-voltage thermostats for the direct control of electric heating make use of liquid-filled elements that respond to both ambient temperature and radiant heat. Also, the design includes slower-moving elements, which can handle direct resistance loads of up to 5000 W of 240 V ac. The differential between the set point and room conditions is reported to be less than that of the snap-action variety.

From an installation viewpoint, wall-mounted line-voltage thermostats must be served by heavy-duty wir-

ing. On commercial work, codes may require that they be run in conduit. Also, where many wires are involved, identification coding is difficult. In conclusion, therefore, line-voltage stats are almost totally limited to space heaters and the room and store type of self-contained air-conditioning products.

The low-voltage thermostat (Fig. A10-5) overcomes the limitations of line-voltage models and is almost universally used in modern central system control circuits. First, the use of spiral-shaped, lightweight bimetallic elements increases the effective length and thus the sensitivity to temperature change. Second, the use of sealed contacts virtually eliminates the problem of dirt and dust. Although there are minor variations among different manufacturers, the contacts are always sealed in a glass tube. Figure A10-6 (on the left) represents a

FIGURE A10-6   Single-action mercury bulb thermostat.

FIGURE A10-7   Thermostat with subbase. (*Courtesy* Robertshaw Controls Company)

Controls   483

single-action mercury bulb design. As the bimetallic strip expands and curves, the mercury fluid moves to the left, completing an electrical circuit between the two electrodes, which carry only 24 V. The differential gap between OFF and ON is very small—$\frac{3}{4}$ to 1°F from the set point. On the right is a sealed tube that has a metal-to-metal contact. The magnet provides the force that closes the contacts.

Combination heating and cooling thermostats may have two sets of contacts—two single-pole, single-throw mercury tubes, or one single-pole, double-throw mercury tube—to provide control of both the heating and cooling systems.

Room thermostats are designed to control room temperatures over a fairly wide range—usually about 50 to 90°F; normal settings are between 68 and 80°F. A dial or arm permits the homeowner to select the desired conditions.

The subbase portion of the thermostat assembly (Fig. A10-7) not only provides a mounting base for the thermostat, but is also used to control the system operation through a series of electrical switches. The system (SYS) switch selects COOL, OFF, or HEAT. The blower operation is controlled by the fan switch, which

is a simple two-position switch. When set in automatic, the fan will cycle with the furnace on heating. If in the ON position, the fan will run continuously.

Somewhat more sophisticated thermostats contain two dials that establish different control points for heating and cooling, with an automatic changeover from heating to cooling. Figure A10-8 represents a two-stage heating and two-stage cooling schematic wiring diagram. Two-stage thermostats are common in heat pumps or rooftop equipment, which use multiple compressors for cooling and two or more stages of heating. Note that seven electrical connections are required; however, M and V terminals are the same power source, so six wires are needed. With color-coded low-voltage wire, however, it is no problem to connect the thermostat to the mechanical equipment.

A10-4
HEAT ANTICIPATION

The sensitivity of room thermostats is affected by both systems lag and operating differential. System lag is the amount of time required for the heating or cooling

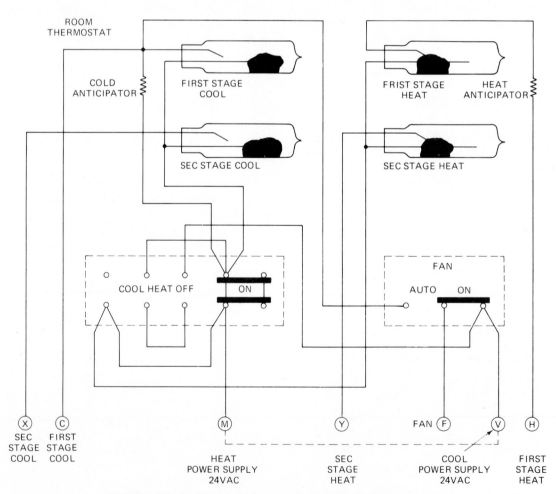

FIGURE A10-8 Two-stage heating and cooling wiring diagram.

system to produce a temperature change that is felt at the thermostat. The operating differential of a thermostat is the change in room air temperature needed to open or close the thermostat contacts.

Heat anticipators are used in low-voltage thermostats to reduce the effects of the system-lag operating differential. They are nothing more than small electrical resistors that generate a small amount of heat when current flows through. If the thermostatic bimetallic strip is exposed to a small amount of "artificial" heat, the sensitivity of the thermostat will increase, because the bimetallic element reacts to changes in surrounding air temperature and will lead to actual changes in room temperature. For example, if the room air temperature increases a certain number of degrees in 5 minutes, the anticipated bimetallic element will feel this change in only 3 minutes, while the unanticipated bimetallic strip will not feel the change of air temperature until 6 minutes has passed. Thus, the anticipated bimetallic strip will shut the furnace burner off before the actual room temperature reaches the thermostat set point. Anticipation results in closer room-temperature control and less overshooting of heating due to residual heat in the furnace heat exchanger that must be dissipated.

Some heating anticipators are fixed, while others are wire-wound variable resistors wired in series with the load. They are rated in amperes or fractions thereof and must be matched to the total load controlled by the thermostat. On the adjustable type, as illustrated in Fig. A10-9, the installer will position the sliding arm to the proper load rating. If on/off cycles are too long or too short, the system operation can be changed to give a faster or slower response. Fixed anticipators obviously cannot do this, and are thus usually used on economy-type heating applications.

A10-5
COOLING ANTICIPATION

A similar technique is used for cold anticipation, except that the cold anticipator is energized only when the cooling is off. The bimetallic element senses the anticipator heat and turns the system on again before the room temperature gets too warm; thus, it leads room air temperature.

Cooling anticipators are fixed resistors wired in parallel with the thermostat contacts; they do not have the capacity to energize the relays, solenoid valves, or contactors normally found in cooling products. These anticipators are factory selected and should not be changed.

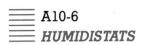

A10-6
HUMIDISTATS

Another popular controller found in residential total comfort applications is the wall-mounted humidistat (Fig. A10-10). It is very similar to the low-voltage thermostat, and it contains both a sensing element and a low-voltage electric switch. The sensing element consists of either an exceptionally thin moisture-sensitive nylon

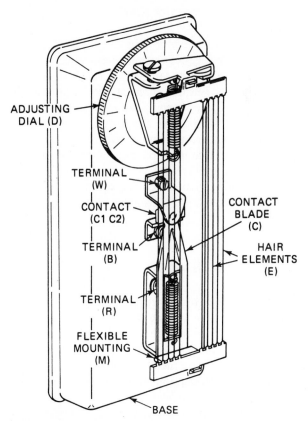

FIGURE A10-10 Construction details of a three-wire wall-mounted humidistat.

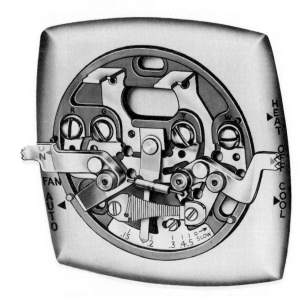

FIGURE A10-9 Low-voltage stat showing heat anticipator. (*Courtesy* Johnson Controls, Inc., Penn Division)

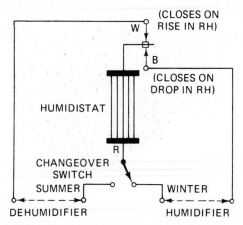

FIGURE A10-11 Schematic diagram of wiring arrangement for year-round control of humidity.

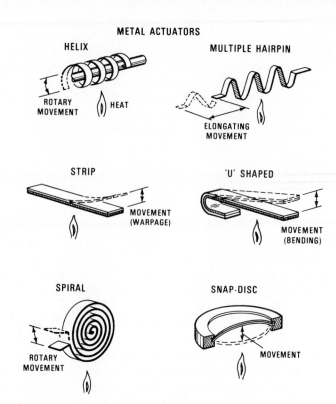

FIGURE A10-12 Bimetallic metal actuators.

ribbon or strands of human hair that react to changes in humidity. The movement of the sensing element is sufficient to make and break electrical contacts directly or when coupled with a mercury-type switch. Humidistats for mounting on ductwork are also available.

Figure A10-11 is a simple schematic wiring arrangement for year-round control of humidity. In winter the elements contract and close the contact to the humidifier. In summer the humidistat will expand in response to a rise in the relative humidity and close the contacts to the dehumidifier (cooling unit) circuit. A summer/winter changeover switch selects the appropriate operation.

The humidistat is not quite as precise a controller as a standard low-voltage room thermostat. Normally, a change of 5% relative humidity is required to actuate switching action. But this is acceptable for normal comfort applications, since people cannot detect, nor do they react to, changes in relative humidity with the same sensitivity as room temperature. The reader should be aware, however, that more precise controls are available for specialized applications, such as computer rooms, libraries, printing plants, etc., where controllers such as wet-bulb and dew-point thermostats would be used.

A10-7
SENSING DEVICES

Before proceeding with other controls, let us examine other types of sensing elements that react to temperature and/or pressure. Bimetallic temperature controls are available in various forms, as illustrated in Fig. A10-12. There are metal actuators that produce rotary movement, elongation, warping, bending, or snap action when exposed to heat. The mechanical movement is relatively large in response to small changes in temperature.

Bellows and diaphragms (Fig. A10-13) react to

changes in pressure. Because of a relatively large internal volume, the bellows will produce greater mechanical movement, but the diaphragm is more accurate; the choice will depend on the amount of movement or sensitivity and the working pressures that must be contained.

The bellows and/or diaphragm can also be adapted to control temperature by the addition of a remote-bulb sensing element (Fig. A10-14) filled with a vapor, a va-

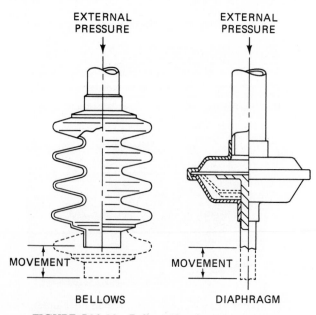

FIGURE A10-13 Bellows/diaphragm actuators.

486 Air-Conditioning

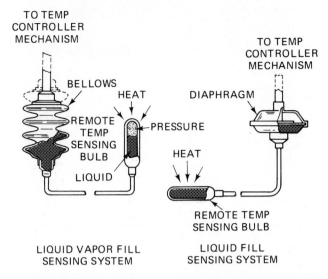

FIGURE A10-14 Bellows/diaphragm with sensing bulb.

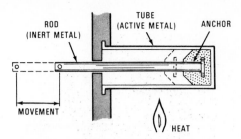

FIGURE A10-16 Rod-and-tube sensors are used in some controllers.

por/liquid, or a solid/liquid. Heat will cause the fill to expand into the bellows or diaphragm cavity, thus creating mechanical movement.

The Bourdon tube (Fig. A10-15) is another sensing element that reacts to changes in pressure. It is an elliptical tube that is sealed and linked to a transducer. As pressure is applied to the pressure connection, the tube will tend to straighten and thus create a mechanical movement. Pressure can come from refrigerant, water, oil, steam, or any fluid that can be pressurized. The Bourdon tube may also be adapted to temperature sensing with a closed bulb filled with liquid or vapor. Bourdon tubes are used most extensively in gauges.

The rod-and-tube sensing element is constructed by placing an inert rod in an active tube (Fig. A10-16). Since the rod is anchored to one end of the tube, a tem-

perature change will cause a pushing or pulling movement that can be made to operate a switch.

Resistance elements are sensing elements designed to vary their electrical resistance when exposed to changes in temperature or light intensity. Thermistors, photoresistance cells, and photocells are typical. A thermistor is an electrical device that will increase its resistance to the flow of current as its temperature decreases (Fig. A10-17). Changes in current flow can be used to activate or deactivate a remote control for various uses.

The cadmium sulfide cell (CdS cells, Fig. A10-18), or photoresistance cell, is a resistance element that reacts to changes in light intensity and is highly resistive to the passage of electrical current when its sensing base is in complete darkness. When exposed to light, the electrical resistance declines proportionally to the intensity of the light.

Some of the more common controls that use the principle of the sensing elements just explained are discussed in the section on residential and light commercial equipment.

In the area of safety the high-limit thermal cutout control is widely used (Fig. A10-19). It employs a snap-disk bimetallic strip to open an electrical circuit. These controls are common in electrical space heaters, electric duct heaters, electric furnaces, heat pumps, and any ap-

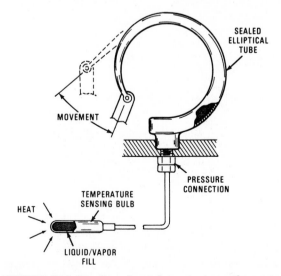

FIGURE A10-15 Bourdon tube actuator with sensing bulb.

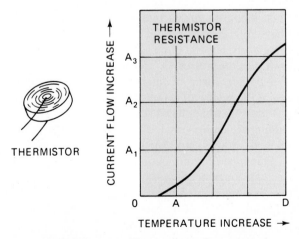

FIGURE A10-17 Thermistor and resistance/temperature curve.

Controls 487

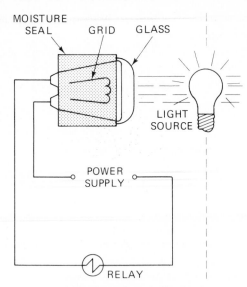

FIGURE A10-18 CdS cell.

FIGURE A10-19 High-limit thermal cutout control.

plication where the small, quarter-sized control can be mounted and exposed to heat sources.

The inherent protector used in hermetic compressors is an excellent example of bimetallic action. The overload (Fig. A10-20) is a current- and heat-sensing device and consists of a bimetallic contact in series with the motor contactor coil control circuit. Whenever the compressor motor is overheated, overloaded, or stalled, the heavy continuous current through the heater warps the bimetallic contact until it opens the motor contactor low-voltage relay coil circuit and stops the flow of current to the compressor.

Another form of high-limit cutout, previously noted in the review of baseboard electric heaters, is the lineal control, which has a sealed vapor/liquid fill tube extending the entire length of the heater. If the heater is blocked by drapes, etc., the safety limit will open.

One type of combination of fan and limit control for a typical gas or oil warm-air furnace (Fig. A10-21) uses the power created by the rotary movement of a helix bimetal. The rotating cam makes or breaks the separate fan and limit electrical contacts.

The combination fan and limit switch illustrated in Fig. A10-22 has a dual function. As a safety limit,

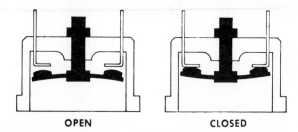

FIGURE A10-20 External type (mounts on housing) snap disk line break overload. Some designs fit inside motor winding.

some have fixed limit temperature settings; others are adjustable (approximately 180 to 200°F is the usual range). This allows a 50 to 60°F rise above normal operation before it opens. The fan control switch is also a temperature-sensing device that is set to turn on the fan after the furnace has warmed up at least 10 to 15°F above room conditions, so that cold drafts are not experienced. It also stops the blower after the burner cuts off, so again there are no uncomfortable drafts. *Note:* We must also recognize that some systems employ constant fan operation and thus override this switch. CAC (constant air circulation) is discussed in later chapters.

Other models may use spiral or flat bimetallic or even liquid-filled elements. Some forms of duct-mounted

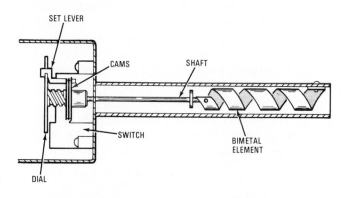

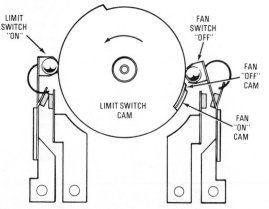

FIGURE A10-21 Details of one combination fan-limit control show how twisting motion of heated helix turns cam to activate fan and limit switches. Arrow on cam indicates counterclockwise rotation on temperature rise.

FIGURE A10-22   Fan/limit control.
(*Courtesy* Honeywell, Inc.)

FIGURE A10-23   Remote bulb temperature controller. (*Courtesy* Honeywell, Inc.)

limit controls use a rod and tube or a liquid-filled bulb to sense the air conditions.

A common control in hydronic heating is the immersion controller, which uses a liquid-filled bulb that operates a snap-acting switch (Fig. A10-23). Immersion controls are inserted directly into the boiler to detect water temperature and can be used as high-limit cutouts and as low-limit and circulator pump controls.

A prime example of the thermistor sensing device is the high-temperature cutout on a hermetic compressor (Fig. A10-24). The thermistor, not much larger than an aspirin tablet, is embedded deep in the motor windings. It senses overheating and will signal a transducer to shut down the compressor until normal temperatures are restored.

The use of the cadmium cell principle is reviewed under oil burner operation. The cadmium cell reacts to the presence or absence of light from the flame and will stop the oil burner if ignition fails before excessive unignited oil can collect in the combustion chamber.

High-pressure and low-pressure controls for compressor and system operation (Fig. A10-25) utilize a diaphragm to sense the refrigeration pressure and react

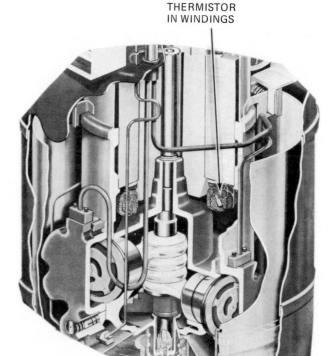

FIGURE A10-24   Motor winding temperature sensor.

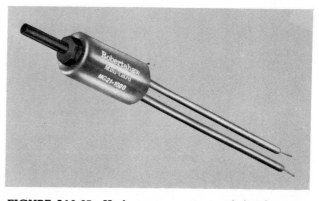

FIGURE A10-25   High-pressure cutout with fixed setting.

FIGURE A10-26   Pressure differential switch.

accordingly through a switch mechanism to shut down the operation if abnormal conditions exist. High-pressure cutout controls frequently must be manually reset in order for the system to operate. This prevents rapid short cycling, which can eventually cause compressor overheating and possible damage.

Pressure controls can also be used on large compressors to detect the buildup of oil pressure. They are equipped with a time-delay mechanism (usually 1 minute or less), and if normal oil pressure is not achieved within the time period, the controller will stop the compressor before damage can occur. It can also be connected to an alarm indicator.

The diaphragm type of air pressure switch (Fig. A10-26) is used in heat pumps to start the defrost cycle. In that case they are used for measuring the differential pressure across the outdoor coil. These devices can also be used to measure positive pressure, as in a central system air duct, and energize an alarm circuit when a change in air pressure is sensed.

Static pressure regulators (diaphragm device) are also air controllers that are used in larger systems to measure static pressure and then, through an electric circuit to a damper motor, to regulate the air damper to maintain a constant static pressure in the system, as illustrated in Fig. A10-27. A static pressure regulator can also be connected to a control that varies the central fan inlet air vane to maintain constant fan performance.

A different form of sensing element not previously described but frequently used in residential and light commercial work is the sail switch (Fig. A10-28). As the name implies, it detects the movement of air by the use

FIGURE A10-28   Sail switch. (*Courtesy* Honeywell, Inc.)

of a sail or paddle attached to an arm; the arm, in turn, is attached to an electrical switch. The illustration is of a combination duct-mounted humidistat and sail switch. The humidistat mounts in the return air duct of a forced-air system. The function of the sail switch is to shut down the humidifier when the supply air is cut off. Sail switches are also used as safety devices to detect airflow or the absence thereof and cut off heating and/or cooling before abnormal system operation occurs.

Most of the controllers reviewed above are of the two-position variety—that is, they are OFF or ON, OPEN or CLOSED—however you wish to describe the action. But even within the scope of residential and light commercial work, there is a need for modulating controls to vary an action somewhere in between fully open or closed or OFF and ON. The proportional controller

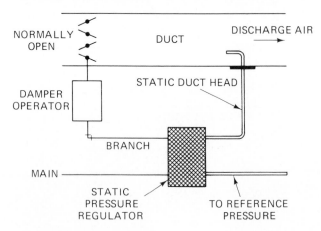

FIGURE A10-27   Pneumatic air damper control.

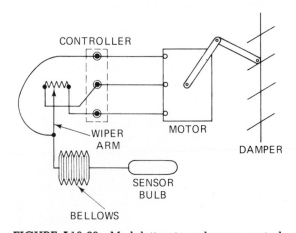

FIGURE A10-29   Modulating-type damper control.

does just that. Proportional or modulating action is achieved by using a potentiometer to vary the position of the controlled device in proportion to any temperature (it could also be pressure or flow) variation felt by the sensing element. A potentiometer (Fig. A10-29) is nothing more than an adjustable three-wire variable resistor in conjunction with a wiper arm connected to the sensor. The wiper arm is the third wire contact. As the arm moves through the complete stroke (throttling range) the current flow changes in proportion to the resistance. This current flow is sent to a motor relay that actuates a damper motor, motorized valve, etc., to bring about the required temperature change in the controlled space.

A10-8
OPERATORS

Operators, both intermediate and final, are on the receiving end of the controller signal. An intermediate operator is one that controls another electrical circuit. A final operator will control or act on a product or final action.

The electric relay (Fig. A10-30) is the best example of an intermediate operator. Its function is to take a low-voltage signal from a controller and, through a magnetic coil, close a set of higher voltage contacts, which in turn will start a fan or compressor or perhaps only supply high voltage to another control circuit. The coil which sets up the magnetic field is installed around a pole piece. The armature moves in response to coil energization to open or close the electrical contact. Return springs are used to pull the armature to the deenergized position.

Relays are used to control electrical loads that draw less than 20 A, such as small motors, valves, ignition transformers, and small compressors. A contactor is a heavy-duty relay designed to handle over 20 A, and in most cases is used on compressor and pump motors. The

volt-ampere draw of the commonly used relay and contactor coils is low; therefore, low-voltage transformers can supply enough power to operate these devices. (*Note:* A starter is a form of contactor or relay that contains overload protection.)

To complete the relay story, it is important to mention briefly a second type of electric relay that has come into use comparatively recently—the so-called thermal relay, which uses a resistance heater and a bimetallic element instead of a magnetic coil and switch. When the low-voltage circuit is energized, perhaps by thermostatic action, the low-voltage heater warms the bimetallic strip, which warps and closes a switch on the high-voltage side. The chief advantage of this relay is its quiet action, and it is therefore well suited to low-capacity residential types of switching needs.

Sealed plug-in relays have been used extensively in commercial work and are becoming popular in residential use to simplify the addition of humidity control and electronic air cleaning. They look much like a square radio tube and plug into a control panel arrangement. Since the electrical contacts are not exposed to dust and dirt, reliability is greatly improved. Also, the plug-in arrangement permits quick replacement in the event of trouble.

Another form of operator is a motor (Fig. A10-31) that responds to a command of a controller and mechanically operates a damper, valve, step controller, etc., that produce a final action. The crank arm is connected to the final apparatus by means by linkage rods. Motors are available for two-position response and may swing through a 160 to 180° arc from start to stop. Others respond to a modulating controller and operate in an

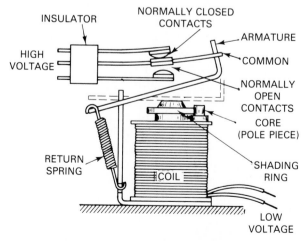

FIGURE A10-30   Relay contacts.

FIGURE A10-31   Damper motor. (*Courtesy* Honeywell, Inc.)

intermediate position to bring the controlled temperature, pressure, etc., back to the set point. Motors vary in arc rotation from 15 seconds to 4 or 5 minutes depending on the application. They are also rated according to the torque force the arm can exert (in inch-pounds or the size in square feet of the damper it can handle).

Some models have a spring return to some normal setting in the event of power failure. Auxiliary switches can be included to perform a variety of other automatic switching combinations for secondary equipment.

The final operators in the example given in the beginning of the chapter are the air-conditioning condensing unit contactor; the blower switching relay, and the fuel or energy control.

The subject of relays and contactors was discussed in Chapter R18. It is suggested that Chapter R18 be reviewed for background for this chapter and succeeding chapters.

A10-9
FUEL OR ENERGY CONTROLS

The fuel controls (gas or oil) and the energy controls (electrical sequences) are classified as final operators. Each is designed to control a specific energy source so as to provide safe as well as the desired results as demanded by the master control—the thermostat.

A10-10
GAS CONTROLS

At one time, the supplying of fuel gas to a heating unit was done by a combination of controls consisting of a gas pressure regulator and a solenoid valve. To meet the exacting demands of the American Gas Association for proper ignition, input control, and quiet cutoff of gas unit operation, these controls were combined into a single valve assembly (Fig. A10-32).

Valves from various manufacturers may differ, but they all accomplish essentially the same functions:

1. Manual control for ignition and normal operation
2. Pilot supply, adjustment, and safety shutoff
3. Pressure regulation of burner gas feed
4. On/off electric solenoid valve controlled by the room thermostat

Figure A10-33 shows a basic valve sketch of these functions for a natural gas valve. The schematic drawing of the valve shows the main diaphragm valve in the open condition that occurs during heat demand. It is assumed that the main diaphragm open schematic applies to a gas heating appliance with the pilot flame

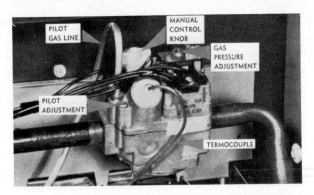

FIGURE A10-32 Combination gas valve/manifold assembly.

burning, that a thermocouple is connected to the automatic pilot magnet operator, and that the lighting operation was previously performed to open the automatic pilot valve.

It is also assumed that the lighting operation was previously performed to open the automatic pilot valve, and that the main gas cock has been turned to the ON position after the pilot is lit. In the application, the 24-V transformer circuit includes the 24-V operator and room thermostat in series. Closure of the room thermostat switch on heat demand has energized the 24-V operator, which causes the armature to be attached to the pole face of the magnet, resulting in a clockwise rotation of the armature, as indicated by the arrows at the end of the armature. This rotation has overcome the valve spring and has pulled the valve stem of the dual operator valve downward, allowing the diaphragm above it to seat on the valve seat encircling the valve stem. The seating of this diaphragm shuts off the bypass porting. Bleed gas can enter the actuator cavity only through the bleed orifice. Bleed gas is allowed to flow from the actuator cavity through the dual-valve operator and the regulator to the outlet through the outlet pressure-sensing port. (The valve stem has a square cross section in a circular guide, allowing gas to pass between the stem and the guide.) The resultant drop in pressure within the actuator cavity and through the main diaphragm allows the inlet pressure above the main diaphragm to open the main diaphragm valve against the force of the main diaphragm valve spring.

After opening, as shown in Fig. A10-33 (main diaphragm valve open), straight-line pressure regulation is secured by the feedback of outlet pressure through the outlet pressure-sensing port to the pressure regulator in the bleed line. A rise in pressure at the control outlet above the set pressure causes a proportional closure of the regulator valve in the bleed line. The proportional closure of the regulator valve causes a proportional rise in pressure in the bleed line ahead of the regulator. This rise in pressure in the actuator cavity acts through the main valve operator port to increase the pressure beneath the main diaphragm and causes a partial closure

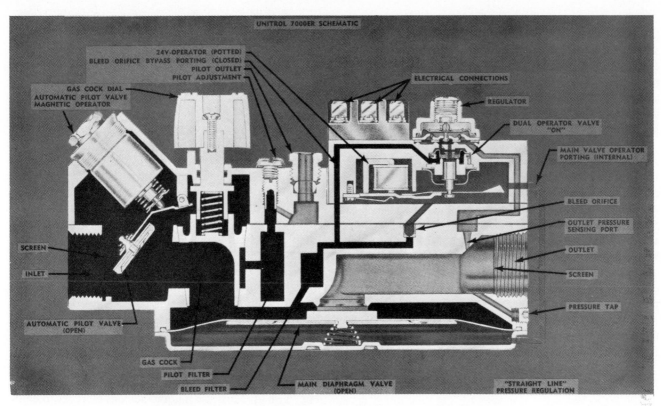

FIGURE A10-33 Combination gas valve cutaway—closed position.

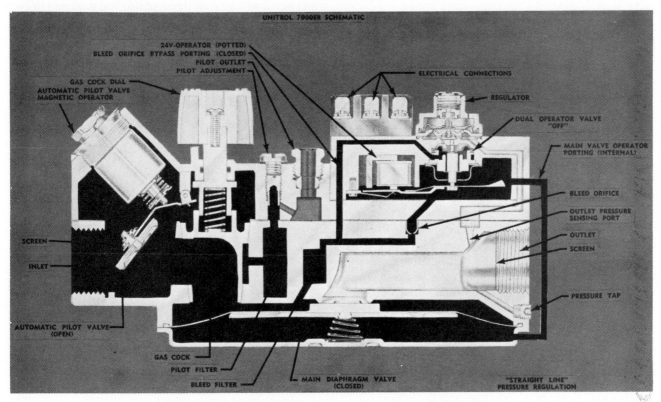

FIGURE A10-34 Combination gas valve cutaway—open position.

of the main diaphragm valve sufficient to lower the control outlet pressure down to the pressure setting. Upon a drop in control outlet pressure, a corresponding decrease in bleed-line pressure transmitted to the underside of the main diaphragm through the action of the bleed-line pressure regulator causes a proportional increase in the main valve opening to bring the delivered outlet pressure back up to the set pressure.

The schematic of the closed main diaphragm (Fig. A10-34) shows the action of the bypass valve in the dual-valve operator during a fast OFF response independent of the bleed orifice. Upon heating "satisfaction," resulting in the opening of the thermostat switch, the 24-V operator is deenergized. A loss of magnetic attraction allows the return spring on the valve stem of the dual valve operator to force the stem upward, closing the center port of the small diaphragm above it and shutting off bleed gas to the bleed regulator and outlet sensing port. The resultant counterclockwise rotation of the armature is indicated by the arrows at the ends of the armature. As the small diaphragm above the valve stem is forced upward, the bleed orifice bypass porting is opened by the diaphragm, leaving the valve seat encircling the valve stem. The actuator cavity, main valve operator port, and the cavity beneath the diaphragm are rapidly exposed to full inlet pressure, which acts to close the main diaphragm valve. The bleed orifice bypass port allows the pressure above and below the mean diaphragm to be equalized rapidly, independently of the bleed orifice. Then this pressure is equalized and the main valve is closed by the main valve spring.

The dual operator valve that provides the bleed orifice bypass porting control for the fast OFF response is an important feature. It helps to prevent flash-back conditions. Flash back can cause pilot outage problems as well as burner and appliance sooting. Relatively slow closing has been a problem on larger-capacity applications. The dual operator valve is an important factor in helping to eliminate this application problem.

The valve has built-in protection against gas-line contaminants through the use of highly effective inlet and outlet screens for the main gas passages, a pilot filter for the pilot line, and a bleed filter for the bleed control line. These means of self-protection are highly effective and protect the control from malfunction due to such contaminants entering the control passages. In the highly unlikely case that the bleed orifice restrictor will become clogged with the main valve in the open or ON position in spite of the bleed filter protection, the dual operator valve will enable the control to operate to shut off the main gas valve when signaled to do so by the room thermostat or limit control.

On an LP furnace, the gas valves do not have the pressure regulation function, but there is still 100% pilot shutoff in the event of pilot outage, which is most important. LP gas is heavier than air, and over a prolonged period enough pilot gas would collect to be a hazard.

The description above is of a method of pilot ignition and gas valve control that is still the most common in residential equipment use. However, the industry has developed and applied electronic ignition devices in rooftop commercial equipment, and we are just beginning to see this technique being offered in residential furnaces.

Figure A10-35 is a picture of an electronic ignition system. Pilot gas is fed to the assembly and burner orifice. The spark electrode is so positioned as to ignite the gas on a signal from the room thermostat. With the pilot ignited and burning, a sensing probe establishes an electrical current sufficient to energize a relay that opens the main gas valve by closing normally open electrical contacts. At the same time the relay's normally closed electrical contacts in the spark ignition circuit open, terminating the spark. As long as the sensing probe recognizes the pilot flame, the relay coil will be energized and normally closed contacts in the spark ignition system will remain open. There are several manufacturers of this type of system using a sensing device to create the changeover from the ignition phase to the operating (gas valve open) phase. This is a single-pole, double-throw switch using a filled actuator such as shown in Fig. A10-14. Most use a liquid such as mercury to cause pressure rise in the switch with a rise in the temperature at the sensing bulb located in the pilot assembly. Figure A10-36 shows the components used to accomplish this marketed by White Rodgers.

Another type of direct spark ignition system uses an electricity-carrying flame sensor rod mounted in such

FIGURE A10-35 Electronic ignition system. (*Courtesy* Bard Manufacturing Company)

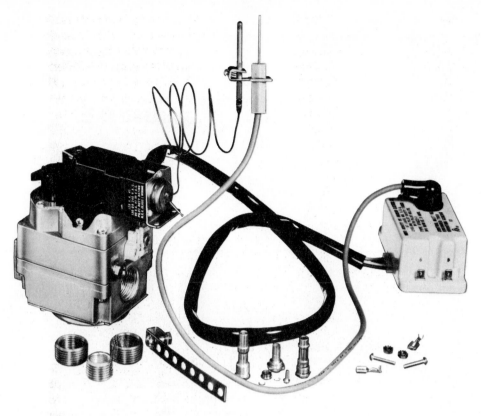

FIGURE A10-36 Electric spark ignition package. (*Courtesy* White-Rodgers, Division of Emerson Electric Company)

a position as to have direct contact by the main burner flame. Because the gas flame will carry electrical energy by means of electrically charging the carbon atoms in the gas before combustion takes place, a current flow can be passed from the burner to a positive-charge flame rod. This current flow controls a circuit in the solid-state control module and keeps the gas valve energized. Figure A10-37 shows the components used to accomplish this. Upon a call from the room thermostat, both the main gas valve and the spark ignitor are activated. Allowing a predetermined time for main flame ignition, the ignition control module will shut down the lockout circuit and maintain burner operation if main flame ignition occurs in that period. The period may be from 4 to 21 seconds, depending on the model of control module used. Usually, the higher the input to the gas unit, the shorter the proving time.

If the main burner flame is not established in the set time, the control module automatically locks out. To reset the circuit, electrical power to the system must be cut off and then back on to start another cycle. Manual reset of such a system is used for maximum safety to the equipment and area sensed.

Because this system uses the main burner assembly as the ground terminal of the spark system, *it is absolutely necessary that the gas burning unit be thoroughly grounded to the electrical supply ground.* It is wise and usually necessary to run a ground (green) wire from the power distribution panel to the unit to provide this positive ground. Because the white or neutral wire is a 120/240-V supply system, it is a current-carrying wire. It is not suitable for use as a unit ground.

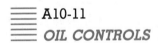

A10-11
OIL CONTROLS

If the heating and air-conditioning control circuit used an oil-fired heating unit instead of a gas-fired unit, a change in the thermostat connection arrangement would be required. The dotted line jumper between terminals *M* and *V* (Fig. A10-8) in the thermostat control setup would be removed and the thermostat would have two separate control circuits: one for cooling between *v* (cooling power) and *c* (first-stage cooling) and *x* (second-stage cooling), the second control circuit between *M* (heat power) and *H* (first-stage heat) and *X* (second-stage heat).

This separation of the two control systems is necessary because the oil burner control has its own power source. Therefore, the heating control system must be isolated from the air-conditioning control system with its own power source. *Never connect two separate power sources together in the same control system. Overloading and burnout will occur.*

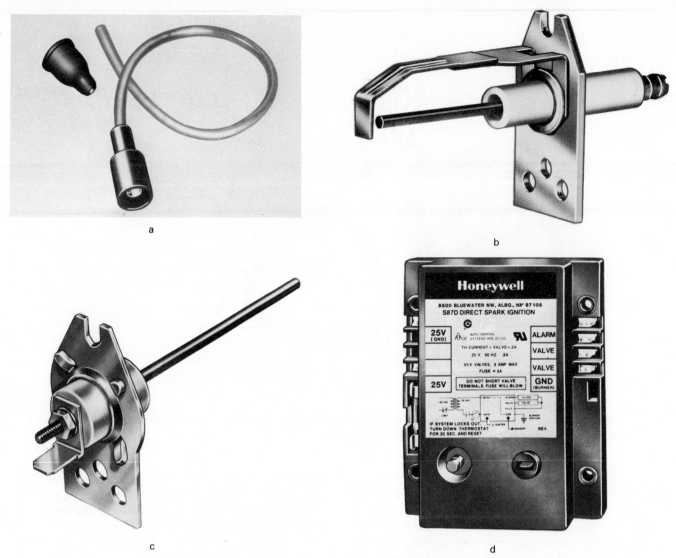

a

b

c

d

FIGURE A10-37   Direct spark ignition kit. (*Courtesy* Honeywell, Inc.)

A10-12
PRIMARY OIL BURNER CONTROLS

The original primary oil burner control, known as a *protecto relay,* was mounted in the smoke pipe. The bimetallic element in the sensing tube would react to a rise in flue product temperature and cause the relay to keep the oil burner operating. Figure A10-38 shows an example of this control with the cover in place and the sensing (stack) element protruding from the rear.

Figure A10-39 shows the same control with front cover removed. In this illustration we see that the control has its own power source transformer as well as temperature-actuation contacts and manual reset safety switch. The operation of the control was by means of a slide clutch operating safety and holding contacts. The drive shaft at the top of the control provided the action

for the slide clutch when moved by the expansion and contraction of the bimetallic element. Two relays are included, one to control the oil burner ignition transformer (relay 1A) and the other to control the oil burner motor (relay 2A). Thermostat low-voltage terminals on the lower left and high voltage (120 V) on the lower right are separated by an insulated barrier. Figure A10-40 shows the internal wiring arrangement of the control as well as the burner motor and ignition transformer circuits, 120-V power supply, and 24-V thermostat circuit.

Following through the control circuit, the action of the cycle would be as follows: The thermostat, calling for heat, closes the circuit from the transformer through the safety switch, through the thermostat, through relay coil 1K, through the right-hand cold contact, the left-hand cold contact, and the safety switch heater to the other side of the transformer.

FIGURE A10-38 Stack-mounted oil burner primary control. (*Courtesy* Honeywell, Inc.)

Relay 1K pulls in closing contact 1K1, powering the ignition transformer; 1K2 energizes relay coil 2K and closing contact 1K3 to the center or common of the cold contacts.

When relay coil 2K is energized, it pulls in, closing contact 2K1, energizing the oil burner motor and oil valve (if used). Contact 2K2 removes the relays from the series circuit through the safety switch heater, reducing the amount of heat produced in the heater. If this action did not occur, the safety switch will open in a very short period of time, possibly 3 to 5 seconds. By removing the

FIGURE A10-39 Stack-mounted oil burner primary control—cover removed. (*Courtesy* Honeywell, Inc.)

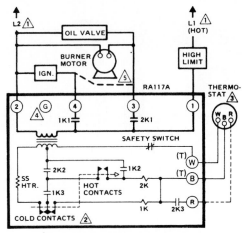

1. POWER SUPPLY. PROVIDE DISCONNECT MEANS AND OVER-LOAD PROTECTION AS REQUIRED.

2. CONTACTS BREAK IN SEQUENCE ON RISE IN TEMP.

3. MAY BE CONTROLLED BY TWO-WIRE THERMOSTAT. CONNECT TO W AND B ONLY. TAPE LOOSE END OF RED WIRE, IF ANY.

4. CONTROL CASE MUST BE CONNECTED TO EARTH GROUND. USE GREEN GROUNDING SCREW PROVIDED.

5. TO REPLACE INTERMITTENT (FORMERLY CALLED CONSTANT) IGNITION DEVICE, WIRE IGNITION LEADWIRE TO TERMINAL 3 ON RA117A.

1112D

FIGURE A10-40 Wiring diagram of oil burner control system. (*Courtesy* Honeywell, Inc.)

relay current from the safety switch circuit, the switch delay time is increased to 60 to 90 seconds.

The start of the oil burner produces hot flue products from the heat exchanger over the bimetallic helix in the sensing tube. This rise in temperature forces the drive shaft forward, opening the left-hand cold contact and removing the safety switch heater from the circuit. A further rise in temperature (forward action of the drive shaft) closes the hot contact; this allows the oil burner motor and oil valve (if used) to continue to operate by passing relay contact 1K2.

Finally, the drive shaft moves forward enough to open the right-hand cold contact, dropping out relay 1K. This stops the ignition and keeps the oil burner operating through the hot contacts. If a flame-out occurs or the burner shuts down from the action of the thermostat, the hot contact opens immediately upon a drop in flue gas temperature. Because contact 1K2 is open, the burner motor cannot operate until the cold contacts close, first the left side to energize the safety switch and then the right side to start the ignition/oil burner motor cycle. By means of these lockout circuits, it is not possible to have a situation of spraying oil vapor into a white hot refractory, resulting in an explosion.

The problem with the stock-mounted protecto relay was extra wiring as well as time, cost, etc., for manufacturing assembly or dependence on the installer to install and wire the device. To overcome this, a flame detection type of relay was developed. Figure A10-41 shows an oil burner protecto relay using a cadmium cell flame detector instead of the bimetallic helix. It is

a

b

FIGURE A10-41   Photo-conductive flame-sensing oil burner relay and cadmium cell. (*Courtesy* Honeywell, Inc.)

FIGURE A10-42   Photo-conductive flame-sensing relay—cover removed. (*Courtesy* Honeywell, Inc.)

mounted directly in the oil burner blast tube directly behind the turbulator plate. The cell sees the light of the burner the instant the flame is established.

Figure A10-42 shows the interior arrangement of the relay using the light-sensitive cadmium cell. Two relays are used: 1K for control of the oil burner and ignition transformer, and 2K, the sensitive relay, controlled by the cadmium cell. Following the circuit action in the schematic in Fig. A10-43, when the thermostat calls for heat and closes the contact between T and T, current flows from the transformer through the thermostat through relay coil 1K, safety switch, timer contact 1, the safety switch heater, and contact 2K1 to the

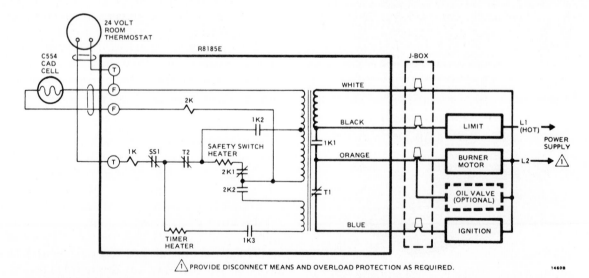

△ PROVIDE DISCONNECT MEANS AND OVERLOAD PROTECTION AS REQUIRED.

14460B

FIGURE A10-43   Typical hook-up and schematic for photo-conductive flame-sensing relay. (*Courtesy* Honeywell, Inc.)

other side of the transformer. This puts a higher voltage through the coil and safety switch heater. If the relay 1K fails to act, the safety switch will open in only a few seconds.

When relay 1K pulls in, however, contact 1K1 pulls in, energizing the oil burner motor, oil valve (if used), and the ignition transformer. Also, contact 1K2 closes, bypassing the safety switch heater, reducing its current draw, and increasing the safety switch action time. If no flame is established, the heater will continue to receive voltage and heat until the safety switch contact (SS1) opens and shuts down the system. The switch is manual reset.

When flame is established, the light strikes the cadmium cell and immediately the resistance of the cell is reduced. The amount of current through relay coil 2K increases and the sensitive relay pulls in. This opens contact 2K1, which deenergizes the safety switch heater and energizes the timer heater through contact 1K3, which closed when relay 1K pulled in. The timer heat now opens contact T1, shutting off the ignition transformer and the holding contact of the sensitive relay. This action will continue as long as burner operation is required.

If flame failure should occur, the cadmium cell resistance will increase, current flow through the sensitive relay will decrease, and the relay will drop out. This opens contact 2K2, which opens the circuit to relay 1K and the burner shuts down.

The burner cannot come on until the timer heater has cooled sufficiently to close the ignition contact T1 and the relay 1K circuit contact T2. This assures the unit sufficient time to vent the furnace heat exchanger of unburned vapors as well as assuring ignition at the next startup.

These are only two of the various types of controls manufactured and used on oil-fired equipment. The service person should collect *and retain* as much information as possible from all manufacturers and all types.

FIGURE A10-44 Delay oil valve.
(*Courtesy* Honeywell, Inc.)

A10-13
OIL VALVES

Allow the oil burner to arrive at operating speed and combustion air volume before oil is supplied and combustion is established. Also, at cutoff of the burner operation, the best method of operation is to have instant cutoff of fuel and combustion.

Figure A10-44 shows a delay-type oil valve that is installed between the oil pump outlet and the burner fixing assembly. The valve, although wired to be energized at the same time as the oil burner motor (wired in parallel), has a delayed opening. This allows full operation of the burner blower and pump before oil flow is allowed. Using a thermistor in the valve coil circuit, the thermistor limits the current to the coil on start. As the thermistor heats, the resistance drops and the current flow increases. After a few seconds, time usually 8 to 10 seconds, the current flow rises sufficiently to cause the coil magnetic pull to open the valve. The valve acts the same as any other type of solenoid valve on cutoff. It closes immediately upon deenergizing the valve and motor. Thus instant cutoff of combustion occurs.

PROBLEMS

A10-1. What are the three major elements in a control system?

A10-2. Give an example of a common controller.

A10-3. Give an example of a controlled load.

A10-4. What is the most common power source in air-conditioning control circuits?

A10-5. Define a power-type transformer.

A10-6. When the voltage out is lower than the voltage in, the transformer is classified as a _____ transformer.

A10-7. The change in voltage is determined by _____ .

A10-8. The change in amperage capacity is in a direct ratio with the change in voltage. True or False?

A10-9. The capacity of a transformer is rated in _____ , which is determined by _____ .

A10-10. What is the bimetallic principle?

A10-11. Why are low-voltage thermostats used instead of high-voltage types?

A10-12. Heating and cooling anticipators are actually small _____ .

A10-13. Is a heating anticipator connected in series with or parallel to the heating contact?

A10-14. Is a cooling anticipator connected in series with or parallel to the cooling contact?

A10-15. Are most heating thermostats fixed or adjustable?

A10-16. By what is the setting of an adjustable heat anticipator determined?

A10-17. What are other types of temperature control sensing elements?

A10-18. A thermistor functions by changing its _____ with a change in temperature.

A10-19. High-pressure controls are made in both _____ and _____ reset types.

A10-20. What are the most common forms of bimetallic configuration used in controls?

A10-21. What are the common forms of power elements used in pressure controls?

A10-22. The control action of controls are of two types. What are they?

A10-23. Define and give an example of an intermittent operator.

A10-24. Explain the difference between a relay, a contactor, and a motor starter.

A10-25. Name the five functions of a natural gas combination gas valve.

A10-26. What is the function of a combination gas valve that is not used with LP gas supply?

A10-27. It is possible to pass an electrical current through a gas flame. True or False?

A10-28. An oil burner is controlled by a device called a _____ .

A10-29. A stack-mounted oil burner control is operated by a _____ , which is actuated by _____ .

A10-30. A burner-mounted oil burner control is operated by a _____ , which is actuated by _____ .

A10-31. A delayed-action oil valve is used to ensure that _____ .

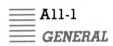

Typical Residential Control Systems

A11-1
GENERAL

With a basic understanding of the mechanical construction of residential heating and air-conditioning as presented in Chapters A5 through A9 and a review of some of the more common electrical controls as discussed in Chapter A10, let's now examine some typical heating and cooling circuits and become familiar with the function and operation of the component parts.

Different manufacturers will have variations in the physical wiring, but the function remains the same in all systems. With a firm understanding of the construction and use of schematic wiring diagrams, the student should be able to wire any manufacturer's product. The industry uses standardized electrical symbols, as shown in Fig. A11-1, to help the technician identify common controls. Additionally, the legend designations, as presented in Fig. A11-2, are related alphabetically to the devices they represent. For example *R* is a general relay, while *CR* is a cooling relay, *HR* is a heating relay, *DR* is a defrost relay, etc. Legends may vary with the particular wiring diagram or the manufacturer's methods of labeling.

Study the symbols and legend designations carefully, for they are all-important to the following discussion, and they will also serve later as a reference.

A11-2
SERVICE ENTRANCE

The starting point of the electrical system is the service entrance. Electricity is distributed through high-power lines by the power companies, and these are reduced down into local residential lines that run on every street. A transformer on the pole will serve several residences, reducing down the high voltage in the transmission lines to normal house entrance voltage. Entrance

service will be 120/240 V single phase, and modern houses will have anywhere from 100- to 200-A service to a circuit breaker or fuse board at the central distribution point. Three wires are brought in by the power company to the house, two of which are the hot wires and one of which is grounded to the board. The serviceperson has no control over the entering house voltage and, therefore, if this is not correct, the only recourse is to call the power company. *But always check the actual voltage value.*

At the house service entrance, power is supplied to various branch circuits to be distributed throughout the house. These branch circuits will be 120-V single-phase two-wire circuits for use in lighting, receptacles, appliances, and a separate circuit for the furnace. The furnace should be the only appliance on that circuit, and its normal fusing is 15 A.

Appliances drawing higher currents, such as electric ranges, clothes dryers, and air-conditioners, will have 240-V service, which is obtained by tapping off each hot side of the entrance circuit. The fuse or circuit-breaker size for these circuits will be in accordance with the current draw of the given appliance; in the case of an air conditioner this might be 30 to 40 A per leg. The circuit breakers for each leg on a 240-V circuit are mechanically interlocked, so that if high current is experienced in one leg, the breaker will trip both legs of the supply service.

A11-3
WIRE SIZE

Wire sizes and fusing for independent appliances will be determined by the manufacturers and specified on their wiring diagrams. Local codes should be examined to determine whether or not conduit is required in bringing the wire from the main circuit breaker to the furnace or air-conditioner. Some codes require conduit

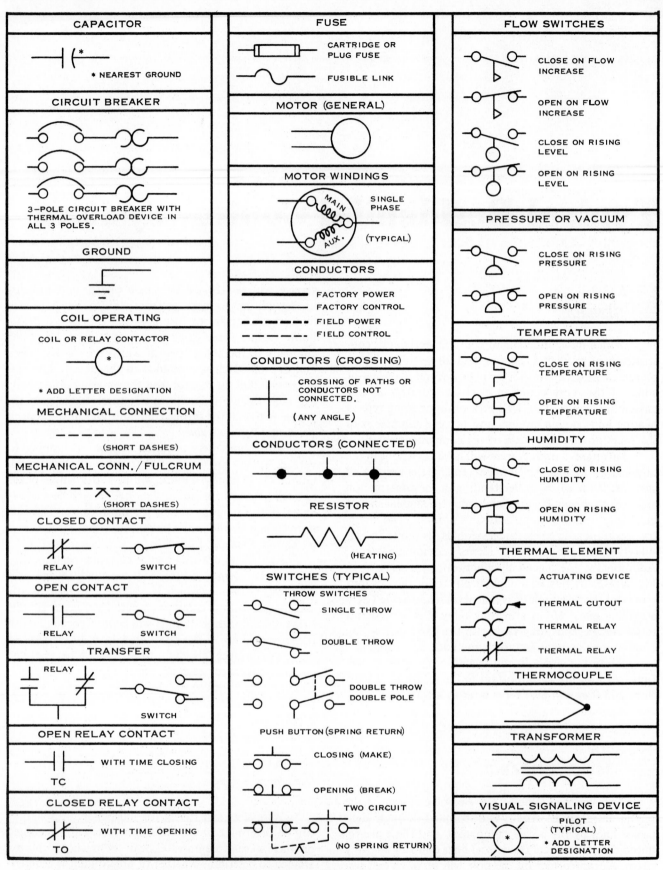

FIGURE A11-1 Electrical symbols. (*Courtesy* Borg-Warner Central Environmental Systems, Inc.)

	Relays			Switches			Miscellaneous
R	Relay General		DI	Defrost Initiation		C-HTR	Crankcase Heater
CR	Cooling Relay		DT	Defrost Termination		RES	Resistor
DR	Defrost Relay		DIT	Defrost Initiation-Defrost		HTR	Heater
FR	Fan Relay			Termination (dual function		PC	Program Control
IFR	Indoor Fan Relay			device)		OL	Overload
OFR	Outdoor Fan Relay		GP	Gas Pressure		L	Indicating Lamp
GR	Guardistor Relay		HP	High Pressure		⊕	Manual Reset Device
HR	Heating Relay		LP	Low Pressure		+	Automatic Reset
LR	Locking Relay (Lock-in or Lockout)		HLP	Combination High-Low Pressure			Device
PR	Protection Relay (Relay in series with protective devices)		OP	Oil Pressure			
VR	Voltage Relay		RM	Reset, Manual			
TD	Time Delay Device		FS	Fan Switch			
THR	Thermal Relay (type)		SS	System Switch			
M	Contactor		HS	Humidity Switch (Humidistat)			
MA	Auxiliary Contact		TA	Thermostat, Ambient			
			TC	Thermostat, Cooling			
	Solenoids		TH	Thermostat, Heating			
			TMA	Thermostat, Mixed Air			
S	Solenoid, General		CT	Thermostat, Compressor Motor			
CS	Capacity Solenoid		HT	High Temperature			
GS	Gas Solenoid		LT	Low Temperature			
RS	Reversing Solenoid		RT	Refrigerant Temperature			
			WT	Water Temperature			

Source: Borg-Warner Central Environmental Systems, Inc.

MAXIMUM LENGTH OF TWO-WIRE RUN*

WIRE SIZE	AMPERES												
	5	10	15	20	25	30	35	40	45	50	55	70	80
14	274	137	91										
12		218	145	109									
10			230	173	138	115							
8				220	182	156	138						
6							219	193	175	159			
4								309	278	253	199		
3									350	319	250	219	
2										402	316	276	
1											399	349	
0											502	439	
00												560	

*To limit voltage drop to 3% at 240V. For other voltages use the following multipliers:

110V	0.458	220V	0.917
115V	0.479	230V	0.966
120V	0.50	250V	1.042
125V	0.521		

EXAMPLE: Find maximum run for #10 wire carrying 30 amp at 120V......
115 x 0.5 = 57 ft.

NOTE: If the required length of run is nearly equal to, or even somewhat more than, the maximum run shown for a given wire size - select the next larger wire. This will provide a margin of safety. The recommended limit on voltage drop is 3 per cent. Something less than the maximum is preferable.

FIGURE A11-3

Typical Residential Control Systems 503

for parts of the circuit even on 120 V. Specification sheets will give the minimum wire and fuse sizes that will meet the requirements of the *National Electric Code®*, and this wire size is based on 125% of the full-load current rating. The length of the wire run will also determine the wire size, and Fig. A11-3 shows the maximum length of two wire runs for various wire sizes. This table is based on holding the voltage drop to 3% at 240 V. For other voltages, the multipliers at the bottom should be used. If the required run is nearly equal to or somewhat more than the maximum run shown by the calculation of the table, the next larger wire size should be used as a safety factor. It is always possible to use larger wire than called for on the specifications, but a smaller wire size should never be used, as this will cause nuisance trips, affect the efficiency of operation, and create a safety hazard. All replacement wire should be the same size and type as the original wire and should be rated at 90 to 105°C.

A11-4
HEATING CIRCUIT

Let's take a quick glance at the simple schematic diagram we will soon be building (Fig. A11-4). It is for a typical gas-fired upflow air furnace. The wire conductors are represented by lines and the other components by means of symbols and letter designations. Notice the very important legend, which identifies the various types of wiring, the symbols, and the letter designations. Although other types of diagrams will be discussed in later discussions, we are interested now in the schematic, since this is the diagram used by service personnel to tell how the system works and why.

There are only five major electrical devices in our diagram: the fan motor, which circulates the heated air; the 24-V automatic gas solenoid valve to control the flow of gas to the burner; the combination blower and limit control, which controls fan operation and governs the flow of current to the low-voltage control circuit; the step-down transformer, which provides 24-V current to operate the automatic gas valve; and the room thermostat, which directly controls the opening and closing of the automatic gas valve.

Now, we'll build this same schematic diagram a step at a time, beginning with the power supply and adding each component as we go along. The wires for the power supply are usually run by the installing technician. They must carry line voltage (single phase, 60 Hz, 120 V) to operate the blower motor, so we will represent them as heavy broken lines to indicate field wiring (Fig. A11-5).

As a protection to our circuit, the power supply must be run through a fused disconnect switch. The disconnect switch is the main switch to the entire system.

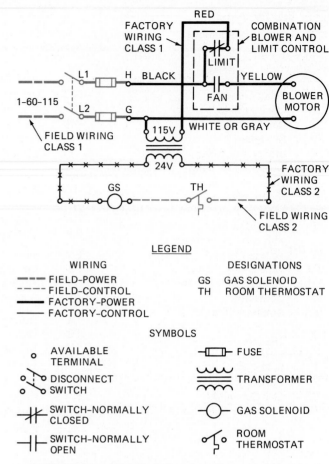

FIGURE A11-4 Heating circuit diagram for gas-fired upflow air furnace. (*Courtesy* Borg-Warner Central Environmental Systems, Inc.)

During normal operation, it must be closed to energize the circuit, but it may be opened or closed manually as necessary for troubleshooting and servicing electrical components. The fuses must be properly sized; large enough to carry the normal flow of current, but small enough to "blow" in case of an overload that might damage the circuit. If the fused disconnect switch is located outside of the building, it must be installed in a weatherproof enclosure.

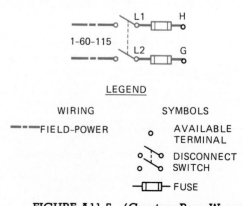

FIGURE A11-5 (*Courtesy* Borg-Warner Central Environmental Systems, Inc.)

504   Air-Conditioning

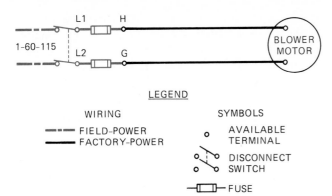

LEGEND

WIRING | SYMBOLS
--- FIELD-POWER | o AVAILABLE TERMINAL
— FACTORY-POWER | DISCONNECT SWITCH
| FUSE

FIGURE A11-6 (*Courtesy* Borg-Warner Central Environmental Systems, Inc.)

In a gas-fired heating system, the fan motor is the heaviest electrical load, so it will be added to the diagram next and connected to the power supply (Fig. A11-6). Since the fan motor is clearly identified on the diagram, we will not add it to the legend. Internal windings are not shown.

Now, close the disconnect switch and see what happens. The fan motor runs, since we have completed a simple circuit from L1 through the windings of the blower motor and back to neutral. But the automatic gas valve is not connected, and we have no means of controlling its operation automatically, so let's keep the disconnect switch open and continue with our schematic diagram.

Since we must have automatic control of both the fan motor and the automatic gas solenoid valve, we will begin to connect the combination fan and limit control (Fig. A11-7). This control, as previously discussed, consists of a sensing element, which feels the temperature

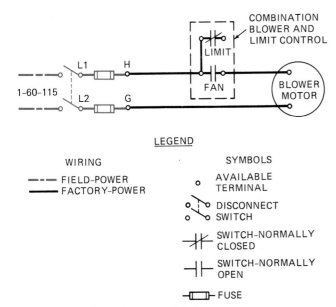

LEGEND

WIRING | SYMBOLS
--- FIELD-POWER | o AVAILABLE TERMINAL
— FACTORY-POWER | DISCONNECT SWITCH
| SWITCH-NORMALLY CLOSED
| SWITCH-NORMALLY OPEN
| FUSE

FIGURE A11-7 (*Courtesy* Borg-Warner Central Environmental Systems, Inc.)

of the heated air inside the furnace, and two adjustable switches: a fan switch and a limit switch.

The fan switch controls the starting and stopping of the blower motor. When the burner starts, the fan switch will not close to start the blower until the air temperature inside the furnace plenum chamber has warmed up to the FAN ON setting. When the burner stops, the fan switch remains closed, keeping the blower in operation until the air temperature inside the furnace cools down to the FAN OFF setting.

The limit switch is a safety control; it prevents the furnace from overheating. As long as the plenum chamber temperature is below the setting of the limit switch, the switch remains closed. If the plenum chamber temperature rises to the switch setting, it opens to deenergize the entire 24-V circuit and close the automatic gas valve.

As illustrated, the combination fan and limit control is partially connected. The fan switch is wired into L_1 in series with the blower motor windings. Even if we close the disconnect switch now, our fan motor will not run because the fan switch in the fan motor circuit is open, and it will remain open until we can warm up the air temperature in the furnace to the switch setting.

We learned earlier that small electrical devices, such as relays and solenoid valves, do very little work and therefore require little current for operation, so a step-down transformer to reduce line voltage to 24 volts for the control circuit is needed.

Figure A11-8 shows the transformer properly installed in the wiring diagram—one side connected to neutral and the other side connected to L_1 through the limit switch. This places the limit switch in control of the entire 24-V circuit.

If we close the disconnect switch now, the blower motor still cannot run, since the fan switch must remain open until the air temperature warms up. However, the transformer and the 24-V power source are energized immediately from L_1 through the closed limit switch and back to L_2. Although we now have a source of 24-V current, nothing can be accomplished until we complete a circuit across the 24-V supply.

Now, with a source of low-voltage current, we are ready to connect the automatic gas solenoid valve and the room thermostat. Essentially, the automatic gas valve is a closed solenoid valve with an adjustable bypass for the pilot flame. During normal operation, the pilot burns continuously to ignite the gas admitted to the burner when the solenoid is energized to open the main gas valve. A warm-air heating system must be properly controlled to maintain a comfortable temperature in the heated space. For this purpose, an adjustable low-voltage room thermostat, which opens on an increase in temperature, is used as the primary operating control. It senses the room temperature and energizes or deenergizes the automatic gas solenoid valve as necessary to maintain the desired temperature.

Typical Residential Control Systems 505

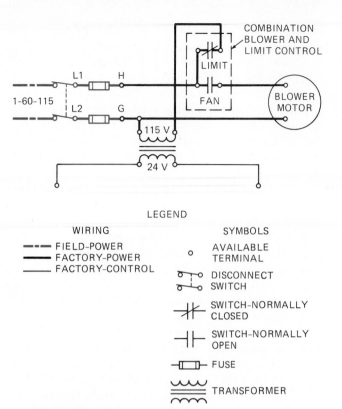

WIRING   SYMBOLS

- - - FIELD-POWER   ○ AVAILABLE
───── FACTORY-POWER   TERMINAL
───── FACTORY-CONTROL

                          ○─⟋○ DISCONNECT
                                SWITCH

                          ─╫─ SWITCH-NORMALLY
                                CLOSED

                          ─╢─ SWITCH-NORMALLY
                                OPEN

                          ─▭─ FUSE

                          ⦚⦚⦚ TRANSFORMER

FIGURE A11-8 (*Courtesy* Borg-Warner Central
Environmental Systems, Inc.)

Now that we have reviewed the functions of the automatic gas solenoid valve and the heating thermostat, let's wire them into our circuit (Fig. A11-9). Since the limit switch is a safety control and not an operating control, the thermostat must control the opening and closing of the gas solenoid, so we will wire them in series across the 24-V circuit. The gas valve is factory wired, so we will use a light, unbroken line to indicate factory control wiring to identify terminals. Normally, the thermostat is located in a room away from the furnace and must be field-wired, as indicated by the light broken lines used to represent control circuit field wiring.

The circuit is now completely wired, so let's close the disconnect switch and see if our system performs as it should. The limit switch should be closed; the thermostat is open. The transformer and low-voltage circuit are immediately energized from L_1 through the closed limit switch back to neutral. The thermostat is open, preventing current from flowing to energize the automatic gas valve, which remains closed. The fan motor cannot operate because the fan switch is open, and it will remain open until the air temperature in the furnace warms up to the FAN ON setting.

When the room becomes cool enough to close the thermostat contacts, the automatic gas valve is energized (Fig. A11-10). This opens the valve to admit gas to the burner, and we now begin to heat the air in the plenum chamber of the furnace. The fan switch would

still be open, so the fan motor cannot run, but as soon as the air temperature in the furnace plenum chamber warms up to the FAN ON setting, the fan switch closes to start the fan motor and circulate warm air to the heated space. The blower will continue to operate as long as the heated air temperature remains above the FAN OFF setting.

When the room temperature warms up to the setting of the thermostat, the thermostat contacts open the 24-V circuit to the automatic gas valve. This closes the valve and extinguishes the burner. However, the fan switch is still closed and the blower is still circulating warm air. With the burner extinguished and the blower running, the air in the furnace plenum chamber becomes cooler and cooler. When the air temperature falls to the FAN OFF setting, the fan switch opens, the blower motor is deenergized and will remain stopped until the air temperature in the furnace plenum is again warmed up to the FAN ON setting.

So far, we have paid very little attention to the all-important limit switch, but it is the overriding system safety control. If, at any time during furnace operation, the plenum chamber tends to overheat, the air temperature soon reaches the setting of the limit switch. This opens the limit switch contacts and immediately de-

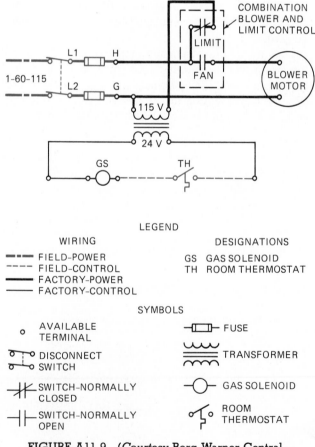

WIRING   DESIGNATIONS

- - - FIELD-POWER   GS GAS SOLENOID
- - - FIELD-CONTROL   TH ROOM THERMOSTAT
───── FACTORY-POWER
───── FACTORY-CONTROL

                          SYMBOLS

○ AVAILABLE   ─▭─ FUSE
    TERMINAL
                          ⦚⦚⦚ TRANSFORMER
○─⟋○ DISCONNECT
      SWITCH
                          ─○─ GAS SOLENOID
─╫─ SWITCH-NORMALLY
      CLOSED
                          ⟋○   ROOM
─╢─ SWITCH-NORMALLY   THERMOSTAT
      OPEN

FIGURE A11-9 (*Courtesy* Borg-Warner Central
Environmental Systems, Inc.)

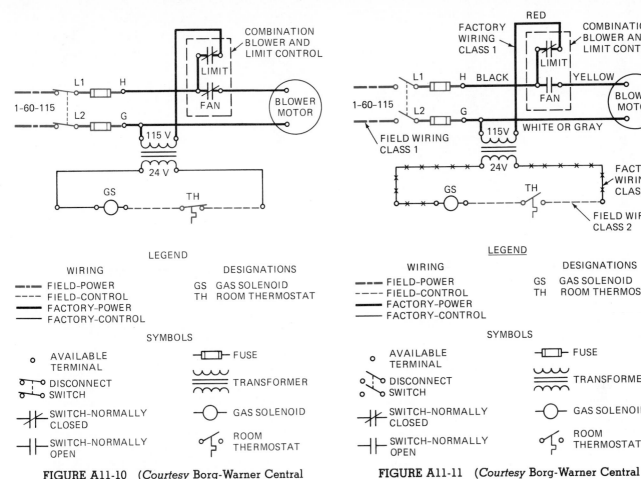

FIGURE A11-10 (*Courtesy* Borg-Warner Central Environmental Systems, Inc.)

FIGURE A11-11 (*Courtesy* Borg-Warner Central Environmental Systems, Inc.)

energizes the entire 24-V circuit. When this happens, the automatic gas valve closes and the burner is extinguished, even though the room thermostat may still be closed and calling for more heat. Although the limit switch can override the room thermostat, it has no effect on fan operation. The fan switch remains closed and the blower continues to run as long as the heated air temperature remains above the FAN OFF setting.

The operating sequence just completed proves that the wiring diagram is correct. The disconnect switch controls the entire electrical system; the blower motor operates independently of the burner, under the control of the fan switch, and the automatic gas valve is cycled on and off by the room thermostat, under the control of the limit switch. However, we have only been concerned with lines and symbols so far. Actually, the finished diagram (Fig. A11-11) contains the complete story on the electrical system just reviewed; the legend clearly identifies all symbol and letter designations used in the diagram, and the notes call attention to important facts relating to the circuit. When you have a thorough knowledge of the simple principles just outlined, the functions of the controls and their relationship to each other in the electrical circuit, you have taken a big step

toward becoming a professional in installation, troubleshooting, and servicing the electrical circuit of a gas-fired heating system—or other heating apparatus.

A11-5
COOLING CIRCUIT

At one time or another, any one just starting a career in air conditioning probably has regarded the formidable cooling system wiring diagram with a feeling of awe and suspicion. We looked at the maze of lines and the dozens of symbols (Fig. A11-12) and wondered what it all meant. Actually, it really is not that bad. When you understand the meaning of a few symbols and their relationship to each other in the electrical circuit, you can tell exactly what is happening in an electrical system: how it happens, when it happens, and why. The ability to read and understand the wiring diagram often makes the difference between an amateur and a well-trained, professional service technician.

Wiring diagrams are made for different purposes, and it is important to recognize each type and know the

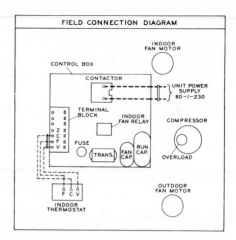

FIGURE A11-12 Cooling circuit diagram. (*Courtesy* Borg-Warner Central Environmental Systems, Inc.)

FIGURE A11-13 Field connection diagram. (*Courtesy* Borg-Warner Central Environmental Systems, Inc.)

intended purpose of each. There are *field connection diagrams* and *schematic diagrams*.

The *field connection diagram* (Fig. A11-13) identifies the various electrical controls in the control box and indicates the necessary field wiring by means of broken lines and shaded areas. The *field connection diagram* is designed to instruct the installing electrician or technician in how to run the proper power supply to the unit and in the correct wiring between sections if the unit consists of more than one section. Internal wiring is not

shown, since we are only interested in installation at this time.

Now let's look at Fig. A11-14, a typical *schematic diagram* for an air-cooled packaged residential air-conditioning unit. One glance tells us that it contains more information than the field connection diagram. This is the diagram used by service personnel that shows how the system works and why. Most manufacturers include a copy of the diagram in the control cover or access panel of the unit. The load devices (motors) and the controls

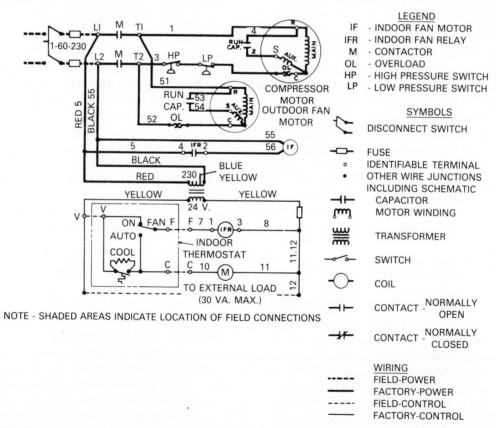

FIGURE A11-14 Schematic diagram. (*Courtesy* Borg-Warner Central Environmental Systems, Inc.)

are represented by symbols, the wires by lines. Notice that some lines are heavy while others are light by comparison. The heavy lines represent wires in the *power circuit,* which supplies higher line-voltage current to the heavy electrical loads, such as motors. The light lines represent the *control circuit,* which supplies low voltage through a step-down transformer to the light load devices, like relay coils that are used only to actuate switches.

Now, start at the beginning and build this cooling electrical circuit step by step, component by component, just as was done for heating. First, the power supply to the unit (Fig. A11-15). Since these wires must be run by the installing electrician or technician in the field, we represent them by heavy broken lines. To protect the circuit, our supply must pass through a properly sized fused disconnect switch. Proper fuse sizes are found in the equipment installation manual. If the fused disconnect switch is located outdoors, it must be installed in a weatherproof enclosure. The power supply is shown as single-phase 60-Hz 240-V current, and the incoming wires are designated as L_1 and L_2. Whenever we complete a path between L_1 and L_2 and install a load in the path, we will have a circuit; current will flow and work will be accomplished.

Since the compressor motor is the heaviest load, it will be placed in the diagram next and connected to the power supply. If we closed the disconnect switch, there would be a completed circuit from L_1 through the compressor motor windings back to L_2 and the compressor would run, as a result of manual switch operation. But the system must run automatically, so we will keep the disconnect open and continue with our diagram.

Before the simplest air-cooled air-conditioning system is complete, we need an outdoor fan and an indoor fan. So let's put them into our diagram next (Fig. A11-16). It is easy to see that both fans are connected across L_1 and L_2, and all three motors are in individual circuits of their own. At this point we have a complete wiring diagram for a simple, manually controlled air-

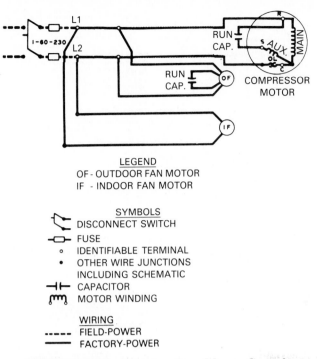

LEGEND
OF - OUTDOOR FAN MOTOR
IF - INDOOR FAN MOTOR

SYMBOLS
DISCONNECT SWITCH
FUSE
IDENTIFIABLE TERMINAL
OTHER WIRE JUNCTIONS
INCLUDING SCHEMATIC
CAPACITOR
MOTOR WINDING

WIRING
----- FIELD-POWER
⸺ FACTORY-POWER

FIGURE A11-16   (*Courtesy* Borg-Warner Central Environmental Systems, Inc.)

conditioning system. If we closed the disconnect switch, the compressor motor would run, the outdoor fan would operate, and the indoor fan would circulate the cooled air to the conditioned space; but it still needs controls to make the system operate automatically—and safely.

Earlier we learned that electrical controls that do very little work require very little current; therefore, we again use a low-voltage (24 V) control circuit. This permits the use of much smaller wire, makes the circuit much safer for residential use and, most important, it provides much closer control of system operation. So we need a source of low-voltage current for our controls and we get this by means of the small step-down transformer. Figure A11-17 shows the transformer in the wiring diagram. In the transformer, 240-V current is reduced to 24 V, all the voltage needed to supply the necessary current to actuate the relays for automatic operation. Remember that the transformer does nothing but provide a source of 24-V power—nothing can happen until we complete a circuit across the 24-V source.

Since an air-conditioning system is designed to maintain a comfortable temperature in the conditioned space, a low-voltage adjustable indoor cooling thermostat is used as the primary system control (Fig. A11-18). It senses the air temperature and tells the secondary controls to start or stop the compressor and fans as necessary. Although there are many variations of indoor thermostats, the one illustrated has two slide switches—a fan switch with two positions, ON and AUTO, and a cooling control switch with positions marked OFF and COOL.

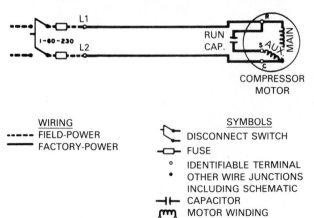

WIRING
----- FIELD-POWER
⸺ FACTORY-POWER

SYMBOLS
DISCONNECT SWITCH
FUSE
IDENTIFIABLE TERMINAL
OTHER WIRE JUNCTIONS
INCLUDING SCHEMATIC
CAPACITOR
MOTOR WINDING

FIGURE A11-15   (*Courtesy* Borg-Warner Central Environmental Systems, Inc.)

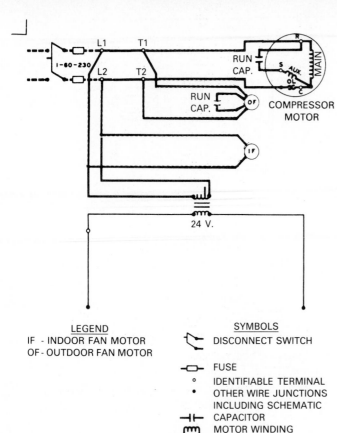

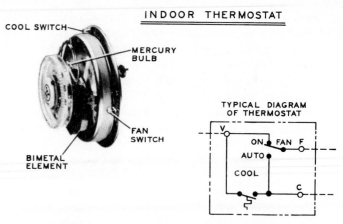

FIGURE A11-18 (*Courtesy* Honeywell, Inc.)

LEGEND
IF - INDOOR FAN MOTOR
OF - OUTDOOR FAN MOTOR

SYMBOLS
⎍ DISCONNECT SWITCH

◻ FUSE
∘ IDENTIFIABLE TERMINAL
• OTHER WIRE JUNCTIONS
  INCLUDING SCHEMATIC
⊣⊢ CAPACITOR
⌇ MOTOR WINDING
⋚ TRANSFORMER

WIRING
----- FIELD-POWER
───── FACTORY-POWER
───── FACTORY-CONTROL

FIGURE A11-17 (*Courtesy* Borg-Warner Central Environmental Systems, Inc.)

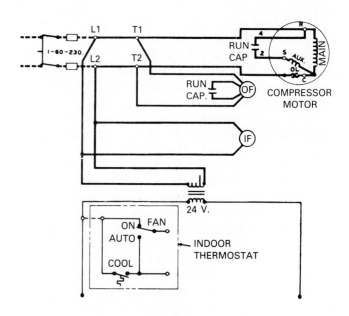

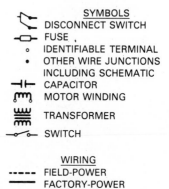

LEGEND
OF - OUTDOOR FAN MOTOR
IF - INDOOR FAN MOTOR

SYMBOLS
⎍ DISCONNECT SWITCH
◻ FUSE
∘ IDENTIFIABLE TERMINAL
• OTHER WIRE JUNCTIONS
  INCLUDING SCHEMATIC
⊣⊢ CAPACITOR
⌇ MOTOR WINDING
⋚ TRANSFORMER
∘⌐∘ SWITCH

WIRING
----- FIELD-POWER
───── FACTORY-POWER

FIGURE A11-19 (*Courtesy* Borg-Warner Central Environmental Systems, Inc.)

Internally, the thermostat consists of two sets of switches. The fan switch controls the indoor fan only. In the ON position, the fan will operate continously; in AUTO, fan operation is actually controlled by the position of the cooling control switch. The contacts of the cooling switch are actually sealed inside a tilting bulb partially filled with mercury. When the bulb is tilted so that mercury covers only one contact, the switch is open. When the bulb is positioned so that mercury covers both contacts, the switch is closed. The bulb is tilted by means of a bimetal element that warps as it feels a change in temperature, to either close or open the mercury switch. The symbol V is the common low-voltage terminal feeding through the fan switch to terminal F and through the mercury bulb to terminal C.

We have just seen how our thermostat works mechanically, so let's put it into our control circuit and see what happens electrically (Fig. A11-19). We have connected our common terminal V to the L_1 side of the low-voltage supply, but nothing new can happen electrically until we complete a circuit to the other side of the 24-V supply.

If we close the main disconnect switch, all three motors will immediately start and continue to operate until the disconnect switch is manually opened. The transformer will be energized and the low-voltage power supply will be available to the thermostat, but no low-voltage circuit is made; we still need a few more controls for automatic operation.

For comfort cooling, an air-cooled residential cooling system must be wired so that both the indoor and outdoor fans operate all the time the compressor is running, the outdoor fan to condense the refrigerant vapor and maintain a safe discharge pressure, and the indoor fan to circulate the cooled air to the conditioned space. In addition, we should be able to operate the indoor fan independently for air circulation without cooling, and this is exactly what we will be able to do when our wiring diagram is complete.

To control the compressor motor automatically, we need a contactor (Fig. A11-20), which, we learned, is nothing more than an oversized relay with a switch large enough to carry the heavy current drawn by motors and a magnetic coil strong enough to actuate the larger switches.

Since we want the cooling switch in the thermostat to govern the action of the contactor in starting and stopping the compressor motor, we will wire the contactor coil M in series with the cooling switch of the thermostat (Fig. A11-21), but we must connect the contactor switches in the power circuit so that they control both the compressor motor and the outdoor fan motor, but not the indoor fan; it must be able to operate in-

FIGURE A11-20 (*Courtesy* Honeywell, Inc.)

dependently. Notice that the contactor coil and both the starter switches are marked M for identification. This simply means that coil M actuates the two switches marked M. Since both switches are normally open, they will close when the coil is energized and open when it is deenergized.

What happens now, when we close our main disconnect switch? Current flows to the indoor fan motor and the 24-V transformer immediately, but the compressor motor and outdoor fan motor cannot start. The thermostat cooling switch is open, so no current can flow to contactor coil M to close its switches marked M in

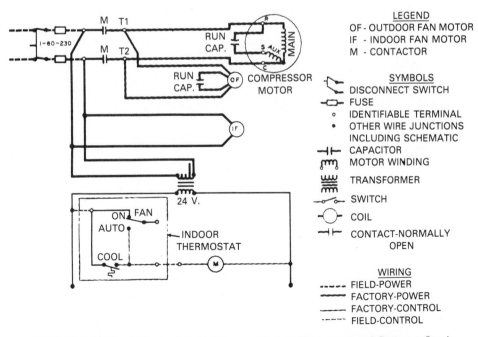

FIGURE A11-21 (*Courtesy* Borg-Warner Central Environmental Systems, Inc.)

the power circuit to the compressor and outdoor fan motors.

What happens when the thermostat cooling switch closes and calls for cooling (Fig. A11-22)? Current immediately flows through starter coil *M*, and the two switches marked *M* in the power circuit close at the same time. This energizes and starts both the compressor motor and the outdoor fan motor at the same time. When the thermostat is satisfied, the cooling switch again moves to the open position. The compressor and outdoor fan motors stop, but our indoor fan motor continues to run, and it will remain in operation as long as the disconnect is closed, regardless of the position of the fan switch.

It seems we still need another control for our indoor fan motor, *IF*. This time we will use a relay with a coil and a normally open switch, both marked *IFR* for indoor fan relay (Fig. A11-23). Notice that the coil is wired in series with the fan switch in the thermostat, and its switch is wired into the indoor fan motor circuit.

With the fan switch in the ON position, low-voltage current now flows through the fan switch to energize relay coil *IFR* and close its switch in the indoor fan power circuit. The indoor fan is now running, and it will continue to run as long as the fan switch remains in the ON position, regardless of the position of the cooling switch. The cooling system—the compressor and the outdoor fan—is free to cycle on and off as the cooling switch dictates with no effect on the indoor fan.

We'll change our wiring diagram once more. Move the fan switch to AUTO (Fig. A11-24). The thermostat cooling switch is open; it is not calling for cooling. Now, close the disconnect. Current immediately flows through the transformer into the 24-V supply, but the fan switch is in AUTO and the cooling switch is open, so we can't get a completed 24-V circuit. Contactor coil *M* and indoor fan relay coil *IFR* cannot be energized, so their switches all remain open, and none of the motors can operate.

Assume that the cooling switch closes and calls for cooling without changing the position of the fan switch or the disconnect—disconnect closed; fan on AUTO (Fig. A11-25). Current immediately flows through the cooling switch and through both the contactor coil *M* and the indoor fan relay coil *IFR* at the same time. This closes both switches marked *M* in the power circuit to the compressor and outdoor fan motors, as well as the switch marked *IFR* in the power circuit to the indoor fan motor.

The compressor and outdoor fan motors are energized and started through closed switches *M*; at the same time, the indoor fan motor is also started through closed switch *IFR*. The entire system is now energized and all motors are running automatically.

When the thermostat is satisfied, the cooling switch opens automatically, and we are back where we started when we closed the disconnect switch; the cooling switch opens the circuit to both the contactor and indoor fan

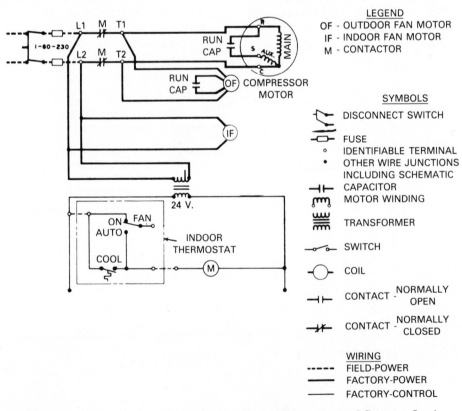

FIGURE A11-22 (*Courtesy* Borg-Warner Central Environmental Systems, Inc.)

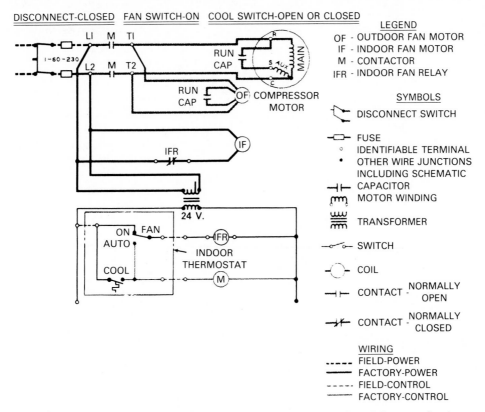

DISCONNECT-CLOSED   FAN SWITCH-ON   COOL SWITCH-OPEN OR CLOSED

LEGEND

OF - OUTDOOR FAN MOTOR
IF - INDOOR FAN MOTOR
M - CONTACTOR
IFR - INDOOR FAN RELAY

SYMBOLS

DISCONNECT SWITCH

FUSE
IDENTIFIABLE TERMINAL
OTHER WIRE JUNCTIONS
INCLUDING SCHEMATIC
CAPACITOR
MOTOR WINDING
TRANSFORMER
SWITCH
COIL
CONTACT - NORMALLY OPEN
CONTACT - NORMALLY CLOSED

WIRING

- - - - FIELD-POWER
——— FACTORY-POWER
- - - - FIELD-CONTROL
——— FACTORY-CONTROL

FIGURE A11-23 (*Courtesy* Borg-Warner Central Environmental Systems, Inc.)

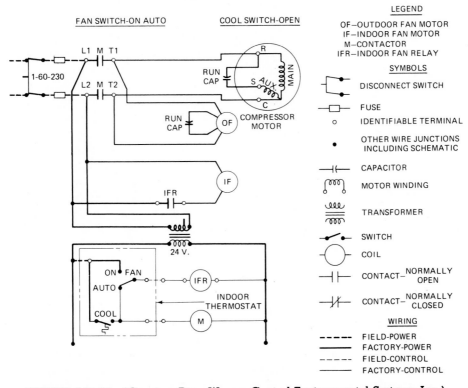

FAN SWITCH-ON AUTO   COOL SWITCH-OPEN

LEGEND

OF—OUTDOOR FAN MOTOR
IF—INDOOR FAN MOTOR
M—CONTACTOR
IFR—INDOOR FAN RELAY

SYMBOLS

DISCONNECT SWITCH

FUSE
IDENTIFIABLE TERMINAL

OTHER WIRE JUNCTIONS
INCLUDING SCHEMATIC
CAPACITOR
MOTOR WINDING
TRANSFORMER
SWITCH
COIL
CONTACT—NORMALLY OPEN
CONTACT—NORMALLY CLOSED

WIRING

- - - - FIELD-POWER
——— FACTORY-POWER
- - - - FIELD-CONTROL
——— FACTORY-CONTROL

FIGURE A11-24 (*Courtesy* Borg-Warner Central Environmental Systems, Inc.)

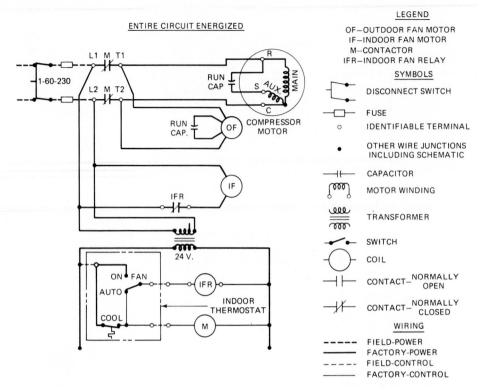

LEGEND
OF—OUTDOOR FAN MOTOR
IF—INDOOR FAN MOTOR
M—CONTACTOR
IFR—INDOOR FAN RELAY

SYMBOLS

DISCONNECT SWITCH

FUSE

IDENTIFIABLE TERMINAL

OTHER WIRE JUNCTIONS
INCLUDING SCHEMATIC

CAPACITOR

MOTOR WINDING

TRANSFORMER

SWITCH

COIL

CONTACT—NORMALLY OPEN

CONTACT—NORMALLY CLOSED

WIRING
FIELD-POWER
FACTORY-POWER
FIELD-CONTROL
FACTORY-CONTROL

FIGURE A11-25 (*Courtesy* Borg-Warner Central Environmental Systems, Inc.)

relay coils, their respective switches immediately snap to their normally open positions, and all motors stop—compressor, outdoor fan, and indoor fan.

Now that we know our wiring diagram is correct from the standpoint of operation, we will add safety controls to protect our system and improve its operating performance (Fig. A11-26).

Overloads *OL* have been added to protect the compressor and outdoor fan motor windings against excessive current flow. Overloads may be installed inside the motor housing as shown for the compressor motor or externally as shown for the outdoor fan motor. Although the windings are not shown for the indoor fan motor, all small single-phase motors are equipped with some type of protection.

As an added precaution against compressor operation under abnormal conditions such as excessive discharge pressure or dangerously low suction pressure, we have wired a high-pressure switch and a low-pressure switch in series with the compressor motor windings. The high-pressure switch, *HP*, is actuated by system discharge pressure to open and stop the compressor if the discharge pressure exceeds the switch setting. The low-pressure switch, *LP*, is actuated by system suction pressure to open and stop the compressor if the suction pressure falls below the switch setting. To protect the relay coils and wiring in the low-voltage control circuit, we have installed a fuse properly sized to open the 24-V circuit in case of excessive current flow.

For closer control of air temperature in the conditioned space, explained in Chap. A10, an anticipator has been wired across the terminals of the thermostat cooling switch. It is a fixed, nonadjustable type.

When we discussed the relay, we learned that one relay coil may actuate as many switches as it is possible to build into one relay, and each switch may be located in a different section of the wiring diagram. As the diagram becomes larger and more relays are used, these switches become more difficult to locate. To solve this problem and make it easy to locate the switches of each relay, the *ladder diagram* system is often used.

On the ladder diagram (Fig. A11-26), each horizontal line is numbered in sequence from top to bottom on the left-hand side for line identification. On the right side of the diagram, opposite each relay coil, are shown the line numbers in which its switches are located. For example, our diagram shows that contactor coil *M* in line 16 has two switches located in lines 1 and 4, while relay coil *IFR* has only one switch located in line 10.

Notice also the factory-installed wiring has been color-coded for ease of identification. (Sometimes there will be a numerical designation instead, depending on the equipment manufacturer.)

Although the diagram we have just built is relatively simple, all wiring diagrams are based upon the same logic. When the technician understands these basic principles, even the most complicated wiring diagram is simplified and much easier to read.

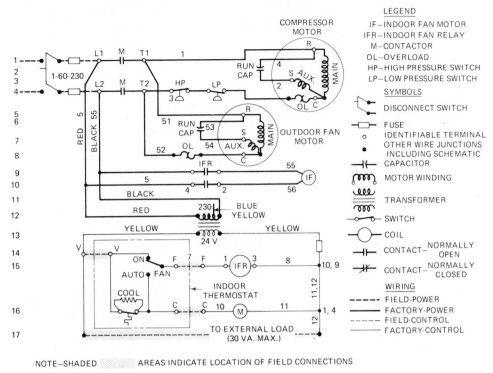

LEGEND
IF—INDOOR FAN MOTOR
IFR—INDOOR FAN RELAY
M—CONTACTOR
OL—OVERLOAD
HP—HIGH PRESSURE SWITCH
LP—LOW PRESSURE SWITCH

SYMBOLS
DISCONNECT SWITCH
FUSE
IDENTIFIABLE TERMINAL
OTHER WIRE JUNCTIONS
INCLUDING SCHEMATIC
CAPACITOR
MOTOR WINDING
TRANSFORMER
SWITCH
COIL
CONTACT—NORMALLY OPEN
CONTACT—NORMALLY CLOSED

WIRING
FIELD-POWER
FACTORY-POWER
FIELD-CONTROL
FACTORY-CONTROL

NOTE—SHADED  AREAS INDICATE LOCATION OF FIELD CONNECTIONS

FIGURE A11-26 (*Courtesy* Borg-Warner Central Environmental Systems, Inc.)

A11-6
TOTAL COMFORT SYSTEMS

It is impossible to review all wiring diagrams that cover the broad range of residential heating and cooling products, along with humidity control and electronic air cleaning that provide a total comfort system. Each manufacturer has its own techniques for combining the various functions and methods of presentation. Most publish diagrams or at least make application recommendations to assist the electrician or the installer in hooking up the systems.

For simplicity of demonstration, this text will review a total comfort system (TCS) using an upflow gas furnace with single-speed fan operation, a split-system air-cooled condensing unit (with indoor coil), a combination heat/cool thermostat, a humidistat, and an electronic air cleaner.

First, the physical wiring and point-to-point connections are illustrated in (Fig. A11-27). This is what the electrical installer will use as a guide during the installation. Note the need for two junction boxes to make the required interlocks. The low-voltage wiring is also detailed as to the number of conductors needed and the terminal or junction connections. Power wiring is black (B) and white (W), and ground conductors are sized according to the load and to meet local codes requirements. The internal wiring of the furnace, condensing unit, etc., is not shown on the installation diagram.

Study the layout for the components and the connections involved.

Figure A11-28 is the schematic ladder diagram of this same system. First locate and relate the major components to the previous drawing; the furnace blower motor, the humidifier, the electronic air cleaner, the condensing unit, etc. Note that the combination thermostat has a HEAT-OFF-COOL system selection and an ON-AUTO fan selection. Anticipators are not shown. The system diagram is shown on the heating mode.

Closing the disconnect switch (line 4) will produce a 120-V line-voltage current flow to the transformer. Trace the black wire through the closed limit (LS) to the transformer primary coil and to the white power return (line 5). A low-voltage potential is created.

Low voltage can now be established through the fan switch ON position and energize the FR (fan relay) coil in line 9. This closes the Fr switch in line 3, allowing line-voltage current to flow to the fan motor and to the humidifier relay (HU) in line 1. Thus the fan runs constantly. Note that the FS (furnace switch) in line 4 may or may not be closed depending on whether there is sufficient heat to close its contacts. The gas valve (GS) operation in line 13 will depend on the activation of the mercury bulb by the room temperature. Note in line 15 that the HEAT position selector switch feeds low voltage to H (the humidistat), and if humidity is needed, it closes its contacts and energizes the humidistat relay coil (HU) also in line 15. The contacts of HU are located in line 1 and will activate the humidifier (fan motor or

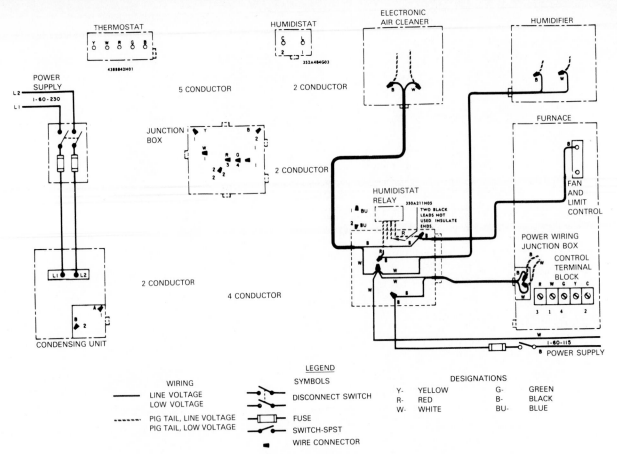

FIGURE A11-27   (*Courtesy* Borg-Warner Central Environmental Systems, Inc.)

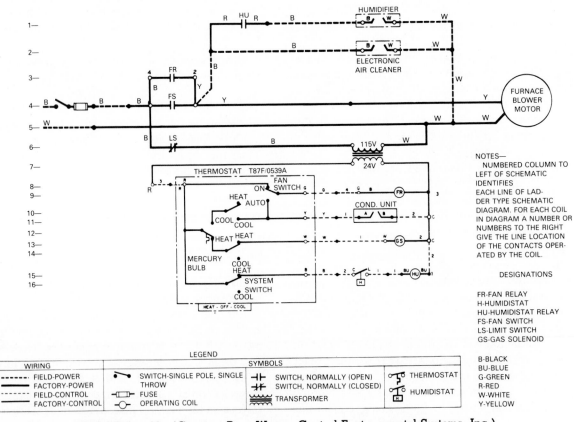

FIGURE A11-28   (*Courtesy* Borg-Warner Central Environmental Systems, Inc.)

water solenoid valve). This is called *permissive humidification* by automatic demand from the humidistat on the heating cycle with the normal cycle of furnace fan. (*Command humidification* would be a system where the humidistat can command fan-only operation, regardless of the heating/cooling mode.) The electronic air cleaner simply parallels the fan operation and is energized any time the fan runs.

If the fan selection is AUTOMATIC, *FR* cannot be energized during heating, but the furnace fan cycles from its fan switch, line 4. On cooling, however, there is a contact to the automatic terminal and the fan cycles with the cooling thermostat. Also, on cooling there is a completed circuit on line 11 to the condensing unit contactor terminals *A* and *B*. This pulls in the compressor

and the outdoor condenser fan. They then cycle off and on with the room temperature mercury switch. Note that the middle system switch on cool drops out the gas valve and the lower switch drops out the humidistat. Note also that on cooling, the furnace fan can operate from the *FR* relay (ON or AUTO) and thus furnish power to the air cleaner whenever it runs.

This is how a simple but common TCS system is installed and wired schematically. The introduction of furnace fan speed control, two-stage cooling and heating thermostats, command humidification, etc., add sophistication and enlarge the diagram, but these items do not necessarily add to its complexity if the schematic is analyzed step by step as to function and operation.

PROBLEMS

A11-1. The electrical service entrance for residential power will be _____ V _____ phase _____ Hz.

A11-2. Main breakers (or fuses) range from _____ to _____ A.

A11-3. The voltage drop in selecting wire sizes should be limited to _____ %.

A11-4. Wiring diagrams require _____ and _____ for proper understanding.

A11-5. Single-phase power is designated by two letters and color codes. Name them.

A11-6. What colors are used for hot wire, neutral wire, and ground?

A11-7. Manufacturers usually furnish two wiring diagrams. What are they?

A11-8. The designation for high- or low-voltage lines are _____ for high voltage and _____ for low voltage.

A11-9. The designations for factory installed or field installed are _____ for factory and _____ for field installed.

A11-10. What is the function of the limit switch in the furnace?

A11-11. What does the furnace fan switch do?

A11-12. The fan cannot operate unless the thermostat is calling for heat. True or False?

A12

Commercial and Engineered Control Systems

The differences between residential control systems and those for commercial uses and engineered systems lie in the increased equipment size, more sophisticated application requirements (low ambient control and economizer cycles, etc.), and in the nature of the space to be controlled.

The unitary type of central heating and air-conditioning (up to 5 tons), which is used in residential work, is all single-phase power, and this same kind of product may also be employed in small commercial situations with no essential changes. In residential work, there is usually only one space thermostat (an exception would be a zone control with two units). Single-point space thermostat control is also common in small and light commercial work where a single zone duct system is used.

It is at the $7\frac{1}{2}$-ton size that the electrical characteristics of the equipment really become three-phase service, although some manufacturers do offer three-phase units in smaller sizes to meet special applications. Above $7\frac{1}{2}$ tons the need for more heavy-duty compressor starting and protection becomes apparent. In the high-tonnage systems, special equipment, such as reduced-voltage starters, open and closed transition starters, etc., become important considerations.

The shift from a single-zone duct system and single-zone space thermostat to the multizone duct systems, induction, and double-duct application, etc., means also a shift to multispace control. That constitutes a change in the point of control for the air-conditioning equipment itself. And as the need for the separation of equipment and space control becomes more pronounced, *control centers* with monitors must be employed to manage the operation, to avoid having personnel running around the entire building checking conditions.

Small and medium-size commercial systems using packaged equipment usually utilize preengineered electrical controls that require a minimum of on-site electrical work. This reduces installation costs, promotes better reliability, and simplifies the problems of maintenance for the operating and service personnel.

On the other hand, *engineered control systems* are almost always tailored to a particular application, and the design is included in the HVAC *plans and specifications* prepared by a *consulting engineer*. It is in this area where the use of pneumatic and electronic types of controls, in conjunction with regular electrical controls, becomes more prevalent.

The installation of *energized control systems* is also a specialized field, and within the HVAC work this job is separately done by control contractors who not only install the systems but do the start-up, balancing, and service. This is a specialized field, and one which offers challenging employment opportunities to the qualified technician.

≣ A12-2
≣ *COMMERCIAL ELECTRICAL*
≣ *CONTROL CIRCUIT*

With this brief introduction of the commercial and engineered system, let's look at the electrical system of a typical commercial rooftop unit with natural gas heating, low ambient control, and an economizer cycle. It is common for the manufacturer to present the information in three parts: (1) the basic refrigeration power circuit; (2) the particular control needed for heating gas, oil, and electric (in this example, natural gas heating is used); and (3) the necessary legend to identify symbols and other items.

Figure A12-1 shows the basic power wiring for compressors and fan motor circuits (note that this is a dual compressor system). Figure A12-2 shows the low-voltage control circuit, and Fig. A12-3 is the legend.

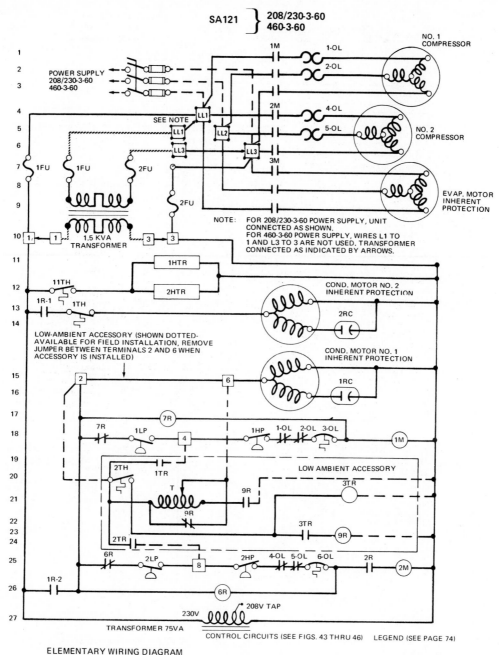

SA121 } 208/230-3-60
         460-3-60

ELEMENTARY WIRING DIAGRAM

FIGURE A12-1 High-voltage wiring diagram. (*Courtesy* Borg-Warner Central Environmental Control Systems, Inc.)

Referring to Fig. A12-1, the power supply is assumed to be 208/240 V, three-phase, 60 Hz, for purposes of sizing fuses and disconnect switch. From the manufacturer's electrical table (Fig. A12-4) on fuses and wire sizes, the model shown at 208/240 V will require a 100-A disconnect and a maximum of 70-A dual-element fuses. The minimum power supply wire size is a function of the distance from the switch to the panel terminals LL_1, LL_2, and LL_3 on the conditioner. Assume that the distance is 125 ft one way. The minimum wire size is No. 4 AWG 60°C wire.

Compressor motors are three phase, with the external overloads (*OL*) in the motor winding circuit. The evaporator motor is also three phase, but it has inherent motor protection. All three motors are started by the contactors marked *1M, 2M,* and *3M,* respectively.

The condenser fan motors and the balance of controls must be single phase so as to place elements in series circuits; this is not possible with three-phase operation. Single-phase line voltage is connected at terminals 1 and 3 on line 10. Ignore the transformer (see the note); its only function is to reduce power from 460

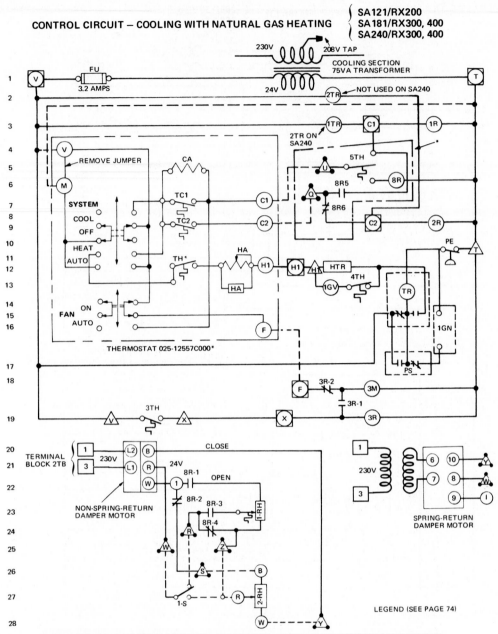

CONTROL CIRCUIT – COOLING WITH NATURAL GAS HEATING
SA121/RX200
SA181/RX300, 400
SA240/RX300, 400

ELEMENTARY WIRING DIAGRAM

*WHEN 100% OUTSIDE AIR OPTION IS NOT USED, CONNECT "C1" AND "C2" OF THERMOSTAT TO "C1" AND "C2", RESPECTIVELY, OF COOLING UNIT.

FIGURE A12-2 Low-voltage control circuit. (*Courtesy* Borg-Warner Central Environmental Control Systems, Inc.)

V to 240 V, and in this example it is not used. Note the added fuse protection *1 FU* and *2 FU* in the single-phase hook-up. Drop down to lines 11 and 12 and note the compressor crankcase heaters (*1 HTR* and *2 HTR*) that come on when the crankcase thermostat closes in response to oil temperature. They can operate as long as power is ON, regardless of the system operation. Therefore, it is important that *the power not be turned off in cold weather.*

Both condenser fans have single-phase motors with external running capacitors (*1 RC* and *2 RC*) and inherent overload protection. As stated, this is a dual circuit unit equipped with low ambient control. Note that in line 13 the No. 2 condenser fan motor is energized by the cooling relay *IR-1* to start and run, but it can also be dropped out by the first low ambient thermostat *1 TH.*

If it were not for the economizer cycle, the No. 1

CA	Anticipator, Cooling	1RC,2RC, 3RC,4RC	Running Capacitor
HA	Anticipator, Heating	S	Switch, Oil Pressure
T	Auto-Transformer, Speed Controller	1-S	Switch, Bypass
FU,1FU,2FU	Fuse	TC1	Thermostat, Cooling 1st Stage
1M,2M	Contactor, Compressor	TC2	Thermostat, Cooling 2nd Stage
3M	Contactor, Blower Motor	TH1	Thermostat, Heating 1st Stage
1GV	Gas Valve	TH2	Thermostat, Heating 2nd Stage
2GV	Gas Valve, Second Stage	1TH,2TH	Thermostat, Low Ambient Control
HTR	Heater in 3TH Fan Thermostat	3TH	Thermostat, Blower (Heat)
1HTR,2HTR	Heater, Compressor Crankcase	4TH	Thermostat, Limit (Heat)
1HP,2HP	High Pressure Cutout, Refrig.	5TH	Thermostat, Mixing Box
IGN	Ignition Trans. (For Pilot Relighter)	11TH	Thermostat, Crankcase Heater
PE	Low Gas Pressure Switch	TR	Time Delay Relay, Pilot Ignition
1LP,2LP	Low Pressure Cutout, Refrig.	VFS	Venter Fan Sail Switch
1-OL,2-OL 3-OL,4-OL 5-OL,6-OL	Overload Protectors, Compressor	10R,11R	Venter Motor Relays
		1RH	Control Damper, Mixed Air
PS	Pilot Safety Switch	2RH	Control Damper Controller, Min. Position
R	Second Stage Gas Valve Relay	⊟	Terminal Block, 1TB
1R,2R	Relay, Control Cooling	☐	Terminal Block, 2TB
3R	Low Voltage Control Relay	◙	24-Volt Terminal Block, 3TB
4R,5R	Relay, Control Electric Heat	△	Identified Connection in Heating Section
8R	Relay, Control, Mixing Box	△	0-100% Outside Air Terminal Block, 6TB (SA121);
9R	Relay, Low Ambient Control		7TB (SA181); 4TB (SA240)

SPECIFIC LEGEND FOR SA121, SA181, SA240 ELEMENTARY WIRING DIAGRAM

Model	SA121		SA181		SA240
6R	Relay, Lockout No. 2 System	6R	Relay, Lockout No. 2 System	6R	Relay, Lockout
7R	Relay, Lockout No. 1 System	7R	Relay, Lockout No. 1 System	MP	Compr. Protection
1TR,2TR	Time Delay Relay (Low Ambient Accessory)	1TR,2TR	Time Delay Relay, Low Ambient Control	1TR	Time Delay Relay, Part Winding Start
3TR	Time Delay Relay, Cond. Fan			2TR	Time Delay Relay, Low Ambient Control
T	Auto-Transformer (Low Ambient Accessory)			3TR	Time Delay Relay, Oil Pressure Switch
				1-SOL	Solenoid, Compr. Unloader
				2-SOL	Solenoid, Evap. Unloader

FIGURE A12-3 Legend. (*Courtesy* Borg-Warner Central Environmental Control Systems, Inc.)

—WIRE, FUSE AND DISCONNECT SWITCH SIZES
COOLING ONLY AND GAS-FIRED HEAT MODELS SA91, SA121, SA181, SA240, SA360

Model	Power Supply	Length Circuit One Way, Ft. Up To	Min. Wire Size Copper AWG 60 C 2% Voltage Drop	Max. Fuse Size Dual Element	Disconnect Switch Size, Amps
SA91-25A	A	100 175 200 250	6 4 3 2	55	60
SA91-45A	A	175 250	10 8	30	30
208/230 VOLT SA121-25B	A	125 175 200 250	4 3 2 1	70	100
SA121-46B	A	200 250	8 6	40	60
SA181-25A	A	150 200 250	1 0 000	100	100
SA181-45A	A	250	4	60	60
SA240-25C	A	150 200 250	00 000 250MCM	175	200

FIGURE A12-4 (*Courtesy* Borg-Warner Central Environmental Control Systems, Inc.)

condenser fan motor would be wired in as shown for the No. 2 fan. But the economizer system works within the same temperature range as the first-stage low ambient, and the two must be electrically interlocked as shown from line 15 to line 26.

IR-2 in line 26 is the No. 2 cooling relay and the key to energizing the circuit that feeds terminals 2 and 6 and the No. 1 condenser fan motor; trace out that circuit. But when the low-ambient accessory kit is used,

the jumper between terminals 2 and 6 is removed and the circuit, shown in dashed lines, is connected. Following line 26, the relay switch *IR-2* power flows to *6R* coil and on to *2R* (cooling contacts) and to the contactor coil *2M*, or it closes the No. 2 compressor contactor. *2R* functions from the second stage of the cooling thermostat. The *6R* coil is the key to furnishing power to line 25 through contacts *6R*. Line 25 has a low-pressure cutout (*2LP*), high-pressure cutout (*2HP*), electrical

Commercial and Engineered Control Systems 521

overload switches *40L* and *50L,* and thermal cutout *60L* . . . all of which protect the No. 2 compressor.

Note that line 18 is a similar protective circuit for the No. 1 compressor. The two lockout relays *6R* and *7R* are normally closed.

Without a complete detailed explanation of the low ambient accessory, it can be noted that *2 TH,* line 20 (low ambient thermostat) is the key to controlling the No. 1 condenser fan motor through a system of the time delay relays *1 TR, 2 TR,* and *3 TR,* eventually feeding to terminal 6 and the condenser fan motor. *T* (line 20) is an auto transformer fan speed control that modulates the fan motor rpm instead of providing a straight OFF-ON. Note that it is put into the circuit when *9R* switches reverse on response to the need for low ambient operation. Otherwise, the fan speed is at full rpm when the current goes around *T,* because *9R* on line 21 is open. Line 27 is the primary line-voltage side of the low-voltage control transformer. It is rated at 75 VA.

Now go to Fig. A12-2, the low-voltage diagram. Line 1 is the 24-V secondary with a fused protection based on the ampere draw. The space thermostat has a COOL-OFF-HEAT-AUTO system selection. In automatic, the system will bring on heating, cooling, or the economizer system without manual selection. *TC1* and *TC2* are first- and second-stage cooling stats. *TH1* is the single-stage heating stat. The indoor fan selection is ON (continuous) or AUTO to cycle with the *3 TH* blower control.

The economizer control thermostat *5 TH* in the mixing box is tied into the *C1* first-stage cooling circuit, so that the No. 1 compressor cannot operate as long as it is open, meaning the temperature of the outside air is sufficient to provide cool air without refrigeration. But when the outside ambient is high enough to call for cooling, *5 TH* closes, furnishing power to relay *8R* and, through contacts *8R5* and *8R6,* allows both cooling relays *R1* and *R2* to close on command of the cooling stats. *1 TR* and *2 TR* are time-delay relays in the cooling circuits that prevent both compressors from coming on at the same time.

On the heating mode, the economizer system is bypassed. Current from the *TH-1* stat (line 12) goes through the heat anticipator *HA* to terminal *H1.* From *H1* it feeds the gas valve *1GV,* provided that the limit control contact *4 TH* is closed. Another circuit on line 17 provides an alternate flow of current to the pilot safety switch, *PS,* and through its contacts feeds the *1GN* pilot ignition transformer. But ignition of the pilot cannot take place unless there is sufficient gas pressure to close the *PE* (pressure/electric) switch (line 11). At the same time the *TR* (time-delay relay) is also energized, the contacts below it are reversed, and the gas valve opens. All of this happens instantly.

If the fan selector is at the ON position, current from terminal *F* goes through normally closed *3R-2*

switch and energizes *3M* (the contactor coil) for the indoor or evaporator fan. In the AUTO position, that circuit is essentially bypassed. Current then (on line 19) flows through *3 TH,* the blower control thermostat (if heat is present to close the switch), and to the *3R* coil, which then closes *3R-1* and opens *3R-2.* Thus *3M* is again energized and the indoor blower comes on. It will remain on until *3 TH* opens to reflect on heat in the plenum.

The power (240 V on lines 20 and 21) to operate the economizer outside air damper is picked up at terminal blocks 1 and 2 from line 10 on Fig. A12-1. Two diagrams are shown. One is of a damper motor with a nonspring return; the other has a spring return. With the spring return the motor will close completely in the event of a power failure in the damper motor system. If there is no spring return, the damper motor will close to some minimum position based on the minimum amount of outside air needed. Note that relay *8R,* coil line 6, is the key to all the switch actions; *8R-1, 8R-2, 8R-3,* and *8R-4* to position the control dampers *1-RH* and *2-RH.*

This review may seem complicated to the beginner, but with a little concentration, it will be seen that its circuits are not that much different than those in residential systems. The difference is the addition of crankcase heaters, low ambient control, economizer dampers, electronic ignition, and dual compressor operation.

Other options that may be found in similar products are two-stage heating, flue venting motors and their control relays, and a ventor fan sail switch to verify air flow.

The control system for this same basic cooling unit but with electric heating instead of gas will be quite different in the low-voltage heating arrangement. The base drawing shown in Fig. A12-1 essentially does not change.

A12-3
ENGINEERED PNEUMATIC CONTROL SYSTEMS

Pneumatic control systems use compressed air to supply energy for the operation of valves, motors, relays, and other pneumatic control equipment. Consequently, the circuits consist of air piping, valves, orifices, and similar mechanical devices.

Pneumatic control systems are reported to offer a number of distinct advantages, especially in commercial and industrial applications:

- Pneumatic equipment is inherently adaptable to modulating operation, yet two-position or positive operation can easily be provided.
- A great variety of control sequences and combi-

nations are available by using relatively simple equipment.

- Pneumatic equipment is said to be relatively free of operating difficulties.
- It is most suitable for controlling explosion hazards.
- Costs may be less than electrical controls if building codes require the total use of electrical conduit.

Pneumatic control systems are made up of the following elements:

1. A source of clean, dry compressed air to provide the operating energy.
2. Air lines, usually copper or plastic, from the air supply to the controlling devices. These are called *mains*.
3. Controlling devices such as thermostats, humidistats, humidity controllers, relays, and switches. These are the *controllers*.
4. Air lines leading from the controlling devices to the controlled devices. These are called *branch lines*.

   Controlled devices such as valves and motors. These are called *operators* or *actuators*.

The air source is an electrically driven compressor (Fig. A12-5), which is connected to the storage tank in which the pressure is maintained between fixed limits (usually between 20 and 35 psi for low-pressure systems). Air leaving the tank is filtered to remove oil and dust, and in many installations a small refrigeration unit is included to condense out any entrained moisture. Pressure-reducing valves control the air pressure to the main that feeds the controller (thermostat).

If suitable compressor air already exists in the building for other purposes, the compressor can be eliminated and a reducing valve station may be installed to clean and reduce the air pressure to required conditions.

The controller function is to regulate the positioning of the controlled device. It does this by taking air from the supply main at a constant pressure and delivering it through the branch line to the controlled device at a pressure that is varied according to the change in the measured conditions.

A bleed type of thermostat control is illustrated in Fig. A12-6. The bimetallic element, which reacts to temperature, positions the vane near or away from the air nozzle, and thus the pressure in the branch line is relative to how much air is bled off. Bleed controls used directly do not have a wide range of control, so they are frequently coupled to a relay that is separately furnished with air for activating purposes, and the bleed thermostat simply controls the relay action. Bleed controls naturally cause a constant drain on the compressed air source.

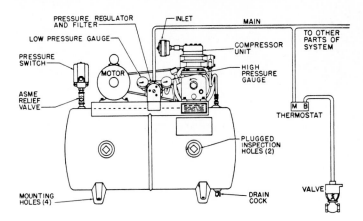

FIGURE A12-5 Basic pneumatic control system.

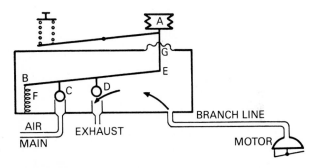

FIGURE A12-6 Bleed thermostat.

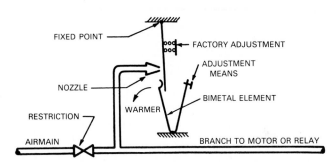

FIGURE A12-7 Nonbleed stat on a decrease in the measured condition.

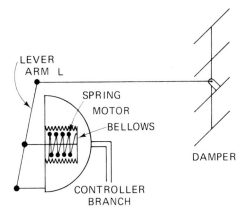

FIGURE A12-8 Pneumatic actuator and normally open damper.

Commercial and Engineered Control Systems   523

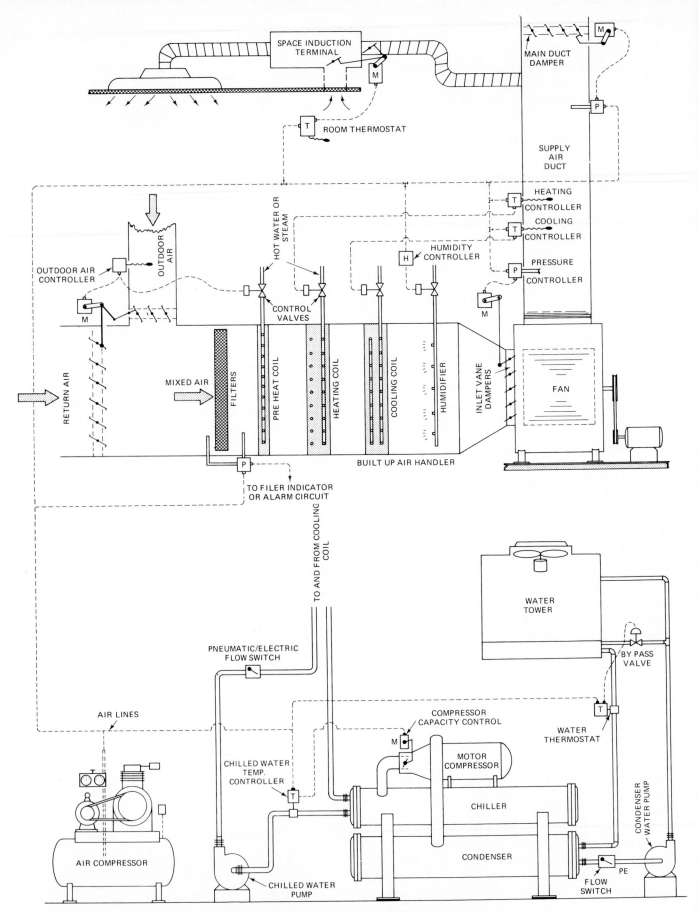

FIGURE A12-9 Pneumatic engineered control system.

Nonbleed-type controllers use air only when the branch line pressure is being increased. The air pressure is regulated by a system of valves (Fig. A12-7), which eliminates the constant bleeding of air that is present in the bleed thermostat. Valves C and D are controlled by the action of the bellows (A) resulting from changes in room temperature. Although the exhaust is in a sense a bleeding action, it is relatively small and occurs only on a pressure increase.

The controlled devices, actuators or operators, are mostly pneumatic damper motor and valves. The principle of operation is the same for both. Figure A12-8 is a diagram of a typical motor. The movement of the bellows as branch-line pressure changes actuates the lever arm or valve stem. The spring exerts an opposing force so that a balanced, controlled position can be stabilized. The motor arm L can be linked to a variety of functions.

Figure A12-9 is a pictorial review of some of the functions that can be involved in a pneumatic engineered control system. Note that there will always be some crossover between the air devices and the electrical system, and the device most widely used is a pneumatic/electric relay.

A12-4
ELECTRONIC CONTROLS

Electronic control systems may also be used effectively in the control of HVAC systems in commercial and industrial work and may become more popular for residential use. Electronic automatic control is the newest of the three types of controls (electric, electronic, and pneumatic). It offers a number of advantages. The sensing elements of electronic controllers are constructed simply; there are no moving parts to interfere with dependable operation. Response is fast. The regulatory element of the controller is usually some distance from the sensing element, and this offers several benefits: (1) all adjustments can be made from a central location, (2) the central area may be cleaner than the place where the sensing element is located, and (3) temperature averaging can be more easily accomplished.

Only simple low-voltage connections are needed between the sensing element and the electric circuit. Flexibility is an important plus, since electronic circuits can be combined with both electric and pneumatic circuits to provide results that ordinarily cannot be achieved separately. Electronic circuits can coordinate temperature changes from several sources (room, outdoor air, fan discharge air, etc.) and program actions accordingly. Another example is a sensing element of perhaps 25 ft in length in a duct to average temperatures as compared to a single-point bulb location.

Electronic controls are based on the Wheatstone bridge concept (Fig. A12-10), which is composed of two

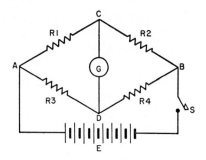

FIGURE A12-10 Wheatstone bridge.

sets of two series resistors (R_1 and R_2; R_3 and R_4) connected in parallel across a dc voltage source. A galvanometer, G (a sensitive indicator of electrical current), is connected across the parallel branches at junctions C and D between the series resistors. If switch S is closed, voltage E (dc battery) is impressed across both branches. If the potential at point C equals that at D, the potential difference is zero and the galvanometer will indicate zero current. When this condition exists, the bridge is said to be *balanced*.

But if the resistance in any one leg is changed, the galvanometer will register a current flow, indicating the existence of a potential difference between points C and D. The bridge is now *unbalanced*.

If that resistance value were changed as a result of temperature reaction, we now have an electronic method of measuring the current flow in relation to the temperature change. With a few minor changes we can create an electronic main bridge circuit such as in Fig. A12-11. A 15-V ac circuit replaces the dc battery. The galvanometer is replaced by a combination voltage amplifier-phase discriminator-switching relay unit. Resistor R_2 is replaced by a sensing element T of an electronic controller.

The purpose of the voltage amplifier is to take the small voltage from the bridge and increase its magnitude by stage amplification to do work. Phase discrimination simply means determining the sensor action. In an electric bimetal thermostat the mechanical movement is directly related to temperature changes. However, the

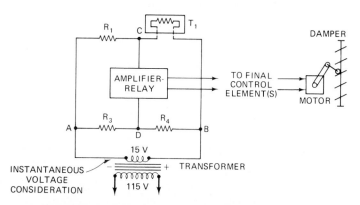

FIGURE A12-11 Electric main bridge circuit.

electronic sensing element is a nonmoving part, and the phase discriminator determines whether the signal will indicate a rise or fall in temperature. The relay then operates the final element action. Phase discrimination can be two-position or, with certain modifications, can be converted to a modulating system.

The crossover point from electronic to electric occurs at the output of the amplifier and relay signal. The motor is a conventional ON-OFF, or a proportional (modulating) electric motorized valve, damper actuator, etc.

Electronic temperature-sensing elements are room thermostats, outdoor thermostats, insertion thermostats for ducts (from several inches to 25 ft or over), insertion thermostats for liquid lines, etc. The typical room thermostat is a coil of fine wire wound on a bobbin. The resistance of the wire varies directly with temperature changes.

Obviously, electronic controls do not sense pressures, but they can detect and control humidity within narrow limits. The sensing element of a humidity controller is gold leaf embossed on a plastic base and coated with a special salt. Its resistance value is much greater than that of the sensing elements used in temperature controllers, but the functional results are the same. Electrical resistance will vary with the change in humidity, and these devices are very sensitive and accurate.

FIGURE A12-12 (*Courtesy* Honeywell, Inc.)

A12-5
CONTROL AUTOMATION

As mentioned at the beginning of this chapter, the complexity of functions and variables to be controlled and monitored in engineered systems have given rise to the use of *building automation techniques* using data control centers (Fig. A12-12). In large buildings, or in a complex of buildings such as a college or university,

it is impossible to have enough operating and maintenance personnel to stop, start and watch each system or component during a 24-hour day.

The control center collects key operating data from the heating, ventilating, and air-conditioning system and incorporates remote-control devices to supervise the system's operation. Although it can be used in preventive maintenance by detecting faulty operation before it can cause serious trouble, the control center's most important contribution is its efficient use and scheduling of personnel to investigate and handle the overall operation. The sophistication of such data centers is, of course, related to the type and size of the installation and economic considerations. Some have continuous scanners with alarm indicators to monitor refrigeration machines, oil and refrigeration pressures, chilled water temperatures, cooling tower water temperatures, air filter conditions, low-water conditions in the boiler, etc., as well as the conventional space temperature and humidity conditions in each zone.

PROBLEMS

A12-1. What is the main electrical distinction between residential and commercial equipment?

A12-2. Commercial system wiring diagrams are usually prepared by _____ .

A12-3. Engineered control systems are usually prepared by _____ .

A12-4. Name three types of commercial and engineered control systems.

A12-5. In cold weather where cooling is not required, power to the unit can be shut off. True or False?

A12-6. What is the advantage of using multiple condenser fans?

A12-7. A 460/230-V 1.5-kVA transformer will provide what rated output amperage?

A12-8. A pneumatic control system is made of five elements. What are they?

A12-9. The air source pressure is usually maintained between _____ and _____ psig.

A12-10. Instead of making and breaking an electrical circuit, a pneumatic controller does what?

A13

Heating, Measuring, and Testing Equipment

A13-1
GENERAL

In Chapter R4 a list of the typical hand tools and their descriptions were given. These tools would be used regardless of the field of endeavor: refrigeration, heating, or air-conditioning.

In Chapter R22 instruments were listed that would generally apply to the refrigeration field. Of course, most could also apply to air-conditioning as well as some to heating. In each field of operation, specific tools and test instruments would also be used. In this chapter we cover those tools and testing instruments that apply to the installation startup and service of heating equipment, covering gas, oil, and electric.

A13-2
GAS

The installation startup and service of a gas furnace requires proper adjustment of temperatures and pressures in several categories: fuel input, combustion efficiency, air quantity, and service trouble diagnosis. The actual procedures of startup, checkout, and operation of heating systems are covered in Chapter A14. This chapter is limited to descriptive material of the test tools and instruments.

A13-2.1
Gas Input

The amount of fuel supplied to a gas furnace in Btu/hr will depend on the Btu content of each cubic foot of the gas as well as the pressure in the manifold supplying the main burners. Two methods measuring this input are by timing the gas meter (the procedure is explained in Chapter A14) and by measuring the manifold pressure.

Timing the meter will require the use of a sweep secondhand on an ordinary watch or, for more accuracy, a stopwatch. The stopwatch can be part of the features in a digital wristwatch or may be a sporting-event stopwatch. Either one is a necessity to accurately check the input to a gas heating unit.

To measure the manifold pressure, we work in units of pressure of less than a pound per square inch gauge. Thus, we have to use a measurement tool called a *manometer* (Fig. A13-1), which is simply a U-tube of glass, or preferably clear plastic, with a sliding inch scale. Fill the tube with water as shown. By connecting a rubber tube to one side, we can blow into the tube, cause the water to rise, and measure the air pressure we exert. If we exert a pound of pressure as seen on the dial, the water would rise 27.71 in., assuming that the glass tube was that long, hence the term "inches of water column" (W.C.). [This is called "inches of water gauge" (W.G.) in the engineering field. The W.C. designation is used throughout this book.] One ounce of pressure equals 1.732 in. of W.C. So when we talk later about adjusting the natural gas manifold pressure to 3.5 in. of W.C., we are talking about 2.02 oz of pressure; we adjust the LP gas to 11 in. of W.C., which equals 6.35 oz.

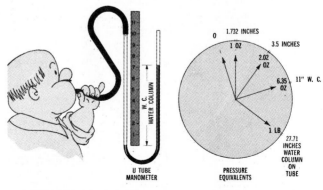

FIGURE A13-1   Measuring pressure. (*Courtesy* Borg-Warner Central Environmental Systems, Inc.)

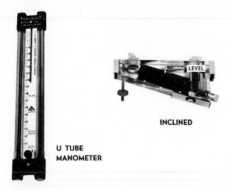

FIGURE A13-2   Instruments for measuring pressure. (*Courtesy* Bacharach Instrument Company and Dwyer Instrument Company)

FIGURE A13-4   Gas service instrument kit. (*Courtesy* Bacharach Instrument Company)

On the left side of Fig. A13-2 is a typical gas pressure manometer that measures up to 15 in. of W.C. When measuring the pressure of air traveling through the air distribution system, pressures of less than 1 in. of W.C. are used. For greater accuracy, when working in tenths of an inch of W.C., an inclined manometer is used, as shown on the right. It is also possible to measure negative or suction pressures in the return air duct. Another type of gauge measuring inches water column and ounces per square inch is a Bourdon-tube type (Fig. A13-3).

A13-2.2
Operation

Once the proper input has been established within plus or minus 10% of the rated capacity of the unit, tests for proper operation should be conducted. The instruments should include:

1. A draft gauge to at least 0.15 in. of W.C.
2. A 0 to 1000°F stack temperature thermometer with a $5\frac{1}{2}$-in. stem

3. A CO_2 indicator with aspirator and pickup tube
4. A fire-efficiency slide rule for gas
5. Two dial-type or pocket thermometers in the range 50 to 200°F.

Usually, these instruments can be purchased in kit form, such as shown in Fig. A13-4. This kit includes a draft gauge (Fig. A13-5), called a neutral pressure point indicator, which is used to determine the amount of negative pressure (draft) in the furnace vent as well as the overfire pressure neutral point in gas conversion burners. With a scale range of $+0.08$ to -0.12 in. of W.C., this instrument has the range requirements to measure draft pressure in most residential and small commercial gas-fired units. For larger units, especially those using power-type burners where draft requirements are higher, the draft gauge shown in Fig. A13-6 would be used. This gauge has a range of $+0.5$ to $+0.25$ in. of W.C. and is of heavier construction. With a remote-type sampling tube, it is also more versatile around larger heating units.

FIGURE A13-3   Gas manifold pressure gauge. (*Courtesy* Marshalltown Instruments, Inc.)

FIGURE A13-5   (*Courtesy* Bacharach Instrument Company)

FIGURE A13-6 (*Courtesy* Bacharach Instrument Company)

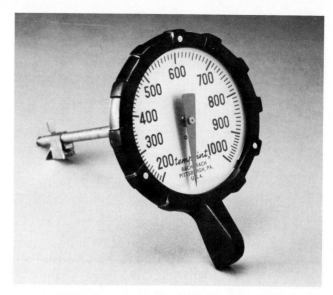

FIGURE A13-8 Flue gas therometer. (*Courtesy* Bacharach Instrument Company)

To check the efficiency of a gas-fired unit, the percent of carbon dioxide (CO_2) in the flue products and the temperature rise of the flue products must both be measured. To measure the CO_2 content, a CO_2 indicator and aspirator bulb are used. Figure A13-7 shows such an instrument. The aspirating tube is inserted in the flue outlet of the heating unit ahead of the draft diverter so as to obtain true flue product samples. The aspirator bulb acts as a gas pump when squeezed, forcing the flue products through the CO_2 indicator. Twenty squeezes or pumps of the bulb are required to purge the tubing, bulb, and indicator and obtain a pure sample. After the

pumping operation, the indicator is inverted several times to mix the flue products and the absorbing liquid to remove the CO_2 from the flue products. The level of liquid in the indicator will rise up the center tube and indicate the percent of CO_2 in the sample. Each instrument has complete instructions that should be read and followed for operation as well as maintenance.

To measure the flue product temperatures, a dial-type thermometer, usually in the range 200 to 1000°F, is used. One with a $11\frac{1}{2}$-in. stem is recommended, as the temperature must be taken ahead of the draft diverter to eliminate temperature loss from mixing with outside air. Figure A13-8 shows the dial thermometer from the set in Fig. A13-4.

With the two measurements taken, CO_2 percentage and stack temperature rise (difference between the temperature of the air entering the burner compartment and the flue product temperature), the fire efficiency finder (Fig. A13-9) is used to determine the combustion efficiency of the unit. To measure the temperature of the air entering the burner compartment or air surrounding

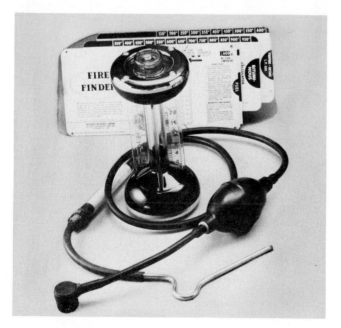

FIGURE A13-7 (*Courtesy* Bacharach Instrument Company)

FIGURE A13-9 (*Courtesy* Bacharach Instrument Company)

Heating, Measuring, and Testing Equipment   529

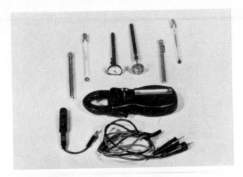

FIGURE A13-10 (*Courtesy* Robinair Division, Sealed Power Corporation)

the unit, dial or pocket-type thermometers such as those shown in Fig. A13-10 can be used. These thermometers would also be used in the return and supply air through the heating unit to determine the temperature rise through the unit. For more about proper temperature rise and CAC (continuous air circulation), see Chapter A14.

A13-2.3
Service

The instruments listed under "input" and "operation" would also be used under the "service" category. In addition, some test instruments would be used primarily for service, as it is assumed that the various parts of the heating unit will perform when installed and started. If malfunction is encountered at the time of startup or after a period of operation, the following instruments are suggested:

1. Clamp-type volt/ammeter
2. Millivoltmeter
3. Dc or micrometer
4. Test drill set

Included in Fig. A13-10 is a clamp-type volt/ammeter used to check the power supply to the heating unit (high voltage of 120 V or possibly 240 V) and the power supply to the control system (usually low voltaged, 24 V). This instrument would also be used to check the amperage draw of the blower motor to determine the blower motor load.

A millivoltmeter, as shown in Fig. A13-11, is required to check the output of thermocouples in the pilot assembly for proper safety dropout as well as operation range. With the introduction of electronic ignition systems, a dc microammeter is required (Fig. A13-12).

To check the main burner orifice size to determine proper input of gas to the unit, a 0-60 number drill set is required. This drill set must be used exclusively for this function and must not be clamped in a drill except

to check the size of the brass orifice. To use for any other purpose will distort the drill sizes and destroy their accuracy.

A13-3
OIL

The installation, startup, and servicing of oil-fired heating units also requires proper adjustment of temperatures and pressures in several categories, fuel input, combustion efficiency, and air quantity, as well as service diagnosis. The step-by-step procedure of startup, operation, and check-out of an oil-fired unit is covered in Chapter A14. Following is a description of the special tools and instruments that would be used in each category of input, operation, and service.

A13-3.1
Input

An oil-fired unit receives fuel in liquid form, and must vaporize it, mix it with the correct amount of air, and burn it for maximum heat generation in the combustion process as well as to provide high efficiency through the heating unit. The first concern is supplying

FIGURE A13-11 (*Courtesy* A. W. Sperry Instruments, Inc.)

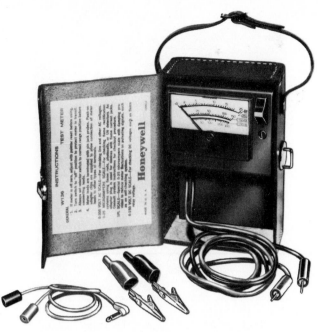

FIGURE A13-12   (*Courtesy* Honeywell, Inc.)

the proper amount of fuel through the burner. This is determined by the pressure the oil burner fuel pump will supply to the spray nozzle in the firing assembly. A compound pressure gauge in the range 0 to 30 in. vacuum and 0 to 150 psig is used on residential and small commercial units. Such a gauge, together with a 12–in. flexible hose as well as several type of fittings, is shown in Fig. A13-13.

The fuel supply may have to come from storage tanks located below the level of the oil burner pump, which means a vacuum to the inlet of the pump. Because of the lift limitation of the pump, it is necessary to know if a vacuum exists, and how much. The gauge is therefore dual ranged (compound type with an overall range of 30 in. vacuum to 150 psig).

A13-3.2
Operation

To check the operation of an oil-fired unit after proper input is established requires the same readings as those for a gas furnace: CO_2, stack temperature, and draft. Although taken in a different manner, the instruments are the same. CO_2 readings in the smoke pipe ahead of any draft control are taken with a CO_2 analyzer (Fig. A13-7). Stack temperature is taken in the same location with the dial-type thermometer (Fig. A13-8). Stack draft and overfire draft are taken with a remote sampling tube type of draft gauge (Fig. A13-6).

In addition to these instruments, a smoke tester is needed to determine the amount of free carbon or "smoke" in the flue products. This is shown in Fig. A13-14. Shown is the smoke tester pump, filter paper, and spot density comparison chart. With the sampling tube inserted into the smoke outlet, and with 10 full strokes of the pump (required by the model RCC-B smoke tester from Bacharach Instrument Company), the correct amount of flue products is drawn through the filter paper. The spot formed on the paper is then inserted behind the test circles for comparison, based on the number of the spot selected for match of the smoke spot colors. The instructions give the following:

Smoke Scale Reading 1   Excellent; little, if any, sooting of furnace or boiler surfaces.

Smoke Scale Reading 2   Good; may be slight sooting with some types of furnace or boiler but little, if any, increase in flue product temperature.

FIGURE A13-13   Oil pressure test set.
(*Courtesy* Robinaire Division, Sealed
Power Corporation)

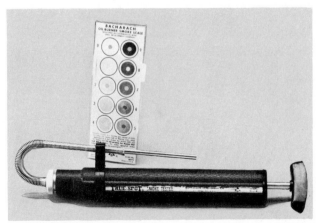

FIGURE A13-14   (*Courtesy* Bacharach Instrument
Company)

Heating, Measuring, and Testing Equipment   531

Smoke Scale Reading 3 Fair; substantial sooting with some types of furnaces or boiler but rarely will require cleaning more than once a year.

Smoke Scale Reading 4 Poor; this is a borderline smoke; some units may soot only moderately, others may soot rapidly.

Smoke Scale Reading 5 Very poor; heavy sooting in all cases; may require cleaning several times during heating season.

Smoke Scale Reading 6 Extremely poor; severe and rapid sooting; may result in damage to stack control and reduce over-fire draft to danger point.

Step-by-step application of the smoke tester and the step information are given in Chapter A14.

Another tool not necessarily required for proper operation but desirable to use is a flame mirror. The flame mirror shown in Fig. A13-15 is a highly polished stainless steel mirror with telescoping handle used to observe flame pattern during operation as well as oil flow after burner cutoff if this exists. It is also useful for checking surfaces and flue product openings in the heat exchanger as well as in inspection for heat exchanger failure.

A13-3.3
Service

Use of the "cad cell" type of oil burner control means that an electrical circuit is the flame detector rather than mechanical action. The cad cell is also located inside the burner blast tube, where operation ob-

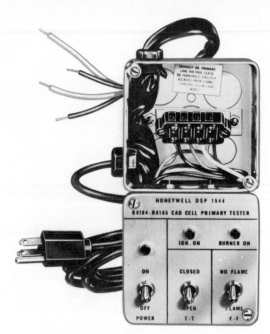

FIGURE A13-16   Cad cell primary tester. (*Courtesy* Honeywell, Inc.)

FIGURE A13-17   Nozzle tool. (*Courtesy* Monarch Manufacturing Works, Inc.)

servation is not possible when the burner is operating. Figure A13-16 shows a cad cell tester that can be wired into the cad cell circuit for testing while the burner is operating.

A tool required in the service of oil burners is a nozzle remover, shown in Fig. A13-17. Using a stainless steel nozzle set into a brass nozzle adapter located between two ceramic-insulated ignition electrodes, use of two adjustable wrenches to loosen the nozzle almost guarantees breakage of the electrode insulators. The nozzle remover moves the action out beyond the electrodes and helps considerably to eliminate electrode damage.

Although there are many tools and pieces of test equipment built and marketed for the heating field, we have attempted to list the most common types. Some products require special testing equipment, which is usually marketed by the product manufacturer involved.

FIGURE A13-15   Flame mirror. (*Courtesy* Monarch Manufacturing Works, Inc.)

A13-1. The input to a gas heating unit can be set by either of two methods. They are _____ , and the instruments required are _____ .

A13-2. What instruments are required to determine the efficiency of a gas heating unit?

A13-3. What must the operator do to operate a CO_2 analyzer properly?

A13-4. When servicing a gas-fired heating unit, a millivolt-meter would be used to check the operation of what device?

A13-5. What is a test drill set used for?

A13-6. The presssure gauge used to check an atomizing oil burner is a compound type, 0 to 30 in. vacuum, 0 to 150 psig. Why?

A13-7. A properly set oil burner will operate with a flue gas of _____ % CO_2, _____ °F stack temperature, and −_____ in. of W.C. over fire draft.

A13-8. To obtain the proper overfire draft, the maximum stack draft would be _____ to _____ in. of W.C.

A13-9. On a pump-type smoke tester, the smoke spot sample should read _____ .

A13-10. The best way to check for possible afterburn would be with a _____ .

A13-11. The cad cell operates by changing the resistance due to the effects of which of the following: heat, light, or air travel over the cell?

A13-12. What is the major advantage of a nozzle remover?

Heating Startup, Checkout, and Operation

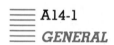

A14-1
GENERAL

In Chapter A5 heating units and component parts were discussed. In Chapter A10 the various types of controls—temperature, pressure, etc.—were discussed. Chapter A11 discussed the heating control systems and Chapter A13 the testing equipment involved. In this chapter we proceed through the process of starting, operating, and checking-out heating units.

A14-2
GAS

To check, test, and adjust a gas-burning unit for highest and safest operating efficiency, the unit must have the proper amount of input for the unit size, the proper adjustment of the burners, as well as the correct amount of combustion air, proper venting, and the correct amount of air through the heat distribution system.

A14-2.1
Input

Before a gas-fired unit can be set for proper burner operation, the *input must be correct*. The amount of gas to be burned in cubic feet of gas per hour for a given Btu/hr input to the unit will depend on two factors:

1. Btu content per cubic foot
2. Specific gravity

The Btu content per cubic foot or *heating value* of the gas is the amount of heat released when the 1 ft³ of gas is mixed with the proper amount of air and burned. The more carbon and hydrogen atoms in the molecules of the gas, the higher will be its heating value number. The largest part of natural gas is methane (CH_4), which is composed of four hydrogen atoms linked to one carbon atom. Ethane (C_2H_6), also found in natural gas but in smaller quantities, is composed of six hydrogen atoms linked to two carbon atoms.

The two gases that are found mixed with liquid petroleum (crude oil) are propane and butane. Propane (C_3H_8) has eight hydrogen atoms linked to three carbon atoms. Butane (C_4H_{10}) has ten hydrogen atoms linked to four carbon atoms.

From this we can see that when burning a cubic foot of each of these gases, a different amount of heat will be released. Because natural gas rarely occurs in a pure state, the mixtures of methane and ethane will vary in each location where the gas is obtained. As a result, the heating value of natural gas is in the range 945 to 1121 Btu/ft³.

Propane and butane are obtained from crude oil wells and do occur in the pure state. Although mixed, they are separatable by condensing them at their respective pressure–temperature relationships. As a result, propane gives a fairly constant 2522 Btu/ft³ and butane yields 3261 Btu/ft³.

The procedures are different for establishing the proper input of propane and butane for natural gas and liquefied petroleum (LP) gas; therefore, we will discuss each procedure separately.

Natural Gas: Several factors affect the input to a natural gas furnace: the size of the supply pipe to the unit, the pressure of the gas in the main burner manifold, and the size of the main burner orifice. All of these are dependent on the amount of gas in cubic feet per hour that must be delivered to the unit as well as the Btu/ft³ and density of the gas.

Supply Piping: To figure the size of the pipe supplying the natural gas to the unit, both the heat content and specific gravity of the gas must be known. The heat source of this information is your local utility. Find out and record this information for each utility in your area.

To explain the method of pipe sizing, we will assume an average Btu content of 1050 per cubic foot and an average specific gravity of 0.62.

Figure A14-1 shows the capacity of iron pipe sizes from $\frac{3}{4}$ to 4 in. in lengths from 15 to 600 ft. The length of the pipe also includes an "ordinary" number of fittings. The assumption is that the equivalent length of the fittings will not be the actual pipe length.

You will note that the table is based on a gas specific gravity of 0.7 and a maximum pressure drop through the pipe of 0.3 in. of W.C.; the average pressure on the outlet of the pressure regulator at the meter is 9 in. of W.C. With a maximum pressure drop in the piping (when the full house load is on—water heater, cooking stove, takeover, and heating unit) of 0.3 in. of W.C., an average pressure of 0.6 in. of W.C. will be supplied to the various operating units.

To use the table, the following conditions are assumed:

Heat content of the gas is 1050 Btu/ft^3

Specific gravity of gas is 0.62

≡ **FIGURE A14-1**
≡ Gas piping table.

LENGTH OF PIPE	DIAMETER OF PIPE—INCHES						
Feet	$\frac{3}{4}$	1	$1\frac{1}{4}$	$1\frac{1}{2}$	2	3	4
15	159	319	694	1130	2296	6019	12853
30	111	223	495	787	1648	4352	8982
45	91	184	403	648	1366	3611	7315
60	80	160	352	565	1195	3195	6297
75	71	143	320	505	1037	2778	5556
90	65	130	287	454	926	2500	5093
105	60	121	264	417	852	2269	4722
120	—	111	250	389	796	2130	4445
150	—	100	224	352	722	1935	4028
180	—	92	208	324	666	1805	3704
210	—	85	190	296	611	1648	3426
240	—	—	176	278	574	1555	3232
270	—	—	165	264	537	1463	3010
300	—	—	157	250	505	1380	2778
450	—	—	130	210	417	1139	2315
600	—	—	110	178	361	954	1972

Capacity of pipe of different diameters and length, in cu. ft. per hr. with press. Drop of 0.3 in. and specific gravity of 0.7.

NOTE: In using this table no allowance for ordinary number of fittings is necessary.

When Specific Gravity *is not* 0.70, multiply by: ★ $\dfrac{0.70}{\text{Sp. of Gas}}$

For Pressure drops other than 0.3, *multiply by:*
★ $\dfrac{\text{Press. Drop, In. of Water}}{0.3}$

Source: Addison Products Company.

Total house load is 260,000 Btu/hr

Heating unit input is 150,000 Btu/hr

House supply pipe is 80 ft long

Furnace branch pipe is 35 ft long

We proceed as follows:

1. Determine the total cubic feet per hour needed by the house.

   $\dfrac{260,000 \text{ Btu/hr}}{1050 \text{ Btu/ft}^3} = 247.6$ ft^3 at 0.7 specific gravity

   $\dfrac{0.7}{0.62} = 1.129$ adjustment factor

   247.6 ft^3 × 1.129 A.F. = 279.5 ft^3 for sizing pipe.

2. Determine the main supply to the house. From the table, using a length of 90 ft, $1\frac{1}{4}$-in. pipe will carry 287 ft^3 per hour; therefore, to carry 279.5 ft^3 per hour for 85 ft, the $1\frac{1}{4}$-in. iron pipe size is required.

3. Determine the total cubic feet per hour needed by the heating unit.

   $\dfrac{150,000 \text{ Btu/hr}}{1010 \text{ Btu/ft}^3} = 142.8$ ft^3 at 0.7 specific gravity

   142.8 × 1.129 A.F. = 161.2 ft^3 for sizing pipe

4. Determine the branch pipe size to the heating unit. From the table, using a length of 35 ft, 1-in. pipe will carry 184 ft^3 for 45 ft. Therefore, to carry 161.2 ft^3 for 35 ft will require 1-in. iron pipe. We also see that the $\frac{3}{4}$-in. iron pipe will carry only 91 ft^3 for 45 ft and 111 ft^3 for 30 ft. Neither condition will supply our example unit.

The carrying capacities given in the table are recommended maximums.

Manifold Pressure: Heating units are the only gas-burning appliance in the home that uses its own pressure regulator. Cook stoves, baking ovens, grills, and water heaters are designed to operate on the 7 in. W.C. average supply-line pressure. They are also not critical of supply-line pressure, which could vary between 5 and 9 in. of W.C.

Heating units are critical between full input and 10% less because of the action of fan and limited controls involved. Therefore, heating units are designed with their own regulators set at an output of $3\frac{1}{2}$ in. of W.C. Burner design will allow operation between 3 and 4 in. of W.C. manifold pressure, but these pressures should not be exceeded.

If the manifold pressure to obtain proper input must be above or below this pressure range, the burner orifice size must be changed. To determine the correct size of orifice for the individual burner, the following information is needed: Btu/hr input per burner, Btu/ft³ of gas, and specific gravity of the gas.

If the heating unit is a four-burner sectional type, each burner would require 25% of the 150,000 Btu/hr input or 37,500 Btu/hr of the 1050-Btu/ft³ gas at 0.62 specific gravity.

The following tables are used with the permission of the American Gas Association and are taken from their Catalog XH0373. It is suggested that this material be referred to for more detailed information.

Figure A14-2 gives the cubic feet per hour capacity for utility gases of 0.60 specific gravity at sea level. If gases of other than 0.60 specific gravity are used and at higher than sea level, adjustments have to be made. These are discussed later.

Because the 0.62 specific gravity of the natural gas

≡ FIGURE A14-2
≡ Utility gases (ft³/hr at sea level).

Orifice Size (Decimal or DMS)	GAS PRESSURE AT ORIFICE - INCHES WATER COLUMN								
	3	3.5	4	5	6	7	8	9	10
.008	.17	.18	.19	.23	.24	.26	.28	.29	.30
.009	.21	.23	.25	.28	.30	.33	.35	.37	.39
.010	.27	.29	.30	.35	.37	.41	.43	.46	.48
.011	.33	.35	.37	.42	.45	.48	.52	.55	.59
.012	.38	.41	.44	.50	.54	.57	.62	.65	.70
80	.48	.52	.55	.63	.69	.73	.79	.83	.88
79	.55	.59	.64	.72	.80	.84	.90	.97	1.01
78	.70	.76	.78	.88	.97	1.04	1.10	1.17	1.24
77	.88	.95	.99	1.11	1.23	1.31	1.38	1.47	1.55
76	1.05	1.13	1.21	1.37	1.52	1.61	1.72	1.83	1.92
75	1.16	1.25	1.34	1.52	1.64	1.79	1.91	2.04	2.14
74	1.33	1.44	1.55	1.74	1.91	2.05	2.18	2.32	2.44
73	1.51	1.63	1.76	1.99	2.17	2.32	2.48	2.64	2.78
72	1.64	1.77	1.90	2.15	2.40	2.52	2.69	2.86	3.00
71	1.82	1.97	2.06	2.33	2.54	2.73	2.91	3.11	3.26
70	2.06	2.22	2.39	2.70	2.97	3.16	3.38	3.59	3.78
69	2.25	2.43	2.61	2.96	3.23	3.47	3.68	3.94	4.14
68	2.52	2.72	2.93	3.26	3.58	3.88	4.14	4.41	4.64
67	2.69	2.91	3.12	3.52	3.87	4.13	4.41	4.69	4.94
66	2.86	3.09	3.32	3.75	4.11	4.39	4.68	4.98	5.24
65	3.14	3.39	3.72	4.28	4.62	4.84	5.16	5.50	5.78
64	3.41	3.68	4.14	4.48	4.91	5.23	5.59	5.95	6.26
63	3.63	3.92	4.19	4.75	5.19	5.55	5.92	6.30	6.63
62	3.78	4.08	4.39	4.96	5.42	5.81	6.20	6.59	6.94
61	4.02	4.34	4.66	5.27	5.77	6.15	6.57	7.00	7.37
60	4.21	4.55	4.89	5.52	5.95	6.47	6.91	7.35	7.74
59	4.41	4.76	5.11	5.78	6.35	6.78	7.25	7.71	8.11
58	4.66	5.03	5.39	6.10	6.68	7.13	7.62	8.11	8.53
57	4.84	5.23	5.63	6.36	6.96	7.44	7.94	8.46	8.90
56	5.68	6.13	6.58	7.35	8.03	8.73	9.32	9.92	10.44
55	7.11	7.68	8.22	9.30	10.18	10.85	11.59	12.34	12.98
54	7.95	8.59	9.23	10.45	11.39	12.25	13.08	13.93	14.65
53	9.30	10.04	10.80	12.20	13.32	14.29	15.27	16.25	17.09
52	10.61	11.46	12.31	13.86	15.26	16.34	17.44	18.57	19.53
51	11.82	12.77	13.69	15.47	16.97	18.16	19.40	20.64	21.71
50	12.89	13.92	14.94	16.86	18.48	19.77	21.12	22.48	23.65
49	14.07	15.20	16.28	18.37	20.20	21.60	23.06	24.56	25.83
48	15.15	16.36	17.62	19.88	21.81	23.31	24.90	26.51	27.89
47	16.22	17.52	18,80	21.27	23.21	24.93	26.62	28.34	29.81
46	17.19	18.57	19.98	22.57	24.72	26.43	28.23	30.05	31.61
45	17.73	19.15	20.52	23.10	25.36	27.18	29.03	30.90	32.51
44	19.45	21.01	22.57	25.57	27.93	29.87	31.89	33.96	35.72
43	20.73	22.39	24.18	27.29	29.87	32.02	34.19	36.41	38.30
42	23.10	24.95	26.50	29.50	32.50	35.24	37.63	40.07	42.14

Orifice Size (Decimal or DMS)	GAS PRESSURE AT ORIFICE - INCHES WATER COLUMN								
	3	3.5	4	5	6	7	8	9	10
41	24.06	25.98	28.15	31.69	34.81	37.17	39.70	42.27	44.46
40	25.03	27.03	29.23	33.09	36.20	38.79	41.42	44.10	46.38
39	26.11	28.20	30.20	34.05	37.38	39.97	42.68	45.44	47.80
38	27.08	29.25	31.38	35.46	38.89	41.58	44.40	47.27	49.73
37	28.36	30.63	32.99	37.07	40.83	43.62	46.59	49.60	52.17
36	29.76	32.14	34.59	39.11	42.76	45.77	48.88	52.04	54.74
35	32.36	34.95	36.86	41.68	45.66	48.78	52.10	55.46	58.34
34	32.45	35.05	37.50	42.44	46.52	49.75	53.12	56.55	59.49
33	33.41	36.08	38.79	43.83	48.03	51.46	54.96	58.62	61.55
32	35.46	38.30	40.94	46.52	50.82	54.26	57.95	61.70	64.89
31	37.82	40.85	43.83	49.64	54.36	58.01	61.96	65.97	69.39
30	43.40	46.87	50.39	57.05	62.09	66.72	71.22	75.86	79.80
29	48.45	52.33	56.19	63.61	69.62	74.45	79.52	84.66	89.04
28	51.78	55.92	59.50	67.00	73.50	79.50	84.92	90.39	95.09
27	54.47	58.83	63.17	71.55	78.32	83.59	89.27	95.04	99.97
26	56.73	61.27	65.86	74.57	81.65	87.24	93.17	99.19	104.57
25	58.87	63.58	68.22	77.14	84.67	90.36	96.50	102.74	108.07
24	60.81	65.67	70.58	79.83	87.56	93.47	99.83	106.28	111.79
23	62.10	67.07	72.20	81.65	89.39	94.55	100.98	107.49	113.07
22	64.89	70.08	75.21	85.10	93.25	99.60	106.39	113.24	119.12
21	66.51	71.83	77.14	87.35	95.63	102.29	109.24	116.29	122.33
20	68.22	73.68	79.08	89.49	97.99	104.75	111.87	119.10	125.28
19	72.20	77.98	83.69	94.76	103.89	110.67	118.55	125.82	132.36
18	75.53	81.57	87.56	97.50	108.52	116.03	123.92	131.93	138.78
17	78.54	84.82	91.10	103.14	112.81	120.33	128.52	136.82	143.91
16	82.19	88.77	95.40	107.98	118.18	126.78	135.39	144.15	151.63
15	85.20	92.02	98.84	111.74	122.48	131.07	139.98	149.03	156.77
14	87.10	94.40	100.78	114.21	124.44	133.22	142.28	151.47	159.33
13	89.92	97.11	104.32	118.18	128.93	138.60	148.02	157.58	165.76
12	93.90	101.41	108.52	123.56	135.37	143.97	153.75	163.69	172.13
11	95.94	103.62	111.31	126.02	137.52	147.20	157.20	167.36	176.03
10	98.30	106.16	114.21	129.25	141.82	151.50	161.81	172.26	181.13
9	100.99	109.07	117.11	132.58	145.05	154.71	165.23	175.91	185.03
8	103.89	112.20	120.65	136.44	149.33	160.08	170.96	182.00	191.44
7	105.93	114.40	123.01	139.23	152.56	163.31	174.38	185.68	195.30
6	109.15	117.88	126.78	142.88	156.83	167.51	178.88	190.46	200.36
5	111.08	119.97	128.93	145.79	160.08	170.82	182.48	194.22	204.30
4	114.75	123.93	133.22	150.41	164.36	176.18	188.16	200.25	210.71
3	119.25	128.79	137.52	156.26	170.78	182.64	195.08	207.66	218.44
2	128.48	138.76	148.61	168.64	184.79	197.66	211.05	224.74	235.58
1	136.35	147.26	158.25	179.33	194.63	209.48	223.65	238.16	250.54
A	145.34	155.48	165.62	189.28	206.18	219.70	236.60	250.12	263.63
B	150.36	160.85	171.33	195.81	213.29	227.25	244.76	258.74	272.73
C	155.45	166.30	177.14	202.44	220.52	234.98	253.06	267.51	281.97
D	160.63	171.84	183.04	209.19	227.87	242.81	261.48	276.44	291.38
E	165.89	177.47	189.05	216.05	235.34	250.77	270.06	285.49	300.93
F	175.43	187.61	199.78	227.75	248.70	265.02	285.40	301.70	318.02
G	180.81	193.43	206.04	235.49	256.50	273.33	294.35	311.18	327.99
H	187.81	200.91	214.01	244.59	266.42	283.89	305.74	323.20	340.67
I	196.38	210.08	223.77	255.75	278.58	296.85	319.68	337.95	356.20
J	203.66	217.87	232.08	265.24	288.92	307.87	331.55	350.49	396.44
K	209.59	224.22	238.84	272.95	297.33	316.82	341.19	360.69	380.18
L	223.23	240.49	257.75	290.71	316.68	337.44	363.40	384.17	404.92

Orifice Size (Decimal or DMS)	GAS PRESSURE AT ORIFICE - INCHES WATER COLUMN								
	3	3.5	4	5	6	7	8	9	10
M	231.00	247.12	263.23	300.83	327.69	349.18	376.03	397.52	419.01
N	242.09	258.98	275.06	315.27	343.42	365.94	394.09	416.61	439.13
O	265.05	283.54	302.03	345.18	376.00	400.66	431.47	456.13	480.79
P	276.92	292.24	315.56	360.64	392.84	418.60	450.80	476.56	502.32
Q	292.57	312.98	333.39	381.03	415.05	442.26	476.28	503.49	530.71
R	305.03	326.33	347.62	397.26	432.73	461.10	496.59	524.95	553.32
S	321.45	344.01	366.57	418.62	456.01	485.91	523.29	553.19	583.09
T	340.19	363.92	387.65	443.04	482.59	514.24	553.79	585.44	617.09
U	359.46	383.54	409.61	468.14	509.93	543.36	585.17	618.60	652.04
V	377.26	403.58	429.90	491.31	535.17	570.27	614.14	649.23	684.33
W	395.48	423.07	450.66	515.05	561.04	597.83	643.82	680.60	717.39
X	418.34	447.53	476.72	544.82	593.47	632.39	681.03	719.94	758.86
Y	433.23	463.45	493.67	564.20	614.58	654.87	705.25	745.56	785.86
Z	452.75	484.33	515.91	589.64	642.26	684.38	737.02	779.14	821.26

Specific Gravity = 0.60
Orifice Coefficient = 0.9
For utility gases of another specific gravity, select factor from Fig. A14-3.
For altitudes above 2,000 feet, first select the equivalent orifice size at sea level from Fig. A14-4.

Source: American Gas Association.

in our example is less than halfway between the 0.60 and 0.65 in Fig. A14-3, we will use the 0.60 factor of 1.000. Figure A14-2 therefore does not need adjustment.

With a burner input of 37,500 Btu/hr of 1050-Btu/ft^3 gas, 35.7 ft^3/hr will be required. We want to keep the manifold pressure between 3.5 and 4 in. of W.C., as close to the 3.5 in. of W.C. as possible. Going down the 3.5 in. of W.C. column, we find that a No. 34 drill size orifice will deliver 35.05 ft^3/hr and a No. 33 drill size orifice will deliver 36.08 ft^3/hr. Better burner operation will develop using the smaller orifice at a slightly higher pressure, so we chose the No. 34 drill. The actual correct input will be set later by timing the meter.

If the specific gravity of the gas is other than the 0.6 on which Fig. A14-2 is based, an adjustment is necessary. Figure A14-3 shows these adjustments. The adjustment is made by dividing the ft^3/hr rate flow needed for the burner by the factor given for the specific gravity of the gas used in order to adjust the rate flow to the 0.60 specific gravity used in Fig. A14-2. For example, if the specific gravity of the gas were 0.75, the correction factor would be 0.894. The 35.7 ft^3/hr of gas at 0.75 specific gravity becomes 35.7 ft^3/hr divided by 0.894, or 39.93 ft^3/hr at 0.60 specific gravity. This figure would be used in Table A14-2 to determine the drill size. At 3.5 in. of W.C. manifold pressure, a No. 32 drill size would deliver 38.3 ft^3/hr and a No. 31 drill size would deliver 40.85 ft^3/hr. We would then use a No. 32 drill size at a slightly higher manifold pressure.

In these examples, we have assumed sea-level op-

FIGURE A14-3

Factors for utility gases of another specific gravity.

Specific Gravity	Factor	Specific Gravity	Factor
0.45	1.155	0.95	0.795
0.50	1.095	1.00	0.775
0.55	1.045	1.05	0.756
0.60	1.000	1.10	0.739
0.65	0.961	1.15	0.722
0.70	0.926	1.20	0.707
0.75	0.894	1.25	0.693
0.80	0.866	1.30	0.679
0.85	0.840	1.35	0.667
0.90	0.817	1.40	0.655

To select an orifice size to provide a desired gas flow rate:
(1) Obtain the factor for the known specific gravity of the gas from above.
(2) *Divide* the desired gas flow rate by this factor, which gives the equivalent flow rate with a 0.60 specific gravity gas.
(3) Using the gas flow rate obtained in (2), use Fig. A14-2 to select orifice size.

To estimate gas flow rate for a given orifice size and gas pressure when it is not possible to meter gas flow:
(1) Use Fig. A14-2 to obtain equivalent gas flow rate with a 0.60 specific gravity gas.
(2) Obtain the factor from Fig. A14-3 for the specific gravity of the gas being burned.
(3) *Multiply* the rate obtained in (1) by the factor from (2) to estimate gas flow rate to the burner.

Source: American Gas Association.

FIGURE A14-4
Equivalent orifice sizes at high altitudes (includes 4% input reduction for each 1000 ft).

ORIFICE SIZE REQUIRED AT OTHER ELEVATIONS

Orifice Size at Sea Level	2000	3000	4000	5000	6000	7000	8000	9000	10000
1	2	2	3	3	4	5	7	8	10
2	3	3	4	5	6	7	9	10	12
3	4	5	7	8	9	10	12	13	15
4	6	7	8	9	11	12	13	14	16
5	7	8	9	10	12	13	14	15	17
6	8	9	10	11	12	13	14	16	17
7	9	10	11	12	13	14	15	16	18
8	10	11	12	13	13	15	16	17	18
9	11	12	13	14	14	16	17	18	19
10	12	13	13	14	15	16	17	18	19
11	13	13	14	15	16	17	18	19	20
12	13	14	15	16	17	17	18	19	20
13	15	15	16	17	18	18	19	20	22
14	16	16	17	18	18	19	20	21	23
15	16	17	17	18	19	20	20	22	24
16	17	18	18	19	19	20	22	23	25
17	18	19	19	20	21	22	23	24	26
18	19	19	20	21	22	23	24	26	27
19	20	20	21	22	23	25	26	27	28
20	22	22	23	24	25	26	27	28	29
21	23	23	24	25	26	27	28	28	29
22	23	24	25	26	27	27	28	29	30
23	25	25	26	27	27	28	29	29	30
24	25	26	27	28	28	29	29	30	30
25	26	27	27	28	29	29	30	30	31
26	27	28	28	29	29	30	30	30	31
27	28	28	29	30	30	30	31	31	32
28	29	29	29	30	30	31	32	33	35
29	29	30	30	31	31	32	36	37	38
30	30	31	31	33	34	35	37	38	40
31	32	32	32	35	36	36	38	40	41
32	33	34	35	36	37	38	39	40	42
33	35	35	36	37	37	38	40	41	43
34	35	36	36	37	38	40	41	42	43
35	36	36	37	39	40	41	42	43	44
36	37	38	38	40	41	42	43	43	43
37	38	39	39	41	42	43	43	44	44
38	39	40	41	41	42	43	43	44	44
39	40	41	41	42	42	43	44	44	44
40	41	42	42	42	43	43	44	44	45

ORIFICE SIZE REQUIRED AT OTHER ELEVATIONS

Orifice Size at Sea Level	2000	3000	4000	5000	6000	7000	8000	9000	10000
41	42	42	42	43	43	44	44	45	46
42	42	43	43	43	44	44	45	46	47
43	44	44	44	45	45	46	47	47	48
44	45	45	45	46	47	47	48	48	49
45	46	47	47	47	48	48	49	49	50
46	47	47	47	48	48	49	50	50	50
47	48	48	49	49	49	50	51	51	51
48	49	49	49	50	50	51	52	51	52
49	49	50	50	51	50	51	52	52	52
50	50	51	51	51	51	52	53	53	53
51	51	52	52	52	52	53	53	53	54
52	51	53	53	53	52	53	54	54	54
53	52	53	53	53	53	54	55	55	55
54	54	54	54	54	54	55	56	56	56
55	54	55	55	55	55	56	56	56	57
56	55	56	57	56	56	58	59	59	60
57	56	59	59	60	60	61	62	63	63
58	58	60	60	61	62	62	63	63	64
59	59	61	61	62	62	63	64	64	65
60	60	61	62	63	63	64	64	65	65
61	61	62	63	63	64	65	65	66	66
62	62	63	63	65	65	66	66	66	67
63	63	64	64	66	66	66	66	67	68
64	64	65	65	66	67	67	67	67	68
65	65	66	66	68	68	69	68	68	69
66	65	67	68	68	69	69	69	69	70
67	67	68	68	69	69	70	70	70	70
68	68	69	69	70	70	71	70	71	71
69	68	70	70	71	71	72	71	72	72
70	70	71	71	73	71	73	72	73	73
71	70	72	72	73	73	74	74	74	74
72	72	73	73	74	74	74	74	74	75
73	73	73	74	75	74	75	75	75	76
74	73	74	75	76	75	76	76	76	76
75	74	75	76	77	76	77	77	77	77
76	75	76	77	78	77	78	77	77	77
77	76	76	77	79	78	79	78	78	78
78	77	78	78	80	79	80	79	80	80
79	79	80	80	80	80	.013	.012	.012	.012
80	80	80	.013	.013	.012	.012	.012	.012	.011

Source: American Gas Association.

eration or operation below 2000 ft altitude. When heating units are operated at higher altitudes, the thinner air encountered reduces the ability of the heating unit to handle the required amount of air for proper combustion. The unit must therefore be derated, depending on the elevation of the unit.

Between sea level and 2000 ft altitude no derating is necessary. Above 2000 ft, however, the unit must be derated 4% per 1000 ft above sea level. For example, at 4000 ft elevation, the unit is derated 4% times 4 (4000 ft above sea level), or 16%. The maximum input at 4000 ft altitude would be 150,000 Btu/hr less 16%, or 126,000 Btu/hr. This new rating can be used for calculating the orifice size or Fig. A14-4 may be used. This table gives the orifice sizes at elevations of 2000 to 10,000 ft based on the orifice calculated at sea level.

Our first example resulted in an orifice drill size of No. 34. Using Fig. A14-4 and No. 34 at sea level, we find a No. 36 at 4000 ft. If we were at 5000 ft (Denver, Colorado), a No. 37 drill size would be required. The higher the altitude, the lower the input and the smaller the orifice size. The final adjustment of the input is accomplished by timing the supply meter.

Figure A14-5 shows the index or dial of a typical domestic gas meter. Included are two test dials, one for $\frac{1}{2}$ ft^3 per revolution and the other for 2 ft^3 per revolution. To determine if the correct amount of gas is being fed to the heating unit, it is only necessary to find the feed rate through the meter. For accuracy, all other appliances must be turned off. *If the pilot lights of the other appliances are turned off, be sure to relight before leaving.* Usually, the requirements of pilot lights are so small that they are ignored.

Gas flow through the meter is determined by the time it takes the test dials to turn one revolution. To determine this time, the following formula can be used:

$$\text{time (sec/ft}^3) = \frac{\text{seconds per hour}}{\text{ft}^3/\text{hr of gas}}$$

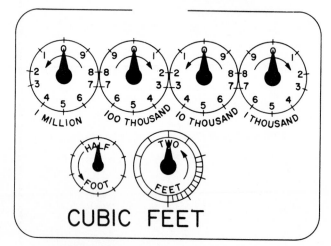

FIGURE A14-5 Typical domestic gas meter index. (*Courtesy* American Gas Association)

In our example, the formula would set up as

$$\text{sec/ft}^3 = \frac{60 \times 60}{142.8} = \frac{3600}{142.8} = 25.2$$

It will require 25.2 seconds for 1 ft^3 of gas to go through the meter. Thus the $\frac{1}{2}$-ft^3 dial would require 12.6 seconds per revolution, and the 2-ft^3 dial, 50.4 seconds per revolution. A good stopwatch is recommended or a digital watch timing function. Rather than use the formula, Fig. A14-6 shows a table of revolution timing for various test dial sizes for various inputs. Using this table, the 142.8 ft^3/hr of gas blown would cause the $\frac{1}{2}$-ft^3 dial to turn between 12 and 13 seconds, the 1-ft^3 dial between 25 and 26 seconds, etc.

Pocket-type charts have been published by various heating unit manufacturers and are usually available at supply houses. A typical chart is shown in Fig. A14-7.

Liquefied Petroleum (LP) Gas: The LP gas industry has established a set manifold pressure for all LP gas-burning appliances—11 in. of W.C. Therefore, it is only necessary to determine that the LP supply system is large enough to maintain 11 in. of W.C. at the units when the total connected load is operating. Actually, the LP supplier will provide the tank installation, pressure-regulating devices, and piping to the furnace, but you should be familiar with the hook-up procedure.

There are two basic systems, as illustrated in Fig. A14-8. The single system on the right, which is used most frequently in residential work, uses only one pressure regulator located at the tank. The two-stage system shown on the left is used mainly in commercial work, where the number of appliances and volume of gas must be greater. Therefore, the line pressure at the outlet of the tank regulator on the two-stage system is 10 to 15 psig, whereas it is 11 in. of W.C. on the single-stage system. The main supply line will carry more ft^3/hr at the higher pressure, and this is smaller. Sometimes it is necessary to connect from a single-stage to a two-stage system if the supply lines are installed undersized.

With the majority of residential and small commercial installations as single-stage systems, let's be more specific about the "single-stage" system. The pressure on the inlet of the regulator will be direct tank pressure and this will vary with fuel and temperature.

To ensure that the gas will flow from the tank into the supply system to the heating unit, the tank pressure must at all times be higher than the line pressure required to supply the system. To maintain a pressure of 11 in. of W.C. at the heating unit, allowing a pressure loss in the line of $\frac{1}{2}$ in. of W.C., the minimum tank pressure would be 2 psig. This means that the minimum outside temperature must be considered when selecting the mixture of LP fuel. Figure A14-9 shows the tank pressure that will occur at temperatures of -30 to $+110°$F for mixtures of propane and butane from 100% pro-

FIGURE A14-6

Gas input to burner (ft³/hr).

CUBIC FEET PER HOUR					CUBIC FEET PER HOUR				
Seconds for One Revolution	Size of Test Meter Dial				Seconds for One Revolution	Size of Test Meter Dial			
	One-Half Cu. Ft.	One Cu. Ft.	Two Cu. Ft.	Five Cu. Ft.		One-Half Cu. Ft.	One Cu. Ft.	Two Cu. Ft.	Five Cu. Ft.
10	180	360	720	1,800	52	35	69	138	346
11	164	327	655	1,636	53	34	68	136	340
12	150	300	600	1,500	54	33	67	133	333
13	138	277	555	1,385	55	33	65	131	327
14	129	257	514	1,286	56	32	64	129	321
15	120	240	480	1,200	57	32	63	126	316
16	112	225	450	1,125	58	31	62	124	310
17	106	212	424	1,059	59	30	61	122	305
18	100	200	400	1,000	60	30	60	120	300
19	95	189	379	947	62	29	58	116	290
20	90	180	360	900	64	29	56	112	281
21	86	171	343	857	66	29	54	109	273
22	82	164	327	818	68	28	53	106	265
23	78	157	313	783	70	26	51	103	257
24	75	150	300	750	72	25	50	100	250
25	72	144	288	720	74	24	48	97	243
26	69	138	277	692	76	24	47	95	237
27	67	133	267	667	78	23	46	92	231
28	64	129	257	643	80	22	45	90	225
29	62	124	248	621	82	22	44	88	220
30	60	120	240	600	84	21	43	86	214
31	58	116	232	581	86	21	42	84	209
32	56	113	225	563	88	20	41	82	205
33	55	109	218	545	90	20	40	80	200
34	53	106	212	529	94	19	38	76	192
35	51	103	206	514	98	18	37	74	184
36	50	100	200	500	100	18	36	72	180
37	49	97	195	486	104	17	35	69	173
38	47	95	189	474	108	17	33	67	167
39	46	92	185	462	112	16	32	64	161
40	45	90	180	450	116	15	31	62	155
41	44	88	176	440	120	15	30	60	150
42	43	86	172	430	130	14	28	55	138
43	42	84	167	420	140	13	26	51	129
44	41	82	164	410	150	12	24	48	120
45	40	80	160	400	160	11	22	45	112
46	39	78	157	391	170	11	21	42	106
47	38	77	153	383	180	10	20	40	100
48	37	75	150	375	190	9.5	19	38	95
49	37	73	147	367	200	9	18	36	90
50	36	72	144	360	210	8.5	17	34	86
51	35	71	141	353	220	8	16	33	82

Note: To convert to Btu per hour multiply by the Btu heating value of gas used.

Source: American Gas Association.

Gas rate (ft³/hr).

Seconds for One Revolution	SIZE OF TEST DIAL					Seconds for One Revolution	SIZE OF TEST DIAL				
	$\frac{1}{4}$ cu. ft.	$\frac{1}{2}$ cu. ft.	1 cu. ft.	2 cu. ft.	5 cu. ft.		$\frac{1}{4}$ cu. ft.	$\frac{1}{2}$ cu. ft.	1 cu. ft.	2 cu. ft.	5 cu. ft.
50	18	36	72	144	360	10	90	180	360	720	1800
51	—	—	—	141	355	11	82	164	327	655	1636
52	—	—	69	138	346	12	75	150	300	600	1500
53	17	34	—	136	340	13	69	138	277	555	1385
54	—	—	67	133	333	14	64	129	257	514	1286
55	—	—	—	131	327	15	60	120	240	480	1200
56	16	32	64	129	321	16	56	113	225	450	1125
57	—	—	—	126	316	17	53	106	212	424	1059
58	—	31	62	124	310	18	50	100	200	400	1000
59	—	—	—	122	305	19	47	95	189	379	947
60	15	30	60	120	300	20	45	90	180	360	900
62	—	—	—	116	290	21	43	86	171	343	857
64	—	—	—	112	281	22	41	82	164	327	818
66	—	—	—	109	273	23	39	78	157	313	783
68	—	—	—	106	265	24	37	75	150	300	750
70	—	—	—	103	257	25	36	72	144	288	720
72	12	25	50	100	250	26	34	69	138	277	692
74	—	—	—	97	243	27	33	67	133	267	667
76	—	—	—	95	237	28	32	64	129	257	643
78	—	—	—	92	231	29	31	62	124	248	621
80	—	—	—	90	225	30	30	60	120	240	600
82	—	—	—	88	220	31	—	—	116	232	581
84	—	—	—	86	214	32	28	56	113	225	563
86	—	—	—	84	209	33	—	—	109	218	545
88	—	—	—	82	205	34	26	53	106	212	529
90	10	20	40	80	200	35	—	—	103	206	514
92	—	—	—	78	196	36	25	50	100	200	500
94	—	—	—	—	192	37	—	—	97	195	486
96	—	—	—	75	188	38	23	47	95	189	474
98	—	—	—	—	184	39	—	—	92	185	462
100	—	—	—	72	180	40	22	45	90	180	450
102	—	—	—	—	178	41	—	—	—	176	439
104	9	17	35	69	173	42	21	43	86	172	429
106	—	—	—	—	170	43	—	—	—	167	419
108	—	—	—	67	167	44	—	41	82	164	409
110	—	—	—	—	164	45	20	40	80	160	400
112	—	—	—	64	161	46	—	—	78	157	391
116	—	—	—	62	155	47	19	38	76	153	383
120	7	15	30	60	150	48	—	—	75	150	375
						49	—	—	—	147	367

Source: Addison Products Company.

pane to 100% butane. From this table we can see that butane is not usable below +40°F, and even propane will not develop sufficient pressures below −30°F. In extremely cold climates, tank heaters are used to ensure adequate fuel supply.

Figure A14-10 shows the Btu/hr capacity for propane of the pipe between the first-stage regulator at the tank and the second-stage regulator at the unit. Given in 1000 Btu/hr, the capacities are shown for line sizes in copper tube and iron pipe in lengths of 10 to 200 ft. The capacities are based on 10-psig line pressure with 2 psig line-pressure loss.

To convert the table in Fig. A14-10 to 15 psig line pressure, the line capacities are increased by multiplying the line capacity by 1.130. If 5-psig line pressure is used, the line capacities are reduced by multiplying the line capacities by 0.879.

Figure A14-11 shows the Btu/hr carrying capacity

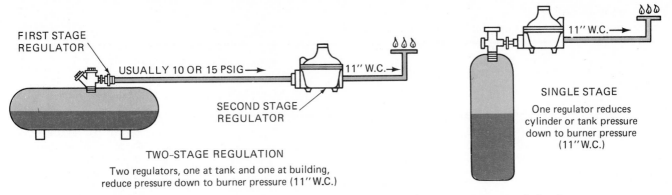

FIGURE A14-8   LP gas supply system. (*Courtesy* Fisher Controls International, Inc.)

FIGURE A14-9
Pressure facts.

| | VAPOR PRESSURE, PSIG | | | | | | | | | | | | | | |
| | Outside Temperature, Degrees Fahrenheit | | | | | | | | | | | | | | |
	−30	−20	−10	0	10	20	30	40	50	60	70	80	90	100	110
100% Propane	6.8	11.5	17.5	24.5	34	42	53	65	78	93	110	128	150	177	204
70% Propane 30% Butane	—	4.7	9	15	20.5	28	36.5	46	56	68	82	96	114	134	158
50% Propane 50% Butane	—	—	3.5	7.6	12.3	17.8	24.5	32.4	41	50	61	74	88	104	122
70% Butane 30% Propane	—	—	—	2.3	5.9	10.2	15.4	21.5	28.5	36.5	45	54	66	79	93
100% Butane	—	—	—	—	—	—	—	3.1	6.9	11.5	17	23	30	38	47

Source: Fisher Controls International, Inc.

of copper tube and iron pipe of various sizes with lengths between 10 and 400 ft. Notice that capacities are not given (their use is not recommended) for copper tube over 250 ft.

In the example of the total house load of 260,000 Btu/hr and 80 ft of supply line, the line size in copper would be $1\frac{1}{8}$ in. OD and in iron pipe would be 1 in. nominal pipe size. In this case it would be better to run a two-stage system with $\frac{3}{8}$-in.-OD copper at 10 psig between the two regulators. For the heating unit load with 35 ft of supply pipe, $\frac{7}{8}$-in.-OD copper tube would be required to handle the fuel supply.

To determine the correct main burner orifice size, a single table is all that is needed. Propane and butane gases are of stable density: 0.51 for propane and 0.58 for butane. Also, the Btu/ft³ is stable, 2500 Btu for propane and 3175 Btu for butane. To determine the orifice size, it is only necessary to know the Btu/hr per burner requirement. In the case of a four-burner 150,000

Btu/hr heating unit or 37,500 Btu/hr per burner, Fig. A14-12 is used to determine the orifice drill size.

The heating unit would use a No. 50 drill at 10.75 in. of W.C. for propane or a No. 51 drill at 10.5 in. of W.C. for butane or butane/propane mixtures. Like natural gas units, LP gas units must be derated at higher altitudes. Once the orifice size is determined from Fig. A14-12, the higher-altitude requirement can be determined from Fig. A14-4. For our example heating unit using a No. 50 drill size on propane at sea level, the drill size would be No. 51 at 4000 ft. For butane the drill size would be No. 51 at 4000 ft.

A14-2.2
Operation

In the operation of a gas-fired heating unit, two major functions occur. The first is the supply of the fuel, natural or LP gas, under the control of a temperature

Pipe sizing between first- and second-stage regulators.

Pipe or Tubing Length, Feet	TUBING SIZE, O.D., TYPE L						NOMINAL PIPE SIZE, SCHEDULE 40								
	$\frac{3}{8}''$	$\frac{1}{2}''$	$\frac{5}{8}''$	$\frac{3}{4}''$	$\frac{7}{8}''$	$1\frac{1}{8}''$	$\frac{1}{2}''$	$\frac{3}{4}''$	$1''$	$1\frac{1}{4}''$	$1\frac{1}{2}''$	$2''$	$2\frac{1}{2}''$	$3''$	$4''$
10	730	1,700	3,200	5,300	8,300	17,000	3,200	7,500	12,800	24,000	40,000	88,000	133,000	237,500	489,000
20	500	1,100	2,200	3,700	5,800	12,000	2,200	4,200	8,800	18,000	33,000	61,000	92,500	165,500	341,000
30	400	920	2,000	2,900	4.700	9,800	1,800	4,000	7,200	14,000	26,000	49,000	76,500	136,500	281,000
40	370	850	1,700	2,700	4,100	8,500	1,600	3,700	6,800	13,500	24,000	46,000	71,000	127,000	262,000
50	330	770	1,500	2,400	3,700	7,600	1,500	3,400	6,300	12,600	22,500	43,000	65,000	116,000	240,000
60	300	700	1,300	2,200	3,300	7,000	1,300	3,100	5,600	12,000	21,700	40,000	61,000	109,000	224,000
80	260	610	1,200	1,900	2,900	6,000	1,200	2,600	4,900	10,000	18,000	34,000	52,000	93,000	192,000
100	220	540	1,000	1,700	2,600	5,400	1,000	2,300	4,300	9,000	15,000	31,000	45,500	81,500	168,000
125	200	490	900	1,400	2,300	4,800	900	2,100	4,000	7,900	13,500	28,000	41,500	74,000	152,500
150	190	430	830	1,300	2,100	4,400	830	1,900	3,600	7,200	12,600	25,000	37,000	66,500	137,000
175	170	400	780	1,200	1,900	4,000	770	1,700	3,300	6,700	11,400	23,500	34,500	61,500	127,000
200	160	380	730	1,100	1,800	3,800	720	1,500	3,100	6,200	10,600	22,000	32,000	57,500	119,000

To convert to capacities at 5 psig settings—Multiply by 0.879.   To convert to capacities at 30 psig settings—Multiply by 1.345

To convert to capacities at 15 psig settings—Multiply by 1.130   To convert to capacities at 40 psig settings—Multiply by 1.488

To convert to capacities at 20 psig settings—Multiply by 1.185   To convert to capacities at 50 psig settings—Multiply by 1.618

Maximum propane capacities listed are based on 2 psig pressure drop at 10 psig setting—Capacities in 1,000 Btu/hr.

Source: Fisher Controls International, Inc.

Pipe sizing between single- or second-stage regulator and heating unit.

Pipe or Tubing Length, Feet	TUBING SIZE, O.D., TYPE L						NOMINAL PIPE SIZE, SCHEDULE, 40								
	$\frac{3}{8}''$	$\frac{1}{2}''$	$\frac{5}{8}''$	$\frac{3}{4}''$	$\frac{7}{8}''$	$1\frac{1}{8}''$	$\frac{1}{2}''$	$\frac{3}{4}''$	$1''$	$1\frac{1}{4}''$	$1\frac{1}{2}''$	$2''$	$2\frac{1}{2}''$	$3''$	$4''$
10	39	92	199	329	501	935	275	567	1,071	2,205	3,307	6,221	10,140	17,990	36,710
20	26	62	131	216	346	630	189	393	732	1,496	2,299	4,331	7,046	12,510	25,520
30	21	50	107	181	277	500	152	315	590	1,212	1,858	3,465	5,695	10,110	20,620
40	19	41	90	145	233	427	129	267	504	1,039	1,559	2,992	4,778	8,481	17,300
50	18	37	79	131	198	376	114	237	448	913	1,417	2,646	4,343	7,708	15,730
60	16	35	72	121	187	340	103	217	409	834	1,275	2,394	3,908	6,936	14,150
80	13	29	62	104	155	289	89	185	346	724	1,086	2,047	3,329	5,908	12,050
100	11	26	55	90	138	255	78	162	307	630	976	1,811	2,991	5,309	10,830
125	10	24	48	81	122	224	69	146	275	567	866	1,606	2,654	4,711	9,613
150	9	21	43	72	109	202	63	132	252	511	787	1,496	2,412	4,218	8,736
200	8	19	39	66	100	187	54	112	209	439	665	1,282	2,038	3,618	7,382
250	8	17	36	60	93	172	48	100	185	390	590	1,138	1,808	3,210	6,549
300	—	—	—	—	—	—	43	90	168	353	534	1,030	1,637	2,905	5,927
350	—	—	—	—	—	—	40	83	155	325	491	947	1,505	2,671	5,450
400	—	—	—	—	—	—	37	77	144	303	458	883	1,404	2,492	5,084

Maximum propane capacities listed are based on $\frac{1}{2}''$ W.C. pressure drop at 11" W.C. setting—Capacities in 1,000 Btu/hr.

Source: Fisher Controls International, Inc.

control (usually a room thermostat) or a pressure control (on a hydronic unit) in conjunction with safety controls: high temperature in a heating unit or high pressure in a hydronic unit. In either case, the control is basically supplied by the operation of a gas valve. In Chapter A10 a typical gas valve was reviewed. In Chapter A5 the gas burner was mentioned, but it would be well to discuss this part in further detail. Proper operation of the burner is absolutely essential for the highest operating efficiency of the heating unit. If insufficient air is supplied for combustion, not all the available heat is obtained from the gas combustion. If too much air is supplied, the flue products pass out of the heat exchanger before the unit has had a chance to extract the maximum

Drill Size (Decimal or DMS)	GAS INPUT, BTU PER HOUR FOR:		Drill Size (Decimal or DMS)	GAS INPUT, BTU PER HOUR FOR:	
	Propane	Butane or Butane-Propane Mixtures		Propane	Butane or Butane-Propane Mixtures
.008	500	554	51	35,330	39,400
.009	641	709	50	38,500	42,800
.010	791	875	49	41,850	45,350
.011	951	1,053	48	45,450	50,300
.012	1,130	1,250	47	48,400	53,550
80	1,430	1,590	46	51,500	57,000
79	1,655	1,830	45	52,900	58,500
78	2,015	2,230	44	58,050	64,350
77	2,545	2,815	43	62,200	69,000
76	3,140	3,480	42	68,700	76,200
75	3,465	3,840	41	72,450	80,200
74	3,985	4,410	40	75,400	83,500
73	4,525	5,010	39	77,850	86,200
72	4,920	5,450	38	81,000	89,550
71	5,320	5,900	37	85,000	94,000
70	6,180	6,830	36	89,200	98,800
69	6,710	7,430	35	95,000	105,300
68	7,560	8,370	34	97,000	107,200
67	8,040	8,910	33	101,000	111,900
66	8,550	9,470	32	105,800	117,000
65	9,630	10,670	31	113,200	125,400
64	10,200	11,300	30	129,700	143,600
63	10,800	11,900	29	145,700	163,400
62	11,360	12,530	28	154,700	171,600
61	11,930	13,280	27	163,100	180,000
60	12,570	13,840	26	169,900	187,900
59	13,220	14,630	25	175,500	194,600
58	13,840	15,300	24	181,700	201,600
57	14,550	16,090	23	186,800	206,400
56	16,990	18,790	22	193,500	214,500
55	21,200	23,510	21	198,600	220,200
54	23,850	26,300	20	203,700	225,000
53	27,790	30,830	19	217,100	241,900
52	31,730	35,100	18	225,600	249,800

	Propane	Butane
Btu per Cubic Foot =	2,500	3,175
Specific Gravity =	1.53	2.00
Pressure at Orifice, Inches Water Column =	11	11
Orifice Coefficient =	0.9	0.9

For altitudes above 2,000 feet, first select the equivalent orifice size at sea level from Figure A14-4.

Source: American Gas Association.

amount of heat. The flue products rise in temperature and more Btu go up the chimney.

A14-2.3
Combustion Air

The portion of the air needed for combustion that is mixed with the gas in the burner, before it passes through the burner port, is called *primary air.* This is approximately 40% of the total air. The rest, or 60%, is called *secondary air.* This air passes around the burner and enters the combustion flame above the burner port. In addition to the air for combustion, air enters the combustion chamber in sufficient quantity to ensure complete combustion. The excess air is usually about 50% of the combustion total, making the heat exchanger handle 1.5 times the amount of air needed for the combustion process. The supplying of excess air ensures complete combustion, but more important, reduces the possibility of the production of carbon monoxide (CO) in the flue products. A properly adjusted burner for proper flame characteristics with the correct amount of secondary and excess air will result in a CO_2 content in the flue products of around 10%. With the standard gas furnace set at correct input and correct amount of air over the heat exchanger to produce an 80° rise in the air temperature (supply air temperature minus return air temperature), the flue product temperature will be about 475°F above the temperature of the air entering the combustion chamber. This rule does not apply to the high-efficiency unit, as some of them operate at flue temperatures as low as 100°F. To properly set a high-efficiency unit, the individual manufacturer's instructions must be followed in detail.

To set the operation of the standard unit, the first adjustment is primary air to the burner. With the heating unit at proper gas input, the burner flame primary air must be adjusted to ensure complete combustion, but not in excess to cause noisy fire and lift-off of flame from the burner. The primary air is controlled by air shutters (Fig. A14-13) as seen from the orifice end of the burner.

Varying the amount of opening controls the airflow into the burner. Opening increases, and closing decreases, the amount of induced primary air. With a fixed gas jet into the burner, we can vary the flame characteristics for the most efficient burning.

The exact flame characteristics depend on the type of burner, but it is recognized that a soft clear blue flame is the most desirable. Closing the air shutters will result in incomplete combustion, resulting in the formation of carbon (soot) deposited on the surfaces of the heat exchanger as well as carbon monoxide (CO) in the flue products.

Primary air shutters opened too wide will allow too much primary air. The flame will be sharp and tend to lift off the burner. This produces noisy fire and forces too much air through the heat exchanger, reducing the efficiency of the heating unit. A sharp, noisy fire increases the heating bill.

FIGURE A14-13 Adjusting primary air. (*Courtesy* Borg-Warner Central Environmental Systems, Inc.)

After checking and setting the Btu/hr input and adjusting the primary air shutters, the vent draft condition should be checked. This has been outlined in Chapter A5.

A14-2.4
Air Temperature Rise

The final adjustment before efficiency testing is the adjustment of the blower for temperature rise of the air through the unit. On heating-only systems, the industry recommendation for heat comfort in the occupied area is CAC—continuous air circulation. This method of fan operation provides as continuous air circulation as possible throughout the heating season. With continuous air circulation, stagnant air discomfort problems, commonly called *cold 70°*, are eliminated and a higher comfort level is achieved in the occupied area.

To obtain CAC, the temperature rise of the air through the heating unit should be as close to 80°F as possible. The fan switch should turn on at 125 to 130°F and should turn off at 95 to 100°F. This will also help to move as much heat as possible out of the heat exchanger compartment of the heating unit before the fan cycles off. On unusually long supply duct runs, where the cool air from the duct on fan startup is objectionable, some gravity or passive effect is desired before the fan cycles ON. The fan switch is set to turn on at 150°F; the turn-off remains the same.

When air conditioning is added to the heating system, allowances are made in the temperature rise used. This is covered in Chapter A17.

To obtain the proper temperature rise of the air through the heating unit, return and supply air temperatures are recorded. Figure A14-14 shows the insertion of dial-type thermometers into the supply and return plenums of the heating unit. Use of a sheet-metal-type scratch owl will allow the formation of a hole large enough to take the $\frac{1}{8}$-in.-diameter stem of the dial thermometer and allow closure of the holes with sheet-metal screws after the test is completed. The supply-air-temperature thermometer should also be located far enough away from the surface of the heat exchanger to reduce the effect on the thermometer of radiant heat from the heat exchanger. This is usually a minimum distance of 12 in. If the thermometer can be located in the main duct off the supply air pressure, the radiant effect will be practically eliminated. After the unit has operated long enough for the supply air thermometer to hold a steady reading (stabilized), the return air temperature should be subtracted from the supply air temperature and the temperature rise recorded. If the temperature rise is below 80°F, the heating unit blower is moving too much air, burning too fast, and the speed should be reduced. If the temperature rise is above 80°F, the blower speed should be increased.

FIGURE A14-14 Measuring air rise. (*Courtesy* Borg-Warner Central Environmental Systems, Inc.

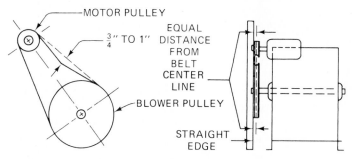

FIGURE A14-15 Align pulleys and tighten belt. (*Courtesy* Borg-Warner Central Environmental Systems, Inc.)

The procedure for this would depend on the type of drive used on the blower. Belt-driven blowers are adjusted by changing the size of the motor or driving pulley, direct-drive blowers by changing the electrical connections to the motor. Belt-driven blowers use a combination of adjustable motor pulley (the driving pulley), blower pulley (the driven pulley), and drive belt. The speed of the blower is adjusted by changing the spread of the flanges of the driving pulley. Opening the pulley by spreading the flanges allows the belt to ride lower in the pulley, thus reducing the effective diameter of the pulley. This in turn reduces the pulley diameter rotation between the two flanges and reduces the blower speed.

Closing the pulley spread increases the drive pulley diameter, resulting in an increase in blower speed. The driven pulley usually has two setscrew flats to allow adjustment of the pulley in one-half-turn increments. *Use these flats—do not drive the setscrew into the pulley adjustment threads.* This malpractice ruins the chance for future adjustment.

After the blower speed has been set, the belt tension and alignment should be checked (see Fig. A14-15). Make sure that the belt tension is correct to avoid slippage but not too tight to cause excessive wear. About $\frac{3}{4}$ to 1 in. of play should be allowed for each 12-in. distance between the motor and blower shafts. Also check the pulley alignment by using a straightedge (yardstick). Misalignment of the pulley and excessive belt tension will cause noise and excessive vibration and wear on the belt and on motor and blower housings.

On direct-drive blowers, the choice of blower speeds is limited to the number of speeds built into the blower motor. On some heating-only units, the blower has only one speed and no choice of temperature rise is available. If air-conditioning is added to the unit, a change in blower motor or blower assembly is required to accommodate the additional pressure loss through the coil.

If the heating unit is equipped with a multispeed blower motor, observe which fan speed lead is connected—high, medium, or low. For the initial startup it

is recommended that the fan be set for medium speed, subject to verification when the temperature rise is obtained.

In either case, increasing the blower speed will increase the ft^3/min of air through the unit and lower the temperature rise. Conversely, lowering the blower speed will decrease the ft^3/min through the unit and increase the temperature rise.

When increasing the blower speed, which increases the load in the motor, the amperage draw of the motor will also increase. A clamp-type ampmeter (such as shown in Figure R19-5) should be used to check the motor operating amperage. If the required ft^3/min causes the motor to draw more than its amperage rating, a motor or blower assembly of larger capacity will have to be substituted.

A14-2.5
Efficiency Testing

With the heating unit input set, the burners properly adjusted, and the unit operating at as close to desirable temperature rise as it is possible to obtain, the unit can be tested for operating efficiency. The instruments used for efficiency testing have been described in Chapter A13. For continuity, however, we will cover those needed in each step.

Basically, the operating efficiency of a heating unit or hydronic unit is the difference between the possible heat obtainable in the combustion process and the actual amount of heat obtained from the unit. A heating unit will put $X\%$ of the heat obtainable from the burning gas into the air as it passes through the unit. A hydronic unit will put $X\%$ of the heat obtainable from the burning gas into the water as it passes through the unit.

The object of adjusting for peak efficiency is to obtain all the heat possible to the conditioned area from the potential amount of heat produced in the combustion process without creating undesirable conditions. To do this, several factors are involved. These factors are

Heating Startup, Checkout, and Operation   547

unit related and any change that affects one will affect the others. These factors are:

1. Amount of fuel to be burned
2. Burner operation
3. Heat exchanger operation

We have discussed the obtaining of proper input and proper burner adjustment as well as a temperature rise through the unit. In this discussion we see how changes in these factors affect the overall efficiency of the unit.

Objective: The objective of efficiency testing is to obtain as high an efficiency rating of the heating unit as possible, taking into account the operating cost, the equipment operating life, and the comfort obtained in the conditioned area. All three factors will be affected by any changes or adjustments made to the heating unit. If the air over the heat exchanger is more than proper for CAC, more heat will be removed from the flue products and the unit operating efficiency will increase. However, with more air circulating through the conditioned area, the greater is the chance for objectionable drafts. Also, if the temperature of the flue products is lowered to the point where the water condenses out of the flue products, water condenses in the chimney or vent pipe and even the heat exchanger itself. This water usually contains a mild acid which can shorten the life of the equipment. On the other hand, reducing the airflow over the unit will reduce the heat extraction from the heat exchanger, increase flue product temperatures, and reduce operating efficiency.

It is necessary, therefore, to reach adjustments that achieve a balance between operating cost (efficiency) and comfort. Certain standards are used for efficiency testing. We have discussed most of them previously but we will list them again.

1. *Input* The unit must be supplied with the correct amount of fuel within 90 to 100% of its rated capacity.
2. *Burner Primary Air Adjustment* Soft blue fire without yellow color or flame lift.
3. *Air Temperature Rise* As close to 80°F temperature rise of the air through the unit as the blower design will allow.
4. *Fan Control Settings* With fan on, at 125 to 130°F; with fan off, at 100 to 105°F. With units having duct runs over 50 ft long, a fan or setting of 150°F is recommended to overcome drafts upon startup.
5. *CO₂* The results of combustion of the gas will produce water (H_2O) when the hydrogen in the gas is burned and carbon dioxide (CO_2) is burned. Water quantity is not of importance and thus is not measured. If stack temperatures get too low, however, this water becomes a corrosive problem. The

amount of CO_2, however, is important. Too low a CO_2 content indicates too much combustion air and an excessive volume of higher-temperature flue products out the chimney or vent. Too high a CO_2 content indicates incomplete combustion and possible production of carbon monoxide (CO). The desirable percentage of CO_2 in the flue products is $8\frac{1}{2}$ to 10%. This will produce a CO_2 level between 70 and 80% of the ultimate CO_2 possible with 100% complete combustion. The combustion efficiency should be between 70 and 80% with less than 0.0005% CO required.

6. *Stack Temperature* The temperature of the flue products leaving the heat exchanger is a direct indication of the efficiency of the heat exchanger operation. With proper input and combustion efficiency, the heat exchanger should have the ability to extract the heat from the flue products at a rate to produce stack temperature in the range 475 to 500°F above the temperature of the combustion air entering the burner compartment. Excessive stack temperatures can be caused by reduced heat exchanger efficiency, such as dirty surfaces, control baffle deterioration, heat exchanger warpage, etc., or excessive combustion air. If the unit is a factory-designed model, the control of the combustion air by baffling is set by the factory. Therefore, if the input, primary air adjustment, and air temperature rise through the unit are correct, the stack temperature rise should not be more than 480°F above the temperature of the air entering the combustion chamber.

When conversion burners are installed in units designed to burn other fuels or in units built up from field-selected components, the method of controlling the amount of total combustion air must be field installed. This is usually done by means of a fixed-adjustment flue restrictor between the flue outlet by the heat exchanger and the draft diverter, or restrictor baffles placed in the secondary flue passages of the heat exchangers. In either case the CO_2 content should be between 7 and 9% and the stack temperature rise should not exceed 580°F.

A14-2.6
Neutral Point

To maintain proper combustion at the flame points of the burner as well as maximum heat recovery from the flue products, the heat exchanger must be able to handle the expanded gas produced by combustion without handling excessive air. When the balance between primary air and total combustion air and the draft at the heat exchanger outlet are correct, the positive pressure produced by the expansion of the hot flue products and the negative pressure produced by the draft at the heat exchanger outlet drawing the combustion air into the combustion chamber balance at what is called the *neutral pressure point*. When the proper balance pressure is obtained, the neutral pressure point (zero pressure) will be at the flame tips of the burner.

Factory-designed units are preset, usually by restrictor baffles located in the outlet of the heat exchanger. These should not be changed or removed. On field-installed units, such as conversion of a solid fuel to gas burning, the firing door is used as a test point.

The American Gas Association code specifies that the installation of gas conversion burners must conform to the Standard for Installation of Domestic Gas Burners contained in the American National Standards Institute Code, Section ANSI Z21.8. This standard requires that the neutral pressure point must be adjusted to between 1 in. below the midpoint and 1 in. below the top of the firing door. Pressure points below this range could seriously disrupt combustion and could produce excessive CO. Pressure points above this range will reduce efficiency and raise operating cost.

A means of adjusting the neutral point must be installed on field-converted units. This is easily done by installing a slide damper in the flue pipe between the heat exchanger flue outlet and the draft diverter. Figure A14-16 shows a step-by-step procedure for doing this.

Step 1 is to cut a slot in the top of the flue pipe wide enough to take a slide damper usually made of No.

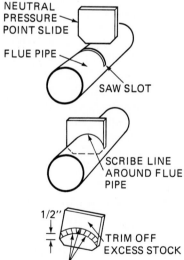
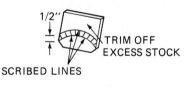

NEUTRAL PRESSURE POINT SLIDE

FLUE PIPE

SAW SLOT

SCRIBE LINE AROUND FLUE PIPE

1/2"

TRIM OFF EXCESS STOCK

SCRIBED LINES

SUGGESTED FORM OF NEUTRAL PRESSURE POINT ADJUSTER. INSERT IN SLOT CUT IN FLUE PIPE.

WHEN ADJUSTMENT IS COMPLETED, SCRIBE A LINE ON ADJUSTER SO IT CAN BE RELOCATED.

REMOVE ADJUSTER, TRIM OFF EXCESS, SLIT REMAINDER VERTICALLY AND BEND SEGMENTS IN ALTERNATE DIRECTIONS.

REPLACE IN FLUE PIPE, RECHECK ADJUSTMENTS TO INSURE UNCHANGED CONDITIONS, AND FASTEN ADJUSTER IN PLACE WITH SHEET METAL SCREWS.

FIGURE A14-16 Neutral pressure point control damper. (*Courtesy*) Bacharach Instrument Company)

20 gauge metal. The slide damper is inserted into the pipe and the restriction to the flue product floor is adjusted until the proper neutral pressure point is obtained.

With the door open only enough to accept the pressure probe of the neutral pressure point meter, located at the 1 in. above center of the door position, the slide damper is moved to restrict the flue if the pressure is negative (draft) or to open the flue if positive (pressure). When the proper position is found, scribe a line on the damper around the surface of the flue pipe. This is step 2.

In step 3, the trim tube can be cut and bent to fasten the restrictor to the flue pipe with minimum slot openings between the tabs by trimming off all the excess material, leaving 1 in. above the scribe line.

In step 4, the restrictor is held in place with sheet-metal screws, ensuring a permanent adjustment.

A14-2.8
Procedure

Efficiency testing requires a "before" and "after" test run to determine if the higher efficiency is obtained as well as the difference in performance of the unit. This information is valuable to a customer relationship. To show the customer an improvement in efficiency or to prove that the unit is operating at its peak efficiency is valuable as a part of the service technician's service selling ability.

A means of recording such information is necessary so that it will become a permanent part of the unit operating and service history. This *efficiency check sheet* should include the following information.

Input

1. Type of gas: Nat. __ Mixed __ Mfg. __ Prop. __ Bu. __
2. Heat content Btu/ft³ _____
3. Specific gravity of the gas _____
4. Main burner orifice drill size: Found __ Left __
5. Manifold pressure (in. W.C.): Found __ Left __
6. Meter test dial size: _____ft³ per rev.
7. Seconds required per revolution of test dial: Found __ Left __

Primary Air Adjustment

1. Flame before adjustment: Sharp blue _____ Soft blue _____ Yellow tips _____ Flame after adjustment: Soft blue _____

Neutral Point Adjustment

1. Factory design—not adjustable _____
2. Conversion burner

Found: Below adjustment range _____
       Above adjustment range _____
       In correct adjustment range _____
Left: In correct adjustment range _____

Air Temperature Rise

	First test	Second test	Left
Supply air temperature			
Return air temperature			
Air temperature rise			

CO_2

First test _____%
Second test _____%
Left _____%

Stack Temperature Rise

	First test	Second test	Left
Stack temperature			
Combustion air temperature			
Stack temperature rise			

Combustion Efficiency

	First test	Second test	Left
% Efficiency			

The first step in any efficiency testing is to establish that proper input exists to the unit and to make adjustments accordingly.

1. Record the Btu/ft³ content and specific gravity of the gas, obtained from the gas supplier or utility.

2. Record the orifice drill size and operating manifold pressure of the unit. This information is needed in case the input is not correct.

3. Test and record the meter operating time. Using the Btu/ft³ and specific gravity of the gas as well as the meter flame rate, the input of natural or mixed gas can be determined. Any adjustment can then be made after the first test is completed. Section A14-2.1 outlines this procedure. However, for comparison of the entire unit performance, complete the first test procedure before any adjustments are made. If the fuel supplied is propane or butane, a meter will not be used; manifold pressure and orifice size will be the only input determining factors.

4. Check the flame conditions of the burner and record.

5. For a conversion burner or field-assembled unit, check for neutral point location (see Fig. A14-17).

6. With thermometers located in the supply and return plenums, record the air temperatures and de-

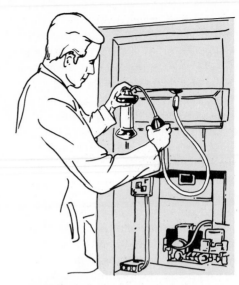

FIGURE A14-17 Using CO_2 analyzer. (*Courtesy* Bacharach Instrument Company)

termine the temperature rise. Figure A14-14 shows the suggested positions of the thermometers for the minimum effect of radiant heat from the heat exchanger and best reading accuracy.

7. Measure and record the CO_2 in the flue products. Each make of CO_2 analyzer has its own procedure for conducting the CO_2 test. The manufacturer's instructions should, therefore, be accurately followed. The CO_2 analyzer shown in Fig. A14-17 specifies the insertion of the test probe into the heat exchanger far enough to eliminate any introduction of outside air through the draft diverter to dilute the sample. In some units this could mean the use of an 18-in. piece of $\frac{1}{4}$-in. copper tube to ensure that sample taking will not get into the heat exchanger. The instructions also specify a minimum of 20 pumps of the sampling bulb to ensure sample purity and a minimum of three inversions of the analyzer to ensure complete absorbtion of the CO_2 into the analyzer fluid. Following the instrument instructions help to promote instrument accuracy.

8. Using the dial stack thermometer and thermometer in the control compartment, measure and record the stack temperature rise. Figure A14-18 shows the location of the thermometer in the flue pipe.

9. After measuring the CO_2 and determining the stack temperature rise (not stack temperature) using the combustion efficiency and stack loss calculator supplied with your CO_2 analyzer, determine the unit efficiency and record. Figure A14-19 shows the use of the calculator supplied with the Bacharach CO_2 analyzer. Slide the net stack temperature slider to position the nearest net stack temperature in the calculator net stack temperature

FIGURE A14-18 Checking flue gas temperature. (*Courtesy* Bacharach Instrument Company)

window. Settings are in 50°F increments and the closest range setting should be used. The vertical slider for CO_2 setting will then reveal the efficiency of the unit as well as the percent of stack loss when the arrow is set at the reading found with the CO_2 analyzer.

Gas-burning equipment of standard design should always be capable of 75 to 80% efficiency. Unless the unit is of a higher-efficiency design (when the manufacturer's instructions and settings must be followed), an efficiency above 80% could adversely affect the draft of the unit as well as cause condensation of moisture in the chimney or flue pipe and on the surfaces of the heat exchanger. If you find the efficiency results to be less than this range, the test should be repeated, checking and setting the proper input. Burner operation and temperature rise until good efficiency is obtained.

More complete detail is obtainable in Bulletin 4006 published by the Bacharach Instrument Company. It is suggested that the bulletin be used as additional study material.

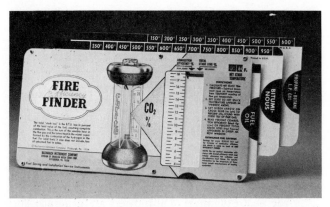

FIGURE A14-19 Combustion efficiency and stack loss calculator. (*Courtesy* Bacharach Instrument Company)

A14-3
OIL

To check, test, and adjust an oil-fired unit for highest and safest operating efficiency, the unit must have the proper input of oil for the unit size, the correct amount of air for combustion and proper venting, and the desired air temperature rise for heat occupany comfort. The air temperature rise for oil-fired units is the same as for gas-fired units, so this subject will not be repeated. Burner operation, as well as draft requirements, are different for oil than for gas and become a separate subject.

A14-3.1
Input

Both gas and oil burners are vapor burners. The difference is that gas is supplied in a vapor state and oil has to be vaporized in the burning process. To understand the vaporizing process, an explanation of how this is accomplished is necessary. Information for this section on oil-fired units has been taken from the publication *The Professional Serviceman's Guide to Oil Heat Savings,* distributed as an industry service by the R.W. Beckett Corporation of Eyria, Ohio, and from information supplied by the Wayne House Equipment Division of Fort Wayne, Indiana.

Gravity or pot-type burners were used in the early years of the oil heating industry. They were individual room-type units. With the introduction of central heating systems, some gravity units were made, but practically all unit manufacturers have changed to the power or pressure-type burner.

A14-3.2
Power Burners

Figure A14-20 shows a high-pressure atomizing-gas-type burner with the various parts identified as seen from the outside. Here we see the motor that drives the blower (inside) and oil or fuel pump, which in turn supplies oil through the oil line to the spray nozzle. Air and oil spray are mixed and burned in front of the flame retention ring within the combustion chamber located in the primary section of the heat exchanger. On top of the unit is the high-voltage transformer used to supply the electric spark to ignite and start the oil combustion process. To the left of the ignition transformer is the oil burner primary control, which in this unit uses the CAD cell for flame detection.

In Figure A14-21 a cross section of a burner is shown. Pictured is the motor-driven blower for force feeding of combustion air, the firing assembly in the air tube, and the ignition system. The firing assembly con-

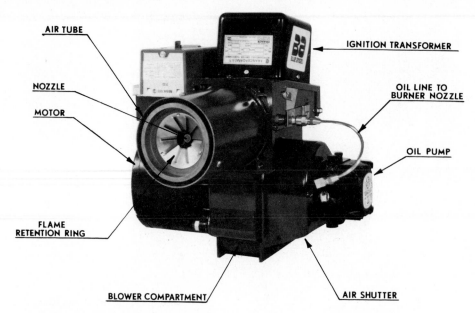

FIGURE A14-20   High-pressure atomizing-gas burner. (*Courtesy* Wayne Home Equipment Company)

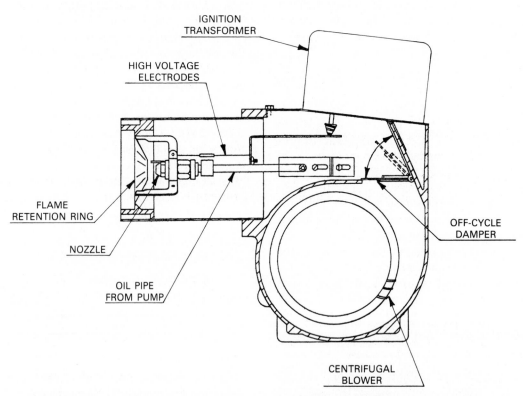

FIGURE A14-21   Cross section of burner. (*Courtesy* Wayne Home Equipment Company)

sists of the oil pipe carrying oil from the fuel pump to the nozzle assembly, the flame retention ring and static disk, which assures the proper flow of combustion air, as well as the desired air patterns for oil/air mixture and the high-voltage electrodes for creation of the ignition spark, supplied by the ignition transformer through the copper ignition leads or bars. This particular model also

has an OFF-cycle damper to stop airflow through the heat exchanger when burner shutdown occurs. This tends to keep the hot flue gases in the heat exchanger longer to promote a longer transfer of heat to the air over the heat exchanger. Oil burners are generally supplied in two mounting configurations: flange mounted where the burner is to be mounted directly onto the

FIGURE A14-22   Direct-mounted oil burner. (*Courtesy R. W. Beckett Company*)

heating unit heat exchanger (Fig. A14-22), or a base-mounted unit for conversion use. This burner (Fig. A14-23) is set up on the floor in front of the heating unit or is pedestal mounted. In all cases the unit is mounted with the firing tube extending to within $\frac{1}{4}$ in. of the inside surface of the combustion chamber or refractory and in a slightly sloped position downward in the direction of airflow. This slope is needed to prevent oil drip from the nozzle (when it occurs) from running backward into the blower compartment. The forward slope keeps any flow of oil after burner shutoff at the fire end of the air tube, where it will burn off without damaging the controls.

FIGURE A14-23   Base-mounted oil burner. (*Courtesy R. W. Beckett Company*)

A14-3.3
Components

Nozzle: In order to burn oil, the oil must first be converted into very fine droplets or mist and vaporized so as to mix with the correct amount of air (oxygen) to obtain as much potential heat release as possible. In this process, for 100% complete combustion, 1 lb of oil (No. 2 distillate heating oil) will mix with 14.36 lb of 20.9% oxygen and 79.1% nitrogen (air) for a mixture weight of 15.36 lb. The combustion process will produce, along with the release of combustion heat, 1.18 lb of water, 11.02 lb of nitrogen, and 3.16 lb of carbon dioxide. This still adds up to the original 15.36 lb of mixture.

However, just as in the gas burner, excess air is required to ensure complete combustion. Therefore, if 1 lb of oil is supplied with 21.54 lb of air for a total of 22.54 lb, this would result in the same amounts of water, nitrogen, and carbon dioxide in the products of combustion. In addition, there would be the 7.15 lb of excess air. The excess air entering the combustion chamber at the burner ambient temperature must be heated even though it does not take part in the combustion process. Therefore, it takes heat from the fire, reducing the amount available to the heat exchanger. We therefore need to keep the excess air to a minimum, yet ensure enough for complete combustion. The desirable operating range of excess air is between 15 and 30%, giving a resulting CO_2 range of 12 to 14%.

The stainless steel nozzle is mounted on the end of the nozzle adapter, which in turn is mounted in the air tube on the end of the oil feed pipe. As shown in Fig. A14-24, the oil flow is into the end of the adapter, through the fine filter (usually made of porous bronze), along the feed groove to the outside of the metering plug, through the swirl slots, and out the nozzle hole. The oil swirled by the action of the swirl slots rotates in a counterclockwise pattern when viewed from the nozzle end of the assembly. In residential oil burners three basic patterns of spray are used. Figure A14-25 shows the solid-core pattern (R), hollow-core pattern (NS), and the special solid-core pattern (AR) that are generally used. In the solid-cone pattern the greater percentage of oil spray is within the center 50% of the area. In the hollow-cone pattern the total oil spray is in the outer ring of the pattern. In the special solid-cone pattern the spray is fairly even throughout the pattern.

After breaking the oil up into fine droplets, the oil must be mixed with the air for combustion. To accomplish the combustion process with as little travel of the oil through the heat exchanger as possible, the air and oil are mixed directly in front of the air tube assembly. Figure A14-26 shows the firing assembly located in the air tube. Fins around the static disk which holds the nozzle adapter assembly and electrodes in place, as well as the turbulator blades in the end cone, force the air to swirl in a clockwise rotation as it leaves the turbulator

Heating Startup, Checkout, and Operation   553

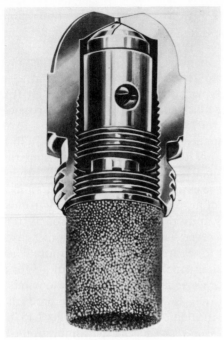

FIGURE A14-24 Nozzle and adapter assembly. (*Courtesy* Monarch Manufacturing Works, Inc.)

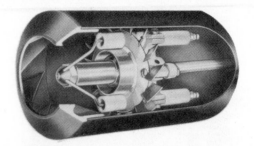

FIGURE A14-26 Air tube assembly cutaway. (*Courtesy* Monarch Manufacturing Works, Inc.)

opening. This clockwise swirl of the air is opposite to the counterclockwise swirl of the oil spray. This results in a fast mixing of oil and air for rapid combustion of the oil. There are two patterns of air discharge cone design: center concentration and ring concentration. It is necessary to match the type of air pattern with the nozzle spray pattern for heat performance. The solid-cone nozzle is used with the center concentration air pattern and the hollow-cone nozzle is used with the ring concentration air pattern. Again, to get rapid complete combustion with the lowest amount of trouble and highest operating efficiency, the oil and air patterns must match. Changing the type of spray pattern of the nozzle only leads to problems. *Do not substitute a solid-cone nozzle for a hollow-cone nozzle, or vice versa.*

Upon ignition of the oil/air mixture, by spark across the electrode mounted above the nozzle (we will discuss the ignition system later), the mixture starts to

SPRAY PATTERNS

R   NS   AR

FIGURE A14-25 Swirl patterns. (*Courtesy* Monarch Manufacturing Works, Inc.)

burn by consuming the hydrogen molecules from the oil first. These are the easiest to mix with the oxygen and burn. The carbon molecules have to be heated before the combustion process will take place. Fortunately, the hydrogen burning provides the heat needed for the carbon-burning process to start.

Refractory or Combustion Chamber: For complete combustion of the carbon, additional heat is necessary. This is provided by the combustion chamber or refractory (Fig. A14-27). Located in the combustion area of the primary section of the heat exchanger, the combustion process takes place inside the refractory. The refractory is made of a high-insulating-value material that will not burn: usually a form of cement, asbestos fiber cement, or the modern higher-temperature ceramics developed by the space program. The purpose of the refractory is to obtain a white-hot surface on the inside as rapidly as possible. This white-hot surface radiates heat back into the flame to provide additional heat for burning the carbon molecules.

When the oil burner first starts, the combustion process is poor and unburned or free carbon flows through the heat exchanger. Most of it passes out of the unit through the flue pipe and chimney or vent. However, some is deposited on these surfaces and held there by the water that condenses out of the flue products onto these cold surfaces. When these surfaces heat up and dry out, the carbon still remains and becomes bonded to the surfaces. Thus, for the first 60 to 90 seconds that an oil burner operates, until the refractory comes up to temperature, soot is deposited in the passages of the heat exchanger, flue pipe, and chimney. To keep this to a minimum, longer ON cycles are required for an oil burner than for a gas burner. The minimum ON cycle recommended is 5 minutes. This is also why an oil-fired unit will soot-up faster in the spring and fall, when the ON cycles are shorter and more frequent than in the dead of winter, when the oil unit operates for longer periods.

The size of the combustion chamber is important. It should not be so large that the reflective surface is too far away from the fire to provide the proper amount of reflected heat. It should not be so small, however, that the oil spray touches the surface before it is burned. If this happens, the carbon will not burn and will build up

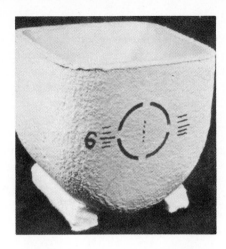

FIGURE A14-27 Soft-fiber refractory combustion chambers. (*Courtesy* R. W. Beckett Corporation)

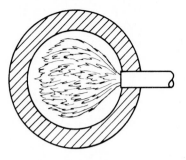

Good Combination

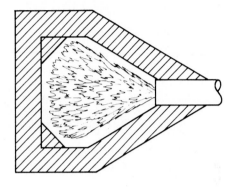

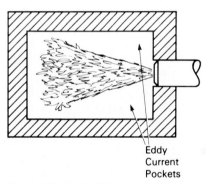

Eddy
Current
Pockets

Corners Should be Filled

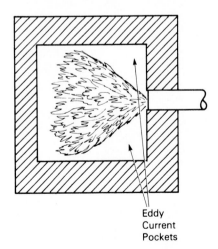

Eddy
Current
Pockets

FIGURE A14-28 Combustion chamber design. (*Courtesy* R. W. Beckett Corporation)

on the surface of the refractory, forming what are called *coke trees*. Generally, the size of the refractory, measured in square inches of the floor area, would be 80 to 90 in.^{2} per gallon of oil burned per hour. Using this sizing standard, it is extremely important that the nozzle capacity is correct and that the nozzle spray pattern and air pattern conform. The most predominate shape of the refractory is round, although other shapes are used depending on the design of the heating or hydronic unit involved.

Figure A14-28 shows the design shape of combustion chambers for both good performance and poor performance. In the square and rectangular refractories, the corners alongside the air tube provide air space that does not take part of the combustion process. This air dilutes the flame and reduces the combustion process and the overall efficiency of the unit.

The height of the combustion chamber must be such that adequate reflective surface is provided for combustion without covering too much of the heat exchanger surfaces. Most refractory applications show the overall height of the refractory to be 2 to $2\frac{1}{2}$ times the height of the nozzle above the floor of the refractory. Figure A14-29 shows combustion chamber sizing data. If, for example, the combustion chamber is sized on an 80-in.²/gal basis, and the burner is to burn 1.35 gal/hr, a 108-in.² combustion chamber will be required. This could be an 11 in. × 11 in. square or $12\frac{3}{8}$-in.-diameter round type. If the nozzle is 5 in. above the floor of the refractory, the height of the combustion chamber would be 5 in. × 2 to $2\frac{1}{4}$ in. or 10 to $12\frac{1}{2}$ in. Sometimes the size of the primary heat exchanger is larger than the refractory required for the unit capacity. This is common where the manufacturer produces the same size unit over a multiple capacity range.

If the unit is fired at less than the highest rating and the combustion chamber is smaller than the primary heat exchanger, the space between the combustion

FIGURE A14-29
Combustion chamber sizing data.

	Oil Consumption G.P.H.	Sq. Inch Area Combustion Chamber	Square Combustion Chamber Inches	Dia. Round Combustion Chamber Inches	Rectangular Combustion Chamber Inches	HEIGHT FROM NOZZLE TO FLOOR INCHES			
						Conventional Burner Wth. × Lgth.	Conventional Burner Single Nozzle	Sunflower Flame Burner Single Nozzle	Sunflower Flame Burner Twin Nozzle
	.75	60	8 × 8	9	—	5	×	5	×
	.85	68	8.5 × 8.5	9	—	5	×	5	×
	1.00	80	9 × 9	$10\frac{1}{8}$	—	5	×	5	×
	1.25	100	10 × 10	$11\frac{1}{4}$	—	5	×	5	×
80 Sq. In.	1.35	108	$10\frac{1}{2} \times 10\frac{1}{2}$	$11\frac{3}{4}$	—	5	×	5	×
Per Gal.	1.50	120	11 × 11	$12\frac{3}{8}$	10 × 12	5	×	6	×
	1.65	132	$11\frac{1}{2} \times 11\frac{1}{2}$	13	10 × 13	5	×	6	×
	2.00	160	$12\frac{5}{8} \times 12\frac{5}{8}$	$14\frac{1}{4}$	6	×	7	×	×
	2.50	200	$14\frac{1}{4} \times 14\frac{1}{4}$	16	$12 \times 16\frac{1}{2}$	6.5	×	7.5	×
	3.00	240	$15\frac{1}{2} \times 15\frac{1}{2}$	$17\frac{1}{2}$	$13 \times 18\frac{1}{2}$	7	5	8	6.5
	3.50	315	$17\frac{3}{4} \times 17\frac{3}{4}$	20	15 × 21	7.5	6	8.5	7
90 Sq. In.	4.00	360	19 × 19	$21\frac{1}{2}$	$16 \times 22\frac{1}{2}$	8	6	9	7
Per Gal.	4.50	405	20 × 20		$17 \times 23\frac{1}{2}$	8.5	6.5	9.5	7.5
	5.00	450	$21\frac{1}{4} \times 21\frac{1}{4}$		18 × 25	9	6.5	10	8
	5.50	550	$23\frac{1}{2} \times 23\frac{1}{2}$		$20 \times 27\frac{1}{2}$	9.5	7	10.5	8
	6.00	600	$24\frac{1}{2} \times 24\frac{1}{2}$		$21 \times 28\frac{1}{2}$	10	7	11	8.5
	6.50	650	$25\frac{1}{2} \times 25\frac{1}{2}$		$22 \times 29\frac{1}{2}$	10.5	7.5	11.5	9
	7.00	700	$26\frac{1}{2} \times 26\frac{1}{2}$		$23 \times 30\frac{1}{2}$	11	7.5	12	9.5
	7.50	750	$27\frac{1}{4} \times 27\frac{1}{4}$		24 × 31	11.5	7.5	12.5	10
	8.00	800	$28\frac{1}{4} \times 28\frac{1}{4}$		25 × 32	12	8	13	10
	8.50	850	$29\frac{1}{4} \times 29\frac{1}{4}$		25 × 34	12.5	8.5	13.5	10.5
	9.00	900	30 × 30		25 × 36	13	8.5	14	11
100 Sq. In.	9.50	950	31 × 31		$26 \times 36\frac{1}{2}$	13.5	9	14.5	11.5
Per Gal.	10.00	1000	$31\frac{3}{4} \times 31\frac{3}{4}$		$26 \times 36\frac{7}{8}$	14	9	15	12
	11.00	1100	$33\frac{1}{4} \times 33\frac{1}{4}$		$28 \times 29\frac{1}{2}$	14.5	9.5	15.5	12.5
	12.00	1200	$34\frac{1}{2} \times 34\frac{1}{2}$		28 × 43	15	10	16	13
	13.00	1300	36 × 36		29 × 45	15.5	10.5	16.5	14
	14.00	1400	$37\frac{1}{2} \times 37\frac{1}{2}$		31 × 45	16	11	17	14.5
	15.00	1500	$38\frac{3}{4} \times 38\frac{3}{4}$		32 × 47	16.5	11.5	17.5	15
	16.00	1600	40 × 40		$33 \times 48\frac{1}{2}$	17	12	18	15
	17.00	1700	$41\frac{1}{4} \times 41\frac{1}{4}$		34 × 50	17.5	12.5	18.5	15.5
	18.00	1800	$42\frac{1}{2} \times 42\frac{1}{2}$		$35 \times 51\frac{1}{2}$	18	13	19	16

Source: R. W. Beckett Company.

chamber and the heat exchanger *must be filled with insulation*. Usually, mica pellets or other high-temperature insulation is used. A poor grade or no fill in this area will shorten the life of the combustion chamber, reduce the efficiency of the unit, and increase combustion noise.

Let's refer back to the discussion of the relationship between the oil spray pattern and air pattern. Any discrepancy will quickly show up on the surfaces of the combustion chamber. If in the round combustion chamber shown in Fig. A14-25 a hollow-cone nozzle is put into a unit designed for a solid-cone nozzle, the large quantity of oil in the outer ring of the mixture pattern will not receive enough combustion, and unburned carbon will be deposited on the sides of the combustion chamber. If a solid-cone nozzle is used in a unit designed for a hollow-cover nozzle, the extra oil supplied to the center of the pattern, where the air pattern is small, will not burn çompletely and coke trees will build in the back of the refractory. In the meantime, in both cases, the unit efficiency is low and soot buildup in the heat exchanger is accelerated. *Do not change the nozzle pattern; use the pattern the burner is designed to require.*

The shape of the refractory will also determine the angle at which the oil and air will be sprayed into the combustion chamber. Round combustion chambers will use a spray angle of 80 to 90° to make maximum use of the combustion chamber reflective surfaces. Long, narrow combustion chambers will use spray angles in the range 30 to 70° depending on the length of the rectangular chamber compared to the width. The longer the chamber, the narrower the spray angle. Again, this is to keep the oil spray from touching the sides of the chamber before the burning process is complete. Before changing the spray angle used in the unit, the unit manufacturer should be consulted. It is not possible to cover all applications of oil burner nozzles in a text of this nature. It is suggested that nozzle manufacturer literature, such as *Bulletin-O* published by the Monarch Manufacturing Works Inc., Philadelphia, Pennsylvania, be used for supplemental information.

The amount of oil that will be supplied by the nozzle is marked on the nozzle in gallons per hour. This rating is at a nozzle oil pressure of 100 psig. Most oil heating unit manufacturers use oil burners operating at 100 psig nozzle pressure. If the burner requires other than the 100 psig nozzle pressure, the nozzle capacity will vary with the operating pressure. Figure A14-30 shows nozzle capacities at various operating pressures.

The size of the nozzle in gallons per hour will be determined by the input requirement of the heating unit. Using the standard figure of 140,000 Btu per gallon of No. 2 fuel oil, it is only necessary to divide the input requirement of the heating unit by 140,000 to get the firing rate required. Once this figure is obtained, the nozzle size in gal/hr is determined. For example, if the unit is rated at 100,000 Btu/hr, a 0.714-gal/hr nozzle is

OIL BURNER NOZZLES

Rate GPH @ 100 PSI	OPERATING PRESSURE: POUNDS PER SQUARE INCH							
	75	100	125	150	175	200	250	300
.40	—	.40	.45	.49	.53	.56	.63	.69
.50	—	.50	.56	.61	.66	.71	.79	.87
.60	—	.60	.67	.74	.79	.85	.95	1.04
.65	—	.65	.73	.80	.86	.92	1.03	1.13
.75	—	.75	.84	.92	.99	1.06	1.19	1.30
.85	—	.85	.95	1.04	1.13	1.20	1.34	1.47
1.00	.87	1.00	1.12	1.23	1.32	1.41	1.58	1.73
1.10	.95	1.10	1.23	1.34	1.45	1.55	1.74	1.91
1.20	1.04	1.20	1.34	1.47	1.59	1.70	1.89	2.08
1.25	1.07	1.25	1.39	1.53	1.65	1.77	1.97	2.16
1.35	1.17	1.35	1.51	1.65	1.79	1.91	2.14	2.34
1.50	1.30	1.50	1.68	1.84	1.98	2.12	2.37	2.60
1.65	1.43	1.65	1.84	2.02	2.18	2.34	2.61	2.86
1.75	1.51	1.75	1.96	2.14	2.32	2.48	2.77	3.03
2.00	1.73	2.00	2.24	2.45	2.65	2.83	3.16	3.46
2.25	1.95	2.25	2.52	2.74	2.98	3.18	3.56	3.90
2.50	2.16	2.50	2.80	3.06	3.30	3.54	3.95	4.34
3.00	2.59	3.00	3.35	3.68	3.97	4.25	4.75	5.20
3.50	3.03	3.50	3.91	4.29	4.63	4.95	5.54	6.06
4.00	3.46	4.00	4.47	4.90	5.30	5.66	6.32	6.94
4.50	3.90	4.50	5.04	5.51	5.95	6.36	7.11	7.80
5.00	4.33	5.00	5.59	6.13	6.61	7.07	7.90	8.66
5.50	4.76	5.50	6.15	6.74	7.27	7.78	8.70	9.54
6.00	5.19	6.00	6.71	7.33	7.94	8.48	9.49	10.40
6.50	5.63	6.50	7.26	7.96	8.60	9.20	10.30	11.25
7.00	6.05	7.00	7.82	8.58	9.25	9.90	11.08	12.12
7.50	6.49	7.50	8.38	9.19	9.91	10.60	11.85	13.00
8.00	6.93	8.00	8.94	9.80	10.58	11.31	12.65	13.85
8.50	7.36	8.50	9.50	10.45	11.27	12.08	13.40	14.70
9.00	7.79	9.00	10.06	11.02	11.91	12.73	14.20	15.60
9.50	8.22	9.50	10.60	11.70	12.60	13.50	15.00	16.45
10.50	9.10	10.50	11.70	12.90	13.90	14.90	16.20	18.20
12.00	10.40	12.00	13.40	14.70	15.90	17.00	19.00	20.80
13.50	11.65	13.50	15.07	16.53	17.90	19.17	21.35	23.40
15.50	13.37	15.50	17.33	18.95	20.56	21.89	24.50	26.85
17.50	15.10	17.50	19.60	21.40	23.20	24.80	27.70	30.40
19.50	16.90	19.50	21.80	23.80	25.80	27.60	30.80	33.80
21.50	18.60	21.50	24.00	26.40	28.40	30.40	34.00	37.30
24.00	20.80	24.00	26.80	29.40	31.80	34.00	38.00	41.60
28.00	24.20	28.00	31.30	34.30	37.00	39.60	44.30	48.50
30.00	26.00	30.00	33.60	36.80	39.70	42.50	47.50	52.00

F-80 NOZZLE CAPACITIES — U.S. GALLONS PER HOUR NO. 2 FUEL OIL

R   NS   AR   PLP   PL   HV

FIGURE A14-30   Nozzle capacities. (*Courtesy* Monarch Manufacturing Works, Inc.)

required. If the exact nozzle size is not available, the size should be reduced to the first available size. *Do not oversize*, as this reduces efficiency and shortens unit life.

According to Fig. A14-30, a 0.65-gal/hr nozzle of 100 psig would furnish 0.65 gal of fuel oil per hour to the unit for an input of 91,000 Btu/hr. This would be within the −10% operating range of the unit. By raising the operating pressure to 110 psig, the input can be raised to the 0.714 gal/hr and stay within the operating characteristics of the average oil burner. Although the following formula is not 100% accurate, it is close enough to use for small burners. For example, the amount of oil required is 0.714 gal/hr. Nozzle selection should not be below 100 psig on nozzles less than 1.00 gal/hr size for best operation. Therefore, a nozzle of smaller than 0.714 gal/hr must be selected. The chart gives a nozzle of 0.65 gal/hr at 100 psig and 0.73 at 125 psig. Somewhere in between would be the pressure for the desired input. Using the following formula, the desired pressure can be obtained.

FIGURE A14-31   Fuel units. (*Courtesy* Suntec Industries, Inc., Rockford, Ill.)

$$\frac{0.65}{100} = \frac{0.714}{X} = \frac{100 \times 0.714}{0.65\,X} = \frac{71.4}{0.65} = 109.85 \text{ psig}$$

where

0.65 = nozzle rating at 100 psig

100 = pressure rating of nozzle

0.714 = desired rating of nozzle

X = required operating pressure, psig

Including this information does not sanction the practice of arbitrarily changing nozzle pressures. Use the pressures stated by the unit manufacturer. To do otherwise could affect the operation and life expectancy of the unit.

Fuel Unit: The oil supplied to the nozzle at the proper pressure is supplied by the oil pump or *fuel unit*. Figure A14-31 shows typical fuel units. Four models are shown. The A2VA-7016 is a single-stage unit normally found on factory-assembled units; A2VA-7116 is a single-stage unit for low-viscosity fuels; A2VA-7416 has a built-in solenoid valve for positive oil flow control; and the B2VA-8216 is a two-stage unit where the oil must be raised from the fuel source as well as supplied to the nozzle.

Figure A14-32 shows a single-stage fuel unit with one set of pump gears. This unit is used where the fuel supply is above the inlet to the fuel unit and the unit must only supply oil to the nozzle at the proper pressure (Fig. A14-33). In the cutaway illustration the upper left horizontal connection is the oil outlet to the oil fuel pipe and nozzle. The upper right horizontal part is the screwdriver-slotted adjustment screw for setting the pump operating pressure. The pressure is measured by connecting a gauge to the upper left vertical pipe tape by means of the appropriate fitting. The upper right vertical connec-

tion has a pipe plug in it. This connection would be used if the pressure-regulating assembly were reversed and the oil outlet were on the right-hand side.

Figure A14-34 shows a two-stage fuel unit with two sets of pump gears. This type of pump is required for installation where the fuel supply is below the level of the pump, such as buried tanks, units suspended at ceiling level, etc. On these installations (see Fig. A14-35), a two-pipe system using an oil supply line from the tank to the pump and an oil return line from the pump to the top of the supply tank should be used. Thus, the pump does not have to carry the direct weight of the oil lift at all times. On startup, both gear sets draw on the oil supply line and air is discharged through the return line. Extremely fast priming occurs. After prime is estab-

FIGURE A14-32   Single-stage fuel unit. (*Courtesy* Suntec Industries, Inc., Rockford, Ill.)

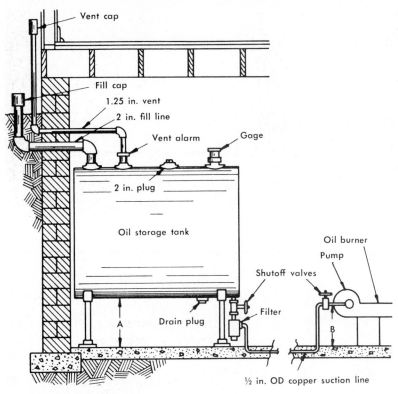

FIGURE A14-33 Typical installation of single-pipe system. (*Courtesy* Addison Products Company)

lished, the first stage continues to discharge back to the tank through the return line. Thus the weight of the oil returning to the tank helps to counteract the weight of the oil rising to the pump by siphon action. The second set of gears receives the oil under a small positive pressure from below the air foam level in the strainer chamber and only needs to raise the oil pressure to nozzle requirement.

The fuel unit must not, however, spray oil into the combustion chamber until the nozzle pressure is high enough to break up the oil properly and must hold the proper pressure while operating. This is accomplished by the use of a piston located at the inlet of the oil outlet connection in the fuel unit. In both Figs. A14-32 and A14-34 the pressure-regulating piston is in the top of the unit. Figure A14-36 shows how this piston operates.

The flow of oil through the fuel unit is from the tank connection, through the strainer, to the pump gears. The relation of the gears forces the oil out and into the upper portion of the pressure-regulating and cut-off valve cylinder. The oil has no outlet and the pressure rises very rapidly. The rise in pressure forces the piston against the regulating spring. When the oil pressure starts to exceed the spring pressure, the piston moves from the outlet orifice and oil flows out the oil feed pipe to the nozzle. Because the nozzle will not allow all the oil to flow out that the pump can deliver, the pressure continues to rise. This causes the piston to move further until the bypass outlet is opened and the excess oil is

returned to the suction side of the unit (the strainer area). Thus, the pump will not feed oil until sufficient oil pressure is created, and yet a maximum oil pressure is maintained. The opening pressure is usually 20 psig below the operating pressure. With a 100 psig operating pressure the opening pressure is approximately 80 psig. In ad-

FIGURE A14-34 Two-stage fuel unit. (*Courtesy* Suntec Industries, Inc.)

Heating Startup, Checkout, and Operation 559

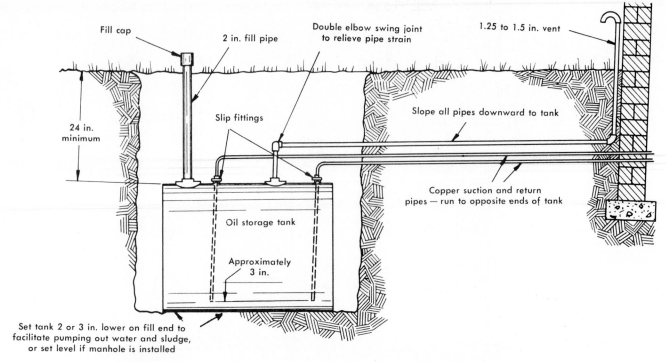

FIGURE A14-35   Typical installation of two-pipe system. (*Courtesy* Addison Products Company)

dition to the advantage of not allowing oil to flow from the nozzle at pressures too low to atomize properly on startup, this feature also cuts off the oil supply on shutdown.

When the pump stops, the oil pressure falls very rapidly, due to the bleed orifice bypass allowing the piston to seat very rapidly. This causes the hydraulic pressure in the nozzle to fall very rapidly, with a minimum amount of flame as the burner operation dies. Therefore, no oil is emitted when the air supply is cut off. If this cutoff does not occur, a flow of oil will occur and produce a very smokey flame because of lack of combustion air.

The "after flow" of oil is caused by the piston not seating properly to cut off the oil pressure. This is usually caused by dirt accumulating in the piston chamber. *Never operate an oil burner without the fuel unit strainer in place.* As a further protection against this problem, a fuel line filter should always be installed on the outlet of the fuel tank and the filter element *replaced at the beginning of each season.*

Blower: In Fig. A14-37 an air shutter is indicated below the burner on the right side. This one is a hinge type. In Fig. A14-38 the air shutter is a rotating band that raises the size of the opening into the blower housing. Both types control the amount of combustion air that the blower can force down the fire tube and into the combustion chamber. The setting or adjustment of this

air shutter is discussed in the check-out section. The burner motor and controls were discussed in previous chapters.

 A14-4
FLUE VENTING

Oil-fired furnaces must have an ample supply of make-up air for combustion, and the methods of introducing makeup air for a gas furnace will also be adequate for oil equipment.

Masonry chimneys used for oil-fired furnaces should be constructed as specified in the National Building Code of the National Board of Fire Underwriters. A masonry chimney shall have a minimum cross-sectional area equivalent to either an 8-in. round or an 8-in. × 8-in. square; it should never be less than the flue outlet of the furnace. Prefabricated lightweight metal chimneys (Fig. A14-39) are also available for use with oil. The double-wall type is filled with insulation. These vents are class A flues. Class B flues, made specifically for gas-fired equipment, are not suitable for use with solid or liquid fuels. Be sure that the class rating states "all fuels."

The terminal vent heights above the roof recommendations previously outlined for gas should also be observed for oil.

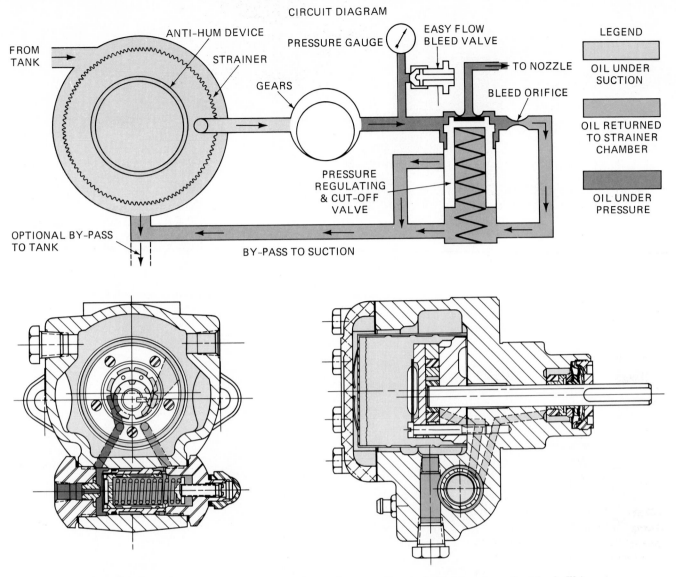

CIRCUIT DIAGRAM

FROM TANK

ANTI-HUM DEVICE

STRAINER

GEARS

PRESSURE GAUGE

EASY FLOW BLEED VALVE

TO NOZZLE

BLEED ORIFICE

PRESSURE REGULATING & CUT-OFF VALVE

OPTIONAL BY-PASS TO TANK

BY-PASS TO SUCTION

LEGEND

OIL UNDER SUCTION

OIL RETURNED TO STRAINER CHAMBER

OIL UNDER PRESSURE

FIGURE A14-36 Fuel unit circuit diagram. (*Courtesy* Suntec Industries, Inc., Rockford, Ill.)

Oil-fired furnaces operate on positive pressure from the burner blower, and it is most important to have a chimney that will develop a minimum draft of 0.01 to 0.02 in. of W.C. as measured at the burner flame inspection port. Consistency or stability of draft is also more critical, and the use of a barometric damper (Fig. A14-40) is required. The damper is usually installed in the horizontal vent pipe between the furnace and chimney. (*Note:* Some manufacturers attach them directly to the furnace flue outlet.) The damper has a movable weight so that it can be set to counterbalance the suction and to maintain reasonably constant flue operation. It is adjusted while the furnace is in operation and the chimney is hot. The overfire draft at the burner should be in accordance with the equipment installation instructions.

A14-5
CLEARANCES

Cabinet temperatures and flue pipe temperatures run warmer for oil-burning equipment, and the clearances from combustible material should be adjusted accordingly. One-inch clearances are common for the sides and rear of the cabinet, as opposed to zero clearances for many gas units. Flue pipe clearances of 9 in. or more are needed, whereas only 6 in. of clearance is needed for gas.

Front clearance is generally determined by the space needed to remove the burner assembly. Combustible floor bases for oil are generally increased as compared to gas. Horizontal oil furnaces are particularly

Heating Startup, Checkout, and Operation **561**

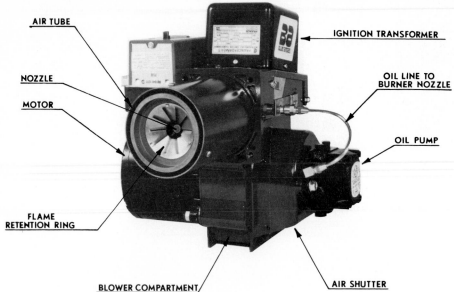

FIGURE A14-37 Oilburner using flap-type air adjuster. (*Courtesy* Wayne Home Equipment Company)

FIGURE A14-38 Oilburner using band-type air adjustment. (*Courtesy* R. W. Beckett Company)

important in respect to attic installations and the installer must check the recommendations carefully.

Oil furnaces are rated and listed by Underwriters Laboratories, and clearances are a vital part of this inspection and compliance procedure along with many other safety considerations.

A14-6
OIL STORAGE

The installation of the fuel oil tank and connecting piping must conform to the standards of the National Fire Protection Association (Standard 31) and/or local code requirements. Regulations and space permitting,

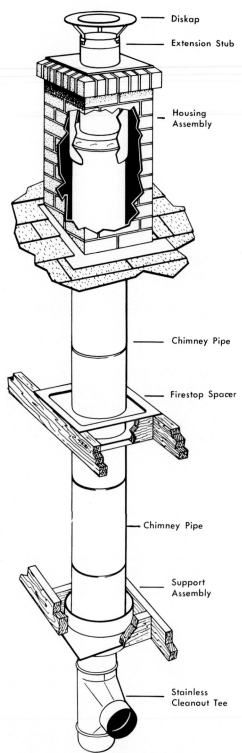

FIGURE A14-39 Factory-built "all fuel" chimney. (*Courtesy* Metalbestos Systems, Wallace Murray Corporation)

FIGURE A14-40   Barometric damper. (*Courtesy* Field Control, Division of Conco, Inc.)

the oil tank can be located indoors. Note the proper use of shutoff valves and an effective filter to catch impurities. If the oil tank is placed outdoors above ground, firm footings must be provided. Exposed tanks and piping above ground are subject to more condensation of water vapor and the possibility of freezing during extremely low temperatures.

If a large tank is installed below ground, it is important to keep it well filled with oil during periods of high water level (i.e., spring rains); otherwise, groundwater may force it to float upward. Extra concrete for added weight over the tank is advised. Flexible copper lines are recommended so as to give with ground movement. Notice the use of suction and return lines; as mentioned previously, it is normal practice to use a two-stage fuel pump with a two-pipe system whenever it is necessary to lift oil from a tank that is below the level of the burner. The oil burner pump suction is measured in terms of inches of mercury vacuum. A two-stage pump should never exceed a 15-in. vacuum. Generally, there is 1 in. of vacuum for each foot of vertical oil lift and 1 in. of vacuum for each 10 ft of horizontal run of supply piping.

A14-7
STARTUP

So far we have discussed the major components of the oil burning system without discussing the sequence of operation. Figure A14-41 shows a typical oil-fired heating-only installation using a heating-only thermostat controlling a cad cell type of primary control with a blower door switch in the low-voltage circuit as an air supply safety switch. In turn, the primary control is con-

trolling the oil burner motor and ignition transformer through the limit control to prevent overheating the unit.

The final function is the operation of the system blower motor by the fan portion of the fan limit control. Properly set, this control will close at 125 to 130°F to start the blower motor and open at 95 to 100°F to stop the blower motor.

A14-7.1
Operating Sequence

The step-by-step operation of the unit would be as follows:

1. With the blower compartment door closed and the blower door switch closed, the room thermostat, upon a call for heat in the conditioned area, will close. This causes current flow from the 40-VA transformer through the control circuits and the burner motor relay. The relay pulls the contact closed.

2. The closed contacts cause a current flow from the "H" or hot side of the supply line through the limit switch to the burner motor and ignition transformer. In addition, a current flow in the 24-V circuit exists through the dark cad cell and the safety switch heater.

3. Immediately, the ignition transformer establishes a spark across the electrodes located above the nozzle (see Fig. A15-2). The spark at this time is only $\frac{1}{8}$ in. long, the distance between the electrode tips.

4. The blower motor is energized at the same time as the ignition transformer. It reaches full-load speed within 1 second. When the full amount of air is

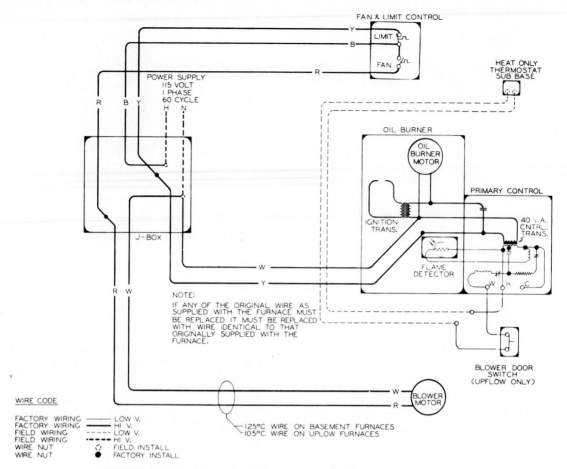

FIGURE A14-41 Typical oil furnace control. (*Courtesy* Luxaire, Inc.)

delivered by the blower, the ignition spark is blown forward into the oil spray and the oil is ignited. This occurs directly in front of the nozzle and quickly expands into a full burst of fire in the combustion chamber.

5. As soon as the light from the fire reaches the cad cell located in the fire tube, the electrical resistance of the cell increases. This increase in resistance reduces the current flow through the safety switch heater and prevents the safety switch from opening.

6. The flame efficiency increases as the refractory temperature rises. As a white-hot surface temperature of the refractory, the burner is operating at peak efficiency. At this point efficiency tests can be taken. To reach these conditions, the burner should be allowed to operate at least 5 minutes.

The initial system startup will require more than just turning on the burner. The first requirement will be to purge the fuel unit of air. This procedure would be different for the single-pipe system, where the fuel supply is above the burner, and the two-pipe system, where the fuel supply is below the burner.

Single-Pipe System: On the single-pipe system with gravity feed (the supply is above the burner) it is only necessary to purge the air from the oil filter, fuel line, and fuel unit. Refer back to Fig. A14-13. The purge valve is located under the valve and the unit is OFF; it is only necessary to open this valve until the air in the system starts to flow out. Usually, this is one-half to one turn of the valve. When the air is out and oil starts to flow out, close the valve to prevent oil leakage.

Two-Pipe System: In a two-pipe system, the pump will force the air down the oil return line to the tank and the system is self-purging. It may take longer than the cycle time of the safety switch to purge the system and the burner may cut off before flame is established. It is then necessary to allow the safety switch to cool to be reset and the burner started again. This should be repeated until the oil lines are clean and fire is established.

A14-7.2
Efficiency Testing

When fire is established and the refractory obtains white heat, efficiency testing can be done on the unit.

As with the gas unit, certain information is necessary. Again, a test sheet is desirable to record the conditions to which the burner were adjusted so that they will be available for future reference. The following information should be included on the test sheet.

1. Make of oil burner _____
2. Model No. _____ Serial No. _____
3. Nozzle: Size (gal/hr) _____ Type _____ Angle _____
4. Refractory: Shape _____

Operation

1. Overfire draft _____
2. Stack draft _____
3. Heat exchanger flow resistance _____
4. $CO_2\%$ _____
5. Net stack temperature _____
6. Efficiency _____
7. Smoke number _____
8. Air temperature—supply plenum _____
9. Air temperature—return plenum _____
10. Air temperature—rise _____

Although it seems unnecessary, it is wise to record on the test sheet the make, model number, and serial number of the burner. This information may be valuable in future situations regarding the burner. Recording the nozzle size, spray angle, and type forces the checking of the unit for proper input as well as spray angle for the shape of the refractory. Round or square refractories will take the 80 or 90° spray angle; a rectangular refractory, the 45 to 60° angle, depending on the length of the refractory. With the unit operating and refractory up to white heat, operating tests can be made.

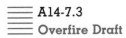
A14-7.3
Overfire Draft

The first test should be overfire draft. Using a draft gauge such as those pictured in Fig. A14-42 and Fig. A14-43 and a $\frac{1}{4}$-in. hole in the pressure relief door over the burner (sometimes called the observation door), insert the test pipe for the gauge through the hole until the end of the pipe is beyond the inner edge of the combustion chamber. To accomplish this, it is sometimes necessary to substitute a longer piece of $\frac{1}{4}$-in. copper tubing for the gauge probe pipe. This test should result in a minimum pressure of 0.02 to −0.04 in. of W.C. to ensure the proper draft for best operation. If the manufacturer of the unit does not specify the coverfire draft, it is best to start at a draft control setting of −0.04 in. of W.C.

FIGURE A14-42 (*Courtesy* Bacharach Instrument Company)

Some manufacturers specify stack draft for a setting. This means that the flow resistance of the heat exchanger must be taken into account for burner operation. By measuring both the overfire draft and stack draft on the new unit, the design draft resistance can be determined. To produce a negative pressure of a given amount over the fire, the stack draft must be a greater negative amount. The difference between the overfire draft and stack draft will be the heat exchanger draft flow resistance. For example, if it is necessary to have a stack draft of −0.06 in. of W.C. to produce an overfire draft of −0.04 in. of W.C, the heat exchanger resistance would be 0.02 in. of W.C. When checking the unit per-

FIGURE A14-43 (*Courtesy* Bacharach Instrument Company)

Heating Startup, Checkout, and Operation 565

formance after a period of operation, if the heat exchanger flow resistance has doubled, it is necessary to mechanically clean the soot from the heat exchanger flue passages.

A negative pressure must be maintained on the heat exchanger to prevent the products of combustion from being forced into the occupied area. Not only do they carry free carbon or soot, but they also contain a high percentage of CO. This is especially true at startup. Therefore, a minimum overfire draft of -0.02 in. of W.C. is usually required for proper operation, with 0.04 in. of W.C. to be on the safe side.

A14-7.4
CO_2

As in the gas furnace, the CO_2 sample is taken far enough ahead of the barometric draft control to obtain a good sample of flue gas without outside air mixture. Again, this may mean a longer sampling tube on the analyzer. Use the CO_2 analyzer, as the one shown in Fig. A14-44. The instrument manufacturer's instructions should be followed for taking the CO_2 sample. The following results should be obtained:

1. **Old-Style Gun Burners** Burners with no special air-handling parts other than an end cone and a stabilizer. A CO_2 reading of 7 to 9% should be obtained unless a CO_2 reading in this range results in more than a No. 2 smoke. If so, the CO_2 reading should be reduced until the smoke test results in below a No. 2 smoke.

2. **Newer-Style Gun Burners** Burners with special air-handling parts should be set in the range 9 to 11%. This should result in less than a No. 2 smoke (closer to a No. 1 smoke).

3. **Flame-Retention Gun Burners** These burners should be set in the range 10 to 12% with the same smoke test results.

4. **Rotary Burners** These burners should be set the same as old-style gun burners.

5. **Rotary Wall Flame Burners** Higher CO_2 settings are available, up to 13.5%, with these burners, but the maximum of No. 2 on the smoke test is required.

A14-7.5
Smoke Testing

Smoke testing is done with a smoke tester, such as the one shown in Fig. A14-45. This instrument uses a pump piston to draw flue products through a filter inserted in the head of the pump between the sample tube connection and the piston body. This particular instrument requires 10 slow strokes of the piston to draw the

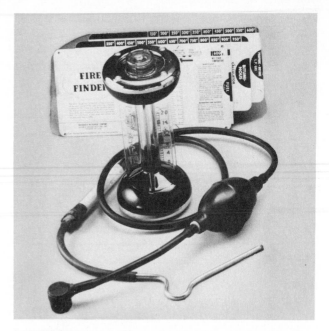

FIGURE A14-44 CO_2 analyzer. (*Courtesy* Bacharach Instrument Company)

required amount of flue products through the filter. After the 10 strokes of the piston, the filtered sample is compared to the numbered rings on the test card. The flue products should not contain any more free carbon (unburned carbon) than that indicated by comparison with the ring marked "2." If the smoke content is higher than No. 2, the burner is receiving insufficient air and is not burning the carbon sufficiently. This will quickly produce carbon deposits and plug heat exchangers passages. If the smoke content is less than No. 1, too much air is being allowed in the burner. The CO_2 content is too low and too much heat is being forced out of the heat exchanger. This results in considerably lower efficiency and much higher operating cost.

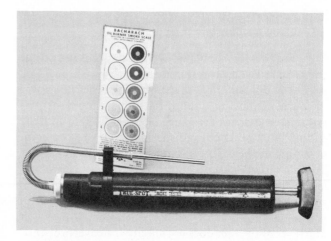

FIGURE A14-45   Smoke tester. (*Courtesy* Bacharach Instrument Company)

A14-7.6
Stack Temperature

Using a high-temperature thermometer, such as that shown in Fig. A14-46, inserted in the same opening in which the stack draft was taken, the stack temperature is taken. The temperature of the flue products will also determine the efficiency of the heating unit. The cleaner the heat exchanger, the better able it is to remove heat from the flue products as they pass through. If the CO_2 content is correct but the stack temperature is too high, this is usually a sign of excessive carbon deposit on the heat exchanger surfaces. To determine the next stack temperature, subtract the temperature of the air entering the burner (usually this is the temperature of the air where the unit is located) from the stack temperature.

A14-7.7
Calculation of Efficiency

Use an efficiency calculation as shown in Fig. A14-47. The horizontal slider is set so that the net stack temperature to the nearest 50°F is shown in the net stack temperature window. The vertical slider is then adjusted so that the tip of the arrow is at the CO_2 determined by the analyzer test. The efficiency of the unit is then shown in the arrow in the vertical slider.

As stated before, burner settings should not result in greater than a No. 2 smoke. Some technicians will start the burner adjustment procedure by adjusting the burner to obtain between a No. 1 and No. 2 smoke with 0.02 in. of W.C. draft overfire and then accept the CO_2 stack temperature and efficiency result. This is permissible if the efficiency rating is within the manufacturer's

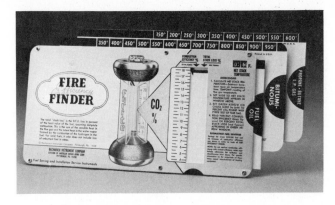

FIGURE A14-47 Fire efficiency finder. (*Courtesy* Bacharach Instrument Company)

specifications. If not, the unit should be examined for the cause of the difference. Usually, a cleaning of the heat exchanger is required.

A14-7.8
Temperature Rise

As stated before for gas units and in the beginning of this chapter, the temperature rise of the air through the unit should be 80°F. The correct rise for CAC is used for testing heating units and should be established before the final test results are accepted.

A14-8
ELECTRIC FURNACES

The typical electrical furnace (Fig. A14-48) is a most flexible and compact heating unit. It consists of a cabinet, blower compartment, filter, and resistance heat-section. The cabinet size is generally more compact than the equivalent gas or oil furnaces, and due to cooler surface temperature, most units enjoy "zero" clearances all around from combustible materials. Thus, they may be located in very small closets. Additionally, since there is no combustion process involved, there is no requirement for venting pipes, chimney, or makeup air, thus simplifying installation and reducing building costs. Service access is the only dimensional consideration. Also, the absence of combustion permits mounting the units for up, down, or horizontal air flow application (Fig. A14-49). Some manufacturers provide space within the furnace cabinet for a cooling coil.

With the panel removed (Fig. A14-50) observe that the blower compartment usually houses a centrifugal, multispeed direct-drive fan or one of a belt-driven design where larger units are involved. Airflow through an electric furnace has less resistance and fan performance is more efficient. Mechanical filters of either the throw-away or cleanable variety are available.

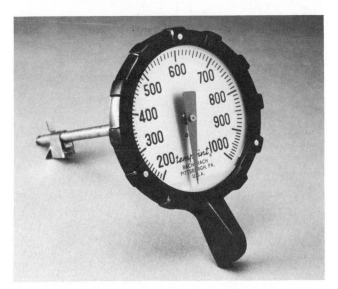

FIGURE A14-46 High temperature stack thermometer. (*Courtesy* Bacharach Instrument Company)

Heating Startup, Checkout, and Operation 567

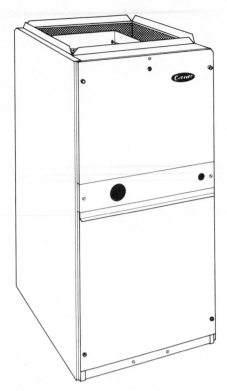

FIGURE A14-48 (*Courtesy* Carrier Air-Conditioning Company)

The heating section consists of banks of resistance heater coils wound from nickel–chrome wire held in place by ceramic spacers. The heater resistance is designed to operate on 208/240-V power with output according to the actual voltage used. The amount of heat per bank is a function of the amperage draw and/or staging. National and local electrical codes control the amount of current that can be put on the line in one surge, so it is necessary for manufacturers to limit this impact by the size (kW) of the heaters in a bank and the sequence of operation. Total furnace output capacities range from a low of 5 kW (17,000 Btu/hr) to 35 kW (119,400 Btu/hr). At approximately 35 kW the amperage draw on 240 V approaches 150 A and, considering that 200 A is the total service to a residence, this leaves only some 50 A for other electrical uses. Seldom are all on continuously, but codes must assume so for safety purposes. Note, however, that a well-insulated, electrically heated home requiring 35 kW will have considerable square feet. Zoning, too, is easier with two electric furnaces since location is not critical.

Electrical heating elements are protected from any overheating that may be caused by fan failure or a blocked filter. These are high-limit switch devices that sense air temperature and when overheated "open" the electrical circuit. Some furnaces also employ fusible links wired in series with the heater; these links melt at approximately 300°F and open the circuit. This is an auxiliary backup for the high-limit switches.

Built-in internal fusing or circuit breakers are provided by some manufacturers and are divided to comply with the National Electric Code. This can be convenient and may also mean installation savings for the contractor; external fuse boxes and associated on-site labor are eliminated.

The sequence control system is an important operation of the electric furnace. On a single-stage thermostat system there is an *electric sequencer control,* which contains a bimetallic strip. On a call for heat, the

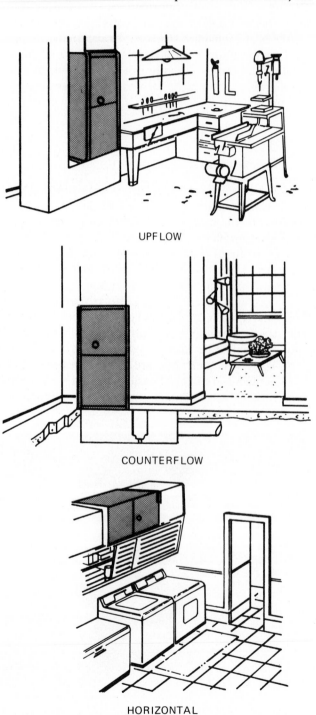

UPFLOW

COUNTERFLOW

HORIZONTAL

FIGURE A14-49 Airflow applications.

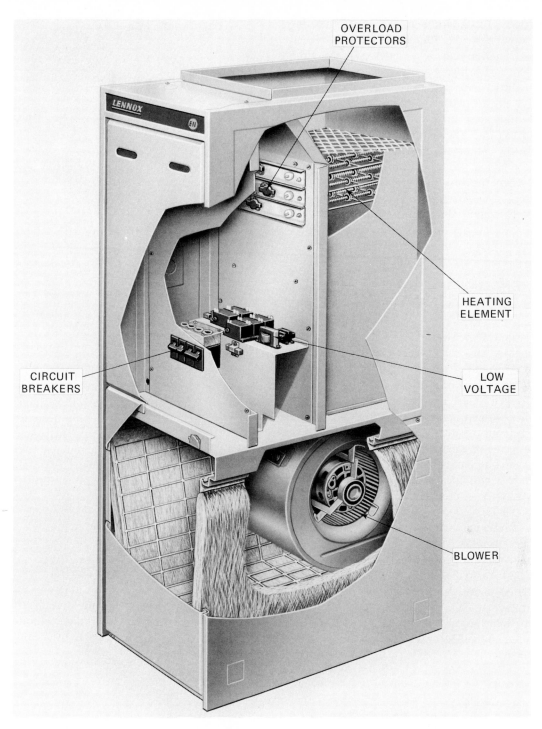

OVERLOAD
PROTECTORS

HEATING
ELEMENT

LOW
VOLTAGE

CIRCUIT
BREAKERS

BLOWER

LENNOX

FIGURE A14-50 Lennox electric E-11 upflow furnace. (*Courtesy* Lennox Industries)

thermostat closes a circuit which applies 24 V across the heater terminals of the sequencer. As the bimetallic element heats up, the blower and first heater come on. The operation characteristics of the sequencer are such that there is a time delay before each additional heater stage is energized. When the thermostat is satisfied, the sequencer is deenergized and the elements are turned off. The time delay (in seconds) is adequate to stagger power inrush and minimize the shock on the power system.

Where larger furnaces are installed, it is common to use two-stage thermostats in connection with the sequencer control. The first stage would operate as previously described and would bring on at least 50% of total capacity. The second stage of the thermostat would respond only when full heating capacity is needed. With this added control, wide variations of indoor temperature are avoided.

Previously, the text emphasized *electric* sequencer

control as opposed to a *motor-driven* variety that may be used on furnaces having four to six elements. It fulfills the same function and, by a series of cams, times out the sequence of bringing heaters ON and OFF.

The air distribution system for an electric furnace should receive extra care due to the normally lower air temperature of heated air coming off the furnace, as compared to gas and oil equipment. Temperatures of 120°F and below can create drafts if improperly introduced into the space. Additional air diffusers are recommended. Also, duct loss through unconditioned areas can be critical, so *well-insulated ducts are a must,* to maintain comfort and reduce operating cost.

A14-9
AIR-HANDLING UNITS
WITH DUCT HEATERS

A variation of the electric furnace is the use of an air-handling unit with duct-type electric heaters (Fig. A14-51). The air handler consists of a blower housed in an insulated cabinet with openings for connections to supply and return ducts. Electric resistance heaters are installed either in the primary supply trunk or in branch ducts leading from the main trunk to the rooms in the dwelling. In terms of sales this system has not been as popular as the complete furnace package concept, primarily because it complicates installation requirements and adds cost. There is more comfort flexibility by zone control when heaters are installed in branch runs; rooms can be individually controlled.

Duct heaters (Fig. A14-52) are made to fit standard sizes and contain overheating protection. Electric duct heaters can also be utilized with other types of ducted heating systems to add heat in remote duct runs or to beef up the system if the house has been expanded. They may be interlocked to come on with the furnace blower and are controlled by a room thermostat.

FIGURE A14-51   Electric duct heater. (*Courtesy* Lennox Industries)

A14-10
SUMMARY

Forced-air heating systems, be they gas, oil, or electric, offer the advantage of mechanical air movement whereby the air is not only heated but is cleaned by filtration, humidified, and "freshened" with an intake of outdoor air, which is circulated to the living area. Thus four of the five elements of total comfort air-conditioning can be performed, with the fifth option of cooling as an easy addition.

A14-11
HYDRONIC SYSTEMS

Another form of heating, and one that has existed for many years, is the hydronic method of conveying heat energy to the point of use by means of hot water

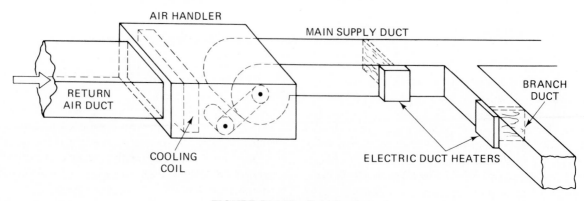

FIGURE A14-52   Duct heaters.

570   Air-Conditioning

or steam. The early steam system (Fig. A14-53) used a boiler partially filled with water. Heat from the fuel converts the water into steam (vapor). The steam passes through piping mains to the terminal units (radiators), where it is condensed upon giving up its heat. The condensate (liquid) returns to the boiler through the piping system, thus maintaining the proper water level within the boiler. The flow of heat is by pressure; no mechanical force is used.

In contrast to the partially filled boiler of the steam-heating system, the boiler, piping, and terminal units of a hot-water heating system (Fig. A14-54) are entirely filled with water. Heat produced by the fuel, be it gas, oil, or electric, is transferred to the water within the boiler or heat exchanger. The heated water is circulated through piping and terminal units of the system. The heat given off as the water passes through the terminal units causes the water to cool. The cooled water returns to the boiler and is reheated and recirculated. (*Note:* A similar but reverse process can be used for cooling by circulating chilled water; this technique will be covered later.)

By definition, hydronics includes all three systems: steam, hot water, and chilled water. However, since steam-heating systems are not frequently installed in residences or small commercial buildings today, this discussion will be confined to hot-water heating.

A14-11.1
Hot-Water Heating

In Fig. A14-54 the flow of water is dependent on gravity circulation due to the difference in the density of water in the supply and return sides. Like steam heating, gravity-flow hot-water systems are pretty much obsolete; much more popular today is the forced circulation system using a pump to circulate the water from the boiler to terminal units and back again. Pipe sizes can be smaller and longer, and terminal units can be smaller and may be located below the boiler if needed. Forced circulation insures better distribution control.

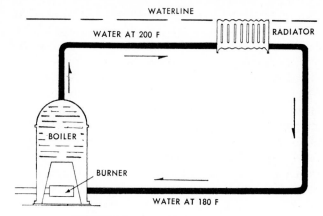

FIGURE A14-54 Hot-water system.

Hot-water systems may be classified according to operating temperatures and pressures:

	Boiler Pressure (psi)	Maximum Hot-Water Temperature (°F)
LTW—low temp.	30	250
MTW—med. temp.	150	350
HTW—high temp.	300	400–500

Medium- and high-temperature systems are limited to large installations and are beyond the scope of this text.

Low-temperature hot-water heating systems are classified according to the piping layout. Four common types of systems are:

1. Series loop
2. One pipe
3. Two pipes
4. Hot-water panel

Piping Systems: The series loop system (Fig. A14-55) is most commonly used in small buildings or in subcir-

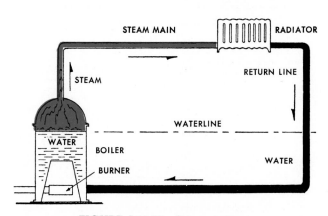

FIGURE A14-53 Steam system.

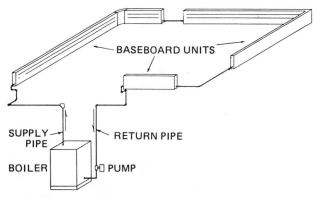

FIGURE A14-55 Series loop baseboard system (single circuit).

cuits of large systems. The terminal units, usually of the baseboard or fin tube radiation, serve as part of the distribution system. Water flows through each consecutive heating element in succession. The water temperature, therefore, is progressively reduced around the circuit, which may require that the length of each heating element be selected and adjusted for its actual operating temperature. The series loop system has the advantage of lower installed costs. Also, interconnecting piping can be above the floor, eliminating the need for pipe trenches, insulation, and furring. Its disadvantage lies in the fact that the water temperature to each unit cannot be regulated, and thus some form of manual air damper on the terminal is desirable. Also, circuit capacity is limited by the size of the tube(s) in the terminal element(s).

The one-pipe system (Fig. A14-56) is a variation of the loop, except that the terminals are connected to the pipe loop by means of branches. Individual terminals can be controlled, but special tees are needed, and these fittings are available only from a few manufacturers. A one-pipe system will generally cost more to install than a comparable series-loop system, because of the extra branch pipe, fittings, and special fittings themselves. The one-pipe system also shows a progressive temperature drop around the circuit with resulting limitations.

Two-pipe systems (Fig. A14-57) use one pipe (supply) to carry hot water to the terminal and a second to return cool water to the boiler. Known as a *two-pipe reverse return,* it is set up so that while terminal 1 is closest to the boiler on the hot water supply, it is the farthest on the return main. The reverse is true for terminal 5; it is farthest away on the hot-water supply and

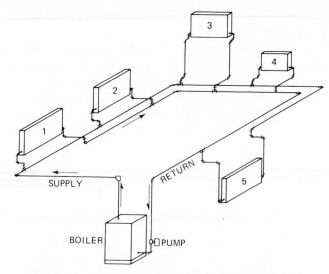

FIGURE A14-57 Two-pipe reverse hot-water heating system.

closest on the return main. This equilization of the distance the water travels through each unit provides even distribution of water throughout the system—and better control. Zones or several circuits may be employed with equal success. The two-pipe system is usually preferred when low pressure-drop terminal units like baseboard are used.

Panel systems (Fig. A14-58) are installed in floors or ceilings and utilize a supply or return header where all of the coils can be brought together. Balancing valves and vents on each of the coils can then be made accessible in a pit or in the ceiling of a closet.

The preceding discussions and illustrations were of single-circuit systems. Any of these hydronic systems may also be installed as multiple circuits where a main or mains form two or more parallel loops between the boiler supply and return.

Multiple circuits are employed for several reasons: to reduce the total length of circuits, to reduce the number of terminal units on one circuit, to reduce pipe size or the quantity of water circulated through a circuit, and

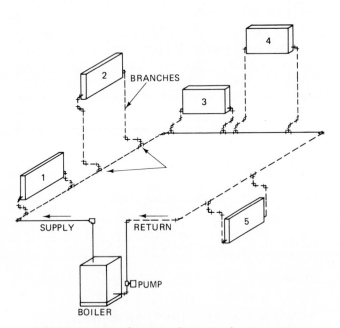

FIGURE A14-56 One-pipe hot-water heating system.

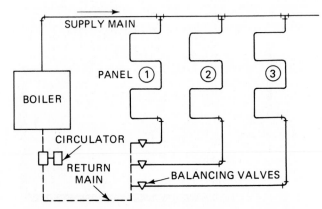

FIGURE A14-58 Forced-circulation hot-water panel system.

for simplification of pipe design in certain types of buildings.

A14-11.2
Boilers

Steam boilers were in existence prior to the mid-1800s, and by the 1870s, hot water was beginning to replace steam. Modern boiler design includes the use of cast iron, steel, and copper materials.

FIGURE A14-59   Gas boiler. (*Courtesy* Weil-McLain, Division of Wylain, Inc.)

Cast-iron boilers are assembled from cast-iron sections (Fig. A14-59). They can burn gas, oil, or coal. Steel boilers are designed with tubes inside a shell. Usually, the system water is in the tubes and the fire and hot gases surround the tubes and thus heat the water. This is called a *water-tube boiler* (Fig. A14-60). However, some boilers function in just the reverse fashion—the fire and hot gases pass through the tubes and heat the water in the shell. These are called *fire-tube boilers*. Residential-size steel boilers are sold as packages with pump, burner, and controls mounted and wired.

Flash boilers for gas or oil burning consist of copper coils which contain the system water surrounded by the fire and hot gases. These are very compact, and some models are designed to be used outdoors.

An electric boiler is essentially a conventional boiler with a large internal volume and electric heaters directly immersed in water. There are also instantaneous types of electric boilers small enough to "hang on the wall" (Fig. A14-61), which are fitted with either immersed electric heaters or electrical heating elements bonded to the outside of the water vessel. The pump and expansion tank are enclosed in the cabinet.

A14-11.3
Typical System

A typical low-temperature hot-water installation (Fig. A14-62) consists of:

FIGURE A14-60   Water-tube boiler.

FIGURE A14-61   Wall-type electric boiler. (*Courtesy* Weil-McLain, Division of Wylain, Inc.)

Heating Startup, Checkout, and Operation   573

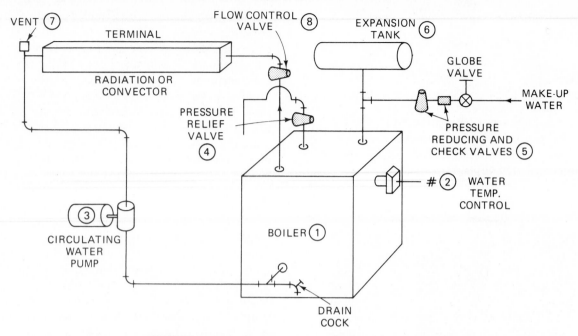

FIGURE A14-62 Typical residential hot-water system.

1. The boiler, be it gas, oil, or electric, as described above and operated at 30 psi, with heating water from 180 to 220°F for normal supply.

2. A boiler water temperature control to cycle the burner or heaters to provide control of hot water temperature.

3. A circulating water pump large enough to allow the flow of sufficient water to heat the space. The control of the pump is from a space or zone thermostat.

4. A relief valve that will ''open'' if the boiler pressure exceeds 30 psi.

5. Pressure-reducing and check valves and an automatic water makeup valve that supplies water to the boiler for initial fill and also if pressure gets too low.

6. A closed air-cushion tank, which, on initial fill, has a pocket of trapped air within the tank. When water is heated it expands and compresses the trapped air as a cushion, thus providing space for extra water without creating excessive pressure.

7. Air vents at the terminals or high point of the piping system to purge unwanted trapped air.

8. A flow-control valve (check valve) to prevent gravity circulation of water when the pump is not in operation.

9. The necessary cocks, purge valves, and drain cocks for balancing, service, and maintenance.

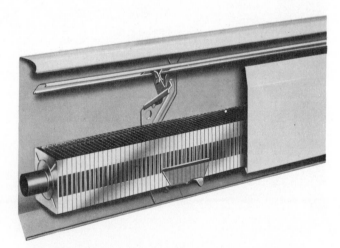

FIGURE A14-63 Finned-tube baseboard radiation. (*Courtesy* Weil-McLain, Division of Wylain, Inc.)

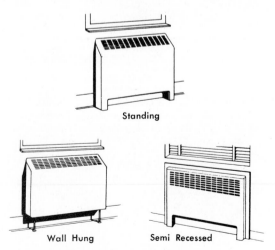

FIGURE A14-64 Convector radiation.

574 Air-Conditioning

Terminal units vary from cast-iron radiators (seldom used today) to finned-tube residential baseboard radiation (Fig. A14-63), to cabinet-type convectors (Fig. A14-64), to hot-water coils for air-handling units or duct mounting (Fig. A14-65). The selection of the type of unit depends on whether it is used in residential, commercial, or institutional buildings, etc.

The selection of the terminal size is essentially a function of the temperature of the water, the flow rate, the surface area of radiation, and the type of space air circulation (natural or fan-forced). Manufacturers' catalogs provide rating tables which prescribe the proper selection.

The overall design of hydronic heating systems is in itself a special art, and we would suggest that those who become involved in this area of heating take advantage of the information available through the Hydronic Institute, 35 Russo Place, Berkeley Heights, New Jersey 07922, and the North American Heating and Air-Conditioning Wholesalers Association, 1661 West Henderson Road, Columbus, Ohio 43220. Both of these associations offer excellent training data on the design, installation, and service of hydronic systems.

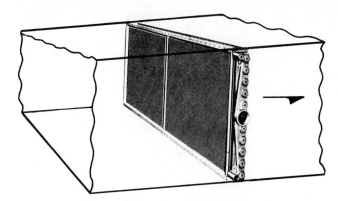

FIGURE A14-65   Hot-water coil in duct.

≡ PROBLEMS ≡

A14-1. To check and test a gas heating unit, the first thing that must be established is _____ .

A14-2. The amount of gas to be burned in a particular unit will depend on _____ and _____ .

A14-3. The heating value of natural gas can vary from _____ to _____ Btu/ft³.

A14-4. The Btu/ft³ of propane is _____ .

A14-5. The Btu/ft³ of butane is _____ .

A14-6. The Btu/ft³ of gas in your area is _____ .

A14-7. The recommended manifold pressure of gas heating units is _____ in. of W.C. with an allowance range from _____ to _____ in. of W.C.

A14-8. If the proper input requires a manifold below or above the allowable range, what must be done to the unit?

A14-9. A gas-fired heating unit is to be installed in Denver, Colorado. How much must the input be derated?

A14-10. What is the formula to find the number of cubic feet of gas to be burned per hour?

A14-11. How many cubic feet of gas would be burned in a 125,000-Btu input unit using 1050-Btu/ft³ gas?

A14-12. When checking the input of a 150,000-Btu/hr unit, using 950-Btu/ft³ gas, by timing the meter using the 1-ft³ dial, how many seconds would it take for the dial to make 1 revolution?

A14-13. The operating manifold pressure of an LP heating unit burning propane or butane is _____ in. of W.C.

A14-14. What is a single-stage LP supply system?

A14-15. What is the main benefit of a two-stage supply system?

A14-16. What are the minimum outside temperatures at which butane and propane can be used as a fuel source without auxiliary devices?

A14-17. In an atmospheric-type burner, the air for the combustion that mixes with the gas inside the burner is called _____ .

A14-18. The air for combustion that passes around the outside of the burner is called _____ .

A14-19. The air through the burner is what percentage of total air for combustion?

A14-20. How much extra air over the amount for combustion must the heat exchanger handle?

A14-21. The best operating flame on an atmospheric-type burner is usually one that is _____ .

A14-22. The correct temperature rise through the heating unit to obtain CAC is _____ °F.

A14-23. For CAC operation, the fan switch should be set to turn the fan on at _____ to _____ °F and off at _____ to _____ °F.

A14-24. Define the "operating efficiency" of a gas-fired heating unit.

A14-25. What are the advantage and disadvantages of increasing the amount of air over the heat exchanger?

A14-26. Define "neutral pressure point."

A14-27. What is the proper operating point for the neutral pressure point?

A14-28. The minimum efficiency requirement of a gas-fired heating unit is _____ to _____ %.

A14-29. Oil burners are installed with a forward downward slope of the firing tube. Why?

A14-30. Name the oil spray patterns of the nozzles used in residential oil burners.

A14-31. Nozzle spray patterns are interchangeable. True or False?

A14-32. The minimum recommended ON cycle for an oil burner is _____ minutes.

A14-33. Carbon deposits on the surface of the refractory are called _____ .

A14-34. What causes carbon deposits on the surface of the refractory?

A14-35. The normal operating pressure of the oil burner fuel supply is _____ psig.

A14-36. Oil burner fuel units are built in two styles. What are they?

A14-37. What style of fuel unit is used in a buried-tank oil burner installation? Why?

A14-38. What causes the "after flow" of oil at burner cutoff?

A14-39. The minimum operating time to conduct an efficiency test on an oil-fired heating unit is _____ .

A14-40. What type of installation permits the installation of a one-pipe system?

A14-41. What type of installation requires the installation of a two-pipe system?

A14-42. If the design resistance of a heat exchanger is 0.10 in. of W.C., what is the maximum resistance permitted before cleaning is required?

A14-43. What is the usual cause of excessive stack temperature in an oil heating unit that has been operating properly?

A14-44. Electric forced-air furnaces use _____ to stagger the heating elements.

A14-45. Elements in an electric furnace use two protective devices in each element. What are they?

A14-46. Hot-water heating systems have been designed in two different circulation systems, _____ and _____ , the most popular being _____ .

A14-47. What are the four most common types of piping layout for hot-water circulation systems?

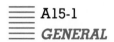 **A15**

Heating Service and Problem Analysis

 **A15-1**
GENERAL

This chapter will acquaint the reader with some of the more common service and maintenance problems that arise in connection with residential and small commercial heating systems, both gas and oil. Once a problem does develop, it will usually be the job of the service technician to handle the complaint, first to diagnose the cause and then to take the necessary remedial action to cure the problems. Both gas- and oil-fired units are covered in this chapter, separated into categories to maintain as much clarity as possible.

The chapter is also arranged in a progressive fashion from the general category of the problem, to the possible symptoms of the problem and the possible cause of the symptom, to an explanation of the cause of the symptom. For example, the cause of "no heat" could be unit A-1, "will not start" because of unit E-2, "season switch is open," possibly because the owner did not know or realize that the season switch was opened last spring at the end of the heating season.

The first tabulation is a list of common complaints under the following categories:

A. Entire system operation
B. Unit operation
C. Burner operation
D. Blower operation
E. Heat exchanger complaints
F. Cost of operation
G. Noise

Under each category is a list of the possible problems in the particular category. As you review the problems you will find duplication because it is possible to have the same problem for more than one complaint.

The number in front of each problem is the sequence in which it appears in the symptom listing. There is no numerical tabulation of problems under each category.

A15-2
GAS PROBLEMS

A. Entire system operation
   1. Will not start
   2. Runs but not enough heat—short cycles
   3. Runs but too high a room temperature
   4. Runs continuously
   5. Blower cycles after thermostat is satisfied
   6. Blows cold air on startup
   7. Startup and/or cool-down noise
   8. Noisy/vibration
   9. Odor
   10. High fuel cost
   11. High electrical cost
B. Unit operation
   1. Will not start
   12. No fuel
   13. Valve won't open
   14. Valve short cycles
   15. Delayed ignition
   16. Pilot outage
C. Burner operation
   15. Delayed ignition
   17. Extinction pop
   18. Burns inside burner
   19. Flame lift
   20. Flame roll-out
   21. Noisy flame

577

22. Yellow fire
23. Carbon deposit
24. Flash back
D. Blower operation
25. Short cycles
E. Heat exchanger complaints
26. Startup and/or cool-down noise
27. Burnout
F. Cost of operation
10. High fuel cost
11. High electrical cost
G. Noise
   a. Combustion noise
     15. Delayed ignition
     21. Noisy flame
     28. Resonance
     29. Noisy pilot
     24. Flash back
     20. Flame roll-out
   b. Mechanical noise
     8. Vibration
     7. Duct system noise
     26. Heat exchanger ticks or bangs

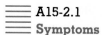

A15-2.1
Symptoms

Following is a list of the symptoms of the problems in the preceding category. Under each symptom is a list of the probable causes. Again, each probable cause is listed in numerical order. In addition, a letter is used to further classify the probable cause as being in the electrical (E) or gas (G) lists of explanations that follow.

1. Will not start
   E1. Season switch open
   E2. Room thermostat set too high
   E3. Disconnect switch open
   E4. Blown fuse
   E5. Limit control open
   E6. Control transformer B.O.
   E7. Open circuit in thermostat and/or sub-base
   E9. Open in control cable
   E11. Open controller in gas valve
   G5. Pilot outage
   G10. Gas valve stuck open or closed
   G8. Safety pilot B.O.
   G1,G4. No fuel

2. Runs but not enough heat—short cycles

E12. Improper heat anticipator setting
E13. Cycling on limit control
G3. Input too low
3. Runs but too high a room temperature
E2. Room thermostat set too high
E12. Improper heat anticipator setting
4. Runs continuously
E8. Short in thermostat and/or subbase
E10. Short in thermostat cable
G10. Gas valve stuck open
5. Blower cycles after thermostat is satisfied
G21. Blower ft³/min adjustment
G22. Fan control setting
6. Blows cold air on startup
G21. Blower ft³/min
G22. Fan control setting
7. Startup and/or cool-down noise
G31. Expansion noise in heat exchanger sections
G32. Expansion noise in heat exchanger restrictors
G33. Duct expansion
8. Noisy/vibrations
G29. Blower wheel unbalanced or out of line
G30. Pulleys unbalanced
G30. Blower belt B.O.
G30. Bearings B.O.
9. Odor
G25. Dust odor at fall startup
G26. Improper venting
G27. Cracked heat exchanger
10. High fuel cost
G20. Improper input
G14. Improper burner adjustment
G35. Improper unit sizing
11. High electrical cost
E14. Improper load on blower motor
12. No fuel
G2. No line pressure
G4. LP tank empty
G3. Regulator B.O.
G1. Supply valve closed
G4. Low outside ambient
13. Gas valve won't open
E6. Control transformer B.O.
E7. Open circuit in thermostat and/or sub-base
E9. Open in control cable
E11. Open controller in gas valve

G10. Gas valve stuck open or closed

G8. Safety pilot B.O.

14. Gas valve short cycles

E12. Improper heat anticipator setting in thermostat

E13. Improper setting of limit control

G23. Dirty air filters

G24. Restrictors in air supply system

15. Delayed ignition

G14. Improper burner adjustment

G16. Improper orifice alignment

G20. Improper input

G14. Improper burner adjustment

G11. Delayed valve opening

G3. Low line pressure

16. Pilot outage

G3. Low line pressure

G7. Pilot orifice B.O.

G6. Thermocouple B.O.

G8. Safety pilot B.O.

15. Delayed ignition

17. Extinction pop

G9. Drafts

17. Extinction pop

G14. Improper primary air adjustment

G15. Improper burner alignment

G16. Improper orifice alignment

G12. Poor valve cutoff

18. Burns inside burner

G14. Improper primary air adjustment

G20. Improper input

17. Extinction pop

G13. Leaking gas valve

19. Flame lift

G20. Improper input

G14. Improper primary air adjustment

20. Flame roll out

G20. Improper input

G18. Restriction in heat exchanger

G26. Improper venting

G19. Improper combustion air supply

21. Noisy flame

G14. Improper primary air adjustment

G20. Improper input

G17. Improper slot or port size

22. Yellow fire

G14. Improper primary air adjustment

G20. Improper input

23. Carbon deposit

G14. Improper primary air adjustment

G20. Improper input

24. Flash back

G20. Improper input

G14. Improper primary air adjustment

G18. Restriction in heat exchanger

G26. Improper venting

25. Intermittent blower operation

G21. Improper ft^3/min adjustment

G22. Fan control settings

G20. Improper input

26. Heat exchanger noise on startup and/or cool down

G31. Stress noises in heat exchanger

G32. Restrictor crawl

27. Heat exchanger burnout

G20. Improper input

G28. Chemical atmosphere

28. Resonance (pipe organ effect)

G20. Improper input

G14. Improper primary air adjustment

G36. Resonance-unit design

A15-2.2
Electrical Causes

In the preceding list, the probable causes of each symptom were given a number beginning with E if in the electrical category and G if in the gas category. The electrical causes are as follows:

E1. Season switch open

E2. Room thermostat set too high

E3. Disconnect switch open

E4. Blown fuse

E5. Limit control open

E6. Control transformer B.O.

E7. Open circuit in thermostat and/or subbase

E8. Short in thermostat and/or subbase

E9. Open in control cable

E10. Short in control cable

E11. Open controller in gas valve

E12. Improper heat anticipator setting

E13. Cycling on limit control

E14. Improper load on blower motor

E1. Season Switch Open: One of the most common causes of "no heat" complaints at the beginning of the heating season is the fact that the season on–off switch located on the unit was put in the OFF position. No one trusted the thermostat to keep the unit off or the owner did not want possible operation if the night happened to get cool. Regardless, the fact that this switch is turned off is easily forgotten over the summer period. When someone calls in a no-heat complaint, it is well to ask if this switch has been turned on. It can save an unnecessary service call and an embarrassed customer.

E2. Room Thermostat Set Too High: The room thermostat may be operating at too high a setting even though the dial pointer is at the desired setting. This is especially true on mercury bulb contact thermostats. Make sure that the thermostat is level before trying to recaliber it. Very seldom does a thermostat go out of calibration, although they are frequently knocked out of level. If the dial setting and the built-in thermometer do not read the same, check the level of the thermostat before recalibrating or adjusting is done.

E3. Disconnect Switch Open: Make sure that the proper voltage is supplied to the unit. Disconnect switches in distribution panels are known to open due to electrical disturbances, electric storms, etc. If there is no voltage to the unit, reset the breakers to check their holding ability.

E4. Blown Fuse: Instant-blow or link-type fuses are sometimes used as branch fuses. Because they will not carry current above their rating for any length of time, they must be sized to carry the maximum starting current of the unit. Time-delay fuses should always be used where a motor is involved. This type will allow closer sizing of the fuse to the running current of the motor for maximum protection.

Checking fuses is best done by means of an ohmmeter. *With the disconnect switch open*, measure the resistance of each fuse. A reading of zero resistance on the ohmmeter indicates that the fuse is good. Infinite resistance or no movement of the needle indicates a blown fuse. A resistance reading indicates a burned fuse that is not completely burned open. This type of fuse should also be replaced.

E5. Limit Control Open: The limit control may be in the high-voltage circuit to the control transformer or it may be in the 24-V or lower-voltage side of the control circuit. In either case, if the control is open, the heat function is shut down. Using a voltmeter, measure if there is voltage across the limit control terminals. Open terminals will indicate full circuit voltage. If no voltage is indicated, the contacts may be closed and some other source of problem is cutting off the voltage. To confirm this, *with the voltage off to the unit*, remove the wires from the control terminals. Using the ohmmeter, measure the resistance of the limit control contact. Infinite resistance or no movement of the needle indicates an open contact. If the temperature of the control is at room temperature, replace the control. If a reading of

zero resistance is found, the contacts are closed and the voltage interruption is elsewhere.

E6. Control Transformer B.O.: The control transformer of the normal heating unit will be a 120- or 240-V primary with a 24-V secondary. To check for transformer operation, use an ac voltmeter to measure the output voltage. If there is 24 V across the output terminals (usually marked R & C) it can be assumed that the transformer is OK. If no voltage is found, the voltage across the primary side should be checked. If there is no voltage in, there will be no voltage out. Using the higher range on the voltmeter, measure the voltage across the primary leads. If no voltage is found, the power supply to the transfer is dead. If voltage is found at the primary leads and no voltage is at the transformer secondary, the transformer should be replaced. Be sure to use a transformer of equal or higher volt-ampere (VA) rating of like input and output voltage as well as the same hertz.

An ohmmeter should be used to determine which winding in the transformer failed. If the primary winding is the open one, the probable cause is high voltage. If the secondary winding is the one that failed, the transformer is overloaded. Either the load should be reduced or a higher-VA-capacity transformer should be substituted.

E7. Open Circuit in Thermostat and/or Subbase: This is easily checked by using a jumper wire across the *R* and *W* terminals of the thermostat subbase. If the gas valve opens, the open is in the thermostat or subbase. These are not field repairable; substitute a new one.

E8. Short in Thermostat and/or Subbase: If the unit runs continuously, that is, the gas valve remains open, in spite of the fact that the thermostat is turned to the lowest setting, remove the thermostat from the subbase. If the valve closes, replace the thermostat. If the valve remains open, remove the red wire from the subbase. If the valve closes, replace the subbase. If the valve still remains open, see E10.

E9. Open in Control Cable: When the jumper wire (in E7) does not open the gas valve, it could be an open lead in the control cable. At the furnace end of the control cable, place a jumper wire connected to the power (*R*) terminal of the transformer and the supply-side terminal of the gas valve. This is usually the red wire at the transformer and the white wire at the gas valve. If the valve opens, there is an open in the circuit. This includes the control cable and may include the limit control. If this control is in the low-voltage circuit, check the control (see E5). If the gas valve opens when the jumper is installed, there is possibly an open in the control cable. To check this an ohmmeter and a coil of insulated wire are necessary. The extra wire should be long enough to reach from the heating unit to the thermostat location.

By connecting the extra wire to each of the control wires in turn and using the ohmmeter to check each control wire, the open wire can be found. Each good wire will indicate a low reading on the ohmmeter. The open wire will read infinitely or no movement of the needle.

All of this is based on the fact that the proper color coding was used to wire the control cable using the red and white wires for the circuit. This is also based on the fact that the cable is one piece from the thermostat to the heating unit. However, this is not always the case. If an open cable is encountered, look for possible splices.

E10. Short in Control Cable: If the gas valve will not close unless the control wires are removed, there may be a short, possibly due to a staple driven through the cable. This can be determined by using the ohmmeter to measure the resistance between the wires in the cable. If a low resistance reading is found between any of the wires, pull in a new cable. It is almost impossible to get at the damaging staple back in the wall.

E11. Open Controller in Gas Valve: The voltmeter says that there is 24 V across the terminals of the gas valve controller and no action. Remove the wires from the controller. Using the ohmmeter, measure the resistance of the circuit through the controller. This reading will be in the range 50 to 5000 Ω. If there is no needle movement, the circuit is open. If the reading is normal, it is possible the controller is stuck in the closed position. Do not try to repair—replace it.

E12. Improper Heat Anticipator Setting: If the heat anticipator setting is too high, too much heat will be generated, causing the thermostat to be satisfied before the area temperature rises to the thermostat setting. If the heat anticipator setting is too low, not enough heat is generated and the area temperature will rise above the thermostat setting before the unit cycles off.

The anticipator should be set at the amount of current traveling in the control circuit when the unit is operating. Do not try to add all the amperage loads in the heating circuit—measure the current. Use a circuit multiplier accessory with your clamp-type ammeter so as to multiply the less-than-1-A current through the circuit to 10 times as high so that it is in the range of your meter. This can also be accomplished by winding the wire around one of the jaws of the meter so that the wire passes through the jaws 10 times. The ends of the wire are then connected across the thermostat leads of the subbase, taking the place of the thermostat. The meter will then read 10 times the actual current through the circuit. The meter reading divided by 10 will be the heat anticipator setting. A very handy instrument for this purpose is the 0- to 1.2-A-range meter marketed by the

T.D. Instrument Company of Rochester, New York 14607.

E13. Cycling on Limit Control: Occasionally, a limit control element will weaken and lower the operating range of the control. The normal range is cut off between 140°F for a counterflow unit to 160 to 220°F on upright and horizontal units. If the control is cycling at a lower range then its setting, replace the control; do not try to recalibrate. If the control is cycling because of high supply plenum temperatures, see Section A15-2.4.

E14. Improper Load on Blower Motor: The amperage draw of the blower motor goes up much faster than an increase in load. Therefore, a small increase in the blower load can cause a larger increase in current consumption. It can also cause higher-than-normal motor temperatures and short motor life.

Measure the amperage draw of the motor under normal load. It should not exceed the amperage rating of the motor. If it does and the load cannot be reduced, a larger motor will have to be substituted. Even with a larger motor handling the load, the wattage used will usually be less than the smaller motor operating in an overloaded condition.

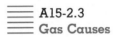

A15-2.3
Gas Causes

The gas causes are as follows:

G1. Supply valve closed
G2. No line pressure
G3. Low line pressure
G4. LP tank empty or low ambient
G5. Pilot outage
G6. Thermocouple B.O.
G7. Pilot orifice B.O.
G8. Safety pilot B.O.
G9. Drafts on pilot
G10. Gas valve stuck open or closed
G11. Delayed valve opening
G12. Poor valve cutoff
G13. Leaking valve
G14. Improper burner adjustment
G15. Improper burner alignment
G16. Improper orifice alignment
G17. Improper slot or port size
G18. Restriction in heat exchanger
G19. Improper combustion air supply
G20. Improper input
G21. Blower ft³/min adjustment
G22. Fan control settings

G23. Dirty air filters

G24. Restrictions in air supply

G25. Dust odor at fall startup

G26. Improper venting

G27. Cracked heat exchanger

G28. Chemical atmosphere

G29. Blower wheel unbalanced or out of line

G30. Blower drive

G31. Expansion noise in heat exchanger sections

G32. Expansion noise in heat exchanger restrictors

G33. Duct expansion

G34. Oil-can effect

G35. Improper unit sizing

G36. Resonance-unit design

G1. Supply Valve Closed: The first place to look when there is no gas supply at the unit is the supply valve on the unit or in the supply line to the unit. Perhaps the gas was turned off at the end of the last heating season and someone forgot to turn it on again. It may also be the main supply into the house that is turned off. *Do not attempt to operate this valve.* This is the responsibility of the gas supply utility and it should be left to them to establish gas supply to the building.

In the case of an LP gas system, contact the gas supplier if the main tank valve is closed off. It had to be closed for a reason that is beyond the responsibility of the service technician.

G2. No Line Pressure: Even though the main valve to the building is open, there can be a situation where a line blockage can cut off the gas supply. In extremely cold climates with exposed gas lines, a line trap full of water can freeze and shut off the gas. Gas meters have been known to accumulate water or dirt to the point that they will not function. This problem should be referred to the gas utility.

G3. Low Line Pressure: This is a difficult situation to analyze unless the unit inspection happens to be during the time that other loads occur. In a residence, the cook stove, oven, and water heater may be on. On complaints of pilot outage or burner problems when the unit tests OK, a test at full load should be conducted.

With a water column gauge connected to the supply line ahead of the heating unit gas valve, the pressure should be noted when the unit and all the others loads are off. This is the idle line pressure. When the unit is turned on, the pressure should be noted. If a drop of more than 0.3 in. of W.C. is indicated, the supply line to the unit is undersized. Refer to Section A14-2.1 for proper sizing.

When the unit operating pressure is recorded, all the other loads connected to this supply system should be turned on. This full operating load should not cause a pressure drop of more than 0.5 in. in W.C. in the gas supply to the heating unit. If higher pressure drop than 0.5 in. in W.C. is encountered, the gas utility should be contacted.

On LP gas systems, low operating pressures are easily encountered because of the smaller lines used. The same procedure should be used to check the line pressure. With the loads off, line pressure should be between 12 and 14 in. of W.C. When the unit is turned on, the pressure should drop to 11 in. of W.C. and hold at this amount. If it drops further than this, it can mean restriction in the supply line or a faulty regulator on the tank. Regulator problems are rare, however, so a search for a line restriction should be the first step.

There may be a kink in the copper tube, a crushed tube, or possibly a dip in the tube that has accumulated water. This problem is more predominant with outside-located packaged units where the supply line is above ground and the winter temperatures drop below 0°F. Water will condense in the main line from the tank to the building. If a branch line is taken off the bottom of the main line to supply the outside unit, the water will enter the branch line and cause obstruction and pressure drop. Always take a branch line off the top of the main line so that any water in the line will continue into the heated building, where it will be gradually dissipated throughout the supply system.

G4. LP Tank Empty or Low Ambient: Figure A15-1 shows tank pressures at various outside ambient temperatures for mixtures of propane and butane from 100% propane to 100% butane. From this you can see that it is not possible to get fuel from a butane tank when its temperature is 30°F or below. The boiling point of propane is much lower (−41.8°F), but when the outside temperature is −40°F or colder, tank pressure is below the $11\frac{1}{2}$ in of W.C. necessary to supply fuel to the system. Special means of maintaining pressure in LP tanks in extremely cold locations are available. The local LP supplier should be contacted if these are required.

G5. Pilot Outage: Pilot outage is the cause of the unit shutoff and there is a no-heat complaint. Pilot outage itself has many causes. The most predominant cause is problems in the pilot assembly itself. This could be improper amount of fire on the thermocouple (see G7), drafts causing the flame to waver away from the thermocouple (see G9), delayed ignition, or extinction pop.

Improper fire on the thermocouple is discussed under G6 and flame waver under G9. Delayed ignition and extinction have several common causes. The most predominant of these are improper burner adjustment (G14), improper burner alignment (G15), or improper orifice alignment; very seldom encountered but a possibility is improper slot or port size (G17). Each of these is discussed under the appropriate title and cause num-

	VAPOR PRESSURE, PSIG														
	Outside Temperature, Degrees Fahrenheit														
	−30	−20	−10	0	10	20	30	40	50	60	70	80	90	100	110
100% Propane	6.8	11.5	17.5	24.5	34	42	53	65	78	93	110	128	150	177	204
70% Propane 30% Butane	—	4.7	9	15	20.5	28	36.5	46	56	68	82	96	114	134	158
50% Propane 50% Butane	—	—	3.5	7.6	12.3	17.8	24.5	32.4	41	50	61	74	88	104	122
70% Butane 30% Propane	—	—	—	2.3	5.9	10.2	15.4	21.5	28.5	36.5	45	54	66	79	93
100% Butane	—	—	—	—	—	—	—	3.1	6.9	11.5	17	23	30	38	47

Source: Fisher Controls International, Inc.

ber. Reference to all sections mentioned should be made when trying to track down the problem of pilot outage. Also keep in mind that low line pressure, discussed in G3, can cause a low pilot flame during periods of high gas usage which can cause a pilot safety to cut off and shut off the heating unit gas supply.

G6. Thermocouple B.O.: The pilot is a small gas burner that causes a flame to impinge on a thermocouple. The thermocouple is made of two dissimilar metals, generally iron and constantan, welded together at the tip of the couple. The iron sheath of the couple is welded to the brass mounting collar. The tip of the couple is the hot junction and the mounting collar is the cold junction. It is the difference in temperature between these two points that generates the output voltage of the thermocouple. This output voltage, carried through the tube and wire conductor and connector tip, is applied to the magnetic coil in the pilot safety of the gas valve, relay, or other pilot control that keeps the valve open, maintains closed contacts, or does whatever it is designed to accomplish to maintain the flow of gas under the control of the thermostat. The more heat that is generated at the tip of the couple as compared to the body temperature, the higher the voltage output.

Three pilot flame sizes are illustrated in Fig. A15-2. In the left illustration, the pilot flame is too small to concentrate the hot tip of the flame on the thermocouple. The center illustration shows the proper flame size with the tip of the combustion cone touching the thermocouple. This gives the highest intensity of heat and the maximum couple output. The illustration on the right shows a typical high-fire situation. The tip of the

thermocouple is actually in the cone of cold unburned gas and the thermocouple output is very low. This condition will be the most predominant, as many people assume that the higher the flame, the higher the output.

Before you adjust anything else, check the flame size to make sure that the pilot flame cone tip just touches the couple. With this flame position, check the thermocouple output. Actually adjusting the flame size for maximum thermocouple output will assure best thermocouple performance. The average thermocouple should develop up to 35 mV. If the maximum output of the thermocouple is less than 20 mV, the thermocouple should be discarded and a new one substituted. However, check the output of the new thermocouple also, as

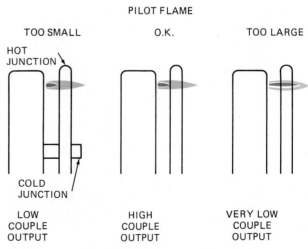

FIGURE A15-2 Pilot flame size.

Heating Service and Problem Analysis 583

it could also be defective. Never trust the part even though it is new; *check it.*

G7. Pilot Orifice B.O.: In situations of low or erratic pilot flame, check the pilot orifice for a plugged orifice hole. Some natural gas pilot orifices have two orifice holes for the desired gas jet spread in the type of pilot burners used. This results in a wide, soft fire for maximum tip impingement as well as burner ignition. Occasionally, one of these holes will become plugged, resulting in only half a flame. Usually, a pilot-dropout situation will develop, but sometimes only delayed main burner ignition will result. In either case, replace the orifice with another of equal total drill size. *Never try to drill out an orifice;* the hole size is critical and it is not worth the time involved. On single-drilling pilot orifices, the result of plugging is complete stoppage of gas flame. This is more easily detected in these cases. *Replace the orifice.*

G8. Safety Pilot B.O.: Even though the thermocouple is capable of putting out the desired millivolts, it may not mean that the safety pilot will hold in the gas valve, pilot-stat, relay, or whatever the couple is connected to. To check the holding power of the electromagnetic oil in the power unit, the milliampere range of your meter is used. Using the battery in the meter as the source of power, the meter is set at maximum milliamperes through the power unit. The pilot-stat is reset and should hold in the operating position. The milliamperes through the power unit are then reduced until the power unit releases. This milliampere release rate should not be less than 150 mA for the average unit. Some special units have different specifications and the manufacturer's specs should be followed. Remember that this is an electrical device and the connection between the thermocouple and the power unit is an electrical connection.

Figure A15-3 shows two power units, both with the thermocouple on the right end. If this connection

has corrosion or dirt on the contact surfaces, the milliamperes through the power unit could be below the dropout range due to the electrical resistance created by the foreign substances on the surfaces. Before testing the output of the thermocouples or the holding power of power units, use sand cloth to clean both the male tip of the thermocouple and the female socket of the power unit.

G9. Drafts on Pilot: The pilot assembly is located in such a position that the normal combustion airflow through the heat exchanger will not cause any flame waver away from the thermocouple, resulting in dropout. However, any unusual amount of air could result in this action.

Wind action is very common on gas packages located on a roof, gas-fired heating units located in closets facing an outside door, or units connected to vents of unusual draw. Anything that will cause a high airflow through the heat exchanger can cause dropout.

Before redesigning the unit by attempting to move the pilot assembly, eliminate the source of unusual air movement—a baffle plate in front of the rooftop unit, a shield in front of the pilot in the unit in the back hall, a draft regulator on that extremely high vent. The situations are many and varied. Each requires a different solution to the high airflow over the pilot.

G10. Gas Valve Stuck Open or Closed: Practically all manufacturers of gas-fired units use gas control valves of component design. Pilot safety devices, controllers, regulators, and main valves are replaceable. They are all a part of the assembly by means of attachment to the pilot body. None are repairable; they are designed to be replaced.

If the main body of the valve becomes inoperative because control passages plug, the main shutoff means is ruptured or sticks, or for any other reason, *replace with a new assembly.* Do not try to field repair. The possibility for liability in the case of damage to unit or surroundings because of faulty repair is too great for the service technician to assume.

G11. Delayed Valve Opening: This was not a problem with the old solenoid-type valves that opened and closed with snap action. The new-type valves using a pressure differential across a diaphragm are slow openers. However, the opening action is rapid enough to give good ignition characteristics. Occasionally, a valve may get water in it that can cause corrosion of the control passages and bleed ports as well as lay on the main diaphragm in the valve. This can cause extremely slow opening action. Ignition may be delayed to the point of minor explosion and roll out of fire from the main burner. This can result in pilot outage, burning of fuel in the main burner, or fire in the control compartment. The valve should be removed and inspected through the

FIGURE A15-3 Power units. (*Courtesy* Honeywell, Inc.)

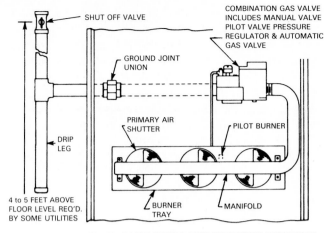

pipe connections for particles or water. If found, replace the valve. Also, blow out the supply lines. To prevent further problems, a drip leg should be installed on the supply line just ahead of the unit.

Figure A15-4 shows the piping arrangement to a gas-fired unit. Note the drip leg in the supply pipe at the left of the unit. This drip leg catches particles and water in the fuel supply before they enter the control valve.

G12. Poor Valve Cutoff: If the valve does not shut off completely, due to particles on the valve seat, it is possible to get flame in the main burner as well as fire in the control compartment of the unit. When the valve closes, .the fire will go out over the burners. If a gas seepage exists through the valve, gas will continue to flow very slowly out of the orifice and up through the burner. Accumulation of gas in the burner will be touched off by the pilot and burner. The gas will then be ignited at the outlet of the orifice and continue to burn. When the valve opens on the next heat-demand cycle, the increase in gas flow will force fire into the main burners and out the primary air openings.

Whenever this unusual fire condition is encountered, turn the gas valve off and, using a small mirror, check the ends of the main burner orifices. Usually, a small flame will be seen. This is only because the gas valve is not closing 100% due to solid particles on the valve seat. Usually, a disassembly of the valve and clearing the valve seat and diaphragm area will cure the problem. At the same time, check for cracks or slits in the seat or diaphragm. When those are found, a new valve is needed.

G13. Leaking Valve: This can easily be confused with G10 or G12. Leaking valves can be either through the valve (ruptured control diaphragm) or external leaks (ruptured gaskets). The most common cause of ruptured

deforms is pressure testing. In some locations the gas supply line must be tested at pressures up to 50 psig and most hold for given periods of time. During these tests, the main shutoff valve to the gas control *must be closed*.

The valve is designed for a maximum inlet pressure of 2 psig. Subjecting the valve to high test pressures can blow holes in the diaphragm. Not only will the valve not shut off, but the unit will operate at full line pressure. Serious fires can result.

External leaks are caused by carelessness by the installer. Use of a pipe wrench on the valve body when installing pipe nipples on the inlet side has caused body distortion, which causes gasket leaks. Valve bodies have wrench pads on the ends. *They should be used.*

G14. Improper Burner Adjustment: A properly adjusted burner will have approximately 40% of the combustion air mixing with 100% of the gas in the burner and 60% of the air mixing with the flame above the burner to complete the combustion process. Figure A15-5 shows a typical burner arrangement. Here we see the gas emitting from the orifice and as it expands it hits the proper place in the throat of the burner venturi. This produces maximum pull of primary air into the burner. Setting the burner for the correct flame condition will mean a minimum opening in the primary air control.

Figure A15-6 shows the size of the opening in an ordinary butterfly-type air control. When the burner and orifice are aligned properly and the burner is working correctly, a small opening in the primary air control will produce the soft blue fire with slightly yellow tips that gives best overall unit performance.

An all-blue sharp fire is receiving too much primary air. This means that there is less radiant heat to heat the lower portion of the heat exchanger. Also, the excess air drives the flame products from the heat exchanger before sufficient time passes to allow good transfer of heat from flue product to heat exchanger. Flue product temperatures rise and unit efficiency drops. If the primary air is reduced too much, heavy yellow tips of improperly burning carbon are produced. These are much lower temperatures and do not produce the heat. Therefore, unit efficiency drops. The carbon can be re-

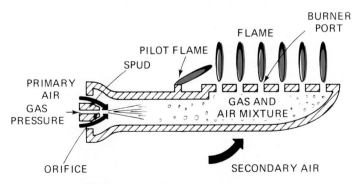

FIGURE A15-5 Gas/air mixtures.

Heating Service and Problem Analysis 585

FIGURE A15-6 Primary air control opening size.

leased from the flame and collected in the heat exchanger to cause sooting and plugging of the flue passages. The proper setting of the primary air quantity is the beginning step in producing high unit operating efficiency.

Improper setting of the primary air shutter can also contribute to pilot outage by producing extinction flash back, called *extinction pop*. To understand this we need to begin with a discussion of flame action of a burner.

When the burner is operating, the gas/air mixture is blowing upward through the burner port at a given speed or velocity. This burner gas velocity is determined by the burner designer. The amount of this velocity will depend on the type of gas to be burned. Natural gas is a slower-burning fuel than propane or butane, so the natural gas burner has different slot or port sizes than for LP gases.

When the gas/air mixture flows from the burner port and burns, the burning action tries to burn down through the port. The velocity of the gas/air mixture prevents this when the two forces are equal. If, however, the burning velocity were to increase due to shutoff of the gas supply (when the gas valve closes), the flame could approach the burner. At this point, the burner body would absorb the heat from the fire, lowering the temperature of the gas/air mixture below the combustion point and extinguish the fire. The trick is to lower the flame down to the burner for extinguishing as rapidly as possible to put the fire out before it has a chance to burn down through the burner port and ignite the mixture in the burner. This ignition produces the extinction pop.

Figure A15-7 shows what happens to the gas flame after the gas valve shuts off. At the moment of shutoff the gas/air mixture inside the burner is at a negative (below atmospheric) pressure. At full fire you have a full cone and a full tail of flame. Immediately after the gas valve closes, the burner pressure collapses. The top sketch shows full fire, full cone, and full tail of flame. In the next sketch you see the partial collapse, where the flame is starting to fall. The third sketch shows the full collapse of the fire down to the burner. The fourth sketch shows the moment of extinction. The flame tries

to burn down through the burner port; the burner absorbs the heat from the fire and extinguishes the flame.

If the speed of gas burning is too high due to too much primary air, the flame does not collapse as rapidly as it should. We do not have sufficient negative pressure in the burner to cause rapid collapse. The flame burns down through the burner port, causing the explosion or extinction pop. This explosion can cause a pressure wave over the pilot that blows the pilot out. A properly adjusted burner greatly reduces the chances of pilot outage.

G15. Improper Burner Alignment: Sometimes it is not possible to obtain good flame characteristics unless

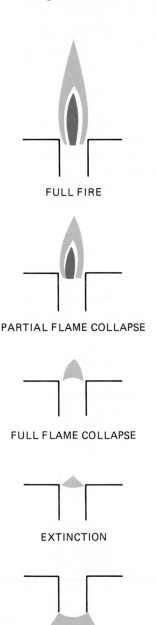

FULL FIRE

PARTIAL FLAME COLLAPSE

FULL FLAME COLLAPSE

EXTINCTION

CARRY THROUGH

FIGURE A15-7 Flame action.

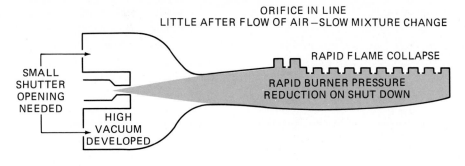

ORIFICE IN LINE
LITTLE AFTER FLOW OF AIR—SLOW MIXTURE CHANGE

SMALL SHUTTER OPENING NEEDED

HIGH VACUUM DEVELOPED

RAPID FLAME COLLAPSE

RAPID BURNER PRESSURE REDUCTION ON SHUT DOWN

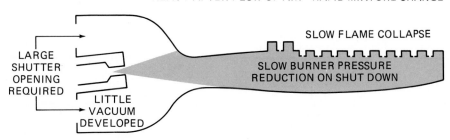

ORIFICE OUT OF LINE
HEAVY AFTER FLOW OF AIR—RAPID MIXTURE CHANGE

LARGE SHUTTER OPENING REQUIRED

LITTLE VACUUM DEVELOPED

SLOW FLAME COLLAPSE

SLOW BURNER PRESSURE REDUCTION ON SHUT DOWN

FIGURE A15-8 Effect of burner alignment.

the air shutter is practically wide open. This tells us that the burner's orifice position is such that the burner venturi is not working properly. Figure A15-8 shows the difference between the action of a properly aligned orifice and one improperly aligned. The top sketch shows the properly aligned orifice with a round impact shape of the gas contacting the throat of the burner venturi. This heavy draw action indicates minimum air shutter opening and good negative operating pressure in the burner. The lower sketch shows an orifice that has been tipped out of alignment. The contact pattern is elliptical, with most of it out of the throat of the venturi. Because of the poor draw action, very little flow resistance to air entering the burner can be tolerated and a wide-open air shutter is needed. As a result, poor burner cutoff occurs, extinction pop is considerable, and this unit has frequent pilot outage.

A common cause of this misalignment is tightening a pipe nipple into the gas valve without a retaining wrench on the valve body. The control manifold assembly takes all the strain and the manifold mounting means gives. It takes a distortion of only 10° out of alignment to affect burner operation. Use two wrenches when installing the supply piping.

G16. Improper Orifice Alignment: Occasionally, only one burner of the assembly will give flash back or extinction pop problems. When this occurs, the problem is usually that one of the orifices is out of alignnment. This could be because the orifice is drilled at an angle or the manifold has been drilled and tapped out of alignment with the other orifices. This is easily checked by rotating the orifice when the burner is operating. If any

change is noted in flame performance, the orifice should be changed. If not, the manifold should be checked for orifice alignment.

The easiest way to check orifice alignment is to remove the manifold from the unit and remove all the orifices from the manifold. Insert pipe nipples in the orifice holes with the nipples at least 6 in. long. With the nipples in place, they should all line up in a row. Any nipples out of alignment will indicate a drilling problem. *Do not try to redrill*; gas leaks will result. Get a new manifold from the manufacturer.

Another source of poor orifice performance is attempting to reduce the input to the heating unit by reducing the orifice drill size. When input has to be reduced it is common practice in the field to peen the end of the orifice with a hammer, reducing the opening, and then redrill for the new size.

Figure A15-9 shows the effect of the reduction of the shank length (the length of the actual hole that determines the spray angle of the gas) when the orifice is pierced and redrilled. When reducing the input of a unit, for example, for high-altitude-capacity reduction, use orifices properly drilled or start with blank orifices if you are going to drill them yourself. Also, use a drill press, not a hand-held drill. No one can drill within the 5° from center required for proper performance.

G17. Improper Slot or Port Size: Occasionally, problems of flame lift or flash back will occur when units are converted to other fuel use. Practically all units using ribbon and slotted burners are designed for an average opening size of the port to accommodate both natural and LP gases. The old-style units as well as

Heating Service and Problem Analysis 587

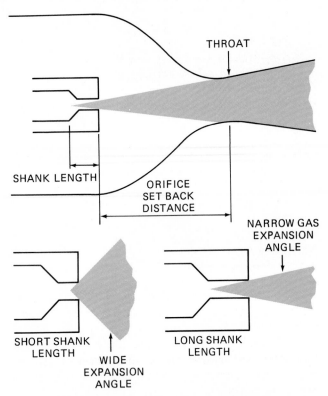

FIGURE A15-9 Effect of orifice drilling.

higher-input units use cast iron or steel burners with drilled ports that were drilled for the specific gas. Natural gas units were drilled with a No. 30 drill and LP units with a No. 32 drill. When an LP unit was converted to natural gas, the smaller port size tended to give a lifting, noisy flame. However, when a natural gas unit was converted to LP gas, the faster-burning LP gas in combination with the larger port opening produced more cases of extinction pop and pilot outage. The manufacturer of the unit should be consulted before conversions are made to be sure that the burner is designed to handle the proposed fuel.

G18. Restriction in Heat Exchanger: Pilot outage, flame roll-out, delayed ignition, and odor in the area surrounding the unit can all be caused by the fact that the heat exchanger cannot carry the flue products fast enough. This would be more noticeable at ignition when the large volume of combustion occurs. The neutral point above the fire is at the middle of the fuel door on conversion units and 3 to 5 in. above the burner pouch opening on designed units.

If there is an outward flow of flame products from the top of the burner pouch opening or flame roll-out, it could indicate restriction in the heat exchanger. Possibly there is sooting of the flue passages, restrictors have shifted, or exchanger suctions have warped closed. Practically all of these causes result from overfiring the unit for a long period of time.

When the trouble has been found and corrected, the unit must be adjusted for proper input. Input tolerance is correct design amount to 10% less. Any over-firing will only produce early unit failure. To set the unit for proper input, refer to Section A14-2.1.

G19. Improper Combustion Air Supply: Excess air is needed for proper combustion even though changes in gas pressure, heat content of the gas, and barometric pressure may occur. Draft conditions also change with barometric pressure and wind conditions.

When a unit encounters insufficient combustion air, the air tends to become hazy and erratic and may even roll over the edge of the burner and out the burner pouch opening. The flame will seek air. Figure A15-10 shows the effect of insufficient secondary air causing floating flame.

To ensure that sufficient combustion air is supplied to the unit, the requirements in the National Fuel Gas Code (ANSI-2223.1) published by the American Gas Association and National Fire Protection Association must be followed. Section A5-3 spells out the method of supplying combustion air under all the various types of equipment locations.

G20. Improper Input: A gas-fired unit is designed to burn efficiently a given amount of fuel. The burner orifice size relationship as well as the burner, heat exchanger, and restrictor relationship are all designed to handle a given amount of fuel and air to produce the correct amount of flue products for the heat exchanger to handle. This means that the gas volume input has to be within design limits. Design limits are from rated input down 10%. All units are rated at the maximum the unit will handle efficiently and with expected unit life. In addition, the controls, fan, and limit, mounted above the burner pouch opening, depend on the radiant heat

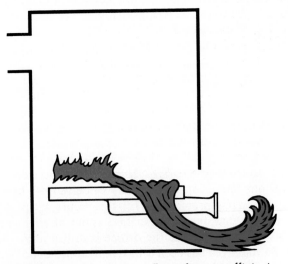

FIGURE A15-10 Floating flame from insufficient secondary air. (*Courtesy* American Gas Association)

of the heat exchanger to operate. The heat exchanger surface must get high enough in temperature to cause the fan control to close without getting so hot that the limit control is actuated.

If the unit is underfired before the heat exchanger gets hot enough to close the fan switch, the supply plenum above the unit will seriously overheat. Very erratic operation such as short cycling of the fan can result.

If the unit is overfired, even though the fan is on, the limit control can be actuated, causing the gas valve to open and close. When erratic control operation is encountered, the first step is to check and make sure that the gas input to the unit is correct. To check and set the input, refer to Section A14-2.1.

G21. Blower CFM Adjustment: The correct adjustment of the air through the unit will result in a temperature rise between return and supply air of 80°F. For the correct method of setting the ft^3/min through the unit, refer to Section A14-2.

G22. Fan Control Settings: Almost all gas-fired units will operate best with fan control settings of 125 to 130°F fan on and 100 to 105°F fan off. On installations with extra-long duct runs, the occupant may encounter movement of cold duct air when the fan first starts. If this is objectionable, it can be partially overcome by increasing the fan-on temperature to 145 to 150°F. The fan-off temperature should remain the same, −100 to 105°F. If this does not handle the problem, the supply duct should be insulated with a minimum of 1 in. of insulation. This will help to keep the air temperature higher and minimize the cold-air effect.

G23. Dirty Air Filters: Dirty air filters cause more erratic operation and waste of fuel than any other factor. Do not wait until the filter is so heavily coated that it looks like a carpet. If you cannot see through it, replace it. The throwaway filter is just that; when it is dirty, throw it away and put in a new one. Air filters should be replaced at the beginning of the operating season: in the fall on heating-only systems and in the fall and spring on year-round systems. If the household includes dogs or cats, the filter should be checked at least once a month and replaced when necessary.

Dirty filters will reduce the ft^3/min of air over the heat exchanger, raising the operating temperature and reducing the operating efficiency. On complaints of unit cycling on and off with little heat or high fuel bills with insufficient heat, the first item is to check the condition of the air filters.

G24. Restrictions in Air Supply: High fuel bills and erratic operation can be encountered because of high temperature rise (lack of air through the unit) even though air filters are clear. People believe that you shutting off the air from or the supply into rooms that get little use will save money. This restriction in the air circulation system can be enough to cause high temperature rise, erratic operation, and could result in high fuel bills. The unit must have sufficient air through the unit to carry the heat from the heat exchanger at the design rate. In these cases it is necessary that the occupant know the negative results of closing supply and return outlets.

G25. Dust Odor at Fall Startup: A common cause of odor complaint at the beginning of the heating season is the burning off of the dust that has accumulated on the heat exchanger during the summer months. If the unit has a summer fan switch and is used for summer ventilation, this problem can be encountered. The odor is temporary and usually lasts one or two cycles of the unit.

Another type of odor is encountered at the startup of a new unit. All unit heat exchangers are treated for rust prevention during storage by an oil coat or even a paint coat. At the first firing, a light smoke will come off the heat exchanger surfaces. Units should be test fired after installation, with heavy ventilation provided in the occupied area. This smoke odor will usually disappear in two or three cycles of the unit.

G26. Improper Venting: As stated in Chapter A14, the mixture of gas and air produces a mixture of water and carbon dioxide that is vented from the heat exchanger. Mixing 1 ft^3 of natural gas with 10 ft^3 of air produces 11 ft^3 of flue product, 1 ft^3 of carbon dioxide, 2 ft^3 of water vapor, and 8 ft^3 of nitrogen. The nitrogen had nothing to do with the process of combustion except to use some of the heat produced by the combustion process to heat it from the room temperature to the flue product temperature. To ensure complete combustion, from 10 to 50% excess air is also added to the process. All of this has be to removed from the heat exchanger. This removal process is called *venting*.

There are two types of venting: *power venting* (now called *active venting*) and atmospheric or *gravity venting* (now called *passive venting*). Power venting uses a mechanical device such as motor-driven blowers to either draw flue products from the heat exchanger or to force combustion air into the heat exchanger. The most popular type is the draw type, where the blower is mounted on the flue outlet and creates the negative pressure in the outlet of the heat exchanger to get the desired combustion efficiency. Because the pressure difference is caused by mechanical power, wind and/or atmospheric conditions have little effect on the venting performance. In atmospheric or gravity venting, hot flue gases pass from the heat exchanger into a flue pipe, chimney, or vent stack. The driving force for a gravity vent is obtained from the fact that hot gases tend to rise in surrounding cooler air. The amount of force will depend on the temperature of hot gases and the height of the gravity vent. The hotter the gases in the vent and/or the higher the vent, the greater the amount of driving force or pull that will be developed. Naturally, the greater the pull on the outlet of the heat exchanger, the more secondary air that will be drawn through the heat exchanger.

The flow of air into the combustion products out of the heat exchanger must be reasonably close to the amounts for which the unit was designed. Enough air must be drawn through to provide complete combustion as well as complete venting of the flue products. However, if too much air is drawn out of the heat exchanger before the correct amount of heat extraction is done, this means higher flue temperatures and reduced unit efficiency.

If the gravity vent pipe were connected directly to the flue outlet, the amount of air drawn through the heat exchanger would vary with the pull of the vent stack, the wind effect on the vent stack, outside temperature, etc. Control of the venting rate on the heat exchanger would be impossible. Further, under some conditions of outside atmosphere it may be possible to have a higher pressure at the outlet of the vent than the combustion process can overcome. This can produce poor combustion with the productions of CO as well as the CO_2 and H_2O.

To overcome the effect of atmospheric conditions, all units use an opening in the venting system called a draft diverter. Figure A15-11 shows four typical heating

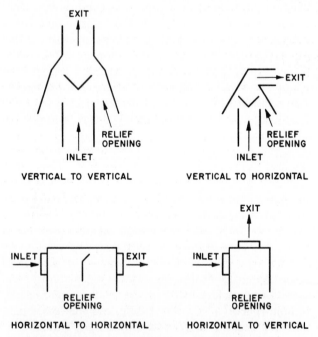

FIGURE A15-11 Typical gas appliance draft hoods. (*Courtesy* American Gas Association)

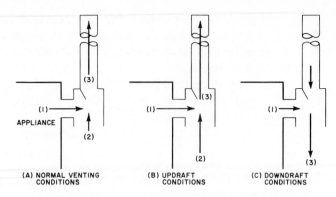

FIGURE A15-12 Operation of a draft diverter under various wind conditions. (*Courtesy* American Gas Association)

unit draft diverters. These consist of an opening from the flue outlet of the heating unit, an opening into the vent pipe, and a relief opening to the surrounding atmosphere.

Figure A15-12 shows the operation of a typical draft diverter under no-wind conditions and with updraft and downdraft conditions. The amount of flue products (1), the dilution air entering the relief opening (2), and the amount of vent gases are indicated by the length of the arrows. With normal venting some air is pulled into the draft diverter by the pull of the gravity vent. The mixture of flue products and surrounding air (called *dilution air*) that blow up the vent is called *vent gas*. The action of the draft diverter is to break the effect of the vent by introducing surrounding air and neutralizing the pull at the flue outlet. The heat exchanger then operates at approximately equal pressure across it from burner pouch opening to flue outlet. The amount of air for combustion is then controlled by the flue restrictors.

If the conditions surrounding the vent stack increase the stack pull, additional air is drawn into the draft diverter to compensate for the increased pull. There is little effect on the heat exchanger performance.

Under conditions where the vent stack pull is reduced or even reversed, creating a downdraft, all combustion flue products are forced into the surrounding area. In addition, the increased pressure in the flue outlet will reduce the flow through the heat exchanger. This can cause incomplete combustion and produce odors carried by the gases and moisture produced by the combustion process. Even though no odors may result, the large amount of moisture produced in the combustion process can accumulate in the occupied area and create adverse living conditions or possibly structural damage.

To check for proper operation of the vent system, use a candle placed below the bottom edge of the diverter opening. With the unit operating and up to temperature, the candle flame should bend in the direction of the opening in the diverter. If the flame is neutral, the draft is on the weak side. Possibly the vent stack is not high enough or large enough. If the candle flame bends outward, a draft problem definitely exists that must be corrected. If the vent stack cannot be lengthened or enlarged, a forced-draft unit must be installed to overcome the problem.

G27. Cracked Heat Exchanger: All heating units must be designed with the heat exchanger on the positive-pressure side of the circulating blower. This is a safety requirement to prevent drawing flue products into the circulating air in the event of heat exchanger failure. In the event of an opening developing in a heat exchanger, the first indication is flame pattern disruption when the blower starts.

Air will be forced through the opening of the heat exchanger and build pressure above the fire, between the burner and the flue restrictors. This increase in pressure will cause the amount of combustion air to be reduced and combustion conditions changed. Products of combustion will be free carbon as well as a higher percentage of CO in the flue products. Disruption can also be great enough to cause flame float and flash back. The way to correct this problem is to replace the heat exchanger.

The cause of this problem may be overfiring or chemical atmosphere (G28). Overfiring to the point where fire contacts the heat exchanger surface will cause high temperatures at the contact spots. These high-temperature spots will crack from the heating and cooling of the unit cycling. Further heating will cause the metal to warp and the cracks to open. Be sure that the input is correct (G20).

G28. Chemical Atmosphere: The most common cause of heat exchanger failure due to "rust out" is due to chlorine and fluorine chemicals in the flue products. When the unit cycles ON, the heat exchanger surfaces are cold. The moisture produced by the combustion process condenses on the heat exchanger draft diverter and vent stack chimney surfaces; the mixture evaporates as soon as the surfaces reach 212°F or higher. The primary surfaces of the heat exchanger dry off first, then the secondary surfaces after the flue restrictors, then the draft diverter, the flue pipe, and finally, the vent stack or chimney.

If there is any chemical in the atmosphere that will introduce chlorine or fluorine into the combustion process, the chemical will combine with the water produced and form hydrocholric or hydrofluoric acid. This acid attacks the metal all the time that the moisture exists. Because the moisture remains there longer on each cycle, the vent stack deteriorates first, then the vent pipe, the draft diverter, and finally the flue outlet suctions of the heat exchanger.

In industrial applications where paint solvents, oil, or any other chlorinated or fluorinated hydrocarbon exists in the atmosphere, heat exchangers have an average life of 3 years. In residences where hair sprays are used extensively, heat exchanger life is normally 7 to 10 years.

If the unit is located in a utility room with an automatic washer, the chlorine from the soap or bleach will shorten the heat exchanger life to 5 to 7 years.

Usually, it is necessary to move the unit to a cleaner atmosphere, possibly outside or in an enclosed area, and provide combustion air from outside the building. Be sure to follow Section 5.3.3 of the National Fuel Gas Code.

G29. Blower Wheel Unbalanced or Out of Line: Usually, the blower wheel is balanced and aligned before being installed in the original unit. However, shipping damage, such as dropping the unit from the tailgate of a truck, can cause the wheel to strike the housing and become warped. It then will not run true and can cause excessive vibration. The only correction is replacement. The way to prevent damage is through careful handling of the unit.

Blower vibration will develop after a unit has been operating for some time. The action of the blower blades will cause dust and dirt to build on the leading surface of the blade. In new construction, sawdust is the chief culprit. If the amount builds up to the point where chunks of the buildup leave the wheel, an imbalance develops and vibration occurs.

The most common cause of wheel vibration is a lazy or incompetent serviceperson. Cleaning the wheel with a brush without disassembling the blower will always lead to excessive vibration. Never touch a blower wheel unless the intent is to dissemble the blower assembly and give the wheel a thorough cleaning. Immersion in detergent and water or jet-type cleaning equipment is required. Then when the unit is assembled, be sure that the blower wheel turns in the right direction. Wheels in backward deliver only approximately 25% of the amount of air that is needed.

G30. Blower Drive: Belt-driven blowers have a higher probability of vibration problems than direct-drive blowers due to the additional parts involved. The most common problem is due to belt tension. Most people believe that the tighter the belt, the better the performance. The opposite is actually the case. The tighter the belt, the harder the motor has to work to get the belt in and out of the pulleys. Therefore, the belt should be as loose as possible without slipping on startup.

Figure A15-13 gives the test for proper belt tension. You should be able to easily depress the belt midway between the motor shaft and blower shafts $\frac{3}{4}$ to 1 in. for each 12 in. of distance between the shafts. Alignment of the motor and blower pulleys is important to keep vibration to a minimum as well as to reduce wear on the sides of the belt.

Finally, each pulley, both motor and blower, should be checked for running true. Any warpage that creates wobble in the pulley requires replacement of the pulley.

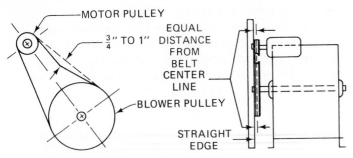

FIGURE A15-13 Align pulleys and tighten belt. (*Courtesy* Borg-Warner Central Environmental Systems, Inc.)

G31. Expansion Noise in Heat Exchanger Sections: Figure A15-14 shows a three-section unit, each section composed of a right-hand and a left-hand drawn steel "clamshell" seam welded together. The sections are then welded into an assembly by fastening to the front mounting plate and rear retainer strap. Sometimes in the welding process, stresses will be setup in the welding area if the two metals are at different temperatures when the bond is made. This results in expansion noises, ticking, and popping as the heat exchanger heats and cools. Most of the time these noises are muffled by the unit casing and duct system to a level where they are not objectionable.

In extreme cases, it is possible to reduce the noises by operating the unit with the blower disconnected, allowing the limit control to turn the unit off and on. This should be done through several cycles of the limit control. Cycling on this extreme heat will cause metal stretch beyond the normal operation range and eliminate the sound. If this does not produce satisfactory results, the only cure is to change the heat exchanger. However, this is no guarantee that the new heat exchanger will be quiet.

G32. Expansion Noise in Heat Exchanger Restrictors: In sectional gas furnaces, most units have a horizontal restrictor located in the top of each heat exchanger section. These restrictors are used to control the amount of secondary air for proper combustion. Most are made in such a shape as to be fastened at the flue outlet end, leaving the other end to move with the expansion and contraction of the metal as the flue products temperature changes.

Occasionally, these restrictors will shift or warp out of line to the point where the end of the restrictor drags on the surface of the heat exchanger. This produces a scratchy or scrubbing noise as the furnace heats up or cools down.

Make sure that the restrictors are located in the center of the heat exchanger section and that any formed

feet that may be punched out of the restrictors have round bearing surfaces instead of sharp points.

G33. Duct Expansion: Any metal duct will have a certain amount of expansion with temperature rise. Aluminum duct has considerably more expansion per degree of temperature rise then does galvanized iron and will be more prone to expansion noise.

The main cause of duct noise is too solid a mounting means. Ductwork mailed directly to floor joists will be extremely noisy from expansion and contraction. It will also be more prone to causing fire because the hot duct is in direct contact with the wooden floor joist.

Properly installed ductwork is suspended 1 in. below the floor joist. Metal straps are used, fastened to the floor joist and extended down under the duct to allow fastening under the duct. The ductwork is hung in the form of a trapeze, which allows free movement of the duct without strain.

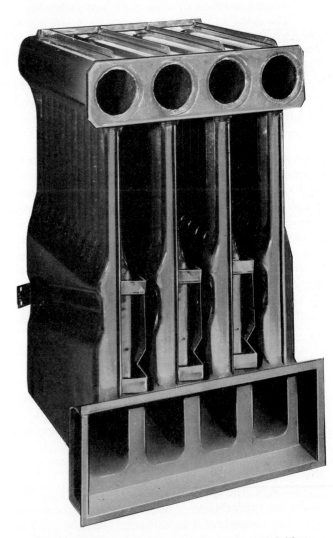

FIGURE A15-14 Heat exchanger. (*Courtesy* Addison Products Company)

FIGURE A15-15   Drum-and-radiator heat exchanger.

G34. Oil-Can Effect: Oil-can effect is the sudden movement of a flat metal surface where a forming stress has been left in the surface. This stress causes the metal to have a slightly concave or convex position rather than a flat plan surface.

Temperature change will cause a stress increase in the material until the metal rapidly changes position to the opposite of its original position. This change will produce a loud "bang" heard throughout the area. Ductwork is very prone to this action and must be cross-broken over any large panel areas. Heat exchangers such as the one shown in Fig. A15-15 will usually have a large flat surface. Unless this surface is cross-broken, it is very subject to the oil-can effect. The best correction is removal of the panel and cross-broking to relieve temperature stresses.

G35. Improper Unit Sizing: For many years if the unit requirement was 100,000 Btu/hr the owner would demand a 150,000-Btu/hr unit in the belief that the larger the unit, the better. The unit did not have to work as hard to do the job. This results in a heavy waste of fuel and high operating cost. We now recognize that the unit should be properly sized for peak efficiency and lowest operating cost.

The actual heat loss of the structure should be properly sized for peak efficiency and lowest operating cost. The actual heat loss should be calculated using the latest manuals in the Environmental Systems Library put out by the Air-Conditioning Contractors of America (ACCA). These manuals are constantly updated and give the most accurate results. The unit should then be sized within ± 10% of this requirement.

G36. Resonance-Unit Design: Occasionally, a gas-fired unit will develop a pipe organ sound that can be severe enough to rattle pictures on the wall. This is a result of the burning frequency of the flame and the pressure wave distance in the burner pouch being in exact synchronization. The sound starts as a low-level hum and builds rapidly to a high-volume sound. There are several solutions available for this problem:

1. Correct the input.
2. Soften the fire.
3. Change the size of the burner pouch.

If the unit is overfired, especially on LP gas, the tendency toward resonance is greater. Therefore, correcting the input to the range of design to minus 10% will usually handle the problem. This also involves adjustment of the primary air to reduce the flame from a sharp cone to a correct lazy blue fire.

If these changes do not correct the problem, change the size of the burner pouch. This is easily done by inserting a vertical sheet of metal alongside the burner or against a side wall. The metal sheet should be fastened with spaces so that there is a $\frac{1}{2}$-in. space between the insert and the side wall of the burner pouch. This changes the beat frequency of the burner pouch so that it is not the same as the flame and the sound buildup is eliminated. This problem was common in units made 15 to 20 years ago but is practically unheard of today.

A15-3
OIL HEATING SYSTEMS

A15-3.1
Problems

The problem tabulation for oil-fired units utilizes the same arrangement as that used for gas-fired units. The first tabulation is a list of common complaints under the following categories:

A. Entire system operation
B. Unit operation
C. Burner operation
D. Blower operation
E. Heat exchanger complaints
F. Cost of operation
G. Noise

Under each category is a list of the possible problems in the particular category. As you review the problems you will find duplications because it is possible to have the same problem for more than one complaint. The number in front of each problem indicates the sequence in which it appears in the symptom listing. There is no numerical tabulation of problems under each category.

A. Entire system operation
1. Will not start
2. Starts but will not continue to run
3. Runs but not enough heat—short cycles
4. Runs but too high a room temperature
5. Runs continuously
6. Blower cycles after thermostat is satisfied
7. Blows cold air on startup
8. Startup and/or cool-down noise
9. Noisy/vibrations
10. Odor
11. High fuel cost
12. High electrical cost

B. Unit operation
1. Will not start
13. Burner short-cycles
14. Smoke and/or odor from observation door
15. Unit pulsates

C. Burner operation
16. Burner cuts off on safety switch
17. Delayed ignition
18. Burns inside burner after cutoff
19. Carbon deposits on refractory
20. Noisy flame
21. Fuel pump "sing"

D. Blower operation
22. Short cycles
24. Noisy/vibration

E. Heat exchanger complaints
1. Startup and/or cool-down noise
22. Burnout

≡ A15-3.2
≡ **Symptoms**

Following is a list of the symptoms of the problems in the preceding category.

1. Will not start
   E1. Room thermostat improperly set
   E2. Season switch open
   E3. System safety switch open/protector relay out on safety switch
   E4. Disconnect switch open
   E5. Blown fuse
   E6. Limit control B.O.
   E8. Protecto-relay transformer B.O.

E10. Open circuit in thermostat and/or subbase
E11. Open in control cable
E9. Protecto relay B.O.

2. Starts but will not continue to run
   A. No fire established
   O1. No fuel
   E13. No ignition
   E9. Protecto Relay B.O.
   O6. Nozzle B.O.
   O7. Fuel Pump B.O.
   B. Fire established
   O6. Nozzle B.O.
   O7. Fuel pump B.O.
   E6. Limit control B.O.
   E9. Protecto relay B.O.
   O6. Nozzle B.O.
   O7. Fuel pump B.O.

3. Runs but not enough heat—short-cycles
   E12. Improper heat anticipator setting
   E6. Cycling on limit control—input too low

4. Runs but too high a room temperature
   E1. High setting of thermostat
   E12. Improper heat anticipator setting

5. Runs continuously
   E10. Short in thermostat and/or subbase
   E11. Short in thermostat cable
   E9. Stuck contacts in protecto relay (protecto relay B.O.)

6. Blower cycles after thermostat is satisfied
   O5. Blower ft^3/min adjustment
   E7. Fan control setting
   O10. Heat exchanger heavy with soot

7. Blows cold air on startup
   O5. Blower ft^3/min adjustment
   E7. Fan control setting

8. Startup and or/cool-down noise
   O17. Expansion noise in heat exchanger
   O18. Duct expansion

9. Noisy/vibration
   O14. Pulsation
   O15. Blower wheel unbalanced or out of line
   O16. Blower drive problems

10. Odor
    O12. Dust odor at fall startup
    O13. Oil odor on unit startup
    O14. Pulsation
    O4. Improper venting
    O11. Cracked heat exchanger

11. High fuel cost
    O2. Improper input

O3. Improper burner adjustment

O20. Improper unit sizing

12. High electric cost

    O19. Improper load on blower motor

13. Burner short-cycles

    E6. Limit control B.O.

    E12. Improper heat anticipator setting

14. Smoke or odor from observation door

    O4. Improper draft or draft control setting

    O4. Inproper venting

    O2. Improper input

    17. Delayed ignition

15. Unit pulsates

    O4. Improper draft

    O4. Improper venting

16. Burner cuts off on safety switch

    A. No fire established

    O1. No fuel

    E13. No ignition

    E9. Protecto relay B.O.

    O6. Nozzle B.O.

    O7. Fuel pump B.O.

    B. Fire established

    O6. Nozzle B.O.

    O7. Fuel pump B.O.

    E6. Limit control B.O.

    E9. Protecto relay B.O.

17. Delayed ignition

    O9. Carbon deposits on firing head

    E14. Improper electrode adjustment

    E15. Cracked electrode insulators

    E16. Ignition leads B.O.

    E17. Ignition transformer B.O.

18. Burns inside burner after cutoff

    O7. Fuel unit B.O.

19. Carbon deposits (coke trees) on refractory

    O6. Nozzle B.O.

20. Noisy flame

    O3. Improper air adjustment

    O6. Improper nozzle size, type, and spray angle

21. Fuel pump sing

    O7. Fuel unit B.O.

22. Heat exchanger burnout

    O2. Input too high

    O6. Nozzle B.O.

23. Blower short-cycles

    O2. Input too low

    O5. Blower ft³/min too high

    E7. Fan control B.O.

A15-3.3
Electrical Causes

In the previous list, the probable causes of each symptom were given a number beginning with an E if electrical and O if in the oil fuel category. The electrical causes are as follows:

E1. Room thermostat improperly set

E2. Season switch open

E3. System safety switch open

E4. Disconnect switch open

E5. Blown fuse

E6. Limit control B.O.

E7. Fan control B.O.

E8. Protecto-relay transformer B.O.

E9. Protecto relay B.O.

E10. Open or short circuit in thermostat and/or sub-base

E11. Open or short in control cable

E12. Improper heat anticipator setting

E13. No ignition

E14. Improper electrode adjustment

E15. Cracked electrode insulators

E16. Ignition leads B.O.

E17. Ignition transformer B.O.

E1. Room Thermostat Improperly Set: The room thermostat may be operating at too high or too low a setting even though the dial pointer is at the desired setting. This is especially true on mercury bulb contact thermostats. Make sure that the thermostat is level before trying to recalibrate it. Very seldom does a thermostat go out of calibration, although they are frequently knocked out of level. If the dial setting and the built-in thermostat do not read the same, check the level of the thermostat before any recalibrating or adjustment is done.

E2. Season Switch Open: One of the most common causes of no-heat complaints, especially at the beginning of the season, is the fact that the season on/off switch on the unit was put in the OFF position. No one trusted the thermostat to keep the unit off, or the owner did not want the unit to come on when the nights got cool. Regardless, the fact that this switch is turned off is easily forgotten over the summertime period. When a no-heat complaint is received, it is well to ask the caller if the switch has been turned on. It can save a service call and an embarrassed customer.

E3. Season Safety Switch Open: Where the unit is located in a basement, a furnace room with only one exit, some localities require that a red switch be located at the exit of the room. This switch is intended to be used in case of emergency only. Some people do not comply with this intent and use it to shut down the system in the nonheating season. Occasionally, someone figures it to be a light switch and turns it off. Again, when a service complaint is received, inquire as to the position of this switch. Make sure that it is in the ON position.

E4. Disconnect Switch Open: Make sure that the proper voltage is supplied to the unit. Disconnect switches in distribution panels are known to open due to electrical disturbances, electric storms, etc. If there is no voltage to the protecto relay on the unit, reset the breakers to check their holding ability.

E5. Blown Fuse: Instant blow- or link-type cartridge fuses are sometimes used as branch fuses. Because they will not carry current above their rating for any length of time, they must be sized to carry the maximum starting current of the oil burner as well as the blower motor in the unit. Time-delay fuses should always be used where motors are involved. This type will allow closer sizing of the fuse to the running current of the unit motors for maximum protection.

Checking of fuses is best done using an ohmmeter. *With the disconnect switch open*, measure the resistance of each fuse. A reading of zero resistance on the ohmmeter indicates that the fuse is good. Infinite resistance or no movement of the instrument needle indicates an open or blown fuse. A resistance reading indicates a burned fuse that is not completely open. This fuse should also be replaced and discarded.

E6. Limit Control B.O.: The limit control on an oil unit is generally in the high-voltage power supply to the protecto relay. If the limit control is open, there will be no main power to the protecto relay. Using the voltmeter across the hot and neutral connections of the relay, full voltage should be indicated. If not, check the season switch (E2), season safety switch (E3), disconnect switch (E4), and any fuses if used (E5) before attempting to check the limit control.

Using the voltmeter, measure the voltage across the terminals of the limit control. If the control is properly closed, voltage will be measured. An open control will indicate full voltage because it is a break in the circuit to the primary side of the control transformer in the protecto relay.

With the voltage off to the unit, remove the wires from the control terminals. Using the ohmmeter, measure the resistance of the limit control contact. Infinite resistance or no movement of the ohmmeter needle indicates an open contact. If the temperature of the control is at room temperature, replace the control. If a reading of zero resistance is found, the contact is closed and the voltage interruption is elsewhere.

E7. Fan Control B.O.: Almost all oil-fired units will operate best with fan control settings of 125°F (51.6°C) to 130°F (54.4°C) with the fan on and 100°F (37.7°C) to 105°F (40.6°C) with the fan off. On installations with extra-long duct runs, the occupant may experience movement of cold air from the supply registers when the fan first starts. If this is objectionable, it can be partially overcome by increasing the fan on temperature to 145°F (62.8°C) to 150°F (65.5°C). The fan-off temperature should remain the same. If this does not handle the problem, the supply duct should be insulated with a minimum of 1 in. of insulation. This will help to keep the supply air temperature higher during the OFF cycle and minimize the cold air effect.

Occasionally, the bimetallic element of the control will weaken and require a higher element temperature to operate the control. This will result in very short ON-time cycles and may cause fan cycling while the burner is on. The best test is substitution of a control that is known to be correct and observing the action of the unit. Contact failure in the control will prevent the blower from operating and the unit will cycle on the limit control.

With a voltmeter or test light across the control, the meter should read zero voltage or the test light should go out when the supply plenum temperature reaches the temperature setting of the control. If not, replace the control.

E8. Protecto-relay Transformer B.O.: It is usually not possible to open the protecto relay to reach the

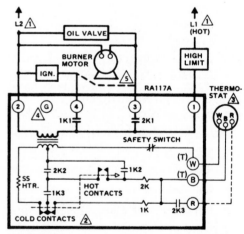

FIGURE A15-16 Internal schematic and typical hook-up. (*Courtesy* Honeywell, Inc.)

transformer to test it. It is not necessary, as corrections to the transformer are done through the T-T terminals on the outside terminal board. Figure A15-16 shows the inside wiring diagram of a typical protecto-relay circuit. With the control wires removed from terminals T-T, a circuit from the top terminal T, through the safety switch (SS), the transformer, normally closed contact (2K1), the safety switch heater, and relay coil (1K) should put 24 V across terminals T-T. With 120 V across the black-and-white leads to the relay, if no voltage is at T-T, replace the protecto relay. If there is voltage at T-T and the thermostat will not operate the relay, check the thermostat and subbase (E10) on the control cable (E11).

E9. Protecto Relay B.O.: Protecto relays are of two different types: a smoke pipe mount using a bimetallic element in the hot flue gases to operate the relay, and a cad cell to operate the relay from the light of the fire. Both types of controls were discussed in Section A10-11.

Bimetallic Controls. As explained in Section A10-11, the bimetallic-type control is clutch operated to move the hot and cold controls through their proper sequence. A sharp blow to the control can release this clutch and throw the control out of sequence. The clutch also wears and can loosen the hold on the contact fingers.

If the control refuses to operate, pull the drive shaft lever (see Fig. A15-17) forward until the stop is reached. Slowly release the drive shaft lever to the cold position. This should close the contacts and the unit should operate. If not, the bimetallic element could be jammed or the contacts defective. To check the bime-

FIGURE A15-18   Light-sensitive cadmium sulfide cell. (*Courtesy* Honeywell, Inc.)

tallic element, remove the protecto relay from the vent stack and check it. Usually, carbon (soot) buildup through the bimetallic helix will be the cause of jamming the drive shaft lever. When cleaning the helix, be careful not to bend or break the helix.

With a clean helix, if the hot and cold contacts still do not hold, the clutch and contact leaves are worn. This requires replacement of the entire control.

Cad Cell. The cad cell protecto relay uses a light-sensitive cadmium sulfide flame detector mounted to the firing tube of the oil burner (see Fig. A15-18).

When the cell is exposed to light, its resistance is very low, which allows current flow through the cell. This current flow is sufficient to pull in the sensitive relay in the protecto relay (see Fig. A15-19).

When the relay pulls in, it opens the circuit to the safety switch heater and prevents cutout of the burner. If the cell is not exposed to the light of the fire or if the

FIGURE A15-17   RA117A with cover removed. (*Courtesy* Honeywell, Inc.)

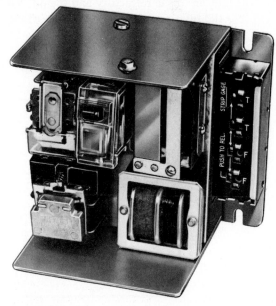

FIGURE A15-19   Internal view of cad cell type of protecto relay. (*Courtesy* Honeywell, Inc.)

Heating Service and Problem Analysis

cell becomes so dirty or covered with soot, it will not allow the current flow to pull in the sensitive relay; the safety switch heater remains in the circuit until the safety switch opens. This breaks the thermostat circuit and the burner stops (see Fig. A15-20).

The safety switch is manually reset. When checking the protecto relay for repeated burner cutoff even though the flame is established, make sure that the cad cell is clean. The cad cell can also be checked using an ohmmeter. Connect the ohmmeter leads across the leads of the cad cell. If the cad cell is exposed to light, the resistance will be less than 10 Ω. Placing a finger over the cell, cutting off the light will raise the resistance to 10,000 Ω or higher.

If the cad cell checks out OK, check the protecto relay by placing a jumper wire across the F-F terminals of the protecto relay. The burner should start and continue to operate. If it cuts off, the timer contacts and heater circuit are defective and the relay should be replaced.

A word of caution: The customer knows that the burner should start if the reset button is pressed. If the relay is cutting off—doing its job—when the problem is poor ignition, it is possible to have a considerable quantity of oil sprayed into the combustion chamber before the owner calls for help. *Before starting the burner, check for liquid oil in the bottom of the refractory.*

When reaching into the refractory, make sure that the power to the burner is off and your arm is covered. If any oil is laying in the bottom of the refractory, it must be removed by soaking it up with sponges or rags. A fire extinguisher must be within reach and the observation door secured open. When flame is established, considerable fire will develop until all the oil is burned out of the bottom of the refractory and the oil that has soaked into the refractory has been burned off. If it is

a large unit of 2.5 gal/hr input or larger, it is advisable to call the fire department for a standby unit before lighting the burner.

E10. Open or Short Circuit in Thermostat and/or Subbase: If the complaint is that the unit will not operate or operates continuously and causes overheating, the thermostat can be at fault. If the unit runs continuously even though the thermostat is turned to its lowest setting, remove the thermostat from the subbase. If the unit stops, replace the thermostat. If the unit continues to operate, remove the red wire from the subbase. If the unit stops, replace the subbase. If the unit continues to operate, check the control cable (E11).

On complaints that the unit will not start even though the thermostat is turned to the highest setting, first check for voltage at the T + T terminals of the protecto relay. With voltage at these terminals, the open is either in the cable (E11) or in the thermostat and/or subbase.

Remove the thermostat from the subbase. Using a jumper wire, jump from R to W on the subbase. If the unit starts, replace the thermostat and subbase. If the unit still does not start, use the voltmeter to measure the voltage across W and R. If there is no voltage, check the control cable (E11). If there is 24 V at this point, use the jumper wire to jump across the T − T terminals at the protecto relay. If the unit does not start, replace the protecto relay (E9).

E11. Open or Short in Control Cable: Where the unit will not operate with the R and W wires at the thermostat fastened together and will operate when the R and W terminals on the protecto relay are jumped, the control cable has an open lead and has to be replaced. When the unit operates continuously with the R and W

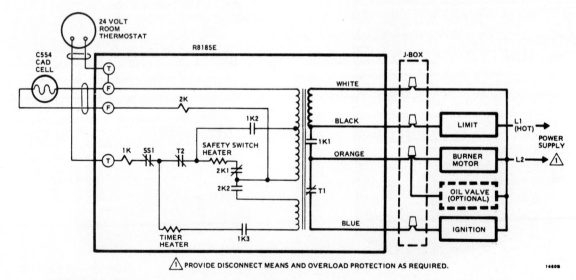

FIGURE A15-20 Internal schematic and hook-up of cad cell type of protecto relay. (*Courtesy* Honeywell, Inc.)

wires removed from the thermostat subbase and the units shuts off when the leads are removed from T – T terminals of the protecto relay, the control cable is internally shorted and must be replaced.

E12. Improper Heat Anticipator Setting: If the heat anticipator setting is too high, too much heat will be generated, causing the thermostat to be satisfied before the area temperature rises to the thermostat setting. If the heat anticipator setting is too low, not enough heat is generated and the area temperature will rise above the setting of the thermostat setting before the unit cycles off.

The anticipator should be set at the amount of current traveling in the control circuit when the unit is operating. Do not try to add all the amperage loads in the control circuit—measure the current. Use a circuit multiplier accessory with your clamp-type ammeter so as to multiply the less-than-1-A current through the circuit to 10 times as high so that it is in the range of your meter. This can also be accomplished by winding a wire around one of the jaws of the meter so that the wire passes through the jaws 10 times. The ends of the wire are then connected across the thermostat leads of the subbase, taking the place of the thermostat. The meter will then read 10 times the actual current through the circuit. The meter reading divided by 10 will be the anticipator setting. A very handy instrument for this purpose is the 0- to 1.2-A-range meter marketed by the T.D. Instrument Company of Rochester, New York 14607.

E13. No Ignition: No ignition means no spark across the electrodes to light the oil spray. If the unit has been operating and develops the no ignition problem, it is usually due to the fact that the unit has developed extensive carbon buildup on the firing assembly. This carbon is a lower-resistance electrical path than the spark gap, and electricity follows the path of least resistance. This problem is an oil-fuel problem—O9. It also could be an electrical problem, covered by E14, E15, E16, or E17.

E14. Improper Electrode Adjustment: To establish the spark necessary to ignite the oil, the electrodes have to be close enough together to present a minimum gap resistance to the 12,000 V supplied by the ignition transformer. To keep the electrodes out of the oil spray and still have the arc flame blown into the oil spray to ignite the oil spray, the electrodes must be high enough above the hole in the nozzle as well as the proper distance ahead of the end of the nozzle. These dimensions will vary with each manufacturer's unit. Figure A15-21 shows the settings of three different burners that have been used on a particular manufacturer's oil-fired unit.

Another dimension is the distance of the end of the nozzle from the end of the firing tube air turbulator.

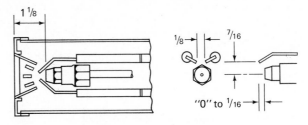

BECKET A-6 BURNER

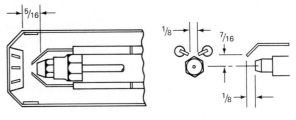

WAYNE MSR-6 BURNER

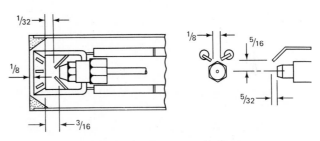

ABC/SUNRAY FC-134 BURNER

FIGURE A15-21 Firing head settings. (*Courtesy* Addison Products Company)

This dimension is important to keep oil spray from impinging on the turbulator and causing carbon buildup and still have the nozzle back far enough to keep the effect of the heat of the fire to a minimum.

You will notice that the three different burners have only one common dimension—the size of the gap between the electrodes—$\frac{1}{8}$ in. The electrode height is the same on the Beckett and Wayne—$\frac{7}{16}$ in.—but only $\frac{5}{10}$ in. on the ABC/Sunray. The forward distance is different on all three. When setting electrode positions, the manufacturer's specifications must be followed.

E15. Cracked Electrode Insulators: Figure A15-22 shows the position of the high-voltage electrodes in the firing assembly. The electrodes are held in a clamp device to ensure remaining in the proper position. Clamped around the ceramic insulator, they hold the electrode and yet insulate the spark voltage from the grounded assembly. The electrode must insulate against 6000 V (one-half the spark voltage) and still stand up against the heat of the burner and combustion chamber when the burner shuts off.

The ceramic insulators are hard and brittle and crack easily. If any twisting or bending pressure is ap-

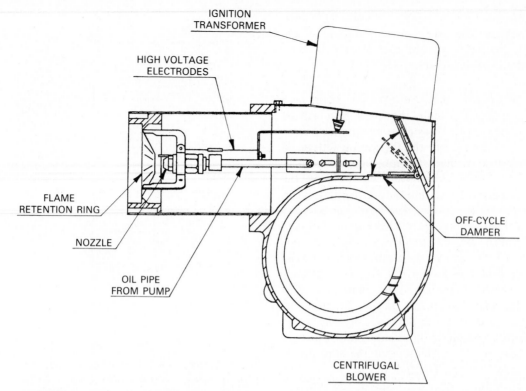

FIGURE A15-22 Oil burner cutaway drawing. (*Courtesy* Wayne Home Equipment Division)

plied to change the position of the electrodes, *loosen the electrode clamps. Do not attempt to bend the electrode wires; they are harder than the ceramic. Also, when tightening the clamps, they should not be overtightened; just drawn up snug.*

If upon examination, any fire cracks or crazing are noted in the surface of the insulators, *replace them;* do not take a chance on the old ones.

E16. Electrode Leads B.O.: If the unit uses copper or brass strips for the electrode leads, check for corrosion at the terminals of the electrodes and where the transformer terminals make contact. If any is noted and cannot be cleaned off, replace the busbars. If high-voltage wire is used, check for cracks in the insulation. Replace if found. If there is any doubt as to the busbar or lead quality, it is better to replace the parts.

E17. Ignition Transformer B.O.: Ignition transformers are 120- to 12,000-V transformers with a grounded center tap on the high-voltage side. With 12,000 V between the terminals and 6000 V from either terminal to ground, attempting to check the transformer by producing a spark with a wire, screwdriver, or other shorting means can be dangerous.

The construction of the transformer is the coils around an iron core in a metal case. The assembly is insulated from the case and the case is sealed with a mastic material, usually tar or petroleum pitch. In time the heat of operation dries this material and cracks develop.

Being in a damp basement, moisture enters the assembly and is absorbed into the windings.

Producing shorts between the windings, the output voltage is lowered to the point of failure of the spark across the ignition gap and faulty ignition results. To check a transformer, the best way is the use of a high-voltage meter in the range 10,000 to 15,000 V.

If this meter is not available, two 120-V voltmeters and a new ignition transformer can be substituted. Figure A15-23 shows the wiring diagram for this test. The secondary of each of the transformers (high-voltage terminals) are connected together to make the new transformer a step-down load of the test transformer. With 120 V applied to the test transformer (measured with one of the voltmeters), the output of the new or testing transformer should be within 10% of the applied voltage. In addition, the output voltage should hold steady. If there is more than a 10% difference or if the output voltage varies, the original transformer has internal shorts and should be replaced.

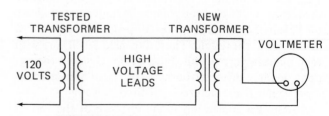

FIGURE A15-23 Transformer test wiring.

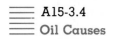
The oil causes are as follows:

O1. No fuel

O2. Improper input

O3. Improper burner adjustment

O4. Improper venting or draft

O5. Blower ft³/min adjustment

O6. Nozzle B.O.

O7. Fuel Unit B.O.

O8. Protecto relay out on safety switch

O9. Carbon deposits on firing head

O10. Heat exchanger heavy with soot

O11. Cracked heat exchanger

O12. Dust odor at fall startup

O13. Oil odor on unit startup

O14. Pulsation

O15. Blower wheel unbalanced or out of line

O16. Blower drive problems

O17. Expansion noises in heat exchanger

O18. Duct expansion and contraction noises

O19. Improper load on blower motor

O20. Improper unit sizing

O1. No Fuel: Do not take the burner apart until you have checked the quantity of oil in the supply tank: the oil gauge on the inside tank or the dip rod for the outside buried tank. The next step is make sure that the tank outlet valve is open.

The third step involves the oil filter in the supply line. Close the tank outlet valve, open the filter cartridge case, and put in a new filter cartridge; 99% of the time one is needed. Be sure to bleed the air from the filter cartridge after assembly.

If this procedure still does not supply oil to the burner and produce fire, remove the oil burner nozzle and check the nozzle filter. If this is plugged, replace the entire nozzle assemble with one of like capacity, spray angle, and cone type.

If these steps do not produce oil flow when the unit runs, check the inlet screen of the fuel unit. Figure A15-24 shows a Sundstrand single-stage fuel unit cut away. Call out No. 3 indicates the fine mesh screen that filters the oil supply before it enters the pump gear assembly. This screen can be removed, cleaned in fresh fuel oil, and replaced. *Do not leave this screen out of the pump.*

After any of the portions of the fuel supply system have been opened, the system must be purged of air. A two-pipe system will automatically purge itself of air. The single-pipe gravity feed type system does not have

FIGURE A15-24 Fuel unit cutaway. (*Courtesy* Suntec Industries, Inc., Rockford, Ill.)

an automatic purge feature. Therefore, purging must be done; call out No. 8 in Fig. A15-23 is the bleed valve for this purpose. With a short piece of plastic hose from the valve to a suitable container and the valve open, operate the unit until a clean stream of fuel oil emits from the hose. It may be necessary to reset the protecto-relay safety switch several times before the supply line is completely purged.

If it is not possible to obtain a flow of clear fuel oil, it is possible that the unit is receiving air through a line or fitting leak. This must be corrected to provide proper burner operation.

O2. Improper Input: Section A14-3 gives a complete discussion of the determination of the Btu/hr input of an oil burner. If the unit is not fired at the proper input, control operation problems, as well as operation odors and heat exchanger failures, can be encountered. The amount of oil burned must produce the Btu/hr input within 90 to 100% of the rating of the unit. *Do not attempt to operate the unit beyond these amounts.*

O3. Improper Burner Adjustment: The correct fuel unit pressure, the correct overfire draft, and the correct CO_2 must be obtained. Efficiency testing is covered in Section A14-3.4. The procedures outlined in this section should be used to adjust the unit for peak efficiency.

O4. Improper Venting or Draft: For the unit to start and operate with the minimum of odor in the unit area as well as operate efficiently, the flue vent or chimney must have sufficient "natural" draft to be able to create a negative pressure of at least −0.06 in. of W.C. in the stack when the unit is operating at an overfire draft or negative pressure of −0.04 in. of W.C. This means that the draft produced by the flue or chimney when the unit is off and the vent is cold must be −0.02 in. of W.C. or better.

Some oil-fired units are connected to short vent stacks through the roof. An example of this is a suspended oil-fired unit hanging from the ceiling, very common in industrial applications. The problem here is that the vent does not produce any negative pressure in the unit. When the unit starts and the first rush of combustion pressure occurs, the gases do not flow through the heat exchanger fast enough to prevent a positive-pressure buildup. When the positive pressure occurs, the combustion air quantity is reduced because of the pressure against the output of the combustion air blower. This reduces combustion and the positive pressure causes oil fumes to emit from the observation door into the unit area.

If the performance of the vent or chimney cannot be improved, a continuous-operating (when the season switch on the thermostat calls for heat) induced-draft fan must be used to ensure proper negative draft upon unit startup.

O5. Blower Ft³/Min Adjustment: The correct adjustment of the air through the unit will result in a temperature rise between return and supply air of 80°F (26.7°C). For the correct method of setting the ft³/min through the unit, refer to Section A14.2.

O.6 Nozzle B.O.: A nozzle that is partially plugged—one or more of the swirl grooves are plugged—will give a one-sided flame that will produce very poor combustion. It could even cause carbon buildup in the refractory. If a cleaning procedure is attempted, the hardest

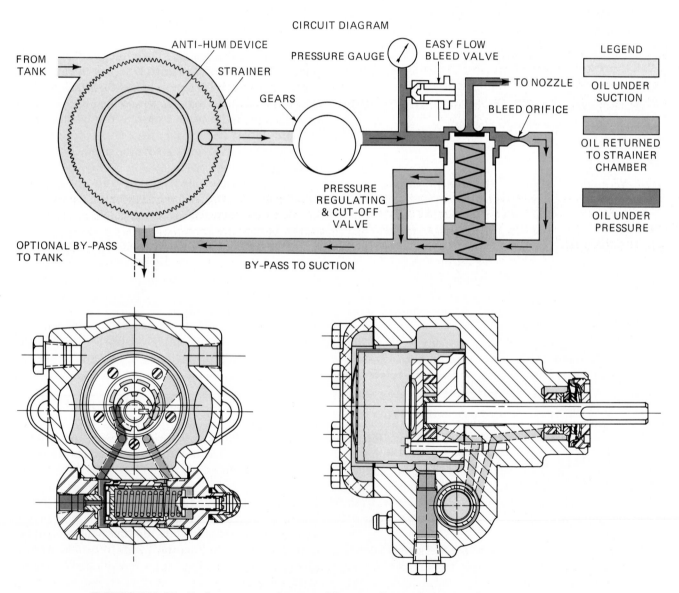

FIGURE A15-25   Fuel unit circuit diagram. (*Courtesy* Suntec Industries, Inc., Rockford, Ill.)

material that can be used to clean the swirl tubes is a wooden toothpick. Better to discard the nozzle and put in a new one.

It is also possible for the nozzle swirl grooves and orifice to wear from the passage of oil through the nozzle. Therefore, the nozzle should be replaced at least every two heating seasons, and in extremely cold climates, at the beginning of each heating season.

The cost of a new nozzle is small compared to the cost of extra fuel if the unit is not operating at peak efficiency. For nozzle selection as well as type, spray angle, and gal/hr capacity, refer to Section A14-3.2.

O7. Fuel Unit B.O.: Section A14.6 has a complete description of the operation of the fuel unit. The most common problem in fuel units is poor cutoff of the oil supply when the unit cycles off. Correct cutoff will provide instantaneous cutoff of the oil to the fuel pipe and nozzle when the pump pressure drops to 80% of the operating pressure. At 100 psig operating pressure the cutoff pressure is 80 psig. This is accomplished by the spring forcing the control piston in the fuel unit against the fuel outlet seat and cutting off the flow of oil. The nozzle pressure drops immediately (see Fig. A15-25).

If a particle buildup occurs on the face of the neoprene seat on the end of the piston, this prevents good full circle contact of the neoprene disk against the seat. Instead of positive cutoff, leakage occurs and the pressure gradually reduces in the nozzle. This gradual reduction of pressure causes oil flow from the nozzle after the unit stops and the air supply disappears. The oil now burns with very little combustion air, producing a very smoky flame.

Carbon builds up on the turbulator end of the firing tube as well as the firing assembly. If the burner is not slanted at least 2° downward toward the combustion chamber, burning oil can flow back toward the blower. This can burn or smoke up the cad cell, with resulting cutoff of the burner on the safety switch. The correction for this is to clean the piston chamber of the fuel unit as well as the intake screen.

A persistent case of this problem due to the quality of fuel oil supply can be reduced by the use of double filtering of the oil before it reaches the unit and the use of a delayed action oil valve (see Fig. A15-26). Located in the fuel line between the outlet of the fuel unit and the firing head, this valve provides a time delay between the time the blower starts and the start of oil spray. This makes sure that air for combustion as well as airflow through the heat exchanger is established before combustion starts. The valve also provides instant cutoff of oil pressure to the nozzle regardless of the action of the fuel unit. This is a highly recommended accessory for any oil-fired unit.

O8. Protecto Relay Out on Safety Switch: The safety switch cut out because the safety switch heater element

FIGURE A15-26 Delayed-action oil valve. (*Courtesy* Suntec Industries, Inc., Rockford, Ill.)

was not deenergized properly. The flame did not come on (O1). Ignition did not take place (E13), the helix element in the stack-mounted relay did not function, or the cad cell is coated with dirt or carbon (O9). Any of these can cause the safety switch to cut off.

When the unit is found to be out on the safety switch, the customer should be questioned as to how many times the switch had been reset. Multiple resetting of the switch could result in considerable oil being sprayed into the refractory. Check and remove as much oil as possible from the bottom of the refractory before attempting to light the burner. Keep a fire extinguisher handy. The firing head should also be removed, cleaned, the orifice removed and checked, and the firing head assembled and properly set before attempting the relight.

O9. Carbon Deposits on Firing Head: Carbon deposits on the firing head can be caused by poor pressure cutoff of the fuel unit (O7), poor draft (O4), overfiring of the unit (O2), or improper positioning of the ignition electrodes (E14).

The first step to correct the problem is to remove and clean the firing head. To do this the firing head must be disassembled. After all parts are cleaned, the firing head is assembled making sure that the electrodes are properly positioned: 1 above the nozzle outlet, 2 ahead of the nozzle, 3 the distance between electrode tips, and 4 the distance of the nozzle from the turbulator. With the firing head in place and a 0- to 150-psig gauge connected to the gauge port of the fuel unit, operate the burner to check and set the operating pressure. Most burners require 100 psig operating pressure. With this pressure set, stop the burner; the gauge pressure should drop immediately to 80 psig and hold there for several seconds. If it does not hold, the fuel unit must be removed and the pressure regulation assembly removed and cleaned. See O7 and Section A14.6.

With the burner operating, check the overfire and stack draft. For proper operation, the overfire draft must be between −0.2 and −0.04 in. of W.C. Any positive pressure will cause poor combustion, high carbon formation, and promote carbon buildup on the firing head (O4).

O10. Heat Exchanger Heavy with Soot: The advantage of checking both overfire and stack draft is that this is an easy way to check the conditon of the heat exchanger for soot buildup. As the flue passages fill with soot, the flow resistance increases. This means that a deeper negative pressure is required in the stack to produce the desired negative pressure over the fire.

Most oil-fired heat exchangers have flue gas flow resistance of 0.015 to 0.025 in. of W.C. This means that with 0-0.04 in. of W.C. overfire draft, the stack draft will be in the range −0.055 to −0.065 in. of W.C. When the draft regulator is set to produce an overfire draft 0.04, if the stack draft has to be–0.09 or higher, the heat exchanger has to be cleaned.

To prevent damage to the heat exchanger, mechanical cleaning of the flue passages is a must. Use a vacuum cleaner and brushes. *Do not use a chemical cleaner on a heat exchanger instead of a vacuum cleaner and brushes.*

After the mechanical cleaning procedure has removed 90% of the carbon, chemical cleaners can be used to finish the job. Chemical cleaners work by causing the carbon to burn and leave the heat exchanger surface; the action starts where the cleaner first encounters the carbon. As the burning process proceeds through the flue passages, the burning increases. By the time it reaches the outlet of the heat exchanger it can be intense enough to melt the heat exchanger. Remove the soot first by means of the vacuum cleaner and then chemically clean.

O11. Cracked Heat Exchanger: All heating units must be designed with the heat exchanger on the positive-pressure side of the circulating blower. This is a safety requirement to prevent drawing flue products into the circulating air in the event of heat exchanger failure.

Heat exchanger failure can be caused by metal failure because of defective metal used in the fabricating process, poor welding procedure, or overfiring. The most common cause is overfiring. When the unit is overfired, the combustion chamber cannot hold all of the flame produced, and contact of the flame to the wall of the heat exchanger above the refractory will result. The flame contact on the metal wall will cause the metal to overheat and burn away, leaving a hole.

Heat exchanger failure will first be detected by an oil fume smell in the heated area when the circulating blower starts. When the burner starts and before the cir-

culating blower starts, the pressure buildup at the time of ignition will force oil fumes into the unit heat exchanger compartment. These will then be carried to the heated area by the circulating air.

In addition, the overfire draft will change. With the blower off, a correct overfire draft can be set. However, when the blower starts, the air forced into the heat exchanger through the hole will reduce the overfire draft and may even cause fumes from the observation door because the pressure builds to a positive amount.

After replacing the heat exchanger, *do not attempt to patch it.* A complete check of input, proper nozzle operation, and efficiency adjustment must be made.

O12. Dust Odor at Fall Startup: A common cause of odor complaint at the beginning of the heating season is the burning off of the dust that has accumulated on the heat exchanger during the summer months. If the unit has a summer fan switch and is used for summer ventilation, this problem can be encountered. The odor is temporary and usually lasts only one or two operating cycles of the unit.

Another type of odor is encountered at the startup of a new unit. All heat exchangers are treated for rust prevention during storage either by an oil coat or a paint coat. At the first firing, a light smoke will come off the heat exchanger surfaces. Units should be test fired after installation with heavy ventilation provided in the occupied area. This odor will usually disappear in two or three cycles of the unit.

O13. Oil Odor on Unit Startup: This has been covered in O4 and O11.

O14. Pulsation: Pulsation of an oil-fired unit is caused by a poor draft when the unit starts. When the unit is standing idle (cold) the vent or chimney should have sufficient pull to maintain air movement through the heat exchanger. On burners with close-off dampers to reduce air movement, the vent or chimney should produce a negative pressure in the heat exchanger. If the vent or chimney does not create this negative pressure, the following will occur:

When the burner starts, the correct amount of air and oil will be supplied for correct combustion. When ignition occurs, a high pressure will be produced in the combustion chamber. This will build to force the flue products through the heat exchanger. The high pressure on the outlet of the burner will reduce the ft³/min of combustion air supplied to the flame. The combustion rate is reduced and oil vapor will rise in the heat exchanger. When the pressure is relieved by the flue products starting through the heat exchanger, the amount of air increases. The extra fuel will then burn, creating a surge of pressure again. This reduces the amount of air, starting the cycle again. This cycling of pressure in the heat exchanger can continue until proper draft is established and can be severe enough to shake the heating unit. The cure for pulsation is proper draft and a correctly set burner.

If the venting system will not maintain the proper OFF-cycle draft conditions, power venting must be used. Power venting uses a mechanical device such as a motor-driven fan or blower to draw the flue products from the heat exchanger. Located between the stack draft control and the vent or chimney, the power ventor creates a negative pressure in the flue pipe. These devices will have more pull then needed, but the draft control will provide the air necessary to satisfy the extra air requirement of the power ventor and keep the heat exchanger pressure in the correct range.

Power ventors should operate continuously during the heating season because their effect is the most important at the time of unit startup and before the heat exchanger has become hot enough to establish its own draft.

Pulsation problems are more prevalent with ceiling-mounted units with short vent pipes and units in utility rooms of one-story-slab or crawl-space homes with the chimney part of a masonry fireplace setup. Because these situations are slow to build temperature for increased draft, they are pulsation prone. Also, pulsation problems are higher during the mild part of the heating season—spring and fall—because the unit does not run long or often enough to establish good vent or chimney temperatures.

O15. Blower Wheel Unbalanced or Out of Line:
Usually, the blower wheel is balanced and aligned before being installed in the original unit. However, shipping damage, such as dropping the unit from the tailgate of a truck, can cause the wheel to strike the housing and become warped. It then will not run true, and excessive vibration will result. The only correction is replacement. The only prevention is careful handling of the unit.

Blower vibration can develop after the unit has been operating for some time. The action of the blower wheel blades will cause dust and dirt to build on the leading surfaces of the blades. In new construction, sawdust is the chief source of this buildup. If the amount of buildup reaches the point where some of the buildup leaves the wheel, an imbalance occurs and vibration results.

The most common cause of wheel vibration is a lazy serviceperson or homeowner who tries to clean the wheel with a brush without disassembling the blower. This will always lead to excessive vibration. Never touch a blower wheel unless the intent is to completely disassemble the blower assembly and gives the wheel a thorough cleaning. Immersion in detergent and water or jet-type cleaning equipment is required. When the unit is assembled, be sure that the blower wheel turns in the right direction. Wheels in backward deliver only approximately 25% of the amount of air that is needed.

O16. Blower Drive Problems: Belt-driven blowers have a higher probability of vibration problems than direct-drive blowers due to the additional parts involved. The most common problem is due to belt tension. Most people believe that the tighter the belt, the better the performance. The opposite is actually the case. The tighter the belt, the harder the motor has to work to get the belt in and out of the pulleys, the slower the blower turns, and the shorter the life of the blower and motor bearings. Therefore, the belt should be as loose as possible without slippage on startup.

Figure A15-27 gives the test for proper belt tension. You should be able to easily depress the belt midway between the motor and blower shafts $\frac{3}{4}$ to 1 in. for each 12 in. of distance between the shafts. Alignment of the motor and blower pulleys is important to keep vibration to a minimum as well as to reduce wear on the sides of the belt.

Finally, each pulley, both motor and blower, should be checked for running true. Any warpage that creates wobble in the pulley requires replacement of the pulley.

O17. Expansion Noises in Heat Exchanger: Figure A15-28 shows a typical oil-fired heat exchanger with large flat surfaces and tubular surfaces welded together. Two types of expansion noises can result. When the seam welding is done, it is possible to produce different rates of expansion in the two surfaces during the welding process. This results in expansion noises, ticking, and popping, as the heat exchanger heats and cools. Most of the time these noises are muffled by the unit casing and duct system to a level where they are not objectionable.

The other expansion noise that can occur is called "oil-can effect," which is the sudden movement of a flat metal surface where a forming stress has been left in the surface during the assembly process. This stress causes the metal to have a slightly concave or convex position rather than a flat surface.

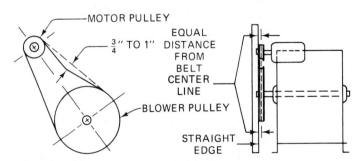

CHECKING BELT TENSION

CHECKING PULLEY ALIGNMENT

FIGURE A15-27 Align pulleys and tighten belt. (*Courtesy* Borg-Warner Central Environmental Systems, Inc.)

FIGURE A15-28   Oil-fired heat exchanger.

Temperature change will cause a stress increase in the material until the metal rapidly changes position to the opposite of its original position. This change will produce a loud "bang" heard throughout the area.

The large flat surfaces in drum-and-radiator heat exchangers should be cross-broken to prevent the oil-can effect. If they can be removed and cross-broken, the problem is usually eliminated. If they cannot be removed, the heat exchanger will have to be replaced.

O18. Duct Expansion and Contraction Noises: Any metal duct will have a certain amount of expansion or contraction with temperature change. Aluminum ducts have considerably more expansion per degree of temperature change than do galvanized iron ducts and will be more prone to expansion and contraction noises.

The main cause of duct noise is too solid a mount-

ing means. Ductwork nailed directly to floor joints will be extremely noisy, as it cannot move with temperature change. It will also be more prone to causing fires because the hot duct is in direct contact with the wood floor joints.

Properly installed ductwork is suspended 1 in. below the floor joints. Metal straps are used, fastened to the floor joints and extended down under the duct. The ductwork is hung in the form of a trapeze, which allows free movement of the ductwork without strain.

O19. Improper Load on Blower Motor: The amperage draw of the blower motor goes up much faster than the increase in load. Therefore, a small increase in the blower load can cause a larger increase in current consumption. It can also cause higher-than-normal motor temperatures and shorter motor life.

Measure the amperage draw of the motor under normal load. It should not exceed the amperage rating of the motor. If it has and the load cannot be reduced, a larger motor will have to be substituted. Even with a larger motor handling the load, the wattage used will usually be less than the smaller motor operating in an overloaded condition.

O20. Improper Unit Sizing: For many years if the unit requirement was 100,000 Btu/hr, the owner would demand a 150,000-Btu/hr unit in the belief that the larger the unit, the better ("the unit did not have to work as hard to do the job"). This results in a heavy waste of fuel and high operating cost. We now recognize that the unit should be properly sized for peak efficiency and lowest operating cost.

The actual heat loss of the structure should be calculated using the latest manuals in the Environmental Systems Library put out by the Air Conditioning Contractors of America (ACCA). These manuals are constantly updated and give the most accurate results. The unit should then be sized within ± 10% of this requirement.

═ PROBLEMS ═

A15-1.   What is the first item to check if the heating unit will not operate?

A15-2.   What is the first item to check if there is no heat because the gas valve will not open?

A15-3.   What would be the easiest way to check for an open circuit in the thermostat or subbase?

A15-4.   There is voltage at the gas valve terminals but the valve will not open. What should the next test be?

A15-5.   The room temperature is much higher or lower than the thermostat setting. What should be checked?

A15-6.   The LP gas branch line supply to an outside heating unit should always be taken off the top of the main-line. Why?

A15-7.   The average thermocouple used in gas-fired heating units should develop between _____ and _____ mV.

A15-8.   To reduce the possibility of water, sediment, rust, etc., from the supply system causing damage, a device called a _____ must be installed in the piping ahead of the unit.

A15-9. Fire in the main burner or control/compartment is usually caused by _____ .

A15-10. An all-blue sharp fire is caused by _____ .

A15-11. The effect of an all-blue sharp fire is _____ .

A15-12. What is the chief cause of burner extinction pop?

A15-13. The input tolerance for a gas-fired heating unit is +_____ % to - _____ %.

A15-14. A major cause of fan short cycling is _____ .

A15-15. Two causes of the gas valve short cycling even though the fan runs steady are _____ and _____ .

A15-16. The proper fan control settings are _____ to°F and _____ to _____ °F off.

A15-17. The first indication of dirty air filters is _____ .

A15-18. What is the easiest way to check for proper draft at the diverter?

A15-19. The main cause of the rusting out of heat exchangers is chlorine or fluorine in the combustion air. What two common products are a source of these?

A15-20. What three methods can be used to cure resonance or pipe organ effect?

A15-21. What kind of control device does a stack-mounted protecto relay use to operate the burner sequence?

A15-22. What kind of control device does a burner-mounted protecto relay use to operate the burner sequence?

A15-23. When called on to fix a burner that fails to keep operating, what is the first item to check?

A15-24. With 120-V power to the oil burner and the thermostat turned as high as it will go, if an oil burner will not run, what is the easiest way to check the thermostat circuit?

A15-25. What is the most common cause of ignition failure?

A15-26. The easiest way to adjust the electrodes is to bend the wires. True or False?

A15-27. What is the most common cause of lack of fuel supply to the oil burner?

A15-28. The input tolerance of an oil-fired heating unit is _____ to _____ % of rated input.

A15-29. When at ambient temperature, a flue or chimney must be able to produce a draft of _____ or better.

A15-30. Oil burner nozzles are cleanable with a wire brush. True or False?

A15-31. What causes carbon buildup in the turbulator end of a firing tube?

A15-32. Chemical cleaners may be used to clean a heat exchanger heavily filled with soot. True or False?

A15-33. If oil fumes are encountered in the occupied area after the unit starts, what is the first thing that should be checked?

A15-34. What is the most common cause of pulsation when an oil-fired unit starts?

A16

Air-Conditioning Measuring and Testing Equipment

A16-1
GENERAL

In Chapter R22, the instruments and procedures that apply to the refrigeration part of the system were covered. Whether the system is applied to the cooling of a dairy case, a frozen-food cabinet, or an air-conditioning system, the refrigeration systems require the same test instruments and equipment to perform service procedures. The air-conditioning system is more dependent on the proper movement of air, and therefore the instruments used in this phase are very important. Some of these instruments were mentioned in Chapter A4. We will, however, cover them in more detail in this chapter.

A16-2
TEMPERATURE MEASUREMENT

To determine if the correct load is being applied to the DX coil in the air-conditioning system, the drop in temperature of the air through the coil must be measured. To determine if the correct amount of refrigerant is in the system, the temperature of the liquid leaving the condenser must be measured. To determine the capacity of the unit, the temperature of the air entering and leaving the condensing unit must be measured. These all require the use of accurate thermometers.

In the past few years, the dial-type thermometer has replaced the glass type for pocket-carrying convenience. It also has a carrying case with a pocket clip (Fig. A16-1). The dial thermometer is more convenient or more practical to use in measuring air temperatures in a duct or strapped to the liquid line of the unit. In air-conditioning service the most popular range is −40 to +160°F. A range of +25 to +125°F is usable for air-conditioning cycle operation, but would be too low for the heating cycle operation.

A more accurate and easier-to-read thermometer is the solid-state or digital type. Using large red LED readout digits for the dial and thermister-type probes for sampling, temperatures can be read within 0.5%, plus or minus one digit. A four-probe scale of this type is pictured in Fig. A16-2. With a range scale of −60 to +300°F, this will cover the range used in most heating and air-conditioning service procedures, with the exception of stack temperatures in the heating unit. A thermocouple type with a range of 0°F (0°C) to 1200°F (340°C) is used for this purpose (see Fig. A16-3).

The dry-bulb and relative-humidity conditions of the air entering the DX coil are required to determine the correct temperature drop of the air through the coil. To measure the dry- and wet-bulb temperatures, a sling psychrometer is used. Two glass-type thermometers, one covered with a wetted cloth sack, are used to take these samples. Figure A16-4 pictures such a sling psychrometer and case. Between the thermometers at the right

FIGURE A16-1 Pocket-type dial thermometer. (*Courtesy* Robinair Manufacturing Corporation)

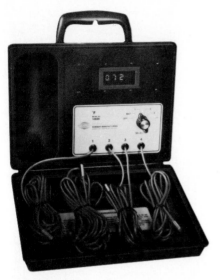

FIGURE A16-2 Four-probe digital thermometer. (*Courtesy* Robinaire Manufacturing Corporation)

FIGURE A16-3 Thermocouple temperature tester. (*Courtesy* Robinaire Manufacturing Corporation)

end is the pivot handle. Using this handle with the sling psychrometer hanging at a right angle to the handle and the sack saturated with water, dry-bulb and wet-bulb samples are obtained. The sling psychrometer is rotated through the air to be tested at approximately 120 rpm. This is 2 revolutions per second, and this rotation is continued until both thermometer readings repeat several times. The thermometers are in balance with the respective temperatures of the air. After the dry- and wet-bulb temperatures have been obtained, refer to Chapter A3 to determine the properties of the air tested.

A16-3
PRESSURE MEASUREMENT

In Chapter A3, the pressures involved in air handling through the supply and return system were discussed. Pressure gauges in the range 0 to 1.5 in. of W.C.,

called manometers, were discussed. Manometers such as that shown in Fig. A16-5 are called incline manometers. The reason for the incline of the pressure-indicating fluid is to spread out the scale into larger spaces, for more accurate reading. This instrument has a range of 0 to 1 in. of W.C. spread out over a scale 8 in. in length. Some instruments use a scale 12 in. in length for extreme accuracy.

In addition to wide range accuracy in the lower-pressure ranges, it may be desirable to measure higher pressures up to 3 in. Figure A16-6 shows a combination range instrument with an inclined range of −0.05 to 3 in. of W.C. and a range of 0.3 to 3 in. of W.C. in a curved tube to vertical. These are applicable to commercial and industrial systems where higher static pressures are used to move more air through smaller ducts.

Because the manometer can be used to read velocity pressure through the use of a pilot tube (see Chapter A3), the gauge can also be calibrated to read air velocity

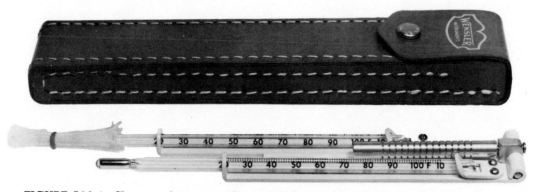

FIGURE A16-4 Sling psychrometer. (*Courtesy* Robinaire Manufacturing, Corporation)

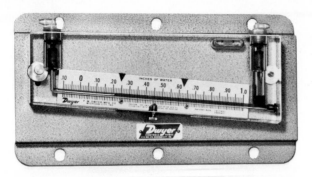

FIGURE A16-5 Inclined manometer. (*Courtesy* Dwyer Instruments, Inc.)

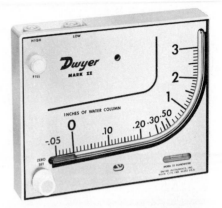

b

FIGURE A16-6 Wide-range manometer. (*Courtesy* Dwyer Instruments, Inc.)

through a duct in feet per minute. This velocity, multiplied by the area of the duct in square feet, will give the ft³/min through the duct.

A16-4
QUANTITY MEASUREMENT

In some instances, the quantity of air from supply outlets or into return grills must be measured in order to balance the system to specifications. The most common of the instruments used for this measurement is the anemometer. Shown in Fig. A16-7, it consists of propellers on a shaft that revolves when held in the airstream from a supply outlet or a return air grill. In the center is a dial that reads in cubic feet. A stopwatch is used for timing. If the instrument is allowed to operate for 1 minute, the total reading on the dial will be the number of cubic feet of air per minute. Some situations

FIGURE A16-7

FIGURE A16-8 Flow hood air quantity meter. (*Courtesy* Alnor Instruments Company)

might require the anemometer to run longer than 1 minute. Longer measurements will provide averages that will be more accurate.

The Flow Hood unit is a direct-reading air quantity instrument. This instrument, shown in Fig. A16-8, reads air quantity directly in cubic feet per minute because it directs all the air via an air-gathering hood through an opening size in the instrument. The meter is then calibrated to read air directly in ft³/min.

A16-5
DIMENSION MEASUREMENT

The size of the duct in square feet, the length of refrigerant lines, the clearance height over the coil, etc., are all necessary in the installation and servicing of air-conditioning equipment. Each service technician must therefore be equipped with some sort of distance device. The most common of these is the steel tape automatic-return case type. It is protected in the case, yet available in easy, flexible steel tape form. The most practical retractable type is the 25-ft length.

The instruments discussed in Chapter R22 should be reviewed before beginning the following chapters on air-conditioning troubleshooting.

PROBLEMS

A16-1. Why is measurement of the temperature drop of the air through the DX coil important?

A16-2. The subcooling of the liquid refrigerant off the condenser must be measured to determine if the _____ _____ is correct.

A16-3. To measure the conditions of the return air to the coil, a _____ is used.

A16-4. The reason for the incline of the pressure-indicating fluid and scale in an inclined manometer is to _____ _____ .

A16-5. The instruments that can be used to measure the air quantity from a register or grill are called an _____ and a _____ .

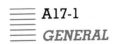

Air-Conditioning Startup, Checkout, and Operation

A17-1
GENERAL

The sole purpose of proper startup, check-out, and operation is to have the air-conditioning unit produce the desired conditions of comfort at the lowest operating cost. This means that the correct amount of air (the cooling load) must be supplied across the DX coil, and the refrigeration system must produce peak efficiency.

An air-conditioning unit only transfers heat from one place and puts it in another place by the change in state of a refrigerant. The load on the coil must be correct depending on the temperature and moisture content of the air to be cooled. This means that the amount of air over the DX coil must be correct.

The refrigeration system must be capable of transferring the desired amount of heat as well as eliminating all added heat from the electricity used. Each component part as well as refrigerant quantity must be correct and functioning properly. With correct ft³/min over the coil and the system operating properly, the unit will then deliver its design capacity. To determine if the unit is operating correctly, each phase of the operation is checked in turn.

Specific information must be obtained at the time of startup, during check-out of the unit, and to determine if the system is operating at peak efficiency. Following is a suggested list of information needed for startup, check-out, and efficiency testing.

1. Condensing unit model no. _____
2. Condensing unit serial no. _____
3. Installation date _____
4. Type of pressure-reducing device
   a. Expansion valve _____
   b. Capillary tube _____
5. Type of fuses _____ Size _____ amperes

6. Wire size to condensing unit _____
7. Line voltage with condensing unit off _____ volts
8. Condition of air filter _____
9. Voltage at condensing unit when trying to start _____
10. Amperage draw of unit at:
    a. Startup _____
    b. Running _____
11. Return air temperature and relative humidity at return air grill _____ °F _____ % RH
12. Return air temperature at unit _____ °F
13. Supply air temperature at unit _____ °F
14. Air temperature drop across DX coil _____ °F
15. Supply air temperature at supply register _____ °F
16. Suction-line temperature at DX coil _____ °F
17. Compressor suction pressure (DX coil evaporating temperature) _____ psig _____ °F
18. DX coil superheat _____ °F
19. Suction-line temperature at compressor _____ °F
20. Temperature of air entering condenser _____ °F
21. Compressor discharge pressure (condensing temperature) _____ °F
22. Condenser split _____
23. Liquid-line temperature _____
24. Liquid subcooling _____ °F
25. Condenser discharge air temperature _____ °F
26. ft³/min through condensing unit _____

TEST BEFORE STARTING UNIT

Items 1 through 8 are used before starting the unit.

1. For the service history file, the model number (1), serial number (2), and installation date (3) should be recorded. Any question of possible warranty, as well as service history, can then be taken into account. Is it a new unit with initial startup, or has it been in operation for a period of time?

4. TK valve systems operate differently regarding pressures and temperatures of the refrigerant. This must be considered.

5. Type of fuse protection—straight fuse or circuit breaker. Time-delay fuses or circuit breakers must be used on high-starting-current motors. If they are not of the time-delay type, they must be changed before the unit is put into operation. If not, unnecessary cutouts will occur. Are the time-delay fuses or circuit breakers the proper size? Condensing units carry a fuse size stamping on the rating plate. Fuses smaller than the rating specified will cause unnecessary cutouts. Fuses or circuit breakers larger than specified will not provide the protection to prevent electrical damage.

6. Wire size to condensing unit. The wire supplying power to the condensing unit must be of sufficient size to prevent excessive voltage drop when the unit attempts to start. If the wire is oversized, usually circuit breaker protection is also oversized, and this reduces unit protection as well as causes an unnecessary increase in installation cost. Undersized wires will cause excessive voltage drop and unit starting problems.

7. Line voltage with condensing unit off. The voltage at the unit with the unit off (no-load voltage) must be within the voltage rating range of the unit. This is plus or minus 10% of the unit rated voltage if the unit is single phase. If the rated voltage is 230 V, the maximum would be 253 V and the minimum would be 207 V. Units rated at 240 V have a range of 264 V on the high side and 216 V on the low side. These voltages are the extremes that the unit will tolerate and still continue to operate. This does not mean that the electrical components will have maximum design life. To accomplish this, the supply voltage must be at unit rated voltage. In three-phase dual-voltage units, the voltage tolerance range is plus 10% minus 5%. This means that on a wide-range 208/230-V unit, the maximum is 230 V plus 10%, or 253 V, and 208 V minus 5%, or 197.6 V.

8. Condition of the air filter. The amount of air through the DX coil seriously affects the performance of the system, as we discuss later in steps 11 through 19. Therefore, the air filter should be checked before startup to eliminate the possibility of test error. Although they may not look like a carpet from the lint and dirt on the surface, it is wise to replace throwaway filters at the beginning of each season—spring for the cooling season and fall for the heating season. If the system has an electronic air filter, it should be cleaned at least twice a year and inspected every month. Some areas, such as the dusty Southwest, require a minimum of six cleanings per year.

TEST ON STARTING UNIT

Two electrical checks should be made on starting the unit. A voltmeter should be connected across the high-voltage terminals on the line side of the contactor and a clamp-type ampmeter connected around one of the wires to the contactor.

9. Voltage at the condensing unit when it is trying to start. When the starting load is cut onto the line, the contactor closes, bringing on the compressor and condenser fan load. The voltage should not

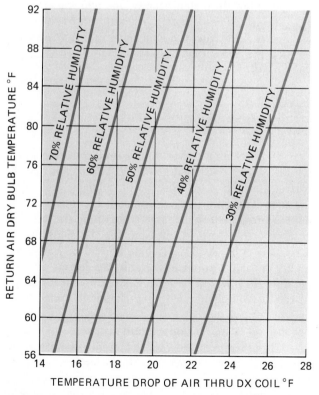

FIGURE A17-1 Air temperature drop for various load ratios.

drop more than 5 V. If it does, either one of the following situations could be the problem:

a. The wire size could be too small for the length installed. Most units will give the wire size in ampacity (amperage capacity) based on a wire length of 60 to 100 ft between the distribution panel and the unit.

b. The building distribution service size could be too small. The startup test should be repeated, measuring the voltage at the circuit breaker in the distribution panel. If the starting voltage drop at the panel is less than 5 V when the drop is more than 5 V at the unit, the branch circuit to the unit is too small. If the drop in starting voltage at the circuit breaker is also more than 5 V, either the service drop to the building is too small or the distribution power transformer is too small or overloaded. For this problem, the utility supplying the power must be contacted.

10. Amperage draw of unit at (a) startup, and (b) running. Any time an induction load (motor) is started, there will be a high inrush current (LPA) and within 1 second this current should drop to the full-load amperes (FLA). The ammeter should be watched during the startup process to make sure that everything is correct and working properly. The unit should draw the proper inrush and running current to remain on the line. If the current does not drop significantly, *the unit must be disconnected immediately.* The unit locked rotor amperes (LRA) and full-load amperes (FLA) are given on the rating plate.

A17-4
DX COIL CFM

To check an air-conditioning unit, *the correct load on the DX coil must be established.* This means that the correct amount of air must go through the coil. With the correct ft³/min through the coil, the unit will produce the desired drop in temperature of the air. It is only necessary, therefore, to measure the ΔT°F of the air across the coils. At 80° dry bulb and 50% relative humidity in the return air to the unit, the correct air temperature drop would be 20°F. Because the heat content of the air will change with any change in temperature and/or relative humidity, the correct amount of air will produce different air temperature drops.

If the temperature and/or humidity increases, the temperature drop will decrease. If the temperature and/or humidity decreases, the air temperature drop will increase. The system will remove only a given amount of heat. Therefore, the greater amount of heat in the air,

the less heat will be removed per cubic foot. Because removal of humidity (latent load) does not affect temperature, the relative humidity change will have the greater affect on the air temperature drop. To determine the correct load on the coil (the correct ft³/min), the correct temperature drop must be determined. This can be determined by the use of a psychrometric chart and a load ratio scale.

The chart in Fig. A17-1 has been plotted for various conditions of air. In this chart a range of relative humidity from 70 to 30% and dry-bulb temperatures from 65 to 92°F are given. Using a sling psychrometer, the dry- and wet-bulb temperatures of the air are taken. The relative humidity is then determined from the psychrometric chart. The psychrometric chart was explained in detail in Chapter A3. Figure A17-2 is the psychrometric chart we will use in our examples.

We will assume that the area to be cooled, at startup of the unit, has a measured dry-bulb temperature of 88°F and a wet-bulb temperature of 79°F. We want to know the correct air temperature drop of the air through the coil. Using the psychrometric chart, we find that the vertical 88°F dry-bulb line and the slope 79°F wet-bulb line intersect at 68% relative humidity.

Using the chart in Fig. A17-1, when we follow from the point where the horizontal 80°F dry-bulb line and the slope 68% relative humidity line cross, down to the temperature drop scale, we find that an air temperature drop of 17°F is desired. The desired conditions we want to maintain would be 80°F dry bulb and 50% relative humidity. These are the conditions at which units are tested at ARI. Air at these conditions would require an air temperature drop of 20°F. Assuming that the conditions of the air are 88°F and 40% relative humidity, the lack of latent load would force the system to cool the air further, taking more sensible heat, and would increase the air temperature drop to 24.1°F.

From this we see that the higher the heat content of the air, especially the latent heat content, the lower the air temperature drop produced. Conversely, the drier the air, the higher the air temperature drop. You must realize that the air temperature drop produced varies with the heat content of the air. A 20°F air temperature drop is correct only at 80° dry bulb and 50% relative humidity. Therefore, before you can properly check an air-conditioning unit, you must set the air quantity for the correct air temperature drop. To determine the correct temperature drop, the sling psychrometer and psychrometric chart must be used. At this stage, any problem can be classified as either an air system or refrigeration system problem.

Assuming that the unit was properly set before the problem developed, if the temperature drop is higher than it should be, the problem is reduction of air quantity; refrigeration systems do not increase in capacity. If the temperature drop is lower than it should be, the problem is in the refrigeration system; blowers do not increase in capacity.

11. Return air temperature and relative humidity at return air grill. Using a sling psychrometer, measure the condition of the air before it enters the return air grill. This measurement should then be compared to the conditions of the air at the return air plenum at the unit.

12. Return air temperature and relative humidity at the unit. Taking the conditions of the return air at both locations will check if the return air system is taking in air from outside the conditioned area. If there is a 3°F or more rise in the return air temperature, the return air system must be sealed and insulated. A 5°F rise can reduce the effective capacity of the system by 20%.

13. Supply air temperature at unit. The supply air temperature must be taken far enough downstream from the coil to get a mixture of the air from each section of the coil. The thermometer must not be located any closer than 6 in. from the coil surface to keep radiant heat transfer to the coil surfaces from affecting the thermometer reading. The temperature of the air leaving each section of the coil should also be checked and compared. If the coil has proper distribution of air and if the refrigerant distribution is correct, the measured air temperatures should be the same. If differences in temperature of more than 2°F are encountered, the reason for the difference should be determined and corrected.

14. Air temperature drop across DX coil. Subtracting the supply air temperature from the return air temperature determines the $\Delta T°F$ of air across the coil. Whether this $\Delta T°F$ is correct will depend on the results determined by the temperature drop graph based on the temperature and relative humidity of the return air (see Section A17-2).

15. Supply air temperatures at supply registers. As in checking the performance of the return air system, the temperature of the air from the various supply registers should be compared to the temperature of the air leaving the DX coil. A maximum of 3°F rise in temperature is permitted. If more than this is encountered, the supply duct must be insulated. A minimum of 2 in. of insulation with a vapor barrier is recommended.

16. Suction-line temperature at DX coil. To determine the operating superheat of the DX coil, the suction-line temperature at a point not more than 6 in. from the coil suction manifold should be measured.

17. Compressor suction pressure. The compressor suction pressure is measured at the condensing unit because this is the pressure tap location provided by most air-conditioning unit manufacturers. This means that the actual pressure in the DX coil is higher than the measured presssure due to pressure drop or vapor flow resistance in the suction line. To promote accuracy, the line pressure drop must be added to the gauge reading. The line pressure drop can be assumed to be 3 psig without too much sacrifice in accuracy.

18. DX coil superheat. Subtracting the temperature equivalent of the coil operating pressure (measured suction pressure plus 3 psig) from the suction-line temperature will determine the operating superheat under the unit operating conditions. Under ARI standard operating conditions, most air-conditioning units will operate with a superheat in the range 7 to 10°F. This is with an outdoor temperature of 95°F.

Capillary Tube Units: As the outdoor temperature varies, however, the superheat will vary. The reason is that the change in outdoor temperature will change the operating head pressure. The change in operating head pressure will change the flow rate of the capillary tubes and the flow rate of refrigerant to the DX coil. As a result, the change in flow rate will change the operating superheat.

Figure A17-3 shows the effect of outdoor temperature, the temperature of the air entering the condenser, and the superheat when the indoor air is 80° DB and 50% RH with 20°FΔT across the DX coil. From this it can be seen that there is no standard superheat setting for capillary tube systems.

TX Valve Units: The TX valve can be set for the proper superheat when the correct load is applied to the DX coil. Using the information in Section A17-2, the ft³/min through the DX coil must be set for the correct amount depending on the dry-bulb temperature and relative humidity of the return air.

19. Suction-line temperature at compressor. The temperature of the suction line at the compressor should be measured and compared to the suction-line temperature at the DX coil. Any rise in gas temperatures at the suction line indicates a heat gain in the suction gas. There will be some gain even though the line is insulated and properly located. A rise of up to 10°F is normal and will not affect compressor cooling. Any rise of more than 10°F or a gas temperature of 80°F or higher could cause compressor overheating and the compressor motor overload to trip off. Any unusual rise in suction gas temperature should be investigated and corrected.

20. Temperature of air entering condenser. The temperature of air entering the condenser, commonly

ASHRAE PSYCHROMETRIC CHART NO. 1

NORMAL TEMPERATURE
BAROMETRIC PRESSURE 29.921 INCHES OF MERCURY
COPYRIGHT 1963
AMERICAN SOCIETY OF HEATING, REFRIGERATING AND AIR-CONDITIONING ENGINEERS, INC.

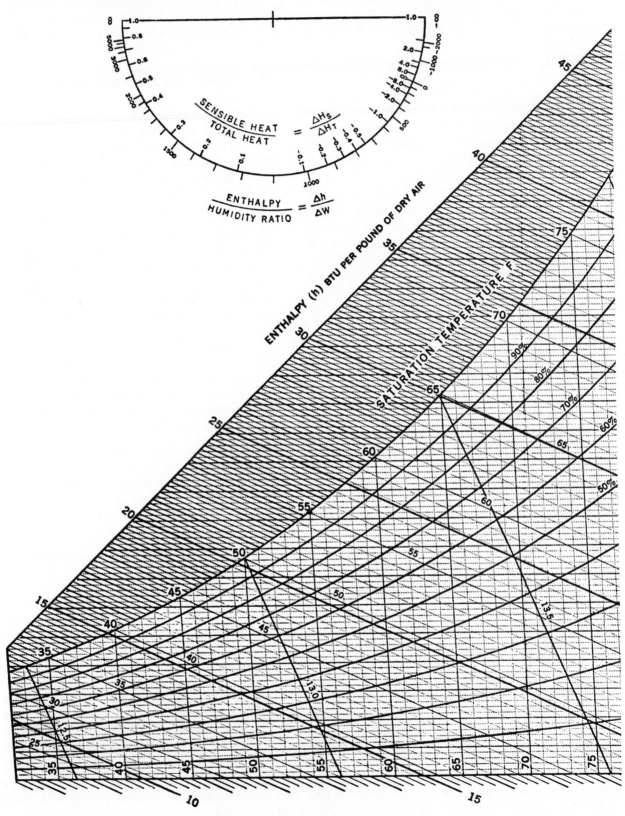

FIGURE A17-2 Psychrometric chart.

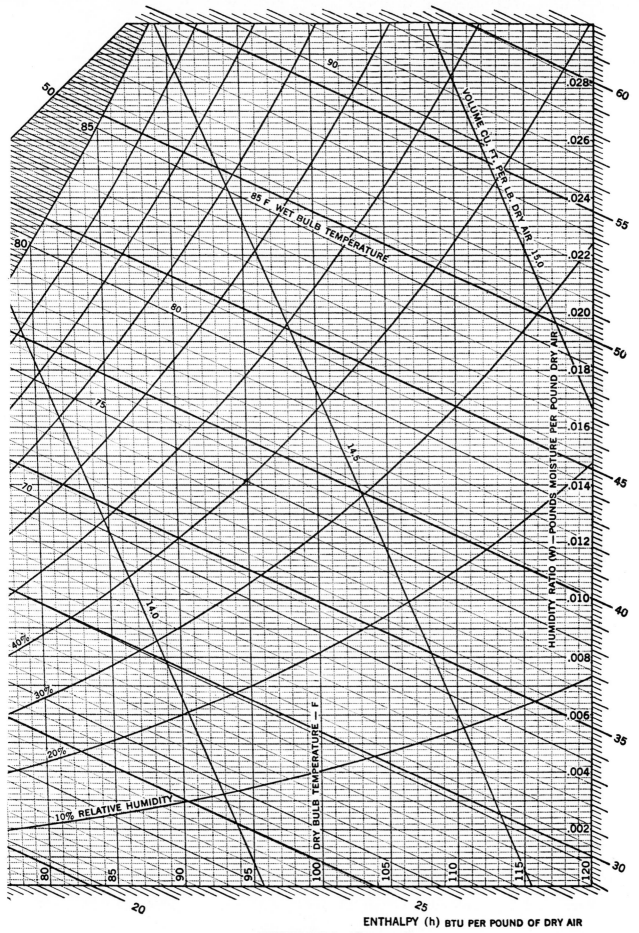

VOLUME CU. FT. PER LB. DRY AIR

85 F. WET BULB TEMPERATURE

HUMIDITY RATIO (W) – POUNDS MOISTURE PER POUND DRY AIR

DRY BULB TEMPERATURE – F

RELATIVE HUMIDITY

ENTHALPY (h) BTU PER POUND OF DRY AIR

FIGURE A17-2 (Continued)

617

Outdoor Air Temperature Entering Condenser Coil (°F)	Superheat (°F)
*65	30
75	25
80	20
85	18
90	15
95	10
105 & above	5

Source: Addison Products Company.

called the condenser ambient temperature, should be taken from a location directly in front of the condenser air inlet but not closer than 6 in. from the condenser surface. On blow-through condensers, this would be at the inlet to the condenser fan. On draw-through units, such as vertical discharge, the temperature should be taken at several locations and averaged. On U-shaped condensers, for example, temperature readings should be taken on both sides as well as the bottom of the U. The average of the three temperatures is used as the entering air temperature.

21. Compressor discharge pressure condensing temperature. The compressor discharge pressure converted to condensing temperature is used to determine the split of the condenser as well as the subcooling. Most units have the pressure tap for measuring discharge pressure on the liquid line at the outlet of the condenser. The difference between this pressure and the actual compressor discharge pressure is the pressure loss through the condenser. Because condenser design flow resistance is between $1\frac{1}{2}$ and 3 psig, this amount of error in the reading is usually ignored. On units having remote condensers, however, the difference could be considerably higher. The actual compressor discharge pressure should be measured and compared to the liquid-line pressure. If this difference is more than 10 psig, corrective measures should be taken.

22. Condenser split. The split of a condenser is the difference between the condensing temperature of the refrigerant and the temperature of the air entering the condenser. Years ago it was common practice to add 30°F to the temperature of the air entering the condenser and use the results as the condensing temperature of the refrigerant. It was a very inaccurate way of determining condensing temperature because the split depends on the amount of heat in the air entering the DX coil.

Figure A17-4 is a graph of the condenser performance of an air-conditioning unit with a 28.5° split at standard ARI conditions—95°F outside temperature, 80°F dry bulb, 50% relative humidity, with 20°ΔT across the coil. From this graph we find that the amount of work the condenser does, and the split produced, will vary from 25° with the entering air at 55°F wet bulb to 35° with 80°F wet bulb entering air. Therefore, unless the exact condenser characteristics of the unit are known, use of a standard condenser split for charging the unit to a specific head pressure (condensing temperature) is not possible.

23. Liquid-line temperature. The temperature of the liquid line taken with 6 in. of the outlet of the condenser must be recorded as part of the test procedure. This temperature and the refrigerant condensing temperature are used to determine the amount of liquid subcooling.

24. Liquid subcooling. When the high-pressure, medium-temperature liquid from in the condenser passes through the pressure-reducing device, the hot liquid is cooled to the DX coil operating temperature by evaporating part of the liquid refrigerant. The resulting vapor, called *flash gas,* must be handled by the compressor. Because the flash gas was not produced by heat absorbtion in the coil, it represents a loss in the system's efficiency. Therefore, the lower the liquid temperature leaving the condenser, the less flash gas produced and the higher the system efficiency. The lower liquid temperature is produced by adding additional refrigerant into the system to produce a buildup of liquid refrigerant in the condenser before it goes into the liquid line. The more refrigerant added, the greater the amount of subcooling. However, as the liquid builds up in the condenser, the effective

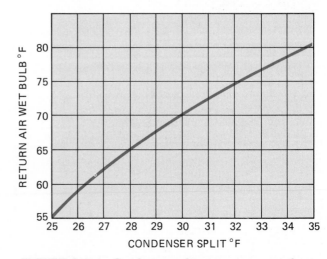

FIGURE A17-4 Condenser split versus return air heat content. (*Courtesy* Addison Products Company)

condensing surface is reduced, causing the remaining amount to work harder, producing a higher split. The condensing temperature rise causes the compressor to work harder and the system efficiency drops. The aim is to charge the system to the proper amount of subcooling, where the rise in efficiency from flash gas reduction is balanced by the drop in compressor efficiency due to the higher split. Practically all standard air-conditioning systems work their best at 18 to 20°F subcooling.

25. Condenser discharge temperature. The temperature of the air off the condenser should be recorded for determining the operating capacity of the unit. This is covered in Section A17-5. When determining the condenser air discharge temperature, an average of readings from several locations should be computed.

On condensing units using blow-through fans (Fig. A17-5), a grid of temperatures must be taken. Because the amount of air passing through the various sections of the condenser will vary depending on the position of the condenser fans, the temperature rise of the air will vary. Temperature-reading location marks are pictured on the condenser out-

Area of placement for thermometers

FIGURE A17-6   Draw-through vertical discharge unit. (*Courtesy* Addison Products Company)

FIGURE A17-5   Blow-through condenser. (*Courtesy* Addison Products Company)

let to obtain as accurate a temperature average as possible. On this particular unit, 15 readings would be used and averaged for the outlet temperature.

On vertical discharge draw-through units, the procedure is considerably easier. Thermometers are located on the discharge grill. A minimum of three should be used, one at midpoint of the condenser and the others placed to divide the circle equally (see Fig. A17-6). Using the three readings, the average is obtained. This temperature, minus the temperature of the air entering the condenser, will give the condensing air temperature rise $-\Delta T °F$.

26. ft³/min through condensing unit. Unless elaborate testing equipment is used, it is very difficult to determine the CFM through the condensing unit. This information, however, is listed in the manufacturers' literature. For example, Fig. A17-7 shows a specification chart of a condensing unit with the condenser for ft³/min given in the third column. This information is necessary to determine unit capacity.

A17-5
OPERATION

After the system has been checked out and adjusted properly, the operation performance of the system can be determined. An air-conditioning unit is rated by the amount of heat, both sensible and latent, that is picked up by the DX coil. If the ft³/min through the coil is known, this can be determined by velocity of air through the duct times and square-foot area of the duct, and the temperature drop of the air through the coil. The latent capacity can be calculated by collecting the amount of water off the coil for 1 hour. This will then

Model Number	Electrical Characteristics (1)	Operating Current (Amps) (2)	Minimum Field Wire (AWG)	Recommended Fuse (Delay Type)	Ampacity (Wire Size Amps)	Compressor L.R. Current (Amps) (3)	Condenser Face Area (Sq. Ft.)	Condenser Tube Dia. (Inches)
QC924-1H	1-60-230/208	12.7/13.2	#12	25A	16/17	54	8.8	$\frac{3}{8}$
QC930-1H	1-60-230/208	15.2/15.2	#12	30A	19/19	65	13	$\frac{3}{8}$
QC936-1H	1-60-230/208	17.2/17.2	#10	35A	22/22	75.8	18.2	$\frac{3}{8}$
QC942-1H	1-60-230/208	21.7/22.2	#10	45A	28/28	93	18.2	$\frac{3}{8}$
QC948-1H	1-60-230/208	24.2/24.2	#10	45A	30/30	95.4	19.7	$\frac{3}{8}$
QC948-3H	3-60-230/208	16.2/16.2	#10	30A	20/20	82	19.7	$\frac{3}{8}$
QC960-1H	1-60-230/208	29.7/32.2	#10	60A	37/40	122	19.7	$\frac{3}{8}$
QC960-3H	3-60-230/208	22.2/19.2	#10	40A	24/28	86	19.7	$\frac{3}{8}$

Model Number	Condenser Fan Dia. (In. O.D.)	Condenser Fan CFM	Fan Motor Type & Size (HP) (4)	Connections (Quick Connect) Suc. & Liq. (5)	Rated Capacity (6) 230V	Rated Capacity (6) 208V	SEER (6) 230V	SEER (6) 208V	Weight (lbs.) Net	Weight (lbs.) Ship
QC924-1H	20	2150	PSC-1/12	$\frac{3}{4}$ - $\frac{1}{4}$	25000	24200	8.95	9.10	164	196
QC930-1H	20	2450	PSC-1/6	$\frac{3}{4}$ - $\frac{1}{4}$	29800	29600	9.20	9.30	186	224
QC936-1H	22	3160	PSC-1/6	$\frac{7}{8}$ - $\frac{3}{8}$	36600	36400	9.30	9.45	228	279
QC942-1H	22	3350	PSC-1/6	$\frac{7}{8}$ - $\frac{3}{8}$	42000	41500	9.05	9.10	231	282
QC948-1H	22	3900	PSC-1/4	$\frac{7}{8}$ - $\frac{3}{8}$	45000	44500	8.90	8.95	261	312
QC948-3H	22	3900	PSC-1/4	$\frac{7}{8}$ - $\frac{3}{8}$	45000(7)	44500(7)	8.8(7)	8.8(7)	261	312
QC960-1H	22	3675	PSC-1/4	$\frac{7}{8}$ - $\frac{3}{8}$	57000	56000	8.45	8.50	326	377
QC960-3H	22	3675	PSC-1/4	$\frac{7}{8}$ - $\frac{3}{8}$	57000(7)	56000(7)	8.7(7)	8.6(7)	326	377

(1) Minimum operating voltage 197V. (2) With matching coil. (3) Compressor—Hermetic PSC, 3450 RPM. (4) Single-speed motor. (5) Refrigerant type R22. (6) With matching coil at DOE test standards. (7) ARI Test Standard.

Source: Addison Products Company.

add up to the unit's total "net" capacity. The net capacity is the unit rating.

By using the information gathered in the check-out procedure, the net capacity can be determined more accurately and more easily. The amount of heat that is ejected from the condenser is made up of the heat energy collected in the coil plus the heat energy produced from the electrical energy supplied to the condenser. This total amount of energy from the condenser is called the *gross capacity* of the unit. If we subtract the motor input energy from the gross capacity, we will determine the coil capacity or unit net capacity.

≡ **A17-5.1**
≡ **Gross Capacity**

The amount of heat from the condenser is determined by using the standard heat content (ft³/min) formula. The basic formula is

$$\text{ft}^3/\text{min} = \frac{\text{Btu/hr}}{\Delta T°\text{F} \times 1.08}$$

where

ft³/min = amount of air traveling per minute through a coil from which sensible heat is being extracted or through a condenser to which sensible heat is being added

Btu/hr = amount of heat that is being extracted or added in a 1-hour period

$\Delta T°$F = temperature change that occurs in the air

1.08 = a constant used to convert the air from cubic feet to pounds and ft³/min to ft³/hr and includes the quantity of heat to change the temperature of each

620 Air-Conditioning

pound of air, derived from the following calculation:

$$1.08 = 60 \text{ min/hr} \times 0.075 \text{ lb/ft}^3 \times 0.24 \text{ Btu/lb/°F}$$

To convert the Btu/hr to Btu/min it is necessary to divide the hour figure by 60. We cannot use the cubic foot of air to determine heat content because the amount of air in a cubic foot changes with temperature. However, a pound of air is constant regardless of how much space it occupies. A standard has to be established in order to use a consistent formula. The standard that is used is air at 70°F and at a barometric pressure of 29.92 in. of mercury. At these conditions, 1 lb of air occupies 0.075 ft³ and has a specific heat of 0.24 Btu/lb/°F temperature change. By multiplying 60 minutes per hour times 0.075 lb/ft³ times 0.24 Btu/lb/°FΔT, the result is 60 $\times$ 0.075 $\times$ 0.24 = 1.08.

To determine the heat ejected from the condenser, the formula is changed to make the Btu/hr the unknown factor:

$$\text{Btu/hr} = \text{ft}^3/\text{min} \times \Delta T°\text{F} \times 1.08$$

Using the known ft³/min value from the manufacturers' literature (Fig. A17-7) and the temperature rise of the air through the condenser determined in Section A17-4, item 25, the numbers are multiplied together along with the constant 1.08, and the result is the amount of heat in Btu/hr ejected from the condenser—or the unit gross capacity.

A17-5.2
Motor Heat Input

The amount of heat pumped into the refrigerant vapor before it enters the condenser is so close to 100% of the electrical input that all the electrical energy is used to calculate the motor input. The formula used to determine the motor Btu/hr input is

motor Btu/hr = volts × amperes × PF
$$\times \text{ 3.413 Btu/W}$$

where

volts = voltage at the line side of the unit contactor with the unit operating. Although the unit is rated at 240 V, the actual operating voltage must be used. If this is a three-phase unit, the voltage across each pair of heads to the motor should be the same. If a voltage difference of more than 3% exists between the sets of leads, the voltage unbalance will cause amperage unbalance and excessive winding temperature in the compressor motor. This will shorten the motor life.

Amps = total amperage that the condensing unit draws, including the condenser fan motor, used in this calculation.

PF = power factor of the condensing unit and must be taken into account. The power factor of condensing units will be between 0.86 and 0.94 depending on the load on the unit. An average of 0.90 is used in the calculation.

Btu/hr/W = heat energy of electrical energy: 3.413 Btu for each watt of electrical energy.

A17-5.3
Unit Net Capacity

As stated before, the motor Btu/hr input subtracted from the condensing unit gross output capacity leaves the amount of heat energy picked up in the DX coil. This quantity is the net capacity of the system, the capacity rating of the system. Comparing the capacity to the manufacturer's rating will indicate the condition of the system. A properly operating system should be within ±10% of the manufacturer's rating.

PROBLEMS

A17-1. What is the purpose of complete and proper startup, check-out, and operation of an air-conditioning unit?

A17-2. To check the operation of the air-conditioning system, the first step is to establish the proper load on the coil. This is done by adjusting for proper _____ .

A17-3. Before starting a unit, the _____ must be checked.

A17-4. Why is it necessary to use a clamp-type ammeter when starting a unit?

A17-5. At standard conditions of air, 80°F DB, and 50% RH, the temperature drop through the air would be _____ °F.

A17-6. The temperature drop for "standard" conditions of air can be used regardless of the conditions of the air encountered. True or False?

A17-7. The temperature drop of the air through the coil varies directly with the temperature of the air. True or False?

A17-8. The temperature drop of the air through the coil varies indirectly with the wet-bulb temperature of the air. True or False?

A17-9. Which has the greater influence on the air temperature drop through the coil—the sensible or latent heat content of the air?

A17-10. Why should the supply and return air temperatures be taken at both the supply and return grills and at the coil?

A17-11. At 80°F DB and 50% RH, a 5°F loss in air temperature represents what percentage of loss in system capacity?

A17-12. The operating superheat of a coil is found by measuring _____ .

A17-13. On a capillary tube system does the superheat go up or down with the outside temperature?

A17-14. With regard to the operating superheat of a TX valve system, within normal outdoor operating temperatures, will the coil superheat go up or down with a rise in outdoor temperatures?

A17-15. Define the "split" or "spread" of a condenser.

A17-16. A split of 30°F can be used to determine the operating head pressure of all standard air-conditioning units. True or False?

A17-17. Define "gross capacity."

A17-18. What is the formula for determining a unit's motor input?

A17-19. Is the rated capacity of the unit the gross capacity or the net capacity?

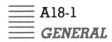

Air-Conditioning Service and Problem Analysis

 A18-1
GENERAL

The process of troubleshooting in the electrical and refrigerant portions of a refrigeration system were covered in Chapters R24 through R26. Chapter R24 listed the most common symptoms of problems as well as possible sources of the problem in both the electrical phase and the refrigeration phase. Chapter R25 outlines the possible sources of problems and possible solutions in the refrigeration circuit. Chapter R26 outlines the possible sources of problems and possible solutions in the electrical circuits.

In this chapter we limit the discussion to those problems and solutions that apply to air-conditioning systems. For those solutions that apply to the refrigeration system, reference will be made to Chapters R24 through R26.

To analyze trouble in air-conditioning systems the same test instruments would be used as those discussed in Chapter A17. The use of the high-pressure gauge, the suction pressure compound gauge, and a minimum of seven thermometers and sling psychrometer are required to properly service the system.

The gauges would be connected to the proper pressure taps: a high-pressure gauge on the compressor discharge valve or liquid-line service tap and a compound gauge on the suction-line service tap. Thermometers are inserted to measure:

1. Supply air temperature
2. Suction-line temperature, either at the coil outlet or at the condensing unit, whichever is more convenient
3. The liquid line at the condenser outlet
4. Condenser entering air
5. Condenser outlet air
6. Return air conditions

 A18-2
PROBLEM ANALYSIS

We are assuming that the unit was originally checked out and put into proper operation according to Chapter A17, or that it is a system that has been working and has developed a problem. The first thing to remember is that the problems in an air-conditioning system are classified in only two categories: air and the refrigerant circuit.

A18-3
AIR

The only thing that can occur in the air category is a reduction in quantity. Air-handling systems do not suddenly increase in capacity, that is increase the amount of air across the coil. On the other hand, the refrigeration system does not suddenly increase in heat-transfer ability. Therefore, the first check is the temperature drop of the air through the DX coil. After measuring the return and supply air temperatures and subtracting to get the temperature drop, is it higher or lower than it should be?

This means that "what it should be" has to be determined first. This is done by using the sling psychrometer to measure and determine the return air wet-bulb temperature and relative humidity. From this the proper temperature drop across the coil can be determined from the chart in Fig. A18-1.

Using the required temperature drop as compared to the actual temperature drop, the problem can be classified as either an air problem or a refrigerant system problem. If the actual temperature drop is greater than the required temperature drop, the air quantity has been reduced; look for problems in the air-handling system. These could be:

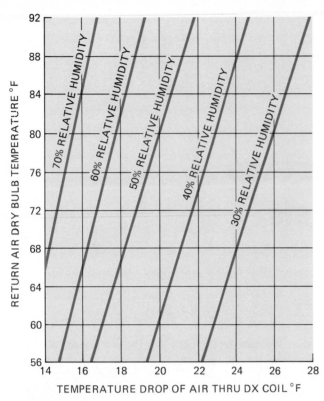

FIGURE A18-1 Air temperature drop for various load ratios.

1. Air filters
2. Blower motor and drive
3. Unusual restrictions in the duct system
4. Failure of the duct system

Air Filters: Air filters of the throwaway type should be replaced at least twice each year, at the beginning of both the cooling and heating seasons. In some areas where dust is high, they may have to be replaced as often as every 30 days. In commercial and industrial applications, a regular schedule of maintenance must be worked out for best performance and longest equipment life. Because this is the most common problem of air failure, *check the filtering system first.*

Blower Motor and Drive: Check the blower motor and drive in the case of belt-driven blowers to make sure that:

1. The blower motor is properly lubricated and operating freely.
2. The blower wheel is clean. The blades could be loaded with dust and dirt or other debris. If the wheel is dirty, it *must be removed and cleaned.* Do not try brushing only because a poor cleaning job will cause an imbalance to occur in the wheel. Extreme vibration in the wheel and noise will result. This could cause deterioration of the wheel.

3. On belt-driven blowers, the blower bearing must be lubricated and operating freely.
4. The blower drive belt must be in good condition and properly adjusted. Cracked or heavily glazed belts must be replaced. Heavy glazing can be caused by too much tension on the belt, driving the belt down into the pulleys. Proper adjustment requires the ability to depress the belt midway between the pulleys approximately 1 in. for each 12 in. between the pulley shaft centers.

Unusual Restrictions in Duct Systems: Placing furniture or carpeting over return air grills reduces the air available for the blower to handle. Shutting off the air to unused areas will reduce the air over the coil. Covering over a return air grill to reduce the noise from the centrally located furnace or air handler may reduce the objectionable noise, but it also drastically affects the operation of the system by reducing the air quantity.

Failure of the Duct System: Collapse of the return air duct system, usually from someone stepping up on it, will affect the entire duct system performance. Air leaks in the return duct will raise the return air temperature and reduce the temperature drop across the coil.

A18-4
REFRIGERATION SYSTEM

When the temperature drop across the coil is less than required, this means that the heat-handling capacity of the system has been reduced. We then look for a possible problem in the refrigeration system.

These problems can be simply divided into two categories: (1) refrigerant quantity, and (2) refrigerant flow rate. If the system has the correct amount of refrigerant charge and it is flowing at the desired rate, the system has to work properly and deliver rated capacity. Any problems in either category will affect the temperatures and pressures that will occur in the unit when the correct amount of air is supplied over the DX coil for the capacity of the unit. Obviously, if the system is empty of refrigerant, a leak has occurred and it must be found and repaired, evacuated thoroughly and then the system recharged with the correct amount of refrigerant.

If the system will not operate, it is obviously an electrical problem that must be found and corrected. Problems of these types have been covered thoroughly in Chapters R24 through R26.

In this chapter the discussion will be confined to those problems that affect the operating capacity of the system. The system will start and run but will not produce satisfactory results. This means that the amount of heat picked up in the coil plus the amount of motor heat added and the total rejected from the condenser is not the total heat quantity the unit is designed to handle. To determine the problem, all the information listed in Section A18-1 must be measured. These results compared to normal operating results will point out the problem. The use of the word "normal" does not imply a fixed set of pressures and temperatures. These will vary with each make and model of the system. There are a few temperatures that are fairly consistent throughout the industry that can be used for comparison:

1. DX coil operating temperature
2. Condensing unit condensing temperature
3. Refrigerant subcooling

These items must also be modified according to the efficiency rating (EER) of the unit. The reason for this is that the amount of evaporation and condensing surface designed into the unit are the main factors in efficiency rating. The larger the condensing surface (the lower the condensing temperature) and the larger the evaporating surface (the higher the suction pressure), the higher the system efficiency (EER) rating.

A18-4.1
DX Coil Operating Temperature

Normal coil operating temperature can be found by subtracting the design coil split from the average of the air temperature going through the coil. The coil split will vary with the system design.

Systems in the EER rating range 7.0 to 8.0 will have design splits in the range 25 to 30°. Systems in the EER rating range 8.0 to 9.0 will have design splits in the range 20 to 25°. Systems with 9.0+ EER ratings will have design splits in the range 15 to 20°. The formula used for determining coil operating temperatures is

$$COT = \frac{EAT + LAT - split}{2}$$

where

COT = coil operating temperature

EAT = temperature of air entering the coil

LAT = temperature of air leaving the coil

The latter two temperatures added together and divided by 2 will give the average air temperature. This is also referred to as the *mean temperature difference* (MTD).

"Split" is the design split according to the EER rating. For example, a unit operating properly with 80° DB, 50% RH air through the coil at a ft³/min to produce 20°FΔT will have a coil operating temperature of:

1. EER rating of 7.0 to 8.0:

$$COT = \frac{80 + 60}{2} \text{ or } 70° - 25 \text{ to } 30° = 40 \text{ to } 45°F$$

2. EER rating of 8.0 to 9.0:

$$COT = \frac{80 + 60}{2} \text{ or } 70° - 20 \text{ to } 25° = 45 \text{ to } 50°F$$

3. EER rating of 9.0+:

$$COT = \frac{80 + 60}{2} \text{ or } 70° - 15 \text{ to } 20°F$$
$$= 50°F \text{ to } 55°F$$

From this you can see that there is no fixed coil operating temperature and suction pressure for air-conditioning systems.

A18-4.2
Condensing Unit Condensing Temperature

This also applies to the condensing temperature of the unit. The amount of surface in the condenser affects the condensing temperature the unit must develop to operate at rated capacity. The variation in the size of the condenser also affects the production cost and price of the unit. The smaller the condenser, the lower the price, but also the lower the efficiency (EER) rating. In the same EER ratings used for the DX coil, at 95°F outside ambient, the 7.0 to 8.0 EER category will operate in the 25 to 30° condenser split range, the 8.0 to 9.0 EER category in the 20 to 25° condenser split range, and the 9.0+ EER category in the 15 to 20° condenser split range.

This means that when the air entering the condenser is at 95°F, the formula for finding the condensing temperature would be

$$EAT + split = RCT$$

where

EAT = temperature of the air entering the condenser

Split = design temperature difference between the entering air temperature and the condensing temperatures of the hot high-pressure vapor from the compressor

RCT = refrigerant condensing temperature

Using the formula with 95° EAT, the split for the various EER systems would be:

1. EER rating of 7.0 to 8.0

$$95° + 25 \text{ to } 30° = 120 \text{ to } 125° RCT$$

2. EER rating of 8.0 to 9.00

$$95° = 20 \text{ to } 25° = 115 \text{ to } 120° \text{ RCT}$$

3. EER rating of 9.0 +

$$95° + 15 \text{ to } 20° = 110 \text{ to } 115° \text{ RCT}$$

Again, from this you can see that operating head pressures vary not only from changes in outdoor temperatures but with the different EER ratings.

A18-4.3
Refrigerant Subcooling

The amount of subcooling produced in the condenser is determined primarily by the quantity of refrigerant in the system. The temperature of the air entering the condenser and the load in the DX coil will have only a small effect on the amount of subcooling produced. The amount of refrigerant in the system has the predominant effect. Therefore, regardless of EER rating, the unit should have, if properly charged, a liquid subcooled to 15 to 20°F. High outdoor temperatures will produce the lower subcooled liquid because of the reduced quantity of refrigerant in the liquid state in the system. More refrigerant will stay in the vapor state to produce the higher pressure and condensing temperatures needed to eject the required amount of heat.

A18-5
REFRIGERANT PROBLEMS

Using the information obtained using the two pressure gauges, a minimum of seven thermometers, the sling psychrometer, and a clamp-type ammeter measuring the condensing unit amperage draw, we can analyze the system problems by using the chart in Fig. A18-2. The figure shows that there are 11 probable causes of trouble in an air-conditioning system. After each probable cause is the reaction that the cause would have on the refrigeration system low-side or suction pressure, the DX coil superheat, the high-side or discharge pressure, the amount of subcooling of the liquid leaving the condenser, and the amperage draw of the condensing unit.

A18-5.1
Insufficient or Unbalanced Load

Insufficient air over the DX coil would be indicated by a greater-than-desired temperature drop through the coil. An unbalanced load on the DX coil would also give the opposite indication; some of the circuits of the DX coil would be overloaded while others would be lightly loaded. This would result in a mixture of air off the coil that would result in reduced temperature drop of the air mixture. The lightly loaded sections of the DX coil would allow liquid refrigerant to leave the coil and enter the suction manifold and suction line.

In TX valve systems the liquid refrigerant passing the feeler bulb of the TX valve would cause the valve to close down. This would reduce the operating temperature and capacity of the DX coil as well as lower the suction pressure. This reduction would be very pronounced. The DX coil operating superheat would be very low, probably zero, because of the liquid leaving some of the sections of the DX coil.

High-side or discharge pressure would be low due to the reduced load on the compressor, reduced amount of refrigerant vapor pumped, and reduced heat load on the condenser. Condenser liquid sub-cooling would be on the high side of the normal range because of the reduction in refrigerant demand by the TX valve. Condensing unit amperage draw would be down due to the reduced load.

In systems using capillary tubes, the unbalanced load would produce a lower temperature drop of the air through the DX coil because the amount of refrigerant supplied by the capillary tubes would not be reduced. Therefore, the system pressure (boiling point) would be approximately the same.

The DX coil superheat would drop to zero with flood-out of the refrigerant into the suction line. Under extreme cases of unbalance, liquid return to the compressor could cause compressor damage. The reduction in heat gathered in the DX coil and the lowering of the refrigerant vapor to the compressor will lower the load on the compressor. The compressor discharge pressure (hot-gas pressure) will be reduced.

The low rate of the refrigerant will be only slightly reduced because of the lower head pressure. The subcooling of the refrigerant will be in the normal range. The amperage draw of the condensing unit will be slightly lower because of the reduced load on the compressor and reduction in head pressure.

A18-5.2
Excessive Load

In this case the opposite effect exists. The temperature drop of the air through the coil will be low, so the unit cannot cool the air as much as it should. Air is moving through the coil at too high a velocity. There is the possibility that the temperature of the air entering the coil is higher than the return air from the conditioned area. This could be from leaks in the return air system drawing air from unconditioned areas.

The excessive load raises the suction pressure. The refrigerant is evaporating at a rate faster than the pump-

Probable Cause	Lowside (Suction) Pressure psig	D.X. Coil Superheat °F	Highside (Hotgas) Pressure psig	Condenser Liquid Subcooling °F	Cond. Unit Amperage Draw Amps.
1 Insufficient or unbalanced load	Low	Low	Low	Normal	Low
2 Excessive load	High	High	High	Normal	High
3 Low ambient (cond. entering air °F)	Low	High	Low	Normal	Low
4 High ambient (cond. entering air °F)	High	High	High	Normal	High
5 Refrigerant undercharge	Low	High	Low	Low	Low
6 Refrigerant overcharge	High	Low	High	High	High
7 Liquid line restriction	Low	High	High	High	Low
8 Plugged capillary tube	Low	High	High	High	Low
9 Suction line restriction	Low	High	Low	Normal	Low
10 Hot gas line restriction	High	High	High	Normal	High
11 Inefficient compressor	High	High	Low	Low	Low

ing rate of the compressor. The superheat developed in the coil will be as follows:

1. If the system uses a TX valve, the superheat will be normal to slightly high. The valve will operate at a higher flow rate to attempt to maintain superheat settings.

2. If the system uses capillary tubes, the superheat will be high. The capillary tubes cannot feed enough increase in refrigerant quantity to keep the DX coil fully active.

The high-side or discharge pressure will be high. The compressor will pump more vapor because of the increase in suction pressure. The condenser must handle more heat and will develop a higher condensing temperature to eject the additional heat. A higher condensing temperature means higher high-side pressure. The quantity of liquid in the system has not changed, nor is the refrigerant flow restricted. The liquid subcooling will be in the normal range. The amperage draw of the unit will be high because of the additional load on the compressor.

A18-5.3
Low Ambient (Condenser Entering
Air Temperature)

In this case, the condenser heat-transfer rate is excessive, producing a discharge pressure that is excessively low. As a result, the suction pressure will be low because the amount of refrigerant through the pressure-reducing device will be reduced. This reduction will re-

duce the amount of liquid refrigerant supplied to the DX coil. The coil will produce less vapor and the suction pressure drops.

The decrease in the flow rate into the coil reduces the amount of active coil and a higher superheat results. In addition, the reduced system capacity will decrease the amount of heat removed from the air. There will be higher temperature and relative humidity in the conditioned area and the high-side pressure will be low. This starts a reduction in system capacity. The amount of subcooling of the liquid will be in the normal range. The quantity of liquid in the condenser will be higher, but the heat-transfer rate of the lower temperatures is less. This will result in a subcooling in the normal range. The amperage draw of the condensing unit will be less. The compressor is doing less work.

The amount of drop in condenser ambient air temperature that the air-conditioning system will tolerate will depend on the type of pressure-reducing device in the system. Systems using capillary tubes will have a gradual reduction in capacity as the outside ambient drops from 95°F. This gradual reduction occurs down to 65°F. Below this temperature the capacity loss is drastic, and some means of maintaining head pressure must be employed. The most reliable means is control of air through the condenser via dampers in the airstream or variable-speed condenser fan motor.

Systems that use TX valves will maintain higher capacity down to an ambient temperature of 35°F. Below this temperature controls must be used. The control of CFM through the condenser using dampers or the condenser fan motor speed control can also be used. In larger TX valve systems, liquid quantity in the condenser is used to control head pressure. These control devices were covered in Section R12-10.

Air-Conditioning Service and Problem Analysis   627

A18-5.4
High Ambient (Condenser Entering Air Temperature)

The higher the temperature of the air entering the condenser, the higher the condensing temperature of the refrigerant vapor to eject the heat in the vapor. The higher the condensing temperature, the higher the head pressure. The suction pressure will be high for two reasons:

1. The pumping efficiency of the compressor will be less.
2. The higher temperature of the liquid will increase the amount of flash gas in the coil, further reducing the system efficiency.

The amount of superheat produced in the coil will be different in a TX valve system and a capillary tube system. In the TX valve system the valve will maintain superheat close to the limits of its adjustment range even though the actual temperatures involved will be higher. In a capillary tube system, the amount of superheat produced in the coil is the reverse of the temperature of the air through the condenser. The flow rate through the capillary tubes is directly affected by the head pressure. The higher the air temperature, the higher the head pressure and the higher the flow rate. As a result of the higher flow rate, the subcooling is lower.

Figure A18-3 shows the superheat that will be developed in a properly charged air-conditioning system using capillary tubes. Do not attempt to change a capillary system below 65°F, as system operating characteristics become very erratic below 65°F.

The head pressure will be high at the higher ambient temperatures because of the higher condensing temperatures required. The condenser liquid subcooling will be in the lower portion of the normal range. The amount of liquid refrigerant in the condenser will be reduced slightly because more will stay in the vapor state to produce the higher pressure and condensing temper-

FIGURE A18-3
Outdoor temperature versus superheat.

Outdoor Air Temperature Entering Condenser Coil (°F)	Superheat (°F)
*65	30
75	25
80	20
85	18
90	15
95	10
105 & above	5

Source: Addison Products Company.

ature. The amperage draw of the condensing unit will be high. The load on the compressor motor will be higher at the higher high-side pressures.

A18-5.5
Refrigerant Undercharge

A shortage of refrigerant in the system means less liquid refrigerant in the DX coil to pick up heat, and lower suction pressure. The smaller quantity of liquid supplied the DX coil means less active surface in the coil for vaporizing the liquid refrigerant and more surface to raise vapor temperature. The superheat will be high. There will be less vapor for the compressor to handle and less heat for the condenser to reject, lower high-side pressure, and lower condensing temperature.

The amount of subcooling will be below normal to none depending on the amount of undercharge. The system operation is usually not affected very seriously until the subcooling is zero and hot gas starts to leave the condenser together with the liquid refrigerant. The amperage draw of the condensing unit will be slightly less than normal.

A18-5.6
Refrigerant Overcharge

An overcharge of refrigerant will affect the system in different ways depending on the pressure-reducing device used in the system and the amount of overcharge.

TX Valve Systems: In systems using a TX valve, there is a different reaction depending on the amount of overcharge. A light overcharge would be up to 50% over the correct charge. A heavy overcharge would be more than 100% over the correct charge.

In TX valve systems with light overcharge, the TX valve will control the refrigerant flow into the coil to maintain the superheat setting of the valve, the only effect on the system would be a reduction in compressor discharge pressure. Under these conditions:

1. The suction pressure would be normal to slightly higher.
2. The DX coil superheat would be in the normal range.
3. The discharge pressure would be higher because the heat rejection ability of the condenser would be reduced due to the large quantity of liquid refrigerant in the condenser.
4. Liquid subcooling would be high. More liquid in the condenser means that the liquid will cool off more before it leaves.
5. The unit amperage draw would be higher because of the higher head pressure.

For TX valve systems with excessive overcharge:

1. The suction pressure will be high. Not only does the reduction in compressor capacity (due to higher head pressure) raise the suction pressure, but the higher pressure will cause the TX valve to overfeed on its opening stroke. This will cause a wider range of "hunt" of the valve.
2. The DX coil superheat will be very erratic from the low normal range to liquid out of the coil.
3. The high-side or discharge pressure will be extremely high.
4. Subcooling of the liquid will also be high because of the excessive liquid in the condenser.
5. The condensing unit amperage draw will be higher because of the extreme load on the compressor motor.

Capillary Tube Systems: The amount of refrigerant in the capillary tube system has a direct effect on the system performance. An overcharge has a greater effect than an undercharge, but both affect system performance, efficiency, and operating cost (EER rating).

Figures A18-4 through A18-6 are performance graphs for a 24,200-Btu/hr split air-conditioning system using an outside ambient temperature of 90°F, inside return air conditions of 75°FDB and 65°FWB (52% relative humidity) and 18.7°FΔT across the DX coil. At 100% of correct charge (55 oz) the unit developed a net capacity of 26,200 Btu/hr. When the amount of charge was varied 5% in either direction, the capacity dropped as the charge varied. Removing 5% (3 oz) of refrigerant reduced the net capacity to 25,000 Btu/hr. Another 5% (2.5 oz) reduced the capacity to 22,000 Btu/hr. From

there on the reduction in capacity became very drastic, 85% (8 oz), 18,000 Btu/hr; 80% (11 oz), 13,000 Btu/hr; and 75% (14 oz), 8000 Btu/hr.

Addition of overcharge had the same effect but at a greater reduction rate. The addition of 3 oz of refrigerant (5%) reduced the next capacity to 24,600 Btu/hr; 6 oz added (10%) reduced the capacity to 19,000 Btu/hr; and 8 oz added (15%) dropped the capacity to 11,000 Btu/hr. This shows that overcharging of a unit has a greater effect per ounce of refrigerant than does undercharging.

Figure A18-5 is a chart showing the amount of electrical energy the unit will demand because of pressure created by the amount of refrigerant in the system, with the only variable being the refrigerant charge. At 100% of charge (55 oz) the unit required 3.06 kWh. As the charge was reduced, the wattage demand also dropped, 3.04 kWh at 95% (3 oz), 2.760 kWh at 90% (6.5 oz), 2.560 kWh at 85% (8 oz), 2.5 kWh at 80% (11 oz), and 2.24 kWh at 75% (14 oz short of correct charge). When the unit was overcharged, the wattage required went up. At 3 oz (5% overcharge) the wattage required was 3.42 kW; at 6 oz (10% overcharge), 3.96 kW; and at 8 oz (15% overcharge), 4.8 kW.

In Fig. A18-6 we have a graph of the efficiency of the unit (EER rating) based on the Btu/hr capacity of the system versus the wattage demand of the condensing unit. At correct charge (55 oz) the efficiency (EER rating) of the unit was 8.49. As the refrigerant was reduced, the EER rating dropped to 8.22 at 95% of charge, 7.97 at 90%, 7.03 at 85%, 5.2 at 80%, and 3.57 at 75% of full refrigerant charge. When refrigerant was added, adding 5% (3 oz) the EER rating dropped to 7.19, at 10% (6 oz) the EER was 4.8, and at 15% overcharge (8

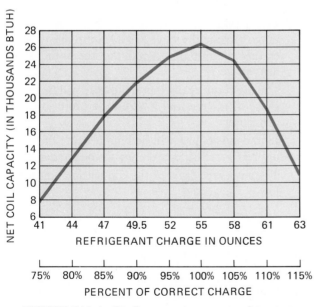

FIGURE A18-4 Btu/hr capacity versus refrigerant charge.

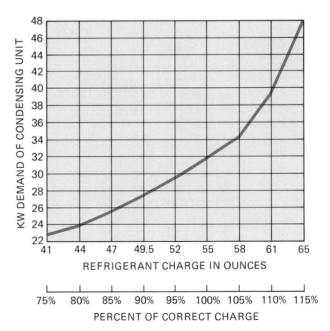

FIGURE A18-5 Refrigerant charge versus RW demand.

Air-Conditioning Service and Problem Analysis 629

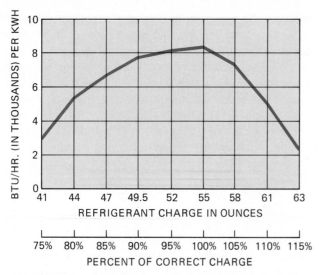

FIGURE A18-6 Refrigerant charge versus Btu/hr/kWh.

oz) the EER was 2.29. From these charts the only conclusion is that the capillary tube systems must be charged to the correct charge with only a −5% tolerance.

The effect of overcharge produces a high suction pressure because the refrigerant flow to the DX coil increases. Suction superheat will decrease because of the additional quantity to the DX coil. At approximately 8 to 10% of overcharge, the suction superheat becomes zero and liquid refrigerant will leave the DX coil. This will cause flooding of the compressor and greatly increases the chance of compressor failure. The high-side or discharge pressure will be high because of the extra refrigerant in the condenser. Liquid subcooling will also be high for the same reason. The wattage draw will increase due to the greater amount of vapor pumped as well as the higher compressor discharge pressure.

A18-5.7
Liquid-Line Restriction

A liquid-line restriction will reduce the amount of refrigerant to the pressure-reducing device. Both TX valve systems and capillary tube systems will then operate with reduced refrigerant flow rate to the DX coil.

1. The suction pressure will be low because of the reduced amount of refrigerant to the DX coil.
2. The suction superheat will be high because of the reduced amount of active portion of the coil, allowing more coil surface for increasing the vapor temperature as well as reducing the refrigerant boiling point.
3. The high-side or discharge pressure will be low because of the reduced load on the compressor.
4. Liquid subcooling will be high. The liquid refrigerant will accumulate in the condenser. It cannot

flow out at the proper rate because of the restriction. As a result, the liquid will cool more than desired.
5. The amperage draw of the condensing unit will be low.

A18-5.8
Plugged Capillary Tube or Feeder Tube

Either plugged capillary tube or plugged feeder tube between the TX valve distributor and the coil will cause part of the coil to be inactive. The system will then be operating with an undersized coil.

1. The suction pressure will be low because the coil capacity has been reduced.
2. The suction superheat will be high in the capillary tube systems. The reduced amount of vapor produced in the coil and resultant reduction in suction pressure will reduce compressor capacity, head pressure, and the flow rate of the remaining active capillary tubes.
3. The high-side or discharge pressure will be low.
4. Liquid subcooling will be high; the liquid refrigerant will accumulate in the condenser.
5. The unit amperage draw will be low.

TX Valve Systems:

1. In TX valve systems, a plugged feeder tube reduces the capacity of the coil. The coil cannot provide enough vapor to satisfy the pumping capacity of the compressor and the suction pressure balances out at a low pressure.
2. The superheat, however, will be in the normal range because the valve will adjust to the lower operating conditions and maintain the setting superheat range.
3. The high-side or discharge pressure will be low because of the reduced load on the compressor and condenser.
4. Liquid subcooling will be high because of the liquid refrigerant accumulating in the condenser.
5. The amperage draw of the condensing unit will be low.

A18-5.9
Suction Line Restriction

A suction-line restriction such as a plugged suction-line strainer, a kink in the suction line, or a solder joint fitted with solder means a high-pressure drop between the DX coil and the compressor.

1. The suction pressure, if measured at the condensing unit end of the suction line, will be low.

2. The superheat, as measured by suction-line temperature at the DX coil and suction pressure (boiling point) at the condensing unit, will be extremely high.

3. The high-side or discharge pressure will be low because of reduced load on the compressor.

4. The low suction and discharge pressure usually indicate a refrigerant shortage. *Warning: The liquid subcooling is normal to slightly above normal.* This indicates a surplus of refrigerant in the condenser. Most of the refrigerant is in the coil where the evaporation rate is low due to the higher operating pressure in the coil.

5. The amperage draw of the condensing unit would be low because of the light load on the compressor.

A18-5.10
Hot-Gas-Line Restriction

With a hot-gas-line restriction, the high-side or compressor discharge pressure will be high if measured at the compressor outlet or low if measured at the condenser outlet or liquid line. In either case the compressor amperage draw will be high. Therefore:

1. The suction pressure is high due to reduced pumping capacity of the compressor.

2. The DX coil superheat is high because the suction pressure is high.

3. The high-side pressure is high when measured at the compressor discharge or low when measured at the liquid line.

4. Liquid subcooling is in high end of normal range.

5. But with all this the compressor amperage draw is above normal. All symptoms point to an extreme restriction in the hot-gas line. This problem is easily found when the discharge pressure is measured at the compressor discharge.

Where the measuring point is the liquid line at the condenser outlet, the facts are very easily misinterpreted. High suction pressure and low discharge pressure will usually be interpreted as an inefficient compressor. *The amperage draw of the compressor must be measured.* The high amperage draw indicates that the compressor is operating against a high discharge pressure. Therefore, there is a restriction between the outlet of the compressor and the pressure measuring point.

A18-5.11
Inefficient Compressor

This problem is last on the list because it is the least likely to be a problem. When the compressor will not pump the required amount of refrigerant vapor:

1. The suction pressure will balance out higher than normal.

2. The DX coil superheat will be high.

3. The high-side or discharge pressure will be extremely low.

4. Liquid subcooling will be low because not much heat will be in the condenser. Therefore, the condensing temperature will be close to the entering air temperature.

5. The amperage draw of the condensing unit will be extremely low, indicating that the compressor is doing very little work.

The analysis of trouble in the air-conditioning system is only a matter of determining if the trouble is in the air quantity or air-supply side of the system or if it is the refrigeration side of the system. If in the refrigeration side, is the problem refrigerant quantity or refrigerant flow rate?

To determine the correct answer, five operating characteristics must be known:

1. Suction pressure
2. DX coil superheat
3. High-side or discharge pressure
4. Liquid subcooling
5. Amperage draw of the condensing unit

PROBLEMS

A18-1. What instruments are needed to properly diagnose problems in an air-conditioning system?

A18-2. At what locations are air temperature measurements required?

A18-3. If the air temperature drop through the coil is higher than when the unit was previously left, is the problem in the refrigeration or the air part of the system?

A18-4. Problems in the refrigeration portion of the system can be divided into two categories. What are they?

A18-5. Define "EER."

A18-6. EER is found by dividing the _____ by the _____ .

A18-7. How many probable causes of trouble are there in an air-conditioning system?

A18-8. If the Btu/hr load is increased on the evaporator, will the superheat of the evaporator increase or decrease on a TX valve coil and a capillary tube coil?

A18-9. What effect does low outside ambient temperature have on the capacity of the air-conditioning unit? Why?

A18-10. What is the minimum outside operating temperature of a capillary tube system and a TX valve system?

A18-11. The minimum outside temperature at which a capillary tube system can be properly charged is _____ .

A18-12. What is the easiest way to determine if a unit does not have the proper amount of refrigerant charge?

A18-13. Which has the greatest adverse effect on the capacity of a system, an overcharge or an undercharge of refrigerant?

A18-14. The change quantity tolerance of a capillary tube system is _____ .

A18-15. To properly diagnose trouble in the refrigeration system, five operating characteristics must be known. What are they?

HEAT PUMPS

Basic Principles

GENERAL

Another source of heat for residential and small commercial applications became very popular with the increase in cost of electrical energy. This source, referred to as the *heat pump,* was actually developed by Lord Kelvin in 1852.

The first practically applied heat pump was installed in Scotland in 1927. Between 1927 and 1950, heat pumps were installed in hundreds of residential and small commercial applications throughout Europe and the southern part of the United States. Many of these installations were merely air-conditioning units converted to heat pumps by the addition of reversing values and applicable controls.

In the early 1950s, heat pumps were built and marketed by companies in the southern portion of the United States for local markets. Unfortunately, these units were also installed in colder climates and in air distribution systems that were totally unsuited for this application. Heat pumps had a high failure rate and therefore acquired a very bad reputation. The bad reputation, together with the relatively low cost of energy, practically destroyed this market.

With the rise in energy cost going into the 1970s, the public again demanded an efficient means of heating with electrical energy and the heat pump was reborn. This time, however, the units marketed were designed for colder climates, with the necessary defrosting equipment as well as cold weather compressor protection. As a result, these "yankee" models perform with the same reliability and life expectancy as straight air-conditioning units.

HP1-2
BASIC PRINCIPLES

The heat pump is a refrigeration system, like any other refrigeration system, that transfers heat from one place to another by the change in state of a liquid. The action has been covered in the previous sections on refrigeration and air-conditioning. In addition, the heat pump is able to reverse the action or direction of heat transfers. It can remove heat from the occupied area for summer cooling and dispose of the heat into outside air, a water supply, or indirectly into earth or other material. By reversing the action, it will also remove heat from the outside air source, the water supply, or from the earth or other material and supply it to the occupied area.

Basically all refrigeration systems are heat pumps in that they transfer heat from a heat source at a low temperature to a heat sink or disposal means of a higher temperature. The title "heat pump" has only been given, however, to the system that is actually a "reverse cycle refrigeration system."

Because basic principles of refrigeration and air-conditioning have already been covered, discussion in this section will be limited to those components, application principles, operation principles, and service procedures that apply to the heat pump. Some of the components will be referred to by using different terms because of the change in their usage. These will be introduced throughout the discussion.

HP1-3
BASIC CYCLE

If we consider the conventional refrigeration cooling cycle (Fig. HP1-1), we know that heat is absorbed by the indoor DX coil or evaporator and discharged by the outside air-cooled condenser. If we can physically reverse these components and absorb heat from the outdoor air, by means of the state changes of the liquid refrigerant, we can discharge this heat into the indoor air. We have now created a method of supplying heat to an occupied area for maintenance of a comfort level of temperature. This is what a heat pump system does, ex-

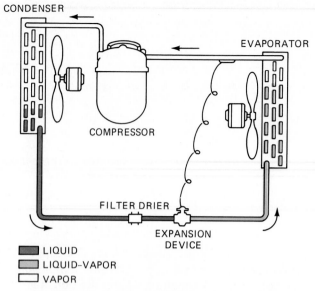

CONDENSER

EVAPORATOR

COMPRESSOR

FILTER DRIER

EXPANSION
DEVICE

■ LIQUID
▨ LIQUID-VAPOR
□ VAPOR

FIGURE HP1-1 (*Courtesy* Borg-Warner Central Environmental Systems, Inc.)

cept that it does not physically reverse the evaporator and condenser. By means of a reversing valve, it can direct the refrigerant flow to make the process provide heating or cooling in the occupied area.

The heat pump cycle is shown in Fig. HP1-2 and HP1-3. We relabel the coils as indoor and outdoor as they are now dual purpose, depending on the usage desired. The outdoor coil is the condenser in the cooling cycle and the evaporator in the heating cycle. The indoor coil is the evaporator in the cooling cycle and the condenser in the heating cycle.

To accomplish the reversing of the refrigerant flow, a reversing valve is used in the suction and discharge lines between the compressor and the two coils. Check valves are also connected in parallel with the pressure-reducing devices to permit removing them from the circuit when they are not to be used.

Referring to the circuit in the cooling phase (Fig. HP1-2), the directional arrows show the high pressure–high temperature compressor discharge gas being directed to the outdoor coil (condenser), where it condenses to the high pressure subcooled liquid. Because we do not want the restriction of the pressure-reducing device connected to the outlet of the outdoor coil, a check valve is connected to open—bypassing the liquid around the pressure-reducing device.

The liquid refrigerant travels to the indoor coil (evaporator) pressure-reducing device. Forced to flow through the device because the check valve, connected in parallel with the pressure-reducing device, has closed, the necessary pressure drop occurs to produce the heat absorbtion in the indoor coil. The refrigerant vapor produced in the indoor coil now travels through the reversing valve and accumulator to the compressor. The cycle is completed.

In the heat cycle (Fig. HP1-3) the reversing valve has changed position, changing the direction of gas flow. The high pressure–high temperature gas from the com-

OUTDOOR
COIL

REVERSING
VALVE

INDOOR
COIL

ACCUMULATOR

COMPRESSOR

HEATING
CYCLE
EXPANSION
DEVICE

COOLING
CYCLE
EXPANSION
DEVICE

CHECK VALVE
#1

CHECK VALVE
#2

■ LIQUID
▨ LIQUID-VAPOR
□ VAPOR

FIGURE HP1-2 (*Courtesy* Borg-Warner Central Environmental Systems, Inc.)

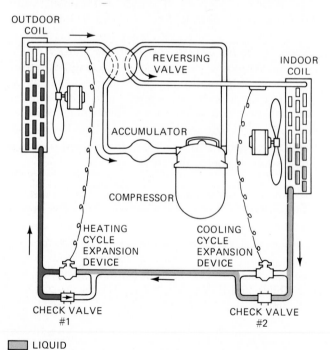

OUTDOOR
COIL

REVERSING
VALVE

INDOOR
COIL

ACCUMULATOR

COMPRESSOR

HEATING
CYCLE
EXPANSION
DEVICE

COOLING
CYCLE
EXPANSION
DEVICE

CHECK VALVE
#1

CHECK VALVE
#2

▨ LIQUID
■ LIQUID-VAPOR
□ VAPOR

FIGURE HP1-3 (*Courtesy* Borg-Warner Central Environmental Systems, Inc.)

636 Heat Pumps

pressor flows through the reversing valve to the indoor coil. This coil now acts as a condenser, ejecting heat into the air from the conditioned area, the vapor is condensed, and the temperature is reduced to produce a high pressure–medium temperature subcooled liquid.

To remove the cooling cycle pressure-reducing valve from the circuit, the check valve opens and allows the liquid refrigerant to flow around the pressure-reducing device. Continuing on through the liquid line, the liquid refrigerant is forced through the outdoor coil pressure-reducing device by the closing of the check valve connected in parallel with the pressure-reducing device. Converted to a low pressure–low temperature liquid, the refrigerant flows into the outdoor coil, which now acts as an evaporator. Heat is picked up from the outdoor air by evaporation of the liquid refrigerant. This refrigerant vapor then flows through the vapor line, reversing valve, and accumulator to the compressor. The cycle is now complete.

Accumulators are recommended for refrigeration and air-conditioning systems; they are a necessity in heat pump systems. This will be covered more thoroughly under the subject of "system defrost."

HP1-3.1
Air Source Systems

The discussion of the basic cycle was based on an air source system referred to as an *air-to-air*-type heat

pump. In this system two air-type tube-and-fin heat exchanges are used. The indoor coil collects heat from (in the cooling cycle) or giving up heat (in the heating cycle) to air that is provided to it by the inside flow and air distribution system. The outdoor coil collects heat from (heating cycle) or gives up heat to (cooling cycle) the outdoor air that is forced through it by the condenser air blower or propeller fan assembly. In each case heat is taken from and given up to "air."

A typical example of an air-to-air system is shown in Fig. HP1-4. The blower assembly in the fan coil unit and the supply and return duct system provide the air through the inside coil. The propeller fan assembly in the outdoor section provides the air through the outdoor coil.

HP1-3.2
Water Source Systems

Water source systems called *water-to-air*-type heat pumps use a liquid heat exchanger in the high-side or "outdoor" section of the heat pump assembly. This section should be located in the heated area to prevent possible freeze-up if the outdoor ambient should drop below 32°F.

In packaged units, of course, this would be automatic, as the entire unit is located in the conditioned area in order to connect to the air distribution system. Figure HP1-5 shows a typical water-to-air unit with the

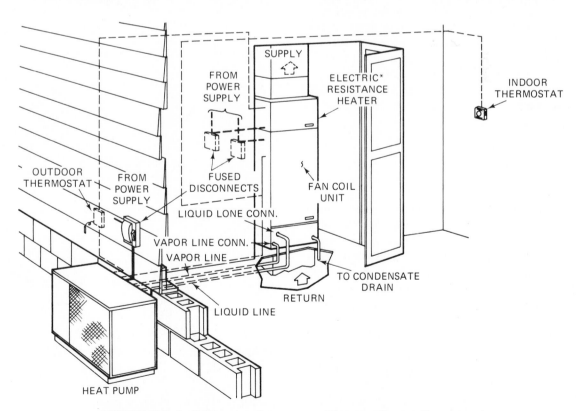

FIGURE HP1-4   Split-system heat pump. (*Courtesy* Carrier Corporation)

FIGURE HP1-5 Water source package heat pump. (*Courtesy* Bard Manufacturing Company)

casing panels removed. In its lower compartment, the double-tube water-to-refrigerant heat exchanger can be seen. Due to the variety of water conditions encountered, these heat exchangers are usually made of cupronickel. For efficiency and compactness, the construction is predominately "tube-in-a-tube" type, called *coaxial* construction. The rest of the refrigeration system, covering compressor, pressure-reducing devices, check valves, and reversing valve, is basically the same as in the air-to-air system.

HP1-3.3
Water-to-Water Systems

Water-to-water heat pump systems have been marketed in small unit sizes for water heating. Mostly in the capacity range 6000 to 12,000 Btu/hr, they are incorporated into a package unit along with a hot-water storage tank or as an add-on unit for existing storage tanks. Figure HP1-6 shows a typical package unit that can be added to a hot-water heater. The unit becomes the primary heat source for domestic or small commercial hot-water use, with the heat source of the water heater as backup. The higher efficiency rating or COP (coefficient of performance) of the water-to-water heat pump will usually supply hot water at a lower energy cost. Installation and service problems are covered in later chapters.

HP1-4
COMPONENT PARTS

The basic heat pump system is composed of two heat exchangers, a compressor for raising the refrigerant boiling point or condensing temperature and pressure-reducing devices for lowering the refrigerant boiling point. Because the system operates at different temperature conditions in the heating and cooling modes, the pressure-reducing devices have to be matched to the coil that will act as the evaporator for the conditions to be encountered. For example, the pressure-reducing device on the inside coil must respond to air conditions entering the coil of 80°F DB and 50% RH as well as 95°F DB entering the outdoor coil (the condenser). The pres-

(a)

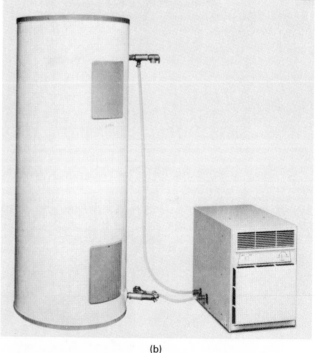

(b)

FIGURE HP1-6 Water-to-water heat pump. (*Courtesy* Borg-Warner Central Environmental Systems, Inc.)

sure-reducing device used in conjunction with the outdoor coil, when it is the evaporator on the heating cycle, must respond to 45°F air entering the coil with 70°F air entering the inside coil, which is now the condenser. From this we can see that the pressure-reducing devices as well as the coils are expected to operate differently than a straight-air conditioning system. Other components are added to the heat pump which are not found in the air-conditioning system. The components, in turn, have their specific function.

HP1-4.1
Heat Exchangers

The inside coil is basically the same in air-to-air and water-to-air systems. Designed to act as a top feed of hot gas and a bottom outlet for liquid when used as a condenser, the coil then operates as a bottom-feed type of evaporator coil. Also, to keep flow resistance down when in the heating mode, these coils usually have more circuits carrying lighter loads than those carried by standard coils.

Their overall size and heat exchanger surface area are larger than standard coils because their primary function is heat rejection as a condenser. Heat pumps designed and marketed since the last surge in popularity have been designed with heating as their prime function and cooling secondary. As a result, field problems have been minimal.

Because of their larger size per 12,000 Btu/hr capacity, the standard heat pump operates like an air-conditioning unit with an oversized coil. The total Btu/hr capacity may be higher, but the latent capacity is greatly reduced. As a result, higher humidity levels in the conditioned area will result from the heat pump operation compared to standard air-conditioning systems.

Figure HP1-7 shows the capillary tubes used for boiling-point reduction when the coil acts as the evaporator on the cooling cycle. Behind the liquid distributor feeding the capillary tubes is the check valve. On the heating cycles, when the coil acts as the condenser and the refrigerant flows in the reverse direction, this valve opens to remove the capillary tube restriction from the circuit.

Air Source Outside Coils: The outside coil is also top feed–bottom outlet coil in order to operate as a condenser on the cooling cycle. Normally, condensers are circuited from top to bottom without regard to circuit loading because the operation is not affected by any circuit load unbalance. However, when used as the evaporator on the heating cycle, it must be circuited properly for highest capacity. Figure HP1-8 shows an outside coil that has been circuited as a stacked double coil. Refrigerant flows from the pressure-reducing device into the bottom row of each section. Its section is then multiple

FIGURE HP1-7 Heat pump "A" coil. (*Courtesy* Bard Manufacturing Company)

circuited to provide the highest heat-absorbing capacity at the lowest possible pressure drop through the coil. In addition to the stacking of two sections, each section is also cross-circuited to promote balanced loading through each section.

Water Source Outside Coils: As stated before, the "outside" coil in water source heat pumps are tube-in-a-tube heat exchangers for transfer of heat from high pressure–high pressure compressor discharge vapor from the compressor to the water in the cooling cycle and from

FIGURE HP1-8 Outdoor heat pump section. (*Courtesy* Bard Manufacturing Company)

Basic Principles **639**

the water to the low pressure–low temperature liquid refrigerant in the heating cycle.

These heat exchangers are usually either a stacked tube in a tube (Fig. HP1-9) or a continuous tube in a tube called a coaxial (Fig. HP1-10). In either case the amount of surface involved is adequate to transfer the required amount of heat.

Figure HP1-9 shows a water-cooled tube-in-tube condenser. The stacked construction with removable head or end plates permits cleaning of the individual water tubes by roding. Pictured is a standard condenser. If it were to be used in a water source heat pump, capillary tubes would be inserted through the liquid-out manifold to supply low pressure–low temperature liquid refrigerant into the bottom of each refrigerant circuit during the heating cycle. Also to provide the maximum capacity, the water flow through the device is reversed, with water entering the top and leaving the bottom. Some sacrifice of capacity is necessary on the cooling cycle to produce peak capacity on the heating cycle.

Continuous tube-in-tube or coaxial heat exchangers are not mechanically cleanable and are limited to a chemical cleansing process. Figure HP1-10 shows a coaxial heat exchanger in the right-hand side of the lower compartment. Female pipe thread connections with water pressure gauge taps are shown to provide means of determining the liming factor of the water tube. Another tube in tube heat exchanger is shown around the base of the compressor. This heat exchanger is used for heating domestic hot water on the cooling cycle.

FIGURE HP1-10   Water source package heat pump. (*Courtesy* Bard Maunufacturing Company)

In water-to-water systems, both heat exchangers would be of the coaxial type, with the necessary controls and water supply pumps to operate properly.

HP1-4.2
Reversing Valve

The ability of the heat pump to reverse its ability to transfer heat is accomplished by reversing the direction of refrigerant vapor flow to the compressor to the heat exchangers. The selection of the heat exchanger function to pick up heat (evaporator) or give up heat (condenser) is determined by the action of the reversing valve. Physically, a typical reversing valve looks like Fig. HP1-11 on the outside. Internally, it is composed of two pistons connected to a sliding block or cylinder with two openings. The illustration shows the position of the piston assembly in the heating mode (Fig. HP1-12) and in the cooling mode (Fig. HP1-13).

The action of the piston is controlled by a solenoid valve that uses high-pressure compressor discharge vapor to move the piston left or right depending on which mode is needed. With the compressor discharge connected directly to the center of the piston chamber, equal pressure is exerted on the internal surfaces of each of the piston ends. To create movement, a pressure difference is produced across the piston by bleeding the cylinder pressure into the suction side of the compressor.

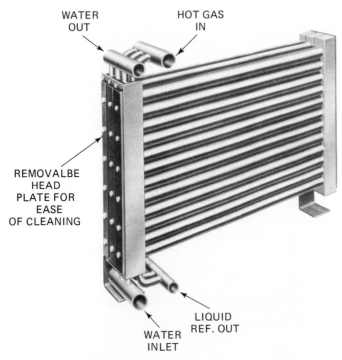

WATER OUT   HOT GAS IN

REMOVALBE HEAD PLATE FOR EASE OF CLEANING

LIQUID REF. OUT

WATER INLET

FIGURE HP1-9   Water-to-refrigerant coil. (*Courtesy* Standard Refrigeration Company)

FIGURE HP1-11 Reversing valve. (*Courtesy* Ranco, Inc.)

which the high-pressure gas can slowly find its way to regulate the speed of the piston travel. In the crossover action, too rapid a change in pressure could result in system shock and excessive noise.

HP1-4.3
Check Valve

The check valves in the heat pump circuit are vital to insure that proper pressures and refrigerant boiling points are maintained to provide the heat absorbtion and transfer at the unit rated capacity. This is easily accomplished, as explained in Section HP1-3, because these valves are either fully open or fully closed, depending

The bleeding action is controlled by the action of the two-way solenoid valve. In Fig. HP1-12 the solenoid valve is deenergized and the control plunger is relaxed in the bottom position. The bottom port is closed and the top port is open to the middle common port. Thus the top vent line is open to the equalizing line. The action that took place was the bleeding off of the pressure from the left cylinder chamber. When the pressure in this chamber was reduced sufficiently to cause the pressure difference across the piston to move the piston (usually 75 to 100 psig), the piston traveled to the left. This caused the valve slider to open the left control port to the center port and the right control port to the piston pressure. In this position the compressor discharge vapor travels to the indoor coil (condenser) and the vapor from the outdoor coil (evaporator) travels to the compressor suction.

In Fig. HP1-13, the solenoid valve coil is energized and the plunger has been lifted. This action has closed the vent line from the left end of the piston and opened the vent line from the right end of the piston. When the pressure in the right end of the piston has dropped sufficiently, the pressure difference across the piston will force the piston to the extreme right end.

This action moves the slide valve, opening the outdoor coil to the compressor discharge and the indoor coil to the compressor suction. The unit is now performing as a standard air-conditioning system. Although they are not shown in the two sketches, a tip on the end of the piston seals into a seat to prevent continuous bypass of the hot vapor from the cylinder through the piston bleed hole into the open vent line.

The piston bleed hole is a small opening through

REVERSING VALVE — COOLING

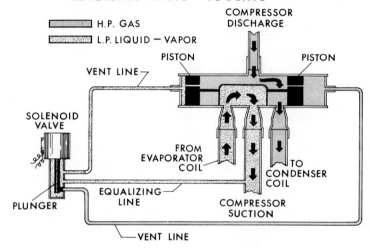

FIGURE HP1-12 Reversing valve—deenergized. (*Courtesy* Ranco, Inc.)

REVERSING VALVE — HEATING

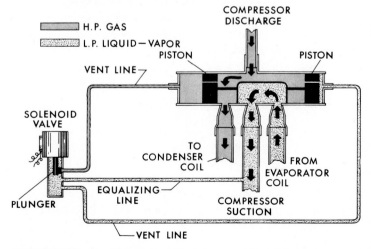

FIGURE HP1-13 Reversing valve—energized. (*Courtesy* Ranco, Inc.)

Basic Principles 641

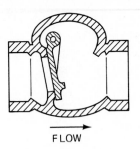

FIGURE HP1-14 Check valve. (*Courtesy Carrier Air-Conditioning Company*)

on the direction of refrigerant flow. Figure HP1-14 shows a swing-type check valve with a directional flow arrow. When the refrigerant flows in the direction of the arrow, the valve opens and allows full flow with minimum resistance.

When the refrigerant is reversed, however, the check plate swings closed and stops the refrigerant flow. Thus, this action forces the refrigerant to flow through the pressure-reducing device and proper operation is obtained.

The first check valves used in residential and small commercial heat pumps were of the disk type. This valve used a flat disk over a circular flat seat with a light pressure spring to aid in closing and still cause very little resistance to flow of the refrigerant. In normal refrigerant flow situations of normal pressure forcing the refrigerant through the check valve and little reverse pressure when the system was idle, the check valve performs adequately. In heat pump systems, where radical pressure differences occur rapidly, the disk would be forced to one side and hang up. This prevents the shut-off of reverse refrigerant flow.

A check valve using a steel ball instead of a flat disk has been developed. This type of check valve will take the heavy reversing of refrigerant flow pressure, especially when the unit comes off the defrost cycle.

Usually, two check valves are incorporated in each system, one connected in parallel with each pressure-reducing device on each of the inside and outside coils. Specific operation and troubleshooting of these devices are covered in Chapter HP6.

HP1-4.4
Pressure-Reducing Devices

As in standard air conditioning systems, the two types of pressure-reducing devices used in heat pumps are thermostatic expansion valves and capillary tubes. Although both types of devices were covered in Chapter R10, an explanation of the difference between usage in straight air-conditioning systems and in heat pump systems is in order.

Capillary Tubes: In heat pump systems the capillary tubes are the same as those in air-conditioning systems

when applied to the indoor coil. Those on the outdoor coil are sized for altogether different conditions of pressures and temperatures. Remember that the heat pump operates, on the heating cycle, with 70°F air entering the condenser and anywhere from −20 to 65°F air entering the evaporator. As in all capillary tube applications, the size and length specified by the manufacturer must be used.

TX Valves: The TX valves used in heat pumps are not interchangeable with those used in standard air-conditioning systems. The feeler bulb of the heat pump TX valve is clamped to the cool suction line in one operating mode and to the hot-gas line in the other mode. Thus the valve must be able to operate when the coil it is connected to is the evaporator and yet stand the high temperatures of the hot-gas line when the coil is the condenser. The power element of the valves have a special pressure-limiting charge to provide this range of operation. For a review of the basic principles of operation of the TX valve, refer to Section R10-4.

HP1-4.5
Accumulators

Accumulators were described in Section R14-6. The application covered was the protection of the compressor from liquid runout from the evaporator during light loads or air reduction problems. Figure HP1-15 shows the top view of an outdoor section showing the accumulator connected between the suction outlet of the reversing valve and the suction inlet of the compressor. In the heat pump, the accumulator has three situations where compressor protection is needed:

1. Flood-out on the cooling cycle: if air restriction causing light load and liquid runout occurs.
2. Flood-out on the heating cycle: when excessive frost buildup on the outdoor coil occurs or if air troubles occur and cause liquid runout.
3. Termination of the defrost cycle: liquid flood-out will always occur when the defrost cycle is terminated.

To melt the frost and ice off the outdoor coil, the heat pump employs the hot-gas method. The system action is reversed, heat is picked up in the inside coil, is raised in pressure and temperature by the compressor, and is forced into the cold outside coil. With the outside blower or fan off, the coil rapidly heats up to melt the frost and/or ice on the outside surface of the coil. While doing this, the coil fills with condensed liquid refrigerant. Before the defrost operation is completed, and the liquid out of the bottom of the condenser reaches the proper termination temperature, usually 55°F, pres-

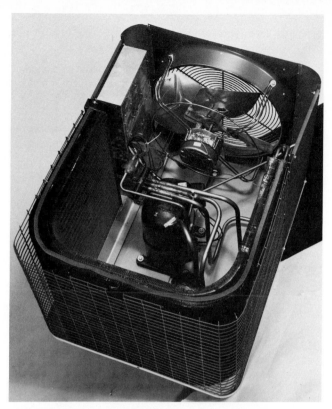

FIGURE HP1-15 (*Courtesy* Amana Refrigeration, Inc.)

sure in the condenser can reach 350 psig and a temperature of 142°F.

When the defrost cycle terminates and the reversing valve switches over, this high pressure is relieved into the suction side of the compressor. Immediately, vapor will form deep in the condenser circuit and force liquid refrigerant out of the coil into the suction line. This action can be compared to removing the cap on the automobile radiator when the engine overheats. Water is blown out by the steam formed in the engine when the pressure is relieved.

Without the accumulator to catch and hold the liquid refrigerant, liquid surge into the compressor can ruin valves. Because of this required protection, the use of an accumulator in the heat pump system does not allow intolerance to refrigerant quantity. The refrigerant change of a heat pump system is very critical.

The refrigerant lines in a heat pump system are not the same as in an air-conditioning or refrigeration system. In these systems, the refrigerant, either as high pressure–medium temperature liquid (liquid line), high pressure–high pressure vapor (hot-gas line), or low pressure–low temperature vapor (suction line), always flows in the same direction: liquid refrigerant from condenser to pressure-reducing device, hot vapor from compressor to condenser, and cold vapor from the evaporator to the compressor.

In heat pumps the refrigerant flow reverses depending on the operating mode. The liquid line is always the liquid line regardless of the operating mode. It carries liquid refrigerant from condenser to evaporator in either mode because the heat exchangers change operating characteristics. Therefore, the liquid pressure in this line is always the high-side liquid operating pressure.

The large vapor line from the inside coil to the outside heat pump section has a dual role. It is the cold vapor line (suction line) in the cooling mode and the hot vapor line (hot gas line) in the heating mode. To check the operating pressures, the service technician must remember to connect a high-pressure gauge to this line to prevent damage to the gauge set.

The only sections of the refrigerant circulating systems that have single duty are the line between the reversing valve and the compressor inlet (suction line) and the line from the compressor outlet to the reversing valve (hot-gas line). To be able to determine problems in the refrigerant system, four gauges are needed: a compound gauge connected to the suction line and standard gauges connected to the hot-gas line, the liquid line, and the vapor line to the outside coil. Use of the gauges is covered in Chapter HP4.

In air-conditioning systems, the dual-purpose vapor line (suction line) is insulated to promote compressor life and reduce line sweating. In heat pumps, insulation on this dual-purpose vapor line is also a necessity, to reduce heat loss between the compressor and the inside condenser. Lack of insulation can produce a loss in system capacity as high as 20%.

PROBLEMS

HP1-1. The heat pump was developed by _____ in _____ .

HP1-2. Describe the difference in operation between an air-conditioning system and a heat pump.

HP1-3. The title "heat pump" has been given to a unit that is actually a _____ .

HP1-4. The proper names for the two coils in the heat pump system are the _____ and _____ coils.

HP1-5. In the cooling mode, the evaporator is the _____ coil.

HP1-6. In the heating mode, the condenser is the _____ coil.

HP1-7. The change from heating mode to cooling mode, and vice versa, is accomplished by a device called a _____.

HP1-8. To protect the compressor from refrigerant surge, an _____ is used.

HP1-9. A tube-in-a-tube heat exchanger is called a _____ type.

HP1-10. Water-to-air heat pumps are always located in the heated area _____.

HP1-11. Generally, water-to-refrigerant heat exchangers used in water-to-air heat pumps are made of _____. Why?

HP1-12. The inside coil in heat pumps is always larger than the coil used in air-conditioners because _____.

HP1-13. What effect does the larger coil have on the results in the cooling mode?

HP1-14. A reversing valve consists of two parts: the _____ and the _____.

HP1-15. The main valve in a reversing valve is moved by the pressure created by the _____.

HP1-16. The minimum pressure required to operate the main valve in a reversing valve is _____ to _____ psig.

HP1-17. The pilot valve operates the main valve by controlling the pressure applied to the valve. True or False?

HP1-18. The assembly consisting of the pressure-reducing device and the check valve is called a _____.

HP1-19. Two types of check valves have been used in heat pumps. What are they?

HP1-20. Which type of check valve has been the least trouble?

HP1-21. Name two types of pressure-reducing devices used in heat pumps.

HP1-22. A TX valve used in heat pumps is not interchangable with air-conditioning system valves because _____.

HP1-23. What are the three situations where accumulators are needed to protect the compressor?

HP1-24. Termination of the defrost cycle occurs when the liquid off the outside coil reaches _____ °F.

HP1-25. During the defrost cycle, condensing pressures and temperatures can reach _____ psig and _____ °F.

HP1-26. In the heat pump system, the lines in which the refrigerant changes direction are the _____ and the _____.

HP1-27. The suction line is connected between the _____ and the _____.

HP1-28. The hot-gas line is always connected between the _____ and the _____.

HP1-29. In a heat pump, the two main purposes for insulation on the vapor line are _____ and _____.

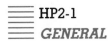

Controls

HP2-1
GENERAL

As in the refrigeration and air-conditioning field, where expansion in application required development of specific controls, this is also true in the heat pump field. Controls used in refrigeration and air-conditioning, such as safety controls and pressure controls that are part of the refrigeration circuit, are covered in chapters R20 and A10.

The discussion of controls in this chapter will be limited to those controls that apply to heat pumps. Such controls are in the following categories:

1. Temperature controls
   a. Thermostats—room
   b. Thermostats—outdoor ambient
2. Defrost controls
   a. Temperature differential
   b. Pressure–temperature
   c. Time–temperature
   d. Pressure–time–temperature

HP2-2
TEMPERATURE CONTROLS

Two basic applications of thermostats exist in the heat pump field: the control of temperature, both heating and cooling in the occupied area, and the control of the auxiliary heat system when other forms of heat are used in conjunction with the heat pump.

HP2-2.1
Thermostats: Room

The basic operation of room thermostats was covered in Chapter A-10. It is necessary, however, to review the functions of the room thermostat as it applies to a heat pump system. To control the heat pump system, the room thermostat performs the following functions:

1. System totally off
2. System on cooling only
3. System on heating only
4. System with two stages of heat
5. System on automatic changeover
6. System on emergency heat
7. Room air circulation only
8. Trouble indication

A typical heat pump thermostat that provides all these functions is pictured in Fig. HP2-1. Shown are the various levers and switches that perform the desired functions together with two signal lights to indicate the unit cutoff and emergency heat situations. On top of the control are the temperature-setting levers heating (A) on the left and cooling (B) on the right. Inside the control is a connecting bar between the two levers that prevents setting the levers any closer than 5°F. The reason for this is explained later.

On the right side of the device are the function control, fan control, emergency heat control, and the safety cutout light. The system switch (D) selects the system mode of operation: "auto" for automatic changeover from heating to cooling, and reverse, to maintain minimum and maximum operating temperature in the occupied area; "heat" for heating-only operation; "off" for complete system shutdown; and "cool" for cooling-only operation. Below this control is the fan switch (E). The "auto" position will cycle the inside blower motor with the compressor on either the heating or cooling operation. In the "on" position the inside blower motor will operate continuously regardless of the position of the function switch.

If the heat pump system should fail, the occupant can substitute the backup heat or auxiliary heat for the

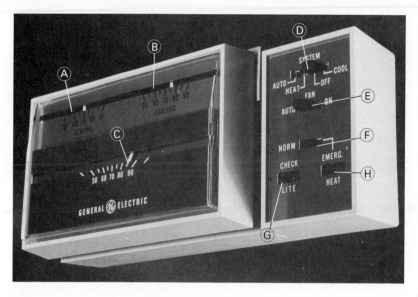

(A) Heating temperature selector
(B) Cooling temperature selector
(C) Temperature indicator
(D) Systems switch
(E) Fan switch
(F) Normal/Emergency heat switch
(G) Check-Lite
(H) Emergency heat light

FIGURE HP2-1   Heat pump thermostat. (*Courtesy* Addison Products Company)

heat pump by moving the "emergency heat" (F) control from "norm" to "emergency heat." This removes the compressor operation from the control and puts the backup heat on the primary or first stage of the thermostat. At the same time, the emergency heat light (H) is energized and remains on as long as the switch is in the "emergency heat" position. This is to constantly remind the occupant that the higher-cost heat source is being used.

If the compressor unit should cut out on high head pressure for whatever reason, the check light (G) will turn on when the compressor lockout relay functions. This serves to remind the occupant that trouble, such as dirty air filters, exists.

In the wiring diagram in Fig. HP2-2, the thermostat is used in conjunction with a heat pump system plus two stages of electric auxiliary heat: three 5-kW elements. At the top of the diagram, the high-voltage wiring of the outdoor unit is shown. At the bottom of the diagram, the high-voltage circuits of the inside electric unit air handler are shown. In the center, all the low-voltage circuits are pictured, the thermostat circuits on the top right and the inside electric air handler on the bottom right. To explain each function, the complete diagram is repeated with the selected circuit in color.

In Position 1: selector switch off-fan "auto," 240-V power is supplied to the control transform primary and 24 V to the power (R) side of the thermostat. All

switches in the thermostat control subbase are open and no action can result (Fig. HP2-3).

In Position 2: selector switch "off"-fan "on," 24-V power is supplied through the R circuit of the thermostat to the "on" terminal of the fan switch, the G terminal of the subbase through the G terminal of the indoor unit, through the 24-V coil of the fan relay to the return or "common" side of the 24-V transformer. The energized relay pulls in and breaks the normally closed contacts 5 and 6 and makes the normally open contacts 2 and 4 of the relay. This supplies 240 V to the high-speed lead and common of the motor. The motor runs on high speed as long as the thermostat subbase fan switch is in the "on" position (Fig. HP2-4).

In Position 3: selector switch on "heat"-fan "auto," with the "heat" selector switch set at 70°F and the room ambient at 69°F, the heating first-stage contact (HTG-1) closes. This provides 24-V power through the fan switch and the G circuit to the fan relay (FR). The fan relay pulls in and the indoor fan operates on high speed. In addition, power is supplied through the normal "norm" position of the emergency switch, the W1 terminal of the subbase, and the Y terminal of the outside unit through the compressor contactor C. Power then flows to the lock out relay (LOR) but cannot energize the relay because the high-pressure control (H.P.C.) and the normally closed contacts (4 and 5) in the lockout relay provide a bypass around the lockout

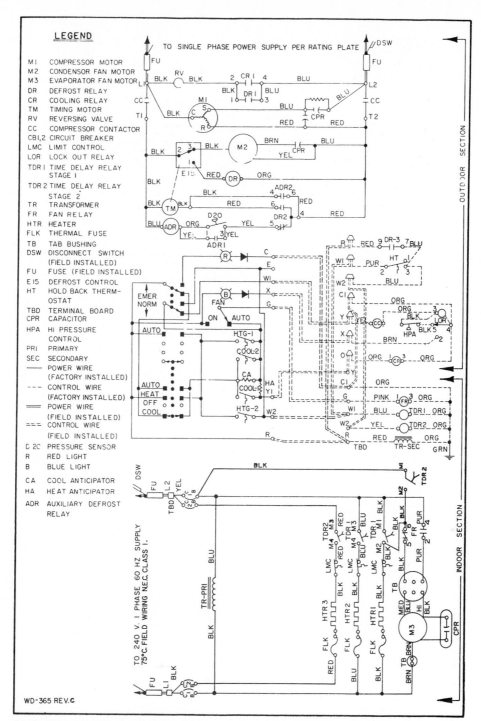

SYSTEM WIRING DIAGRAM—2 THRU 3 TON WITH 15 KW

FIGURE HP2-2   Heat pump system wiring diagram. (*Courtesy* Addison Products Company)

relay. The circuit from the lockout relay is back through common (Cl) to the return or common side of the power transformer. The energizing of the compressor contactor closes the contacts (CC) in the high-voltage circuit, energizing both the condenser fan and compressor. These continue to operate as long as the thermostat first stage calls for heat (Fig. HP2-5).

If the condition area temperature should continue to drop because the heating load is greater than the ca-

pacity of the heat pump, the second-stage contact (HTG-2) in the thermostat closes (see Fig. HP2-6). This supplies 24-V power through the W2 terminal of the subbase and the W1 terminal of the electric air handler to the first-stage electric heat time-delay relay (TDR-1). This energized time-delay relay closes its contacts in step sequence with 20 to 30 seconds between the steps. Step 1 closes contacts M1 and M2, providing 24-V power across heater 1 (HTR1). The second set of contacts (M3

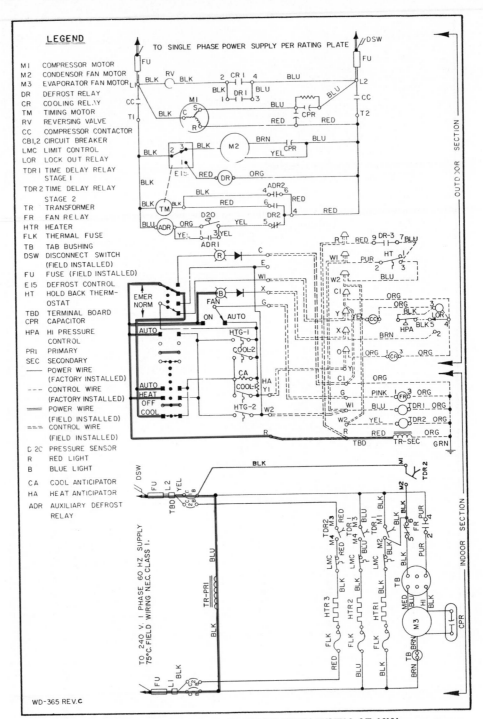

SYSTEM WIRING DIAGRAM—2 THRU 3 TON WITH 15 KW

FIGURE HP2-3 Position 1: selector switch "off"—fan "auto." (*Courtesy* Addison Products Company)

and M4) then close to provide power across heater 2 (HTR2).

When the area temperature rises, the second stage of the thermostat opens to cut off the auxiliary heat before the heat pump is cut off. There is a 2° difference between the first and second steps to prevent the reverse action of stopping the heat pump before stopping the auxiliary heat.

In Position 4: selector switch on cooling–fan "auto," when the area temperature rises above the heat setting of the thermostat, the thermostat automatically prepares the circuits for cooling operation. A 3° rise in occupied area temperature causes the final-stage cooling thermostat (COOL-1) to close. This energizes the cooling relay, through the Y1 and Y terminals and circuits.

When the cooling relay is energized and pulls in,

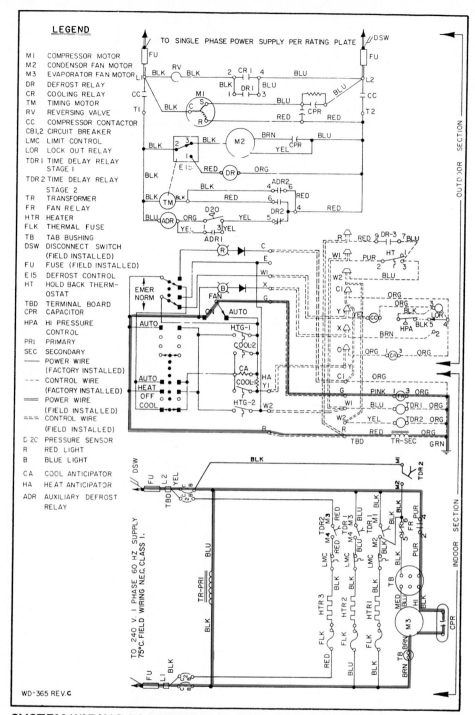

SYSTEM WIRING DIAGRAM—2 THRU 3 TON WITH 15 KW

FIGURE HP2-4   Position 2: selector switch "off"—fan "on." (*Courtesy* Addison Products Company)

normally open contacts 2 and 4 close and put 240-V power across the reversing valve coil (RV). The reversing valve pilot valve operates to put the valve in the cooling position should the compressor start because cooling operation is needed (Fig. HP2-7).

A further rise of 2°F in the conditioned area temperature will cause the second-stage cooling thermostat (COOL-2) to close to operate the compressor.

Close inspection of the circuits will show that thermostat contacts HTG-1 and COOL-2 are in parallel and closing of either one will provide compressor operation. Whether heating or cooling is provided will depend on the position of the first-stage cooling thermostat (COOL-1), which controls the position of the reversing valve. To provide a temperature range for this operation is the reason for the 5° minimum differential between

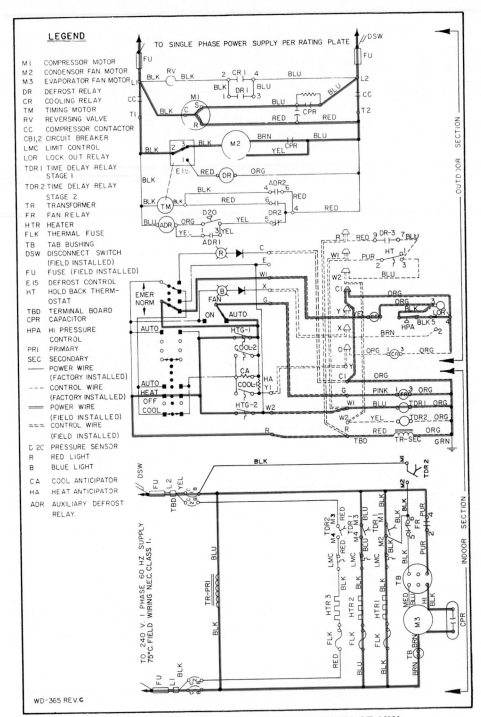

SYSTEM WIRING DIAGRAM—2 THRU 3 TON WITH 15 KW

FIGURE HP2-5 Position 3: selector switch on "heat"—fan "auto." (*Courtesy* Addison Products Company)

the heating and cooling setting of the thermostat.

In Position 5: emergency heat switch on emergency (EMER)–fan "auto," (Fig. HP2-8) when the emergency heat switch is moved to the "EMER" position, two things happen:

1. Power is supplied directly to the "emergency heat" light (R) and it burns continuously as long as the switch is in the "EMER" position.

2. The circuit from the first-stage heating thermostat (HTG-1) is switched from W1 to the E circuit, which feeds down to W2 through W1 of the electric air handler and operates the first-stage heat through time-delay relay (TDR-1). This provides an occupied area temperature of the first-stage setting of the thermostat if the heat pump should fail.

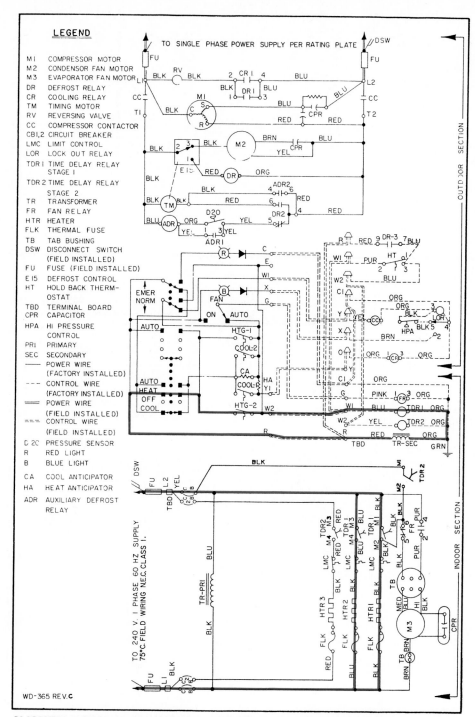

SYSTEM WIRING DIAGRAM—2 THRU 3 TON WITH 15 KW

FIGURE HP2-6   (*Courtesy* Addison Products Company)

With the selector switch on automatic–fan "auto," in the heat or cool positions, the thermostat will operate only in the range selected by the individual selector areas on the top of the thermostat. When in the automatic (auto) position, all four stages of the thermostat are energized and can provide operation if the area temperature requires it (Fig. HP2-9). The sequence of operation would be:

1. Room temperature below the second stage heating thermostat means the compressor operating along with the auxiliary heat. When the room temperature reaches 2°F below the heat setting, HTG-2 opens and the auxiliary heat shuts off.
2. When the occupied area temperature rises the 2°F to the setting of the thermostat, (HTG-1) the compressor cycles off.

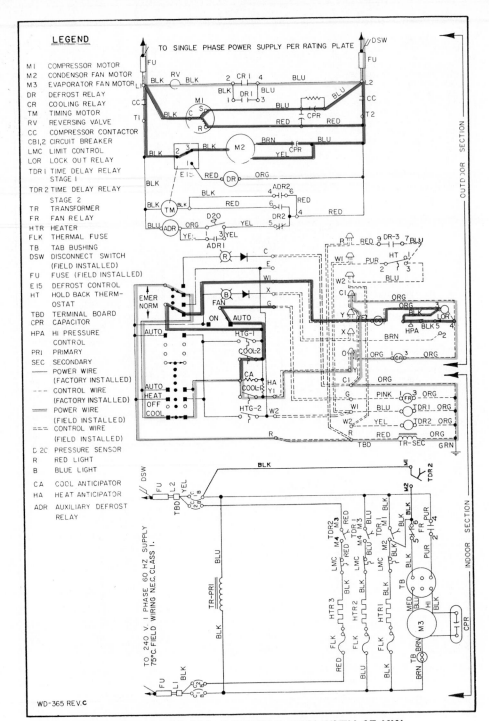

SYSTEM WIRING DIAGRAM—2 THRU 3 TON WITH 15 KW

FIGURE HP2-7   Position 4: selector switch on cooling—fan "auto." (*Courtesy* Addison Products Company)

3. A further rise of 3°F causes the first-stage cooling thermostat (COOL-1) to close, energizing the reversing valve.

4. Another 2°F rise closes the second-stage thermostat (COOL-2) and the compressor cycles on.

The span of the operating temperatures can be extended above the 5°F but cannot be less than the 5°F because of the operating differentials needed for each thermostat element.

HP2-2.2

Thermostat: Outdoor Ambient

When the second stage heat is required, it is possible to lower the operating cost by limiting the opera-

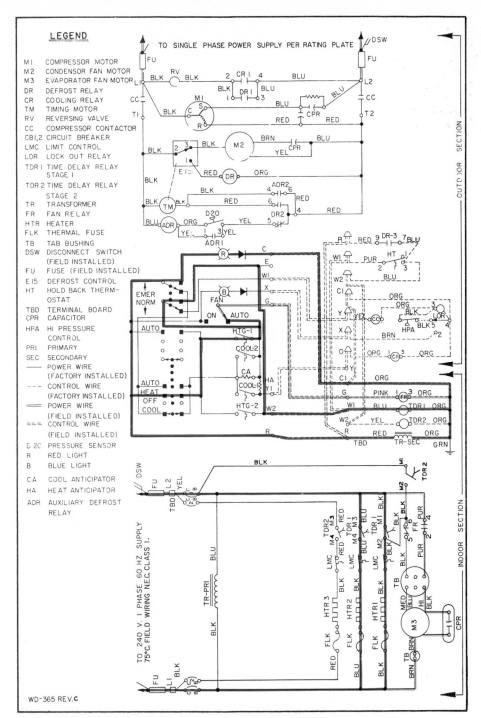

SYSTEM WIRING DIAGRAM—2 THRU 3 TON WITH 15 KW

FIGURE HP2-8   Position 5: emergency heat switch on emergency (EMER)—fan "auto." (*Courtesy* Addison Products Company)

tion of the electric heat elements until the outdoor temperature (the building heat loss) requires it. To prevent element operation above the point of heat loss where they are not required, outdoor ambient thermostats are used. In Fig. HP2-10 the circuit through the hold-back thermostat (outdoor ambient thermostat) is outlined. It is a temperature-operated switch located in

the outdoor unit that is connected between the first- and second-stage connections (W1 and W2) to prevent current flow to the second-stage time-delay relay (TDR 2) when the outside temperature is above the setting of the thermostat. The selection of the setting of the thermostat is covered in Chapter HP5, Heat Pump Startup, Checkout, and Operation.

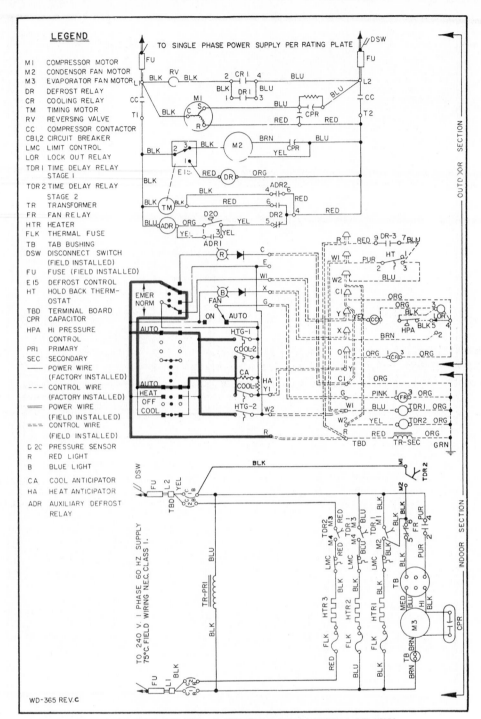

SYSTEM WIRING DIAGRAM—2 THRU 3 TON WITH 15 KW

FIGURE HP2-9   Selector switch on automatic—fan "auto." (*Courtesy* Addison Products Company)

≡ **HP2-3**
≡ *DEFROST CONTROLS*

When operating on the heating cycle, if the outdoor temperature and humidity are such that the air is cooled below the dew point, moisture will condense on the outside coil. If the coil temperature is below 32°, the moisture turns to frost and sometimes ice. Because the frost forms between the coil fins, it adds resistance to the airflow, reduces the amount of air through the coil, and reduces the unit capacity. Therefore, when the frost has accumulated enough to seriously affect the unit performance, it has to be removed.

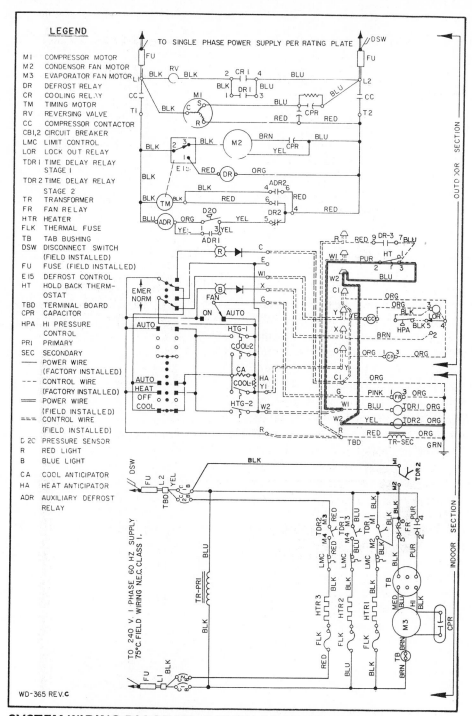

SYSTEM WIRING DIAGRAM—2 THRU 3 TON WITH 15 KW

FIGURE HP2-10 (*Courtesy* Addison Products Company)

Although many systems of warm air, electric heaters, etc., have been tried, the only successful way has been by the hot-gas method or reversing the system to the cooling cycle with the outdoor blower off. This means that the unit is operated as a cooling unit, taking heat from the indoor air (the occupied area during the defrost cycle). This, of course, causes a drop in area temperature during this operation. To overcome this, part of the auxiliary heat was turned on to prevent the air temperature drop.

This operation means that three separate functions have to take place in three separate locations in the control system. This is done by means of a multiswitch type of relay. Regardless of the means of starting the defrost cycle (initiation) and stopping the defrost cycle (termination), some type of defrost relay is involved.

Temperature Initiation-Temperature Termination

The first defrost control was a two-temperature control produced by Ranco, Inc. that compared the temperature of the outside coil with the temperature of the air entering the outside coil (Fig. HP2-11). The outside air temperature was measured by means of a straight capillary type of sensing element connected to a pressure diaphram in the control. The coil temperature was measured by a bulb-type sensing element attached to the return bend of the final refrigerant circuit of the coil. This circuit would be the last portion to reach the required temperature, to assure the coil staying free of frost and ice (Fig. HP2-12).

Under normal clear conditions, the coil operating as a DX coil operated with design spread of temperature between coil temperature and air temperature. On most heat pumps, this spread varied from 15 at 40°F air temperature to 10 at 10°F air temperature. Performance tests also revealed that the highest rate of frost formation usually occurred between 45 and 35°F air temperature. Above 45 to 48°F, frost formation is practically nonexistent. Therefore, these controls had the air temperature capillary, called the ambient airstream reference, purposely charged to prevent defrost operation above 45 to 48°F air temperature.

The unfortunate part of this control was that it depended on coil temperatures following the correct temperature change according to the amount of air through its coil. This meant that at all times the operating discharge and suction pressure had to be correct. Unfortunately, at the time that this control was used, heat pumps were connected to a varying quality of inside duct systems, from undersized to totally inadequate, to handle the required amount of air. As a result, operating head pressures were higher than normal, forcing

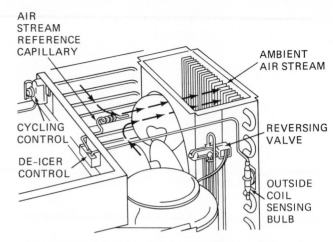

FIGURE HP2-12   Defrost control. (*Courtesy* Ranco, Inc.)

coil operating pressures higher than normal. As a result, the high coil operating temperatures forced the formation of larger quantities of frost to form before the control could sense the required temperatures to start the defrost cycle. After many attempts to adjust a control that was not field adjustable, the control usage was discouraged.

Pressure Initiation-Temperature Termination

The pressure initiated–temperature terminated defrost control (Fig. HP2-13) works on the principle of measuring the airflow resistance through the outside coil. A clean coil at full rated airflow has a resistance of 0.15 to 0.25 in. of W.C. At the time these controls were used, horizontal airflow units were the design. The pressure taps were connected, one to the inside of the

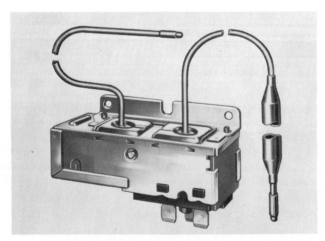

FIGURE HP2-11   (*Courtesy* Ranco, Inc.)

FIGURE HP2-13   Defrost control. (*Courtesy* Ranco, Inc.)

cabinet to measure the negative pressure of the air downstream of the coil and the other connected outside the cabinet to measure atmospheric pressure. All the units were of the draw-through type with the fan located downstream from the coil.

Two requirements had to be met in order for the unit to go into defrost:

1. The coil temperature at the inlet on the heating cycle (outlet on the cooling cycle) had to be below 26°F.

2. The static pressure drop through the coil had to reach 0.65 in. of W.C.

When these two conditions were encountered, the contacts would close and the defrost relay would be energized and pull the unit into the defrost cycle. When the temperature bulb reached 55°F, the coil should be free of frost and ice and the contacts opened. This terminated the defrost cycle and the unit was back in the heating cycle.

The major problem with the control was wind. If the wind was against the coil discharge, it prevented the development of the necessary pressure drop until an excessive amount of frost gathered on the coil. Sometimes the unit would not go into the defrost cycle. If the control was adjusted for the heavy wind pressure, it would go into unnecessary or "nuisance" defrost cycles during periods of little or no wind.

HP2-3.3
Time Initiation-Temperature Termination

To remove the problem of wind effect, the industry moved to the theory that the heat pump will require defrosting every 30 minutes of compressor operation. In dryer climates the time period may be 90 minutes. This time period, also, is with the terminating thermostat closed, indicating that the coil inlet is below 26°F, cold enough to build frost.

Figure HP2-14 shows a typical timer installation in a unit electrical control panel. The defrost timer in the upper right-hand corner (item 2) is used in conjunction with the defrost relay (item 3) and the terminating thermostat (Fig. HP2-15) fastened to the outside coil inlet circuit. In order to determine the cycle time of the timer, two drive cams were used. Because the predominate time period was the shorter one, most of the factories shipped the timer with the 30-minute cam installed and the 90-minute cam as a loose spare item.

Figure HP2-16 shows the 30-minute cam arrangement in the top picture and the double-cam 90-minute arrangement in the bottom section. In both arrangements, the action of the cam caused the contact point to close when the cam, turning clockwise, caused the bottom follower to drop into the cam notch. This will

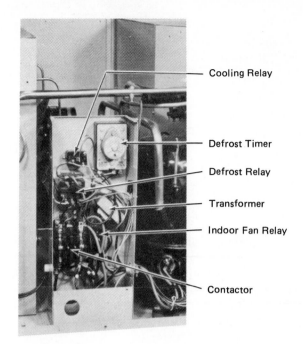

FIGURE HP2-14 (*Courtesy* Addison Products Company)

close the contact, energizing the relay. Approximately 30 seconds later the second follower drops, opening the circuit to the relay. The relay has a holding contact that holds it in the energized condition when the timer contact opens.

Figure HP2-17 outlines the defrost control circuit of the timer contacts as well as the defrost relay contacts. The defrost relay controls the action of the outside fan motor through the normally closed contact of terminal 4 (common) and terminal 5. When the defrost relay is energized, this contact opens and the outdoor fan stops.

The terminal combination of 1 and 3 (normally open) is the holding contact around the initiating contacts, terminals 3 and 4 of the defrost timer. When the defrost timer completes the required amount of running timer of the compressor—either 30 or 90 minutes, depending on the cams used—and the contact between terminals 3 and 4 closes, the relay is energized. This closes the contact between terminals 1 and 3. This contact is in parallel with the timer contact. Thirty seconds later, when the timer contact opens, the relay circuit is still closed and the relay remains energized.

FIGURE HP2-15 (*Courtesy* Addison Products Company)

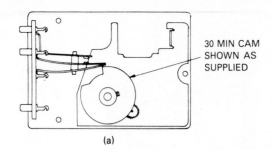

30 MIN CAM
SHOWN AS
SUPPLIED

(a)

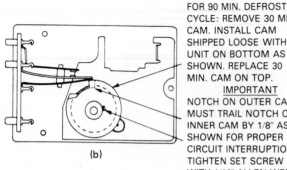

FOR 90 MIN. DEFROST
CYCLE: REMOVE 30 MIN.
CAM. INSTALL CAM
SHIPPED LOOSE WITH
UNIT ON BOTTOM AS
SHOWN. REPLACE 30
MIN. CAM ON TOP.
IMPORTANT
NOTCH ON OUTER CAM
MUST TRAIL NOTCH ON
INNER CAM BY 1/8" AS
SHOWN FOR PROPER
CIRCUIT INTERRUPTION
TIGHTEN SET SCREW
WITH 1/16" ALLEN WRENCH.

(b)

FIGURE HP2-16 (a) The timer with cover removed and 30-minute cam installed (as shipped from the factory). (b) The timer with cover removed and 90-minute cam installed. (*Courtesy* Addison Products Company)

In addition to the relay stopping the outdoor fan motor and holding the circuit, the relay also energizes the reversing valve through terminals 2 and 4 of the second defrost relay. This places the unit in the cooling mode to provide the hot gas for defrost. (On later units these two relays are combined into one assembly.) This action continues until the defrost thermostat (Fig. HP2-15) opens because the liquid refrigerant from the bottom outlet of the coil (now acting as a condenser) reaches 55°F, signifying that the coil is free of frost and ice.

If, however, the power to the compressor circuit or the 24-V control circuit were to be interrupted, the defrost control system would assume the defrost process were complete and the defrost cycle would be terminated. This is covered in Chapter HP6. The major problem with this control system was the potential for unnecessary defrost periods, which would tend to raise operating cost.

HP2-3.4
Pressure-Time Initiation-Temperature Termination

The hold this false cost to a minimum, a pressure switch measuring the coil airflow resistance has been combined with the time clock and termination bulb to provide the following action:

1. No action until the coil temperature is below 26°F

and the coil has reached 80% blockage from ice formation

2. A means of adjusting the time between defrost cycles to allow a closer relationship between the length of the defrost cycle required and the frequency of the defrost cycle

3. A means of terminating the defrost cycle in the case of high wind and cold outside ambient temperatures

Figure HP2-18 shows the wiring diagram that is used in the thermostat section with the defrost control circuit as well as auxiliary relay system. The auxiliary relay system has been added by some manufacturers to eliminate high-frequency interference from TV sets and radios. This interference was caused by the rapid make and break of the high-voltage pressure switch (D20) before the pressure buildup became enough to close the switch securely.

The action of this type of system occurs in the following steps:

1. The frost buildup is sufficient to cause the contacts in the pressure switch (D20) to close.

2. This contact closure energizes the auxiliary defrost relay (ADR). Energizing the relay closes the contact between terminals 1 and 3, which shorts out the pressure switch and keeps the circuit active.

3. In addition, the contacts between terminals 4 and 6 of the auxiliary defrost relay close and energize the defrost timer motor. After a suitable period, if the temperature of the terminating bulb is below 26°F, the mechanical switch portion of the timer motor actuates, shuts off the condenser fan motor, and energizes the defrost relay.

4. The defrost relay performs the following functions:

   a. Energizes the reversing valve to put it in the cooling mode.

   b. Brings on the auxiliary heat to keep the supply air to the room above the comfort level.

   c. Disconnects the auxiliary defrost relay circuit by opening the contact between terminals 4 and 5 and closing the holding contact between terminals 4 and 6. This holds the power to the timer motor because the original making contact between terminals 4 and 6 in the auxiliary defrost realy will open when this relay drops out.

5. When the temperature of the terminating thermostat bulb reaches 55°F, the mechanical portion of the defrost control forces the contacts between terminals 2 and 1 to open, deenergizing the defrost relay and causing the contacts between terminals 2 and 3 to close, energizing the outside fan motor.

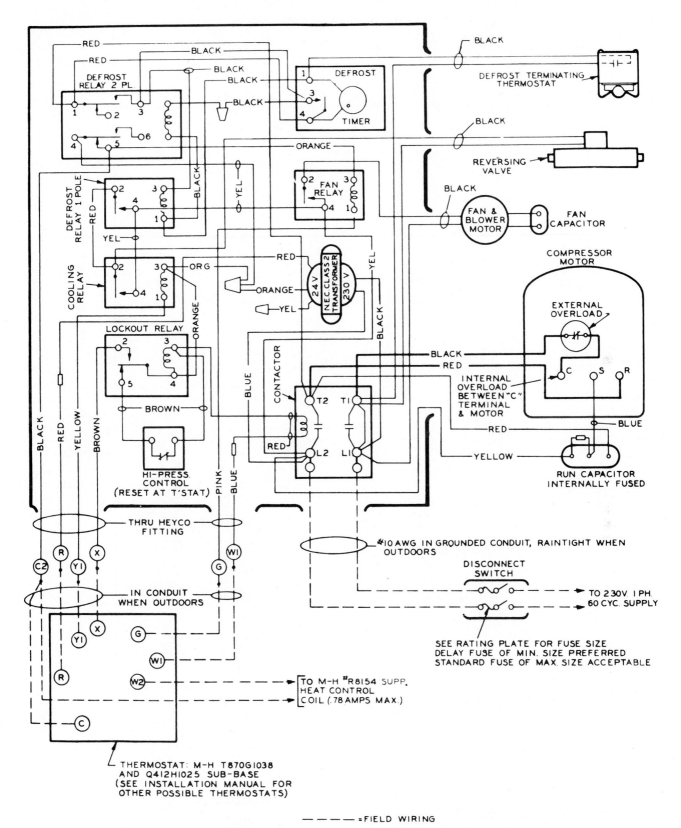

FIGURE HP2-17 (*Courtesy* Addison Products Company)

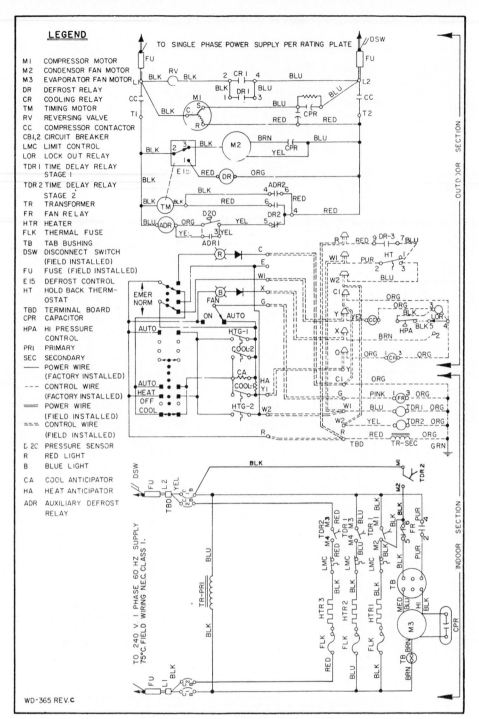

SYSTEM WIRING DIAGRAM—2 THRU 3 TON WITH 15 KW

FIGURE HP2-18 (*Courtesy* Addison Products Company)

The deenergized relay returns the contacts to their original position. The reversing valve returns to the heat mode, the auxiliary heat is turned off, and the contact controlling the ADR is closed, making the system ready for the next defrost period. This defrost control is that of model E15 manufactured by Ranco Controls of Plain City, Ohio.

HP2-4
HIGH-PRESSURE LOCKOUT

Item G in the description of the thermostat is the "check lite" to indicate that the outdoor unit has cut out due to excessive condensing pressure. Something has caused excessive head pressure to the setting of the "hi-

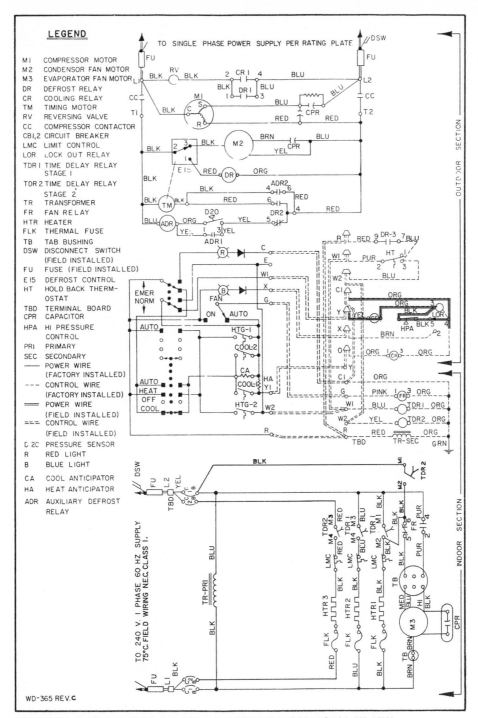

SYSTEM WIRING DIAGRAM—2 THRU 3 TON WITH 15 KW

FIGURE HP2-19 (*Courtesy* Addison Products Company)

pressure control'' (HPA). To provide manual reset of this control as required by UL and not to require the removal of the outside unit panels, a reset means is provided through the thermostat.

Figure HP2-19 shows the unit wiring diagram with the lockout circuit emphasized. In the compressor contactor coil circuit, the ''hi-pressure control'' (HPA) is connected across the coil of the lockout relay. This is the normal operating position of the control. The lockout coil, which has a very high resistance, is not in the circuit. When the thermostat calls for compressor operation by closing HTG-1 in the heating mode or COOL-2 in the cooling mode, the full 24 V of the control circuit is placed across the compressor contactor (CC) coil. The

Controls **661**

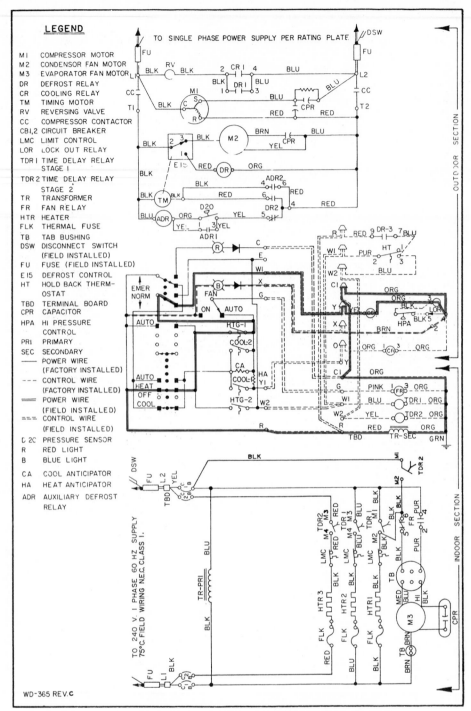

SYSTEM WIRING DIAGRAM—2 THRU 3 TON WITH 15 KW

FIGURE HP2-20   (*Courtesy* Addison Products Company)

coil is fully energized and exerts enough power to close the contacts.

If the "hi-pressure switch" opens, as shown in the marked circuit in Fig. HP2-20, the coil of the lockout relay is placed in series with the coil of the compressor contactor. The addition of the lockout relay coil increases the total load resistance and reduces the current flow through the circuit. The resulting amperage is

sufficient to pull the lockout relay in, but the compressor contactor drops out, stopping the compressor.

With the compressor off, the unit will cool down, reducing the head pressure and closing the "hi-pressure control." The lockout relay, however, has broken the circuit through the control by opening the normally closed contacts between terminals 4 and 5. At the same time, the normally open contacts between terminals 4 and 2 are closed. This provides a completed circuit through the blue light (B) in the thermostat and it glows, indicating that the unit is "off" on "lockout."

The unit will stay in this condition as long as power is supplied through the thermostat. To reset the control, only power interruption is needed. This can be provided by placing the function control switch in the "off" position and then returning it to the original operating mode. If the action of the lockout system should repeat, the system should be placed on emergency heat and a service technician called in to check the reason for lockout.

All the other controls have been reviewed in the refrigeration and air-conditioning sections.

PROBLEMS

HP2-1. Two types of temperature controls are used in heat pumps. What are they?

HP2-2. Four types of defrost control systems have been used in heat pumps. What are they?

HP2-3. Normal heat pump system room thermostats have eight functions. What are they?

HP2-4. In two-stage mechanical thermostats, the difference in temperatures held by the two stages is usually _____ °F.

HP2-5. A minimum of 5°F is used between the heating and cooling levers of the standard mechanical room thermostat. Why?

HP2-6. This 5°F temperature period is called the _____ .

HP2-7. What is the main purpose of an outdoor ambient thermostat?

HP2-8. The defrost relay has three separate functions. What are they?

HP2-9. The temperature initiation–temperature termination defrost control used two temperatures. What are they?

HP2-10. The highest rate of frost formation usually occurs at outside ambient temperatures between _____ and _____ °F.

HP2-11. What pressure is used by the pressure initiated–temperature terminated defrost system to initiate the defrost cycle?

HP2-12. What temperature is used by the pressure initiated–temperature terminated defrost system to terminate the defrost cycle?

HP2-13. In a time initiation–temperature termination defrost system, on what is the time the clock motor operates on based?

HP2-14. In a pressure-time actuated–temperature terminated defrost system, on what is the clock operating time based?

HP2-15. Manual reset of a automatic reset high-pressure control is provided by a device called a _____.

HP2-16. When the lockout relay is actuated in series with the contactor, why does the contactor drop out and the relay stay in?

HP2-17. To reset a lockout relay, it is only necessary to __ .

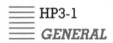

HP3

Heat Pump Equipment

HP3-1
GENERAL

Heat pump equipment is marketed in both unitary (complete package) and split systems (portion outdoors and portion indoors). The appearance of the equipment is exactly the same as that of its air-conditioning counterpart. The difference, of course, is in the design of the two coil assemblies involved, as well as the extra refrigerant flow controls needed for the reverse action.

The predominant product has been the package type, but the split or remote type has rapidly gained popularity due to the ease of installation as well as its versatility. In both designs, the types of heat source are duplicated. The heat source used as well as reversing ability determines the general classification of the particular unit.

Heat pumps are classified as follows:

1. Air to air–dual operation
2. Liquid to air–dual operation
3. Air to air–single operation
4. Liquid to air–single operation
5. Air to liquid–single operation

HP3-2
AIR TO AIR-DUAL OPERATION

The air to air–dual operation heat pump is the most predominant type of unit sold. Its capability to both heat and cool makes it a year-round application. Because of its energy and operating cost savings when applied with resistance electric heat, its popularity increased as energy costs increased. The air-to-air heat pump is marketed in both a package type, with all components in one package, and a split type, with an air handler in the

area and the compressor/heat exchanger assembly outdoors.

Figure HP3–1 shows a typical self-contained or package heat pump, including a complete cabinet view as well as an interior view of the refrigerant circuit and electrical components. Included in the package view are the optional electric heat strips located in the blower discharge area. Some type of auxiliary heat is required except in semitropical climates such as that in southern Florida. The easiest to incorporate into the heat pump assembly are electric strip heaters.

Figure HP3-2 includes both the air handler with an electric strip heat and outdoor section. Although not visible, the electric strip heat is located in the upper section in the blower discharge area, behind the control assembly and enclosing panel.

Both types of equipment have their advantages and disadvantages. The package unit has no refrigerant line installation cost but is limited in application situations. The split system can be used in a wide range of vertical, horizontal, or downflow applications, but installation cost is higher. However, both types of systems have their place in the market.

HP3-3
LIQUID TO AIR-DUAL OPERATION

Where groundwater is available, the liquid-to-air unit is rising in popularity. In addition to the use of groundwater as a heat source, this type of system is used on indirect liquid systems. In these systems, a liquid is pumped through coils buried in the ground or other source of heat. The advantages of the application are the control of the water characteristics through the coil in the heat pump, as well as the use of heat storage means, such as a solar-heated storage system. The liquid-to-air heat pump is marketed in both configurations: package and split.

(a)

(b)

FIGURE HP3-1 (*Courtesy* Bard Manufacturing Company)

An upright-type self-contained unit is shown in Fig. HP3-3 and a horizontal type of Fig. HP3-4. This type of unit is located totally within the conditional area. The only connection to the outside area would be the water supply and disposal.

Figure HP3-5 shows the interior parts arrangement of the vertical model shown in Fig. HP3-3. The upper section of the unit contains the inside coil with the necessary refrigerant controls as well as the blower and motor assembly for the inside air movement. The bottom compartment contains the motor-compressor assembly, the water-to-refrigerant heat exchanger, reversing valve, and the operating controls. A liquid-to-air unit does not contain the defrost control system, as it is not designed to operate with water-to-refrigerant heat exchanger surfaces below 32°F. This unit employs a tube-in-a-tube or coaxial type of heat exchanger as well as a water regulating valve to control the flow of water

FIGURE HP3-2 (*Courtesy* Addison Products Company)

and a condensate pump for disposal of the inside coil condensate to drains elevated above the unit.

Figure HP3-6 shows the interior parts arrangement of the horizontal unit shown in Figure HP3-4.

HP3-4
AIR TO AIR–SINGLE OPERATION

Several companies, especially in the North-Central area, have marketed a single-operation heat pump. This unit provides a heating-only cycle. Therefore, it does not

FIGURE HP3-3 (*Courtesy* Koldwave Division, Heat Exchangers, Inc.)

Heat Pump Equipment   665

FIGURE HP3-4 (*Courtesy* Koldwave Division, Heat Exchangers, Inc.)

FIGURE HP3-6 (*Courtesy* Koldwave Division, Heat Exchangers, Inc.)

contain a reversing valve or the pressure-reducing device and check valve combination on the inside coil. It also does not have a check valve in conjunction with the pressure-reducing device on the outside coil. Located in areas where the cooling load is slight or nonexistent, this unit has a market.

HP3-5
LIQUID TO AIR-SINGLE OPERATION

Like the air to air–single operation system, this unit is designed for heating only in extreme northern locations. Designed to operate with water temperatures slightly above the freezing point, the water-to-refrigerant heat exchanger is larger with more surface to provide the necessary heat transfer with very little temperature drop of the water through the heat exchange. Liquid to air units–double operation operate with an 8 to 10°F ΔT of the water through the heat ex-

changer. Liquid to air units–single operation operate with only a 2 to 3°F ΔT in the water through the heat exchanger. Considerably more water is circulated, but the unit operates in the colder-water zones.

HP3-6
AIR TO LIQUID-SINGLE OPERATION

Most air-to-liquid applications have been for heating domestic hot water. The increase in the use of heated swimming pools has opened a market for large Btu/hr capacities in this type of heat pump. Figure HP3-7 shows a smaller domestic water heating unit. This unit is designed to sit on top of a standard electric water heater. This provides hot water by taking heat from the surrounding air and heating the water with an EER factor of 2.9—thus more hot water for less energy cost.

Some units are designed to be connected into the water circulating system of the water heater. Shown in Fig. HP3-8 is a remote-type unit. In this application, plastic hose (copper or galvanized pipe where required

FIGURE HP3-5 (*Courtesy* Koldwave Division, Heat Exchangers, Inc.)

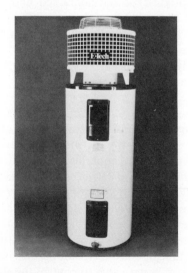

FIGURE HP3-7 (*Courtesy* E-Tech, Inc.)

by code) is used to connect the heat pump to the hot water tank. The heat pump is set to maintain a higher water temperature than the settings of the thermostats of the gas or electric water heater. This means that the heat pump is the primary means of heating the water. The auxiliary electric or gas heat source cuts in only if the heat pump cannot provide sufficient heating capacity for the amount of water drawn at the particular time.

Usually, the thermostat settings are 130°F for the heat pump and 120°F for the auxiliary heat source. The exact settings will vary with each type of equipment, and the manufacturer's instructions should be followed.

FIELD PIPING

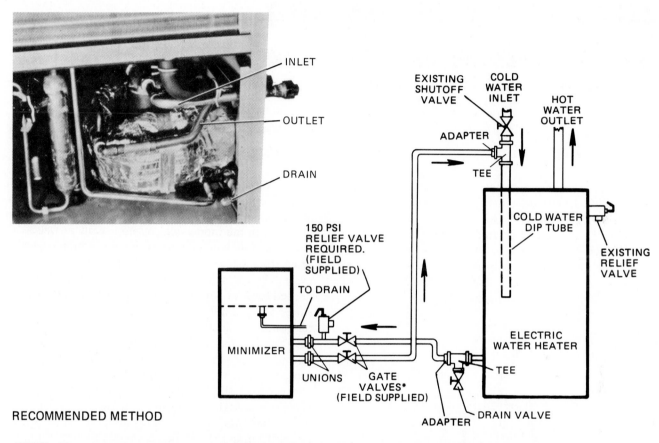

RECOMMENDED METHOD

*NOTE: These shutoff valves will allow the MiniMizer to be serviced and/or cleaned without draining the water heater tank.

FIGURE HP3-8   Field piping. (*Courtesy* Borg-Warner Environmental Systems, Inc.)

PROBLEMS

HP3-1. Heat pumps are produced in five classifications. What are they?

HP3-2. Air-to-air units are marketed in two types. What are they?

HP3-3. The most popular form of auxiliary heat is _____ .

HP3-4. The advantage of a package-type unit is _____ .

HP3-5. The advantage of the split-type system is _____ .

HP3-6. Liquid-to-air units are manufactured in the same configurations as air-to-air with the same advantages. True or False?

HP3-7. Which type of unit is the easiest to use in a solar heating application?

HP3-8. Liquid-to-air units must be located in the conditioned area. Why?

HP3-9. On a liquid-to-air unit, two water regulating valves are used. What is their purpose?

HP3-10. Liquid to air–single operation units operate with a 2 to 3°F ΔT of the water through the evaporator. Why is this?

HP3-11. In what market are air to liquid–single operation units primarily used?

Heat Pump Measuring and Testing Equipment

HP4-1
GENERAL

The use of thermometers and pressure gauges, as well as electrical characteristics, have been covered in Chapters R22, A13, and A16. To repeat that coverage would add a considerable number of pages to the book. The reader should refer to these chapters for information on this subject. There are, however, some special instruments and accessories that have been popularized by the growth of the heat pump industry.

FIGURE HP4-1 (*Courtesy* Robinaire Manufacturing Corporation, Division of Kent Moore)

HP4-2
TEMPERATURE MEASUREMENT

Measurement of the performance temperatures in a heat pump system are similar to that in the refrigeration or air-conditioning systems. Air temperature across both the indoor and outdoor coils as well as refrigerant liquid temperatures are important regardless of whether the heat pump is in the heating or the cooling cycle. Thermometers that cover a wider range would be needed for heat pump applications, although the normal range of the dial-type thermometer usually covers both air-conditioning and heat pumps. Refrigeration systems may operate at such temperatures that are below the normal range and would require a different-range thermometer. On the other hand, thermometers used to check out fossil-fuel heating units are expected to operate over a higher range of temperatures.

Wide-range thermometers are desirable if it is desirable to keep the number involved to a minimum. Accuracy, however, is sacrificed if too wide a range is involved. Actual temperatures encountered with heat pumps are discussed in Chapter HP5.

HP4-3
PRESSURE MEASUREMENT

The same pressure measurements as in an air-conditioning unit, suction and discharge, are required for proper analysis of problems. In addition, a third pressure reading is required. This reading is the pressure of the vapor line between the outside coil and the reversing valve. This line, called the *vapor line,* will have suction pressure during the cooling cycle and discharge pressure during the heating cycle.

Figure HP4-1 shows a three-gauge set that is used for heat pump service. The lower left-hand gauge is the compound gauge to measure suction pressure. To obtain

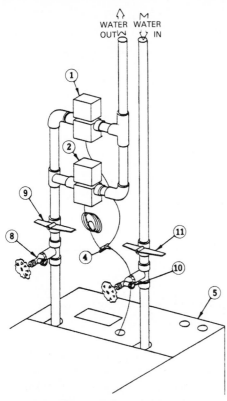

FIGURE HP4-2 Piping arrangement.
(*Courtesy* Bard Manufacturing Company)

the reading, a gauge tap is placed in the line between the centerline of the reversing valve and the compressor inlet. This line contains suction pressure regardless of the mode in which the unit is operating.

The lower right-hand gauge is a standard 0 to 500-psig gauge used to measure liquid-line pressure. Connected to the liquid line, it measures the pressure downstream from the condenser regardless of whether the inside or outside coil is performing the condensing function.

The third gauge, the top one, is connected to the vapor line. This one, in the cooling mode, will read suction pressure, and in the heating mode, discharge pressure. The difference between the readings on this gauge and the others is part of the information used in diagnosing trouble.

FIGURE HP4-3

HP4-4
PRESSURE MEASUREMENT—WATER SOURCE

On water source heat pumps, the fouling factor (lime buildup) in the outdoor coil is important for troubleshooting. Therefore, pressure gauges are used on the inlet and outlet sides of this coil. Figure HP4-2 shows the piping arrangement of the water supply and return to a typical water heat source unit.

In addition to shutoff valves (gate or ball type) in the main supply and return lines, shutoff valves are located in the branch connections of tees in these lines for connection to pressure gauges such as that shown in Figure HP4-3. This gauge set, with a flexible hose connector, will measure -30 to 150 psig and will operate very satisfactorily for measuring water pressure. If a constant flow valve (No. 7) is used, three gauge sets are required to get a complete pressure condition. *Do not use a refrigeration gauge set. Water in a refrigeration system means trouble.* Operating pressures are covered in Chapter HP5.

PROBLEMS

HP4-1. A gauge set for measuring pressures in a heat pump must be able to read what pressures?

HP4-2. In the heating mode, the vapor-line pressure will be close to the _____ pressure.

HP4-3. On liquid-to-air units, what is meant by the "fouling factor"?

HP4-4. How can the fouling factor be determined?

HP4-5. What types of valves are recommended for use in the liquid circuit?

Heat Pump Startup, Checkout, and Operation

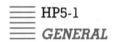

HP5-1
GENERAL

The proper startup, checkout, and operation are desirable from both the cooling and heating cycle standpoint in dual-operation units and in the heating-only operation in single-operation units. In all cases the correct amount of air or liquid must be supplied through the inside coil as well as through the outside coil to have the system operate at peak efficiency. To determine if the system is operating at peak efficiency at the time of startup, during checkout of the unit, and during operation, specific information is needed. Because this is different for each type of unit, each type is considered separately.

HP5-2
AIR TO AIR-DUAL OPERATION

Following is a suggested list of information needed for the startup, checkout, and efficiency testing of the air to air–dual operation heat pump.

1. Outdoor unit model no. _____
2. Outdoor unit serial no. _____
3. Indoor unit model no. _____
4. Indoor unit serial no. _____
5. Type of pressure-reducing device
   a. Outdoor unit
      (1) Expansion valve
      (2) Capillary tube
   b. Indoor unit
      (1) Expansion valve
      (2) Capillary tube
6. Type of fuse _____ Size _____ amperes
7. Wire size to outdoor unit _____

8. Wire size to indoor unit _____
9. Line voltage with unit off _____ volts
10. Condition of air filter

With a dual-operation heat pump, two separate sets of test conditions exist. The unit has operational limits. Therefore, it cannot be operated and checked out in both phases of operation unless the outside temperature is in the narrow overlapping range of temperatures for each phase of operation.

In the cooling cycle, operation is possible down to 65°F. In the heating cycle, operation is possible up to 70°F. Therefore, the initial checkout of the unit will be in either the heating or cooling cycle.

11. Voltage at condensing unit when trying to start _____ volts
12. Amperage draw of unit at:
    a. Startup _____ amperes
    b. Running _____ amperes
13. Return air temperature and relative humidity at return air grill _____ °F _____ %RH
14. Air temperature and humidity at coil _____ °F
15. Supply air temperature at unit _____ °F
16. Supply air temperature at supply register _____ °F
17. Suction-line temperature at inside coil _____ °F
18. Compressor suction pressure _____ psig
19. Refrigerant boiling point—inside coil _____ °F
20. Inside coil superheat _____ °F
21. Suction-line temperature at compressor _____ °F
22. Temperature of air entering outside coil _____ °F
23. Compressor discharge pressure _____ psig
24. Condensing temperature in outside coil _____ °F
25. Outside coil—ambient temperature–condensing temperature split _____ °F
26. Liquid subcooling _____ °F

27. Outside coil discharge temperature _____°F
28. Ft³/min through outside unit

HP5-2.1
Test Before Starting Unit

Items 1 through 10 are used before starting the unit in either the cooling or the heating mode.

1. For the service history file, the model number (1) and (3) and the serial number (2) and (4) of both the outdoor and indoor sections of the unit should be recorded. Any questions of possible warranty as well as service history can be taken into account.

5. The type of pressure-reducing devices used in both the outdoor and indoor unit should be recorded. This can be very important in checking and diagnosing possible problems. The action of the system is different if TX valves are used rather than capillary tubes.

6. Type of fuse protection—straight fuse or circuit breaker. Time-delay fuses or circuit breakers must be used on high-starting current motors. If they are not of time-delay type, they must be changed before the unit is placed into operation. If this is not done, unnecessary cutouts will occur.

    Are the time-delay fuses or circuit breakers of the proper size? Both outdoor and indoor units carry a fuse size stamping on the rating plate. Fuses smaller than the rating specified will cause unnecessary cutouts. Fuses or circuit breakers larger than specified will not provide the protection to prevent electrical damage.

7. Wire size to outdoor unit. The wires supplying power to the outdoor unit must be of sufficient size to prevent excessive voltage drop when the unit attempts to start. If the wire is oversized, usually the circuit breaker protection is also oversized, and this reduces unit protection as well as causing an unnecessary increase in installation cost. Undersized wire will cause excessive voltage drop and unit starting problems.

8. Wire size to indoor unit. The wire to the indoor unit must carry the control and blower electrical load as well as the auxiliary heat load. This equipment also carries a listing of the required wire size stamped on the rating plate. This listing should be followed.

9. Line voltage with unit off. The voltage at the unit, both the inside and outside sections (no-load voltage), must be within the voltage range of the unit. This is ±10% of the unit rated voltage if the unit is single phase. If the unit-rated voltage is 240 V, this means a maximum 264 V and a minimum of 216 V. These voltages are the extremes that the unit will tolerate and still continue to operate. This does not mean that the electrical system will have maximum design life. To accomplish this, the supply voltage must be at the unit rated voltage.

    In three-phase dual-voltage units, the voltage tolerance is +10%, −5%. This means that in a wide-range 208/230-V three-phase unit, the maximum is 230 V plus 10%, or 253 V, and 208 V minus 5%, or 198 V.

10. Condition of air filter. The amount of air through the indoor coil will seriously affect the performance of the system in both the cooling and the heating mode. In the cooling mode the reduction in ft³/min will cause coil icing and possible liquid return to the compressor. In the heating mode, the reduction in ft³/min will raise the head pressure and cutout on the lockout relay. In heat pump systems, throwaway filters should be changed once a month. Electronic filters should be cleaned once a month. Remember that the indoor blower is used throughout the year.

HP5-2.2
Test Upon Starting Unit

Two electrical checks should be made upon starting the unit. A voltmeter should be connected across the high-voltage terminals on the live side of the contactor and a clamp-type ammeter (preferably a digital type that has a high-amperage lock) around one of the wires to the contactor.

11. Voltage at the outdoor unit when it is trying to start. When the starting load is cut onto the line, the contactor closes, bringing on the compressor and condenser fan load. The voltage should not drop more than 5 V. If it does, either one of the following situations could be the problem:

    a. The wire size could be too small for the length installed. Most units will give the wire size in ampacity (amperage capacity) based on 60 to 100 ft of wire length between the distribution panel and the unit.

    b. The building distribution service size could be too small. The starting test should be repeated measuring the voltage at the circuit breaker in the distribution panel.

    If the starting voltage drop at the distribution panel is less than 5 V when the drop is more than 5 V at the unit, the branch circuit to the unit is too small. If the drop in starting voltage at the distribution panel is more than 5 V, either the service drop to the building is too small or the distribution power transformer is too small or overloaded. For this problem, the utility supplying the power should be contacted.

12. Amperage draw of the unit, both starting and running amps. Any time an induction load (motor) is started, there will be a high inrush current (LPA) and within 1 second this current should drop to the full-load amperes (FLA). The ammeters should be watched during the startup process to make sure that everything is correct and working properly. The unit should draw the proper inrush and running current to remain on the line. If the current does not drop below the rated FLA, *the unit must be disconnected immediately.*

FIGURE HP5-1

HP5-2.3
Test Upon Start—Cooling Cycle

The heat pump is sized to the cooling load rather than to the heating load. Any extra heating load above the cooling load is handled by auxiliary heat equipment. The most popular auxiliary heat devices are electric resistance elements. However, gas- or oil-fired units are sometimes used when the heat pump can no longer handle the load.

Because the unit is selected for the cooling load, the amount of ft³/min over the inside coil during the cooling cycle must be correct for the local weather design conditions. With the correct ft³/min through the coil, the unit will produce the desired temperature drop of the air. It is only necessary to measure the T °F of the air through the coil. Figure HP5-1 shows the suggested location of dial-type thermometers used to measure these temperatures.

At 80°F dry bulb and 50% relative humidity in the air entering the inside coil, the correct air temperature drop would be 20°F. Because the heat content of the air will change with any change in temperature and/or relative humidity, the correct amount of air will produce different air temperature drops.

If the temperature and/or humidity increases, the temperature drop will decrease. If the temperature and/or humidity decrease, the air temperature drop will increase. The system will remove only a given amount of heat. Therefore, the greater amount of heat in the air, the less heat will be removed per cubic foot. Because removal of humidity (latent load) does not affect temperature, the relative humidity change will have a greater effect on the air temperature drop. To determine the correct load on the inside coil (the correct ft³/min), the correct temperature drop must be determined. This is determined by the use of a psychrometric chart and a load ratio chart. The chart in Fig. HP5-2 has been plotted for the various conditions of air within the operating range of the unit. In this chart a range of relative humidity from 70 to 30% and dry-bulb temperatures from 65 to 92°F are given. Using a sling psychrometer, the

dry- and wet-bulb temperatures of the air to the coil are measured; the relative humidity is then determined from the psychrometric chart. The psychrometric chart was explained in detail in Chapter A3. Figure HP5-3 is the psychrometric chart we will use in our examples.

We will assume that the area to be cooled, at startup of the unit, has a measured dry-bulb temperature of 86°F and a wet-bulb temperature of 79°F. We want to know the correct air temperature drop of the air through the inside coil. Using the psychrometric chart (Fig. HP5-3), we find that the vertical 88°F dry-bulb line and the sloped 79°F wet-bulb line intersect at 68% relative humidity.

Using the air temperature drop of various load ratios chart (Fig. HP5-2), when we follow from the point

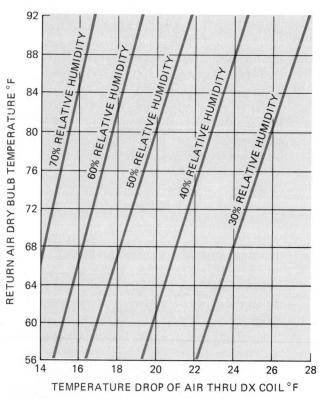

FIGURE HP5-2 Air temperature drops for various load ratios.

where the horizontal 80°F dry-bulb line and the sloped 68% relative humidity line cross down to the temperature drop scale, we find that an air temperature drop of 17°F is desired. The desired conditions we want to maintain, the same conditions used in ARI testing, would require a temperature drop of 20°F.

If we assume that the starting load conditions of the air were 88°F and 40% relative humidity, the lack of latent load would force the unit to cool the air further, taking more sensible heat, and the air temperature drop would increase to 24.1°F.

13. Return air temperature and relative humidity at return air grill. Using the sling psychrometer (Fig. A16-4), the dry-bulb and wet-bulb temperatures of the air at the return air grill as well as the test item 14.

14. Air temperature and relative humidity of the air at the indoor coil. Will show any air leakage in the return air system. If a temperature rise between the return air grill and the coil of more than 2°F is read, the return duct system should be checked for leaks and insulation.

15. Supply air temperature at coil. This temperature was taken to obtain the correct temperature drop information. This reading should be compared to the temperature of the air at the supply register.

16. Supply air temperature at register. The temperature of the air from the supply register at the end of the duct system, farthest from the coil, should be read and compared to the temperature off the coil. If a rise in temperature of more than 3°F is found, the supply duct system must be insulated. With a 20°F temperature drop of the air through the coil, every 1°F rise in the supply air temperature is a 5% loss in unit capacity.

17. Suction line temperature at inside coil suction outlet. To determine the operating superheat of the inside coil on the cooling cycle, the suction-line temperature at a point not more than 6 in. from the coil suction manifold should be measured.

18. Compressor suction pressure. The compressor suction pressure is measured at the tap located between the reversing valve and the compressor suction inlet. This is the true suction line. It remains the suction line in both the cooling and heating modes.

19. Refrigerant boiling point. Using a pressure–temperature chart, the compressor suction pressure is converted to a refrigerant boiling point. Because an accumulator and reversing valve as well as suction-line pressure drop (especially on split systems) are between the coil and the compressor, a pressure drop of 5 to 15 psig can be in the circuit, depending on the load. Therefore, measuring superheat is not very accurate on heat pump systems.

20. Inside coil superheat. Subtracting the temperature equivalent of the coil operating pressure (the refrigerant boiling point) from the suction line temperature will determine the operating range of the superheat under the unit operating conditions. Under ARI standard operating conditions, most heat pumps on the cooling cycle will operate with superheat of 10 to 15°F. This is with an outdoor temperature of 95°F. The superheat will vary widely if capillary tubes are used as the pressure-reducing devices. As explained in item 18 of Section A17-4, the coil operating characteristics using capillary tubes or thermostatic expansion valves in straight air-conditioning units and air-to-air heat pumps in the cooling mode are very similar.

21. Suction-line temperature at compressor. The temperature of the suction line at the compressor should be measured and compared to the suction-line temperature at the inside coil. Any rise in line temperature indicates a heat gain in the suction gas. There will be some gain even if the line is insulated and properly located. The gain should be kept to an absolute minimum. A temperature rise of more than 10°F should be investigated. This line is the hot-gas line on the heating cycle and any part of the line exposed without insulation is a costly and unnecessary loss of unit capacity.

22. Temperature of air entering outside coil. The temperature of the air entering the outside coil should be taken from a location directly in front of the coil inlet but not closer than 6 in. to the coil surface. On blow-through units, this would be at the inlet to the fan or blower. On draw-through units, such as vertical discharge, the temperature should be taken at several locations and averaged. On U-shaped coils, for example, temperature readings should be taken on both sides as well as the midpoint of the U. The average of three temperatures is used as the entering air temperature.

23. Compressor discharge pressure. After the unit has run for a sufficient period of time for the operating pressures to stabilize, the compressor discharge pressure is recorded. If the unit has a pressure tap in the compressor discharge line between the compressor and the reversing valve, this tap should be used to record the discharge pressure. If the pressure tap in the liquid line is used, trouble with the check valves not functioning properly can give false head pressure readings. Occasionally, it is necessary to put a gauge tap in this line. Some manufacturers have started to supply such a gauge connection.

24. Condensing temperature in outside coil. Converting the pressure to temperature (some gauges do this by having a pressure–temperature chart as part of the scale) will give the condensing temperature

ASHRAE PSYCHROMETRIC CHART NO. 1

NORMAL TEMPERATURE
BAROMETRIC PRESSURE 29.921 INCHES OF MERCURY
COPYRIGHT 1963
AMERICAN SOCIETY OF HEATING, REFRIGERATING AND AIR-CONDITIONING ENGINEERS, INC.

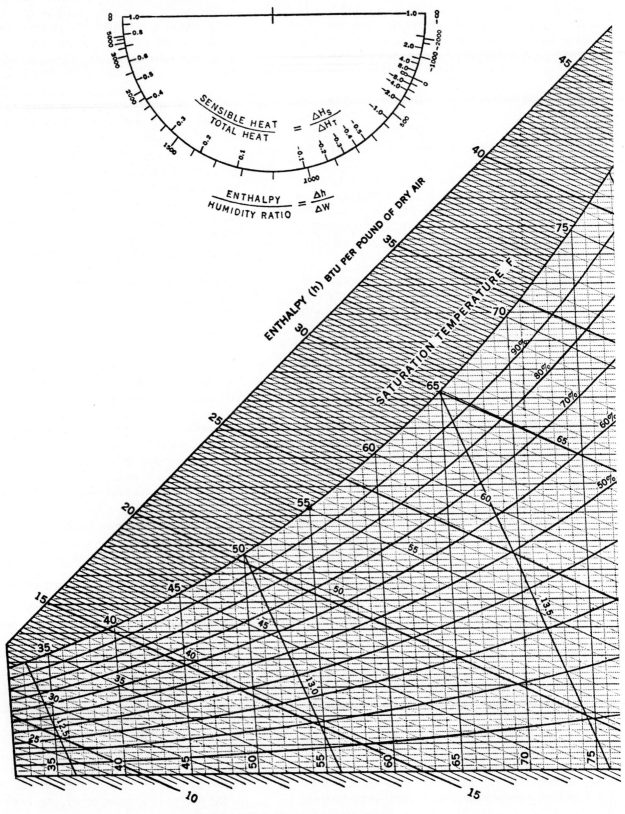

FIGURE HP5-3   Psychrometric chart.

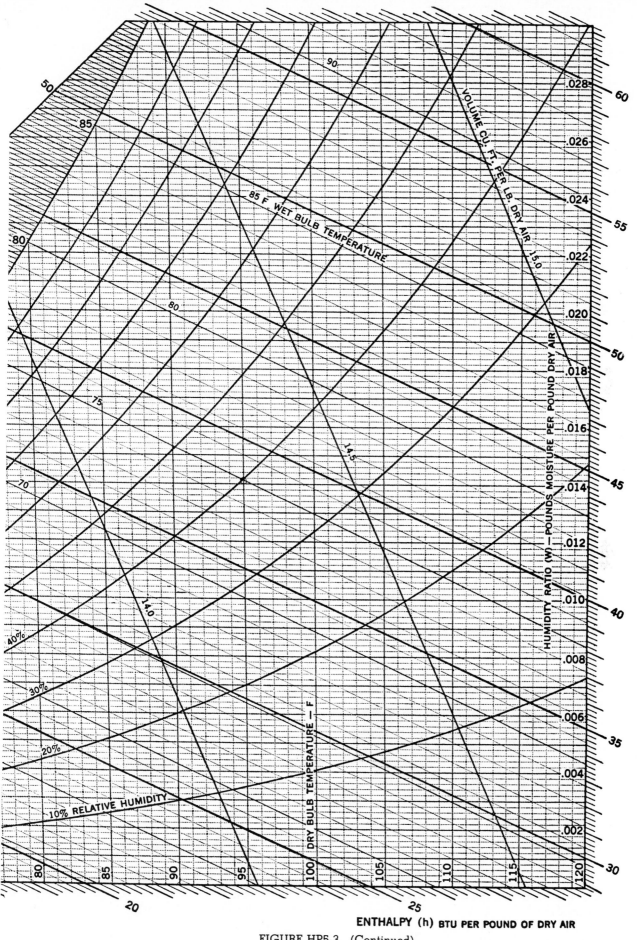

ENTHALPY (h) BTU PER POUND OF DRY AIR

FIGURE HP5-3 (Continued)

of the refrigerant in the outside coil. This temperature minus the temperature of the air entering the outside coil will give the "split" of the coil at the operating conditions. To judge the accuracy of the split, you must have the manufacturer's specifications for the particular unit.

25. Outside coil ambient temperature. As explained in item 24, the condensing temperature minus the outside coil ambient temperature will give the split of the coil.

26. Liquid subcooling. To obtain the amount of subcooling of the liquid refrigerant, the temperature of the liquid is measured. This measurement must be obtained by thermometer located not more than 6 in. from the liquid outlet of the outside coil. The liquid temperature subtracted from the condensing temperature will give the subcooling. Like straight air-conditioning units, most air-to-air heat pumps work their best at between 18 and 20°F subcooling.

27. Outside coil discharge temperature. The temperature of the air off the outside coil should be recorded for determining the capacity of the unit on the cooling cycle. This is covered in Section HP5-2.4. When determining the outside coil air discharge temperature, an average of reading from several locations should be done.

On outside units using blow-through fans (Fig. HP5-4), a grid of temperatures must be taken. Because the amount of air passing through the various sections of the outside coil will vary depending on the position of the fan, the temperature rise of the air will vary. Temperature-reading location marks are pictured on the coil outlet to obtain as accurate a temperature average as possible. On this particular unit, 15 readings would be used and averaged for the outlet temperature.

On vertical discharge draw-through units, thermometers are located on the discharge grills.

FIGURE HP5-5 Draw-through vertical discharge unit. (*Courtesy* Addison Products Company)

A minimum of three should be used, with one always located at the midpoint of the outside coil. The others are then placed to divide the circle equally (see Fig. HP5-5). Using the three or more readings, the average is obtained. This temperature minus the temperature of the air entering the coil, the coil ambient, will give the air temperature rise through the coil.

28. Ft³/min through the outside coil. Unless elaborate testing equipment is used, it is very difficult to determine the ft³/min through the outdoor unit. This information is listed in the manufacturer's literature. Figure HP5-6 shows a specification table of an outdoor unit. The outdoor (condenser) fan ft³/min is given in the third column from the right. This information is used to determine the unit capacity to be explained in HP5-2.5.

≡ HP5-2.4
≡ Operation—Cooling Cycle

After the unit has been checked out and adjusted properly, the operation performance of the system can be determined. Operating as an air-conditioning unit, in the cooling cycle, the net capacity can be determined by the same method as outlined in Section A17-5. Chapter A17 should be reviewed in its entirety for a clearer understanding of an air-to-air heat pump operating on the cooling cycle.

≡ HP5-2.5
≡ Test Upon Start—Heating Cycle

Because the unit is sized for the cooling load, it may not handle the heating load. If this is the case, an auxiliary heat supply is necessary to provide the comfort

FIGURE HP5-4 Blow-through unit. (*Courtesy* Addison Products Company)

Model	Electrical Charac-teristics	Operating Current (Amps) (1)	Minimum Fieldwire (AWG) (2)	Maximum Wire Run (Ft.) (3)	Minimum Fuse (Delay)	Compres-sor L.R. Current	Condenser Face Area (Sq. Ft.)	Condenser Fan CFM	Fan Motor	WEIGHT	
										Net	Ship
QH818-1D	1-60-230/208	11.1/11.8	14	33	20A	42	9.1	2670	PSC-$\frac{1}{4}$	166	192
QH824-1D	1-60-230/208	13.9/13.9	12	42	20A	53	9.1	2670	PSC-$\frac{1}{4}$	166	192
QH830-1D	1-60-230/208	17.2/17.2	10	48	25A	75	9.1	2670	PSC-$\frac{1}{4}$	189	215
QH836-1D	1-60-230/208	19.5/21.0	10	43	30A	81	9.1	3680	PSC-$\frac{1}{3}$	197	228
QH842-1D	1-60-230/208	23.0/23.0	10	35	35A	97	13.0	3620	PSC-$\frac{1}{3}$	242	297
QH848-1D	1-60-230/208	26.5/26.5	8	50	40A	118	13.0	3920	PSC-$\frac{1}{3}$	267	322
QH848-3D	3-60-230/208	19.0/19.0	10	40	30A	90	13.0	3950	PSC-$\frac{1}{3}$	267	322
QH860-1D	1-60-230/208	32.9/32.9	6	66	50A	139	13.0	4450	PSC-$\frac{1}{2}$	293	348
QH860-3D	3-60-230/208	20.4/20.4	10	34	30A	104	13.0	4450	PSC-$\frac{1}{2}$	293	348

(1) Per ARI Standard 240 Rating Conditions. (2) Copper wire only. (3) For longer run, increase wire gauge one size.

Source: Addison Products Company.

level required in the occupied area. For calculation of the heating requirement as well as the proper application of heat pumps, refer to Manual J, *Load Calculation for Winter and Summer Air Conditioning,* and Manual H, *Heat Pump Equipment Selection and Application,* published by the Air-Conditioning Contractors of America (ACCA), 1228 17th Street N.W., Washington, D.C. 20036.

It is necessary, however, to be sure that the heat pump is delivering the rated capacity so as to use its operating cost advantage as much as possible. The unit must be checked and information recorded even though some of the information applies to a different part of the system.

The part of the information listing that applies to both the cooling and heating cycles, items 1 through 12, will not be repeated. Other items in the remaining portion of the listing that do not apply to the heating cycle or are limited in their application will be so marked.

13. Return air temperature at return air grill. In the heating cycle the temperature rise across the inside coil, now the condenser, is the important item. Relative humidity has so little effect on the indoor coil operation that it is ignored.

14. Return air temperature at the indoor coil. Measuring the return air temperature at both the return air grill and the indoor coil is a method of determining the tightness as well as the amount of insulation on the return air system. A rise of more than 3°F means that the system should be checked and corrected.

15. Supply air temperature off the indoor coil. This temperature measurement and the temperature measurement of the air of the indoor coil will give an indication of the performance of the system. Because the output of a heat pump varies with outdoor temperature, it goes down as outdoor temperature drops. A single temperature rise cannot be used to grade performance. The amount of air over this coil is also a factor in determining the temperature rise. Usually, the leaving air temperature is ignored unless it is too high, which could cause high head pressure and unit cutoff on a high-pressure cutout.

Most heat pumps operate with an indoor coil leaving air temperature of 100 to 105°F at 70°F outdoors to 80 to 85°F at −10°F outdoors. If the temperature rise through the coil is within the range specified by the manufacturer and the unit is delivering the rated Btu/hr, it can be assumed that the unit is operating properly. The final check on air quantity should be done during the next cooling season.

16. Supply air temperature at register. As in cooling, any heat loss in the supply system means higher operating cost. A 3°F loss in a fossil-fuel system (80°F temperature rise) is only a loss of 3.75%. However, a 3°F loss in the heat pump system is 10% of 70°F and 20% at −10°F outside. Tight duct, well insulated, is an absolute necessity.

17. Suction-line temperature at outside coil suction outlet. The outside coil is now the evaporator. To determine the operating superheat on the heating

cycle, the temperature of the line leaving the outdoor coil is measured. The temperature reading should be taken within 6 in. of the coil outlet.

18. Compressor suction pressure. With the compound gauge connected to the gauge connection located in the suction line between the reversing valve and the compressor, the operating suction pressure is read. Because an accumulator and reversing valve are in the circuit between the outdoor coil (the evaporator on the heating cycle) and the compressor, the suction pressure reading can be anywhere from 5 to 15 psig below the actual coil pressure. Therefore, this reading is not accurate unless a wide difference is found. This is then used to analyze problems in the system.

19. Refrigerant boiling point. Converting the suction pressure to a boiling-point temperature will give the coil operative boiling point. This is done after the range in pressure drop is taken into consideration.

20. Outside coil superheat. Subtracting the temperature equivalent of the coil operating pressure from the coil outlet temperature will determine the operating superheat range under the operating conditions. Because the pressure drop through the accumulator and reversing valve will be higher at 60°F outside than at 0°F outside, the error in superheat will be higher at the 60°F.

21. Suction-line temperature at compressor. The suction-line temperature at the compressor is not important, as it is above the temperature of the coil outlet only due to the heat gained in the accumulator and reversing valve. The only importance is the minimum difference. If the two temperatures are very close, the superheat of the coil is low or nonexistent because of liquid feedout from the coil. This feedout could be of sufficient quantity to fill the accumulator and flood through the reversing valve.

22. Temperature of air entering outside coil. The ambient temperature of the air entering the outside coil is important in determining the rated capacity of the unit. Figure HP5-7 shows unit rated capacities at the various outdoor temperatures between 70 and −10°F. Both the Btu/hr heating capacity and COP (coefficient of performance) rating are given. This is used to determine unit size when applying the unit to a particular job as well as comparing the results of a performance check.

23. Compressor discharge pressure. The compressor discharge pressure reflects the condition of the filter in the inside unit as well as the condition of the coil and the air distribution system. Anything that affects the amount of air through the inside coil will be reflected in the compressor discharge pressure.

If the unit has a gauge pressure tap between the compressor discharge outlet and the reversing valve (some manufacturers have provided this), the discharge pressure reading at this point helps in diagnosing problems in the inside coil check valve assembly.

If the compressor discharge pressure is measured at the liquid line, the motor amperage draw must be measured at the same time. This is discussed further in Chapter HP6.

24. Condensing temperature inside coil. To find the amount of subcooling on the heating cycle, the compressor discharge pressure is converted to condensing temperature. This temperature is also used in conjunction with the temperature of the air entering the inside coil to determine the split of the coil.

25. Inside coil ambient temperature. In this mode of operation, the inside coil ambient temperature and the temperature of the return air are the same.

26. Liquid subcooling. The refrigerant condensing temperature minus the temperature of the refrigerant liquid leaving the inside coil is the liquid subcooling. The temperature of the liquid must be measured within 6 in. of the coil outlet. This is usually the inside coil connection immediately outside the inside coil cabinet. Most heat pumps with an 18 to 20°F subcooling on the cooling cycle will operate with a 13 to 15°F subcooling on the heating cycle.

27. Inside coil discharge temperature. This is the supply air temperature from the unit into the conditioned area. This temperature reading, together with the inside coil ambient temperature, give the temperature rise of the air through the inside coil. (This information is useful to check operating air temperatures and also to determine air quantity through the coil.)

28. Ft³/min through inside coil. The ft³/min through the inside coil can be determined by means of static pressure measurements or by means of auxiliary heat test. If the auxiliary heat is electric resistance elements, the process is very simple. If the auxiliary heat is a fossil-fuel (gas or oil) supply, the process is more complicated and the static pressure method is advised.

   a. To use the static pressure method, the performing characteristics of the unit must be known. All manufacturers publish this information in their specifications because it is needed to satisfactorily apply the unit to the air distribution duct system. Figure HP5-8 shows the amount of ft³/min a unit will deliver against the total static pressure external to the unit. By measuring the supply pressure and return pressure added together, the external static pressure is

FIGURE HP5-7
Application ratings—heating.

Model	OUTDOOR AMBIENT (°F DB)																	
	−10°		0°		10°		20°		30°		40°		50°		60°		70°	
	BTUH	COP	BTUH	COP	BTUH	COP	BTUH	COP	BTUH	COP	BTUH	COP	BTUH	COP	BTUH	COP	BTUH	COP
QH824/AH65HD	7200	1.27	9100	1.46	11200	1.68	13500	1.92	17000	2.20	21800	2.44	25900	2.73	29200	2.82	30900	2.98
QH830/AH65HE	10200	1.3	12200	1.5	16300	1.8	19900	2.1	24000	2.35	29500	2.65	34000	2.9	36900	3.1	38500	3.2
QH386/AH68HF	16200	1.35	18700	1.6	21400	1.85	25000	2.08	29500	2.35	35000	2.6	40000	2.85	43500	3.0	46000	3.1
QH842/AH68HG	18000	1.3	20500	1.55	24000	1.8	28000	2.05	33000	2.35	39000	2.65	45500	2.9	50500	3.05	53000	3.2
QH848/AH68HH	19500	1.3	23000	1.55	27000	1.75	31500	2.0	37000	2.25	43500	2.5	50000	2.75	56000	2.95	60000	3.05
QH860/AH68HK	2200	1.2	27000	1.4	3300	1.6	40000	1.9	50000	2.2	60500	2.55	71500	2.85	79000	3.05	83500	3.15

(1) Above at standard rating airflow with 70° entering air temperature.

Source: Addison Products Company.

FIGURE HP5-8
Indoor section specifications.

Model	Blower Motor	EVAPORATOR COIL CFM (1) @ EXTERNAL STATIC PRESSURES (INS. OF WATER)							WEIGHT	
		.10	.15	.20	.25	.30	.40	.50	Net	Ship
AH65HD	1/4	870	840	810	780	750	690	620	113	128
AH65HE	1/2	1260	1240	1220	1195	1170	1100	1025	121	136
AH68HF	1/2	1410	1380	1350	1310	1290	1220	1150	158	173
AH68HG	3/4	1760	1740	1720	1685	1650	1575	1450	155	170
AH68HH	3/4	1870	1845	1830	1790	1750	1690	1590	160	175
AH68HK	3/4	1870	1845	1830	1790	1750	1690	1590	170	186

(1) Since filter furnished with unit, total E.S.P. available to ducts and grills.

Source: Addison Products Company.

determined. The cabinet model number gives the blower size and motor horsepower. Using the horizontal line of the cabinet model number and the vertical line of the external static pressure, the ft³/min through the coil is listed where the two lines cross. For example, if an AH68HF unit has an external static pressure of 0.25 in. of W.C., the air quantity through the unit is 1310 ft³/min.

b. If the characteristic table on ft³/min is not available from the manufacturer, the temperature-rise-on-auxiliary-heat method can be used.

(1) Operate the system on the second stage of the thermostat by shutting off the power to the outside unit and turning the thermostat as high as it will go.

(2) With thermometers in the supply and return, operate the unit until these thermometers stabilize.

(3) While waiting for the thermometers to stabilize, measure the voltage at the auxiliary unit electrical terminals and the total amperage draw of the heater deck.

(4) After determining the volts, amperage, and air temperature rise, the following formula is used to determine the ft³/min:

$$ft^3/min = \frac{volts \times amperes \times 3.414 \; Btu/W}{\Delta T \; °F \; air \; through \; coil \times 1.08}$$

The volts multiplied by the amperes is the watts the elements are drawing. Because this is a resistance load, no power factor is involved. The watts times 3.414 Btu/W is the Btu/hr the unit is putting into the air. This then evolves into the standard air-handling formula of

$$ft^3/min = \frac{Btu/hr}{\Delta T \; °F \times 1.08}$$

This results in the ft³/min through the inside coil on both heating and cooling.

HP5-2.6
Net Capacity—Cooling

To calculate the net capacity on the cooling cycle, the same method is used as that explained in Section A17-5. The heat pump on the cooling cycle is rated the same as for air-conditioning units under the ASHRAE Standard 210/240–84, "Unitary Air-Conditioning and Air Source Heat Pump Equipment."

HP5-2.7
Gross Capacity—Heating

On the heating cycle, the amount of heat from the inside coil is the rated capacity of the unit. This heat quantity consists of the heat picked up in the outside coil (the net capacity) plus the heat from the electrical energy in the compressor and blower motor (the motor heat input).

To calculate the gross capacity, it is only necessary to know the ft³/min through the inside coil and the air temperature rise created by operation of the system. The ft³/min is obtained by following the process outlined in item 29. After the ft³/min is obtained and the unit operates on the heating cycle until the supply and return air temperature thermometers have stabilized, the standard air formula is used to find the unit heating capacity:

$$Btu/hr = ft^3/min \times \Delta T \; °F \times 1.08$$

HP5-3
LIQUID TO AIR—DUAL OPERATION

The checklist for a liquid to air–dual operation is the same as air to air–dual operation, with the exception of the following items:

22. Temperature of liquid entering liquid to refrigerant coil
27. Temperature of liquid leaving the liquid to refrigerant coil
28. Weight of liquid through liquid to refrigerant coil
29. Specific heat of liquid through liquid to refrigerant coil (Btu/lb)

The checkout process on the liquid-to-air unit is exactly the same as for the air-to-air system regarding the operation of the inside coil and the refrigerant flow system. The change is in the adjustment and testing of the liquid-to-refrigerant coil.

HP5-3.1
Liquid to Air-Cooling Cycle

On the cooling cycle, the liquid-to-refrigerant coil is the liquid-cooled condenser. As most of these use water as a heat sink for cooling or as a heat source for heating, we will use water in the discussion of this type of unit. This does not imply limitation of this type of unit to water only. These have been used with glycol–water solutions from solar-heated storage as well as oil and liquid chemicals from waste heat sources. The

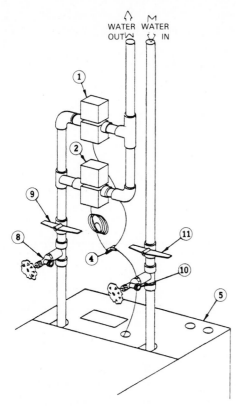

FIGURE HP5-9 (*Courtesy* Bard Manufacturing Company)

air temperature drop of various load ratios chart (Fig. HP5-2).

The compressor discharge pressure is set by regulating the amount of liquid through the liquid-to-refrigerant coil. The most common setting is the same as for water-cooled air-conditioning units, 105°F condensing temperature. At this condensing temperature, the compressor discharge pressure would be 126.6 psig for R-11, 210.8 psig for R-22, and 231.7 psig for R-502.

HP5-3.2
Net Capacity—Cooling

To find the net capacity on the cooling cycle, the process is the same as for the air-to-air unit except for the calculation of the unit gross capacity.

HP5-3.3
Gross Capacity—Cooling

To find the unit gross capacity, the following formula is used:

$$\text{gross Btu/hr} = \text{gal/hr} \times \text{lb/gal} \times \text{specific heat} \times \Delta T \text{ °F}$$

where

gross Btu/hr = total amount of heat being rejected from the system; this capacity consists of the heat picked up in the inside coil plus the motor input heat

gal/min = amount of liquid per hour through the coil; usually measured for a 1-minute period and the quantity multiplied by 60

lb/gal = pounds per gallon of the liquid used; lb/gal of water is 8.33

specific heat = amount of heat in Btu needed to change the temperature of 1 lb of the liquid 1°F; the specific heat of water is 1

$\Delta T°F$ = amount of temperature change in the liquid between entering and leaving the coil

For example, if the flow rate of the water is 4 gal/min and the temperature rise is 16°F, the gross capacity would be

$$\text{Btu/hr} = 4 \times 60 \times 8.33 \times 1 \times 16 = 31,987 \text{ Btu/hr}$$

only thing necessary to know is the specific heat (Btu/lb) to make these discussions apply.

On liquid-to-air units the flow of the liquid through the liquid-to-refrigerant coil is controlled by two liquid flow regulating valves. These valves are connected to the vapor line between the coil and the reversing valve.

One of the valves is a maximum pressure valve that opens on pressure rise and the other is a minimum pressure valve that opens on pressure fall. Figure HP5-9 shows a typical piping arrangement of the valves, shutoff valves, and test valves on a unit. The water regulating valves (Nos. 1 and 2) are connected to a tee fitting (4) and then to the vapor line in the unit (5) to isolate the unit for test and maintenance purposes. Hand shutoff valves are installed in the entering side of the coil (11) and the leaving side (9) ahead of the regulating valves.

Additional hand valves are installed between the hand valves and the unit in the branch of a tee fitting. These valves (8 and 10) are used to install pressure gauges for checking the flow conditions of the liquid coil. This is discussed in Chapter HP6.

To adjust the unit on the cooling cycle, the air temperature drop through the indoor coil is the same as for the air-to-air unit. Find the dry-bulb temperature and relative humidity of the air to the inside coil, the return air, and set the ΔT °F across the coil according to the

Heat Pump Startup, Checkout, and Operation 681

HP5-3.4
Motor Input

As for the air-to-air unit or straight air-conditioning unit, the motor input is the amount of electrical energy the condensing section of the unit uses. The voltage measured at the compressor contactor times the amperage draw times the power factor times 3.414 Btu/W is the motor input.

$$\text{motor input Btu/hr} = \text{volts} \times \text{amperes} \times \text{PF} \times 3.414 \text{ Btu/W}$$

The power factor of a capacitor run motor is assumed to be 90% or 0.9. Assuming the applied voltage to be 232 V and the compressor draw to be 16 A, the motor Btu/hr input would be

$$\text{motor input Btu/hr} = 232\text{V} \times 12\text{A} \times 0.9 \text{ PF} \times 3.414 \text{ Btu/W}$$

In this case the motor input would be 11,405 Btu/hr.

HP5-3.5
Unit Net Capacity

The unit gross capacity minus the motor input is the unit net capacity. In the example:

Unit gross capacity	33,754 Btu/hr
less motor input	8,554 Btu/hr
Unit net capacity	25,200 Btu/hr

As long as the unit net capacity is within $\pm 5\%$ of the rated capacity, the unit performance is satisfactory.

HP5-3.6
Gross Capacity—Heating

The heating output, the rated output, is the gross capacity of the unit. This heating output is equal to the amount of heat removed from the liquid through the outside coil plus the motor heat. Therefore, this heating capacity can be measured in either of two ways.

1. Directly by using the ft³/min through the inside coil and the temperature rise of the air through the inside coil (see Section HP5-3).
2. Indirectly by measuring the net capacity of the liquid-to-refrigerant coil and adding the motor input heat.

To obtain the amount of heat absorbed from the liquid through the outside coil, the formula used in Section HP5-3.3 would be used, the only difference being that a temperature drop would be used instead of a temperature rise.

$$\text{net capacity Btu/hr} = \text{gal/hr} \times \text{lb/gal} \times \text{specific heat} \times \Delta T \text{ °F}$$

where

net capacity Btu/hr = amount of heat picked up by the liquid through the liquid-to-refrigerant coil

gal/hr = amount of liquid in gallons per hour; this quantity is usually measured for a 1-minute period and multiplied by 60; if a container and scale are used, the quantity in pounds per hour is measured directly

lb/gal = quantity in gallons per hour multiplied by the weight of the liquid per gallon; the weight of water per gallon is 8.33 lb

specific heat = amount of heat needed to change the temperature of 1 lb of the liquid 1°F; the specific heat of water is 1

ΔT °F = temperature drop of the water as it flows through the coil; entering water temperature minus leaving water temperature is the ΔT °F

If, for example, the flow rate of the unit on the heating cycle were 5 gal/min × 60 or 300 gal/hr with a drop in temperature of 9°F, the unit net capacity would be 22,491 Btu/hr.

$$22,491 \text{ Btu/hr} = 5 \text{ gal/min} \times 60 \text{ min} \times 8.33 \text{ lb/gal} \times 9 \Delta T \text{ °F}$$

Motor input is measured and calculated as described in Section HP5-5.2. If the unit is our example had a current draw of 11 A at 232 V, the motor input would be 7841 Btu/hr. Added together, the total of the net capacity (22,491 Btu/hr) and the motor input (7841 Btu/hr) would be 30,332 Btu/hr, the gross capacity or rated capacity of the unit.

HP5-4
AIR TO AIR–SINGLE OPERATION

Several manufacturers market an air to air–single operation unit with a heating-only cycle. These units do not have reversing valves but do have defrost control systems. The startup, checkout, and operation of this type of system would be the same as for a dual operation unit on the heating cycle. The only difference is that the large vapor line from the compressor to the inside coil is permanently the hot-gas line and the pressure drop through the reversing valve is eliminated. For an explanation of the startup, checkout, and operation, review the heating cycle operation of the dual operation unit in Section HP5-2.

HP5-5
LIQUID TO AIR–SINGLE OPERATION

Like the unit described in Section HP5-4, the liquid-to-air unit is a heating-only system. No reversing valve is included and defrost controls are not needed. The startup, checkout, and operation of these systems are covered in the heating portions of Section HP5-3.

HP5-6
AIR TO LIQUID–SINGLE OPERATION

These heating-only heat pump systems are found predominantly in the domestic and small commercial water heating market. For example, an add-on type of unit is pictured in Fig. HP5-10. It is designed to be directly connected to the domestic hot water system. Using flexible plastic lines (where local codes permit) the system uses the normal pressure relief valve and drain valve location on the water heater unit. In some locations, solid piping of copper or galvanized pipe is required. Figure HP5-11 pictures an installation using flexible hose supplies with the unit. These units are built using a completely sealed refrigeration system along the same lines as a window-type air-conditioner. It is not possible to conveniently measure suction and discharge pressures, nor is it necessary to do so. As the output (the gross capacity) is the rate at which it heats water, it is only necessary to measure the time it takes to raise the temperature of a given quantity of water. For example, if the unit will heat the water in a 30-gal water heater tank from 55°F to 120°F in 2 hours and 20 minutes, the capacity of the unit is 6961 Btu/hr. This should come within ±10% of the nameplate rating. The capacity would be figured using the following formula:

$$\text{gross capacity} = \frac{\text{quantity} \times 8.33 \times (\text{OFF } T°\text{F} - \text{ON } T°\text{F}) \times 60}{\text{time (min)}}$$

where

gross capacity = amount of total heat the unit is putting into the water in 1 hour's time; this is the rated capacity of the unit

quantity = amount of water to be heated; care must be taken to prevent withdrawal of water from the tank during the test period

8.33 = pounds of water per gallon

OFF T = temperature of the water at the supply outlet when the unit cycles off on the built-in control

ON T = temperature of the water before the test started at the supply outlet

60 = minutes per hour; it is necessary to find the capacity of the unit in Btu/min and then convert to Btu/hr

time (min) = time it takes the unit to raise the water temperature

FIGURE HP5-10 Water heating heat pump. (*Courtesy* Borg-Warner Central Environmental Systems, Inc.)

Heat Pump Startup, Checkout, and Operation   683

FIELD PIPING

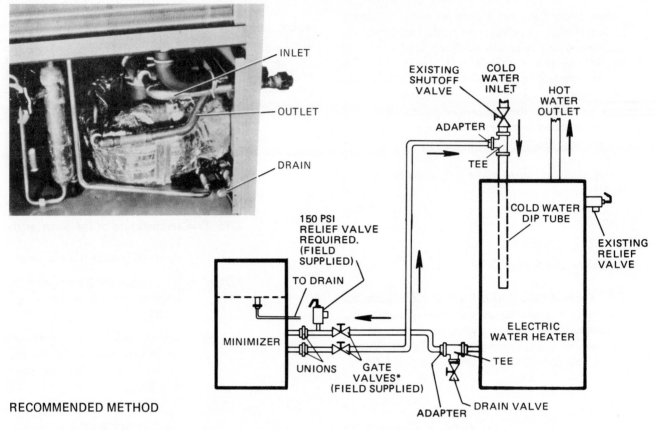

RECOMMENDED METHOD

*NOTE: These shutoff valves will allow the MiniMizer to be serviced and/or cleaned without draining the water heater tank.

FIGURE HP5-11   Field piping. (*Courtesy* Borg-Warner Central Environmental Systems, Inc.)

The calculation procedure would be as follows: 30 gallons (tank capacity) times 8.33 equals 249.9 lb of water in the tank. The off temperature of 120°F minus the off temperature of 55°F equals a 65° temperature rise. The total capacity of the unit is 16,243.5 Btu.

To find the Btu/hr capacity, the total capacity is divided by the number of minutes it takes to deliver the 16.243.5 Btu. This will determine the Btu/min.

$$\frac{16,243.5 \text{ Btu}}{2 \text{ hr} + 20 \text{ min (140 min)}} = 116.025 \text{ Btu/min}$$

The Btu/hr is then determined by multiplying the Btu/min × 60:

gross Btu/hr = 116.025 Btu/hr × 60 = 6961.5 Btu/hr

Any problem that is encountered in units of this type that reduce the gross capacity will result in a long running time to do the rated capacity. This is discussed in Chapter HP6.

PROBLEMS

HP5-1. In order to check the operating capacity of an air-to-air or liquid-to-air unit, what must be determined first?

HP5-2. The applied voltage of a heat pump is plus or minus _____ %.

HP5-3. An air-to-air heat pump can be checked in both the heating and cooling modes regardless of the outside conditions. True or False?

HP5-4. A liquid-to-air heat pump can be checked in both the

heating and cooling modes regardless of the outside conditions. True or False?

HP5-5. How often should throwaway filters be changed in heat pump systems?

HP5-6. How often should electronic filters be cleaned?

HP5-7. When starting the unit, the voltage at the unit drops 12 V and at the distribution panel it drops 9 V. What is the problem?

HP5-8. When setting the temperature drop through the unit in the cooling mode, are different temperature drops required for heat pumps than for air-conditioning systems?

HP5-9. The maximum allowable temperature rise of the air in the return duct is ———————— °F.

HP5-10. To determine the superheat of the evaporator in the heating mode, where would the suction-line temperature thermometer be fastened?

HP5-11. The rise in suction gas temperature in the cooling mode should not be over 10°F. Why?

HP5-12. Where is the best place to obtain the ft³/min of air through the outdoor portion of an air-to-air unit?

HP5-13. Regardless of the outdoor temperature on an air-to-air unit, the supply air temperature to the conditioned area will remain constant. True or False?

HP5-14. Regardless of the outdoor temperature on a liquid-to-air unit, the supply air temperature to the conditioned area will remain constant. True or False?

HP5-15. In the heating mode, the liquid subcooling would be measured on which coil?

HP5-16. For a unit with electric auxiliary heat, what is the formula for determining the ft³/min through the inside coil?

HP5-17. In the heating mode, is the rated capacity of the unit the gross or the net capacity?

HP5-18. The rated capacity on heating is made up of what heat source?

HP5-19. What is the cooling capacity of a liquid-to-air unit using 8 gal/min of water with a 12° rise in temperature if the unit draws 16 A at 240 V?

HP5-20. Find the heating capacity of an air-to-liquid heat pump heating 40 gal of water from 70°F to 135°F in 3 hours and 20 minutes.

HP6

Heat Pump Service and Problem Analysis

HP6-1
GENERAL

The majority of heat pump problems pertain to the refrigeration and air circulation systems. The refrigeration system problems other than the items found only in heat pumps are covered in Chapters R24 through R26. The air system problems are covered in Chapters A16 through A18. In these six chapters, the instruments and test equipment used in all three categories are outlined. Heat pumps are included in these lists because they apply to the refrigeration and air-handling portions of the system, with some exceptions. These exceptions would be:

1. Pressure gauge and manifold set. Where we only need two gauges for an air-conditioning or refrigeration unit, a minimum of three gauges are required for a heat pump. These include a compound gauge for suction pressure and two straight O- to 500-psig gauges for discharge and vapor-line pressure. Some manufacturers have added a gauge pressure tap in the hot-gas line off the compressor, requiring a third straight pressure gauge. This would mean that we need four gauges. The fourth tap makes trouble analysis much easier and should be used.

2. Eight thermometers plus a sling psychrometer for temperature measurement of refrigerant and air.

   a. Air temperature leaving inside coil.

   b. Suction-line temperature at center tube of reversing valve.

   c. Suction-line temperature on tube between the coil acting as the evaporator and the reversing valve. This location will change with the change from the heating to the cooling cycle, and vice versa.

   d. Liquid-line temperature at the outlet of the coil

acting as the condenser. Like the thermometer on the reversing valve suction-line inlet, this thermometer will alternate between the indoor and outdoor coil outlet depending on operation in the heating or cooling cycle. In either case, the thermometer must be located within 6 in. of the coil outlet and be well insulated.

   e. Temperature of the air entering the outside coil. This will be the air entering the condenser on the cooling cycle or the evaporator on the heating cycle. The sling psychrometer can be used to measure the temperature.

   f. Three or more thermometers placed on the air discharge side of the outdoor coil to measure the overage of the leaving air temperature. The purpose is to measure the temperature rise on the cooling or temperature drop on the heating cycle.

   g. A sling psychrometer to measure the dry-bulb and wet-bulb temperatures of the return air to the inside coil during the cooling cycle or the dry-bulb temperature only on the heating cycle.

Unlike an air-conditioning unit, which can develop problems in either the refrigeration system or the air system, the heat pump can develop problems in the cooling cycle, the heating cycle, or both.

HP6-2
AIR SYSTEM PROBLEMS

When in the cooling cycle, the inside coil must be able to produce the correct amount of sensible and latent capacity to produce the desired room conditions. Assuming that the system has previously been set to the required temperature drop, the first check is the temperature drop through the evaporator (see Section A18-

3). If the temperature drop has increased, the amount of air through the coil has decreased. The capacity of the refrigeration system will not increase; therefore, a reduction in the ft³/min has occurred.

If the unit is in the heating cycle, the temperature rise of the air through the inside coil, now the condenser, indicates a reduction of air through the coil. The reduction can be severe enough to cause the unit to cut out on the high head pressure lockout relay system. Repetitive cutout bringing on the lockout light on the thermostat should be investigated by a competent service technician before damage to the compressor results.

Air filters. The heat pump operates on a year-round basis and the amount of air through the inside coil is more critical then in a heating–air conditioning system. Throwaway filters should be replaced every 30 days. It is advisable to use electronic air cleaners with heat pumps due to their low static resistance even when dirty.

Blower motor and drive. Inspection of the blower and drive should be done at least once a year. Both blower motor and blower bearings should be lubricated with no more than 10 drops of No. 20 electric motor oil. This oil is a detergent-free oil. Use of automobile oil is discouraged because it has detergent (soap) which coats the outer surface of the sintered bronze bearing and prevents oil passage through the bearing.

Unusual restrictions in the duct system such as closing off rooms that are not used as well as placement of furniture or carpeting over supply and/or return grills will cause coil frosting in the cooling cycle and unit cutoff in the heating cycle. This practice has a greater effect in the heating cycle than in the cooling cycle.

Failure of the duct system also has a greater effect in the heating cycle than in the cooling cycle. With 70°F in the occupied area the supply air temperature will only be in the range 100 to 105°F when the outdoor temperature is 60 to 65°F down to the 85°F range when it is −10°F outside. Therefore, any leakage in the duct system will seriously reduce the capacity of the unit to handle the heating load. As a result, the operating cost will be higher than normal.

HP6-3
REFRIGERATION SYSTEM PROBLEMS

When measurement of the temperature drop through the inside coil is less than it should be on either the cooling cycle or the heating cycle, the refrigeration system should be suspected. As in an air-conditioning system, it can be classified into (1) refrigerant quantity or (2) refrigerant flow rate. If the system has the proper amount of refrigerant and it is flowing at the desired rate, the system has to work properly and deliver rated capacity. Any problem in either category will affect the

temperatures and pressures that will occur in the unit when the correct amount of air is supplied over the inside coil for the capacity of the unit.

Obviously, if the system is low on refrigerant, a leak has occurred. It must be found and repaired, then the system evacuated thoroughly and recharged with the correct amount of refrigerant. If the system will not operate, it is obviously an electrical problem that must be found and corrected.

Those problems that are common for refrigeration, air-conditioning, and heat pump systems are covered in Chapters R24 through R26 as well as Chapters A16 through A18. Because of the additional control systems used in heat pumps, some of these problems will be repeated in this chapter. The chapters listed above should be reviewed, however, together with the discussion in this chapter.

To compare the air-conditioning system with the heat pump system, the two systems are shown in Fig. HP6-1, HP6-2 and HP6-3. In Fig. HP6-1 a conventional air-conditioning system is shown. This system shows the refrigerant flow from the discharge of the compressor to the condenser, the outside coil. The refrigerant condenses and flows from the outside coil through the liquid line, the filter drier, and pressure-reducing device to the inside coil. The refrigerant expands, picking up heat in the evaporator, in the inside coil. From the evaporator it flows as a vapor to the compressor.

In Fig. HP6-2 the same action takes place except that some devices have been added. A reversing valve has been added to the suction and discharge lines to enable the system to reverse the flow of refrigerant in the system. The position of the reversing valve still directs the hot gas from the compressor to the outside coil and

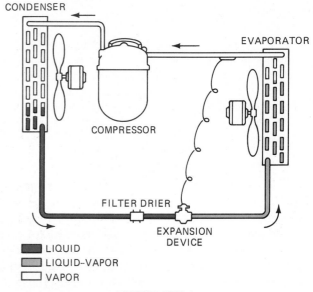

LIQUID
LIQUID-VAPOR
VAPOR

FIGURE HP6-1

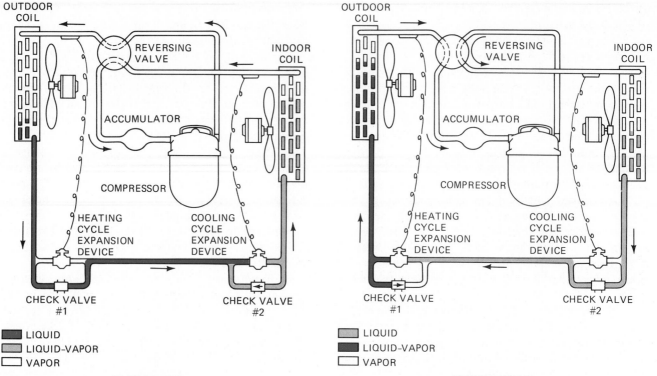

LIQUID
LIQUID-VAPOR
VAPOR

FIGURE HP6-2

LIQUID
LIQUID-VAPOR
VAPOR

FIGURE HP6-3

the vapor from the inside coil to the compressor. In addition, a check valve has been installed around the pressure-reducing device, feeding refrigerant to the inside coil. This check valve is connected so as to prevent flow around the pressure-reducing device during the cooling operation. It opens during the heating operation to eliminate the restriction of the pressure-reducing device.

A pressure-reducing device has also been added to the outside coil for this coil to operate as an evaporator during the heating cycle. A check valve is also connected around this pressure-reducing device to remove its pressure drop during the cooling cycle.

Following the refrigerant flow we see that it is the same as for the air-conditioning unit. Hot vapor flows from the compressor to the outdoor coil, the condenser. After cooling and condensing to a liquid, it flows around the pressure-reducing device connected to the outdoor coil and to the check valve–pressure reducing device assembly connected to the inside coil. The check valve in this assembly, called a *trombone,* closes and forces the liquid refrigerant to flow through the pressure-reducing device. This results in the required pressure reduction to produce the desired boiling point of the refrigerant. From here it vaporizes in the evaporator and flows through the reversing valve to the compressor.

In Fig. HP6-3 the reversing valve has changed position. The hot refrigerant vapor flows to the inside coil, the condenser. Giving up heat to the air supplied to the occupied area, the hot refrigerant vapor cools, condenses, and flows out of the bottom of the condenser. With

the flow reversed, the check valve opens and allows the liquid refrigerant to flow around the pressure-reducing device on the inside coil to eliminate any pressure loss. The liquid refrigerant flows in the reverse direction to the trombone on the outside coil. The refrigerant is forced to flow through the pressure-reducing device on the outside coil (now the evaporator). In the outside coil (the evaporator) the liquid refrigerant, at a lower pressure and boiling point, vaporizes and picks up heat from the outside air. The vaporized refrigerant flows through the reversing valve and the accumulator to the compressor.

From this discussion we can see that the only reversing of the refrigerant flow is from the reversing valve through the coils, pressure-reducing devices, and check valves. The refrigerant vapor always flows from the reversing valve, accumulator, and compressor back to the reversing valve.

HP6-4
REFRIGERANT FLOW PROBLEMS

From the preceding discussion we see that the only difference between the heat pump and the air-conditioning system is the addition of a reversing valve, two check valves, and a second pressure-reducing device. We will limit the refrigerant problem discussion to these devices. The problem discussions presented in Chapter A18 should be reviewed before covering this section.

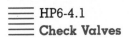

Check valves have the capability of sticking in either the open or closed position. Therefore, the system will work correctly in any of the cycles depending on the valve's location in the system. If it is doing what it is supposed to be doing, the system will operate properly. If not, the problem will show up. For example,

1. *The check valve on the inside coil sticks in the open position.* The system will operate properly in the heating cycle. The valve is supposed to be open in this cycle. In the cooling cycle, however, there is no pressure-reducing device in the circuit, so the refrigerant will flood through the evaporator, suction pressure will be high, discharge pressure will be low, and the accumulator will be filled with liquid refrigerant. It may be flooding back to the compressor if the liquid line is short or in a packaged heat pump.

2. *The check valve on the inside coil sticks in a closed position.* The system will operate properly on the cooling cycle. The valve is supposed to be closed. In the heating cycle, the suction pressure will be much lower than normal. In units using capillary tubes, the suction pressure will be the result of two capillary tubes in series. In systems using TX valves, the valve will close and the compressor could pull the suction pressure into a vacuum. At the same time, the discharge pressure will be low due to little vapor for the compressor to pump. *Before adding gas to the system,* check the system by switching to the opposite cycle.

3. *The check valve on the outside coil sticks in a closed position.* The system will operate properly on heating. The valve is supposed to be closed. On cooling, low suction pressure and discharge pressure will result.

4. *The check valve on the outside coil sticks open.* The system will operate on cooling but not on heating. The outside check valve is supposed to be open on the cooling cycle.

Check Valve Repair: Usually, a check valve can be released from the stuck-open position by means of a magnet placed against the outlet end of the valve. Moving the magnet toward the center will force the ball or flapper to move to the seat. If in the stuck-closed position, place the magnet at the middle of the valve and move it to the outlet end. If you have no success, replace the valve. To reduce the possibility of future problems, use a ball-type check. However, before installing the new valve, shake it. If it rattles, install it. If it does not rattle, take it back to the supplier; it is already stuck and it will not function properly.

Figure HP6-4 shows the exterior view of a reversing valve. The tube connection is always connected to the compressor discharge. The bottom middle connection is always connected to the compressor suction connection. When an accumulator is in the circuit, this connection is to the accumulator inlet. The accumulator outlet is then connected to the compressor suction. This puts the accumulator upstream from the compressor and provides surge protection in either the heating or cooling mode.

The right and left connections are the connections to the outlets of the inside and outside coils. Which goes to where depends on whether the operating coil is energized on heating or cooling. Most present-day heat pumps are designed to operate in colder climates where the heating operating hours are more than the cooling operating hours. To reduce the operating coil on time, the coil is energized during the cooling season. In this case, the flow through the coil would be from the left connection to the middle and from the compressor discharge through the right-hand connection. The left-hand connection would be to the inside coil (the condensor) and the right-hand connection to the outside coil (the evaporator).

With the coil energized, the valve slide assembly would be in the opposite position. The hot gas from the compressor would flow through the left-hand connection to the outside coil, now the condenser. The cold gas from the inside coil (now the evaporator) flows through the right-hand connection to the middle connection to the accumulator and on to the compressor.

If the cycling of the valve is reversed, the coil is energized during the heating cycle and the outside connections would be reversed. Before checking the position of the valve operation, the electrical control system

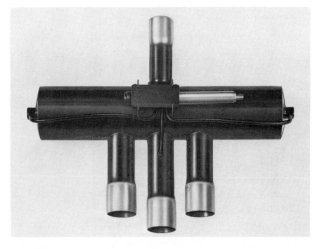

FIGURE HP6-4   Reversing valve. (*Courtesy* Ranco, Inc.)

Heat Pump Service and Problem Analysis   689

should be checked to determine the operating requirements of the valve.

Operation: Reviewing Fig. HP6-5, the solenoid valve is actually a single port, double throw valve. The center connection is connected to the suction line and relieves pressure out of the valve. The bottom port has been closed. This port is considered "normally closed." Because this port is closed, compressor discharge pressure builds up behind the main valve piston and in this line.

The top port is open and the pressure in this line has drained into the suction line. Because discharge pressure is on the right side of the left piston in the main valve and suction pressure on the left side, the piston has been forced to the left side of the main cylinder. This moved the bypass valve to cover the middle and left-hand outlets. Not shown is the V point on the end of the piston that seats in the outlet sent to shut off gas flow through the piston port into the vent line.

When the pilot valve (solenoid valve) is energized, (Figure HP6-6) the ports in the pilot valve reverse; the lower valve opens and the upper valve closes. This lowers the pressure in the right end of the main cylinder to suction pressure. The pressure difference that develops across the right-hand piston forces the piston to the right end of the main cylinder. The bypass valve is moved to cover the center and right-hand outlets. When the piston stroke is completed, the gas flow is shut off by the V point entering the outlet valve seat.

From this we see that the main valve piston is controlled by the action of the pilot valve draining the pressure off the ends of the main valve piston to the suction side of the compressor. The valve works on the compressor differential pressure. The minimum pressure required to operate the valve is 75 to 100 psig. This means that the refrigeration system must be fully charged with refrigerant and operating long enough to develop a 75-psig difference between suction and discharge pressure.

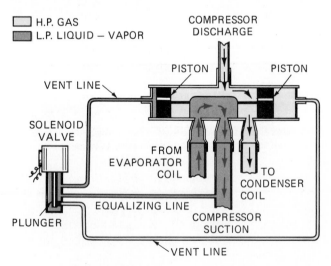

FIGURE HP6-5   (*Courtesy* Ranco, Inc.)

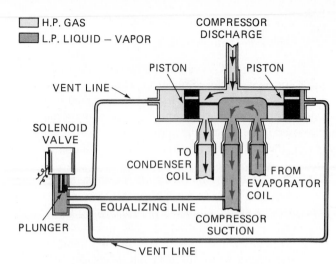

FIGURE HP6-6   (*Courtesy* Ranco, Inc.)

Testing: Problems in reversing valves are either electrical or mechanical. Electrical problems are confined to the solenoid coil on the pilot valve. When the solenoid coil is supposedly energized and nothing happen, test the coil as follows:

1. Make sure that voltage is applied to the coil. Some units have the coil in the 240-V portion of the system, while others use it in the 24-V portion. Check the wiring diagram for coil voltage before applying the leads of your voltmeter. Also, start the test with the voltmeter set to the higher range. This helps to reduce meter burnout.

2. With voltage applied to the coil, remove the coil-holding nut and attempt to pull the coil off the pilot valve plunger casing. If you feel a resistance to removing the coil, the coil is active. If no pull is felt, the coil is dead. Shut off the power, remove the coil heads, and check for continuity with the ohmmeter. If open, replace the coil. If a circuit exists, check the leads and the connections for continuity.

    If the coil is active, when removing the coil a "click" should be heard when the pilot plunger returns to the normally closed position. When replacing the coil, a click should also be heard when the pilot plunger is lifted to the open position. If no clicks are heard, the pilot valve is stuck. The only repair is to replace the reversing valve.

3. When the pilot valve checks out and the main valve does not shift, make sure that suction and discharge show more than a 100-psig difference. If the valve is in the cooling position, block off the air to the condenser with plastic on the inlet face of the coil. Allow the unit to operate until the head pressure reaches 130°F condensing temperature. With the unit operating, cycle the valve on and off several times. This will usually free the main valve

to operate again. If you get no results, change the valve.

When changing a reversing valve, after completely removing the refrigerant from the system, be sure to read the installation instructions supplied with the valve. Replace the valve with one of a comparable size. Always position the valve so that the main piston is in a horizontal position and the pilot valve is higher than the main valve. This is to keep oil from gathering in the pilot valve and affecting its operation.

The valve body *must always be protected from heat* by wrapping the body with some type of thermoplastic material. The maximum temperature the valve body will tolerate is 250°F.

HP6-5
PROBLEM ANALYSIS

Like refrigeration and air-conditioning systems, heat pumps have problems develop in the refrigeration system as well as the basic electrical control systems. These problems were covered in Chapters R24 through R26 as well as Chapters A17 and A18. Therefore, only those problems characteristic to heat pumps will be discussed in this chapter.

HP6-5.1
Thermostats

A list of possible problems in this category might consist of the following:

1. *The outdoor unit will not start on either heating or cooling regardless of the setting of the operation selector switch.* With a problem of this type the system will operate on the first stage of the auxiliary electric heat. The problem encountered here is the matching of the terminal numbers on split-system heat pumps. The terminal designation on thermostat, inside air handler, and outside compressor unit do not match. On packaged heat pumps, the manufacturer does the intercomponent wiring, so this problem does not come up.

The solution to this problem is close scrutiny of the wiring diagram regarding terminal use for the system. The standard terminal designation for the first-stage heating in the case of an electric furnace would be the first-stage electric heat (W1). When a heat pump is added to the electric furnace, the first stage of thermostat energizes the Y circuit of the compressor contactor to operate the heat pump compressor.

Figure HP6-7 shows a typical field wiring arrangement. The only mismatch that occurs is in the control of the outdoor unit compressor and reversing valve. In this case, the unit uses a high-voltage coil controlled by a 24-V relay. This relay is connected through the out-

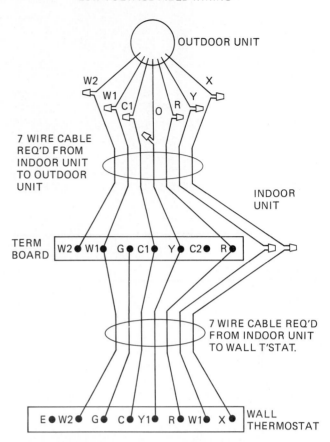

FIGURE HP6-7 Low voltage field wiring. (*Courtesy* Addison Products Company)

door unit O lead and indoor unit dummy terminal Y to the Y1 terminal of the thermostat. The W1 terminal of the thermostat is connected directly to the Y lead that controls the compressor contactor. Further inspection shows that the W2 terminal of the thermostat, the second stage, is connected to the W1 or first stage of the auxiliary electric heat. A circuit also comes off this W1 to the W1 in the outside unit and from the W2 in the outside unit back to the W2, a second stage of heat in the inside unit. The outside unit has the outdoor ambient thermostat that controls the linking together of the two stages of electric heat.

2. *The trouble light keeps coming on.* The trouble light in the thermostat is controlled by the lockout relay, which in turn is controlled by the head pressure control. The unit is cutting out on high head pressure. This is a refrigeration problem (see Section R25-4).

3. *The room temperature drops at certain outside temperatures.* In mechanical thermostats, such as that shown in Fig. HP6-8, there is a 2°F difference between the closing of the first- and second-stage contacts. Because of the difference, when the heat loss of the conditioned area reaches the capacity of the heat pump, the temperature of the occupied area has to drop 2°F to

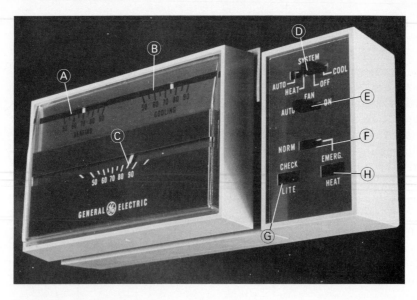

(A) Heating temperature selector
(B) Cooling temperature selector
(C) Temperature indicator
(D) Systems switch
(E) Fan switch
(F) Normal/Emergency heat switch
(G) Check-Lite
(H) Emergency heat light

FIGURE HP6-8 (*Courtesy* Addison Products Company)

bring in the second stage of the thermostat. The temperature outside that produces the heat loss resulting in continuous running of the unit is called the *initial balance point.* For an explanation of how to determine the balance point, see Manual H, *Heat Pump Equipment Selection and Application,* published by the Air-Conditioning Contractors Association (ACCA), 1228 17th Street N.W., Washington, D.C. 20036.

Because of the 2°F mechanical difference in the thermostat, the temperature of the occupied area will drop 2°F at the balance point to bring in the second stage. The area temperature will be maintained 2°F below the heat setting of the thermostat.

The customer has three alternatives in this case:

1. Tolerate the 2°F temperature drop.
2. Raise the thermostat setting the 2°F difference.
3. Replace the thermostat with a solid-state thermostat that compensates for this situation by measuring outside temperature and incorporating this action into the performance of the control.

HP6-5.2
Refrigeration System

1. *The unit will not change cycle, heating to cooling, or vice versa, regardless of thermostat switch settings.* The changeover of the system is controlled by the action of the reversing valve. No action of this control can be either electrical or mechanical.

Electrical. Either a high-voltage or low-voltage valve must have applied voltage with ±10% of the design voltage range in order for the control to operate properly. The operating coil should also be able to produce the proper magnetic pull to operate the pilot valve (see Section HP6-9).

Mechanical. The main valve piston may be stuck in one end of the cylinder because it is not getting the proper pressure changes form the pilot valve. This can be determined by touching the coil connection tubes with the coil energized and deenergized. Any sticking of the pilot valve, plugs in the small tubes between the pilot valve and the ends of the main valve, or seizure of the main valve requires replacement of the valve.

Repair of the valve is not justified because of the time cost and high percentage of failure repeat.

2. *The unit will work fine on heating but very poorly on cooling.* This problem is usually found in the action of the check valves. If the suction pressure is higher than is reasonable for the temperature of the air leaving the coil (Section A18-4), it is very possible that the check valve on the inside coil is stuck open. This removes the pressure-reducing device from the circuit and refrigerant is flooding through the coil. Superheat

will be low or nonexistent and the temperature of the accumulator, reversing valve tubes, and possibly the compressor shell, will be very low.

If the suction pressure is lower than normal, the check valve on the outside coil is stuck closed. This keeps the pressure-reducing device in the circuit. If it is a TX valve, the valve will close and shut off the refrigerant flow to the liquid line. The suction pressure will drop extremely low, possibly into a vacuum.

If the pressure-reducing devices are capillary tubes, refrigerant flow will take place but much less than normal. The unit will have two capillary tube sets in series with the higher-pressure reduction of both sets. On some units, the head pressure is measured on the liquid line between the check valve–pressure reducing valve sets. With TX valves, the head pressure gauge will indicate a loss of refrigerant. With capillary tubes, the head pressure will be low along with the low suction pressure. In either case, *do not add refrigerant* until the subcooling has been checked. The subcooling will be found to be extremely high because the refrigerant has accumulated in the condenser.

3. *The unit will work fine on cooling but will not work on heating.* The same action will take place as in problem 2 except that the check valve on the inside coil (the condenser) may be stuck closed or the valve on the outside coil (the evaporator) may be stuck open. Suction and discharge pressures will act the same.

Other refrigeration problems are covered in Chapters R24 through R26, A17, and A18.

HP6-5.3
Defrost System

To diagnose problems in the defrost system, it is necessary to determine the type of defrost system involved. The four most predominant systems are covered here. On those systems that are used by individual manufacturers, the manufacturer should be contacted for service information.

Temperature Differential Defrost System:

1. *Unit will not go into defrost.* This is the most common complaint on this control. The control was operated by the temperature difference between the entering air temperature of the outside coil (the evaporator) and the coil operating temperature (see Section HP2-3.1). The predominant reason that the unit would not go into defrost was that the coil would not reach a low-enough temperature to provide an increase in the control differential to activate the defrost cycle. The major reason for this was that the head pressure was too high, which forced the suction pressure and coil boiling point up. Only a 2°F rise in boiling point was necessary to

require complete coverage of the coil with frost and ice before initiating the defrost cycle. Rather than correct the air problem—the duct system was inadequate or the occupants did not like the low-temperature discharge air—the service technician would replace the control. Worse yet, he or she would try to adjust the control, which was not possible to adjust without closely controlled water baths.

2. *Unit goes into defrost with very little frost on bottom of coil.* The location of the coil temperature sensing bulb was critical. If the bulb was located too far to the bottom of the coil, too close to the entering low-pressure liquid refrigerant, the lower temperature of this portion of the coil would promote premature initiation of the defrost cycle. The cure was to move the coil temperature bulb up one return bend at a time until the correct location was obtained. Once the defrost cycle was initiated, the coil had to reach a preset temperature to terminate the defrost cycle. Usually, this temperature was 50 to 60°F.

3. *Unit does not defrost entire coil; ice ring buildup around the bottom of coil.* The ice buildup indicates that the coil temperature bulb is located too high on the coil. The bulb is reaching termination temperature before the coil is completely defrosted. The bulb should be lowered one return bend at a time until complete defrosting takes place.

Pressure–Temperature Defrost System: In this system, the pressure differential across the coil initiates the defrost cycle and the temperature of the liquid leaving the coil terminates it. The liquid-line thermostat also controls the possibility of a defrost cycle (see Section HP2-3.2).

A thermostat fastened to the liquid line at the bottom outlet of the coil is exposed to the expanding liquid refrigerant when the unit is operating in the heating cycle. If the evaporator coil is operating with an entering liquid of 26°F or higher, there is little chance that frost will form on the coil. If this temperature is below 26°F, the thermostat closes and completes the circuit through the defrost relay to the differential pressure switch.

At such time as the frost buildup causes enough resistance to close the differential pressure switch, this initiates the defrost cycle. When the liquid leaving the coil reaches 55°F, the circuit is broken, the defrost relay circuit opens, and the relay drops out. This opens the holding circuit of the relay and completes the switchover to the heating cycle.

1. *The unit will not go into defrost.* Two possible causes can prevent the unit from going into defrost. The most probable cause is that the termination thermostat has come loose from the coil outlet pipe. The thermostat cannot reach the 29°F or below to close and allow the defrost circuit to initiate. The other probable cause is plugged tubes to the pressure differential switch. Insects sometimes plug these pipes.

2. *The unit goes into defrost with very little frost on the bottom of the coil.* Because the pressure drop of the air through the coil is the determining factor for initiating the defrost cycle, both the cleanliness of the coil and the frost buildup make up this pressure drop. Therefore, the more dirt buildup on the coil, the less frost has to form to reach the required pressure drop to initiate the defrost cycle. When the unit starts defrosting with little frost buildup, clean the coil.

3. *The unit does not defrost the entire coil; an ice ring forms at the bottom of the coil.* The fact that the ice ring forms indicates that the defrost cycle is interrupted before the defrost cycle is completed. The most common cause of this occurrence is too much reheat when the unit goes into defrost, which brings on the reheat as part of the defrost cycle. The amount of reheat should not exceed the sensible capacity of the cooling cycle of the heat pump.

When the coil is defrosted, the termination thermostat breaks the power to the defrost relay. The holding contact in the relay that keeps the relay energized is broken. The system must now develop the necessary frost buildup to close the pressure switch.

If the power supply to the defrost circuit is interrupted, the same action occurs. Therefore, if the reheat is great enough to warm the occupied area and cause the room thermostat to open before the defrost cycle is completed, the incomplete defrost cycle will leave an ice buildup at the bottom of the coil. The alternate thawing and freezing of the water held on the coil by the ice coating will cause collapse of the coil tubes and loss of refrigerant.

Time-Temperature Defrost System:

1. *The unit will not go into defrost.* This system uses a timer to control the defrost cycle under the control of the termination thermostat. The power to the timer is supplied from the compressor circuit. Therefore, the timer operates when even the compressor is operating and the termination thermostat is below 26°F.

The most common cause of this problem is a loose termination thermostat. A poor contact between the thermostat and the coil tube prevents the thermostat temperature dropping below 26°F. If contact is made, the thermostat should be checked to determine if it closes at a temperature below 26°F. Immersion in an ice and saltwater bath using an immersion thermometer will accomplish this. The timer motor should also be checked for proper motor operation.

2. *The unit goes into defrost with very little frost on the bottom of the coil.* The timer may be operating on too short a cycle time. Most timers have 30-minute and 90-minute cycle cams. Convert the timer from the 30-minute cam to the 90-minute cam.

3. *The unit does not defrost the entire coil; there is ice ring buildup around the bottom of the outside coil.* The correction to this problem is to reduce the amount of reheat used in the defrost cycle to an amount not more than the sensible capacity of the unit on the cooling cycle.

Pressure-Time-Temperature Defrost System: This system uses a differential pressure switch to measure the amount of frost on the outside coil. When the pressure drop through the coil reaches a preset amount, the switch closes the timer circuit.

If the termination thermostat is below 26°F, the timer will operate. This timer usually requires 5 minutes of pressure switch closure time to initiate the defrost cycle by energizing the defrost relay. When the termination thermostat reaches 55°F, the power to the defrost relay is interrupted and the unit changes over to the heating cycle.

1. *The unit will not go into defrost.* The most common cause of failure to go into the defrost cycle is a loose or poorly connected termination bulb of the defrost control. It is very important that the bulb be securely fastened with heat-transfer compound on the joint, insulated, and sealed from moisture. Ice can build up at this location and loosen the bulb fastening. The second cause can be failure of the pressure switch due to dirt and/or insects in the pressure switch tube connections. There is always the possibility of failure of the timer control, but this is remote. Check the previous two items before replacing any parts.

2. *The unit goes into defrost with very little frost on the bottom of the coil.* The major cause of this action is a dirty outside coil. Its frost-free resistance is so high that it takes very little increase in pressure drop to start the defrost cycle. A thorough coil cleaning is in order.

3. *The unit does not defrost entire coil.* This system has an automatic defrost cycle termination anywhere from 5 to 12 minutes, depending on the frost accumulation rate of the outside weather conditions.

Because it is not possible to cover all defrost systems in a single book, it is recommended that manufacturers' literature be obtained and retained for complete information.

PROBLEMS

HP6-1. Fastening a thermometer to the center tube from the reversing valve to the compressor and from the evaporator coil to the reversing valve would be a possible check for what type of problem?

HP6-2. The term "trombone" applies to what assembly?

HP6-3. The only lines in a heat pump in which the refrigerant flows in one direction regardless of the operating mode are the _____ and the _____ .

HP6-4. Before condemning a part or parts of a heat pump system, what must be done to the system?

HP6-5. When operating the dual operation unit in the cooling mode, which check valve will be open and which closed?

HP6-6. The check valve on the inside coil sticks open. In which mode will be problem show up?

HP6-7. The check valve on the outside coil sticks closed. In which mode will the problem show up?

HP6-8. In the heating mode, the head pressure reading taken on the liquid line is low, the suction pressure is low, but the compressor amperage draw is high. Performance in the cooling mode is good. What is the trouble in the system?

HP6-9. Placing your hand on the center tube of the reversing valve, you find the temperature higher than that on the line from the evaporator. What is the problem?

HP6-10. How much refrigerant must be in the system to check the operation of the check valve?

HP6-11. What is the minimum pressure difference across a reversing valve for it to operate?

HP6-12. When changing a reversing valve, what is the maximum temperature the valve body will take?

HP6-13. Upon looking up the low-voltage control system and turning the unit on, the thermostat will bring on the first-stage electric heat instead of the compressor. What is the problem?

HP6-14. The control that controls the operation of the second stage of the auxiliary heat is called the _____ .

HP6-15. The outdoor temperature at which the heat loss of the building equals the capacity of the heat pump is called the _____ .

HP6-16. The air-to-air heat pump operates with an ice ring at the bottom of the outside coil. What is the most likely problem?

HP6-17. Most termination thermostats open at what temperature?

HP6-18. On temperature termination defrost controls, what is the most likely cause of failure to go into the defrost cycle?

HP6-19. On a pressure-initiation defrost system, the unit goes into defrost with very little frost on the coil. What is the problem?

HP6-20. What is the maximum amount of reheat that should be used?

Appendices

TABLE 3 ... PROPERTIES OF LIQUID AND SATURATED VAPOR (continued)

TEMP F	PRESSURE lb per sq in		VOLUME cu ft per lb		DENSITY lb per cu ft		ENTHALPY ** Btu per lb			ENTROPY ** Btu per (lb) (°R)		TEMP F
t	Absolute P	Gage p	Liquid v_f	Vapor v_g	Liquid $1/v_f$	Vapor $1/v_g$	Liquid h_f	Latent h_{fg}	Vapor h_g	Liquid s_f	Vapor s_g	t
-25	13.556	2.320*	0.010730	2.7295	93.197	0.36636	3.1724	71.391	74.563	0.007407	0.17164	-25
-24	13.886	1.649*	.010741	2.6691	93.098	0.37466	3.3848	71.288	74.673	.007894	.17151	-24
-23	14.222	0.966*	.010753	2.6102	92.999	0.38311	3.5973	71.185	74.782	.008379	.17139	-23
-22	14.564	0.270*	.010764	2.5529	92.899	0.39171	3.8100	71.081	74.891	.008864	.17126	-22
-21	14.912	0.216	.010776	2.4972	92.799	0.40045	4.0228	70.978	75.001	.009348	.17114	-21
-20	15.267	0.571	0.010788	2.4429	92.699	0.40934	4.2357	70.874	75.110	0.009831	0.17102	-20
-19	15.628	0.932	.010799	2.3901	92.599	0.41839	4.4487	70.770	75.219	.010314	.17090	-19
-18	15.996	1.300	.010811	2.3387	92.499	0.42758	4.6618	70.666	75.328	.010795	.17078	-18
-17	16.371	1.675	.010823	2.2886	92.399	0.43694	4.8751	70.561	75.436	.011276	.17066	-17
-16	16.753	2.057	.010834	2.2399	92.298	0.44645	5.0885	70.456	75.545	.011755	.17055	-16
-15	17.141	2.445	0.010846	2.1924	92.197	0.45612	5.3020	70.352	75.654	0.012234	0.17043	-15
-14	17.536	2.840	.010858	2.1461	92.096	0.46595	5.5157	70.246	75.762	.012712	.17032	-14
-13	17.939	3.243	.010870	2.1011	91.995	0.47595	5.7295	70.141	75.871	.013190	.17021	-13
-12	18.348	3.652	.010882	2.0572	91.893	0.48611	5.9434	70.036	75.979	.013666	.17010	-12
-11	18.765	4.069	.010894	2.0144	91.791	0.49643	6.1574	69.930	76.087	.014142	.16999	-11
-10	19.189	4.493	0.010906	1.9727	91.689	0.50693	6.3716	69.824	76.196	0.014617	0.16989	-10
-9	19.621	4.925	.010919	1.9320	91.587	0.51759	6.5859	69.718	76.304	.015091	.16978	-9
-8	20.059	5.363	.010931	1.8924	91.485	0.52843	6.8003	69.611	76.411	.015564	.16967	-8
-7	20.506	5.810	.010943	1.8538	91.382	0.53944	7.0149	69.505	76.520	.016037	.16957	-7
-6	20.960	6.264	.010955	1.8161	91.280	0.55063	7.2296	69.397	76.627	.016508	.16947	-6
-5	21.422	6.726	0.010968	1.7794	91.177	0.56199	7.4444	69.291	76.735	0.016979	0.16937	-5
-4	21.891	7.195	.010980	1.7436	91.074	0.57354	7.6594	69.183	76.842	.017449	.16927	-4
-3	22.369	7.673	.010993	1.7086	90.970	0.58526	7.8745	69.075	76.950	.017919	.16917	-3
-2	22.854	8.158	.011005	1.6745	90.867	0.59718	8.0898	68.967	77.057	.018388	.16907	-2
-1	23.348	8.652	.011018	1.6413	90.763	0.60927	8.3052	68.859	77.164	.018855	.16897	-1
0	23.849	9.153	0.011030	1.6089	90.659	0.62156	8.5207	68.750	77.271	0.019323	0.16888	0
1	24.359	9.663	.011043	1.5772	90.554	0.63404	8.7364	68.642	77.378	.019789	.16878	1
2	24.878	10.182	.011056	1.5463	90.450	0.64670	8.9522	68.533	77.485	.020255	.16869	2
3	25.404	10.708	.011069	1.5161	90.345	0.65957	9.1682	68.424	77.592	.020719	.16860	3
4	25.939	11.243	.011082	1.4867	90.240	0.67263	9.3843	68.314	77.698	.021184	.16851	4
5	26.483	11.787	0.011094	1.4580	90.135	0.68588	9.6005	68.204	77.805	0.021647	0.16842	5
6	27.036	12.340	.011107	1.4299	90.030	0.69934	9.8169	68.094	77.911	.022110	.16833	6
7	27.597	12.901	.011121	1.4025	89.924	0.71300	10.033	67.984	78.017	.022572	.16824	7
8	28.167	13.471	.011134	1.3758	89.818	0.72687	10.250	67.873	78.123	.023033	.16815	8
9	28.747	14.051	.011147	1.3496	89.712	0.74094	10.467	67.762	78.229	.023494	.16807	9
10	29.335	14.639	0.011160	1.3241	89.606	0.75523	10.684	67.651	78.335	0.023954	0.16798	10
11	29.932	15.236	.011173	1.2992	89.499	0.76972	10.901	67.539	78.440	.024413	.16790	11
12	30.539	15.843	.011187	1.2748	89.392	0.78443	11.118	67.428	78.546	.024871	.16782	12
13	31.155	16.459	.011200	1.2510	89.285	0.79935	11.336	67.315	78.651	.025329	.16774	13
14	31.780	17.084	.011214	1.2278	89.178	0.81449	11.554	67.203	78.757	.025786	.16765	14
15	32.415	17.719	0.011227	1.2050	89.070	0.82986	11.771	67.090	78.861	0.026243	0.16758	15
16	33.060	18.364	.011241	1.1828	88.962	0.84544	11.989	66.977	78.966	.026699	.16750	16
17	33.714	19.018	.011254	1.1611	88.854	0.86125	12.207	66.864	79.071	.027154	.16742	17
18	34.378	19.682	.011268	1.1399	88.746	0.87729	12.426	66.750	79.176	.027608	.16734	18
19	35.052	20.356	.011282	1.1191	88.637	0.89356	12.644	66.636	79.280	.028062	.16727	19
20	35.736	21.040	0.011296	1.0988	88.529	0.91006	12.863	66.522	79.385	0.028515	0.16719	20
21	36.430	21.734	.011310	1.0790	88.419	0.92679	13.081	66.407	79.488	.028968	.16712	21
22	37.135	22.439	.011324	1.0596	88.310	0.94377	13.300	66.293	79.593	.029420	.16704	22
23	37.849	23.153	.011338	1.0406	88.201	0.96098	13.520	66.177	79.697	.029871	.16697	23
24	38.574	23.878	.011352	1.0220	88.091	0.97843	13.739	66.061	79.800	.030322	.16690	24
25	39.310	24.614	0.011366	1.0039	87.981	0.99613	13.958	65.946	79.904	0.030772	0.16683	25
26	40.056	25.360	.011380	0.98612	87.870	1.0141	14.178	65.829	80.007	.031221	.16676	26
27	40.813	26.117	.011395	0.96874	87.760	1.0323	14.398	65.713	80.111	.031670	.16669	27
28	41.580	26.884	.011409	0.95173	87.649	1.0507	14.618	65.596	80.214	.032118	.16662	28
29	42.359	27.663	.011424	0.93509	87.537	1.0694	14.838	65.478	80.316	.032566	.16655	29
30	43.148	28.452	0.011438	0.91880	87.426	1.0884	15.058	65.361	80.419	0.033013	0.16648	30
31	43.948	29.252	.011453	0.90286	87.314	1.1076	15.279	65.243	80.522	.033460	.16642	31
32	44.760	30.064	.011468	0.88725	87.202	1.1271	15.500	65.124	80.624	.033905	.16635	32
33	45.583	30.887	.011482	0.87197	87.090	1.1468	15.720	65.006	80.726	.034351	.16629	33
34	46.417	31.721	.011497	0.85702	86.977	1.1668	15.942	64.886	80.828	.034796	.16622	34
35	47.263	32.567	0.011512	0.84237	86.865	1.1871	16.163	64.767	80.930	0.035240	0.16616	35
36	48.120	33.424	.011527	0.82803	86.751	1.2077	16.384	64.647	81.031	.035683	.16610	36
37	48.989	34.293	.011542	0.81399	86.638	1.2285	16.606	64.527	81.133	.036126	.16604	37
38	49.870	35.174	.011557	0.80023	86.524	1.2496	16.828	64.406	81.234	.036569	.16598	38
39	50.763	36.067	0.011573	0.78676	86.410	1.2710	17.050	64.285	81.335	0.037011	0.16592	39

*Inches of mercury below one standard atmosphere.

TABLE 3 ... PROPERTIES OF LIQUID AND SATURATED VAPOR (continued)

TEMP F	PRESSURE lb per sq in		VOLUME cu ft per lb		DENSITY lb per cu ft		ENTHALPY ** Btu per lb			ENTROPY ** Btu per (lb) (°R)		TEMP F
	Absolute P	Gage p	Liquid v_f	Vapor v_g	Liquid $1/v_f$	Vapor $1/v_g$	Liquid h_f	Latent h_{fg}	Vapor h_g	Liquid s_f	Vapor s_g	
t												t
40	51.667	36.971	0.011588	0.77357	86.296	1.2927	17.273	64.163	81.436	0.037453	0.16586	40
41	52.584	37.888	.011603	.76064	86.181	1.3147	17.495	64.042	81.537	.037893	.16580	41
42	53.513	38.817	.011619	.74798	86.066	1.3369	17.718	63.919	81.637	.038334	.16574	42
43	54.454	39.758	.011635	.73557	85.951	1.3595	17.941	63.796	81.737	.038774	.16568	43
44	55.407	40.711	.011650	.72341	85.836	1.3823	18.164	63.673	81.837	.039213	.16562	44
45	56.373	41.677	0.011666	0.71149	85.720	1.4055	18.387	63.550	81.937	0.039652	0.16557	45
46	57.352	42.656	.011682	.69982	85.604	1.4289	18.611	63.426	82.037	.040091	.16551	46
47	58.343	43.647	.011698	.68837	85.487	1.4527	18.835	63.301	82.136	.040529	.16546	47
48	59.347	44.651	.011714	.67715	85.371	1.4768	19.059	63.177	82.236	.040966	.16540	48
49	60.364	45.668	.011730	.66616	85.254	1.5012	19.283	63.051	82.334	.041403	.16535	49
50	61.394	46.698	0.011746	0.65537	85.136	1.5258	19.507	62.926	82.433	0.041839	0.16530	50
51	62.437	47.741	.011762	.64480	85.018	1.5509	19.732	62.800	82.532	.042276	.16524	51
52	63.494	48.798	.011779	.63444	84.900	1.5762	19.957	62.673	82.630	.042711	.16519	52
53	64.563	49.867	.011795	.62428	84.782	1.6019	20.182	62.546	82.728	.043146	.16514	53
54	65.646	50.950	.011811	.61431	84.663	1.6278	20.408	62.418	82.826	.043581	.16509	54
55	66.743	52.047	0.011828	0.60453	84.544	1.6542	20.634	62.290	82.924	0.044015	0.16504	55
56	67.853	53.157	.011845	.59495	84.425	1.6808	20.859	62.162	83.021	.044449	.16499	56
57	68.977	54.281	.011862	.58554	84.305	1.7078	21.086	62.033	83.119	.044883	.16494	57
58	70.115	55.419	.011879	.57632	84.185	1.7352	21.312	61.903	83.215	.045316	.16489	58
59	71.267	56.571	.011896	.56727	84.065	1.7628	21.539	61.773	83.312	.045748	.16484	59
60	72.433	57.737	0.011913	0.55839	83.944	1.7909	21.766	61.643	83.409	0.046180	0.16479	60
61	73.613	58.917	.011930	.54967	83.823	1.8193	21.993	61.512	83.505	.046612	.16474	61
62	74.807	60.111	.011947	.54112	83.701	1.8480	22.221	61.380	83.601	.047044	.16470	62
63	76.016	61.320	.011965	.53273	83.580	1.8771	22.448	61.248	83.696	.047475	.16465	63
64	77.239	62.543	.011982	.52450	83.457	1.9066	22.676	61.116	83.792	.047905	.16460	64
65	78.477	63.781	0.012000	0.51642	83.335	1.9364	22.905	60.982	83.887	0.048336	0.16456	65
66	79.729	65.033	.012017	.50848	83.212	1.9666	23.133	60.849	83.982	.048765	.16451	66
67	80.996	66.300	.012035	.50070	83.089	1.9972	23.362	60.715	84.077	.049195	.16447	67
68	82.279	67.583	.012053	.49305	82.965	2.0282	23.591	60.580	84.171	.049624	.16442	68
69	83.576	68.880	.012071	.48555	82.841	2.0595	23.821	60.445	84.266	.050053	.16438	69
70	84.888	70.192	0.012089	0.47818	82.717	2.0913	24.050	60.309	84.359	0.050482	0.16434	70
71	86.216	71.520	.012108	.47094	82.592	2.1234	24.281	60.172	84.453	.050910	.16429	71
72	87.559	72.863	.012126	.46383	82.467	2.1559	24.511	60.035	84.546	.051338	.16425	72
73	88.918	74.222	.012145	.45686	82.341	2.1889	24.741	59.898	84.639	.051766	.16421	73
74	90.292	75.596	.012163	.45000	82.215	2.2222	24.973	59.759	84.732	.052193	.16417	74
75	91.682	76.986	0.012182	0.44327	82.089	2.2560	25.204	59.621	84.825	0.052620	0.16412	75
76	93.087	78.391	.012201	.43666	81.962	2.2901	25.435	59.481	84.916	.053047	.16408	76
77	94.509	79.813	.012220	.43016	81.835	2.3247	25.667	59.341	85.008	.053473	.16404	77
78	95.946	81.250	.012239	.42378	81.707	2.3597	25.899	59.201	85.100	.053900	.16400	78
79	97.400	82.704	.012258	.41751	81.579	2.3951	26.132	59.059	85.191	.054326	.16396	79
80	98.870	84.174	0.012277	0.41135	81.450	2.4310	26.365	58.917	85.282	0.054751	0.16392	80
81	100.36	85.66	.012297	.40530	81.322	2.4673	26.598	58.775	85.373	.055177	.16388	81
82	101.86	87.16	.012316	.39935	81.192	2.5041	26.832	58.631	85.463	.055602	.16384	82
83	103.38	88.68	.012336	.39351	81.063	2.5413	27.065	58.488	85.553	.056027	.16380	83
84	104.92	90.22	.012356	.38776	80.932	2.5789	27.300	58.343	85.643	.056452	.16376	84
85	106.47	91.77	0.012376	0.38212	80.802	2.6170	27.534	58.198	85.732	0.056877	0.16372	85
86	108.04	93.34	.012396	.37657	80.671	2.6556	27.769	58.052	85.821	.057301	.16368	86
87	109.63	94.93	.012416	.37111	80.539	2.6946	28.005	57.905	85.910	.057725	.16364	87
88	111.23	96.53	.012437	.36575	80.407	2.7341	28.241	57.757	85.998	.058149	.16360	88
89	112.85	98.15	.012457	.36047	80.275	2.7741	28.477	57.609	86.086	.058573	.16357	89
90	114.49	99.79	0.012478	0.35529	80.142	2.8146	28.713	57.461	86.174	0.058997	0.16353	90
91	116.15	101.45	.012499	.35019	80.008	2.8556	28.950	57.311	86.261	.059420	.16349	91
92	117.82	103.12	.012520	.34518	79.874	2.8970	29.187	57.161	86.348	.059844	.16345	92
93	119.51	104.81	.012541	.34025	79.740	2.9390	29.425	57.009	86.434	.060267	.16341	93
94	121.22	106.52	.012562	.33540	79.605	2.9815	29.663	56.858	86.521	.060690	.16338	94
95	122.95	108.25	0.012583	0.33063	79.470	3.0245	29.901	56.705	86.606	0.061113	0.16334	95
96	124.70	110.00	.012605	.32594	79.334	3.0680	30.140	56.551	86.691	.061536	.16330	96
97	126.46	111.76	.012627	.32133	79.198	3.1120	30.380	56.397	86.777	.061959	.16326	97
98	128.24	113.54	.012649	.31679	79.061	3.1566	30.619	56.242	86.861	.062381	.16323	98
99	130.04	115.34	.012671	.31233	78.923	3.2017	30.859	56.086	86.945	.062804	.16319	99
100	131.86	117.16	0.012693	0.30794	78.785	3.2474	31.100	55.929	87.029	0.063227	0.16315	100
101	133.70	119.00	.012715	.30362	78.647	3.2936	31.341	55.772	87.113	.063649	.16312	101
102	135.56	120.86	.012738	.29937	78.508	3.3404	31.583	55.613	87.196	.064072	.16308	102
103	137.44	122.74	.012760	.29518	78.368	3.3877	31.824	55.454	87.278	.064494	.16304	103
104	139.33	124.63	0.012783	0.29106	78.228	3.4357	32.067	55.293	87.360	0.064916	0.16301	104

TABLE 3... PROPERTIES OF LIQUID AND SATURATED VAPOR (continued)

TEMP F	PRESSURE lb per sq in		VOLUME cu ft per lb		DENSITY lb per cu ft		ENTHALPY ** Btu per lb			ENTROPY ** Btu per (lb) (^{o}R)		TEMP F
t	Absolute P	Gage p	Liquid v_f	Vapor v_g	Liquid $1/v_f$	Vapor $1/v_g$	Liquid h_f	Latent h_{fg}	Vapor h_g	Liquid s_f	Vapor s_g	t
105	141.25	126.55	0.012806	0.28701	78.088	3.4842	32.310	55.132	87.442	0.065339	0.16297	105
106	143.18	128.48	.012829	.28303	77.946	3.5333	32.553	54.970	87.523	.065761	.16293	106
107	145.13	130.43	.012853	.27910	77.804	3.5829	32.797	54.807	87.604	.066184	.16290	107
108	147.11	132.41	.012876	.27524	77.662	3.6332	33.041	54.643	87.634	.066606	.16286	108
109	149.10	134.40	.012900	.27143	77.519	3.6841	33.286	54.478	87.764	.067028	.16282	109
110	151.11	136.41	0.012924	0.26769	77.376	3.7357	33.531	54.313	87.844	0.067451	0.16279	110
111	153.14	138.44	.012948	.26400	77.231	3.7878	33.777	54.146	87.923	.067873	.16275	111
112	155.19	140.49	.012972	.26037	77.087	3.8406	34.023	53.978	88.001	.068296	.16271	112
113	157.27	142.57	.012997	.25680	76.941	3.8941	34.270	53.809	88.079	.068719	.16268	113
114	159.36	144.66	.013022	.25328	76.795	3.9482	34.517	53.639	88.156	.069141	.16264	114
115	161.47	146.77	0.013047	0.24982	76.649	4.0029	34.765	53.468	88.233	0.069564	0.16260	115
116	163.61	148.91	.013072	.24641	76.501	4.0584	35.014	53.296	88.310	.069987	.16256	116
117	165.76	151.06	.013097	.24304	76.353	4.1145	35.263	53.123	88.386	.070410	.16253	117
118	167.94	153.24	.013123	.23974	76.205	4.1713	35.512	52.949	88.461	.070833	.16249	118
119	170.13	155.43	.013148	.23647	76.056	4.2288	35.762	52.774	88.536	.071257	.16245	119
120	172.35	157.65	0.013174	0.23326	75.906	4.2870	36.013	52.597	88.610	0.071680	0.16241	120
121	174.59	159.89	.013200	.23010	75.755	4.3459	36.264	52.420	88.684	.072104	.16237	121
122	176.85	162.15	.013227	.22698	75.604	4.4056	36.516	52.241	88.757	.072528	.16234	122
123	179.13	164.43	.013254	.22391	75.452	4.4660	36.768	52.062	88.830	.072952	.16230	123
124	181.43	166.73	.013280	.22089	75.299	4.5272	37.021	51.881	88.902	.073376	.16226	124
125	183.76	169.06	0.013308	0.21791	75.145	4.5891	37.275	51.698	88.973	0.073800	0.16222	125
126	186.10	171.40	.013335	.21497	74.991	4.6518	37.529	51.515	89.044	.074225	.16218	126
127	188.47	173.77	.013363	.21207	74.836	4.7153	37.785	51.330	89.115	.074650	.16214	127
128	190.86	176.16	.013390	.20922	74.680	4.7796	38.040	51.144	89.184	.075075	.16210	128
129	193.27	178.57	.013419	.20641	74.524	4.8448	38.296	50.957	89.253	.075501	.16206	129
130	195.71	181.01	0.013447	0.20364	74.367	4.9107	38.553	50.768	89.321	0.075927	0.16202	130
131	198.16	183.46	.013476	.20091	74.209	4.9775	38.811	50.578	89.389	.076353	.16198	131
132	200.64	185.94	.013504	.19821	74.050	5.0451	39.069	50.387	89.456	.076779	.16194	132
133	203.15	188.45	.013534	.19556	73.890	5.1136	39.328	50.194	89.522	.077206	.16189	133
134	205.67	190.97	.013563	.19294	73.729	5.1829	39.588	50.000	89.588	.077633	.16185	134
135	208.22	193.52	0.013593	0.19036	73.568	5.2532	39.848	49.805	89.653	0.078061	0.16181	135
136	210.79	196.09	.013623	.18782	73.406	5.3244	40.110	49.608	89.718	.078489	.16177	136
137	213.39	198.69	.013653	.18531	73.243	5.3965	40.372	49.409	89.781	.078917	.16172	137
138	216.01	201.31	.013684	.18283	73.079	5.4695	40.634	49.210	89.844	.079346	.16168	138
139	218.65	203.95	.013715	.18039	72.914	5.5435	40.898	49.008	89.906	.079775	.16163	139
140	221.32	206.62	0.013746	0.17799	72.748	5.6184	41.162	48.805	89.967	0.080205	0.16159	140
141	224.00	209.30	.013778	.17561	72.581	5.6944	41.427	48.601	90.028	.080635	.16154	141
142	226.72	212.02	.013810	.17327	72.413	5.7713	41.693	48.394	90.087	.081065	.16150	142
143	229.46	214.76	.013842	.17096	72.244	5.8493	41.959	48.187	90.146	.081497	.16145	143
144	232.22	217.52	.013874	.16868	72.075	5.9283	42.227	47.977	90.204	.081928	.16140	144
145	235.00	220.30	0.013907	0.16644	71.904	6.0083	42.495	47.766	90.261	0.082361	0.16135	145
146	237.82	223.12	.013941	.16422	71.732	6.0895	42.765	47.553	90.318	.082794	.16130	146
147	240.65	225.95	.013974	.16203	71.559	6.1717	43.035	47.338	90.373	.083227	.16125	147
148	243.51	228.81	.014008	.15987	71.386	6.2551	43.306	47.122	90.428	.083661	.16120	148
149	246.40	231.70	.014043	.15774	71.211	6.3395	43.578	46.904	90.482	.084096	.16115	149
150	249.31	234.61	0.014078	0.15564	71.035	6.4252	43.850	46.684	90.534	0.084531	0.16110	150
151	252.24	237.54	.014113	.15356	70.857	6.5120	44.124	46.462	90.586	.084967	.16105	151
152	255.20	240.50	.014148	.15151	70.679	6.6001	44.399	46.238	90.637	.085404	.16099	152
153	258.19	243.49	.014184	.14949	70.500	6.6893	44.675	46.012	90.687	.085842	.16094	153
154	261.20	246.50	.014221	.14750	70.319	6.7799	44.951	45.784	90.735	.086280	.16088	154
155	264.24	249.54	0.014258	0.14552	70.137	6.8717	45.229	45.554	90.783	0.086719	0.16083	155
156	267.30	252.60	.014295	.14358	69.954	6.9648	45.508	45.322	90.830	.087159	.16077	156
157	270.39	255.69	.014333	.14166	69.770	7.0592	45.787	45.088	90.875	.087600	.16071	157
158	273.51	258.81	.014371	.13976	69.584	7.1551	46.068	44.852	90.920	.088041	.16065	158
159	276.65	261.95	.014410	.13789	69.397	7.2523	46.350	44.614	90.964	.088484	.16059	159
160	279.82	265.12	0.014449	0.13604	69.209	7.3509	46.633	44.373	91.006	0.088927	0.16053	160
161	283.02	268.32	.014489	.13421	69.019	7.4510	46.917	44.130	91.047	.089371	.16047	161
162	286.24	271.54	.014529	.13241	68.828	7.5525	47.202	43.885	91.087	.089817	.16040	162
163	289.49	274.79	.014570	.13062	68.635	7.6556	47.489	43.637	91.126	.090263	.16034	163
164	292.77	278.07	.014611	.12886	68.441	7.7602	47.777	43.386	91.163	.090710	.16027	164
165	296.07	281.37	0.014653	0.12712	68.245	7.8665	48.065	43.134	91.199	0.091159	0.16021	165
166	299.40	284.70	.014695	.12540	68.048	7.9743	48.355	42.879	91.234	.091608	.16014	166
167	302.76	288.06	.014738	.12370	67.850	8.0838	48.647	42.620	91.267	.092059	.16007	167
168	306.15	291.45	.014782	.12202	67.649	8.1950	48.939	42.360	91.299	.092511	.16000	168
169	309.56	294.86	0.014826	0.12037	67.447	8.3080	49.233	42.097	91.330	0.092964	0.15992	169

TABLE 7... PROPERTIES OF LIQUID AND SATURATED VAPOR (continued)

TEMP F	PRESSURE lb per sq in		VOLUME cu ft per lb		DENSITY lb per cu ft		ENTHALPY ** Btu per lb			ENTROPY ** Btu per (lb) (°R)		Temp F
	Absolute P	Gage p	Liquid v_f	Vapor v_g	Liquid $1/v_f$	Vapor $1/v_g$	Liquid h_f	Latent h_{fg}	Vapor h_g	Liquid s_f	Vapor s_g	
t												t
-50	11.674	6.154*	0.011235	4.2224	89.004	0.23683	- 2.511	101.656	99.144	-0.00604	0.24209	-50
-49	11.996	5.498*	.011248	4.1166	88.905	.24292	- 2.262	101.519	99.257	- .00543	.24176	-49
-48	12.324	4.829*	.011261	4.0140	88.806	.24913	- 2.012	101.381	99.369	- .00483	.24143	-48
-47	12.660	4.144*	.011273	3.9145	88.707	.25546	- 1.762	101.242	99.480	- .00422	.24110	-47
-46	13.004	3.445*	.011286	3.8179	88.607	.26192	- 1.511	101.103	99.592	- .00361	.24078	-46
-45	13.354	2.732*	0.011298	3.7243	88.507	0.26851	- 1.260	100.963	99.703	-0.00301	0.24046	-45
-44	13.712	2.002*	.011311	3.6334	88.407	.27523	- 1.009	100.823	99.814	- .00241	.24014	-44
-43	14.078	1.258*	.011324	3.5452	88.307	.28207	- 0.757	100.683	99.925	- .00181	.23982	-43
-42	14.451	0.498*	.011337	3.4596	88.207	.28905	- 0.505	100.541	100.036	- .00120	.23951	-42
-41	14.833	0.137	.011350	3.3764	88.107	.29617	- 0.253	100.399	100.147	- .00060	.23919	-41
-40	15.222	0.526	0.011363	3.2957	88.006	0.30342	0.000	100.257	100.257	0.00000	0.23888	-40
-39	15.619	0.923	.011376	3.2173	87.905	.31082	0.253	100.114	100.367	.00060	.23858	-39
-38	16.024	1.328	.011389	3.1412	87.805	.31835	0.506	99.971	100.477	.00120	.23827	-38
-37	16.437	1.741	.011402	3.0673	87.703	.32602	0.760	99.826	100.587	.00180	.23797	-37
-36	16.859	2.163	.011415	2.9954	87.602	.33384	1.014	99.682	100.696	.00240	.23767	-36
-35	17.290	2.594	0.011428	2.9256	87.501	0.34181	1.269	99.536	100.805	0.00300	0.23737	-35
-34	17.728	3.032	.011442	2.8578	87.399	.34992	1.524	99.391	100.914	.00359	.23707	-34
-33	18.176	3.480	.011455	2.7919	87.297	.35818	1.779	99.244	101.023	.00419	.23678	-33
-32	18.633	3.937	.011469	2.7278	87.195	.36660	2.035	99.097	101.132	.00479	.23649	-32
·31	19.098	4.402	.011482	2.6655	87.093	.37517	2.291	98.949	101.240	.00538	.23620	-31
-30	19.573	4.877	0.011495	2.6049	86.991	0.38389	2.547	98.801	101.348	0.00598	0.23591	-30
-29	20.056	5.360	.011509	2.5460	86.888	.39278	2.804	98.652	101.456	.00657	.23563	-29
-28	20.549	5.853	.011523	2.4887	86.785	.40182	3.061	98.503	101.564	.00716	.23534	-28
-27	21.052	6.536	.011536	2.4329	86.682	.41103	3.318	98.353	101.671	.00776	.23506	-27
-26	21.564	6.868	.011550	2.3787	86.579	.42040	3.576	98.202	101.778	.00835	.23478	-26
-25	22.086	7.390	0.011564	2.3260	86.476	0.42993	3.834	98.051	101.885	0.00894	0.23451	-25
-24	22.617	7.921	.011578	2.2746	86.372	.43964	4.093	97.899	101.992	.00953	.23423	-24
-23	23.159	8.463	.011592	2.2246	86.269	.44951	4.352	97.746	102.098	.01013	.23396	-23
-22	23.711	9.015	.011606	2.1760	86.165	.45956	4.611	97.593	102.204	.01072	.23369	-22
-21	24.272	9.576	.011620	2.1287	86.061	.46978	4.871	97.439	102.310	.01131	.23342	-21
-20	24.845	10.149	0.011634	2.0826	85.956	0.48018	5.131	97.285	102.415	0.01189	0.23315	-20
-19	25.427	10.731	.011648	2.0377	85.852	.49075	5.391	97.129	102.521	.01248	.23289	-19
-18	26.020	11.324	.011662	1.9940	85.747	.50151	5.652	96.974	102.626	.01307	.23262	-18
-17	26.624	11.928	.011677	1.9514	85.642	.51245	5.913	96.817	102.730	.01366	.23236	-17
-16	27.239	12.543	.011691	1.9099	85.537	.52358	6.175	96.660	102.835	.01425	.23210	-16
-15	27.865	13.169	0.011705	1.8695	85.431	0.53489	6.436	96.502	102.939	0.01483	0.23184	-15
-14	28.501	13.805	.011720	1.8302	85.326	.54640	6.699	96.344	103.043	.01542	.23159	-14
-13	29.149	14.453	.011734	1.7918	85.220	.55810	6.961	96.185	103.146	.01600	.23133	-13
-12	29.809	15.113	.011749	1.7544	85.114	.56999	7.224	96.025	103.250	.01659	.23108	-12
-11	30.480	15.784	.011764	1.7180	85.008	.58207	7.488	95.865	103.353	.01717	.23083	-11
-10	31.162	16.466	0.011778	1.6825	84.901	0.59436	7.751	95.704	103.455	0.01776	0.23058	-10
- 9	31.856	17.160	.011793	1.6479	84.795	.60685	8.015	95.542	103.558	.01834	.23033	- 9
- 8	32.563	17.867	.011808	1.6141	84.688	.61954	8.280	95.380	103.660	.01892	.23008	- 8
- 7	33.281	18.585	.011823	1.5812	84.581	.63244	8.545	95.217	103.762	.01950	.22984	- 7
- 6	34.011	19.315	.011838	1.5491	84.473	.64555	8.810	95.053	103.863	.02009	.22960	- 6
- 5	34.754	20.058	0.011853	1.5177	84.366	0.65887	9.075	94.889	103.964	0.02067	0.22936	- 5
- 4	35.509	20.813	.011868	1.4872	84.258	.67240	9.341	94.724	104.065	.02125	.22912	- 4
- 3	36.277	21.581	.011884	1.4574	84.150	.68615	9.608	94.558	104.166	.02183	.22888	- 3
- 2	37.057	22.361	.011899	1.4283	84.042	.70012	9.874	94.391	104.266	.02241	.22864	- 2
- 1	37.850	23.154	.011914	1.4000	83.933	.71431	10.142	94.224	104.366	.02299	.22841	- 1
0	38.657	23.961	0.011930	1.3723	83.825	0.72872	10.409	94.056	104.465	0.02357	0.22817	0
1	39.476	24.780	.011945	1.3453	83.716	.74336	10.677	93.888	104.565	.02414	.22794	1
2	40.309	25.613	.011961	1.3189	83.606	.75822	10.945	93.718	104.663	.02472	.22771	2
3	41.155	26.459	.011976	1.2931	83.497	.77332	11.214	93.548	104.762	.02530	.22748	3
4	42.014	27.318	.011992	1.2680	83.387	.78865	11.483	93.378	104.860	.02587	.22725	4
5	42.888	28.192	0.012008	1.2434	83.277	0.80422	11.752	93.206	104.958	0.02645	0.22703	5
6	43.775	29.079	.012024	1.2195	83.167	.82003	12.022	93.034	105.056	.02703	.22680	6
7	44.676	29.980	.012040	1.1961	83.057	.83608	12.292	92.861	105.153	.02760	.22658	7
8	45.591	30.895	.012056	1.1732	82.946	.85237	12.562	92.688	105.250	.02818	.22636	8
9	46.521	31.825	.012072	1.1509	82.835	.86892	12.833	92.513	105.346	.02875	.22614	9
10	47.464	32.768	0.012088	1.1290	82.724	0.88571	13.104	92.338	105.442	0.02932	0.22592	10
11	48.423	33.727	.012105	1.1077	82.612	.90275	13.376	92.162	105.538	.02990	.22570	11
12	49.396	34.700	.012121	1.0869	82.501	.92005	13.648	91.986	105.633	.03047	.22548	12
13	50.384	35.688	.012138	1.0665	82.389	.93761	13.920	91.808	105.728	.03104	.22527	13
14	51.387	36.691	.012154	1.0466	82.276	.95544	14.193	91.630	105.823	.03161	.22505	14

*Inches of mercury below one standard atmosphere.

TABLE 7 ... PROPERTIES OF LIQUID AND SATURATED VAPOR (continued)

TEMP F	PRESSURE lb per sq in		VOLUME cu ft per lb		DENSITY lb per cu ft		ENTHALPY ** Btu per lb			ENTROPY ** Btu per (lb) (°R)		TEMP F
t	Absolute P	Gage p	Liquid v_f	Vapor v_g	Liquid $1/v_f$	Vapor $1/v_g$	Liquid h_f	Latent h_{fg}	Vapor h_g	Liquid s_f	Vapor s_g	t
15	52.405	37.709	0.012171	1.0272	82.164	0.97352	14.466	91.451	105.917	0.03218	0.22484	15
16	53.438	38.742	.012188	1.0082	82.051	0.99188	14.739	91.272	106.011	.03275	.22463	16
17	54.487	39.791	.012204	0.98961	81.938	1.0105	15.013	91.091	106.105	.03332	.22442	17
18	55.551	40.855	.012221	0.97144	81.825	1.0294	15.288	90.910	106.198	.03389	.22421	18
19	56.631	41.935	.012238	0.95368	81.711	1.0486	15.562	90.728	106.290	.03446	.22400	19
20	57.727	43.031	0.012255	0.93631	81.597	1.0680	15.837	90.545	106.383	0.03503	0.22379	20
21	58.839	44.143	.012273	.91932	81.483	1.0878	16.113	90.362	106.475	.03560	.22358	21
22	59.967	45.271	.012290	.90270	81.368	1.1078	16.389	90.178	106.566	.03617	.22338	22
23	61.111	46.415	.012307	.88645	81.253	1.1281	16.665	89.993	106.657	.03674	.22318	23
24	62.272	47.576	.012325	.87055	81.138	1.1487	16.942	89.807	106.748	.03730	.22297	24
25	63.450	48.754	0.012342	0.85500	81.023	1.1696	17.219	89.620	106.839	0.03787	0.22277	25
26	64.644	49.948	.012360	.83978	80.907	1.1908	17.496	89.433	106.928	.03844	.22257	26
27	65.855	51.159	.012378	.82488	80.791	1.2123	17.774	89.244	107.018	.03900	.22237	27
28	67.083	52.387	.012395	.81031	80.675	1.2341	18.052	89.055	107.107	.03958	.22217	28
29	68.328	53.632	.012413	.79604	80.558	1.2562	18.330	88.865	107.196	.04013	.22198	29
30	69.591	54.895	0.012431	0.78208	80.441	1.2786	18.609	88.674	107.284	0.04070	0.22178	30
31	70.871	56.175	.012450	.76842	80.324	1.3014	18.889	88.483	107.372	.04126	.22158	31
32	72.169	57.473	.012468	.75503	80.207	1.3244	19.169	88.290	107.459	.04182	.22139	32
33	73.485	58.789	.012486	.74194	80.089	1.3478	19.449	88.097	107.546	.04239	.22119	33
34	74.818	60.122	.012505	.72911	79.971	1.3715	19.729	87.903	107.632	.04295	.22100	34
35	76.170	61.474	0.012523	0.71655	79.852	1.3956	20.010	87.708	107.719	0.04351	0.22081	35
36	77.540	62.844	.012542	.70425	79.733	1.4199	20.292	87.512	107.804	.04407	.22062	36
37	78.929	64.233	.012561	.69221	79.614	1.4447	20.574	87.316	107.889	.04464	.22043	37
38	80.336	65.640	.012579	.68041	79.495	1.4697	20.856	87.118	107.974	.04520	.22024	38
39	81.761	67.065	.012598	.66885	79.375	1.4951	21.138	86.920	108.058	.04576	.22005	39
40	83.206	68.510	0.012618	0.65753	79.255	1.5208	21.422	86.720	108.142	0.04632	0.21986	40
41	84.670	69.974	.012637	.64643	79.134	1.5469	21.705	86.520	108.225	.04688	.21968	41
42	86.153	71.457	.012656	.63557	79.013	1.5734	21.989	86.319	108.308	.04744	.21949	42
43	87.655	72.959	.012676	.62492	78.892	1.6002	22.273	86.117	108.390	.04800	.21931	43
44	89.177	74.481	.012695	.61448	78.770	1.6274	22.558	85.914	108.472	.04855	.21912	44
45	90.719	76.023	0.012715	0.60425	78.648	1.6549	22.843	85.710	108.553	0.04911	0.21894	45
46	92.280	77.584	.012735	.59422	78.526	1.6829	23.129	85.506	108.634	.04967	.21876	46
47	93.861	79.165	.012755	.58440	78.403	1.7112	23.415	85.300	108.715	.05023	.21858	47
48	95.463	80.767	.012775	.57476	78.280	1.7398	23.701	85.094	108.795	.05079	.21839	48
49	97.085	82.389	.012795	.56532	78.157	1.7689	23.988	84.886	108.874	.05134	.21821	49
50	98.727	84.031	0.012815	0.55606	78.033	1.7984	24.275	84.678	108.953	0.05190	0.21803	50
51	100.39	85.69	.012836	.54698	77.909	1.8282	24.563	84.468	109.031	.05245	.21785	51
52	102.07	87.38	.012856	.53808	77.784	1.8585	24.851	84.258	109.109	.05301	.21768	52
53	103.78	89.08	.012877	.52934	77.659	1.8891	25.139	84.047	109.186	.05357	.21750	53
54	105.50	90.81	.012898	.52078	77.534	1.9202	25.429	83.834	109.263	.05412	.21732	54
55	107.25	92.56	0.012919	0.51238	77.408	1.9517	25.718	83.621	109.339	0.05468	0.21714	55
56	109.02	94.32	.012940	.50414	77.282	1.9836	26.008	83.407	109.415	.05523	.21697	56
57	110.81	96.11	.012961	.49606	77.155	2.0159	26.298	83.191	109.490	.05579	.21679	57
58	112.62	97.93	.012982	.48813	77.028	2.0486	26.589	82.975	109.564	.05634	.21662	58
59	114.46	99.76	.013004	.48035	76.900	2.0818	26.880	82.758	109.638	.05689	.21644	59
60	116.31	101.62	0.013025	0.47272	76.773	2.1154	27.172	82.540	109.712	0.05745	0.21627	60
61	118.19	103.49	.013047	.46523	76.644	2.1495	27.464	82.320	109.785	.05800	.21610	61
62	120.09	105.39	.013069	.45788	76.515	2.1840	27.757	82.100	109.857	.05855	.21592	62
63	122.01	107.32	.013091	.45066	76.386	2.2190	28.050	81.878	109.929	.05910	.21575	63
64	123.96	109.26	.013114	.44358	76.257	2.2544	28.344	81.656	110.000	.05966	.21558	64
65	125.93	111.23	0.013136	0.43663	76.126	2.2903	28.638	81.432	110.070	0.06021	0.21541	65
66	127.92	113.22	.013159	.42981	75.996	2.3266	28.932	81.208	110.140	.06076	.21524	66
67	129.94	115.24	.013181	.42311	75.865	2.3635	29.228	80.982	110.209	.06131	.21507	67
68	131.97	117.28	.013204	.41653	75.733	2.4008	29.523	80.755	110.278	.06186	.21490	68
69	134.04	119.34	.013227	.41007	75.601	2.4386	29.819	80.527	110.346	.06241	.21473	69
70	136.12	121.43	0.013251	0.40373	75.469	2.4769	30.116	80.298	110.414	0.06296	0.21456	70
71	138.23	123.54	.013274	.39751	75.336	2.5157	30.413	80.068	110.480	.06351	.21439	71
72	140.37	125.67	.013297	.39139	75.202	2.5550	30.710	79.836	110.547	.06406	.21422	72
73	142.52	127.83	.013321	.38539	75.068	2.5948	31.008	79.604	110.612	.06461	.21405	73
74	144.71	130.01	.013345	.37949	74.934	2.6351	31.307	79.370	110.677	.06516	.21388	74
75	146.91	132.22	0.013369	0.37369	74.799	2.6760	31.606	79.135	110.741	0.06571	0.21372	75
76	149.15	134.45	.013393	.36800	74.664	2.7174	31.906	78.899	110.805	.06626	.21355	76
77	151.40	136.71	.013418	.36241	74.528	2.7593	32.206	78.662	110.868	.06681	.21338	77
78	153.69	138.99	.013442	.35691	74.391	2.8018	32.506	78.423	110.930	.06736	.21321	78
79	155.99	141.30	.013467	.35151	74.254	2.8449	32.808	78.184	110.991	.06791	.21305	79

TABLE 7... PROPERTIES OF LIQUID AND SATURATED VAPOR (continued)

TEMP F	PRESSURE lb per sq in		VOLUME cu ft per lb		DENSITY lb per cu ft		ENTHALPY ** Btu per lb			ENTROPY ** Btu per (lb) (°R)		TEMP F
t	Absolute P	Gage p	Liquid v_f	Vapor v_g	Liquid $1/v_f$	Vapor $1/v_g$	Liquid h_f	Latent h_{fg}	Vapor h_g	Liquid s_f	Vapor s_g	t
80	158.33	143.63	0.013492	0.34621	74.116	2.8885	33.109	77.943	111.052	0.06846	0.21288	80
81	160.68	145.99	.013518	.34099	73.978	2.9326	33.412	77.701	111.112	.06901	.21271	81
82	163.07	148.37	.013543	.33587	73.839	2.9774	33.714	77.457	111.171	.06956	.21255	82
83	165.48	150.78	.013569	.33083	73.700	3.0227	34.018	77.212	111.230	.07011	.21238	83
84	167.92	153.22	.013594	.32588	73.560	3.0686	34.322	76.966	111.288	.07065	.21222	84
85	170.38	155.68	0.013620	0.32101	73.420	3.1151	34.626	76.719	111.345	0.07120	0.21205	85
86	172.87	158.17	.013647	.31623	73.278	3.1622	34.931	76.470	111.401	.07175	.21188	86
87	175.38	160.69	.013673	.31153	73.137	3.2100	35.237	76.220	111.457	.07230	.21172	87
88	177.93	163.23	.013700	.30690	72.994	3.2583	35.543	75.968	111.512	.07285	.21155	88
89	180.50	165.80	.013727	.30236	72.851	3.3073	35.850	75.716	111.566	.07339	.21139	89
90	183.09	168.40	0.013754	0.29789	72.708	3.3570	36.158	75.461	111.619	0.07394	0.21122	90
91	185.72	171.02	.013781	.29349	72.564	3.4073	36.466	75.206	111.671	.07449	.21106	91
92	188.37	173.67	.013809	.28917	72.419	3.4582	36.774	74.949	111.723	.07504	.21089	92
93	191.05	176.35	.013836	.28491	72.273	3.5098	37.084	74.690	111.774	.07559	.21072	93
94	193.76	179.06	.013864	.28073	72.127	3.5621	37.394	74.430	111.824	.07613	.21056	94
95	196.50	181.80	0.013893	0.27662	71.980	3.6151	37.704	74.168	111.873	0.07668	0.21039	95
96	199.26	184.56	.013921	.27257	71.833	3.6688	38.016	73.905	111.921	.07723	.21023	96
97	202.05	187.36	.013950	.26859	71.685	3.7232	38.328	73.641	111.968	.07778	.21006	97
98	204.87	190.18	.013979	.26467	71.536	3.7783	38.640	73.375	112.015	.07832	.20989	98
99	207.72	193.03	.014008	.26081	71.386	3.8341	38.953	73.107	112.060	.07887	.20973	99
100	210.60	195.91	0.014038	0.25702	71.236	3.8907	39.267	72.838	112.105	0.07942	0.20956	100
101	213.51	198.82	.014068	.25329	71.084	3.9481	39.582	72.567	112.149	.07997	.20939	101
102	216.45	201.76	.014098	.24962	70.933	4.0062	39.897	72.294	112.192	.08052	.20923	102
103	219.42	204.72	.014128	.24600	70.780	4.0651	40.213	72.020	112.233	.08107	.20906	103
104	222.42	207.72	.014159	.24244	70.626	4.1247	40.530	71.744	112.274	.08161	.20889	104
105	225.45	210.75	0.014190	0.23894	70.472	4.1852	40.847	71.467	112.314	0.08216	0.20872	105
106	228.50	213.81	.014221	.23549	70.317	4.2465	41.166	71.187	112.353	.08271	.20855	106
107	231.59	216.90	.014253	.23209	70.161	4.3086	41.485	70.906	112.391	.08326	.20838	107
108	234.71	220.02	.014285	.22875	70.005	4.3715	41.804	70.623	112.427	.08381	.20821	108
109	237.86	223.17	.014317	.22546	69.847	4.4354	42.125	70.338	112.463	.08436	.20804	109
110	241.04	226.35	0.014350	0.22222	69.689	4.5000	42.446	70.052	112.498	0.08491	0.20787	110
111	244.25	229.56	.014382	.21903	69.529	4.5656	42.768	69.763	112.531	.08546	.20770	111
112	247.50	232.80	.014416	.21589	69.369	4.6321	43.091	69.473	112.564	.08601	.20753	112
113	250.77	236.08	.014449	.21279	69.208	4.6994	43.415	69.180	112.595	.08656	.20736	113
114	254.08	239.38	.014483	.20974	69.046	4.7677	43.739	68.886	112.626	.08711	.20718	114
115	257.42	242.72	0.014517	0.20674	68.883	4.8370	44.065	68.590	112.655	0.08766	0.20701	115
116	260.79	246.10	.014552	.20378	68.719	4.9072	44.391	68.291	112.682	.08821	.20684	116
117	264.20	249.50	.014587	.20087	68.554	4.9784	44.718	67.991	112.709	.08876	.20666	117
118	267.63	252.94	.014622	.19800	68.388	5.0506	45.046	67.688	112.735	.08932	.20649	118
119	271.10	256.41	.014658	.19517	68.221	5.1238	45.375	67.384	112.759	.08987	.20631	119
120	274.60	259.91	0.014694	0.19238	68.054	5.1981	45.705	67.077	112.782	0.09042	0.20613	120
121	278.14	263.44	.014731	.18963	67.885	5.2734	46.036	66.767	112.803	.09098	.20595	121
122	281.71	267.01	.014768	.18692	67.714	5.3498	46.368	66.456	112.824	.09153	.20578	122
123	285.31	270.62	.014805	.18426	67.543	5.4272	46.701	66.142	112.843	.09208	.20560	123
124	288.95	274.25	.014843	.18163	67.371	5.5058	47.034	65.826	112.860	.09264	.20542	124
125	292.62	277.92	0.014882	0.17903	67.197	5.5856	47.369	65.507	112.877	0.09320	0.20523	125
126	296.33	281.63	.014920	.17648	67.023	5.6665	47.705	65.186	112.891	.09375	.20505	126
127	300.07	285.37	.014960	.17396	66.847	5.7486	48.042	64.863	112.905	.09431	.20487	127
128	303.84	289.14	.014999	.17147	66.670	5.8319	48.380	64.537	112.917	.09487	.20468	128
129	307.65	292.95	.015039	.16902	66.492	5.9164	48.719	64.208	112.927	.09543	.20449	129
130	311.50	296.80	0.015080	0.16661	66.312	6.0022	49.059	63.877	112.936	0.09598	0.20431	130
131	315.38	300.68	.015121	.16422	66.131	6.0893	49.400	63.543	112.943	.09654	.20412	131
132	319.29	304.60	.015163	.16187	65.949	6.1777	49.743	63.206	112.949	.09711	.20393	132
133	323.25	308.55	.015206	.15956	65.766	6.2674	50.087	62.866	112.953	.09767	.20374	133
134	327.23	312.54	.015248	.15727	65.581	6.3585	50.432	62.523	112.955	.09823	.20354	134
135	331.26	316.56	0.015292	0.15501	65.394	6.4510	50.778	62.178	112.956	0.09879	0.20335	135
136	335.32	320.63	.015336	.15279	65.207	6.5450	51.125	61.829	112.954	.09936	.20315	136
137	339.42	324.73	.015381	.15059	65.017	6.6405	51.474	61.477	112.951	.09992	.20295	137
138	343.56	328.86	.015426	.14843	64.826	6.7374	51.824	61.123	112.947	.10049	.20275	138
139	347.73	333.04	.015472	.14629	64.634	6.8359	52.175	60.764	112.940	.10106	.20255	139
140	351.94	337.25	0.015518	0.14418	64.440	6.9360	52.528	60.403	112.931	0.10163	0.20235	140
141	356.19	341.50	.015566	.14209	64.244	7.0377	52.883	60.038	112.921	.10220	.20214	141
142	360.48	345.79	.015613	.14004	64.047	7.1410	53.238	59.670	112.908	.10277	.20194	142
143	364.81	350.11	.015662	.13801	63.848	7.2461	53.596	59.298	112.893	.10334	.20173	143
144	369.17	354.48	.015712	.13600	63.647	7.3529	53.955	58.922	112.877	.10391	.20152	144

TABLE 15... PROPERTIES OF LIQUID AND SATURATED VAPOR (continued)

TEMP F	PRESSURE lb per sq in		VOLUME cu ft per lb		DENSITY lb per cu ft		ENTHALPY ** Btu per lb			ENTROPY ** Btu per (lb) (°R)		TEMP F
	Absolute P	Gage p	Liquid v_f	Vapor v_g	Liquid $1/v_f$	Vapor $1/v_g$	Liquid h_f	Latent h_{fg}	Vapor h_g	Liquid s_f	Vapor s_g	
t												t
- 35	21.42	6.72	0.01089	1.8637	91.85	0.5366	1.30	74.50	75.80	0.0031	0.1785	- 35
- 34	21.93	7.24	.01090	1.8226	91.74	.5487	1.56	74.37	75.93	.0037	.1784	- 34
- 33	22.46	7.76	.01091	1.7825	91.63	.5610	1.82	74.24	76.06	.0043	.1783	- 33
- 32	23.00	8.30	.01093	1.7436	91.52	.5735	2.07	74.11	76.18	.0049	.1781	- 32
- 31	23.54	8.84	.01094	1.7056	91.41	.5863	2.34	73.97	76.31	.0055	.1780	- 31
- 30	24.10	9.40	0.01095	1.6687	91.30	0.5993	2.60	73.84	76.44	0.0061	0.1779	- 30
- 29	24.66	9.97	.01097	1.6328	91.19	.6124	2.85	73.71	76.56	.0067	.1778	- 29
- 28	25.24	10.54	.01098	1.5978	91.08	.6259	3.12	73.57	76.69	.0073	.1777	- 28
- 27	25.82	11.13	.01099	1.5637	90.97	.6395	3.38	73.44	76.82	.0079	.1776	- 27
- 26	26.42	11.72	.01101	1.5305	90.85	.6534	3.64	73.30	76.94	.0085	.1775	- 26
- 25	27.02	12.33	0.01102	1.4982	90.74	0.6675	3.90	73.17	77.07	0.0091	0.1774	- 25
- 24	27.64	12.95	.01103	1.4667	90.63	.6818	4.16	73.03	77.19	.0097	.1773	- 24
- 23	28.27	13.57	.01105	1.4360	90.52	.6964	4.43	72.89	77.32	.0103	.1772	- 23
- 22	28.91	14.21	.01106	1.4061	90.40	.7112	4.69	72.75	77.44	.0109	.1771	- 22
- 21	29.56	14.86	.01108	1.3770	90.29	.7262	4.95	72.62	77.57	.0115	.1770	- 21
- 20	30.22	15.52	0.01109	1.3486	90.18	0.7415	5.21	72.48	77.69	0.0121	0.1769	- 20
- 19	30.89	16.19	.01110	1.3209	90.06	.7571	5.48	72.34	77.82	.0127	.1768	- 19
- 18	31.57	16.88	.01112	1.2939	89.95	.7729	5.74	72.20	77.94	.0133	.1767	- 18
- 17	32.27	17.57	.01113	1.2676	89.83	.7889	6.01	72.06	78.07	.0139	.1766	- 17
- 16	32.97	18.28	.01115	1.2419	89.72	.8052	6.27	71.92	78.19	.0145	.1766	- 16
- 15	33.69	18.99	0.01116	1.2169	89.60	0.8218	6.54	71.78	78.32	0.0151	0.1765	- 15
- 14	34.42	19.72	.01117	1.1925	89.49	.8386	6.80	71.64	78.44	.0156	.1764	- 14
- 13	35.16	20.46	.01119	1.1686	89.37	.8557	7.06	71.50	78.56	.0162	.1763	- 13
- 12	35.91	21.22	.01120	1.1454	89.26	.8731	7.33	71.36	78.69	.0168	.1762	- 12
- 11	36.68	21.98	.01122	1.1227	89.14	.8907	7.60	71.21	78.81	.0174	.1761	- 11
- 10	37.46	22.76	0.01123	1.1006	89.02	0.9086	7.86	71.07	78.93	0.0180	0.1760	- 10
- 9	38.25	23.55	.01125	1.0790	88.91	.9268	8.13	70.93	79.06	.0186	.1760	- 9
- 8	39.05	24.35	.01126	1.0579	88.79	.9453	8.40	70.78	79.18	.0192	.1759	- 8
- 7	39.86	25.17	.01128	1.0373	88.67	.9640	8.66	70.64	79.30	.0198	.1758	- 7
- 6	40.69	26.00	.01129	1.0172	88.55	.9831	8.93	70.49	79.42	.0204	.1757	- 6
- 5	41.53	26.84	0.01131	0.9976	88.44	1.0024	9.20	70.34	79.54	0.0209	0.1756	- 5
- 4	42.39	27.69	.01132	.9784	88.32	1.0220	9.47	70.20	79.67	.0215	.1756	- 4
- 3	43.26	28.56	.01134	.9597	88.20	1.0420	9.74	70.05	79.79	.0221	.1755	- 3
- 2	44.14	29.44	.01135	.9414	88.08	1.0622	10.00	69.91	79.91	.0227	.1754	- 2
- 1	45.03	30.33	.01137	.9236	87.96	1.0828	10.27	69.76	80.03	.0233	.1753	- 1
0	45.94	31.24	0.01138	0.9061	87.84	1.1036	10.54	69.61	80.15	0.0239	0.1753	0
1	46.86	32.16	.01140	.8891	87.72	1.1248	10.81	69.46	80.27	.0244	.1752	1
2	47.79	33.10	.01142	.8724	87.60	1.1463	11.08	69.31	80.39	.0250	.1751	2
3	48.74	34.05	.01143	.8561	87.48	1.1681	11.35	69.16	80.51	.0256	.1751	3
4	49.71	35.01	.01145	.8402	87.36	1.1902	11.62	69.01	80.63	.0262	.1750	4
5	50.68	35.99	0.01146	0.8247	87.24	1.2126	11.89	68.86	80.75	0.0268	0.1749	5
6	51.68	36.98	.01148	.8094	87.12	1.2354	12.16	68.70	80.86	.0273	.1749	6
7	52.68	37.99	.01149	.7946	87.00	1.2585	12.43	68.55	80.98	.0279	.1748	7
8	53.70	39.01	.01151	.7800	86.88	1.2820	12.70	68.40	81.10	.0285	.1747	8
9	54.74	40.04	.01153	.7658	86.76	1.3058	12.98	68.24	81.22	.0291	.1747	9
10	55.79	41.09	0.01154	0.7519	86.63	1.3300	13.25	68.08	81.33	0.0296	0.1746	10
11	56.86	42.16	.01156	.7383	86.51	1.3545	13.52	67.93	81.45	.0302	.1745	11
12	57.94	43.24	.01158	.7250	86.39	1.3793	13.80	67.77	81.57	.0308	.1745	12
13	59.03	44.34	.01159	.7120	86.26	1.4045	14.06	67.62	81.68	.0314	.1744	13
14	60.14	45.45	.01161	.6992	86.14	1.4301	14.34	67.46	81.80	.0319	.1743	14
15	61.27	46.57	0.01163	0.6868	86.02	1.4561	14.62	67.30	81.92	0.0325	0.1743	15
16	62.41	47.72	.01164	.6746	85.89	1.4824	14.89	67.14	82.03	.0331	.1742	16
17	63.57	48.86	.01166	.6626	85.77	1.5091	15.16	66.98	82.14	.0336	.1742	17
18	64.75	50.05	.01168	.6510	85.64	1.5362	15.44	66.82	82.26	.0342	.1741	18
19	65.94	51.24	.01169	.6395	85.52	1.5637	15.71	66.66	82.37	.0348	.1740	19
20	67.14	52.45	0.01171	0.6283	85.39	1.5915	15.99	66.50	82.49	0.0354	0.1740	20
21	68.37	53.67	.01173	.6174	85.26	1.6198	16.26	66.34	82.60	.0359	.1739	21
22	69.61	54.91	.01175	.6066	85.14	1.6485	16.54	66.17	82.71	.0365	.1739	22
23	70.86	56.17	.01176	.5961	85.01	1.6775	16.81	66.01	82.82	.0371	.1738	23
24	72.13	57.44	.01178	.5858	84.88	1.7070	17.10	65.84	82.94	.0376	.1738	24
25	73.42	58.73	0.01180	0.5757	84.76	1.7369	17.37	65.68	83.05	0.0382	0.1737	25
26	74.73	60.04	.01182	.5659	84.63	1.7672	17.65	65.51	83.16	.0388	.1736	26
27	76.06	61.36	.01183	.5562	84.50	1.7980	17.93	65.34	83.27	.0393	.1736	27
28	77.40	62.70	.01185	.5467	84.37	1.8292	18.21	65.17	83.38	.0399	.1735	28
29	78.76	64.06	.01187	.5374	84.24	1.8608	18.48	65.01	83.49	.0405	.1735	29

TABLE 15... PROPERTIES OF LIQUID AND SATURATED VAPOR (continued)

TEMP F	PRESSURE lb per sq in		VOLUME cu ft per lb		DENSITY lb per cu ft		ENTHALPY ** Btu per lb			ENTROPY ** Btu per (lb) (°R)		TEMP F
t	Absolute P	Gage p	Liquid v_f	Vapor v_g	Liquid $1/v_f$	Vapor $1/v_g$	Liquid h_f	Latent h_{fg}	Vapor h_g	Liquid s_f	Vapor s_g	t
30	80.13	65.44	0.01189	0.5283	84.11	1.8928	18.76	64.84	83.60	0.0410	0.1734	30
31	81.53	66.83	.01191	.5194	83.98	1.9253	19.04	64.67	83.71	.0416	.1734	31
32	82.94	68.24	.01193	.5106	83.85	1.9583	19.32	64.49	83.81	.0422	.1733	32
33	84.37	69.67	.01194	.5021	83.72	1.9917	19.60	64.32	83.92	.0427	.1733	33
34	85.82	71.12	.01196	.4937	83.59	2.0256	19.88	64.15	84.03	.0433	.1732	34
35	87.28	72.59	0.01198	0.4854	83.46	2.0600	20.17	63.97	84.14	0.0438	0.1732	35
36	88.77	74.07	.01200	.4774	83.33	2.0948	20.44	63.80	84.24	.0444	.1731	36
37	90.27	75.58	.01202	.4695	83.20	2.1301	20.73	63.62	84.35	.0450	.1730	37
38	91.80	77.10	.01204	.4617	83.07	2.1659	21.01	63.44	84.45	.0455	.1730	38
39	93.34	78.64	.01206	.4541	82.93	2.2022	21.29	63.27	84.56	.0461	.1729	39
40	94.90	80.20	0.01208	0.4466	82.80	2.2390	21.57	63.09	84.66	0.0466	0.1729	40
41	96.48	81.78	.01210	.4393	82.67	2.2763	21.86	62.91	84.77	.0472	.1728	41
42	98.08	83.38	.01212	.4321	82.53	2.3142	22.14	62.73	84.87	.0478	.1728	42
43	99.70	85.00	.01214	.4251	82.40	2.3525	22.42	62.55	84.97	.0483	.1727	43
44	101.3	86.64	.01216	.4182	82.26	2.3914	22.71	62.36	85.07	.0489	.1727	44
45	103.0	88.30	0.01218	0.4114	82.13	2.4308	22.99	62.18	85.17	0.0494	0.1726	45
46	104.7	89.97	.01220	.4047	81.99	2.4708	23.28	61.99	85.27	.0500	.1726	46
47	106.4	91.67	.01222	.3982	81.86	2.5113	23.57	61.81	85.38	.0505	.1725	47
48	108.1	93.39	.01224	.3918	81.72	2.5524	23.85	61.62	85.47	.0511	.1725	48
49	109.8	95.13	.01226	.3855	81.58	2.5940	24.14	61.43	85.57	.0517	.1724	49
50	111.6	96.89	0.01228	0.3793	81.44	2.6362	24.42	61.25	85.67	0.0522	0.1724	50
51	113.4	98.66	.01230	.3733	81.31	2.6790	24.71	61.06	85.77	.0528	.1723	51
52	115.2	100.5	.01232	.3673	81.17	2.7224	25.00	60.87	85.87	.0533	.1723	52
53	117.0	102.3	.01234	.3615	81.03	2.7664	25.29	60.67	85.96	.0539	.1722	53
54	118.8	104.1	.01236	.3557	80.89	2.8110	25.58	60.48	86.06	.0544	.1722	54
55	120.7	106.0	0.01238	0.3501	80.75	2.8562	25.87	60.28	86.15	0.0550	0.1721	55
56	122.6	107.9	.01241	.3446	80.61	2.9020	26.16	60.09	86.25	.0555	.1721	56
57	124.5	109.8	.01243	.3392	80.47	2.9485	26.44	59.90	86.34	.0561	.1720	57
58	126.4	111.7	.01245	.3338	80.33	2.9956	26.73	59.70	86.43	.0566	.1720	58
59	128.4	113.7	.01247	.3286	80.18	3.0434	27.02	59.50	86.52	.0572	.1719	59
60	130.3	115.6	0.01249	0.3234	80.04	3.0918	27.32	59.30	86.62	0.0578	0.1719	60
61	132.3	117.6	.01252	.3184	79.90	3.1409	27.61	59.10	86.71	.0583	.1718	61
62	134.3	119.6	.01254	.3134	79.76	3.1907	27.91	58.89	86.80	.0589	.1717	62
63	136.4	121.7	.01256	.3085	79.61	3.2411	28.19	58.69	86.88	.0594	.1717	63
64	138.4	123.7	.01258	.3037	79.47	3.2923	28.48	58.49	86.97	.0600	.1716	64
65	140.5	125.8	0.01261	0.2990	79.32	3.3442	28.78	58.28	87.06	0.0605	0.1716	65
66	142.6	127.9	.01263	.2944	79.18	3.3968	29.08	58.07	87.15	.0611	.1715	66
67	144.8	130.1	.01265	.2898	79.03	3.4502	29.37	57.86	87.23	.0616	.1715	67
68	146.9	132.2	.01268	.2854	78.88	3.5043	29.67	57.65	87.32	.0622	.1714	68
69	149.1	134.4	.01270	.2810	78.74	3.5591	29.96	57.44	87.40	.0627	.1714	69
70	151.3	136.6	0.01272	0.2766	78.59	3.6147	30.25	57.23	87.48	0.0633	0.1713	70
71	153.5	138.8	.01275	.2724	78.44	3.6712	30.55	57.01	87.56	.0638	.1712	71
72	155.8	141.1	.01277	.2682	78.29	3.7284	30.85	56.80	87.65	.0644	.1712	72
73	158.0	143.3	.01280	.2641	78.14	3.7864	31.15	56.58	87.73	.0649	.1711	73
74	160.3	145.6	.01282	.2601	77.99	3.8452	31.45	56.36	87.81	.0655	.1711	74
75	162.7	148.0	0.01285	0.2561	77.84	3.9049	31.74	56.14	87.88	0.0660	0.1710	75
76	165.0	150.3	.01287	.2522	77.68	3.9654	32.04	55.92	87.96	.0665	.1709	76
77	167.4	152.7	.01290	.2483	77.53	4.0268	32.34	55.70	88.04	.0671	.1709	77
78	169.8	155.1	.01292	.2446	77.38	4.0890	32.64	55.47	88.11	.0676	.1708	78
79	172.2	157.5	.01295	.2408	77.22	4.1522	32.94	55.25	88.19	.0682	.1707	79
80	174.6	159.9	0.01298	0.2372	77.07	4.2162	33.24	55.02	88.26	0.0687	0.1707	80
81	177.1	162.4	.01300	.2336	76.91	4.2812	33.54	54.79	88.33	.0693	.1706	81
82	179.6	164.9	.01303	.2300	76.76	4.3471	33.84	54.56	88.40	.0698	.1706	82
83	182.1	167.4	.01305	.2266	76.60	4.4140	34.14	54.33	88.47	.0704	.1705	83
84	184.7	170.0	.01308	.2231	76.44	4.4819	34.45	54.09	88.54	.0709	.1704	84
85	187.2	172.5	0.01311	0.2197	76.29	4.5507	34.75	53.86	88.61	0.0715	0.1703	85
86	189.8	175.1	.01314	.2164	76.13	4.6206	35.06	53.62	88.68	.0720	.1703	86
87	192.5	177.8	.01316	.2132	75.97	4.6915	35.36	53.38	88.74	.0726	.1702	87
88	195.1	180.4	.01319	.2099	75.80	4.7634	35.67	53.14	88.81	.0731	.1701	88
89	197.8	183.1	.01322	.2068	75.64	4.8364	35.97	52.90	88.87	.0737	.1701	89
90	200.5	185.8	0.01325	0.2036	75.48	4.9105	36.28	52.65	88.93	0.0742	0.1700	90
91	203.2	188.5	.01328	.2006	75.32	4.9856	36.59	52.40	88.99	.0747	.1699	91
92	206.0	191.3	.01331	.1976	75.15	5.0619	36.89	52.16	89.05	.0753	.1698	92
93	208.8	194.1	.01334	.1946	74.99	5.1394	37.20	51.91	89.11	.0758	.1697	93
94	211.6	196.9	.01337	.1916	74.82	5.2180	37.51	51.65	89.16	.0764	.1697	94

TABLE 15 ... PROPERTIES OF LIQUID AND SATURATED VAPOR (continued)

TEMP F	PRESSURE lb per sq in		VOLUME cu ft per lb		DENSITY lb per cu ft		ENTHALPY ** Btu per lb			ENTROPY ** Btu per (lb) (°R)		TEMP F
t	Absolute P	Gage p	Liquid v_f	Vapor v_g	Liquid $1/v_f$	Vapor $1/v_g$	Liquid h_f	Latent h_{fg}	Vapor h_g	Liquid s_f	Vapor s_g	t
95	214.4	199.7	0.01340	0.1888	74.65	5.2979	37.82	51.40	89.22	0.0769	0.1696	95
96	217.3	202.6	.01343	.1859	74.48	5.3789	38.13	51.14	89.27	.0775	.1695	96
97	220.2	205.5	.01346	.1831	74.32	5.4612	38.44	50.88	89.32	.0780	.1694	97
98	223.1	208.4	.01349	.1804	74.15	5.5447	38.75	50.62	89.37	.0786	.1693	98
99	226.1	211.4	.01352	.1776	73.97	5.6296	39.06	50.36	89.42	.0791	.1692	99
100	229.1	214.4	0.01355	0.1750	73.80	5.7157	39.37	50.10	89.47	0.0796	0.1692	100
101	232.1	217.4	.01358	.1723	73.63	5.8033	39.68	49.83	89.51	.0802	.1691	101
102	235.1	220.4	.01361	.1697	73.45	5.8921	40.00	49.56	89.56	.0807	.1690	102
103	238.2	223.5	.01365	.1672	73.28	5.9824	40.31	49.29	89.60	.0813	.1689	103
104	241.3	226.6	.01368	.1646	73.10	6.0741	40.62	49.02	89.64	.0818	.1688	104
105	244.4	229.7	0.01371	0.1621	72.92	6.1673	40.94	48.74	89.68	0.0824	0.1687	105
106	247.6	232.9	.01375	.1597	72.74	6.2620	41.25	48.47	89.72	.0829	.1686	106
107	250.7	236.0	.01378	.1573	72.56	6.3582	41.57	48.18	89.75	.0834	.1685	107
108	254.0	239.3	.01382	.1549	72.38	6.4560	41.88	47.90	89.78	.0840	.1684	108
109	257.2	242.5	.01385	.1525	72.20	6.5554	42.20	47.62	89.82	.0845	.1683	109
110	260.5	245.8	0.01389	0.1502	72.01	6.6564	42.52	47.33	89.85	0.0851	0.1682	110
111	263.8	249.1	.01392	.1480	71.83	6.7590	42.83	47.04	89.87	.0856	.1680	111
112	267.1	252.4	.01396	.1457	71.64	6.8634	43.15	46.75	89.90	.0862	.1679	112
113	270.5	255.8	.01400	.1435	71.45	6.9695	43.47	46.45	89.92	.0867	.1678	113
114	273.9	259.2	.01403	.1413	71.26	7.0775	43.79	46.15	89.94	.0872	.1677	114
115	277.3	262.6	0.01407	0.1391	71.07	7.1872	44.11	45.85	89.96	0.0878	0.1676	115
116	280.8	266.1	.01411	.1370	70.87	7.2988	44.43	45.55	89.98	.0883	.1674	116
117	284.3	269.6	.01415	.1349	70.68	7.4124	44.75	45.24	89.99	.0889	.1673	117
118	287.8	273.1	.01419	.1328	70.48	7.5279	45.07	44.93	90.00	.0894	.1672	118
119	291.4	276.7	.01423	.1308	70.28	7.6454	45.39	44.62	90.01	.0899	.1671	119
120	295.0	280.3	0.01427	0.1288	70.08	7.7649	45.71	44.31	90.02	0.0905	0.1669	120
121	298.6	283.9	.01431	.1268	69.88	7.8866	46.04	43.99	90.03	.0910	.1668	121
122	302.2	287.5	.01435	.1248	69.68	8.0105	46.36	43.67	90.03	.0916	.1666	122
123	305.9	291.2	.01439	.1229	69.47	8.1365	46.68	43.35	90.03	.0921	.1665	123
124	309.7	295.0	.01444	.1210	69.26	8.2648	47.00	43.02	90.02	.0926	.1663	124
125	313.4	298.7	0.01448	0.1191	69.05	8.3955	47.33	42.69	90.02	0.0932	0.1662	125
126	317.2	302.5	.01453	.1173	68.84	8.5285	47.65	42.36	90.01	.0937	.1660	126
127	321.0	306.3	.01457	.1154	68.62	8.6639	47.97	42.03	90.00	.0942	.1659	127
128	324.9	310.2	.01462	.1136	68.41	8.8019	48.29	41.69	89.98	.0948	.1657	128
129	328.8	314.1	.01467	.1118	68.19	8.9424	48.62	41.35	89.97	.0953	.1655	129
130	332.7	318.0	0.01471	0.1101	67.96	9.0855	48.95	41.00	89.95	0.0958	0.1654	130
131	336.6	321.9	.01476	.1083	67.74	9.2313	49.27	40.65	89.92	.0964	.1652	131
132	340.6	325.9	.01481	.1066	67.51	9.3798	49.59	40.30	89.89	.0969	.1650	132
133	344.7	330.0	.01486	.1049	67.28	9.5312	49.91	39.95	89.86	.0974	.1648	133
134	348.7	334.0	.01491	.1032	67.05	9.6854	50.24	39.59	89.83	.0979	.1646	134
135	352.8	338.1	0.01497	0.1016	66.81	9.8425	50.56	39.23	89.79	0.0985	0.1644	135
136	357.0	342.3	.01502	.09997	66.58	10.003	50.88	38.87	89.75	.0990	.1642	136
137	361.1	346.4	.01508	.09837	66.33	10.166	51.21	38.50	89.71	.0995	.1640	137
138	365.3	350.6	.01513	.09679	66.09	10.332	51.53	38.13	89.66	.1000	.1638	138
139	369.6	354.9	.01519	.09522	65.84	10.502	51.86	37.75	89.61	.1006	.1636	139
140	373.8	359.1	0.01525	0.09368	65.59	10.674	52.17	37.38	89.55	0.1011	0.1634	140
141	378.2	363.5	.01531	.09216	65.33	10.850	52.49	37.00	89.49	.1016	.1632	141
142	382.5	367.8	.01537	.09067	65.07	11.030	52.81	36.62	89.43	.1021	.1630	142
143	386.9	372.2	.01543	.08919	64.81	11.212	53.13	36.23	89.36	.1026	.1627	143
144	391.3	376.6	.01549	.08773	64.54	11.398	53.45	35.84	89.29	.1031	.1625	144
145	395.8	381.1	0.01556	0.08630	64.27	11.588	53.77	35.45	89.22	0.1036	0.1622	145
146	400.3	385.6	.01563	.08489	63.99	11.780	54.08	35.06	89.14	.1041	.1620	146
147	404.8	390.1	.01570	.08350	63.71	11.977	54.39	34.67	89.06	.1046	.1617	147
148	409.4	394.7	.01577	.08213	63.42	12.176	54.70	34.27	88.97	.1051	.1615	148
149	414.0	399.3	.01584	.08078	63.13	12.379	55.01	33.87	88.88	.1056	.1612	149
150	418.6	403.9	0.01591	0.07946	62.84	12.585	55.32	33.47	88.79	0.1061	0.1610	150
151	423.3	408.6	.01599	.07816	62.53	12.794	55.62	33.08	88.70	.1065	.1607	151
152	428.1	413.4	.01607	.07688	62.22	13.007	55.92	32.68	88.60	.1070	.1604	152
153	432.8	418.1	.01615	.07563	61.91	13.222	56.22	32.28	88.50	.1075	.1601	153
154	437.6	422.9	.01624	.07441	61.59	13.439	56.51	31.88	88.39	.1079	.1599	154
155	442.5	427.8	0.01632	0.07321	61.26	13.659	56.80	31.48	88.28	0.1084	0.1596	155
156	447.4	432.7	.01641	.07204	60.92	13.882	57.10	31.08	88.18	.1088	.1593	156
157	452.3	437.6	.01651	.07089	60.58	14.106	57.38	30.69	88.07	.1093	.1590	157
158	457.2	442.5	.01661	.06978	60.22	14.331	57.66	30.30	87.96	.1097	.1587	158
159	462.3	447.6	.01671	.06869	59.86	14.557	57.94	29.91	87.85	.1101	.1585	159
160	467.3	452.6	0.01681	0.06764	59.49	14.784	58.21	29.53	87.74	0.1105	0.1582	160

MISCELLANEOUS DATA

Metric Conversions

The factors for converting a number of refrigeration properties from English to Metric units and vice versa, are given in Table A.

Temperature Conversion

Table B gives the corresponding Fahrenheit and Centigrade temperatures for a wide range of temperature.

Table A. Metric Conversions

To Convert From	To	Multiply By
Atmospheres of pressure	Bars	1.013
	Inches of mercury	29.92
	Kilograms/sq centimeter	1.033
	Pounds/sq inch	14.696
Bars	Atmospheres	0.987
	Inches of mercury	29.53
	Kilograms/sq centimeter	1.020
	Pounds/sq inch	14.504
Btu	Calories	252
	Horsepower-minutes	0.01757
	Watt-hours	0.293
	Joules	1054
Btu/cubic foot	Calories/cubic centimeter	0.00890
	Calories/liter	8.90
	Kilocalories/cubic meter	8.90
Btu/minute	Calories/minute	252
	Horsepower	0.02357
	Kilowatts	0.01757
Btu/lb	Calories/gram	0.556
	Watt-hours/gram	0.000646
	Joules/gram	2.324
Btu/(lb) (°F)	Calories/(gram) (°C)	1
Calories	Btu	0.00397
	Watt-hours	0.00116
	Joules	4.184
Calories/cubic centimeter	Btu/cubic foot	112.4
	Calories/liter	0.001
	Kilocalories/cubic meter	0.001
Calories/gram	Btu/lb	1.8
	Watt-hours/gram	0.00116
	Joules/gram	4.184

To Convert From	To	Multiply By
Calories/(gram) (°C)	Btu/(lb) (°F)	1
Calories/liter	Btu/cubic foot	0.112
	Calories/cubic centimeter	1000
	Kilocalories/cubic meter	1
Calories/minute	Btu/minute	0.00397
	Horsepower	9.35×10^{-5}
	Kilowatts	69.7
Cubic centimeters	Cubic feet	3.53×10^{-5}
	Cubic meters	1×10^{-6}
	Gallons (U.S. Liq.)	0.000264
	Liters	0.001
Cubic centimeters/gram	Cubic feet/pound	0.0160
	Gallons (U.S. Liq.)/pound	0.120
	Liters/gram	0.001
Cubic feet	Cubic centimeters	28,320
	Cubic meters	0.0283
	Gallons (U.S. Liq.)	7.48
	Liters	28.3
Cubic feet/minute	Cubic meters/minute	0.0283
	Liters/second	0.472
Cubic feet/pound	Cubic centimeters/gram	62.43
	Cubic meters/kilogram	0.0624
	Gallons (US. Liq.)/pound	7.481
	Liters/kilogram	62.43
Cubic meters	Cubic centimeters	1×10^{6}
	Cubic feet	35.3
	Gallons (U.S. Liq.)	264
	Liters	1000
Cubic meters/kilogram	Cubic centimeters/gram	1000
	Cubic feet/pound	16.0
	Gallons (U.S. Liq.)/pound	120
	Liters/kilogram	1000
Cubic meters/minute	Cubic feet/minute	35.3
	Liters/second	16.7
Degrees K	Degrees R	1.8
Gallons	Cubic centimeters	3790
	Cubic feet	0.134
	Cubic meters	0.00379
	Liters	3.79
Grams	Pounds	0.00220
Grams/cubic centimeter	Kilograms/liter	1
	Pounds/cubic foot	62.43
	Pounds/gallon	8.345

To Convert From	To	Multiply By
Horsepower	Btu/minute	42.44
	Calories/minute	10,690
	Kilocalories/minute	10.7
	Kilowatts	0.746
Inches of Mercury	Atmospheres	0.0334
	Bars	0.0339
	Kilograms/sq centimeter	0.0345
	Pounds/sq inch	0.491
Joules	Btu	0.000948
	Calories	0.239
	Watt-hours	0.000278
Joules/gram	Btu/lb	0.430
	Calories/gram	0.239
Kilocalories	Calories	1000
Kilocalories/cubic meter	Btu/cubic foot	0.112
	Calories/cubic centimeter	1000
	Calories/liter	1
Kilograms	Grams	1000
Kilograms/cubic meter	Grams/cubic centimeter	0.001
	Kilograms/liter	0.001
	Pounds/cubic foot	0.0624
	Pounds/gallon (U.S. Liq.)	0.00835
Kilograms/sq centimeter	Atmospheres	0.968
	Bars	0.981
	Inches of mercury	28.96
	Pounds/sq inch	14.22
Kilowatts	Btu/minute	56.9
	Calories/minute	14,340
	Horsepower	1.34
Liters	Cubic centimeters	1000
	Cubic feet	0.0353
	Cubic meters	0.001
	Gallons (U.S. Liq.)	0.264
Liters/kilogram	Cubic centimeters/gram	1
	Cubic feet/pound	0.0160
	Cubic meters/kilogram	0.001
	Gallons (U.S. Liq.)/pound	0.120
Liters/second	Cubic feet/minute	2.12
	Cubic meters/minute	0.060
Pounds	Grams	453.6

To Convert From	To	Multiply By
Pounds/cubic foot	Grams/cubic centimeter	0.0160
	Kilograms/cubic meter	16.0
	Kilograms/liter	0.0160
	Pounds/gallon (U.S. Liq.)	0.134
Pounds/gallon (U.S. Liq.)	Grams/cubic centimeter	0.120
	Kilograms/cubic meter	120
	Kilograms/liter	0.120
	Pounds/cubic foot	7.48
Pounds/square inch	Atmospheres	0.0680
	Bars	0.0689
	Inches of mercury	2.036
	Kilograms/sq centimeter	0.0703
Tons of Refrigeration	Btu/minute	200
	Horsepower	4.716
	Kilocalories/minute	50.4

Temperature Conversions

$$°K = °C + 273.15 = \frac{°F + 459.67}{1.8}$$

$$°R = °F + 459.67 = (°C + 273.15)\ 1.8$$

$$°C = (°F - 32)\frac{5}{9}$$

$$°F = (°C \times \frac{9}{5}) + 32$$

Vacuum Conversions

Units of vacuum are frequently used in refrigeration and are expressed as:

Inches of mercury vacuum
Inches of mercury below one atmosphere

These units can be converted to pressure units as follows:

Inches of mercury vacuum = 29.921 − inches of mercury pressure.

Inches of mercury vacuum = 29.921 − (29.921) (atmospheres)

Inches of mercury vacuum = 29.921 − (2.036) (pounds/sq inch, abs.)

Inches of mercury vacuum = 29.921 − (28.96) (kilograms/sq centimeter)

$$\text{Atmospheres} = \frac{29.921 - \text{inches of mercury vacuum}}{29.921}$$

Pounds/sq inch, abs = 0.491 (29.921 − inches of mercury vacuum)

Kilograms/sq centimeter = 0.0345 (29.921 − inches of mercury vacuum)

$$\text{Inches of mercury} = \frac{\text{centimeters of mercury}}{2.54}$$

Centimeters of mercury vacuum = 76 − centimeters of mercury pressure

$$\text{Atmospheres} = \frac{76 - \text{centimeters of mercury}}{76}$$

Pounds/sq inch, abs = 0.193 (76 − centimeters of mercury vacuum)

Kilograms/sq centimeter = 0.0136 (76 − centimeters of mercury vacuum)

Table B. Temperature Conversion

°F	Temp. to be Converted	°C	°F	Temp. to be Converted	°C	°F	Temp. to be Converted	°C
−112.0	−80	−62.2	−31.0	−35	−37.2	50.0	10	−12.2
−110.2	−79	−61.7	−29.2	−34	−36.7	51.8	11	−11.7
−108.4	−78	−61.1	−27.4	−33	−36.1	53.6	12	−11.1
−106.6	−77	−60.6	−25.6	−32	−35.6	55.4	13	−10.6
−104.8	−76	−60.0	−23.8	−31	−35.0	57.2	14	−10.0
−103.0	−75	−59.4	−22.0	−30	−34.4	59.0	15	− 9.4
−101.2	−74	−58.9	−20.2	−29	−33.9	60.8	16	− 8.9
− 99.4	−73	−58.3	−18.4	−28	−33.3	62.6	17	− 8.3
− 97.6	−72	−57.8	−16.6	−27	−32.8	64.4	18	− 7.8
− 95.8	−71	−57.2	−14.8	−26	−32.2	66.2	19	− 7.2
− 94.0	−70	−56.7	−13.0	−25	−31.7	68.0	20	− 6.7
− 92.2	−69	−56.1	−11.2	−24	−31.1	69.8	21	− 6.1
− 90.4	−68	−55.6	− 9.4	−23	−30.6	71.6	22	− 5.6
− 88.6	−67	−55.0	− 7.6	−22	−30.0	73.4	23	− 5.0
− 86.8	−66	−54.4	− 5.8	−21	−29.4	75.2	24	− 4.4
− 85.0	−65	−53.9	− 4.0	−20	−28.9	77.0	25	− 3.9
− 83.2	−64	−53.3	− 2.2	−19	−28.3	78.8	26	− 3.3
− 81.4	−63	−52.8	− 0.4	−18	−27.8	80.6	27	− 2.8
− 79.6	−62	−52.2	1.4	−17	−27.2	82.4	28	− 2.2
− 77.8	−61	−51.7	3.2	−16	−26.7	84.2	29	− 1.7
− 76.0	−60	−51.1	5.0	−15	−26.1	86.0	30	− 1.1
− 74.2	−59	−50.6	6.8	−14	−25.6	87.8	31	− 0.6
− 72.4	−58	−50.0	8.6	−13	−25.0	89.6	32	0.0
− 70.6	−57	−49.4	10.4	−12	−24.4	91.4	33	0.6
− 68.8	−56	−48.9	12.2	−11	−23.9	93.2	34	1.1
− 67.0	−55	−48.3	14.0	−10	−23.3	95.0	35	1.7
− 65.2	−54	−47.8	15.8	− 9	−22.8	96.8	36	2.2
− 63.4	−53	−47.2	17.6	− 8	−22.2	98.6	37	2.8
− 61.6	−52	−46.7	19.4	− 7	−21.7	100.4	38	3.3
− 59.8	−51	−46.1	21.2	− 6	−21.1	102.2	39	3.9
− 58.0	−50	−45.6	23.0	− 5	−20.6	104.0	40	4.4
− 56.2	−49	−45.0	24.8	− 4	−20.0	105.8	41	5.0
− 54.4	−48	−44.4	26.6	− 3	−19.4	107.6	42	5.6
− 52.6	−47	−43.9	28.4	− 2	−18.9	109.4	43	6.1
− 50.8	−46	−43.3	30.2	− 1	−18.3	111.2	44	6.7
− 49.0	−45	−42.8	32.0	0	−17.8	113.0	45	7.2
− 47.2	−44	−42.2	33.8	1	−17.2	114.8	46	7.8
− 45.4	−43	−41.7	35.6	2	−16.7	116.6	47	8.3
− 43.6	−42	−41.1	37.4	3	−16.1	118.4	48	8.9
− 41.8	−41	−40.6	39.2	4	−15.6	120.2	49	9.4
− 40.0	−40	−40.0	41.0	5	−15.0	122.0	50	10.0
− 38.2	−39	−39.4	42.8	6	−14.4	123.8	51	10.6
− 36.4	−38	−38.9	44.6	7	−13.9	125.6	52	11.1
− 34.6	−37	−38.3	46.4	8	−13.3	127.4	53	11.7
− 32.8	−36	−37.8	48.2	9	−12.8	129.2	54	12.2

°F	Temp. to be Converted	°C	°F	Temp. to be Converted	°C	°F	Temp. to be Converted	°C
131.0	55	12.8	167.0	75	23.9	203.0	95	35.0
132.8	56	13.3	168.8	76	24.4	204.8	96	35.6
134.6	57	13.9	170.6	77	25.0	206.6	97	36.1
136.4	58	14.4	172.4	78	25.6	208.4	98	36.7
138.2	59	15.0	174.2	79	26.1	210.2	99	37.2
140.0	60	15.6	176.0	80	26.7	212.0	100	37.8
141.8	61	16.1	177.8	81	27.2	213.8	101	38.3
143.6	62	16.7	179.6	82	27.8	215.6	102	38.9
145.4	63	17.2	181.4	83	28.3	217.4	103	39.4
147.2	64	17.8	183.2	84	28.9	219.2	104	40.0
149.0	65	18.3	185.0	85	29.4	221.0	105	40.6
150.8	66	18.9	186.8	86	30.0	222.8	106	41.1
152.6	67	19.4	188.6	87	30.6	224.6	107	41.7
154.4	68	20.0	190.4	88	31.1	226.4	108	42.2
156.2	69	20.6	192.2	89	31.7	228.2	109	42.8
158.0	70	21.1	194.0	90	32.2	230.0	110	43.3
159.8	71	21.7	195.8	91	32.8	231.8	111	43.9
161.8	72	22.2	197.6	92	33.3	233.6	112	44.4
163.4	73	22.8	199.4	93	33.9	235.4	113	45.0
165.2	74	23.3	201.2	94	34.4	237.2	114	45.6

Pressure-temperature Relationships

TEMP F	"Freon" Refrigerants							
	11	12	13	22	113	114	500	502
−50	28.9	15.4	57.0	6.2	—	27.1	13.1	0.0
−48	28.8	14.6	60.0	4.8	—	26.9	12.1	0.7
−46	28.7	13.8	63.1	3.4	—	26.7	11.1	1.5
−44	28.6	12.9	66.2	2.0	—	26.5	10.1	2.3
−42	28.5	11.9	69.4	0.5	—	26.3	9.0	3.2
−40	28.4	11.0	72.7	0.5	—	26.0	7.9	4.1
−38	28.3	10.0	76.1	1.3	—	25.8	6.7	5.1
−36	28.2	8.9	79.7	2.2	—	25.5	5.4	6.0
−34	28.1	7.8	83.3	3.0	—	25.2	4.2	7.0
−32	27.9	6.7	87.0	3.9	—	25.0	2.8	8.1
−30	27.8	5.5	90.9	4.9	29.3	24.6	1.4	9.2
−28	27.7	4.3	94.9	5.9	29.3	24.3	0.0	10.3
−26	27.5	3.0	98.9	6.9	29.2	24.0	0.8	11.5
−24	27.4	1.6	103.0	7.9	29.2	23.6	1.5	12.7
−22	27.2	0.3	107.3	9.0	29.1	23.2	2.3	14.0
−20	27.0	0.6	111.7	10.1	29.1	22.9	3.1	15.3
−18	26.8	1.3	116.2	11.3	29.0	22.4	4.0	16.7
−16	26.6	2.1	120.8	12.5	28.9	22.0	4.9	18.1
−14	26.4	2.8	125.5	13.8	28.9	21.6	5.8	19.5
−12	26.2	3.7	130.4	15.1	28.8	21.1	6.8	21.0
−10	26.0	4.5	135.4	16.5	28.7	20.6	7.8	22.6
− 8	25.8	5.4	140.5	17.9	28.6	20.1	8.8	24.2
− 6	25.5	6.3	145.8	19.3	28.5	19.6	9.9	25.8
− 4	25.3	7.2	151.1	20.8	28.4	19.0	11.0	27.5
− 2	25.0	8.2	156.5	22.4	28.3	18.4	12.1	29.3
0	24.7	9.2	162.1	24.0	28.2	17.8	13.3	31.1
2	24.4	10.2	167.8	25.6	28.1	17.2	14.5	32.9
4	24.1	11.2	173.7	27.3	28.0	16.5	15.7	34.8
6	23.8	12.3	179.7	29.1	27.9	15.8	17.0	36.9
8	23.4	13.5	185.8	30.9	27.7	15.1	18.4	38.9
10	23.1	14.6	192.1	32.8	27.6	14.4	19.8	41.0
12	22.7	15.8	198.5	34.7	27.5	13.6	21.2	43.2
14	22.3	17.1	205.7	36.7	27.3	12.8	22.7	45.4
16	21.9	18.4	211.9	38.7	27.1	12.0	24.2	47.7
18	21.5	19.7	218.7	40.9	27.0	11.1	25.7	50.0
20	21.1	21.0	225.7	43.0	26.8	10.2	27.3	52.5
22	20.6	22.4	232.9	45.3	26.6	9.3	29.0	54.9
24	20.1	23.9	240.2	47.6	26.4	8.3	30.7	57.5
26	19.7	25.4	247.7	49.9	26.2	7.3	32.5	60.1
28	19.1	26.9	255.4	52.4	26.0	6.3	34.3	62.8
30	18.6	28.5	263.2	54.9	25.8	5.2	36.1	65.6
32	18.1	30.1	271.2	57.5	25.6	4.1	38.0	68.4
34	17.5	31.7	279.4	60.1	25.3	2.9	40.0	71.3
36	16.9	33.4	287.7	62.8	25.1	1.7	42.0	74.3
38	16.3	35.2	296.2	65.6	24.8	0.6	44.1	77.4

Pressure-temperature Relationships

"Freon" Refrigerants

TEMP F	11	12	13	22	113	114	500	502
40	15.6	37.0	304.9	68.5	24.5	0.4	46.2	80.5
42	15.0	38.8	313.9	71.5	24.2	1.0	46.4	83.8
44	14.3	40.7	322.9	74.5	23.9	1.7	50.7	87.0
46	13.6	42.7	332.2	77.6	23.6	2.4	53.0	90.4
48	12.8	44.7	341.5	80.8	23.3	3.1	55.4	93.9
50	12.0	46.7	351.2	84.0	22.9	3.8	57.8	97.4
52	11.2	48.8	360.9	87.4	22.6	4.6	60.3	101.1
54	10.4	51.0	371.0	90.8	22.2	5.4	62.9	104.8
56	9.6	53.2	381.2	94.3	21.8	6.2	65.5	108.6
58	8.7	55.4	391.6	97.9	21.4	7.0	68.2	112.4
60	7.8	57.7	402.3	101.6	21.0	7.9	71.0	116.4
62	6.8	60.1	413.3	105.4	20.6	8.8	73.8	120.5
64	5.9	62.5	424.3	109.3	20.1	9.7	76.7	124.6
66	4.9	65.0	435.4	113.2	19.7	10.6	79.7	128.9
68	3.8	67.6	446.9	117.3	19.2	11.6	82.7	133.2
70	2.8	70.2	458.7	121.4	18.7	12.6	85.8	137.6
72	1.6	72.9	470.6	125.7	18.2	13.6	89.0	142.2
74	0.5	75.6	482.9	130.0	17.6	14.6	92.3	146.8
76	0.3	78.4	495.3	134.5	17.1	15.7	95.6	151.5
78	0.9	81.3	508.0	139.0	16.5	16.8	99.0	156.3
80	1.5	84.2	520.8	143.6	15.9	18.0	102.5	161.2
82	2.2	87.2	534.0	148.4	15.3	19.1	106.1	166.2
84	2.8	90.2	—	153.2	14.6	20.3	109.7	171.4
86	3.5	93.3	—	158.2	13.9	21.6	113.4	176.6
88	4.2	96.5	—	163.2	13.2	22.8	117.3	181.9
90	4.9	99.8	—	168.4	12.5	24.1	121.2	187.4
92	5.6	103.1	—	173.7	11.8	25.5	125.1	192.9
94	6.4	106.5	—	179.1	11.0	26.8	129.2	198.6
96	7.1	110.0	—	184.6	10.2	28.2	133.3	204.3
98	7.9	113.5	—	190.2	9.4	29.7	137.6	210.2
100	8.8	117.2	—	195.9	8.6	31.2	141.9	216.2
102	9.6	120.9	—	201.8	7.7	32.7	146.3	222.3
104	10.5	124.6	—	207.7	6.8	34.2	150.9	228.5
106	11.3	128.5	—	213.8	5.9	35.8	155.4	234.9
108	12.3	132.4	—	220.0	4.9	37.4	160.1	241.3
110	13.1	136.4	—	226.4	4.0	39.1	164.9	247.9
112	14.2	140.5	—	232.8	3.0	40.8	169.8	254.6
114	15.1	144.7	—	239.4	1.9	42.5	174.8	261.5
116	16.1	148.9	—	246.1	0.8	44.3	179.9	268.4
118	17.2	153.2	—	252.9	0.1	46.1	185.0	275.5
120	18.2	157.7	—	259.9	0.7	48.0	190.3	282.7
122	19.3	162.2	—	267.0	1.3	49.9	195.7	290.1
124	20.5	166.7	—	274.3	1.9	51.9	201.2	297.6
126	21.6	171.4	—	281.6	2.5	53.8	206.7	305.2
128	22.8	176.2	—	289.1	3.1	55.9	212.4	312.9
130	24.0	181.0	—	296.8	3.7	58.0	218.2	320.8
132	25.2	185.9	—	304.6	4.4	60.1	224.1	328.9
134	26.5	191.0	—	312.5	5.1	62.3	230.1	337.1
136	27.8	196.1	—	320.6	5.8	64.5	236.3	345.4
138	29.1	201.3	—	328.9	6.5	66.7	242.5	353.9

Glossary of Technical Terms As Applied to Refrigeration, Air-Conditioning, and Heat Pumps

A

Absolute Humidity: Amount of moisture in the air, indicated in grains per cubic foot.

Absolute Pressure: Gauge pressure plus atmospheric pressure (14.7 lb per in²).

Absolute Temperature: Temperature measured from absolute zero.

Absolute Zero Temperature: Temperature at which all molecular motion ceases (−460°F and −275°C).

Absorbent: Substance which has the ability to take up or absorb another substance.

Absorber: A device containing liquid for absorbing refrigerant vapor or other vapors. In an absorption system, that part of the low side used for absorbing refrigerant vapor.

Absorption Refrigerator: Refrigerator which creates low temperatures by using the cooling effect formed when a refrigerant is absorbed by chemical substance.

Accelerate: To add to speed; hasten progress of development.

Accumulator: Storage tank which receives liquid refrigerant from evaporator and prevents it from flowing into suction line.

Acid Condition in System: Condition in which refrigerant or oil in system is mixed with fluids which are acid in nature.

ACR Tubing: Tubing used in refrigeration which has ends sealed to keep tubing clean and dry.

Activated Alumina: Chemical used as a drier or desiccant.

Activated Carbon: Specially processed carbon used as a filter-drier; commonly used to clean air.

Acoustical: Pertaining to sound.

Acoustical Duct Lining: Duct with a lining designed to control or absorb sound and prevent transmission of sound from one room to another.

Adiabatic Compression: Compressing refrigerant gas without removing or adding heat.

Adsorbent: Substance which has property to hold molecules of fluids without causing a chemical or physical change.

Aerosol: An assemblage of small particles, solid or liquid, suspended in air. The diameters of the particles may vary from 100 microns down to 0.01 micron or less, e.g., dust, fog, smoke.

Agitator: Device used to cause motion in confined fluid.

Air (Saturated): A mixture of dry air and saturated water vapor all at the same dry-bulb temperature.

Air (Specific Heat of): The quantity of heat absorbed by a unit weight of air per unit temperature rise.

Air (Standard): Air with a density of 0.075 lb per ft³ and an absolute viscosity of 0.0379×10^{-5} lb mass per (ft) (sec). This is substantially equivalent to dry air at 70°F and 29.92 in. Hg barometric pressure.

Air Blast: Forced air circulation.

Air Changes: A method of expressing the amount of air leakage into or out of a building or room in terms of the number of building volumes or room volumes exchanged.

Air Cleaner: Device used for removal of airborne impurities.

Air Coil: Coil used with some types of heat pumps which may be used either as an evaporator or as a condenser.

Air Conditioner: Device used to control temperature, humidity, cleanliness, and movement of air in conditioned space.

Air-Conditioning: The simultaneous control of all or at least the first three, of the following factors affecting the physical and chemical conditions of the atmosphere within a structure: Temperature, humidity, motion, distribution, dust, bacteria, odors, toxic gases and ionization—most of which affect in greater or lesser degree human health or comfort.

Air-Cooled Condenser: Heat of compression is transferred from condensing coils to surrounding air. This may be done either by convection or by a fan or blower.

Air Cooler: Mechanism designed to lower temperature of air passing through it.

Air Cycle, Air-Conditioning: System which removes heat from air and transfers this heat to air.

Air Diffuser: Air distribution outlet designed to direct airflow into desired patterns.

Air Return: Air returned from conditioned or refrigerated space.

Air Sensing Thermostat: Thermostat unit in which sensing element is located in refrigerated space.

Air Spill-Over: Refrigerating effect formed by cold air from freezing compartment in refrigerator spilling over, or flowing into normal storage area of refrigerator.

Air Washer: Device used to clean air, which may increase or decrease humidity.

Alcohol Brine: Water and alcohol solution which remains a liquid at below 32°F.

Allen-Type Screw: Screw with recessed head designed to be turned with hex shaped wrench.

Altitude Adjustment: Adjusting refrigerator controls so unit will operate efficiently at altitude in which it is to be used.

Ambient Temperature: Temperature of fluid (usually air) which surrounds object on all sides.

Ammeter: An electric meter used to measure current, calibrated in amperes.

Ammonia: Chemical combination of nitrogen and hydrogen (NH_3). Ammonia refrigerant is identified by R-117.

Amperage: Electron or current flow of one coulomb per second past given point in circuit.

Ampere: Unit of electric current equivalent to flow of one coulomb per second.

Ampere Turns: Term used to measure magnetic force. Represents product of amperes times number of turns in coil of electromagnet.

Amplifier: Electrical device which increases electron flow in a circuit.

Anemometer: Instrument for measuring the rate of flow of air.

Anhydrous Calcium Sulphate: Dry chemical made of calcium, sulphur and oxygen ($CaSO_4$).

Annealing: Process of heat treating metal to obtain desired properties of softness and ductility (easy to form into new shape).

Anode: Positive terminal of electrolytic cell.

Anticipating Control: One which is artificially forced to cut in or cut out before it otherwise would, thus starting the cooling before needed or stopping the heating before control point is reached, to reduce the temperature fluctuation or override.

A.S.A.: Formerly, abbreviation for American Standards Association. Now known as United States of America Standards Institute.

Aspect Ratio: Ratio of length to width of rectangular air grille or duct.

Aspirating Psychrometer: A device which draws sample of air through it for humidity measurement purposes.

Aspiration: Movement produced in a fluid by suction.

Atmospheric Pressure: Pressure that gases in air exert upon the earth; measured in pounds per square inch.

Atom: Smallest particle of element that can exist alone or in combination.

Atomize: Process of changing a liquid to minute particles, or a fine spray.

Attenuate: Decrease or lessen in intensity.

Attic Fan: An exhaust fan to discharge air near the top of a building while cooler air is forced (drawn) in at a lower level.

Automatic Defrost: System of removing ice and frost from evaporators automatically.

Automatic Expansion Valve (AEV): Pressure controlled valve which reduces high pressure liquid refrigerant to low pressure liquid refrigerant.

Automatic Ice Cube Maker: Refrigerating mechanism designed to produce ice cubes in quantity automatically.

Autotransformer: A transformer in which both primary and secondary coils have turns in common. Step-up or stepdown of voltage is accomplished by taps on common winding.

Automatic Refrigerating System: One which regulates itself to maintain a definite set of conditions by means of automatic controls and valves usually responsive to temperature or pressure.

Azeotropic Mixture: Example of azeotropic mixture—refrigerant R-502 is mixture consisting of 48.8% refrigerant R-22, and 51.2% R-115. The refrigerants do not combine chemically, yet azeotropic mixture provides refrigerant characteristics desired.

B

Back Pressure: Pressure in low side of refrigerating system; also called suction pressure or low side pressure.

Baffle: Plate or vane used to direct or control movement of fluid or air within confined area.

Balance Point: The outdoor temperature at which the output of the heat pump in a specific application is equal to the heat requirement of the structure.

Ball Check Valve: Valve assembly call which permits flow of fluid in one section only.

Balloon Type Gasket: Flexible refrigerator door gasket having a large cross section.

Barometer: Instrument for measuring atmospheric pressure. It may be calibrated in pounds per square inch or in inches of mercury in column.

Bath: A liquid solution used for cleaning, plating, or maintaining a specified temperature.

Baudelot Cooler: Heat exchanger in which water flows by gravity over the outside of the tubes or plates.

Bearing: Low friction device for supporting and aligning a moving part.

Bellows: Corrugated cylindrical container which moves as pressures change, or provides a seal during movement of parts.

Bending Spring: Coil spring which is mounted on inside or outside to keep tube from collapsing while bending it.

Bernoulli's Theorem: In stream of liquid, sum of elevation head, pressure head and velocity remains constant along any line of flow provided no work is done by or upon liquid in course of its flow, and decreases in proportion to energy lost in flow.

Bimetal Strip: Temperature regulating or indicating device which works on principle that two dissimilar metals with unequal expansion rates, welded together, will bend as temperatures change.

Blast Heater: A set of heat-transfer coils or sections used to heat air which is drawn or forced through it by a fan.

Bleed off: The continuous or intermittent wasting of a small fraction of the circulating water to a cooling tower to prevent the buildup and concentration of scale-forming chemicals in the water.

Bleed-Valve: Valve with small opening inside which permits a minimum fluid flow when valve is closed.

Blow (Throw): The distance an air stream travels from an outlet to a position at which air motion along the axis is reduced to a velocity of 50 ft per minute.

Blower: A fan used to force air under pressure.

Boiling Point: (See Boiling Temperature)

Boiling Temperature: Temperature at which a fluid changes from a liquid to a gas.

Bore: Inside diameter of a cylindrical hole.

Bourdon Tube: As used in pressure gauges. Thin walled tube of elastic metal flattened and bent into circular shape, which tends to straighten as pressure inside is increased.

Bowden Cable: Tube containing a wire used to regulate a valve or control from a remote point.

Boyle's Law: Law of physics—volume of a gas varies as pressure varies, if temperature remains the same. Examples: If pressure is doubled on quantity of gas, volume becomes one half. If volume becomes doubled, gas has its pressure reduced by one half.

Brazing: Method of joining metals with nonferrous filler (without iron) using heat between 800°F and melting point of base metals.

Breaker Strip: Strip of wood or plastic used to cover joint between outside case and inside liner of refrigerator.

Brine: Water saturated with chemical such as salt.

British Thermal Unit (Btu): Quantity of heat required to change temperature of one pound of water one degree F.

Bulb, Sensitive: Part of sealed fluid device which reacts to temperature to be measured, or which will control a mechanism.

Bunker: In commercial installations, space in which ice or cooling element is installed.

Bypass: Passage at one side of, or around, regular passage.

C

Cadmium Plated: Parts coated with thin corrosion-resistant covering of cadmium metal.

Calcium Sulfate: Chemical compound ($CaSO_4$) which is used as a drying agent or desiccant in liquid line driers.

Calibrate: To determine position indicators as required to obtain accurate measurements.

Calorie: Heat required to raise temperature of one gram of water one degree centigrade.

Calorimeter: Device used to measure quantities of heat or determine specific heats.

Capacitance (C): Property of nonconductor (condenser or capacitor) that permits storage of electrical energy in an electrostatic field.

Capacitor: Type of electrical storage device used in starting and/or running circuits on many electric motors.

Capacitor-Start Motor: Motor which has a capacitor in the starting circuit.

Capacity-Gross Heating: The gross capacity of a heat pump on the heating cycle is the total heat ejected into the inside air or material. This is made up of the heat picked up from the outside heat source plus the heat energy equivalent of the electrical energy used.

Capacity-Net Cooling: The cooling capacity of an air-conditioning system or heat pump on the cooling cycle is the amount of Sensible and Latent heat (total heat) removed from the inside air.

Capacity, Refrigerating: The ability of a refrigerating system, or part thereof, to remove heat expressed as a rate of heat removal, usually measured in Btu/hr or tons/24 hr.

Capacity Reducer: In a compressor, a device such as a clearance pocket, movable cylinder head.

Capillary Tube: A type of refrigerant control. Usually consists of several feet of tubing having small inside diameter. Friction

of liquid refrigerant and bubbles of vaporized refrigerant within tube serve to restrict flow so that correct high side and low side pressures are maintained while the compressor is operating. A capillary tube refrigerant control allows high side and low side pressures to balance during off cycle. Also, a small diameter tubing used to connect temperature control bulbs to control mechanisms.

Carbon Dioxide (CO_2): Compound of carbon and oxygen which is sometimes used as a refrigerant. Refrigerant number is R-744.

Carbon Filter: Air filter using activated carbon as air cleansing agent.

Carbon Tetrachloride: A colorless nonflammable liquid used as solvent and in fire extinguishers. Very toxic. Should never be allowed to touch skin, or fumes inhaled.

Carnot Cycle: A sequence of operations forming the reversible working cycle of an ideal heat engine of maximum thermal efficiency. It consists of isothermal expansion, adiabatic expansion, isothermal compression, and adiabatic compression to the initial state.

Carrene: A refrigerant in group 1 (R-11). Chemical combination of carbon, chlorine, and fluorine.

Cascade System: One having two or more refrigerant circuits, each with a pressure imposing element, condenser, and evaporator, where the evaporator of one circuit cools the condenser of the other (lower-temperature) circuit.

Cascade Systems: Arrangement in which two or more refrigerating systems are used in series; uses cooling coil of one machine to cool condenser of other machine. Produces ultra-low temperatures.

Casehardened: Heat treating ferrous metals (iron) so surface layer is harder than interior.

Cathode: Negative terminal of an electrical device. Electrons leave the device at this terminal.

Celsius: German language word for centigrade, the metric system temperature scale.

Centigrade Scale: Temperature scale used in metric system. Freezing point of water is 0; boiling point 100.

Centimeter: Metric unit of linear measurement which equals 0.3937 in.

Centrifugal Compressor: Compressor which compresses gaseous refrigerants by centrifugal force.

Central Fan System: A mechanical indirect system of heating, ventilating, or air-conditioning, in which the air is treated or handled by equipment located outside the rooms served, usually at a central location, and is conveyed to and from the rooms by means of a fan and a system of distributing ducts.

Charge: The amount of refrigerant in a system.

Charging: Putting in a charge.

Charging Board: Specially designed panel or cabinet fitted with gauges, valves and refrigerant cylinders used for charging refrigerant and oil into refrigerating mechanisms.

Charles' Law: The volume of a given mass of gas at a constant pressure varies according to its temperature.

Check Valve: A device which permits fluid flow only in one direction.

Chemical Refrigeration: A system of cooling using a disposable refrigerant.

Chimney Effect: The tendency of air or gas in a duct or other vertical passage to rise when heated due to its lower density compared with that of the surrounding air or gas. In buildings, the tendency toward displacement (caused by the difference in temperature) of internal heated air by unheated outside air due to the difference in density of outside and inside air.

Choke Tube: Throttling device used to maintain correct pressure difference between high side and low side in refrigerating mechanism. Capillary tubes are sometimes called choke tubes.

Circuit: A tubing, piping or electrical wire installation which permits flow from the energy source back to energy source.

Clearance: Space in cylinder not occupied by piston at end of compression stroke, or volume of gas remaining in cylinder at same point. Measured in percentage of piston displacement.

Clearance Pocket Compressor: A small space in cylinder from which compressed gas is not completely expelled. This space is called the compressor clearance space or pocket. For effective operation, compressors are designed to have as small clearance space as possible.

Code Installation: A refrigeration or air-conditioning installation which conforms to the local code and/or the national code for safe and efficient installations.

Closed Cycle: Any cycle in which the primary medium is always enclosed and repeats the same sequence of events.

Coefficient of Conductivity: The measure of the relative rate at which different materials conduct heat. Copper is a good conductor of heat and therefore has a high coefficient of conductivity.

Coefficient of Expansion: The change in length per unit length or the change in volume per unit volume per degree change in temperature.

Coefficience of Performance (COP): The efficiency of a heat pump is the amount of heat available to the inside load compared to the heat energy equivalent of the electrical energy used by the unit.

Coil: Any heating or cooling element made of pipe or tubing connected in series.

Coil-Inside: The coil located in the inside portion of the heat pump system. Used for evaporation on the cooling cycle and condensing on the heating cycle.

Coil-Outside: The coil located in the outside section of the heat pump system. Used for the condenser on the cooling cycle and evaporator on the heating cycle.

Cold: Cold is the absence of heat; a temperature considerably below normal.

Cold Junction: That part of the thermoelectric system which absorbs heat as the system operates.

Cold Wall: Refrigerator construction which has the inner lining of refrigerator serving as the cooling surface.

Colloids: Miniature cells in meat, fish and poultry.

Comfort Air-Conditioning: The simultaneous control of all, or at least the first three, of the following factors affecting the physical and chemical conditions of the atmosphere within a structure for the purpose of human comfort; temperature, humidity, motion, distribution, dust, bacteria, odors, toxic gases, and ionization, most of which affect in greater or lesser degree human health or comfort.

Comfort Chart: Chart used in air-conditioning to show the dry bulb temperature and humidity for human comfort conditons.

Comfort Cooler: A system used to reduce the temperature in the living space in homes. These systems are not complete air conditioners as they do not provide complete control of heating, humidifying, dehumidification, and air circulation.

Comfort Zone: Area on psychrometric chart which shows conditions of temperature, humidity, and sometimes air movement, in which most people are comfortable.

Commutator: Part of electric motor rotor which conveys electric current to rotor windings.

Compound Gauge: Instrument for measuring pressures both above and below atmospheric pressure.

Compound Refrigerating Systems: System which has several compressors or compressor cylinders in series. The system is used to pump low pressure vapors to condensing pressures.

Compression: Term used to denote increase of pressure on a fluid by using mechanical energy.

Compression Gauge: Instrument used to measure positive pressures (pressures above atmospheric pressures) only. These gauges are usually calibrated from 0 to 300 pounds per square inch of pressure, gauge, (psig).

Compressor: The pump of a refrigerating mechanism which draws a vacuum or low pressure on cooling side of refrigerant cycle and squeezes or compresses the gas into the high pressure or condensing side of the cycle.

Compressor-Actual Capacity: A function of the ideal capacity and the overall volumetric efficiency.

Compressor-Brake Horsepower: A function of the power input to the ideal compressor and to the compression, mechanical, and volumetric efficiencies of the compressor.

Compressor Efficiency: A measure of the deviation of the actual compression from the perfect compression cycle. Is defined as the work done within the cylinders.

Compressor, Hermetic: Compressor in which driving motor is sealed in the same dome or housing that contains the compressor.

Compressor, Open-Type: Compressor in which the crankshaft extends through the crankcase and is driven by an outside motor.

Compressor, Reciprocating: Compressor which uses a piston and cylinder mechanism to provide pumping action.

Compressor, Rotary: A compressor which uses vanes, eccentric mechanisms, or other rotating devices to provide pumping action.

Compressor Seal: Leakproof seal between crankshaft and compressor body.

Condensate: Fluid which forms on an evaporator.

Condensate Pump: Device used to remove fluid condensate that collects beneath an evaporator.

Condensation: Liquid or droplets which form when a gas or vapor is cooled below its dew point.

Condense: Action of changing a gas or vapor to a liquid.

Condenser: The part of refrigeration mechanism which receives hot, high pressure refrigerant gas from compressor and cools gaseous refrigerant until it returns to liquid state.

Condenser, Air-Cooled: A heat exchanger which transfers heat to surrounding air.

Condenser Comb: Comb-like device, metal or plastic, which is used to straighten the metal fins on condensers or evaporators.

Condenser-Evaporative: A condenser in which heat is absorbed from the surface by the evaporation of water sprayed or flooded over the surface.

Condenser Fan: Forced air device used to move air through air-cooled condenser.

Condenser, Water-Cooled: Heat exchanger which is designed to transfer heat from hot gaseous refrigerant to water.

Condenser Water Pump: Forced water moving device used to move water through condenser.

Condensing Unit: That part of refrigerating mechanism which pumps vaporized refrigerant from evaporator, compresses it, liquefies it in the condenser, and returns the liquid refrigerant to refrigerant control.

Condensing Unit Service Valves: Shutoff hand valves mounted on condensing unit to enable serviceman to install and/or service unit.

Conductance (Surface Film): The time rate of heat flow per unit area under steady conditions between a surface and the ambient fluid for a unit temperature difference between the surface and the fluid. In English units its value is usually expressed in Btu per (hour) (square foot) (Fahrenheit degree temperature difference between surface and fluid).

Conductance (Thermal): "C" factor—The time rate of heat flow per unit area under steady conditions through a body from one of its bounding surfaces to the other for a unit temperature difference between the two surfaces. In English units its value is usually expressed in Btu per (hour) (square foot) (Fahrenheit degree). The term is applied to specific bodies or constructions as used, either homogeneous or heterogeneous.

Conduction (Thermal): The process of heat transfer through a material medium in which kinetic energy is transmitted by the particles of the material from particle to particle without gross displacement of the particles.

Conductivity (Thermal): "k" factor—The time rate of heat flow through unit area of a homogeneous material under steady conditions when a unit temperature gradient is maintained in the direction perpendicular to the area. In English units its value is usually expressed in Btu per (hour) (square foot) (Fahrenheit degree per inch of thickness) Materials are considered homogeneous when the value of "k" is not affected by variation in thickness or in size of sample within the range normally used in construction.

Conductor (Thermal): A material which readily transmits heat by means of conduction.

Connecting Rod: That part of compressor mechanism which connects piston to crankshaft.

Constrictor: Tube or orifice used to restrict flow of a gas or a liquid.

Contaminant: A substance (dirt, moisture, etc.) foreign to refrigerant or refrigerant oil in system.

Continuous Cycle Absorption System: System which has a continuous flow of energy input.

Control: Automatic or manual device used to stop, start and/or regulate flow of gas, liquid, and /or electricity.

Control, Compressor: (See Motor Control)

Control, Defrosting: Device to automatically defrost evaporator. It may operate by means of a clock, door cycling mechanism, or during "off" portion of refrigerating cycle.

Control, Low Pressure: Cycling device connected to low pressure side of system.

Control, Motor: A temperature of pressure operated device used to control running of motor.

Control, Pressure Motor: A high or low pressure control which is connected into the electrical circuit and used to start and stop motor when there is need for refrigeration or for safety purposes.

Control, Refrigerant: Device used to regulate flow of liquid refrigerant into evaporator, such as capillary tube, expansion valves, high and low side float valves.

Control, Temperature: A thermostatic device which automatically stops and starts motor, operation of which is based on temperature changes.

Controlled Evaporator Pressure: Controlled system which maintains definite pressure or range of pressures in evaporator.

Convection: Transfer of heat by means of movement or flow of a fluid or gas.

Convection, Forced: Transfer of heat resulting from forced movement of liquid or gas by means of fan or pump.

Convection, Natural: Circulation of a gas or liquid due to difference in density resulting from temperature differences.

Conversion Factors: Force and power may be expressed in more than one way. A horse power is equivalent to 33,000 ft lbs of work per minute, 746 W, or 2,546 Btu per hour. These values can be used for changing horsepower into foot pounds, Btu, or watts.

Cooling Coil: (See Evaporator)

Cooling Tower: Device which cools water by water evaporation in air. Water is cooled to wet bulb temperature of air.

Copper Plating: Condition developing in some units in which copper is electrolytically deposited on compressor part surfaces.

Counterflow: Flow in opposite direction.

"Cracking" a Valve: Opening valve a small amount.

Crankshaft Seal: Leakproof joint between crankshaft and compressor body.

Crank Throw: Distance between center line of main bearing journal and center line of the crankpin or eccentric.

Crisper: Drawer or compartment in refrigerator designed to provide high humidity along with low temperature to keep vegetables, especially leafy vegetables, cold and crisp.

Critical Pressure: Condition of refrigerant at which liquid and gas have same properties.

Critical Temperature: Temperature at which vapor and liquid have same properties.

Critical Vibration: Vibration which is noticeable and harmful to structure.

Cross Charged: Sealed container containing two fluids which together create a desired pressure-temperature curve.

Cryogenic Fluid: Substance which exists as a liquid or gas at ultra-low temperatures ($-250°F$ or lower).

Cryogenics: Refrigeration which deals with producing temperatures at 250°F below zero and lower.

Cut-In: Temperature or pressure valve which closes control circuit.

Cut-Out: Temperature or pressure valve which opens control circuit.

Cycle: Series of events which have tendency to repeat same events in same order.

Cylinder Head: Part which encloses compression end of compressor cylinder.

Cylinder, Refrigerant: Cylinder in which refrigerant is purchased and dispensed. Color code painted on cylinder indicates kind of refrigerant cylinder contains.

Cylindrical Commutator: Commutator with contact surfaces parallel to the rotor shaft.

D

Dalton's Law: Vapor pressure exerted on container by a mixture of gases is equal to sum of individual vapor pressures of gases contained in mixture.

Damper: Valve for controlling airflow.

Decibel: Unit used for measuring relative loudness of sounds. One decibel is equal to approximate difference of loudness ordinarily detectible by human ear, the range of which is about 130 decibels on scale beginning with one for faintest audible sound.

Defrost Control: A control system used to detect buildup of frost on the outside coil during the heating cycle and cause system reversal for hot gas defrost of the coil.

Defrost Cycle: Refrigerating cycle in which evaporator frost and ice accumulation is melted.

Defrost Timer: Device connnected into electrical circuit which shuts unit off long enough to permit ice and frost accumulation on evaporator to melt.

Defrosting: Process of removing frost accumulation from evaporators.

Defrosting Type Evaporator: An evaporator operating at such temperatures that ice and frost on surface melt during off part of operating cycle.

Degreasing: Solution or solvent used to remove oil or grease from refrigerator parts.

Degree-Day: Unit that represents one degree of difference from given point in average outdoor temperature of one day and is often used in estimating fuel requirements for a building. Degree-days are based on average temperature over a 24 hour period. As an example, if an average temperature for a day is 50°F, the number of degree-day for that day would be equal to 65°F minus 50°F or 15 degree-days (65 − 50 = 15). Degree-days are useful when calculating requirements for heating purposes.

Dehumidify: To remove water vapor from the atmosphere. To remove water or liquid from stored goods.

Dehumidifier: Device used to remove moisture from air in enclosed space.

Dehumidifier (Surface): An air-conditioning unit, designed primarily for cooling and dehumidifying air through the action of passing the air over wet cooling coils.

Dehumidifying Effect: The difference between the moisture contents, in pounds per hour, of the entering and leaving air, multiplied by 1.060.

Dehydrate: To remove water in all forms from matter. Liquid water, hygroscopic water, and water of crystallization or water of hydration are included.

Dehydrated Oil: Lubricant which has had most of water content removed (a dry oil).

Dehydration: The removal of water vapor from air by the use of absorbing or adsorbing materials; the removal of water from stored goods.

Dehydrator: (See Drier)

Dehydrator-Receiver: A small tank which serves as liquid refrigerant reservoir and which also contains a desiccant to remove moisture. Used on most automobile air-conditioning installations.

Deice Control: Device used to operate refrigerating system in such a way as to provide melting of the accumulated ice and frost.

Delta Transformer: A three-phase electrical transformer which has ends of each of three windings electrically connected.

Demand Meter: An instrument used to measure kilowatt-hour consumption of a particular circuit or group of circuits.

Density: Closeness of texture or consistency.

Deodorizer: Device which absorbs various odors, usually by principle of absorption. Activated charcoal is a common substance used.

Desiccant: Substance used to collect and hold moisture in refrigerating system. A drying agent. Common desiccants are activated alumina, silica gel.

Detector, Leak: Device used to detect and locate refrigerant leaks.

Dew Point: Temperature at which vapor (at 100 percent humidity) begins to condense and deposit as liquid.

Dialectric Fluid: Fluid with high electrical resistance.

Diaphragm: Flexible membrane usually made of thin metal, rubber, or plastic.

Dichlorodifluoromethane: Refrigerant commonly known as R-12. Chemical formula is CCl_2F_2. Cylinder color code is white. Boiling point at atmospheric pressure is −21.62°F.

Die Cast: A process of moulding low melting temperature metals in accurately shaped metal moulds.

Die Stock: Tool used to hold dies with external threads.

Dies (Thread): Tool used to cut external threads.

Differential: As applied to refrigeration and heating: difference between "cut-in" and "cut-out" temperature or pressure of a control.

Direct Expansion Evaporator: An evaporator coil using a pressure-reducing device to supply liquid refrigerant at the correct boiling point for heat absorbtion into the refrigerant.

Displacement, Piston: Volume obtained by multiplying area of cylinder bore by length of piston stroke.

Distilling Apparatus: Fluid reclaiming device used to reclaim used refrigerants. Reclaiming is usually done by vaporizing and then recondensing refrigerant.

Dome-Hat: Sealed metal container for the motor-compressor of a refrigerating unit.

Double Duty Case: Commercial refrigerator which has part of it for refrigerated storage and part equipped with glass windows for display purposes.

Double Thickness Flare: Copper, aluminum or steel tubing end which has been formed into two-wall thickness, 37 to 45 deg. bell mouth or flare.

Dowel Pin: Accurately dimensioned pin pressed into one assembly part and slipped into another assembly part to insure accurate alignment.

Draft: A current of air. Usually refers to the pressure difference which causes a current of air or gases to flow through a flue, chimney, heater, or space.

Draft Gauge: Instrument used to measure air movement.

Draft Indicator: An instrument used to indicate or measure chimney draft or combustion gas movement. Draft is measured in units of 0.1 in. of water column.

Drier: A substance or device used to remove moisture from a refrigeration system.

Drift: Entrained water carried from a cooling tower by air movement.

Drip Pan: Pan-shaped panel or trough used to collect condensate from evaporator coil.

Dry Bulb: An instrument with sensitive element which measures ambient (moving) air temperature.

Dry Bulb Temperature: Air temperature as indicated by ordinary thermometer.

Dry Ice: A refrigerating substance made of solid carbon dioxide which changes directly from a solid to a gas (sublimates). Its subliming temperature is 109°F below zero.

Dry System: A refrigeration system which has the evaporator liquid refrigerant mainly in the atomized or droplet condition.

Duct: Heating and air-conditioning. A tube or channel through which air is conveyed or moved.

Dust: An air suspension (aerosol) of particles of any solid material, usually with particle size less than 100 microns.

Dynamometer: Device for measuring power output or power input of a mechanism.

E

Ebulator: A pointed or sharp edged solid substance inserted in flooded type evaporators to improve evaporation (boiling) of refrigerant in coil.

Eccentric: A circle or disk mounted off center. Eccentrics are used to adjust controls and connect compressor driveshafts to pistons.

Eductor: A device used in absorption refrigeration systems to increase the ability of the absorbing solution.

Effective Area: Actual flow area of an air inlet or outlet. Gross area minus area of vanes or grille bars.

Effective Temperature: Overall effect on a human of air temperature, humidity, and air movement.

Effective Temperature Difference: The difference between the room air temperature and the supply air temperature at the outlet to the room.

Ejector: Device which uses high fluid velocity, such as a venturi, to create low pressure or vacuum at its throat to draw in fluid from another source.

Electric Defrosting: Use of electric resistance heating coils to melt ice and frost off evaporators during defrosting.

Electric Heating: House heating system in which heat from electrical resistance units is used to heat rooms.

Electric Water Valve: Solenoid type (electrically operated) valve used to turn water flow on and off.

Electronics: Field of science dealing with electron devices and their uses.

Electronic Leak Detector: Electronic instrument which measures electronic flow across gas gap. Electronic flow changes indicate presence of refrigerant gas molecules.

Electronic Sound Tracer: Instrument used to detect leaks by locating source of high frequency sound caused by leak.

Electrostatic Filter: Type of filter which gives particles of dust electric charge. This causes particles to be attracted to plate so they can be removed from air stream or atmosphere.

End Bell: End structure of electric motor which usually holds motor bearings.

End Play: Slight movement of shaft along center line.

Energy Efficiency Ratio (EER): The heat transfer ability of the refrigeration system, expressed in Btu/h, compared to the watts of electrical energy necessary to accomplish the heat transfer. This comparison is expressed in Btu/h./Watt of electrical energy.

Enthalpy: Total amount of heat in one pound of a substance calculated from accepted temperature base. Temperature of 32°F is accepted base for water vapor calculation. For refrigerator calculations, accepted base is −40°F.

Entropy: Mathematical factor used in engineering calculations. Energy in a system.

Enzyme: A complex organic substance originating from living cells that speeds up chemical changes in foods. Enzyme action is slowed by cooling.

Epoxy (Resins): A synthetic plastic adhesive.

Equalizer Tube: Device used to maintain equal pressure or equal liquid levels between two containers.

Eutectic Mixture or Solution: A mixture which melts or freezes completely at constant temperature and with constant composition. Its melting point is the lowest possible for mixtures of the given substances.

Evaporation: A term applied to the changing of a liquid to a gas. Heat is absorbed in this process.

Evaporative Condenser: A device which uses open spray or spill water to cool a condenser. Evaporation of some of the water cools the condenser water and reduces water consumption.

Evaporator: Part of a refrigerating mechanism in which the refrigerant evaporizes and absorbs heat.

Evaporator Coil: Device made of a coil of tubing which functions as a refrigerant evaporator.

Evaporator, Dry Type: An evaporator into which refrigerant is fed from a pressure reducing device. Little or no liquid refrigerant collects in the evaporator.

Evaporator Fan: Fan which cools extended heat exchange surface of evaporator.

Evaporator, Flooded: An evaporator containing liquid refrigerant at all times.

Exfiltration: Air flow outward through a wall, leak, membrane, etc.

Exhaust Opening: Any opening through which air is removed from a space which is being heated or cooled, or humidified or dehumidified, or ventilated.

Expansion Valve: A device in refrigerating system which maintains a pressure difference between the high side and low side and is operated by pressure.

Expendable Refrigerant System: System which discards the refrigerant after it has evaporated.

Extended Surface: Heat transfer surface, one side of which is increased in area by the use of fins, ribs, pins, etc.

External Equalizer: Tube connected to low pressure side of an expansion valve diaphragm and to exit of evaporator.

F

Fahrenheit Scale: On a Fahrenheit thermometer, under standard atmospheric pressure, boiling point of water is 212° and freezing point is 32° above zero on its scale.

Fail Safe Control: Device which opens circuit when sensing element fails to operate.

Fan: A radial or axial flow device used for moving or producing artificial currents of air.

Fan (Centrifugal): A fan rotor or wheel within a scroll type of housing and including driving mechanism supports for either belt drive or direct connection.

Fan (Propeller): A propeller or disc-type wheel within a mounting ring or plate and including driving mechanism supports for either belt drive or direct connection.

Fan (Tubeaxial): A propeller or disc-type wheel within a cylinder and including driving mechanism supports for either belt drive or direct connection.

Fan (Vaneaxial): A disc-type wheel within a cylinder, a set of air guide vanes located either before or after the wheel and including driving mechanism supports for either belt drive or direct connection.

Faraday Experiment: Silver chloride absorbs ammonia when cool and releases ammonia when heated. This is basis on which some absorption refrigerators operate.

Field Pole: Part of stator of motor which concentrates magnetic field of field winding.

File Card: Tool used to clean metal files.

Filter: Device for removing small particles from a fluid.

Fin: An extended surface to increase the heat transfer area, as metal sheets attached to tubes.

Finned Tubes: Heat transfer tube or pipe with extended surface in the form of fins, discs, or ribs.

Flammability: The ability of a vapor to combine with oxygen in the combustion process.

Flame Test for Leaks: Tool which is principally a torch and when an air-refrigerant mixture is fed to flame, this flame will change color in presence of heated copper.

Flapper Valve: The type of valve used in refrigeration compressors which allows gaseous refrigerants to flow in only one direction.

Flare: Copper tubing is often connected to parts of refrigerating system by use of flared fittings. These fittings require that the end of tube be expanded at about 45° angle. This flare is firmly gripped by fittings to make a strong leakproof seal.

Flare Nut: Fitting used to clamp tubing flare against another fitting.

Flared Single Thickness Connection: Tube ending formed into $37\frac{1}{2}°$ or 45° bell mouth or flare.

Flash Gas: This is the instantaneous evaporation of some liquid refrigerant in evaporator which cools remaining liquid refrigerant to desired evaporation temperature.

Flash Point: Temperature at which an oil will give off sufficient vapor to support a flash flame but will not support continuous combustion.

Flash Weld: A resistance type weld in which mating parts are brought together under considerable pressure and a heavy electrical current is passed through the joint to be welded.

Float Valve: Type of valve which is operated by sphere or pan which floats on liquid surface and controls level of liquid.

Floc Point: The temperature at which the wax in oil will start to separate from the oil.

Flooded System: Type of refrigerating system in which liquid refrigerant fills evaporator.

Flooded System, Low-Side Float: Refrigerating sytem which has a low side float refrigerant control.

Flooding: Act of filling a space with a liquid.

Flow Meter: Instrument used to measure velocity or volume of fluid movement.

Fluid: Substance in a liquid or gaseous state; substance containing particles which move and change position without separation of the mass.

Fluid Coupling: Device which transmits drive energy to energy absorber through a fluid.

Fluid Flow: The movement of a fluid by a presssure difference created by mechanical means or difference in density created by the addition or removal of heat energy.

Flush: An operation to remove any material or fluids from refrigeration system parts by purging them to the atmosphere using refrigerant or other fluids.

Flux-Brazing, Soldering: Substance applied to surfaces to be joined by brazing or soldering to free them from oxides and facilitate good joint.

Flux, Magnetic: Lines of force of a magnet.

Foam Leak Detector: A system of soap bubbles or special foaming liquids brushed over joints and connections to locate leaks.

Foaming: Formation of a foam in an oil-refrigerant mixture due to rapid evaporation of refrigerant dissolved in the oil. This is most likely to occur when the compressor starts and the pressure is suddenly reduced.

Foot Pound: A unit of work. A foot pound is the amount of work done in lifting one pound one foot.

Force: Force is accumulated pressure and is expressed in pounds. If the pressure is 10 psi on a plate of 10 sq. in. area, the force is 100 lbs.

Forced Convection: Movement of fluid by mechanical force such as fans or pumps.

Force-Feed Oiling: A lubrication system which uses a pump to force oil to surfaces of moving parts.

Free Area: The total minimum area of the openings in a grille, face, or register through which air can pass.

Freezer Alarm: Device used in many freezers which sounds an alarm (bell or buzzer) when freezer temperature rises above safe limit.

Freezer Burn: A condition applied to food which has not been properly wrapped and that has become hard, dry, and discolored.

Freeze-Up: 1-The Formation of ice in the refrigerant control device which may stop the flow of refrigerant into the evaporator. 2-Frost formation on a coil may stop the airflow through the coil.

Freezing: Change of state from liquid to solid.

Freezing Point: The temperature at which a liquid will solidify upon removal of heat. The freezing temperature for water is 32°F at atmospheric pressure.

Freon: Trade name for a family of synthetic chemical refrigerants manufactured by DuPont, Inc.

Frost Back: Condition in which liquid refrigerant flows from evaporator into suction line; indicated by frost formation on suction line.

Frost Control, Automatic: A control which automatically cycles refrigerating system based on frost formation on evaporator.

Frost Control, Manual: A manual control used to change refrigerating system to produce defrosting conditions.

Frost Control, Semiautomatic: A control which starts defrost part of a cycle manually and then returns system to normal operation automatically.

Frost Free Refrigerator: A refrigerated cabinet which operates with an automatic defrost during each cycle.

Frosting Type Evaporator: A refrigerating system which maintains the evaporator at frosting temperatures during phases of cycle.

Full Floating: A mechanism construction in which a shaft is free to turn in all the parts in which it is inserted.

Fumes: Smoke; aromatic smoke; odor emitted, as of flowers; a smoky or vaporous exhalation, usually odorous, as that from concentrated nitric acid. The word fumes is so broad and inclusive that its usefulness as a technical term is very limited. Its principal definitive characteristic is that it implies an odor. The terms vapor, smoke, fog, etc., which can be more strictly defined, should be used whenever possible. Also defined as solid particles generated by condensation from the gaseous state, generally, after volatilization from molten metals, etc., and often accompanied by a chemical reaction such as oxidation. Fumes flocculate and sometimes coalesce.

Fuse: Electrical safety device consisting of strip of fusible metal in circuit which melts when current is overloaded.

Fusible Plug: A plug or fitting mode with a metal of a known low melting temperature, used as safety device to release pressures in case of fire.

G

Galvanic Action: Corrosion action between two metals of different electronic activity. The action is increased in the presence of moisture.

Gas: Vapor phase or state of a substance.

Gasket: A resilient or flexible material used between mating surfaces of refrigerating unit parts or of refrigerator doors to provide a leakproof seal.

Gasket, Foam: A joint sealing device made of rubber or plastic foam strips.

Gas—Noncondensible: A gas which will not form into a liquid under pressure-temperature conditions.

Gas Valve: Device for controlling flow of gas.

Gauge, Compound: Instrument for measuring pressures both below and above atmospheric pressure.

Gauge, High Pressure: Instrument for measuring pressures in range of 0 psig to 500 psig.

Gauge, Low Pressure: Instrument for measuring pressures in range of 0 psig and 50 psig.

Gauge Manifold: A device constructed to hold compound and high pressure gauges and valved to control flow of fluids through it.

Gauge, Vacuum: Instrument used to measure pressures below atmospheric pressure.

Generator: A device used in absorption-type refrigeration systems to heat the absorbing liquid to drive off the refrigerant vapor for condensing to a liquid before entering the evaporator.

Grain: A unit of weight and equal to one 7000th of a pound. It is used to indicate the amount of moisture in the air.

Gravity (Specific): The specific gravity of a solid or liquid is the ratio of the mass of the body to the mass of an equal volume of water at some standard temperature. At the present time a temperature of 4 C (39°F) is commonly used by physicists, but the engineer uses 60°F. The specific gravity of a gas is usually expressed in terms of dry air at the same temperature and pressure as the gas.

Grille: An ornamental or louvered opening placed at the end of an air passageway.

Grommet: A plastic metal or rubber doughnut-shaped protector for wires or tubing as they pass through hole in object.

Ground Coil: A heat exchanger buried in the ground which may be used either as an evaporator or as a condenser.

Ground, Short Circuit: A fault in an electrical circuit allowing electricity to flow into the metal parts of the structure.

Ground Wire: An electrical wire which will safely conduct electricity from a structure into the ground.

H

Halide Refrigerants: Family of refrigerants containing halogen chemicals.

Halide Torch: Type of torch used to detect halogen refrigerant leaks.

Hastelloy: Trade name for a hard, noncorroding metal alloy.

Head (Total): In flowing fluid, the sum of the static and velocity pressures at the point of measurement.

Head Pressure: Pressure which exists in condensing side of refrigerating system.

Head-Pressure Control: Pressure operated control which opens electrical circuit if high side pressure becomes excessive.

Head, Static: Pressure of fluid expressed in terms of height of column of the fluid, such as water or mercury.

Head, Velocity: In flowing fluid, height of fluid equivalent to its velocity pressure.

Heat: Form of energy the addition of which causes substances to rise in temperature; energy associated with random motion of molecules.

Heat (Latent): Heat characterized by a change of state of the substance concerned, for a given pressure and always at a constant temperature for a pure substance, i.e., heat of vaporization or of fusion.

Heat (Sensible): A term used in heating and cooling to indicate any portion of heat which changes only the temperature of the substances involved.

Heat (Specific): The heat absorbed (or given up) by a unit mass of a substance when its temperature is increased (or decreased) by 1-degree Common Units: Btu per (pound) (Fahrenheit degree), calories per (gram) (Centigrade degree). For gases, both specific heat at constant pressure (c_p) and specific heat at constant volume (c_v) are frequently used. In air-conditioning, c_p is usually used.

Heat Exchanger: Device used to transfer heat from a warm or hot surface to a cold or cooler surface. Evaporators and condensers are heat exchangers.

Heat Lag: When a substance is heated on one side, it takes time for the heat to travel through the substance. The time is called heat lag.

Heat Leakage: Flow of heat through a substance is called heat leakage.

Heat Load: Amount of heat, measured in Btu, which is removed during a period of 24 hrs.

Heat of Compression: Mechanical energy of pressure transformed into energy of heat.

Heat of Fusion: The heat released in changing a substance from a liquid state to a solid state. The heat of fusion of ice is 144 Btu per pound.

Heat of Respiration: The process by which oxygen and carbohydrates are assimilated by a substance; also when carbon dioxide and water are given off by a substance.

Heat Pump: A name given to an air-conditioning system that is reversible so as to be able to remove heat from or add heat to a given space or material upon demand.

Heat Pump-Air Source: A device that transfers heat between two different air quantities, in either direction, upon demand.

Heat Pump-Water Source: A device that uses a water supply as a source of heat or for disposal of heat depending upon the operational demand.

Heat Sink: The material in which the refrigeration system puts the heat picked up as well as the electrical energy used.

Heat Source: The material from which the refrigeration system extracts heat.

Heat Transfer: Movement of heat from one body or substance to another. Heat may be transferred by radiation, conduction, convection, or a combination of these three methods.

Heating Coil: A heat transfer device which releases heat.

Heating Control: Device which controls temperature of heat transfer unit which releases heat.

Heating Value: Amount of heat which may be obtained by burning a fuel. It is usually expressed in Btu per pound or Btu per gallon.

Heavy Ends, Hydocarbon Oils: The heavy molecules or larger molecules of hydrocarbon oils.

Hermetic Motor: Compressor drive motor sealed within same casing which contains compressor.

Hermetic System: Refrigeration system which has a compressor driven by a motor contained in compressor dome or housing.

Hermetically Sealed Unit: A sealed hermetic-type condensing unit is a mechanical condensing unit in which the compressor and compressor motor are enclosed in the same housing with no external shaft or shaft seal, the compressor motor operating in the refrigerant atmosphere. The compressor and compressor motor housing may be of either the fully welded or brazed type, or of the service-sealed type. In the fully welded or brazed type, the housing is permanently sealed and is not provided with means of access for servicing internal parts in the field. In the service-sealed type, the housing is provided with some means of access for servicing internal parts in the field.

Hg (Mercury): Heavy silver-white metallic element; only metal that is liquid at ordinary room temperature. Symbol, Hg.

High Pressure Cut-Out: Electrical control switch operated by the high side pressure which automatically opens electrical circuit if too high head pressure or condensing pressure is reached.

High Side: Parts of a refrigerating system which are under condensing or high side pressure.

High Side Float: Refrigerant control mechanism which controls the level of the liquid refrigerant in the high pressure side of mechanism.

High Vacuum Pump: Mechanism which can create vacuum in 1000 to 1 micron range.

High Velocity System: Usually large commercial or industrial air distribution systems designed to operate with static pressures of 6 to 9 inches water gauge.

Hollow-Tube Gasket: Sealing device made of rubber or plastic with tubular cross section.

Hone: Fine-grit stone used for precision sharpening.

Horsepower: A unit of power equal to 33,000 foot pounds of work per minute. One electrical horsepower equals 746 watts.

Hot Gas Bypass: Piping system in refrigerating unit which moves hot refrigerant gas from condenser into low pressure side.

Hot Gas Defrost: A defrosting system in which hot refrigerant gas from the high side is directed through evaporator for short period of time and at predetermined intervals in order to remove frost from evaporator.

Hot Gas Line: The line that carries the hot discharge gas from the compressor to the condenser.

Hot Junction: That part of thermoelectric circuit which releases heat.

Hot Wire: A resistance wire in an electrical relay which expands when heated and contracts when cooled.

Humidifiers: Device used to add to and control the humidity in a confined space.

Humidistat: An electrical control which is operated by changing humidity.

Humidity: Moisture; dampness. Relative humidity is ratio of quantity of vapor present in air to greatest amount possible at given temperature.

Hydrolen—Tar: A hydrocarbon by-product of oil industry. Used as a low melting temperature, waterproof sealing compound.

Hydrometer: Floating instrument used to measure specific gravity of a liquid. Specific gravity is ratio of weight of any volume of a substance to weight of equal volume of substance used as a standard.

Hygrometer: An instrument used to measure degree of moisture in the atmosphere.

Hygroscopic: Ability of a substance to absorb and retain moisture and change physical dimensions as its moisture content changes.

I

ICC—Interstate Commerce Commission: A government body which controls the design and construction of pressure containers.

Ice Cream Cabinet: Commercial refrigerator which operates at approximately 0°F and is used for storage of ice cream.

Ice Melting Equivalent (I.M.E.) (Ice Melting Effect): Amount of heat absorbed by melting ice at 32°F is 144 Btu per pound of ice or 288,000 Btu per ton.

Idler: A pulley used on some belt drives to provide the proper belt tension and to eliminate belt vibration.

Ignition Transformer: A transformer designed to provide a high voltage current. Used in many heating systems to ignite fuel.

Impeller: Rotating part of a centrifugal pump.

Induction Motor: An AC motor which operates on principle of rotating magnetic field. Rotor has no electrical connection, but receives electrical energy by transformer action from field windings.

Induction Units (Low-Pressure Type): Essentially induction-type convectors. They use a jet of conditioned air (or primary air) to induce into the unit a flow of room or secondary air which mixes with the primary air. The mixture is discharged into the room through a grille at the top of the unit. Heating coils are located in the secondary air stream for use in heating.

Industrial Air-Conditioning: Air-conditioning for other uses than comfort.

Infrared Lamp: An electrical device which emits infrared rays; invisible rays just beyond red in the visible spectrum.

Insulation, Thermal: Substance used to retard or slow flow of heat through wall or partition.

Isothermal: Changes of volume or pressure under conditions of constant temperature.

Isothermal Expansion and Contraction: An action which takes place without a temperature change.

J

Joint (Brazed, High-Temperature): A gas tight joint obtained by the joining of metal parts with metallic mixtures or alloys which melt at temperatures below 1500°F but above 1000°F.

Joint (Soldered): A gas-tight joint obtained by the joining of metal parts with metallic mixtures or alloys which melt at temperatures below 1000°F.

Joint (Welded): A gas-tight joint obtained by the joining of metal parts in the plastic or molten state.

Joule-Thomson Effect: Change in temperature of a gas on expansion through a porous plug from a high pressure to a lower pressure.

Journal, Crankshaft: Part of shaft which contacts the bearing.

Junction Box: Group of electrical terminals housed in protective box or container.

K

Kata Thermometer: Large bulb alcohol thermometer used to measure air velocities or atmospheric conditions by means of cooling effect.

Kelvin Scale (K): Thermometer scale on which unit of measurement equals the centigrade degree and according to which absolute zero is 0°, the equivalent of −273.16° C. Water freezes at 273.16° and boils at 373.16°.

Kilometer: A metric unit of liner measurement = 1000 meters.

Kilowatt: Unit of electrical power, equal to 1000 watts.

L

Lacquer: A protective coating or finish which dries to form a film by evaporation of a volatile constituent.

Lamps, Steri: A lamp which gives forth a high intensity ultraviolet ray and is used to kill bacteria. It is often used in food storage cabinets.

Lapping: Smoothing a metal surface to high degree of refinement or accuracy using a fine abrasive.

Latent Heat: Heat energy absorbed in process of changing form of substance (melting, vaporization, fusion) without change in temperature or pressure.

Leak Detector: Device or instrument such as a halide torch, an electronic sniffer; or soap solution used to detect leaks.

Limit Control: Control used to open or close electrical circuits as temperature or pressure limits are reached.

Liquid Absorbent: A chemical in liquid form which has the property to "take on" or absorb moisture.

Liquid Indicator: Device located in liquid line which provides a glass window through which liquid flow may be observed.

Liquid Line: The tube which carries liquid refrigerant from the condenser or liquid receiver to the pressure reducing device.

Liquid Nitrogen: Nitrogen in liquid form which is used as a low temperature refrigerant in chemical (or expendable) refrigerating systems.

Liquid Receiver: Cylinder connected to condenser outlet for storage of liquid refrigerant in a system.

Liquid Return Line: The line that carries the liquid refrigerant from the outlet of the condenser to the inlet of the receiver (when used).

Liquid-Vapor Valve Refrigerant Cylinder: A dual hand valve on refrigerant cylinders which is used to release either gas or liquid refrigerant from the cylinder.

Litharge: Lead powder mixed with glycerine to seal pipe thread joints.

Liquor: Solution used in absorption refrigeration.

Liter: Metric unit of volume which equals 61.03 in.^{3}.

Load: The amount of heat per unit time imposed on a refrigerating system, or the required rate of heat removal.

Louvers: Sloping, overlapping boards or metal plates intended to permit ventilation and shed falling water.

Low Side: That portion of a refrigerating system which is under the lowest evaporating pressure.

Low Side Float Valve: Refrigerant control valve operated by level of liquid refrigerant in low pressure side of system.

Low Side Pressure: Pressure in cooling side of refrigerating cycle.

Low Side Pressure Control: Device used to keep low side evaporating pressure from dropping below certain pressure.

Low Velocity System: An air distribution system designed to operate at a static pressure of 0.25 inch water gauge or less.

M

Makeup Water: The water required to replace the water lost from a cooling tower by evaporation, drift, and bleedoff.

Manifold, Service: A device equipped with gauges and manual valves, used by serviceman to service refrigerating systems.

Manometer: Instrument for measuring pressure of gases and vapors. Gas pressure is balanced against column of liquid such as mercury, in U-shaped tube.

Mass: A quantity of matter cohering together to make one body which is usually of indefinite shape.

Mean Effective Pressure (M.E.P.): Average pressure on a surface when a changing pressure condition exists.

Mean Temperature Difference: The average temperature between the temperature before process begins and the temperature after process is completed.

Mechanical Cycle: Cycle which is a repetitive series of mechanical events.

Medium Velocity System: An air distribution system designed to operate at a design static pressure of 2.0 to 3.0 inches water gauge.

Melting Point: Temperature at atmospheric pressure, at which a substance will melt.

Mercoid Bulb: An electrical circuit switch which uses a small quantity of mercury in a sealed glass tube to make or break electrical contact with terminals within the tube.

Meter: Metric unit of linear measurement equal to 39.37 in.

Methanol Drier: Alcohol type chemical used to change water in refrigerating system into a nonfreezing solution.

Methyl Chloride (R-40): A chemical once commonly used as a refrigerant. The chemical formula is CH_3Cl. Cylinder color code is orange. The boiling point at atmospheric pressure is $-10.4°F$.

Metric System: A decimal system of measures and weights, based on the meter and gram. Length of one meter, 39.37 in.

Micrometer: A precision measuring instrument used for making measurements accurate to 0.001 to 0.0001 in.

Micron: Unit of length in metric system; a thousandth part of one millimeter.

Micron Gauge: Instrument for measuring vacuums very close to a perfect vacuum.

Milli: A combining form denoting one thousandth; example, millivolt, one thousandth of a volt.

Modulating: A type of device or control which tends to adjust by increments (minute changes) rather than by either full on or full off operation.

Modulating Refrigeration Cycle: Refrigerating system of variable capacity.

Moisture Determination: An action using instruments and calculations to measure the relative or absolute moisture in an air conditioned space.

Moisture Indicator: Instrument used to measure moisture content of a refrigerant.

Molecule: Smallest portion of an element or compound that retains chemical identity with the substance in mass.

Molliers Diagram: Graph of refrigerant pressure, heat, and temperature properties.

Monel: A trademark name for metal alloy consisting chiefly of copper and nickel.

Monitor Top: Unit built by General Electric which had a cylindrical condenser surrounding the motor-compressor, mounted on top of the cabinet.

Monochlorodifluoromethane: A refrigerant better known as Freon 12 or R-22. Chemical formula is $CHClF_2$. Cylinder color code is green.

Motor—2-Pole: A 3600 rpm electric motor (synchronous speed).

Motor—4-Pole: An 1800 rpm electric motor (synchronous speed).

Motor, Capacitor: A single-phase induction motor with an auxiliary starting winding connected in series with a condenser (capacitor) for better starting characteristics.

Motor Burnout: Condition in which the insulation of electric motor has deteriorated by overheating.

Motor Control: Device to start and/or stop a motor at certain temperature or pressure conditions.

Motor Starter: High capacity electric switches usually operated by electromagnets.

Muffler, Compressor: Sound absorber chamber in refrigeration system used to reduce sound of gas pulsations.

Mullion: Stationary part of a structure between two doors.

Multiple Evaporator System: Refrigerating system with two or more evaporators connected in parallel.

Multiple System: Refrigerating mechanism in which several evaporators are connected to one condensing unit.

N

Natural Convection: Movement of a fluid caused by temperature differences (density changes).

Neoprene: A synthetic rubber which is resistant to hydrocarbon oil and gas.

Neutralizer: Substance used to counteract acids, in refrigeration system.

No-Frost Freezer: A low temperature refrigerator-cabinet in which no frost or ice collects on produce stored in cabinet.

Nominal Size Tubing: Tubing measurement which has an inside diameter the same as iron pipe of the same stated size.

Non-Code Installation: A functional refrigerating system installed where there are no local, state, or national refrigeration codes in force.

Noncondensable Gas: Gas which does not change into a liquid at operating temperatures and pressures.

Nonferrous: Group of metals and metal alloys which contain no iron.

Nonfrosting Evaporator: An evaporator which never collects frost or ice on its surface.

Normal Charge: The thermal element charge which is part liquid and part gas under all operating conditions.

North Pole, Magnetic: End of magnet from which magnetic lines of force flow.

O

Off Cycle: That part of a refrigeration cycle when the system is not operating.

Oil Binding: Physical condition when an oil layer on top of refrigerant liquid hinders it from evaporating at its normal pressure-temperature condition.

Oil, Refrigeration: Specially prepared oil used in refrigerator mechanism circulates to same extent with refrigerant. The oil must be dry (entirely free of moisture), otherwise, moisture will condense out and freeze in the refrigerant control and may cause refrigerant mechanism to fail. An oil classified as a refrigerant oil must be free of moisture and other contaminants.

Oil Rings: Expanding rings mounted in grooves and piston; designed to prevent oil from moving into compression chamber.

Oil Separator: Device used to remove oil from gaseous refrigerant.

Open Circuit: An interrupted electrical circuit which stops flow of electricity.

Open Display Case: Commercial refrigerator designed to maintain its contents at refrigerating temperatures even though the contents are in an open case.

Open Type System: A refrigerating system which uses a belt-driven compressor or a coupling-driven compressor.

Orifice: Accurate size opening for controlling fluid flow.

Oscilloscope: A fluorescent coated tube which visually shows an electrical wave.

Outside Air: External air; atmosphere exterior to refrigerated or conditioned space; ambient (surrounding) air.

Overload: Load greater than load for which system or mechanism was intended.

Overload Protector: A device, either temperature, pressure, or current operated, which will stop operation of unit if dangerous conditions arise.

Ozone: A gaseous form of oxygen usually obtained by silent discharge of electricity in oxygen or air.

P

Partial Pressures: Condition where two or more gases occupy a space and each one creates part of the total pressure.

Pascal's Law: A pressure imposed upon a fluid is transmitted equally in all directions.

Perm: The unit of permeance. A perm is equal to 1 grain per (sq ft) (hr) (inch of mercury vapor pressure difference).

Permeance: The water vapor permeance of a sheet of any thickness (or assembly between parallel surfaces) is the ratio of water vapor flow to the vapor pressure difference between the surfaces. Permeance is measured in perms.

Permanent Magnet: A material which has its molecules aligned and has its own magnetic field; bar of metal which has been permanently magnetized.

Photoelectricity: A physical action wherein an electrical flow is generated by light waves.

Pinch-Off Tool: Device used to press walls of a tubing together until fluid flow ceases.

Piston: Close fitting part which moves up and down in a cylinder.

Piston Displacement: Volume displaced by piston as it travels length of stroke.

Pitch: Pipe slope, in direction of vapor flow, to promote the transfer of oil through the pipe.

Pitot Tube: Tube used to measure air velocities.

Plenum Chamber: Chamber or container for moving air or other gas under a slight positive pressure.

Polyphase Motor: Electrical motor designed to be used with three-phase electrical circuit.

Polystyrene: Plastic used as an insulation in some refrigerator cabinet structures.

Ponded Roof: Flat roof designed to hold quantity of water which acts as a cooling device.

Porcelain: Ceramic china-like coating applied to steel surfaces.

Pour Point (Oil): Lowest temperature at which oil will pour or flow.

Power: Time rate at which work is done or energy emitted; source or means of supplying energy.

Power Element: Sensitive element of a temperature operated control.

Pressure: An energy impact on a unit area; force or thrust exerted on a surface.

Pressure Drop: The pressure difference at two ends of a circuit, or part of a circuit, the two sides of a filter, or the pressure difference between the high side and low side in a refrigerator mechanism.

Pressure Limiter: Device which remains closed until a certain pressure is reached and then opens and releases fluid to another part of system.

Pressure-Heat Diagram: Graph of refrigerant pressure, heat, and temperature properties (Mollier's diagram).

Pressure Motor Control: A device which opens and closes on electrical circuit as pressures change to desired pressures.

Pressure-Operated Altitude (POA) Valve: Device which maintains a constant low side pressure independent of altitude of operation.

Pressure-Reducing Device: The device used to produce a reduction in pressure and corresponding boiling point before the refrigerant is introduced into the evaporator.

Pressure Regulator, Evaporator: An automatic pressure regulating valve. Mounted in suction line between evaporator outlet and compressor inlet. Its purpose is to maintain a predetermined pressure and temperature in the evaporator.

Pressure Suction: Pressure in low pressure side of a refrigerating system.

Pressure Water Valve: Device used to control water flow which is responsive to head pressure of refrigerating system.

Primary Control: Device which directly controls operation of heating system.

Process Tube: Length of tubing fastened to hermetic unit dome, used for servicing unit.

Protector, Circuit: An electrical device which will open an electrical circuit if excessive electrical conditions occur.

psi: A symbol or initials used to indicate pressure measured in pounds per square inch.

psia: A symbol or initials used to indicate pressure measured in pounds per square inch absolute. Absolute pressure equals gauge pressure plus atmospheric pressure.

psig: A symbol or initials used to indicate pressure in pounds per square inch gauge. The "g" indicates that it is gauge pressure and not absolute pressure.

Psychrometer or Wet Bulb Hygrometer: An instrument for measuring the relative humidity of atmospheric air.

Psychrometric Chart: A chart that shows relationship between the temperature, pressure, and moisture content of the air.

Psychrometric Measurement: Measurement of temperature pressure, and humidity using a psychrometric chart.

Pull Down: An expression indicating action of removing refrigerant from all or a part of refrigerating system.

Pump Down: The act of using a compressor or a pump to reduce the pressure in a container or a system.

Purging: Releasing compressed gas to atmosphere through some part or parts for the purpose of removing contaminants from that part or parts.

Pyrometer: Instrument for measuring high temperatures.

Q

Quenching: Submerging hot solid object in cooling fluid.

Quick Connect Coupling: A device which permits easy, fast, connecting of two fluid lines.

R

R-11, Trichloromonofluoromethane: Low pressure, synthetic chemical refrigerant which is also used as a cleaning fluid.

R-12, Dichlorodifluoromethane: A popular refrigerant known as Freon 12.

R-22, Monochlorodifluoromethane: Synthetic chemical refrigerant.

R-40, Methyl Chloride: Refrigerant which was used extensively in the 1920s and 1930s.

R-113, Trichlorotrifluoroethane: Synthetic chemical refrigerant.

R-160, Ethyl Chloride: Refrigerant which is seldom used at present time.

R-170, Ethane: Low temperature application refrigerant.

R-290, Propane: Low temperature application refrigerant.

R-500: Refrigerant which is azeotropic mixture of R-12 and R-152a.

R-502: Refrigerant which is azeotropic mixture of R-22 and R-115.

R-503: Refrigerant which is azeotropic mixture of R-23 and R-13.

R-504: Refrigerant which is azeotropic mixture of R-32 and R-115.

R-600, Butane: Low temperature application refrigerant; also used as a fuel.

R-611, Methyl Formate: Low pressure refrigerant.

R-717, Ammonia: Popular refrigerant for industrial refrigerating systems; also a popular absorption system refrigerant.

R-764, Sulphur Dioxide: Low pressure refrigerant used extensively in 1920s and 1930s. Not in use at present; chemical is often used as an industrial bleaching agent.

Radial Commutator: Electrical contact surface on a rotor which is perpendicular or at right angles to the shaft center line.

Radiation: Transfer of heat by heat rays.

Range: Pressure or temperature settings of a control; change within limits.

Rankin Scale: Name given the absolute (Fahrenheit) scale. Zero on this scale is −460°F.

Receiver-Drier: A cylinder in a refrigerating system for storing liquid refrigerant and which also holds a quantity of desiccant.

Receiver Heating Element: Electrical resistance mounted in or around liquid receiver, used to maintain head pressures when ambient temperature is at freezing or below freezing.

Reciprocating: Action in which the motion is back and forth in a straight line.

Recording Ammeter: Electrical instrument which uses a pen to record amount of current flow on a moving paper chart.

Recording Thermometer: Temperature measuring instrument which has a pen marking a moving chart.

Rectifier, Electric: An electrical device for converting ac into dc.

Reed Valve: Thin flat tempered steel plate fastened at one end.

Refrigerant: Substance used in refrigerating mechanism to absorb heat in evaporator coil by change of state from a liquid to a gas, and to release its heat in a condenser as the substance returns from the gaseous state back to a liquid state.

Refrigerant Charge: Quantity of refrigerant in a system.

Refrigerant Control: Device which meters refrigerant and maintains pressure difference between high pressure and low pressure side of mechanical refrigerating system while unit is running.

Refrigerating Effect: The amount of heat in Btu/h or cal/hr the system is capable of transferring.

Refrigeration: The process of transferring heat from one place to another by the change in state of a liquid.

Refrigeration-Absorption: Refrigerating effect produced by the change in pressure in the system produced by the changes in the ability of a substance to retain a liquid dependent upon the temperature of the substance.

Refrigeration-Mechanical: Refrigerating effect produced by the changes in pressure in the system produced by mechanical action of a compressor.

Refrigeration System: A system composed of parts necessary to produce the changes in boiling point necessary to accomplish heat transfer by the change in state of the refrigerant.

Register: Combination grille and damper assembly covering on an air opening or end of an air duct.

Relative Humidity: Ratio of amount of water vapor present in air to greatest amount possible at same temperature.

Relay: Electrical mechanism which uses small current in control circuit to operate a valve switch in operating circuit.

Relay-Lock Out: A relay used in conjunction with an automatic reset high pressure control to cause manual reset interruption of the unit operation when high head pressure is encountered. Reset of the system is from a remote location by interruption of power to the circuit.

Relief Valve: Safety device designed to open before dangerous pressure is reached.

Remote Power Element Control: Device with sensing element located apart from operating mechanism.

Remote System: Refrigerating system which has condensing unit located outside and separate from refrigerator cabinet.

Repulsion-Start Induction Motor: Type of motor which has an electrical winding on the rotor for starting-purposes.

Reverse Cycle Defrost: Method of heating evaporator for defrosting purposes by using valves to move hot gas from compressor into evaporator.

Reversing Valve: Device used to reverse direction of the refrigerant flow depending upon whether heating or cooling is desired.

Ringelmann Scale: Measuring device for determining smoke density.

Riser Valve: Device used to manually control flow of refrigerant in vertical piping.

Rotary Blade Compressor: Mechanism for pumping fluid by revolving blades inside cylindrical housing.

Rotary Compressor: Mechanism which pumps fluid by using rotating motion.

Rotor: Rotating part of a mechanism.

Running Winding: Electrical winding of motor which has current flowing through it during normal operation of motor.

S

Saddle Valve (Tap-A-Line): Valve body shaped so it may be silver brazed to refrigerant tubing surface.

Safety Control: Device which will stop the refrigerating unit if unsafe pressures and/or temperatures are reached.

Safety Motor Control: Electrical device used to open circuit if the temperature, pressure, and/or the current flow exceed safe conditions.

Safety Plug: Device which will release the contents of a container above normal pressure conditions and before rupture pressures are reached.

Saturation: A condition existing when a substance contains maximum of another substance for that temperature and pressure.

Scavenger Pump: Mechanism used to remove fluid from sump or container.

Schrader Valve: Spring loaded device which permits fluid flow when a center pin is depressed.

Scotch Yoke: Mechanism used to change reciprocating motion into rotary motion or vice-versa. Used to connect crankshaft to piston in refrigeration compressor.

Sealed Unit: (See Hermetic System) A motor-compressor assembly in which motor and compressor operate inside sealed dome or housing.

Seal Leak: Escape of oil and/or refrigerant at the junction where shaft enters housing.

Seal, Shaft: A device used to prevent leakage between shaft and housing.

Secondary Refrigerating System: Refrigerating system in which condenser is cooled by evaporator of another or primary refrigerating system.

Second Law of Thermodynamics: Heat will flow only from material at certain temperature to material at lower temperature.

Sensible Heat: Heat which causes a change in temperature of a substance.

Sensor: A material or device which goes through a physical change or an electronic characteristic change as the conditions change.

Separator, Oil: A device used to separate refrigerant oil from refrigerant gas and return the oil to crankcase of compressor.

Sequence Controls: Group of devices which act in series or in time order.

Servel System: One type of continuous operation absorption refrigerating system.

Serviceable Hermetic: Hermetic unit housing containing motor and compressor asembled by use of bolts or threads.

Service Valve: A device to be attached to system which provides opening for gauges and/or charging lines. Also provides means of shutting off or opening gauge and charging ports, and controlling refrigerant flow in system.

Shaded Pole Motor: A small ac motor used for light start loads. Has no brushes or commutator.

Sharp Freezing: Refrigeration at temperature slightly below freezing, with moderate air circulation.

Shell-and-Tube Flooded Evaporator: Device which flows water through tubes built into cylindrical evaporator or vice-versa.

Shell Type Condenser: Cylinder or receiver which contains condensing water coils or tubes.

Short Cycling: Refrigerating system that starts and stops more frequently than it should.

Shroud: Housing over condenser or evaporator.

Sight Glass: Glass tube or glass window in refrigerating mechanism which shows amount of refrigerant, or oil in system; or, pressure of gas bubbles in liquid line.

Silica Gel: Chemical compound used as a drier, which has ability to absorb moisture when heated. Moisture is released and compound may be reused.

Silver Brazing: Brazing process in which brazing alloy contains some silver as part of joining alloy.

Sintered Oil Bearing: Porous bearing metal, usually bronze, and which has oil in pores of bearing metal.

Sling Psychrometer: Humidity measuring device with wet and dry bulb thermometers, which is moved rapidly through air when measuring humidity.

Slug: A unit of mass equal to the weight (English units) of object divided by 32.2 (acceleration due to the force of gravity).

Smoke: An air suspension (aerosol) of particles, usually but not necessarily solid, often originating in a solid nucleus, formed from combustion or sublimation. Also defined as carbon or soot particles less than 0.1 micron in size which result from the incomplete combustion of carbonaceous materials such as coal, oil, tar, and tobacco.

Smoke Test: Test made to determine completeness of combustion.

Solar Heat: Heat from visible and invisible energy waves from the sun.

Soldering: Joining two metals by adhesion of a low melting temperature metal (less than 800°F).

Solenoid Valve: Electromagnet with a moving core which serves as a valve, or operates a valve.

Solid Absorbent Refrigeration: Refrigerating system which uses solid substance as absorber of the refrigerant during cooling part of cycle and releases refrigerant when heated during generating part of cycle.

South Pole, Magnetic: That part of magnet into which magnetic flux lines flow.

Specific Gravity: Weight of liquid compared to water which is assigned value of 1.0.

Specific Heat: Ratio of quantity of heat required to raise temperature of a body one degree to that required to raise temperature of equal mass of water one degree.

Specific Volume: Volume per unit mass of a substance.

Splash System, Oiling: Method of lubricating moving parts by agitating or splashing oil.

Split-Phase Motor: Motor with two stator windings. Winding in use while starting is disconnected by centrifugal switch after motor attains speed, then motor operates on other winding.

Split System: Refrigeration or air-conditioning installation which places condensing unit outside or remote from evaporator. Also applicable to heat pump installations.

Spray Cooling: Method of refrigerating by spraying refrigerant inside of evaporator or by spraying refrigerated water.

Squirrel Cage: Fan which has blades parallel to fan axis and moves air at right angles or perpendicular to fan axis.

Standard Atmosphere: Condition when air is at 14.7 psia pressure, at 68°F temperature.

Standard Conditions: Used as a basis for air-conditioning calculations. Temperature of 68°F, pressure of 29.92 in. of Hg and relative humidity of 30 percent.

Static Pressure: The pressure exerted against the inside of a duct in all directions. Roughly defined as *bursting pressure*.

Starting Relay: An electrical device which connects and/ or disconnects starting winding of electric motor.

Starting Winding: Winding in electric motor used only during brief period when motor is starting.

Glossary of Technical Terms (Refrigeration, Air-Conditioning, and Heat Pumps)   729

Stationary Blade Compressor: A rotary pump which uses blade inside pump to separate intake chamber from exhaust chamber.

Stator, Motor: Stationary part of electric motor.

Steam: Water in vapor state.

Steam-Heating: Heating system in which steam from a boiler is conducted to radiators in space to be heated.

Steam Jet Refrigeration: Refrigerating system which uses a steam venturi to create high vacuum (low pressure) on a water container causing water to evaporate at low temperature.

Stethoscope: Instrument used to detect sounds.

Stoker: Machine used to supply a furnace with coal.

Strainer: Device such as a screen or filter used to retain solid particles while liquid passes through.

Stratification of Air: Condition in which there is little or no air movement in room; air lies in temperature layers.

Strike: Door part of a door latch.

Subcooling: Cooling of liquid refrigerant below its condensing temperature.

Sublimation: Condition where a substance changes from a solid to a gas without becoming a liquid.

Suction Line: Tube or pipe used to carry refrigerant gas from evaporator to compressor.

Suction Pressure Control Valve: Device located in the suction line which maintains constant pressure in evaporator during running portion of cycle.

Suction Service Valve: A two-way manual-operated valve located at the inlet to compressor, which controls suction gas flow and is used to service unit.

Sulfur Dioxide: Gas once commonly used as a refrigerant. Refrigerant number is R-764; chemical formula is SO_2. Cylinder color code, black; boiling point at atmospheric pressure 14°F.

Sulphur Dioxide: (See Sulfur Dioxide)

Superheat: Temperature of vapor above boiling temperature of its liquid at that pressure.

Superheater: Heat exchanger arranged to cool liquid going to evaporator using this heat to superheat vapor leaving evaporator.

Surface Plate: Tool with a very accurate flat surface, used for measuring purposes, and for lapping flat surfaces.

Surge: Modulating action of temperature or pressure before it reaches is final value or setting.

Surge Tank: Container connected to a refrigerating system which increases gas volume and reduces rate of pressure change.

Swaging: Enlarging one tube end so end of other tube of same size will fit within.

Swash Plate-Wobble Plate: Device used to change rotary motion to reciprocating motion, used in some refrigeration compressors.

Sweating: This term is used two different ways in refrigeration work: 1—Condensation of moisture from air on cold surface. 2—Method of soldering in which the parts to be joined are first coated with a thin layer of solder.

Sweet Water: Term sometimes used to describe tap water.

Sylphon Seal: Corrugated metal tubing used to hold seal ring and provide leak-proof connection between seal ring and compressor body or shaft.

Synthetic Rubber, Neoprene: Soft resilient material made of a synthetic chemical compound.

T

Tap-A-Line: Device used to puncture or tap a line where there are no service valves available; sometimes called a saddle valve.

Tap Drill: Drill used to form hole prior to placing threads in hole. The drill is the size of the root diameter of tap threads.

Tap (Screw Thread): Tool used to cut internal threads.

Teflon: Synthetic rubber material often used for O rings.

Temperature: Degree of hotness or coldness as measured by a thermometer; measurement of speed of motion of molecules.

Temperature Humidity Index: Actual temperature and humidity of sample of air, compared to air at standard conditions.

Test Light: Light provided with test leads, used to test or probe electrical circuits to determine if they are alive.

Thermal Relay (Hot Wire Relay): Electrical control used to actuate a refrigeration system. This system uses a wire to convert electrical energy into heat energy.

Thermistor: Material called a semiconductor, which is between a conductor and an insulator, which has electrical resistance that varies with temperature.

Thermocouple: Device which generates electricity, using principle that if two dissimilar metals are welded together and junction is heated, a voltage will develop across open ends.

Thermocouple Thermometer: Electrical instrument using thermocouple as source of electrical flow, connected to milliammeter calibrated in temperature degrees.

Thermodisk Defrost Control: Electrical switch with bimetal disk which is controlled by electrical energy.

Thermodynamics:

1st law of: energy can neither be created nor destroyed—it can only be changed from one form to another.

2nd law of: to cause heat energy to travel, a temperature (heat intensity) difference must be created and maintained.

Thermoelectric Refrigeration: A refrigerator mechanism which depends on Peletier effect. Direct current flowing through electrical junction between dissimilar metals provides heating or cooling effect depending on direction of flow of current.

Thermometer: Device for measuring temperatures.

Thermomodule: Number of thermocouples used in parallel to achieve low temperatures.

Thermostat: Device responsive to ambient temperature conditions.

Thermostat-Outdoor Ambient: A control used to limit the amount of auxiliary electric heat according to outside ambient temperature to reduce electrical surge and operating cost.

Thermostatic Control: Device which operates system or part of system based on temperature changes.

Thermostatic Expansion Valve: A control valve operated by temperature and pressure within evaporator coil, which controls flow of refrigerant. Control bulb is attached to outlet of coil.

Thermostatic Motor Control: Device used to control cycling of unit through use of control bulb attached to evaporator.

Thermostatic Valve: Valve controlled by thermostatic elements.

Thermostatic Water Valve: Valve used to control flow of water through system, actuated by temperature difference. Used in units such as water-cooled compressor or condenser.

Throttling: Expansion of gas through orifice or controlled opening without gas performing any work in expansion process.

Timers: Mechanism used to control on and off times of an electrical circuit.

Timer-Thermostat: Thermostat control which includes a clock mechanism. Unit automatically controls room temperature and changes it according to time of day.

Ton of refrigeration: Refrigerating effect equal to the melting of one ton of ice in 24 hours. This may be expressed as follows:

288,000 Btu/24 hr

12,000 Btu/1 hr

200 Btu/min

Ton Refrigeration Unit: Unit which removes same amount of heat in 24 hours as melting of one ton of ice.

Torque: Turning or twisting force.

Torque Wrenches: Wrenches which may be used to measure torque or pressure applied to a nut or bolt.

Total Pressure: The sum of static pressure and velocity pressure at the point of measurement.

Toxicity: The number classification where concentrations of a vapor become harmful to humans.

Transducer: Device actuated by power from one system and supplies power to another form to second system.

Trichlorotrifluoroethane: Complete name of refrigerant R-113. Group 1 refrigerant in rather common use. Chemical compounds which make up this refrigerant are chlorine, fluorine, and ethane.

Triple Point: Pressure temperature condition in which a substance is in equilibrium in solid, liquid and vapor states.

Trombone: The assembly portion of a heat pump refrigeration system consisting of the pressure-reducing device and check valve connected in parallel.

Truck, Refrigerated: Commercial vehicle equipped to maintain below atmospheric temperatures.

Tube, Constricted: Tubing that is reduced in diameter.

Tube-Within-A-Tube: A water cooled condensing unit in which a small tube is placed inside large unit. Refrigerant passes through one tube; water through the other.

Tubing: Fluid carrying pipe which has a thin wall.

Turbulent Flow: The movement of a liquid or vapor in a pipe in a constantly churning and mixing fashion.

Two-Temperature Valve: Pressure opened valve used in suction line on multiple refrigerator installations which maintains evaporators in system at different temperatures.

U

"U" Factor: The amount of heat energy in Btu/h that will be absorbed by one square foot of surface for each degree of mean temperature difference through the surface material.

Ultraviolet: Invisible radiation waves with frequencies shorter than wave lengths of visible light and longer than X-Ray.

Universal Motor: Electric motor which will operate on both ac and dc.

Urethane Foam: Type of insulation which is foamed in between inner and outer walls of display case.

V

Vacuum: Reduction in pressure below atmospheric pressure.

Vacuum Pump: Special high efficiency compressor used for creating high vacuums for testing or drying purposes.

Valve: Device used for controlling fluid flow.

Valve-Check: A device used to bypass the refrigerant flow around or force the refrigerant flow through the pressure-reducing device on each coil depending upon the flow direction of the refrigerant.

Valve, Expansion: Type of refrigerant control which maintains pressure difference between high side and low side pressure in refrigerating mechanism. Valve is caused to operate by pressure in low or suction side. Often referred to as an automatic expansion valve or AEV.

Valve Plate: Part of compressor located between top of compressor body and head which contains compressor valves.

Valve-Reversing: The control used to regulate the flow of hot high pressure vapor from the compressor and the cool low pressure vapor to the compressor from the coils depending upon the direction of heat flow desired.

Valve, Service: Device used by service technicians to check pressures and charge refrigerating units.

Valve, Solenoid: Valve actuated by magnetic action by means of an electrically energized coil.

Valve, Suction: Valve in refrigeration compressor which allows vaporized refrigerant to enter cylinder from suction line and prevents its return.

Valves, Water: Most water cooling units are supplied with water valves. These valves provide a flow of water to cool the system while it is running. Most water valves are controlled by solenoids.

Vapor: Word usually used to denote vaporized refrigerant rather than the word gas.

Vapor Barrier: Thin plastic or metal foil sheet used in air-conditioned structures to prevent water vapor from penetrating insulating material.

Vapor Charged: Lines and component parts of system which are charged at the factory.

Vapor Line: The large line from the inside coil to the outside portion of the heat pump is dual purpose—suction line on cooling and hot gas line on heating. Only a high pressure gauge must be used to measure pressure in this line.

Vapor Lock: Condition where liquid is trapped in line because of bend or improper installation which prevents the vapor from flowing.

Vapor Pressure: Pressure imposed by either a vapor or gas.

Vapor Pressure Curve: Graphic presentation of various pressures produced by refrigerant under various temperatures.

Vapor, Saturated: A vapor condition which will result in condensation into droplets of liquid as vapor temperature is reduced.

Variable Pitch Pulley: Pulley which can be adjusted to provide different pulley ratios.

V-Belt: Type of belt that is commonly used in refrigeration work. It has a contact surface which is in the shape of letter V.

V-Block: V-shaped groove in metal block used to hold shaft.

Velocimeter: Instrument used to measure air velocities using a direct reading air speed indicating dial.

Velocity: A vector quantity which denotes at once the time rate and the direction of a motion. $V = ds/dt$. For uniform linear motion $V = s/t$. Common units are feet per second or feet per minute.

Velocity Pressure: The pressure exerted in direction of flow.

Viscosity: Term used to describe resistance of flow of fluids.

Volatile Liquid: Liquid which evaporates at low temperature and pressure.

Voltage Control: It is necessary to provide some electrical circuits with uniform or constant voltage. Electronic devices used for this purpose are called voltage controls.

Voltmeter: Instrument for measuring voltage action in electrical circuit.

Volume (Specific): The volume of a substance per unit mass; the reciprocal of density. Units: cubic feet per pound, cubic centimeters per gram, etc.

Volumetric Efficiency: Term used to express the relationship between the actual performance of a compressor or of a vacuum pump and calculated performance of the pump based on its displacement versus its actual pumping ability.

Vortex Tube: Mechanism for cooling or refrigerating which accomplishes cooling effect by releasing compressed air through specially designed opening. Air expands in rapidly spiraling column of air which separates slow moving molecules (cool) from fast moving molecules (hot).

Vortex Tube Refrigeration: Refrigerating or cooling devices using principle of vortex tube, as in mining suits.

W

Walk-In Cooler: Large commercial refrigerated space kept below room temperature. Often found in large supermarkets or wholesale meat distribution centers.

Water-Cooled Condenser: Condensing unit which is cooled through use of water.

Water Defrosting: Use of water to melt ice and frost from evaporator during off-cycle.

Wax: Ingredient in many lubricating oils which may separate out if cooled sufficiently.

Wet Bulb: Device used in measurement of relative humidity. Evaporation of moisture lowers temperature of wet bulb compared to dry bulb temperature in same area.

Wet Cell Battery: Cell or connected group of cells that converts chemical energy into electrical energy by reversible chemical reactions.

Window Unit: Commonly used when referring to air conditioners which are placed in a window. Normally a domestic application.

Glossary of Technical Terms As Applied to Heating

A

Air Binding or Air Bound: A condition in which a bubble or other pocket of air is present in a pipeline or item of equipment and, by its presence, prevents or reduces the desired flow or movement of the liquid or gas in the pipeline or equipment.

Air Cushion Tank: A closed tank, generally located above the boiler and connected to a hydronic system in such a manner that when the system is initially filled with water, air is trapped within the tank. When the water in the system is heated it expands and compresses the air trapped within the air cushion tank, thus providing space for the extra volume of water without creating excessive pressure. Also called expansion tank.

Air-Gas Ratio: The ratio of combustion air supply flow rate to the fuel gas supply flow rate.

Air Shutter: An adjustable shutter on the primary air openings of a burner, which is used to control the amount of combustion air introduced into the burner body.

Air Vent: A valve installed at the high points in a hot water system to permit the elimination of air from the system.

Aldehyde: A class of compounds, which can be produced during incomplete combustion of a fuel gas. They have a pungent distinct odor.

Ambient Temperature: The temperature of the air in the area of study or consideration.

Available Head: The difference in pressure which can be used to circulate water in the system. The difference in pressure which may be used to overcome friction within the system. (See Pump Head, Head)

Atmospheric Burner: (See Burner)

Atmospheric Pressure: The pressure exerted upon the earth's surface by the weight of atmosphere above it.

Atom: The smallest unit of an element which retains the particular properties of that element.

Automatic Gas Pilot Device: A gas pilot incorporating a device, which acts to automatically shut off the gas supply to the appliance burner if the pilot flame is extinguished.

B

Backfire Protection: (See Flashback Arrestor)

Baffle: A surface used for deflecting fluids, usually in the form of a plate or wall.

Balancing Fit: (See Balance Fitting)

Balance Fitting: A pipe fitting or valve designed so that its resistance to flow may be varied. These are used to balance the pressure drop in parallel circuits.

Balancing Valve: (See Balance Fitting)

Baseboard: A terminal unit resembling the base trim of a house. These units are the most popular terminal unit for residential systems.

Boiler, Heating: That part of a hydronic heating system in which heat is transferred from the fuel to the water. If steam is generated it is a steam boiler. If the temperature of the water is raised without boiling, it is classed as a hot-water boiler.

Boiler Horsepower: The equivalent evaporation of 34.5 lb of water per hr from and at 212°F. This is equal to a heat output of $970.3 \times 34.5 = 33,475$ Btu/h.

Bonnet: The part of the furnace casing which forms a plenum chamber from where supply ducts receive warmed air. Also called supply plenum.

Branch: That portion of the piping system which connects a terminal unit to the circuit.

Btu or British Thermal Unit: The quantity of heat required to change the temperature of one pound of water one degree Fahrenheit.

Bull Head: The installation of a pipe tee in such a way that water enters (or leaves) the tee at both ends of the run (the straight through section of the tee) and leaves (or enters) through the side connection only.

Bunsen-Type Burner: A gas burner in which combustion air is injected into the burner by the gas jet emerging from the gas orifice, and this air is premixed with the gas supply within the burner body before the gas burns on the burner port.

Burner: A device for the final conveyance of gas, or a mixture of gas and air, to the combustion zone. (See also specific type of burner)

(1) Injection Burner. A burner employing the energy of a jet of gas to inject air for combustion into the burner and mix it with gas.

(a) Atmospheric Injection Burner. A burner in which the air injected into the burner by a jet of gas is supplied to the burner at atmospheric pressure.

(2) Power Burner. (See also Forced Draft Burner, Induced Draft Burner, Premixing Burner, and Pressure Burner) A burner in which either gas or air or both are supplied at pressure exceeding, for gas, the line pressure, and for air, atmospheric pressure.

(3) Yellow-Flame Burner. A burner in which secondary air only is depended on for the combustion of the gas.

Burner Flexibility: The degree at which a burner can operate with reasonable characteristics with a variety of fuel gases and/or variations in input rate (gas pressure).

Burner Head: That portion of a burner beyond the outlet of the mixer tube which contains the burner ports.

Burner Port: (See Port)

Burning Speed: (See Flame Velocity)

Butane: A hydrocarbon fuel gas heavier than methane and propane and a major constituent of liquefied petroleum gases.

C

Cadmium Cell: A device that controls the current flow in an electrical circuit according to the amount of light falling upon the active face of the cell.

Calorimeter: Device for measuring heat quantities, such as machine capacity, heat of combustion, specific heat, vital heat, heat leakage, etc.; also device for measuring quality (or moisture content) of steam or other vapor.

Cfm: Cubic feet per minute.

Chap.: Chapter

Chimney Effect: The upward movement of warm air or gas, compared with the ambient air or gas, due to the lesser density of the warmed air or gas.

Circuit: The piping extending from the boiler supply tapping to the boiler return tapping.

Circuit Main: The portion of the main in a multiple circuit system that carries only a part of the total capacity of the system.

Circulator: A motor driven device used to mechanically circulate water in the system. Also called Pump.

Coke Trees: The accumulation of carbon on the walls of the refractory due to poor mixing of oil spray and air during the combustion process.

Colorimetric Detection Device: A device for detecting the presence of a particular substance, such as carbon monoxide, in which the presence of that substance will cause a color change in a material in the detector.

Combustion: The rapid oxidation of fuel gases accompanied by the production of heat or heat and light.

Combustion Air: Air supplied in an appliance specifically for the combustion of a fuel gas.

Combustion Chamber: The portion of an appliance within which combustion normally occurs.

Combustion Products: Constituents resulting from the combustion of a fuel gas with the oxygen in air, including the inerts, but excluding excess air.

Commercial Buildings: Such buildings as stores, shops, restaurants, motels, and large apartment buildings.

Compression Tank: (See Air Cushion Tank)

Compound: A distinct substance formed by the chemical combination of two or more elements in definite proportions.

Condensable: A gas which can be easily converted to liquid form, usually by lowering the temperature and/or increasing pressure.

Connected Load: The total load in Btu/h attached to the boiler. It is the sum of the outputs of all terminal units and all heat to be supplied by the boiler for process applications.

Controls: Devices designed to regulate the gas, air, water or electricity supplied to a gas appliance. They may be manual, semi-automatic or automatic.

Control Valves: Any valve used to control the flow of water in a hydronic system.

Convection: The movement of a fluid set up by a combination of differences in density and the force of gravity. For example, warm water at the bottom of a vertical tank will rise and displace cooler water at the top. The cooler water will sink to the bottom as the result of its greater density.

Convector: A terminal unit surrounded on all sides by an enclosure having an air outlet at the top or upper front. Convectors operate by gravity recirculated room air.

Converter: A heat exchange unit designed to transfer heat from one distributing system to another. These may be either steam to water or water to water units. They are usually of shell and tube design.

Counterflow: In heat exchange between two fluids, opposite direction of flow, coldest portion of one meeting coldest portion of the other.

Cubic Foot of Gas: (Standard Conditions) The amount of gas which will occupy 1 cubic foot when at a temperature of 60°F, and under a pressure equivalent to that of 30 in. of mercury.

D

Damper: A valve or plate which is installed in the cold and warm air ductwork and used to regulate the amount of air flowing through the duct. A damper may also be used in the flue of a furnace.

Dead Space: The short distance between a burner port and the base of a flame.

Degree Day: A unit used to estimate fuel consumption and to specify the heating load in winter, based on temperature difference and time. There are as many degree days for any one day as there are degrees F difference in temperature between the mean temperature for the day and 65°F.

Density: The weight of a substance per unit volume. As applied to gases, the weight in pounds of a cubic foot of gas at standard pressure and temperature.

Design Heat Loss: The heat loss of a building or room at design indoor-outdoor temperature difference.

Design Load: The design heat loss plus all other heating requirements to be provided by the boiler.

Design Temperature Difference: The difference between the design indoor and outdoor temperatures.

Design Water Temperature: The average of the temperature of the water entering and leaving the boiler (or sub-circuit) when the system is operating at design conditions.

Design Water Temperature Drop: The difference between the temperature of the water leaving the boiler and returning to the boiler when the system is operating at design conditions. In large systems employing sub-circuits the design temperature drop is usually taken as the difference in the temperature of the water entering and leaving each sub-circuit.

Dilution Air: Air which enters a draft hood and mixes with the flue gases.

Direct-Indirect Heating Unit: A heating unit located in the room or space to be heated and partially enclosed, the enclosed portion being used to heat air which enters from outside the room.

Direct Return: A two pipe system in which the first terminal unit taken off the supply main is the first unit connected to the return main.

Discharge Coefficient: The ratio of the actual flow rate of a gas from the orifice or port to the theoretical, calculated flow rate. Always less than 1.0.

Distillation: Removal of gaseous substances from solids or liquids by applying heat.

D.M.S.: Drill Manufacturer's Standard—equivalent to Standard Twist Drill or Steel Wire Gauge Numbers.

Domestic Hot Water: The heated water used for domestic or household purposes such as laundry, dishes, bathing, etc.

Double Heat Transfer: The transfer of heat from the plant to the heated medium (usually liquid) and from the liquid to the air in the conditioned space.

Down-feed One-pipe Riser (Steam): A pipe which carries steam downward to the heating units and into which condensate from the heating units drains.

Down Feed System: A hydronic system in which the main is located above the level of the terminal units.

Down-Feed System (Steam): A steam heating system in which the supply mains are above the level of the heating units which they serve.

Drain Cock: A valve installed in the lowest point of a boiler or at low points of a heating system to provide for complete drainage of water from the system.

Draft: A current of air, usually referring to the difference in pressure which causes air or gases to flow through a chimney flue, heating unit or space.

Draft Hood: (Draft Diverter) A device built into an appliance, or made part of a vent connector from an appliance, which is designed to: (1) assure the ready escape of the products of combustion in the event of no draft, backdraft, or stoppage beyond the draft hood; (2) prevent a backdraft from entering the appliance; and (3) neutralize the effect of stack action of a chimney or gas vent upon the operation of the appliance.

Downdraft: Excessive high pressure existing at the outlet of chimney or stack which tends to make gases flow downward in the stack.

Drilled Port Burner: A burner in which the ports have been formed by drilled holes in a thick section in the burner head or by a manufacturing method which results in holes similar in size, shape and depth.

Duct: Round or rectangular sheet metal pipes through which heat is carried from the furnace to the various rooms in the building.

E

Eccentric Reducer: A pipe fitting designed to change from one pipe size to another and to keep one edge of both pipes in line. These fittings should be installed so that the ''in line'' section of pipe is at the top.

Effective Heat Allowance: An allowance added to the test output of certain designs of radiation to compensate for a better distribution of heat within the heated space. Some agencies do not permit the use of effective heat allowance.

Electric Heating Element: A unit assembly consisting of a resistor, insulated supports, and terminals for connecting the resistor to electric power.

Element: One of the 96 or more basic substances of which all matter is composed.

Excess Air: Air which passes through an appliance and the appliance flues in excess of that which is required for complete combustion of the gas. Usually expressed as a percentage of the air required for complete combustion of the gas.

Expansion Tank: (See Air Cushion Tank)

Exposed Area: The area of any wall, window, ceiling, floor, or partition separating a heated room from the out-of-doors or from an unheated space.

Extinction Pop: (See Flashback)

F

Fahrenheit: The common scale of temperature measurement in the English system of units. It is based on the freezing point of water being 32°F and the boiling point of water being 212°F at standard pressure conditions.

Fan-Coil: A terminal unit consisting of a finned-tube coil and a fan in a single enclosure. These units may be designed for heating, cooling, or a combination of the two. Some fan-coil units are designed to receive duct work so that the unit may serve more than one room.

Ferrous: As used in this course, ferrous relates to objects made of iron or steel.

Filter: A porous material (fiberglass or foam plastic) which is installed in the air circulation system of a furnace to remove dust particles and pollen. Some are disposable, whereas some may be cleaned and re-used.

Fig: Figure.

Finned-Tube: A heat exchange device consisting of a metal tube through which water or steam may be circulated. Metal plates or fins are attached to the outside of the tube to increase the heat transfer surface. Finned tube or fin tube, may consist of one, two, or three tiers and are designed for installation bare, or with open type grilles, covers, or enclosures having top, front, or inclined outlets. Usually finned-tube units are for use in other than residential buildings.

Fire Tube Boiler: A steel boiler in which the hot gases from combustion are circulated through tubes which are surrounded by boiler water which fills the space between the boiler shell and the tubes.

Firing Device: The burner, either oil, gas, or coal.

Fixed Orifice: (See Orifice Spud)

Flame Arrestor: (See Flashback Arrestor)

Flame Retention Device: A device added to a burner which aids in holding the flame base close to the burner ports.

Flame Rollout: A condition where flame rolls out of a combustion chamber when the burner is turned on.

Flame Velocity: The speed at which a flame moves through a fuel-air mixture.

Flammability Limits: The maximum percentages of a fuel in an air-fuel mixture which will burn.

Flashback: The movement of the gas flame down through the burner port upon shutdown of the gas supply. Usually caused by excessive primary air.

Flashback Arrestor: A gauze, grid, or any other portion of a burner assembly used to avert flashback.

Flashtube: An ignition device, commonly used for igniting gas on range top burners. An air-gas mixture from the burner body is injected into the end of a short tube. The mixture moves along the tube, is ignited by a standing pilot flame at the other open end of the tube and the flame travels back through the mixture in the flashtube to ignite the gas at the burner ports.

Flash Boiler: A boiler with very limited water capacity. Usually about one gal. of water per 1000 Btu/h net rating.

Floating Flames: An undesirable burner operating condition, usually indicating incomplete combustion in which flames leave the burner ports to "reach" for combustion air.

Flow Control Valve: A specially designed check valve, usually installed in the supply pipe, to prevent gravity circulation of hot water within the heating system when the pump is not in operation.

Flue: An enclosed passage in the chimney to carry exhaust smoke and fumes of the heating plant to escape to the outer air.

Flue Gases, Flue Products: Products of combustion and excess air in appliance flues or heat exchangers before the draft hood.

Flue Loss: The heat lost in flue products exiting from the flue outlet of an appliance.

Flue Outlet: The opening provided in an appliance for the escape of flue gases.

Fluid: A gas or liquid, as opposed to a solid.

Foot of Water: A measure of pressure. One foot of water is the pressure created by a column of water one foot in height. It is equivalent ot 0.433 lb/in.^{2}.

Forced Hot Water: Or forced circulation hot water. Hot water heating systems in which a pump is used to create the necessary flow of water.

Forced Draft Burner: A burner in which combustion air is supplied by a fan or blower.

Friction Head: In a hydronic system the friction head is the loss in pressure resulting from the flow of water in the piping system.

ft: Foot or feet.

Fuel: Any substance used for combustion.

Fuel Gas: Any substance in gaseous form when used for combustion.

Fuel-oil Burner, Pressure Atomizing or Gun Type: A burner designed to atomize the oil for combustion under an oil supply pressure of 100 psig.

Fuel-oil Burner, Rotary Type: A burner employing a thrower ring that mixes the oil and the air.

Fuel-oil Burner, Vaporizing or Pot Type: These burners use the heat of combustion to vaporize the oil in a pool beneath the vaporizing ring, and this vapor rising through the ring ignites and maintains combustion in the burner.

Fuel Unit: A gear-type pump used to supply oil to the nozzle at the proper operating pressure. A second function is to supply instant cutoff of the oil supply at shutdown.

Furnace: That part of a warm air heating system in which combustion takes place.

G

Gal: Gallon or gallons.

Gate Valve: A valve designed in such a way that the opening for flow, when the valve is fully open, is essentially the same as the pipe and the direction of flow through the valve is in a straight line.

gpm: The abbreviation for "gallons per minute" which is a measure of rate of flow.

Grate Area: Grate surface area measured in square feet, used in estimating the fuel burning rate.

Gravity Warm Air Heating System: (See Warm Air Heating System)

Gravity Hot Water: Hot water heating systems in which the circulation of water through the system is due to the difference in the density of the water in the supply and return sides of the system.

Gross Output: A rating applied to boilers. It is the total quantity of heat which the boiler will deliver and at the same time meet all limitations of applicable testing and rating codes.

H

Hard Flame: A flame with a hot, tight, well-defined inner cone.

Head: As used in this course, head refers to a pressure difference. See pressure head, pump head, available head.

Header: A piping arrangement for inter-connecting two or more supply or return tappings of a boiler. Also a section of pipe, usually short in length, to which a number of branch circuits are attached.

Heat: A form of energy.

Heat Distributing Units: (See Terminal Units)

Heating Element: (See Terminal Unit)

Heat Flow: (See Heat Loss)

Heating Effect Factor: An arbitrary allowance added to the test output of some types of terminal units when establishing the catalog ratings. This allowance is intended to give credit for improved heat distribution obtained from the terminal unit.

Heat, Latent: The heat which changes the form of a substance without changing its temperature.

Heat, Sensible: Heat which changes the temperature of a substance without changing its form.

Heat Exchanger: Any device for transferring heat from one fluid to another.

Heat Loss: As used in this course, the term applies to the rate of heat transfer from a heated building to the outdoors.

Heat Loss Factor: A number assigned to a material or construction indicating the rate of heat transmission through that material or construction for a one degree temperature difference.

Heating Element (Electric): A unit assembly consisting of a resistor, insulated supports, and terminals for connecting the resistor to electric power.

Heat Transmission: Any time-rate of heat flow; usually refers to conduction, convection, and radiation combined.

Heat Transmission Coefficient: Any one of a number of coefficients used in calculating heat transmission through different materials and structures, by conduction, convection, and radiation.

Heating Surface: All surfaces which transmit heat from flames or flue gases to the medium being heated.

Heating Unit (Electric): A structure containing one or more heating elements, electrical terminals or leads, electric insulation, and a frame or casing, all assembled together in one unit.

Heating Value: The number of British thermal units produced by the complete combustion at constant pressure of one cubic foot of gas. Total heating value includes heat obtained from cooling the products to the initial temperature of the gas and air and condensing the water vapor formed during combustion.

High Limit Control: A switch controlled by the temperature of the water in the boiler and used to limit burner operation whenever the boiler water temperature reaches the maximum to be permitted. A safety control.

High-Temperature Water System (HTW): A hot water system operating at temperatures over 350°F and usual pressures of about 300 psi.

High Voltage Controls: Also called "line voltage controls." Controls designed to operate at normal line voltage, usually 115 V.

Hot Water Heating Systems: Hydronic systems in which heated water is circulated through the terminal units.

Humidistat: An instrument that is used to regulate the operation of a humidifier to control the amount of humidity in the conditioned air.

Humidity, Absolute: The amount of moisture actually in a given unit volume of air.

Humidity, Relative: A ratio of the weight of moisture that air actually contains at a certain temperature as compared to the amount that it could contain if it were saturated.

Hydrocarbon: Any of a number of compounds composed of carbon and hydrogen.

Hydronics: Pertaining to heating or cooling with water or vapor.

I

Ignition: The act of starting combustion.

Ignition Temperature: The minimum temperature at which combustion can be started.

Ignition Velocity: (See Flame Velocity)

Impingement Target Burner: A burner consisting simply of a gas orifice and a target, with the gas jet from the orifice entraining combustion air in the open and the mixture striking and burning on the target surface. No usual burner body is used.

Inches of Mercury Column: A unit used in measuring pressures. One inch of mercury column equals a pressure of 0.491 $lb/in.^2$.

Inches of Water Column: A unit used in measuring pressures. One inch of water column equals a pressure of 0.578 $oz/in.^2$. One inch of mercury column equals about 13.6 in. water column.

Incomplete Combustion: Combustion in which the fuel is only partially burned.

Indirect Water Heater: A coil or bundle of tubes, usually copper, surrounded by hot boiler water. The domestic water is within the tube and is heated by transfer of heat from the hot boiler water surrounding the tube.

Indoor Design Temperature: The indoor air temperature used when calculating the design heat loss. The indoor design temperature is usually assumed to be 70°F.

Indoor-Outdoor Temperature Difference: The temperature of the indoor air minus the temperature of the outdoor air.

Industrial Buildings: Such buildings as small manufacturing plants, garage, and storehouses.

Induced Draft Burner: A burner which depends on draft induced by a fan or blower at the flue outlet to draw in combustion air and vent flue gases.

Inerts: Non-combustible substances in a fuel, or in flue gases, such as nitrogen or carbon dioxide.

Infiltration: Air leakage into a building from the out-of-doors as a result of wind and indoor-outdoor temperature difference.

Infrared Burner: (Radiant Burner). A burner which is designed to operate with a hot, glowing surface. A substantial amount of its energy output is in the form of infrared radiant energy.

Injection: Drawing primary air into a gas burner by means of a flow of fuel gas.

Input Rate: The quantity of heat or fuel supplied to an appliance, expressed in volume or heat units per unit time, such as cubic feet per hour or Btu per hour.

Input Rating: The gas-burning capacity of an appliance in Btu per hour as specified by the manufacturer. Appliance input ratings are based on sea level operation up to 2,000 feet elevation. For operation at elevations above 2,000 ft, input ratings should be reduced at the rate of 4 percent for each 1,000 ft above sea level.

Instantaneous Water Heater: (See Tankless Water Heater)

J

Joint, Expansion, Bellows: An item of equipment used to compensate for the expansion and contraction of a run of pipe. The device is built with a flexible bellows that stretches or is compressed as necessary to accept the movement of the piping.

Joint, Expansion, Slip: A joint in which the provision for expansion and contraction consists of a cylinder that moves in and out of the main body of the device.

Jet Burner: A burner in which streams of gas or air-gas mixtures collide in air at some point above the burner ports and burn there.

L

Lean Mixture: An air-gas mixture which contains more air than the amount needed for complete combustion of the gas.

Lifting Flames: An unstable burner flame condition in which flames lift or blow off the burner port(s).

Liquefied Petroleum Gases: The terms "Liquefied Petroleum Gases," "LPG," and "LP Gas" mean and include any fuel gas which is composed predominantly of any of the following hydrocarbons, or mixtures of them: propane, propylene, normal butane or isobutane and butylenes.

LNG: Liquefied natural gas. Natural gas which has been cooled until it becomes a liquid.

Low Link Control: A switch operated by the temperature of the water in the boiler and used to start the burner at any time the water temperature drops to some prescribed minimum. This control is used if the boiler is supplying domestic hot water as well as heat for the building.

Low-Temperature Water System (LTW): A hot water heating system operating at design water temperatures of 250°F or less and a maximum working pressure of 160 psi.

Low Voltage Control: Controls designed to operate at voltages of 20 to 30 V.

LP Gas-Air Mixtures: Liquefied petroleum gases distributed at relatively low pressures and normal atmospheric temperatures which have been diluted with air to produce desired heating value and utilization characteristics.

Luminous Flame Burner: (See Burner, Yellow Flame)

M

Main: The pipe used to carry water between the boiler and the branches of the terminal units.

Make-up Air: The air which is supplied to a building to replace air that has been removed by an exhaust system.

Make Up Water Line: The water connection to the boiler or system for filling or adding water when necessary.

Manifold: The conduit of an appliance which supplies gas to the individual burners.

Manifold Pressure: The gas pressure in an appliance manifold, upstream of burner orifices.

Manufactured Gas: A fuel gas which is artificially produced by some process, as opposed to natural gas, which is found in the earth. Sometimes called town gas.

Medium, Heating: A substance used to convey heat from the heat source to the point of use. It is usually air, water, or steam.

Medium-Temperature Water System: A hot water system operating at temperatures of 350°F or less, with pressures not exceeding 150 psi.

Methane: A hydrocarbon gas with the formula CH_4, the principal component of natural gases.

Mixed Gas: A gas in which the heating value of manufactured gas is raised by mixing with natural or LPG (except where natural gas or LPG is used only for "enriching" or "reforming").

Mixer: That portion of a burner where air and gas are mixed before delivery to the burner ports.

1. Mixer Face. The air inlet end of the mixer head.
2. Mixer Head. That portion of an injection type burner, usually enlarged, into which primary air flows to mix with the gas stream.
3. Mixer Throat (Venturi throat). That portion of the mixer which has the smallest cross-sectional area, and which lies between the mixer head and the mixer tube.
4. Mixer Tube. That portion of the mixer which lies between the throat and the burner head.

Molecule: The smallest portion of an element or compound which retains the identity and characteristics of the element or compound.

Multiple Circuit: A system in which the main, or mains, form two or more parallel loops between the boiler supply and the boiler return.

Multiple Zone: A system controlled by two or more thermostats.

N

Natural Draft: The motion of flue products through an appliance generated by hot flue gases rising in a vent connected to the furnace flue outlet.

Natural Gas: Any gas found in the earth, as opposed to gases which are manufactured.

Needle, Adjustable: A tapered projection, coaxial with and movable with respect to a fixed orifice, used to regulate the flow of gas.

Needle, Fixed: A tapered projection, the position of which is fixed, coaxial with an orifice which can be moved with respect to the needle to regulate flow of gas.

Net Rating: A rating applied to boilers. It is the quantity of heat available in Btu/h for the connected load.

Neutral Point: The position in the heat exchanger where the positive pressure produced by combustion of the gas/air mixture and the negative pressure produced by the draft on the outlet of the heat exchanger balance out. The desirable position of the "Neutral Point" is at the tips of the flame on the burner.

Non-Ferrous: Metals other than iron or steel. In heating systems the principal non-ferrous metals are copper and aluminium.

Nozzle: The device on the end of the oil fuel pipe used to form the oil into fine droplets by forcing the oil through a small hole to cause the oil to breakup. The oil is also forced into definite swirl patterns to mix with air for complete combustion.

O

Odorant: A substance added to an otherwise odorless, colorless and tasteless gas to give warning of gas leakage and to aid in leak detection.

Oil Burner Relay: A special, multi-purpose control used with oil burners. The device controls the operation of the oil burner and also acts as a safety to prevent operation in the event of malfunction.

One-Pipe Fitting: A specially designed tee for use in a one-pipe system to connect the supply or return branch into a circuit. These fittings cause a portion of the water flowing through the circuit to pass through the terminal unit.

One-Pipe System: A forced hot-water system using one continuous pipe or main from the boiler supply to the boiler return. The terminal units are connected to this pipe by two smaller pipes known as supply and return branches.

Orifice: An opening in an orifice cap (hood), orifice spud or other device through which gas is discharged, and whereby the flow of gas is limited and/or controlled. (See also Universal Orifice)

Orifice Cap (Hood): A movable fitting having an orifice which permits adjustment of the flow of gas by changing its position with respect to a fixed needle or other device extending into the orifice.

Orifice Discharge Coefficient: (See Discharge Coefficient)

Orifice Spud: A removable plug or cap containing an orifice which permits adjustment of the gas flow either by substitution with a spud having different sized orifices (fixed orifice) or by motion of an adjustable needle into or out of the orifice (adjustable orifice).

Outdoor Design Temperature: The outdoor temperature on which design heat losses are based.

Overrating: Operation of a gas burner at a greater rate than it was designed for.

Oxygen: An elemental gas that comprises approximately 21 percent of the atmosphere by volume. Oxygen is one of the elements required for combustion.

P

Packaged Boiler: A boiler having all components, including burner, boiler, controls, and auxiliary equipment, assembled as a unit.

Panel Heating: A heating system in which heat is transmitted by both radiation and convection from panel surfaces to both air and surrounding surfaces.

Panel Radiator: A heating unit placed on or flush with a flat wall surface, and intended to function essentially as a radiator.

Panel Systems: Or radiant system. A heating system in which the ceiling or floor serves as the terminal unit.

Peak Load: The maximum load carried by a system or a unit of equipment over a designated period of time.

pH or pH Value: A term based on the hydrogen ion concentration in water, which denotes whether the water is acid, alkaline, or neutral. A pH value of 8 or more indicates a condition of alkalinity: of 6 or less, acidity. A pH of 7 means the water is neutral.

Pilot: A small flame which is used to ignite the gas at the main burner.

Pilot-Intermittent: A gas ignition pilot that cycles with the main burner.

Pilot-Standing: A gas ignition pilot that is continuously operating.

Pilot Switch: A control used in conjunction with gas burners. Its function is to prevent operation of the burner in the event of pilot failure.

Piping and Pick-up Allowance: That portion of the gross boiler output that is allowed for warming up the heating system and for taking care of the heat emission from a normal amount of piping.

Pitch: The amount of slope given to a horizontal pipe when it is installed in a heating system.

Plenum Chamber: An air compartment maintained under pressure, and connected to one or more distributing ducts.

Pressure Head: The force available to cause circulation of water or vapor in a hydronic system. (See Head, Pump Head, Available Head)

Port: Any opening in a burner head through which gas or an air-gas mixture is discharged for ignition.

Port Loading: The input rate of a gas burner per unit of port area, obtained by dividing input rate by total port area. Usually expressed in terms of Btu per hour per square inch of port area.

Power Burner: (See Burner)

Premixing Burner: A burner in which all, or nearly all, combustion air is mixed with the gas as primary air.

Pressure Burner: A burner in which an air and gas mixture under pressure is supplied, usually at 0.5 to 14 in. water column.

Pressure Regulator: A device for controlling and maintaining a uniform outlet gas pressure.

Pressure Reducing Valve: A diaphragm operated valve installed in the make-up water line of a hot water heating system to introduce water into the system and to prevent the system from possible exposure to city water pressures higher than the working pressure of the boiler.

Pressure Relief Valve: A device for protecting a hot water boiler (or a hot water storage tank) from excessive pressure by opening at a pre-determined pressure and discharging water, or steam, at a rate sufficient to prevent further build-up of pressure.

Primary Air: The combustion air introduced into a burner which mixes with the gas before it reaches the port. Usually expressed as a percentage of air required for complete combustion of the gas.

Primary Air Inlet: The opening or openings through which primary air is admitted into a burner.

Propane: A hydrocarbon gas heavier than methane but lighter than butane. It is used as a fuel gas alone, mixed with air or as a major constituent of liquefied petroleum gases.

psig: Pounds per square inch gauge pressure.

Pulsation: The heavy combustion vibrations in an oil-fired unit at startup due to poor or nonexistent draft when the unit starts.

Pump: A motor driven device used to mechanically circulate water in the system. Also called a circulator.

Pump Head: The difference in pressure on the supply and intake sides of the pump created by the operation of the pump.

Q

Quenching: A reduction in temperature whereby a combustion process is retarded or stopped.

R

Radiant Burner: (See Infrared Burner)

Radiant Heating: A heating system in which only the heat radiated from panels is effective in providing the heating requirements. The term radiant heating is frequently used to include both panel and radiant heating.

Radiation: The transmission of energy by means of electromagnetic waves.

Radiator: A heating unit exposed to view within the room or space to be heated. A radiator transfers heat by radiation to objects within visible range, and by conduction to the surrounding air which in turn is circulated by natural convection; a so-called radiator is also a convector, but the term radiator has been established by long usage.

Radiator (Concealed): A heating device located within, adjacent to, or exterior to the room being heated, but so covered or enclosed or concealed that the heat transfer surface of the device, which may be either a radiator or a convector, is not visible from the room. Such a device transfers its heat to the room largely by convection air currents.

Radiator Valve: A valve installed on a terminal unit to manually control the flow of water through the unit.

Rate: (See Input)

Recirculated Air: Return air passed through the conditioner before being again supplied to the conditioned space.

Reducing Fitting: A pipe fitting designed to change from one pipe size to another.

Refractory: A device made of high insulating type of noncombustible material needed to supply additional heat into the combustion area for complete burning of the oil and air mixture.

Regulator: (See Pressure Regulator)

Relay: An electrically operated switch. Usually the control circuit of the switch uses low voltage while the switch makes and breaks a line voltage circuit. However, both the control and load circuits are of the same voltage in some instances.

Relief Opening: The opening in a draft hood to permit ready escape to the atmosphere of flue products from the draft hood in event of no draft, back draft or stoppage beyond the draft hood, and to permit inspiration of air into the draft hood in the event of a strong chimney updraft.

Residential Buildings: Single family homes, duplexes, apartment buildings.

Resonance: The pipe organ effect produced by a gas furnace when the frequency of the burner flame combustion and the pressure wave distance in the burner pouch are in exact synchronization.

Return Branch: The piping used to return water from a terminal unit to the main, circuit main, or trunk.

Return Main: The pipe used to carry water from the return branches of the terminal units to the boiler.

Return Mains: Pipes or conduits which return the heating or cooling medium from the heat transfer unit to the source of heat or refrigeration.

Return Piping: That portion of the piping system that carries water from the terminal units back to the boiler.

Return Tapping: The opening in a boiler into which the pipe used for returning condensate or water to the boiler is connected.

Reverse Acting Control: A switch controlled by temperature and designed to open on temperature drop and close on temperature rise.

Reverse Return: A two-pipe system in which the return connections from the terminal units into the return main are made in the reverse order from that in which the supply connections are made in the supply main.

Rich Mixture: A mixture of gas and air containing too much fuel or too little air for complete combustion of the gas.

Riser: This generally refers to the vertical portion of the supply or return branches. However, any vertical piping in the heating system might be termed a riser.

Run-Out: This term generally applies to the horizontal portion of branch circuits.

S

Safety Valve: A device for protecting a steam boiler from excessive pressure by opening at a predetermined pressure setting and allowing steam to escape at a rate equal to or greater than the steam generating capacity of the boiler.

Secondary Air: Combustion air externally supplied to a burner flame at the point of combustion.

Sequencer: A time-delay-type relay, usually operated by a bimetallic-type motor.

Series Loop: A forced hot water heating system with the terminal units connected so that all the water flowing through the circuit passes through each series-connected unit in the circuit.

Single Circuit System: A hydronic system composed of only one circuit.

Single Port Burner: A burner in which the entire air-gas mixture issues from a single port.

Soft Flame: A flame partially deprived of primary air such that the combustion zone is extended and inner cone is ill-defined.

Soot: A black substance, mostly consisting of small particles of carbon, which can result from incomplete combustion and appear as smoke.

SNG: Supplementary natural gas. Gases which are manufactured to duplicate natural gas.

Spud: (See Orifice)

Square Foot (Steam): A term used to express the output of boilers and radiation. When applied to boilers, it is 240 Btu/h; when applied to terminal units, it represents the amount of radiation which will emit 240 Btu/h when supplied with steam at 215°F and air at 65°F.

Square Head Cock: A type of valve often used as a balancing valve. In place of the valve handle, the stem is made square. A wrench is used to adjust the valve setting.

Specific Gravity: Specific gravity is the ratio of the weight of a given volume of gas to that of the same volume of air, both measured at the same temperature and pressure.

Spoiler Screw: (Breaker Bolt) A screw or bolt moved in or out of the gas jet in a burner to control primary air injection.

Standard Conditions: Pressure and temperature conditions selected for expressing properties of gases on a common basis. In gas appliance work, these are normally 30 in. of mercury and 60°F.

Static Pressure: The normal force per unit area at a small hole in a wall of the pipe through which the fluid (water) flows.

Steam Heating System: A hydronic system in which steam is circulated through the terminal units.

Sub-circuits: A term applied to circuits taken off of the primary distribution loop of a complex hydronic system.

Supply Branch: The piping used to supply heated water from a main, circuit main, or trunk to the terminal unit.

Supply Main: The pipe used to distribute water from the boiler to the supply branches of the terminal units.

Supply Piping: That portion of the piping system that carries water from the boiler to the terminal units or to the point of use.

Supply Tapping: The opening in a boiler into which the supply main is connected.

System Temperature: The average of the temperatures of the water leaving the boiler and returning to the boiler.

T

Tankless Water Heater: An indirect water heater designed to operate without a hot water storage tank in the system. Also called an instantaneous heater.

Tee: A pipe fitting designed to connect three sections of pipe together. Two of the connections are in line, the third is at right angles to the other two.

Terminal Units: That part of a hydronic system in which heat is transferred from the water to the air in the air-conditioned space. Common terminal units include radiators, convectors, baseboard, unit heaters, finned tube, etc.

Thermal Conductivity: A term indicating the ability of a material to transmit heat. Thermal conductivity is the reciprocal of thermal resistance.

Thermal Head: The head produced by the difference in weight of the heated water in the supply side of the system and the cooler water in the return side. This is the only head available to cause circulation of water in a gravity system.

Thermal Radiation: The transmission of heat from a hot surface to a cooler one in the form of invisible electromagnetic waves, which on being absorbed by the cooler surface, raise the temperature of that surface.

Thermal Resistance: The resistance a material offers to the transmission of heat. Insulating materials have high thermal resistance. Materials such as metals have low thermal resistance.

Therm: A unit of heat having a value of 100,000 Btu.

Thermostat: A control (switch) which is operated by the temperature of the air.

Throat: (See Venturi)

Tie Rod: The sections of cast-iron sectional boilers are held in tight contact by means of tie rods that pass entirely through the sections.

Transformer: A device designed to change voltage. In heating controls the transformer usually converts line voltage (115 V) to low voltage (24 V).

Trunk: Or trunk main. The section of the main in a multiple circuit system that carries the combined capacity of two or more of the circuits.

Two-Pipe System: A hot-water heating system using one pipe from the boiler to supply heated water to the terminal units, and a second pipe to return the water from the terminal units back to the boiler.

Total Air: The total amount of air supplied to a burner. It is the sum of primary, secondary, and excess air.

Total Pressure: Also called impact pressure. The pressure measured in a moving fluid by an impact tube. It is the sum of the velocity pressure and the static pressure.

Town Gas: (See Manufactured Gas)

Turndown: The ratio of maximum to minimum input rates.

U

Ultimate CO_2: The percentage of carbon dioxide in dry combustion products when a fuel (gas) is completely burned with exactly the amount of air needed for complete combustion. This is the theoretical maximum CO_2 which can be obtained for a given gas in burning the gas in air.

Unit Heater: Also see fan coil. The term applies to a terminal unit designed to heat a given space. It consists of a fan and motor, a heating element, and an enclosure.

Universal Orifice: A combination fixed and adjustable orifice designed for the use of two different gases, such as LPG and natural gas.

Updraft: Excessively low air pressure existing at the outlet of a chimney or stack which tends to increase the velocity and volume of gases passing up the stack.

Up-Feed System: A hydronic system in which the supply main is located below the level of the terminal units.

Unit Ventilator: A terminal unit in which a fan is used to mechanically circulate air over the heating coil. These units are so constructed that both outdoor and room air may be circulated so as to provide ventilation as well as heat. These units may contain a cooling coil for summer operation.

Utility Gases: Natural gas, manufactured gas, liquefied petroleum gas-air mixtures or mixtures of any of these gases.

V

Vapor: The gaseous form of a substance that, under other conditions of pressure, temperature, or both, is a solid or a liquid.

Vapor Barrier: A material that is impervious to the passage of water vapor through it.

Valve-Oil-Delayed Actions: A device between the fuel unit outlet and the nozzle used to delay the delivery of the oil to the nozzle until complete combustion air pattern has been established. Also provides instant cutoff on shutdown.

Velocity Pressure: Pressure exerted by a flowing gas by virtue of its movement in the direction of its motion. It is the difference between total pressure and static pressure.

Vent: A device, such as a pipe, to transmit flue products from an appliance to the outdoors. This term also is used to designate a small hole or opening for the escape of a fluid (such as in a gas control).

Vents: (See Air Vents)

Ventilation: The introduction of outdoor air into a building by mechanical means.

Vent Gases: Products of combustion from gas appliances plus excess air, plus dilution air in the venting system above a draft hood.

Venturi: A section in a pipe or a burner body that narrows down and then flares out again.

Viscosity: The property of a fluid to resist flow.

W

Water Column: Abbreviated as W.C. A unit used for expressing pressure. One inch water column equals a pressure of 0.578 oz/in.^{2}.

Water Tube Boiler: A steel, hot-water boiler in which the water is circulated through the tubes and the hot gases from combustion of the fuel are circulated around the tubes inside the shell.

Y

Yellow Flame Burner: (See Burner)

Yellow Tips: (Yellow Tipping) The appearance of yellow tips in an otherwise blue flame, indicating the need for additional primary air.

Z

Zone: That portion of a hydronic system, the operation of which is controlled by a single thermostat.

Zoned System: A hydronic system in which more than one thermostat is used. This permits independent control of room air temperature at more than one location.

Zone Valve: A valve, the operation of which is controlled by a thermostat. They are used in hydronic systems to control the flow of water in localized parts of the system, thus making it possible to independently control the temperature in different zones, or areas, of the building.

Glossary of Technical Terms As Applied to Electricity and Electrical Data

A

Alternating Current: Abbreviated ac. Current that reverses polarity or direction periodically. It rises from zero to maximum strength and returns to zero in one direction then goes to similar variation in the opposite direction. This is a cycle which is repeated at a fixed frequency. It can be single phase, two phase, three phase, and poly phase. Its advantage over direct or undirectional current is that its voltage can be stepped up by transformers to the high values which reduce transmission costs.

Alternator: A machine which converts mechanical energy into alternating current.

American Wire Gauge: Abbreviated AWG. A system of numbers which designate cross-sectional area of wire. As the diameter gets smaller, the number gets larger, e.g., AWG # 14 = 0.0641 in. AWG #12 = 0.0808.

Ammeter: An instrument for measuring the quality of electron flow in amperes.

Ampere: Abbreviated amp. A basic unit which designates the amount of electricity passing a certain point at a specific time.

Ampere-Turn: Abbreviated AT or NI. Unit of magnetizing force produced by a current flow of one ampere through one turn of wire in a coil.

Amplitude: The maximum instantaneous value of alternating current or voltage. It can be in either a positive or negative direction.

Angle of Lag or Lead: Phase angle difference between two sinusoidal wave forms having the same frequency.

Anticipator: A heater used to adjust thermostat operation to produce a closer temperature differential than the mechanical capability of the control.

Armature: The moving or rotating component of a motor, generator, relay or other electromagnetic device.

Atom: The smallest particle of matter which exhibits the properties of an element.

Atomic Weight: The number of protons in an atom of a material is classified as its *atomic weight*.

B

Battery: Two or more primary or secondary electrically interconnected cells.

Break: Electrical discontinuity in the circuit generally resulting from the operation of a switch or circuit breaker.

Breaker Points: Metal contacts that open and close a circuit at timed intervals.

Bridge Circuit: A circuit for determining an unknown value or resistance. A power source and three known values or resistors are interconnected in a series-parallel network along with the unknown value or resistor. If all resistors are equal, a galvanometer bridged across the parallel legs and connected between each set of resistors will give no reading. If the unknown resistor is lesser or greater than the three known resistors, the galvanometer will indicate the degree and direction of the imbalance.

Brush: A conducting material, usually of carbon or graphite, which makes continuous contact with a moving commutator through which current can flow for the remainder of the circuit.

Bus Bar: A primary power distribution source which is connected to a main power source—usually a heavy conductor consisting of rigid bar or strap.

Burnout: Accidental passage of high voltage through an electrical circuit or device which causes damage.

C

Cable: A stranded single-conductor cable or a combination multiple-conductor cable. Cable refers to larger sizes. Small cable is called *stranded wire* or *cord*. Cable can be bare or insulated. Insulated cable may be armored with lead or with steel wire or bands.

Capacitor: Two conductors or electrodes in the form of plates separated from each other by a dielectric (insulator). The capacitor blocks the flow of direct current and partially impedes the flow of alternating current. This impedance decreases as the current frequency increases.

Cation: A positively charged ion, which is attracted to the cathode during electrolysis.

Charge: The electrostatic charge is the quantity of electricity held by a capacitor or insulated object. The charge is negative if it has excess electrons, and positive if it has less electrons than normal.

Choke Coil: A coil which has low ohmic resistance and high impedance to alternating current flow. It allows direct current to pass, but limits the flow of alternating current.

Circuit: A closed circuit provides a complete path for current flow. It is an open circuit if current flow is interrupted. It may contain a number of electrical components designed to perform one or more specific functions.

Circuit Breaker: A device that is actuated by electromagnetic or thermal energy to open a circuit when current exceeds a predetermined setting. A reset method is usually provided.

Coil: A helically or spirally wound conductor, which creates a strong magnetic field with current passage.

Commutator: A ring of copper segments insulated from each other and connecting the armature and brushes of a motor or generator. It passes power into or from the brushes.

Contact: The part of a switch or relay that carries current. Current is controlled by touching or separating contacts.

Conductance: The ability of material to carry electrical current. It is the reciprocal of resistance—therefore, it is expressed in mhos (ohms spelled backward).

Conductivity: The ease of electrical transmission through a substance.

Conductor: Material or substance which readily passes electricity.

Connected Load: The sum of the capacities or continuous ratings of the load-consuming apparatus connected to a supplying system.

Contactor: A device for making or breaking load-carrying contacts by a pilot circuit through a magnetic coil.

Controller: Measures the difference between sensed output and desired output and initiates a response to correct the difference.

Core: A magnetizable portion of a device, which affords an easy path for the magnetic flux lines of a coil.

Counter EMF: Counter electromotive force; the EMF induced in an armature or coil, which opposes applied voltage.

Coulomb: An electrical unit of charge, one coulomb per second equals one ampere or 6.25×10^{18} electrons past a given point in one second.

Covalent Bonding: The combining of atoms that have a high number of electrons in the outer shell. Bonding results in the sharing of electrons via extended electron paths around two or more atoms.

Current: The transfer of electrical charge through a conductor between points of different voltage potential.

Current Limiter: A protective device in a high amperage circuit.

Cut-In: Switch action in a conducting mode.

Cut-Out: Switch action to the OFF position.

Cycle: A complete positive and a complete negative alterna-

tion of voltage or current. Cycles per seconds (CPS) or hertz (Hz) denotes frequency.

D

Delta Connection: The connection in a three-phase system in which terminal connections are triangular similar to the Greek letter delta. With the inductors of one phase opposite the middle of the poles, and assuming maximum current to be induced at this moment, only one half the same current value will be induced at the same moment in the other two phases.

Delta Transformer Connection: Connection with both primaries and secondaries connected in delta grouping.

Delta-Y Connection: Primaries connected in delta grouping and secondaries in star grouping.

Demand: The size of any load generally averaged over a specified interval of time but occasionally instantaneously. Demand is expressed in kilowatts, kilovolt-amperes, or other suitable units. Occasionally used interchangeably with load.

Demand (Billing): The demand upon which billing to a customer is based, as specified in a rate schedule or contract. The billing demand need not coincide with the actual measured demand of the billing period.

Demand Charge: The specified charge to be billed on the basis of the billing demand, under an applicable rate schedule or contract.

Demand Meter: A device which indicates or records the demand or maximum demand. (NOTE: Since demand involves both an electrical factor and a time factor, mechanisms responsive to each of these factors are required as well as an indicating or recording mechanism. These mechanisms may be either separate from or structurally combined with one another. Demand meters may be classified as follows: Class 1—Curve-drawing meters, Class 2—Integrated-demand meters, Class 3—Lagged-demand meters.)

Device: A component that primarily carries, but does not utilize, electrical energy, does not perform useful work; e.g., controller or switch.

Dielectric: An insulator-nonconductor. The material between plates of a capacitor, which may be used to store electrostatic energy.

Dielectric Strength: Maximum voltage a dielectric can take without a breakdown.

Diode: A device that will carry current in one direction but not the reverse direction.

Direct Current: Abbreviated dc. Electric current that flows only in one direction—undirectional. The varying current in one direction is a pulsating direct current, which may be derived from rectified alternating current.

Disconnecting Switch: A knife switch that opens a circuit after the load has been disconnected by some other means.

Doping: The improving of the energy-carrying capacity of a semiconductor by mixing foreign atoms with the semiconductor atoms.

Double-Pole Switch: Simultaneously opens and closes two wires of a circuit. Completes a live circuit on each side of the OFF position.

Dynamic Electricity: Electrons in motion—the movement of the electrons is called *current*.

E

E: Symbol for volts.

Eddy Currents: The induced circulating currents in a conducting material that are erected by a varying magnetic field.

Efficiency: A percentage value denoting the ratio of power output to power input.

Electrical Angle: The method specifying the exact instant in an alternating current cycle. Each cycle is 360°; therefore, 180° would indicate one-half cycle, 45° one-eighth cycle.

Electric Circuit: The complete path for electric current.

Electric Field: A magnetic region in space which has force and direction.

Electric Filament Lamp: A light source consisting of a glass bulb containing a filament electrically maintained at incandescence. (NOTE: A lighting unit, consisting of an electric filament lamp with shade, reflector, enclosing globe, housing, or other accessories, is also commonly called a lamp. In such cases, in order to distinguish between the assembled lighting unit and the light source within it, the latter is often called a bulb.)

Electrical Degree: One-360th of an alternating current or voltage cycle.

Electricity: The effect created by interaction of positive and negative electrical charges. Electrostatic attraction and repulsion cause motion and movement of current carriers, which, when given force and direction, become electrical current flow.

Electrodes: The plates in a bath to accomplish electrolysis. Current enters the solution through the anode and leaves through the cathode. Also, each plate in a storage battery.

Electrolysis: Changing the chemical characteristics of a compound through application of an electrical current.

Electrolyte: A solution of a substance (liquid or paste) that is capable of conducting electricity.

Electromagnet: A temporary magnet that creates a magnetic field only during current flow.

Electromotive Force: Abbreviated EMF. The force or electrical pressure (voltage) that produces current flow. The difference in potential electrical energy between two points.

Electron: The light portion of the basic nucleus of an atom that carries the negative charge.

Electrostatic: Electricity at rest—static electricity.

Electrostatic Charge: The electrical energy located on the surface of insulating material or in a capacitor because of an excess or deficiency of electrons.

Electro-valent Bond: The combining of atoms into common paths of the electrons to produce molecules of various types of atoms.

Energy Levels: The energy states of electrons in an atom.

F

Farad: The unit of capacitance. The storage in a capacitor of one coulomb when one volt is applied.

Feedback: The transfer of energy output back to input.

Field: The space involving the magnetic lines of force.

Fluorescent Lamp: An electric discharge lamp in which the radiant energy from the electric discharge is transferred by suitable materials (phosphors) into wave lengths giving higher luminous efficiency.

Flux: The electric or magnetic lines of force in a region.

Frequency: The number of complete cycles or vibrations in a unit of time.

Four-Wire, Three-Phase Transformer: The secondaries of three transformers are star connected, the fourth wire is connected from a neutral point. The voltage between any main wire and the neutral will be 57% of the voltage between any two main wires.

Full-load Amperage (FLA): The amount of amperage an inductive load (motor) will draw at its full design load.

Fuse: An element designed to melt or dissipate at a predetermined current value, and intended to protect against abnormal conditions of current.

G

Galvanic: Current generated chemically, which results when two dissimilar conductors are immersed in an electrolyte. In the closed electrical circuit, current will flow, which will oxidize the most easily oxidized conductor.

Gang: Mechanical connection of two or more circuit devices so they may be varied simultaneously.

Galvanometer: An instrument for measuring small dc currents.

Gauss: Electromagnetic unit of flux density: one maxwell per square centimeter.

Generator: A machine that converts mechanical energy into electrical energy. See *alternator*.

Ground: Intentional or accidental connection between an electrical circuit and earth onto a common point of zero potential.

H

Henry: Abbreviated H. The basic unit of inductance, one henry results from the current variation of one ampere per second and produces one volt.

Holes: The portion of the outer circle (shell) of an atom that is not occupied by an electron.

Horsepower: Abbreviated hp. A unit of power equivalent to raising 33,000 lbs one foot in one minute, or 550 ft. lbs per second, or 776 W. of electrical power.

Humidistat: A control for electrical circuits actuated by changes in humidity surrounding the control.

Hysteresis: A lag—subjecting ferromagnetic material to a varying magnetic field causes a lagging of the magnetic flux behind the magnetizing force.

I

Illumination: The density of the luminous flux on a surface; it is the quotient of the flux by the area of the surface when the latter is uniformly illuminated. (NOTE: The term illumination is also commonly used in a qualitative or general sense to designate the act of illuminating or the state of being illuminated. Usually the context will indicate which meaning is intended, but occasionally it is desirable to use the expression, level of illumination, to indicate that the quantitative meaning is intended.)

Impedance: Total opposition to alternating current flow. It can consist of any combination of inductive reactance, capacitive reactance, and resistance. Z is the symbol for impedance and the ohm its unit.

Induced: Current or voltage in a conductor which results when the conductor is moved perpendicularly to a magnetic field or subjected to a varying magnetic field.

Inductance: The characteristic of an alternating current circuit to oppose a change in current flow. The change in current through the circuit causes a change in the magnetic field. The change in magnetic field induces a counter-voltage, which tends to oppose the change in current. It is a fly-wheel effect in which the countervoltage causes current retardation with current buildup and prolongs current flow when current decreases. It is symbolized by the letter L.

Induction: The act that produces electrification, magnetization, or induced voltage in an object by exposure to a magnetic field.

Inductive Load: A device that uses electrical energy to produce motion.

Inductive Reactance: Opposition, measured in ohms, to an alternating or pulsating current.

In Phase: The condition existing when two waves of the same frequency have their maximum and minimum values of like polarity at the same instant.

Interlock: A safety device that allows power to a circuit only after a predetermined function has taken place.

Ion: An atom or molecule that has more or less electrons than normal. The negative ion has more electrons than normal and the positive ion less.

IR Drop: Voltage drop resulting from current flow through a resistor.

J

Joule: Abbreviated J. The unit of energy or work resulting from one ampere of current flowing through one ohm of resistance in one second.

Jumper: A conductor used to bypass a switch or a break in a circuit or to make electrical connection between terminals.

Junction: A place in the circuit where two or more wires are joined.

Junction Box: A metal box into which cables or wires are inserted and joined.

K

Kilowatt: A unit for measuring electrical power. One kilowatt is equal to 1,000 W. One kilowatt is equal to 3,413 Btu per hour.

Kilowatt-Hour: A unit for measuring electrical energy. One kilowatt-hour is equal to 3,413 Btu.

Kilovolt-Ampere: Abbreviated kva. The unit of apparent power as differentiated from kilowatts or true power.

Kirchhoff's Current Law: A basic electrical law stating that the sum of all the currents flowing into a point in the circuit must be equal to the sum of the currents flowing away from the point.

Kirchhoff's Voltage Law: States that the sum of all the voltage rises in a complete circuit must equal the sum of the circuit's voltage drops.

L

Lag: The time one wave is behind another of the same frequency—usually expressed in electrical degrees.

Laws of Electric Charge: Like charges repel, unlike charges attract.

Laws of Magnetism: Like poles repel, unlike poles attract.

Lead: The phase of one alternating quantity that is ahead of the other measured in angular degrees. Also a wire connection.

Left-Hand Rule: Used to determine the direction of magnetic lines of force around a current-carrying wire. If the fingers of the left hand are placed so the thumb points in the direction of electron flow, the fingers will point in the direction of the magnetic field.

Linear: A proportional ratio of change in two related quantities. Output that varies in direct proportion to input.

Line Drop: The voltage drop between two points on a power or transmission line due to leakage, resistance, or reactance.

Line of Force: A line in an electric or magnetic field that shows direction of force.

Line Voltage: Voltage existing at wall outlets or terminals of a power line system above 30 V.

Load: A device that converts electrical energy to another form of usable energy.

Load Factor: The ratio of the average load in kilowatts supplied during a designated period to the peak or maximum load occurring in that period. Typical formula:

$$\text{Load factor} = \frac{\text{Kwhr supplied in period}}{\text{Peak kw in period} \times \text{hours in period}}$$

The kilowatt-hours and peak generally are on a net output basis. The peak generally is for a 60-minute demand interval.

Locked Rotor Amperage (LRA): The amount of energy a motor will draw under stalled conditions.

M

Magnetic Circuit: The closed path of magnetic lines of force.

Magnetic Field: The space in which a magnetic force exists.

Magnetic Flux: The sum of the lines of force issuing from the magnetic pole.

Magnetic Hysteresis: Internal friction resulting from subjecting a ferromagnetic substance to a varying magnetic field.

Magnetic Poles: Field concentrations in a magnet consisting of a north and south pole of equal strength.

Magnetize: Molecular rearrangement in a material, which converts it into a magnet.

Mega: Prefix for one million; e.g., megohm—one million ohms.

Metallic Bonding: The combining of atoms that have a minimum of electrons in their outer shell.

Micro: Prefix for one-millionth; e.g., microampere—one-millionth of an amp; microfarad—one-millionth of a farad.

Micron: A unit of length. One-thousandth of a millimeter or one-millionth of a meter.

Milli: A prefix for one-thousandth; e.g., milliampere—one-thousandth of an ampere.

Modulate: Output response, which varies proportionally with the input signal.

Module: A combination of components in a circuit that are in a unitized package and capable of performing a complete function.

Molecule: The smallest particle of a substance that retains all the characteristics of the substance.

Mutual Inductance: A circuit condition that exists when the relative positions of two inductors cause magnetic lines of force from one to link with the turns of the other.

N

"N" Type Semiconductor: Conductors doped for an excess of free electrons with predominately negative current carriers.

National Electrical Code: Abbreviated NEC. A code of electrical rules based on fire underwriters' requirements for interior electric wiring. Wiring must be in conformance to NEC rules to meet insurance and municipality requirements.

Negative: The terminal or electrode with excess electrons.

Network: Two or more interconnected electrical circuits; an electrical distribution system.

Neutron: An atomic particle in the nucleus having the weight of a proton but having no electrical charge.

Normally Closed: Switch contacts closed with the circuit de-energized.

Normally Open: Switch contacts open with the circuit de-energized.

Nuclear Action: The production of electrical energy by the use of heat energy produced from nuclear energy released from nuclear reaction.

Null: Zero.

O

Ohm: A unit of resistance; the resistance that allows one ampere to flow with the potential of one volt.

Ohm's Law: Current is directly proportional to voltage and inversely proportional to resistance.

Open Circuit: Noncontinuous circuit.

Oscillator: An electrical component that generates an alternating voltage.

Overload Protection: A component to interrupt current when it reaches a predetermined high.

P

"P" Type Semiconductor: Conductors doped for an excess of holes for heavier positive current carrying capacity.

Panelboard: A single panel, or a group of panel units, designed for assembly in the form of a single panel; including buses, and with or without switches and/or automatic overcurrent protective devices for the control of light, heat, or power circuits of small individual as well as aggregate capacity; designed to be placed in a cabinet or cutout box placed in or against a wall or partition and accessible only from the front.

Parallel: Circuit connected so current has two or more paths to follow.

Parasitic Load: A load that is designed to be operated with a primary load that can limit the current flow.

Permeability: The ease with which magnetic lines of force can flow through a material as compared with air, which is considered unity.

Phase: Angular relationship between waves and expressed in degrees.

Photovoltaic: The generation of electrical potential across a cell by the conversion of light energy to electrical energy.

Piezoelectric Effect: The generation of electrical potential across a given material by application of pressure across the material.

Polarity: The condition denoting direction of current flow; the condition of being positive or negative, or having a magnetic north and south pole.

Pole: The part of a magnet where flux lines are concentrated or where they enter and leave the magnet; also, the electrode of a battery.

Polyphase: Two or more alternating currents that differ in phase by a predetermined number of degrees.

Positive: A point of attraction for electrons.

Potential: The amount of voltage or charge between points of a circuit.

Potentiometer: Abbreviated Pot. A resistor with three contacts. One contact is on a movable arm, which can move across the resistor. This adds resistance to one leg of a Wheatstone bridge circuit while simultaneously decreasing resistance in the other, thereby varying current to each leg of the bridge.

Power: The rate of doing work.

Power Factor: The rate of actual power as measured by a wattmeter in an alternating circuit to the apparent power determined by multiplying amperes by volts.

Power Source: A source of electrical energy, be it a battery, generator, piezo-crystal, or thermo-couple, from which an electrical load draws electrical energy.

Primary Load: A load that is capable of absorbing the full voltage of the power source.

Primary Voltage: The voltage of the circuit supplying power to a transformer is called the primary voltage, as opposed to the output voltage of load-supply voltage which is called secondary voltage. In power supply practice the primary is almost always the high-voltage side and the secondary, the low-voltage side of a transformer.

Proton: The portion of the basic nucleus of an atom that carries the positive charge.

Pull-In Voltage The voltage value that causes the relay armature to seat on the pole face.

Pump Down: A method of controlling the operation of a refrigeration system by controlling the flow of refrigerant to the evaporator.

Q

Quick Disconnect: Quick attaching and releasing connecter halves.

R

Reactance: Opposition to alternating current by either inductance or capacitance or both. Symbolized by X and measured in ohms.

Reed Switch: Two or more highly conductive reeds encapsulated in a glass enclosure. Energy from a magnetic field will cause the reed contacts to open or close simultaneously.

Relay (Contactor): A two-circuit device. A pilot duty circuit controls a load-carrying circuit.

Reluctance: The opposition a material offers to magnetic lines of force. Equals the magnetomotive force divided by the magnetic flux.

Resistance: The opposition to current flow by a physical conductor when voltage is applied.

Resistor: A circuit element that offers resistance to current flow.

Retentivity: A material's ability to hold magnetism.

Rheostat: An adjustable or variable resistor.

Rotor: A rotating armature or member of an electrical motor or generator.

S

Secondary Voltage: The output, or load-supply voltage, of a transformer or substation is called the secondary voltage.

Self-Induction: The process by which an EMF is induced in the circuit by its own magnetic field.

Series: A circuit with one continuous path for current flow.

Series Wound: A motor or generator with armature and field windings wired in series.

Service: The conductors and equipment for delivering energy from the electricity supply system to the wiring system of the premises served.

Service Conductors: That portion of the supply conductors which extends from the street main or duct or from transformers to the service equipment of the premises supplied. For overhead conductors this includes the conductors between the last pole or other aerial support and the service equipment.

Service Drop: That portion of overhead service conductors between the last pole or other aerial support and the first point of attachment to the building.

Service Equipment: The necessary equipment, usually consisting of circuit breaker or switch and fuses, and their accessories, located near point of entrance of supply conductors to a building and intended to constitute the main control and means of cutoff for the supply to that building.

Service Raceway: The rigid metal conduit, electrical metallic tubing, or other raceway, that encloses service entrance conductors.

Servo: A control mechanism that converts a small force into a greater force.

Servo Mechanism A closed-loop system which initiates an input signal with deviation from a desired condition; the signal is fed back into the control system until a continued response eliminates the signal.

Shell: The path that the electrons take as they travel around the proton.

Short Circuit: A low-resistance connection (usually accidental and undesirable) between two parts of an electrical circuit.

Shunt: A part connected in parallel with another part.

Sine: In a right-angle triangle the ratio of the opposite of any angle to the hypotenuse.

Single Phase A single alternating voltage.

Single-Pole Switch: A switch with movement from an OFF position to a live contact.

Sinusoidal: Current that varies in proportion to the sine of an angle or time function; e.g., ordinary alternating current.

Snap Switch: One with contacts that make and break quickly through mechanical linkage.

Solenoid: A movable plunge activated by an electromagnetic coil.

Space Charge The electrical space charge resulting from ions and electrons; the cloud of electrons around a hot vacuum-tube cathode.

Stable Atom: An atom whose outer shell contains more than 50% of the potential to contain electrons.

Stator: The stationary part of the magnetic circuit of a motor or generator.

Switch: A mechanical device for making or breaking an electrical circuit.

Switchboard: A large single panel, frame, or assembly of panels, on which are mounted, on the face or back or both, switches, overcurrent and other protective devices, buses, and usually instruments. Switchboards are generally accessible from the rear as well as from the front and are not intended to be installed in cabinets.

T

Terminal: A point of connection for electrical conductors.

Thermal Cutout: An overcurrent protective device which contains a heater element in addition to and effecting a renewable fusible member which opens the circuit. It is not designed to interrupt short circuits.

Thermocouple: A junction of two dissimilar metals that develops a voltage when heated.

Thermoelectric: Conversion of heat to electrical energy or vice versa.

Thermostat: A control for electrical circuits activated by changes in the surrounding air temperatures.

Three-Phase Delta-Connected System: In wiring, this arrangement employs three wires. Assuming 100 amps. and 1,000

Vs in each phase winding, the pressure between any two conductors is the same as the pressure in the winding, and the current in any conductor is equal to the current in the winding multiplied by the square root of 3, that is, $100 \times 1.723 = 173.2$ amps., or, disregarding the fraction, 173 amp.

Three-Phase Delta Grouping: A method of grouping a three-phase winding, in which the three circuits are connected together in the form of a triangle, and the three-corners are connected to the three terminals.

Three-Phase Four-Wire System: A three-phase, three-wire system having a fourth wire connected to the netural point of the source, which may be grounded.

Three-Phase Y(Star)-Connected Systems: In wiring, there are systems with star connections employing three wires or four wires. Assuming winding, the pressure between any two conductors is equal to the pressure in one winding multiplied by $\sqrt{3}$, that is, $1,000 \times 1.732 = 1,732$ Vs. The current in each conductor is equal to the current in the winding, or 100 amps.

Three-Phase Y (Star) Grouping: A method of grouping a three-phase winding in which one end of each of the three circuits is brought to a common junction, usually insulated, and the three other ends are connected to three terminals. It is commonly called a Y connection or grouping, owing to the resemblance of its diagrammatic representation to the letter Y.

Three-Phase System Wiring: There are various ways of arranging the circuit for three-phase current giving numerous three-phase systems, which may be classed:

1. With respect to the number of wires used as: (a) three wire; (b) four wire.
2. With respect to the connections as: (a) star; (b) delta; (c) star delta; (d) delta star.

Torque: Torque is a force or combination of forces that produces or tends to produce rotation.

Total Interruption: A control device that controls by 100% make or break of the electrical circuit.

Transformer: An electromagnetic device, with two or more coils linked by magnetic lines of force, used to increase or decrease ac voltage. With voltage increase, current is decreased, and vice versa.

Transmission Lines: A conductor or system of conductors for the purpose of carrying electrical energy from source to load.

U

Underwriters Laboratories: Abbreviated UL. Underwriters Laboratories, Inc. maintain and operate laboratories for the examination and testing of devices, systems, and materials.

Universal Motor: A motor that can operate on ac or dc current.

Unstable Atom: An atom whose outer shell contains less than 50% of the potential to contain electrons.

Utility Rate Structure: A utility's approved schedules of charges to be made in billing for utility service rendered to various classes of customers.

V

Valence Shell: The outside shell of the path the electrons take that are the farthest from the center proton.

Vector: A line that represents the direction and magnitude of a quantity.

Volt: The unit of electrical potential or pressure. The electromotive force (EMF) that will send one ampere through one ohm of resistance.

Voltage Relay: One that functions at a predetermined voltage value.

Volt-Ampere: Abbreviated VA. Volts times amperes.

W

Watt: The electrical unit of power or rate of doing work. In its simplest terms, it is the rate of energy transfer equivalent to one ampere flowing under the pressure of one volt at unity power factor. It is analogous to horsepower or foot-pounds per minute of mechanical power. One horsepower equals 746 watts.

Watt-hour Meter: An electricity meter which measures and registers the integral, with respect to time, of the active power of the circuit in which it is connected. This power integral is the energy delivered to the circuit during the interval over which the integration extends, and the unit in which it is measured is usually the kilowatt-hour.

Waveform: The graphical representation of an electromagnetic wave, which shows variations in amplitude plotted against time.

Wheatstone Bridge: A bridge with four main arms containing a resistor in each arm. With known values for three resistors, the value for the unknown resistor can be computed.

Winding: One or more wire turns, which form a continuous coil or conductive path.

Answers to Problems in Refrigeration

CHAPTER R1

R1-1. China.   R1-2. 1900.   R1-3. True.   R1-4. 13.   R1-5. False.   R1-6. False.   R1-7. Compressor, condenser, evaporator, and pressure-reducing device.   R1-8. Liquid line, suction line, hot-gas line, and liquid return line. R1-9. Transfer of heat by the change in state of a liquid.   R1-10. Suction line.   R1-11. Hot-gas line.   R1-12. Liquid line.   R1-13. Liquid return line.   R1-14. Bare pipe and finned coils.

CHAPTER R2

R2-1. Solid, liquid, and gas.   R2-2. Molecule.   R2-3. Intensity.   R2-4. The point at which, theoretically, there is a complete absence of heat and molecular activity. This is believed to occur at $-460°$F or $-273°$C.   R2-5. Energy cannot be created or destroyed, it can only be converted from one form to another.   R2-6. To cause heat to travel a temperature difference must be established and maintained.   R2-7. The amount of heat needed to change the temperature of 1 pound of water 1 degree fahrenheit at sea level.   R2-8. $°C = \frac{5}{9}(F - 32) = \frac{5}{9}(68 - 32) = 20°C$. R2-9. Btu $= W \times \Delta T = 100(120° - 70°) = 100 \times 50° = 5000$ Btu.

R2-10. Btu $= W \times \Delta T; \Delta T = \dfrac{Btu}{W} = \dfrac{750}{15} = 50°F$.

R2-11. 1.0.   R2-12. False.   R2-13. Conduction, convection, and radiation.   R2-14. Sensible heat.   R2-15. Latent heat.   R2-16. Sublime.   R2-17. Latent; vaporization.   R2-18. The Btu needed to melt 1 ton of ice in 24 hours. R2-19. 12,000 Btu/hr.   R2-20. Poor.

CHAPTER R3

R3-1. Fluid pressure is the force per unit area that is exerted by a gas or liquid.   R3-2. Pounds per square inch (psi). R3-3. The total weight of the substance. R3-4. Total force = weight of water = 3 ft $\times$ 3 ft $\times$ 1 ft = 9 ft³
$= 9$ ft³ $\times$ 62.4 lb/ft³ = 561.6 lb

Unit pressure per square inch $= \dfrac{561.6 \text{ lb}}{3 \text{ ft} \times 3 \text{ ft}} = \dfrac{561.6 \text{ lb}}{36 \times 36} = 0.433$ psi.

R3-5. The depth of a body of water is called the "head."   R3-6. Psi = 0.433 psi $\times$ 8 ft of head = 3.464 psi. R3-7. One-half of 3.464 or 7.732 psi.

R3-8. $\dfrac{1580 \text{ lb}}{12 \text{ ft}^2} = 134.67$ lb/ft² or 0.914 psi.

R3-9. $\dfrac{600 \text{ lb}}{16 \text{ in.}^2} \times 2 \text{ in.}^2 = 75 \text{ psi.}$

R3-10. $\dfrac{75 \text{ psi.}}{2 \text{ in.}^2} = 37.5 \text{ psi.}$

R3-11. Weight of a substance per unit of volume. Most density is expressed in weight per cubic foot. **R3-12.** 62.4 lb/ft³. **R3-13.** The weight of a substance as compared to a standard substance. **R3-14.** The number of cubic feet occupied by 1 pound of the substance. **R3-15.** Temperature. **R3-16.** Temperature and pressure. **R3-17.** The volume of a gas varies inversely as its pressure if the temperature of the gas remains constant. **R3-18.** 183.8 psi. **R3-19.** 382.3 psi. **R3-20.** The volume of gas is in direct proportion to its absolute temperature provided that its pressure is kept constant; and the absolute pressure of a gas is in direct proportion to its absolute temperature, provided that the volume is kept constant. **R3-21.** 5.46 ft³.

R3-22. $\dfrac{P_1 V_1}{T_1} = \dfrac{P_2 V_2}{T_2}$ or $P_1 V_1 T_2 = P_2 V_2 T_1.$

R3-23. $P_2 = \dfrac{P_1 V_1 T_2}{V_2 T_1} = \dfrac{(57.7 + 14.7) \times 10 \times (126.6 + 460)}{5 \times (60 + 460)} = \dfrac{72.4 \times 10 \times 586.6}{5 \times 520}$

$= \dfrac{424{,}698.4}{2600} = 163.4 \text{ psia} - 14.7 = 148.6 \text{ psig.}$

R3-24. $V_2 = \dfrac{P_1 V_1 T_2}{P_2 \times T_1} = \dfrac{(70.2 + 14.7) \times 20 \times (115 + 460)}{(146.8 + 14.7) \times (70 + 460)} = \dfrac{84.9 \times 20 \times 575}{161.5 \times 530}$

$= \dfrac{976{,}350}{85{,}595} = 11.4 \text{ ft}^3$

R3-25. Added heat that raises the temperature of a vapor above the boiling point. **R3-26.** Lowering the temperature of a liquid below its condensing temperature. **R3-27.** Saturated liquid is at the boiling point before latent heat of vaporization has been added. Saturated vapor is at the same boiling point after latent heat of vaporization has been added.

CHAPTER R4

R4-1. Flare nut wrench. **R4-2.** True. **R4-3.** 6 and 8. **R4-4.** False. **R4-5.** Allen. **R4-6.** Flat blade tip and Phillips tip. **R4-7.** Single cut and double cut. **R4-8.** False. **R4-9.** 1/1000 (0.001 in.). **R4-10.** False. **R4-11.** Occupational safety and health act.

CHAPTER R5

R5-1. Copper. **R5-2.** False. **R5-3.** K, L, and M. **R5-4.** K and L. **R5-5.** Soft drawn and hard drawn. **R5-6.** Flared, compression, and mechanical bonding. **R5-7.** 45° angle. **R5-8.** Five times the diameter of the tube. **R5-9.** Open end. **R5-10.** Swedging. **R5-11.** Proper cutting, cleaning, proper fluxing, and correct support. **R5-12.** Neutral.

CHAPTER R6

R6-1. Evaporator, compressor, condenser, and pressure-reducing device. **R6-2.** Increase the surface area of the coil by adding fins or use a fan or blower to increase movement of air across the coil. **R6-3.** When the evaporator must operate in a space where high humidity must be maintained. **R6-4.** Submerged coils, shell-in-tube and tube-in-tube configurations, and Baudelot coolers. **R6-5.** Reciprocating, rotary, and centrifugal. **R6-6.** Reed and ring. **R6-7.** The compressor and motor are sealed in the same housing in the hermetic. The open compressor is belt driven by use of a pulley-and-belt drive with a separate motor. **R6-8.** Rolling piston and rotating blade or vane. **R6-9.** None. **R6-10.** To dissipate the heat picked up by the refrigerant in the evaporator. **R6-11.** Air cooled, liquid cooled, and evaporative cooled.

R6-12. Automatic expansion valve, thermostatic expansion valve, capillary tube, low-side float, and high-side float.
R6-13. The capillary tube, because it has no moving parts to wear and require replacement.   **R6-14.** To permit the flow of a liquid or vapor through a line or pipe in one direction only.   **R6-15.** To permit control of a liquid or vapor flow under the control of an electrical circuit or some other means of actuation.

CHAPTER R7

R7-1. It absorbs heat from the place where it is not wanted and expels it elsewhere.   **R7-2.** Latent heat of vaporization.
R7-3. Chemical and physical.   **R7-4.** Green.   **R7-5.** Electronic leak detector or soap solution.   **R7-6.** 80.
R7-7. 125.   **R7-8.** R-12, white; R-22, green; R-502, orchid.   **R7-9.** 76 psig.   **R7-10.** 115°F.

CHAPTER R8

R8-1. The amount of heat in Btu that each pound of refrigerant absorbs in the evaporator.   **R8-2.** The heat content of the liquid entering the pressure-reducing device and the total heat content of the vapor leaving the evaporator.

R8-3.
$$W = \frac{200}{NRE}$$
W = weight of refrigerant circulated per minute
200 = 200 Btu/min = 1 ton of refrigerating effect
NRE = net refrigerating effect (Btu/lb) of the
refrigerant used at the particular operating conditions.

R8-4. The heat added to a vapor that raises its temperature above its boiling point.   **R8-5.** Saturated vapor is vapor at the boiling point—superheated vapor has had sensible heat added to raise its temperature above the boiling point.   **R8-6.** Its total displacement and its volumetric efficiency.   **R8-7.** It is a graphic representation on a Mollier diagram of the various processes within a refrigeration cycle.   **R8-8.** In the central area are shown the changes in state of the refrigerant; in the left-hand area are shown the subcooled liquid factors; and in the right-hand area are shown the factors for superheated vapor.
R8-9. Heat control (enthalpy), entrophy, temperature, pressure, and volume.

CHAPTER R9

R9-1. Bare tube, finned tube, and plate.   **R9-2.** Fins provide a higher heat-transfer rate per square inch of tube surface, thus increasing the coil capacity. Also, a smaller coil may be used for a given load.   **R9-3.** Conduction.   **R9-4.** When natural convection of air through the coil will not provide adequate air circulation to handle the heat load.   **R9-5.** Amount of total surface in the coil; heat-transfer rate (U factor) in Btu per square foot per degree temperature difference; and mean temperature difference (MTD) of air traveling through the coil.   **R9-6.** In low applications where heavy frosting of the coil occurs.   **R9-7.** Frost acts as an insulator, reducing the heat transfer between the air and the refrigerant in the coil, thus reducing the efficiency of the system.   **R9-8.** A hot-gas defrost removes frost from the coil while the compressor is running. When the hot gas gives up its heat in the process and condenses to a liquid, a large quantity (called a *slug*) of liquid can flow to the compressor and cause damage.   **R9-9.** For assurance of proper compressor lubrication.   **R9-10.** Oil in the evaporator reduces the heat-transfer rate of the coil as well as reducing the volume left for the handling of refrigerant.

CHAPTER R10

R10-1. Automatic expansion valve, thermostatic expansion valve, capillary tube, low-side float, and high-side float.
R10-2. Must be used on a constant load.   **R10-3.** Superheat.   **R10-4.** Spring pressure, body pressure, and bulb pressure.
R10-5. Spring and body pressure oppose the bulb pressure.   **R10-6.** False.   **R10-7.** Initial cost and system pressure balance during the OFF cycle, resulting in lower starting torque required.   **R10-8.** They do not have the ability to regulate refrigerant flow to change in load and they require a critical refrigerant change.

CHAPTER R11

R11-1. Reciprocating, rotary, helical, and centrifugal. R11-2. False. R11-3. Open drive and hermetic. R11-4. True. R11-5. Field serviceable or bolted and full hermetic or welded. R11-6. Induction. R11-7. Positive return of oil to the compressor crankcase. R11-8. It is a way of stating how "thick" an oil is. R11-9. Splash feed system and positive or forced-feed system. R11-10. It must remain fluid at low temperatures; it must remain stable at high temperatures; it must not react chemically with refrigerants, metals, motor insulation, air or other contaminants; it must not decompose into carbon under expected operating conditions; it must not deposit wax at extremely low temperature; and it must be dry and free of moisture. R11-11. The lowest temperature at which the oil will flow. R11-12. When hermetic compressors are used, the oil returning to the compressor passes over the motor windings, terminals, and power leads. R11-13. Heat. R11-14. 3%. R11-15. 10; 10. R11-16. 10; 5. R11-17. Energy efficiency ratio. R11-18. Capacity. R11-19. High ratio. R11-20. Life expectancy.

CHAPTER R12

R12-1. Air cooled, water cooled, and evaporative. R12-2. Double pipe, open vertical shell and tube, horizontal shell and tube, and shell and coil. R12-3. Active (forced draft) and passive (gravity). R12-4. Latent heat. R12-5. Wet bulb. R12-6. Natural draft and forced draft. R12-7. Scale-forming concentrates. R12-8. Evaporation, bleed-off, and drift. R12-9. The number of degrees fahrenheit the water is cooled in the tower and the difference in temperature of water entering and leaving the tower. R12-10. The temperature difference between the temperature of the water leaving the tower and the wet-bulb temperature of the air entering the tower. R12-11. 3; 4. R12-12. 5. R12-13. 65. R12-14. False. R12-15. 80%.

CHAPTER R13

R13-1. Provide a passage for the refrigerant and provide a passage for the oil to return to the compressor. R13-2. Keep the pipe clean, proper sizing, use as few fittings as possible, take special precautions in making connections, and pitch horizontal lines in direction of refrigerant flow. R13-3. Poor. R13-4. 500 ft/min; 1000 ft/min. R13-5. 2 psig for R-12 and 3 psig for R-22 and R-502. R13-6. True. R13-7. When the condensing unit is installed more than 15 ft above the 2x coil. R13-8. Those in which the system has compressor capacity control. R13-9. False. R13-10. True. R13-11. 6. R13-12. Yes. R13-13. 4 psig. R13-14. Gas formed in the liquid line ahead of the pressure-reducing device. R13-15. Lowering of liquid pressure and boiling point of liquid refrigerant below the temperature surrounding the liquid line or raising the temperature surrounding the liquid line above the boiling point of the liquid. R13-16. When the liquid line is at a higher temperature than the condensing temperature of the liquid.

CHAPTER R14

R14-1. False. R14-2. Reciprocating. R14-3. Vertical (down gas flow), and horizontal. R14-4. To reduce the temperature or subcool the liquid refrigerant and add heat to the suction gas to prevent liquid flow to the compressor. R14-5. To reduce the amount of moisture circulating in the system and filter out any solid particles from the liquid refrigerant. R14-6. Liquid and vapor. R14-7. To prevent liquid refrigerant from entering the compressor. R14-8. Contact, insertion, and wraparound. R14-9. Prevent liquid refrigerant accumulations in the compressor crankcase during the OFF cycle. R14-10. Color. R14-11. The refrigerant pressure in the discharge or high side of the system. R14-12. False. R14-13. D. R14-14. They are one and the same. R14-15. To relieve pressure from the system, to prevent formation of pressures that could be destructive. R14-16. To relieve pressures from the system, to prevent formation of pressures that could be destructive. R14-17. The pressure relief valve operates on pressure buildup, and the fusible plug operates on temperature buildup. R14-18. A pressure relief valve resets with a drop in pressure; a rupture disk is a one-time valve that does not close.

CHAPTER R15

R15-1. Heat. R15-2. Lower. R15-3. Water. R15-4. Lithium bromide. R15-5. Generator, absorber, evaporator, and condenser. R15-6. The evaporator. R15-7. Generator. R15-8. Noncondensible. R15-9. Manual. R15-10. True.

CHAPTER R16

R16-1. Protons; electrons. R16-2. Negative. R16-3. True. R16-4. The condition when the electrons are at rest but have the potential to move. R16-5. Electrons in motion. R16-6. Unlike; like. R16-7. Primary: flashlight battery, not rechargeable; and secondary: automobile battery, rechargeable. R16-8. Dc. R16-9. Chemical. R16-10. Thermocouple. R16-11. 0.15 to 0.35 V. R16-12. Thermopile. R16-13. Photovoltaic cell. R16-14. Piezoelectric effect. R16-15. Electromechanical. R16-16. Electromotive force. R16-17. Voltage. R16-18. True. R16-19. Ohm. R16-20. Ampere. R16-21. Series and parallel. R16-22. Power source, load, and conductors.

R16-23. $I \text{ (amperage)} = \dfrac{V \text{ (voltage)}}{R \text{ (resistance)}}.$

R16-24. 12 Ω. R16-25. True. R16-26. False. R16-27. Inductive reactance. R16-28. Henry. R16-29. Voltage. R16-30. Capacitor. R16-31. Microfarad. R16-32. The ratio of consumed power to supplied power. R16-33. 90.

CHAPTER R17

R17-1. True. R17-2. Sine curve. R17-3. False—60 cycles in 1 second. R17-4. True. R17-5. 120 and 240 V, single phase. R17-6. Commercial and industrial customers. R17-7. Conductor. R17-8. Circular mil. R17-9. *National Electrical Code*. R17-10. Raceway. R17-11. Conductors. R17-12. Time-delay.

CHAPTER R18

R18-1. Power source, load, and conductors. R18-2. Controls. R18-3. Step-up and step-down. R18-4. In a step-up transformer, the secondary (output) has a greater number of turns than the primary coil. In a step-down transformer, the primary (input) has the greater number of turns. R18-5. The volts times the amperes is practically the same in both windings. If voltage is doubled in the secondary over the primary, the amperage will be one-half, ignoring the transformer efficiency. If the voltage change is one-tenth (240–24 V), the amperage output will be 10 times the voltage in (ignoring transformer efficiency). R18-6. 2400 V. R18-7. Primary type—capable of being placed across the applied voltage without damage; and parasitic type—must have a primary load connected in series to limit current flow. R18-8. Thermostat heat anticipator. R18-9. Magnetic and thermal. R18-10. Total interruption and characteristic change. R18-11. Bimetallic strips. R18-12. Bourdon. R18-13. The voltage of the control device and the load are not the same, the control cannot carry the current demand of the load, and more than one circuit must be controlled by a single control device. R18-14. False. R18-15. Two plates separated by an insulator where, when a positive voltage is applied to one plate, a negative voltage will be created on the other. R18-16. Manual; automatic. R18-17. Starting and running. R18-18. Starting: high capacitance, electrolytic filled for short-time duty; running: low capacitance, oil-filled metal can for continuous duty. R18-19. A force, or combination of forces, that produces or tends to produce rotation. R18-20. Split phase, capacitor start, permanent split capacitor, capacitor start–capacitor run, and shaded pole. R18-21. False. R18-22. To prevent damage to the motor from overheating.

CHAPTER R19

R19-1. Electrical circuit. R19-2. (a) Voltmeter. (b) Megger. (c) Ammeter. (d) Ohmmeter. (e) Wattmeter. (f) Ohmmeter. (g) Ohmmeter. R19-3. Highest. R19-4. False. R19-5. Minimum. R19-6. Open.

R19-7. High voltage contact 1 Ω, low voltage contacts 0.5 Ω.   R19-8. To check out an electrical circuit having low voltage and current measurements.   R19-9. Thermocouple circuits on gas appliances.   R19-10. 500V.

CHAPTER R20

R20-1. Normally open.   R20-2. Normally closed.   R20-3. (a) 70°F or 21.1°C. (b) Atmospheric pressure—0 psig. (c) Power to circuit is off.   R20-4. Normally open.   R20-5. Normally closed.   R20-6. DPST.   R20-7. DPST. R20-8. One of each.   R20-9. Triple pole, single throw; normally open.   R20-10. Yes—by hooking two contacts in parallel to double the capacity of each set of contacts.   R20-11. Normally closed.

CHAPTER R21

R21-1. Protons; electrons.   R21-2. Positive.   R21-3. Negative.   R21-4. The number of protons in the nucleus. R21-5. 29.   R21-6. Shell.   R21-7. 2.   R21-8. Valance shell.   R21-9. 8.   R21-10. Outer.   R21-11. An electron with enough additional energy to force it from the valance shell.   R21-12. The outer shell contains fewer than five electrons.   R21-13. Two copper atoms combined with one oxygen atom produce a copper oxide molecule with eight valence electrons.   R21-14. It has an equal number of protons and electrons.   R21-15. Ion.   R21-16. Positive ion. R21-17. Negative ion.   R21-18. The process of combining rows of opposite charges.   R21-19. The ability of the valence electron to move freely from electron to electron.   R21-20. When stable atoms are combined, they share electrons rather than give them up.   R21-21. They do not have free electrons to carry energy from atom to atom.   R21-22. The space left in the valence shell when an electron leaves the atom.   R21-23. The amount of electron transfer becomes higher with temperature rise; therefore, the higher the temperature, the lower the resistance to current flow.   R21-24. Negative free electron current and positive hole current.   R21-25. Free electron flow.   R21-26. Doping.   R21-27. Pentavalent. R21-28. Trivalent.   R21-29. N.   R21-30. P.   R21-31. A doped semiconductor, so that the current will pass through in one direction but not the reverse direction.   R21-32. Metallic rectifiers and semiconductor diodes.   R21-33. Copper oxide; selenium.   R21-34. The region where the P and N semiconductors are joined to form an unbalanced condition of electrons and holes between semiconductors.   R21-35. Half-wave.   R21-36. The voltage applied across the nonconducting direction of a diode that destroys its ability to hold back current flow.   R21-37. 4.   R21-38. When the desired voltage is within 10% of the power supply voltage.   R21-39. To limit the current when the capacitors are charging to protect the diode.   R21-40. Choke coil.   R21-41. L section.   R21-42. Pi section.   R21-43. Voltage-doubling circuit.   R21-44. The diode in the primary of the pulse transformer.   R21-45. Follow the manufacturer's instructions.

CHAPTER R22

R22-1. Glass and dial type.   R22-2. Thermister.   R22-3. Above; below.   R22-4. Saturated temperature. R22-5. Testing manifold.   R22-6. False.   R22-7. Purging.   R22-8. 2.   R22-9. Explosion and fire. R22-10. Deep evacuation and triple evacuation.   R22-11. 1/25,400.   R22-12. 500 or less.   R22-13. False. R22-14. Charging bottle.   R22-15. Vapor.

CHAPTER R23

R23-1. National Fire Protection Association.   R23-2. False.   R23-3. Codes are rules to be followed; standards are recommended guidelines.   R23-4. Serviceability.   R23-5. False.   R23-6. Scale inside copper tubing. R23-7. Sweating.   R23-8. Canvas.   R23-9. Pinch-off tube.   R23-10. False.   R23-11. False—must be within the range of the operating charge.   R23-12. Total evacuation and weighing in the factory-listed charge. R23-13. Overcharging.   R23-14. False.   R23-15. False.

CHAPTER R24

R24-1. Pressure.   R24-2. Above.   R24-3. Spread.   R24-4. Bottom.   R24-5. 2 to 3.   R24-6. Air; refrigerant system.   R24-7. High.   R24-8. 65°F.   R24-9. Refrigerant quantity; refrigerant flow rate.

CHAPTER R25

R25-1. Control; boiling point.   R25-2. Low.   R25-3. Stay the same.   R25-4. Decrease.   R25-5. Fall.
R25-6. False.   R25-7. Clogged condenser.   R25-8. Clogged air filters.   R25-9. Shut off the air supply with the unit running and check the coil circuits for frost-up.   R25-10. $\frac{1}{3}$ to $\frac{1}{2}$.   R25-11. Ambient temperature.   R25-12. Excessive refrigerant charge; restricted liquid line.

CHAPTER R26

R26-1. 10; 10.   R26-2. 10; 5.   R26-3. 3.   R26-4. 1; 0.5.   R26-5. Time-delay.   R26-6. Loose fuse-holding clips.
R26-7. Heat-assisted externally mounted, internal thermostat/main line heater element, and internal line break.
R26-8. Start capacitor and capacitor relay.   R26-9. High, 264 V; low, 216 V.   R26-10. High, 253 V; low, 198 V.
R26-11. At the contactor terminals.   R26-12. Low applied voltage.   R26-13. Increase.   R26-14. Decrease.
R26-15. False.   R26-16. False—they contain soap.   R26-17. Current and potential.   R26-18. Simply, 0; the relay depends on amperage quantity only.   R26-19. False—they are not field adjustable.   R26-20. 10; 10.   R26-21. 0; 10.

R26-22. $\text{MFD} = \dfrac{\text{amperage} \times 2650}{\text{voltage across capacitor}}$.

R26-23. True.

CHAPTER R27

R27-1. Heat transmission, air infiltration, product loads, and supplementary heat.   R27-2. Sensible; latent.
R27-3. High.   R27-4. 875 Btu/hr.   R27-5. Loose fill, flexible, ragid, reflective, and formed in place.   R27-6. The temperature heat is exceeded only 1% of the time during the summer season.   R27-7. Air infiltration.   R27-8. Specific heat.   R27-9. Above.   R27-10. Moisture or water.

CHAPTER R28

R28-1. Sulfur dioxide (SO_2).   R28-2. By using larger evaporators to reduce humidity removal and maintain high relative humidity in the cabinet.   R28-3. 0 to 10°F.   R28-4. Destructive and preservative.   R28-5. Any delay in refrigeration allows the ripening process to continue and reduces the allowable storage time.   R28-6. Ice in end bunkers of the cars with an air circulation system between the ice and the product.   R28-7. Devices made of a high-specific-heat material that are cooled and provide long-term heat absorbtion without the refrigeration system attached.   R28-8. The amount of surface of the ice. Flake ice will cool faster than cube ice.   R28-9. Because products have different moisture and temperature requirements.   R28-10. Mixed storage.

Answers to Problems in Air-Conditioning

CHAPTER A1

A1-1. Fire, clothing, caves.   A1-2. John Garrie.   A1-3. Willis Carrier.   A1-4. Freon.   A1-5. 1935.   A1-6. Air-Conditioning and Refrigeration Institute.   A1-7. Air-conditioning and refrigeration equipment manufacturers.   A1-8. Air-Conditioning Contractors Association.   A1-9. Contractors in the heating, air-conditioning, and environmental phase of the refrigeration and air-conditioning industry.   A1-10. Trained personnel available.   A1-11. 11.

CHAPTER A2

A2-1. Temperature, moisture, air movement, cleanliness, and ventilation.   A2-2. 98.6°F.   A2-3. Humidity.
A2-4. The combinations of temperature and humidity in which the majority of people feel comfortable.   A2-5. 68 to 70°F.
A2-6. 78 to 80°F and 50% relative humidity.   A2-7. Feet per minute.   A2-8. 50 ft/min.   A2-9. Dry-bulb.
A2-10. Wet-bulb.   A2-11. Sling psychrometer.   A2-12. Wet-bulb depression.   A2-13. Comfort zone.   A2-14. 36.
A2-15. Dust, pollen, smoke, fumes, and chemicals.   A2-16. 10; 20.   A2-17. 95.

CHAPTER A3

A3-1. The science and understanding of the properties of air.   A3-2. 77% nitrogen, 23% oxygen, with traces of other rare gases.   A3-3. Grains; pounds.   A3-4. Ordinary thermometer.   A3-5. Evaporation rate.   A3-6. Psychrometer.
A3-7. Wet-bulb depression.   A3-8. The temperature at which water vapor will condense to a liquid.   A3-9. The actual weight of water vapor in the air measured in pounds or grains per pound of dry air.   A3-10. 7000.   A3-11. The percent of actual water vapor in the air as compared to the total amount it could hold at a given temperature.   A3-12. 100.
A3-13. Sensible heat; latent heat.   A3-14. The number of cubic feet occupied by 1 pound of the mixture of air and water vapor.   A3-15. The total heat content of air and water vapor.   A3-16. In Btu per pound of air.   A3-17. Sensible temperature.   A3-18. Humidity ratio expressed in pounds of moisture per pound of dry air.   A3-19. Saturated.
A3-20. Saturation line.   A3-21. Wet-bulb temperatures.   A3-22. Relative humidity.   A3-23. Sensible.
A3-24. Latent.

CHAPTER A4

A4-1. Mechanical pressure or temperature change.   A4-2. Turbulence, changes in shape or direction, and air control devices.   A4-3. Static pressure (P_S); velocity pressure (P_V).   A4-4. Manometer.   A4-5. Anemometer.   A4-6. Pitot tube.   A4-7. Air friction chart.   A4-8. Equivalent length of straight duct.   A4-9. Main duct: supply, 1000 ft/min;

return, 800 ft/min. Branch: supply, 600 ft/min; return, 600 ft/min.   A4-10. Forward-curved multiblade.
A4-11. Velocity reduction; equal friction; and static regain.

CHAPTER A5

A5-1. Coal, oil, gas, and electricity.   A5-2. No. 1 and No. 2; approximate 140,000 Btu/gal.   A5-3. Methane.
A5-4. Liquefied petroleum; propane and butane.   A5-5. Heavier.   A5-6. 10 parts of air to 1 part of gas.
A5-7. More than 24 ft³ of air to 1 ft³ of gas.   A5-8. Carbon dioxide; water.   A5-9. Carbon monoxide (CO).
A5-10. 34 lb.   A5-11. American Gas Association (AGA).   A5-12. Underwriters' Laboratories (UL).   A5-13. Baseboard.
A5-14. Highboy, lowboy, horizontal, and counterflow (downflow).   A4-15. Highboy has blower under the heat exchanger
and lowboy has the blower alongside the heat exchanger.   A5-16. 96,000 Btu/hr; 80%.   A5-17. Combustible materials.
A5-18. To provide safe travel of flue products to outside atmosphere and produce and maintain a draft on the unit.
A5-19. Minimize the effect of draft variation on the operation of the furnace.   A5-20. All air from inside building; all air
from outdoors; all air from ventilated attic; and air in from crawl space, out into attic.   A5-21. No—a very dangerous
practice.   A5-22. Sectional and drum and radiator.   A5-23. Primary.   A5-24. Secondary.   A5-25. Refractory.
A5-26. High-pressure atomizing gun.   A5-27. Fuel unit; air blower; electric motor; ignition transformer; and blast tube and
firing assembly.   A5-28. 5.   A5-29. High-limit switches and fusible links.   A5-30. Sequencer.   A5-31. To stagger
the power inrush and minimize the shock on the power supply system.   A5-32. Steam, hot water, and chilled water.
A5-33. 30; 250.   A5-34. Series loop; one-pipe; two-pipe; and hot-water panel.   A5-35. A boiler in which the water is in
the boiler tubes and the hot flue products travel around the tubes.   A5-36. Plate type; atomizing type; and wetted element
type.   A5-37. Humidistat.

CHAPTER A6

A6-1. Throwaway.   A6-2. 75.   A6-3. False—it can cause most service problems.   A6-4. Ionizing types and charged
media types.   A6-5. 12,000 V dc.   A6-6. Ft³/min capacity.   A6-7. A unit of measure—1/25,400 of an inch.
A6-8. 50.   A6-9. True—except in the case of the automatic wash type.   A6-10. Soaking in tub or sink with detergent
cleaner.

CHAPTER A7

A7-1. Association of Home Appliance Manufacturers (AHAM).   A7-2. 80°F dry bulb and 67°F wet bulb inside air and
95°F dry bulb outside air.   A7-3. 75% (0.75).   A7-4. 115.   A7-5. Mild-weather operation.   A7-6. Low-ambient
operation.   A7-7. Controlling the speed of the condenser fan.   A7-8. Parallel.   A7-9. Antifreeze and water solution.
A7-10. "A" type, flat, slant, and horizontal.   A7-11. Horizontal and vertical airflow.   A7-12. Quick-connect,
compession, and flare.   A7-13. Vertical lift.   A7-14. 25.   A7-15. Air-handling unit.   A7-16. Cylinder unloading
and hot-gas bypass.   A7-17. To ensure proper oil return in small quantities to protect the compressor.   A7-18. Air lock.

CHAPTER A8

A8-1. Incremental systems.   A8-2. Single-zone and multizone.   A8-3. Econonomizer.   A8-4. Electronic.
A8-5. All external electrical power to a single connection on the unit.   A8-6. Supply and return air through a combination
diffuser.   A8-7. 3.

CHAPTER A9

A9-1. Applied systems, applied equipment, applied machinery, and engineered systems.   A9-2. Field.   A9-3. Air; liquid;
refrigerant.   A9-4. Water chiller, boiler, air-handling unit, water cooling tower, and control system.   A9-5. 40; 45.

A9-6. 180; 200.   A9-7. To control air temperatures by regulating the amount of air through the coil and bypass area.
A9-8. Reciprocating.   A9-9. The number of times the water will travel through the length of the shell from inlet to outlet.
A9-10. Fouling factor is an assumed safety margin anticipating scale buildup on water tubing.   A9-11. 4.
A9-12. Cupro-nickel steel.   A9-13. Direct; gear.   A9-14. Two-, three-, and four-pipe.   A9-15. Water spray, steam pan, and steam grid.   A9-16. Low, medium, and high velocity.   A9-17. Multizone.   A9-18. Maximum flexibility.

CHAPTER A10

A10-1. Power source, load or loads, and controllers.   A10-2. Thermostat, humidistat, or pressurestat.   A10-3. Relay, contactor, valve, or damper.   A10-4. 24-V output transformer.   A10-5. Two unconnected coils of insulated wire wound around an iron core.   A10-6. Step-down type.   A10-7. The ratio of turns of wire in each coil.   A10-8. False.
A10-9. Volt-amperes (VA); multiplying the output voltage by the rated amperage capacity.   A10-10. Two metals of different expansion rates welded together that have different rates of expansion causing a change in shape with temperature change.   A10-11. The closer contact opening allowed with low voltage produces closer temperature control.
A10-12. Resistors.   A10-13. In series with.   A10-14. Parallel to.   A10-15. Adjustable.   A10-16. The total operating current flow through the heating contact.   A10-17. Remote bulb, rod and tube, and thermisters.   A10-18. Resistance.
A10-19. Automatic; manual.   A10-20. Helix, multiple hairpin, strip, U-shaped, spiral, and snap disk.   A10-21. Bellows, diaphragm, and bourdon tube.   A10-22. Total interruption (on or off) and modulating.   A10-23. One that controls another electrical circuit—a relay or contactor.   A10-24. Relay—has a contact rating of less than 20 A; contactor—has a contact rating of 20 A or more; motor starter—relay or contactor with built-in overload protection.   A10-25. Manual control for ignition and normal operation; pilot supply, adjustment, and safety shutoff; pressure regulation of gas burner feed; and on/off operation, controlled by external controller such as room thermostat.   A10-26. Pressure regulation of gas burner feed.   A10-27. True.   A10-28. Primary control or protecto relay.   A10-29. A bimetallic element; flue gas temperature.   A10-30. A cadmium cell; light produced by combustion of the oil.   A10-31. Full air supply is attained before oil is supplied for combustion.

CHAPTER A11

A11-1. 120/240; 1; 60.   A11-2. 100; 200 (200 A in an all-electric home).   A11-3. 3.   A11-4. Legends; symbols.
A11-5. B (black) and W (white).   A11-6. Hot, black; neutral, white; ground, green; neutral gray or bare wire.
A11-7. Pictorial and schematic.   A11-8. Heavy; light.   A11-9. Continuous; broken.   A11-10. To stop the heating process if air gets too hot.   A11-11. Cycles the fan on and off in response to air temperature.   A11-12. False—the fan is a separate circuit from the thermostat circuit.

CHAPTER A12

A12-1. Single-phase and three-phase power.   A12-2. The equipment manufacturer.   A12-3. Consulting engineer or control system coordinator.   A12-4. Electric, electronic, and pneumatic.   A12-5. False—compressors always need power for crankcase heat protection.   A12-6. Cycling of the fan motors can be used for low-ambient control.   A12-7. 6.52 A.
A12-8. Source of clean dry air; main air lines; controllers; branch lines; and operators or actuators.   A12-9. 20; 35.
A12-10. Regulates the flow of air through a controlled bleed port.

CHAPTER A13

A13-1. Manifold pressure using a manometer; timing the meter using a stopwatch.   A13-2. Draft gauge, stack temperature thermometer, CO_2 indicator, fire efficiency slide rule, and two air temperature thermometers.   A13-3. Follow the instrument manufacturer's instructions completely.   A13-4. Pilot thermocouple.   A13-5. Checking the size of the opening in the main burner orifice.   A13-6. Pressure side is used to check setting of nozzle pressure and vacuum side used to check the lift of oil to the pump.   A13-7. 9 to 12; 600; −04.   A13-8. −0.6; −0.8.   A13-9. No. 2.
A13-10. Flame mirror.   A13-11. Light.   A13-12. Ability to remove the nozzle without endangering the electrodes.

A14-1. Proper input.   A14-2. Btu content per cubic foot; specific gravity.   A14-3. 945; 1121.   A14-4. 2522.   A14-5. 3261.   A14-6. You must find this out from your local gas utility.   A14-7. $3\frac{1}{2}$; 3; 4.   A14-8. The size of the hole in the orifice must be changed.   A14-9. Denver is 5280 ft above sea level—5 times 4% = 20%.

A14-10. $\text{Cubic feet per hour} = \dfrac{\text{Btu per hour}}{\text{Btu per cubic foot}}.$

A14-11. 119.05 ft³/hr.

A14-12. $\text{Cubic feet per hour} = \dfrac{150{,}000 \text{ Btu/hr input}}{970 \text{ Btu/ft}^3 \text{ gas}} = 154.64 \text{ ft}^3.$

$\text{Seconds per cubic foot} = \dfrac{60 \times 60}{154.64} = 23.3 \text{ seconds per revolution.}$

A14-13. 11.   A14-14. One regulator is used to drop the pressure from tank pressure to unit pressure.   A14-15. The main supply line can carry more ft³/hr of gas or may be a smaller pipe.   A14-16. Butane 140°F, propane 30°F.   A14-17. Primary air.   A14-18. Secondary air.   A14-19. 40.   A14-20. 50%.   A14-21. Soft, clear blue.   A14-22. 80.   A14-23. 125; 130; 95; 100.   A14-24. The difference between the heat obtainable from the combustion process compared to the actual amount of heat obtained from the unit.   A14-25. The increased air will extract the heat faster and increase the operating efficiency; the increased air can also cause drafts in the conditioned area as well as cause water to condense from the flue products and shorten heat exchanger life.   A14-26. The point at which the positive pressure caused by the expansion of the hot flue products and the negative pressure produced by the unit draft balance.   A14-27. At the flame tips of the burner.   A14-28. 75; 80.   A14-29. In case of oil after the flow from the nozzle the oil will stay at the refractory until burned.   A14-30. Solid core, hollow core, and special solid.   A14-31. False.   A14-32. 5.   A14-33. Coke trees.   A14-34. Unburned carbon from incomplete combustion of the oil.   A14-35. 100.   A14-36. One-stage and two-stage.   A14-37. A two-stage pump because the fuel must be lifted as well as nozzle pressure developed.   A14-38. The fuel unit pressure regulating piston not seating properly.   A14-39. 5 minutes.   A14-40. Where the fuel supply is higher than the burner—oil flows by gravity to the burner.   A14-41. Where the oil supply is lower than or a distance away from the oil burner.   A14-42. 0.20 in. of W.C. double the design resistance.   A14-43. Excessive carbon deposits in the heat exchanger.   A14-44. Sequence controls.   A14-45. High-limit cutout switch and fusible link.   A14-46. Gravity; forced circulation; forced circulation.   A14-47. Series loop, one-pipe, two-pipe, and hot-water panel.

CHAPTER A15

A15-1. The season switch could be on "off" or "cool."   A15-2. If voltage is at the terminals of the valve.   A15-3. Jumper between the power terminal and heating terminal—unit should operate.   A15-4. Check the coil for continuity.   A15-5. The heating anticipator setting.   A15-6. To prevent water in the main line from flowing into the branch line, freezing, and shutting off the gas supply.   A15-7. 20; 30.   A15-8. Drip leg.   A15-9. Failure of the gas valve to shut off completely.   A15-10. Receiving too much primary air.   A15-11. Reduction in operating efficiency of the unit.   A15-12. Too much primary air.   A15-13. 0; 10.   A15-14. Unit underfired.   A15-15. Unit overfired; dirty filters.   A15-16. 125; 130; 100; 105.   A15-17. Short cycling of the gas valve.   A15-18. Use of a candle flame which should bend into the diverter.   A15-19. Hair spray and washing bleach.   A15-20. Correct the input; soften the fire; and change the size of the burner pouch.   A15-21. A clutch driven by a V-type bimetallic element.   A15-22. A light-sensitive cadmium cell.   A15-23. Check for liquid oil in the bottom of the refractory.   A15-24. Jumper the T-T terminals of the protecto relay.   A15-25. Carbon buildup on the electrodes.   A15-26. False—bending the wires will crack the insulators.   A15-27. Empty fuel tank.   A15-28. 90; 100.   A15-29. −0.02 in. of W.C.   A15-30. False—better to discard and replace with new one.   A15-31. Poor oil cutoff of the fuel unit.   A15-32. False.   A15-33. The heat exchanger, for cracks or other openings.   A15-34. Lack of OFF-cycle flue draft.

CHAPTER A16

A16-1. The correct Btu/hr load must be established in order to check the system performance.   A16-2. Refrigerant charge.   A16-3. Sling psychrometer.   A16-4. Spread out the scale for larger spaces on the scale for more accurate readings.   A16-5. Anemometer; stopwatch.

CHAPTER A17

A17-1. The desired comfort conditions at the lowest operating cost.   **A17-2.** Air temperature drop across the coil. **A17-3.** Line voltage.   **A17-4.** To make sure that the amperage draw falls properly from LRA to FLA.   **A17-5.** 20. **A17-6.** False.   **A17-7.** False.   **A17-8.** True.   **A17-9.** Latent heat.   **A17-10.** To check the heat loss or gain in the duct distribution system.   **A17-11.** 25%.   **A17-12.** The suction pressure and leaving gas temperature.   **A17-13.** Down. **A17-14.** Neither; it will remain constant.   **A17-15.** The difference between the temperature of the air entering the condenser and the condensing temperature of the refrigerant.   **A17-16.** False.   **A17-17.** The heat picked up in the coil plus the heat produced from the electrical energy the unit consumes.   **A17-18.** Motor Btu/hr = volts × amperes × PF × 3.413 Btu/W. **A17-19.** Net capacity.

CHAPTER A18

A18-1. High-pressure gauge; low-pressure gauge; sling psychrometer; seven thermometers; volt meter; and clamp-type ammeter.   **A18-2.** Supply air temperature; return air temperature and relative humidity; liquid-line temperature; suction-line temperature; condenser inlet air temperature; and average of condenser outlet air temperature.   **A18-3.** The air part of the system.   **A18-4.** Refrigerant quantity and refrigerant flow rate.   **A18-5.** The energy efficiency rating of the unit; the number of Btu/hr the unit removed per each watt of electricity consumer.   **A18-6.** Btu/hr net capacity of the unit; watts draw of the condensing unit.   **A18-7.** 11.   **A18-8.** Will stay the same on a TX valve coil and will increase on a capillary tube coil.   **A18-9.** Will reduce the capacity—reduces the head pressure, which reduces the refrigerant flow. **A18-10.** Capillary tube system 65°F, and TX value system 35°F.   **A18-11.** 65°F.   **A18-12.** Measure the amount of subcooling.   **A18-13.** An overcharge.   **A18-14.** 1 oz.   **A18-15.** Suction pressure; DX coil superheat; discharge pressure; liquid subcooling; and amperage draw of the condensing unit.

Answers to Problems in Heat Pumps

HP1-1. Lord Kelvin; 1852.   **HP1-2.** An air-conditioning unit transfers heat in one direction only; a heat pump can reverse the direction of heat transfer.   **HP1-3.** Reverse cycle refrigeration system.   **HP1-4.** Indoor; outdoor.   **HP1-5.** Indoor.   **HP1-6.** Indoor.   **HP1-7.** Reversing valve.   **HP1-8.** Accumulator.   **HP1-9.** Coaxial.   **HP1-10.** To prevent freeze-up of the water coil.   **HP1-11.** Cupro-nickel—because of the variety of groundwater encountered.   **HP1-12.** The coil is sized as a condenser in the heating mode, which requires more surface.   **HP1-13.** The latent capacity of the unit is reduced, which results in higher humidity in the conditioned area.   **HP1-14.** Main valve; pilot valve.   **HP1-15.** Compressor.   **HP1-16.** 75; 100.   **HP1-17.** False—it controls the pressure bled off the end of the valve.   **HP1-18.** Trombone.   **HP1-19.** Disk type and ball type.   **HP1-20.** The ball type.   **HP1-21.** TX valves and capillary tubes.   **HP1-22.** The power element has a special charge to operate when on the cooling mode and yet take the vapor-line temperature in the heating mode.   **HP1-23.** Flood-out on the cooling cycle—dirty filters, blower trouble, etc.; flood-out on the heating cycle—frost buildup on outside coil; and termination of the defrost cycle—flood-out occurs each time.   **HP1-24.** 55.   **HP1-25.** 350; 142.   **HP1-26.** Liquid line; vapor line.   **HP1-27.** Compressor; reversing valve.   **HP1-28.** Compressor; reversing valve.   **HP1-29.** To reduce heat loss; to keep system efficiency high.

HP2-1. Room and outdoor ambient.   **HP2-2.** Temperature differential; pressure temperature; time temperature; and pressure–time temperature.   **HP2-3.** System totally off; cooling only; heating only; two-stage heat; automatic changeover; emergency heat; room air circulation only; and trouble indication.   **HP2-4.** 2.   **HP2-5.** To allow 3°F difference between heating and reversing valve change and 2°F difference between reversing valve change and cooling.   **HP2-6.** Dead period.   **HP2-7.** Limit the amount of auxiliary heat that can be brought on when it is not required; limit inrush current to the equipment.   **HP2-8.** Reverse the reversing valve; cut off the outdoor fan; and provide holding contact around the defrost initiating device.   **HP2-9.** Outside coil temperature and temperature of air entering the outside coil.   **HP2-10.** 35; 45.   **HP2-11.** The airflow resistance or pressure drop through the outside coil.   **HP2-12.** The temperature of the liquid refrigerant in the bottom line of the outside coil (the condenser).   **HP2-13.** The operating time of the compressor.   **HP2-14.** The contact closing time of the pressure switch.   **HP2-15.** Lockout relay.   **HP2-16.** The current flow through the circuit is too low to keep the heavier contactor pulled in.   **HP2-17.** Break the power to the lockout circuit.

HP3-1. Air to air–dual operation; liquid to air–dual operation; air to air–single operation; liquid to air–single operation; and air to liquid–single operation.   **HP3-2.** Package and split.   **HP3-3.** Electric strip.   **HP3-4.** Lower installation cost.   **HP3-5.** More versatile installation—one unit can be vertical, horizontal, or in downflow application.   **HP3-6.** True.   **HP3-7.** Liquid to air.   **HP3-8.** To prevent freezing of the outdoor coil.   **HP3-9.** One to regulate head pressure in the cooling mode, and the other to regulate suction pressure on the heating mode.   **HP3-10.** So that they can operate with water temperatures close to freezing.   **HP3-11.** Heating of domestic water.

CHAPTER HP4

HP4-1. Suction pressure; discharge pressure; and vapor-line pressure.   **HP4-2.** Discharge.   **HP4-3.** The lime buildup in the liquid-to-refrigerant coil that affects performance.   **HP4-4.** By measuring the flow resistance through the water coil and comparing it to the manufacturer's specs.   **HP4-5.** Gate or ball.

CHAPTER HP5

HP5-1. The correct amount of air through the inside coil.   **HP5-2.** 10.   **HP5-3.** False.   **HP5-4.** True.
HP5-5. Once a month.   **HP5-6.** Once a month.   **HP5-7.** Service drop to the building and/or distribution power transformer is too small.   **HP5-8.** No—both respond the same to the conditions of the air.   **HP5-9.** 2.   **HP5-10.** At the outlet of the outside coil.   **HP5-11.** The heat loss in the heating mode will greatly reduce the capacity of the system.
HP5-12. From the manufacturer's literature.   **HP5-13.** False.   **HP5-14.** True.   **HP5-15.** The inside coil.

HP5-16. $\text{ft}^3/\text{min} = \dfrac{\text{volts} \times \text{amperes} \times 3.414 \text{ Btu/W}}{\Delta T{}^\circ\text{F through coil} \times 1.08}$.

HP5-17. Gross capacity.   **HP5-18.** Net capacity (heat picked up by air or liquid) plus motor heat.   **HP5-19.** Gross capacity = 47,980.8 Btu/hr; motor input = 11.798.8 Btu/hr; net capacity = 36.182 Btu/hr.

HP5-20. 6497 Btu/hr heating capacity $= \dfrac{\text{quantity of water} \times 8.33 \times \Delta T{}^\circ\text{F} \times 60}{\text{time}}$

$$= \frac{40 \times 8.33 \times (135^\circ - 70^\circ) \times 60}{200 \text{ minutes}}$$

$$= \frac{40 \times 8.33 \times 65 \times 6}{200} = 6497 \text{ Btu/hr.}$$

CHAPTER HP6

HP6-1. Excessive leakage from the high side to the low side through the reversing valve.   **HP6-2.** The pressure-reducing device and the check valve in parallel.   **HP6-3.** Suction line from reversing valve to compressor; discharge line from the compressor to the reversing valve.   **HP6-4.** The system must be operated in both modes.   **HP6-5.** The check valve on the outside coil will be open and the one on the inside coil will be closed.   **HP6-6.** The cooling mode.   **HP6-7.** The cooling mode.   **HP6-8.** The check valve on the inside coil is stuck closed.   **HP6-9.** The valve is stuck in the middle position.   **HP6-10.** Full system charge.   **HP6-11.** 75 psig.   **HP6-12.** 250°F.   **HP6-13.** W1 on the thermostat is connected to W1 on the electric heat instead of the Y lead to the compressor.   **HP6-14.** Outdoor ambient control.
HP6-15. Initial balance point.   **HP6-16.** Too much reheat on the defrost cycle, causing the room thermostat to be satisfied before the defrost cycle is completed.   **HP6-17.** 55°F.   **HP6-18.** The termination thermostat is loose and not cooling down properly.   **HP6-19.** The coil needs cleaning.   **HP6-20.** The sensible capacity of the unit in the cooling mode.

Index